THE CAMBRIDGE HANDBOOK OF RESEARCH METHODS IN CLINICAL PSYCHOLOGY

Edited by

Aidan G. C. Wright
University of Pittsburgh

Michael N. Hallquist
Pennsylvania State University

CAMBRIDGE UNIVERSITY PRESS

CAMBRIDGE
UNIVERSITY PRESS

University Printing House, Cambridge CB2 8BS, United Kingdom

One Liberty Plaza, 20th Floor, New York, NY 10006, USA

477 Williamstown Road, Port Melbourne, VIC 3207, Australia

314–321, 3rd Floor, Plot 3, Splendor Forum, Jasola District Centre, New Delhi – 110025, India

79 Anson Road, #06–04/06, Singapore 079906

Cambridge University Press is part of the University of Cambridge.

It furthers the University's mission by disseminating knowledge in the pursuit of education, learning, and research at the highest international levels of excellence.

www.cambridge.org
Information on this title: www.cambridge.org/9781107189843
DOI: 10.1017/9781316995808

© Cambridge University Press 2020

This publication is in copyright. Subject to statutory exception and to the provisions of relevant collective licensing agreements, no reproduction of any part may take place without the written permission of Cambridge University Press.

First published 2020

Printed in the United Kingdom by TJ International Ltd. Padstow Cornwall

A catalogue record for this publication is available from the British Library.

Library of Congress Cataloging-in-Publication Data
Names: Wright, Aidan G. C., 1980– editor. | Hallquist, Michael N., 1981– editor.
Title: The Cambridge handbook of research methods in clinical psychology / edited by Aidan G. C. Wright, University of Pittsburgh, Michael N. Hallquist, Pennsylvania State University.
Description: Cambridge ; New York, NY : Cambridge University Press, 2020. | Series: Cambridge handbooks in psychology | Includes bibliographical references and index.
Identifiers: LCCN 2019037660 (print) | LCCN 2019037661 (ebook) | ISBN 9781107189843 (hardback) | ISBN 9781316639528 (paperback) | ISBN 9781316995808 (epub)
Subjects: LCSH: Clinical psychology–Research–Methodology–Handbooks, manuals, etc.
Classification: LCC RC467.2 .C36 2020 (print) | LCC RC467.2 (ebook) | DDC 616.89–dc23
LC record available at https://lccn.loc.gov/2019037660
LC ebook record available at https://lccn.loc.gov/2019037661

ISBN 978-1-107-18984-3 Hardback
ISBN 978-1-316-63952-8 Paperback

Cambridge University Press has no responsibility for the persistence or accuracy of URLs for external or third-party internet websites referred to in this publication and does not guarantee that any content on such websites is, or will remain, accurate or appropriate.

THE CAMBRIDGE HANDBOOK OF RESEARCH METHODS IN CLINICAL PSYCHOLOGY

This book integrates the philosophy of science, data acquisition methods, and statistical modeling techniques to present readers with a forward-thinking perspective on clinical science. It reviews modern research practices in clinical psychology that support the goals of psychological science, study designs that promote good research, and quantitative methods that can test specific scientific questions. It covers new themes in research including intensive longitudinal designs, neurobiology, developmental psychopathology, and advanced computational methods such as machine learning. Core chapters examine significant statistical topics, for example missing data, causality, meta-analysis, latent variable analysis, and dyadic data analysis. A balanced overview of observational and experimental designs is also supplied, including preclinical research and intervention science. This is a foundational resource that supports the methodological training of the current and future generations of clinical psychological scientists.

AIDAN G. C. WRIGHT is Associate Professor of Psychology at the University of Pittsburgh, USA, where he primarily teaches graduate statistics. His work has been recognized by several awards, including the Society for a Science of Clinical Psychology's Susan Nolen-Hoeksema Early Career Research Award and the American Psychological Association's David Shakow Early Career Award for Contributions to Clinical Psychology.

MICHAEL N. HALLQUIST is Assistant Professor in the Department of Psychology at Penn State University, USA. His research is supported by the National Institute of Mental Health and in 2019 he was recognized as an International Society for the Study of Personality Disorders Young Investigator.

Contents

List of Figures	*page* viii
List of Tables	x
List of Contributors	xi
Acknowledgments	xiii

PART I CLINICAL PSYCHOLOGICAL SCIENCE: AN EVOLVING FIELD

1 Trends in the Evolving Discipline of Clinical Psychology 3
   Aidan G. C. Wright and Michael N. Hallquist

2 Defining and Redefining Phenotypes: Operational Definitions as
   Open Concepts 5
   Peter Zachar, Eric Turkheimer, and Kenneth F. Schaffner

3 Building Models of Psychopathology Spanning Multiple Modalities of
   Measurement 18
   Noah C. Venables and Christopher J. Patrick

PART II OBSERVATIONAL APPROACHES

4 The Conceptual Foundations of Descriptive Psychopathology 33
   Trevor F. Williams and Leonard J. Simms

5 Survey and Interview Methods 45
   Douglas B. Samuel, Meredith A. Bucher, and Takakuni Suzuki

6 Psychometrics in Clinical Psychological Research 54
   R. Michael Furr

7 Latent Variable Models in Clinical Psychology 66
   Aidan G. C. Wright

8 Psychiatric Epidemiology Methods 80
   David S. Fink, Hyunsik Kim, and Nicholas R. Eaton

PART III EXPERIMENTAL AND BIOLOGICAL APPROACHES

9 Conceptual Foundations of Experimental Psychopathology: Historical
   Context, Scientific Posture, and Reflections on Substantive
   and Method Matters 93
   Mark F. Lenzenweger

10	A Practical Guide for Designing and Conducting Cognitive Studies in Child Psychopathology	105
	Cynthia Huang-Pollock, Zvi Shapiro, Hilary Galloway-Long, and Jason Feldman	
11	Peripheral Psychophysiology	118
	Amanda Levinson and Greg Hajcak	
12	Behavioral and Molecular Genetics	136
	Gretchen Saunders and Matt McGue	
13	Concepts and Principles of Clinical Functional Magnetic Resonance Imaging	153
	Avram J. Holmes and Angus W. MacDonald III	
14	Clinical Computational Neuroscience	168
	Michael N. Hallquist and Alexandre Y. Dombrovski	

PART IV DEVELOPMENTAL PSYCHOPATHOLOGY AND LONGITUDINAL METHODS

15	Studying Psychopathology in Early Life: Foundations of Developmental Psychopathology	193
	Dante Cicchetti	
16	Adolescence and Puberty: Understanding the Emergence of Psychopathology	205
	Sarah D. Lynne, Allison S. Metz, and Julia A. Graber	
17	Quantitative Genetic Research Strategies for Studying Gene-Environment Interplay in the Development of Child and Adolescent Psychopathology	219
	Isabell Brikell and Henrik Larsson	
18	Designing and Managing Longitudinal Studies	230
	M. Brent Donnellan and Deborah A. Kashy	
19	Measurement and Comorbidity Models for Longitudinal Data	241
	Phillip K. Wood	

PART V INTERVENTION APPROACHES

20	The Multiphase Optimization Strategy for Developing and Evaluating Behavioral Interventions	267
	Kate Guastaferro, Chad E. Shenk, and Linda M. Collins	
21	Future Directions in Developing and Evaluating Psychological Interventions	279
	Shannon Sauer-Zavala	
22	Health Psychology and Behavioral Medicine: Methodological Issues in the Study of Psychosocial Influences on Disease	287
	Timothy W. Smith	

PART VI INTENSIVE LONGITUDINAL DESIGNS

23	Ambulatory Assessment	301
	Stuart G. Ferguson, Tina Jahnel, Katherine Elliston, and Saul Shiffman	
24	Modeling Intensive Longitudinal Data	312
	Marlies Houben, Eva Ceulemans, and Peter Kuppens	

25 Modeling the Individual: Bridging Nomothetic and Idiographic
    Levels of Analysis 327
    Peter C. M. Molenaar and Adriene M. Beltz

26 Social Processes and Dyadic Designs 337
    Robert D. Vlisides-Henry, Sheila E. Crowell, Erin A. Kaufman,
    and Betty Lin

27 Models for Dyadic Data 350
    Eduardo Estrada, David A. Sbarra, and Emilio Ferrer

PART VII GENERAL ANALYTIC CONSIDERATIONS

28 Reproducibility in Clinical Psychology 371
    Christopher J. Hopwood and Simine Vazire

29 Meta-Analysis: Integration of Empirical Findings through
    Quantitative Modeling 383
    Kristian E. Markon

30 Mediation, Moderation, and Conditional Process Analysis:
    Regression-Based Approaches for Clinical Research 396
    Nicholas J. Rockwood and Andrew F. Hayes

31 Statistical Inference for Causal Effects in Clinical Psychology:
    Fundamental Concepts and Analytical Approaches 415
    Reagan Mozer, Donald B. Rubin, and Jose Zubizarreta

32 Analyzing Nested Data: Multilevel Modeling and Alternative
    Approaches 426
    Daniel J. Bauer, Daniel M. McNeish, Scott A. Baldwin, and
    Patrick J. Curran

33 Missing Data Analyses 444
    Amanda N. Baraldi and Craig K. Enders

34 Machine Learning for Clinical Psychology and Clinical Neuroscience 467
    Marc N. Coutanche and Lauren S. Hallion

Index   483

Figures

2.1	Two-dimensional NMDS for NEO-PI facet scales	*page* 12
3.1	Results from a three-factor with one higher-order factor confirmatory factor analytic model of inhibitory control and disinhibition (DIS) as assessed by differing measurement modalities and a *multimodal* factor corresponding to the covariance across measurement modalities	23
6.1	Steps in exploratory factor analysis	55
6.2	Steps in confirmatory factory analysis	56
7.1	Examples of common factor models	69
7.2	Two alternative representations of a hierarchical factor model	74
7.3	Graphical depiction of latent distributions associated with various factor mixture models	76
10.1	Statistical model of the planned missingness approach	109
10.2	Diffusion model of a numerical decision task	111
11.1	Selected overview of the organ systems indexed with peripheral psychophysiological measures and their links to ANA activity	119
11.2	Electrocardiography (EKG) and impedance cardiography (ICG) waveforms for one heartbeat	123
11.3	Recommended electrode placement for EDA measurement	125
11.4	Grand average SCR waveform and its commonly scored components	126
11.5	Electrode placement for startle eye-blink and postauricular	128
12.1	Multifactorial threshold (MFT) model	138
12.2	Univariate ACE model	138
12.3	Examples of gene-by-environment interactions	147
13.1	Different task conditions and the corresponding convolution with a prototypical hemodynamic response function	155
13.2	A summary of intrinsic connectivity analyses methods and cortical parcellations based in-vivo brain imaging	156
14.1	Neural network models	170
14.2	Learned action value estimates for cup and bowl according to a Q-learning reinforcement learning model	173
14.3	The model-predicted probability of choosing the cup (versus bowl) on each of 100 trials	175
14.4	Trust game experiment	181
14.5	Comparison of RL models	182
14.6	Meta-analytic maps of prediction error and learned value signals from human fMRI studies	185
19.1	Parsimonious factor pattern models	245
19.2	Commonly used growth curve models	247
19.3	BSI general severity index (BSI-GSI) observed and model-predicted means	249
19.4	Growth curve model for BSI General Severity Index (BSI-GSI)	251

19.5	Scatterplot of log-likelihood and leverage for general psychological distress model	253
19.6	Observed and predicted means for alcohol consumption variables	253
19.7	Growth curve model for number of times "high" or "lightheaded"	255
19.8	Growth curve model for five or more drinks in a sitting	256
19.9	Growth curve model for number of times "drunk"	257
19.10	Comorbidity subdiagram by sex	258
19.11	Model-predicted curves for BSI-GSI and general problematic alcohol use by sex	261
20.1	The multiphase optimization strategy (MOST) process	269
20.2	Conceptual model for the hypothetical intervention targeting PTSD symptoms in children	270
22.1	Systems hierarchy in the biopsychosocial model	288
22.2	Effects of stress exposure, reactivity, recovery, and restoration on physiological response, across levels of psychosocial risk and resilience	293
24.1	Graphs showing simulated time series data for two fictive subjects	317
24.2	Diagram showing which models to use for several types of research questions	323
25.1	Cattell's (1952) data box with two orthogonal slices corresponding to interindividual variation and intraindividual variation	328
26.1	Schematic of the actor-partner interdependence model	344
27.1	Path diagram of an actor partner independence model (APIM)	353
27.2	Bivariate (dyadic) latent growth curve (LGC) model	356
27.3	Bivariate (dyadic) latent change score (LCS) model	358
27.4	Bivariate (dyadic) autoregressive cross-lagged model	360
27.5	Bivariate (dyadic) dynamic factor analysis (DFA) model	362
27.6	Bivariate (dyadic) differential equation model (DEM)	363
29.1	Funnel plots for fixed-effect meta-analyses of 100 hypothetical studies reporting Cohen's d values	387
30.1	A simple mediation model, parallel multiple mediator model, serial multiple mediator model, and a conceptual representation of moderation	397
30.2	Four examples of conditional process models with a single moderator W	403
30.3	A first-stage moderated parallel multiple mediator model in conceptual form and in the form of a path diagram	405
32.1	Schematic representation of the therapy arms of Dimidjian et al. (2006)	427
32.2	Plot of hypothetical data showing the relation between patient outcomes and patient-therapist alliance strength	430
32.3	Same data as Figure 32.2 with nested structure shown	430
32.4	Plot of the psychotherapist means and the effect of patient-therapist alliance strength on patient outcomes between psychotherapists	431
32.5	Depiction of the total effect, the between effect, and the within effects	431
32.6	Visual depiction of two levels of variability represented with a multilevel growth model with random intercepts and slopes	436
33.1	A path diagram for the multiple regression model in Equation 33.1	446
33.2	A path diagram for the multiple regression model in Equation 33.1 recast as a latent variable model	453
33.3	A path diagram demonstrating use of the saturated correlates approach to include auxiliary variables in the analysis model	453
33.4	A path diagram demonstrating use of the extra dependent variable approach to include auxiliary variables in the analysis model	454
34.1	An example confusion matrix	477

Tables

4.1	Summary of descriptive psychopathology approaches	*page* 34
6.1	Five issues in evaluating validity	61
8.1	Fourfold table	85
8.2	A hypothetical study of 10 individuals with a risk factor followed up for eight years	86
12.1	Individual and meta-analyses of twin studies of psychopathology	140
12.2	Genome-wide association studies (GWAS) of psychopathology	144
13.1	Experimental approaches using functional magnetic resonance imaging	154
13.2	Analytic techniques to examine task-evoked datasets	158
13.3	Analytic techniques of intrinsic brain function	159
16.1	Adolescent prevalence of major depressive episodes and substance use	206
17.1	Methodological considerations and assumptions in quantitative genetic research	220
18.1	Critical design decisions in longitudinal research	231
19.1	Multigroup factor loadings, means, and manifest variable intercepts	252
19.2	Factor correlations/covariances by sex	259
19.3	Alcohol subfactor loadings and conditional variances (disturbances)	260
19.4	Manifest variable error variances	260
20.1	Experimental conditions and per-condition n in 2^5 factorial design	273
23.1	Types of sampling used in AA studies	302
24.1	Overview of some major dynamic properties of single or multiple symptoms (and their interrelations) and how they can be calculated or modeled	314
25.1	Person-oriented principles and methods adapted from Sterba and Bauer (2010) and updated by Molenaar (2010)	331
27.1	Overview of dyadic models	352
30.1	Regression coefficients, standard errors, p-values, and confidence intervals from the motivational interviewing conditional process analysis	406
30.2	Conditional indirect effects of motivational interviewing on alcohol use and the index of moderated mediation	407
31.1	An example of the observed data for N units	416
31.2	An example of the "Science" for N units under SUTVA	416
31.3	Artificial example of a confounded assignment mechanism	417
33.1	Summary of variables in motivating example	445
33.2	Correlations among study variables	445
33.3	Pooled estimates and standard errors from multiple imputation analysis example	450
33.4	Estimates and standard errors from maximum likelihood analysis example	454

Contributors

Scott A. Baldwin, *Brigham Young University*
Amanda N. Baraldi, *Oklahoma State University*
Daniel J. Bauer, *University of North Carolina at Chapel Hill*
Adriene M. Beltz, *University of Michigan*
Isabell Brikell, *Karolinska Institutet & Aarhus University*
Meredith A. Bucher, *Purdue University*
Eva Ceulemans, *KU Leuven*
Dante Cicchetti, *University of Minnesota*
Linda M. Collins, *Pennsylvania State University*
Marc N. Coutanche, *University of Pittsburgh*
Sheila E. Crowell, *University of Utah*
Patrick J. Curran, *University of North Carolina at Chapel Hill*
Alexandre Dombrovksi, *University of Pittsburgh*
M. Brent Donnellan, *Michigan State University*
Nicholas R. Eaton, *Stony Brook University*
Katherine Elliston, *University of Tasmania*
Craig K. Enders, *University of California, Los Angeles*
Eduardo Estrada, *University of California, Davis*
Jason Feldman, *Pennsylvania State University*
Stuart G. Ferguson, *University of Tasmania*
Emilio Ferrer, *University of California, Davis*
David S. Fink, *Columbia University*
R. Michael Furr, *Wake Forest University*
Hilary Galloway-Long, *Pennsylvania State University*
Julia A. Graber, *University of Florida*
Kate Guastaferro, *Pennsylvania State University*
Greg Hajcak, *Florida State University*
Lauren S. Hallion, *University of Pittsburgh*
Michael N. Hallquist, *Pennsylvania State University*
Andrew F. Hayes, *Ohio State University*
Avram J. Holmes, *Yale University & Massachusetts General Hospital, Boston*
Christopher J. Hopwood, *University of California, Davis*
Marlies Houben, *KU Leuven*
Cynthia Huang-Pollock, *Pennsylvania State University*
Tina Jahnel, *University of Tasmania*
Deborah A. Kashy, *Michigan State University*
Erin A. Kaufman, *University of Pittsburgh Medical Center*
Hyunsik Kim, *Stony Brook University*
Peter Kuppens, *KU Leuven*
Henrik Larsson, *Karolinska Institutet & Örebro University*
Mark F. Lenzenweger, *State University of New York at Binghamton & Weill Cornell Medical College*

LIST OF CONTRIBUTORS

Amanda Levinson, *Stony Brook University*
Betty Lin, *State University of New York at Albany*
Sarah D. Lynne, *University of Florida*
Angus W. MacDonald, III, *University of Minnesota*
Kristian E. Markon, *University of Iowa*
Matt McGue, *University of Minnesota*
Daniel M. McNeish, *Arizona State University*
Allison S. Metz, *University of Florida*
Peter C. M. Molenaar, *Pennsylvania State University*
Reagan Mozer, *Bentley University*
Christopher J. Patrick, *Florida State University*
Nicholas J. Rockwood, *Loma Linda University*
Donald B. Rubin, *Harvard University*
Douglas B. Samuel, *Purdue University*
Shannon Sauer-Zavala, *University of Kentucky*
Gretchen Saunders, *University of Minnesota*
David A. Sbarra, *University of Arizona*
Kenneth F. Schaffner, *University of Pittsburgh*
Zvi Shapiro, *Pennsylvania State University*
Chad E. Shenk, *Pennsylvania State University*
Saul Shiffman, *University of Pittsburgh*
Leonard J. Simms, *University at Buffalo, the State University of New York*
Timothy W. Smith, *University of Utah*
Takakuni Suzuki, *Purdue University*
Eric Turkheimer, *University of Virginia*
Simine Vazire, *University of California, Davis*
Noah C. Venables, *University of Minnesota – Twin Cities*
Robert D. Vlisides-Henry, *University of Utah*
Trevor F. Williams, *University at Buffalo, State University of New York*
Phillip K. Wood, *University of Missouri*
Aidan G. C. Wright, *University of Pittsburgh*
Peter Zachar, *Auburn University at Montgomery*
Jose Zubizarreta, *Harvard University*

Acknowledgments

We are deeply indebted to all of the authors who each contributed exceptional chapters that are outstanding in their depth and breadth of thought. In addition, we reserve special thanks for the efforts of Brinkley M. Sharpe, whose editorial contributions made this book possible.

PART I CLINICAL PSYCHOLOGICAL SCIENCE

An Evolving Field

1 Trends in the Evolving Discipline of Clinical Psychology

AIDAN G. C. WRIGHT AND MICHAEL N. HALLQUIST

Clinical psychology sits at the confluence of many basic and applied sciences including psychometrics, measurement, biology, neuroscience, cognition, psychotherapy, development, and epidemiology. These merging scientific streams have supported the ascendancy of clinical psychology as a broad and fast-moving discipline that has the potential to deliver important findings about the etiology and treatment of psychopathology. However, navigating these waters can be challenging due to the field's complexity, made even more difficult by rapid changes in the dominant methods and thinking in recent decades. Some of the most marked changes have come from continued debates about the fundamental nature of the constructs we seek to study. For instance, is psychopathology composed of kinds or continua? Although the field continues to be organized in part by a diagnostic nosology based on legacy categories, the cumulative evidence now clearly tilts in favor of dimensions of pathology, with little compelling evidence in support of discrete mental disorders. This shift in a longstanding debate has major implications that ripple across the field. For example, who should be included in research studies if there is no clear cutoff between normality and pathology? How should they be assessed, and along which dimensions? If psychopathology is predominantly dimensional, how can we develop treatments that are empirically supported for a given person? And what sorts of statistical models can accommodate the complexity of multidimensionality?

In addition to the question of categories or dimensions, there are many other paradigm shifts occurring in clinical psychology. For instance, findings from genetics and neuroscience are increasingly being infused into the conceptualization of psychopathology. Although exciting, this shift comes with the challenge of pursuing translational research that bridges wide divides in levels of analysis. Some trends are more methodological, such as the rise of technologies that provide unprecedented access to human experiences in daily life. For example, intensive sampling methods such as ecological momentary assessment have the potential to measure dynamic psychological processes at the timescale of greatest relevance, but unlocking the potential of such approaches depends on advanced quantitative analyses of time-varying processes. Likewise, many modern studies collect an immense amount of multimodal data on each participant, in some cases covering much of the lifespan. Thus, we must address both the promises and pitfalls of "big data" that require one to synthesize across methods and timescales (e.g., ecological momentary assessment, fMRI, and self-report). Some approaches to "big data" in clinical psychology have begun to emphasize the value of predicting clinical outcomes using machine learning methods that can cope with an overwhelming volume of data, even if they at times lack a strong theoretical basis. Finally, as in many other disciplines, clinical psychology is becoming increasingly aware of the value of examining how we conduct our science. This shift has promoted a greater emphasis on staying true to best practices and avoiding approaches that can lead to problems with replicability and reproducibility or to questionable research and interpretive practices.

Clinical scientists of today can no longer equip themselves for a successful program of research by selecting a preferred diagnosis, picking a level of analysis (e.g., behavior, physiology), and looking at an F-table to find the critical values for an ANOVA of a hypothesized clinical group x experimental condition effect. Rather, to succeed, clinical psychologists increasingly have to grapple with difficult philosophical, technological, and methodological issues. We believe that progress in the field will be best supported by a tight synthesis between theory and method founded on a conceptual continuity in the way clinical psychologists think about the phenomena of interest, how they sample data, and how they model the data quantitatively. The overarching aim of this book is to provide a map of the major tributaries of modern clinical science and to look downstream toward where the field will flow in the future.

Relative to other similar handbooks, we have solicited contributions that emphasize philosophy of science, paradigms of inquiry, data acquisition apparatuses, and quantitative methods. In the first part, the field's larger questions are addressed, including the definition of clinical phenotypes (Chapter 2) and models of psychopathology that transcend multiple levels of analysis (Chapter 3). This transitions naturally into a part on observational designs, headlined by a defining treatise on the foundations of descriptive psychopathology (Chapter 4). This part also covers important fundamentals, including chapters focused on survey and interview methods (Chapter 5), psychometrics (Chapter 6), latent variable models (Chapter 7), and epidemiology (Chapter 8). Although we do not intend to suggest that experimental and observational designs are incompatible, the next part focuses on experimental designs as an important and distinct approach to understanding psychopathology. This part also covers biological approaches, as they have so often been incorporated within the experimental paradigm. A background on the field of experimental psychopathology (Chapter 9) is followed by a detailed look at strategies and issues in the area (Chapter 10). The coverage then turns to biological approaches, including deep dives into the well-established but continually developing fields of peripheral psychopathology (Chapter 11), molecular and behavioral genetics (Chapter 12), and functional neuroimaging (Chapter 13). This part ends with an overview of the emerging field of computational clinical neuroscience (Chapter 14).

In the next few parts, the focus shifts to longitudinal research designs. The first of these parts emphasizes a developmental perspective, beginning with an overview of developmental psychopathology (Chapter 15). This introductory chapter is followed by a review of how and why to study pubertal changes in psychopathology (Chapter 16). The focus then turns to quantitative genetic strategies for studying gene-environment interactions (Chapter 17). This part concludes by examining more general issues in longitudinal research designs (Chapter 18) and tuning statistical models of longitudinal data to specific research questions (Chapter 19).

At their heart, treatment studies are also longitudinal research, and the first chapter in the next part covers contemporary strategies in developing and evaluating psychological interventions (Chapter 20). As the field moves away from treating specific diagnoses as unique disorders, so too do treatment designs need to accommodate broad dimensions of dysfunction that encapsulate specific impairments. The next chapter describes progress and future directions in developing psychological treatments from a transdiagnostic dimensional perspective (Chapter 21). This part concludes with a chapter on behavioral medicine and health psychology, which integrate the diverse areas of behavioral science and traditional medicine (Chapter 22).

Traditional longitudinal designs have been built on sampling relatively few observations per person (e.g., fewer than 10), but many processes of theoretical interest in clinical psychology occur at a finer timescale and may necessitate more intensive measurement. The past couple of decades have seen rapid advances in the methodology for sampling and modeling intensive longitudinal data, and the next part focuses on these developments. Ambulatory assessment, perhaps the methodology most associated with intensive longitudinal designs, is covered in the first chapter of the part (Chapter 23), followed by analytic strategies for such data (Chapter 24). When a study collects many observations per participant, a model can be developed for each individual. Although idiographic designs have been around for a long time, they have seen a major resurgence alongside the recent increased interest in personalized medicine. This has ushered in a need for methodology for person-specific modeling and generalizing to the nomothetic from the idiographic. This perspective and methodology are covered in Chapter 25. Social process and dyadic research have long focused on micro-processes that require intensive longitudinal designs. Thus, an overview of study design (Chapter 26) and appropriate statistical models (Chapter 27) round out this section.

In the final part, we have enlisted world leaders in philosophical and methodological topics that span specific research designs and are relevant for all earlier parts of the volume. These are foundational topics we believe all clinical scientists should understand because they are fundamental to the research enterprise. The topics covered in this part include reproducible science (Chapter 28), reviews and meta-analyses (Chapter 29), mediation, moderation, and conditional process analyses (Chapter 30), causal reasoning and inference (Chapter 31), nested data structures (e.g., observations nested within persons; Chapter 32), accommodating and modeling missing data (Chapter 33), and finally the value of machine learning in clinical psychology (Chapter 34).

Altogether, these chapters cumulatively reflect our view of current and emerging research themes in clinical psychology. We hope that this volume aids the training and continuing education of clinical psychologists whose research employs interdisciplinary approaches that can support a more mechanistic approach to clinical science. Many of the scientific questions addressed in clinical psychology are complex, and key variables such as genetic vulnerability, environmental exposure to stress, and social support are often outside of our (experimental) control. Thus, we believe that better methods can potentially help us to characterize dynamic psychological processes in observational studies. Likewise, more nuanced experimental manipulations and biological approaches can potentially help us test the role of specific processes in psychopathology. Furthermore, if mental illness can be conceptualized in terms of the configuration of many dimensions of pathology, our approaches to developing treatments must carefully examine which components are crucial to positive change and what are the best targets.

Finally, we would like to express our deep gratitude to the many authors of this volume whose contributions reflect clear, contemporary thinking on key methodological topics in clinical psychology.

2 Defining and Redefining Phenotypes

Operational Definitions as Open Concepts

PETER ZACHAR, ERIC TURKHEIMER, AND KENNETH F. SCHAFFNER[*]

It is remarkable how long a period elapsed before it definitively dawned upon inquirers that the designation of *thermal states* by *numbers* reposed on a *convention*. Thermal states exist in nature, but the concept of temperature exists only by virtue of our arbitrary *definition*, which could very well have taken another form. Yet until very recently inquirers in this field appear more or less unconsciously to have sought after a *natural* measure of temperature, a real temperature, a sort of Platonic Idea of temperature, of which the temperatures read from the thermometric scales were only the imperfect and inexact expression. (Mach, 1903, p. 154)

In the early years of the twenty-first century, scientific psychology entered a crisis period in response to the non-replicability of many of its results (Munafò et al., 2017). In a few cases, scientific misconduct led to the publication of spurious findings that could not be verified, but in other cases different factors were involved, including a variety of acceptable practices that cumulatively resulted in capitalizing on chance findings. Several solutions to this problem have been offered, many of them involving the introduction of new methodological standards.

This was not the first time psychologists attempted to solve a scientific crisis of confidence by introducing new methodological standards. In the 1930s, Stanley Smith Stevens (1935, p. 517) advocated for the use of operational definitions to rid psychology of the "hazy ambiguities which result in ceaseless argument and dissention." Operational definitions had been introduced a few years earlier by the physicist Percy Bridgman (1927) in order to regiment the meaning of concepts in science. In an operational definition, one defines a scientific concept by specifying how it is measured – thus making its meaning clear. For example, an operational definition of depression is:

Little interest or pleasure in doing things
Feeling down, depressed or hopeless
Trouble falling or staying asleep, or sleeping too much
Feeling tired or having little energy
Poor appetite or overeating
Feeling bad about yourself – or that you are a failure or have let yourself or your family down.
(Kroenke et al., 2009, p. 171)

Someone who has experienced five or more of these symptoms for at least eight days in the past two weeks would be considered depressed.

Stevens hoped that introducing *operational definitions* would help psychologists reach agreement on the meaning of their important concepts, but the way operational definitions came to be conceptualized in psychology may have made the problem worse. When researchers across labs use different measures of a concept, such as working memory, they often assume that these differences are inconsequential and that they are studying the same thing. This assumption might not be scientifically justified (Poldrack & Yarkoni, 2016; Sullivan, 2017). They may be studying different things under the same name.

Focusing on complex psychological traits such as phenotypes, this chapter will expound upon an important philosophical interpretation of operational definitions – their status as open concepts. An open concept is one that is revisable in the light of new information. To provide some background, we will review the roots of open concepts in the philosophy of science – in both physics and in the more advanced stages of logical positivism. In addition to elucidating the openness of concepts for psychological phenotypes such as schizophrenia and the structure of personality, we will also elucidate the currently open status of the endophenotype concept.

We would like to alert readers that we are advocating for a reading of what Bridgman intended in proposing the operational point of view, which is different from how he is typically understood. In 1927's *The Logic of Modern Physics*, sentences such as "the concept is synonymous with the corresponding set of operations" (Bridgman, 1927, p. 5), "If we have more than one set of operations, we have more than once concept," and "strictly there should be a separate name to correspond to each different set of operations" (Bridgman, 1927, p. 10) are

---

[*] Thanks to Lucas Mathews, Hasok Chang, Hanna van Loo, Denny Borsboom, Randy Russell, Jan-Willem Romeijn, and Ken Kendler for insightful comments on an earlier version of this chapter.

understandably taken as explicating a theory of meaning. Hasok Chang (2009) argues that statements such as this were not intended to specify a universal and precise theory of meaning for scientific concepts. Bridgman had something more limited in mind. He was opposed to treating any concept as having a fixed, essential meaning that remains the same in every situation where the concept is used. Based on our reading of Bridgman, this chapter emphasizes the partialness of operational definitions and how concepts so defined are modifiable in the light of new information.

We will also explore open concepts by tentatively reintroducing a feature of philosophical empiricism that was more prominent in the early days of American psychology: *conventionalism*. The primary example we use to elucidate conventionalism is factor analysis. Calling attention to the conventional features of factor-analytic models, such as the five-factor model of personality, illustrates their intrinsically provisional nature.

To set the stage, however, we will begin our analysis by giving a historical example of an open concept in the domain of psychopathology. The name of that concept is hysteria.

THE HYSTERIA PHENOTYPE: AN OPEN CONCEPT

The seventeenth-century physician Thomas Sydenham is considered the father of modern medical classification. Many of his ideas about classification were developed in response to his experience with epidemics, but Sydenham claimed that the chronic disease he encountered most often was hysterio-hypochondriasis (Trillat, 1995). Hysteria was typically diagnosed in women, hypochondriasis in men. In the eighteenth century they both became fashionable diagnoses, representing an increasing recognition of "psychiatric" maladies or "nervous distempers" that do not denote "insanity."

A large number of symptoms were included under the description of hysteria. "In its French fin-de-siècle incarnation, the hysteria diagnosis encompassed nearly the entire field of what twentieth-century psychiatry would call the neuroses, spilling over into other collateral clinical areas now claimed by the psychoses and various nonpsychiatric maladies" (Micale, 2008, pp. 147–148). The extended hysteria phenotype includes convulsions and fits, mutism, numbness and tingling, paralysis, tics, sensory hypersensitivity, vertigo, joint pain, swelling, gastrointestinal complaints, headaches, shortness of breath, anorexia, sexual problems, and fatigue. More psychologically there are intrusive thoughts, worries about health, anxiety, feelings of being overwhelmed, dysphoria, attention, concentration, and memory problems, fugue, cognitive diffuseness, suggestibility, emotional overreactivity, emotional lability, lack of emotional depth, egocentrism, dependency, deficient self-control, indiscriminate seductiveness, impulsivity, and manipulativeness.

As an explanatory construct, hysteria has a sorry history, long used to demean women and justify their mistreatment women (Chodoff, 1982; Kaplan, 1983; Showalter, 1985). Indeed, Micale (2008) notes that disorders commonly diagnosed in men, such as melancholia, hypochondria, and neurasthenia, were less stigmatized. Beginning in the seventeenth century, however, those who were inclined to define hysteria as a neurological disease tended to see these male-oriented syndromes as part of a larger hysteria family, and they were often in conflict with those who defined hysteria as a woman's disease.

The abandonment of the explanatory machinery associated with the hysteria concept (such as wandering wombs, vapors, and failed repression) is consistent with Paul Meehl's (1978) remarks about the instability of many psychological theories over the long term. According to Meehl, in every generation psychological topics become fashionable and spawn considerable scholarly activity – but over time the phenomena turn out to be more complicated than envisioned and competing theoretical perspectives proliferate. Meehl's concern was that eventually the topic falls out of fashion and psychologists move on to something else, but the something else does not typically represent cumulative progress on what came before.

Despite the abandonment of previous explanatory hypotheses for hysteria, if one examines the cluster of symptoms encompassed under the hysteria phenotype, most of them remain a part of the psychopathological landscape. Indeed, the phenotypic descriptions of thinkers such as Sydenham are still familiar today.

> Nor are they unhappy (due to physical symptoms only) for their minds are worse affected than their bodies, for an incurable desperation is mixed with the very nature of the disease ... fear, anger, jealousies, suspicions and worse passions of the mind, if any can be worse ... They adhere to no mean in anything, their only consistency is inconsistency ... Sometimes they love above measure, and presently they hate the same without any reason. (Sydenham, 1734, p. 306)

Commenting on the disappearance of hysteria in the late twentieth century, Micale (1993) claims that some of what was diagnosed as hysteria was redefined as epilepsy, or untreated syphilis, or prodromal psychosis. As these new syndromes were demarcated more explicitly, the clinician's working concepts of hysteria changed – indeed some of them dissipated like the morning dew. The introduction of panic disorder, generalized anxiety disorder, dissociative disorders, and trauma-related disorders –all capturing some part of the hysteria phenotype – altered the concept even further (Zachar, 2014). In the *Diagnostic and Statistical Manual of Mental Disorders*-III (DSM-III) of 1980, what was left of hysteria was fragmented into histrionic personality disorder and somatization disorder. In the DSM-III-R of 1987 and the DSM-IV of 1994 some diagnostic criteria were deleted or modified to better differentiate histrionic personality disorder from borderline and narcissistic personality disorder, which restricted the construct even further.

Blashfield, Reynolds, and Stennet (2012) observe that borderline personality disorder has become the new fashionable diagnosis among clinicians, supplanting histrionic personality in the clinic and the lab. Sydenham's description of hysteria, however, likely struck many readers as more like borderline personality disorder than hysteria, indicating that borderline traits were part of the larger hysteria phenotype before being redescribed as "borderline."

The persistence of the descriptive phenotype(s) in these diagnostic reformulations calls to mind another important idea promulgated by Meehl – that of open concepts. Hysteria certainly seems to have been an exemplary open concept. With its supposed chameleon-like nature and fluidity of symptoms over historical time, hysteria encompasses a large portion of mild and moderate psychiatric distress – a notion also reflected in concepts such as the general distress factor of the Minnesota Multiphasic Personality Inventory (MMPI; Welsh, 1956), trait neuroticism (Clark, 2005), and the general psychopathology factor or the "p" factor (Caspi et al., 2014).

OPEN CONCEPTS: DEFINITIONS THAT ARE NOT FIXED

According to Meehl (1986) in his extended autobiography, he learned about open concepts from his encounters with Arthur Pap in the early years of the Minnesota Center for the Philosophy of Science. The idea of open concepts, described as "vagueness in our constructs" (Meehl, 1986, p. 294), was introduced to psychology by Cronbach and Meehl (1955) in their groundbreaking article on construct validity – an extension of their earlier work as members of an American Psychological Association Committee on Psychological Test Standards (American Psychological Association, 1954). Meehl reports that he "initially fought an uphill fight against Cronbach and all (or almost all) the members of the committee to convince them about the acceptability of open concepts and the incorrectness of the strict operationist philosophy of science" (Meehl, 1986, p. 42).

Carnap's Partial Definitions

One way to understand an open concept is to examine a contrast concept, in this case a closed concept. An example of a "closed concept" is a triangle. A triangle is a two-dimensional figure with three straight sides and three angles. Having three straight sides and three angles is necessary and sufficient for triangularity. These are identity-determining features. Anything possessing them is a triangle. Anything lacking them is not a triangle. Definitions based on such identity-determining features are called essentialist definitions. With such a definition we can identify triangles with perfect accuracy. Any concept that can be so defined is considered *closed*.

What kind of definition is an *operational definition*? An example is "a substance is soluble if it dissolves when placed in water." Does this definition close off what is meant by soluble? No, for instance, if a substance is never placed in water, it still my be soluble. This particular operational definition defines solubility only for objects placed in water.

According to Rudolf Carnap (1936, 1937) dispositional terms such as "soluble" cannot be completely defined by such an operational definition. Other substances are "soluble" only in different solutes, such as alcohol or fat. Carnap referred to the operational definition of solubility as a test condition. A test condition for solubility is at best an incomplete or *partial definition*.

Carnap believed that because it was possible to propose new test conditions, operationally defined concepts remain open. Rather than specifying the essence of a concept, operational definitions are provisional and putatively extendable.

Pap's Open Concepts

Arthur Pap's (1953, 1958) work on open concepts was inspired by Carnap, but Pap extended Carnap's thinking in two ways. First, Pap said that any test condition can be seen as a probability indicator of the target hypothetical concept. Some tests are better indicators than others.

Second, he claimed that a strictly necessary versus contingent distinction cannot be applied to open concepts. In simple terms this means that, for open concepts, what we treat as a defining (essential, necessary) property rather than a contingent property is provisional. By essential property we mean identity-determining. It is necessary for anything to which the concept applies that the essential properties be present. For example, it is necessary that a triangle has three straight sides and three angles. By contingent property we mean a property that can occur, but need not occur.

For open concepts Pap rejected fixing defining properties as "necessary," favoring instead potential correction by future facts. For instance, the definition "bachelors are unmarried men" is a convention – a rule about what the term bachelor means. It fixes the meaning of bachelor, putting limits on what can be included in the concept. You cannot call a married man or an unmarried woman a bachelor.

Do the following statements have definitional roles that fix the meaning of terms analogous to "bachelors are unmarried men"? Intelligence is what is measured by the Wechsler Adult Intelligence Scale-IV (WAIS-IV); schizophrenia has a deteriorating course; and depression is a disorder of mood. Pap argued that these empirical definitions can be treated as fixed only on a provisional basis. Very importantly, the meanings of these concepts are subject to future correction in the light of new facts and are therefore open.

Next, we illustrate open concepts further by exploring schizophrenia and its diagnostic phenotype. Emil Kraepelin's definition of dementia praecox as a syndrome with a deteriorating course was the forerunner to schizophrenia. His definition was revived by Eli Robins and Samuel Guze (1970) when they proposed diagnostic criteria for schizophrenia formulated as a psychotic disorder with poor prognosis.

Assume that psychiatric nosologists adopt a convention in which they fix poor prognosis as definitional of schizophrenia. If poor prognosis is treated as a postulate, it places limits on the phenotype. For example, assume a young man is diagnosed with schizophrenia and later makes a full recovery. Someone who used deteriorating course as a definition would say that the young man's recovery demonstrates that he did not actually have schizophrenia. Recovery lies outside the limits of the concept for the phenotype. If this sounds fantastical, keep in mind that the concepts of schizophreniform disorder, brief psychotic disorder, and schizotypal and borderline personality disorder each split off different clusters of cases from the schizophrenia spectrum, in part to preserve the definitional role of deteriorating course (Zanderson, Henrikson, & Parnas, 2019).

In Robins and Guze's framework, deteriorating course is called a validator. Validators are properties correlated with a symptom cluster that are given a definitional role. Is this schizophrenia? Confirm that there is a deteriorating course. Is this major depressive disorder? Confirm that the symptoms occurred out of the blue or are out of proportion to the precipitants. Is this hysteria? Confirm that the patient uses repression as a primary defense.

For open concepts, however, definitions can always be rethought in the light of new evidence. For instance, by the early twenty-first century it became clear that people who met diagnostic criteria for Robins and Guze's poor prognosis syndrome had a broad range of outcomes, including full recovery (Bentall, 1990; Jablensky, 2010). If we treat poor prognosis as definitional of schizophrenia, then the conclusion is that the Robins and Guze criteria generate false positive diagnoses. If we do not treat poor prognosis as definitional, but rather see it as a contingent feature, then it is possible to modify the schizophrenia construct to allow a variable prognosis.

Interestingly, echoing claims about the disappearance of hysteria (Micale, 1993) and the death of histrionic personality disorder (Blashfield et al., 2012), several thinkers have recently been writing about the disappearance and/or slow death of schizophrenia (Bergsholm, 2016; Guloksuz & van Os, 2017). One of their core ideas is that the restricted, poor prognosis definition is a postulate and, in combination with other Kraepelinian postulates, it places overly stringent limits on the phenotype. As we have seen, the deteriorating course postulate potentially forbids recovery.

Another common postulate is "schizophrenia is a syndrome distinct from other psychiatric disorders." Under this postulate, the complex symptom pictures that include depressive episodes, manic episodes, and panic attacks that can be observed in some cases of schizophrenia are unexpected and therefore confusing.

Rather than keeping these two postulates fixed, another option is to incorporate the facts about variable outcomes and complex symptom pictures and revise the description of phenotype. Van Os and colleagues would also rename the redescribed phenotype *psychosis spectrum disorder*. Under this description, new postulates are introduced such as "psychotic disorders have variable outcomes."

Waismann's Open Textures

Carnap and Pap's writings are difficult to decipher for anyone without a solid background in modern logic. A more accessible articulation of open concepts – under the auspices of *the open texture of empirical concepts* – was offered by Friedrich Waismann in 1945. Waismann was also a member of the famous Vienna Circle that spawned logical positivism.

By open textures, Waismann meant that when scientists propose a method for measuring a concept (e.g., schizophrenia is measured by scale 8 of the MMPI) or discover a fact about a concept (e.g., diagnosed cases of schizophrenia vary in outcome), they are specifying through elaboration what the concept means. Because scientists can never exhaust either methods of measurement or correlated facts, the meaning of empirical concepts remains open for further elaboration. Although our focus in this chapter is on operational definitions, open textures more broadly apply to any concept encompassing facts about the world (i.e., empirical concepts).

Try as we may, no concept is limited in such a way that there is no room for any doubt. We introduce a concept and limit it in *some* directions; for instance, we define gold in contrast to some other metals such as alloys. This suffices for our present needs, and we do not probe any farther. We tend to *overlook* the fact that there are always other directions in which the concept has not been defined. And if we did, we could easily imagine conditions which would necessitate new limitations. (Mackinnon, Waismann, & Kneale, 1945, p. 123)

As, in fact, we can never eliminate the possibility of some unforeseen factor emerging, we can never be quite sure that we have included in our definition everything that should be included, and thus the process of defining and refining an idea will go on without ever reaching a final stage. In other words, every definition stretches into an open horizon. Try as we may, the situation will always remain the same: no definition of an empirical term will cover all possibilities. (Mackinnon et al., 1945, pp. 125–126)

From a Waismannian open-textures perspective, defining and redefining phenotypes is potentially open-ended. At any one time, fixing definitions may facilitate empirical progress, but that can always change.

Proposing definitions and holding to them come what may is a small-scale version of staying committed to a Kuhnian "paradigm" in the face of troubling anomalies. In the philosophy of science, a paradigm is a field-wide model such as the classification strategy adopted in the *Diagnostic and Statistical Manual of Mental Disorders*. Steven Hyman (2010) argues that the diagnostic categories of the DSM such as schizophrenia have become so reified (taken as fixed, unalterable matters of fact) that they are impeding empirical progress.

Borrowing some ideas from Imre Lakatos (1970), it turns out that holding to definitions may be a rational strategy because one should not always, especially in the short run, capitulate to critique. As befits the self-correcting process of science, however, it is also rational for researchers to exploit open textures in order to make progress. These two strategies of embracing and elaborating on concepts and replacing or revising concepts to accommodate new facts are in continual tension with each other.

RECOGNIZING SCIENTIFIC CONVENTIONS ILLUMINATES OPENNESS IN CONCEPTS

In the previous section we discussed the fixing of definitions as establishing conventions. There is a long history of conventionalism in the philosophy of science, pertaining largely to physics. Conventions are what resides in the decisions we make, not exclusively in the facts discovered.

One of the most respected scientists in history, Albert Einstein, is famous for proposing the theory of relativity. As both Fine (1986) and Norton (2010) argue, Einstein was able to work out the principles of relativity because he was not committed to the fixity of widely accepted definitions of scientific concepts such as space and time. Inspired by his reading of the philosophical empiricists David Hume and Ernst Mach, Einstein understood the sometimes arbitrary nature of such concepts.

The key insight that led Einstein to introduce the special theory of relativity in 1905 was related to his proposal about how to measure whether two very distant events occur at the same time. For everyday distances between the relatively slow-moving objects we observe on the Earth (where the velocity of light is irrelevant), every accurate observation leads to the same judgment about whether two events occur at the same time. Measuring simultaneity for events that are light years apart, however, alters our concept of simultaneity.

Einstein's measure of simultaneity was based on synchronizing two clocks, A and B, by sending a light signal between them. Simply put, if the light leaves A at noon and takes one hour to reach B, if you set B to 1PM when the light arrives, then the two clocks are synchronized. Einstein realized that for people in different relative states of motion, A and B may read 1PM at the same time (be simultaneous), A may read 1PM before B does, or B may read 1PM after A does. As a result, there is no absolute fact about whether two events occur at the same time for people in different frames of reference.

This realization on Einstein's part was an important event in the history of physics. His insight that whether two distant events occur at the same time was relative to one's inertial frame of reference meant that in order for the velocity of light to be constant, space and time themselves cannot be absolute, they must be relative to one's inertial frame of reference, i.e., the first part of his theory of relativity.

As discussed by Chang (2004), Einstein's measurement-based definition of simultaneity was what inspired Percy Bridgman to introduce operational definitions into our scientific methodology. Bridgman's own research program involved experimentally creating very high pressures. Because previously invented pressure gauges failed to work in these extremely high-pressure conditions, Bridgman had to invent new measures to determine how high his pressures were. As a result, Bridgman was attuned to the close relationship that exists between methods of measurement and scientific concepts. In the world of physics, Einstein's redescription of simultaneity and the nature of space and time was revolutionary, but Bridgman proposed that if concepts were defined only by the specific practices whereby they are "measured," new theories like Einstein's would not radically alter our understanding of scientific concepts.

For complicated reasons, Einstein claimed that his measure of simultaneity rested on a convention because it involved synchronizing two clocks by assuming a constant value for the velocity of light, but that value cannot be known unless two clocks were already synchronized. In Einstein's view, no possible discovery of fact could ever get us out of this vicious circle. Although no physicist questions the relativity of simultaneity, its conventionality is a matter of debate.

In promulgating his ideas about conventionality, Einstein harkened back to the earlier writings of Hume and Mach, as noted, but also Henri Poincaré. Poincaré proposed that scientists often rely on conventions that are taken to be factual descriptions of the world, but actually work more like definitions (Heinzmann & Stump, 2017; Poincaré, 1905/2007). Poincaré's example was Euclidean geometry, which seems to be a straightforward factual description of space, but once non-Euclidean geometries were proposed that were also empirically adequate, our ideas about geometry changed. An axiom such as "A straight line is the shortest distance between two points" is not dictated by *the facts*, it is a definition (postulate, convention) that fixes what we mean when we use the words straight line (Sklar, 2000). When using the Euclidean system, this convention facilitates the discovery of empirical facts about distances.

Our previous example regarding the role of deteriorating course in schizophrenia also touched on conventions. As a convention, deteriorating course plays multiple roles:

deteriorating course defines what schizophrenia is and deteriorating course is used to diagnose, measure, and validate schizophrenia. The use of diagnostic criteria that are sensitive and specific and the setting of diagnostic thresholds are also conventional in this way; they play both definitional roles and "measurement" roles. As Hanna van Loo and Jan-Willem Romejin (2015) argue, it is through such conventions that we are able to coordinate theoretical concepts such as schizophrenia with the factual structure of the world.

With respect to deteriorating course, there are persuasive reasons for embracing the alternative definition that allows for variable outcomes, but no privileged set of facts or crucial experiments compel us to decide it one way or the other. Each definition is underdetermined by the direct evidence. If so, it would be a mistake to conclude that it is an affront to science to hold to the deteriorating course definition rather than to embrace the variable outcomes definition, or vice versa.

The point we wish to emphasize in this chapter is that if conventions are recognized *as* conventions, they draw attention to the inherently open textures of empirical concepts. If certain conventions and other auxiliary assumptions are modified, other facts can become more salient. The facts thus discovered are perfectly good facts, but they are conditionalized, depending in part on the conventions adopted (Schaffner, 2012).

The scientific importance of recognizing openness in our constructs was nicely expressed in Einstein's 1916 obituary for Ernst Mach:

concepts that have proved useful for ordering things easily assume so great an authority over us, that we forget their terrestrial origin and accept them as unalterable facts. They then become labeled as "conceptual necessities," "a priori situations," etc. The road of scientific progress is frequently blocked for long periods by such errors. (Cited in Fine, 1986, pp. 15–16)

RETURNING TO PSYCHOLOGY WITH LESSONS FROM PHYSICS

Writing about the influential five-factor model of personality, McCrae and John (1992) claim that just as it is a fact that there are seven continents and eight classes of vertebrates, it is a fact that there are five personality factors: neuroticism (N), extroversion (E) openness to experience (O), agreeableness (A), and conscientiousness (C). In their view, the five-factor model has been replicated so many different times that it has been verified to be the correct description of structure of the personality phenotype.

Given that there are also empirically adequate models with both less than and more than five factors (Almagor, Tellegen, & Waller, 1995; Cloninger, 1998; De Raad et al., 2010), or alternative five-factor models that carve the space up differently (Jackson, 1994), many scientific psychologists are reluctant to accept such a metaphysical comparison between reality of factors and reality of continents. In this section we intend to argue that, even for the data sets that reliably produce the five-factor model as conceptualized by the *NEO Personality Inventory*, important details of that model partly depend on conventions. Although the model is highly meaningful, rather than corresponding to the correct structure of personality, it is best seen as scientifically expedient.

Before continuing, let us illustrate what we mean by meaningful but expedient. Turkheimer, Ford, and Oltmanns (2008) observe that a typical geographical map of the state of Virginia is divided into five geographical regions – Tidewater, Piedmont, Blue Ridge Mountain, Valley and Ridge, and Appalachian Plateau. There is a meaningful distinction to be made between the peaks of the Blue Ridge and the rolling plains of the Piedmont, but the precise line separating them is arbitrary. Within each region, the geography is very similar, but there is no geographical fact determining where the Blue Ridge region ends and the Piedmont begins any more than there is a geographical fact determining where China ends and Tibet begins or where "Palestine" ends and Israel begins. They are all practical categories – *meaningful but expedient* (and revisable).

As we saw earlier, Poincaré argued that mathematical models such as Euclidean geometry include conventions that are taken to be factual descriptions of the world, but are not. Some readers may infer that conventions and postulates are highly pertinent to purely logical, deductive disciplines such as geometry and theoretical physics but less relevant to empirical disciplines such as psychology. Yet as both methodology and statistical procedures are also largely logical-deductive, this is not necessarily the case.

An argument regarding the conventionality of multivariate statistical procedures for the description of complex psychological data, including phenotypes, was articulated by Louis Guttman (1954). Guttman's key insight was that different aspects of systems for data classification and description are constrained by observational data to different degrees and in different ways. The methods that interested Guttman – factor analysis and the related methods of multidimensional scaling and smallest space analysis – begin with a data matrix consisting of pairwise measurements of the similarity or dissimilarity of objects. The similarities might come from anywhere. They could be correlation coefficients between personality scales, clinician ratings of the similarity of pairs of patients, or experimentally obtained ratings of the similarity of visual stimuli. The goal of the multivariate statistical procedure is to model the pairwise similarities in a lower-dimensional space than the one in which they were originally observed.

The similarity matrix represents what is given to the scientist by the data, and its structure cannot be contradicted without violating a realist view of the world being described. The mountainous region of Virginia is closer to the Piedmont than it is to the ocean, and a geographic model that denies this fact is incorrect on realist grounds. The binding of the model to realism, however, is loosened as one proceeds.

The first decision (as opposed to discovery) to be made in representing the data involves how complex the model will be, which is specified in terms of its dimensionality. Here there is a tradeoff between simpler models with fewer dimensions that are more approximate and highly complex models with many dimensions that provide a closer fit the data. The decision made by the scientist is guided by the fit of the models (which can be estimated numerically) and one's preference for more or less measurement error (simple models liberally treat a lot of variance as error variance, complex models treat less variance as error but encounter the problem of overfitting in which error variance is incorporated into the model). Often, the choice of a particular balance of complexity and simplicity is guided by the practicalities of the task at hand. A flat map of Virginia is fine for many purposes, but for a geologist or an aviator a complex three-dimensional model might be more practically worthwhile – and even life-saving for a pilot.

Once the similarities have been established and a dimensionality selected, the model must be oriented. At this decision point, the constraints imposed by the data are loosened almost entirely. For instance, south-up maps work fine, and are even politically preferable to some groups living in the southern hemisphere. In the same way, when the whole Earth is mapped as a grid, geographers use parallel gridlines running from north to south (longitude) and parallel gridlines running east to west (latitude). This convention for mapping longitude and latitude in continuous space does not possess greater fidelity than would grids of the Earth oriented running from southeast to northwest and northeast to southwest. There is nothing about the continuous nature of the Earth that fixes north-south longitude and east-west latitude as the factually correct gridlines.

In factor analysis and related methods, this is the step of factor rotation, and it is not meaningfully constrained by the data: all factor rotations fit the data equally well. In factor analysis, it has become commonplace to simplify the choices involved in orienting a model by applying a conventional criterion called simple structure. Rotation to simple structure transforms the results of the factor analysis so all of the factor loadings lie as close as possible to one of the axes. The result is that indicators correlate highly with one factor and have very low correlations with all the other factors.

There are two practical advantages to this convention that make it useful. The first advantage is that it makes the multivariate structure more interpretable for humans. It is easier to decide what a factor measures when you have indicators that either correlate highly with it or not at all. A second advantage is that indicators with high correlations (or factor loadings) are expedient for constructing psychological test scales. You select those indicators with high correlations as items on the scale. The five-factor model of the personality is a trait model that has been rotated, expediently and conventionally, to conform as closely as possible to simple structure and the *NEO Personality Inventory* has psychological test scales representing each of those factors.

Here is the thing. The conventional aspects of factor analysis were obvious in the 1940s when factor analyses were computed by hand because the people doing the computing could see that once dimensions were extracted, they could be rotated in many ways. After the introduction of desktop computer programs such as SPSS, most psychologists' relationship with statistics changed. With drop-down menus, after the data file has been opened, with one click you select the data reduction family, with the next click you select the family of factor procedures, then you click on type of extraction and next type of rotation. With one last click, the results are churned out. This way of doing statistics makes rotation look like an inherent feature of the data, but it is not. With statistical tools for describing a large multidimensional space, these decision points (executed by the hands of the scientist not the procedure itself) become rather important.

Guttman's analysis of the relationship between theory (in the form of a statistical model) and empirical data produced one final insight that was crucial to the metatheory he eventually developed. Once a set of similarities is modeled in a space of a given dimensionality, but before any conventional structure is imposed on it, the scientist can observe one final data-driven aspect of the objects of study: their structure. Suppose one has decided on a two-dimensional model. When the objects of interest are located in the two-dimensional space, the shape of their relation to each other is fixed by the data, even though their orientation is not. So, for example, the model of the similarities might form a ring, which is a particular representation of how objects are like and unlike other objects. Such a ring, which Guttman called a circumplex, would represent a real characteristic of the data. The well-known color circle, a model of human color perception, is the most famous example of such a structure. In the personality domain, the interpersonal circumplex is a well-known alternative to the five-factor model as a means of modeling relations among personality traits (Wiggins, 1968).

Other two-dimensional structures, like the filled-in disk called a radex, model the structure of similarity in different ways. We will describe a radex structure for the five personality factors shortly.

Having established that the building of abstract models of real-world phenomena entailed activities that ranged from purely empirical to completely expedient, Guttman worked to develop a scientific methodology that would encompass both the discovery-driven and decision-driven components of model building. The result of his efforts was called facet theory, a metatheory that was designed to focus on the parts of modeling that were constrained by

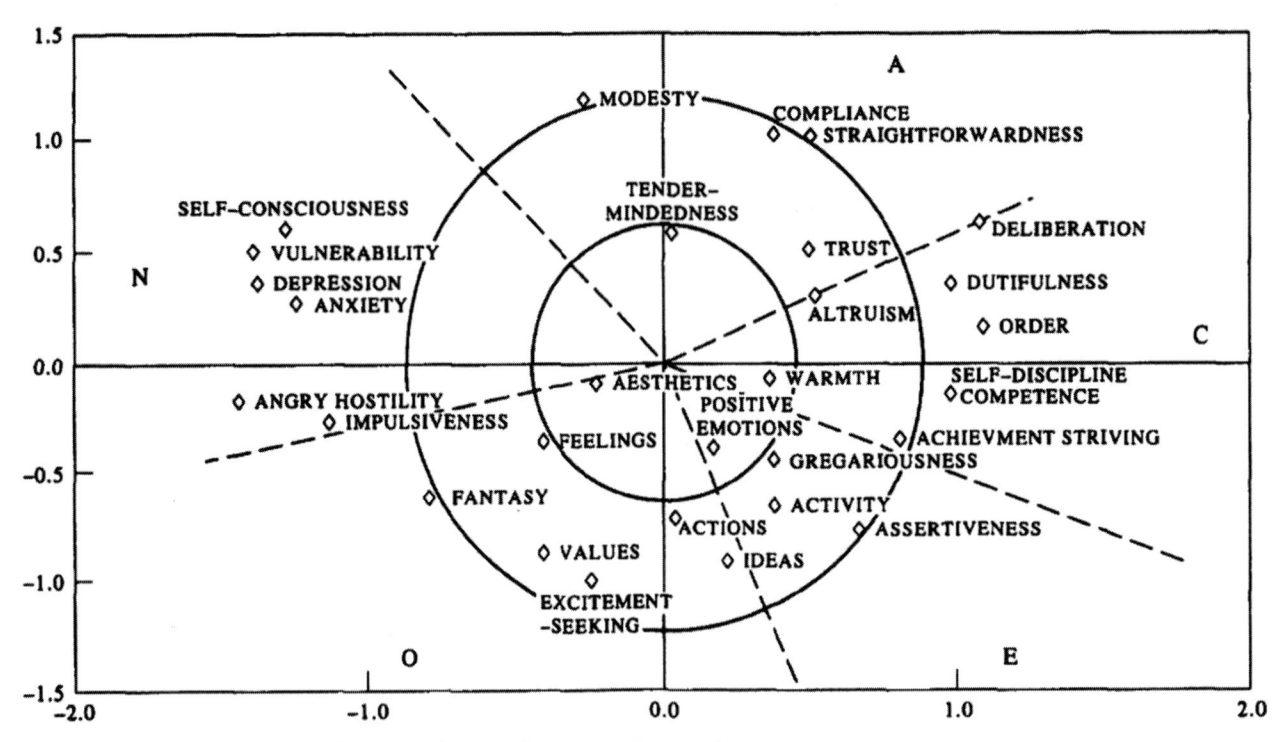

Figure 2.1 Two-dimensional NMDS for NEO-PI facet scales

data, while allowing the conventional aspects to vary according to the needs of the researcher. In facet theory, the investigator defines a domain of objects whose structure is to be understood, obtains similarity data for them, decides about an appropriate dimensionality for the modeled space, and observes the structure of the objects in the space once they are located there. No rotation or boundary drawing is involved.

In the prescribed schema for theory building (which has not become widely accepted) the scientist constructs "mapping sentences" that express hypotheses about the structure of the space and the relative proximity of the objects within the structure. For instance, hypothesizing that the personality domain is organized into five broad traits would be a mapping sentence. As most psychologists learn, the subscales that make up the five factors are called facets. For example, the facets of the extroversion factor include warmth, excitement-seeking, and positive emotionality. Hypothesizing that some of the facets constituting a trait have higher than average correlations with the other facets (centrality) is another mapping sentence. In this tradition, Maraun (1997) has shown that the facet scales of the five-factor model have a radex structure. His radex of the NEO Personality Inventory (NEO-PI) facet scales is presented in Figure 2.1.

In the radex, going clockwise, regions for the five factors are labeled A, C, E, O, and N. The factors occupy neighboring regions of the space, but rather than being represented by uncorrelated dimensions, factors that are more similar to each other lie next to each other within the circle. For instance, extraversion lies next to openness on one side and conscientiousness on the other, and is more similar to each of them than it is to neuroticism. Interestingly, in this structure, rather than belonging to extraversion, the excitement-seeking facet lies in the openness region and the warmth facet lies in the conscientiousness region. Also, the central facets of aesthetics and positive emotions have a higher than average number of correlations with the other facets.

In this radex, the relationships between the domains and the facets are not quite how they are portrayed in the standard five-factor model. The radex model captures a characteristic of personality that is lost in models based on uncorrelated dimensions – traits blend into each other continuously, with no up and down orientation and no boundaries between traits determined by data.

The radex is, accordingly, an empirically adequate description of the phenotypic space of personality. Therefore, any claim that five orthogonal factors represent the inherent structure of personality is treating a discovery partly dependent on conventions as a privileged factual description of the world. By recognizing that rotational indeterminacy signals a convention, psychologists can see the personality phenotype as having an open texture and not be bound by the limitations imposed by one meaningful, but expedient model. Indeed, in order to better coordinate theories and facts, we may decide that it is better to rely on alternative conventions.

ENDOPHENOTYPE: ALSO AN OPEN CONCEPT

Much of what we have been discussing would seem counterintuitive with respect to how people are taught about science from grade school onward. Measuring the speed of light from point A to point B rests on a convention. Schizophrenia necessarily involves an irreversible deterioration of functioning, unless we decide otherwise. The replicated structure of the five-factor model looks like a factual description of the personality phenotype, but incorporates arbitrary decision points. As the theory of open concepts might have counterintuitive implications, we will illustrate open concepts with one more example, this one taken from behavioral genetics.

Turkheimer and Gottesman's (1991) first law of behavioral genetics states that all human behavioral traits are heritable. A complex psychological trait is partly a result of many genes of small effect, but these genes are distributed across the genome. According to Turkheimer (2017), polygenes do not cluster together into a coherent unit that can be called a genetic mechanism. Genes are a part of the causal picture for complex traits, but not in the sense of constituting a *specific mechanism* for such traits.

This outcome has been dubbed a form of the gloomy prospect. The phrase "gloomy prospect" was initially used by Plomin and Daniels (1987) in reference to their dawning realization that in behavioral genetics, the variance associated with the nonshared environment (anything that makes two identical twins differ) might refer to unsystematic and idiosyncratic events. If so, there may be no environmental regularities (or laws) to discover.

In the early days of the human genome project, Turkheimer (2000) predicted that the gloomy prospect might also be applicable to the genetic variance. This would mean that no gene or set of genes are definitional of complex traits, i.e., necessary and sufficient for the presence of the trait. In 2016 he suggested that the gloomy prospect looked to have been fulfilled (Turkheimer, 2016). The results of genome-wide association studies (GWAS) indicate that genes are at best probability indicators (in Pap's sense).

Such causal complexity is another source of openness in concepts for complex psychological phenotypes. With so many varied inputs, a complex phenotype in the population is a family of similar but not identical behavioral patterns. The use of family resemblance models, prototypes, and radial categories in psychopathology are often responses to this openness (Parnas & Bovet, 2015; Schaffner, 2012; Schwartz, Wiggins, & Norko, 1995).

Interestingly, endophenotypes were introduced to the science of psychopathology by Gottesman and Shields (1972) in response to their early insights about polygenic determination, which suggested that complex phenotypes such as schizophrenia might be too complex to be defined by an identity-determining gene or fixed set of genes. However, they reasoned, perhaps a complex phenotype could be partitioned into underlying facets that are themselves the results of genes of large effect. With this idea in mind, Gottesman and Shields defined the endophenotype as a biological process that was correlated with the phenotype, but could be expected to be more genetically tractable, i.e., an intermediate between genes and phenotype that has a stronger genetic signal than the phenotype. For instance, increased cortisol release is a biological marker for stress. If also heritable, it would be an endophenotype in behavioral genetics.

It may surprise some, but the endophenotype concept is also an exemplary open concept. For instance, Gottesman and Shields (1972), Gottesman and Gould (2003), Cannon and Keller (2006), Kendler and Neale (2010), Lenzenweger (2013), Miller and Rockstroh (2013), and Gottesman and McGue (2015) proposed many different definitional criteria for endophenotypes. Our sample of these criteria include:

A biological marker associated with a phenotype that is that is heritable
Less genetically complex than the phenotype
Manifests in a person whether the disorder is present or not
Imperceptible to the naked eye (i.e., latent)
Shares a common genetic diathesis with associated disorders
Shares a common genetic diathesis with causal substrates of associated disorders
Associated with causes rather than effects

Not all of these definitions/validators align. According to the first two definitions, the endophenotype is a biological marker of the phenotype that has a stronger genetic signal than the phenotype. As noted, the GWAS paradigm has found that genetic variants each explain a very small percentage of variance for any complex psychological trait – less than 1 percent usually. More surprisingly, putative endophenotypes are often nearly as genetically complex as the phenotypes (Bilder, Howe, & Sabb, 2013).

So the closer the endophenotype is to the phenotype, the more genetically complex it is. As endophenotypes become more genetically tractable, they are often too remote from the phenotype to be diagnostically useful. This suggests that a diagnostically useful biomarker (definition 1) and a less genetically complex biomarker (definition 2) may be different things.

Consider also the criterion that the endophenotype manifests in the person whether the disorder is present or not. This defines the endophenotype as a risk factor. Kendler and Neale (2010) point out that the endophenotype as a liability risk factor for psychopathology versus the endophenotype as a mediator of psychopathology lead to different predictions about the effects of interventions on the endophenotype. In a risk-factor model, the endophenotype is correlated with the phenotype because it shares a common genetic diathesis, but it is not itself a causal factor in the manifestation of the target phenotype. Thus, intervening on the endophenotype should not alter the phenotype. In the mediator model, the endophenotype lies in the causal path between genes and phenotype, thus intervening on it could be expected to alter the phenotype.

It may even be possible, claim Kendler and Neale, that the term endophenotype could be applied to a biomarker that is somewhat heritable, but that more directly indexes an environmental event such as toxic exposure. Thinking along similar lines, Miller and Rockstroh (2013) claim that more advanced models of gene action do not require that the genetic contributions to the endophenotype be the most prominent causal factors.

Finally, one of the more promising endophenotypes is neuroticism (Goldstein & Klein, 2014). Neuroticism, however, is not a biomarker by any stretch of the imagination. Is this an extension too far? That is a matter of decision, not discovery. Neuroticism shares enough features with other heritable biomarkers that, for consistency's sake, the endophenotype concept has been extended to encompass it. This also illustrates both the conventional and open nature of the endophenotype concept.

Such openness is also illustrated in the following quote from Gottesman and McGue (2015, p. 1069): "The rules for defining an endophenotype are not yet ready to be carved into marble, and we expect them to evolve as the fields of psychology and of neuroscience evolve, always based on the best evidence available at the time." The vagueness in the endophenotype concept could be reduced by the promulgation of an official operational definition, but doing so would prematurely fix the meaning of endophenotype and constrain its scientific refinement.

DOES THE THEORY OF OPEN CONCEPTS UNDERMINE PSYCHOLOGY'S SCIENTIFIC ASPIRATIONS?

In the philosophy of science, the theory of open concepts represented a rethinking of operational definitions in light of the complexity and limitations of scientific practice. In psychology, open concepts were part of a critique of a strict operationalism, not of operationalism per se. Indeed, the theory of open concepts recovers two metatheoretical features of operational definitions that were deemphasized as they became a part of standard scientific practice in psychology: their conventional and provisional nature. Consider this quote by S. S. Stevens (1935) in his one of his articles introducing operational definitions to psychology:

Properly, a definition is the sum total of the criteria (operations) by which we determine the applicability of a term in any particular instance. These criteria are nothing more than social conventions which take into account the state of factual knowledge at a given time. It is for this reason that the discovery of [a] new related fact may make a revision of the criteria necessary so that we may include or exclude the new observation from the class denoted by the original definition. (Stevens, 1935, p. 519)

Writers have long discussed the differences between Bridgman's and Stevens' understanding of operationalism (Israel & Goldstein, 1944; Ribes-Iñesta, 2003; Walter, 1990). For Bridgman, operationalism was a turning away from absolutes and essences, toward practices and activities. For Stevens, operationalism was a method for establishing a public, intersubjective meaning for a concept. For both of them, however, *the operationalist attitude* was inherently contrary to reification. As noted by Stevens (1935, p. 19): "No concept can ever be allowed to congeal: operational procedures insure against fixation."

The problem is that once operational definitions became institutionalized in psychological research, they were taken literally, so that, for example, by schizophrenia we mean nothing other than meeting the six required DSM-5 diagnostic criteria. As a result, operational definitions became reified.

This was a mistake. As Schaffner and Tabb (2015) argue, operationalism does not require being resolutely atheoretical, or renouncing the organization of observable features using abstract principles. Understanding that no partial definition will be a content-valid representation of a concept and that concepts are the vehicles on which scientific theories ride are important auxiliaries to the use of operational definitions.

As we saw earlier, Pap held that some test conditions are better indicators of the target concept than others. He also held that consistency of results across different operational definitions provided evidence for treating those tests as independent measures of the same thing. The philosopher Herbert Feigl (1950) and his onetime student Paul Meehl took consistency of results as evidence that those tests are measuring something "real."

One of Meehl's (1990) favorite examples of "triangulating" toward reality was Avogadro's number – which refers to the approximate number of atoms in 12 grams of carbon, i.e., 6.02×10^{23} particles. No human can see atoms, nor can we ever count such a high number of anything in our lifetimes. However, a variety of independent formulae all triangulate on the same estimate for the number of atoms in a sample, suggesting that they are estimating the same thing.

By analogy, consistent estimates of an unobserved psychological construct can also be taken as measures of the same thing. For example, if the *MMPI-2* depression scale, the *Beck Depression Inventory*, and the *Hamilton Rating Scale for Depression* are highly correlated with each other and show a similar pattern of correlation with other measures, we might presume that these three scales are slightly different measures of the same thing.

Meehl's work was an understandable response to the problem of being overly literal about operational definitions, but one negative consequence of adopting such a scientific realist view of psychological measurement was that it initiated a slight shift in emphasis from the provisional and extendable to inferences about approximating a fixed reality, and the operationalist attitude's inherent resistance to reification was weakened in a different way. Once you have convinced yourself that you have a valid enough measure of the construct, you are likely to begin closing it off.

As we alluded to at in the opening pages of this chapter, the philosopher Jackie Sullivan (2014, 2017) and the psychologists Russel Poldrack and Tal Yarkoni (2016) argue that psychologists too readily assume that when different operational definitions putatively measure the same construct, variations between the measures are inconsequential. They each show how seemingly subtle variations in the measure can make results across labs hard to compare. In addition to the shared variance between measures, the unshared variance can be important as well – and what Chang (2009) calls the operationalist conscience would motivate us to conceptually understand the systematic differences between distinct measures. Finding better ways to uncover these systematic differences and understand their implications might be a worthwhile research goal.

A more recently proposed methodological solution to the proliferation of different ways of measuring the putatively same construct can be found in the variety of "ontologies" that have been proposed by those working in the area of informatics. Some names for these ontologies are the cognitive atlas (Poldrack et al., 2011), the cognitive paradigm ontology (Turner & Laird, 2012), and the human phenotype ontology (Köhler et al., 2017). An ontology is defined as a formal representation of the types of entities that exist in a domain, including their properties and their interrelationships.

Although such ontologies could be interpreted as *metaphysical ontologies*, we suspect that clinical scientists will be more inclined to view them as practical solutions to the massive proliferation of partial definitions that has ensued in a field committed to operationalism. The computer databases in which these ontologies are stored specify how the entity was measured, and thus can track differences that exist between distinct operational definitions that share the same construct name – and learn more about where the differences between operational definitions matter and where they do not. The hope is that, rather than putting constraints on the proliferations of operational definitions, large-scale data analysis can detect patterns that have thus far escaped our notice, and thus promote progress.

Paul Meehl believed that open concepts are more intrinsic to the subject matter of psychology than to the natural sciences – and this is one reason why defining and redefining seemingly goes on without end. It is not realistic to ever think that constructs for complex psychological phenotypes can be as well-demarcated as the fundamental particles of physics – they are too intertwined with each other. But that does not mean we cannot improve on past knowledge by finding order in the complexity, and transmit that improvement to the next generation.

We would like to conclude the chapter by proposing that open concepts are not a problem to be solved but that recognizing openness is part of a solution to the problem of an overly metaphysical embracement of concepts. The theory of open concepts with its emphasis on convention and open-textured alterability in the light of new facts is a counterweight to both less and more vigorous inclinations to fix the meaning of concepts for complex psychological phenotypes – in both its strict operationalist and scientific realist versions. As Chang (2004, p. 147) says about Bridgman's operationalism: "Bridgman was interested in advancing science, not in carping against it. Operational analysis is an excellent diagnostic tool for revealing where our knowledge is weak, in order to guide our efforts in strengthening it."

REFERENCES

Almagor, M., Tellegen, A., & Waller, N. G. (1995). The Big Seven Model: A Cross-Cultural Replication of the Basic Dimensions of Natural Language Trait Descriptions. *Journal of Personality and Social Psychology, 69*, 300–307.

American Psychological Association. (1954). Technical Recommendations for Psychological Tests and Diagnostic Techniques. *Psychological Bulletin, 51*(2, Pt. 2), 1–38.

Bentall, R. P. (Ed.) (1990). *Reconstructing Schizophrenia*. New York: Routledge.

Bergsholm, P. (2016). Is Schizophrenia Disappearing? The Rise and Fall of the Diagnosis of Functional Psychoses: An Essay. *BMC Psychiatry, 16*, 387. Retrieved from https://bmcpsychiatry.biomedcentral.com/track/pdf/10.1186/s12888-016-1101-5

Bilder, R. M., Howe, A. G., & Sabb, F. W. (2013). Multilevel Models from Biology to Psychology: Mission Impossible? *Journal of Abnormal Psychology, 122*, 917–927.

Blashfield, R. K., Reynolds, S. M., & Stennett, B. (2012). The Death of Histrionic Personality Disorder. In T. A. Widiger (Ed.), *The Oxford Handbook of Personality Disorders* (pp. 603–627). New York: Oxford University Press.

Bridgman, P. W. (1927). *The Logic of Modern Physics*. New York: Macmillan.

Cannon, T. D., & Keller, M. C. (2006). Endophenotypes in the Genetic Analyses of Mental Disorders. *Annual Review of Clinical Psychology, 2*, 267–290.

Carnap, R. (1936). Testability and Meaning. *Philosophy of Science, 3*, 419–471.

Carnap, R. (1937). Testability and Meaning – Continued. *Philosophy of Science, 4*, 1–40.

Caspi, A., Houts, R. M., Belsky, D. W., Goldman-Mellor, S. J., Harrington, H., Israel, S., … Moffitt, T. E. (2014). The P Factor: One General Psychopathology Factor in the Structure of Psychiatric Disorders? *Clinical Psychological Science, 2*(2), 119–137.

Chang, H. (2004). *Inventing Temperature: Measurement and Scientific Progress*. New York: Oxford University Press.

Chang, H. (2009). Operationalism. In E. N. Zalta (Ed.), *The Stanford Encylopedia of Philosophy*. Retrieved from https://plato.stanford.edu/archives/fall2009/entries/operationalism/

Chodoff, P. (1982). Hysteria and Women. *The American Journal of Psychiatry, 139*(5), 545–551.

Clark, L. A. (2005). Temperament as a Unifying Basis for Personality and Psychopathology. *Journal of Abnormal Psychology, 114*(4), 505–521.

Cloninger, C. R. (1998). The Genetics and Psychobiology of the Seven-Factor Model of Personality. In K. R. Silk (Ed.), *Biology of Personality Disorders* (pp. 63–92). Washington, DC: American Psychiatric Association.

Cronbach, L. J., & Meehl, P. E. (1955). Construct Validity in Psychological Tests. *Psychological Bulletin, 52*(4), 281–302.

De Raad, B., Barelds, D. P. H., Levert, E., Ostendorf, F., Mlačić, B., Blas, L. D., ... Katigbak, M. S. (2010). Only Three Factors of Personality Description Are Fully Replicable across Languages: A Comparison of 14 Trait Taxonomies. *Journal of Personality and Social Psychology, 98*(1), 160–173.

Feigl, H. (1950). Existential Hypotheses. *Philosophy of Science, 17*, 35–62.

Fine, A. (1986). *The Shaky Game: Einstein, Realism, and the Quantum Theory*. Chicago: University of Chicago Press.

Goldstein, B. L., & Klein, D. N. (2014). A Review of Selected Candidate Endophenotypes for Depression. *Clinical Psychology Review, 34*(5), 417–427.

Gottesman, I. I., & Gould, T. D. (2003). The Endophenotype Concept in Psychiatry: Etymology and Strategic Intentions. *American Journal of Psychiatry, 160*, 636–645.

Gottesman, I. I., & McGue, M. (2015). Endophenotypes. In R. L. Cautin & S. O. Lilienfeld (Eds.), *The Encyclopedia of Clinical Psychology* (Vol. II, pp. 1068–1075). Chichester: John Wiley & Sons.

Gottesman, I. I., & Sheilds, J. (1972). *Schizophrenia Genetics: A Twin Study Vantage Point*. New York: Academic Press.

Guloksuz, S., & van Os, J. (2017). The Slow Death of the Concept of Schizophrenia and the Painful Birth of the Psychosis Spectrum. *Psychological Medicine, 48*, 229–244.

Guttman, L. (1954). A New Approach to Factor Analysis: The Radex. In P. S. Laxarsfeld (Ed.), *Mathematical Thinking in the Social Sciences* (pp. 258–348). New York: The Free Press.

Heinzmann, G., & Stump, D. (2017). Henri Poincaré. In E. N. Zalta (Ed.), *The Stanford Encyclopedia of Philosophy*. Retrieved from https://plato.stanford.edu/cgi-bin/encyclopedia/archinfo.cgi?entry=poincare

Hyman, S. E. (2010). The Diagnosis of Mental Disorders: The Problem of Reification. *Annual Review of Clinical Psychology, 6*, 155–179.

Israel, H., & Goldstein, B. (1944). Operationism in Psychology. *Psychological Review, 51*(3), 177–188.

Jablensky, A. (2010). The Diagnostic Concept of Schizophrenia: Its History, Evolution, and Future Prospects. *Dialogues in Clinical Neuroscience, 12*(3), 271–287.

Jackson, D. N. (1994). *Jackson Personality Inventory – Revised*. London, Ontario: Research Psychologists Press.

Kaplan, M. (1983). A Woman's View of DSM-III. *American Psychologist, 38*(7), 786–792.

Kendler, K. S., & Neale, M. C. (2010). Endophenotype: A Conceptual Analysis. *Molecular Psychiatry, 15*(8), 789–797.

Köhler, S., Vasilevsky, N. A., Engelstad, M., Foster, E., McMurry, J., Aymé, S., ... Robinson, P. N. (2017). The Human Phenotype Ontology in 2017. *Nucleic Acids Research, 45*(Database issue), D865–D876.

Kroenke, K., Strine, T. W., Spitzer, R. L., Williams, J. B. W., Berry, J. T., & Mokdad, A. H. (2009). The PHQ-8 as a Measure of Current Depression in the General Population. *Journal of Affective Disorders, 114*(1), 163–173.

Lakatos, I. (1970). Falsification and the Methodology of Scientific Research Programmes. In I. Lakatos & A. Musgrave (Eds.), *Criticism and the Growth of Knowledge* (pp. 91–196). New York: Cambridge University Press.

Lenzenweger, M. F. (2013). Endophenotype, Intermediate Phenotype, Biomarker: Definitions, Concept Comparisons, Clarifications. *Depression and Anxiety, 30*(3), 185–189.

Mach, E. (1903). Critique of the Concept of Temperature (T. J. McCormack, Trans.). In P. Carus (Ed.), *The Open Court: A Monthly Magazine Volume XVII* (pp. 154–161). Chicago: The Open Court Publishing Company.

Mackinnon, D. M., Waismann, F., & Kneale, W. C. (1945). Symposium: Verifiability. Proceedings of the Aristotelian Society, Supplementary Volumes, 19, 101–164.

Maraun, M. D. (1997). Appearance and Reality: Is the Big Five the Structure of Trait Descriptors. *Personality and Individual Differences, 22*(5), 629–647.

McCrae, R. R., & John, O. P. (1992). An Introduction to the Five Factor Model and Its Implications. *Journal of Personality, 60*, 175–215.

Meehl, P. E. (1978). Theoretical Risks and Tabular Asterisks: Sir Karl, Sir Ronald, and the Slow Progress of Soft Psychology. *Journal of Consulting and Clinical Psychology, 46*(4), 806–834.

Meehl, P. E. (1986). *Paul E. Meel Extended Autobiography*. Retrieved from http://meehl.umn.edu/sites/g/files/pua1696/f/139autoextended.pdf

Meehl, P. E. (1990). Appraising and Amending Theories: The Strategy of Lakatosian Defense and Two Principles That Warrant It. *Psychological Inquiry, 1*(2), 108–141.

Micale, M. S. (1993). On the "Disappearance" of Hysteria. *Isis, 84*, 496–526.

Micale, M. S. (2008). *Hysterical Men*. Cambridge, MA: Harvard University Press.

Miller, G. A., & Rockstroh, B. (2013). Endophenotypes in Psychopathology Research: Where Do We Stand? *Annual Review of Clinical Psychology, 9*, 177–213.

Munafò, M. R., Nosek, B. A., Bishop, D. V., Button, K. S., Chambers, C. D., du Sert, N. P., ... Ioannidis, J. P. (2017). A Manifesto for Reproducible Science. *Nature Human Behaviour, 1*, 0021. Retrieved from https://doi.org/10.1038/s41562-016-0021

Norton, J. D. (2010). How Hume and Mach Helped Einstein Find Special Relativity. In M. Dickson & M. Domski (Eds.), *Discourse on New Method: Reinvigorating the Marriage of History and Philosophy of Science* (pp. 359–386). La Salle, IL: Open Court.

Pap, A. (1953). Reduction Sentences and Open Concepts. *Methodos, 5*, 3–30.

Pap, A. (1958). *Semantics and Necessary Truth*. New Haven, CT: Yale University Press.

Parnas, J., & Bovet, P. (2015). Psychiatry Made Easy: Operation (al)ism and Some of Its Consequences. In K. S. Kendler & J. Parnas (Eds.), *Philosophical Issues in Psychiatry III: The Nature and Sources of Historical Change* (pp. 190–212). New York: Oxford University Press.

Plomin, R., & Daniels, D. (1987). Why Are Children in the Same Family so Different from One Another. *Behavioral and Brain Sciences, 10*, 1–16.

Poincaré, H. (1905/2007). Non-Euclidean Geometries. *MacTutor History of Mathematics Archive*. Retrieved from www-history.mcs.st-andrews.ac.uk/Extras/Poincare_non-Euclidean.html

Poldrack, R., Kittur, A., Kalar, D., Miller, E., Seppa, C., Gil, Y., ... Bilder, R. (2011). The Cognitive Atlas: Toward a Knowledge Foundation for Cognitive Neuroscience. *Frontiers in Neuroinformatics, 5*(17). Retrieved from www.frontiersin.org/article/10.3389/fninf.2011.00017

Poldrack, R. A., & Yarkoni, T. (2016). From Brain Maps to Cognitive Ontologies: Informatics and the Search for Mental Structure. *Annual Review of Psychology, 67,* 587–612.

Ribes-Iñesta, E. (2003). What Is Defined in Operational Definitions? The Case of Operant Psychology. *Behavior and Philosophy, 31,* 111–126.

Robins, E., & Guze, S. B. (1970). Establishment of Diagnostic Validity in Psychiatric Illness: Its Application to Schizophrenia. *American Journal of Psychiatry, 126*(7), 983–986.

Schaffner, K. F. (2012). A Philosophical Overview of the Problems of Validity for Psychiatric Disorders. In K. S. Kendler & J. Parnas (Eds.), *Philosophical Issues in Psychiatry II: Nosology* (pp. 169–189). Oxford: Oxford University Press.

Schaffner, K. F., & Tabb, K. (2015). Hempel as a Critic of Bridgman's Operationalism: Lesson for Psychiatry from the History of Science. In K. S. Kendler & J. Parnas (Eds.), *Philosophical Issues in Psychiatry III: The Nature and Sources of Historical Change* (pp. 213–220). New York: Oxford University Press.

Schwartz, M. A., Wiggins, O. P., & Norko, M. A. (1995). Prototypes, Ideal Types, and Personality Disorders: The Return to Classical Phenomenology. In W. J. Livesley (Ed.), *The DSM-IV Personality Disorders* (pp. 417–432). New York: The Guilford Press.

Showalter, E. (1985). *The Female Malady.* New York: Pantheon Books.

Sklar, L. (2000). Convention, Role of. In W. H. Newton-Smith (Ed.), *A Companion to the Philosophy of Science* (pp. 56–64). Malden, MA: Blackwell.

Stevens, S. S. (1935). The Operational Definition of Psychological Concepts. *Psychological Review, 42*(6), 517–527.

Sullivan, J. A. (2014). Stabilizing Mental Disorders: Prospects and Promises. In H. Kincaid & J. A. Sullevin (Eds.), *Classifying Psychopathology: Mental Kinds and Natural Kinds* (pp. 257–281). Cambridge, MA: The MIT Press.

Sullivan, J. A. (2017). Coordinated Pluralism as a Means to Facilitate Integrative Taxonomies of Cognition. *Philosophical Explorations, 20*(2), 129–145.

Sydenham, T. (1734). *The Whole Works of That Excellent Practical Physician, Dr. Thomas Sydenham* (J. Pechey, Trans. 10 edn.). London: W. Feales.

Trillat, E. (1995). Conversion Disorder and Hysteria: Clinical Section. In G. E. Berrios & R. Porter (Eds.), *A History of Clinical Psychiatry* (pp. 433–441). London: The Athlone Press.

Turkheimer, E. (2000). Three Laws of Behavior Genetics and What They Mean. *Current Directions in Psychological Science, 9*(5), 160–164.

Turkheimer, E. (2016). Weak Genetic Explanation 20 Years Later: Reply to Plomin et al. (2016). *Perspectives on Psychological Science, 11,* 24–28.

Turkheimer, E. (2017). The Hard Question in Psychiatric Nosology. In K. S. Kendler & J. Parnas (Eds.), *Philosophical Issues in Psychiatry IV* (pp. 27–44). Oxford: Oxford University Press.

Turkheimer, E., Ford, D. C., & Oltmanns, T. F. (2008). Regional Analysis of Self-Reported Personality Disorder Criteria. *Journal of Personality, 76*(6), 1587–1622.

Turkheimer, E., & Gottesman, I. I. (1991). Is $H^2 = 0$ a Null Hypothesis Anymore. *Behavioral and Brain Sciences, 14,* 410–411.

Turner, J. A., & Laird, A. R. (2012). The Cognitive Paradigm Ontology: Design and Application. *Neuroinformatics, 10*(1), 57–66.

Van Loo, H. M., & Romejin, J.-W. (2015). Psychiatric Comorbidity: Fact or Artifact. *Theoretical Medicine and Bioethics, 36,* 41–60.

Walter, M. L. (1990). *Science and Cultural Crisis: An Intellectual Biography of Percy Williams Bridgman (1882–1961).* Stanford, CA: Stanford University Press.

Welsh, G. S. (1956). Factor Dimensions A and R. In G. S. Welsh & W. G. Dahlstrom (Eds.), *Basic Reading on the MMPI in Psychology and Medicine* (pp. 264–281). Minneapolis: University of Minnesota Press.

Wiggins, J. S. (1968). Personality Structure. *Annual Review of Psychology, 19,* 293–350.

Zachar, P. (2014). *A Metaphysics of Psychopathology.* Cambridge, MA: The MIT Press.

Zanderson, M., Henrikson, M. G., & Parnas, J. (2019). A Recurrent Question: What Is Borderline. *Journal of Personality Disorders, 33,* 341–369.

3 Building Models of Psychopathology Spanning Multiple Modalities of Measurement

NOAH C. VENABLES AND CHRISTOPHER J. PATRICK

Psychopathology research and clinical practice stand to benefit greatly from effective use of neurobiological methods and measures in efforts to understand the organization and bases of mental health problems. Such approaches can also aid in optimizing procedures for characterizing, preventing, and treating them. The field's growing interest in incorporating neurobiology into conceptualizations and assessments of psychological disorders is evidenced by major federal initiatives. For example, the National Institute of Mental Health's (NIMH) Research Domain Criteria initiative (RDoC; Insel et al., 2010; Kozak & Cuthbert, 2016) advocates for greater use of neuroscience methods and concepts in reshaping nosology of mental health and illness. Specifically, RDoC calls for an investigative focus on constructs such as acute threat, reward valuation, and response inhibition across multiple "units of analysis" (ranging from genes to brain circuitry to overt behavior) to advance biobehavioral conceptions and assessments of mental health problems. Parallel efforts are being advocated by the National Institute on Alcohol Abuse and Alcoholism (NIAAA; e.g., Kwako et al., 2016) and the National Institute of Drug Abuse (NIDA; e.g., Ramey, 2017) to help in explicating mechanisms of addition, addressing heterogeneity of substance use disorders (SUDs), and delineating individualized targets for treatment (i.e., precision medicine). Along similar lines, a recent report of the National Research Council (NRC, 2015) highlighted the need for incorporating data from brain and behavioral modalities into assessments of human capabilities in applied contexts (e.g., employment selection, military duty assignment). The shared focus of these initiatives is on characterizing individual differences in terms of cognitive, affective, regulatory, and social processes, assessed using measures from behavioral and physiological modalities along with report-based measures.

However, there is also growing concern about the "rigor and replicability" of neuroscientific findings for clinical phenotypes and associated trait constructs (Button et al., 2013; Ioannidis, 2011; Lilienfeld, 2017; Patrick & Hajcak, 2016; Tackett et al., 2017). Given this tension, a concrete methodological strategy is needed to systematize and coordinate efforts in the field directed at integrating neuroscientific methods and measures into psychopathology research and practice. In this chapter, we describe a *psychoneurometric* strategy for establishing models of core trait liabilities for psychopathology that span multiple modalities of measurement. This strategy uses experimental psychopathology and individual-difference assessment methods in combination to identify psychopathology-relevant variables from different modalities that covary across individuals, and clarify the nature of attributes underlying their covariance. Its focus is on producing reliable multimodal assessments of clinically relevant trait constructs with well-delineated nomological networks (i.e., patterns of convergent and discriminant relations with meaningful measures of other types), that have viable utility in applied settings – for example, in a precision medicine framework for psychopathology (see Insel & Cuthbert, 2015).

As referents for bridging between psychological report and neurobiological measurement modalities, and with measures from the modality of behavioral performance, the psychoneurometric approach focuses on *neurobehavioral (NB) trait* constructs: latent attributes with different observable manifestations, including self-reported proclivities, behavioral performance on neurocognitive or emotional task procedures, and measures of brain or other physiological reactivity in such tasks or at rest. NB traits differ importantly from traditional personality trait concepts in that they are informed by and incorporate data from modalities other than self-report. For example, in studies seeking to link impulsive traits to deficits in behavioral control, the traditional investigative approach has been to correlate self-reported traits with performance on tasks designed to index behavioral control within the laboratory. In such studies, impulsivity is typically assessed via self-report – either using a purpose-built measure of impulsivity and its facets (e.g., Whiteside & Lynam, 2001), or through use of traits from general personality frameworks such as the five-factor model (McRae & John, 1992) or the multidimensional personality model (Tellegen & Waller, 2008).

Studies focusing on associations between individual self-report measures of impulsivity and individual task performance measures of behavioral control or related capacities have generally reported weak and inconsistent convergence between the two (e.g., Creswell et al., 2019; MacKillop et al., 2016). In a review of findings from studies reporting correlations between individual measures of these two types, Sharma and colleagues (2014, p. 390) reported that "[t]he great majority of correlations [were] low; only six [were] above r = |0.30|, and only one was above r = |0.40|." The authors also reported that to the extent self-report and lab-task measures of impulsivity each predicted impulse-related clinical problems (i.e., substance use, aggression, delinquency, and gambling), their predictive relations were generally separate rather than overlapping. Sharma et al. concluded that self-report and laboratory task measures of impulsivity index differing aspects of impulse control problems.

However, a different picture of the intersection between impulsive-disinhibitory proclivities and lab-task measures of control capacity emerges from research on general proneness to externalizing problems (Krueger et al., 2007). Empirical findings have shown that this dispositional propensity is appreciably heritable (Krueger et al., 2002; Young et al., 2000) and that it relates to both impulsive and negative affective personality traits (e.g., Krueger, 1999; Venables & Patrick, 2012). Other research has shown that general externalizing proneness is associated with impaired cognitive processing of visual task stimuli as indexed by P3 brain response (e.g., Iacono et al., 2002; Patrick et al., 2006), and that this negative association is attributable mostly (if not entirely) to heritable variance in common between externalizing proneness and P3 responding (Hicks et al., 2007). Importantly, a large-sample twin analysis by Young et al. (2009) demonstrated that general externalizing proneness also relates negatively to scores on the common performance factor derived from task measures of inhibitory control – interpretable as indexing a common executive function (EF; see Miyake & Friedman, 2000) – due in large part to shared heritable variance. These converging lines of evidence suggest that an effective self-report index of general externalizing proneness, combining elements of impulsiveness and negative affectivity, could be expected to show negative associations with both P3 brain response and task-assessed EF function – potentially as a function of shared genetic influence. As we discuss later in this chapter, there is indeed compelling evidence for this (e.g., Brennan & Baskin-Sommers, 2018; Nelson, Patrick, & Bernat, 2011; Venables et al., 2018a; Yancey et al., 2013).

POTENTIAL ADVANTAGES OF A MULTIMODAL ASSESSMENT APPROACH

Of relevance to the current discussion is emerging evidence indicating that measures of a clinical construct from different measurement modalities can provide distinctive information at different levels of an underlying latent trait. Balsis et al. (2018) illustrated this using brain-structural, task-performance, and observer-rating indicators of Alzheimer's disease progression – showing that measures of adaptive functioning along with brain volume and metabolic activity variables provided enhanced precision in quantifying a latent dimension of Alzheimer's disease beyond commonly used cognitive performance measures alone. The implication is that effective quantification of a clinically relevant latent attribute may require indicators that provide nonredundant information. In this respect, multimodal assessments may be uniquely well-suited to characterizing continua of mental illness progression, from dispositional risk to incipient dysfunction to mild and more severe expressions of illness.

It seems likely that an important part of the distinctive information provided by nonreport-based (e.g., neurophysiological, task-behavioral) measures is their ability to index characteristics not accessible via self-report (e.g., structural brain anomalies indicative of an incipient disease state, in Balsis et al.'s [2018] illustrative example) or subject to response bias (e.g., socially undesirable propensities or attitudes). Information of these types provided by non-report-based indicators can be especially valuable in certain contexts. For example, given evidence that P3 amplitude and inhibitory task-performance measures each index genetic liability for externalizing problems, multimodal assessment batteries could be developed to quantify risk for problems of this type. Such an approach would be akin to routine medical screening assessments for heart disease that integrate lab test results (e.g., blood pressure and cholesterol levels) with family history and self-reported behaviors and lifestyle. Another application of assessments that include measures from non-report modalities would be in forensic or disability evaluations where strong motivation often exists to either deny or to feign or exaggerate symptoms.

Conversely, it can be useful to relate physiological or task-performance measures to well-established self-report scale measures of NB target constructs (e.g., inhibitory control), in order to clarify their psychological meaning and evaluate their specificity to a particular domain of clinical problems. For example, single neurobiological measures can include distinct components of variance related to different clinical problems or clinically relevant traits. This was demonstrated in recent work by Perkins and colleagues (2017), which illustrated how a single brain response measure, amplitude of P3 response to noise probes during viewing of affective pictures, contained separate portions of variance related in opposing directions to self-report-assessed NB traits of threat sensitivity and inhibitory control (inhibition–disinhibition). Results from this study provide evidence that report-based measures of NB traits can help to partition variance in neurobiological measures (e.g., via simple regression analysis, in this case) that relates to one clinical problem domain versus another (e.g., variance in noise probe P3 response related to fear fulness vs. externalizing). The psychoneurometric approach, described in more detail below, facilitates partitioning of variance in neurobiological or task-performance measures as related to different clinical problem domains by systematically mapping their patterns of associations.

CHALLENGES TO INTEGRATING NEURAL AND BEHAVIORAL MEASURES INTO CLINICAL CONCEPTUALIZATIONS AND ASSESSMENTS

Given the modest correspondence between measures from different assessment modalities, along with other conceptual and methodological challenges highlighted next, it is little wonder that limited progress has been made to date in linking psychological trait and clinical problem variables to neurobiological systems and measures, particularly for more applied uses. While initiatives that advocate for the integration of brain, physiological, and behavioral measures offer tremendous potential for advancing psychopathology research and practice beyond current diagnostic classification systems (see Clark et al., 2017), there are key issues needing to be addressed before neurobiological data can effectively shape how we characterize psychopathology or be utilized in practical clinical assessments for remediation or prevention purposes. One issue concerns the questionable or unknown psychometric properties of non-report, lab-task measures. The dominant experimental paradigm for studying biobehavioral aspects of psychopathology is to test for associations between clinical phenotypes (either trait liabilities or symptoms of mental health problems) and individual measures of physiological or behavioral response derived from laboratory tasks. However, lab-task measures have a number of documented weaknesses, including weak reliability relative to report-based measures, unknown or assumed construct validity, measurement error inherent to single measures, and problems in replicating small-N neuroscientific findings of associations between neurobiological measures and trait or clinical phenotypes (e.g., Lilienfeld & Treadway, 2016; Patrick et al., 2013).

Method variance, defined as variance unique to a particular modality of assessment, poses another challenge. In particular, method variance constrains the magnitude of associations between measures of a target construct from different measurement modalities. This is due to the fact that measures from a particular response modality contain variance *unrelated* to the target construct (i.e., method-specific variance). As a result, relations between measures from different modalities tend to be systematically lower than those between measures from the same modality (Campbell & Fiske, 1959). Measures of the same construct assessed in *different modalities* are expected to correlate moderately (0.3–0.5) rather than strongly (0.6 or higher). For example, reported fear of snakes might be expected to relate at this level to behavioral approach toward snakes. Critically, correlations are expected to be even lower (0.1–0.3 range) for measures of *only somewhat related* constructs from different assessment modalities. Given that lab-task-derived behavioral and brain measures index constructs only loosely related to those assessed by self- or other-report, correlations between report-based and lab-task measures are expected to fall within this modest (0.1–0.3) range. Indeed, effect sizes for relations between lab-task measures and report-based traits or clinical phenotypes in large-N investigations tend to be small in magnitude (e.g., Castellanos-Ryan et al., 2014). In general, the lack of attention to basic psychometric properties of neurobiological measures likely accounts to a substantial degree for the lack of progress made to date in identifying dependable non-report indicators – including biological indicators, or biomarkers – of psychological disorders or risk for such disorders.

PSYCHONEUROMETRIC RESEARCH STRATEGY

One research strategy that has been proposed for integrating measures from different modalities in clinical assessments is the *psychoneurometric* (PNM) approach (Patrick et al., 2013; Patrick, Iacono, & Venables, 2019; see also Patrick, Durbin, & Moser, 2012; Yancey, Venables, & Patrick, 2016). This research strategy conceives of NB traits as core biobehavioral dispositions that affect measurable responses in different modalities (e.g., scale-report, behavioral performance, and neurophysiology). As an illustration of this, Nelson et al. (2011) reported convergence among different brain-potential indicators of externalizing problems and, in turn, a common factor reflecting their shared variance was more robustly related to a scale measure of trait disinhibition than were individual neural indicators. Furthermore, when the scale measure was entered into a factor analysis along with the brain measures, a single common factor emerged on which all measures loaded, and scores on this factor showed robust associations with externalizing symptomatology as well as with separate neurophysiological criterion measures (see also Patrick et al., 2013). In addition to variance in common among a set of indicators, there is also potential for different subsets of neural or behavioral performance measures to account for different portions of variance in self-reported traits such as inhibitory control. For example, there is evidence that distinguishable control capacities are involved in regulating affective states versus inhibiting prepotent responses to accomplish task goals (see Nigg, 2017). Evidence of this kind can highlight the need for refinement of the *neurobehavioral construct* to include distinct thematic subcomponents (facets). The PNM approach can readily be extended to other measurement modalities including behavioral task performance to identify sets of converging indicators and characterize their associations with neural and self-report indicators of a target trait such as inhibitory control. This process of empirical mapping and integration provides a means for establishing multimodal assessment models for NB traits, which can be viewed as the *products* of the PNM research strategy. In the next section, we describe a multimodal measurement model for the construct of inhibitory control (inhibition–disinhibition) resulting from application of this research strategy, and describe evidence for its validity in relation to clinical and lab-task criterion measures.

MULTIMODAL, NEUROBEHAVIORAL TRAIT ASSESSMENT: THE EXAMPLE OF INHIBITORY CONTROL AS RELATED TO EXTERNALIZING PSYCHOPATHOLOGY

Given the aforementioned issues concerning multimodal assessments, we have advocated for a model-based analytic approach to integrating measures of NB traits from neurophysiological, behavioral, and report-based modalities (Patrick et al., 2012, 2013; Venables et al., 2018a; Yancey et al., 2016). Quantifying NB traits using multiple indicators from different modalities addresses the issue of method variance and reduces measurement error in scores, and allows for reshaping of the target individual-difference construct on the basis of observed indicators and their interrelations. The remainder of this chapter describes in detail our development of a model of inhibition–disinhibition which conceives of liability to externalizing problems in terms of variations along a continuum of inhibitory control capacity.

It is noteworthy that this dimensional liability conceptualization is compatible with movements in the field toward a dimensional classification system for psychopathology – as exemplified by the recently proposed Hierarchical Taxonomy of Psychopathology (HiTOP) model (Kotov et al., 2017). Current psychiatric nosologies that rely on consensus-based categorical diagnostic entities (i.e., the *Diagnostic and Statistical Manual of Mental Disorders* [DSM] and the International Classification of Disease [ICD]) have greatly impeded progress toward understanding mental health problems in neurobiological terms and establishing neurobiologically informed interventions (Krueger et al., 2018). By contrast, efforts directed at reconceptualizing psychopathology in neural-systems terms (e.g., RDoC) currently lack a systematic methodological strategy for linking process-oriented constructs such as acute threat or response inhibition to broad psychopathology liabilities and specific clinical expressions (Patrick & Hajcak, 2016). A multimodal NB trait assessment approach provides a means for addressing these issues by considering neurobiological and other non-report sources of information in explicating boundaries between normal and aberrant psychological functioning.

To illustrate our proposed strategy for building models of psychopathology that incorporate indicators from multiple modalities, we describe work directed at establishing a measurement model for the NB trait of inhibitory control. A key point of reference for this work was factor-analytic studies of the systematic comorbidity among common mental disorders that revealed evidence for a coherent *externalizing* factor, representing the systematic shared variance among impulsive antisocial behavior problems and substance-related disorders (e.g., Kendler et al., 2003; Krueger, 1999), as a counterpart to an *internalizing* comorbidity factor for fear, anxiety, and unipolar mood disorders. As noted earlier, behavior genetic research on the externalizing factor has provided compelling evidence for its substantial heritability (Kendler et al., 2003; Krueger et al., 2002; Young et al., 2000), leading to the idea of a broad latent liability for externalizing problems (Krueger et al., 2002; Patrick et al., 2012; see also Yancey et al., 2013).

As a basis for systematizing the assessment of externalizing problems and affiliated traits and quantifying dispositional liability for such problems in self-report terms, Krueger and colleagues (2007) developed the Externalizing Spectrum Inventory (ESI). The ESI includes 23 facet scales that load together onto a general disinhibition factor, with loadings strongest for facets of problematic impulsivity, irresponsibility/lack of dependability, thievery, and alienation. Scores on the ESI's general factor can be indexed effectively using a disinhibition (DIS) scale consisting of items from these highly loading facets, and ESI-DIS scores in turn correlate highly with interview-based assessments of externalizing symptomatology, in large part due to shared genetic influence (Yancey et al., 2013). The implication is that the ESI-DIS scale indexes the heritable liability for externalizing problems, which has been theorized to reflect brain-based impairments in inhibitory control (Iacono et al., 2008; Patrick et al., 2012).

NEUROBEHAVIORAL INDICATORS OF EXTERNALIZING PRONENESS

Besides contributing to impulsive proclivities and clinical problems such as antisocial behavior and substance abuse (Krueger et al., 2002), the latent externalizing liability is hypothesized to influence patterns of behavioral and brain responding on laboratory tasks. In this subsection, we review findings demonstrating that behavioral-performance measures from inhibitory control (or general executive function) task procedures and brain response measures index this same latent externalizing liability.

Individual differences in cognitive control capacity or executive function involve variations in the ability to regulate behavior in new or dynamic situations on the basis of internal representations of motivational and goal states (e.g., Miller & Cohen, 2001). In a review of task measures for indexing executive function, Miyake and Friedman (2012) presented evidence that scores from tasks of this type (i.e., memory updating, set shifting, and response inhibition) load together onto a common "executive function" factor – with some tasks (i.e., inhibitory control tasks such as Stroop, antisaccade, and stop-signal) loading exclusively on the EF factor, and others (i.e., memory updating and set-shifting tasks) loading onto both the EF factor and task-specific subfactors (see also Miyake et al., 2000). These findings indicate that response-inhibition tasks operate as relatively "pure" indicators of general EF, whereas memory-updating and set-shifting tasks contain subfactor-specific variance along with general EF variance.

The general EF factor of the Miyake and Friedman (2012) model of cognitive performance was linked to disinhibitory problems and traits by Young et al. (2009). Using two samples consisting of monozygotic and dizygotic twin pairs, these investigators showed that scores on a factor defined by clinical and trait indicators of disinhibition (akin to the general factor of the ESI) robustly predicted scores on an EF factor defined by performance on the abovementioned inhibitory control tasks (Stroop, antisaccade, stop-signal). Twin concordance (biometric) analyses revealed that scores on the disinhibition factor reflected both additive genetic and shared environmental influences, whereas scores on the lab-task-based EF factor were almost exclusively attributable to additive genetic influences (~98 percent). Importantly, the genetic correlation between the clinical-disinhibition and task-EF factors was similarly high in two separate adolescent cohorts (–0.60 and –0.61), indicating substantial overlap in genetic influences contributing to disinhibitory problems and lower EF capacity.

As noted earlier, other research has identified reduced amplitude of P3 brain response to visual task stimuli as a dependable neurophysiological indicator of proneness to externalizing problems. Phenotypically, reduced P3 response has been demonstrated in relation to both externalizing symptom scores (Patrick et al., 2006) and scores on the ESI-DIS scale (Yancey et al., 2013). Etiologically, P3 amplitude has been shown to prospectively predict externalizing problems (e.g., Iacono et al., 2002) and has been shown to relate to externalizing symptomatology and trait disinhibition predominantly as a function of shared genetic influences (Hicks et al., 2007; Yancey et al., 2013). Considered along with the findings of Young et al. (2009), these results provide compelling evidence that reduced P3 amplitude and EF-task performance operate as different-modality indicators of heritable disinhibitory liability, and suggested that brain and behavioral factors defined by variants of P3 and inhibition task scores, respectively, should covary with one another – and, in turn, with a self-report factor defined by scores on scale measures of trait disinhibition. In the next section, we report findings from work designed to test these hypotheses and establish a multimodal assessment framework for the construct of inhibitory control (inhibition-disinhibition).

Toward a Multimodal Multivariate Model of Inhibition–Disinhibition

Drawing on the foregoing lines of evidence and utilizing a PNM approach, we (Venables et al., 2018a) recently reported on the development of a multimodal measurement model for quantifying externalizing proneness in terms of variations in inhibitory control (inhibition-disinhibition). In doing so, we hypothesized that relevant measures from modalities of self-report, neurophysiology, and behavioral performance would operate as overlapping indicators of proneness toward externalizing problems. Specifically, we evaluated interrelations among 12 different indicators of inhibition–disinhibition that included four measures from each of three response modalities: self-report, behavioral performance, and brain response. Self-report indicators consisted of four scale measures of trait disinhibition consisting of a 20-item scale indexing the general factor of the ESI externalizing inventory, along with two measures of disinhibition composed of relevant items from other personality inventories, and a well-established measure of unsocialized-delinquent proclivities. The behavioral performance indicators included reaction time and accuracy measures from the following inhibitory control tasks shown to operate as indicators of general EF (Miyake & Friedman, 2012; Miyake et al., 2000): antisaccade, stop-signal, Stroop, and Go/No-Go. The brain-response indicators consisted of variants of P3 from three separate cognitive task procedures: a rotated-heads oddball task that has been used in many of the previously cited work on associations with externalizing, a flanker discrimination task, and a pseudo-gambling task that required choosing between two options on each of a series of trials.

We used factor-analytic techniques (both exploratory and confirmatory) characterize for patterns of covariance within each measurement modality, and across different modalities. First, we conducted exploratory factor analyses (EFAs) to establish that measures would cohere within each measurement modality to form a single factor reflecting the expression of inhibitory control in that modality. In each analysis, we found evidence for a single dominant factor. Loadings of scale indicators on the self-report modality factor were uniformly high (range = |0.67–0.88|, median = 0.81), whereas (consistent with prior work) loadings of behavioral and brain measures on their respective lab-task factors were more modest: behavioral performance measures (range = |0.27–0.59|, median = 0.48) and brain-response measures (range = |0.37–0.70|, median = 0.54).[1] Notably, *covariance across measurement modalities* was more reliable at the aggregate (i.e., factor score) level than at the level of individual measures. Specifically, the associations between factor scores

---

[1] Some indicators were coded such that higher scores reflected better inhibitory control, whereas others were coded in the direction of poorer inhibitory control; thus, correlations between some indicators were positive whereas others were negative.

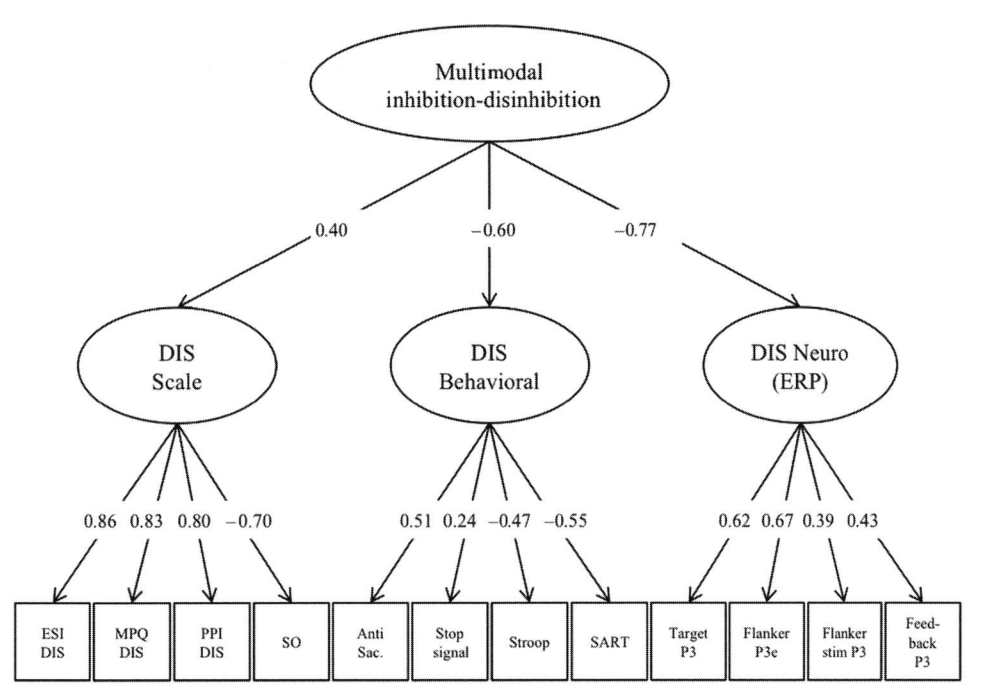

Figure 3.1 Results from a three-factor with one higher-order factor confirmatory structural model of inhibitory control versus disinhibition (DIS) as assessed by differing measurement modalities and a *multimodal* factor corresponding to the covariance across measurement modalities. Standardized parameter estimates are provided. The Scale factor loads positively onto the higher-order (Multimodal) factor because higher scores on this scale factor reflect higher disinhibition; the Behavioral and Neuro factors load negatively because higher scores on these factors reflect lower disinhibition (higher inhibitory capacity). ESI-DIS = Externalizing Spectrum Inventory Disinhibition scale; MPQ-DIS = Multidimensional Personality Questionnaire Disinhibition scale; PPI-DIS = Psychopathic Personality Inventory Disinhibition scale; SO = Gough's Socialization scale; AntiSac. = Antisaccade accuracy; Stop Signal = proactive inhibition from the Stop Signal task; Stroop = reaction time to incongruent stimuli from the Stroop task; SART = reaction time variability from the sustained attention to response (Go/No-Go) task; ERP = event-related potential; Target P3 = amplitude of P3 ERP to targets in the Oddball task; Flanker P3e = amplitude of P3 ERP following errors in the Flanker task; Flanker stim. P3 = amplitude of P3 ERP to arrow stimuli in the Flanker task; and Feedback P3 = amplitude of P3 ERP to feedback cues in the pseudo-gambling task
Adapted with permission from Venables et al. (2018a).

corresponding to the latent variable from each EFA were higher ($rs = |0.28$ to $0.21|$) than the magnitudes of bivariate correlations between individual indicators across modalities (median $rs = |0.10$ to $0.14|$). These results indicate that quantifying the covariance among indicators within each modality in terms of a common factor results in more robust and reliable associations across different modalities. Results from the initial series of EFAs highlight how stronger associations were observed across measurement modalities at the latent variable level as compared to the pattern of bivariate associations. This is likely due to enhanced construct validity at the latent variable level as a result of convergence among alternative relevant indicators and reduced measurement error.

Next, we utilized a confirmatory factor analytic (CFA) approach to specify a measurement model that included a higher-order factor reflecting individual differences in inhibitory control as assessed across all measurement modalities. A benefit of CFA techniques is the availability of full-information maximum-likelihood estimation to accommodate missing data for certain indicators arising from technical errors, lack of understanding, or poor task compliance on the part of subjects (e.g., as indicated by below-chance responding). The structural model depicted in Figure 3.1 is a higher-order correlated-factors model in which indicators from each response modality (self-report scales, behavioral performance, and neurophysiology) defined a lower-order factor within each measurement modality. These lower-order measurement modality specific factors then loaded onto a general inhibition-disinhibition higher-order factor, thereby quantifying degree of covariance observed across the three methods of measurement. This model fit the data well according to multiple quantitative indices and fit substantially better than a model that specified all 12 indicators as loading onto a single factor.

Our demonstration of a broad factor that accounted for covariance across assessment modalities accords with the view of inhibition–disinhibition as an individual difference characteristic that influences responses in different modes of measurement. More broadly, these findings have important implications for models of psychopathology that span multiple measurement modalities. One notable technical finding was that the single-factor model did not fit the data well. This highlights the need for operationalizing modality-specific variance as a first step toward building measurement models that explicitly span multiple assessment methods. While it has long been known (Campbell & Fiske, 1959) that measures of distinct types contain method-specific variance unrelated to the target of an assessment protocol (e.g., a clinical phenotype or psychological trait), analytic techniques for parsing construct-specific versus method-specific variance have not been systematically applied in research directed at integrating psychological and neurobiological data. Given that CFA approaches for specifying a multitrait-multimethod model suffer from a number of problems including underidentification (Kenny & Kashy, 1992), it will be important for quantitative modeling methods to be developed for quantifying method-specific variance in non-report-based measures and partitioning it from psychological construct variance. Given the current lack of analytic procedures for isolating method variance, one limitation of the Venables et al. (2018a) model is that the lower-order modality-specific factors include both modality-specific variance along with psychological construct (i.e., inhibition–disinhibition) related variance. It is only at the high-order level, where a general factor is specified that reflects variance shared across modalities, that a purer construct-specific representation emerges.

One conceptual implication of the Venables et al. (2018a) model is that the higher-order factor was weighted more heavily toward behavioral and brain-response modalities (–0.60 and –0.77, respectively) as compared to self-report (0.40). This suggests a tighter coupling of variance between behavioral performance and brain response than was observed for either behavior or brain response in relation to self-reported psychological experiences and attributes. As a result, scores corresponding to the general factor were only moderately associated with self-reported acts of crime/delinquency, substance use, and Cluster B Personality Disorder traits, but more robustly associated with neurophysiological criteria. However, an alternative model was fit in which the loadings of the lower-order factors onto the higher-order factor were constrained to be equivalent across the three modality factors. The model in which the higher-order factor was weighted equally across self-report, behavior, and brain response evidenced comparable fit to the model in which the loadings onto the higher-order factor were allowed to vary freely. Interestingly, scores from the constrained alternative model demonstrated higher predictive associations with the self-report clinical criterion measures.

A key implication is that alternative representations of inhibition–disinhibition can be specified depending on the particular assessment and prediction purpose. In the case of the unconstrained model, higher-order factor scores would have great utility in studies seeking to elaborate on the specific neural circuitry (e.g., through brain imaging techniques) of inhibitory control that is implicated in externalizing psychopathology. However, if the intended purpose is to aid in clinical diagnosis (e.g., in an applied clinical setting), scores from the constrained model would be expected to exhibit enhanced validity for predicting clinical criteria. As such, when building models of psychopathology that integrate across assessment modalities, it is critical at the planning stages of the investigation to identify an explicit goal tied to a specific prediction purpose (Messick, 1989).

Furthermore, there are interesting implications when examining results from each EFA model and the omnibus CFA model. The loadings for indicators on the self-report factor were uniformly high, indicating that these measures were largely interchangeable and that the scale factor provided good representation of what was shared among the four scale indicators. This high level of convergence can be attributed to the strong psychometric properties of these scales, each of which consisted of multiple items selected to cohere together around a clearly defined psychological attribute. As for the comparatively weaker loadings of task-derived measures on the behavioral performance factor, it appears that, despite their modest interrelations, they formed a coherent factor with good construct validity, as evidenced by predicted associations with both the self-report and brain-response factors. The fact that loadings for the brain-response indicators on their factor were somewhat higher than those for the task-performance indicators, but still appreciably weaker than for the scale indicators, likely reflects the fact that they were variants of a common brain response component (P3) but from different tasks.

In light of these findings, it may be useful to consider behavioral performance and brain response measures from different tasks as "items," akin to individual questions on self-report scales, rather than as "tests." To the extent that individual behavioral and brain indicators ("items") are less reliable, exhibit greater measurement error, and reflect multiple sources of variance, it can be expected that some level of aggregation will be required to concentrate their psychological construct "signal." Just as individual items from a self-report scale would not be used to index a target trait, and single symptoms would not be used to diagnose a clinical disorder, we believe that single brain or lab-task measures are best viewed as items in a psychological assessment context, and should be aggregated (e.g., through averaging, or latent-variable modeling) to index psychological attributes. Establishing reliable neurobiological "scales" composed of multiple converging indicators would facilitate subsequent investigations of cognitive and affective processes in psychopathology research and practice.

FUTURE DIRECTIONS AND CHALLENGES

In this chapter, we have described a psychoneurometric research strategy for building multimodal models of psychopathology that use neurobehavioral trait constructs to interface reported clinical problems with non-report-based (physiological response, task performance) measures. We have illustrated initial efforts toward building such multimodal models in the domain of externalizing psychopathology, in terms of a measurement model for the NB trait construct of inhibitory control capacity (inhibition–disinhibition). This work aligns with scientific initiatives that call for incorporation of neurobiological concepts and measures into novel models of psychopathology (e.g., RDoC; Kozak & Cuthbert, 2016), utilizing measures of different types to account for heterogeneity in clinical problems such as substance addiction (Kwako et al., 2016), and integrating data of different types into assessments of performance potential (NRC, 2015). We have recently detailed the procedural steps for undertaking a PNM research approach and for implementing multimethod NB trait-assessment protocols into applied clinical assessments (Patrick et al., 2019). Thus, the goal of the final subsections that follow is to highlight remaining conceptual issues and future avenues for research that can help to advance multimodal models of psychopathology and address challenges to implementing such models into applied practice.

Neurobehavioral Traits as Liabilities for Psychopathology

The development of a multimodal model of inhibition-disinhibition emerged out of research demonstrating a shared liability across externalizing disorders and relevant dispositional traits (Krueger et al., 2002), and other work indicating that P3 amplitude (Iacono, Malone, & McGue, 2003) and performance on inhibitory control-task procedures (Young et al., 2009) index this shared liability. The approach we used to develop this model of inhibition-disinhibition (Venables et al., 2018a) differs somewhat from the approach used to develop a counterpart model for the NB trait of threat sensitivity. The NB conception of threat sensitivity emerged out of work examining individual difference correlates of a distinct physiological index of defensive "fear" reactivity – namely, enhancement (potentiation) of the protective startle reflex during processing of aversive visual foreground stimuli. For example, Vaidyanathan and colleagues (2009) presented evidence that various fear and fearlessness scales previously shown to correlate with startle potentiation during aversive picture viewing covary with one another, and that it is this common component of variance that accounts for their relations with aversive startle potentiation (see also Kramer et al., 2012). Building on this work, Yancey and colleagues (2016) utilized a PNM approach to quantify threat sensitivity as the common factor emerging from a structural analysis of four indicators of defensive (fear) reactivity – a scale measure designed to index the common component of variance among different fear/fearlessness scales identified by Vaidyanathan et al., and three physiological indicators of negative emotional reactivity to aversive picture stimuli (startle potentiation, heart-rate acceleration, corrugator-muscle activation). Thus, the beginnings of a multimodal, NB trait model of threat sensitivity emerged out of work identifying a common component of variance among scale measures known to relate in common with a physiological index of defensive (fear) reactivity – aversive startle potentiation.

One key question regarding neurobiological investigations of psychopathology is the extent to which observed deviations in neural systems reflect liability versus manifestation of psychopathology. Given the evidence that disinhibition (Iacono et al., 2003; Krueger et al., 2002; Yancey et al., 2013) and threat sensitivity (Kramer et al., 2012) are heritable dispositions that interface with neurobiological systems, we theorize that these NB traits primarily reflect liabilities for externalizing and fear disorders, respectively. To test this hypothesis, Venables et al. (2017) examined the etiological overlap (genetic and environmental influences) among disinhibition and threat sensitivity with DSM-based fear, distress, and substance use symptom dimensions. NB traits were quantified as latent variable scores that included both self-report and psychophysiological indicators. Venables et al. (2017) found evidence for robust genetic overlap between disinhibition and substance-use disorders, and for threat sensitivity with fear disorders. Each NB trait also showed modest positive associations with distress-related clinical problems, as a function of independent genetic contributions. Furthermore, the NB traits synergistically (interactively) predicted distress problems such that individuals scoring concurrently high on both disinhibition and threat sensitivity exhibited greater endorsement of depressive mood and anxiety symptoms (see Carver & Johnson, 2018). As a follow-up to this study, Venables et al. (2018b) investigated the phenotypic and etiological overlap of disinhibition and threat sensitivity (using the same sample and measures as in the previous study) with suicidal behavior. Both NB traits were independently predictive of suicidal behavior (a composite of prior attempts and ideation), in each case largely as a function of genetic variance in common with suicidality. The NB traits also interactively predicted suicidal behavior, with individuals scoring high on both trait dimensions reporting the greatest levels of suicidal behavior.

Taken together, these results suggest that (a) NB traits of disinhibition and threat sensitivity reflect genetic liabilities for psychopathology and (b) these traits operate synergistically in predicting clinical problems involving heightened distress proneness, such as major depression and suicidality. It will be critical in future work to identify the mechanisms by which NB traits relate to different expressions of psychopathology. In the case of suicidal behavior, there is emerging evidence to suggest that disinhibition and threat sensitivity constitute distal risk factors that operate to enhance suicide risk by contributing to the emergence of promotive processes that are proximal and specific to deliberate self-harm (Buchman-Schmitt et al., 2017). Future work along these lines could aid in determining how trait liabilities ultimately manifest in different forms of psychopathology, and in identifying factors that enhance resilience to psychopathology when trait liabilities are present.

Biological Reductionism

A critical consideration when building multimodal models of psychopathology is to avoid the problem of biological reductionism (see Miller, 2010). Initiatives including NIMH's RDoC have been criticized for being overly reliant on neurobiological "units of analysis" without clearly articulating how or why neurobiological process constructs should relate to clinical problems, and for failing to consider other variables important to the development of psychopathology such as environmental and cultural influences (Lilienfeld, 2014). Others have argued against privileging either psychological or biological explanations in accounting for the emergence and maintenance of psychopathology, in light of evidence that most psychopathological conditions reflect the dynamic interplay of biological and psychological factors in the context of environmental and cultural influences (see, e.g., Borsboom, Cramer, & Kalis, 2019).

While RDoC shows caution in this respect by using the term "units of analysis" in place of "levels of analysis," thereby avoiding the notion of a hierarchical organization with "more versus less basic" factors, we advocate instead for the term *assessment modality* when referring to the use of different methods of measurement for operationalizing constructs of interest. In addition, our multimodal approach conceives of neurobehavioral traits as latent attributes with indicators in different modalities including self- (or other-) report, neurophysiology, and task performance. Rather than considering measures of one type as criteria and others as predictors, the multimodal approach utilizes measures of various types as indicators of a target construct. Once a multimodal measurement model has been established for a trait construct of interest, subsequent work can be undertaken to delineate other facets of the construct that lie outside the measurement model as specified, but that can be connected with it. For example, measures from the neuroimaging modality could be incorporated into the model to detail the neural circuitry implicated in a trait dimension such as inhibition-disinhibition (e.g., Bari & Robbins, 2013) or further refine it through the incorporation of measures of this type. As another example, a multimodal model could be revised to accommodate evidence for distinct facets of a target trait construct (e.g., distinguishable forms of inhibitory control including motoric, affective, and attentional; Nigg, 2017) by incorporating additional indicators to index these facets as subfactors within the model.

Score Estimation and Prediction Purpose

Future implementations of the PNM approach could move toward estimating a given individual's level of a latent trait (i.e., their theta level in Item-Response Theory [IRT] terms) based on the administration of a partial rather than full set of the indicators used in the model for that trait. This could facilitate applied clinical or personnel assessment, or serve as the basis for subject recruitment in resource-intensive efforts such as neuroimaging protocols, detailed ambulatory assessments, or large-scale etiologically informative studies (i.e., longitudinal, twin, adoption, etc.).

For example, results from the Venables et al. (2018a) model suggest that indicators from the modality of self-report are largely interchangeable, based on the uniformly high loadings (>0.8) of alternative scale measures onto the scale factor of the model. It would be quite tractable for large-scale projects to routinely include brief self-report measures of NB traits, to provide for a common data element across studies that vary in their focus (e.g., neuroimaging, etiologically informative designs, clinical samples, etc.). Once large normative datasets are established, a model-based latent-variable framework could allow for estimation of an individual's level along the latent trait dimension (e.g., inhibition–disinhibition) using a reduced set of indicators (e.g., self-report disinhibition scale along with oddball P3 response and antisaccade task performance) referenced to normative data for the full model. Furthermore, techniques such as multidimensional IRT could be used to characterize the extent to which measures from different modalities provide overlapping versus nonredundant information across the latent trait continuum (see, e.g., Balsis et al., 2018). Such analytical methods would be helpful for determining whether scores on a multimodal factor can be estimated most effectively based on data for a subset of indicators, and the degree to which multiple indicators from a particular modality (e.g., task performance or brain response) are necessary for accurate assessments due to provision of nonredundant information across differing levels of the latent trait.

Given that the goal of clinical science and practice is fundamentally to predict future outcomes and behaviors (e.g., likelihood of developing a disorder, expected course of illness, potential to respond to treatment, etc.), machine learning techniques (see Chapter 20) may prove useful as a complement to the analytic approach we have described in this chapter. For example, machine learning approaches could help to refine multimodal assessment batteries by optimizing the indicators (features) for specific prediction purposes (e.g., likelihood to respond to a given treatment).

Conclusion

In this chapter, we have described a psychoneurometric approach to building models of psychopathology that span multiple assessment modalities including self-reports, behavioral performance measures on cognitive and affective tasks, and brain response in task procedures. As a means for bridging between clinical problem and neurobiological assessment domains, we advocate for a focus on neurobehavioral traits – i.e., individual difference characteristics that systematically affect responses across different modalities of measurement. While our main focus in this chapter has been on one such trait, inhibition-disinhibition, and its relevance to externalizing psychopathology, this approach can readily be extended to other neurobehavioral traits and related psychopathologies. For example, threat sensitivity is relevant to focal fear disorders, reward sensitivity to mood disorders and substance misuse, and affiliative capacity to clinical problems involving callous disregard of others. Furthermore, the PNM approach we have described can be expanded to incorporate other assessment sources including molecular genetics and neuroimaging, in order to extend and refine multimodal psychopathology models for these and other NB traits.

The multimethod strategy we have described is intended to address current tensions in the field between initiatives that seek to establish multimodal models of psychopathology and concerns regarding basic measurement issues. It is our perspective that until the various issues we have highlighted are addressed, neurobiological measures will continue to have limited utility for practical clinical purposes, including the development of novel treatments. Systematic progress toward a neurobiologically informed science of psychopathology, and its application to clinical practice, will depend upon the development of novel strategies for integrating measures across different response modalities. We hope this chapter provides some useful pointers in this direction.

REFERENCES

Balsis, S., Choudhury, B. S., Geraci, L., & Patrick, C. J. (2018). Alzheimer's Disease Assessment: A Review and Illustrations Focusing on Item Response Theory Techniques. *Assessment*, 25, 360–373.

Bari, A., & Robbins, T. W. (2013). Inhibition and Impulsivity: Behavioral and Neural Basis of Response Control. *Progress in Neurobiology*, 108, 44–79.

Borsboom, D., Cramer, A., & Kalis, A. (2019). Brain Disorders? Not Really: Why Network Structures Block Reductionism in Psychopathology Research. *Behavioral and Brain Sciences*, 42, e2.

Brennan, G. M., & Baskin-Sommers, A. R. (2018). Brain-Behavior Relationships in Externalizing: P3 Amplitude Reduction Reflects Deficient Inhibitory Control. *Behavioural Brain Research*, 337, 70–79.

Buchman-Schmitt, J. M., Brislin, S. J., Venables, N. C., Joiner, T. E., & Patrick, C. J. (2017). Trait Liabilities and Specific Promotive Processes in Psychopathology: The Example of Suicidal Behavior. *Journal of Affective Disorders*, 216, 100–108.

Button, K. S., Ioannidis, J. P. A, Mokrysz, C., Nosek, B. A., Flint, J., Robinson, E. S. J., & Munafo, M. R. (2013). Power Failure: Why Small Sample Size Undermines the Reliability of Neuroscience. *Nature Reviews Neuroscience*, 14, 365–376.

Campbell, D. T., & Fiske, D. W. (1959). Convergent and Discriminant Validation by the Multitrait-Multimethod Matrix. *Psychological Bulletin*, 56(2), 81–105.

Carver, C. S., & Johnson, S. L. (2018). Impulsive Reactivity to Emotion and Vulnerability to Psychopathology. *American Psychologist*, 73, 1067–1078.

Castellanos-Ryan, N., Struve, M., Whelan, R., Banaschewski, T., Barker, G. J., Bokde, A. L., ..., & the IMAGEN Consortium (2014). Neural and Cognitive Correlates of the Common and Specific Variance across Externalizing Problems in Young Adolescence. *American Journal of Psychiatry*, 171(12), 1310–1319.

Clark, L. A., Cuthbert, B., Lewis-Fernández, R., Narrow, W. E., & Reed, G. M. (2017). Three Approaches to Understanding and Classifying Mental Disorder: ICD-11, DSM-5, and the National Institute of Mental Health's Research Domain Criteria (RDoC). *Psychological Science in the Public Interest*, 18, 72–145.

Creswell, K. G., Wright, A. G., Flory, J. D., Skrzynski, C. J., & Manuck, S. B. (2019). Multidimensional Assessment of Impulsivity-Related Measures in Relation to Externalizing Behaviors. *Psychological Medicine*, 49, 1678–1690.

Hicks, B. M., Bernat, E. M., Malone, S. M., Iacono, W. G., Patrick, C. J., Krueger, R. F., & McGue, M. (2007). Genes Mediate the Association between P3 Amplitude and Externalizing Disorders. *Psychophysiology*, 44, 98–105.

Iacono, W. G., Carlson, S. R., Malone, S. M., & McGue, M. (2002). P3 Event-Related Potential Amplitude and the Risk for Disinhibitory Disorders in Adolescent Boys. *Archives of General Psychiatry*, 59, 750–757.

Iacono, W. J., Malone, S. M., & McGue, M. (2003). Substance Use Disorders, Externalizing Psychopathology, and P300 Event-Related Potential Amplitude. *International Journal of Psychophysiology*, 48, 147–178.

Iacono, W. G., Malone, S. M., & McGue, M. (2008). Behavioral Disinhibition and the Development of Early-Onset Addiction: Common and Specific Influences. *Annual Review of Clinical Psychology*, 4, 325–348.

Insel, T. R., & Cuthbert, B. N. (2015). Brain Disorders? Precisely. *Science*, 348, 499–500.

Insel, T., Cuthbert, B., Garvey, M., Heinssen, R., Pine, D. S., Quinn, K., ... & Wang, P. (2010). Research Domain Criteria (RDoC): Toward a New Classification Framework for Research on Mental Disorders. *American Journal of Psychiatry*, 167, 748–750.

Ioannidis, J. P. A. (2011). Excess Significance Bias in the Literature on Brain Volume Abnormalities. *Archives of General Psychiatry*, 68, 773–780.

Kendler, K. S., Prescott, C. A., Myers, J., & Neale, M. C. (2003). The Structure of Genetic and Environmental Risk Factors for Common Psychiatric and Substance Use Disorders in Men and Women. *Archives of General Psychiatry*, 60, 929–937.

Kenny, D. A., & Kashy, D. A. (1992). Analysis of the Multitrait-Multimethod Matrix by Confirmatory Factor Analysis. *Psychological Bulletin*, 112, 165–172.

Kotov, R., Krueger, R. F., Watson, D., Achenbach, T. M., Althoff, R. R., Bagby, M., ..., & Zimmerman, M. (2017). The Hierarchical Taxonomy of Psychopathology (HiTOP): A Dimensional Alternative to Traditional Nosologies. *Journal of Abnormal Psychology*, 126, 454–477.

Kozak, M. J., & Cuthbert, B. N. (2016). The NIMH Research Domain Criteria Initiative: Background, Issues, and Pragmatics. *Psychophysiology, 53*(3), 286–297.

Kramer, M. D., Patrick, C. J., Krueger, R. F., & Gasperi, M. (2012). Delineating Physiologic Defensive Reactivity in the Domain of Self-Report: Phenotypic and Etiologic Structure of Dispositional Fear. *Psychological Medicine, 42*(6), 1305–1320.

Krueger, R. F. (1999). The Structure of Common Mental Disorders. *Archives of General Psychiatry, 56*, 921–926.

Krueger, R. F., Hicks, B. M., Patrick, C. J., Carlson, S. R., Iacono, W. G., & McGue, M. (2002). Etiologic Connections among Substance Dependence, Antisocial Behavior and Personality: Modeling the Externalizing Spectrum. *Journal of Abnormal Psychology, 111*, 411–424.

Krueger, R. F., Kotov, R., Watson, D., Forbes, M. K., Eaton, N. R., Ruggero, C. J., ... & Bagby, R. M. (2018). Progress in Achieving Quantitative Classification of Psychopathology. *World Psychiatry, 17*(3), 282–293.

Krueger, R. F., Markon, K. E., Patrick, C. J., Benning, S. D., & Kramer, M. D. (2007). Linking Antisocial Behavior, Substance Use, and Personality: An Integrative Quantitative Model of the Adult Externalizing Spectrum. *Journal of Abnormal Psychology, 116*(4), 645–666.

Kwako, L. E., Momenan, R., Litten, R. Z., Koob, G. F., & Goldman, D. (2016). Addictions Neuroclinical Assessment: A Neuroscience-Based Framework for Addictive Disorders. *Biological Psychiatry, 80*(3), 179–189.

Lilienfeld, S. O. (2014). The Research Domain Criteria (RDoC): An Analysis of Methodological and Conceptual Challenges. *Behaviour Research and Therapy, 62*, 129–139.

Lilienfeld, S. O. (2017). Psychology's Replicability Crisis and the Grant Culture: Righting the Ship. *Perspectives in Psychological Science, 12*, 660–664.

Lilienfeld, S. O., & Treadway, M. T. (2016). Clashing Diagnostic Approaches: DSM-ICD versus RDoC. *Annual Review of Clinical Psychology, 12*, 435–463.

McCrae, R. R., & John, O. P. (1992). An Introduction to the Five-Factor Model and Its Applications. *Journal of Personality, 60*, 175–215.

MacKillop, J., Weafer, J., Gray, J. C., Oshri, A., Palmer, A., & de Wit, H. (2016). The Latent Structure of Impulsivity: Impulsive Choice, Impulsive Action, and Impulsive Personality Traits. *Psychopharmacology, 233*, 3361–3370.

Messick, S. (1989). Validity. In R. L. Linn (Ed.), *The American Council on Education/Macmillan Series on Higher Education. Educational measurement*, (pp. 13–103). New York: American Council on Education & Macmillan Publishing Co.

Miller, E. K., & Cohen, J. D. (2001). An Integrative Theory of Prefrontal Cortex Function. *Annual Review of Neuroscience, 24*, 167–202.

Miller, G. A. (2010). Mistreating Psychology in the Decades of the Brain. *Perspectives on Psychological Science, 5*, 716–743.

Miyake, A., & Friedman, N. P. (2012). The Nature and Organization of Individual Differences in Executive Functions: Four General Conclusions. *Current Directions in Psychological Science, 21*(1), 8–14.

Miyake, A., Friedman, N. P., Emerson, M. J., Witzki, A. H., Howerter, A., & Wager, T. (2000). The Unity and Diversity of Executive Functions and Their Contributions to Complex "Frontal Lobe" Tasks: A Latent Variable Analysis. *Cognitive Psychology, 41*, 49–100.

National Research Council (NRC) (2015). *Measuring Human Capabilities: An Agenda for Basic Research on the Assessment of Individual and Group Performance Potential for Military Accession.* Committee on Measuring Human Capabilities, Division of Behavioral and Social Sciences and Education. Washington, DC: National Academy of Sciences.

Nelson, L. D., Patrick, C. J., & Bernat, E. M. (2011). Operationalizing Proneness to Externalizing Psychopathology as a Multivariate Psychophysiological Phenotype. *Psychophysiology, 48*, 64–72.

Nigg, J. T. (2017). Annual Research Review: On the Relations among Self-Regulation, Self-Control, Executive Functioning, Effortful Control, Cognitive Control, Impulsivity, Risk-Taking, and Inhibition for Developmental Psychopathology. *Journal of Child Psychology and Psychiatry, 58*, 361–383.

Patrick, C. J., Bernat, E., Malone, S. M., Iacono, W. G., Krueger, R. F., & McGue, M. K. (2006). P300 Amplitude as an Indicator of Externalizing in Adolescent Males. *Psychophysiology, 43*, 84–92.

Patrick, C. J., Durbin, C. E., & Moser, J. S. (2012). Conceptualizing Proneness to Antisocial Deviance in Neurobehavioral Terms. *Development and Psychopathology, 24*, 1047–1071.

Patrick, C. J., & Hajcak, G. (2016). RDoC: Translating Promise into Progress. *Psychophysiology, 53*, 415–424.

Patrick, C. J., Foell, J., Venables, N. C., & Worthy, D. A. (2015). Substance Use Disorders as Externalizing Outcomes. In T. Beauchaine & S. Hinshaw (Eds.), *Oxford Handbook of Externalizing Spectrum Disorders* (pp. 38–60). New York: Oxford University Press.

Patrick, C. J., Iacono, W. G., & Venables, N. C. (2019). Incorporating Neurophysiological Measures into Clinical Assessments: Fundamental Challenges and a Strategy for Addressing Them. *Psychological Assessment*. Retrieved from https://psycnet.apa.org/doiLanding?doi=10.1037%2Fpas0000713

Patrick, C. J., Venables, N. C., Yancey, J. R., Hicks, B. M., Nelson, L. D., & Kramer, M. D. (2013). A Construct-Network Approach to Bridging Diagnostic and Physiological Domains: Application to Assessment of Externalizing Psychopathology. *Journal of Abnormal Psychology, 122*, 902–916.

Perkins, E. R., Yancey, J. R., Drislane, L. D., Venables, N. C., Balsis, S., & Patrick, C. J. (2017). Methodological Issues in the Use of Individual Brain Measures to Index Trait Liabilities: The Example of Noise-Probe P3. *International Journal of Psychophysiology, 111*, 245–255.

Ramey, T. (2017). NIDA Phenotyping Battery [Webinar]. In *Clinical Trials Network Dissemination Library*, April 21. Retrieved from https://ctndisseminationlibrary.wordpress.com/2017/04/14/ctn-webinar-nida-phenotyping-battery-april-21-1-2pm-edt/

Sharma, L., Markon, K. E., & Clark, L. A. (2014). Toward a Theory of Distinct Types of "Impulsive" Behaviors: A Meta-Analysis of Self-Report and Behavioral Measures. *Psychological Bulletin, 140*, 374–408.

Tackett, J. L., Lilienfeld, S. O., Patrick, C. J., Johnson, S. L., Krueger, R. F., Miller, J. D., ... & Shrout, P. E. (2017). It's Time to Broaden the Replicability Conversation: Thoughts for and from Clinical Psychological Science. *Perspectives on Psychological Science, 12*(5), 742–756.

Tellegen, A., & Waller, N. (2008). Exploring Personality through Test Construction: Development of the Multidimensional Personality Questionnaire. In G. J. Boyle, G. Matthews, & D. H. Saklofske (Eds.), *Handbook of Personality Theory and Testing:*

Vaidyanathan, U., Patrick, C. J., & Bernat, E. M. (2009). Startle Reflex Potentiation during Aversive Picture Viewing as an Indicator of Trait Fear. *Psychophysiology, 46*, 75–85.

Venables, N. C., Foell, J., Yancey, J. R., Kane, M. J., Engle, R. W., & Patrick, C. J. (2018a). Quantifying Inhibitory Control as Externalizing Proneness: A Cross-Domain Model. *Clinical Psychological Science, 6*(4). Retrieved from https://doi.org/10.1177%2F2167702618757690

Venables, N. C., Hicks, B. M., Yancey, J. R., Kramer, M. D., Nelson, L. D., Strickland, C. M., … & Patrick, C. J. (2017). Evidence of a Prominent Genetic Basis for Associations between Psychoneurometric Traits and Common Mental Disorders. *International Journal of Psychophysiology, 115*, 4–12.

Venables, N. C., & Patrick, C. J. (2012). Validity of the Externalizing Spectrum Inventory in a Criminal Offender Sample: Relations with Disinhibitory Psychopathology, Personality, and Psychopathic Features. *Psychological Assessment, 24*, 88–100.

Venables, N. C., Yancey, J. R., Kramer, M. D., Hicks, B. M., Krueger, R. F., Iacono, W. G., Joiner, T. J., & Patrick, C. J. (2018b). Psychoneurometric Assessment of Dispositional Liabilities for Suicidal Behavior: Phenotypic and Etiological Associations. *Psychological Medicine, 48*, 463–472.

Whiteside, S. P., & Lynam, D. R. (2001). The Five Factor Model and Impulsivity: Using a Structural Model of Personality to Understand Impulsivity. *Personality and Individual Differences, 30*, 669–689.

Yancey, J. R., Venables, N. C., Hicks, B. M., & Patrick, C. J. (2013). Evidence for a Heritable Brain Basis to Deviance-Promoting Deficits in Self-Control. *Journal of Criminal Justice, 41*, 309–317.

Yancey, J. R., Venables, N. C., & Patrick, C. J. (2016). Psychoneurometric Operationalization of Threat Sensitivity: Relations with Clinical Symptom and Physiological Response Criteria. *Psychophysiology, 53*(3), 393–405.

Young, S. E., Friedman, N. P., Miyake, A., Willcutt, E. G., Corley, R. P., Haberstick, B. C., & Hewitt, J. K. (2009). Behavioral Disinhibition: Liability for Externalizing Spectrum Disorders and Its Genetic and Environmental Relation to Response Inhibition across Adolescence. *Journal of Abnormal Psychology, 118*, 117–130.

Young, S. E., Stallings, M. C., Corley, R. P., Krauter, K. S., & Hewitt, J. K. (2000). Genetic and Environmental Influences on Behavioral Disinhibition. *American Journal of Medical Genetics, 96*, 684–695.

PART II OBSERVATIONAL APPROACHES

4 The Conceptual Foundations of Descriptive Psychopathology

TREVOR F. WILLIAMS AND LEONARD J. SIMMS

When the answer cannot be put into words, neither can the question be put into words.
The riddle does not exist.
If a question can be framed at all, it is also *possible* to answer it.
– Tractatus Logicus-Philosophicus, Ludwig Wittgenstein (1922, proposition 6.5)

The act of description allows for language to serve as an intermediary between empirical observations and conceptual models, making it a critical process for science, clinical psychology, and humans more generally (Adams, Luscher, & Bernat, 2001; Sokal, 1974). Formal systems for describing the natural world emerged in ancient civilizations (e.g., Greece) and included psychological phenomena (e.g., Hippocrates' four humors); however, descriptive models became substantially more refined with the advent of science (Engel & Kristensen, 2013; Millon, 2004, pp. 16–17). For example, in the eighteenth century Carl Linnaeus iteratively constructed a taxonomy of the natural world, based on emerging botany and zoology research (e.g., species discoveries; Stearn, 1959). Importantly, other researchers continued to refine, test, and react to Linnaeus' taxonomy many years after his death (Hull, 1965), illustrating descriptive model research as a scientific enterprise (Hempel, 1965).

Despite the creation of scientific descriptive models, the value of such work has been underestimated at times. Scientists often prefer research that provides insights into causation and process-based mechanisms, which *seemingly* contrasts with descriptive science's focus on primarily observing phenomena (Grimaldi & Engel, 2007). Nonetheless, descriptive models have served important roles in many sciences. The work of Linnaeus and others in primarily descriptive sciences (e.g., anatomy) provided the foundation for early and contemporary advances in evolutionary theory (Stearn, 1959). Similarly, Mendeleev's periodic table of elements not only provided a systematic organizational framework, but instantiated natural laws and gave rise to predictions and discoveries in both chemistry and physics (Scerri, 2012). More abstractly, Sokal (1974) outlined three contributions descriptive models make in science: (a) portraying naturally existing systems, (b) allowing efficient communication and information retrieval, and (c) explaining interrelations among entities (e.g., individuals) such that general statements can be made about groups of entities. Although descriptive models vary considerably, both across and within scientific disciplines, it seems likely that scientific descriptive models can be broadly defined by their attempt to make at least one of these contributions, through both conceptual and empirical analysis.

Consistent with their value in other sciences, psychopathology researchers have also argued for scientific descriptive models (e.g., Clark, Watson, & Reynolds, 1995; Eaton et al., 2013; Millon, 1991). Blashfield and Draguns (1976) provide a particularly thorough discussion of the need for descriptive psychopathology models, citing varied functions that they serve: (a) communication between scientists and clinicians, (b) organization of theory and research to facilitate information retrieval, (c) clear articulation of similarities and differences among psychiatric patients, (d) allowing predictions about etiology, prognosis, and treatment based on existing knowledge, and (e) providing a foundation for theories of psychopathology (e.g., organizing explanatory principles). Additionally, such models also serve administrative (e.g., insurance) and socio-political (advocacy groups, public health initiatives, etc.) purposes (Langenbucher & Nathan, 2006). Thus, descriptive psychopathology models play an integral role in psychopathology research, applied work, and public policy.

Paralleling biological taxonomy, descriptive psychopathology models often develop iteratively; however, psychopathology taxonomy more strongly depends on measurement as an intermediary between the theorized model and evidence that might lead to its revision (Kline, 1998). In psychopathology research, theorized models often include an entity that cannot be directly observed or definitively operationalized, typically referred to as a "hypothetical construct," and instead must be measured by an instrument whose meaning is inferred from a network of relations to direct observations (e.g., behavior) and other hypothetical constructs (e.g., building a

Table 4.1 Summary of descriptive psychopathology approaches

Question	Clinical theory	Descriptive psychiatry	Quantitative models	Biological models
What is the model's purpose?	Facilitate psychosocial treatments	Coordinate scientists, clinicians, and society	Map observed covariation	Provide biological explanation
What are the most basic observed units?	Observed behavior and subject reports	Observed behavior and subject reports	Observed behavior and subject reports	Physical observations of subjects
How are observed units typically assessed?	Unstructured interviewing and some standardized assessments	Clinicians: unstructured interviewing; Researchers: structured interviewing and questionnaires	Structured interviewing and questionnaires	External and internal quantitative-physical assessments
How are observations organized?	Theory-specific psychological structures and processes	Atheoretical diagnostic categories	Latent variables	Physical structures and processes
How was the organization originally developed?	Specified a priori based on clinical experience	Specified a priori based on clinical experience	Exploratory latent variable modeling	Observations of individual physical structures and processes
How has the model been evaluated?	Therapy efficacy, studies of individual theoretical concepts	Expert consensus	Statistical modeling and construct validation	Varied levels of physical granularity

nomological network; Cronbach & Meehl, 1955). Notably, the very process of developing such an instrument and elaborating its nomological network (e.g., construct validation) may alter an investigator's understanding of the phenomenon that they wished to measure in the first place (Loevinger, 1957; Patrick et al., 2013). The process of construct validation often has been described in the context of developing psychological tests (e.g., Simms & Watson, 2007); however, Skinner (1981) used the logic of construct validation to create a metatheory of descriptive psychopathology model development, noting three steps: (a) *theory formulation* (delineate content domain, define classification model, and specify relations to external variables), (b) *internal validation* (operationalize constructs and statistically model their internal structure), and (c) *external validation* (predictive validity and clinical utility). Skinner (1981) noted that this process may be iterative, in the sense that empirical results at any stage may lead to the modification of either the operationalized taxonomy (e.g., adding questionnaire items; Simms & Watson, 2007) or the underlying theoretical model.

Skinner's (1981) metatheory is useful in that it explains how descriptive psychopathology models are developed based on the explicit and implicit decisions made by theorists and researchers, potentially providing insight into why models differ (see Table 4.1). To start, theorists and researchers establish goals for their models, whether it be predicting outcomes (e.g., recidivism), guiding treatment, or informing theories of psychopathology (Dowden & Brown, 2002; Ramnerö & Törneke, 2008). Beyond this, decisions also must be made regarding what counts as an observation in the model (e.g., symptoms; Millon, 1991) and how such observations should be ideally operationalized (e.g., self-report). More substantively, it must be decided how observations are interrelated or organized (e.g., latent variables) and the extent to which this framework (e.g., structure) is specified a priori or later determined by data-driven analyses. Finally, scientific descriptive models need to be empirically evaluated; however, the developers of the model may choose to weight one or more methods of evaluation (e.g., factor analysis) more than others (Wright & Zimmerman, 2015).

Ideally, the myriad decisions described in the preceding paragraph and in Table 4.1 should be grounded in basic science, methodological research, and a sound philosophy of science. However, historical, professional, and political factors likely also affect the choices made by individual researchers. Specifically, a theorist or researcher's training lineage or involvement in a specific science or practice community directs their conceptual focus and methodological preferences (Kuhn, 1977, pp. 294–297; Wiggins, 2003). An essential thesis of this chapter is that communities of psychopathology theorists and researchers use diverging descriptive concepts (core beliefs, symptom dimensions, etc.), assessment methods (self-report, neuroimaging, etc.), and data analysis techniques (*t*-tests, factor analysis, etc.), complicating communication among communities and slowing scientific progress. We aim to illustrate this by reviewing four prominent approaches to descriptive psychopathology: Clinical Theory, Descriptive Psychiatry, Quantitative Models, and Biological Models. We further compare methods among these approaches, using the diagnosis of borderline personality disorder as an example to demonstrate similarities and differences. We conclude by considering strategies for integrative descriptive psychopathology practice and research.

APPROACHES TO DESCRIPTIVE PSYCHOPATHOLOGY

Although early descriptive psychopathology models existed in ancient civilizations and endured for centuries, the rapid growth of science and emergence of rigorous biological taxonomies in the seventeenth and eighteenth centuries set the stage for their replacement (Kendler, 2009; Millon, 2004). In this section of the chapter, we review the modern descriptive psychopathology approaches that followed. Specifically, we begin with the Clinical Theory and Descriptive Psychiatry approaches, which emerged in the eighteenth and nineteenth centuries, then move to the more recently developed Quantitative and Biological modeling approaches (Hothersall, 2004, pp. 277–297; Pilgrim, 2007). These four approaches and the communities that adopted them are neither perfectly homogenous nor distinct from one another; however, through focusing on core features, we hope to illustrate important distinctions between them.

Clinical Theory

All major schools of psychotherapy address clinical description to some extent, as this is necessary for understanding clients' presenting problems, consulting with colleagues, and communicating about the therapy process (Eells, 2007; Langenbucher & Nathan, 2006). Clinical descriptive models typically share the following qualities: (a) detailed description of within-person patterns in thoughts, feelings, and behaviors over time; (b) a focus on client reports and observed in-session behavior, describable in first-person language (e.g., "*I* had the thought that..."), as the base units of description; (c) theoretical concepts (e.g., schemas) that often are idiosyncratic to specific therapies and derived from a mixture of clinical observation and basic psychological science; and (d) person-specific case conceptualizations that integrate observations and theoretical concepts (Fisher, Newman, & Molenaar, 2011; Kendler, 2014). Clinical theorists describe individual differences through comparing individuals in terms of single theoretical concepts (e.g., core beliefs; Fournier, DeRubeis, & Beck, 2012) or case conceptualizations (e.g., Linehan, 1993). Single clinical theoretical concepts, however, often focus on one source of individual differences (e.g., thoughts vs. behavior) and, in contrast, case conceptualizations may be too complex to systematically compare. Thus, although clinical theorists offer rich and clinically useful descriptive models, they tend to rely on other approaches (e.g., Descriptive Psychiatry's diagnoses) to describe individual differences in psychopathology comprehensively and parsimoniously, which are useful for connecting clinical practice to relevant research (Dudley, Kuyken, & Padesky, 2011).

Practitioners typically assess clinical descriptive models through unstructured observations and interviewing, though structured formats for integrating this information do exist (e.g., Beck, 2011). Beyond this, there are self-report instruments and a limited number of structured interviews that assess components of clinical descriptive models; however, many of these instruments are relatively new and used inconsistently (e.g., Baer et al., 2006; Lee et al., 2017). The reliance on less standardized methods of assessment may in part be due to some clinical theory constructs (e.g., core beliefs) and processes (e.g., psychological flexibility) being difficult to operationalize (Hayes, Strosahl, & Wilson, 2012). Notably, methods in experimental psychopathology and intensive repeated measures studies may offer methodological advances that allow more standardized assessments of such constructs in the future (Lazarus et al., 2014; see also Parts III and VI on these topics). General evidence for clinical theory descriptive models may be gleaned from treatment efficacy and treatment mechanism research, which can suggest the overall utility of these models. Beyond this, basic research is sometimes used to more directly examine clinical theory descriptive models; however, such studies typically only examine one aspect of a descriptive model (e.g., Fournier et al., 2012). These methods of evaluation are arguably limited, in that rigorous examinations of the full descriptive model, or multiple components of the model, are rare. Nonetheless, indirect and direct evidence often can be found for varied *individual* components of clinical theory descriptive models in laboratory studies using simple univariate tests, ecological momentary assessment studies using multilevel modeling, and cross-sectional factor-analytic studies.

Clinical theories of borderline personality disorder (BPD) illustrate many of the aforementioned qualities. Kernberg (1984), Linehan (1993), Young and colleagues (2003), and Bateman and Fonagy (2013) all provide rich descriptions of interacting structures (e.g., schemas) and processes (e.g., hypermentalizing) that exist within individuals and that are the focus of clinical interventions (e.g., interpersonal effectiveness skills). Oftentimes, these descriptive models initially were developed based on clinical observations and by importing theories from other areas of psychology (e.g., cognitive; Linehan, 1993; Young et al., 2003); however, the constructs and processes within these models also have begun to accrue their own literatures (e.g., Crowell, Beauchaine, & Linehan, 2009). Despite this, the full descriptive models implied by these theorists are not typically examined, likely due to model complexity and difficult-to-operationalize constructs (e.g., interacting schemas, operations, modes, and coping styles; Young et al., 2003). Furthermore, clinical theories of BPD inherently rely upon psychiatric diagnoses to describe between-person differences parsimoniously and contextualize the theory in relation to other forms of psychopathology (e.g., Linehan, 1993; Young et al., 2003; though see Kernberg, 1984).

Clinical theories of BPD illustrate the ability of this approach to provide detailed individual descriptions with direct connections to therapeutic action, qualities that are important to frontline practitioners. Despite this utility, clinical BPD theories also demonstrate how theorized structures and processes may be idiosyncratic to specific models, difficult to compare (e.g., splitting vs. self-invalidation; Kernberg, 1984; Linehan, 1993), and ultimately supplemented by other theorized entities (e.g., diagnoses) to describe individual differences. As will be discussed in the sections that follow, other approaches have more strongly emphasized the ability to compare individuals' psychopathology in a way that can be used by competing theories.

Descriptive Psychiatry

In contrast to clinical theories, psychiatric nosology often has served a more extensive array of scientific, practical, and political purposes (e.g., Blashfield & Draguns, 1976; Langenbucher & Nathan, 2006). Modern descriptive psychiatry began with late-eighteenth-century psychiatrists (e.g., de Sauvages; Pilgrim, 2007) and came to prominence during the neo-Kraeplinian revolution of the 1970s, which led descriptive psychiatry to dominate official nosology thereafter (e.g., *Diagnostic and Statistical Manuals of Mental Disorders* [DSM]; American Psychiatric Association [APA], 1980; World Health Organization, 1993). Defining features of this approach include: (a) an emphasis on observed symptoms, (b) an assumption of underlying biological disease processes, (c) viewing mental illness as categorically distinct from normative functioning, (d) a basis in clinical observations, often tracing back to nineteenth-century psychiatrists, and (e) the influence of consensus-building politics (e.g., Decker, 2013; Markon, 2013; Pilgrim, 2007). In comparison with clinical theories, descriptive psychiatry aims to communicate between-patient differences in psychopathology efficiently (i.e., through diagnoses). Additionally, although early DSMs (e.g., DSM-I; APA, 1952) were steeped in psychoanalytic theory, modern descriptive psychiatry is largely atheoretical, meaning that researchers and practitioners often need to apply theory to diagnostic categories to make them practically useful.

Assessment in descriptive psychiatry tends to vary considerably between researchers and clinicians. Clinicians often simply use unstructured interviewing to assess the criteria of DSM diagnoses (Blashfield et al., 2014; Hunsley & Mash, 2007) and then rely on the precise, though often complex, guidelines in the DSM to determine whether a specific disorder is present or absent. Researchers interested in psychiatric epidemiology, psychopathology structure, mechanisms, and intervention efficacy often rely on structured or semistructured interviews as gold standards for assessment (e.g., Heeringa et al., 2004; Wright & Simms, 2015), though well-validated self-report instruments also are used (see Chapter 5, by Samuel). Data analyses based on these methods and guided by the principles of descriptive psychiatry often focus on diagnoses, which are treated as dichotomous (i.e., present or absent). The specific analytic techniques vary considerably, however, given the multitude of purposes that these models serve. A subset of this research, focusing on diagnostic reliability (e.g., interrater), cognitive processes within clinicians (e.g., prototyping), and the latent structure of psychopathology, has directly aimed to investigate the validity of the DSM (Blashfield et al., 2014); however, changes to the DSM related to the results of such studies often have been limited, a point which has led to questions about the DSM's scientific validity (e.g., Kotov et al., 2017; Markon, 2013).

BPD, which first appeared in psychiatric nosology in DSM-III (APA, 1980), exemplifies a number of these points. Spitzer, Endicott, and Gibbon (1979) note that, pre-DSM-III, there was considerable clinical interest in BPD, with "borderline" typically referring to clients presenting with symptoms related to either (a) "instability and vulnerability" or (b) schizophrenia spectrum features. Based on a previous review (Gunderson & Singer, 1975) and consultations with expert clinicians, Spitzer and colleagues (1979) developed diagnostic criteria for BPD and schizotypal personality disorder (PD) and then examined their reliability, the discriminability of the two diagnoses, and the diagnostic threshold for deeming BPD "present" or "absent." Despite the addition of the "transient, stress-related dissociation" criterion and refined language for several criteria (e.g., identity disturbance), later DSMs largely have been consistent with Spitzer and colleagues' definition of BPD (Skodol et al., 2002). Notably, DSM-III and later DSMs have placed BPD and schizotypal PD among a number of other PD diagnoses (e.g., eight in DSM-5; APA, 2013), which were not part of the criteria development process for BPD. Research often indicates that BPD overlaps considerably with these other diagnoses and calls into question the meaning of BPD as a distinct diagnosis (Skodol et al., 2002; Williams, Scalco, & Simms, 2018); however, likely in part due to political pressures within psychiatry and clinical psychology, such research has failed to affect psychiatric nosology (Markon, 2013). Thus, DSM BPD can be seen as an attempt to provide an atheoretical operationalization of a purportedly distinct medical disorder, based largely on clinician observations.

Descriptive psychiatry, with its behaviorally focused criterion sets that aim to define psychiatric syndromes comprehensively, has both strengths and limitations. Although descriptive psychiatry models have improved reliability and served as a sociopolitical interface for the field, validity concerns such as those with BPD abound in the literature. As will be reviewed below, the limitations of descriptive psychiatry have motivated alternative descriptive approaches.

Quantitative Models

The quantitative modeling approach to describing psychopathology has emerged in part as a reaction to limitations of current descriptive psychiatry models with the aim of providing descriptive models that accurately reflect the empirical organization (e.g., structure) of psychopathology (Kotov et al., 2017). In particular, descriptive psychiatry models suffer from extensive comorbidity, within-diagnosis heterogeneity, poor coverage (e.g., "Not Otherwise Specified" diagnoses), and a falsely assumed normal-abnormal discontinuity (Markon & Krueger, 2005; Ormel et al., 2015; Verheul, Bartak, & Widiger, 2007). Defining features of the quantitative modeling approach include (a) data-driven latent variable modeling that uses observed covariation among psychopathology indicators (symptoms, items, etc.) to identify underlying hypothetical descriptive units (dimensions, disorders, etc.), (b) a hierarchical focus with varying levels of construct breadth (e.g., dysphoria, fear, internalizing), (c) an emphasis on construct validity, and (d) rigorous measure development (Kotov et al., 2017; Simms & Watson, 2007. This approach is similar to the descriptive psychiatry approach, in that it aims to describe between-person differences in psychopathology comprehensively and does not explicitly concern itself with theories of within-person processes. However, it differs from descriptive psychiatry approaches in that statistical modeling and construct validity evidence are preferred to clinical intuition and committee consensus in adjudicating the structure of psychopathology (e.g., Williams et al., 2018; also see Chapter 9, by Wright and colleagues).

Unfortunately, quantitatively based descriptive models tend not to be used by clinicians, and researchers vary in their integration of such models (e.g., Rodriguez-Seijas, Eaton, & Krueger, 2015). Specifically, treatment, laboratory, and epidemiology research generally focuses on psychiatric nosology (Barlow et al., 2014; Heeringa et al., 2004; Lazarus et al., 2014), whereas quantitative models have been more heavily adopted as the basis (and output) of modern psychological assessment research (Kotov et al., 2017).[1] Regardless of the context, quantitative models tend to be assessed by self-report questionnaires (e.g., Simms et al., 2011) or structured interviews (e.g., Watson et al., 2012). Quantitative model researchers often place great emphasis on the internal validity of their models and use a variety of latent variable modeling techniques to test structural hypotheses (e.g., exploratory factor analysis; Wright & Zimmerman, 2015), though network analysis is an emerging yet controversial approach (Forbes et al., 2017) Finally, quantitative model researchers often examine the construct validity of their models, using multimethod longitudinal data collection to establish nomological networks (e.g., Watson et al., 2012); however, it is rarer that quantitative models are connected to within-person processes and causal mechanisms, which can lead to stronger and richer theories (though see Patrick et al., 2013).

A quantitative modeling perspective on BPD can be informed by a vast literature that suggests BPD is: (a) an exceptionally heterogeneous category (e.g., 256 diagnostic criteria configurations), (b) multidimensional, (c) highly comorbid with other PDs, and (d) best represented as a dimension (or multiple dimensions) continuous with normative functioning (Samuel et al., 2013; Sanislow et al., 2002; Wright & Zimmerman, 2015). Recent work suggests that BPD can be best conceptualized as either (a) the presence of general (i.e., non-specific) impairments common to *all* PDs (Williams et al., 2018) or (b) extreme positions on normal personality dimensions (e.g., five-factor model; Samuel et al., 2013). Quantitatively guided BPD models thus imply that it is inefficient to focus on BPD as a unified and distinct construct and that more homogenous PD dimensions hold promise for mechanism and treatment research (e.g., Crowell et al., 2009; Skodol et al., 2002).

Quantitative modeling approaches to descriptive psychopathology are well-suited to developing descriptive systems based on observed covariation and for evaluating the diagnostic assumptions of any classification system. Despite this strength, quantitative models frequently are atheoretical and disconnected from work on psychological and biological mechanisms, which often guide treatment and provide more satisfying explanatory accounts of psychopathology.

Biological Models

Previously described approaches focused on observed behavior and subjective experience; in contrast, biological models seek to provide explanatory accounts of psychopathology based on physical structures and neural processes within organisms (Schwartz et al., 2016; Walter, 2013). Typically, this approach has involved articulating models of DSM diagnoses in terms of neurobiological concepts (genes, hormones, neural networks, etc.; Ruocco & Carcone, 2016); however, this strategy has been increasingly questioned, with calls for a shift in focus to endophenotypes and narrower psychological processes (e.g., Gottesman & Gould, 2003; Insel et al., 2010). An example of this shift is the National Institute of Mental Health's

---

[1] Arguably, this may be caused by the need for dichotomous decisions (cutoffs for inclusion in treatment studies, prevalence estimates, etc.) in these research contexts; however, in theory, the dimensions in quantitative models could be dichotomized for these purposes.

Research Domain Criteria (RDoC) initiative, which articulates five broad domains of psychological processes that have been previously linked to neurobiological measures (i.e., negative valence, positive valence, cognitive systems, social processes, and arousal systems; Cuthbert, 2014). Models such as RDoC may have considerable promise for understanding the neurobiology of specific maladaptive processes; however, researchers have questioned whether RDoC – or any primarily biological model – will be effective for describing individual differences in psychopathology (Berenbaum, 2013; Kendler, 2014; Lilienfeld & Treadway, 2016). For instance, providing a complete description of an individual's psychopathology would likely require conceptualizing an exceptional number of neurobiological systems as interconnecting and interacting, which may impede between-person comparisons and clinical decisions; understanding neurobiological systems in terms of higher-order psychological constructs (disorders, dimensions, etc.; Patrick et al., 2013) and translating neurobiological findings into mental terms would appear necessary (Kendler, 2014).

Relative to other approaches, biological models of psychopathology are examined using exceptionally diverse data collection and analytic methods. RDoC provides one example of how biological models can organize such diverse methods, through defining "units of analysis" (i.e., genes, molecules, cells, circuits, physiology, behavior, and self-reports) and the measurement methods associated with each unit type (Cuthbert, 2014). For instance, physiology might be measured through heart rate, electromyography, skin conductance, eye blink, startle, and salivary assays, among other methods. Neuroimaging methods (e.g., functional magnetic resonance imaging) – which can provide information about functional relations among brain regions – are particularly central to biological approaches, given the role that these methods have played in the rise of neuroscience (Walter, 2013). It is difficult to broadly characterize the statistical methods used in biological psychopathology research, largely because its methods result in such varied forms of data; however, at a conceptual level, this might hint at an important analytic consideration. Due to the methodological and analytic diversity within and across units of analysis, it can be challenging to integrate the results of such studies effectively (e.g., Ruocco & Carcone, 2016) and connect them to broader psychopathology models (Schwartz et al., 2016).

Numerous reviews have considered the neurobiology of BPD (e.g., Ruocco & Carcone, 2016; Schmahl et al., 2014; van Zutphen et al., 2015). Despite evidence for BPD's heritability, initial research has failed to connect BPD to specific genes (Ruocco & Carcone, 2016; Schmahl et al., 2014), leading researchers to suggest gene-environment interactions and narrower psychological processes as future directions of promise. In contrast, research indicates that negative life events may influence the neurobiology of BPD (e.g., trauma; Kuhlmann et al., 2013). In particular, Ruocco and Carcone (2016) note that trauma or cumulative stress may affect neuroendocrine function, which may have downstream effects on cognitive and emotional control (e.g., regulation). Furthermore, regarding more immediate processes, van Zutphen and colleagues (2015) review evidence suggesting that emotional sensitivity in BPD might be related to greater amygdala activity in response to emotion-provoking stimuli, whereas decreased top-down emotion regulation might be associated with decreased anterior cingulate cortex activity. Although there are multiple ways to interpret and integrate research on BPD's neurobiology, the above findings illustrate how multiple sources of biological data can be used to build conceptual models of a disorder.

This brief review of neurobiological research on BPD demonstrates general strengths of biological models, such as their ability to inform theory and use of multiple methods; however, it is unclear how to interpret such findings in the context of other disorders (e.g., BPD with another comorbid disorder) or the clinical utility of such results (e.g., Ruocco & Carcone, 2016). Although movement away from diagnosis-focused research (e.g., Insel et al., 2010) may address some issues in this approach, it seems likely that it will still need to remain tied to other descriptive approaches that invoke higher-order psychological constructs and psychosocial interventions (Kendler, 2014; Patrick & Hajcak, 2016).

Summary

In surveying modern descriptive psychopathology approaches, points of convergence and divergence naturally arise. Although all approaches address the metagoal of "describing psychopathology," they clearly diverge in their more specific goals for description. Despite having unique aims, these approaches intersect in various ways. For instance, clinical theory, descriptive psychiatry, and quantitative models all use observed behavior and subjective reports as primary observations. Greater complexity emerges when examining how observations are assessed and translated into theoretical concepts. Clinical theorists often use unstandardized assessment methods, focus on within-person variation, and employ idiosyncratic theoretical concepts that form dynamic systems. Descriptive psychiatry and quantitative model researchers often use standardized assessment instruments, organize observations based on theoretical constructs that describe between-person variation, and are relatively agnostic regarding dynamic processes and etiology. In further contrast, biological models use observable physical phenomena to describe psychopathology in terms of neurobiological structures and processes.

These complex discrepancies have implications for understanding the ability to communicate findings across approaches. First, theorists and researchers are starting with widely differing data, because approaches privilege different methods of assessment (unstructured observations, structured questions, brain activity, etc.). Differences in data sources, metrics, and measurement complicate the integration of findings across multiple approaches, because (a) discrepant findings across methods may be difficult to resolve and (b) individual researchers may lack the background knowledge to critically evaluate research from other approaches. Second, theorists and researchers often use theoretical language that is meaningful within the context of an individual approach and its goals, but does not always intuitively fit within other approaches. For example, the dimensions of quantitative models are rarely a part of the dynamic portions of clinical theory descriptive models and thus clinicians may struggle to use quantitative models to formulate an intervention. Both of these points of discrepancy – assessment methods and theoretical language – would appear to highlight the need for careful thought regarding translating between descriptive approaches.

The fact that approaches have diverging goals may explain these difficulties in communication; however, it is also likely that some of these difficulties are related to the history of the approaches. Approaches vary in the extent to which clinical impressions were influential in shaping initial models (e.g., descriptive psychiatry vs. biological models; e.g., Kendler, 2009), whether their full descriptive models were evaluated (e.g., clinical theory vs. quantitative models), and their dependence on technological advances (e.g., Walter, 2013). However, despite the influence of diverging goals and histories, the past decade has brought these approaches into closer contact in ways that promise to shape the future of descriptive psychopathology.

THE CURRENT STATE OF DESCRIPTIVE PSYCHOPATHOLOGY

Dialogue, competition, and convergence among the four descriptive approaches have recently accelerated. The most salient trend is the increasingly strong criticism of descriptive psychiatry from clinical theorists (e.g., Barlow et al., 2014), quantitative model researchers (e.g., Kotov et al., 2017), and biological model researchers (e.g., Cuthbert & Kozak, 2013). Critiques across approaches vary in emphasis, but most consistently note the heterogeneity within and comorbidity among DSM-based diagnostic constructs. Although these limitations most directly suggest that DSM diagnoses have poor internal validity (e.g., Skinner, 1981), they also argue against the DSM's utility for (a) understanding treatment effects, (b) guiding intervention decisions, (c) communicating meaningfully among professionals and researchers, (d) developing measurement instruments, and (e) understanding the pathophysiology of psychopathology. The publication of DSM-5 (APA, 2013) was seen by many as a failure to address these limitations, thus potentially setting the stage for a dramatic shift in descriptive psychopathology in the near future (e.g., Lilienfeld & Treadway, 2016; Markon, 2013).

Emerging Alternatives

Clinical, quantitative, and biological researchers have each offered alternatives to relying on DSM diagnostic categories for guiding research and practice. Clinical theorists and researchers have offered an array of transdiagnostic processes (Mansell & McEvoy, 2017; Sauer-Zavala et al., 2017) and therapies (e.g., Barlow et al., 2014; Hayes et al., 2012), which emphasize descriptive accounts wherein a common psychological mechanism (e.g., negative repetitive thinking) explains the symptoms experienced by individuals with varying DSM diagnoses. Recently, a large and growing number of prominent quantitative researchers have joined the Hierarchical Taxonomy of Psychopathology (HiTOP) consortium, which has recently published a model that synthesizes decades of quantitative descriptive psychopathology research and aims to increase the influence of such research on clinical practice and psychiatric classification (Kotov et al., 2017). Finally, as discussed above, RDoC attempts to integrate diverse psychological processes within a multidimensional neurobiological framework (e.g., Kozak & Cuthbert, 2016). Given that HiTOP, RDoC, and transdiagnostic clinical processes reflect promising contemporary alternatives to the predominant descriptive psychiatry approach (i.e., DSM-5), it is important to consider their relative strengths and weaknesses.

First, HiTOP has the best-articulated and most evidence-based organizational structure; half a century of research has directly examined hierarchical-dimensional models of symptoms, with increasing comprehensiveness (e.g., Achenbach, 1966; Wright & Simms, 2015). Despite this, much like other quantitative models, HiTOP dimensions (a) lack clearly explicated theories of within-person processes (psychological or biological) and (b) there is limited work connecting these dimensions to well-validated interventions (Barlow et al., 2014; Kotov et al., 2017).

Second, and in contrast, transdiagnostic clinical processes largely have been studied within the context of psychotherapy research and directly consider psychological processes, as well as altering them (Barlow et al., 2014; Klemanski et al., 2017). Despite this promise, research on transdiagnostic clinical processes is relatively new, it is unclear how distinct specific processes are (e.g., rumination vs. negative repetitive thinking), their measurement remains complex, and there is no comprehensive organizing framework for these processes (Mansell & McEvoy, 2017; Sauer-Zavala et al., 2017).

Third, RDoC provides a model for articulating and connecting distinct psychological-biological processes relevant to a broad understanding of psychopathology, where the measurement of processes is clearly explicated, and a preliminary higher-order organizational framework is offered to assemble related processes (Cuthbert, 2014). These strengths, however, are tempered by the untested nature of RDoC's domain-level structure, as well as its unclear connection to clinical problems and interventions (Berenbaum, 2013; Patrick & Hajcak, 2016).

These alternatives to DSM-5 share a general strength of being responsive to contemporary research; however, further evaluation is needed to determine their ultimate validity and utility. One aspect of this evaluation likely will involve establishing the convergence and divergence of these alternative models. For instance, some research indicates that HiTOP's internalizing dimension (Kotov et al., 2017), RDoC's negative valence domain (Cuthbert, 2014), and varied transdiagnostic foci (e.g., neuroticism; Barlow et al., 2014) substantially overlap; however, additional work is needed to clarify such associations (e.g., Watson, Stanton, & Clark, 2017). Along with establishing points of convergence between these alternative models, it will be necessary to identify their unique strengths and weaknesses. The preceding paragraphs suggest that such divergences are likely, raising the possibility that adopting only one of these models would be insufficient for the purposes of current research and clinical practice. Additionally, it may also be argued that these models have not fully subsumed the functions of descriptive psychiatry, which has included the provision of clearly defined public standards for mental disorder diagnosis that can be used for legal, legislative, insurance, and advocacy purposes (e.g., Langenbucher & Nathan, 2006). Thus, integrating across descriptive approaches and these particular models, to build upon the unique advantages of each, would have substantial value; however, doing so will require overcoming several practical and conceptual barriers.

Challenges to Integrating Descriptive Psychopathology Approaches

In considering practical and theoretical challenges that might impede integrating descriptive psychopathology approaches in their modern forms, it is helpful to think of these approaches as embedded within scientific communities in which members share similar technical skills and conceptual vocabularies (e.g., Kuhn, 1977). With this in mind, a first challenge to integrating descriptive approaches is myopia, or nearsightedness, in which scholars struggle to see the importance or possibility of expanding beyond their descriptive approach. Many researchers are trained primarily in the methods, models, and important findings of one descriptive psychopathology approach and may fail to appreciate the strengths of alternative approaches or struggle to see the relevance of such approaches for their work. For example, most quantitative model researchers do not provide neurobiological accounts of dimensions within their models, either because they have limited knowledge of basic neurobiology or perhaps because they undervalue the potential for neurobiological methods to contribute to their research. Such myopia is regrettable, since it impedes scientific progress and the impact of this progress on clinical practice and public policy.

Second, many researchers lack the technical skills associated with other approaches. For instance, researchers knowledgeable about the biological units of analysis in RDoC may not have the appropriate statistical foundations to understand or implement latent variable modeling, a foundational technique within the quantitative modeling approach. Linked with these skill deficits is a third issue: individual researchers often lack the resources to examine more than one descriptive approach. Such a situation might occur when a psychotherapy researcher wishes to use the RDoC framework, as they may lack the equipment for behavioral tasks or neurobiological measures. Skill and resource deficits likely are intertwined, in that graduate students often are trained within one laboratory, whose resources strongly influence which skills a student acquires.

A final issue emerges when considering the functions of descriptive psychiatry, which through the DSM has been able to forge a communicative pathway to legislators, lawyers, insurance agencies, and advocacy groups. Proponents of HiTOP, transdiagnostic clinical processes, and RDoC have not put forth plans to assume this burden; however, as these models gain traction, they will likely have to perform these functions. Addressing this issue may prove difficult, due to the inertia of bureaucratic systems (e.g., Markon, 2013) and the fact that researchers may lack the training (e.g., policy advocacy), time, or incentive to press for necessary changes. Positioning a model to serve societal functions – along with motivational, educational, and resources limitations – presents an important barrier to a more integrative science of psychopathology. In the next section, we present recommendations for how such barriers may be surmounted.

CONCLUSIONS AND RECOMMENDATIONS

Approaches to descriptive psychopathology are diverse and currently in flux. Descriptive models serve a range of purposes (communication, prediction, explanation, etc.) and use a variety of concepts and methods to achieve these ends. As described above, HiTOP, transdiagnostic clinical processes, and RDoC have emerged as promising alternatives to the traditional descriptive psychiatry approach embodied by the DSM and International Classification of Diseases (ICD). Notably, however, each of these models has weaknesses that would appear to be complemented by the qualities of one (or more) of the other models. Additionally, although current descriptive psychiatry models (e.g., DSM-5) may have limited empirical support, they serve important functions that other approaches need to consider. Thus, integrating descriptive psychopathology approaches is a compelling solution; however, as illustrated in the preceding section, doing so will be complex. In addressing these issues, we have several recommendations for how individual researchers and the field as a whole can strive for an integrative descriptive approach.

First, researchers within each of the descriptive psychopathology perspectives highlighted in this chapter must recognize the historical context of their work and acknowledge the strengths of alternative approaches to avoid a myopic approach to research. Ideally, this should also involve meaningfully extending their work to interface with alternative descriptive approaches. For example, quantitative model researchers might aim to (a) develop simple, efficient, and reliable diagnostic criteria and associated measures (i.e., similar to descriptive psychiatry), (b) provide theoretical accounts of their quantitatively informed diagnostic constructs that describe within-person process and translate into theory-based interventions (i.e., similar to clinical theory), and (c) study the neurobiological underpinnings of their models. Adopting such a mindset is a significant first step; however, it is important to acknowledge the limitations of individual researchers, in terms of their expertise and resources.

Thus, as a second recommendation, psychopathology researchers need to move toward models of cross-specialty collaboration (e.g., team science) and data sharing (Munafò et al., 2017; Patrick & Hajcak, 2016). Ideally, this would connect researchers who have training in varied descriptive approaches and have access to unique and complementary resources. Notably this is common and effective in other fields (e.g., biology), in which teams produce some of the most impactful work (Wuchty, Jones, & Uzzi, 2007). Furthermore, there are examples of this within psychopathology research, such as multisite longitudinal research (e.g., Sanislow et al., 2002), research on the reproducibility of psychological science (e.g., Munafò et al., 2017), and work integrating multiple descriptive approaches (e.g., Patrick et al., 2013). Additionally, it is worth noting that the HiTOP consortium has recently evolved to include specific work groups, comprised of members with diverse expertise, which focus on clinical translation and neurobiology in the context of the HiTOP model. Collaboration between researchers of diverse expertise represents a promising solution to issues intractable for individual scientists.

Third, descriptive psychopathology scholars must interface with mental health practitioners, policy makers, advocacy groups, and perhaps even patients themselves. This may take the form of including these individuals on research teams or fostering channels of communication with them. Including such stakeholders in the research process should increase the relevance and impact of the work, improve the research conducted, and hasten its translation into clinical practice. Steps such as these will facilitate the translation of insights from the clinical theory, quantitative, and biological perspectives into informing practice and policy in ways traditionally monopolized by descriptive psychiatry models, such as the DSM.

Fourth, laboratories and training programs should emphasize training across descriptive psychopathology approaches rather than adopt silo models that emphasize expertise in only a single approach. One way to do this is to ensure coursework meaningfully covers each descriptive approach. A second, more compelling way that a program might accomplish this is through requiring students to conduct cross-laboratory research, where they gain experience by working in the laboratory of a second advisor. Such immersive experiences are a part of some doctoral psychology programs and are more common in natural science graduate training. Ideally, this training will lead students to develop skills relevant to a broader range of descriptive approaches and enhance their appreciation of the value of each approach as well.

These four recommendations – contextualizing individual research programs, collaboration and open science, community and political engagement, and integrative graduate training – represent important steps toward integrating the strengths of multiple approaches to describing psychopathology. This chapter has aimed to illustrate the importance and complexity of such an integration, in that the four approaches examined have diverging goals, histories, methods, and theoretical organization; such diversity maximizes the incremental value of individual descriptive approaches, but can make learning the language of an alternative approach daunting to individual researchers. Overcoming the barriers to communication across these descriptive approaches is an imperative task for the academic community to face and resolve in the service of a more complete science of psychopathology.

REFERENCES

Achenbach, T. M. (1966). The Classification of Children's Psychiatric Symptoms: A Factor-Analytic Study. *Psychological Monographs*, *80*, 1–37.

Adams, H. E., Luscher, K. A., & Bernat, J. A. (2001). The Classification of Abnormal Behavior: An Overview. In P. B. Sutker & H. E. Adams (Eds.), *Comprehensive Handbook of Psychopathology* (3rd edn., pp. 3–28). New York: Kluwer Academic.

American Psychiatric Association (APA). (1952). *Diagnostic and Statistical Manual: Mental Disorders*. Washington, DC: Author.

American Psychiatric Association (APA). (1980). *Diagnostic and Statistical Manual of Mental Disorders* (3rd edn.). Washington, DC: Author.

American Psychiatric Association (APA). (2013). *Diagnostic and Statistical Manual of Mental Disorders* (5th edn.). Washington, DC: Author.

Baer, R. A., Smith, G. T., Hopkins, J., Krietemeyer, J., & Toney, L. (2006). Using Self-Report Assessment Methods to Explore Facets of Mindfulness. *Assessment*, *13*(1), 27–45.

Barlow, D. H., Sauer-Zavala, S., Carl, J. R., Bullis, J. R., & Ellard, K. K. (2014). The Nature, Diagnosis, and Treatment of

Neuroticism: Back to the Future. *Clinical Psychological Science*, *2*(3), 344–365.

Bateman, A., & Fonagy, P. (2013). Mentalization-Based Treatment. *Psychoanalytic Inquiry*, 33, 595–613.

Beck, J. S. (2011). *Cognitive Behavior Therapy: Basics and Beyond* (2nd edn.). New York: Guilford Press.

Berenbaum, H. (2013). Classification and Psychopathology Research. *Journal of Abnormal Psychology*, *122*(3), 894–901.

Blashfield, R. K., & Draguns, J. G. (1976). Toward a Taxonomy of Psychopathology: The Purpose of Psychiatric Classification. *British Journal of Psychiatry*, *129*, 574–583.

Blashfield, R. K., Keeley, J. W., Flanagan, E. H., & Miles, S. R. (2014). The Cycle of Classification: *DSM-I* through *DSM-5*. *Annual Review of Clinical Psychology*, *10*, 25–51.

Clark, L. A., Watson, D., & Reynolds, S. (1995). Diagnosis and Classification of Psychopathology: Challenges to the Current System and Future Directions. *Annual Review of Psychology*, *46*(1), 121–153.

Crowell, S. E., Beauchaine, T. P., & Linehan, M. M. (2009). A Biosocial Developmental Model of Borderline Personality: Elaborating and Extending Linehan's Theory. *Psychological Bulletin*, *135*(3), 495–510.

Cronbach, L. J., & Meehl, P. E. (1955). Construct Validity in Psychological Tests. *Psychological Bulletin*, *52*(4), 281–302.

Cuthbert, B. N. (2014). The RDoC Framework: Facilitating Transition from ICD/DSM to Dimensional Approaches that Integrate Neuroscience and Psychopathology. *World Psychiatry*, *13*(1), 28–35.

Cuthbert, B. N., & Kozak, M. J. (2013). Constructing Constructs for Psychopathology. *Journal of Abnormal Psychology*, *122*(3), 928–937.

Decker, H. P. (2013). *The Making of DSM-III*. Oxford: Oxford University Press.

Dowden, C., & Brown, S. L. (2002). The Role of Substance Abuse Factors in Predicting Recidivism: A Meta-Analysis. *Psychology, Crime and Law*, *8*(3), 243–264.

Dudley, R., Kuyken, W., & Padesky, C. A. (2011). Disorder Specific and Trans-Diagnostic Case Conceptualisation. *Clinical Psychology Review*, *31*(2), 213–224.

Eaton, N. R., Krueger, R. F., Docherty, A. R., & Sponheim, S. R. (2013). Toward a Model-Based Approach to the Clinical Assessment of Personality Psychopathology. *Journal of Personality Assessment*, *96*(3), 1–10.

Eells, T. D. (2007). History and Current Status of Psychotherapy Case Formulation. In T. D. Eells (Ed.) *Handbook of Psychotherapy Case Formulation* (2nd edn., pp. 3–32). New York: Guilford Press.

Engel, M. S., & Kristensen, N. P. (2013). A History of Entomological Classification. *Annual Review of Entomology*, *58*, 585–607.

Fisher, A. J., Newman, M. G., & Molenaar, P. C. M. (2011). A Quantitative Method for the Analysis of Nomothetic Relationships between Idiographic Structures: Dynamic Patterns Create Attractor States for Sustained Posttreatment Change. *Journal of Consulting and Clinical Psychology*, *79*(4), 552–563.

Forbes, M. K., Wright, A. G. C., Markon, K. E., & Krueger, R. F. (2017). Evidence That Psychopathology Symptom Networks Have Limited Replicability. *Journal of Abnormal Psychology*, *126*(7), 969–988.

Fournier, J. C., DeRubeis, R. J., & Beck, A. T. (2012). Dysfunctional Cognitions in Personality Pathology: The Structure and Validity of the Personality Belief Questionnaire. *Psychological Medicine*, *42*(4), 795–805.

Gottesman, I. I., & Gould, T. D. (2003). The Endophenotype Concept in Psychiatry: Etymology and Strategic Intentions. *American Journal of Psychiatry*, *160*(4), 636–645.

Gunderson, J. G., & Singer, M. T. (1975). Defining Borderline Patients: An Overview. *American Journal of Psychiatry*, *132*(1), 1–10.

Grimaldi, D. A., & Engel, M. S. (2007). Why Descriptive Science Still Matters. *Bioscience*, *57*(8), 646–647.

Hayes, S. C., Strosahl, K. D., & Wilson, K. G. (2012). *Acceptance and Commitment Therapy: The Process and Practice of Mindful Change*. New York: Guilford Press.

Heeringa, S. G., Wagner, J., Torres, M., Duan, N., Adams, T., & Berglund, P. (2004). Sample Designs and Sampling Methods for the Collaborative Psychiatric Epidemiology Studies (CPES). *International Journal of Methods in Psychiatric Research*, *13*(4), 221–240.

Hempel, C. G. (1965). *Aspects of Scientific Explanation and Other Essays in the Philosophy of Science*. New York: Free Press.

Hothersall, D. (2004). *History of Psychology* (4th edn.). New York: McGraw-Hill.

Hull, D. L. (1965). The Effect of Essentialism on Taxonomy – Two Thousand Years of Stasis (I). *British Journal for the Philosophy of Science*, *15*(60), 314–326.

Hunsley, J., & Mash, E. J. (2007). Evidence-Based Assessment. *Annual Review of Clinical Psychology*, *3*, 29–51.

Insel, T., Cuthbert, B., Garvey, M., Heinssen, R., Pine, D. S., Quinn, K., Sanislow, C., & Wang, P. (2010). Research Domain Criteria (RDoC): Toward a New Classification Framework for Research on Mental Disorders. *American Journal of Psychiatry*, *167*(7), 748–751.

Kendler, K. S. (2009). A Historical Framework for Psychiatric Nosology. *Psychological Medicine*, *39*, 1935–1941.

Kendler, K. S. (2014). The Structure of Psychiatric Science. *American Journal of Psychiatry*, *171*(9), 931–938.

Kernberg, O. F. (1984). *Severe Personality Disorders: Psychotherapeutic Strategies*. New Haven, CT: Yale University.

Klemanski, D. H., Curtiss, J., McLaughlin, K. A., & Nolen-Hoeksema, S. (2017). Emotion Regulation and the Transdiagnostic Role of Repetitive Negative Thinking in Adolescents with Social Anxiety and Depression. *Cognitive Therapy and Research*, *41*(2), 206–219.

Kline, P. (1998). *The New Psychometrics: Science, Psychology and Measurement*. Philadelphia, PA: Routledge.

Kotov, R., Krueger, R. F., Watson, D., Achenbach, T. M., Althoff, R. R., Bagby, R. M., ... & Eaton, N. R. (2017). The Hierarchical Taxonomy of Psychopathology (HiTOP): A Dimensional Alternative to Traditional Nosologies. *Journal of Abnormal Psychology*, *126*(4), 454–477.

Kozak, M. J., & Cuthbert, B. N. (2016). The NIMH Research Domain Criteria Initiative: Background, Issues, and Pragmatics. *Psychophysiology*, *53*(3), 286–297.

Kuhn, T. S. (1977). *The Essential Tension: Selected Studies in Scientific Tradition and Change*. Chicago, IL: University of Chicago Press.

Kuhlmann, A., Bertsch, K., Schmidinger, I., Thomann, P. A., & Herpertz, S. C. (2013). Morphometric Differences in Central Stress-Regulating Structures between Women with and without Borderline Personality Disorder. *Journal of Psychiatry & Neuroscience*, *38*(2), 129–137.

Langenbucher, J., & Nathan, P. E. (2006). Diagnosis and Classification. In M. Hersen & J. C. Thomas (Eds.), *Comprehensive Handbook of Personality and Psychopathology* (2nd edn., pp. 3–20). Hoboken, NJ: Wiley.

Lazarus, S. A., Cheavens, J. S., Festa, F., & Rosenthal, M. Z. (2014). Interpersonal Functioning in Borderline Personality Disorder: A Systematic Review of Behavioral and Laboratory-Based Assessments. *Clinical Psychology Review*, 34(3), 193–205.

Lee, D. J., Weather, F. W., Sloan, D. M., Davis, M. T., & Domino, J. L. (2017). Development and Initial Psychometric Evaluation of the Semi-Structured Emotion Regulation Interview. *Journal of Personality Assessment*, 99(1), 56–66.

Lilienfeld, S. O., & Treadway, M. T. (2016). Clashing Diagnostic Approaches: DSM-ICD versus RDoC. *Annual Review of Clinical Psychology*, 12, 435–463.

Linehan, M. (1993). *Cognitive-Behavioral Treatment of Borderline Personality Disorder*. New York: Guilford.

Loevinger, J. (1957). Objective Tests as Instruments in Psychological Theory. *Psychological Reports*, 3, 635–694.

Mansell, W., & McEvoy, P. M. (2017). A Test of the Core Process Account of Psychopathology in a Heterogenous Clinical Sample of Anxiety and Depression: A Case of the Blind Men and the Elephant. *Journal of Anxiety Disorders*, 46, 4–10.

Markon, K. E. (2013). Epistemological Pluralism and Scientific Development: An Argument against Authoritative Nosologies. *Journal of Personality Disorders*, 27(5), 554–579.

Markon, K. E., & Krueger, R. F. (2005). Categorical and Continuous Models of Liability to Externalizing Disorders: A Direct Comparison in NESARC. *Archives of General Psychiatry*, 62(12), 1352–1359.

Millon, T. (1991). Classification in Psychopathology: Rationale, Alternatives, and Standards. *Journal of Abnormal Psychology*, 100(3), 245–261.

Millon, T. (2004). *Masters of the Mind: Exploring the Story of Mental Illness from Ancient Times to the New Millennium*. Hoboken, NJ: Wiley.

Munafò, M. R., Nosek, B. A., Bishop, D. V., Button, K. S., Chambers, C. D., du Sert, N. P., ... & Ioannidis, J. P. (2017). A Manifesto for Reproducible Science. *Nature Human Behaviour*, 1, 0021.

Ormel, J., Raven, D., van Oort, F., Hartman, C. A., Reijneveld, S. A., Veenstra, R., ... & Oldehinkel, A. J. (2015). Mental Health in Dutch Adolescents: A TRAILS Report on Prevalence, Severity, Age of Onset, Continuity and Co-morbidity of DSM Disorders. *Psychological Medicine*, 45(2), 345–360.

Patrick, C. J., & Hajcak, G. (2016). RDoC: Translating Promise into Progress. *Psychophysiology*, 53(3), 415–424.

Patrick, C. J., Venables, N. C., Yancey, J. R., Hicks, B. M., Nelson, L. D., & Kramer, M. D. (2013). A Construct-Network Approach to Bridging Diagnostic and Physiological Domains: Application to Assessment of Externalizing Psychopathology. *Journal of Abnormal Psychology*, 122(3), 902–916.

Pilgrim, D. (2007). The Survival of Psychiatric Diagnosis. *Social Science & Medicine*, 65, 536–547.

Ramnerö, J., & Törneke, N. (2008). *The ABCs of Human Behavior: Behavioral Principles for the Practicing Clinician*. Oakland, CA: New Harbinger.

Rodriguez-Seijas, C., Eaton, N. R., & Krueger, R. F. (2015). How Transdiagnostic Factors of Personality and Psychopathology can Inform Clinical Assessment and Intervention. *Journal of Personality Assessment*, 97(5), 425–435.

Ruocco, A. C., & Carcone, D. (2016). A Neurobiological Model of Borderline Personality Disorder: Systematic and Integrative Review. *Harvard Review of Psychiatry*, 24(5), 311–329.

Samuel, D. B., Carroll, K. M., Rounsaville, B. J., & Ball, S. A. (2013). Personality Disorders as Maladaptive, Extreme Variants of Normal Personality: Borderline Personality Disorder and Neuroticism in a Substance Using Sample. *Journal of Personality Disorders*, 27(5), 625–635.

Sanislow, C. A., Grilo, C. M., Morey, L. C., Bender, D. S., Skodol, A. E., Gunderson, J. G., ... & McGlashan, T. H. (2002). Confirmatory Factor Analysis of DSM-IV Criteria for Borderline Personality Disorder: Findings from the Collaborative Longitudinal Personality Disorders Study. *American Journal of Psychiatry*, 159(2), 284–290.

Sauer-Zavala, S., Gutner, C. A., Farchione, T. J., Boettcher, H. T., Bullis, J. R., & Barlow, D. H. (2017). Current Definitions of "Transdiagnostic" in Treatment Development: A Search for Consensus. *Behavior Therapy*, 48(1), 128–138.

Scerri, E. R. (2012). The Periodic Table. In R. F. Hendry, P. Needham, & A. I. Woody (Eds.), *Handbook of the Philosophy of Science (Volume 6: Philosophy of Chemistry)*. Oxford: Oxford University Press.

Schmahl, C., Herpertz, S. C., Bertsch, K., Ende, G., Flor, H., Kirsch, P., ... & Bohus, M. (2014). Mechanisms of Disturbed Emotion Processing and Social Interaction in Borderline Personality Disorder: State of Knowledge and Research Agenda of the German Clinical Research Unit. *Borderline Personality Disorder and Emotion Dysregulation*, 1, 12.

Schwartz, S. J., Lilienfeld, S. O., Meca, A., & Sauvigné, K. C. (2016). The Role of Neuroscience within Psychology: A Call for Inclusiveness over Exclusiveness. *American Psychologist*, 71(1), 52–70.

Simms, L. J., & Watson, D. (2007). The Construct Validation Approach to Personality Scale Construction. In R. W. Robins, R. C. Fraley, & R. F. Krueger (Eds.), *Handbook of Research Methods in Personality Psychology* (pp. 240–258). New York: Guilford Press.

Simms, L. J., Goldberg, L. R., Roberts, J. E., Watson, D., Welte, J., & Rotterman, J. H. (2011). Computerized Adaptive Assessment of Personality Disorder: Introducing the CAT-PD Project. *Journal of Personality Assessment*, 93(4), 380–389.

Skinner, H. A. (1981). Toward the Integration of Classification Theory and Methods. *Journal of Abnormal Psychology*, 90(1), 68–87.

Skodol, A. E., Gunderson, J. G., Pfohl, B., Widiger, T. A., Livesley, W. J., & Siever, L. J. (2002). The Borderline Diagnosis I: Psychopathology, Comorbidity, and Personality Structure. *Biological Psychiatry*, 51(12), 936–950.

Sokal, R. (1974). Classification: Purposes, Principles, Progress, Prospects. *Science*, 185(4157), 1115–1123.

Spitzer, R. L., Endicott, J., & Gibbon, M. (1979). Crossing the Border into Borderline Personality and Borderline Schizophrenia. *Archives of General Psychiatry*, 36, 17–24.

Stearn, W. T. (1959). The Background of Linnaeus's Contributions to the Nomenclature and Methods of Systematic Biology. *Systematic Zoology*, 8(1), 4–22.

Van Zutphen, L., Siep, N., Jacob, G. A., Goebel, R., & Arntz, A. (2015). Emotional Sensitivity, Emotion Regulation and Impulsivity in Borderline Personality Disorder: A Critical Review of fMRI Studies. *Neuroscience & Biobehavioral Reviews*, 51, 64–76.

Verheul, R., Bartak, A., & Widiger, T. (2007). Prevalence and Construct Validity of Personality Disorder not Otherwise Specified (PDNOS). *Journal of Personality Disorders, 21*(4), 359–370.

Walter, H. (2013). The Third Wave of Biological Psychiatry. *Frontiers in Psychology, 4*, 1–8.

Watson, D., O'Hara, M. W., Naragon-Gainey, K., Koffel, E., Chmielewski, M., Kotov, R., Stasik, S. M. & Ruggero, C. J. (2012). Development and Validation of New Anxiety and Bipolar Symptom Scales for an Expanded Version of the IDAS (the IDAS-II). *Assessment, 19*(4), 399–420.

Watson, D., Stanton, K., & Clark, L. A. (2016). Self-Report Indicators of Negative Valence Constructs within the Research Domain Criteria (RDoC): A Critical Review. *Journal of Affective Disorders, 216*, 58–69.

Wiggins, J. S. (2003). *Paradigms of Personality Assessment.* New York: Guilford.

Williams, T. F., Scalco, M. D., & Simms, L. J. (2018). The Construct Validity of General and Specific Dimensions of Personality Pathology. *Psychological Medicine, 48*(5), 834–848.

Wittgenstein, L. (1922). *Tractatus Logicus-Philosophicus.* London: Kegan.

World Health Organization. (1993). *The ICD-10 Classification of Mental and Behavioral Disorders: Diagnostic Criteria for Research.* Geneva: WHO.

Wright, A. G., & Simms, L. J. (2015). A Metastructural Model of Mental Disorders and Pathological Personality Traits. *Psychological Medicine, 45*(11), 2309–2319.

Wright, A. G. C., & Zimmerman, J. (2015). At the Nexus of Science and Practice: Answering Basic Clinical Questions in Personality Disorder Assessment and Diagnosis with Quantitative Modeling Techniques. In S. Huprich (Ed.), *Personality Disorders: Toward Theoretical and Empirical Integration in Diagnosis and Assessment* (pp. 109–144). Washington, DC: American Psychological Association.

Wuchty, S., Jones, B. F., & Uzzi, B. (2007). The Increasing Dominance of Teams in Production of Knowledge. *Science, 316* (5827), 1036–1039.

Young, J. E., Klosko, J. S., & Weishaar, M. E. (2003). *Schema Therapy: A Practitioner's Guide.* New York: Guilford Press.

5 Survey and Interview Methods

DOUGLAS B. SAMUEL, MEREDITH A. BUCHER, AND TAKAKUNI SUZUKI

As discussed in the preceding chapter, the measurement of a psychological construct is a *sine qua non* of clinical psychological science. Quite literally, without an adequate measurement of diagnostic, outcome, or functional constructs there is no way to conduct reliable and valid research. This volume highlights and expands on a number of methods and approaches to measuring those constructs, but none have as lengthy a history as simply asking an individual to report on their experiences, thoughts, feelings, and behaviors. The ubiquitous use of survey or interview methods has, at times, garnered criticism (Haeffel & Howard, 2010; Nisbett & Wilson, 1977). Many a reviewer or author has lamented the reliance on questionnaire methods. Indeed, these methods – like any other – do have limitations. Nonetheless, there are reasons, beyond simple convenience, that they have proven so popular. Survey and interview methods also represent the measurement approach that has, by far, the largest empirical background and typically offers well-articulated psychometric properties and available evidence for construct validity.

In this chapter, we articulate the choices that go into choosing a method and highlight those factors that make surveys and/or interviews well suited for a given question. We also detail those research questions or constructs for which these methods are not recommended. Next, we highlight the similarities and differences between the survey and interview methods and make suggestions about how a researcher might choose between the two. Finally, we recommend best practices for scale development and a series of decision points in the use of these methods.

Throughout this chapter it is important to note that we consider the merit of survey and interview *methods* separately from the *source* from whom the information is gathered. Although in many cases surveys (and interviews) are completed by the participant on their own experiences or symptoms, there is also a rich history of using these methods to assess the opinions of an informant (i.e., a spouse, co-worker, parent, or even a clinician) about the target. Once we have covered the general pros and cons of the methods, we will briefly turn to the question of to whom the instruments are administered.

WHEN ARE SURVEYS AND INTERVIEWS USEFUL?

The central hypothesis underlying the use of any method that asks a participant to provide a linguistic answer to a question (whether a survey or interview), is that the information being sought is available to conscious awareness. Thus, a central point of consideration in using a survey or interview is whether the construct is one for which the participant is reasonably able to have knowledge and articulate it verbally. For example, a research question aimed at understanding neural connectivity or any other explicitly internal biological process is not particularly amenable to survey or interview methods. Similarly, a psychological construct that is defined as being outside of conscious awareness, such as a defense mechanism, would be less suitable for a survey (e.g., the source of the ratings was knowledgeable and well positioned to have access to the relevant information). Nonetheless, there remains a multitude of psychological constructs relevant to clinical research that fit comfortably within these desired boundaries, making them amenable to assessment via survey or interview. Particularly in clinical diagnosis, the conscious, subjective experience, such as distress, is often the central construct of interest (*Diagnostic and Statistical Manual of Mental Disorders*, Fifth Edition – DSM-5). Such constructs, by definition, are assessed readily via surveys or interviews.

Another central consideration in choosing a method is the physical distance between the researcher and the participant. In this regard, surveys and interviews are particularly well suited to efficiently and effectively collect a sample from across a broad geographic area, regardless of the location of the assessor. Although rapid advances in technology have closed the gap (e.g., wearable technology; Trull & Ebner-Priemer, 2013), surveys and interviews have the distinct advantage of being able to be administered to the participant from anywhere on the globe due to the availability to complete surveys via mail or online and conduct interviews by phone or video call. In contrast, many other methods (e.g., neurobiological assessments or experimental paradigms) require a physical visit to a laboratory. This produces a significant burden on the participant (and likely limits the geographic range from which a study can recruit participants) as well as the experimenter. Furthermore, the intensive nature of other methods often limits participation to only a single individual at a given time, making it inefficient to collect large samples rapidly.

Differences between Survey and Interview Methods

Having outlined the benefits shared by surveys and interviews, choosing among those two becomes the next pivotal question. The clear similarity between surveys and interviews is that they both ask direct questions of the subject, who then provides an answer. In this way, both methods are predominantly scored on the basis of subjects' responses. Surveys often (but not necessarily) instruct a participant to select from among a list of potential response options whereas interviews will often allow for a more elaborated or open-ended response. Nonetheless, there is no a priori reason why either should be restricted in this way. A survey *could* be crafted so that the subject provides an open-ended written response (e.g., Sentence Completion Test; Loevinger, 1979) just as easily as an interview item could ultimately be answered as true versus false, although this is not typical.

The key distinctions between these methods are with regard to how the test stimuli are presented as well as whose opinion determines the score. In the case of a self-report questionnaire survey, the individual is free to answer each item as they desire and the researcher has limited ability to alter the scoring. This makes a survey a direct and unfiltered report from the target. In contrast, items from an interview are scored by the researcher/interviewer based on their opinion of the individual's response as well as additional information. For example, consider an item asking about a person's humility. In an interview the interviewee could reply to the item by saying: "Oh yes, I am extremely humble. Probably the most humble person you will ever meet. I've had some amazing successes in my life, but I've never let it go to my head. People always comment on how remarkably humble I am given my impressive accomplishments." In such a case the interviewer would be free to score the item as low on humility despite the stated answer, whereas the interviewee would likely endorse a survey item as indicating high humility. Therefore, a careful consideration of the constructs and the method of administration is crucial in deciding between these two methods.

The other primary difference is in the presentation of stimuli. Survey questions are presented to the respondent in written format, typically in a standardized order. The result is an equivalent stimulus for all participants given the required level of reading comprehension. The hallmark of the interview method is that a member of the research team administers the items to the respondent. There is considerably more variability in the sequencing and standardization of the items on interviews than on surveys, depending primarily on the nature of the interview. Interviews can range from completely structured, where the interviewer reads items verbatim and asks specific follow-up questions depending on the answer in a way that resembles the survey format (e.g., the Wechsler intelligence scales), to completely unstructured that allows the interviewer to ask whatever questions they see fit to ascertain the scoring or diagnosis desired. The latter are standard in clinical practice (Perry, 1992) but only rarely employed in research settings, which may complicate the translation of research into practice (Samuel, Suzuki, & Griffin, 2016). The most typical interview methods are "semistructured" in that they provide a set of standardized prompts that are asked of all participants, but then also permit the interviewer to follow up on the responses or ask more questions as they see fit (First et al., 2014).

SELECTING AN INTERVIEW OR SURVEY METHOD

In light of the considerable differences between interview and survey methods, it is useful to examine the costs and benefits of each method so that researchers can determine which is the best fit for a given study.

Pros of Surveys

A major benefit of surveys – and indeed likely a major reason they are commonly used in psychological research – is their incredible efficiency. Surveys allow large samples to be collected with little researcher effort. Historically this meant the use of "mass testing" situations where numbers of participants would take a paper-and-pencil survey at the same time in a given room or mail out copies of the survey for participants to complete on their own time, greatly reducing the ratio of experimenter effort per data point. However, advances in the implementation of surveys via the web (e.g., Qualtrics) have exponentially decreased the effort even further, such that it is now possible to collect samples that are limited in size only by the researcher's budget or the population of interest. Additionally, advances in overall computer technology allow for the use of computerized adaptive testing in which the test adapts items administered based on participants' responses (Zetin & Glenn, 1999). This can further reduce administration time by removing unnecessary, or extra, questions without sacrificing test validity.

The standardized format and direct ratings by the subject also makes surveys very easy to use. The use of existing surveys requires only that researchers administer them in accordance with the instructions, leaving very little chance for administration errors or variation across participants. This feature also makes the assessment process easy for the participant as they are entirely self-paced and thus the survey need take only as long as the individual completing it requires. Taken together, this level of standardization provides some assurance that the items completed (and the responses to those items) are comparable across participants.

Surveys also have the potential benefit of anonymity for the respondent that may maximize honesty for illegal or sensitive topics (e.g., substance use, sexual behaviors). For example, Ong and Weiss (2000) found that the rate of admitted cheating among undergraduate students was 74 percent when participants were guaranteed anonymity, but only 25 percent when the survey was confidential. Even if not anonymous, many people will find it much easier to report personal matters, undesirable characteristics, or embarrassing behaviors when it involves clicking a button on a keyboard rather than detailing it to another person in a face-to-face setting (Newman et al., 2002) – although some exceptions have been noted (e.g., Poulin, 2010).

A final benefit of surveys, as commonly used, is the elimination of scoring errors on the part of the researcher. Because most surveys – although not all (e.g., patient satisfaction surveys) – rely on closed-ended questions with limited response options and a specified scoring algorithm, the scoring process can be automated to the point that scoring errors can only stem from incorrectly entering the items or scoring metrics. This eliminates the need for calculation of interrater reliability.

Cons of Surveys

One frequently articulated concern of survey research is that the answering process is a black box, in that researchers have no way of knowing *why* a respondent chose a particular answer. This yields additional concerns about a myriad of threats to the validity of the observed score on a given item or scale. These threats include lack of reading ability or comprehension, lack of insight/knowledge, attentional limitations, response styles, demographic biases, or even deliberate faking on the part of the respondent. It goes without saying that a participant who lacks the ability to read and comprehend the items on a survey cannot provide an answer that accurately reflects the true level of the construct that the items represent. Further, even if they are able to comprehend the items, there is also the possibility that the individual lacks the insight or knowledge to accurately report on a given item. For example, on a self-report questionnaire, an individual asked about relationship conflicts may be prone to underestimate their role in these conflicts due to lack of insight. Further, a spouse or other informant who is asked to infer the mental state of a target may simply lack the requisite information to make such a rating (e.g., Vazire, 2010). Even in situations where the target has the knowledge and insight to make valid ratings on a survey, there may still be a host of factors that threaten the validity of their scores.

The attention span of the respondent is one such factor that can limit validity in situations where surveys are lengthy. There is also a broad literature on how response styles can impact the validity of surveys (Van Vaerenbergh & Thomas, 2013). The specifics of the response styles are beyond the scope of this chapter, but are thought of as systematic tendencies toward answering items in a given way. These styles, or biases, can include nay-saying (denying attributes), acquiescence (answering affirmatively), extreme responding, midpoint responding, or social desirability and are considered mostly nondeliberate. In contrast, an additional issue that can be particularly relevant in clinical research is deliberate faking whereby participants deliberately distort their responses to make the target seem better than they are (i.e., "faking good") or to appear more impaired than is accurate (i.e., "faking bad"). Although there are ways to minimize such biases or detect such attempts, which we will detail later, researchers and clinicians should continue to be aware of the possibility of response biases.

In addition to the validity threat, there are also a few issues that are tradeoffs to the way in which surveys are typically used. A downside to the standardized stimuli that characterize surveys is that there is no flexibility in following up on responses to ask in another way or clarify the participant's response. While such a strategy is possible by using survey branching or contingent item sets within computerized surveys, there are still limits to the range of follow-up questions that can be done via survey. A final potential downside of surveys is again a tradeoff for their efficiency in that the researcher cedes control of the administration setting. This is not an inherent limitation of surveys, as it is possible for them to be completed in the laboratory. However, practically this is rarely done unless a visit is required for the completion of other methods. The result is that, practically, survey respondents can complete items in any number of contexts (home, work, school, movie theater, etc.) or states (happy, angry, distracted, intoxicated, etc.) that may confound score validity.

Pros of Interviews

It should be clarified again that there is considerable variability in the amount of structure built into interviews. At one extreme, a fully structured interview (one that allows no probes) is, in reality, simply a survey that is administered verbally. Such a case is likely quite infrequent as even structured interviews typically allow for follow-up questions – or probes – to better understand the nature of the response and flesh out details. Thus, a real strength of interviews is the ability of both the interviewer and interviewee to seek clarification. Participants in an interview are able to address issues of comprehension by asking for items to be restated or reworded. Further, they are able to elaborate on their responses in order to communicate nuances and/or indicate the rationale behind their answers.

The ability of the interviewer to probe responses and seek elaboration on answers is the primary advantage of the interview method. In doing so, the interviewer has the ability to build from existing items (in the case of a semi-structured interview) or ask whatever questions he or she sees fit (in the case of an unstructured interview) to garner the information needed to arrive at a score for each item. In this way, the score on an interview represents the integration of multiple points of information (e.g., nonverbal behavior, pace of speech, affective intensity, etc.) that go well beyond the interviewee's stated response.

In addition to the opportunity to seek additional information, the scores on an interview are usually provided by a researcher who has been well trained in the scoring protocol and is a (presumably) neutral third party. This makes interview scores potentially less vulnerable to the response biases, lack of insight, or other validity threats that can hamper surveys. Finally, the administration setting for an interview can be standardized such that each interviewee completes the interview in a laboratory setting that is removed from everyday life in a way that can free them from distractions and potentially affective interference.

Cons of Interviews

We hope it has been clear to this point that a major feature of interviews is the intensive oversight of the information being gathered on the part of the trained interviewer. Of course, this also represents its greater downside as such a procedure is incredibly costly in terms of researcher time since it requires an interviewer to be present one-on-one with each research participant. Not only does this greatly impede the efficiency of data collection, but it also results in a much greater per-participant financial cost. As such, the advantages of interviews need to be considered carefully in light of the significant cost – perhaps even to the point that some demonstrable benefit must recommend interviews, rather than treating them as the default gold standard.

Another factor to consider is that the one-on-one nature of interviews also makes it all the more challenging to find a suitable time for an individual to participate. Scheduling a participant necessitates the researcher and participant agreeing on a mutually workable time and then coordinating any logistical details that might go along with it (e.g., a physical space for an in-person interview or the calling details for a phone interview). To manage this successfully, a great degree of flexibility is required on the part of a research team in order to accept all interested subjects who might have nontraditional work schedules or caretaking responsibilities. Regardless of the level of flexibility on the part of the assessor, it still remains probable that the nature of some pathologies may actually be related to greater difficulty in scheduling and completing an interview, making missing data points more problematic.

For any interview to be valid all interviewers must be interchangeable with one another. To achieve (or approximate) this goal, extensive training is often necessary before a research staff member is ready to conduct an interview. This training can be quite time-consuming and expensive. For example, the training listed on the website for the Structured Clinical Interview for DSM-IV Axis I (First et al., 1996), suggested each interviewer first watch an 11-hour DVD training program before embarking on any onsite training. Even once initial training has been successful (i.e., each interviewer has been trained to some preestablished criterion), ongoing training is necessary to guard against rater drift (Rogers, 2001). This often requires regularly scheduled meetings in which all interviewers discuss difficult coding decisions and receive ongoing fidelity assessments (Widiger & Samuel, 2005).

Even when interview training is thorough, a routine best practice for the valid administration of interviews is estimating the reliability across raters. This is because, as noted above, the interviewer introduces a new source of error in that two different interviewers may yield different scores for the same interviewee. This can occur for a variety of reasons but can basically be distilled into three major categories: idiosyncratic administration, interviewer stimulus effects, or unreliable scoring (Rogers, 2001). Aside from fully structured interviews, a feature of this method is the ability for the interviewer to use follow-up probes or ask additional questions to clarify an answer for scoring purposes. Although this is designed to increase validity, it can introduce unreliability. For example, if two interviewers differ in terms of the frequency, style, or type of question asked then this idiosyncratic administration can elicit different information from the interviewee. It may also be the case that idiosyncratic features of the interviewer (or the similarity with the interviewee) may differentially elicit information from the participant. These can take the form of demographic or physical features of the interviewer (e.g., a female participant may be more willing to disclose details of their sexual history to another female than to a male) as well as more state-based features (e.g., an interviewer who is particularly chipper on a given day versus tired on another may elicit different responses from the participant). Finally, even if interviewers elicit the same responses or information from participants they still might assign different scores. In such a case, this idiosyncratic scoring might be due to biases on the part of the interviewer (e.g., gender biases) or inaccurate encoding of all information to formulate a final score (e.g., Morey & Benson, 2016). There may also be halo effects in which scores on specific items bleed over and color the interviewer-assigned ratings more broadly.

In sum, a central concern about interviews – beyond their intensive nature – is the possibility of unreliability across raters (Samuel, 2015). Researchers often take a number of steps to minimize unreliability, including training, as well as analyses that examine interrater reliability. The typical form of interrater reliability utilized in most diagnostic interview settings is to have another interviewer listen to an audio or video recording of the session and make ratings, which can be compared to those of the original interviewer. Importantly, this method can only examine the reliability of scoring assignments. As Chmielewski and colleagues (2015) have noted, this is a less than perfect approach as it does not take into account potential differences related to the administration of the interview or stimulus value of the interviewer. To do this, the interview should be independently administered by two separate researchers to see how well scores match.

CHOOSING AN EXISTING MEASURE OR DEVELOPING A NEW ONE

Once the researcher has determined that either the survey or interview methods will likely suit their needs, the next step is to determine if there is an existing measure that will assess the construct of interest or if a new one must be developed. In either case, a thorough review of the literature is an imperative first step. This will reveal measures of similar/same constructs and the researcher can determine if any meet the needs of the particular study. Ideally, any existing measure chosen would have been vetted so that its psychometric properties and validity are known, as opposed to using a measure that is developed and used ad hoc in a given study. If the researcher determines that no existing measure provides the type of assessment they seek, then a new measure can be created.

Principles of Scale Construction

A number of wonderful resources exist on the topic of scale construction (e.g., Clark & Watson, 1995) and it is well beyond the scope of this chapter to reproduce those in their entirety. Nonetheless, it is worth summarizing some of the basics here. Most notable is that this process is iterative and self-correcting, in that results from one step provide feedback on prior and subsequent steps. Following a thorough literature review to determine that no suitable assessment exists, the researcher should create an operational definition of the construct of interest. This definition should specify both the breadth of the construct (i.e., what it does include) as well as its boundaries (i.e., what it does not include). An operational definition should also specify any lower-order components of that construct. Moreover, this operational definition should offer a conceptual and theoretical account of what this construct is and how it fits into the larger context (Loevinger, 1957), see Chapter 2, by Zachar and colleagues, for more details.

Following the literature review and construct operationalization the scholar next creates a large list of items that strategically samples all the content within the defined construct (including subconstructs), as well as content that is at the edges of, and even past, the boundaries. The idea behind this process is that the irrelevant items will be identified in subsequent validation steps and discarded, but relevant content cannot be added back in at a later point. Thus, the goal of this item development process is one of overinclusiveness (Clark & Watson, 1995). In addition, items should be worded as simply as possible to capture the intended meaning so as to improve readability and avoid idiosyncratic interpretations. Items are also best suited to detect a construct when they reflect *only* that construct. That is, it is advisable to avoid items that represent a blend of two different constructs.

In the next stage, this large pool of items is administered to a sample of the population the scale is intended to measure. This point is both obvious and crucial, as it would be of little value to examine item performance in a sample irrelevant to the population of interest. It is worth mentioning here that this step can look quite different in questionnaires versus interviews. Indeed, developing a large (500+) item pool and administering it to several hundred participants is readily achievable for questionnaire development (e.g., Simms et al., 2013), but it would be quite onerous to do this in the development of a structured interview. This provides some inherent psychometric advantages for questionnaires over surveys as they are typically developed.

The properties of items in this development pool can be evaluated to eliminate items that show uniform endorsement (all participants answer the question the same way), redundancy (typically items correlating above 0.5), or that show poor internal consistency (i.e., those that do not correlate with the others). Most of these procedures stem from classical test theory (CTT), which emphasizes psychometric indicators of observed items. A related alternative is item response theory (IRT; Embretson & Reise, 2000), which examines latent item properties such as the discrimination parameter (i.e., how well the item differentiates individuals at various levels of the latent trait) and the difficulty parameter (i.e., the point along the latent trait continuum where the item best discriminates those individuals). Although sometimes pitched as competitors, these approaches are similar and are perhaps best thought of as complementary and overlapping. One area where they may well differ, though, is in their treatment of infrequently endorsed items. As noted above, a typical approach within CTT is to remove infrequent items as they have less information value. However, within IRT, the test developer may actually prefer such items as they will provide information at different levels of the latent trait and, as such, will maximize the coverage of that trait by the scale. This property has proven quite useful on high-stakes tests predicting academic achievement.

Regardless of the method of deriving the item pool, there is general consensus that the resulting instrument should have homogenous scales (Smith, McCarthy, & Zapolski, 2009). Although an instrument can have multiple scales, each individual one should measure a unitary construct, rather than some amalgam. This again emphasizes the need to have items that are themselves univocal.

A central step in any scale's construction is the examination of construct validity. Construct validation is again a very broad concept and has been the source of many seminal works (Cronbach & Meehl, 1955), prior reviews (Strauss & Smith, 2009), and is discussed by Furr in Chapter 6. For all those reasons, we will not repeat this in depth here, but rather summarize briefly the key issues of construct validation. A baseline requirement for construct validation is reliability. This term can connote somewhat different meanings, but typically represents the degree to which the scale is internally consistent. This should not rely exclusively on Cronbach's alpha, as it varies considerably based on the number of items (Lance, Butts, & Michels, 2006). Instead, one should prefer metrics like average interitem correlation or McDonald's omega (McDonald, 1999), which detects the degree to which variance in the items are attributable to a single factor. In cases where the construct is conceptualized as traitlike, such as personality, the degree of stability over brief intervals (i.e., dependability; Watson, 2004) can also be an indicator of transient error, or unreliability. The idea behind each of these metrics is that the reliability of a measure places a cap on its validity.

Validity is itself a multifaceted idea, with a variety of specific types of validity subsumed under the umbrella of construct validity, which indicates the degree to which a given scale actually measures the latent construct it intends to operationalize. This can include content validity, which indicates the degree to which the item/scale is deemed to represent the content of the operationalized construct definition and is typically assessed by consulting expert raters, but may also be evidenced by simple face validity (i.e., the item appears to measure the intended construct). Most typically, this includes an evaluation of criterion-related validity, which can either be concurrent or predictive. In a typical construct validation study, the researcher will choose a set of constructs assessed by existing measures that range in their degree of conceptual linkage with the target construct. For example, a novel measure of intelligence might reference existing measures of intelligence as well as proximal constructs, such as academic achievement, and more distally related (or even potentially unrelated) concepts like openness to experience. The goal is to utilize these measures to situate the newly created measure in the "nomological network" (Cronbach & Meehl, 1955). Ideally the new measure should correlate most highly with the most conceptually related constructs (i.e., convergent validity) and fail to correlate with conceptually unrelated constructs (i.e., discriminant validity).

During the construct validation, the test developer is also encouraged to attend to the issue of shared method variance. It is well known that measures from the same method (e.g., self-report questionnaire) will often correlate more highly than measures of a different method (Campbell & Fiske, 1959). As such, it is advised to select not only an array of related and unrelated constructs, but also to vary the methods in order to create the multitrait, multimethod matrix.

Finally, the newly developed measure should be administered to a new sample to examine the psychometric properties and construct validity. Often the newly constructed measure is optimized for the initial sample. However, due to sampling error, the properties of the measure should be examined in an independent sample to ensure the findings are replicable and generalizable.

The goal of many assessment measures is often to make predictions about future states or behaviors (e.g., risk of violence, job or academic performance, illness trajectories) and so another important aspect is predictive validity. Their longitudinal nature makes these tests less common in the psychometric literature, but all the more informative when available. No test can be said to be validated, or valid in an absolute sense; rather a test has validity support for a given purpose.

OTHER CONSIDERATIONS FOR CONSTRUCTING AND EVALUATING SURVEYS OR INTERVIEWS

Construct Specification

Within the overarching theme of construct specification, much attention is rightly drawn to the issue of operationally defining the construct and specifying how it should relate to other measures. However, there are other features of construct specification that are less commonly considered. A necessary step in defining a construct is to hypothesize about the temporal features of the construct explicitly; that is, should the target be considered a state that will shift over time, or be more traitlike and durable across time and situation. This has major implications for the proposed nomological network as well as for the type of items that are written and the timeframe specified by the question. It would be ideal for the construct specification phase to hypothesize the expected test-retest stability over various intervals (e.g., a measure of the emotion of surprise would be expected to have very low stability even over short intervals).

As noted earlier, it is preferable for a given scale to be homogenous and assess a single latent construct. That said, it is quite routine for an instrument to include multiple scales, including some that may be conceptually orthogonal and others that are highly intercorrelated, but separable (e.g., facets within personality models). For this reason, test developers should specify a priori how they believe the scale and overarching instrument will be structured. There are a variety of possibilities (unidimensional, correlated higher-order factors, bifactor, etc.), which might all be reasonable, but it is incumbent upon the developer to offer a falsifiable hypothesis about how it will be structured. Furthermore, the iterative process of scale construction and validation will often feed back on itself such that subsequent examinations of structure may well result in modifications to the theory. For example, two scales that were thought to be distinct may be combined based on tests of latent structure, or a scale thought to load on one higher-order domain may shift to another. In other cases, however, the departure of latent structure analyses from a priori expectation may suggest the need to revise the scale's content (Clark & Watson, 1995).

A final point, which in our estimation is often neglected, is the issue of continuum specification. Continuum specification refers to the hypothesized nature of a construct's distribution as well as the number and nature of its end poles. For example, there has been a great deal of work concerning whether the emotions of sadness and happiness represent opposite poles of the same continuum (i.e., bipolar) versus two distinct (i.e., unipolar) concepts (Tay & Kuykendall, 2017). When defining and operationalizing a construct the developer should specify the ends of the continuum and note whether a low standing reflects simply the absence of the construct or an elevated level of an opposite construct. This property of constructs has major implications for a test developer as it provides the breadth of item content that should be included. To be clear, not every measure should aim to assess all possible levels of the continuum. For example, a scale intended to measure depression will reasonably focus its assessment on the extreme levels of sadness, but it would also be helpful to specify whether low scores reflect a lack of depression (e.g., neutral mood), happiness, or even risk for mania.

Practical Considerations in Scale Development

The preceding sections have focused primarily on the conceptual issues in developing survey and interview instruments. This final section concerns primarily practical considerations in developing an instrument. A broad issue that we return to at this point is the *source* of the information, or the person who will actually be completing the measure. In many cases, particularly with surveys, there is an expectation that the questionnaire will be completed by the individual who is being assessed (i.e., the target). Indeed, in many cases the target is the best source of information about their own thoughts, feelings, behaviors, and abilities. Yet there is also a broad literature demonstrating the value of considering informant sources for assessing psychopathology (Rescorla et al., 2016), job performance (Barrick, Mount, & Judge, 2001), and minors (Tackett et al., 2013). These informants might be those with any variety of relation to the target, including parents, peers, spouses, teachers, supervisors, observers, or healthcare providers. Test developers should carefully consider the source, or sources, that might provide the most useful information about the target construct and create the instrument accordingly (Alexander et al., 2017). For example, a self-report scale assessing advanced dementia would seem less than ideal in isolation, as would an informant report of a target's internal mental states.

Another practical consideration is the tendency for people to answer items (whether survey or interviews) in ways that may reflect an idiosyncratic response style, response bias, or even intentional distortion that might subvert the validity of their scores. There are a number of methods that have been implemented to deal with these possibilities. At the most extreme level, many instruments – particularly for psychopathology – contain a set of validity scales to assess for these response styles (e.g., the Minnesota Multiphasic Personality Inventory [MMPI]; Ben-Porath, 2012). These typically include extremely rare or impossible items (e.g., "I was born on the moon") to detect lapses in attention, random responding, excessive endorsement, defensiveness, denial, faking, or variable responding across the instrument. In each case, these scales are solely used to screen out invalid responses, as attempts to utilize validity metrics to "correct" scores on substantive scales has been shown to reduce, rather than improve, score validity (e.g., the "K-scale correction" on the MMPI).

It is notable here that response bias concerns and validity scales have been exclusive to self-report sources, yet there is no compelling evidence that whatever problem exists is restricted to self-report. Indeed, most interviews or informant survey measures that are completed by spouses, parents, peers, or clinicians do not have validity scales, yet we are unaware of any evidence to suggest these sources have any fewer issues in this regard. Obviously, test developers should consider carefully how they might handle response biases and future research should more clearly articulate the nature and magnitude of concerns from multiple sources.

There are also more subtle techniques designed to account for some of these response styles. For example, utilizing reverse-keyed items can be used to account for individuals who simply endorse all items affirmatively. That said, it has not always been clear that reverse-keyed items actually solve this problem in practice. It is not atypical for reversed items to hang together and perhaps even form their own (sub)factor and generally have inferior psychometric properties to positively worded items (Rodebaugh, Woods, & Heimberg, 2007; van Sonderen, Sanderman, & Coyne, 2013). This appears to be the case due to the increased comprehension demands for the test-taker, potentially amplifying problems with groups that have lower reading levels, cognitive decline, or for translated measures. Thus, although their usage remains fairly common – we have used them ourselves (Samuel et al., 2012) – test developers should carefully consider their pros and cons when developing new surveys or interviews.

Another central consideration that often gets less attention from test developers is carefully choosing the response options/anchors to items as well as item probes. This entails the instructions to examinees about what they are rating. Most typically, mental health questionnaires utilize an agreement metric, but there are also a number of examples of tests where the answer is based on the frequency with which a certain symptom is exhibited. This can create confusion, particularly when writing items about relatively infrequent, but intense, affective states. An item such as "Please rate the extent to which you have experienced an urge to self-harm in the past week" can be quite confusing for respondents. Considering a person who has had a single episode of intense desire to self-harm, should they respond by saying "a little," based on the fact that it only happened once in the past week? Or should they respond "quite a bit," based on how intensely they felt in that one instance? Clear instructions on how to rate the frequency or intensity (or perhaps even having them rate both, if that information is valuable) is key.

The number and type of response options are also areas that rely primarily on anecdote and tradition rather than empirical findings. Typically, most scales utilize either a true-false format or a Likert-type scale. Although Likert-type scales typically use between four and five response options (with intense debate about the utility and meaning of a "middle" option), there is a wide variety used, including visual analog scales that have a nearly infinite number. Unfortunately, the amount of research on these considerations is quite slim (e.g., Froman, 2014; Preston & Colman, 2000). Ideally, a test developer would both consider this issue carefully and pilot test a variety of candidates before settling on the set of response options that maximizes the reliability, validity, and utility of the scale.

A final consideration is the length of a scale. These run the gamut from several hundred items to even some "instruments" that claim to assess complex constructs with only a single item. In practice, the advice we provide is to start with the absolute minimum of three items per construct and move up from there based on the cost-benefit analysis of informational value versus time cost. The minimum of three items provides the ability to call it a "scale" and makes it amenable to latent variable modeling. Ideally, any number above that point will be based on a number of factors including construct breadth and depth, assessment time, and precision.

In closing, we hope that this chapter has provided a summary of the pros and cons of the survey and interview methods of clinical research for the reader to make an informed decision regarding the measurement they use. Despite criticisms, these remain – by far – the most common methods of assessing clinical constructs in the research literature. This is driven in no small part by their efficiency and ease of use. Nonetheless, this ease of use has also allowed these methods to develop a massive literature on their psychometric properties. One can only hope that alternative methods of assessment are ultimately held to the same psychometric standards as these methods.

REFERENCES

Alexander, L. A., McKnight, P. E., Disabato, D. J., & Kashdan, T. B. (2017). When and How to Use Multiple Informants to Improve Clinical Assessments. *Journal of Psychopathology and Behavioral Assessment*, 39(4), 669–679.

Barrick, M. R., Mount, M. K., & Judge, T. A. (2001). Personality and Performance at the Beginning of the New Millennium: What Do We Know and Where Do We Go Next? *International Journal of Selection and Assessment*, 9(1–2), 9–30.

Ben-Porath, Y. S. (2012). *Interpreting the MMPI-2-RF*. Minneapolis, MN: University of Minnesota Press.

Campbell, D. T., & Fiske, D. W. (1959). Convergent and Discriminant Validation by the Multitrait-Multimethod Matrix. *Psychological Bulletin*, 56(2), 81–105.

Chmielewski, M., Clark, L. A., Bagby, R. M., & Watson, D. (2015). Method Matters: Understanding Diagnostic Reliability in DSM-IV and DSM-5. *Journal of Abnormal Psychology*, 124(3), 764–769.

Clark, L. A., & Watson, D. (1995). Constructing Validity: Basic Issues in Objective Scale Development. *Psychological Assessment*, 7(3), 309–319.

Cronbach, L. J., & Meehl, P. E. (1955). Construct Validity in Psychological Tests. *Psychological Bulletin*, 52(4), 281–302.

Embretson, S. E., & Reise, S. P. (2000). *Item Response Theory for Psychologists*. Mahwah, NJ: Lawrence Erlbaum.

First, M. B., Bhat, V., Adler, D., Dixon, L., Goldman, B., Koh, S., ... Siris, S. (2014). How Do Clinicians Actually Use the Diagnostic and Statistical Manual of Mental Disorders in Clinical Practice and Why We Need to Know More. *Journal of Nervous and Mental Disease*, 202(12), 841–844.

First, M. B., Spitzer, R. L., Gibbon, M., & Williams, J. B. W. (1996). *Structured Clinical Interview for DSM-IV Axis I Disorders*. Washington, DC: American Psychiatric Press.

Froman, R. D. (2014). The Ins and Outs of Self-Report Response Options and Scales. *Research in Nursing & Health*, 37(6), 447–451.

Haeffel, G. J., & Howard, G. S. (2010). Self-Report: Psychology's Four-Letter Word. *American Journal of Psychology*, 123(2), 181–188.

Lance, C. E., Butts, M. M., & Michels, L. C. (2006). The Sources of Four Commonly Reported Cutoff Criteria: What Did They Really Say? *Organizational Research Methods*, 9(2), 202–220.

Loevinger, J. (1957). Objective Tests as Instruments of Psychological Theory. *Psychological Reports*, 3(4), 635–694.

Loevinger, J. (1979). Construct Validity of the Sentence Completion Test of Ego Development. *Applied Psychological Measurement*, 3, 281–311.

McDonald, R. P. (1999). *Test Theory: A Unified Treatment*. Mahwah, NJ: Lawrence Earlbaum.

Morey, L. C., & Benson, K. T. (2016). An Investigation of Adherence to Diagnostic Criteria, Revisited: Clinical Diagnosis of the DSM-IV/DSM-5 SECTION II Personality Disorders. *Journal of Personality Disorders*, 30(1), 130–144.

Newman, J. C., Jarlais, D., Turner, C. F., Gribble, J., Cooley, P., & Paone, D. (2002). The Differential Effects of Face-to-Face and Computer Interview Modes. *American Journal of Public Health, 92*(2), 294–297.

Nisbett, R. E., & Wilson, T. D. (1977). Telling More Than We Can Know – Verbal Reports on Mental Processes. *Psychological Review, 84*(3), 231–259.

Ong, A. D., & Weiss, D. J. (2000). The Impact of Anonymity on Responses to Sensitive Questions. *Journal of Applied Social Psychology, 30*(8), 1691–1708.

Perry, J. C. (1992). Problems and Considerations in the Valid Assessment of Personality Disorders. *American Journal of Psychiatry, 149*(12), 1645–1653.

Poulin, M. (2010). Reporting on First Sexual Experience: The Importance of Interviewer-Respondent Interaction. *Demographic Research, 22*, 237–287.

Preston, C. C., & Colman, A. M. (2000). Optimal Number of Response Categories in Rating Scales: Reliability, Validity, Discriminating Power, and Respondent Preferences. *Acta Psychologica, 104*(1), 1–15.

Rescorla, L. A., Achenbach, T. M., Ivanova, M. Y., Turner, L. V., Arnadottir, H., Au, A., … Zasepa, E. (2016). Collateral Reports and Cross-Informant Agreement about Adult Psychopathology in 14 Societies. *Journal of Psychopathology and Behavioral Assessment, 38*(3), 381–397.

Rodebaugh, T. L., Woods, C. M., & Heimberg, R. G. (2007). The Reverse of Social Anxiety Is Not Always the Opposite: The Reverse-Scored Items of the Social Interaction Anxiety Scale Do Not Belong. *Behavior Therapy, 38*(2), 192–206.

Rogers, R. (2001). *Handbook of Diagnostic and Structured Interviewing*. New York: Guilford Press.

Samuel, D. B. (2015). A Review of the Agreement Between Clinicians' Personality Disorder Diagnoses and Those From Other Methods and Sources. *Clinical Psychology-Science and Practice, 22*(1), 1–19.

Samuel, D. B., Riddell, A. D. B., Lynam, D. R., Miller, J. D., & Widiger, T. A. (2012). A Five-Factor Measure of Obsessive-Compulsive Personality Traits. *Journal of Personality Assessment, 94*(5), 456–465.

Samuel, D. B., Suzuki, T., & Griffin, S. A. (2016). Clinicians and Clients Disagree: Five Implications for Clinical Science. *Journal of Abnormal Psychology, 125*(7), 1001–1010.

Simms, L. J., Goldberg, L. R., Watson, D., Roberts, J., & Welte, J. (2013). The CAT-PD Project: Introducing an Integrative Model & Efficient Measure of Personality Disorder Traits. Paper presented at the Society for Research in Psychopathology, Oakland, CA.

Smith, G. T., McCarthy, D. M., & Zapolski, T. C. B. (2009). On the Value of Homogeneous Constructs for Construct Validation, Theory Testing, and the Description of Psychopathology. *Psychological Assessment, 21*(3), 272–284.

Strauss, M. E., & Smith, G. T. (2009). Construct Validity: Advances in Theory and Methodology. *Annual Review of Clinical Psychology, 5*, 1–25.

Tackett, J. L., Herzhoff, K., Reardon, K. W., Smack, A. J., & Kushner, S. C. (2013). The Relevance of Informant Discrepancies for the Assessment of Adolescent Personality Pathology. *Clinical Psychology-Science and Practice, 20*(4), 378–392.

Tay, L., & Kuykendall, L. (2017). Why Self-Reports of Happiness and Sadness May Not Necessarily Contradict Bipolarity: A Psychometric Review and Proposal. *Emotion Review, 9*, 146–154.

Trull, T. J., & Ebner-Priemer, U. (2013). Ambulatory Assessment. *Annual Review of Clinical Psychology 9*(9), 151–176.

Van Sonderen, E., Sanderman, R., & Coyne, J. C. (2013). Ineffectiveness of Reverse Wording of Questionnaire Items: Let's Learn from Cows in the Rain. *Plos One, 8*(7), e68967.

Van Vaerenbergh, Y., & Thomas, T. D. (2013). Response Styles in Survey Research: A Literature Review of Antecedents, Consequences, and Remedies. *International Journal of Public Opinion Research, 25*(2), 195–217.

Vazire, S. (2010). Who Knows What About a Person? The Self-Other Knowledge Asymmetry (SOKA) Model. *Journal of Personality and Social Psychology, 98*(2), 281–300.

Watson, D. (2004). Stability versus Change, Dependability versus Error: Issues in the Assessment of Personality over Time. *Journal of Research in Personality, 38*(4), 319–350.

Widiger, T. A., & Samuel, D. B. (2005). Evidence-Based Assessment of Personality Disorders. *Psychological Assessment, 17*(3), 278–287.

Zetin, M., & Glenn, T. (1999). Development of a Computerized Psychiatric Diagnostic Interview for Use by Mental Health and Primary Care Clinicians. *CyberPsychology & Behavior, 2*(3), 223–229.

6 Psychometrics in Clinical Psychological Research

R. MICHAEL FURR

Measurement is at the heart of science, and good science depends on good measurement. Clinical psychological science is no exception – well-grounded empirical, theoretical, and practical clinical work hinges on high-quality measurement of clinical constructs. The purpose of this chapter is to outline core psychometric principles and practices, and to highlight recent psychometric trends relevant to clinical psychological science. The chapter's ultimate goal is to enhance readers' ability to produce high-quality clinical research by providing an integrated, up-to-date overview of fundamental psychometric principles and practices.

THE IMPORTANCE OF GOOD PSYCHOMETRIC QUALITY IN CLINICAL PSYCHOLOGICAL RESEARCH

Good measurement facilitates scientific progress. In terms of statistical results, poor psychometric quality can lead to underestimation of important effects (e.g., inaccurately weak correlations or small differences between groups). However, it can also lead researchers to *over*estimate trivial effects (e.g., obtain inaccurately strong correlations or large mean differences). More conceptually, poor measurement can create ambiguity or inaccuracy when interpreting scores derived from psychological measures. That is, poor measurement might lead researchers to misinterpret the psychological meaning of a set of test (or inventory or survey) scores.

Considering such problems, poor psychometric quality can lead to inaccurate conclusions about one's overarching psychological questions or hypotheses. If one's statistical results are inaccurate or if one's psychological interpretation of the data's meaning is incorrect, then the ensuing conclusions are very likely inaccurate.

Such problems underscore the importance of measurement quality. This applies to both observational and experimental research, and to all types of measures, including self-report inventories, state-like measures of emotion, clinical diagnoses, physiological measures, cognitive assessments, and ability or knowledge tests. To minimize problems arising from poor measurement quality, researchers should use measures with strength in three areas of psychometric quality – well-understood dimensionality (i.e., factorial structure), good reliability, and well-supported construct validity.

DIMENSIONALITY

Meaning

Dimensionality is the number and nature of the psychological constructs affecting responses to a measure (Furr, 2018, chapter 4). Some measures might reflect a single psychological construct, with only that construct systematically driving responses. However, many clinical measures have more complex dimensionality, with two or more constructs driving responses.

Consider the Generalized Anxiety Disorder Questionnaire-IV (GAD-Q-IV; Newman et al., 2002). The GAD-Q-IV is a nine-item self-report measure intended to reflect Generalized Anxiety Disorder (GAD). Participants' responses to the items are assumed to be affected by one psychological construct – GAD. A participant with a high level of GAD is expected to respond to all items in a way that reveals GAD (e.g., responding "yes" to the item "Do you experience excessive worry?"). But is this assumption of unidimensionality warranted? Perhaps multiple constructs systematically affect participants' responses. For example, perhaps some items represent more cognitive facets of anxiety, whereas others reflect more affective facets. If cognitive symptoms do not go hand in hand with affective symptoms, then the GAD-Q-IV might reflect two distinguishable psychological constructs, rather than one. If so, then how strongly connected are those constructs? Cognitive symptoms might be strongly correlated with affective symptoms; however, a participant might report a high level of cognitively oriented symptoms and a low level of affective symptoms, suggesting a weak correlation. The number and nature of the constructs reflected in a measure are issues of dimensionality.

IMPORTANCE

Dimensionality is important for at least two reasons. First and most practically, it has implications for scoring a measure, as each score should reflect a single dimension. That is, each score should be based upon a set of items with a unidimensional structure (Zeigler & Hagemann, 2015). If a set of items is *not* unidimensional, then it is inadvisable to create a single score for that set. Indeed, a score based upon a multidimensional set of items might have no clear psychological meaning. That said, a set of items might be close enough to unidimensional to be scored as such (Reise, Waller, & Comrey, 2000). For example, a set of items might reflect two dimensions that are highly correlated with each other. This might allow for three scores – one for each dimension, and one total score combining the dimensions. The key point is that researchers must understand a measure's dimensionality in order to obtain psychologically meaningful scores from that measure.

Second, and of greater theoretical importance, dimensionality informs an understanding of the construct(s) reflected by a measure (Floyd & Widaman, 1995). For example, if the GAD-Q-IV reflected cognitive and affective dimensions, then this might affect researchers' understanding of the nature of GAD more generally. Perhaps even the diagnosis would shift with the recognition that GAD is realized via different pathways. Thus, for both practical and theoretic reasons, researchers should evaluate dimensionality when developing, evaluating, and potentially revising a measure.

Evaluation via Exploratory Factor Analysis

Dimensionality is typically evaluated through factor analysis – either exploratory factor analysis (EFA) or confirmatory factor analysis (CFA). EFA is most appropriate when researchers have no robust hypotheses about a measure's dimensionality, whereas CFA is appropriate when there are clear hypotheses.

Figure 6.1 summarizes the key steps in EFA. The first step is selecting an extraction method, which is the mathematical procedure to use. For most psychometric purposes, a true factor analysis method (e.g., principal axis factoring, PAF) is preferable to a principal components analysis (Fabrigar et al., 1999; Floyd & Widaman, 1995; see Widaman 1993 for details on the differences), though results are often similar (but cf. Widaman, 1993). Indeed, PAF is common and reasonable, particularly when data are continuous and normally distributed. However, additional methods are available, including maximum likelihood, weighted least squares, and unweighted least squares. Although the choice often makes little practical difference, the optimal choice can depend on the nature of the data – e.g., are the data binary/ordinal or continuous; are responses normally distributed or non-normally distributed? For example, Mplus (Muthén & Muthén, 1998–2017) offers different options and defaults depending on

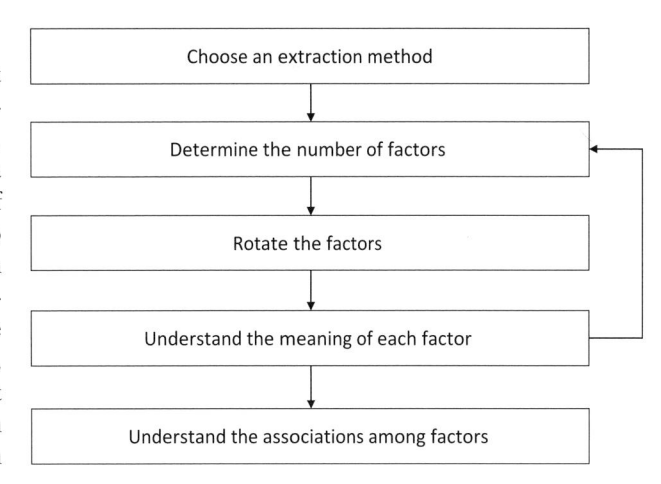

Figure 6.1 Steps in exploratory factor analysis

such issues. Similarly, Revelle (2017) recommends using polychoric or tetrachoric correlations (rather than Pearson correlations) when factor analyzing binary or ordinal data. Using such correlations is easily done via the *psych* package in R (Revelle, 2017), and possible (though not as easily accomplished) in SAS and SPSS (e.g., the HETCOR extension in SPSS, the CORR procedure in SAS).

A second step is identifying the appropriate number of dimensions (or factors), and there are several methods for evaluating this. Perhaps the best known is the problematic "eigenvalue greater than one" rule, which should generally be avoided partly because it often suggests more factors than is reasonable (Fabrigar et al., 1999). Another well-known method is to evaluate how many factors are needed to explain a targeted amount of common variance (e.g., Floyd and Widaman recommend 80 percent). Others have recommended the "scree plot" in which one plots the ordered eigenvalues of the factor solution and looks for an "elbow" to determine how many factors to retain. Although decisions based upon scree plots are subjective and clear decisions are not always possible, most statistical packages provide this information and it is a relatively good method (Fabrigar et al., 1999). Parallel analysis (Hayton, Allen, & Scarpello, 2004) is yet another method for identifying the number of dimensions, but despite receiving endorsements, it does not yet enjoy widespread use. Fortunately, several software packages, including Mplus, SAS, SPSS, and R, now include procedures for conducting parallel analysis (Muthén & Muthén, 1998–2017; O'Connor, 2000; Revelle, 2017). The recently developed Comparison Data method (CD; Ruscio & Roche, 2012) is yet another promising method, but it is currently available only via R (Ruscio, n.d.). A variety of additional good methods have been developed. These include the Minimum Average Partial test (MAP; Velicer, 1976), the Very Simple Structure method (VSS; Revelle & Rocklin, 1979), and the use of fit indices when analyses are based upon Maximum Likelihood extraction. The MAP and VSS methods are not widely available, though they can be conducted via SAS, SPSS, or R (O'Connor, 2000;

Revelle, 2017). Considering their current availability in statistical software, their relative accuracy, and readers' likely familiarity with the procedures, a combination of scree plots and parallel analysis may be the best approach to evaluating the number of dimensions in an EFA context.

Although there are many statistically oriented methods for identifying the appropriate number of dimensions reflected in a measure, theory and conceptual clarity are also crucial considerations. That is, the psychological meaningfulness of any given dimensionality should be weighed heavily alongside any statistical indices. Moreover, the optimal approach is to examine several relevant statistical indices, and consider their results in light of conceptual clarity (Fabrigar et al., 1999). For example, a parallel analysis of the GAD-IV might suggest three factors. However, inspection of the scree plot might strongly indicate only two factors, and inspection of the factors' content (see step 4, below) might suggest that the third factor is conceptually unclear or uninterpretable. Balancing such statistical and conceptual considerations might lead toward the two-factor structure.

Rotation is a third step in EFA. Its purpose is, roughly, to produce results that are relatively likely to be clear and interpretable (i.e., that achieve "simple structure," see step 4). Thus, rotation is typical, though different types of rotation are possible. One broad distinction is between orthogonal and oblique rotations. Whereas orthogonal rotations (e.g., varimax) force factors to be uncorrelated, oblique rotations (e.g., promax, oblimin) allow factors to be correlated. Most statistical software offers several options for each of these. Despite the apparent popularity of orthogonal rotations, oblique rotations are likely to be more useful and reasonable in psychometric work (Reise et al., 2000). In many testing-focused applications of EFA, latent constructs are likely correlated with each other (at least to some degree). Therefore, rotations that allow and reveal such correlations are preferable to those that assume and impose orthogonality.

A fourth step is examining connections between factors and items in order to understand the factors' psychological meaning. Ideally, a simple structure is revealed, in which each item loads strongly on only one factor, and each factor has multiple items loading strongly on it. A factor's meaning is then defined by the items loading strongly on it. Practically, the items loading strongly on a factor would typically be aggregated (e.g., averaged) to form a scale score. This score would then be interpreted in terms of the psychological dimension reflected in those items. Importantly, EFA might reveal items that do not have simple structure. For example, some items might load strongly on multiple factors, some items might load weakly on all factors, or some factors might include only one or two items that load strongly. As shown in Figure 6.1, such results might motivate reexamination of the number of factors (e.g., moving from two to three factors). Alternatively, researchers might drop problematic items. For example, an item that loads weakly on all factors likely does not belong on the measure.

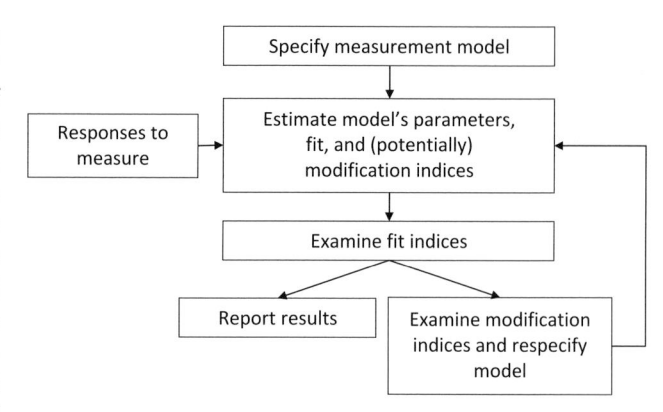

Figure 6.2 Steps in confirmatory factory analysis

If there are multiple factors (i.e., the measure is multidimensional) and if an oblique rotation was chosen at step 3, then a fifth step is to examine the correlations among the factors. This informs the advisability of creating a total score for the measure. Researchers are likely to create a score for each dimension of a multidimensional measure. However, they might also consider creating a "total score" reflecting all dimensions. For example, if steps 1–4 suggest that the GAD-Q-IV has two dimensions, then we would likely create a score for each dimension – say a cognitive anxiety score and an affective anxiety score (depending on what was revealed about the factors in step 4). However, we might also consider a "total GAD" score comprising both cognitive and affective facets of anxiety. Such a total score is meaningful only if the cognitive and affective factors themselves are sufficiently correlated (say $r > 0.60$). If factors are weakly correlated, then a total score is not meaningful and is not warranted.

Evaluation via Confirmatory Factor Analysis

CFA is inherently connected to psychometric theory and practice. No longer simply an analytic tool for examining psychometric issues, CFA's concepts are now used to conceptualize some key psychometric issues. Thus, a sophisticated perspective on psychometrics requires familiarity with CFA (see Brown's [2015] thorough introduction).

Figure 6.2 presents broad steps in CFA. The process begins with clear views about a measure's dimensionality, as reflected in a coherent measurement model. A model is defined by several fundamental aspects, including the number of factors, the connections between factors and items (at least in terms of which items are, and are not, associated with each factor), and whether there are non-zero associations among factors (for multidimensional measurement models). Additional specifications can be integrated into measurement models for more complex designs.

After specifying the measure's model, responses to the measure are analyzed and several key pieces of information are obtained. Statistical software such as LISREL (Diamantopoulos & Siguaw, 2000), AMOS (from SPSS,

Byrne, 2016), SAS (O'Rourke & Hatcher, 2013), Mplus (Byrne, 2012), and R (Rosseel, 2012) can be used to conduct these analyses. Key information includes fit indices, parameter estimates (e.g., factor loadings, interfactor correlations), and modification indices. Fit indices (Chi square, Root Mean Square Error of Approximation, Tucker-Lewis Index, etc.) reflect the overall degree to which the actual responses to the measure are consistent with the specified measurement model. For example, if responses to the GAD-Q-IV are driven by two psychological dimensions, then a unidimensional model would be inconsistent with those responses, and analyses will produce poor fit indices. In CFA, parameter estimates are provided for a model specified by the user. For example, if a researcher specifies a unidimensional model for the 9 GAD-Q-IV items, then analyses would estimate one factor loading (and one error variance) for each item. Modification indices provide information about the way in which the model could be made more consistent with the actual responses.

The next step is usually to examine fit indices, which can lead in two directions. If fit indices are good, then researchers usually conclude that their hypothesized measurement model is consistent with the actual responses. They then report those indices and the parameter estimates. However, if fit indices are poor, then researchers conclude that the measurement model is inconsistent with actual responses. They often then examine modification indices, identify ways in which the model is inconsistent with the responses, respecify the model to enhance this consistency, and estimate the parameters for the modified model. This iterative "modify-and-respecify" procedure can continue until model fit becomes good, hopefully with very few modifications (readers are encouraged to consult Chapter 28 by Hopwood and Vazire for further discussion of these procedures as related to reproducibility in clinical psychology). This procedure, though common, blurs the distinction between confirmatory and exploratory research (Wright, 2017). Moreover, any modifications should be psychologically reasonable.

Through EFA or CFA, researchers hopefully achieve deeper understanding of a measure's dimensionality, and produce better measures. This is important not only for reasons mentioned earlier, but also because factor analytic results (factor loadings, error variances, etc.) are the basis for examining additional psychometric issues. These include, as described later, examinations of reliability and test bias.

It is worth noting that Exploratory Structural Equation Modeling (ESEM) is a relatively new procedure integrating EFA and CFA. CFA models typically allow each item to load (at a nonzero level) on only one factor, but ESEM relaxes this constraint. ESEM allows researchers, for example, to hypothesize that some specified factor loadings will be close to zero (though not exactly zero), while other loadings might not be close to zero. In this way, all loadings are estimated, thus reflecting an EFA approach; however, estimations are guided by a priori expectations, thus reflecting a CFA perspective. Since it does not impose the hypothesis that many loadings are exactly zero, ESEM provides an arguably more realistic representation of psychometric reality than the perhaps too-restrictive typical CFA model (Marsh et al., 2014). Currently, ESEM can be performed via Mplus or R's "psych" package (Revelle, 2017).

RELIABILITY

Meaning

The second key psychometric issue is reliability, the precision with which scores represent a psychological construct. There are two approaches to the conceptualization and evaluation of reliability – Classical Test Theory (CTT) and Item Response Theory (IRT).

CTT is the most well-known perspective on reliability, viewing an individual's score on a measure (i.e., the observed score, X_o) as determined by two components:

$$X_o = X_t + X_e \quad (6.1)$$

The first component is the individual's true score (X_t), which can be seen as the individual's average score across an infinite number of measurements. Perhaps more intuitively, it is (roughly) the person's level on the latent construct that drives responses to the test, though this conceptualization is not technically ideal. The second component is random error (X_e), which reflects random, transient factors that artificially inflate or deflate an individual's score. When researchers administer a measure, they are interested in their participants' true scores. If observed scores are good approximations of the true score, then they are reliable. However, if observed scores are poor approximations (i.e., if they are heavily affected by measurement error), then they are unreliable.

In CTT, reliability (r_{xx}) can be defined in several ways (see Furr, 2018, chapter 5). The most familiar is a signal-to-noise ratio:

$$r_{xx} = \frac{\sigma_t^2}{\sigma_t^2 + \sigma_e^2} = \frac{\sigma_t^2}{\sigma_o^2} \quad (6.2)$$

Where σ_t^2 is the variance of a set of true scores (i.e., the signal that researchers wish to detect), σ_e^2 is the variance of error scores (i.e., the noise masking the signal), and σ_o^2 is the variance of observed scores. Note that true scores and error scores are not directly observable. As discussed later, the theoretical basis of reliability, being based upon true and error scores, is accompanied by practical methods for estimating reliability. From a theoretical perspective, reliability can also be defined as the (squared) correlation between observed scores and true scores (r_{to}^2):

$$r_{xx} = r_{to}^2 \quad (6.3)$$

This highlights reliability as the degree to which differences in respondents' observed scores on a measure are consistent with differences in their true scores.

Reliability ranges from 0 to 1.0, with higher values representing greater precision (i.e., greater consistency between observed and true scores). Although there is not complete consensus and lower reliability is common, reliability ≥0.80 has been cited as satisfactory for research (Raykov & Marcoulides, 2011), with even higher reliability (≥0.90) desirable when tests are used to inform decisions about individuals.

Importance

As the precision with which a measure's scores reflect true psychological characteristics, reliability affects the estimation of statistical effects. That is, reliability affects the accuracy with which the associations (e.g., correlations, regression slopes) or differences obtained when analyzing *measures* reflect *true* psychological associations or differences. Poor reliability produces underestimation (i.e., attenuation) of observed effects in comparison to true psychological effects.

Imagine that we correlate GAD-Q-IV scores with scores from a measure of anxious attachment. Further, imagine that the true psychological connection between GAD and anxious attachment is $r'_{xy} = 0.30$, and that the measures' reliabilities are $r_{xx} = 0.80$ and $r_{yy} = 0.70$ for the GAD-Q-IV and the attachment measure, respectively. If CTT's core assumptions are met (see Furr, 2018, chapter 5) then the observed correlation between our measures is only $r_{xy} = 0.22$.

$$\begin{aligned} r_{xy} &= r'_{xy}\sqrt{r_{xx}r_{yy}} \\ r_{xy} &= 0.30\sqrt{(0.80)(0.70)} \\ r_{xy} &= 0.30(0.75) \\ r_{xy} &= 0.22 \end{aligned} \quad (6.4)$$

Thus, according to CTT, measures with imperfect reliability (which all are) underestimate true psychological effects ($r_{xy} = 0.22$ vs. $r'_{xy} = 0.30$). This is true for correlations in observational research, and for group/condition differences in experimental research (see Furr, 2018, chapter 6). In all research, better reliability produces less attenuation and results in greater psychological accuracy.

Moreover, because effect sizes affect statistical significance, reliability affects the likelihood of obtaining significant results. That is, poor reliability attenuates the effect sizes obtained in a study, which reduces the magnitude of inferential statistics, which in turn reduces the likelihood of statistical significance. To the degree that researchers care about obtaining accurate effect sizes and avoiding Type II errors (i.e., incorrectly failing to reject a null hypothesis), reliability matters.

Evaluation

Because researchers do not know individuals' true or error scores, they can never truly know the reliability of a measure's observed scores. Fortunately, there are methods for estimating reliability, including alternate forms, test-retest, and internal consistency. Alternate forms are used infrequently in many areas of research, including clinical research, so we focus on other methods (but see Furr, 2018, chapter 6, for details on this method).

A crucial but underappreciated fact is that the accuracy of any estimate of reliability hinges on a set of assumptions. If those assumptions are not met, then one's estimated reliability might be higher or lower than the measure's true reliability. Moreover, different methods hinge on different assumptions – some methods hinge on assumptions that are quite strict and not often valid. The various sets of assumptions are reflected in four CTT-defined measurement models – parallel, tau-equivalent, essentially tau-equivalent, and congeneric (in decreasing order of restrictiveness).

The test-retest method is fairly simple, and its accuracy hinges on the parallel tests model. Researchers administer the measure to the same respondents at two occasions, and the correlation between the two occasions' scores is an estimate of the measure's reliability. The parallel tests model's assumptions include: (a) respondents' true scores have not changed between occasions ($X_{t\_occasion1} = X_{t\_occasion2}$), (b) error variances are equal across the two occasions ($\sigma^2_{e\_occasion1} = \sigma^2_{e\_occasion2}$), and (c) error scores affecting one occasion are uncorrelated with error scores affecting the other occasion ($r_{e\_occasion1,e\_occasion2} = 0$) and with the true scores from each occasion ($r_{e\_occasion1,t\_occasion2} = 0$, $r_{t\_occasion1,e\_occasion2} = 0$). If these assumptions hold true, then the test-retest correlation accurately estimates reliability.

The internal consistency method is applicable for scores based on multiple parts (e.g., multiple items). Whereas the test-retest method estimates reliability through the consistency between a measure and itself over time, internal consistency methods estimate reliability through the consistency between a measure's different parts. Within this general method, there are many specific indices of reliability, including split-half and alpha (i.e., Cronbach's alpha). Because it is easily obtained (both in terms of requiring only one measurement occasion and in terms of availability in statistical software), alpha is the most widely used method for estimating reliability. It is based upon two factors: (a) the degree to which a measure's items are consistent with each other and (b) the number of items (see Furr, 2018, chapter 6, for details).

Unfortunately, alpha may provide inaccurate estimates of reliability in many cases (Green & Yang, 2009; but cf. Raykov & Marcoulides, 2019). Alpha's accuracy requires an essentially tau-equivalent (or more restrictive) model (Osburn, 2000). This model's assumptions include: (a) respondents' true scores differ across items only by a constant ($X_{t\_item1} = a + X_{t\_item2}$) and (b) error scores affecting one item are uncorrelated with other items' error scores ($r_{e1e2} = 0$, $r_{e1e3} = 0$, etc.) and their true scores ($r_{e\_item1,\ t\_item2} = 0$, $r_{t\_item1,\ e\_item2} = 0$, $r_{t\_item1,\ e\_item3} = 0$). Unfortunately, these assumptions may not hold true in many situations

(McNeish, 2018, but cf. Raykov & Marcoulides, 2019), and alpha might either underestimate or overestimate reliability (Raykov & Marcoulides, 2011). Thus, despite alpha's popularity and ease, researchers should examine the viability of its assumptions when considering its use, and they might often select a method with more realistic assumptions.

Fortunately, such methods exist (Bentler, 2017; McNeish, 2018; Revelle & Zinbarg, 2009), though they are unfamiliar to many researchers. One particularly useful approach, known as omega or composite reliability (McDonald, 1999; Raykov, 2004), is based upon factor analysis. It parallels reliability's conceptual meaning as a signal-to-noise ratio (Equation 6.2, above). Here, signal, or true score variance, is operationalized as items' connections to a common latent variable (i.e., factor loadings), and noise is operationalized as items' error variances. That is:

$$\hat{r}_{xx} = \frac{(\sum \lambda_i)^2}{(\sum \lambda_i)^2 + \sum \theta_{ii} + 2\sum \theta_{ij}}, \quad (6.5)$$

where $\hat{r}_{xx}$ is the estimated reliability of a score derived from a set of items, λ_i is an item's factor loading, θ_{ii} is an item's error variance, and θ_{ij} is a covariance between two items' errors. These components are obtainable via CFA procedures available in many statistical software packages (see the earlier discussion of CFA). This index ranges from 0 to 1.0, it is relatively straightforward for researchers familiar with CFA, and it is illustrated in several sources (e.g., Furr, 2018, chapter 12; Raykov, 2004).

This method estimates reliability accurately in a wider range of situations than many other methods, including alpha. This is because the assumptions underlying the method are less restrictive than many other indices. Based upon a congeneric measurement model, its main assumption is that respondents' true scores differ across items by a constant and a linear slope that can take any values ($X_{t1} = a + b(X_{t2})$), which is less restrictive than the similar assumption for alpha (which can be seen as assuming that $b = 1$). This index requires no assumptions about correlations among error scores, as alpha does.

Because any reliability estimates' accuracy hinges on assumptions about the data being examined, researchers should evaluate those assumptions before having much confidence in a reliability estimate. Fortunately, CFA can be used in this way, with descriptions of this process available in Brown (2015, pp. 207–221), Furr (2018, chapter 12), and Raykov and Marcoulides (2011, pp. 130–136).

Item Response Theory

Thus far, discussion has highlighted CTT, but IRT is an alternative framework for conceptualizing and evaluating reliability. IRT is used increasingly in developing, validating, and revising measures in many areas of psychological, including clinical psychology (e.g., Huprich, Paggeot, & Samuel, 2015).

IRT is based on mathematical models articulating the probability that a particular person will respond a particular way to a particular item. All models include a "person" component reflecting a person's level of a relevant underlying construct (e.g., anxiety). Models also include one or more "item" components, such as difficulty or discrimination (i.e., the degree to which an item differentiates between people with high versus low levels of the underlying construct). Models differ in several ways, including the number of item parameters, and the number of response options (two versus more than two). To conduct an IRT analysis, researchers select and apply a relevant model to participants' responses.

Based upon the selected model, IRT provides information about the psychometric properties of each item and about the entire measure. Some information parallels information from CTT (e.g., IRT's item discrimination is akin to factor loadings or item-total correlations from CTT); however, there are crucial differences. For example, whereas CTT conceptualizes reliability as a property of a *set* of scores (i.e., a set of scores has a single reliability value, see Equation 6.5, above), IRT views reliability as potentially varying across a set of scores. That is, an item or scale might effectively differentiate among people who have low levels of the relevant construct, but it might be ineffective at differentiating among people who have high levels. Moreover, from an IRT perspective, each item has a maximally informative point on the underlying construct – easy items effectively differentiate at low levels of the construct, whereas more difficult items effectively differentiate at mid or high levels of the construct.

Based upon such differences, IRT may offer benefits over CTT. For example, IRT might allow researchers to produce measures that are shorter, more reliable, or both (e.g., Pilkonis et al., 2011). Furthermore, IRT facilitates computerized adaptive testing in which different items are presented to different participants, presenting only items that sharply target each participant's level of the relevant psychological construct (e.g., Sunderland et al., 2019). In addition, IRT encourages and allows researchers to consider the range of the psychological construct that is effectively assessed by a measure. That is, IRT reveals whether items are better calibrated to differentiate, for example, highly depressed from *very* highly depressed participants, or to differentiate non-depressed from moderately depressed participants. With such information, researchers can develop and use measures effectively targeted to the population of interest. For example, it might motivate researchers to develop new items that cover a wider range of the construct.

Given its potential benefits, researchers are encouraged to learn more about an IRT approach. Good introductory sources are available for IRT in general and with clinical applications in particular (Furr, 2018, chapter 14; Reise & Waller, 2009; Thomas, 2011). Moreover, statistical software provides increasing user-friendly access to sophisticated IRT procedures (e.g., proc IRT in SAS, ltm and mirt packages in R).

Additional Comments and Advances

At least three additional issues merit comment. First, and practically, reliability estimates obtained from many methods (including alpha) can be negative, though reliability is conceptually bounded between 0 and 1.0. Often, this occurs due to errors in reverse scoring.

Second, procedures are available for conceptualizing and estimating reliability for multidimensional measures. Indeed, many clinical measures are multidimensional, often with correlations among dimensions. Reliability can be conceptualized and estimated in various ways for such measures, depending on their intended use (for details and illustrations, see Revelle & Zinbarg, 2009; Rodriguez, Reise, & Haviland, 2016).

Third, clinical researchers are increasingly interested in experience sampling methods, in which participants report experiences multiple times throughout their daily lives (e.g., Law et al., 2015). Such methods offer insight into important phenomena, but they present challenges when estimating reliability – e.g., the reliability of within-person variability or of within-person associations. There are specific tools for estimating reliability in experience sampling data, and this step is just as important as in traditional measurement methods (Fuller-Tyszkiewicz et al., 2017; Nezlek, 2017).

VALIDITY

Meaning

Validity is the third and ultimately most crucial psychometric issue. It is "the degree to which evidence and theory support the interpretations of test scores for proposed uses" (AERA, APA, & NCME, 2014, p. 11).

This definition has at least two important points. First, validity is a property of the interpretation of scores rather than a property of a measure. That is, a measure is neither valid nor invalid; instead, one's interpretation of the measure's scores is more or less valid. For example, a measure's scores might be validly interpreted as reflecting depression, but not validly interpreted as reflecting anxiety. The definition's focus on interpretation highlights the importance of psychological constructs in validity. A second important point is that validity hinges on solid empirical evidence (and theory) regarding the psychological meaning of a measure's scores. Without solid empirical evidence, researchers should avoid a measure.

It is worth noting that the concept of validity has evolved and that the definition above is not universally accepted. For example, some psychometricians object to the emphasis on interpretation rather than scores, while others accept the importance of interpretation but reject the relevance of 'proposed uses' (e.g., Borsboom, Mellenbergh, & Van Heerden, 2004; Cizek, 2012). However, most core points in the view above are generally accepted (Kane, 2013; Messick, 1989).

Importance

Validity's importance is probably obvious: if researchers fail to interpret measures correctly, then any conclusions based upon those measures are fundamentally flawed. Without due attention to validity, researchers might mistakenly draw conclusions about one construct, when they have, in fact, measured an entirely different construct.

Moreover, when using measures to guide recommendations or decisions about individuals, psychologists should attend to validity as well. Whether used for diagnostic purposes or for tracking psychological change, measures must be interpreted correctly. Otherwise, recommendations and decisions may be misguided or even harmful. For example, imagine again that the GAD-Q-IV truly reflects two separable factors (i.e., cognitive and affective facets of anxiety), but a clinician interprets it solely as reflecting an anxiety-related affect. If that interpretation guides treatment decisions for an individual scoring high on the GAD-Q-IV (interpreted as a single unidimensional measure), then treatment might mistakenly minimize or bypass altogether attention to anxiety-related cognition. In this case, a more valid interpretation of the test as a multidimensional measure reflecting both affect *and* cognition might lead to more appropriate treatment foci.

Evaluation

Table 6.1 summarizes five issues to consider when evaluating validity (AERA et al., 2014; Furr, 2018, chapters 8 and 9). Most efforts to establish validity seem to focus on only two of these – internal structure and associative validity. Efforts have focused less frequently on issues such as response process and consequential validity.

Content Validity

The first issue is the degree to which a measure's actual content corresponds with the content that the measure should have. Of course, the proper content for a measure depends upon the intended interpretation of the measure.

For many measures, content validity is often conceived in terms of items. Items should sample from all content relevant to the intended construct, no more and no less. For example, a measure of depression should reflect core phenomena of depression (e.g., patterns of mood and cognition), but not phenomena related to anxiety, unless such content is relevant to depression. A measure that fails to reflect core phenomena suffers from construct underrepresentation, while one that includes irrelevant phenomena may lack discriminant validity.

Table 6.1 Five issues in evaluating validity

Issue	Definition – The match between...	Typical relevant evidence
Content	A measure's *actual* content and the content that *should be* in the measure	Expert ratings of test content
Internal structure	A measure's *actual* internal structure and the structure that the measure *should* possess	Factor analysis of responses (EFA and/or CFA)
Response process	The psychological processes that respondents *actually* use when completing a measure and the processes that they *should* use	Interviewing respondents about response processes. Experimental manipulation of psychological states and processes. Indicators such as response times, eye tracking
Associations	A measure's *actual* associations with other measures and the associations that the test *should* have with the other measures	Correlations with other measures, including relevant group variables. Convergent, discriminant, concurrent, predictive correlations
Consequences	The *actual* consequences of using a measure and the consequences that *should be* seen.	Evaluation of intended consequences, differential impact, and systemic changes

Content validity is typically evaluated through expert ratings of a measure's content. Experts review a measure's content and judge it against a particular construct. Relevant procedures have been developed, including a statistical index of a measure's content validity (Haynes, Richard, & Kubany, 1995; Koller, Levenson, & Glück, 2017; Newman et al., 2013).

Internal Structure Validity

The second issue in evaluating validity is the degree to which a measure's actual dimensionality (i.e., internal structure) corresponds with the dimensionality that it should have. Of course, the appropriate dimensionality depends upon the dimensionality of the intended construct. A measure intended to reflect a single unidimensional construct should have a unidimensional structure; further, a measure intended to reflect a multidimensional construct should exhibit the particular form of multidimensionality implied by the construct (e.g., correlated vs. uncorrelated dimensions).

Even further, internal structure also refers to whether a test's underlying construct is either dimensional (e.g., a personality trait) or categorical (e.g., a set of personality types). Internal structure validity also hinges on the correspondence between a test's actual and intended structure in terms of this distinction (Clark & Watson, 1995).

Internal structure validity is typically evaluated via factor analysis (see this chapter's earlier discussion) and related procedures. Both EFA and CFA are often used (e.g., Siefert et al., 2018).

Response Process Validity

The third issue is the degree to which the psychological processes that participants actually use when responding to a measure correspond to the processes that they should use (Bornstein, 2011; Borsboom et al., 2004; Embretson, 2016; Hubley & Zumbo, 2017). This issue is based upon the idea that psychological constructs affect responses by shaping the processes "that underlie what people do, think, or feel when interacting with and responding to, the item or task" (Hubley & Zumbo, 2017). If these processes occur in a way that makes theoretical sense in terms of the construct being measured, then scores are valid.

As Borsboom and his colleagues suggest (2004, p. 1062), to evaluate response process validity, researchers must articulate and test "a theory of response behavior." Although done relatively infrequently in clinical research, such theories can be tested in a variety of ways. IRT models can be used to examine specific processes (Embretson, 2016; Zettler et al., 2016). In addition, researchers can interview participants while (or immediately after) completing a measure, to understand the processes used (Morell & Tan, 2009). Researchers have also used techniques such as eye-tracking and recording mouse movements in computerized administration of cognitive assessments to reveal the processes participants use when experiencing and responding to a measure (Ivie & Embretson, 2010). Finally, Bornstein (2011) presents a framework for using experimental manipulations of participants' psychological states and processes to identify those shaping participants' responses to a measure. Recent work has presented examples of the latter procedure in examining response processes underlying the Rorschach (Mihura et al., 2019).

Associative Validity

A fourth issue in evaluating validity is the degree to which a measure's actual association with other measures corresponds with the associations that it should have. Ideally, this involves testing predictions about the associations between a measure's scores and scores on other measures. The predictions should arise from a solid conceptualization of the construct being assessed by a measure, in terms of its meaning and its hypothesized connections to other constructs (i.e., its nomological network; Cronbach & Meehl, 1955). If a measure is correlated with other measures with which it should be correlated, it has convergent validity. If it is *un*correlated (or weakly correlated) with measures with which it should be uncorrelated, it has discriminant validity.

Associative validity is a frequent focus of validity work in clinical psychology, with at least four general ways of proceeding (Furr, 2018, chapter 9). First, some measures have a single gold-standard criterion against which they can be evaluated (e.g., a diagnostic interview). If a measure's scores are robustly associated with that criterion, then its associative validity is supported. In clinical research, such associations are sometimes examined via sensitivity/specificity analyses. For example, a recent study examined the degree to which a particular test reflected cognitive deficits linked to schizophrenia (e.g., De la Torre et al., 2016). Test scores were compared to the results of (independently conducted) diagnostic interviews. Results revealed that participants with high test scores were significantly more likely to be diagnosed with schizophrenia as compared to participants with low test scores. Such results provide compelling evidence supporting the proposed interpretation/use of the test's scores.

A second method of examining associative validity is via patterns of associations. Researchers administer the focal measure along with a set of other measures, "eyeball" the pattern of associations between measures, and gauge the pattern's consistency with expectations. This approach is quite common (e.g., Hill et al., 2004) and often relies upon researchers' subjective judgments of the degree of associative validity.

The third and fourth approaches also involve patterns of associations, but they reduce subjectivity in judging evidence of validity. The multitrait-multimethod approach requires researchers to examine a set of constructs, all of which are measured via a common set of methods. By examining the pattern of associations across all constructs and methods according to well-specified criteria (Cambell & Fiske, 1959; Eid et al., 2008), researchers can gauge the evidence of validity relatively objectively. Finally, Westen and Rosenthal (2003) outlined a "Quantifying Construct Validity" method in which researchers produce specific quantitative predictions about the correlations to be observed, and statistically gauge the correspondence between the actual pattern of correlations and the predicted pattern. Despite this approach's potential value (Furr & Heuckeroth, 2019; but cf. Smith, 2005), it is rarely used in in current psychometric practice (e.g., Poythress et al., 2010). The recent availability of an R package dedicated to QCV analysis might increase its use (Furr & Heuckeroth, 2018).

Consequential Validity

The final key issue in evaluating validity is the degree to which the actual consequences of using a measure correspond with the intended consequences. Certainly, researchers and practitioners would hope that using a measure produces desirable consequences (e.g., improving healthcare, selecting appropriate candidates) with minimal undesirable consequences.

Kane (2013) suggests three facets to this issue. Specifically, researchers should evaluate the degree to which using a measure: (a) accomplishes its intended consequences, (b) has unintended differential impact on groups of people (e.g., based on gender or race), and (c) has unintended systemic effects (e.g., affecting healthcare delivery).

Consequential validity is controversial in terms of its relevance to validity (Cizek, 2012; Kane, 2013; Sireci, 2016). Some researchers argue that validity should hinge solely on the interpretation of a measure's scores, while others argue that both interpretation and use are crucial to evaluating a measure.

DIFFERENCES IN PSYCHOMETRIC QUALITY

Psychologists are often interested in detecting and understanding group differences, or differences over time. Whether comparing groups or time points in terms of mean levels of a construct or in terms of associations among constructs (e.g., group differences in correlations), valid interpretations require that the constructs are measured equally well in each group or at each time point.

Thus, before interpreting such differences, researchers should evaluate the nature of any group/time differences in psychometric properties of the measures. Such group differences are sometimes known as test bias, and can be conceptualized in terms of construct bias and predictive bias (Reynolds & Suzuki, 2013). Construct bias exists when a measure reflects a construct differently in different groups. This is primarily examined in terms of group differences in internal structure (i.e., "measurement invariance"; Vandenberg & Lance, 2000), differential item functioning (via IRT; Walker, 2011), or reliability (Feldt, Woodruff, & Salih, 1987; Raykov, 2002), though additional methods are available as well (see Furr, 2018, chapter 11). Predictive bias exists when a measure is related to important outcomes in different ways for different groups (Aguinis, Culpepper, & Pierce, 2010), and this seems to be a concern in areas such as employee selection, educational testing, and college admissions more than in clinical research. Indeed, clinical researchers seem to be more concerned about construct bias than predictive bias (e.g., Garcia et al., 2018).

CREATING AND REFINING CLINICAL MEASURES

Given the conceptual, statistical, and empirical consequences of psychometric quality, researchers should use measures having clear dimensionality, strong reliability, and robust validity. To accomplish this goal, researchers should develop measures with these qualities in mind, should evaluate these qualities when they are not already known, and should seek to refine measures when necessary.

There are a variety of sources outlining procedures for developing and refining measures (Clark & Watson, 1995, 2019; Danner et al., 2016; Furr, 2011, chapter 2; Markus & Borsboom, 2013, chapter 12). Briefly, several issues – practical, conceptual, psychometric, and statistical – should guide scale development. The process should begin by clarifying the nature of the construct(s) to be assessed, the intended purposes of the measure, and the intended respondents. These issues, along with practical guidelines (e.g., regarding double-barreled items), should guide item writing and initial selection of items to ensure content validity and item clarity. Ideally, empirical evidence is then gathered in a sequence of studies in which dimensionality is tested and clarified, alongside reliability estimation. These initial empirical steps often result in additional item development or refinement (e.g., based upon item-level reliability or dimensionality indices – see Furr, 2018, chapters 4 and 7). Subsequent steps involve evaluating the range of issues in validity, often including assessment of group differences in psychometric quality (see Furr, 2018, chapter 11, for an overview).

It is worth noting that these steps are based upon the common deductive approach in which scale development begins with a fairly clear idea of the construct(s) to be measured (Ozer, 1989). From a more exploratory, inductive approach to scale development, the nature of the relevant constructs emerges as the scale development process unfolds (Tellegen & Waller, 2008).

CONCLUSIONS

Good psychometric quality is fundamental to measurement, which is fundamental to science. When interpreting results of psychological tests, careful attention to psychometric quality facilitates appropriate conclusions or, when necessary, produces appropriate skepticism. When developing or refining clinical measures, attending to the issues outlined in this chapter should produce good measures with strong conceptual and psychometric foundations. Ultimately, the use of measures with strong conceptual and psychometric foundations will advance clinical psychological science.

REFERENCES

Aguinis, H., Culpepper, S. A., & Pierce, C. A. (2010). Revival of Test Bias Research in Preemployment Testing. *Journal of Applied Psychology*, 95, 648–680.

American Educational Research Association (AERA), American Psychological Association(APA), & National Council on Measurement in Education (NCME). (2014). *Standards for Educational and Psychological Testing*. Washington, DC: American Educational Research Association.

Bentler, P. M. (2017). Specificity-Enhanced Reliability Coefficients. *Psychological Methods*, 22, 527–540.

Bornstein, R. F. (2011). Toward a Process-Focused Model of Test Score Validity: Improving Psychological Assessment in Science and Practice. *Psychological Assessment*, 23, 532–544.

Borsboom, D., Mellenbergh, G. J., & Van Heerden, J. (2004). The Concept of Validity. *Psychological Review*, 111, 1061–1071.

Brown, T. A. (2015). *Confirmatory Factor Analysis for Applied Research* (2nd edn.). New York: Guilford Press.

Byrne, B. M. (2012). *Structural Equation Modeling with Mplus: Basic Concepts, Applications, and Programming*. New York: Routledge.

Byrne, B. M. (2016). *Structural Equation Modeling with AMOS: Basic Concepts, applications, and Programming* (3rd edn.). New York: Routledge.

Campbell, D. T., & Fiske, D. W. (1959). Convergent and Discriminant Validation by the Multitrait-Multimethod Matrix. *Psychological Bulletin*, 56, 81–104.

Cizek, G. J. (2012). Defining and Distinguishing Validity: Interpretations of Score Meaning and Justifications of Test Use. *Psychological methods*, 17, 31–43.

Clark, L. A., & Watson, D. (1995). Constructing Validity: Basic Issues in Objective Scale Development. *Psychological Assessment*, 7, 309–319.

Clark, L. A., & Watson, D. (2019). Constructing Validity: New Developments in Creating Objective Measuring Instruments. *Psychological Assessment*. Retrieved from http://dx.doi.org/10.1037/pas0000626

Cronbach, L. J., & Meehl, P. E. (1955). Construct Validity in Psychological Tests. *Psychological Bulletin*, 51, 281–302.

Danner, D., Blasius, J., Breyer, B., Eifler, S., Menold, N., Paulhus, D. L., ... & Ziegler, M. (2016). Current Challenges, New Developments, and Future Directions in Scale Construction. *European Journal of Psychological Assessment*, 32, 175–180.

De la Torre, G. G., Perez, M. J., Ramallo, M. A., Randolph, C., & González-Villegas, M. B. (2016). Screening of Cognitive Impairment in Schizophrenia: Reliability, Sensitivity, and Specificity of the Repeatable Battery for the Assessment of Neuropsychological Status in a Spanish Sample. *Assessment*, 23, 221–231.

Diamantopoulos, A., & Siguaw, J. A. (2000). *Introducing LISREL: A Guide for the Uninitiated*. London: Sage.

Eid, M., Nussbeck, F. W., Geiser, C., Cole, D. A., Gollwitzer, M., & Lischetzke, T. (2008). Structural Equation Modeling of Multitrait-Multimethod Data: Different Models for Different Types of Methods. *Psychological Methods*, 13, 230–253.

Embretson, S. E. (2016). Understanding Examinees' Responses to Items: Implications for Measurement. *Educational Measurement: Issues and Practice*, 35, 6–22.

Fabrigar, L. R., Wegener, D. T., MacCallum, R. C., & Strahan, E. J. (1999). Evaluating the Use of Exploratory Factor Analysis in Psychological Research. *Psychological Methods*, 4, 272–299.

Feldt, L. S., Woodruff, D. J., & Salih, F. A. (1987). Statistical Inference for Coefficient Alpha. *Applied Psychological Measurement*, 11, 93–103.

Floyd, F. J., & Widaman, K. F. (1995). Factor Analysis in the Development and Refinement of Clinical Assessment Instruments. *Psychological Assessment*, 7(3), 286–299.

Fuller-Tyszkiewicz, M., Hartley-Clark, L., Cummins, R. A., Tomyn, A. J., Weinberg, M. K., & Richardson, B. (2017). Using

Dynamic Factor Analysis to Provide Insights into Data Reliability in Experience Sampling Studies. *Psychological Assessment*, 29, 1120–1128.

Furr, R. M. (2011). *Scale Construction and Psychometrics for Social and Personality Psychology*. London: Sage.

Furr, R. M. (2018). *Psychometrics: An Introduction* (3rd edn.). Thousand Oaks, CA: Sage.

Furr, R. M., & Heuckeroth, S. A. (2018). qcv: Quantifying Construct Validity. R Package Version 1.0. Retrieved from https://cran.r-project.org/package=qcv

Furr, R. M., & Heuckeroth, S. A. (2019). The "Quantifying Construct Validity" Procedure: Its Role, Value, Interpretations, and Computation. *Assessment*, 26, 555–566.

Garcia, A. F., Berzins, T., Acosta, M., Pirani, S., & Osman, A. (2018). The Anxiety Depression Distress Inventory-27 (ADDI-27): New Evidence of Factor Structure, Item-Level Measurement Invariance, and Validity. *Journal of Personality Assessment*, 100, 321–332.

Green, S. B., & Yang, Y. (2009). Commentary on Coefficient Alpha: A Cautionary Tale. *Psychometrika*, 74, 121–135.

Haynes, S. N., Richard, D. C. S., & Kubany, E. S. (1995). Content Validity in Psychological Assessment: A Functional Approach to Concepts and Methods. *Psychological Assessment*, 7, 238–247.

Hayton, J. C., Allen, D. G., & Scarpello, V. (2004). Factor Retention Decisions in Exploratory Factor Analysis: A Tutorial on Parallel Analysis. *Organizational Research Methods*, 7, 191–205.

Hill, R. W., Huelsman, T. J., Furr, R. M., Kibler, J., Vicente, B. B., & Kennedy, C. (2004). A New Measure of Perfectionism: The Perfectionism Inventory (PI). *Journal of Personality Assessment*, 82, 80–91.

Hubley, A.M., & Zumbo, B.D. (2017). Response Processes in the Context of Validity: Setting the Stage. In B. D. Zumbo and A.M. Hubley (Eds.), *Understanding and Investigating Response Processes in Validation Research* (pp. 1–12). New York: Springer.

Huprich, S. K., Paggeot, A. V., & Samuel, D. B. (2015) Comparing the Personality Disorder Interview for DSM-IV (PDI-IV) and SCID-II Borderline Personality Disorder Scales: An Item-Response Theory Analysis. *Journal of Personality Assessment*, 97, 13–21.

Ivie, J. L., & Embretson, S. E. (2010). Cognitive Process Modeling of Spatial Ability: The Assembling Objects Task. *Intelligence*, 38, 324–335.

Kane, M. T. (2013). Validating the Interpretations and Uses of Test Scores. *Journal of Educational Measurement*, 50, 1–73.

Koller, I., Levenson, M. R., Glück, J. (2017). What Do You Think You Are Measuring? A Mixed-Methods Procedure for Assessing the Content Validity of Test Items and Theory-Based Scaling. *Frontiers in Psychology*, 8, 1–20.

Law, M. K., Furr, R. M., Arnold, E. M., Mneimne, M., Jaquett, C., & Fleeson, W. (2015). Does Asking Frequently and Repeatedly about Suicide Cause Harm? A Randomized Control Study. *Psychological Assessment*, 27, 1171–1181.

Markus, K. A., & Borsboom, D. (2013). *Frontiers of Test Validity Theory: Measurement, Causation, and Meaning*. New York: Routledge.

Marsh, H. W., Morin, A. J. S., Parker, P. D., Kaur, G. (2014) Exploratory Structural Equation Modeling: An Integration of the Best Features of Exploratory and Confirmatory Factor Analysis. *Annual Review of Clinical Psychology*, 10, 85–110

McDonald, R. P. (1999) *Test Theory: A Unified Treatment*. Mahwah, NJ: Lawrence Erlbaum.

McNeish, D. (2018). Thanks Coefficient Alpha, We'll Take It from Here. *Psychological Methods*, 23, 412–433.

Messick, S. (1989). Validity. In R. L. Linn (Ed.), *Educational Measurement* (3rd edn., pp. 13–103). New York: Macmillan.

Mihura, J. L., Dumitrascu, N., Roy, M., & Meyer, G. J. (2019). The Centrality of the Response Process in Construct Validity: An Illustration via the Rorschach Space Response. *Journal of Personality Assessment*, 101, 374–392.

Morell, L., & Tan, R. J. B. (2009). Validating for Use and Interpretation: A Mixed Methods Contribution Illustrated. *Journal of Mixed Methods Research*, 3, 242–264.

Muthén, L. K., & Muthén, B. O. (1998–2017). *Mplus User's Guide* (8th edn.). Los Angeles, CA: Muthén & Muthén.

Newman, I., Lim, J., & Pineda., F. (2013). Content Validity Using a Mixed Methods Approach: Its Application and Development through the Use of a Table of Specifications Methodology. *Journal of Mixed Methods Research*, 7, 243–260.

Newman, M. G., Zuellig, A. R., Kachin, K. E., Constantino, M. J., Przeworski, A., Erickson, T., & Cashman-McGrath, L. (2002). Preliminary Reliability and Validity of the Generalized Anxiety Disorder Questionnaire – IV: A Revised Self-Report Diagnostic Measure of Generalized Anxiety Disorder. *Behavior Therapy*, 33, 215–233.

Nezlek, J. B. (2017). A Practical Guide to Understanding Reliability in Studies of Within-Person Variability. *Journal of Research in Personality*, 69, 149–155.

O'Connor, B. P. (2000). SPSS and SAS Programs for Determining the Number of Components using Parallel Analysis and Velicer's MAP Test. *Behavior Research Methods, Instrumentation, and Computers*, 32, 396–402.

O'Rourke, N., & Hatcher, L. (2013). *A Step-by-Step Approach to Using the SAS System for Factor Analysis and Structural Equation Modeling* (2nd edn.). Cary, NC: SAS Institute.

Osburn, H. G. (2000). Coefficient Alpha and Related Internal Consistency Reliability Coefficients. *Psychological Methods*, 5, 343–355.

Ozer, D. J. (1989). Construct Validity in Personality Assessment. In D. Buss & N. Cantor (Eds.), *Personality Psychology: Recent Trends and Emerging Directions* (pp. 225–234). New York: Springer-Verlag.

Pilkonis, P. A., Choi, S. W., Reise, S. P., Stover, A. M., Riley, W. T., & Cella, D. (2011). Item Banks for Measuring Emotional Distress from the Patient-Reported Outcomes Measurement Information System (PROMIS®): Depression, Anxiety, and Anger. *Assessment*, 18, 263–283.

Poythress, N. G., Lilienfeld, S. O., Skeem, J. L., Douglas, K. S., Edens, J. F., Epstein, M., & Patrick, C. J. (2010). Using the PCL-R to Help Estimate the Validity of Two Self-Report Measures of Psychopathy with Offenders. *Assessment*, 17, 206–219

Raykov, T. (2002). Examining Group Differences in Reliability of Multiple-Component Instruments. *British Journal of Mathematical and Statistical Psychology*, 55, 145–158.

Raykov, T. (2004). Behavioral Scale Reliability and Measurement Invariance Evaluation using Latent Variable Modeling. *Behavior Therapy*, 35, 299–331.

Raykov, T., & Marcoulides, G. A. (2011). *Introduction to Psychometric Theory*. New York: Routledge, Taylor & Francis Publishers.

Raykov, T., & Marcoulides, G. A. (2019). Thanks, Coefficient Alpha – We Still Need You! *Educational and Psychological Measurement*, 79(1), 200–210.

Reise, S. P., & Waller, N. G. (2009). Item Response Theory and Clinical Measurement. *Annual Review of Clinical Psychology, 5,* 25–46.

Reise, S. P., Waller, N. G., & Comrey, A. L. (2000). Factor Analysis and Scale Revision. *Psychological Assessment, 12,* 287–297.

Revelle, W. (2017) *Psych: Procedures for Personality and Psychological Research,* Northwestern University, Evanston, Illinois, USA. Retrieved from https://CRAN.R-project.org/package=psych Version = 1.7.5.

Revelle, W., & Rocklin, T. (1979). Very Simple Structure: An Alternative Procedure for Estimating the Optimal Number of Interpretable Factors. *Multivariate Behavioral Research, 14,* 403–414.

Revelle, W., & Zinbarg, R. E. (2009). Coefficients Alpha, Beta, Omega and the Glb: Comments on Sijtsma. *Psychometrika, 74*(1), 145–154.

Reynolds, C. R., & Suzuki, L. (2013). Bias in Psychological Assessment: An Empirical Review and Recommendations. In J. R. Graham, J. A. Naglieri, & I. B. Weiner (Eds.), *Handbook of Psychology, Volume 10: Assessment Psychology* (2nd edn., pp. 82–113). Hoboken, NJ: Wiley.

Rodriguez, A., Reise, S. P., & Haviland, M. G. (2016). Evaluating Bifactor Models: Calculating and Interpreting Statistical Indices. *Psychological Methods, 21*(2), 137–150.

Rosseel, Y. (2012). Lavaan: An R Package for Structural Equation Modeling. *Journal of Statistical Software, 48,* 1–36.

Ruscio, J. (n.d.). Retrieved from http://ruscio.pages.tcnj.edu/quantitative-methods-program-code/.

Ruscio, J., & Roche, B. (2012). Determining the Number of Factors to Retain in an Exploratory Factor Analysis Using Comparison Data of Known Factorial Structure. *Psychological Assessment, 24,* 282–292.

Siefert, C. J., Stein, M., Slavin-Mulford, J., Haggerty, G., Sinclair, S. J., Funke, D., & Blais, M. A. (2018). Exploring the Factor Structure of the Social Cognition and Object Relations – Global Rating Method: Support for Two- and Three-Factor Models. *Journal of Personality Assessment, 100,* 122–134.

Sireci, S. G. (2016). On the Validity of Useless Tests. *Assessment in Education: Principles, Policy & Practice, 23,* 226–235.

Smith, G. T. (2005). On Construct Validity: Issues of Method and Measurement. *Psychological Assessment, 17,* 396–408.

Sunderland, M., Batterham, P., Carragher, N., Calear, A., & Slade, T. (2019). Developing and Validating a Computerized Adaptive Test to Measure Broad and Specific Factors of Internalizing in a Community Sample. *Assessment, 25,* 1030–1045.

Tellegen, A., & Waller, N. G. (2008). Exploring Personality through Test Construction: Development of the Multidimensional Personality Questionnaire. In G. J. Boyle, G. Matthews, & D. H. Saklofske (Eds.), *The SAGE Handbook of Personality Theory and Assessment: Vol. 2. Personality Measurement and Testing* (pp. 261–292). London: Sage.

Thomas, M. (2011). The Value of Item Response Theory in Clinical Assessment: A Review. *Assessment, 18,* 291–307

Vandenberg, R. J., & Lance, C. E. (2000). A Review and Synthesis of the Measurement Invariance Literature: Suggestions, Practices, and Recommendations for Organizational Research. *Organizational Research Methods, 3,* 4–70.

Velicer, W. F. (1976). Determining the Number of Components from the Matrix of Partial Correlations. *Psychometrika, 41,* 321–327.

Walker, C. (2011). What's the DIF? Why Differential Item Functioning Analyses Are an Important Part of Instrument Development and Validation. *Journal of Psychoeducational Assessment, 29,* 364–376.

Westen, D., & Rosenthal, R. (2003). Quantifying Construct Validity: Two Simple Measures. *Journal of Personality and Social Psychology, 84,* 608–618.

Widaman, K. F. (1993). Common Factor Analysis versus Principal Component Analysis: Differential Bias in Representing Model Parameters? *Multivariate Behavioral Research, 28,* 263–311.

Wright, A. G. C. (2017). The Current State and Future of Factor Analysis in Personality Disorder Research. *Personality Disorders: Theory, Research, and Treatment, 8,* 14–25.

Zettler, I., Lang, J. W. B., Hülsheger, U. R., & Hilbig, B. E. (2016). Dissociating Indifferent, Directional, and Extreme Responding in Personality Data: Applying the Three-Process Model to Self- and Observer Reports. *Journal of Personality, 84,* 461–472.

Ziegler, M., & Hagemann, D. (2015). Testing the Unidimensionality of Items. *European Journal of Psychological Assessment, 31,* 231–237.

7 Latent Variable Models in Clinical Psychology

AIDAN G. C. WRIGHT

Most of what clinical psychology concerns itself with is directly unobservable. Concepts like neuroticism and depression, but also learning and development, represent dispositions, states, or processes that must be inferred and cannot (currently) be directly measured. Latent variable modeling encompasses a range of techniques that involve estimating the presence and effect of unobserved variables from observed data. In other words, latent variable models are a class of statistical techniques that allow investigators to work at the construct level, even when the constructs in question elude direct measurement. They can also be used to directly compare distinct and often competing conceptualizations of mental disorder. The basic logic underpinning these approaches is similar to the diagnostician's task of inferring an inaccessible disease state from outwardly available signs and symptoms. For instance, if an individual patient were to complain of lack of interest and/or pleasure, persistent low mood, guilt, as well as a number of vegetative symptoms like fatigue and appetite disturbance, the clinician might infer that they are suffering from depression. In this example, depression is a clinical concept that is presumed to drive the manifestation of these debilitating mental and physical states. If one wants to study depression, one would ideally have access not just to those observable features, all of which have other potential causes, but to what they share in the form of the unobserved depression episode – latent variable models provide investigators direct access to the inferred construct. This is the crux of this set of techniques, although as I discuss below, they can be leveraged in creative ways to understand the very nature of psychopathology.

This chapter provides a nontechnical introduction of latent variable models with an emphasis on how they can be used to answer challenging questions relevant to clinical psychology. Therefore, the emphasis is largely conceptual, and the reader is directed elsewhere for technical treatments and detailed instructions for applications (e.g., Bollen, 1989; Brown, 2015; Collins & Lanza, 2010; Loehlin, 2004; Mulaik, 2010). I begin with a discussion of the definition of latent variables, then follow with an elaboration of several exemplar models including factor analytic techniques and mixture modeling. Portions of the model overviews borrow heavily from Wright (2017) and Wright and Zimmerman (2015), which can be consulted for additional detail. Throughout, examples from the clinical literature are provided.

DEFINITIONS AND CONCEPTUAL UNDERPINNINGS

How should we think about latent variables? What are their defining features? What role can they play in applied clinical research programs? Scholars and methodologists have answered these questions in different ways over the years, with some providing formal but narrow definitions in specific quantitative terms and others providing informal and loosely defined criteria. In a review on the use of latent variable models in social science, Bollen (2002) summarized several of the most common definitions, and in so doing contrasted informal and formal definitions. On the informal side, Bollen (2002) noted that several authors have argued that latent variables are "hypothetical variables" or "hypothetical constructs" (Edwards & Bagozzi, 2000; Harman, 1960; Nunnally, 1978). Although this definition is certainly accurate, it is only conceptual, and does not link back to the formal statistical models in any clear way. Another common informal definition that Bollen (2002) identified is that latent variables are "unobservable or unmeasurable." The major concern with this definition is that it presupposes advances in technology that may make what currently defies measurement feasible to measure some day in the future. Finally, some have argued that latent variables are merely summaries of the observed variables serving little more than a descriptive function. Adopting this perspective unencumbers the researcher of making any challenging assumptions about the true nature of latent variables, but it also defangs the models and reduces them to their weakest form.

Although Bollen (2002) also reviewed a number of formal definitions that have been offered for latent variables over the years (e.g., local independence definition), he noted that each of these are overly restrictive or too narrow for various reasons. Ultimately, he offered a new definition that is simultaneously formal yet nonrestrictive – a latent variable is a variable (i.e., not a constant) for which there is no sample realization in a given sample. What this means is that any variable that is not directly observed in (at least some portion of) a sample is latent. One attractive feature of this definition is that it acknowledges that a variable may be latent in one sample (because it is unmeasured), but observed in another, and that this may be something that changes over time (e.g., as new technology is developed). This approach is also useful for accommodating certain technical aspects of latent variable models, such as their use in handling missing data, allowing for correlated residuals, and treating residuals in regressions and random effects in mixed effects models as latent variables. Finally, although very abstract, this definition does provide the necessary link between the unobserved and the observed data. Thus, this definition encompasses all of the informal definitions listed above, and other formal definitions often serve as special cases of this more general definition.

The "sample realization" definition is useful for establishing when a variable is latent, but it does not speak to their ontology. That is, how should we conceive of latent variables? Borsboom, Mellenbergh, and van Heerden (2003) took up this question in what has become a classic treatment on the conceptualization of latent variables. Borsboom and colleagues raise a number of technical points to motivate this question, but the fundamental issue they consider is whether latent variables exist independent of the data used to estimate them. That is to say, given any set of data, one can run and estimate a latent variable model and if it fits well, the latent variable can be interpreted. However, that's not to say that anything meaningfully independent of the data has been ascertained. One needs to make an ontological assumption to link the operational latent variable estimated from the observed data to the formal latent variable of theoretical interest. Borsboom and colleagues describe two distinct ontological stances, although one of these has layers of stringency. The first stance is the realist stance, which assumes that the latent variable in question is something real in nature distinct from the data that are used to measure it. This can be contrasted with the constructivist stance, which regards latent variables as constructed by the human mind and therefore not existing independent of their measurement. An extreme variation on this latter perspective is that the latent variable is just a data reduction method, much like Bollen (2002) discussed as one informal definition of latent variables.

Borsboom and colleagues (2003) argue that to take latent variable modeling seriously, one must adopt a realist stance. Which is to say, one must assume that the latent variable is something that exists in nature independent of the data and causes the patterns in observed data. In contrast, an operationalist perspective is fundamentally at odds with latent variable theory, and assumes nothing more than a summary of the available data. That is, any latent variable estimated from the data is not independent from it, and it is just a construction of our minds that does not otherwise exist in reality.

Borsboom and colleagues' arguments are well conceived and articulated, but they would seem to place clinical psychologists in an uncomfortable position by drawing a crisp distinction between realism and constructivism. Either you must assert that our constructs like depression, narcissism, obsessive-compulsivity, and the like are real and exist in nature per se, or you must adopt a constructivist stance that treats our theories and constructs as little more than summaries of what are often unsatisfyingly imprecise data. Given the current state of our science, I believe that a pseudo-realism (but probably more realistic) perspective is justified. Pseudo-realism, as I define it here, argues that latent variables represent the sum of the shared processes generating covariation among some set of observed variables. In this perspective, one assumes that the latent data-generating processes are real, but our understanding and labeling of these processes are only approximations for what the real processes are due our limited knowledge and poor measurement tools. Several implications follow from this perspective. First, a single latent variable need not represent a single process, and can (and likely does in clinical psychology) represent several processes at once. These can include processes of substantive interest (e.g., neuroticism) as well as artifacts of measurement (e.g., response acquiescence). Second, as with Bollen's (2002) sample realization definition, the latent variable is assumed to exist separate from the data used to estimate it. Third, because there is no expectation that a latent variable be fixed and real as currently conceived, they require construct validation (Cronbach & Meehl, 1955) and further explication through any variety of research programs (see Chapter 4 by Williams and Simms and Chapter 9 by Lenzenweger).

To illustrate this perspective, I will borrow and expand on the thermometer example used by Bollen (2002). A thermometer offers a classic example of measuring and estimating a latent variable. In this case heat is the latent variable and volume of mercury is the observed data. As heat increases mercury expands, and as heat decreases mercury contracts. Thus, latent heat, an unobserved variable, can be inferred based on the observed volume of mercury. It would seem that adopting a fully realist perspective here is warranted, as what could be more real than heat? But what is heat exactly? It should be clear that the concept of heat is a mere abstraction, and a more precise description of the link between the latent variable and the observed data requires an understanding of a process – namely, what has long been referred to as heat is now understood to be the transfer of energy among matter. This understanding is predated by the earliest thermometers by centuries or even millennia, depending on how you define a thermometer. In a similar fashion, in clinical psychology, we may treat as real what we can only assume to be approximations of the true underlying processes, and which we hope are better understood and described, if not replaced, in the years, decades, and centuries to come. In this way, psychopathologists may comfortably adopt a pseudorealist stance, assuming that latent variables allow them to access our current best approximations of clinically relevant processes, even if they also hold the assumption that how we understand these variables will evolve over time.

Assuming that a latent variable exists separate from the observed data does not mean that all measures (i.e., data) will allow for an equivalent and equally good estimate of the latent variable, even if they are designed to do so. Common paradigms for data collection in clinical psychology research, including survey methodology (Chapter 5 by Samuel, Bucher, and Suzuki), functional magnetic resonance imaging (fMRI; Chapter 13 by Holmes and MacDonald), and peripheral psychophysiology (Chapter 11 by Levinson and Hajcak), are all based on implicit latent variable models, at least as they pertain to use in clinical psychology. Survey responses, blood oxygenation in the limbic system, and galvanic skin conductance have all been used as indicators of an underlying emotional state, where the underlying emotional state is presumed to drive the measured variables. At the same time, each of these is a very different type of data, and so it would be imprudent to assume that there are not large methodological processes at work driving the data in addition to the latent variable of interest. Thus, all of these could be used to estimate the same latent variable, but in doing so care must be taken to isolate that process.

Some texts provide coverage of both reflective and formative latent variables. Reflective latent variable models are those in which the latent variable causes the data (i.e., the variable is reflected in the data), whereas formative models are those where the data generate (i.e., form) the latent variable. As Borsboom and colleagues (2003) note, only the reflective latent variable assumes a realist interpretation. Therefore, formative variables will not be considered here, because they are something "different" in that they assume no shared processes in what generates them, although they can be useful to consider. For instance, Bollen (2002) refers to the concept of "exposure to media violence," which might be estimated from the amount of violent television and movies watched and violent videogames played. It would be illogical to assume that the exposure causes the viewing and gaming, but rather the other way around. Other classic examples of putative formative latent variables, like socioeconomic status, are debatable. It may seem to some that an amalgam of variables like educational attainment, employment, neighborhood crime, and the like are distinct enough that they could only be combined arbitrarily. However, strong arguments can be made for social and institutional processes that do, in fact, contribute to positive associations among these variables that are distinct from the data. Whether psychopathology and other constructs relevant for clinical psychology are reflective or formative is no doubt a matter worthy of debate. If it is the latter, it is not clear that latent variable models are the best approach for interrogating their nature. Without foreclosing on this argument, this chapter is intended for those who are interested in using reflective latent variables to study clinical constructs.

MODEL ESTIMATION AND EVALUATION

The basic logic of estimation underlying latent variable models is to find values for a proposed model's parameters that would reproduce the observed data. Traditionally latent variable modeling has used summary data (e.g., sample variances, covariances, and means), although contemporary methods make use of the raw data because it allows for certain attractive features (e.g., accommodating missingness and severely nonnormal data).

Latent variables are estimated from the observed data through an iterative process, wherein a software-based algorithm auditions and repeatedly modifies parameter coefficient values in an effort to arrive at the best possible match to the data given the model's structure. Although the exact mathematical procedures differ as a function of the underlying estimator, maximum likelihood estimation is the most widely used estimation technique and therefore will be used as an example. To further keep this example simple, we can consider a basic confirmatory factor analysis (CFA; e.g., as depicted in Figure 7.1B). Underlying a CFA is a basic formula that defines how the observed variables are related to each other. In the model shown in Figure 7.1B, circles refer to latent variables and boxes refer to observed variables. The two observed variables Y1 and Y2 are related to each other because they each are indicators of (i.e., are caused by, note that the arrows emerge from the latent variable toward the observed variables) the latent variable F1. Thus, their observed correlation is explained by their associations with the latent variable. In contrast, Y2 and Y5 are related to each other not by shared variance with the same latent variable, but rather because they are each an indicator for latent variables that are correlated, as well as sharing an additional direct association depicted by the double-headed arrow between them. Without going into the details of the matrix algebra that can be used to calculate their predicted association given the model, it is important to understand that such a formula exists and can be used to generate an implied covariance matrix from the model's parameter values. In maximum-likelihood estimation, initially each parameter in the user-defined model is assigned a starting value, which can be used to calculate an initial implied covariance matrix. This matrix is then compared to the observed data, although presumably it is far off the observed values. However, this comparison can be used to adjust the parameter values so as to reduce the difference

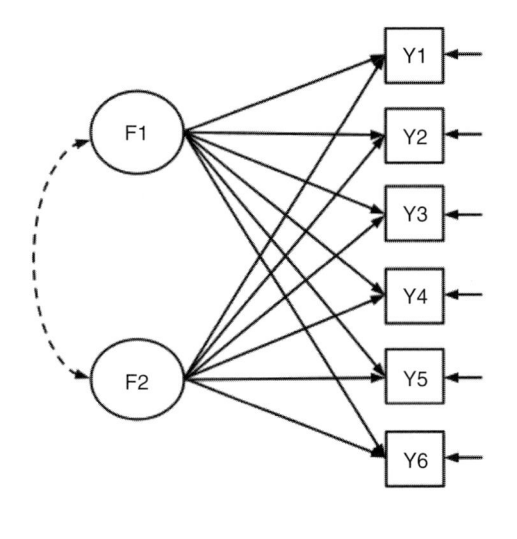

A. Exploratory factor analysis

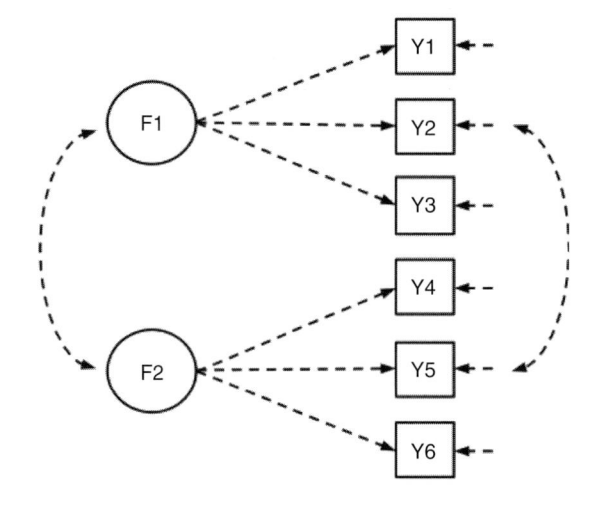

B. Confirmatory factor analysis

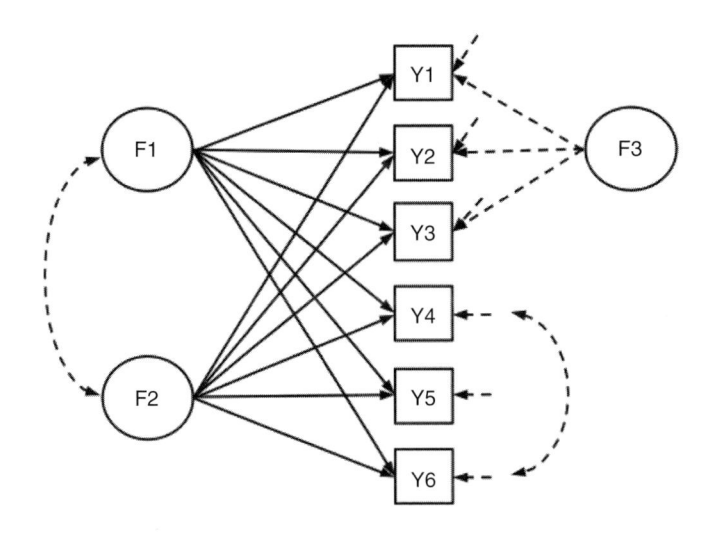

C. Exploratory structural equation modeling

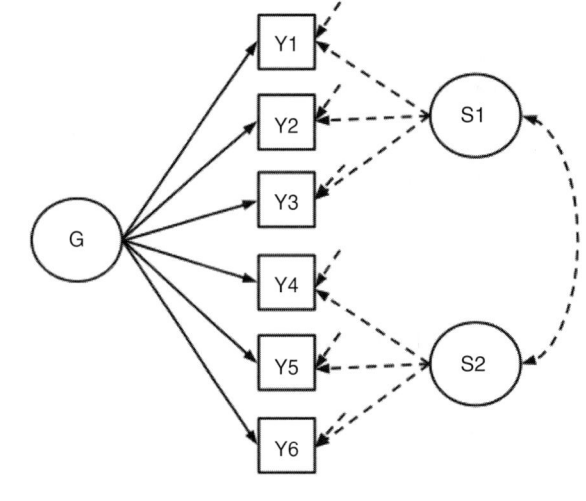

D. Bifactor modeling

Figure 7.1 Examples of common factor models

between the implied and observed data, and the amount of overall change in this step can be quantified. This step is repeated iteratively. Initially the amount of change is large, but through successive adjustments the amount of change between one set of values and the next is small. Once this discrepancy is small enough (i.e., reaches an established threshold), the model is said to have converged on a set of values. In the case of maximum likelihood estimation, these should be the set of values that *maximize the likelihood* of the data given the model. Importantly, these values do not necessarily provide a good match between the model implied and the observed data – rather, they provide the best match possible. That is, the model converges when it is no longer improving beyond a certain point at each iteration, not when the match between it and the data are below some certain discrepancy. This has attractive features; namely the ability to falsify models by testing absolute fit as well as comparing models to each other to ascertain relative fit.

Once the model estimation steps arrive at the closest match between the model implied and the observed data, the set of resulting values can be evaluated. There are several ways in which models can be evaluated, including global tests or indices of how well the data fit the model (usually some form of chi-square test), tests of local strain that are more circumscribed areas of misfit (e.g., residuals between the model implied covariances and the observed covariances), whether the model has generated plausible values (e.g., no values that are out of bounds like a negative variance), and whether the model in question fits better than alternative models under consideration.

MAPPING THE LATENT VARIABLE LANDSCAPE

Latent variable models come in different shapes and sizes. I mean this quite literally, in that one of the major continua along which models can be organized is whether they assume the latent variable to be dimensional, categorical, or somewhere in the middle (Masyn, Henderson, & Greenbaum, 2010). Additionally, latent variables can be distinguished from each other by whether they are exploratory or confirmatory in nature. Comprehensive coverage of all possible instantiations of latent variables would extend far beyond the scope of this chapter. Coverage here is selective based on those models that have received the most usage and support in clinical psychology. In particular, dimensional latent variable models are emphasized, including exploratory factor analysis (EFA), confirmatory factor analysis (CFA), exploratory structural equation modeling (ESEM), and bifactor models. Following this, factor mixture models (i.e., categorical and hybrid latent variable models) are introduced briefly, and research is reviewed that has found consistent support favoring latent dimensional structures over these mixture structures. Finally, latent variable models can also be enlisted in the study of processes of change over time. Indeed, a latent growth curve model is just a CFA with factor loadings fixed by the user to specific values. However, due to space limitations longitudinal models will not be covered here, but the interested reader is directed to Bollen and Curran (2006), Preacher and colleagues (2008), and Newsom (2015) for more detailed coverage, as well as Wood (Chapter 19) for advanced applications.

EXPLORATORY FACTOR ANALYSIS

The earliest latent variable models were exploratory factor analyses, first developed by Spearman (1904) as a quantitative approach to test his general theory of intelligence. He had observed that those who did well on one mental test tended to perform well on others, thereby perhaps reflecting a shared underlying cause. Initially, factor-analytic approaches were limited to EFA, which are termed exploratory because the investigator does not specify the patterning of items loading on factors, and instead all associations between latent and observed variables are freely estimated (the reader is referred to Mulaik [2010] for detailed treatment of EFA). Figure 7.1A provides a graphical representation of EFA. In this diagram square boxes represent observed variables, circles represent latent variables or factors, straight arrows connecting circles and squares represent factor loadings (i.e., the regression of the observed variable on the latent variable), arrows only pointing toward squares represent observed variable uniqueness (i.e., variability not accounted for by the latent factors, which includes both unique variance and error variance), and curved arrows represent covariances/correlations. Additionally, solid lines are used to represent model-specified parameters, whereas dashed lines represent parameters that can be specified by the investigator. In this example there are six observed variables and two correlated factors (i.e., an oblique model), and each of the observed variables loads on each of the two factors.

In EFA the investigator does not assign observed variables to factors; rather the relationship between each is estimated and the pattern of loadings is evaluated or "interpreted" after the analysis is run. Because of this, EFA has sometimes been called an atheoretical analytic approach, which is unfortunate as many aspects of EFA are, in fact, theoretically driven. For one, the latent variables are assumed to be dimensional and normally distributed. Second, it is frequently the case that the investigator has some hypothesis about how many factors are needed to account for the observed variables. Consequently, usually there is a theory, or perhaps this is better construed as an expectation, about which observed variables serve as significant markers for the same factors. More generally, the key modeling decisions in EFA (e.g., selecting which items to include, number of factors to retain, etc.) should ideally be made based on substantive theory. For instance, factors must be interpreted and labeled, and the emergence of a factor that is uninterpretable may prompt one to select fewer factors, drop items, or collect more data. EFA can be a very interactive technique, in the sense that several models are often run under different conditions and compared before settling on a final solution.

An investigator should consider three core questions when conducting an EFA: (1) *Which observed variables should be included in order to arrive at a valid latent structure* (this is true of statistical modeling in general)? One concern here is that if too few indicators for a specific construct are included, a corresponding factor is unlikely to emerge and will not be well determined if it does. An example of this can be found in Wright and colleagues (2013), which examined the latent structure of a number of symptoms of mental disorders in an Australian epidemiological sample. There were only two markers of manic episodes, which is insufficient to determine a distinct mania factor. On the flip side of that coin, an over-representation of content from a particular construct will almost guarantee a separate factor, even if the construct is subordinate to another domain (i.e., a bloated specific; see Oltmanns & Widiger, 2016 for a relevant example). (2) *How many factors should be retained?* Contemporary best practices for selecting the number of factors to retain involve the consultation of quantitative criteria like Horn's (1965) parallel analysis, Velicer's (1976) minimum average partial test, Ruscio and Roche's (2012) comparison data technique, and model fit criteria (e.g., chi-square, Root Mean Square Error of Approximation [RMSEA]) when available based on the estimator (e.g., maximum likelihood) to *inform* the number to retain. However, regardless of which methods are used, these are fallible tools that should be weighed in the decision but not blindly followed. The investigator should still make careful choices based on all pertinent information, especially theory. (3) *How should these factors be rotated?* Factor rotation involves adjusting the relationship between the factors and the indicators so that they are more interpretable. Usually this involves using an algorithm (e.g., Varimax, Oblimin, Geomin) to try and achieve something that approximates simple structure (i.e., each indicator loads on only one factor). Despite the many options available for factor rotation (see Sass & Schmitt, 2010 for a review), the most important distinction is between *orthogonal* or *oblique* factors. In an orthogonal rotation, the factors are forced to be unrelated to each other, whereas in an oblique rotation factors are allowed to correlate. Oblique rotation methods are generally preferable because they do not preclude an orthogonal solution from emerging, but allow for substantial factor correlations when indicated. This is a key consideration in psychopathology research, given that there are theoretical and empirical rationales for why factors might be expected to correlate substantially (Caspi & Moffitt, 2018; Sharp et al., 2015). However, factor rotation will potentially have nonnegligible effects on factor interpretation, and therefore it should be given thorough consideration. This is especially the case in psychopathology data, where indicators tend to be positively correlated. Recall that the model will attempt to reproduce the observed correlations with the model parameters. Therefore, in an orthogonal model where factors are uncorrelated, the only way that indicators can be associated is "through" the factors, by increasing the factor loadings (particularly secondary or cross-loadings). In an oblique model, because the factors are allowed to correlate, thereby accounting for some of the indicator associations, the secondary loadings are often relatively smaller (see Sharp et al., 2015 for an example of this phenomenon). Conceptually and practically this becomes a question of where you want to have the model complexity, at the level of the items or at the level of latent constructs.

In clinical psychology EFA is often used in the development of assessment tools and measurement inventories (Clark & Watson, 1995; see also Chapter 6 by Furr). Often a large number of items are auditioned as potential indicators for latent constructs of interest, and items are retained or excluded based on their pattern and strength of factor loadings. It has also been put to good use in studying the structure of psychopathology, particularly as quantitative structural models of psychopathology have started to incorporate large numbers of indicators without precise a priori hypotheses about how they should relate to each other (e.g., Forbes et al., 2017; Wright & Simms, 2015).

CONFIRMATORY FACTOR ANALYSIS

As the name indicates, unlike EFA, CFA is intended to serve primarily as a hypothesis *testing* analytic approach (the reader is referred to Brown [2015] for detailed discussion of CFA techniques). The confirmatory aspects are that (a) the user may specify any of the model parameters, and (b) the fit (or, more specifically, the lack of fit) of the observed data to the specified model is tested. Figure 7.1B illustrates a hypothetical typical two-factor CFA. In this model the observed variables Y1–Y3 serve as indicators of latent factor F1 only, and Y4–Y6 serve as indicators of F2 only. Please note that the CFA in Figure 7.1B differs from the EFA in Figure 7.1A in that the presence of each factor loading was user specified, and not all items load on each factor. Much like the EFA model the factors are allowed to correlate, making it an oblique model. However, there is no rotation to choose; in CFA factors are either correlated (oblique) or uncorrelated (orthogonal). This is because in CFA the investigator has the ability to impose true simple structure (i.e., indicators load on only one factor), which rotation algorithms are designed to approximate. Further, each observed variable has a residual variance, reflecting unique variability unaccounted for by the factor plus measurement error. Finally, notice the curved arrow between Y2 and Y5. This reflects a residual covariance, indicating that there is shared variance in items Y2 and Y5 unaccounted for by the modeled factors.

Recall when testing this model, the statistical package would first optimize the values of the parameters in an effort to match the data set, then it would compare the fit of the model-implied covariance matrix to the observed covariance matrix and generate goodness-of-fit indices based on the degree of match. Each modeling decision has implications for the implied pattern of covariation. For instance, in the case where there are no free residual covariances, the factors must account for *all* of the covariation among the observed variables (i.e., conditional independence). Any unaccounted-for residual covariation in the actual data will contribute to worse fit.

CFA does allow for deviation from the assumption of conditional independence. Factor models are usually specified such that there is no covariance among the indicator residuals, the assumption being that the observed variables are independent from each other once the factors are accounted for (i.e., conditional on the factors). Although reasonable given the goal of factor analysis, relaxing this assumption has legitimate uses. For instance, it can be used to account for method variance between specific item sets (e.g., scales from the same instrument, scales completed by the same reporter). However, unprincipled use of residual covariances is discouraged, as it can capitalize on chance in any given data set, especially when sample size is large, and result in nonreplicable model complexity.

CFA has a long track record of use in studying the latent structure of psychopathology (e.g., Brown, Chorpita, & Barlow, 1998; Girard et al., 2017; Kotov et al., 2011; Krueger, 1999; Wright et al., 2013). The observation that certain patterns of diagnostic covariation occur at rates much higher than chance beckons for latent variable hypotheses. Specifically, that there might be underlying processes that generate these patterns of observed covariation. In the adult psychopathology literature, the seminal study by Krueger (1999) used the National Comorbidity Study data to show that patterns of diagnostic covariation among the common mental disorders could be accounted for by two broad factors of internalizing (mood and anxiety disorders) and externalizing (substance use and antisocial behavior). CFA also forms the measurement model in full structural equation models that allow for structural (i.e., regression paths) among latent variables. So, CFAs are frequently used to develop error-free measures of latent constructs that are then related in more complex patterns of associations in structural equations.

THE EXPLORATORY-CONFIRMATORY SPECTRUM

Often exploration and confirmation are presented as if they are discrete and mutually exclusive modes of inquiry. In practice, though, the boundary between them is much fuzzier. For instance, it is hard to imagine someone conducting an EFA on a set of items with no intuition or expectations about what they might find. Similarly, it is hard to imagine testing a CFA with not only all latent to observed relations articulated, but also their precise values specified a priori. Indeed, either of these scenarios might represent unlikely extremes along a continuum of exploratory to confirmatory. In practice, latent variable modeling contains some mixture of both – in EFAs the investigator usually has some sense about the general structure that might emerge, although the details are quite uncertain, whereas in CFA the investigator often enjoys much greater confidence about many aspects of the model, but some degree of uncertainty remains about the exact values that might emerge. Furthermore, there is likely a willingness to make alterations to noncentral features of the model without doing extreme violence to the theoretical structure being tested (e.g., there may be a strong theory about the number of factors but a weaker theory about whether there are cross-loadings or not). This led me to propose the Exploratory-Confirmatory Spectrum (Wright, 2017) in factor analysis, which is intended to better contextualize one's degree of a priori knowledge and expectations about the results. Typical EFA modeling scenarios would fall toward the exploratory end of the spectrum and typical CFA estimation would fall toward the confirmatory end, although both could be pushed further out with procedures of the sort I alluded to above. Falling in the middle between the two are variations in certainty about the expected structure in EFA and willingness to make modifications in CFA. The two poles of the spectrum are brought together with ESEM (Asparouhov & Muthén, 2009).

ESEM blends the core features of EFA (i.e., exploratory factors, range of rotations) and CFA (i.e., the ability to specify parameters, user-specified factors, multiple group analysis) allowing for near-total flexibility in modeling. Numerous advantages are gained by this innovation. These include the ability to estimate method factors in EFA analyses of multiple scales from different measures, correlated residuals, and adding parameter equality constraints across two or more scientifically interesting groups (e.g., genders, patients vs. nonpatients). Figure 7.1C provides a hypothetical example of an ESEM model. In this diagram, in addition to two obliquely rotated EFA factors (F1 and F2), there is a third, investigator-specified factor (F3) that is orthogonal to the other two. F3 could perhaps represent shared method variance for observed variables Y1–Y3, or that they are markers for more than one construct. Finally, the residuals for Y4 and Y6 are allowed to correlate. In the modeling of complex personality data that has large item sets, ESEM benefits from the efficiencies of the EFA framework, while allowing the investigator control over specific modeling features that are afforded with CFA.

Similar to CFA, ESEM relies on estimation methods that ultimately result in in an implied covariance matrix that can be compared to an observed matrix in various ways to generate goodness-of-fit indices. The fact that the EFA portion of the structure can model a large number of potentially conceptually negligible but statistically significant cross-loadings generally results in considerable improvement in fit over a strict (and implausible) simple structure imposed by many CFAs. However, it is worth emphasizing that factor-analytic techniques are largely separable from the estimation approach. While certain estimation methods (e.g., principal factor analysis) are reserved for EFA, estimators like maximum likelihood and weighted least squares can be applied to EFA, CFA, or ESEM. This underappreciated fact often results in models erroneously labeled as ESEMs, when in reality only a maximum likelihood EFA has been conducted. Although this produces fit criteria, no additional user-specified parameters have been included. EFA is a very useful technique, and the objection to labeling a maximum-likelihood EFA an ESEM is that it creates the perception that there are user-specified parameters without the user having specified any beyond a standard EFA. Alternatively, a maximum-likelihood EFA can be considered a special case of ESEM, and the same can be said for CFA.

Given that well-validated personality inventories often fit poorly in CFA models (Hopwood & Donnellan, 2010), personality researchers generally have been early adopters of ESEM. And as such, ESEM in clinical psychology was quickly adopted by personality disorder researchers who deal with similar issues. For instance, Gore and Widiger (2013) estimated the joint factor structure of four personality inventories to examine whether the normal range and pathological scales combined to indicate the same five factors. An initial maximum-likelihood EFA resulted in poor fit to the data, so an ESEM was estimated allowing the residuals of indicators from the same personality inventory to correlate across factors. This ultimately resulted in an excellently fitting model, and a theoretically expected five-factor structure. In this case an ESEM allowed Gore and Widiger (2013) to account for the dependency among scales from the same inventory within an otherwise exploratory analytic framework. Other examples include Wright and Simms (2014, 2015), who dealt with the same issue in a similar but distinct fashion by estimating method factors for each inventory used in an otherwise exploratory model. For example, Wright and Simms (2015) tested whether the joint structure of clinical syndromes, DSM personality disorder dimensions, and maladaptive personality trait scales would conform to a recognizable five-factor structure (Negative Affectivity, Detachment, Antagonism, Disinhibition, and Psychoticism). In addition to estimating correlated substantive exploratory factors on which all items loaded, orthogonal measure-specific factors were included for the clinical syndrome interview, the personality disorder interview, and the trait scales, on which all indicators from each measure loaded. This served to isolate shared method variance while retaining substantive variance in each indicator.

The advent of ESEM offers investigators considerably more flexibility than EFA or CFA alone. Researchers are encouraged to think of models not as either exploratory or confirmatory, but as falling somewhere along a continuum between those two poles. Thought should be given to whether any parameters can be specified and tested, even if parts of the model will be determined via exploratory techniques.

BIFACTOR MODELS

In recent years, the bifactor model has made a considerable impact on our conceptualization of the structure of psychopathology (see Greene et al., 2019, for a review). Although it was first described in the 1930s (Holzinger & Swineford, 1937) the bifactor model received little attention from the applied literature until the mid-to-late 2000s. But since then it has had an outside effect on the field; so much so that it deserves its own section to unpack the merits and demerits of the model. First, it is important to understand that a bifactor model is just an ordinary factor model (it can be estimated as either an EFA or CFA), but with a specific structure (see Figure 7.1D). The key feature is that a general factor, on which all indicators load, is estimated as orthogonal to some number of "group" or "specific" factors, on each of which only a subset of items loads (or in the case of EFA that load strongly). The specific factors may be either orthogonal or oblique, although orthogonal is more common. In Figure 7.1D, a bifactor model is depicted with a general factor and two oblique specific factors.

The principal motivation for estimating a bifactor model is to partition the common variance in each indicator into what it shares with all other indicators (i.e., general variance) and into what it shares with only a smaller subset of indicators (i.e., specific or group variance). Conceptually, this can be understood to isolate those processes that are driving scores in all indicators from those processes that are driving scores on specific groupings. For instance, recently Fournier and colleagues (2019) estimated a bifactor model of the NEO-PI-Rs (Costa & McCrae, 1992) neuroticism scales. The NEO-PI-R is structured hierarchically by design, such that eight individual questions are combined to form facet scales, and six facet scales are combined to form a domain, like neuroticism. Neuroticism facets include depression, anxiety, angry hostility, impulsiveness, self-consciousness, and vulnerability. Fournier and colleagues' (2019) motivation was to examine whether what was specific to the depression facet was incrementally related to external variables beyond the general factor (i.e., what was shared across all indicators). Thus, a bifactor model was estimated to partition the variance into neuroticism-general and depression-specific processes. Findings demonstrated, across a community and a clinical sample, that although the standard NEO-PI-R depression scale correlated significantly and strongly with other depression measures, the specific depression facet factor in a bifactor model was unassociated with other depression measures. In other words, whatever processes account for the association between the NEO-PI-R depression facet and other measures of depression, these are processes that are general to all neuroticism items and not specific to the depression items.

This sort of theoretically driven analysis, which seeks to partition and separately examine distinct sources of variance in observed indicators, is something the bifactor model is uniquely situated to address (see also Sharp et al., 2015 for another example). However, the bifactor's popularity has also been driven by its use as an estimate of a higher-order factor in a hierarchical structure while also providing exceptionally good fit to the data. To elaborate, factors estimated from observed indicators can also be used as indicators for higher-order factors, assuming there are sufficient lower-order factors present to identify the higher-order factor (i.e., at least three). See Figure 7.2B for an example. The bifactor is an alternative approach to estimating a higher-order factor, in that the general factor in the bifactor would capture shared variance in all indicators, and group or specific factors could be estimated to capture what is unique in the indicators for the aforementioned first-order factors. See Figure 7.2A for a bifactor hierarchical model in the same data. The general factor and the higher-order factor in these models will be very similar, in that the general factor captures what all the indicators share and the higher-order factor captures what is shared in the lower-order factors, which themselves are based on what groups of indicators share. At the same time, they are not identical, because in the higher-order factor model the shared variance is "funneled" through the

A. Bifactor model

B. Higher-order factor analysis

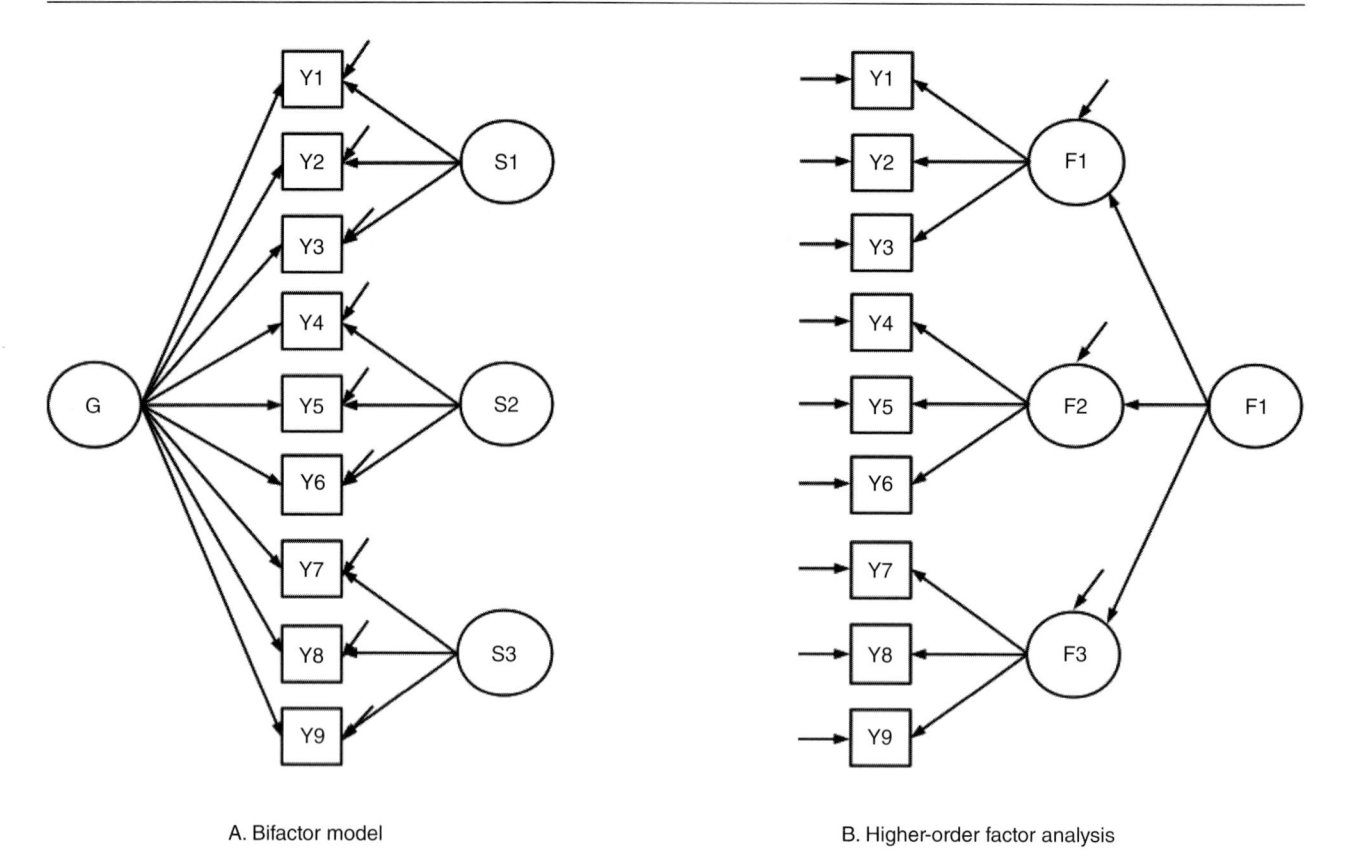

Figure 7.2 Two alternative representations of a hierarchical factor model

lower-order factors on its way up to the higher-order model. The implication is that any additional shared variance among the indicators that is not captured by the group factors would not be explained by the higher-order factor, but would be explained by the general factor in the bifactor model. Importantly, the specific factors will not be identical to, and in many cases in clinical research will be quite distinct from, the first-order factors in a standard higher-order factor model, because the specific factors will only be estimated from the variance in a subset of indicators net of the variance shared with all items. As I will describe below, this last issue has become a point of criticism of the bifactor approach, but those criticisms often neglect an important reality of higher-order models.

In the domain of psychopathology structure, although conceptually distinct domains like internalizing and externalizing have been estimated in many data sets, they often correlate strongly (e.g., population sample $r \sim 0.5$). This has led some to hypothesize a higher-order general factor of psychopathology, or the "p-factor" (Caspi et al., 2014; Lahey et al., 2012, 2015). In their seminal paper on the topic, Caspi and colleagues (2014) referred to the p-factor as the propensity or liability to develop psychopathology. They compared models with three correlated factors (internalizing, externalizing, and thought disorder) explaining the covariation of mental disorder diagnoses over the lifetime with several other structures, including a bifactor. In this early work, the bifactor model was chosen over the correlated factors model and alternatives because of its markedly superior relative fit in the data. Note that a three-factor oblique model would have identical statistical fit to a higher-order model with one factor explaining the covariation among the three factors. Although papers examining a general factor of psychopathology (e.g., Lahey et al., 2012) predate the work by Caspi et al. (2014), this was the paper that ignited an explosion of literature in the past half a decade examining the p-factor and often comparing its fit to a correlated factor model (Greene et al., 2019).

As with any highly visible and potentially transformative research program, the p-factor research has met with detractors, and critics have often taken aim at the bifactor model specifically. Concerns have been raised on both quantitative and conceptual grounds, and these extend well beyond the domain of clinical psychology and p-factors. Nevertheless, they are important to cover here because of the outsized impact this model has had on the field and because bifactors are now used extensively in clinical psychology, not just in the p-factor literature. The statistical concerns largely center on the fact that bifactor models are exceptionally good at matching the empirical structure of the data – perhaps too good, in that some have raised concerns about bias and ill-performing fit statistics in comparisons of bifactor models to other models

(Bonifay, 2017; Gignac, 2016; Greene et al., 2019; Mansolf & Reise, 2017; Morgan et al., 2015; Murray & Johnson, 2013; Reise et al., 2016). To summarize a technically complex and nuanced literature in a couple of brief points, what these studies have shown is that the bifactor model is better able to accommodate departures from the expected pattern of effects generated by other models, most notably the higher-order model described above. In other words, when factor models are not a faithful match to the true data-generating mechanism, which is always the case to some degree, the features of the bifactor model allow it to better approximate the data than comparable models, even if it is not the true data-generating mechanism. The types of misspecifications that would lead fit statistics to favor the bifactor model over the higher-order factor model, even when the higher-order factor model is the true model, include unmodeled cross-loadings of items in the first-order factors and residual correlations. An added issue is that these sorts of model features can arise spuriously by chance in any given data set, thereby favoring the bifactor spuriously.

However, in applied settings the true data-generating mechanism is unknown, and whether effects are spurious or not may not be known either. As Mansolf and Reise (2017) put it, this leaves applied researchers with a conundrum. If the fit statistics favor the bifactor model, that could be because it is a better representation of the data, or it could be because it has accommodated unmodeled (and possibly spurious) complexities in the structure from something like the higher-order model. They and Gignac (2016) also point out that there are conditions under which the bifactor model will have identical fit to the higher-order factor model, even when the bifactor model is the true data-generating mechanism. Ultimately, what this boils down to is that selecting the bifactor model or a comparator based solely on fit is dubious practice, and instead conceptual arguments need to be made in favor of one or the other. As noted above, quantitative indices like fit statistics are useful but fallible guides.

Concerns have also been raised about the bifactor model on conceptual grounds. One of the major issues has been that partitioning variance into general and specific in this way is something of a false exercise because individuals only produce one set of scores, not multiple sources of scores. Or, alternatively, one cannot determine from a single set of scores what underlying processes generated their manifestation. This criticism fundamentally misunderstands the utility of latent variable modeling as providing access to something distinct from the observed data. A second criticism has been that bifactor models can often generate "nonsensical" specific factor loadings. For instance, it is not uncommon to find null or even negative loadings of indicators on a specific factor, when those same indicators would positively and strongly load on a first-order factor. An example of this can be found in the item loadings depression facet in the community sample of the aforementioned Fournier et al. (2019).

The concern is that the specific factors in a bifactor model "are not the same thing" as the first-order factors in a higher-order model. This is most assuredly the case, but whether it is a reasonable comparison is another matter. The correct comparison would not be to the first-order factor in a higher-order factor model, but rather to the residuals (sometimes referred to as a disturbance) of those first-order factors after partialing out the higher-order factor (see Figure 7.2). Often relatively little variance remains in the first-order factor after accounting for the higher-order factor, and these are rarely if ever evaluated independently. One of the reasons this is the case is that it can be difficult to estimate identified structural models to evaluate the effect of these residuals.

In sum, the bifactor model can be a useful tool for partitioning variance into general and specific components when theoretically justified. Yet it is likely that the field has placed too much weight and confidence in the model due to its ability to provide a better match to what is often complex observed data. Moving forward, applied researchers should understand the model for what it is and use it when it makes strong conceptual sense, placing less emphasis on its statistical fit.

FACTOR MIXTURE MODELING AND COMPARING LATENT STRUCTURES

To this point, only dimensional latent variable models have been considered, with the shared assumption being that the latent variables are continuous and normally distributed. This distributional assumption is not a requirement and, as introduced above, models exist that assume nonnormality and discontinuities in the latent space. Although initially developed as distinct models, categorical (e.g., latent class and latent profile analysis) and dimensional latent variable models can now be subsumed within the broader framework of factor mixture modeling. Factor mixtures are not limited to simplistic categorical vs. dimensional dichotomies, but rather encompass a *Categorical-Dimensional Spectrum* (Masyn et al., 2010). Thus, the modeled structures can range from the fully dimensional (i.e., factor analyses) to the fully categorical (i.e., latent class analysis), with variations that combine aspects of the two in between (see also Hallquist & Wright, 2014). To illustrate some of the latent structures that are possible in this framework, Figure 7.3 provides graphical depictions of many, but not all, of the possible structures. Factor analysis (Figure 7.3A) represents one pole of the categorical-dimensional spectrum, and assumes a fully dimensional latent structure, such that individuals vary continuously along a normally distributed latent trait (or traits). At the other end of the spectrum is latent class analysis (Figure 7.3F) that assumes a fully categorical latent structure, such that individuals differ discretely from each other exclusively in terms of a pattern of features shared among a homogenous subgroup. In terms of

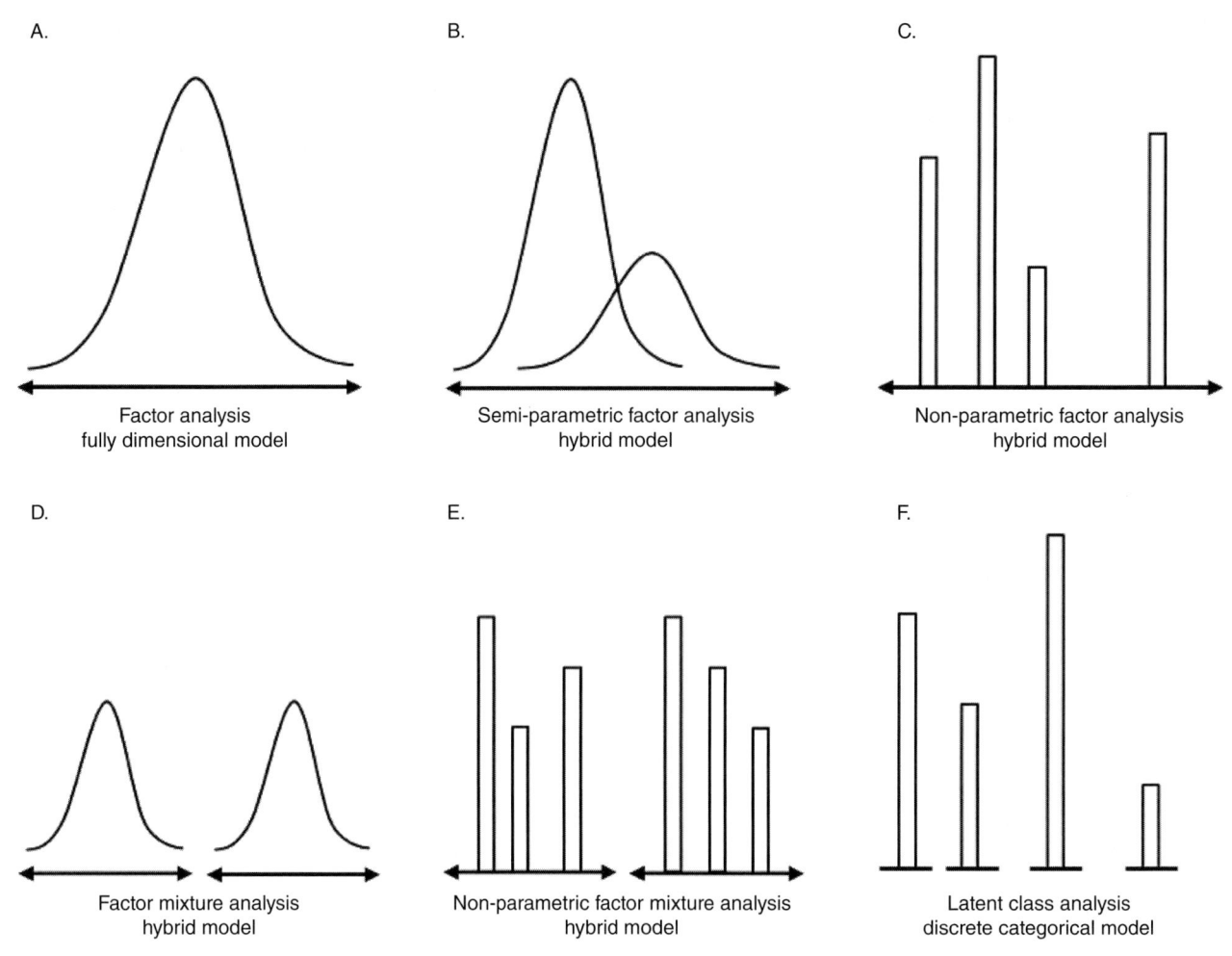

Figure 7.3 Graphical depiction of latent distributions associated with various factor mixture models. Lines at the base with arrows represent continuous dimensions. Columns represent groupings of individuals with no variance in latent scores

hybrid models, Semi-Parametric Factor Analysis (Figure 7.3B) estimates a mixture of normally distributed groups along a common dimension to model a nonnormal, but continuous, distribution. Thus, individuals vary along the same trait, but it allows for an extreme tail, or other nonnormal (e.g., bimodal) distributions. Alternatively, Non-Parametric Factor Analysis (also referred to as Located Latent Class Analysis; Figure 7.3C) models discrete latent groups along a shared dimension. In this case, there are defined "gaps" between latent groups of individuals along the same latent trait. It is also possible to model factor structures that differ across groups (Figure 7.3D), which imply different latent dimensions or that the questions or symptoms have different meanings, or function differently across groups. This can even be extended to include discrete disjunctions in those factors (Figure 7.3E). This is not an exhaustive catalogue of these models, but rather a sampling to encourage researchers and practitioners to think in more nuanced ways about the possible latent structure of clinical constructs beyond the typical categories and continua. Importantly, these can all be estimated and compared with each other in real data to test theoretical assumptions about the actual latent structure of pathology. This is made possible by advances in maximum-likelihood estimation that place all of these models on comparable quantitative footing.

Adjudicating debates about the latent structure of psychopathology is one fruitful use of factor mixture modeling in clinical psychology. Specifically, numerous studies have now accumulated that have compared dimensional, categorical, and hybrid structures to determine which structure the empirical data fit better. In a typical example of this literature, Conway, Hammen, and Brennan (2012) examined the latent structure of the DSM's nine borderline personality disorder criteria in a large community sample of young adults at risk for psychopathology. They compared dimensional (factor analytic), categorical (latent class analysis), and hybrid models (non-parametric factor analyses) finding that the data best fit a fully dimensional latent structure. In a more recent study, Aslinger and colleagues (2018) performed similar analyses on the DSM's narcissistic personality disorder criteria, but extended the typical approach by including semiperimetric factor analyses as an alternative hybrid model as well as attempting to replicate initial exploratory results across five samples. The consistent finding favored the dimensional model (CFA) over hybrids or fully categorical models. Expanding the lens beyond single disorders to broad domains of psychopathology (e.g., internalizing, externalizing, psychosis), the finding has been consistent, with fit criteria favoring latent dimensional models over categorical or hybrid models (e.g., Eaton et al., 2013; Markon & Krueger, 2005; Walton, Ormel, & Krueger, 2011; Wendt et al., in press; Witkiewitz et al., 2013; Wright et al., 2013).

Although the broad picture is quite consistent, that the latent structure of psychopathology is likely dimensional in nature, several other aspects of this modeling space are worth noting. For one, factor mixture models can quickly become complex, because they allow basically any parameter to vary across classes (e.g., factor loadings, variances, residuals, intercepts). Thus, exhaustive testing of structures is impractical and runs the risk of capitalizing on chance. As a result, most applied use of factor mixtures in clinical science has been restricted to these coarse comparisons as outlined in the preceding paragraph (cf. Hallquist & Pilkonis, 2012). It may be that judicious and targeted selection of hybrid models will prove useful for some questions. One example is in the modeling of non-normality in a dimensional latent space (Masyn et al., 2010). A further consideration is that even when a particularly latent structure is estimated it implicitly provides support for another. For instance, Bornovalova and colleagues (2010) estimated a latent class model of borderline personality disorder symptoms, but the four retained classes differed from each *only* in severity, *not* in the configuration of symptom endorsements (i.e., quantitative, not qualitative differences between classes), suggesting a dimensional latent structure. Just because a particular structure is estimated and is well fitting and provides sensible values does not mean it is the most appropriate fit. In many respects this is similar to the issues revolving around bifactor modeling, and serves as a reminder that considerations of interpretability and conceptualization must be close bedfellows of quantitative indices.

CONCLUSION

Latent variable models play an important role in contemporary clinical psychology. Here I have provided a conceptual overview of the techniques, in part because any technical treatment limited to a single chapter would be woefully incomplete. More important than that, though, is that the utility of these rests most strongly on how we think about them and what they can tell us. At the start, I argued that it is important to take the ontology of latent variable models seriously, but that does not mean one has to adopt an extreme position. The pseudorealist perspective asserts that the latent variables estimated are meaningful and distinct from the data, but hedges on whether they are identifying one true underlying cause. In fact, as I think will be clear to readers, much of what clinical psychology is working with is approximations. This seems reasonable for a science as young as ours, and I believe it will continue to improve with the help of latent variable models. Consistent with this theme, throughout the chapter I have noted that many of the ongoing thorny issues in latent variable modeling (e.g., the overreliance on fit statistics when evaluating bifactor models) require deep conceptual thinking, not careful adherence to statistics.

The review presented here strongly favored factor analysis or latent dimensional models. This is in large part because these models have provided more replicable and interpretable results. However, the use of mixture models has been and continues to be (Aslinger et al., 2018; Wendt et al., in press) useful in the adjudication of latent structure. Another limitation is that all of the examples I provided were based on self-report or diagnostic interview data. There is no reason to limit these models to psychometric scale data of this sort, and one can just as easily use biological or observational behavior data as well. Chapter 3 by Venables and Patrick gives examples of how this might proceed fruitfully. More broadly, I hope readers take away the wider perspective, that anytime a measure of something observable stands in for some underlying construct, the researcher is invoking a latent variable model conceptualization.

REFERENCES

Aslinger, E. N., Manuck, S. B., Pilkonis, P. A., Simms, L. J., & Wright, A. G. C. (2018). Narcissist or Narcissistic? Evaluation of the Latent Structure of Narcissistic Personality Disorder. *Journal of Abnormal Psychology*, 127(5), 496–502.

Asparouhov, T., & Muthén, B. (2009). Exploratory Structural Equation Modeling. *Structural Equation Modeling*, 16, 397–438.

Bollen, K. A. (1989). *Structural Equations with Latent Variables*. New York: Wiley.

Bollen, K. A. (2002). Latent Variables in Psychology and the Social Sciences. *Annual Review of Psychology*, 53(1), 605–634.

Bollen, K. A., & Curran, P. J. (2006). *Latent Curve Models: A Structural Equation Perspective*. New York: Wiley.

Bonifay, W. (2017). On the Complexity of Item Response Theory Models. *Multivariate Behavioral Research*, 52(4), 1–20.

Bornovalova, M. A., Levy, R., Gratz, K. L., & Lejuez, C. W. (2010). Understanding the Heterogeneity of BPD Symptoms through Latent Class Analysis: Initial Results and Clinical Correlates among Inner-City Substance Users. *Psychological Assessment*, 22(2), 233–245.

Borsboom, D., Mellenbergh, G. J., & van Heerden, J. (2003). The Theoretical Status of Latent Variables. *Psychological Review*, 110(2), 203–219.

Brown, T. A. (2015). *Confirmatory Factor Analysis for Applied Research* (2nd edn.). New York: Guilford Press.

Brown, T. A., Chorpita, B. F., & Barlow, D. H. (1998). Structural Relationships among Dimensions of the DSM-IV Anxiety and

Mood Disorders and Dimensions of Negative Affect, Positive Affect, and Autonomic Arousal. *Journal of Abnormal Psychology*, 107(2), 179–192.

Caspi, A., Houts, R. M., Belsky, D. W., Goldman-Mellor, S. J., Harrington, H., Israel, S., ... Poulton, R. (2014). The P Factor One General Psychopathology Factor in the Structure of Psychiatric Disorders? *Clinical Psychological Science*, 2(2), 119–137.

Caspi, A., & Moffitt, T. E. (2018). All for One and One for All: Mental Disorders in One Dimension. *American Journal of Psychiatry*, 175(9), 831–844.

Clark, L. A., & Watson, D. (1995). Constructing Validity: Basic Issues in Objective Scale Development. *Psychological Assessment*, 7(3), 309–319.

Collins, L. M., & Lanza, S. T. (2010). *Latent Class and Latent Transition Analysis: With Applications in the Social Behavioral, and Health Sciences*. Hoboken, NJ: Wiley.

Conway, C., Hammen, C., & Brennan, P. A. (2012). Comparison of Latent Class, Latent Trait, and Factor Mixture Models of DSM-IV Borderline Personality Disorder Criteria in a Community Setting: Implications for DSM-5. *Journal of Personality Disorders*, 26, 793–803.

Costa, P. T., Jr., & McCrae, R. R. (1992). *Revised NEO Personality Inventory (NEO-PI-R) and NEO Five-Factor Inventory (NEO-FFI) Professional Manual*. Odessa, FL: Psychological Assessment Resources.

Cronbach, L., & Meehl, P. (1955). Construct Validity in Psychological Tests. *Psychological Bulletin*, 52(4), 281–302.

Eaton, N. R., Krueger, R. F., Markon, K. E., Keyes, K. M., Skodol, A. E., Wall, M., Hasin, D. S., ... Grant, B. F. (2013). The Structure and Predictive Validity of the Internalizing Disorders. *Journal of Abnormal Psychology*, 122, 86–92.

Edwards, J.R., & Bagozzi, R.P. (2000). On the Nature and Direction of Relationships between Constructs and Measures. *Psychological Methods*, 5, 155–174.

Forbes, M. K., Kotov, R., Ruggero, C. J., Watson, D., Zimmerman, M., & Krueger, R. F. (2017). Delineating the Joint Hierarchical Structure of Clinical and Personality Disorders in an Outpatient Psychiatric Sample. *Comprehensive Psychiatry*, 79, 19–30.

Fournier, J. C., Wright, A. G. C., Tackett, J. L., Uliaszek, A., Pilkonis, P. A., Manuck, S. B., & Bagby, R. M. (2019). Decoupling Personality and Acute Psychiatric Symptoms in a Depressed and a Community Sample. *Clinical Psychological Science*, 7(3), 566–581.

Gignac, G. E. (2016). The Higher-Order Model Imposes a Proportionality Constraint: That Is Why the Bifactor Model Tends to Fit Better. *Intelligence*, 55, 57–68.

Girard, J. M., Wright, A. G. C., Beeney, J. E., Lazarus, S., Scott, L. N., Stepp, S. D., & Pilkonis, P. A. (2017). Interpersonal Problems across Levels of the Psychopathology Hierarchy. *Comprehensive Psychiatry*, 79, 53–69.

Gore, W. L., & Widiger, T. A. (2013). The DSM-5 Dimensional Trait Model and Five-Factor Models of Personality. *Journal of Abnormal Psychology*, 122(3), 816–821.

Greene, A. L., Eaton, N. R., Li, K., Forbes, M. K., Krueger, R. F., Markon, K. E., ... Kotov, R. (2019). Are Fit Indices Used to Test Psychopathology Structure Biased? A Simulation Study. *Journal of Abnormal Psychology*, 128(7), 740–764.

Hallquist, M., & Pilkonis, P. (2012). Refining the Phenotype of Borderline Personality Disorder. *Personality Disorders: Theory, Research, and Treatment*, 3(3), 228–246.

Hallquist, M. N. & Wright, A. G. C. (2014). Mixture Modeling Methods for the Assessment of Normal and Abnormal Personality Part I: Cross-Sectional Models. *Journal of Personality Assessment*, 96(3), 256–268.

Harman, H. H. (1960). *Modern Factor Analysis*. Chicago: University of Chicago Press

Holzinger, K. J., & Swineford, F. (1937). The Bi-Factor Method. *Psychometrika*, 2(1), 41–54.

Hopwood, C. J., & Donnellan, M. B. (2010). How Should the Internal Structure of Personality Inventories Be Evaluated? *Personality and Social Psychology Review*, 14, 332–346.

Horn, J. L. (1965). A Rationale and Test for the Number of Factors in Factor Analysis. *Psychometrika*, 30(2), 179–185.

Kotov, R., Ruggero, C. J., Krueger, R. F., Watson, D., Yuan, Q., & Zimmerman, M. (2011). New Dimensions in the Quantitative Classification of Mental Illness. *Archives of General Psychiatry*, 68, 1003–1011.

Krueger, R. F. (1999). The Structure of Common Mental Disorders. *Archives of General Psychiatry*, 56, 921–926.

Lahey, B. B., Applegate, B., Hakes, J. K., Zald, D. H., Hariri, A. R., & Rathouz, P. J. (2012). Is There a General Factor of Prevalent Psychopathology during Adulthood? *Journal of Abnormal Psychology*, 121(4), 971–977.

Lahey, B. B., Rathouz, P. J., Keenan, K., Stepp, S. D., Loeber, R., & Hipwell, A. E. (2015). Criterion Validity of the General Factor of Psychopathology in a Prospective Study of Girls. *Journal of Child Psychology and Psychiatry*, 56(4), 415–422.

Loehlin, J. C. (2004). *Latent Variable Models: An Introduction to Factor, Path, and Structural Equation Analysis*. Abingdon: Taylor & Francis.

Mansolf, M., & Reise, S. P. (2017). When and Why the Second-Order and Bifactor Models Are Distinguishable. *Intelligence*, 61, 120–129.

Markon, K. E., & Krueger, R. F. (2005). Categorical and Continuous Models of Liability to Externalizing Disorders: A Direct Comparison in NESARC. *Archives of General Psychiatry*, 62, 1352–1359.

Masyn, K. E., Henderson, C. E., & Greenbaum, P. E. (2010). Exploring the Latent Structures of Psychological Constructs in Social Development Using the Dimensional-Categorical Spectrum. *Social Development*, 19(3), 470–493.

Morgan, G. B., Hodge, K. J., Wells, K. E., & Watkins, M. W. (2015). Are Fit Indices Biased in Favor of Bi-Factor Models in Cognitive Ability Research? A Comparison of Fit in Correlated Factors, Higher-Order, and Bi-Factor Models via Monte Carlo Simulations. *Journal of Intelligence*, 3(1), 2–20.

Mulaik, S. (2010). *Foundations of factor analysis* (2nd edn.). Boca Raton, FL: Chapman & Hall.

Murray, A. L., & Johnson, W. (2013). The Limitations of Model Fit in Comparing the Bi-Factor versus Higher-Order Models of Human Cognitive Ability Structure. *Intelligence*, 41(5), 407–422.

Newsom, J. T. (2015). *Longitudinal Structural Equation Modeling: A Comprehensive Introduction*. New York: Routledge.

Nunnally, J. C. (1978). *Psychometric Theory*. New York: McGraw-Hill

Oltmanns, J. R., & Widiger, T. A. (2016). Self-Pathology, the Five-Factor Model, and Bloated Specific Factors: A Cautionary Tale. *Journal of Abnormal Psychology*, 125(3), 423–434.

Preacher, K. J., Wichman, A. L., Briggs, N. E., & MacCallum, R. C. (2008). *Latent Growth Curve Modeling*. Newbury Park, CA: Sage.

Reise, S. P., Kim, D. S., Mansolf, M., & Widaman, K. F. (2016). Is the Bifactor Model a Better Model or Is It Just Better at

Modeling Implausible Responses? Application of Iteratively Reweighted Least Squares to the Rosenberg Self-Esteem Scale. *Multivariate Behavioral Research, 51*(6), 818–838.

Ruscio, J., & Roche, B. (2012). Determining the Number of Factors to Retain in an Exploratory Factor Analysis Using Comparison Data of Known Factorial Structure. *Psychological Assessment, 24*(2), 282–292.

Sass, D. A., & Schmitt, T. A. (2010). A Comparative Investigation of Rotation Criteria within Exploratory Factor Analysis. *Multivariate Behavioral Research, 45*, 73–103.

Sharp, C., Wright, A. G. C., Fowler, J. C., Freuh, C., Allen, J. G., Oldham, J., & Clark, L. A. (2015). The Structure of Personality Pathology: Both General ("g") and Specific ("s") Factors? *Journal of Abnormal Psychology, 124*(2), 387–398.

Spearman, C. (1904). General Intelligence, Objectively Determined and Measured. *American Journal of Psychology, 15*, 201–293.

Velicer, W. F. (1976). Determining the Number of Components from the Matrix of Partial Correlations. *Psychometrika, 41*, 321–327.

Walton, K. E., Ormel, J., & Krueger, R. F. (2011). The Dimensional Nature of Externalizing Behaviors in Adolescence: Evidence from a Direct Comparison of Categorical, Dimensional, and Hybrid Models. *Journal of Abnormal Child Psychology, 39*, 553–561.

Wendt, L. P., Wright, A. G. C., Pilkonis, P. A., Nolte, T, Fonagy, P., Montague, R. P., ... Zimmermann, J. (in press). Evaluating the Latent Structure of Interpersonal Problems: Validity of Dimensions and Classes. *Journal of Abnormal Psychology*.

Witkiewitz, K., King, K., McMahon, R. J., Wu, J., Luk, J., Bierman, K. L., ... Conduct Problems Prevention Research Group (2013). Evidence for a Multi-Dimensional Latent Structural Model of Externalizing Disorders. *Journal of Abnormal Child Psychology, 41*, 223–237.

Wright, A. G. C. (2017). The Current State and Future of Factor Analysis in Personality Disorder Research. *Personality Disorders: Theory, Research, and Treatment, 8*(1), 14–25.

Wright, A. G. C., & Simms, L. J. (2014). On the Structure of Personality Disorder Traits: Conjoint Analyses of the CAT-PD, PID-5, and NEO-PI-3 Trait Models. *Personality Disorders: Theory, Research, and Treatment, 5*(1), 43–54.

Wright, A. G. C., & Simms, L. J. (2015). A Metastructural Model of Mental Disorders and Pathological Personality Traits. *Psychological Medicine, 45*(11), 2309–2319.

Wright, A.G.C., & Zimmermann, J. (2015). At the Nexus of Science and Practice: Answering Basic Clinical Questions in Personality Disorder Assessment and Diagnosis with Quantitative Modeling Techniques. In S. Huprich (Ed.), *Personality Disorders: Toward Theoretical and Empirical Integration in Diagnosis and Assessment* (pp. 109–144). Washington, DC: American Psychological Association.

Wright, A. G. C., Krueger, R. F., Hobbs, M. J., Markon, K. E., Eaton, N. R., & Slade, T. (2013). The Structure of Psychopathology: Toward an Expanded Quantitative Empirical Model. *Journal of Abnormal Psychology, 122*(1), 281–294.

8 Psychiatric Epidemiology Methods

DAVID S. FINK,[*] HYUNSIK KIM,[*] AND NICHOLAS R. EATON

Psychiatric epidemiology is the study of the distributions and determinants of psychiatric conditions in human populations, and the application of this knowledge to prevent disease and improve population health and wellbeing. This definition articulates the two fundamental goals of the field: to (1) study the distributions of psychiatric disorders within and between populations and (2) identify the causes of psychiatric disease and wellbeing. Moreover, this definition serves to motivate and concretize the ultimate goal of the field: the study of psychiatric conditions in human populations with the purpose of informing interventions to improve population health. In this chapter, we focus on the two central activities of epidemiology, study design and data analysis, with regard to how they can inform clinical psychology research. First, we present an overview of study designs. Next, we review the statistical methods and analytic approaches commonly apply to psychiatric epidemiology to describe the distribution of disease and identify the causes of disease.

STUDY DESIGNS

In this section we provide an introduction to methodology concerning study design for psychiatric epidemiology, with a focus on cohort, experimental, and case-control study designs. Decisions researchers make during the early stages of study design go on to influence their ability to accurately answer the research questions that drive their inquiry.

Cohort Studies

A cohort study is a longitudinal study that follows one or more groups of individuals (i.e., cohorts) who are, have been, or in the future may be exposed to a hypothesized risk factor. The cohort study design compares the proportion of individuals within each group who develop a disease over a specified period of time. While cohort studies can be *prospective* or *retrospective*, depending on when outcomes of interest occurred in relation to recruitment of cohorts (Rothman, Greenland, & Lash, 2008), the defining feature of the cohort study is that individuals are disease-free at the start of the risk period. By determining the exposure status before the onset of disease, investigators can minimize uncertainty about the temporal sequence of exposure and disease.

The primary aim of a cohort study design is to estimate incidence (i.e., the frequency of new cases/onsets) of the outcome and establish the extent to which risk factors observed prior to or during follow-up contribute to increased incidence. Furthermore, the measurement of covariates at multiple time points allows investigators to minimize the potential effect of confounding by either controlling for confounders in (a) the design phase (e.g., match subjects in cohorts by relevant individual differences or covariates, such as age, gender) or (b) the analytical phase (e.g., multiple regression, stratification, standardization, propensity score). In this section, we examine the three stages of the cohort study that allow investigators to estimate the relationship between an exposure and a disease outcome: (1) planning the study; (2) defining and assessing the exposure; and (3) following the cohort and identifying the outcome.

Planning a Cohort Study

The planning stage of the cohort study begins with articulating a specific empirical research question. Questions that are best suited for the cohort study aim to determine the causal relationship between an exposure and a subsequent psychiatric outcome. Unlike descriptive questions that aim to describe a variable of interest (e.g., "what is the prevalence of depression in population A?") or comparative questions that examine the differences between two or more groups on one or more variables (e.g., "what is the prevalence of depression in men and women?"), causal questions aim to determine whether one or more factors' causes or effects change in one or more outcomes. Although observational studies most often estimate

---

[*] Fink and Kim contributed equally to this document and are considered joint first authors.

associations between exposure and outcome variables, it is useful to begin by asking a causal question. Some important causal questions that have been addressed by psychiatric epidemiology involve: Does early adversity (e.g., childhood maltreatment) predict later development of mental disorders? (Björkenstam et al., 2017; Elliott et al., 2016; Kim et al., 2014; Meyers et al., 2015); Can prenatal exposure to stress increase risk of schizophrenia? (Jenkins, 2013); and Does psychological treatment or pharmacological treatment reduce symptoms of a given mental disorder (Abramowitz, 1997; Tran et al., 1997; Van Etten & Taylor, 1998). However, some questions aimed to determine the causal relationship between an exposure and disease outcome is not necessarily well suited for a cohort study. Indeed, a cohort study is poorly suited for the investigation of ubiquitous exposures, rare outcomes, and diseases that have long latency periods. Ubiquitous exposures (e.g., television viewing in the United States, public education among low-income youth) limit the variability necessary to acquire individuals with different levels of exposure. Rare outcomes (e.g., suicide, schizophrenia) make it difficult to accumulate a large enough number of cases. Long latency periods require a great deal of resources and time to study. Despite these limitations, the cohort study is well suited to answer many important causal questions of interest to psychiatric epidemiology.

After articulating a specific research question and considering the limitations of the cohort study design, the next step of the planning stage is to define the population of interest or target population. Without specifying a target population of interest, it is impossible to determine whether the data used are appropriate for answering our research question. One method of identifying the target population is to ask: what are the characteristics of the population in which we would like to know the answer to our research question? Do we want to know whether childhood trauma affects psychopathology in adolescents? Among adults? The elderly? Or over the entire life course? Do we want to know about people in college? Unemployed? Retired? Each answer will inform how we identify our study data.

A target population can be defined by any number of variables, ranging from a geographic location (e.g., city, state, worksite) to a specific time or era, or by a shared characteristic (e.g., gender, occupation) or exposure to a particular event of interest (e.g., post-9/11 rescue worker, hurricane victim). For example, the target population of the Dunedin Multidisciplinary Health and Development Study (Silva, 1990) was newborn babies in Dunedin, New Zealand, in 1972, whereas the target population of the 9/11 cohort study was people who lived, worked, or went to school in the area of the World Trade Center disaster, or were involved in rescue or recovery efforts. Furthermore, the individuals who compose the population may remain consistent or vary over time. Whereas a static population will comprise the same individuals over time (e.g., 9/11 cohort study), people may migrate into and out of dynamic populations (e.g., United States residents, West Virginia coal miners). The careful definition of the target population is a prerequisite for identifying study data.

Survey and administrative data are two important resources for epidemiologic research. Epidemiologists use surveys to collect information on the health status and behaviors of individuals sampled from a larger source population (i.e., the group of people from which a study sample is selected). Study samples can be drawn using either probability- or nonprobability-based sampling methods. Whereas probability sampling uses a random process to select a representative sample of the source population, nonprobability sampling uses a nonrandom process to select a specific subset of the source population or a convenient sample of the population. Probability-based sampling can take many forms, including: simple random sampling, stratified random sampling, systematic sampling, cluster random sampling, and multistage random sampling. Although the process for each probability sampling technique differs slightly from the next, the basis is the same: every member of the population must have a known, nonzero chance of being selected.

Whereas survey data are explicitly collected for research purposes, administrative data are collected by governments or other organizations for recordkeeping or billing purposes. Administrative data contain consistent elements, collected over time, on large proportions of the population. Common sources of administrative data include hospital discharge records, pharmaceutical claims, arrests and convictions, and vital records, such as records of birth, marriage, and death.

Defining and Measuring Exposures

Defining and measuring the exposure(s) of interest begins after articulating a research question and before the start of follow-up. An exposure is a variable, hypothesized by the investigator and dictated by the research question that may cause a health outcome. Exposures in psychiatric epidemiology can vary across several dimensions, including type, chronicity, severity, and timing. Type of exposure, for example, can range from individual (e.g., nutrition, smoking cigarettes) to macrosocial (e.g., political economy, culture). Additional dimensions of exposures include the persistence of the exposure ranging from acute to chronic, exposure severity from mild to severe, and when the exposure occurred during the lifespan – from gestation to death and across generations. These are the factors whose causal effect is to be estimated. As such, the exposure must be accurately defined and measured at baseline and over time.

In a cohort study, an exposed group represents individuals who have been exposed to a risk factor, while a non-exposed group represents individuals who have not. Thus, reliable and valid assessment of the primary exposure of interest is critical. For example, suppose that we would like to study whether individuals who experience childhood trauma develop depression in adulthood more frequently than individuals who do not experience childhood trauma. In this example, childhood trauma is the primary exposure of interest, the factor whose causal effect is to be estimated. Once identified, the next step is to determine how to measure this construct. Sometimes valid and reliable measures of an exposure exist, such as, in this example, the Childhood Trauma Questionnaire (Bernstein et al., 2003) or the Adverse Childhood Experience Questionnaire (Anda et al., 2010; Bruskas & Tessin, 2013), whereas other times the investigator may need to create a new tool to capture the underlying construct of interest.

A good measure should be both reliable and valid. First, a reliable measure is one that acts in a consistent and predictable manner (DeVellis, 2016). Test-retest reliability is one way that we can estimate the reliability of a measure. Using this method, an investigator will have a respondent complete a measure at two or more time points and, assuming that the construct is stable over time, the respondent should complete the measure in the same manner. Other methods often used to assess the reliability of a measure include using an Analysis of Variance (ANOVA) to estimate an interclass correlation coefficient (ICC) and calculating Cronbach's alpha to estimate the internal consistency of a measure (DeVellis, 2016). Second, a valid measure is one that accurately captures the underlying construct of interest or which permits appropriate inferences. For a measure to demonstrate validity, it must also be reliable. One way to estimate the validity of a measure is to compare the measure with a known and accepted "gold standard" instrument. For example, a new measure for a mental disorder is often compared to a diagnosis arising from a clinician-administered interview to determine its validity. A measure that performs well against a "gold standard" instrument is said to have criterion-related validity. In addition to criterion-related validity, a measure is said to have construct validity if it acts in a way that is consistent with theory (Cronbach & Meehl, 1955). For example, a measure of childhood trauma should be positively associated with factors associated with childhood trauma (e.g., reported case referrals to social services, conduct disorder) and negatively associated with factors unrelated to childhood trauma (e.g., allergies). The reliability and validity of a study's measures and associated inferences are often evaluated using a pilot study, prior to the start of the cohort study, to ensure that each measure possesses the psychometric properties necessary to answer the research question.

Following the Cohort and Identifying the Outcome

In the final stage of the cohort study, investigators follow the cohort until participants either develop the disease, leave the cohort for some reason (e.g., death), or reach the end of the study. During the follow-up period, investigators maintain contact with the participants at predetermined intervals using any number of methods. The timing of follow-up assessments can range from a few minutes to years. For example, a study investigating the efficacy of a major depression treatment after cancer surgery might administer follow-up assessments every four weeks for 16 weeks, whereas a study investigating whether or not frequent marijuana use during adolescence is associated with psychosis might assess a cohort every five years for 20 years.

The methods used to identify new cases can vary considerably, ranging from postal mail or online questionnaires to phone or in-person interviews, laboratory tests, or registries containing detailed information about each variable of interest. In theory, any decisions about the timing of follow-up assessments and the identification of new cases should be driven by the research question. However, investigators must balance a confluence of factors including: cost of assessment, risk associated with assessment, access to alternative sources of data (e.g., medical records, death records), and willingness of individuals to participate in future assessments.

Finally, investigators use information about the exposure of interest and other risk factors, measured at an earlier time, to examine how these risk factors differently predicted the development of a disorder of interest between the cohorts with and without exposure. For example, a prospective cohort study investigated how peer victimization assessed at age of 13 predicted anxiety disorder diagnoses assessed at age of 18, indicating that victimized adolescents had a higher risk of developing anxiety disorders than did nonvictimized adolescents (Stapinski et al., 2014). Given that cohort study designs enable researchers to examine the incidence of a given mental disorder of interest in exposure groups, this design usually involves calculating risk of a disorder, such as absolute risk, relative risk, and attributable risk, which are discussed below.

Epidemiologic Experiments

Epidemiologic experiments include field trials and randomized controlled trials (RCTs). In field trials, treatment is randomly assigned to different groups of participants, attempting to investigate the initial occurrence of an illness of interest (Rothman et al., 2008). For example, in an RCT (Balance Investigators, 2010), 330 individuals with bipolar I disorder were randomly assigned to three different treatment conditions (i.e., lithium plus valproate combination therapy, lithium monotherapy, and valproate monotherapy) after receiving 4–8 weeks of treatment with lithium carbonate and valproate semi-sodium during a run-in phase. Then, the patients were followed up for 24 months to compare relapse rates across different treatment conditions during the follow-up period.

RCTs can be considered as a special form of the cohort study where investigators assign participants to an exposure at random, thereby balancing the distribution of potential confounders (i.e., extraneous variables, which are associated with the exposure and outcome of interest, that can distort the effect measure of interest) among the exposure groups. What makes RCTs different from cohort studies, which are observational in nature, is that the investigator has control over who is assigned to which exposure group in RCTs. Much like observational cohort studies, investigators employing an RCT must still articulate a specific research question, define the source population, follow participants over time, and identify new cases of disease. Nonetheless, investigator control over the assignment mechanism, through randomization, confers a substantial benefit over observational cohort studies.

Randomization is intended to balance the distribution of extraneous factors across the exposure groups and therefore minimize third-variable confounds. By creating two or more groups of people, equal on all factors except the exposure of interest, investigators can increase their confidence that any differences in the proportion of new cases observed between the exposure groups is most likely the result of the exposure itself and not some other factor.

Randomization is not without limitations. While randomization can enable accurate inference about a specific exposure-disease relationship, some important research questions in psychiatric epidemiology have risk factors that are not suited for randomization. Indeed, it is unethical to expose participants to certain deleterious factors, such as potentially traumatic events (e.g., physical assault, natural disasters), harmful toxins (e.g., lead, streptococcus), diseases (e.g., Lyme disease, HIV), or social conditions (e.g., poverty, unemployment). In addition, the results of RCTs might not be generalizable to other populations. Inclusion criteria, often created to minimize risk of and increase adherence to the exposure under study, can create unique study samples that do not represent the population at large. A recent review (Moberg & Humphreys, 2017) of 22 substance use disorder treatment studies found that between 64 and 96 percent of potential study participants were excluded from substance use disorder treatment studies. Removal of such a large proportion of the population has been found to cause an upward bias that can increase estimates of treatment effectiveness (Storbjörk, 2014). Finally, identifying causal mechanisms in RCTs can be difficult. Experimental studies randomly assign participants to levels of the exposure and measure the mediating and outcome variables, assuming the mediator causally affects the outcome variable. However, non-randomized assignment of the mediator introduces risk of bias from confounding (Bullock, Green, & Ha, 2010). While new methods of experimentally manipulating the mediator have been proposed to overcome this chance of bias, it is not always feasible to randomize the exposure and mediator (Bullock et al., 2010; Pirlott & MacKinnon, 2016). Future research is needed to improve estimation of causal mechanisms using RCTs.

Despite these restrictions, the RCT has proved to be a powerful tool for detecting causes. As such, RCTs are being applied in an increasingly wide range of circumstances. A study exposing individuals with chronic, combat-related post-traumatic stress disorder (PTSD) to a one-hour video of either trauma-related footage or a neutral topic can generate information about the biological mechanisms that underpin PTSD – information that might help inform future behavioral or pharmaceutical interventions (Geracioti et al., 2008, 2013). Moving to Opportunity, a social experiment that randomized families from high-poverty to low-poverty neighborhoods in five United States cities from 1994 to 2006 found that males and females experienced a qualitatively different effect of moving to a low-poverty neighborhood, such that males tended to experience increased rates of mental illness and females experienced decreased rates (Kessler et al., 2014; Ludwig et al., 2008). The gendered effects of moving to a low-poverty neighborhood on mental health provided new information about the effects of disturbing familial and social ties on adolescent wellbeing. With its ability to control the assignment mechanism, thereby balancing the distribution of potential confounders among the exposure groups, the RCT can be a useful form of the cohort design to identify causes.

Case-Control Studies

In contrast to a cohort study in which investigators compare groups of people differing in exposure status with regard to their risk of developing the disease outcome, investigators conducting a case-control study compare groups of people differing in outcome status. These case-control study groups are compared with regard to their odds of exposure to some hypothesized risk factor. Thus, case-control studies group individuals by outcome status, while cohort studies group individuals by exposure status. The primary advantage to using a case-control study over a cohort study is its efficiency when investigating very rare diseases. Typically, a case-control study will require a fraction of the resources necessary to conduct a comparable cohort study. The primary challenge to using a case-control study design is to recruit controls that represent the distribution of experiences of the population that gave rise to the cases. In this section, we explain the process for conducting a case-control study. We do not aim to explore all the variants of the case-control design but rather to illustrate how a case-control study can be used for psychiatric epidemiology. A large body of literature has comprehensively discussed several aspects of the case-control study, including control selection (Vandenbroucke & Pearce, 2012; Wacholder et al., 1992a), variants of the case-control design (Gatto et al., 2004; Vandenbroucke & Pearce, 2012; Wacholder & Boivin, 1987; Wacholder et al., 1992b), and analysis of case-control studies (Greenland & Thomas, 1982; Vandenbroucke & Pearce, 2012).

Case-control studies, much like cohort studies, begin with articulating a specific question and specifying a source population. Suppose that we would like to investigate the relationship between cannabis use and risk of developing psychosis, using a case-control study design. The initial step would be to articulate in very specific terms the eligibility criteria for cases (i.e., a group known to have the outcome) to be enrolled in the study. For example, Di Forti and colleagues (2009, 2015) defined cases as "all patients aged 18–65 years who presented with first-episode psychosis ... at the inpatient units of the South London and Maudsley (SLaM) NHS Foundation Trust." With the cases well defined, the study base will be defined as the population experience that generated those cases. Di Forti and colleagues' study base would be composed of any person aged 18–65 years who would have presented at the inpatient units of the South London and Maudsley (SLaM) NHS Foundation Trust had they developed first-episode psychosis. To identify such people, those authors used internet and newspaper advertisements and distributed leaflets at train stations, shops, and job centers in the area. In this example, they used a secondary base (i.e., a study base comprised of the population that gave rise to cases).

A secondary base represents individuals who would have been cases if they developed the disease of interest during the study period. In a study with a secondary base, investigators define the population from a case series, specify the source population, and sample controls from the well-defined source population. For example, Harrison and colleagues (2001) used a secondary base to study the association between schizophrenia and social inequality at birth. In this study, the cases were defined as every first episode case of psychosis presenting to specialist mental health services, within a defined catchment area, over a two-year period. The base for this case series is comprised of all people who, had they developed psychosis, would have been diagnosed at specialist mental health services, within this catchment, during this two-year period. What is crucial in a secondary base study is to accurately define the study base in order to sample controls validly. Defining the study base for a community hospital might be fairly straightforward; however, it can be particularly challenging to define the study base for large research or specialty hospitals (e.g., Mayo Clinic) that attract people from all over the world. Whereas a secondary base starts with the cases and then attempts to identify a hypothetical cohort that gave rise to them, a primary study base can be defined before cases appear.

A primary study base is a well-defined population that gives rise to cases before cases appear, typically defined by a geographic area or existing cohort study. Studies examining the predictors of suicide tend to use a primary study base defined by a geographic region or area. For example, Waern and colleagues (2002) aimed to study the association between physical illness and suicide in elderly people. Prior to identifying cases, this study defined the study base as people aged 65 years and older living in the city of Gothenburg and two adjacent counties. In this study, cases were obtained by identifying "cases of suicide in Scandinavian born people aged 65 years and older who underwent necropsy at the Gothenburg Institute of Forensic Medicine" and controls were obtained by identifying Scandinavian-born people aged 65 or older living in the same area. This is an example of a population-based case-control study, which defines the study base by area of residents. A nested case-control study defines the study base as participants in a cohort study. In a nested case-control study, investigators identify cases either that have already occurred or as they occur, and then randomly sample, for each case, a specified number of controls from those remaining in the cohort. Whether a study uses a primary or secondary study base, the chosen controls need to be similar to a random sample from the study base that generated the cases.

The primary advantage to using a case-control study, over a cohort study, is its efficiency when investigating rare outcomes. As described above, the cohort study determines the exposure status before the onset of disease. It follows, then, that a substantial number of persons would be needed in each exposure group to accumulate sufficient cases. For example, approximately 20 per 100,000 males in the United States die by suicide each year (Curtin, Warner, & Hedegaard, 2016). Therefore, to accumulate 100 cases of suicides among American males, a cohort study would have to follow either 500,000 males for 1 year, 100,000 males for 5 years, or 5,000 males over 100 years. In contrast, a systematic review found the median sample size of 27 extant studies investigating the relationship between psychiatric diagnoses and suicide was 148 participants, with studies ranging from 42 to 1,055 (Arsenault-Lapierre, Kim, & Turecki, 2004). Thus, case-control studies can save substantial resources, measured in both money and time, compared to a comparable cohort study.

Nonetheless, the case-control design is not without limitations. There are three central challenges to implementing a case-control study. First, case-control studies are poorly suited for investigating the effect of rare exposures. Second, selecting cases and controls from different source populations can introduce bias. Hospital-based controls can be particularly risky for this reason. Suppose, for example, that Di Forti and colleagues (2009, 2015) selected controls from inpatient units at the same hospital used to identify cases with first-episode psychosis. Although it would stand to reason that patients at the same hospital would arise from the same source population, hospitals can draw patients from multiple different source populations. For example, because persons with multiple diseases have a higher probability of being hospitalized than persons with only one disease, it is likely that controls selected from inpatient units at the same hospital used to identify cases with first-episode psychosis will be selected from a more unwell source population than the cases. This is an example of Berkson's bias (Berkson, 1946). In 1946, Joseph Berkson showed that two diseases that are unrelated in the general population can become

Table 8.1 Fourfold table

	Diseased	Nondiseased	Risk of outcome
Exposed	A	B	A/(A + B)
Unexposed	C	D	C/(C + D)
Column totals	A + C	B + D	A + B + C + D

"spuriously associated" in hospital-based case-control studies. For this reason and others, special care needs to be taken when conducting hospital-based case-control studies.

Finally, accurate measurement of past exposures can be challenging to assess. For instance, the experience of being diagnosed with a disease can change the perception of past experiences. Further, differences in the recall of past exposures between cases and controls can bias estimates; however, validation studies can be used to correct for differential measurement of exposures (Espeland & Hui, 1987; Flanders, Drews, & Kosinski, 1995; Greenland & Kleinbaum, 1983). Even with these limitations, a case-control study, conducted with rigor, can yield accurate data for causal inference.

EPIDEMIOLOGIC ANALYSES

Epidemiologic research moves through phases, each phase building on the last, with the ultimate aim of quantifying the causal relation between exposure and disease. In the previous section, we provided guidance on how to develop and conduct different types of epidemiologic studies. In this section, we introduce the types of epidemiologic analyses that make it possible to describe the distribution of disease within a population and identify its causes. While there is a substantial body of literature on the theory of and methods for causal inference using epidemiologic data, we focus our attention on the most common quantitative measures that describe the strength of association between an exposure and disease.

The Fourfold Table: The Risk Ratio, Risk Difference, and Odds Ratio

Epidemiologic studies often investigate the association between a binary exposure (exposed and unexposed) and a binary outcome (diseased and nondiseased). In this classic scenario, with both a binary exposure and a binary outcome, data can be displayed and investigated using a 2×2 contingency table or fourfold table. A fourfold table, composed of four cells (A, B, C, D), displays the count of individuals within the study sample with different combinations of exposure and disease. In Table 8.1, A is the count of study participants who were both exposed and diseased, B is the count of participants exposed and nondiseased, C is the count of participants unexposed and diseased, and D is the count of participants unexposed and nondiseased.

Organization of the data in a fourfold table facilitates the analysis of three measures of association: the risk ratio, the risk difference, and the odds ratio. In a cohort study in which all participants are followed for the same amount of time, data from the fourfold table can be used to calculate the cumulative incidence (conditional risk) of disease among the exposed ($A/(A+B)$) and among the unexposed ($C/(C+D)$). The risk ratio is therefore $[A/(A+B)/C/(C+D)]$ and the risk difference is $[A/(A+B) - C/(C+D)]$. Likewise, in a case-control study, the data from the fourfold table is used to calculate the odds of disease among the exposed (A/B) and the unexposed (C/D), and the odds ratio is $[(A/B)/(C/D)]$. In instances when the disease affects less than 10 percent of the population, the case-control odds ratio will approximate the risk ratio that would arise from a corresponding population-based cohort study (Greenland & Thomas, 1982; Pearce, 1993).

These measures of association provide two different perspectives on the same data. In each of these three measures of association, the exposure-disease relation among the unexposed group is used to represent the expected experience that the exposed group would have had if they had not been exposed. Therefore, whereas the measures of relative risk (i.e., risk ratio, odds ratio) indicate the strength of the exposure-disease relation, the risk difference indicates the excess risk of disease conferred by the exposure. For example, suppose the risk of posttraumatic stress disorder among deployed service members is 15 percent and among nondeployed service members it is 5 percent. The relative risk is 3, indicating that the risk of PTSD is three times greater among deployed service members than nondeployed service members. The risk difference is 10 percent, indicating deployed service members had 10 percent greater risk of PTSD than nondeployed service members or, if the exposure-disease relation is causal, 10 percent of PTSD cases among deployed service members can be attributed to being deployed.

The fourfold table is the simplest scenario employed for calculating measures of association; more complex scenarios might require the use of statistical modeling. In particular, two scenarios require further discussion. First, it is common for the exposure or outcome variable to take a different form than a two-level binary variable. For example, suppose an investigator researching the effect of depressive symptoms on suicide used the Patient Health Questionnaire (PHQ) to assess depressive symptoms, a nine-item scale that asks participants to rate the severity of The *Diagnostic and Statistical Manual of Mental Disorders*, Fifth Edition (DSM-5) criteria from 0 (not at all) to 3 (nearly every day). For the purposes of analysis, the investigator could create a dichotomous exposure for each symptom (Not at all/Several days vs. More than half the days/Nearly every day), a four-level categorical variable (Not at all, Several days, More than half the days, Nearly

Table 8.2 A hypothetical study of 10 individuals with a risk factor followed up for eight years

Subjects	2010	2011	2012	2013	2014	2015	2016	2017	Time (years) at risk
1	O	O	O	D					3
2	O	O	O	D					3
3	O	O	O	O	O	X			5
4	O	O	O	O	O	X			5
5	O	O	O	O	O	O	O	O	8
6	O	O	O	O	O	O	O	O	8
7	O	O	O	O	O	O	O	O	8
8	O	O	O	O	O	O	O	O	8
9	O	O	O	O	O	O	D		6
10	O	O	O	O	O	O	D		6
								Total years at risk =	60

Note: O = disease free year; D = disease; X = death.

every day), or a continuous variable (a total score of the nine items). Whether the exposure variable is dichotomous, multilevel categorical, or continuous will influence whether the exposure-disease relation is measured using a 2 × 2 contingency table, a $k \times 2$ contingency table (where k is the number of distinct exposure levels), or a more advanced statistical model.

Second, it is common to be interested in the exposure-disease relation, conditional on one or more potentially confounding variables (i.e., a factor associated with both the exposure and the outcome variable that can bias measures of association). In this scenario, suppose the same investigator researching the effect of depressive symptoms on suicide is concerned that age might be a potential confounder. Much like the above scenario, a contingency table cannot accommodate the introduction of a third variable to the analysis. Therefore, whether the exposure variable is continuous or the analysis requires three or more variables, a more advanced statistical modeling technique is needed. Although outside the scope of this chapter, a number of outstanding textbooks provide a complete discussion of these advanced statistical modeling techniques (Agresti & Kateri, 2011; Hosmer Jr, Lemeshow, & Sturdivant, 2013; Kleinbaum et al., 2013). However, in the next section, we discuss a general set of methods for analyzing the time until disease.

Incidence Rate and Incidence Rate Ratio

In the previous section, the fourfold table was introduced for calculating and comparing the cumulative incidence (average risk) of disease among the exposed and unexposed groups. When each participant is observed from the beginning of the study to the end, the cumulative incidence, calculated as the number of new cases over the population at risk for becoming a case, is an accurate representation of a sample's experience of health and disease. However, prospective cohort studies that extend over long durations of time may be subject to losses to follow-up. In a cohort study in which some people are lost over time, a different set of statistical procedures are needed.

When a cohort study has losses to follow up, the incidence rate, calculated as the number of new cases per observed person-time, is used to estimate the experience of health and disease within the sample. Observed person-time is a measure of the time-at-risk that each participant contributed to the study follow-up before either developing the disease or dropping out of the study. Imagine a hypothetical study that follows a cohort of 10 people over eight years (Table 8.2). Two individuals developed the disease in the fourth year of follow-up, two individuals died in the sixth year, two individuals developed the disease in the seventh year, and four individuals did not develop the disease during the follow-up period. If we assume that both disease and death occurred at the beginning of a given year for the calculation convenience, the incidence rate for this study is four per 60 person-years (i.e., number of new cases per observed person-time year): 32 person-years (4 × 8; for the four disease-free individuals for the entire eight-year period) plus six person-years (2 × 3; for the two who developed the disease at the fourth year) plus 10 person-years (2 × 5; for the two who died in the sixth year) and plus 12 person-years (2 × 6; for the two who developed the disease in the seventh year), which yields a total of 60 person-years. Much like a risk ratio is the cumulative incidence among the exposed divided by the cumulative incidence among those unexposed, the incidence rate ratio is the incidence rate among the exposed divided by the incidence rate among those unexposed.

WHAT IS A CAUSE?

Identifying the causes of disease is the central aim of epidemiology. We have explained the steps to study design, from identifying the source populations to measuring exposures of interest and collecting cases, all with the aim of identifying cause of disease and informing interventions to mitigate harmful exposures and improve population health. In this next section, we will define a cause and consider whether the effect estimate above, such as the risk ratio or incidence rate ratio, is causal.

Causal inference begins with a model, is informed by data, but relies on the triangulation of evidence from multiple studies. Although many definitions of a cause exist, an early definition by John Locke from 1690 still proves helpful: "that which produces any simple or complex idea, we denote by the general name cause', and that which is produced, 'effect'." More formally, the cause of a specific disease event is "an antecedent event, condition, or characteristic that was necessary for the occurrence of the disease at the moment it occurred, given that other conditions are fixed" (Rothman, Greenland, & Lash, 2008, p. 8). Thus, a cause has two definitive properties: direction and time order – that is, it occurs before the outcome and it changes the outcome (Susser, 1973). The temporal direction arises from substantive knowledge or a model that explains a cause (e.g., smoking causes lung cancer but not vice versa). Alternatively, it can be ensured by study design, either a cohort study measuring the exposure before the outcome occurs or an experiment that manipulates the exposure through randomization. With the temporal direction ensured, the causal effect can be measured.

A causal effect is the difference in a subject (e.g., a person, a clinic, a school) under two different states. For example, a person would have a particular level of posttraumatic stress symptoms had they been exposed to a combat zone during the Vietnam War, whereas they would have a different level of symptoms had they been stationed stateside during the Vietnam War. To measure the causal effect of exposure to the Vietnam War versus being stateside, the investigator would compare the posttraumatic stress symptoms for the same individual in both alternative futures. Since it is impossible to observe the outcome for the same individual under two different exposures, one of the outcomes is observable (factual) and one of the outcomes is always missing (counterfactual). This is the fundamental problem of causal inference (Holland, 1986): An object can only be observed under one exposure. To overcome this problem, epidemiologists use groups of people to estimate causal effects, comparing the posttraumatic stress symptoms of people deployed to Vietnam to people stateside during this time. Experiments can create two groups of people, equal on all factors other than the assigned exposure, through randomization. When this is achieved (i.e., the two groups are equal on all other factors that act or interact with the exposure to cause the outcome), the exposed and unexposed groups are said to be exchangeable with each other. Whereas some exposures can be randomized (e.g., medication, schools), epidemiologists deal primarily with exposures that cannot be assigned (e.g., military deployments).

Causal inference is challenging when data are observational and collected outside of the lab. Central to the complexities involved is the noncomparability between groups at different levels of an exposure, that is to say, the risk of an outcome in the unexposed group is different than the risk of the outcome in the exposed group had subjects in the unexposed group not received the exposure. Randomization is used in experiments to induce exchangeability of the exposure groups, implying that the average risk of the outcome in the unexposed group is equal to the average risk in the exposed group had the exposed group not received the exposure. Therefore, in observational studies, exchangeability becomes a necessary assumption that is required for an association to equal causation.

The exchangeability of exposure groups can be assessed using directed acyclical graphs (DAGs). Epidemiologists use DAGs to identify the minimum set of variables sufficient for confounding adjustment to assume exchangeability between exposure groups. This is done through mapping out the causal relations between the exposure and the outcome under study. In a DAG, an arrow or line (also called an edge) is used to represent a direct link from a cause to an effect. To create a DAG, the first step is to graph the exposure-outcome relation. Next, all common causes of the exposure and the outcome (i.e., confounders) are charted. After all common causes are charted, the final step is to identify the minimum set of variables sufficient for confounding adjustment. This is done through closing all unblocked paths, tracing out routes along or against arrows that connect the exposure and the disease. A path is said to be blocked if the path enters and exits a variable through arrowheads (i.e., a collider variable) or a variable is adjusted for in the model. For more information (see Pearl, 1988, 1995).

Causal inference is difficult in epidemiology. The counterfactual is unobservable and exposures are rarely randomized. Thus, epidemiologists rely on informed theoretical models, thoughtful study design, and triangulation of evidence and study replication to identify causes.

Considerations for Clinical Psychologists Using Epidemiologic Methods

We will briefly discuss two considerations commonly faced by clinical psychologists who undertake psychiatric epidemiological projects. The first consideration is availability of data. While some studies in clinical psychology have recruited, for instance, representative samples, the majority of clinical psychology research has focused on nonrepresentative clinic, university, or community samples. These studies that rely on nonrandom and non-representative convenience samples can be highly informative, but they often lack the design-based rigor to support strong generalizations and inferences about larger populations. For example, a study by Pruchno and colleagues (2008) compared statistical differences in outcome measures frequently included in caregiver research (e.g., caregiver mental health, care recipients' cognitive and mental problems, caregiver-care recipient relationship) between a convenience sample and a sample recruited by use of a random digital dial method. Their results indicated that researchers should be cautious when using convenience samples, given that convenience samples can lead to different outcomes and conclusions. Also, a study by Sauver and colleagues (2012) compared standard variables included in epidemiologic research (e.g., demographic, ethnic, and socioeconomic status information) across five different levels of populations (i.e., a county in Minnesota, the state of Minnesota, the upper Midwest, white individuals in the United States, and the entire United States population), suggesting that (a) no single community can be entirely representative of the United States and that (b) researchers should be cautious when attempting to generalize findings from a study conducted on a selective population. Thus, for clinical psychologists who wish to collect representative data, we strongly recommend the early incorporation of an epidemiologist into study design and implementation, because sampling is a complex topic.

Many psychiatric epidemiologic questions can be addressed without collection of new data sets, however. Various large and representative psychiatric epidemiologic surveys have been conducted, and many are freely available. While there are benefits associated with using archival data, such as freely available data from the National Epidemiologic Survey on Alcohol and Related Conditions (NESARC) – the largest psychiatric epidemiologic study conducted in the United States to date – there are also limitations. Most obviously, archival data has already been collected, limiting the questions that researchers can answer as well as the methods they may use to answer these questions. Skip-out rules in diagnostic assessments are common, for instance, meaning that investigation of symptom counts may be problematic, because some symptoms would not be assessed in individuals who did not screen positive on gateway questions.

A second major issue for clinical psychologists analyzing epidemiologic data is that many of these studies employ what is known as complex sampling designs. That is, many epidemiological studies are designed and conducted with particular sampling methodologies in mind. Perhaps sampling was conducted primarily in certain geographic areas to reduce travel costs of interviewers, introducing geographic clustering into the data. Perhaps the data were intended to be representative with regard to some population characteristics; for instance, the first two waves of the NESARC were representative of the age, gender, and race/ethnic distributions of the United States in the 2000 US Census. These complex sampling design features require incorporation into all analyses, which is not a topic commonly discussed in clinical psychology's methodological training. Specialized software, such as Stata, Mplus, or the *survey* package in R is required to incorporate design features such as sampling weights, stratification, and primary sampling units (PSUs). The incorporation of these design feature variables into analyses allows for parameter estimates (e.g., regression coefficients, prevalence rates) that are representative of a given population and statistical tests with accurate standard errors and *p*-values (e.g., corrected for clustered observations). Finally, it is worth noting that data sets are best not considered "representative," but rather they are representative with regard to certain population characteristics.

SUMMARY

Psychiatric epidemiologic methods and data allow for analysis of numerous critical questions in clinical psychology. From the documentation of prevalence rates of disorders that are generalizable to a broader population, to the estimation of causal processes that may underlie the development and maintenance of psychopathology and substance use, psychiatric epidemiology holds great promise to inform our understanding of mental health. Researchers, however, should be aware that the designs and statistical analyses employed within epidemiology differ – sometimes notably – from those employed in clinical psychology, and they often require specialized training, experience, and software to be conducted appropriately. When conducted and interpreted correctly, the methods of psychiatric epidemiology can address critical public health questions, and their broader incorporation into clinical psychology research represents a key future direction for the field.

REFERENCES

Abramowitz, J. S. (1997). Effectiveness of Psychological and Pharmacological Treatments for Obsessive-Compulsive Disorder: A Quantitative Review. *Journal of Consulting and Clinical Psychology, 65*(1), 44–52.

Agresti, A., & Kateri, M. (2011). *Categorical Data Analysis: International Encyclopedia of Statistical Science* (pp. 206–208). Berlin: Springer.

Anda, R., Tietjen, G., Schulman, E., Felitti, V., & Croft, J. (2010). Adverse Childhood Experiences and Frequent Headaches in Adults. *Headache: The Journal of Head and Face Pain, 50*(9), 1473–1481.

Arsenault-Lapierre, G., Kim, C., & Turecki, G. (2004). Psychiatric Diagnoses in 3275 Suicides: A Meta-Analysis. *BMC Psychiatry, 4* (37), 1–11.

Balance Investigators. (2010). Lithium plus Valproate Combination Therapy versus Monotherapy for Relapse Prevention in Bipolar I Disorder (BALANCE): A Randomised Open-Label Trial. *The Lancet, 375*(9712), 385–395.

Berkson, J. (1946). Limitations of the Application of Fourfold Table Analysis to Hospital Data. *Biometrics Bulletin, 2*(3), 47–53.

Bernstein, D. P., Stein, J. A., Newcomb, M. D., Walker, E., Pogge, D., Ahluvalia, T., ... Desmond, D. (2003). Development and Validation of a Brief Screening Version of the Childhood Trauma Questionnaire. *Child Abuse & Neglect, 27*(2), 169–190.

Björkenstam, E., Ekselius, L., Burström, B., Kosidou, K., & Björkenstam, C. (2017). Association between Childhood Adversity and a Diagnosis of Personality Disorder in Young Adulthood: A Cohort Study of 107,287 Individuals in Stockholm County. *European Journal of Epidemiology, 32*(8), 721–731.

Bruskas, D., & Tessin, D. H. (2013). Adverse Childhood Experiences and Psychosocial Well-Being of Women Who Were in Foster Care as Children. *The Permanente Journal, 17*(3), 131–141.

Bullock, J. G., Green, D. P., & Ha, S. E. (2010). Yes, but What's the Mechanism?(Don't Expect an Easy Answer). *Journal of Personality and Social Psychology, 98*(4), 550–558.

Cronbach, L. J., & Meehl, P. E. (1955). Construct Validity in Psychological Tests. *Psychological Bulletin, 52*(4), 281–309.

Curtin, S. C., Warner, M., & Hedegaard, H. (2016). *Increase in Suicide in the United States, 1999–2014*. US Department of Health and Human Services, Centers for Disease Control and Prevention, National Center for Health Statistics.

DeVellis, R. F. (2016). *Scale Development: Theory and Applications* (Vol. 26). Thousand Oaks, CA: Sage.

Di Forti, M., Marconi, A., Carra, E., Fraietta, S., Trotta, A., Bonomo, M., ... Russo, M. (2015). Proportion of Patients in South London with First-Episode Psychosis Attributable to Use of High Potency Cannabis: A Case-Control Study. *The Lancet Psychiatry, 2*(3), 233–238.

Di Forti, M., Morgan, C., Dazzan, P., Pariante, C., Mondelli, V., Marques, T. R., ... Paparelli, A. (2009). High-Potency Cannabis and the Risk of Psychosis. *The British Journal of Psychiatry, 195* (6), 488–491.

Elliott, J. C., Stohl, M., Wall, M. M., Keyes, K. M., Skodol, A. E., Eaton, N. R., ... Hasin, D. S. (2016). Childhood Maltreatment, Personality Disorders and 3-Year Persistence of Adult Alcohol and Nicotine Dependence in a National Sample. *Addiction, 111* (5), 913–923.

Espeland, M. A., & Hui, S. L. (1987). A General Approach to Analyzing Epidemiologic Data that Contain Misclassification Errors. *Biometrics, 43*(4) 1001–1012.

Flanders, W. D., Drews, C. D., & Kosinski, A. S. (1995). Methodology to Correct for Differential Misclassification. *Epidemiology, 6*(2), 152–156.

Gatto, N. M., Campbell, U. B., Rundle, A. G., & Ahsan, H. (2004). Further Development of the Case-Only Design for Assessing Gene-Environment Interaction: Evaluation of and Adjustment for Bias. *International Journal of Epidemiology, 33*(5), 1014–1024.

Geracioti, T. D., Baker, D. G., Kasckow, J. W., Strawn, J. R., Mulchahey, J. J., Dashevsky, B. A., ... Ekhator, N. N. (2008). Effects of Trauma-Related Audiovisual Stimulation on Cerebrospinal Fluid Norepinephrine and Corticotropin-Releasing Hormone Concentrations in Post-Traumatic Stress Disorder. *Psychoneuroendocrinology, 33*(4), 416–424.

Geracioti, T. D., Jefferson-Wilson, L., Strawn, J. R., Baker, D. G., Dashevsky, B. A., Horn, P. S., & Ekhator, N. N. (2013). Effect of Traumatic Imagery on Cerebrospinal Fluid Dopamine and Serotonin Metabolites in Posttraumatic Stress Disorder. *Journal of Psychiatric Research, 47*(7), 995–998.

Greenland, S., & Kleinbaum, D. G. (1983). Correcting for Misclassification in Two-Way Tables and Matched-Pair Studies. *International Journal of Epidemiology, 12*(1), 93–97.

Greenland, S., & Thomas, D. C. (1982). On the Need for the Rare Disease Assumption in Case-Control Studies. *American Journal of Epidemiology, 116*(3), 547–553.

Harrison, G., Gunnell, D., Glazebrook, C., Page, K., & Kwiecinski, R. (2001). Association between Schizophrenia and Social Inequality at Birth: Case-Control Study. *British Journal of Psychiatry, 179*(4), 346–350.

Holland, P. W. (1986). Statistics and Causal Inference. *Journal of the American Statistical Association, 81*(396), 945–960.

Hosmer Jr, D. W., Lemeshow, S., & Sturdivant, R. X. (2013). *Applied Logistic Regression* (Vol. 398). Hoboken, NJ: John Wiley & Sons.

Jenkins, T. A. (2013). Perinatal Complications and Schizophrenia: Involvement of the Immune System. *Frontiers in Neuroscience, 7,* 110–118.

Kessler, R. C., Duncan, G. J., Gennetian, L. A., Katz, L. F., Kling, J. R., Sampson, N. A., ... Ludwig, J. (2014). Associations of Housing Mobility Interventions for Children in High-Poverty Neighborhoods with Subsequent Mental Disorders during Adolescence. *Jama, 311*(9), 937–947.

Kim, J. H., Martins, S. S., Shmulewitz, D., Santaella, J., Wall, M. M., Keyes, K. M., ... Hasin, D. S. (2014). Childhood Maltreatment, Stressful Life Events, and Alcohol Craving in Adult Drinkers. *Alcoholism: Clinical and Experimental Research, 38* (7), 2048–2055.

Kleinbaum, D., Kupper, L., Nizam, A., & Rosenberg, E. (2013). *Applied Regression Analysis and Other Multivariable Methods*. Ontario: Nelson Education.

Ludwig, J., Liebman, J. B., Kling, J. R., Duncan, G. J., Katz, L. F., Kessler, R. C., & Sanbonmatsu, L. (2008). What Can We Learn about Neighborhood Effects from the Moving to Opportunity Experiment? *American Journal of Sociology, 114*(1), 144–188.

Meyers, J. L., Lowe, S. R., Eaton, N. R., Krueger, R., Grant, B. F., & Hasin, D. (2015). Childhood Maltreatment, 9/11 Exposure, and Latent Dimensions of Psychopathology: A Test of Stress Sensitization. *Journal of Psychiatric Research, 68,* 337–345.

Moberg, C. A., & Humphreys, K. (2017). Exclusion Criteria in Treatment Research on Alcohol, Tobacco and Illicit Drug Use Disorders: A Review and Critical Analysis. *Drug and Alcohol Review, 36*(3), 378–388.

Pearce, N. (1993). What Does the Odds Ratio Estimate in a Case-Control Study? *International Journal of Epidemiology, 22*(6), 1189–1192.

Pearl, J. (1988). *Probabilistic Reasoning in Intelligent Systems: Networks of Plausible Inference*. Burlington, MA: Morgan Kaufmann.

Pearl, J. (1995). Causal Diagrams for Empirical Research. *Biometrika, 82*(4), 669–688.

Pirlott, A. G., & MacKinnon, D. P. (2016). Design Approaches to Experimental Mediation. *Journal of Experimental Social Psychology, 66,* 29–38.

Pruchno, R. A., Brill, J. E., Shands, Y., Gordon, J. R., Genderson, M. W., Rose, M., & Cartwright, F. (2008). Convenience Samples and Caregiving Research: How Generalizable Are the Findings? *The Gerontologist, 48*(6), 820–827.

Rothman, K. J., Greenland, S., & Lash, T. L. (2008). *Modern Epidemiology* (3rd edn.). Philadelphia, PA: Lippincott Williams & Wilkins.

Sauver, J. L. S., Grossardt, B. R., Leibson, C. L., Yawn, B. P., Melton, L. J., & Rocca, W. A. (2012). Generalizability of Epidemiological Findings and Public Health Decisions: An Illustration from the Rochester Epidemiology Project. Paper presented at the Mayo Clinic Proceedings, *87*(2), 151–160.

Silva, P. A. (1990). The Dunedin Multidisciplinary Health and Development Study: A 15 Year Longitudinal Study. *Paediatric and Perinatal Epidemiology, 4*(1), 76–107.

Stapinski, L. A., Bowes, L., Wolke, D., Pearson, R. M., Mahedy, L., Button, K. S., ... Araya, R. (2014). Peer Victimization during Adolescence and Risk for Anxiety Disorders in Adulthood: A Prospective Cohort Study. *Depression and Anxiety, 31*(7), 574–582.

Storbjörk, J. (2014). Implications of Enrolment Eligibility Criteria in Alcohol Treatment Outcome Research: Generalisability and Potential Bias in 1-and 6-Year Outcomes. *Drug and Alcohol Review, 33*(6), 604–611.

Susser, M. (1973). *Causal Thinking in the Health Sciences: Concepts and Strategies of Epidemiology*. New York: Oxford University Press.

Tran, P. V., Hamilton, S. H., Kuntz, A. J., Potvin, J. H., Andersen, S. W., & Tollefson, G. D. (1997). Double-Blind Comparison of Olanzapine versus Risperidone in the Treatment of Schizophrenia and Other Psychotic Disorders. *Journal of Clinical Psychopharmacology, 17*(5), 407–418.

Van Etten, M. L., & Taylor, S. (1998). Comparative Efficacy of Treatments for Post-Traumatic Stress Disorder: A Meta-Analysis. *Clinical Psychology & Psychotherapy: An International Journal of Theory and Practice, 5*(3), 126–144.

Vandenbroucke, J. P., & Pearce, N. (2012). Case-Control Studies: Basic Concepts. *International Journal of Epidemiology, 41*(5), 1480–1489.

Wacholder, S., & Boivin, J.-F. (1987). External Comparisons with the Case-Cohort Design. *American Journal of Epidemiology, 126*(6), 1198–1209.

Wacholder, S., Silverman, D. T., McLaughlin, J. K., & Mandel, J. S. (1992a). Selection of Controls in Case-Control Studies: II. Types of Controls. *American Journal of Epidemiology, 135*(9), 1029–1041.

Wacholder, S., Silverman, D. T., McLaughlin, J. K., & Mandel, J. S. (1992b). Selection of Controls in Case-Control Studies: III. Design Options. *American Journal of Epidemiology, 135*(9), 1042–1050.

Waern, M., Rubenowitz, E., Runeson, B., Skoog, I., Wilhelmson, K., & Allebeck, P. (2002). Burden of Illness and Suicide in Elderly People: Case-Control Study. *British Medical Journal, 324*(7350), 1355–1358.

PART III EXPERIMENTAL AND BIOLOGICAL APPROACHES

9 Conceptual Foundations of Experimental Psychopathology

Historical Context, Scientific Posture, and Reflections on Substantive and Method Matters

MARK F. LENZENWEGER

The nature of the hypothesis to be tested should determine the procedure which is used.

– Brendan A. Maher, *Principles of Psychopathology: An Experimental Approach* (1966)

When Professors Wright and Hallquist invited me to contribute to this handbook on the topic of conceptual foundations of experimental psychopathology, I thought the foundations could be easily contained in the following brief definition that I had presented earlier:

Experimental psychopathology is the psychological science discipline that uses the methods of the experimental psychology laboratory in conjunction with quantitative analytic approaches to gain leverage on etiology and pathogenesis of psychopathology, within a brain-based (genomic, endophenotype, neurobiological) diathesis-stressor matrix. (Lenzenweger, 2010, p. 19)

Unpacking this definition to specify foundations, however, is something altogether different as embedded in it are any number of assumptions, approaches, and avenues for exploration that would extend far beyond a single chapter (see Lenzenweger, 2010). I have therefore sought to selectively interpret this definition in the form of conceptual guidelines and a general vantage point on experimental psychopathology placed within a brief historical context. As I was taught, experimental psychopathology is something of a hybrid psychological science subdiscipline supported by clinical psychological science, experimental psychology, cognitive and affective (neuro)science, psychometrics and measurement, genetics/genomics, neurobehavioral systems, statistical analysis, and taxonomic science. Importantly, each conceptual pillar supporting experimental psychopathology as a subdiscipline ultimately serves a common mission. That common mission, which I often discussed with my late colleague, Professor Brendan Maher, is to understand the etiology, underlying processes, and pathogenesis of psychopathology. In short, the task of experimental psychopathology is to illuminate the causes and unfolding of mental illness, in all its representations, and across levels of analysis. In this chapter, I briefly describe the conceptual foundations, with some treatment of important historical cornerstones, of experimental psychopathology[1] and then review a necessarily selective assortment of conceptual issues that should be kept in mind when investigating the territory of mental disorder from the experimental psychopathology vantage point.[2] This chapter is not about research design; it is more about important things to bear in mind when pursuing the goals of experimental psychopathology – the foundation stones on which the discipline rests. The other chapters within this section of the present volume also illuminate the complex identity of experimental psychopathology and, implicitly, its supporting assumptions from different angles.

CONCEPTUAL ROOTS: THE CENTRAL ROLE OF SCHIZOPHRENIA AS A HISTORICAL FOCUS

Experimental psychopathology emerged largely out of the efforts of several psychologists that began to probe the nature of mental disorders using the methods of the experimental psychology laboratory. Whereas one can

---

[1] The interested reader can find historical detail on the development and growth of experimental psychopathology from different scientific and professional perspectives in Abramson and Seligman (1977), Forsyth and Zvolensky (2002), Kimmel (1971), Korolenko et al. (1966), and Zvolensky, Forsyth, and Johnson (2013).

[2] In this discussion, the term psychopathology means expressed forms of well-established and recognized (diagnosable or otherwise measureable) psychopathology, or detectable/measureable forms of subclinical forms of psychopathology, or plausible, well-reasoned manifestations of liability for psychopathology. In this context, experimental psychopathology is not viewed as a descriptive enterprise, though rich clinical description clearly has a role in the study of psychopathology writ large. Experimental psychopathology as intended here should not be taken to mean analog research or the laboratory creation of psychopathology mimics or models. Analogs or laboratory models may have some utility; however, they are necessarily constrained with respect to ecological validity and lack the full array of causal factors conferred by nature in the determination of recognized psychopathology. The question of animal models of psychopathology is complex and vexing as the vast majority of human psychopathology cannot find a ready imitation in laboratory animals and inferences drawn from the study of animals vis-à-vis psychopathology entail, minimally, grand inferential leaps.

easily see the relevance of the work of psychological pioneers (James, Pavlov) in setting the stage for laboratory research on psychopathology, experimental psychopathology as an active approach is seen most clearly in early work seeking to dissect genuine or bona fide, expressed (diagnosable) mental disorders, not *analogue* conditions. Forgoing an extended history, it is fair to say that the disorder that most powerfully captured the attention of early psychopathologists was clearly schizophrenia, and early efforts to study schizophrenia shaped the nascent discipline. If one consults the *Journal of Abnormal Psychology* (known as *Journal of Abnormal and Social Psychology* from 1921 until 1965) from 1906 through 1960, from its very beginnings, a clear predominance of research reports focused on schizophrenia is apparent. This is not to say there were not studies of manic depression (as it was known in the early years), depression, and sociopathic personality (the defunct term that previously described what we now know as psychopathy) appearing in those early issues – there were indeed such studies. But the overwhelming presence of experimental laboratory studies of schizophrenia was palpable. The benefits of laboratory study of schizophrenia, such as controlled conditions, precision measurement apparatuses, and proper experimental methods (i.e., control groups, experimenter blinds, and so on), clearly appealed to early workers in contrast to clinical approaches that were largely descriptive and geared toward theory, assessment, and/or treatment.

Two figures that were particularly prominent in the early years of experimental psychopathology were David Shakow, PhD, and Joseph Zubin, PhD, both of whom did extensive laboratory research on schizophrenia. Shakow conducted classic experimental studies focused on cognitive deficits and decline as well as "segmental set" in schizophrenia, making use of reaction time metrics and other measures to document findings. Joseph Zubin was Chief of Psychiatric Research in Biometrics at the renowned New York State Psychiatric Institute and became well known for his emphasis on experimental psychopathology methods and his quest to understand the etiology of schizophrenia. In fact, he coedited one of the first formal collections of scholarly papers devoted expressly to experimental psychopathology (with Paul Hoch) in 1957 (Hoch & Zubin, 1957).

Following the trailblazing efforts of Shakow and Zubin came other researchers, largely focused on schizophrenia, who would embody the experimental psychopathology approach, such as Brendan A. Maher, Philip S. Holzman, Sarnoff Mednick, Rue Cromwell, and Loren and Jean Chapman. Importantly, other psychologists, bringing along substantive and methodological rigor from related specialties such as genetics (Irving I. Gottesman, David Rosenthal), theory development and philosophy of science (Paul E. Meehl), and developmental psychology (Norman Garmezy), would also impart their imprint on the developing discipline of experimental psychopathology. One development in the emergence of experimental psychopathology is particularly salient and often noted amongst psychopathologists as pivotal, a true watershed event. That was the appearance of Brendan A. Maher's (1966) monograph *Principles of Psychopathology: An Experimental Approach*. Maher's *Principles*, at one fell swoop, placed experimental psychopathology squarely on the psychological science disciplinary map and provided a blueprint that continues to exert its influence to this day (Lenzenweger & Hooley, 2003). Common features of the work and teaching of Shakow, Zubin, and (later) Maher as well as the others noted above were experimental rigor, conceptual clarity, and advancement of quantification, and careful data collection and hypothesis testing using laboratory methods. This collection of features constitutes the essence of experimental psychopathology today and continues to define its conceptual foundations and scientific values. In short, experimental psychopathology is grounded in the rigorous collection and testing of empirical data in the laboratory, not (as a Professor Maher once quipped to me) in "overstuffed leather chairs, sipping glasses of sherry, and discussing the world of theory devoid of data."[3] Moreover, in experimental psychopathology, hypothesis development and the selection of measurement and proper laboratory procedures are inextricably intertwined.

Although developed primarily by psychologists, the approach was also embraced by many research-oriented psychiatrists during the neo-Kraepelinian revolution of the late 1970s and the 1980s. It clearly influenced the sea change in article content that took place in what had long been treatment/clinically oriented outlets such as the *Archives of General Psychiatry* (now *JAMA Psychiatry*) and *American Journal of Psychiatry* during the 1970s and 1980s. Unraveling the nature of schizophrenia has remained a vigorous focus for experimental psychopathology, but the general approach in recent decades has been brought to bear directly and fruitfully on a wide variety of psychopathology, including depression, bipolar illness, anxiety-related pathologies, personality disorders, and more. Finally, a scientific society expressly dedicated to experimental psychopathology, the Society for Research in Psychopathology (SRP) (inaugural meeting in 1986 at Harvard University), was born from a small meeting of schizophrenia-focused experimental psychopathologists that gathered at Cornell University in 1985 to discuss recent experimental research on positive and negative symptoms in schizophrenia. SRP, though initially founded to foster discussion of experimental psychopathology research on schizophrenia, remains a forum dedicated to the experimental psychopathology approach but now welcomes scientists who examine a range of psychopathologies.

---

[3] This is not to say that other areas of psychological science do not share the same ethos. Rather, I seek to differentiate an experimental approach to psychopathology from traditional clinical and descriptive psychopathology approaches in which laboratory testing and empirical data collection were less central.

BROAD CONCEPTUAL FOUNDATIONS OF EXPERIMENTAL PSYCHOPATHOLOGY: A BRIEF OVERVIEW OF SUBSTANTIVE, METHODOLOGICAL, AND MATTERS OF APPROACH

I would like to selectively review a number of issues that I think any student of experimental psychopathology should keep in mind. The issues and reflections are in no particular order of importance; rather, the overall gestalt is the point. Each of the matters I address could be developed at far greater length, and many other critical issues could be addressed, but space constraints prevent an exhaustive review. The interested student is referred to Lenzenweger (2004, 2010, 2013) for more extended discussion. Finally, when illustrating a particular point, I will refer to examples from the schizophrenia and schizotypy literature, though one could substitute other disorders *mutatis mutandis* to provide comparable illustrations.

Importance of Clinical Observation

Careful clinical observation is essential for psychopathology research in many ways. Consider just two of these. First, clinical observation is necessary for the careful enumeration of psychopathology manifestations found in research subjects. If the diagnostic work characterizing subjects has not been done well, then a study is hobbled from the start. To that end, genuine training in descriptive psychopathology is essential and to be encouraged. I note this rather axiomatic point because, ironically, some of the most "research-oriented" training programs in psychopathology today have come to deemphasize clinical exposure in their training models. Furthermore, it strikes me that the most seasoned and clinically skilled members of a research team should be centrally involved in the diagnostic and symptom assessments of study patients. Assessments should *not* be delegated to the most junior members of study teams as is done in some settings. Clinical presentations are often complex and the elicitation of symptomatology requires skill, insight, and considerable exposure to many patients. Second, clinical observation remains a goldmine for hypothesis generation for psychopathology studies. Many research vectors across different laboratories have their roots or genesis in clinical observations of actual patients. Clinical observation is ideally suited to the "context of discovery" that is discussed below. Although the experimental and/or laboratory procedures that we use in our studies often receive the greatest attention in developing a protocol, it is critical to not underestimate the importance of rigorous clinical observation, phenomenology, and diagnosis.

Determinants versus Correlates of Psychopathology

The overall goal of experimental psychopathology is to discover etiologic factors in mental illness, stated succinctly. However, in the context of the foregoing discussion about clinical observation, it is essential to note that the subjects of our studies (human beings affected with impairing and limiting mental disorders) present us with a complex picture of behavior, cognition, affect, and so on. This complex picture is necessarily a mixture of the results of their primary pathology as well as behaviors or strategies that have emerged to adapt to or to cope with the distress that the primary pathology produces (following Maher, 2003). This is a critical distinction and is necessarily related to the phenomena we end up studying in psychopathology. Namely, we could be examining some aspect of primary pathology or we might be unknowingly distracted by a correlate of psychopathology. For example, in schizotypic individuals one often sees a diminished capacity for the experience of pleasure (i.e., low hedonic experience or hypohedonia). This observation supports the following question: "is this hypohedonia part of the primary pathology of the schizotype or is it the negative affective tone and experience that emerges from enduring years of adverse experience in social relations, occupational efforts, and attempts at intimacy?" Experimental psychopathologists should always pause to consider whether they are concerned with primary pathology or the correlates of pathology, and how to tell them apart. In this context, I encourage psychopathologists to remember that patients suffering from some form of psychopathology, particularly severe mental illness, are likely to do worse on almost any task set before them in comparison to controls. Again, do the observed differences stem from primary pathology or merely correlates of pathology (possibly distracting artifacts)? There are times, however, when a correlate is not an artifact but an interesting aspect of the development of a psychopathology. For example, many schizotypic individuals develop a hypohedonic state over time owing to attempting to pursue life while they harbor the diathesis for schizophrenia, which impacts their neurocognitive and interpersonal functioning.

Rating and Counting – When Possible, Don't Rate, Count

The role of rating in psychopathology is at once central and ubiquitous. Yet the nature of ratings is often overlooked or not considered fully in terms of its profound implications for some of the most fundamental data that are collected. Although experimental psychopathology relies in many instances on apparatus-based laboratory tasks (e.g., computerized or behavioral tasks), many of the most basic data collected in psychopathology critically rely upon self-rating and observer-rating techniques. For example, the diagnosis of a patient (e.g., schizophrenia vs. panic disorder vs. depression) is still made on the basis of a patient's self-report (symptoms; self-rating) and the observations of a diagnostician (signs; observer rating). It is a *rating* enterprise. When a researcher assesses the level of thought disorder in a patient's speech, a *rating* is made. When a symptom of schizotypic personality disorder is diagnosed (e.g., "Does the patient have an odd or eccentric appearance?"), a *rating* is made. When a patient completes an inventory regarding personality disorder features, such as self-concept clarity or impulsivity, a self-*rating* is made. The process of *rating* plays a major role in psychopathology research, even with the rise of experimental paradigms and laboratory science. This problem has long been known in personality science, and the proponents of experimental psychopathology have drawn attention to it for decades.

To the extent we can move away from rating approaches in our work (being mindful that some research objectives necessarily preclude this, for now) and move more toward a *counting* methodology for collecting our basic data, we will enhance the precision of our work and realize fully the benefits of quantification. Counting implies discrete numerical aggregation of data, typically involving a rational zero value (e.g., Kelvin temperature measurement; number of milliseconds required for reaction time; number of murders committed). Remember, clinical ratings just serve as rough "estimates" of a property or construct we seek to measure. Ratings have nebulous underlying scaling attributes (ordinal, at best), yet are treated in analyses as if they embody the sought-after characteristics of ratio-based scaling. To illustrate this general point, consider this anecdote: "When you check out at a supermarket, you don't eyeball the heap of purchases and say to the clerk, 'Well it looks to me as it it's about $17.00 worth; what do you think?' The clerk adds it up" (Meehl, 1986, p. 372). In psychopathology ratings we rely upon descriptors such as "severe," "most of the time," "clinically significant," "strongly disagree," "true," or "false" and really have no method of validly "adding things up," because they cannot truly be added together.

NULL HYPOTHESIS STATISTICAL TESTING: PLEASE REMEMBER THAT THE NULL HYPOTHESIS IS QUASI-NEARLY ALWAYS FALSE

The problem with the entire null hypothesis statistical testing (NHST) enterprise is that the null hypothesis, particularly as framed in psychological science applications, is *quasi-nearly always false*. What does this mean (Cohen, 1994)? In short, psychologists rarely do more than test the presence or absence of differences between two or more groups, or test the significance of a correlation coefficient deviating from a zero value. They sometimes do little more than wait for a nondirectional statistically significant result to arrive (i.e., a *p*-value that allows one to reject the null hypothesis in good conscience). The difficulty with this approach is that one will nearly always be able to reject the null hypothesis, especially as one increases the size of one's sample (which increases power). Meehl (1967) and many others have discussed this issue at length (even *ad nauseam*) but to little avail. The problem here is not with the mathematics or any computational issue, the problem resides in the fact that we are attempting to evaluate hypotheses (or hunches derived from models) using an approach that will nearly always deliver the same result, assuming large enough samples, namely a "significant" finding and rejection of the null hypothesis. Aside from the sample size issue, following Meehl (1990), there are other reasons the NHST approach is flawed as typically applied in psychology, such as (a) the relevance of NHST to testing theory, (b) the lack of well-articulated and reasonable statistical expectations for most applications in psychology, (c) the usual focus on naturally assigned grouping distinctions (i.e., nonexperimental designs), and (d) while "there is a negligible difference between the substantive theory of interest and the counter null hypothesis in agronomy, ... in theoretical soft psychology they are distinctly different and frequently separated by what one could call a large 'logical distance'" (Meehl, 1978). To better understand the latter point by Meehl, consider that in agronomy one might posit that a particular fertilizer will make corn grow taller based on a consideration of the plant physiology or genetics vs. positing that the time a child waits to take a candy from a dish in a laboratory room derives from their capacity to delay gratification that hails from their level of psychosexual development. In short, as noted above and worth stating again, in psychological research the null hypothesis is quasi-nearly always false.

Unfortunately, psychopathology research (and psychological science as a whole) still very much relies on the NHST approach, the *p*-value rules the day in the eyes of most investigators and journal editors, and alternative approaches to evaluate data are looked upon askance with a mixture of anxiety and resistance to change. Psychopathology research (and the field writ large) should move more in the direction of evaluating point predictions (as is done in physics, for example), make use of confidence intervals more often, and keep effect sizes in mind (as opposed to simply paying attention to *p*-values) (consider, for example, Bayesian approaches, see Wagenmakers, Morey, & Lee, 2016). The interested student is referred to Meehl (1998/2006) and references therein as well as Cohen (1994) and Wilkinson (1999) for additional discussion of this issue.

Stay Close to Your Data (and Experience Matters)

Students sometimes get the mistaken impression that as one gets on in one's career in psychological science, one should spend more and more time farther and farther away from the patients and the laboratory where data are collected. Indeed, some students who are completing psychopathology research apprenticeships at laboratories around the country will remark to me, "yes, it is Professor So and So's lab, but I've never seen him/her in the lab, ever!" I recommend that to maintain an intimate grasp of one's data, irrespective of one's career development, one must stay close to the laboratory and the subjects of study.

In experimental psychopathology, this means: be sure to test and calibrate all equipment carefully and continuously (especially true in neuroimaging studies), observe your students processing subjects, review data routinely for anomalies or irregularities, and constantly review laboratory protocols for any evidence of lack of precision. However, the odd thing in psychopathology research (and psychological science as a whole) is that, unless one makes a concerted effort to do otherwise, one can often move farther away from the data collection process as one moves up the tenure ranks. This partly reflects the demands on many advanced researchers, who spend more time in managing budgets, grant applications and renewals, and staff than they do in collecting or analyzing data. Data cleaning and analysis are often delegated to graduate students, who often assign the most basic (but most critical) task of data collection to the undergraduate psychology major. Many of you reading this chapter will have collected data in a laboratory under the guidance of a graduate student and might have only vague memories of a professor even being involved in the process. This is a highly unfortunate development in our field, one that quietly works against creativity and risk taking (as well as working against the goals of quality training/education in science). In short, my recommendation is to stay close to your data – do not become so distant that you could not detect some anomaly that could reveal a crucial error or miss some feature of the data that suggests a fascinating new direction for further research. Make sure you know how subjects are moving through a protocol and know the rhythm and flow of incoming data.

Post-Hoc Data Analysis: Context of Discovery versus Context of Justification

One of the strangest things that I have encountered in graduate students in experimental psychopathology, as well as in some psychopathologists, is the quasi-unshakeable belief that one is only allowed to test hypotheses in a data set that were formulated *before* data collection. Of course, one tests all a priori hypotheses. However, I think strict reliance on only those hypotheses that have been formulated a priori is misguided. Rather, data often harbor fascinating relationships waiting to be discovered and, if appropriate, followed up in further empirical studies. Such fascinating relationships, even if not specified a priori, can be found in what I term "mindful" exploration – note the term "mindful" – of data after the a priori hypotheses have been examined, or mindful exploratory post-hoc analysis. There seems to be an unprincipled objection to post-hoc analysis burned deeply into the cortex of many psychologists. Where does this objection come from? It comes from the mistaken notion that everything that can be possibly known about a phenomenon will be known beforehand and, therefore, only analyses linked that that (pseudo-)understanding of a phenomenon are allowed. Let me be rather straightforward about this – the data we collect are gathered with a great expenditure of time, effort, money, and the kind cooperation of innumerable research subjects. It makes no sense to believe that one can only tap a data set for a few correlation coefficients, a couple of t-tests, or a single contrast specified beforehand and then pack it up, place it on the shelf, and allow it to gather dust in copy-paper boxes.

Hans Reichenbach (1891–1953) wrote in his 1938 monograph *Experience and Prediction* on the distinction between what he called the *context of discovery* and the *context of justification* (Reichenbach, 1938). The context of discovery refers to the thinking processes of the scientist involved in the research process. The discovery process need not be particularly orderly, grounded in theory or a model, or cleanly linear in the unfolding of new insights or knowledge. This is the place where we ponder a puzzle, consider various angles of a problem, and allow our ideas to develop freely. In terms of data analysis, this is where we probe the data, come to know it, ask questions of it, look at it from different angles and use it to illuminate a problem of interest. Theory may play a small role here, for example it might give you some ideas about potentially interesting effects in the data or how to explore the data to gain some leverage, but it need not constrain the enterprise (e.g., what variables to pick, how to configure them). Experienced researchers, for example, know that the material that ends up in scientific reports are highly ordered and coherent, whereas the nature of the investigative and creative processes leading up to the final published result is typically far from such a pristine ideal. But this is really quite acceptable, as much work is done within the context of discovery and one need not be preoccupied with confirming relationships in some definitive manner or presenting a highly polished linear rendition of the discovery process.

The scientist who discovers a theory is usually guided to his discovery by guesses; he cannot name a method by means of which he found the theory and can only say that it appeared plausible to him, that he had the right hunch, or that he saw intuitively which assumption would fit the facts ... The act of discovery escapes logical analysis; there are no logical rules in terms of which a "discovery machine" could be constructed to take over the creative function of the genius. But it is not the logician's task to account for scientific discoveries. (Reichenbach, 1956, pp. 230, 231)

The *context of justification*, on the other hand, refers to the process of evaluating a formally presented theory or model (typically in some written form, such as a book or journal article). The idea here is the effort after confirmation or disconfirmation, the seeking of definitive evaluation, subjecting the model or theory to a genuine test or attempt at falsification. In other words, this is where the "rubber meets the road," and we subject our claim, thesis, model, or hypothesis to some form of evaluation or test.

At this juncture some readers might be rather perplexed. Much of the foregoing material probably flies in the face of what they believe is the process of science (and may have even been taught is the process of science). The illusion that science and the process of discovery is necessarily a highly linear, step-by-step process specified a priori that moves from hypotheses to study to analysis to research report is really rather something of a fiction. The main point here is simply this: discovery and exploration are *both* important modes in which the psychopathologist functions rather routinely in the quest to understand psychopathology and these modes are greatly enhanced by post-hoc analysis and sifting through data, based on the assumption that interesting discoveries will be followed up with later studies. Post-hoc data analysis alone (or what some might call "fishing expeditions"), importantly, is not a substitute for carefully designed subsequent studies to follow up on what might have been discovered post hoc.

How We Think About Our Predictors, *p*(S|D) versus *p*(D|S)

Many psychopathologists spend a good deal of time comparing the means (averages) of two or more groups (typically a psychopathology group vs. normal control group) in terms of any variety of outcomes or dependent variables (e.g., IQ, anxiety levels, depression levels). These same psychopathologists also spend a good deal of time discussing the selection of the risk factors or putative liability indicators for, say, schizophrenia. Typically, discussions regarding such liability indicators take the form of a discussion of "predictors" of schizophrenia liability or "indicators" that tap schizophrenia liability. Such theoretical discussions have been buttressed by empirical results that have shown the presence of deficits in the processes or constructs of interest in schizophrenia patients, first-degree relatives of schizophrenia patients, and/or schizotypic patients otherwise defined. Thus, one learns that schizophrenia patients show lower accuracy scores than normal individuals (i.e., a control group) on a measure of eye tracking. Likewise, first-degree relatives of schizophrenia patients display poorer average performance than normal subjects on a measure of sustained attention. Such studies often include a psychiatric control group as well in order to evaluate if an observed group difference is reasonably specific to schizophrenia or it is merely reflective of general psychopathology.

This all sounds well and good – one learns about group differences in the constructs (or processes) of interest and discusses them within a predictive framework. The problem is that the conceptual framework used to discuss the so-called predictors or indicators is usually wrongly configured. What I mean by this is that many such discussions presume we know something about the ability of a predictor or indicator to actually predict or identify the form of psychopathology (or psychopathology liability) in question, but we normally do not have this sort of empirical (i.e., predictive power, probability) information at hand.

What we do have at hand are data that come from studies where we already know that a liability exists in our experimental group and, therefore, we really are evaluating the extent to which deviance is found on an indicator or predictor given the presence of liability. Thus, in more formal terms, although we discuss putative predictors and liability indicators in the following manner p (Schizophrenia| Deviance on indicator), what we actually know is only p (Deviance on indicator|Schizophrenia). Experimental psychopathology can offer a great deal to illuminate predictors that actually map on to the model we espouse, namely p (Schizophrenia|Deviance on indicator), but we need to recognize that much of our research is not configured in this manner. Thus, we need to remind ourselves that we need to understand empirical data in a manner that actually is consistent with the nature of the argument (e.g., predicting liability) that one wishes to make.

Heterogeneity in Our Patients and Data – A Reality We Must Embrace

In most experimental psychopathology studies, we nearly always begin with clinically defined pathology in a subject group, whether rated according to the *Diagnostic and Statistical Manual of Mental Disorders* (DSM) or some other rubric (see rating discussion above), as well as some form of control group. The diagnostic categories we employ, however, are highly heterogeneous – that is, not all patients within the category possess the same symptom profiles. Given the extensive reliance upon subjectively rated phenomenology in the diagnosis of psychopathology, it is likely that heterogeneity will characterize any group of patients despite being assigned to the same diagnostic category. For example, one schizophrenia patient may have prominent persecutory delusions (e.g., the CIA has placed a small Martian in their brain to spy on the British Intelligence group MI6), bizarre behavior (e.g., carrying a dead chicken around in a lunch bag), and diminished volition (e.g., no desire to get out of bed, bath, or go to a rehabilitation training program), whereas another schizophrenia patient will display marked auditory hallucinations (e.g., hearing voices outside their head – when alone – commenting on their behavior), severe thought disorder (e.g., incoherent speech), and flattened affect (i.e., showing no emotion on their face). Now, both of these people suffer from schizophrenia, yet they seem so different with respect to phenomenology.

This is so for several reasons. First, the current diagnostic systems in use (the DSM-5 and the International Classification of Diseases [ICD-10]) define disorders using a polythetic criteria set and diagnostic thresholds are set primarily in terms of the *number* of criteria met (i.e., not which criteria are met). Thus, there are usually any number of ways in which a patient can be seen as fulfilling the diagnostic criteria for a disorder and, therefore, two patients that have been diagnosed with the same illness can display remarkable variation in their clinical phenomenology. Second, it is well known, particularly with respect to schizophrenia, that the illness itself displays considerable variation with respect to its clinical presentation (an attempt to capture this feature of the illness is reflected, in part, in the polythetic approach to definition noted already). Many other forms of psychopathology display remarkable variation in clinical presentation as well. Variation in the clinical presentation of patients in concert with a system that essentially encourages vast numbers of symptom combinations leads inexorably to a high degree of heterogeneity at the clinical or phenotypic level even within well-described categories of psychopathology. Third, diagnostic practice, assessment instruments, and diagnostic standards vary across sites, though this is less a problem than it was 40–50 years ago, and combine to increase heterogeneity in patient samples even though all patients have received the same diagnosis. (In this context, it is important to note that use of dimensional measures does not somehow get one out of the heterogeneity conundrum; a multidimensional space can harbor immense heterogeneity reflected in densifications and phenotypic clumpings of patients.)

More vexing is the additional likelihood that not all patients within the category share the same causal pathway from etiology to expressed psychopathology (common diagnosis notwithstanding), a principle referred to as *equifinality*. This raises the unsettling prospect of not only heterogeneity in phenotypic expression, but the nontrivial probability of heterogeneity in etiologic processes (i.e., causal factors). This is perhaps more interesting from the standpoint of theory and model building. Could it be that the heterogeneity that appears in a clinical sample of similarly diagnosed patients – say, schizophrenia – is really just an umbrella term to cover similar-appearing clinical manifestations of *multiple forms* of psychopathology or multiple pathological processes? In other words, perhaps there are multiple *types* or *forms* of schizophrenia, which would probably imply multiple etiological pathways to a similar-appearing class of phenomenology. One could easily apply this same scenario to conditions such as borderline personality disorder or psychopathy. I refer to the source of such variation as *etiological heterogeneity*.

Finally, when one ponders the methods of experimental psychopathology, for example laboratory measures of sustained attention, working memory, context processing, executive functioning, or smooth-pursuit eye movements (or some other current focus of attention in the literature), one must also confront the issue of heterogeneity in performance on such laboratory measures. Simply stated, it is rarely the case (if at all) that all schizophrenia patients (or schizotypes) in a given study sample will display deficits on any single measure. It is much more likely the case that, for the sake of illustration, 40 percent of an adult sample of schizophrenia patients tested for sustained attention will actually display deficits in attention in their performance, or 60 percent of an adult sample will display evidence of eye-tracking dysfunction. It is surely not the case that all schizophrenia patients display comparable/similar deficits across multiple laboratory measures. We speak of this type of variation as *laboratory task performance heterogeneity*. The notion of heterogeneity in performance on laboratory tasks is a central reality and important challenge in experimental psychopathology research (Lenzenweger, Jensen, & Rubin, 2003; Maher, 2003; see also Silverstein, 2008). Finally, an important related concept concerns what could be called task impurity, namely a given task (e.g., a working memory task) in one study could be quite different from a task used in another study, with both putatively studying the same process. When variation in a common process arises across studies, do not always assume that such differences are the result of differences in a sample – they often reflect differences in the tasks or measures used (or how they were administered).

Many psychopathology researchers seek to diminish the impact of heterogeneity when analyzing their data by suggesting it does not account for much in observed relationships. My recommendation is in the opposite direction, namely embrace heterogeneity and explore it, surely do not wish it away. For example, considering sex differences, many researchers will conduct analyses to suggest that there are none in their data and then assume they are "good to go," so to speak (i.e., they then collapse their data across sexes and move on with the analyses). I suggest that we always inspect the sources of heterogeneity with the hope of finding something interesting – it is not error, it may be the source of new hypotheses and insights (recall the *context of discovery*). What is the message here? Perhaps one of the most important things one must know about heterogeneity in psychopathology research is simply *to expect it*. Heterogeneity in phenomenology and laboratory task performance should be assumed to be a given in any study of schizotypy or schizophrenia.

Levels of Analysis

As argued by Kosslyn and Rosenberg (2005), any psychological phenomenon can be studied from multiple *levels of analysis*. This is an important concept to commit to memory. Particularly in the age of neuroscience and neuroimaging, it is relatively easy for students to err in their thinking and come to believe that only neuroscience (particularly neuroimaging) investigations of psychological phenomena are worthy of attention. Or, worse yet, students might come to believe (wrongly) that only evidence directly related to brain functioning is worth studying (such evidence can be falsely persuasive, see Weisberg et al., 2008). The "levels of analysis" approach advocated by Kosslyn and Rosenberg (2005, p. 78) represents a way of thinking that helps to place the brain in a broader context, which allows "facts about the brain to illuminate facets of phenomena at other levels of analysis." The approach encourages scientists to make distinctions among at least three levels of analysis: (a) level of the brain, (b) level of the person, and (c) level of the group. One can readily see how this dissection allows one to place the brain within the person and the person within the group, thereby capturing the essence of modern psychological science investigations and theory. The level of the brain really refers to biological mechanisms – think brain structures, brain circuits and networks, neurobiological pathways and interacting neurobehavioral systems, hormones, and so on. By implication, the level of the brain is also concerned with the effects of genes on the brain, neural development, neurobiology and hormones. The "level of the person," which is not the same as the "whole person," refers to the content of mental processes (such as beliefs, feelings). The "level of the group" refers to the physical (including natural and human-made) and social worlds. A key idea in the Kosslyn-Rosenberg proposal is that "events at each of the levels affect events at the other levels" (Kosslyn & Rosenberg, 2005, p. 78).

Consider this example to gain an appreciation for how this organizing scheme works: For example, right this minute your brain is processing these squiggly black lines in front of you, interpreting them as conveying meaning (level of the brain). These lines were created by others (specifically, we authors – level of the group) in order to convey specific ideas (level of the person). If we have successfully conveyed that information, your beliefs will change (level of the person), which in turn will affect not only how your brain organizes and stores information in the future, but possibly even how you interact with other people. (Kosslyn & Rosenberg, 2005, p. 78)

All psychological phenomena emerge from *interactions* that occur between events at the three different levels of analysis proposed by Kosslyn and Rosenberg (2005). The "whole person," so to speak, is the consequence of *all* such interactions. In this context, one might ask about individual differences in personality, temperament, or even psychopathology – how do these things come about within the Kosslyn-Rosenberg framework? In short, these variations in personality and temperament arise from interactions between events in the brain, content (the results of experience/learning), and social context. A critical foundation of experimental psychopathology is to consistently entertain multiple levels of analysis in one's investigations and to seek to integrate findings across levels of analysis.

Specific Etiology

In psychopathology research, as noted above, our goal is to illuminate etiological factors that are uniquely or most powerfully responsible for pathological development for particular forms of psychopathology. We want to understand pathogenesis from beginning to end, as it were. Of critical importance in this quest is the identification of those etiological factors that have a high degree of specificity for the type of psychopathology under investigation. In general, the more specific a causal or etiological input is to the determination of psychopathology, the better. Although a relatively straightforward idea, the definition of *specific etiology* requires clear conceptual thinking as well as innovative research strategies. A thorough treatment of this topic cannot be attempted here. Some of the most careful thinking on this topic has been done by Meehl (in particular, see Part II: Specific Etiology of *A Paul Meehl Reader*: Waller et al., 2006). To provide a sense of the issues in play when discussing specific etiology, consider this illustrative sampling of Meehl's discourse on this topic: Meehl articulated several "distinguishable and equally defensible meanings" (Meehl, 1972, p. 22) of specific etiology ranging from a *qualitative* (i.e., present vs. absent) factor that is both necessary and sufficient for the illness to occur, through progressively weaker meanings of the term. For example, one could have a necessary qualitative causal factor that is not sufficient to cause an illness on its own unless it interacts with a variety of nonspecific contributory factors. Moving away from a purely qualitatively structured causal factor, a dimensionally (or quantitatively) structured underlying specific liability could exert its influence with a threshold effect. In such a circumstance an illness would be likely to develop only when values above a certain threshold on the liability dimension were present, and only then would contributory factors play a role in the emergence of the illness. A fourth meaning of specific etiology, albeit a weaker form of specificity, is what Meehl termed a *uniformly most potent* factor. There might be a given variable or factor that has the greatest impact everywhere in the multivariate array or space of variables known to influence the development of an illness –in this conceptualization, small differences in this causal factor are more potent than differences in other factors in the array. Meehl lists several other forms of specific etiology including "specific step function" etiology and "uniquely nonfungible factor" etiology (Waller et al., 2006, p. 173). Consideration of what the term specific etiology means is foundational to experimental psychopathology (see Meehl, 1977). This seminal paper is useful to consult when designing studies to determine what form of etiology one might be studying through a particular protocol or how one's conceptual model of some aspect of psychopathology maps on to the different forms of specific etiology.

Reliability and Validity

Most experimental psychopathology students learn about reliability and validity as separate concepts with corresponding methods used to establish them. However, it is always important to keep in mind that reliability and validity have some very intimate relations to one another that are worth committing to memory (for a more detailed treatment of this topic, see Chapter 6, by Furr). First, reliability of measurement does *not* ensure validity. One can generate highly repeatable measurements, although that is no guarantee that the measurements possess validity. Second, when measurements are made validly, however, they will be, by definition, reliable. Third, one can impact the validity of measurement through alterations in the reliability of measurement. One can even squeeze validity out of measurements by increasing reliability to very high levels, which is somewhat counterintuitive. For example, one could narrow the definition of schizophrenia to such an extent (albeit yielding reliable ratings) that the illness appears no longer to be susceptible to genetic influences (a known criterion of validity for the illness) (see Lenzenweger, 2010, for further examples). Fourth, reliability sets an upper bound on validity (i.e., the square root of the reliability sets an upper bound on the validity of a measure). Fifth, excessively high reliability can be a telltale sign that one is dealing with an interesting state of affairs in terms of commingled latent structures harbored within the data (Waller, 2008). For example, one might discover internal consistency levels for a measure in the mid-1990s, which represents a level that is somewhat atypical in psychometrics, and closer inspection reveals that one is not sampling from a single population, rather the data set is composed of scores that are drawn from multiple populations. The presence of commingled populations often serves to bias measures of internal consistency upward.

The DSM System/Architecture Has Not Cornered the Market on Truth

Many an instructor in Abnormal Psychology (undergraduate) and Psychopathology (graduate) courses spends a good deal of time teaching students the DSM approach to the classification of mental disorders. The explicit approach of the system is clearly categorical; that is, one is either diagnosed as having or not having a disorder. Such an approach has utility, but one should not come to believe that the diagnostic scheme presented in the DSM amounts to the "truth." Indeed, a particularly dangerous combination can be found in (a) thinking that the DSM amounts to truth and (b) having excessively high levels of confidence in one's ability to apply the *DSM* system (see Berner & Graber, 2008, for an excellent discussion of the effects of overconfidence on diagnostic error in medicine).

To explore this issue, let us consider one of the more engaging sections of Meehl's (1986, p. 220) seminal paper on nosological revision focused on what he described as three different intellectual views of the DSM constructs as: (a) "admirable," (b) "criticizable, but fairly harmless," and (c) "scientifically malignant." The "admirable" view sees the diagnostic constructs explicated in the DSM as representing symptomatically defined syndromes that hang together (descriptively or statistically), have some unknown or unspecified etiology (that is likely shared by those diagnosed with the conditions), and have some communicative value. This reveals a reasonable intellectual approach to the thorny problem represented by the classification and treatment challenges represented by psychopathology. The "fairly harmless" view is one that holds that the constructs defined in the DSM system are somehow correct and the best we can do for now, with an implied openness to revision down the line. This is an OK view, so to speak, just as long as those who think the DSM is the best we can do for now do not get in the way of those seeking to explore other entities or models of psychopathology (e.g., consider the Alternative Model for Personality Disorders in section III of the DSM-5) (see Zachar, Krueger, & Kendler, 2016, for an illuminating oral history of the development of this model).

The mistaken and intellectually "malignant" indefensible view is that the DSM constructs represent the *truth* and this is indeed bad news. Not only can such a misguided view get in the way of progress, it is also grounded in a view of entities and operational definitions that has long been abandoned by informed philosophers of science and psychological science researchers (for further considerations, see Chapter 2, by Zachar and colleagues). What is meant here? If someone believes that the *meaning* of the conditions in the DSM is defined by the list of signs and symptoms (or DSM-defined diagnostic rules), then one is implicitly subscribing to an *operationism* (so-called operational definitions) that has long been discounted in philosophy and psychology. However, the DSM does not specify any operations to be followed for the definition of conditions, it provides no formalism linking signs/symptoms to an underlying disease entity, and there is no statement of procedures to be followed (or that were followed) in defining the process to determine the nature of the pathology under consideration. Moreover, conditions (or diseases) in traditional organic medicine are defined not merely by signs and symptoms; rather they represent (implicitly) information regarding etiology, pathophysiology, and so on. If information about etiology and pathophysiology are absent from such constructs in organic medicine, then, at a minimum, the construct in question is viewed as in need of further research and has not attained "disease" status.

In this context, I would urge students of experimental psychopathology to "think outside the DSM boxes," as it were. Experimental psychopathology and clinical psychological science, it seems, have been more comfortable exploring the manifestations of severe psychopathology that fall outside what appears in print on the pages of the DSM. For example, concepts such as schizotypy (the liability for schizophrenia), psychopathy (as distinct from the sociodemographically defined antisocial personality disorder), or borderline personality organization (e.g., using Kernberg's [1984] phenomenological organizing framework for personality pathology) serve as but a few examples. Each of these theoretical concepts has spawned rich research literatures that have illuminated important aspects of more traditionally defined psychiatric conditions. For example, the laboratory study of schizotypy has generated a remarkable corpus of findings that support schizotypic psychopathology as an alternative manifestation of schizophrenia liability. Frankly, the work of psychopathologists and psychologists over the past 4–5 decades, working outside of strictly categorical models of pathology, has generated the corpus that has supported increasing interest in transdiagnostic dimensional models such as the current National Institute of Mental Health (NIMH) Research Domain Criteria (RDoC) approach. To work outside the realm of DSM disorders in psychopathology research, however, requires persistence and tenacity because many reviewers and funding agencies prefer papers and proposals seeking to investigate "established" entities.

Genetic Influences, Emergence, and Causality in a Correlational World

In closing this essay, I wish to stress, if briefly, the importance of three additional perspectives on experimental psychopathology. The role of genetic influences in the determination of psychopathology is now well established as scientific fact and few, if any, informed psychopathologists would dispute this. An implication of this fact is the reality that, in shorthand, genes will always make a difference if one is doing a study properly – they are always in the mix, so to speak. Beyond advances in molecular genetics and related research designs (e.g., genome-wide association studies), I would strongly recommend that all experimental psychopathologists consider the endophenotype concept, proposed originally by Gottesman and Shield, worthy of detailed study (Gottesman & Gould, 2003; Shields & Gottesman, 1973; see also Lenzenweger, 2013). Always remember, nature assigns genes, and that process is not under experimenter control.

The second issue concerns the concept of emergence, which argues that complex end products cannot be simply reduced to their constituent parts. The concept of *emergence* (Bickhard & Campbell, 2000; Meehl & Sellars, 1956) is highly relevant to psychopathology as most disorders represent complex configural outcomes of multiple interacting systems. Emergence plays a critical role in many areas of science (e.g., condensed matter and material physics, see Kivelson & Kivelson, 2016) and, closer to home, contemporary cognitive neuroscience (Rumelhart, 1984), which views most psychological phenomena as having emergent properties (e.g., consider mental imagery). Many psychopathological phenomena, in my view, represent the final interactive product of multiple systems *and* they cannot be reduced to a mere collection of constituent parts – they are emergent constructs reflecting emergent properties. For example, consider the topic of personality disorders, we (see Lenzenweger & Depue, 2016, for details) have long argued that such disorders are emergent phenomena best understood within a framework of multiple, interacting neurobehavioral systems, which are themselves influenced by genetic factors, and are acted upon by social and physical environmental inputs. In our view, incorporating the emergence viewpoint, one cannot simply parse a personality disorder into its component parts – rather the underlying component parts have come together interactively to form a distinctly rich phenotype. Moreover, interactions *across levels of analysis* within the human being conspire to generate a phenotype that can be thought of as an emergent property or manifestation.

The third issue that I wish to emphasize as a foundational element in experimental psychopathology is the concept of causality. Recall that the working definition of experimental psychopathology speaks to etiology, which implicitly embraces the concept of causality. What causes schizophrenia? What causes borderline personality disorder? This is the underlying question that drives experimental psychopathology, but it is not easily addressed when one is working principally within a correlational framework wherein nature has assigned group membership (i.e., nature, not the investigator, decides who develops an illness and who does not). Experimental psychopathology has no choice but to recognize and embrace the correlational paradigm while simultaneously seeking to realize the ultimate goal, which is to explain etiology and pathogenesis. One thing is for certain: causality is not illuminated simply by conducting structural equation modeling or path analysis. That is not to say such techniques cannot help to illuminate matters to some extent. The issue of causality is discussed richly by Professor Rubin (Chapter 31, by Mozer, Rubin, and Zubizaretta) and the reader is directed to his thinking on the matter and his creative statistical approach known as the Rubin Causal Model (Imbens & Rubin, 2010; see also the work of Pearl, 2009).

SUMMARY AND FINAL COMMENTS

This overview of the historical and conceptual foundations of experimental psychopathology, albeit necessarily selective, points to the many rich considerations that one must bear in mind when using the methods of the laboratory to probe the nature and causes of psychopathology. The value placed on clinical observation cannot be overemphasized, both in terms of generating hypotheses as well as providing high-quality characterizations of study samples. The focus of the discipline on using precise laboratory methods to examine processes in psychopathology, rather than reliance on subjective or impressionistic ratings (no matter how reliable) or empirically ungrounded models is central. Every step of the experimental psychopathology research process, from conceptualization of hypotheses for testing in the laboratory, through data collection, on to analysis, and interpretation of the results merit careful thought and engagement if the discipline is to reach its potential in the study of psychopathology. Reliance on traditional null hypothesis statistical testing in one's work must always receive scrutiny and other approaches to the statistical analysis of data should be considered. Finally, bearing in mind the multiple levels of analysis that can be pursued in experimental psychopathology investigations will likely bear fruit in the dissection of one's data as well as the design of future studies, all the while seeking to illuminate causal processes deriving from etiology and directing pathogenesis in psychopathology.

REFERENCES

Abramson, L. Y., & Seligman, M. E. P. (1977). Modeling Psychopathology in the Laboratory: History and Rationale. In J. P. Maser & M. E. P. Seligman (Eds.), *Psychopathology: Experimental Models* (pp. 1–26). San Francisco: Freeman.

Berner, E. S., & Graber, M. L. (2008). Overconfidence as a Cause of Diagnostic Error in Medicine. *American Journal of Medicine*, *121*(Suppl. 5), s2–s23.

Bickhard, M. H., & Campbell, D. T. (2000). Emergence. In P. B. Andersen, C. Emmeche, N. O. Finnemann, & P. V. Christiansen (Eds.), *Downward Causation: Minds, Body, Matter* (pp. 322–348). Århus: Århus University Press.

Cohen, J. (1994). The Earth is Round ($p < 0.05$). *American Psychologist*, *49*, 997–1003.

Forsyth, J. P., & Zvolensky, M. J. (2002). Experimental Psychopathology, Clinical Science, and Practice: An Irrelevant or Indispensable Alliance? *Applied and Preventive Psychology: Current Scientific Perspectives*, *10*, 243–264.

Gottesman, I. I., & Gould, T. D. (2003). The Endophenotype Concept in Psychiatry: Etymology and Strategic Intentions. *American Journal of Psychiatry*, *160*, 636–645.

Hoch, P. H. & Zubin, J. (Eds.) (1957). *Experimental Psychopathology*. New York: Grune & Stratton.

Imbens, G. W., & Rubin, D. B. (2010). *Causal Inference in Statistics and the Medical and Social Sciences*. New York: Cambridge University Press.

Kernberg, O. F. (1984). *Severe Personality Disorders*. New Haven, CT: Yale University Press.

Kimmel, H. D. (1971). Introduction. In H. D. Kimmel (Ed.), *Experimental Psychopathology: Recent Research and Theory* (pp. 1–10). New York: Academic Press.

Kivelson, S., & Kivelson, S. A. (2016). Defining Emergence in Physics. *NPJ Quantum Materials*, *1*, 1–2.

Korolenko, C. P., Volkov, P. P., Evceeva, T. A., & Shmatko, N. S. (1966). Experimental Psychopathology and Its Significance for the Clinical Study of Exogenous Psychoses. *L' Evolution psychiatrique*, *31*, 777–785.

Kosslyn, S. M., & Rosenberg, R. S. (2005). The Brain and Your Students: How to Explain Why Neuroscience Is Relevant to Psychology. In B. Perlman, L. I. McCann, & W. Buskist (Eds.), *Voices of Experience: Memorable Talks from the National Institute on the Teaching of Psychology Volume One* (pp. 71–82). Washington, DC: American Psychological Society.

Lenzenweger, M. F. (2004). Consideration of the Challenges, Complications, and Pitfalls of Taxometric Analysis. *Journal of Abnormal Psychology*, *113*, 10–23.

Lenzenweger, M. F. (2010). *Schizotypy and Schizophrenia: The View from Experimental Psychopathology*. New York: Guilford Press.

Lenzenweger, M. F. (2013). Thinking Clearly about the Endophenotype vs. Intermediate Phenotype vs. Biomarker Distinctions in Developmental Psychopathology Research. Invited Essay for 25th Anniversary Issue. *Development & Psychopathology*, *25*, 1347–1357.

Lenzenweger, M. F., & Depue, R. A. (2016). Toward a Developmental Psychopathology of Personality Disturbance: A Neurobehavioral Dimensional Model Incorporating Genetic, Environmental, and Epigenetic Factors (pp. 1079–1110). In D. Cicchetti (Ed.), *Developmental Psychopathology, Volume 3, Maladaptation and Psychopathology* (3rd edn.). New York: Wiley.

Lenzenweger, M. F., & Hooley, J. M. (Eds.) (2003). *Principles of Experimental Psychopathology: Essays in Honor of Brendan A. Maher*. Washington, DC: American Psychological Association.

Lenzenweger, M. F., Jensen, S., & Rubin, D. B. (2003). Finding the "Genuine" Schizotype: A Model and Method for Resolving Heterogeneity in Performance on Laboratory Measures in Experimental Psychopathology Research. *Journal of Abnormal Psychology*, *112*, 457–468.

Maher, B. A. (1966). *Principles of Psychopathology: An Experimental Approach*. Oxford: McGraw-Hill.

Maher, B. A. (2003). Psychopathology and Delusions: Reflections on Methods and Models. In M. F. Lenzenweger & J. M. Hooley (Eds.), *Principles of Experimental Psychopathology: Essays in Honor of Brendan A. Maher* (pp. 9–28). Washington, DC: American Psychological Association.

Meehl, P. E. (1967). (1967). Theory-Testing in Psychology and Physics: A Methodological Paradox *Philosophy of Science*, *34*, 103–115.

Meehl, P. E. (1972). Specific Genetic Etiology, Psychodynamics and Therapeutic Nihilism. *International Journal of Mental Health*, *1*, 10–27.

Meehl, P. E. (1977). Specific Etiology and Other Forms of Strong Influence: Some Quantitative Meanings. *Journal of Medicine and Philosophy*, *2*, 33–53.

Meehl, P. E. (1978). Theoretical Risks and Tabular Asterisks: Sir Karl, Sir Ronald, and the Slow Progress of Soft Psychology. *Journal of Consulting and Clinical Psychology*, *46*, 806–834.

Meehl, P. E. (1986). Diagnostic Taxa as Open Concepts: Metatheoretical and Statistical Questions about Reliability and Construct Validity in the Grand Strategy of Nosological Revision. In T. Millon & G. L. Klerman (Eds.), *In Contemporary Directions in Psychopathology* (pp. 215–231). New York: Guilford.

Meehl, P. E. (1990). Toward an Integrated Theory of Schizotaxia, Schizotypy, and Schizophrenia. *Journal of Personality Disorders*, *4*, 1–99.

Meehl, P. E. (1998/2006). The Power of Quantitative Thinking. In N. G. Waller, L. J. Yonce, W. M. Grove, D. Faust, & M. F. Lenzenweger (Eds.). *A Paul Meehl Reader: Essays on the Practice of Scientific Psychology* (pp. 433–444). Mahwah, NJ: Erlbaum.

Meehl, P. E., & Sellars, W. (1956). The Concept of Emergence. In H. Feigl & M. Scriven (Eds.), *Minnesota Studies in the Philosophy of Science: Vol. I. The Foundations of Science and the Concepts of Psychology and Psychoanalysis* (pp. 239–252). Minneapolis: University of Minnesota Press.

Pearl, J. (2009). *Causality: Models, Reasoning and Inference* (2nd edn.). New York: Cambridge University Press.

Reichenbach, H. (1938). *Experience and Prediction*. Chicago, IL: University of Chicago Press.

Reichenbach, H. (1956). *The Rise of Scientific Discovery*. Berkeley, CA: University of California Press.

Rumelhart, D. E. (1984). The Emergence of Cognitive Phenomena from Sub-Symbolic Processes. In *Proceedings of the Sixth Annual Conference of the Cognitive Science Society Colorado, 1984*. Hillsdale, NJ: Erlbaum.

Shields, J., & Gottesman, I. I. (1973). Genetic Studies of Schizophrenia as Signposts to Biochemistry. In L. L. Iversen & S. P. R. Rose (Eds.), *Biochemistry and Mental Illness* (pp. 165–174). London: Biochemical Society.

Silverstein, S. M. (2008). Measuring Specific, rather than Generalized, Cognitive Deficits and Maximizing Between-Group Effect Size in Studies of Cognition and Cognitive Change. *Schizophrenia Bulletin, 34*(4), 645–655.

Wagenmakers, E.-J., Morey, R. D., & Lee, M. D. (2016). Bayesian Benefits for the Pragmatic Researcher. *Current Directions in Psychological Science, 25*, 169–176.

Waller, N. G. (2008). Commingled Samples: A Neglected Source of Bias in Reliability Analysis. *Applied Psychological Measurement, 32*, 211–223.

Waller, N. G., Yonce, L. J., Grove, W. M., Faust, D. A., & Lenzenweger, M. F. (2006). *A Paul Meehl Reader: Essays on the Practice of Scientific Psychology*. Mahwah, NJ: Lawrence Erlbaum.

Weisberg, D. S., Keil, F. C., Goodstein, J., Rawson, E., & Gray, J. R. (2008). The Seductive Allure of Neuroscience Explanations. *Journal of Cognitive Neuroscience, 20*(3), 470–477.

Wilkinson, L., & The Task Force on Statistical Inference. (1999). Statistical Methods in Psychology Journals: Guidelines and Explanations. *American Psychologist, 54*, 594–604.

Zachar, P., Krueger, R. F., & Kendler, K. S. (2016). Personality Disorder in DSM-5: An Oral History. *Psychological Medicine, 46*, 1–10.

Zvolensky, M. J., Forsyth, J. P., & Johnson, K. (2013). Laboratory Methods in Experimental Psychopathology. In J. S. Comer & P. C. Kendall (Eds.), *Oxford Library of Psychology: The Oxford Handbook of Research Strategies for Clinical Psychology* (pp. 7–23). New York: Oxford University Press.

10 A Practical Guide for Designing and Conducting Cognitive Studies in Child Psychopathology

CYNTHIA HUANG-POLLOCK, ZVI SHAPIRO, HILARY GALLOWAY-LONG, AND JASON FELDMAN

The intent of this chapter is to provide a practical guide to planning and designing laboratory-based cognitive research studies in children with behavioral or emotional disorders. Because of its centrality as an organizing construct, cognition has long been integrated into multiple broad fields of inquiry including clinical neuropsychology, cognitive psychopathology, cognitive neuroscience, cognitive genomics, and cognitive psychopharmacology. This integration has been facilitated by the explosion in the number of portable, well-designed cognitive paradigms that have been developed for behavioral research. Many of these paradigms and the theoretical models upon which they have been built, have been critical tools in the identification of the brain networks that govern complex behavior.

Cognitive psychopathology is a branch of study which applies the concepts of cognitive psychology to understand how cognitive dysfunctions contribute to the development/maintenance of psychopathology and to measure/capture functional outcomes of psychopathology (Bentall, 1996; Haywood & Raffard, 2017). Studies of cognition generally fall into two broad design categories: experimental and descriptive. Experimental designs (also called Type II designs: Zvolensky, Forsyth, & Johnson, 2013) evaluate the effects of an experimental manipulation within a clinical/subclinical population to understand the processes that cause, maintain, or exacerbate psychopathology. For example, in their seminal work on theory of mind in autism, Baron-Cohen and colleagues asked children with autism, Down's syndrome, and typically developing controls to arrange a four-panel sequence of pictures into a predetermined sequence (Baron-Cohen, Leslie, & Frith, 1986). The storylines depicted (a) object sequences (e.g., a ball rolling down a hill), (b) overt behavior between people that did not require consideration of mental states (e.g., a person or people acting in a routine social or nonsocial activity), or required (c) understanding of mental states (e.g., false belief tasks). Whereas children with autism performed well on both the mechanical and behavioral conditions, they had considerable difficulty in their ability to successfully identify the storyline in the false belief condition compared to children with Down's syndrome and typically developing controls. The implication of those findings were that the social communication deficits which are a hallmark of autism, can be explained, or are caused by, an inability to understand the minds of others. As noted in this example, the emphasis in the design of experimental studies is to maximize the internal (vs. external) validity of the manipulation. Here, two experimental control conditions controlled for, or evaluated, basic understanding of cause and effect (between objects and people in social and nonsocial situations). In this way, the interpretation of a specific deficit in theory of mind was better supported.

In contrast, descriptive designs (also called Type IV designs: Zvolensky et al., 2013) provide information on cognitive correlates of psychopathology. The search for cognitive endophenotypes is perhaps the best broad example of this approach. In the context of clinical science, endophenotypes refer to "intermediate" and heritable traitlike phenotypes that are (a) associated with psychopathology, but are also (b) simpler in their genetic underpinnings than observable behavior, (c) co-segregate with illness within families, and (d) are present in nonaffected family members at a higher rate than the general population (Gottesman & Gould, 2003). The hope is that the identification of endophenotypes will bridge the gap in our understanding of how genetic risk factors are ultimately translated into psychopathology. This in turn could alter modern psychiatric taxonomy from a system that is based on behavior to one that is based on etiology. In fact, the aim of the National Institute of Mental Health (NIMH) Research Domain Criteria (RDoC) project is to establish new ways to classify mental illness based on cross-diagnostic processes. Many of the RDoC processes are putative endophenotypes, including various aspects of executive function and cognitive control (Kozak & Cuthbert, 2016; Sanislow et al., 2010). Descriptive designs may utilize a cognitive paradigm such as a complex span task as an index of working memory (Conway et al., 2005), or the stop signal reaction time task as an index of inhibitory control (Logan et al., 2014), but they differ from experimental designs by the lack of a within-subject manipulation.

Importantly, because participants cannot randomly be assigned to "pathologic" vs. "normative" groups, both "experimental" and "descriptive" designs are in fact quasiexperimental. Group differences in performance cannot be unequivocally interpreted as due to the presence or absence of psychopathology despite the temptation to do so. Both designs are critically dependent on the manner in which groups are recruited for study, the definition and measurement of psychopathology, and the manner in which specific aspects of cognition are indexed.

With this in mind, we begin with a discussion of the selection and recruitment of target and control groups, followed by the assessment of psychopathology and cognition, and end with a review of pragmatic concerns when working with children and families. Of course, the constructs and design considerations that will be discussed also apply to the study of adult psychopathology. We leave the discussion of neuroimaging and psychophysiological approaches, many of which utilize cognitive paradigms, to other chapters. Similarly, studies whose main goal is to evaluate the effectiveness of a treatment or prevention program are also covered in other chapters.

SELECTION OF TARGET GROUP

Depending on the age and sex of the child, 4–20 percent of children between the ages of 9 and 16 meet criteria for a psychiatric disorder at any given time (Costello et al., 2003), with worldwide prevalence rates estimated at 13 percent (Polanczyk et al., 2015). This makes recruiting an adequately powered sample of participants with a specific clinical presentation or condition a challenge. For studies intending to evaluate the effects of an experimental manipulation or degree of association of a given risk factor with psychopathology *among treatment-seeking individuals*, the appeal and convenience of sampling from a psychiatric or psychological clinic is self-evident. However, the manner in which participants are recruited is one of the most important boundary conditions for the generalizability of findings. Whether or not any individual study should utilize such a strategy needs to be carefully considered.

The primary limitation of such an approach is that children recruited via a clinic or hospital-based recruitment strategy are not representative of the vast majority of children or families who have a disorder or condition, but who are not seeking treatment. This distinction is not just academic. As a rule, children and families who are seeking or currently receiving services express more chronic and severe forms of illness, earlier onset, greater incidence of comorbidity, and reduced social support (Cohen & Cohen, 1984; Goodman et al., 1997). The clinician's illusion and Berkson's bias are two well-known types of selection bias that can arise when participants are sampled from secondary or tertiary care clinics. The clinician's illusion refers to a distortion in the data in which the chronicity of mental health problems is overestimated, and the positive long-term outcomes among affected individuals are underestimated (Cohen & Cohen, 1984). This occurs because of the census nature of the clinical setting (i.e., children who meet current criteria for a disease) as opposed to an incidence sample (i.e., those who have ever met criteria, even if no longer current and therefore not being seen in a clinic). Berkson's bias (Berkson, 1946) arises because the probability of being seen in a clinical setting is greater for those who have two disorders than for those who have one disorder. When samples are taken from a hospital setting, there can then appear to be a greater-than-chance covariance between the two disorders (or a disorder and a risk factor), suggesting the existence of a shared risk factor, even when no such relationship exists. Those with only a single disorder are more likely to remain in the community than to present at a hospital or clinic, and thus be missed by the use of a hospital-based recruitment strategy.

Programs of research that measure cognition within the context of disease often seek to understand the mechanisms by which psychopathology arises or is maintained. Their interest is generally not restricted to those currently seeking treatment. Thus, the vast majority of cognitive psychopathology studies require a community-based sampling strategy. A community-based sampling strategy ensures that children have an equal probability of being recruited, regardless of whether their families have sought or are engaged in services. Recruiting from schools, where almost all children are enrolled, is both a valid and viable community-based sampling strategy. But, being granted that permission is incumbent in part on the research scientist maintaining a positive, collaborative, and bidirectional working relationship with the school districts, principals, and teachers (and parents/guardians who ultimately provide consent). Volunteering to present at teacher in-service events, to parent groups organized within the school, hosting educational tables at school functions, and other similar "give-back" opportunities can help form such a collaborative relationship. Increasingly, schools and school districts maintain a formal approval process, in which approval is based partially on burden of request (e.g. distributing flyers advertising for the study vs. allowing students to be pulled out of classes or after-school programs for data collection) and partially on direct benefit to the school and to the children in return for their assistance. For example, many research programs provide monetary compensation to parents, teachers, and children for their participation. However, offering free screening/assessment and feedback on relevant clinical findings is often the greater incentive for parents to consent for their children to participate, particularly given the cost and wait time of such services in the private sector.

One of the most efficient and cost-effective methods is to request permission to send flyers home in student folders.

Other effective community-based recruitment approaches include social media/TV/radio ad buys, health system referrals (e.g. pamphlets or brochures placed in general pediatric or health clinics), and paid mass mailings. Advertisements should be engaging, as well as clear and free of research jargon. Despite federal mandates to increase representation of ethnic and racial minorities in research (Hohmann & Parron, 1996), successfully recruiting historically underrepresented minorities is complicated by distrust of research within those communities following a long history of racial discrimination and exploitation by universities and medical centers in the US (Clay et al., 2003; Corbie-Smith, Thomas, & St. George, 2002). As with school recruitment, active participation in community outreach (e.g., presentations/meetings, contact with community leaders, participation at community events) is essential to building positive collaborative relationships within local communities (Clay et al., 2003; Farmer et al., 2007; UyBico, Pavel, & Gross, 2007).

SELECTION OF CONTROL GROUP

The appropriate selection of a control group is as important to the internal validity of a study as the selection of a target group. Furthermore, deciding how to define and recruit a control group can be among the most complicated of design decisions. In fact, an entire field of study has blossomed around this question (Grobbee & Hoes, 2015; Ibrahim & Spitzer, 1979; Kopec & Esdaile, 1990; Rothman, 1986). Each of the various types of controls (clinic/hospital, population, neighborhood, friend-based, etc.) have their own benefits and limitations, and the appropriateness of each type is study-dependent (Dantas et al., 2007; Grimes & Schulz, 2005; Ma et al., 2004). Whereas the relatively low base rate of specific mental health conditions creates challenges to recruiting a large sample of target participants, lack of motivation to participate is the primary challenge to recruiting an adequate sample of controls (Wacholder et al., 1992a).

A common misconception is to assume that a control group should be free of *any* disorder when, in fact, a control group should be free of the disorder(s), conditions, or specific behaviors/concerns under investigation, but *be otherwise representative of the sampling frame* (including the presence of other disorders, behaviors, or conditions). This is because the purpose of a control group is to index the background frequency of risk or exposure (Grimes & Schulz, 2005). A control group that is free of any and all disorders does not do this, and would instead represent a supranormative group of individuals. In cognitive psychopathology studies, a control group provides normative information for the cognitive process under investigation. Therefore, how a control group is recruited or sampled is fully dependent on the population from which the target cases have been sampled (Kopec & Esdaile, 1990; Stavraky & Clarke, 1983; Wacholder et al., 1992a). If target cases have been sampled from a particular source (e.g., school roster), then controls should also be sampled from that same source.

If controls are *not* representative of the population of targets, then measures of association or effect size are distorted (i.e., biased). For example, despite the fact that Attention Deficit Hyperactivity Disorder (ADHD) has become virtually synonymous with the presence of executive dysfunction within the lay press, there is in fact substantial overlap in performance distributions among those with and without ADHD. The moderate effect sizes reported in the literature (Boonstra et al., 2005; Huang-Pollock & Nigg, 2003; Huang-Pollock et al., 2012; Lijffijt et al., 2005; Oosterlaan, Logan, & Sergeant, 1998; Pauli-Pott & Becker, 2011; Schwartz & Verhaeghen, 2008; Willcutt et al., 2005) has prompted the field to move toward specifying *how much* and *what types* of neurocognitive heterogeneity might exist among children with ADHD, as well as the moderators of that heterogeneity.

A critical first step in answering how much heterogeneity exists is to obtain accurate estimates of the size of the ADHD-executive function (EF) association. But the size of the association is dependent in part on the nature of the control group. As a demonstration, in a 2005 review, two large research sites utilizing community- or school-based recruitment strategies reported that ~35 percent of children with ADHD were impaired on the Stroop or Continuous Performance Test, when impairment was defined as performing worse than the 90th percentile based on performance of comparison controls (Nigg et al., 2005). However, a third large research site using clinic-based recruitment reported only 20 percent of children as impaired. The underestimation of the effect derives from the selection bias inherent when a clinic-based recruitment strategy is used. The risk factor (in this case, executive dysfunction) is likely not independent of the selection process (i.e. seeking psychiatric care) because treatment-seeking individuals are likely to be more impaired across domains. Ultimately obtaining a true population estimate of effect size (and, perhaps more importantly, the population-based variance around that effect size), will require both first- and second-order meta-analyses, as well as correction for sampling error and other artifacts (Schmidt, 2010). The key point, however, is simply that if target children who are recruited from a clinic are not representative of the population of children with a disorder, then clinic-recruited controls without a disorder are similarly (or even less) representative of the population of children without the disorder.

Keeping in mind the job of a control group is to minimize threats to internal validity through indexing background levels of risk, care should be taken to ensure target and controls are equivalent on basic demographics including gender, age, socioeconomic status, and ethnic/racial distributions (Wacholder et al., 1992b). Friend- and neighborhood-based controls are often utilized for this reason, though few studies exist that empirically test the equivalence of different types of controls (Dantas et al., 2007; Lund, 1989; Ma et al., 2004). In our own experience, potential controls tend to be of high average to above average IQ based on national norms, which may speak to systematic biases that motivate different types of families to participate in cognition research. Intellectual functioning is negatively associated with psychological functioning (Caspi et al., 2014; Masten et al., 1999), and is also correlated with other indices of cognition including processing speed, executive function, and academic achievement (Deary et al., 2007; Engle et al., 1999; Geary, 2011). Thus, studies of cognitive performance should take care in matching or otherwise equating samples for IQ to ensure that unintended group differences in intellectual ability do not confound interpretation of findings. When studying disorders for which a co-occurring intellectual disability is common, including a clinical control group (not to be confused with a clinic-based sampling frame) can also help determine whether findings reflect a specific domain of impairment (as opposed to impairment in broad intellectual functioning). For example, approximately 68 percent of children with autism also present with an intellectual disability (Fombonne, 2009; Yeargin-Allsopp et al., 2003). It is therefore not uncommon for studies of autism to include two types of controls: a typically developing and an intellectually disabled nonautistic group. If the cognitive atypicality is absent from the clinical controls, then it suggests that the target group is not more impaired across all domains.

ASSESSMENT OF PSYCHOPATHOLOGY

The identification of target and control participants assumes a systematic and valid method to assess for the presence, absence, or degree of psychopathology. Much has already been written about the (a) low to moderate correlations between multiple informant reports of behavior (i.e., parent and teacher; parent and child; child and peer), (b) degree to (and the contexts within) which such discrepancies reflect true variation in behavior vs. measurement error, and (c) lack of an agreed-upon method to handle multiple informant reports in a research or clinical context (Achenbach, McConaughy, & Howell, 1987; Achenbach et al., 2005; Bird, Gould, & Staghezza, 1992; De Los Reyes et al., 2013, 2015; Jensen et al., 1999; Kraemer et al., 2003). That being said, cross-informant studies have repeatedly found that consistency of informant report across context (i.e., parent and teacher) tends to be better for externalizing/observable concerns than internalizing/less observable concerns, and among informants who observe the child in the same setting (i.e., mother and father vs. mother and teacher) (De Los Reyes et al., 2015). Furthermore, the predictive validity of adult informant ratings of child externalizing behavior tends to be greater than child self-report (Bird et al., 1992; Hart et al., 1994; Johnston & Murray, 2003).

During the DSM field trials for ADHD, Oppositional Defiant Disorder (ODD), and Conduct Disorder (CD), which were conducted to identify and provide evidence-based recommendations on what diagnostic thresholds to adopt, a symptom was counted as present if either the parent or teacher (or child, in the case of CD) reported its presence (Lahey et al., 1994a, 1994b). Thus, for these particular clinical disorders, there is explicit direction on how multiple informant reports should be combined, at least with respect to providing formal DSM diagnoses for those disorders (which does not, of course, preclude ongoing investigation, e.g. Martel et al., 2015; Shemmassian & Lee, 2015). There is less of a consensus within the empirical literature with respect to the predictive or incremental validity of adult raters vs. child self-report of internalizing experiences. In those cases, there are strong theoretical arguments that both adult and child reports provide unique information on child psychological adjustment and functioning (Klein, Dougherty, & Olino, 2005; Silverman & Ollendick, 2005).

Setting aside the question of how information from multiple informants should ultimately be combined, there is also the question of how data on psychopathology can be collected most efficiently. Age-normed standardized behavioral rating scales and structured or semistructured clinical interviews are the two most common methods of obtaining information on psychiatric symptomology. There are benefits and drawbacks to both. Rating scales provide a brief and inexpensive method to screen for a variety of potential risk factors or conditions. They can be administered and completed online at the time and location of the respondents' choosing and are commonly scored using age-based norms. However, questions and response options can be easily misinterpreted (Hinshaw & Nigg, 1999; Schwarz, 1999), subscales may show clinical elevations even when core symptoms are absent (e.g., elevated depression scores without a persistent depressed mood), and response biases are often present (i.e., halo effects, presentation bias, leniency or severity, central tendency or range restrictions; Abikoff et al., 1993; Hartung et al., 2006; Martin, Hooper, & Snow, 1986; Schachar, Sandberg, & Rutter, 1986). All of this reduces the accuracy of measurement.

In contrast, structured interviewing allows for the systematic collection of diagnostic information and the opportunity to improve accuracy by clarifying participants' responses. However, clinical interviews lack normative data, are time-intensive, and valid administration/scoring requires intensive training, a process that is time consuming and subsequently increases the cost of assessment (Basco et al., 2000; Jensen-Doss & Hawley, 2010). For participants, attending in-person assessments often requires overcoming major obstacles (e.g., missed work, finding childcare), as well as additional monetary compensation for the time committed to participation.

To address the individual strengths and weaknesses of each of these approaches, both rating scales and

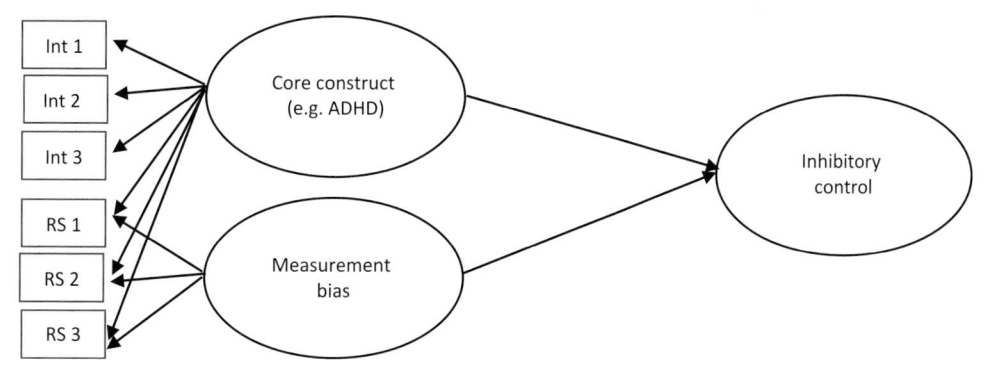

Figure 10.1 Statistical model of the planned missingness approach. Int = interview, RS = rating scale

interviews are commonly collected and reported within a single study to validate target and control groups. However, there is relatively little empirical evidence to suggest substantive gains in diagnostic validity when information is obtained from a single informant, given that correlation between methods is frequently high (Brunshaw & Szatmari, 1988; Conners, Erhardt, & Sparrow, 1999; Doyle et al., 1997; Levin-Aspenson & Watson, 2018; Longwell & Truax, 2005; Magnússon et al., 2006; Martin et al., 2006; Pelham, Fabiano, & Massetti, 2005; Reynolds & Kobak, 1995; Richman et al., 1999; Vaughn & Hoza, 2013). Thus, the degree to which each measure adds unique predictive value to clinical assessment is not clear. It makes intuitive sense that briefer and less expensive tests should be used for clinical assessment if they perform as well as longer and more expensive ones (Hunsley & Meyer, 2003; Johnston & Murray, 2003). Optimal symptom assessment procedure would therefore combine the validation of clinical severity provided by diagnostic interviews with the speed and cost efficiency of clinical rating scales.

Recently, two-method measurement (TMM), a planned missingness research design, has been proposed to improve data-collection efficiency in the context of a scientific research study (Graham, 2012; Graham et al., 2006; Shapiro et al., in press). This design takes advantage of situations in which tradeoffs exist between the cost and validity of two different measurement tools of the same construct. In this design, the expensive measure is purposefully not collected from a randomly selected subset of participants, and the savings are allocated toward collecting the less expensive measure from an additional number of participants who could not otherwise be recruited due to limited funds. Statistical power increases as the overall N increases, and the expensive measure is used to correct for the measurement error from the less expensive measure. Alternatively, this design can also be used to decrease costs while keeping power constant.

Using structural equation modeling, a latent variable representing the construct of interest is created, composed of both the expensive/valid as well as less expensive measure (see Figure 10.1). A second latent variable comprised of the shared variance of the less expensive measurement is also modeled, which represents unwanted measurement bias. The potential utility of this design has been demonstrated for diagnoses of adult attention deficit hyperactivity disorder (Shapiro et al., in press), studies of heart disease (Zawadzki, Graham, & Gerin, 2012), and could also be applied to a wide range of research questions including smoking behavior, classroom behavior, and cognitive ability (Garnier-Villarreal, Rhemtulla, & Little, 2014; Graham et al., 2006; Little & Rhemtulla, 2013; Rhemtulla & Little, 2012). In the context of assessment of psychopathology, the more expensive measure would be structured clinical interviewing, and the less expensive measure would be responses on behavioral rating scales. In the context of cognitive measurement, performance data collected in a laboratory setting would be the more expensive/valid measure (because compliance to task and environmental context could be monitored), with performance on a mobile app as the less expensive/less valid measure.

Importantly, in a TMM design, assignment to the complete case or planned missingness group must be determined at random or using a predetermined algorithm, and unlike computerized adaptive testing (Gibbons et al., 2016), is not applicable to the assessment of individuals. Because TMM is specific to Structural Equation Modelling (SEM), it is most appropriate for descriptive designs (i.e. Type IV designs, described earlier).

ASSESSMENT OF COGNITION

Cognitive psychopathology has been at the forefront of a movement to shift the assessment of psychopathology away from behavioral manifestation toward a more mechanism-based nosology. This movement has been spurred in large part by decades of research finding cognitive atypicalities in various psychiatric conditions, including (but not limited to) executive function deficits, attentional biases to threat, verbal deficits, and impaired theory of mind (e.g., Dudeney, Sharpe, & Hunt, 2015; Klin,

2000; Moffitt, 1993; Willcutt et al., 2005; Yirmiya et al., 1998).

Because of its dominance within the cognitive psychopathology tradition, we focus our discussion in this section on modern approaches to conceptualizing, evaluating, and improving the measurement of executive function. Executive dysfunction has been identified in a range of mental health disorders including autism, ADHD, schizophrenia, and bipolar disorder (Barkley, 1997; Hill, 2004; Nieuwenstein, Aleman, & de Haan, 2001; Quraishi & Frangou, 2002) and is arguably one of the most promising potential transdiagnostic features of psychopathology (Huang-Pollock et al., 2017b; McTeague et al., 2017; Nigg et al., 2017; Nolen-Hoeksema & Watkins, 2011; Pennington & Ozonoff, 1996; White et al., 2017). However, despite its promise, the empirical evidence that executive dysfunction might be an endophenotype per se remains relatively weak. Working memory and other aspects of cognitive control have so far not shown themselves to have a simpler or more direct series of genetic associations than behavioral symptomology (Bilder et al., 2011), and the evidence that such deficits are present in unaffected family members remains tenuous (Nigg et al., 2004; Seidman et al., 2000). Recent but as yet unreplicated efforts in ADHD have found that the relationship between polygenetic risk scores and ADHD is mediated by aspects of EF (Nigg et al., 2018), but some unresolved challenges many need to be addressed first before the status of EF in psychopathology is formally clarified.

First, EF continues to refer to a frustratingly broad construct (Halperin, 2016), in part due to the manner in which it has historically been measured. Traditional neuropsychological tests of EF were originally designed to detect frank brain injury (Golden et al., 1981; Vega & Parsons, 1967). To that end, they tend to tap multiple component processes (visual search, fine motor control, receptive/expressive language, etc.) so that difficulty at any step of the flow of information processing as a result of brain injury could be detected. This of course undermines the cognitive specificity of an EF assessment. No task is process-pure (Jacoby, 1991), but the adoption of empirically supported models and paradigms of cognitive control from the cognitive sciences is essential if psychopathologists are to move closer to specifying the cognitive processes, supporting neural networks, and areas of localization involved in disorder.

That being said, the most frequently used indices of performance on the vast majority of even the most well-validated measures (e.g., go-no-go, continuous performance tasks, working memory span tasks, flanker tasks, or Stroop) use mean reaction time (RT) or mean accuracy as the primary outcome variable. This reliance on mean RT or mean accuracy is problematic for three reasons. First, as two descriptors of a single response, RT and accuracy are produced simultaneously and are not independent. However, based on history or tradition, they are often analyzed separately, even if this leads to important differences in interpretation. For example, a meta-analysis of the continuous performance test (CPT, a measure of sustained attention) among children with ADHD found no reliable evidence of impaired performance when RT was the dependent measure, but very large effect sizes when accuracy was used (Huang-Pollock et al., 2012). This paradox is hardly limited to the CPT. A recent empirical study found that children with ADHD performed more poorly on four out of five EF tests (Choice RT, Attention Network Test, go-no-go, stop signal RT task, and N-back) when accuracy was used as a dependent variable, one out of five when RT was used, and five out of five when standard deviation of RT was used (Epstein et al., 2011). With few exceptions (see Halperin et al., 1991; Huang-Pollock, Nigg, & Halperin, 2006), RT and accuracy are generally not incorporated into a single variable to describe performance (e.g., slow hits as evidence of inattention, not just omission errors). Furthermore, how should rank order (i.e., individual differences) be determined given two different individuals, one who responds more accurately but slowly, and another who responds more quickly but less accurately? Considering RT and accuracy separately provides at best an incomplete understanding of performance, and at worst erroneous interpretations of data.

Second, RTs are not normally distributed. Rather, they are positively skewed, bound by the fastest possible response of zero milliseconds, and a potentially infinite number of slow responses. Therefore, the most accurate description of RTs is not given by an index of central tendency (i.e., mean, standard deviation). Performance would be better described by the parameters associated with an ex-Gaussian distribution, which combines an exponential distribution with a normal distribution (Dawson, 1988). Parameters of ex-Gaussian distributions include: mu (μ) and sigma (σ), which represent the mean and standard deviation of the normal portion of the distribution; and tau (τ), which characterizes the exponential portion of the distribution (Lacouture & Cousineau, 2008).

Third, RT is multiply influenced. It is of course influenced by the efficiency with which information is accumulated to make a decision, which is typically the construct of interest (e.g., "do I go or not go?," "is this a threat or non-threat?"). But it is also influenced by the amount of time needed to encode a stimulus; to prepare and execute a motor response; and whether one tends (or has been instructed) to emphasize speed over accuracy, or vice versa. When individual differences are detected in mean RT, are they due to the core construct of interest, or to difficulties in associated processes? Rather than invoking a homuncular "executive function" or "cognitive control" construct, some have even argued that it would be better to reframe the control construct with these more basic, easily measured, and (as we describe below) computationally defined processes (Verbruggen, McLaren, & Chambers, 2014).

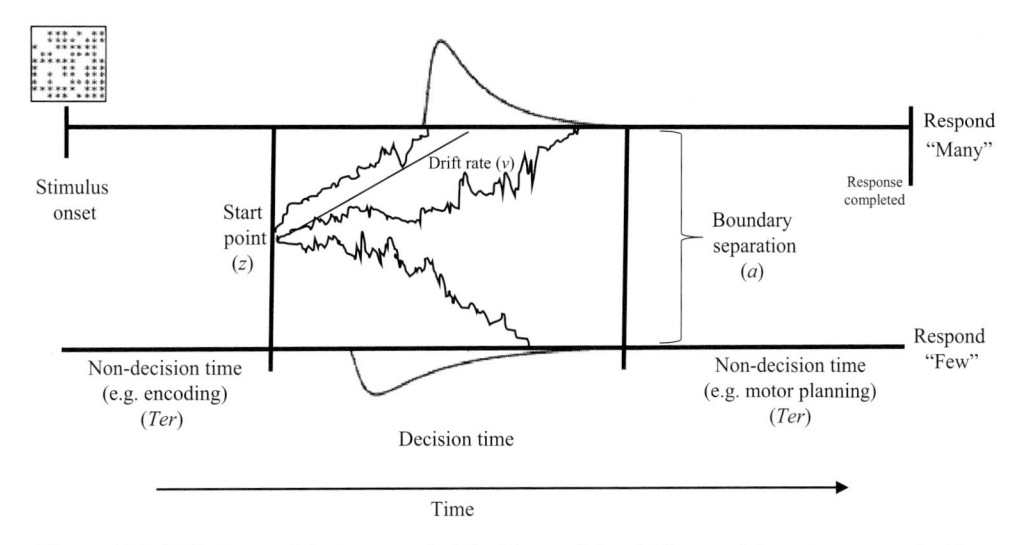

Figure 10.2 Diffusion model of a numerical decision task in which a participant is presented with a 10×10 array of asterisks and asked to judge whether there are "many" or "few" stars. An individual's response time (RT) is influenced by the time it takes to encode a stimulus; the amount of time it takes for information to accumulate toward one or the other choice boundary (e.g., "Many" or "Few"); the amount of time needed to prepare and execute a response; and the degree to which speed is emphasized over accuracy (i.e., boundary separation).

Therefore, an empirically supported method is needed that can (a) incorporate both error rate and RT into a single set of performance indicators, (b) capture the shape of the distribution of performance, and (c) better parse information processing into its component parts. A well-known, well-validated, and reliable computational method called diffusion modeling (DM) presents a solution (Ratcliff & McKoon, 2008; Voss, Nagler, & Lerche, 2013). Commonly used in cognitive research in college-aged adults, the DM has also been validated in typically developing children (Cohen-Gilbert et al., 2014; Ratcliff et al., 2012), healthy aging (Ratcliff, Thapar, & McKoon, 2001, 2006), and clinical populations, most notably childhood ADHD (Huang-Pollock et al., 2017a, 2012; Karalunas & Huang-Pollock, 2013; Karalunas, Huang-Pollock, & Nigg, 2012; Mulder et al., 2010; Salum et al., 2014; Weigard & Huang-Pollock, 2014, 2017) and anxiety disorders (Cohen-Gilbert et al., 2014; White et al., 2010, 2015).

The DM parameters used to describe performance are based on the *shape* of the RT distributions for correct *and* incorrect responses. The primary model parameters are drift rate, or v, which is an index of the rate at which information is accumulated to make a decision; boundary separation, or a, which represents the degree of speed-accuracy tradeoff; start point, or z, which represents response bias toward either of the response options; and "non-decision time," or Ter, which represents stimulus encoding and motor preparation time (Figure 10.2).

By making full use of the data, these parameters provide insight into how different cognitive mechanisms might combine to yield distinct patterns of performance that are otherwise obscured by a molar RT. For example, the DM has been used to show that slow RTs in healthy aging are not due to a slowing of drift rate, as might be assumed, but are instead due to slower nondecision times (i.e., motor preparation) and to wider boundaries (i.e., valuing accuracy over speed) (Ratcliff et al., 2001). Our own work has shown that slow/variable RTs in childhood ADHD can be attributed to slow drift rate (Karalunas et al., 2014; Weigard et al., 2018), and that slow drift rate is also responsible for impairments in working memory, motor disinhibition, sustained attention, and error monitoring in affected children (Huang-Pollock et al., 2012; 2017a; Karalunas & Huang-Pollock, 2013; Weigard & Huang-Pollock, 2017; Weigard, Huang-Pollock, & Brown, 2016). It has also allowed us to test competing theories of ADHD that assume deficits in different processes, but predict similar accuracy rates and response-time distributions (Weigard et al., 2018).

The diffusion model was developed to model simple two-alternative forced-choice perceptual decisions. More recent iterations have validated it in reinforcement learning as well as classic inhibitory and interference-control tasks (Logan et al., 2014; Matzke et al., 2013; Ratcliff & Frank, 2012; White, Ratcliff, & Starns, 2011). And, the related linear ballistic accumulator model can be applied to multiple-choice tasks (Brown & Heathcote, 2008), or even tasks with continuous outcomes (Smith, 2016). But of course, even with these innovations, not all cognitive paradigms or questions can be designed to meet the requirements or assumptions of the diffusion or other formal RT models. In those cases, it remains incumbent upon clinical scientists to consider the accuracy and completeness of the performance outcome variables, ensuring that they are comprehensive enough to fully capture and describe performance.

We end this section with a reminder that selecting an experimental measure or model of cognition needs to carefully consider both its construct validity and its relevance to the real-world behaviors of interest. The ultimate benefit and impact of a study's findings to the field of cognitive psychopathology is the extent to which that measure taps into latent cognitive processes, and the degree to which it contributes to a given symptom or form of psychopathology.

PRAGMATIC CONCERNS

Finally, we turn to pragmatic considerations in data collection. Working with children who have emotional or behavioral concerns (and other special populations, including adults with dementia, intellectual, or other developmental disabilities) requires an additional level of study planning and awareness. We note first that children are not adults, and that the cognitive paradigms that have been developed for adults are often not suitable for children. Use of age- and grade-appropriate language and instructions, including additional example or practice rounds with physical or onscreen visual aids, are often necessary. Gamification of common paradigms may be necessary to address issues of motivation and compliance that are less common when working with adults (Lumsden et al., 2016).

In our laboratory, we have developed an extensive training program for potential research assistants. During this process, RAs are trained on and practice how to handle difficult situations such as: when a child is non-compliant, grabs at materials, is talking off-topic, attempts to gain feedback on performance, is rolling on the floor, attempts to bolt out of (or succeeds in leaving) the room, or starts to cry. The majority of behavioral concerns that arise during testing are mild, can be caught before escalation, and are easily handled/deescalated with standard behavioral management techniques. Planned snack/play/wiggle breaks and/or brief consultation or reassurance from the attending parent (who is required to be onsite at all times) can also be helpful.

Some of the more effective behavioral techniques include *specific* praise (for effort, for looking at all the choices, for continuing to work when the task is hard/boring); statements that convey warmth and approachability (talking about hobbies, music, empathic statements validating the difficulty of the task, or telling jokes prior to or between tasks); minor token or marking systems to manage expectations (marking the number of tasks to be completed, and choosing stickers to check off each task); and the opportunity to pick a small "prize" at the end. Of course, careful consideration about the frequency, form, and situations in which behavioral management should be applied without violating the intent or internal validity of the study are needed on a study-by-study basis. However, without adequate rapport and active assent, any data produced is likely to be invalid.

Implementing these behavioral strategies in real time while simultaneously maintaining standardized procedures and rapport is not trivial. To help manage these competing requirements, the testing room and materials are both organized and arranged to ensure efficiency of process as well as child safety (including positioning the RA and testing table between the child and door). If materials (e.g. blocks, paper, pencil, keyboard, response box) are no longer needed for a new task, they should be removed from the child's reach until such time as they are needed again. Prior to being allowed to work directly with participants, new RAs are required to pass a standardized "test out" protocol in which an experienced RA role-plays a potential child participant, beginning with the introduction and assenting process, and ending after the entire test battery has been run. This allows us to evaluate the new RA's comfort and ability to gain/maintain rapport; their real-time command of the experimental protocol; and the basal, reverse, ceiling, and query rules.

Paying particular attention to providing RAs with extensive and ongoing training protects the validity of data, decreases the potential for missing data, and helps ensure positive participant experience. To protect against procedural drift, all testing visits and interviews are video recorded to allow for review and supervision at weekly laboratory meetings. Our particular program of research does not require behavioral or facial coding, so to protect privacy, recordings are taken from behind the child and focus on the RA and the testing materials. Some have recently argued that recordings of experimental procedures should be made public to improve and accelerate the progress in reproducibility science (Gilmore & Adolph, 2017; Gilmore et al., 2017). Databrary (databrary.org) is a digital library designed to store and share video for research purposes, and has an explicit policy framework to protect and share video with participant consent to address privacy risks.

SUMMARY

Cognitive psychopathology seeks to identify the cognitive causes, correlates, and outcomes of clinical disorders. It sits at the intersection of multiple broad fields of inquiry, and has been central to recent attempts to inform an improved taxonomy of mental health disorders. However, the quasi-experimental nature of this form of research poses numerous challenges to study design. Care must first be taken to identify the appropriate sampling frame for targets and controls, and in deciding how to most validly and efficiently measure psychopathology. In turn, the internal validity of any experimental manipulation (or of the cognitive construct under investigation) must be maximized, while simultaneously recognizing that no task is process pure and the specificity of the adopted paradigms will always be imperfect. That being said, advances in cognitive modeling now allow for a more accurate, nuanced, and complete description of performance than traditional indices of performance, and are poised to further close the gap in knowledge of the links that exist between neurobiology and behavior. Combined with increased attention to data-collection pragmatics and reproducibility, addressing these design challenges will help ensure that the significance of cognitive psychopathological insights to research endure.

REFERENCES

Abikoff, H., Courtney, M., Pelham, W. E., Jr., & Koplewicz, H. S. (1993). Teachers' Ratings of Disruptive Behaviors: The Influence of Halo Effects. *Journal of Abnormal Child Psychology, 21*(5), 519–533.

Achenbach, T. M., McConaughy, S. H., & Howell, C. T. (1987). Child Adolescent Behavioral and Emotional Problems: Implications of Cross-Informant Correlations for Situational Specificity. *Psychological Bulletin, 101*(2), 213–232.

Achenbach, T. M., Krukowski, R. A., Dumenci, L., & Ivanova, M. Y. (2005). Assessment of Adult Psychopathology: Meta-Analyses and Implications of Cross-Informant Correlations. *Psychological Bulletin, 131*(3), 361–382.

Barkley, R. A. (1997). Behavioral Inhibition, Sustained Attention, and Executive Functions: Constructing a Unifying Theory of ADHD. *Psychological Bulletin, 121*(1), 65–94.

Baron-Cohen, S., Leslie, A. M., & Frith, U. (1986). Mechanical, Behavioral, and Intentional Understanding of Picture Stories in Autistic Children. *British Journal of Developmental Psychology, 4*, 113–125.

Basco, M. R., Bostic, J. Q., Davies, D., Rush, A. J., Witte, B., Hendrickse, W., & Barnett, V. (2000). Methods to Improve Diagnostic Accuracy in a Community Mental Health Setting. *American Journal of Psychiatry, 157*(10), 1599–1605.

Bentall, R. P. (1996). At the Centre of a Science of Psychopathology? Characteristics and Limitations of Cognitive Research. *Cognitive Neuropsychiatry, 1*(4), 265–273.

Berkson, J. (1946). Limitations of the Application of Fourfold Table Analysis to Hospital Data. *Biometrics Bulletin, 2*(3), 47–53.

Bilder, R. M., Howe, A., Novak, N., Sabb, F. W., & Parker, D. S. (2011). The Genetics of Cognitive Impairment in Schizophrenia: A Phenomic Perspective. *Trends in Cognitive Sciences, 15*(9), 428–435.

Bird, H. R., Gould, M. S., & Staghezza, B. (1992). Aggregating Data from Multiple Informants in Child Psychiatry Epidemiologic Research. *Journal of the American Academy of Child and Adolescent Psychiatry, 31*(1), 78–85.

Boonstra, A. M., Oosterlaan, J., Sergeant, J. A., & Buitelaar, J. K. (2005). Executive Functioning in Adult ADHD: A Meta-Analytic Review. *Psychological Medicine, 35*(8), 1097–1108.

Brown, S. D., & Heathcote, A. (2008). The Simplest Complete Model of Choice Response Time: Linear Ballistic Accumulation. *Cognitive Psychology, 57*(3), 153–178.

Brunshaw, J. M., & Szatmari, P. (1988). The Agreement between Behaviour Checklists and Structured Psychiatric Interviews for Children. *Canadian Journal of Psychiatry, 33*(6), 474–481.

Caspi, A., Houts, R. M., Belsky, D. W., Goldman-Mellor, S. J., Harrington, H., Israel, S., ... Moffitt, T. E. (2014). The p Factor: One General Psychopathology Factor in the Structure of Psychiatric Disorders? *Clinical Psychological Science, 2*(2), 119–137.

Clay, C., Ellis, M. A., Amodeo, M., Fassler, I., & Griffin, M. L. (2003). Recruiting a Community Sample of African American Subjects: The Nuts and Bolts of a Successful Effort. *Families in Society – The Journal of Contemporary Human Services, 84*(3), 396–404.

Cohen, P., & Cohen, J. (1984). The Clinician's Illusion. *Archives of General Psychiatry, 41*(12), 1178–1182.

Cohen-Gilbert, J. E., Killgore, W. D. S., White, C. N., Schwab, Z. J., Crowley, D. J., Covell, M. J., ... Silveri, M. M. (2014). Differential Influence of Safe versus Threatening Facial Expressions on Decision-Making during an Inhibitory Control Task in Adolescence and Adulthood. *Developmental Science, 17*(2), 212–223.

Conners, C. K., Erhardt, D., & Sparrow, E. P. (1999). *Conners' Adult ADHD Rating Scales (CAARS): Technical Manual*. Toronto: Multi-Health Systems.

Conway, A., Kane, M. J., Bunting, M. F., Hambrick, D. Z., Wilhelm, O., & Engle, R. W. (2005). Working Memory Span Tasks: A Methodological Review and User's Guide. *Psychonomic Bulletin & Review, 12*(5), 769–786.

Corbie-Smith, G., Thomas, S. B., & St. George, D. M. M. (2002). Distrust, Race, and Research. *Archives of Internal Medicine, 162*(21), 2458–2463.

Costello, E. J., Mustillo, S., Erkanli, A., Keeler, G., & Angold, A. (2003). Prevalence and Development of Psychiatric Disorders in Childhood and Adolescence. *Archives of General Psychiatry, 60*(8), 837–844.

Dantas, O. M., Ximenes, R. A., de Albuquerque, M. d. F. P., Montarroyos, U. R., de Souza, W. V., Varejão, P., & Rodrigues, L. C. (2007). Selection Bias: Neighbourhood Controls and Controls Selected from Those Presenting to a Health Unit in a Case Control Study of Efficacy of BCG Revaccination. *BMC Medical Research Methodology, 7*(1), 11.

Dawson, M. R. W. (1988). Fitting the Ex-Gaussian Equation to Reaction Time Distributions. *Behavior Research Methods Instruments & Computers, 20*(1), 54–57.

De Los Reyes, A., Thomas, S. A., Goodman, K. L., & Kundey, S. M. A. (2013). Principles Underlying the Use of Multiple Informants' Reports. *Annual Review of Clinical Psychology, 9*(9), 123–149.

De Los Reyes, A., Augenstein, T. M., Wang, M., Thomas, S. A., Drabick, D. A. G., Burgers, D. E., & Rabinowitz, J. (2015). The Validity of the Multi-Informant Approach to Assessing Child and Adolescent Mental Health. *Psychological Bulletin, 141*(4), 858–900.

Deary, I. J., Strand, S., Smith, P., & Fernandes, C. (2007). Intelligence and Educational Achievement. *Intelligence, 35*(1), 13–21.

Doyle, A., Ostrander, R., Skare, S., Crosby, R. D., & August, G. J. (1997). Convergent and Criterion-Related Validity of the Behavior Assessment System for Children-Parent Rating Scale. *Journal of Clinical Child Psychology, 26*(3), 276–284.

Dudeney, J., Sharpe, L., & Hunt, C. (2015). Attentional Bias towards Threatening Stimuli in Children with Anxiety: A Meta-Analysis. *Clinical Psychology Review, 40*, 66–75.

Engle, R. W., Tuholski, S. W., Laughlin, J. E., & Conway, A. (1999). Working Memory, Short-Term Memory, and General Fluid Intelligence: A Latent-Variable Approach. *Journal of Experimental Psychology-General, 128*(3), 309–331.

Epstein, J. N., Langberg, J. M., Rosen, P. J., Graham, A., Narad, M. E., Antonini, T. N., ... Altaye, M. (2011). Evidence for Higher Reaction Time Variability for Children with ADHD on a Range of Cognitive Tasks including Reward and Event Rate Manipulations. *Neuropsychology, 25*(4), 427–441.

Farmer, D. F., Jackson, S. A., Camacho, F., & Hall, M. A. (2007). Attitudes of African American and Low Socioeconomic Status White Women toward Medical Research. *Journal of Health Care for the Poor and Underserved, 18*(1), 85–99.

Fombonne, E. (2009). Epidemiology of Pervasive Developmental Disorders. *Pediatric Research, 65*(6), 591–598.

Garnier-Villarreal, M., Rhemtulla, M., & Little, T. D. (2014). Two-Method Planned Missing Designs for Longitudinal Research. *International Journal of Behavioral Development, 38*(5), 411–422.

Geary, D. C. (2011). Cognitive Predictors of Achievement Growth in Mathematics: A 5-Year Longitudinal Study. *Developmental Psychology, 47*(6), 1539–1552.

Gibbons, R. D., Weiss, D. J., Frank, E., & Kupfer, D. (2016). Computerized Adaptive Diagnosis and Testing of Mental Health Disorders. *Annual Review of Clinical Psychology, 12*(1), 83–104.

Gilmore, R. O., & Adolph, K. E. (2017). Video Can Make Science More Open, Transparent, Robust, and Reproducible. Retrieved from https://psyarxiv.com/tcfqf/

Gilmore, R. O., Diaz, M. T., Wyble, B. A., & Yarkoni, T. (2017). Progress toward Openness, Transparency, and Reproducibility in Cognitive Neuroscience. *Annals of the New York Academy of Sciences, 1396*(1), 5–18.

Golden, C. J., Fishburne, F. J., Lewis, G. P., Conley, F. K., Moses, J. A., Engum, E., … Graber, B. (1981). Cross-Validation of the Luria-Nebraska Neuropsychological Battery for the Presence, Lateralization, and Localization of Brain-Damage. *Journal of Consulting and Clinical Psychology, 49*(4), 491–507.

Goodman, S. H., Lahey, B. B., Fielding, B., Dulcan, M., Narrow, W., & Regier, D. (1997). Representativeness of Clinical Samples of Youths with Mental Disorders: A Preliminary Population-Based Study. *Journal of Abnormal Psychology, 106*(1), 3–14.

Gottesman, I. I., & Gould, T. D. (2003). The Endophenotype Concept in Psychiatry: Etymology and Strategic Intentions. *American Journal of Psychiatry, 160*(4), 636–645.

Graham, J. W. (2012). *Missing Data: Analysis and Design*. New York: Springer.

Graham, J. W., Taylor, B. J., Olchowski, A. E., & Cumsille, P. E. (2006). Planned Missing Data Designs in Psychological Research. *Psychological Methods, 11*(4), 323–343.

Grimes, D. A., & Schulz, K. F. (2005). Compared to What? Finding Controls for Case-Control Studies. *Lancet, 365*(9468), 1429–1433.

Grobbee, D., & Hoes, A. (2015). *Clinical Epidemiology: Principles, Methods, and Applications for Clinical Research*. Burlington, MA: Jones and Bartlett Learning, LLC.

Halperin, J. M. (2016). Executive Functioning – A Key Construct for Understanding Developmental Psychopathology or a "Catch-All" Term in Need of Some Rethinking? *Journal of Child Psychology and Psychiatry, 57*(4), 443–445.

Halperin, J. M., Wolf, L., Greenblatt, E. R., & Young, G. (1991). Subtype Analysis of Commission Errors on the Continuous Performance Test in Children. *Developmental Neuropsychology, 7*(2), 207–217.

Hart, E. L., Lahey, B. B., Loeber, R., & Hanson, K. S. (1994). Criterion Validity of Informants in the Diagnosis of Disruptive Behavior Disorders in Children: A Preliminary Study. *Journal of Consulting and Clinical Psychology, 62*(2), 410–414.

Hartung, C. M., Van Pelt, J. C., Armendariz, M. L., & Knight, L. A. (2006). Biases in Ratings of Disruptive Behavior in Children: Effects of Sex and Negative Halos. *Journal of Attention Disorders, 9*(4), 620–630.

Haywood, H. C., & Raffard, S. (2017). Cognition and Psychopathology: Overview. *Journal of Cognitive Education and Psychology, 16*(1), 3–8.

Hill, E. L. (2004). Executive Dysfunction in Autism. *Trends in Cognitive Sciences, 8*(1), 26–32.

Hinshaw, S. P., & Nigg, J. T. (1999). Behavior Rating Scales in the Assessment of Disruptive Behavior Problems in childhood. In D. Shaffer, C. P. Lucas, & J. E. Richters (Eds.), *Diagnostic Assessment in Child and Adolescent Psychopathology* (pp. 91–126). New York: Guilford Press.

Hohmann, A. A., & Parron, D. L. (1996). How the New NIH Guidelines on Inclusion of Women and Minorities Apply: Efficacy Trials, Effectiveness Trials, and Validity. *Journal of Consulting and Clinical Psychology, 64*(5), 851–855.

Huang-Pollock, C., & Nigg, J. T. (2003). Searching for the Attention Deficit in Attention Deficit Hyperactivity Disorder: The Case of Visuospatial Orienting. *Clinical Psychology Review, 23*(6), 801–830.

Huang-Pollock, C., Nigg, J. T., & Halperin, J. M. (2006). Single Dissociation Findings of ADHD Deficits in Vigilance but not Anterior or Posterior Attention Systems. *Neuropsychology, 20*(4), 420–429.

Huang-Pollock, C. L., Karalunas, S. L., Tam, H., & Moore, A. N. (2012). Evaluating Vigilance Deficits in ADHD: A Meta-Analysis of CPT Performance. *Journal of Abnormal Psychology, 121*(2), 360–371.

Huang-Pollock, C., Ratcliff, R., McKoon, G., Shapiro, Z., Weigard, A., & Galloway-Long, H. (2017a). Using the Diffusion Model to Explain Cognitive Deficits in Attention Deficit Hyperactivity Disorder. *Journal of Abnormal Child Psychology, 45*(1), 57–68.

Huang-Pollock, C., Shapiro, Z., Galloway-Long, H., & Weigard, A. (2017b). Is Poor Working Memory a Transdiagnostic Risk Factor for Psychopathology? *Journal of Abnormal Child Psychology, 45*(8), 1477–1490.

Hunsley, J., & Meyer, G. J. (2003). The Incremental Validity of Psychological Testing and Assessment: Conceptual, Methodological, and Statistical Issues. *Psychological Assessment, 15*(4), 446–455.

Ibrahim, M. A., & Spitzer, W. O. (1979). Case Control Study: Problem and the Prospect. *Journal of Chronic Diseases, 32*(1–2), 139–144.

Jacoby, L. L. (1991). A Process Dissociation Framework – Separating Automatic from Intentional Uses of Memory. *Journal of Memory and Language, 30*(5), 513–541.

Jensen-Doss, A., & Hawley, K. M. (2010). Understanding Barriers to Evidence-Based Assessment: Clinician Attitudes toward Standardized Assessment Tools. *Journal of Clinical Child and Adolescent Psychology, 39*(6), 885–896.

Jensen, P. S., Rubio-Stipec, M., Canino, G., Bird, H. R., Dulcan, M. K., Schwab-Stone, M. E., & Lahey, B. B. (1999). Parent and Child Contributions to Diagnosis of Mental Disorder: Are Both Informants Always Necessary? *Journal of the American Academy of Child and Adolescent Psychiatry, 38*(12), 1569–1579.

Johnston, C., & Murray, C. (2003). Incremental Validity in the Psychological Assessment of Children and Adolescents. *Psychological Assessment, 15*(4), 496–507.

Karalunas, S. L., & Huang-Pollock, C. L. (2013). Integrating Impairments in Reaction Time and Executive Function Using a Diffusion Model Framework. *Journal of Abnormal Child Psychology, 41*(5), 837–850.

Karalunas, S. L., Huang-Pollock, C. L., & Nigg, J. T. (2012). Decomposing Attention-Deficit/Hyperactivity Disorder (ADHD)-Related

Effects in Response Speed and Variability. *Neuropsychology, 26*(6), 684–694.

Karalunas, S. L., Geurts, H. M., Konrad, K., Bender, S., & Nigg, J. T. (2014). Annual Research Review: Reaction Time Variability in ADHD and Autism Spectrum Disorders: Measurement and Mechanisms of a Proposed Trans-Diagnostic Phenotype. *Journal of Child Psychology and Psychiatry, 55*(6), 685–710.

Klein, D. N., Dougherty, L. R., & Olino, T. M. (2005). Toward Guidelines for Evidence-Based Assessment of Depression in Children and Adolescents. *Journal of Clinical Child & Adolescent Psychology, 34*(3), 412–432.

Klin, A. (2000). Attributing Social Meaning to Ambiguous Visual Stimuli in Higher-Functioning Autism and Asperger Syndrome: The Social Attribution Task. *Journal of Child Psychology and Psychiatry and Allied Disciplines, 41*(7), 831–846.

Kopec, J. A., & Esdaile, J. M. (1990). Bias in Case Control Studies, A Review. *Journal of Epidemiology and Community Health, 44*(3), 179–186.

Kozak, M. J., & Cuthbert, B. N. (2016). The NIMH Research Domain Criteria Initiative: Background, Issues, and Pragmatics. *Psychophysiology, 53*(3), 286–297.

Kraemer, H. C., Measelle, J. R., Ablow, J. C., Essex, M. J., Boyce, W. T., & Kupfer, D. J. (2003). A New Approach to Integrating Data from Multiple Informants in Psychiatric Assessment and Research: Mixing and Matching Contexts and Perspectives. *American Journal of Psychiatry, 160*(9), 1566–1577.

Lacouture, Y., & Cousineau, D. (2008). How to Use MATLAB to Fit the Ex-Gaussian and Other Probability Functions to a Distribution of Response Times. *Tutorials in Quantitative Methods for Psychology, 4*(1), 35–45.

Lahey, B. B., Applegate, B., Barkley, R. A., Garfinkel, B., McBurnett, K., Kerdyk, L., ... Shaffer, D. (1994a). DSM-IV Field Trials for Oppositional Defiant Disorder and Conduct Disorder in Children and Adolescents. *American Journal of Psychiatry, 151*(8), 1163–1171.

Lahey, B. B., Applegate, B., McBurnett, K., Biederman, J., Greenhill, L., Hynd, G. W., ... Shaffer, D. (1994b). DSM-IV Field Trials for Attention-Deficit Hyperactivity Disorder in Children and Adolescents. *American Journal of Psychiatry, 151*(11), 1673–1685.

Levin-Aspenson, H. F., & Watson, D. (2018). Mode of Administration Effects in Psychopathology Assessment: Analyses of Gender, Age, and Education Differences in Self-Rated versus Interview-Based Depression. *Psychological Assessment, 30*(3), 287–295.

Lijffijt, M., Kenemans, J. L., Verbaten, M. N., & van Engeland, H. (2005). A Meta-Analytic Review of Stopping Performance in Attention-Deficit/Hyperactivity Disorder: Deficient Inhibitory Motor Control? *Journal of Abnormal Psychology, 114*(2), 216–222.

Little, T. D., & Rhemtulla, M. (2013). Planned Missing Data Designs for Developmental Researchers. *Child Development Perspectives, 7*(4), 199–204.

Logan, G. D., Van Zandt, T., Verbruggen, F., & Wagenmakers, E.-J. (2014). On the Ability to Inhibit Thought and Action: General and Special Theories of an Act of Control. *Psychological Review, 121*(1), 66–95.

Longwell, B. T., & Truax, P. (2005). The Differential Effects of Weekly, Monthly, and Bimonthly Administrations of the Beck Depression Inventory-II: Psychometric Properties and Clinical Implications. *Behavior Therapy, 36*(3), 265–275.

Lumsden, J., Edwards, E. A., Lawrence, N. S., Coyle, D., & Munafo, M. R. (2016). Gamification of Cognitive Assessment and Cognitive Training: A Systematic Review of Applications and Efficacy. *Journal of Medical Internet Research Serious Games, 4*(2), 14.

Lund, E. (1989). The Validity of Different Control Groups in a Case-Control Study: Oral Contraceptive Use and Breast Cancer in Young Women. *Journal of Clinical Epidemiology, 42*(10), 987–993.

Ma, X. M., Buffler, P. A., Layefsky, M., Does, M. B., & Reynolds, P. (2004). Control Selection Strategies in Case-Control Studies of Childhood Diseases. *American Journal of Epidemiology, 159*(10), 915–921.

Magnússon, P., Smári, J., Sigurðardóttir, D., Baldursson, G., Sigmundsson, J., Kristjánsson, K., ... Guðmundsson, Ó. Ó. (2006). Validity of Self-Report and Informant Rating Scales of Adult ADHD Symptoms in Comparison with a Semistructured Diagnostic Interview. *Journal of Attention Disorders, 9*(3), 494–503.

Martel, M. M., Schimmack, U., Nikolas, M., & Nigg, J. T. (2015). Integration of Symptom Ratings from Multiple Informants in ADHD Diagnosis: A Psychometric Model with Clinical Utility. *Psychological Assessment, 27*(3), 1060–1071.

Martin, A., Rief, W., Klaiberg, A., & Braehler, E. (2006). Validity of the Brief Patient Health Questionnaire Mood Scale (PHQ-9) in the General Population. *General Hospital Psychiatry, 28*(1), 71–77.

Martin, R. P., Hooper, S., & Snow, J. (1986). Behavior Rating Scale Approaches to Personality Assessment in Children and Adolescents. In H. M. Knoff (Ed.), *The Assessment of Child and Adolescent Personality* (pp. 309–348). New York: Guilford Press.

Masten, A. S., Hubbard, J. J., Gest, S. D., Tellegen, A., Garmezy, N., & Ramirez, M. (1999). Competence in the Context of Adversity: Pathways to Resilience and Maladaptation from Childhood to Late Adolescence. *Development and Psychopathology, 11*(1), 143–169.

Matzke, D., Love, J., Wiecki, T. V., Brown, S. D., Logan, G. D., & Wagenmakers, E. J. (2013). Release the BEESTS: Bayesian Estimation of Ex-Gaussian Stop Signal Reaction Time Distributions. *Frontiers in Psychology, 4*, 918.

McTeague, L. M., Huemer, J., Carreon, D. M., Jiang, Y., Eickhoff, S. B., & Etkin, A. (2017). Identification of Common Neural Circuit Disruptions in Cognitive Control across Psychiatric Disorders. *American Journal of Psychiatry, 174*(7), 676–685.

Moffitt, T. E. (1993). Adolescence-Limited and Life-Course Persistent Antisocial Behavior: A Developmental Taxonomy. *Psychological Review, 100*(4), 674–701.

Mulder, M. J., Bos, D., Weusten, J. M. H., van Belle, J., van Dijk, S. C., Simen, P., ... Durston, S. (2010). Basic Impairments in Regulating the Speed-Accuracy Tradeoff Predict Symptoms of Attention-Deficit/Hyperactivity Disorder. *Biological Psychiatry, 68*(12), 1114–1119.

Nieuwenstein, M. R., Aleman, A., & de Haan, E. H. F. (2001). Relationship between Symptom Dimensions and Neurocognitive Functioning in Schizophrenia: A Meta-Analysis of WCST and CPT Studies. *Journal of Psychiatric Research, 35*(2), 119–125.

Nigg, J. T., Blaskey, L. G., Stawicki, J. A., & Sachek, J. (2004). Evaluating the Endophenotype Model of ADHD Neuropsychological Deficit: Results for Parents and Siblings of Children with ADHD Combined and Inattentive Subtypes. *Journal of Abnormal Psychology, 113*(4), 614–625.

Nigg, J. T., Willcutt, E. G., Doyle, A., & Sonuga-Barke, E. J. S. (2005). Causal Heterogeneity in Attention-Deficit/Hyperactivity

Disorder: Do We Need Neuropsychologically Impaired Subtypes? *Biological Psychiatry, 57*(11), 1224–1230.

Nigg, J. T., Jester, J. M., Stavro, G. M., Ip, K. I., Puttler, L. I., & Zucker, R. A. (2017). Specificity of Executive Functioning and Processing Speed Problems in Common Psychopathology. *Neuropsychology, 31*(4), 448–466.

Nigg, J. T., Gustafsson, H. C., Karalunas, S. L., Ryabinin, P., McWeeney, S. K., Faraone, S. V., ... Wilmot, B. (2018). Working Memory and Vigilance as Multivariate Endophenotypes Related to Common Genetic Risk for Attention-Deficit/Hyperactivity Disorder. *Journal of the American Academy of Child and Adolescent Psychiatry, 57*(3), 175–182.

Nolen-Hoeksema, S., & Watkins, E. R. (2011). A Heuristic for Developing Transdiagnostic Models of Psychopathology: Explaining Multifinality and Divergent Trajectories. *Perspectives on Psychological Science, 6*(6), 589–609.

Oosterlaan, J., Logan, G., & Sergeant, J. A. (1998). Response Inhibition in AD/HD, CD, Comorbid AD/HD+CD, Anxious, and Control Children: A Meta-Analysis of Studies with the Stop Task. *Journal of Child Psychology and Psychiatry, 39*(3), 411–425.

Pauli-Pott, U., & Becker, K. (2011). Neuropsychological Basic Deficits in Preschoolers at Risk for ADHD: A Meta-Analysis. *Clinical Psychology Review, 31*(4), 626–637.

Pelham, W. E., Jr., Fabiano, G. A., & Massetti, G. M. (2005). Evidence-Based Assessment of Attention Deficit Hyperactivity Disorder in Children and Adolescents. *Journal of Clinical Child & Adolescent Psychology, 34*(3), 449–476.

Pennington, B. F., & Ozonoff, S. (1996). Executive Functions and Developmental Psychopathology. *Journal of Child Psychology and Psychiatry, 37*(1), 51–87.

Polanczyk, G. V., Salum, G. A., Sugaya, L. S., Caye, A., & Rohde, L. A. (2015). Annual Research Review: A Meta-Analysis of the Worldwide Prevalence of Mental Disorders in Children and Adolescents. *Journal of Child Psychology and Psychiatry, 56*(3), 345–365.

Quraishi, S., & Frangou, S. (2002). Neuropsychology of Bipolar Disorder: A Review. *Journal of Affective Disorders, 72*(3), 209–226.

Ratcliff, R., & Frank, M. J. (2012). Reinforcement-Based Decision Making in Corticostriatal Circuits: Mutual Constraints by Neurocomputational and Diffusion Models. *Neural Computation, 24*(5), 1186–1229.

Ratcliff, R., & McKoon, G. (2008). The Diffusion Decision Model: Theory and Data for Two-Choice Decision Tasks. *Neural Computation, 20*(4), 873–922.

Ratcliff, R., Thapar, A., & McKoon, G. (2001). The Effects of Aging on Reaction Time in a Signal Detection Task. *Psychology and Aging, 16*(2), 323–341.

Ratcliff, R., Thapar, A., & McKoon, G. (2006). Aging, Practice, and Perceptual Tasks: A Diffusion Model Analysis. *Psychology and Aging, 21*(2), 353–371.

Ratcliff, R., Love, J., Thompson, C. A., & Opfer, J. E. (2012). Children Are Not Like Older Adults: A Diffusion Model Analysis of Developmental Changes in Speeded Responses. *Child Development, 83*(1), 367–381.

Reynolds, W. M., & Kobak, K. A. (1995). Reliability and Validity of the Hamilton Depression Inventory: A Paper-and-Pencil Version of the Hamilton Depression Rating Scale Clinical Interview. *Psychological Assessment, 7*(4), 472–483.

Rhemtulla, M., & Little, T. (2012). Tools of the Trade: Planned Missing Data Designs for Research in Cognitive Development. *Journal of Cognition and Development: Official Journal of the Cognitive Development Society, 13*(4), 425–438.

Richman, W. L., Kiesler, S., Weisb, S., & Drasgow, F. (1999). A Meta-Analytic Study of Social Desirability Distortion in Computer-Administered Questionnaires, Traditional Questionnaires, and Interviews. *Journal of Applied Psychology, 84*(5), 754–775.

Rothman, K. (1986). *Modern Epidemiology*. Boston: Little, Brown, and Company.

Salum, G. A., Sergeant, J., Sonuga-Barke, E., Vandekerckhove, J., Gadelha, A., Pan, P. M., ... Rohde, L. A. P. (2014). Specificity of Basic Information Processing and Inhibitory Control in Attention Deficit Hyperactivity Disorder. *Psychological Medicine, 44*(3), 617–631.

Sanislow, C. A., Pine, D. S., Quinn, K. J., Kozak, M. J., Garvey, M. A., Heinssen, R. K., ... Cuthbert, B. N. (2010). Developing Constructs for Psychopathology Research: Research Domain Criteria. *Journal of Abnormal Psychology, 119*(4), 631–639.

Schachar, R., Sandberg, S., & Rutter, M. (1986). Agreement between Teachers' Ratings and Observations of Hyperactivity, Inattentiveness, and Defiance. *Journal of Abnormal Child Psychology, 14*(2), 331–345.

Schmidt, F. (2010). Detecting and Correcting the Lies that Data Tell. *Perspectives on Psychological Science, 5*(3), 233–242.

Schwartz, K., & Verhaeghen, P. (2008). ADHD and Stroop Interference from Age 9 to Age 41 Years: A Meta-Analysis of Developmental Effects. *Psychological Medicine, 38*(11), 1607–1616.

Schwarz, N. (1999). Self-Reports: How the Questions Shape the Answers. *American Psychologist, 54*(2), 93–105.

Seidman, L. J., Biederman, J., Monuteaux, M. C., Weber, W., & Faraone, S. V. (2000). Neuropsychological Functioning in Nonreferred Siblings of Children with Attention Deficit/Hyperactivity Disorder. *Journal of Abnormal Psychology, 109*(2), 252–265.

Shapiro, Z., Huang-Pollock, C. L., Graham, J., & Neely, K. (in press). *Making the Most of It: Application of Planned Missingness Design to Increase the Efficiency of Diagnostic Assessment*.

Shemmassian, S. K., & Lee, S. S. (2015). Predictive Utility of Four Methods of Incorporating Parent and Teacher Symptom Ratings of ADHD for Longitudinal Outcomes. *Journal of Clinical Child & Adolescent Psychology, 45*(2), 1–12.

Silverman, W. K., & Ollendick, T. H. (2005). Evidence-Based Assessment of Anxiety and Its Disorders in Children and Adolescents. *Journal of Clinical Child and Adolescent Psychology, 34*(3), 380–411.

Smith, P. L. (2016). Diffusion Theory of Decision Making in Continuous Report. *Psychological Review, 123*(4), 425–451.

Stavraky, K. M., & Clarke, E. A. (1983). Hospital or Population Controls: An Unanswered Question. *Journal of Chronic Diseases, 36*(4), 301–307.

UyBico, S. J., Pavel, S., & Gross, C. P. (2007). Recruiting Vulnerable Populations into Research: A Systematic Review of Recruitment Interventions. *Journal of General Internal Medicine, 22*(6), 852–863.

Vaughn, A. J., & Hoza, B. (2013). The Incremental Utility of Behavioral Rating Scales and a Structured Diagnostic Interview in the Assessment of Attention-Deficit/Hyperactivity Disorder. *Journal of Emotional and Behavioral Disorders, 21*(4), 227–239.

Vega, A., & Parsons, O. A. (1967). Cross-Validation of Halstead-Reitan Tests for Brain Damage. *Journal of Consulting Psychology, 31*(6), 619–625.

Verbruggen, F., McLaren, I. P. L., & Chambers, C. D. (2014). Banishing the Control Homunculi in Studies of Action Control and Behavior Change. *Perspectives on Psychological Science, 9*(5), 497–524.

Voss, A., Nagler, M., & Lerche, V. (2013). Diffusion Models in Experimental Psychology: A Practical Introduction. *Experimental Psychology, 60*(6), 385–402.

Wacholder, S., Silverman, D. T., McLaughlin, J. K., & Mandel, J. S. (1992a). Selection of Controls in case-Control Studies 2: Types of Controls. *American Journal of Epidemiology, 135*(9), 1029–1041.

Wacholder, S., Silverman, D. T., McLaughlin, J. K., & Mandel, J. S. (1992b). Selection of Controls in Case-Control Studies 3: Design Options. *American Journal of Epidemiology, 135*(9), 1042–1050.

Weigard, A., & Huang-Pollock, C. L. (2014). A Diffusion Modeling Approach to Understanding Contextual Cueing Effects in Children with ADHD. *Journal of Child Psychology and Psychiatry, 55*(12), 1336–1344.

Weigard, A., & Huang-Pollock, C. (2017). The Role of Speed in ADHD-Related Working Memory Deficits. *Clinical Psychological Science, 5*(2), 195–211.

Weigard, A., Huang-Pollock, C., & Brown, S. (2016). Evaluating the Consequences of Impaired Monitoring of Learned Behavior in Attention-Deficit/Hyperactivity Disorder Using a Bayesian Hierarchical Model of Choice Response Time. *Neuropsychology, 30*(4), 502–515.

Weigard, A., Huang-Pollock, C., Brown, S., & Heathcote, A. (2018). Testing Formal Predictions of Neuroscientific Theories of ADHD with a Cognitive Model-Based Approach. *Journal of Abnormal Psychology, 127*(5), 529–539.

White, C. N., Ratcliff, R., & Starns, J. J. (2011). Diffusion Models of the Flanker Task: Discrete versus Gradual Attentional Selection. *Cognitive Psychology, 63*(4), 210–238.

White, C. N., Ratcliff, R., Vasey, M. W., & McKoon, G. (2010). Anxiety Enhances Threat Processing without Competition among Multiple Inputs: A Diffusion Model Analysis. *Emotion, 10*(5), 662–677.

White, C. N., Skokin, K., Carlos, B., & Weaver, A. (2015). Using Decision Models to Decompose Anxiety-Related Bias in Threat Classification. *Emotion,* 16(2), 196–207.

White, L. K., Moore, T. M., Calkins, M. E., Wolf, D. H., Satterthwaite, T. D., Leibenluft, E., . . . Gur, R. E. (2017). An Evaluation of the Specificity of Executive Function Impairment in Developmental Psychopathology. *Journal of the American Academy of Child & Adolescent Psychiatry, 56*(11), 975–982.

Willcutt, E. G., Doyle, A., Nigg, J. T., Faraone, S. V., & Pennington, B. F. (2005). Validity of the Executive Function Theory of Attention-Deficit/Hyperactivity Disorder: A Meta-Analytic Review. *Biological Psychiatry, 57*(11), 1336–1346.

Yeargin-Allsopp, M., Rice, C., Karapurkar, T., Doernberg, N., Boyle, C., & Murphy, C. (2003). Prevalence of Autism in a US Metropolitan Area. *Journal of the American Medical Association, 289*(1), 49–55.

Yirmiya, N., Erel, O., Shaked, M., & Solomonica-Levi, D. (1998). Meta-Analyses Comparing Theory of Mind Abilities of Individuals with Autism, Individuals with Mental Retardation, and Normally Developing Individuals. *Psychological Bulletin, 124*(3), 283–307.

Zawadzki, M. J., Graham, J. W., & Gerin, W. (2012). Increasing the Validity and Efficiency of Blood Pressure Estimates Using Ambulatory and Clinic Measurements and Modern Missing Data Methods. *American Journal of Hypertension, 25*(7), 764–769.

Zvolensky, M., Forsyth, J., & Johnson, K. (2013). Laboratory Methods in Experimental Psychopathology. In J. S. Comer & P. C. Kendall (Eds.), *The Oxford Handbook of Research Strategies for Clinical Psychology*. New York: Oxford University Press.

11 Peripheral Psychophysiology

AMANDA LEVINSON AND GREG HAJCAK

For over a century, peripheral psychophysiology has played an important role in clinical psychology research. Historically, nosologies in clinical psychology have used expert consensus to delineate diagnostic categories (e.g., the *Diagnostic and Statistical Manual of Mental Disorders*, or DSM). These categories have been criticized for being highly heterogeneous within category, having significant overlap across categories, and for their basis in subjective interpretation rather than empirical evidence (Kendler, 2009). Contemporary nosological efforts, such as the Research Domain Criteria (RDoC) initiative, have therefore turned toward dimensional measures with strong reliability and validity (Insel et al., 2010). These efforts have particularly emphasized the integration of biological measures into clinical psychology research for their objectivity and more direct associations with the functioning of key brain systems. Peripheral psychophysiological measures are ideally suited to this mission. Among their many strengths, peripheral psychophysiology measures are objective, inexpensive, and noninvasive. Peripheral psychophysiology measures are also some of the best-characterized biological indices in clinical psychology research, which facilitates their interpretation. This chapter is by no means a comprehensive account of peripheral psychophysiology, a topic which can fill (and has filled) many tomes. Rather, we begin with an orientation to influential theories in peripheral psychophysiology and some broad factors to consider when conducting physiological research. We then discuss a handful of specific measures, including current recommendations for standardized measurement and study design considerations, and at least one clinical application of each measure.

The term "peripheral psychophysiology" refers both to an extensive set of methods and to a distinct field of study with its own complex theoretical models for understanding psychopathology. For the purposes of this chapter, we define peripheral psychophysiology as the use of noninvasive measures to study autonomic nervous system (ANS) functioning as it relates to psychological processes. The ANS is subdivided into three distinct systems: the sympathetic nervous system (SNS), the parasympathetic nervous system (PSNS), and the enteric system (Jänig, 2008). Figure 11.1 presents an overview of the various organ systems that are indexed in peripheral psychophysiology research and their links to the ANS.

The PSNS and SNS are of greatest interest to psychological research. The SNS can be thought of as an accelerator that supports mobilization of energetic resources to respond to acute threat. Complementarily, the PSNS can be thought of as the brakes, inhibiting SNS-mediated activation. Together, the SNS and PSNS often act as opposing forces, interacting in manifold complex ways to optimize physiological response to environmental demands. The accentuated antagonism hypothesis of cardiovascular control (Levy, 1971), for example, suggests that SNS and PSNS activation are tethered such that as activity in one system increases, activity in the other decreases. Peripheral psychophysiological indices are regulated to differing degrees by the SNS and PSNS, and patterns of activation across multiple measures can provide useful information about the mechanisms underlying associated psychological phenomena. For example, skin conductance responses (SCRs) are determined almost entirely by the SNS (Beauchaine, 2001; Fowles, 1980; Zisner & Beauchaine, 2016), whereas respiratory sinus arrhythmia (RSA), when respiration rate is controlled, is determined almost entirely by the PSNS. Used in tandem, these measures may be able to partially dissociate SNS and PSNS correlates of clinical phenomena.

THEORETICAL MODELS

Early models linking physiology to psychopathology drew from the principle of *homeostasis*, which states that healthy systems use feedback loops to maintain equilibrium at set points (Cannon, 1932). For a simple example, the human body generally maintains a temperature of 98.6°F (37°C). When the body is too hot, it sweats to cool down. When it is too cold, it shivers to warm up. Applied to clinical psychology, the homeostatic principle predicts that psychopathology should be linked to either (a)

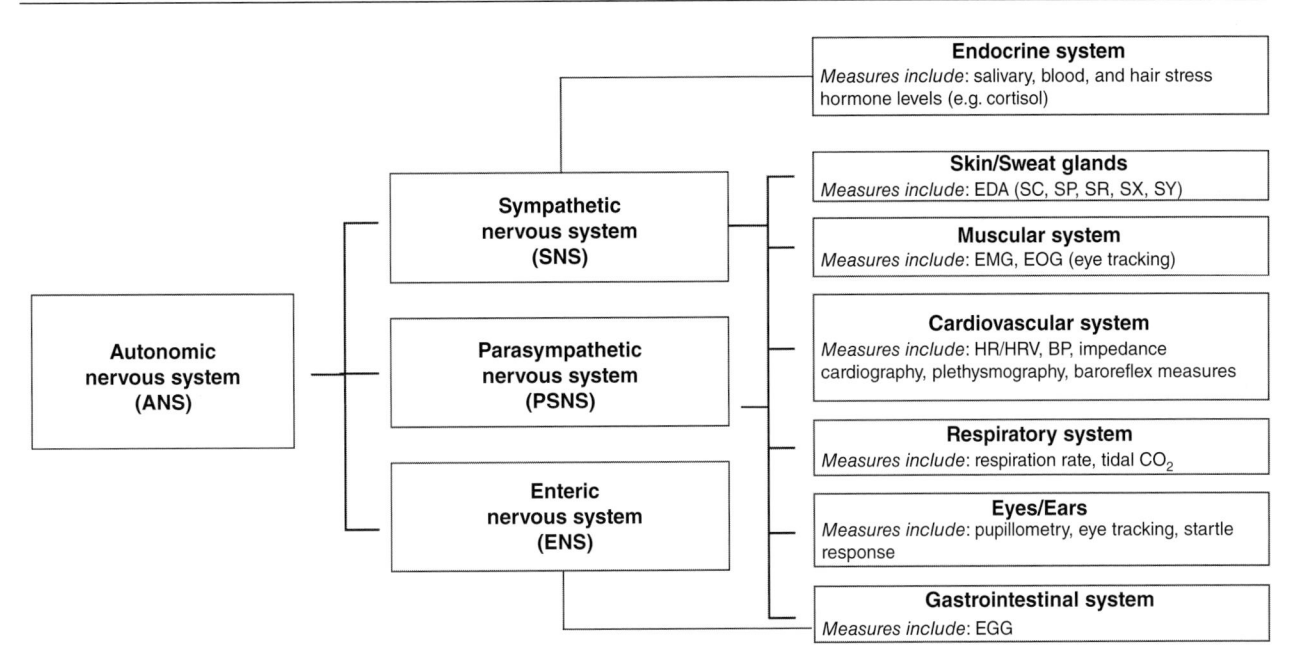

Figure 11.1 Selected overview of the organ systems indexed with peripheral psychophysiological measures and their links to ANA activity

maintenance of a maladaptive set point or (b) excessive lability away from the set point. For example, early models of anxiety physiology suggested that the disproportionate experience of fear could be attributed to tonically elevated SNS activity (Cannon, 1929; Roth et al., 1986) or excessive and dysregulated SNS fluctuations (Friedman, 1945; Stein, Tancer, & Uhde, 1992).

While the principle of homeostasis remains a mainstay in biological research, many indices of physiological health do not comport with the "homeostatic ideal." Rather, a sign of healthy functioning is often *more* physiological variability. Several models have been proposed which collectively describe how adaptive physiological variability can contribute to overall organismic stability. The *heterostatic* model, for instance, describes the stabilization of one physiological variable through strategic fluctuations in at least one other physiological system (Davis, 1958). Returning to the example of body temperature, to maintain the internal temperature of 98.6°F in the face of hot external temperatures, the body upregulates sweat gland activity. The *homeorhetic* model, meaning "similar flow," describes the tendency of natural systems to stabilize along a developmental trajectory (Waddington, 1957). Waddington (1957) first used this framework to describe the canalized pathway that cells follow as they develop from undifferentiated, embryonic cells to specialized cells within the human body. The homeorhetic model has been helpful in epigenetic models to conceptualize how genotypic and phenotypic changes are not one-to-one. Specifically, decades after Waddington's work, we now know that when one allele becomes mutated, the rest of the genome buffers against that mutation causing change in the phenotype (Debat & David, 2001). In psychophysiology, the homeorhetic model would suggest that healthy development of physiological systems should be stable along a typical developmental trajectory. Finally, the *allostatic* model argues that physiological set points are adaptable (Sterling, 2004; Sterling & Eyer, 1988). The three tenets of allostasis are (1) set points are not stable but change with environmental demands; (2) regulation is most efficiently accomplished by a central command center (i.e., central nervous system); and (3) the ANS incorporates past experiences to predict environmental demands and it adapts physiological functioning preemptively (Ramsay & Woods, 2014; Sterling, 2004; Sterling & Eyer, 1988). In sum, these models illustrate the complexity of physiological dynamics, in which numerous systems must coordinate to achieve systemic equilibrium in the face of changing environmental and developmental conditions.

Two influential theories have made significant strides in explaining the psychological link to physiological dynamics: the polyvagal theory (Porges, 1995) and the neurovisceral integration (NVI) model (Thayer & Lane, 2000). While the research supporting these models has primarily examined cardiac outcome variables, each model has implications for all physiological systems. The polyvagal theory concerns the vagus nerve (tenth cranial nerve), which mediates autonomic control of visceral organs including the heart, lungs, and digestive tract (Berthoud & Neuhuber, 2000). "Vagal tone" is a theoretical construct, which is the power of the vagus to regulate the viscera such that greater vagal tone corresponds to better regulated peripheral systems. While studying newborn infants, Porges published a paper suggesting that strong vagal tone was a marker of resilience to stress (Porges, 1992). He then received a letter from a physician describing a counterexample: a potentially fatal condition called fetal bradycardia which can result from excessive vagus activation. Porges called these conflicting findings the "vagal paradox" (i.e., "How can the vagus be both good and bad?"), and his study of this paradox culminated in the polyvagal theory (Porges, 1995).

The polyvagal theory states that the vagus, previously thought to convey one uniform signal, actually contains two different types of efferent fibers: the "smart vagus" and the "vegetative vagus" (Porges, 1995). The smart vagus is responsible for mediating the adaptive responses that Porges had reported. The smart vagus is phylogenetically newer, originates at the nucleus ambiguous (NA), and conveys regulatory signals to two systems: (1) the social communication system and (2) the defensive mobilization system (fight or flight). Under most circumstances, it is optimal for one of these two systems to be dominant, and strong vagal tone allows the individual to shift fluidly between these two systems as the environment demands. When the environment is safe, the vagus shifts the organism into "social communication" mode, dampening the SNS responses and allowing the individual to relax and engage with the people around them. When a threat is present, the vagus shifts into "defensive mobilization" mode, increasing SNS activity and preparing the individual to fight the danger back or flee to safety. Poor functioning of the smart vagus may lead to difficulty shifting between social communication and fight or flight, which in turn can result in disorders of emotion dysregulation. The vegetative vagus is phylogenetically older, originates at the dorsal motor nucleus (DMNX) of the vagus, and mediates involuntary immobilization (freezing) behaviors. The freezing system comes online as a last resort, when the life of the organism is in danger, and there is no chance for fighting or fleeing. While this system should be activated infrequently, exposure to trauma or chronic stress can lead to an overactive freezing system.

Beauchaine (2001) proposed a model integrating the polyvagal theory with Gray's motivational theory, which is well established in the clinical literature (Beauchaine, 2001). Gray's motivational theory outlines three distinct interconnected neuropsychological systems: (1) the fight-or-flight system; (2) the behavioral activation system (BAS), also called the reward system; and (3) the behavioral inhibition system (BIS), also called punishment system (Gray, 1975). In Beauchaine's (2001) integrated model, the polyvagal theory delineates a biological pathway to generalized emotion dysregulation, and the specific manifestation of that dysregulation is determined by Gray's other systems (BIS and BAS). Specifically, weak vagal tone causes insufficient vagally mediated PSNS regulation of emotions. As the SNS increases physiological arousal (i.e. fight-or-flight response), a dominant BIS may lead to an anxious or depressed presentation, while a dominant BAS would lead to an impulsive or aggressive presentation.

The neurovisceral integration model makes two key contributions to the polyvagal theory: (1) the integration of nonlinear dynamical systems theory to explain psychophysiological patterns and (2) the delineation of a neural network responsible for psychophysiological regulation (Thayer & Lane, 2000). Nonlinear dynamical systems theory is a branch of mathematics used to describe the behavior of systems in which the parts relate nonlinearly. Chaos theory, a subdiscipline of nonlinear dynamics, is used to identify patterns in apparently chaotic behavior that emerge in deterministic, nonrandom systems due to complex interactions of the system's constituent parts. Nonlinear dynamics and chaos define many natural systems (e.g., weather patterns), and the NVI model suggests that the integrated system of brain and body is no different. In the NVI model, physiological systems act as "a set of loosely coupled bio-oscillators" (Thayer & Lane, 2000, p. 203). While each physiological system has numerous degrees of freedom, the system is drawn to a limited number of configurations where internal consistency is maximized, called *attractor basins*. These attractor basins represent the different emotions: coordinated repertoires of physiological, attentional, and affective behaviors which help the organism enact goal-directed behavior. In this model, emotions are not distinct categories, but rather represent different points in the same state-space to which physiological systems are more likely to be drawn. Changes in environmental demands can shift goals, in response to which the healthy organism can flexibly move from one emotional state or attractor basin to another. Pathology arises, however, when lack of flexibility prevents movement between attractors. For example, anxiety disorders are characterized physiologically by elevated heart rate and limited heart rate variability, maintaining a constant state of physiological hyperarousal. Behaviorally, this corresponds to difficulty shifting out of a fearful state after a danger has passed. The NVI model calls this adaptability "neurovisceral integration," to indicate the coordination of the brain and body.

Neurovisceral integration can be indexed by heart rate variability (HRV), where greater variability in resting heart rate (HR) is associated with better integration. The NVI model posits that psychophysiological regulation depends on the central autonomic network (CAN), which supports autonomic regulation and structurally overlaps with networks responsible for attention and affect (Benarroch, 1993). The network includes structures throughout the brain from lower regions such as brainstem nuclei, the periaqueductal gray, hypothalamus, amygdala, and basal forebrain to cortical regions responsible for higher-level cognitive processes including the cingulate, insula, somatosensory cortex, and the ventromedial and dorsolateral prefrontal cortices. Recent updates to the NVI model argue that the neural network is structured hierarchically, such that lower levels integrate afferent signals from the body about metabolic needs in the moment, while higher levels integrate information from memory and external cues in the environment (Smith et al., 2017).

METHODOLOGICAL CONSIDERATIONS

There are several broad considerations for research using peripheral psychophysiological measures. Here, we discuss: (1) baselines, (2) tonic vs. phasic measures (with a case for phasic measures), (3) the selection of task and stimuli, and (4) population-level confounds.

Baselines

A person's physiological activity fluctuates throughout the day as a function of several variables (e.g., mood, activity level). To reduce the risk of mistaking these chance variations for scientifically relevant inter- or intra-individual variability, psychophysiological research uses a variety of techniques to establish a "true baseline." Most commonly, time is set aside at the start of the experimental session to induce a relatively neutral emotional state. Some researchers recommend a baseline rest period during which the participant is asked to do nothing and may even be directed to look at a bland stimulus such as a blank wall. Recommendations for the duration of these rest periods have varied from 5 to 20 minutes, with the shorter intervals used in children with attentional deficits (Hastrup, 1986; Jennings et al., 1992; Zisner & Beauchaine, 2016). The goal is to have a baseline rest period that is long enough for the participant to reach a calm, wakeful state, but not so long that the participant becomes fatigued and unfocused. Other researchers argue that there is too much variability in what a participant may be thinking about during such baseline rest periods. They therefore recommend the use of a "vanilla" baseline, in which participants are given a boring activity such as watching a quiet video or completing a repetitive and undemanding task (Gavin & Davies, 2008; Piferi et al., 2000). Some measures are so sensitive to stimulation, however, that even a vanilla activity could influence their magnitude (e.g., RSA; Beauchaine et al., 2001).

Tonic versus Phasic Measures (and an Argument for Phasic)

Tonic, or basal, measures assess physiological states at rest, ideally following a quiet baseline period, while phasic measures assess change in response to a stimulus. Note that variability measures can be taken both as tonic measures (i.e., variability during a period of rest) or phasic measures (i.e., variability in response to a stimulus). In past decades, tonic measures were preponderant in research, and some of the best-replicated findings have used tonic physiology. For example, antisocial behavior has a well-replicated association with low resting HR (for a recent review, see Portnoy & Farrington, 2015). Today, phasic measures have largely supplanted tonic measures, and for some good reasons. First, phasic measures are less subject to baseline noise because they are calculated by subtracting values at baseline from the stimulus-evoked values. As noted, baseline/resting physiology is difficult to interpret because it may reflect a genuine traitlike baseline or it may reflect a phasic reaction to one's thoughts wandering free. For example, Portnoy and Farrington's (2015) large-scale meta-analysis of 114 studies showed that the association between antisocial behavior and low resting HR is significant, but the effect was small ($d = -0.20$; Cohen, 1988) and 30 percent of studies found an association in the opposite direction (Portnoy & Farrington, 2015). One possible explanation for this variability may be the baseline noise inherent in tonic measures. Of note, phasic measures may not completely escape the effects of baseline noise. Per the law of initial value (LIV; Wilder, 1958), pre-stimulus baseline magnitude is negatively associated with the magnitude of change in response to a stimulus. The LIV applies both for statistical reasons (i.e., ceiling effects) and because of biomechanical limitations (e.g., the physical limits of stroke volume create boundaries for possible cardiac output). Phasic measures can also account for some of the noise we see in studies of tonic measures alone by testing for context-specific physiology-psychopathology associations. For example, startle eyeblink responses are potentiated in anxious individuals relative to controls during worry-related tasks but there are no group differences at baseline (Ray et al., 2009). Using tonic measures alone, those group differences in physiology would be obscured.

As a final note on the topic, the term "vagal tone," while useful, has perhaps added some confusion to the literature on the correlates of tonic vs. phasic physiology. Vagal tone, as described above, is a construct that describes the adaptive response of vagally mediated cardiac activity to environmental demands. While the name suggests that vagal tone is indexed with tonic measures, there are important tonic *and* phasic correlates of vagal tone. Namely, strong vagal tone is broadly associated with greater HRV at baseline and reduced HRV in response to a stimulus. Thus, readers of vagal tone literature should attend to whether the measures used are tonic or phasic when interpreting the findings.

Population-Level Considerations

As is the case for all human subjects research, lack of diverse representation in psychophysiological research limits generalizability (Gatzke-Kopp, 2016). Normative values of physiological indices can vary as a function of age, sex, education, and race. For example, larger differential change from rest in skin conductance during fear acquisition is associated with more education, younger age, and being female (Rosenbaum et al., 2015). Additionally, there are moderators of physiological functioning that are population-specific. For example, multiple studies have demonstrated a robust effect of menstrual cycle phase on cardiovascular measures, which, of course, represents a confound only relevant to female participants

(Bai et al., 2009; Hirshoren et al., 2002; Mckinley et al., 2009; Teixeira et al., 2015). It is critical, therefore, to (1) accurately report demographic variables to help with later interpretation; (2) consider findings in the context of appropriate population norms, where such normative data is available; and (3) strive for more diverse, representative samples in future research.

OVERVIEW OF COMMON PERIPHERAL MEASURES

There are far more psychophysiological methods and indices than can be adequately covered in this chapter, so we instead summarize a selection of indices that are popular and promising in clinical research. These indices fall into three categories: cardiovascular measures (e.g., HR, HRV, RSA), electrodermal activity (e.g., skin conductance), and facial measures (e.g., startle response, pupillary changes, and eye tracking).

Cardiovascular Measures

Overview and History

"Go to your bosom; / Knock there, and ask your heart what it doth know."
– William Shakespeare, *Measure for Measure*, 2:2:164–165

The heart and mind have been linked to mental states in art, literature, and cultural beliefs throughout human history. In ancient Egypt, for instance, the heart and not the brain was believed to be the seat of the mind. In his writings about the James–Lange theory of emotion, William James, founder of American psychology, spoke of cardiac measures as one of the visceral cues that lead to emotional experiences (James, 1884). While James' ideas were based on subjective interoceptive observations, cardiovascular measurement technologies have been readily adopted by psychology research as they become available. The electrocardiogram (EKG) had only recently been invented when Eppinger and Hess proposed the vagotonia hypothesis, which conceptualized cardiac tendencies as indicators of personality traits (Eppinger et al., 1915). In the 1950s, after being described in several case studies, HRV became a measure of interest in clinical psychological research, with the observation that increased HRV was linked to impulsivity (Lacey & Lacey, 1958). Today, the leading unifying theories of clinical peripheral psychophysiology, such as the polyvagal theory and NVI model, use indices of HRV as their primary measures of interest (Porges, 1995; Thayer & Lane, 2000). In sum, the history of cardiac psychophysiology is almost as long as the history of psychology research itself.

Biological Mechanisms

The cardiovascular system, responsible for circulation of blood throughout the body, consists of a large central pump (i.e., the heart) and a system of channels for conveying blood throughout the body (i.e., vasculature). The sinoatrial (SA) node, called the "pacemaker of the heart," sets heart rate by controlling the firing rate of the autorhythmic cells, distributed throughout the heart (Gordan, Gwathmey, & Xie, 2015). Unobstructed, the SA node sets an accelerated heart rate of 100 beats per minute (bpm), which would wear out the heart over time. Thankfully, the SA node receives inputs from both the PSNS and SNS, which regulate heart rate based on metabolic needs. At rest, the PSNS cholinergic "brakes" directly signal the SA node via the vagus, slowing the heart rate to roughly 60–75 bpm and reducing conduction velocity of cardiac electrical signals. During physical exertion or emotional excitation, norepinephrine signals from the SNS increase heart rate and cardiac signal conduction. Unlike PSNS signals, SNS signals also influence the contractility of cardiac muscle, so the heart not only beats faster, but also harder.

Methodology and Measures

Preparation for Testing. A recent recommended premeasurement protocol (Laborde, Mosley, & Thayer, 2017), based on review of the literature, suggests that participants should: (1) have a normal night's sleep the night before testing; (2) within 24 hours of testing, refrain from excessive exercise and alcohol; and (3) within 2 hours of testing, refrain from eating meat and consuming caffeine.

EKG. The electrocardiogram (abbreviated as ECG or EKG) assesses cardiac activity by measuring the electrical output of the heart. Electrodes are placed on the surface of the skin in a strategic arrangement relative to the position of the heart based on Einthoven's triangle hypothesis (Einthoven, Fahr, & de Waart, 1913). This arrangement is important because the different electrical patterns at each lead provide information needed to isolate and adjust for artifacts created by respiration (i.e., that HR accelerates on the inhale and slows down on the exhale; Goldberger, 1945). While an EKG conducted for cardiological assessment uses 12 leads, psychophysiological research generally uses only 3 leads – an anode (positive lead), a cathode (negative lead), and a ground (Tassinary, Hess, & Carcoba, 2012). Lead arrangement 1 places the positive lead on the inside of the left wrist, and negative lead on the inside of the right wrist. Lead arrangement 2 places the positive lead on the left ankle and the negative lead on the right wrist. Finally, lead arrangement 3 places the positive lead on the left ankle and the negative lead on the left wrist. The ground can be placed anywhere that is sufficiently distant from the bipolar lead, often on the lower abdomen or leg. Of these arrangements, 2 is most commonly used in psychophysiology and it is also often modified to reduce movement-related noise by moving the positive lead from the leg to under the left ribcage and the negative lead to under the right side of the sternum (Tassinary et al., 2012).

Under normal conditions, each heartbeat produces a predictable EKG waveform with several distinct and meaningful components (see Figure 11.2). From the

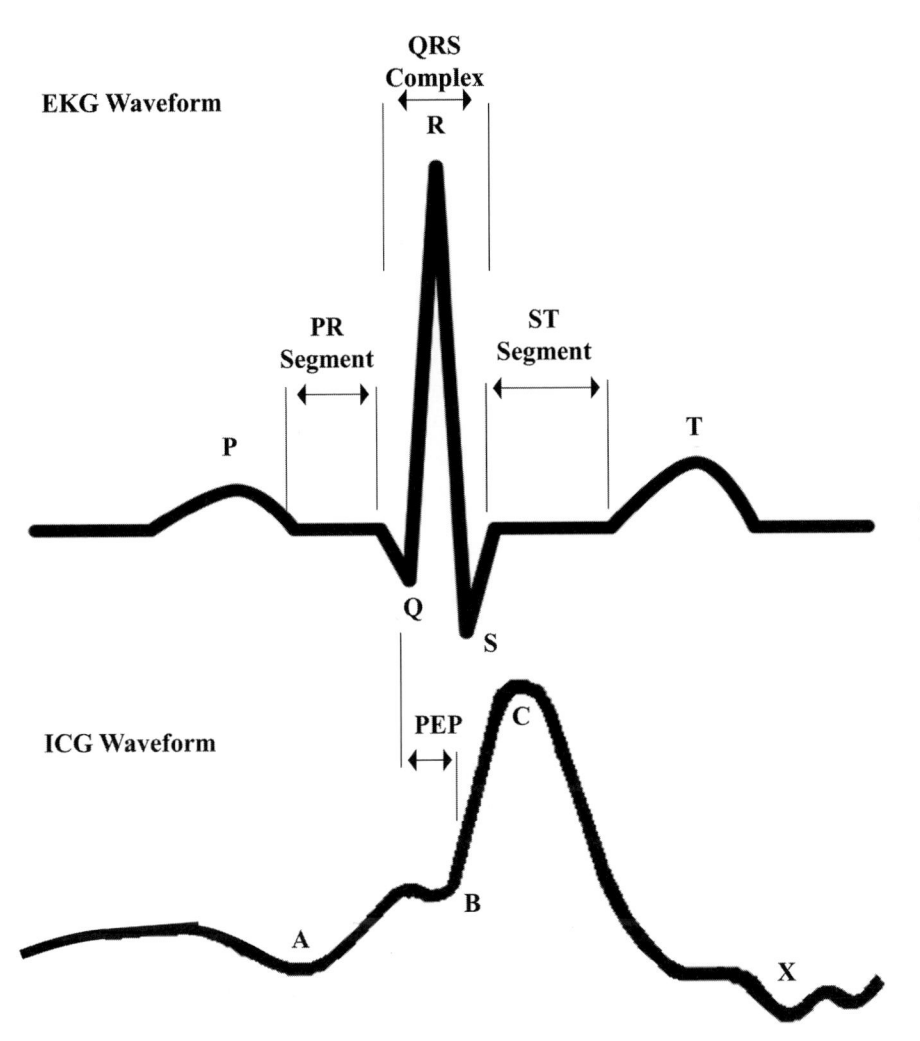

Figure 11.2 Electrocardiography (EKG) and impedance cardiography (ICG) waveforms for one heartbeat

timing, amplitude, and frequencies of these components we can derive several common measures used in psychophysiological research (Cacioppo, Tassinary, & Berntson, 2007). Heart period is calculated from EKG data as the time (in milliseconds) between R peaks (i.e., the duration of the R-R interval). Heart rate is the reciprocal of heart period, calculated in beats per minute by dividing 60,000 by the heart period. Heart rate variability, often a proxy for vagal tone, is calculated several different ways (see Laborde et al., 2017, for review). To summarize just a few, in the time domain the simplest HRV index is the standard deviation of the normal-normal beat interval (SDNN), calculated as the standard deviation of the heart periods in a set time interval (5 minutes are recommended; Camm et al., 1996). Unlike other HRV measures, SDNN is not thought to index vagal tone. Another commonly used formula is the root mean square of successive differences (RMSSD). RMSSD is calculated by first computing the duration between all successive heartbeats. These values are then squared and averaged, and the square root of this average is the RMSSD. RMSSD is thought to index vagal tone, and is robust to respiratory confounds. Measures in the frequency domain are typically obtained through spectral analysis in which the EKG signal during the R-R interval is converted to a spectral density function, which can then be broken up into different frequency bands. Vagal tone is indexed with high-frequency HRV (abbreviated simply, HF) in the range 0.15–0.40Hz. HF is highly correlated with RMSSD, but is linked most closely to the breathing-related respiratory sinus arrhythmia. RSA is a type of HRV which indexes fluctuations of HR as a function of the respiratory cycle. As mentioned above, HR accelerates on the inhale when there is more oxygenated blood in need of circulation, and slows down on the exhale. The question of whether to control for respiration is a source of some debate. Some experts argue that measures adjusted for respiration rate (e.g., HF and RSA) better control for the confounding effect of respiration and more purely capture PSNS vagal tone (Beauchaine, 2015). Multiple studies, however, find that normal respiration has little impact on HRV (Bertsch et al., 2012; Larsen et al., 2010) and even without controlling for respiration, HRV has robust demonstrated clinical implications (Beauchaine & Thayer, 2015).

Preliminary reports of psychometric properties of cardiac measures are mixed. In one study of cardiac indices across a range of tasks, within-task internal consistency reliabilities (e.g., Cronbach's α) were in the acceptable to very good range but reliability was notably weaker across tasks (Kelsey, Ornduff, & Alpert, 2007). In another small study of women, one-year test-retest correlations of cardiac measures (HR, BP) and cortisol were strong and highly significant (Burleson et al., 2003). Finally in a study specifically of HR and HRV measures, only 3 of 28 variables (RR interval length in the N2 and N3 stages of slow-wave sleep and ln HF HRV measured during presleep wakefulness) measured during sleep reached minimal criteria for acceptable stability (Cellini et al., 2016). These data suggest that HF HRV may the best psychometric properties of vagally mediated HRV measures.

Impedance Cardiography. Impedance cardiography (ICG) is another technique used in conjunction with EKG. In brief, ICG involves continuously running an alternating current around the thorax. Because blood is a good conductor, changes in impedance (i.e., opposition to an electrical current passing through a circuit) as the current travels around the thorax can index blood flow. A committee from the Society for Psychophysiological Research (Sherwood et al., 1990) summarized ICG methods as follows: ICG is typically measured using band electrodes encircling the thorax, but can also be conducted with spot electrodes. The band method employs a tetrapolar set of four band electrodes, each made of Mylar coated in aluminum and attached to the skin with adhesive tape. Two of the four bands, called the "voltage electrodes," are placed around the neck and the chest and are used to detect impedance. The remaining two bands, called the "current electrodes," have the alternating current running through them. One current electrode is placed 3 centimeters above the neck voltage electrode and the other is 3 centimeters below the chest electrode. Sherwood and colleagues (1990) recommend a prestimulus baseline period of 10 minutes for impedance measures. The most popular ICG measure in psychophysiological research is the pre-ejection period (PEP; Cacioppo et al., 2007). PEP is the time interval beginning with left ventricle depolarization (Q-wave on the EKG) and concluding when blood is ejected from the heart into the aorta (B-wave on the ICG; see Figure 11.3). Whereas HRV indexes PSNS functioning, phasic changes in PEP are regulated by SNS adrenergic signaling such that greater sympathetic arousal produces greater reductions in PEP duration (Schächinger et al., 2001).

Selected Clinical Findings using Cardiac Measures
HR. Meta-analytic work and large-scale longitudinal studies find a robust link between decreased resting HR and antisocial behavior (Ortiz & Raine, 2004; Portnoy et al., 2014; Raine, 2002; Sijtsema et al., 2010). This association has been replicated both with concurrent symptoms and predicting the onset of future symptoms. The sensation-seeking theory of behavioral psychopathology suggests that low-tonic HR reflects low overall arousal, which the individual experiences as a reduced sense of excitement (Ortiz & Raine, 2004; Portnoy et al., 2014). Antisocial behaviors (e.g., aggression, substance abuse, criminality) are perpetrated to create a feeling of excitement artificially. A direct test of this model in 335 adolescent boys measured HR, self-reported impulsive sensation seeking, and self-reported antisocial behavior (Portnoy et al., 2014). This study found that sensation-seeking tendencies did statistically mediate the relationship between low resting HR and antisocial behavior.

HRV/RSA. Low HRV, specifically measures linked to vagal tone (RMSSD, RSA, HF), is considered a transdiagnostic biomarker of emotion dysregulation. One well-replicated finding is that low vagally mediated HRV is associated with increased anxiety symptoms including panic, generalized anxiety disorder (GAD)/worry, social anxiety, and post-traumatic stress disorder (PTSD), though findings have been inconsistent for specific phobias and obsessive-compulsive disorder (OCD; Chalmers et al., 2014). A possible reason for the mixed effects for specific phobia and OCD may be due to the use of tonic, rather than phasic, measures. For example, one study found that OCD patients did not display irregularities in HRV measured while relaxed, but did show group differences when HRV was measured in response to a hyperventilation task (Pittig et al., 2013). Biofeedback interventions designed to increase HRV, a popular research intervention for performance enhancement, have been found to alleviate anxiety symptoms (Lehrer & Gevirtz, 2014). This research suggests that attenuated HRV may be necessary for the maintenance of anxiety, if not a causal mechanism in anxiety development.

PEP. PEP is commonly used as a marker of reward responsiveness such that greater reward responsiveness corresponds to greater shortening of PEP beginning approximately 20 seconds after reward feedback (Zisner & Beauchaine, 2016). PEP reactivity to rewards is strongly correlated with CNS indices of reward-related mesolimbic dopaminergic activity (Zisner & Beauchaine, 2016). Blunted PEP reactivity to reward differentiates externalizing populations (e.g., attention-deficit/hyperactivity disorder [ADHD], oppositional defiant disorder [ODD], conduct disorder [CD], antisocial personality disorder [ASPD]) from controls (Beauchaine & Gatzke-Kopp, 2012) and prospectively predicts early substance use in middle-schoolers (Brenner & Beauchaine, 2011). Additionally, blunted PEP reactivity to reward has been found to be associated with dysphoria symptoms (Franzen & Brinkmann, 2015). Thus, PEP may be a useful biomarker in disorders characterized by deficits in reward processing, however more data is needed to assess how robust these effects are across studies.

Electrodermal Activity

Overview and History

Research using electrodermal activity (EDA) makes use of the relationships between psychological phenomena and electrical potentials measured on the surface of the skin. The earliest psychological research using EDA measures has been attributed to Romain Vigouroux (1879), who found that changes in psychological state during magnetic therapy altered basal skin resistance. In the following years, seminal works established a variety of techniques by which EDA measures may be obtained. Féré (1888) reported on the exosomatic method and Tarchanoff (1890) described the endosomatic method (each described below). In the latter half of the twentieth century, EDA measures were well-established enough to begin standardizing research methods. Brown (1967) proposed a standardized nomenclature and Lykken and Venables (1971) proposed an influential set of guidelines for standardized EDA measurement which remain largely unchanged (for updated guidelines see Roth, Dawson, & Filion, 2012).

Biological Mechanisms

In early EDA research, two competing theories emerged to explain the biological mechanism underlying psychologically induced changes in EDA. The vascular theory, supported by Vigouroux (1879), proposed that psychologically induced changes in EDA were driven by changes in blood circulation. By contrast, the secretory theory, often attributed to Tarchanoff (1890) and now broadly accepted as correct (Dawson, Schell, & Filion, 2007; Roy et al., 2012), argues that psychologically induced changes in EDA are driven by sweat-gland activity. This sweat-gland activity, dubbed "psychological sweating," produces detectable EDA responses before excreted sweat reaches the surface of the skin, suggesting that EDA measures are responsive not to sweat itself, but another type of sweat-gland activity. Of the two types of sweat glands, eccrine and apocrine, EDA measures have generally been linked to eccrine gland activity, though there is some evidence that apocrine sweat glands may also contribute to the EDA signal (Harker, 2013). Eccrine sweat-gland activity is regulated by SNS cholinergic pathways (Harker, 2013), so EDA is used to index sympathetic arousal.

Methodology and Measures

Recommended methods are summarized below. For a comprehensive text on EDA methods, see Boucsein (2012).

Exosomatic versus Endosomatic Measurement.
There are three core methods for EDA measurement: (1) endosomatic, (2) exosomatic with direct current, and (3) exosomatic with alternating current (Boucsein, 2012; Roth et al., 2012). The endosomatic method simply measures

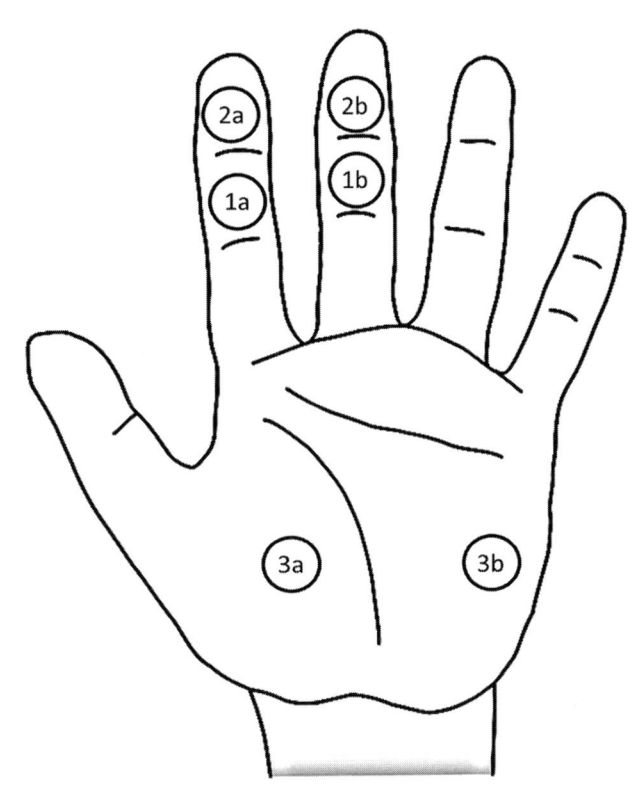

Figure 11.3 Recommended electrode placement for EDA measurement

the difference in charge between two electrodes placed on the skin's surface, and the resulting output signals are called skin potentials (SP). The exosomatic method, so named because it employs an exogenous electrical current, involves passing of an electrical current from one electrode to another along the surface of the skin. When a direct current (DC) is applied to the skin, the resulting output is skin conductance (SC) or its reciprocal, skin resistance (SR). When an alternating current (AC) is applied, the resulting output is skin impedance (SZ) and skin admittance (SY). A recent comparison of exosomatic EDA measures found that AC and DC measures are strongly correlated, suggesting that measure choice may not have a large impact on study outcome (Pabst et al., 2017). Nevertheless, the most-studied measure is skin conductance. Tonic EDA measure are given the suffix "level" (i.e., skin conductance level or SCL), while phasic measures are given the suffix "response" (i.e., skin conductance response or SCR). Note that before standardized naming was introduced SCR was called the *galvanic skin response* or GSR.

Placement of Electrodes.
EDA measures are obtained using sintered silver/silver chloride electrodes placed on the surface of the skin. The greatest densities of eccrine sweat glands are on the palmar (palms of the hands), plantar (soles of the feet), axillary (armpits), and facial surfaces (Harker, 2013). Most research uses the palmar surface for its combination of high eccrine sweat-gland

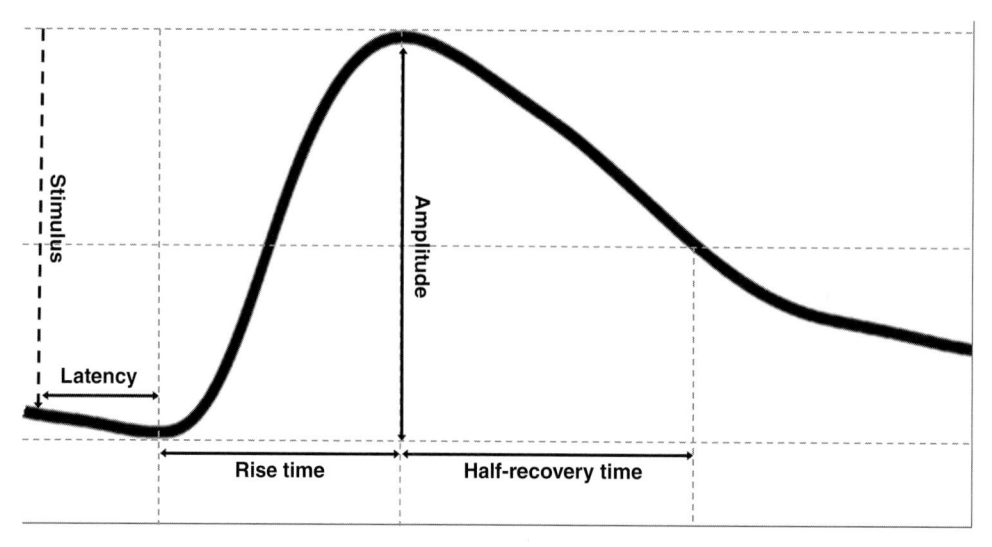

Figure 11.4 Grand average SCR waveform and its commonly scored components

densities and ease of access. A recent comparison across several sites found that the palmar surface produces the largest EDA signals, supporting this traditional placement (Payne, Schell, & Dawson, 2016). Figure 11.3 presents three common electrode arrangements on the palmar surface. In practice, arrangement 1 on the medial phalanges of the index and middle fingers is often preferred to arrangement 2 (distal phalanges) because the attachment of the sensor is more secure. Two small studies, however, suggest that distal phalanges produced up to 3.5 times greater skin conductance amplitudes (Freedman et al., 1994; Scerbo et al., 1992). Thus, arrangement 2 may be psychometrically beneficial because of the superior signal-to-noise ratio. Arrangement 3 is less popular, particularly for tasks requiring movement, because the electrodes on the palm can be obtrusive.

Ambient Temperature and Humidity. Previous research suggests that the effects of psychological sweating may only be detectable at normal room temperature. As a result, EDA methodology guidelines include instructions to closely control the temperature and humidity of the testing room. Venables and Christie (1980) recommend a temperature of 21°C (69.8°F), while Boucsein (2012) recommends a temperature of 23°C (73.4°F) and 50 percent humidity.

Quantification of Skin Conductance. Skin conductance magnitudes are expressed in microSiemens (μS) and can be graphed as μS/time. Tonic SCLs are calculated as the average μS over a given period, often a period of rest. Because SCLs are often subject to significant negative skew and kurtosis in the sample distribution, some methodological guides recommend log-transforming the raw averages (Venables & Christie, 1980). SCRs appear as positive deflections in the SCL waveform (Figure 11.4). SCRs occur both spontaneously (called nonspecific SCRs, or NS-SCRs) and in response to stimuli. To quantify stimulus-linked SCRs, data are segmented into epochs, beginning at stimulus onset and lasting typically 1–3 seconds poststimulus. Within each epoch, SCRs are scored if they meet a minimum response amplitude cutoff (usually between 0.01μS and 0.05μS), to avoid including nonmeaningful fluctuations. SCRs are then scored as the change in amplitude from the bottom of the SCR curve to its apex and averaged across trials. SCRs can also be quantified using other components of the waveform (e.g., latency, rise time, half-recovery time; Figure 11.4). These alternative waveform components, however, have been shown to be unreliable across time whereas amplitude measures have relatively strong psychometric properties (Lang et al., 1993; Schell et al., 2002). Alternatively, model-based computational approaches to SCR computation have gained in popularity in the last two decades, and a variety of model-based calculations exist. In brief, whereas peak scoring calculation methods rely on the assumption that SCRs are the result of SNS activity, model-based calculations fit the observed EDA data to a model, incorporating a priori information (e.g., about the hypothesized neural and physiological processes involved) to better capture the SNS-driven SCR. These SCR values can then be plugged back into the model and the model can be "inverted" to obtain a value that represents SNS activity. In a psychphsyiological model (calculable using the open-source Matlab suite PsPM; Staib, Castegnetti, & Bach, 2015), the model includes information both about the behavior of sudomotor nerves involved in generating SCRs and the timing by which SNS activity can lead to SCRs. PsPM has been found to have better positive predictive value than both traditional peak-detection scoring and alternative model-based calculations that use only information about sudomotor nerve function for detecting the presence of a fear-conditioned stimulus (Staib et al., 2015).

Selected Clinical Findings

One well-replicated clinical finding in EDA literature is that depression is associated with attenuation of skin conductance measures, both tonic SCL and phasic SCRs. A recent systematic review of 77 studies reported that compared to healthy controls, depressed individuals have lower overall SCL and blunted SCR magnitudes (Sarchiapone et al., 2018). One included study found that SCR distinguished between depressed and healthy groups with 80 percent efficiency (Dawson et al., 1985). Furthermore, among depressed individuals, those who were suicidal demonstrated further attenuation of SCRs compared to non-suicidal depressed individuals. The authors noted that group differences in SCR are sensitive to the type of elicitation stimulus. SCRs to neutral stimuli (e.g., the "go" of a reaction time task or neutral words/tones) reliably produced the expected group differences while affective stimuli produced mixed results. In most studies, depressed individuals had attenuated SCRs to both pleasant and unpleasant emotional stimuli, but some studies found the opposite effect, particularly when fear-inducing stimuli were used.

Facial Measures

The Startle Response – Eyes and Ears

Overview and History. The startle response is a coordinated, multicomponent, physiological response to a surprising stimulus (e.g., a loud noise). The startle response was identified in the late 1800s (Exner, 1874) but was repeatedly lost and rediscovered in the literature for several decades before it resurfaced and became the popular topic it is today (for a thorough history, see Dawson, Schell, & Bohmelt, 2008). Similarly, discovery of the "fear potentiated startle" has been attributed to multiple authors, though the earliest example we found was in the early 1950s (Brown, Kalish, & Farber, 1951). Thus, while the history of the startle response is somewhat muddled, we know that it has played an important role in clinical research for many decades.

Although there are several measurable components of the startle response, by far the most studied measure is the *startle eye-blink*. This response is a cued contraction of the muscle that forms a circle around the eye opening, called the orbicularis oculi. Another increasingly popular index of the startle response is the post-auricular muscle response (PAMR), which is a flexion of the muscles behind the ears. In the clinical literature, research has focused particularly on the degree to which the startle response is modulated by contextual factors (i.e., phasic startle measures are more common than basal measures). Startle eye-blink response is potentiated in the presence of threatening or unpleasant stimuli, a phenomenon called fear-potentiated startle (FPS). By contrast, PAMR is potentiated in the presence of pleasant stimuli (Aaron & Benning, 2016). Together, these two startle indices provide useful probes of both defensive (FPS) and appetitive (PAMR) motivational systems.

Biological Mechanisms. Given the short latency of the startle response (as little as 20 milliseconds; Blumenthal et al., 2005), it is unsurprising that the neural network which generates this response is relatively parsimonious. Based largely on evidence from the animal literature, the primary neural pathway responsible for the acoustic startle response includes only three core synapses: (1) from the auditory nerve fibers to the cochlear root neurons; (2) from the cochlear root neurons to a portion of the reticular formation called the nucleus reticularis pontis caudalis (RPC); and (3) from the RPC to facial and spinal motor neurons (Davis, 2006; Eaton, 1984). Secondary pathways have also been found to modulate the startle response. For example, fear potentiation of the startle response is facilitated by projections from the amygdala to the RPC (Davis, 2006).

Methodology and Measures. The following method is based on guidelines for the startle eye-blink, proposed by a committee from the Society for Psychophysiological Research (Blumenthal et al., 2005). To measure the startle response, two small Ag/AgCl electrodes are placed on the area of interest (Figure 11.5). To facilitate conduction of the electrical signal, the skin is first cleaned with an alcohol swab and a conductive gel is applied between the skin surface and the electrode. For the eye-blink response, sensors are placed on the bottom lid of the eye, one directly below the pupil and the second 1 cm lateral to the first (Figure 11.5, black triangles). For the PAMR, the ear is gently folded forward and sensors are placed approximately halfway up the back of the ear (black circles), 1 cm apart, along the fold where the ear meets the head.

To elicit a startle response effectively, probe stimuli must be of sufficient intensity (e.g., loud volume) and have a near-instantaneous rise time (i.e., quickly reach maximum intensity). Acoustic startle probes (most often a burst of white noise) range from 50 to 100 decibels in volume, which balances competing goals of maximizing intensity while minimizing participant discomfort. An electronic switch is used to reduce frequency splatter at the onset of the probe, minimizing "rise time" to allow for a clean, sudden start to the sound. Probe duration also influences startle response such that longer acoustic probes (up to 50 milliseconds) elicit greater responses. While less common, startle eye-blinks are also elicited using (1) a puff of air to the eye, (2) a burst of light, and (3) mild electric shocks to the surface of the skin. PAMRs are also elicited with a "click," an acoustic probe that is softer and shorter than the traditional startle probe. The benefit of the click probes is that the resulting PAMRs do not habituate across trials (Yoshie & Okudaira, 1969), whereas traditional startle probe PAMRs habituate quickly.

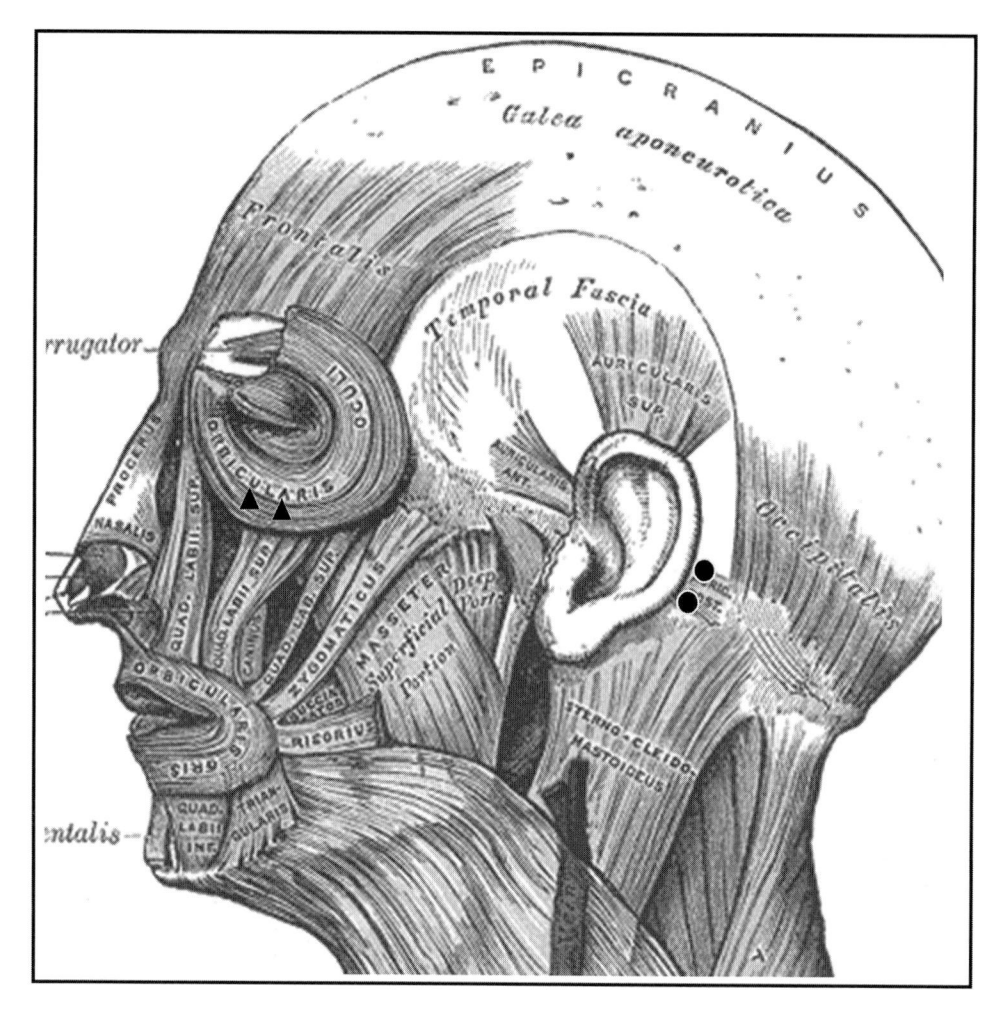

Figure 11.5 Electrode placement for startle eye-blink (triangles) and postauricular (circles). Two markers are shown at each location to indicate that bipolar leads are used

Reliable startle eye-blink responses have been obtained with as few as six trials in one promising study (Lieberman et al., 2017). Nevertheless, psychometric properties of startle responses vary as a function of task. For the startle eye-blink response, reliability indices are excellent in resting-state tasks, acceptable-to-excellent in a fear-conditioning paradigm called the NPU task (see below), and poor-to-fair during affective picture viewing (Aaron & Benning, 2016; Kaye, Bradford, & Curtin, 2016; Lieberman et al., 2017; Nelson, Hajcak, & Shankman, 2015). There is relatively less information on the psychometric properties of the PAMR, but one study found poor psychometric properties when elicited with an affective picture-viewing task (Aaron & Benning, 2016). Additionally, scoring method affects reliability such that untransformed scores are psychometrically superior to other common scoring strategies, including standardized scores and percentage change (Bradford et al., 2015; Kaye et al., 2016). In sum, non-transformed startle response measures elicited in resting-state or NPU tasks are psychometrically preferable to affective picture viewing.

Selected Clinical Findings

FPS. The startle eye-blink response is most commonly used in the clinical literature to study FPS as it relates to anxiety. Exaggerated FPS is a transdiagnostic biomarker of anxiety (Lang & McTeague, 2009), and is thought to reflect elevated threat sensitivity, a core feature of anxiety pathology (Richards et al., 2002). There is, however, some nuance to the FPS and anxiety literature that merits discussion. First, anxiety effects on FPS vary as a function of diagnosis, such that the fear disorders (e.g., specific phobia, social phobia, panic) show greater FPS than the distress disorders (e.g., GAD, OCD, PTSD; Watson, 2005). Interestingly, in a direct comparison of several anxiety diagnoses, individuals with a single-trauma PTSD diagnosis exhibit the largest FPS responses, and multiple trauma PTSD the smallest (McTeague & Lang, 2012). This suggests that exposure to trauma initially sensitizes the defensive responding indexed by FPS, but when this defensive system is overexerted with "multiple hits," it weakens. The NPU task, which separately assesses startle potentiation to no threat (N), predictable threat (P), and unpredictable threat (U), offers clarification to these heterogeneous results. Large-scale studies suggest that FPS to unpredictable but not predictable threat drives the association between anxiety and FPS. Furthermore, FPS to unpredictable threat prospectively predicts the development of anxiety symptoms (Gorka et al., 2017; Nelson & Hajcak, 2017).

PAMR. While PAMR has been relatively less studied in the clinical literature, preliminary research suggests that, like other indices of appetitive motivation, the PAMR to pleasant/rewarding stimuli is attenuated in disorders associated with reward-processing deficits such as substance use (Lubman et al., 2009) and depressive disorders (Benning & Ait Oumeziane, 2017; Sloan & Sandt, 2010).

Oculometry

Overview and History. As Delabarre wrote: "Many problems suggest themselves to the psychologist whose solution would be greatly furthered by an accurate method of recording the movements of the eye" (Delabarre, 1898, p. 572). Oculometry, or eye tracking, addresses this methodological need (for a review of oculometry methods, see Duchowski, 2007). Early oculometry methods were introduced around the turn of the twentieth century. These methods often required an object (e.g., suction ring, plaster mold) placed directly on the surface of the eye, with the obvious disadvantage of creating participant discomfort. To address this limitation, in the 1970s electro-oculography (EOG) became popular. EOG uses electrodes placed on skin around the eye to measure electrical potentials of the muscles that direct eye movement (Young & Sheena, 1975). EOG only assesses eye movements relative to head position, and therefore EOG is not useful for identifying what the participants is looking at (i.e., *point of regard*). Also in the 1970s, photo-oculography (POG) methods were introduced into the literature (Lambert, Monty, & Hall, 1974), which assess eye position based on the reflection of light off the eye's surface. This basic strategy of measuring light on the eye is the foundation of modern video-based eye tracking, which is the most commonly used measure today.

Biological Mechanisms. Six extraocular muscles facilitate eye movements along all axes: horizontal movements are controlled by the medial and lateral recti, vertical movements by the superior and inferior recti, and rotation by the superior and inferior obliques. These muscles receive inputs via cranial nerves III, IV, and VI, stemming from multiple neural pathways collectively known as the oculomotor system. *Smooth pursuit* eye movements are facilitated by a cerebro-ponto-cerebellar pathway (Thier & Ilg, 2005). Smooth pursuit signals originate in the cerebral cortex, the most notable area V5 of the occipital cortex, and the middle superior temporal area. They are then relayed by the dorsal pontine nucleus in the pons, and arrive in the cerebellum, where the necessary motor control signals are coordinated. Within the cerebellum, the flocculus-paraflocculus coordinates movements in simultaneous eye-head smooth pursuit, whereas the posterior vermis mediates the early portion of smooth pursuit while the trajectory of the target may not yet be predictable.

Point of regard measures are determined by a rather more complex set of neural mechanisms. Nevertheless, regions of interest include the superior colliculus, involved in translating auditory cues to the visual space (i.e., knowing where to look for a sound) and programming eye movements when attention is drawn to a new visual location. Motor signals that direct eye movements are informed by directional selectivity cells in the occipital cortex, which perceive motion (Duchowski, 2007). Additionally, nonvisual systems influence point of regard through top-down processes (i.e., motivation to visually search for a stimulus of interest). The top-down direction of selective visual attention occurs in part through the increased firing rate of neurons in the occipital cortex that code for the attended stimulus (Moore & Zirnsak, 2017).

Methodology and Measures. Modern oculometry uses video-based eye tracking, using either a table-mounted or a head-mounted apparatus, with the latter allowing for free head movement. Eye tracking systems often use multiple cameras to capture both images of the eye(s) and images from the participant's point of view. Video-based eye tracking can assess *point of regard* by holding the head steady and then measuring gaze direction relative to the known head position or by measuring two different features of the eye, such as the center of the pupil and corneal light reflections (Purkinje images). Each eye tracking session begins with a calibration procedure during which the participant is directed to look at various fixation points to insure a match between actual and calculated point of regard. While eye tracking systems can be expensive, several low-cost systems (approximately $100–$500) have been developed by individual research groups with good results (e.g. Coyne & Sibley, 2016).

In clinical research, there are two broad measures of interest in oculometry: *point of regard* and *smooth pursuit* eye movements. *Point of regard* measures are often used to assess attentional bias, as indexed by a tendency to spend disproportionate time looking at or away from certain stimuli (e.g., more time spent looking at frightening images suggests an attention bias toward threat). Attention bias is typical quantified in units of time (e.g., milliseconds) by calculating the duration of the time spent looking at the object of interest. *Smooth pursuit* eye movements refer to the fluid motion of the eye while visually tracking a moving object. Excessive saccades or other deviations from the trajectory of the moving object suggests deficits in basic attention and the ability to predict the path the object will take (Franco et al., 2014). There are several ways to quantify smooth pursuit data, two common metrics being root mean squared of the error (RMSE) and the number of saccades over time (e.g., saccades per minute or per trial). While number of saccades represents a specific irregularity in eye movements, RMSE represents a global measure of the deviation of the participant's gaze from the trajectory traveled by the target object. RMSE is calculated by squaring the difference between eye position and target position in each slice of time (e.g., trial), and then averaging the square root of these numbers. Across multiple studies, psychometric properties were strong for RMSE and saccade frequency measures, using both internal consistency and test-retest reliability indices (Ettinger et al., 2003; Gooding, Iacono, & Beiser, 1994; Iacono & Lykken, 1981; Kumra et al., 2001).

Selected Clinical Findings

Smooth Pursuit. Both the oldest and most common clinical use of *smooth pursuit* eye movements is in the study of schizophrenia. Just after the turn of the century, a case series noted deficits in smooth pursuit eye movements in those with schizophrenia, such that there were unusually high numbers of jerky eye movements (i.e., saccades) while tracking a moving object (Diefendorf & Dodge, 1908). Since then, these deficits have been repeatedly found to distinguish schizophrenic samples from healthy controls with a large meta-analytic effect size (d's between 0.7 and 0.8; O'Driscoll & Callahan, 2008). Similar deficits are found in nonpsychotic first-degree relatives of schizophrenic patients at about half the level of severity as is seen in the proband, suggesting that smooth pursuit measures may tap a schizophrenia endophenotype (Kathmann et al., 2003; Keefe et al., 1997).

Point of Regard. Point-of-regard measures have been used extensively to study attention biases in two clinical populations: autism spectrum disorders (ASD) and affective disorders (i.e., anxiety and depression). In ASD, there is a well-documented attentional bias away from social stimuli. Two recent meta-analyses found that ASD individuals spend significantly less time looking at a variety of social stimuli, including direct social engagement (e.g., eye-contact; Papagiannopoulou et al., 2014) and indirect social engagement (e.g., pictures of people; Chita-Tegmark, 2016). In affective disorders, there are well-documented attentional biases to emotionally valenced stimuli. Specifically, a meta-analysis of 33 eye tracking studies found that greater anxiety corresponded to longer gaze duration toward threatening stimuli, while depression was associated with longer gaze toward sad stimuli and shorter gaze toward happy stimuli (Armstrong & Olatunji, 2012).

Pupillometry

Overview and History. Pupillometry is the measurement of the diameter of the pupil, the aperture at the center of the eye through which light reaches the photoreceptors on the retina. The pupil dilates and constricts as a function of ambient light (i.e., pupils are more open in the dark) and, to a lesser extent, psychological events. A handful of psychology studies in the early 1900s reported pupillometric findings, but their methods were generally considered too crude to produce meaningful results (Loewenfeld, 1999). Thus, credit for introducing pupillometry to psychological literature is generally given to Eckhard Hess for studies conducted in the 1960s. These studies demonstrated pupil dilations as a function of subjective interest (Hess & Polt, 1960) and as a function of greater cognitive load (Hess & Polt, 1964). Today, pupillometry is still most commonly used as an index of cognitive load and, by extension, cognitive limits. Similarly, uncertainty is associated with pupil dilation, suggesting that we invest more cognitive effort into assessing uncertain than certain events (Nassar et al., 2012). Additionally, in recent affective neuroscience work, the pupils have been found to dilate to both pleasant and unpleasant emotional stimuli, which allows pupillometry to be used as a measure of emotional responsiveness (e.g. Bradley et al., 2008).

Biological Mechanisms. Pupil aperture is regulated by two opposing muscle groups in the iris (the colored part of the eye): the dilator muscles and the sphincter (a.k.a. constrictor) muscles (for a full description, see Sirois & Brisson, 2014). The dilator muscles are controlled by the adrenergic receptors linked to the SNS while constrictor muscles are linked to the PSNS. Thus, pupil diameter reflects both sympathetic and parasympathetic forces. In normal light conditions, the pupil is approximately 3 mm in width, but can range from about 1 to 9 mm. Fluctuations due to psychological factors, however, produce variations of approximately 0.5 mm, and are therefore not detectable by the naked eye. Task-related pupil dilations are linked to the locus coeruleus (LC), a midbrain structure that acts as the relay station for all norepinephrine (NE) to the brain. The LC-NE system has been dubbed the "pacemaker of attention," and its activity is thought of as a biological index of the Yerkes-Dodson curve. In other words, moderate LC-NE activity is linked to a productive level of attention/arousal, while high and low activity of the LC-NE system are linked to poor attention due to hypo- or hyperarousal (Bast, Poustka, & Freitag, 2018).

Methodology and Measures. Contemporary pupillometry is conducted using video-based eye tracking technology (see oculometry methods for details). To reduce luminance confounds, these eye trackers may use infrared illumination that does not elicit a pupillary light reflex, and study sessions are conducted in moderately lit rooms. Using these eye tracking systems, pupil diameter can be measured continuously. Because clinical psychology research is primarily interested in stimulus-evoked pupillary changes, continuous data is segmented into stimulus-locked epochs. Trials that are >50 percent blinks are typically removed, and blinks are then interpolated linearly out of the remaining trials. Pupil diameter in response to the stimulus (measured in millimeters) can then be quantified in a number of ways, such as baseline corrected and plotted over time, calculated as an average change from baseline within a given time window, or calculated as the average difference between two conditions (Sirois & Brisson, 2014).

To our knowledge, all published psychometric studies of pupillometry to date have been conducted in samples with neurological or ophthalmologic disorders, and thus may not be representative of the general population. Nevertheless, these studies find moderate to high test-retest reliability for pupil diameter (e.g., Farzin et al., 2011). Additionally, one study in a healthy sample found that phasic pupillometry measures are robust to the luminance confounds seen with tonic measures (Peysakhovich, Vachon, & Dehais, 2017). Thus, phasic measures may be more reliable than measures taken at rest due to better signal-to-noise ratios.

Selected Clinical Findings. Like the smooth pursuit oculometry indices described above, pupillometry has historically been used in clinical research to study cognitive deficits in schizophrenia (Zahn, Frith, & Steinhauer, 1991). In brief, these studies find that individuals with schizophrenia exhibit attenuated pupil dilations and irregular latencies to pupil constrictions in a variety of cognitive tasks, as well as stress-induction tasks such as cold pressor application. Furthermore, because this pupillary pattern does not resolve with medication or with symptom remission, pupil responses may represent a trait-like biomarker of schizophrenia. One conceptualization of these deficits is that they reflect broad autonomic hyporesponsiveness in schizophrenia. Functionally, one study also found that within schizophrenic samples, greater deficits in pupillary responses were related to "defeatist attitudes," which represent a negative prognostic factor in schizophrenia (Granholm et al., 2016).

Recently, using tasks that target narrower sources of impairment, other clinical uses for pupillometry have emerged. For example, increased pupil dilation specifically to sad stimuli has been documented in depression and depression risk (Burkhouse, Siegle, & Gibb, 2014; Burkhouse et al., 2015). Similarly, one recent study in adolescent girls found that pupil dilation to maternal criticism prospectively predicted increased borderline personality disorder symptoms (Scott et al., 2017). Together, these applications illustrate how pupillometry can be used to assess both broad cognitive deficits and deficits that emerge in symptom-specific contexts.

SUMMARY AND CONCLUSIONS

We hope that this chapter has piqued your curiosity about peripheral psychophysiology, though we could only cover what is very much the tip of the proverbial iceberg. Psychophysiological biomarkers are among the most promising indices in clinical research, both because of their strong replicability and the insights they can provide about the mechanisms that undergird psychopathology. For those interested in learning more, in addition to the texts we have recommended throughout the chapter, we direct the reader to Cacioppo and colleagues' *Handbook of Psychophysiology* (the 4th edition of which is in press) and to the Society for Psychophysiological Research, which is dedicated to fostering the advancement of psychophysiological work and which places a strong emphasis on the education of interested researchers.

REFERENCES

Aaron, R. V., & Benning, S. D. (2016). Postauricular Reflexes Elicited by Soft Acoustic Clicks and Loud Noise Probes: Reliability, Prepulse Facilitation, and Sensitivity to Picture Contents. *Psychophysiology, 53*(12), 1900–1908.

Armstrong, T., & Olatunji, B. O. (2012). Eye Tracking of Attention in the Affective Disorders: A Meta-Analytic Review and Synthesis. *Clinical Psychology Review, 32*(8), 704–723.

Bai, X., Li, J., Zhou, L., & Li, X. (2009). Influence of the Menstrual Cycle on Nonlinear Properties of Heart Rate Variability in Young Women. *American Journal of Physiology – Heart and Circulatory Physiology, 297*(2), H765–H774.

Bast, N., Poustka, L., & Freitag, C. M. (2018). The Locus Coeruleus–Norepinephrine System as Pacemaker of Attention – A Developmental Mechanism of Derailed Attentional Function in Autism Spectrum Disorder. *European Journal of Neuroscience, 47*(2), 115–125.

Beauchaine, T. (2001). Vagal Tone, Development, and Gray's Motivational Theory: Toward an Integrated Model of Autonomic Nervous System Functioning in Psychopathology. *Development and Psychopathology, 13*(2), 183–214.

Beauchaine, T. P. (2015). Respiratory Sinus Arrhythmia: A Transdiagnostic Biomarker of Emotion Dysregulation and Psychopathology. *Current Opinion in Psychology, 3*, 43–47.

Beauchaine, T. P., & Gatzke-Kopp, L. M. (2012). Instantiating the Multiple Levels of Analysis Perspective in a Program of Study on Externalizing Behavior. *Development and Psychopathology, 24*(3), 1003–1018.

Beauchaine, T. P., Katkin, E. S., Strassberg, Z., & Snarr, J. (2001). Disinhibitory Psychopathology in Male Adolescents: Discriminating Conduct Disorder from Attention-Deficit/Hyperactivity Disorder through Concurrent Assessment of Multiple Autonomic States. *Journal of Abnormal Psychology, 110*(4), 610–624.

Beauchaine, T. P., & Thayer, J. F. (2015). Heart Rate Variability as a Transdiagnostic Biomarker of Psychopathology. *International Journal of Psychophysiology, 98*(2), 338–350.

Benarroch, E. E. (1993). The Central Autonomic Network: Functional Organization, Dysfunction, and Perspective. In *Mayo Clinic Proceedings* (Vol. 68, No. 10, pp. 988–1001). Rochester, MN: Elsevier.

Benning, S. D., & Ait Oumeziane, B. (2017). Reduced Positive Emotion and Underarousal are Uniquely Associated with Subclinical Depression Symptoms: Evidence from Psychophysiology, Self-Report, and Symptom Clusters. *Psychophysiology, 54*(7), 1010–1030.

Berthoud, H.-R., & Neuhuber, W. L. (2000). Functional and Chemical Anatomy of the Afferent Vagal System. *Autonomic Neuroscience, 85*(1), 1–17.

Bertsch, K., Hagemann, D., Naumann, E., Schächinger, H., & Schulz, A. (2012). Stability of Heart Rate Variability Indices Reflecting Parasympathetic Activity. *Psychophysiology, 49*(5), 672–682.

Blumenthal, T. D., Cuthbert, B. N., Filion, D. L., Hackley, S., Lipp, O. V., & Van Boxtel, A. (2005). Committee Report: Guidelines for Human Startle Eyeblink Electromyographic Studies. *Psychophysiology, 42*(1), 1–15.

Boucsein, W. (2012). *Electrodermal Activity*. New York: Springer Science & Business Media.

Bradford, D. E., Starr, M. J., Shackman, A. J., & Curtin, J. J. (2015). Empirically Based Comparisons of the Reliability and Validity of Common Quantification Approaches for Eyeblink Startle Potentiation in Humans. *Psychophysiology, 52*(12), 1669–1681.

Bradley, M. M., Miccoli, L., Escrig, M. A., & Lang, P. J. (2008). The Pupil as a Measure of Emotional Arousal and Autonomic Activation. *Psychophysiology, 45*(4), 602–607.

Brenner, S. L., & Beauchaine, T. P. (2011). Pre-Ejection Period Reactivity and Psychiatric Comorbidity Prospectively Predict Substance Use Initiation among Middle-Schoolers: A Pilot Study. *Psychophysiology, 48*(11), 1588–1596.

Brown, C. C. (1967). A Proposed Standard Nomenclature for Psychophysiologic Measures. *Psychophysiology, 4*(2), 260–264.

Brown, J. S., Kalish, H. I., & Farber, I. (1951). Conditioned Fear as Revealed by Magnitude of Startle Response to an Auditory Stimulus. *Journal of Experimental Psychology, 41*(5), 317–328.

Burkhouse, K. L., Siegle, G. J., & Gibb, B. E. (2014). Pupillary Reactivity to Emotional Stimuli in Children of Depressed and Anxious Mothers. *Journal of Child Psychology and Psychiatry, 55*(9), 1009–1016.

Burkhouse, K. L., Siegle, G. J., Woody, M. L., Kudinova, A. Y., & Gibb, B. E. (2015). Pupillary Reactivity to Sad Stimuli as a Biomarker of Depression Risk: Evidence from a Prospective Study of Children. *Journal of Abnormal Psychology, 124*(3), 498–506.

Burleson, M. H., Poehlmann, K. M., Hawkley, L. C., Ernst, J. M., Berntson, G. G., Malarkey, W. B., ... Cacioppo, J. T. (2003). Neuroendocrine and Cardiovascular Reactivity to Stress in Mid-Aged and Older Women: Long-Term Temporal Consistency of Individual Differences. *Psychophysiology, 40*(3), 358–369.

Cacioppo, J. T., Tassinary, L. G., & Berntson, G. (2007). *Handbook of Psychophysiology*. Cambridge: Cambridge University Press.

Camm, A. J., Malik, M., Bigger, J., Breithardt, G., Cerutti, S., Cohen, R., ... Kleiger, R. (1996). Heart Rate Variability: Standards of Measurement, Physiological Interpretation and Clinical Use. Task Force of the European Society of Cardiology and the North American Society of Pacing and Electrophysiology. *Circulation, 93*(5), 1043–1065.

Cannon, W. B. (1929). *Bodily Changes in Pain, Fear, Hunger, and Rage*. New York: Appleton.

Cannon, W. B. (1932). *The Wisdom of the Body*. New York: W. W. Norton.

Cellini, N., Whitehurst, L. N., McDevitt, E. A., & Mednick, S. C. (2016). Heart Rate Variability during Daytime Naps in Healthy Adults: Autonomic Profile and Short-Term Reliability. *Psychophysiology, 53*(4), 473–481.

Chalmers, J. A., Quintana, D. S., Maree, J., Abbott, A., & Kemp, A. H. (2014). Anxiety Disorders are Associated with Reduced Heart Rate Variability: A Meta-Analysis. *Frontiers in Psychiatry, 5*, 1–11.

Chita-Tegmark, M. (2016). Social Attention in ASD: A Review and Meta-Analysis of Eye-Tracking Studies. *Research in Developmental Disabilities, 48*, 79–93.

Cohen, J. (1988). *Statistical Power Analysis for the Behavioral Sciences* (2nd edn.). Hillsdale, NJ: Lawrence Erlbaum Associates.

Coyne, J., & Sibley, C. (2016). Investigating the Use of Two Low Cost Eye Tracking Systems for Detecting Pupillary Response to Changes in Mental Workload. In *Proceedings of the Human Factors and Ergonomics Society Annual Meeting* (Vol. 60, No. 1, pp. 37–41). Los Angeles, CA: Sage'

Davis, M. (2006). Neural Systems Involved in Fear and Anxiety Measured with Fear-Potentiated Startle. *American Psychologist, 61*(8), 741–756.

Davis, R. C. (1958). The Domain of Homeostasis. *Psychological Review, 65*(1), 8–13.

Dawson, M. E., Schell, A. M., & Bohmelt, A. H. (2008). *Startle Modification: Implications for Neuroscience, Cognitive Science, and Clinical Science*. Cambridge: Cambridge University Press.

Dawson, M. E., Schell, A. M., Braaten, J. R., & Catania, J. J. (1985). Diagnostic Utility of Autonomic Measures for Major Depressive Disorders. *Psychiatry Research, 15*(4), 261–270.

Dawson, M. E., Schell, A. M., & Filion, D. L. (2007). The Electrodermal System. *Handbook of Psychophysiology, 2*, 200–223.

Debat, V., & David, P. (2001). Mapping Phenotypes: Canalization, Plasticity and Developmental Stability. *Trends in Ecology & Evolution, 16*(10), 555–561.

Delabarre, E. B. (1898). A Method of Recording Eye-Movements. *The American Journal of Psychology, 9*(4), 572–574.

Diefendorf, A. R., & Dodge, R. (1908). An Experimental Study of the Ocular Reactions of the Insane from Photographic Records. *Brain, 31*(3), 451–489.

Duchowski, A. T. (2007). *Eye Tracking Methodology. Theory and Practice* (3rd edn.). Basel: Springer International.

Eaton, R. C. (1984). *Neural Mechanisms of Startle Behavior*. New York: Springer Science & Business Media.

Einthoven, W., Fahr, G., & de Waart, A. (1913). Ueber die Rechtung und die Manifeste Grösse der Potential schwankungon im menschlechen Herzen und ueber den Einfluss der Herzglage auf die Form des Electrokardiogramms. *Pflügers Archives European Journal of Physiology, 150*, 275–315.

Eppinger, H., Hess, L., Kraus, W. M., & Jelliffe, S. E. (1915). *Vagotonia: A Clinical Study in Vegetative Neurology*. New York: Nervous and Mental Disease Publishing Company.

Ettinger, U., Kumari, V., Crawford, T. J., Davis, R. E., Sharma, T., & Corr, P. J. (2003). Reliability of Smooth Pursuit, Fixation, and Saccadic Eye Movements. *Psychophysiology, 40*(4), 620–628.

Exner, S. (1874). Experimental Investigation of the Simplest Mental Process: First Article. *Pflugers Archiv:* European Journal of Physiology, 7, 601–660.

Farzin, F., Scaggs, F., Hervey, C., Berry-Kravis, E., & Hessl, D. (2011). Reliability of Eye Tracking and Pupillometry Measures in Individuals with Fragile X Syndrome. *Journal of Autism and Developmental Disorders, 41*(11), 1515–1522.

Fere, C. (1888). Note on Changes in Electrical Resistance under the Effect of Sensory Stimulation and Emotion. *Comptes rendus des Seancs de la Societe de Biologie, 5*, 28–33.

Fowles, D. C. (1980). The Three Arousal Model: Implications of Gray's Two-Factor Learning Theory for Heart Rate, Electrodermal Activity, and Psychopathy. *Psychophysiology, 17*(2), 87–104.

Franco, J., De Pablo, J., Gaviria, A., Sepulveda, E., & Vilella, E. (2014). Smooth Pursuit Eye Movements and Schizophrenia: Literature Review. *Archivos de la Sociedad Española de Oftalmología (English Edition), 89*(9), 361–367.

Franzen, J., & Brinkmann, K. (2015). Blunted Cardiovascular Reactivity in Dysphoria during Reward and Punishment Anticipation. *International Journal of Psychophysiology, 95*(3), 270–277.

Freedman, L. W., Scerbo, A. S., Dawson, M. E., Raine, A., McClure, W. O., & Venables, P. H. (1994). The Relationship of Sweat Gland Count to Electrodermal Activity. *Psychophysiology, 31*(2), 196–200.

Friedman, M. (1945). Studies Concerning the Etiology and Pathogenesis of Neurocirculatory Asthenia: III. The Cardiovascular Manifestations of Neurocirculatory Asthenia. *American Heart Journal, 30*(5), 478–491.

Gatzke-Kopp, L. M. (2016). Diversity and Representation: Key Issues for Psychophysiological Science. *Psychophysiology, 53*(1), 3–13.

Gavin, W., & Davies, P. (2008). Obtaining Reliable Psychophysiological Data with Child Participants: Methodological Considerations. In *Developmental Psychophysiology: Theory, Systems, and Methods* (pp. 424–447). New York: Cambridge University Press.

Goldberger, E. (1945). The Validity of the Einthoven Triangle Hypothesis. *American Heart Journal, 29*(3), 369–377.

Gooding, D. C., Iacono, W. G., & Beiser, M. (1994). Temporal Stability of Smooth-Pursuit Eye Tracking in First-Episode Psychosis. *Psychophysiology, 31*(1), 62–67.

Gordan, R., Gwathmey, J. K., & Xie, L.-H. (2015). Autonomic and Endocrine Control of Cardiovascular Function. *World Journal of Cardiology, 7*(4), 204–214.

Gorka, S. M., Lieberman, L., Shankman, S. A., & Phan, K. L. (2017). Startle Potentiation to Uncertain Threat as a Psychophysiological Indicator of Fear-Based Psychopathology: An Examination across Multiple Internalizing Disorders. *Journal of Abnormal Psychology, 126*(1), 8–18.

Granholm, E., Ruiz, I., Gallegos-Rodriguez, Y., Holden, J., & Link, P. C. (2016). Pupillary Responses as a Biomarker of Diminished Effort Associated with Defeatist Attitudes and Negative Symptoms in Schizophrenia. *Biological Psychiatry, 80*(8), 581–588.

Gray, J. A. (1975). *Elements of a Two-Process Theory of Learning.* Oxford: Academic Press.

Harker, M. (2013). Psychological Sweating: A Systematic Review Focused on Aetiology and Cutaneous Response. *Skin Pharmacology and Physiology, 26*(2), 92–100.

Hastrup, J. L. (1986). Duration of Initial Heart Rate Assessment in Psychophysiology: Current Practices and Implications. *Psychophysiology, 23*(1), 15–18.

Hess, E. H., & Polt, J. M. (1960). Pupil Size as Related to Interest Value of Visual Stimuli. *Science, 132*(3423), 349–350.

Hess, E. H., & Polt, J. M. (1964). Pupil Size in Relation to Mental Activity during Simple Problem-Solving. *Science, 143*(3611), 1190–1192.

Hirshoren, N., Tzoran, I., Makrienko, I., Edoute, Y., Plawner, M. M., Itskovitz-Eldor, J., & Jacob, G. (2002). Menstrual Cycle Effects on the Neurohumoral and Autonomic Nervous Systems Regulating the Cardiovascular System. *Journal of Clinical Endocrinology & Metabolism, 87*(4), 1569–1575.

Iacono, W. G., & Lykken, D. T. (1981). Two-Year Retest Stability of Eye Tracking Performance and a Comparison of Electro-Oculographic and Infrared Recording Techniques: Evidence of EEG in the Electro-Oculogram. *Psychophysiology, 18*(1), 49–55.

Insel, T., Cuthbert, B., Garvey, M., Heinssen, R., Pine, D. S., Quinn, K., ... Wang, P. (2010). Research Domain Criteria (RDoC): Toward a New Classification Framework for Research on Mental Disorders. *American Journal of Psychiatry, 167*(7), 748–751.

James, W. (1884). What Is an Emotion? *Mind, 9*(34), 188–205.

Jänig, W. (2008). *Integrative Action of the Autonomic Nervous System: Neurobiology of Homeostasis.* Cambridge: Cambridge University Press.

Jennings, J. R., Kamarck, T., Stewart, C., Eddy, M., & Johnson, P. (1992). Alternate Cardiovascular Baseline Assessment Techniques: Vanilla or Resting Baseline. *Psychophysiology, 29*(6), 742–750.

Kathmann, N., Hochrein, A., Uwer, R., & Bondy, B. (2003). Deficits in Gain of Smooth Pursuit Eye Movements in Schizophrenia and Affective Disorder Patients and Their Unaffected Relatives. *American Journal of Psychiatry, 160*(4), 696–702.

Kaye, J. T., Bradford, D. E., & Curtin, J. J. (2016). Psychometric Properties of Startle and Corrugator Response in NPU, Affective Picture Viewing, and Resting State Tasks. *Psychophysiology, 53*(8), 1241–1255.

Keefe, R. S., Silverman, J. M., Mohs, R. C., Siever, L. J., Harvey, P. D., Friedman, L., ... Schmeidler, J. (1997). Eye Tracking, Attention, and Schizotypal Symptoms in Nonpsychotic Relatives of Patients with Schizophrenia. *Archives of General Psychiatry, 54*(2), 169–176.

Kelsey, R. M., Ornduff, S. R., & Alpert, B. S. (2007). Reliability of Cardiovascular Reactivity to Stress: Internal Consistency. *Psychophysiology, 44*(2), 216–225.

Kendler, K. S. (2009). An Historical Framework for Psychiatric Nosology. *Psychological Medicine, 39*(12), 1935–1941.

Kumra, S., Sporn, A., Hommer, D. W., Nicolson, R., Thaker, G., Israel, E., ... Gochman, P. (2001). Smooth Pursuit Eye-Tracking Impairment in Childhood-Onset Psychotic Disorders. *American Journal of Psychiatry, 158*(8), 1291–1298.

Laborde, S., Mosley, E., & Thayer, J. F. (2017). Heart Rate Variability and Cardiac Vagal Tone in Psychophysiological Research–Recommendations for Experiment Planning, Data Analysis, and Data Reporting. *Frontiers in Psychology, 8*, 1–18.

Lacey, J. I., & Lacey, B. C. (1958). The Relationship of Resting Autonomic Activity to Motor Impulsivity. *Research Publications of the Association for Research in Nervous & Mental Disease, 36*, 144–209.

Lambert, R. H., Monty, R. A., & Hall, R. J. (1974). High-Speed Data Processing and Unobtrusive Monitoring of Eye Movements. *Behavior Research Methods & Instrumentation, 6*(6), 525–530.

Lang, P. J., Greenwald, M. K., Bradley, M. M., & Hamm, A. O. (1993). Looking at Pictures: Affective, Facial, Visceral, and Behavioral Reactions. *Psychophysiology, 30*(3), 261–273.

Lang, P. J., & McTeague, L. M. (2009). The Anxiety Disorder Spectrum: Fear Imagery, Physiological Reactivity, and Differential Diagnosis. *Anxiety, Stress, & Coping, 22*(1), 5–25.

Larsen, P., Tzeng, Y., Sin, P., & Galletly, D. (2010). Respiratory Sinus Arrhythmia in Conscious Humans during Spontaneous Respiration. *Respiratory Physiology & Neurobiology, 174*(1), 111–118.

Lehrer, P. M., & Gevirtz, R. (2014). Heart Rate Variability Biofeedback: How and Why Does It Work? *Frontiers in Psychology*, 5, 1–9.

Levy, M. N. (1971). Brief Reviews: Sympathetic-Parasympathetic Interactions in the Heart. *Circulation Research*, 29(5), 437–445.

Lieberman, L., Stevens, E. S., Funkhouser, C. J., Weinberg, A., Sarapas, C., Huggins, A. A., & Shankman, S. A. (2017). How Many Blinks are Necessary for a Reliable Startle Response? A Test Using the NPU-Threat Task. *International Journal of Psychophysiology*, 114, 24–30.

Loewenfeld, I. E. (1999). *The Pupil: Anatomy, Physiology, and Clinical Applications*. Oxford: Butterworth-Heinemann.

Lubman, D. I., Yücel, M., Kettle, J. W., Scaffidi, A., MacKenzie, T., Simmons, J. G., & Allen, N. B. (2009). Responsiveness to Drug Cues and Natural Rewards in Opiate Addiction: Associations with Later Heroin Use. *Archives of General Psychiatry*, 66(2), 205–212.

Lykken, D. T., & Venables, P. H. (1971). Direct Measurement of Skin Conductance: A Proposal for Standardization. *Psychophysiology*, 8(5), 656–672.

Mckinley, P. S., King, A. R., Shapiro, P. A., Slavov, I., Fang, Y., Chen, I. S., ... Sloan, R. P. (2009). The Impact of Menstrual Cycle Phase on Cardiac Autonomic Regulation. *Psychophysiology*, 46(4), 904–911.

McTeague, L. M., & Lang, P. J. (2012). The Anxiety Spectrum and the Reflex Physiology of Defense: From Circumscribed Fear to Broad Distress. *Depression and Anxiety*, 29(4), 264–281.

Moore, T., & Zirnsak, M. (2017). Neural Mechanisms of Selective Visual Attention. *Annual Review of Psychology*, 68, 47–72.

Nassar, M. R., Rumsey, K. M., Wilson, R. C., Parikh, K., Heasly, B., & Gold, J. I. (2012). Rational Regulation of Learning Dynamics by Pupil-Linked Arousal Systems. *Nature Neuroscience*, 15(7), 1040–1046.

Nelson, B. D., & Hajcak, G. (2017). Anxiety and Depression Symptom Dimensions Demonstrate Unique Relationships with the Startle Reflex in Anticipation of Unpredictable Threat in 8 to 14 Year-Old Girls. *Journal of Abnormal Child Psychology*, 45(2), 397–410.

Nelson, B. D., Hajcak, G., & Shankman, S. A. (2015). Event-Related Potentials to Acoustic Startle Probes during the Anticipation of Predictable and Unpredictable Threat. *Psychophysiology*, 52(7), 887–894.

O'Driscoll, G. A., & Callahan, B. L. (2008). Smooth Pursuit in Schizophrenia: A Meta-Analytic Review of Research since 1993. *Brain and Cognition*, 68(3), 359–370.

Ortiz, J., & Raine, A. (2004). Heart Rate Level and Antisocial Behavior in Children and Adolescents: A Meta-Analysis. *Journal of the American Academy of Child & Adolescent Psychiatry*, 43(2), 154–162.

Pabst, O., Tronstad, C., Grimnes, S., Fowles, D., & Martinsen, Ø. G. (2017). Comparison between the AC and DC Measurement of Electrodermal Activity. *Psychophysiology*, 54(3), 374–385.

Papagiannopoulou, E. A., Chitty, K. M., Hermens, D. F., Hickie, I. B., & Lagopoulos, J. (2014). A Systematic Review and Meta-Analysis of Eye-Tracking Studies in Children with Autism Spectrum Disorders. *Social Neuroscience*, 9(6), 610–632.

Payne, A. F., Schell, A. M., & Dawson, M. E. (2016). Lapses in Skin Conductance Responding across Anatomical Sites: Comparison of Fingers, Feet, Forehead, and Wrist. *Psychophysiology*, 53(7), 1084–1092.

Peysakhovich, V., Vachon, F., & Dehais, F. (2017). The Impact of Luminance on Tonic and Phasic Pupillary Responses to Sustained Cognitive Load. *International Journal of Psychophysiology*, 112, 40–45.

Piferi, R. L., Kline, K. A., Younger, J., & Lawler, K. A. (2000). An Alternative Approach for Achieving Cardiovascular Baseline: Viewing an Aquatic Video. *International Journal of Psychophysiology*, 37(2), 207–217.

Pittig, A., Arch, J. J., Lam, C. W., & Craske, M. G. (2013). Heart Rate and Heart Rate Variability in Panic, Social Anxiety, Obsessive-Compulsive, and Generalized Anxiety Disorders at Baseline and in Response to Relaxation and Hyperventilation. *International Journal of Psychophysiology*, 87(1), 19–27.

Porges, S. W. (1992). Vagal Tone: A Physiologic Marker of Stress Vulnerability. *Pediatrics*, 90(3), 498–504.

Porges, S. W. (1995). Orienting in a Defensive World: Mammalian Modifications of Our Evolutionary Heritage. A Polyvagal Theory. *Psychophysiology*, 32(4), 301–318.

Portnoy, J., & Farrington, D. P. (2015). Resting Heart Rate and Antisocial Behavior: An Updated Systematic Review and Meta-Analysis. *Aggression and Violent Behavior*, 22, 33–45.

Portnoy, J., Raine, A., Chen, F. R., Pardini, D., Loeber, R., & Jennings, J. R. (2014). Heart Rate and Antisocial Behavior: The Mediating Role of Impulsive Sensation Seeking. *Criminology*, 52(2), 292–311.

Raine, A. (2002). Biosocial Studies of Antisocial and Violent Behavior in Children and Adults: A Review. *Journal of Abnormal Child Psychology*, 30(4), 311–326.

Ramsay, D. S., & Woods, S. C. (2014). Clarifying the Roles of Homeostasis and Allostasis in Physiological Regulation. *Psychological Review*, 121(2), 225–247.

Ray, W. J., Molnar, C., Aikins, D., Yamasaki, A., Newman, M. G., Castonguay, L., & Borkovec, T. D. (2009). Startle Response in Generalized Anxiety Disorder. *Depression and Anxiety*, 26(2), 147–154.

Richards, A., French, C. C., Calder, A. J., Webb, B., Fox, R., & Young, A. W. (2002). Anxiety-Related Bias in the Classification of Emotionally Ambiguous Facial Expressions. *Emotion*, 2(3), 273–287.

Rosenbaum, B. L., Bui, E., Marin, M.-F., Holt, D. J., Lasko, N. B., Pitman, R. K., ... Milad, M. R. (2015). Demographic Factors Predict Magnitude of Conditioned Fear. *International Journal of Psychophysiology*, 98(1), 59–64.

Roth, W. T., Telch, M. J., Taylor, C. B., Sachitano, J. A., Gallen, C. C., Kopell, M. L., ... Pfefferbaum, A. (1986). Autonomic Characteristics of Agoraphobia with Panic Attacks. *Biological Psychiatry*, 21(12), 1133–1154.

Roth, W. T., Dawson, M. E., & Filion, D. L. (2012). Publication Recommendations for Electrodermal Measurements. *Psychophysiology*, 49(8), 1017–1034.

Roy, J.-C., Boucsein, W., Fowles, D. C., & Gruzelier, J. (2012). *Progress in Electrodermal Research* (Vol. 249). New York: Springer Science & Business Media.

Sarchiapone, M., Gramaglia, C., Iosue, M., Carli, V., Mandelli, L., Serretti, A., ... Zeppegno, P. (2018). The Association between Electrodermal Activity (EDA), Depression and Suicidal Behaviour: A Systematic Review and Narrative Synthesis. *BMC Psychiatry*, 18(1), 1–27.

Scerbo, A. S., Freedman, L. W., Raine, A., Dawson, M. E., & Venables, P. H. (1992). A Major Effect of Recording Site on Measurement of Electrodermal Activity. *Psychophysiology*, 29(2), 241–246.

Schächinger, H., Weinbacher, M., Kiss, A., Ritz, R., & Langewitz, W. (2001). Cardiovascular Indices of Peripheral and Central

Sympathetic Activation. *Psychosomatic Medicine, 63*(5), 788–796.

Schell, A. M., Dawson, M. E., Nuechterlein, K. H., Subotnik, K. L., & Ventura, J. (2002). The Temporal Stability of Electrodermal Variables Over a One-Year Period in Patients with Recent-Onset Schizophrenia and in Normal Subjects. *Psychophysiology, 39*(2), 124–132.

Scott, L. N., Zalewski, M., Beeney, J. E., Jones, N. P., & Stepp, S. D. (2017). Pupillary and Affective Responses to Maternal Feedback and the Development of Borderline Personality Disorder Symptoms. *Development and Psychopathology, 29*(3), 1089–1104.

Sherwood, A., Allen, M. T., Fahrenberg, J., Kelsey, R. M., Lovallo, W. R., & Doornen, L. J. (1990). Methodological Guidelines for Impedance Cardiography. *Psychophysiology, 27*(1), 1–23.

Sijtsema, J. J., Veenstra, R., Lindenberg, S., van Roon, A. M., Verhulst, F. C., Ormel, J., & Riese, H. (2010). Mediation of Sensation Seeking and Behavioral Inhibition on the Relationship between Heart Rate and Antisocial Behavior: The TRAILS Study. *Journal of the American Academy of Child & Adolescent Psychiatry, 49*(5), 493–502.

Sirois, S., & Brisson, J. (2014). Pupillometry. *Wiley Interdisciplinary Reviews: Cognitive Science, 5*(6), 679–692.

Sloan, D. M., & Sandt, A. R. (2010). Depressed Mood and Emotional Responding. *Biological Psychology, 84*(2), 368–374.

Smith, R., Thayer, J. F., Khalsa, S. S., & Lane, R. D. (2017). The Hierarchical Basis of Neurovisceral Integration. *Neuroscience & Biobehavioral Reviews, 75*, 274–296.

Staib, M., Castegnetti, G., & Bach, D. R. (2015). Optimising a Model-Based Approach to Inferring Fear Learning from Skin Conductance Responses. *Journal of Neuroscience Methods, 255*, 131–138.

Stein, M. B., Tancer, M. E., & Uhde, T. W. (1992). Heart Rate and Plasma Norepinephrine Responsivity to Orthostatic Challenge in Anxiety Disorders: Comparison of Patients with Panic Disorder and Social Phobia and Normal Control Subjects. *Archives of General Psychiatry, 49*(4), 311–317.

Sterling, P. (2004). Principles of Allostasis: Optimal Design, Predictive Regulation, Pathophysiology, and Rational. In J. Schulkin (Ed.), *Allostasis, Homeostasis, and the Costs of Physiological Adaptation* (pp. 17–64). Cambridge: Cambridge University Press.

Sterling, P., & Eyer, J. (1988). Allostasis: A New Paradigm to Explain Arousal Pathology. In S. Fisher & J. Reason (Eds.), *Handbook of Life Stress, Cognition and Health* (pp. 629–649). Oxford: John Wiley.

Tarchanoff, J. (1890). Galvanic Phenomena in the Human Skin during Stimulation of the Sensory Organs and during Various Forms of Mental Activity. *Pflügers Archiv für die gesammte Physiologie des Menschen und der Tiere, 46*, 46–55.

Tassinary, L. G., Hess, U., & Carcoba, L. M. (2012). Peripheral Physiological Measures of Psychological Constructs. *APA Handbook of Research Methods in Psychology, 1*, 461–488.

Teixeira, A. L., Ramos, P. S., Vianna, L. C., & Ricardo, D. R. (2015). Heart Rate Variability across the Menstrual Cycle in Young Women Taking Oral Contraceptives. *Psychophysiology, 52*(11), 1451–1455.

Thayer, J. F., & Lane, R. D. (2000). A Model of Neurovisceral Integration in Emotion Regulation and Dysregulation. *Journal of Affective Disorders, 61*(3), 201–216.

Thier, P., & Ilg, U. J. (2005). The Neural Basis of Smooth-Pursuit Eye Movements. *Current Opinion in Neurobiology, 15*(6), 645–652.

Venables, P. H., & Christie, M. J. (1980). Electrodermal Activity. In I. Martin & P. H. Venables (Eds.), *Techniques in Psychophysiology* (pp. 3–67). New York: John Wiley.

Vigouroux, R. (1879). Sur le role de la resistance electrique des tissues dans l'electro-diagnostic. *Comptes Rendus Societe de Biologie, 31*, 336–339.

Waddington, C. H. (1957). *The Strategy of the Genes: A Discussion of Some Aspects of Theoretical Biology. With an Appendix by H. Kacser*. London: Allen & Unwin

Watson, D. (2005). Rethinking the Mood and Anxiety Disorders: A Quantitative Hierarchical Model for DSM-5. *Journal of Abnormal Psychology, 114*(4), 522–536.

Wilder, J. (1958). Modern Psychophysiology and the Law of Initial Value. *American Journal of Psychotherapy, 12*, 199–221.

Yoshie, N., & Okudaira, T. (1969). Myogenic Evoked Potential Responses to Clicks in Man. *Acta Oto-laryngologica, 67* (sup252), 89–103.

Young, L. R., & Sheena, D. (1975). Survey of Eye Movement Recording Methods. *Behavior Research Methods & Instrumentation, 7*(5), 397–429.

Zahn, T. P., Frith, C. D., & Steinhauer, S. R. (1991). Autonomic Functioning in Schizophrenia: Electrodermal Activity, Heart Rate, Pupillography. In S. R. Steinhauer, J. H. Gruzelier, & J. Zubin (Eds.), *Handbook of Schizophrenia, Vol. 5. Neuropsychology, Psychophysiology, and Information Processing* (pp. 185–224). New York: Elsevier Science.

Zisner, A. R., & Beauchaine, T. P. (2016). Psychophysiological Methods and Developmental Psychopathology. In D. Cicchetti (Ed.), *Developmental Psychopathology* (pp. 832–884). Hoboken, NJ: John Wiley.

12 Behavioral and Molecular Genetics

GRETCHEN SAUNDERS AND MATT MCGUE

Existing on the fringes of psychopathology research throughout much of the twentieth century, genetics has been fully embraced by the field in the twenty-first century. It is no longer possible to open a textbook in abnormal or clinical psychology and not find discussion of twin studies, heritability, and recent molecular genetic findings. The importance of genetic approaches was established by twin and adoption studies documenting the heritability of behavior. These behavioral genetic studies also provided a foundation for the molecular transformation of the field, as researchers seek to leverage findings from the Human Genome Project to identify the specific genetic variants that underlie psychopathology risk. In this chapter we will summarize the logic and findings of twin and adoption studies, as well as explore our current understanding of genetic influences on psychopathology risk through molecular genetic studies. The chapter concludes with a discussion of the most exciting yet challenging research in this area: the exploration of how genetic and environmental factors combine to influence complex behavioral phenotypes, and where the field may go from here.

TWIN AND ADOPTION STUDIES

Overview

The importance of genetic factors to psychopathology is one of the most robust behavior genetic findings to date. From Mendel's experiments showing simple Mendelian inheritance, to R. A. Fisher's biometric derivation of the inheritance patterns for complex quantitative phenotypes, it has long been known that traits follow a familial pattern of inheritance. Family history of disease has served as an important predictor of individual risk for much of this time. Determining whether this risk is conferred through genetic factors transmitted from parent to offspring or through environmental factors of being raised in the same household, however, has been the major goal of biometric analyses.

Twin and adoption studies aim to uncover the sources of variation that exist in nearly all human traits (Plomin et al., 2013). Variation between individuals is due to some combination of genetic and environmental factors as well as their interaction. Genetic variation is a consequence of the differences in the sequence of DNA bases that exist across people. Environmental variation refers to experiential and contextual differences that occur among individuals. Behavioral geneticists typically distinguish the effect of environmental factors that contribute to phenotypic similarity among reared-together relatives (e.g., socioeconomic status of the rearing home), which they designate the shared environment, from the effect of environmental factors that contribute to phenotypic differences (e.g., accidental traumatic experiences), which they designate the nonshared environment (Plomin et al., 2013).

Biometric Modeling and Heritability

Biometric modeling utilizes a genetically informative sample, like twins reared together or apart, or adoptive families, to decompose the total variance in a phenotype into variance due to genetic and environmental factors (Neale & Cardon, 2013),

$$V_P = V_{genetic} + V_{environment} \qquad (12.1)$$

The total environmental variance is the sum of the shared (designated as C) and nonshared environmental (E) variance,

$$V_{environment} = V_C + V_E \qquad (12.2)$$

while the total genetic variance is the sum of additive (A), dominant (D), and epistatic (I) genetic effects,

$$V_{genetic} = V_A + V_D + V_I \qquad (12.3)$$

A represents the additive effects of individual genetic alleles summed over the multiple alleles affecting the phenotype. D and I refer to nonadditive genetic effects due to the interactive effect of genetic variants at the same loci or at different loci, respectively. So, the total phenotypic variance can be rewritten as,

$$V_P = V_A + V_D + V_I + V_C + V_E \qquad (12.4)$$

It is important to note that E contains the effect of both nonshared environmental factors and measurement error. Also, in practice, epistatic and dominance variance are typically assumed to be negligible as these nonadditive genetic effects are difficult to estimate in practice (Carlborg, Haley, & Carlborg, 2004; Phillips, 2008) and there is empirical and theoretical evidence suggesting that these effects are not likely to be large (Hill, Goddard, & Visscher, 2008).

Heritability (h^2) is defined as the proportion of the phenotypic variance accounted for by genetic variance, or $h^2 = \frac{(V_A+V_D+V_I)}{V_P}$. Heritability consequently varies between 0, i.e., genetic factors do not contribute to individual differences, and 100 percent, i.e., individual differences on the trait are totally accounted for by genetic factors. Heritability is a ratio of variances, and thus applies to differences among individuals rather than a specific individual's trait level. If there is no phenotypic variance, for instance the trait of having a head, the heritability estimate will be undefined despite the fact that having a head is clearly influenced by genetic factors. Alternatively, if there is no genetic variance, the heritability estimate will be zero. While heritability estimates are often misinterpreted, they are important indicators of the likelihood that we will be able to identify genetic variants associated with a phenotype of interest.

Misinterpretation of heritability has led to considerable controversy within psychology (Tenesa & Haley, 2013). Importantly, heritability is not an indicator of genetic determinism, nor does it indicate whether the trait is malleable through intervention. There are many examples of environmental change and intervention that have a marked effect on phenotypes that are highly heritable. Using height as an example, the heritability of height is approximately 0.80. This means that 80 percent of the variance in height can be accounted for by genetic differences among individuals. Nonetheless, the average height of humans has increased considerably over time, approximately 11cm in Western Europe from the 1850s to the 1980s or 20cm for Dutch males from the eighteenth to twentieth centuries (Stulp & Barrett, 2016). This increase is assuredly not due to a change in the genetic factors influencing height, as evolutionarily these do not change so quickly, and must be due to environmental effects. Additionally, the difference in height between North and South Koreans is approximately 7 cm (Schwekendiek, 2009). This difference as well is most likely due to differences in nutrition and not genetic factors. Moreover, heritability estimates are not expected to be fixed constants. As environmental variance changes, so does heritability. An increase in environmental variance leads to a reduction in the estimate of heritability, implying that heritability estimates may differ considerably across groups or cultures, as well as over time and age.

Heritability, as described so far, applies to quantitative phenotypes. Psychopathology research, on the other hand, is often concerned with dichotomous phenotypes such as diagnoses. The multifactorial threshold (MFT; also called the liability-threshold model) model provides a basis for understanding the inheritance of discrete phenotypes within the Fisherian biometric tradition (Falconer, 1965; Tenesa & Haley, 2013). The MFT model assumes that discrete phenotypes follow an underlying continuous distribution of risk, termed liability. A diagnosed condition occurs when an individual's total liability exceeds a fixed threshold on the hypothesized liability continuum (Figure 12.1). Genetic and environmental factors that influence risk apply to the variance of the unobserved continuous liability. As such, heritability estimates of discrete phenotypes refer to the variance in this underlying liability.

Twin Studies

Sir Francis Galton was one of the first people to use twins to explore the differential contributions of nature and nurture. He postulated about different types of twins as far back as the 1870s, even though he had no method to distinguish them (Burbridge, 2001). Monozygotic (MZ) twins share all of their genes, meaning the raw DNA sequence of MZ twins within a pair is effectively identical. Dizygotic (DZ) twins share on average 50 percent of their segregating alleles, as is the case with non-twin siblings. Both MZ and DZ twins, reared together, share all of their common environment. These known relationships can be exploited to estimate the relative contributions of genetic and environmental factors on a given phenotype.

Historically, variation due to genetic and environment factors was derived using Falconer's formulas by comparing the correlations within MZ twin pairs to the correlations within DZ twin pairs (Falconer & Mackay, 1996). The Falconer model assumes all genetic effects are additive (for reasons specified above), so that the following variance components can be estimated: $\frac{V_a}{V_p} = a^2 = 2(r_{MZ} - r_{DZ})$, $\frac{V_c}{V_p} = c^2 = 2r_{DZ} - r_{MZ}$, and $\frac{V_e}{V_p} = e^2 = 1 - a^2 - c^2$, where r is the twin correlation for a quantitative phenotype (or tetrachoric correlation for a categorical phenotype), a^2 is the proportion of the total phenotypic variance attributable to additive genetic factors (additive heritability), c^2 is the proportion of total variance attributable to shared environmental factors, and e^2 is the proportion of total variance attributable to nonshared environmental factors. While Falconer's estimates were useful in a historical sense, they can result in negative estimates of variance parameters. Modern estimation of biometric variance components uses structural equation modeling to decompose the genetic and environmental influences on a given phenotype, often referred to as ACE modeling (Boomsma, Busjahn, & Peltonen, 2002; Neale & Cardon, 2013). An example of the univariate ACE decomposition is shown in Figure 12.2.

Multifactorial threshold model

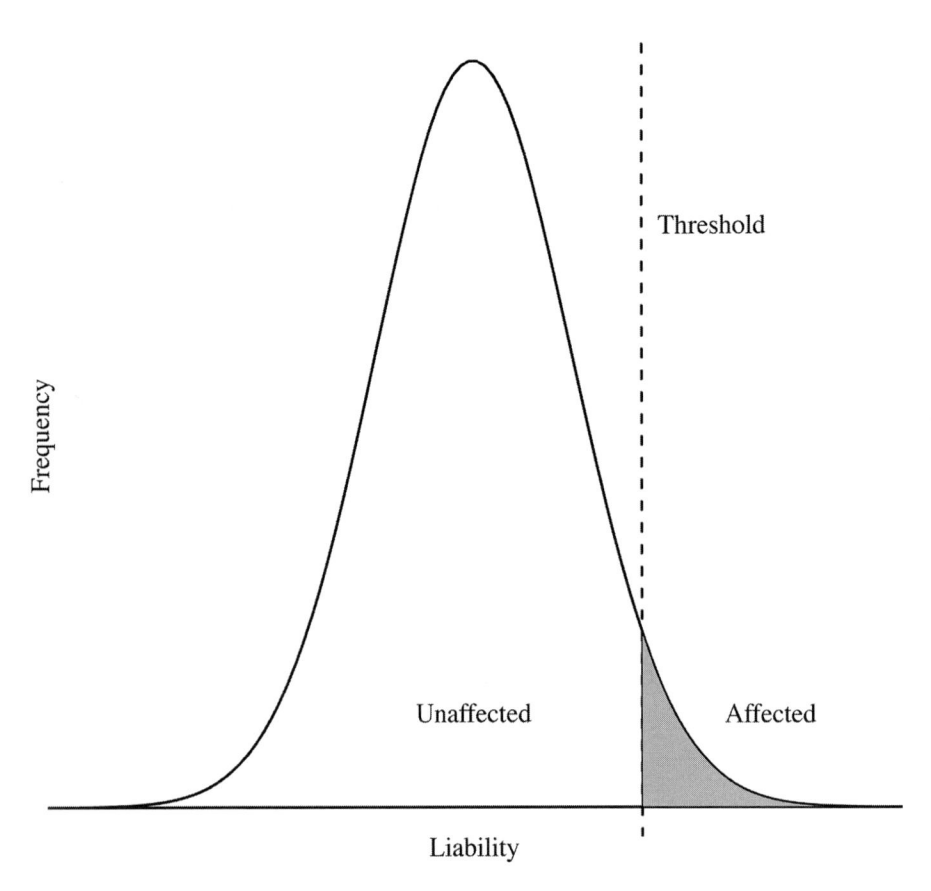

Figure 12.1 Multifactorial threshold (MFT) model. The MFT assumes that a categorical phenotype (e.g., psychiatric diagnosis) can be accounted for by the existence of an underlying quantitative variable, termed liability, such that an individual is affected if their liability exceeds some threshold along the liability continuum

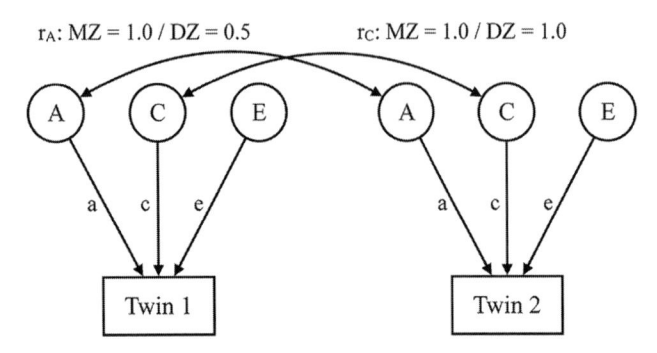

Figure 12.2 Univariate ACE model. The typical model used in analyzing data from monozygotic (MZ) and dizygotic (DZ) twins assumes that phenotypic variance is a function of additive genetic effects (A), shared environmental effects (C) and non-shared environmental effects (E)

Twin studies inherently include several assumptions when estimating heritability. The equal environments assumption (EEA) states that MZ and DZ twins do not differ on environmental similarity (Kendler et al., 1993), so that environmental factors do not lead to greater similarity in MZ than DZ twins. There have been numerous studies testing the EEA assumption and it appears that the assumption holds in most cases (Felson, 2014). A second assumption is that of random mating or that individuals do not select mates with phenotypic values similar to their own. For most psychological traits the degree of spousal similarity is minimal (Maes et al., 1998) and failure to meet this assumption, which would lead to an underestimation of heritability, is not thought to have much of an effect. Standard analysis of twin data additionally assumes that only additive genetic effects contribute to heritability estimates. Any interaction between loci (dominance or epistasis) or interaction between genes and the environmental is not considered. Violation of this assumption could lead to an overestimation of heritability and a corresponding underestimation of the effect of the shared environment, although there is not much evidence for the existence of nonadditive genetic effects. Finally, twin studies rely on the notion that twins are representative of the general population. Despite early claims that twins were at greater risk for psychopathology, evidence suggests that there is little, if any, difference in the rates of psychopathology between twins and singletons (Andrew et al., 2001). When taken in aggregate, the assumptions underlying the basic twin study are reasonable or likely to be approximately true. Nonetheless, most researchers recognize that the approach can only provide approximate estimates of biometric variance components and they look for other study designs to confirm twin study findings.

Adoption Studies

Adoption studies provide an alternative way of decomposing the phenotypic variance into genetic and environmental components (Plomin, Owen, & McGuffin, 1994; Plomin et al., 2013). In adoption studies, the resemblance between adopted individuals is compared to both their adoptive and biological relatives. Since the adopted individual and their adoptive relatives share only their environment, their similarity implicates the importance of environmental factors. Adopted individuals and their biological relatives only share genetic factors, and no environmental factors, so similarity between them must be due to these genetic factors.

Adoption studies rely on a couple of key assumptions as well. One is that adoptive families are representative of the general population. It is known, however, that adoptive families often have greater socioeconomic status than the general population (McGue et al., 2007), which can result in an underestimation of environmental effects (Stoolmiller, 1999). A second assumption is a lack of selective placement. Selective placement refers to placing an individual in an adoptive home that is similar to their biological background. The presence of selective placement will inflate estimates of heritability in adoption studies (Neale & Cardon, 2013). As with a twin study, there are limitations to the adoption study approach, but the limitations differ in the two types of studies. Consequently, convergent findings across the two designs are thought to reflect real effects rather than a failure to meet underlying assumptions.

Findings from Twin and Adoption Studies

Findings from biometric studies of twins have been quite robust. Large registries of twins have been collected and followed over time (Boomsma et al., 2002; Hur & Craig, 2013). Meta-analyses combining these large twin studies with many adoption studies have also helped to robustly estimate heritabilities for a multitude of phenotypes (Polderman et al., 2015). Table 12.1 displays heritability estimates for various forms of psychopathology from twin studies. Results are from large-scale studies or meta-analyses, when available.

Consistent with what Eric Turkheimer has termed the "three laws of behavior genetics" (Turkheimer, 2000), several patterns emerge from twin studies of psychopathology. First, nearly all psychopathology is to some degree heritable. MZ twin concordance is almost always larger than DZ concordance, and additive genetic variance (a^2) is moderate for most phenotypes (between 30 and 60 percent). Second, genetic factors account for more of the total variance than shared environmental factors (estimates of a^2 are larger than c^2). This suggests that the effect of genes has a larger impact than the effect of the home (shared) environment. Lastly, after accounting for the variance attributable to genetic and shared environmental factors, there is a substantial portion of the variance left over. The nonshared environmental estimates (e^2) are substantially greater than zero across the board, indicating that the environment has a relatively large effect on psychopathology but that the key environmental factors are those that are not shared within the same family. Consistent with this, MZ twins are never perfectly concordant, further emphasizing the role of environmental factors.

Patterns also emerge in adoption studies. For most psychopathology there is greater similarity between adopted individuals and their biological relatives, than with their adoptive families. This further emphasizes the importance of genetic factors on psychopathology. Adoption studies are better able to assess shared environmental variance, unconfounded by genetic factors, than twin studies. For several phenotypes, there is evidence that shared environmental factors play an important role in risk for psychopathology, including depression and anxiety disorders, as well as conduct and oppositional defiant disorders (Burt, 2009). This is supported by similarity between adopted individuals and their adoptive families, despite not sharing any genetic factors.

To date, most biometric studies of relatively common psychopathology are generally well powered, with adequate sample sizes, though there is some evidence of studies that are underpowered to detect small, but significant, effects of the shared environment (Burt, 2009). Meta-analyses, in which sample sizes are pooled, are relatively prevalent now and may provide more reliable estimates of variance components for psychological disorders than individual studies. The low prevalence of some forms of psychopathology, however, makes it difficult to achieve sample sizes large enough to reliably estimate genetic and environmental influences. Estimated ACE components based on small sample sizes should be corroborated with larger studies, meta-analyses in which sample sizes are pooled, or other biometric methodologies. Adoption studies are less common than twin studies due to greater difficulty in obtaining the sample. In general, however, both twin and adoption studies have produced similar heritability estimates, despite their different assumptions, providing robust evidence for the importance of genetic factors in many disorders (Nikolas & Alexandra, 2010; Sullivan et al., 2000; Verhulst et al., 2015).

Identifying Genetic Variants

A nonzero heritability estimate implies that genetic factors contribute to the variance in a phenotype. The next logical step is to identify the specific genetic variants that contribute to a given phenotype. Doing so would not only enhance our understanding of the causes of psychopathology, but would also ultimately help in prevention and more effective treatment of these disorders. The public health impact could be enormous in terms of targeted preventative therapy and treatment.

Table 12.1 Individual and meta-analyses of twin studies of psychopathology

	First author (year)	Type of study	Number of twin pairs (sex)	Genetic a^2	95% CI	Shared environment[a] c^2	95% CI	Nonshared environment e^2	95% CI
Major depression									
	Sullivan et al. (2000)	Meta-analysis	8,485 (M&F)	0.37	(0.33, 0.42)	–		0.63	(0.58, 0.67)
	Kendler et al. (2006)	Twin study	4,358 (M)	0.29	(0.19, 0.38)	–		0.71	(0.62, 0.81)
			5,502 (F)	0.42	(0.36, 0.47)	–		0.58	(0.53, 0.64)
Anxiety disorders									
Generalized anxiety disorder									
	Hettema et al. (2001)	Meta-analysis	4,561 (M)	0.32	(0.24, 0.39)	–		0.68	(0.61, 0.76)
			832 (F)	0.32	(0.24, 0.39)	0.17	(0.03, 0.29)	0.51	(0.41, 0.64)
Panic disorder									
	Hettema et al. (2001)	Meta-analysis	9,007 (M&F)	0.43	(0.32, 0.53)	–		0.57	(0.47, 0.68)
Obsessive-compulsive disorder									
	Taylor (2011)	Meta-analysis	22,690 (M&F)	0.38	(0.30, 0.46)	–		0.52	(0.50, 0.55)
Bipolar disorder									
	McGuffin et al. (2003)	Twin study	67 (M&F)	0.85	(0.73, 0.93)	–		0.15	(0.07, 0.27)
	Kieseppa et al. (2004)	Twin study	25 (M&F)	0.93	(0.69, 1.00)	–		0.07	(0.00, 0.31)
Schizophrenia									
	Sullivan et al. (2003)	Meta-analysis	2,106 (M&F)	0.81	(0.73, 0.90)	0.11	(0.03, 0.19)	0.08	NR
ADHD									
	Faraone et al. (2005)	Meta-analysis		0.76	NR	NR		NR	
Autism spectrum disorder									
	Tick et al. (2016)	Meta-analysis	8,451 (M&F)	0.74	(0.70, 0.87)	0.25	(0.12, 0.37)	0.01	(0.01, 0.03)

Table 12.1 (cont.)

	First author (year)	Type of study	Number of twin pairs (sex)	Genetic a^2	95% CI	Shared environment[a] c^2	95% CI	Nonshared environment e^2	95% CI
Eating disorders									
Anorexia nervosa									
	Bulick et al. (2006)	Twin study	382 (F)	0.31	(0.00–0.62)	0.00	(0.00–0.44)	0.68	(0.37–1.00)
Bulimia nervosa									
	Mazzeo et al. (2010)	Twin study	1,024 (F)	0.63	(0.26, 0.74)	–		0.37	(0.24, 0.56)
Substance use disorders									
Alcohol									
	Verhulst et al. (2015)	Meta-analysis	48,491 (M&F)	0.51	(0.45, 0.56)	0.10	(0.03, 0.16)	0.39	(0.38, 0.42)
Cannabis									
	Verweij et al. (2010)	Meta-analysis	7,539 (M)	0.51	(0.38, 0.65)	0.20	(0.11, 0.28)	0.29	(0.22, 0.35)
			4,483 (F)	0.59	(0.44, 0.73)	0.15	(0.01, 0.30)	0.26	(0.23, 0.30)
Tobacco									
	Li et al. (2003)	Meta-analysis	34,973 (M&F)	0.52	(0.54, 0.63)	0.13	(0.06, 0.22)	0.35	(0.34, 0.41)
Externalizing behavior									
	Bezdjian et al. (2011)	Meta-analysis	12,623 (M&F)	0.49	NR	–		0.51	NR

[a] Dash indicates that this parameter was constrained to be zero in the best fitting model; M = male, F = female, a^2 = additive genetic heritability, c^2 = shared environmental component of variance, e^2 = nonshared environmental component of variance, CI = confidence interval, NR = not reported.

For Mendelian disorders, in which the presence or absence of the disorder is determined by the status of one gene at a specific location in the genome (Antonarakis & Beckmann, 2006), identifying the gene involved has been very successful through linkage analysis followed by candidate-gene studies (Manolio et al., 2009). These disorders are often referred to as single-gene (monogenic) disorders because they are caused by the allele status of a particular gene. An example of a Mendelian disorder is phenylketonuria where mutations in the phenylalanine hydroxylase (PAH) gene result in the inability to break down phenylalanine (Scriver, 2007). Other examples of Mendelian disorders include Huntington's disease and cystic fibrosis (Chial, 2008).

In terms of psychopathology, single-gene disorders are essentially nonexistent. Psychological disorders are almost exclusively considered polygenic, in which multiple genetic variants contribute to the phenotype (Buckholtz & Meyer-Lindenberg, 2012). These complex phenotypes are influenced not only by multiple genetic variants, but also by a multitude of environmental factors. Given this complexity, it is not surprising that identifying the underlying genes that contribute to polygenic phenotypes has been much more difficult than for single-gene disorders and much more difficult than originally anticipated (Wilson & Nicholls, 2015).

Candidate-Gene Studies

Historically, the search for genetic variants began with candidate-gene studies. In this approach, a gene of interest was identified as being related to a phenotype (Kwon & Goate, 2000). The allele status (one of the alternative forms a gene can take) at this particular genetic locus was then compared in cases, individuals with the phenotype, and controls, those without it. If there was a significant difference in allele frequency between cases and controls, one would conclude that this genetic variant is associated with the phenotype. Candidate-gene studies require that a gene, or genes, with a possible association to a phenotype be specified in advance. Doing so requires hypotheses about how and why the gene is related to the phenotype.

Overall, candidate-gene studies have not been particularly successful, largely due to lack of replication of their findings (Munafò, 2006). They display poor replicability for several reasons. Original candidate-gene studies often had small samples sizes. These studies, at the time, were thought to be adequately powered. We now know, however, that the effect sizes of individual variants are exceedingly small, requiring very large samples for detection (Hardy & Singleton, 2009; Robinson, Wray, & Visscher, 2014).

The necessity of a hypothesis-driven approach also limited the applicability of candidate-gene studies. Specifying candidate genes in advance was difficult for phenotypes in which little was known about the underlying biology, as is the case with most forms of psychopathology. What was needed was an approach that did not depend on biological knowledge, a hypothesis-free approach to identifying variants that were not already known to be related to a phenotype.

Genome-Wide Association Studies

Technological advances soon allowed for such an approach, called a genome-wide association study (GWAS). A GWAS is a hypothesis-free approach that scans the entire genome for variants that might be associated with a particular phenotype (Hardy & Singleton, 2009; Hirschhorn & Daly, 2005). In GWAS, several hundred thousand, to over a million, single nucleotide polymorphisms (SNPs) distributed across the whole genome are genotyped. A SNP is a single base pair location in the genome that commonly varies within a population. Envisioning DNA as a double helical structure, a SNP is a location on that structure where the nucleotide base (i.e., A, C, T, or G) varies among humans. Theoretically, humans can vary at any base pair location in the genome, but in most of these cases the variation would be considered rare. SNPs get at common genetic variants in which the rarer of the possible alleles (alternate forms of the possible nucleotide bases at each SNP location) occurs in greater than 1 percent of the population.

There are millions of catalogued SNPS in the human genome, although in practice only a subset of these SNPs needs to be genotyped to adequately cover the whole genome. This is due to what is called linkage disequilibrium (LD). LD refers to the phenomenon whereby SNPs that are located near each other, within a couple of thousand base pairs, tend to be highly correlated with each other (Hirschhorn & Daly, 2005). If we know the allele status at one SNP, we can infer the allele status at a nearby SNP with high accuracy. This allows us to genotype only a portion of SNPs in order to provide coverage for most SNP variation (Jorde, 2000).

The logic behind GWAS is exceedingly simple. At each genotyped SNP, cases and controls are compared on the frequency of the base alleles. If there is a significant difference in allele frequency between cases and controls, then this SNP would be identified as being related to the phenotype of interest. For quantitative phenotypes, the analysis involves regressing the outcome on a count of the number of one of the alleles for each SNP (i.e., 0, 1, or 2) (Bush & Moore, 2012). In both cases, the comparison of each SNP is a separate hypothesis test. Assuming a type I error rate of 5 percent with one million genotyped SNPs, this would result in 50,000 false positives under the null hypothesis (i.e., 50,000 of the one million tests would incorrectly suggest a significant SNP effect due to chance). To control for this unacceptably high rate of expected false positive findings, the significance threshold in a GWAS is usually lowered to 5×10^{-8} (Johnson et al., 2010), equivalent to a Bonferroni correction for multiple testing of one million independent tests. Additionally, the vast majority of GWAS studies control for population structure (Yang et al., 2011b). Allelic differences can occur among human population groups as a function of their differing evolution rather than due to disease risk. Failure to account for this structure can result in spurious relationships between the phenotype and a SNP.

When adequately powered, GWAS typically results in the identification of multiple SNPs associated with the phenotype, but these results should be interpreted with caution. A SNP identified as significantly associated with a phenotype is not necessarily a causal variant (Hardy & Singleton, 2009). In other words, it may not be the identified SNP that increases/decreases risk for a phenotype, but rather a SNP that is nearby and so in LD with the identified one that is causal. In some cases, the identified SNP will fall within a gene, which would implicate this gene in the causal pathway, but in other cases the true causal gene may lie either upstream or downstream of the SNP in the genome. Regardless, identification of a significantly associated SNP allows for a more focused follow-up genomic search for the true causal variant.

GWAS studies were originally designed around the common-disease/common-variant (CDCV) model, which is thought to be particularly applicable to psychopathology because they are relatively common disorders. In the CDCV model, it is assumed that the genetic variants that affect disease risk are common within a population but have a very small effect on the phenotype (Reich & Lander, 2001; Zondervan & Cardon, 2004). This helps explain why some psychiatric disorders persist in a population despite being selected against (low fecundity rates), as many unaffected individuals will carry the risk alleles. Because the effects are assumed to be very small, this model implies that very large samples sizes are required to detect these genetic variants (Robinson et al., 2014). Researchers consequently pool studies and undertake cumulative meta-analyses in order to achieve the large samples needed. The SNPs that are identified in initial studies will have the largest effect; their signal is the strongest and easiest to detect. As sample size grows, each subsequent SNP that is identified is expected to have a smaller and smaller effect. Increasing sample sizes acts to boost the signal of these tiny effects, making them more likely to be detected.

Schizophrenia provides a good example of this process. Several schizophrenia consortiums have pooled numerous samples of cases and controls to increase the total sample size for schizophrenia GWAS studies. The first results from these efforts were published in 2011 with 9,394 cases and 12,462 controls (Schizophrenia Psychiatric Genome-Wide Association Study Consortium, 2011). Seven SNPs were found to be significantly associated with schizophrenia, with five of these SNPs being newly identified. Following up in 2013 with 21,246 cases and 38,072 controls identified 13 SNPs significantly associated with schizophrenia (Ripke et al., 2013). Finally, in 2014, with 36,989 cases and 113,075 controls, 128 SNPs were found to be associated with schizophrenia (Schizophrenia Working Group of the Psychiatric Genomics Consortium, 2014). The increasing sample size corresponding to increases in identification of SNPs highlights the importance of large samples and pooled consortiums to further uncover genetic variants associated with psychopathology.

GWAS Findings

A summary of current GWAS findings of common psychopathology is shown in Table 12.2. While GWAS studies are still building adequate sample sizes, particularly with psychological disorders, one feature has emerged in nearly every large-scale GWAS undertaken: the effect of any individual SNP on a phenotype is very small. In the most recent GWAS of schizophrenia, the SNP with the smallest p-value had an associated odds ratio of 1.21 on schizophrenia risk (Schizophrenia Working Group of the Psychiatric Genomics Consortium, 2014) This implies that hundreds, if not thousands, of SNPs contribute to the variation in each specific form of psychopathology. A current estimate of the total number of SNPs contributing to schizophrenia risk is upwards of 8,000 (Ripke et al., 2013). Given that only 128 variants have been identified thus far, we clearly have a long way to go.

Major Findings and Controversies

Overall, findings from twin and adoption studies show that there is a nontrivial heritability associated with nearly all behavioral phenotypes. Building upon this, GWAS, when adequately powered, has begun to detect genetic variants associated with several behavioral phenotypes and psychological disorders. GWAS are still in their relative infancy and almost assuredly a wealth of genetic variants will be identified as sample sizes increase further. Vast quantities of genetic and phenotypic data exist globally and are now commonly being pooled to create GWAS consortiums.

While the prospect of identifying gene variants associated with psychopathology appears promising, the slow process of uncovering these genetic variants currently hinders the field. The idea of missing heritability is also a current controversy in the area.

Missing Heritability

Missing heritability refers to difference in heritability estimates from twin/adoption studies and the heritability identified through GWAS findings. There appears to be a sizable amount of missing heritability across all measured phenotypes, including height. The heritability of height estimated from twin studies is 0.80. The 697 variants identified by GWAS in total account for 20 percent of the variance in height (Wood et al., 2014). This means that the total phenotypic variance of height that can be explained by the combined effect of the 697 significantly associated SNPs is 20 percent. The difference between 80 and 20 percent is the missing heritability for height (a more recent preprint, not yet subject to peer review, has now identified 3,290 significantly associated SNPs that explain almost 25 percent of the variance in height; Yengo et al., 2018). The picture is less favorable for behavioral phenotypes. The heritability of schizophrenia, one of the most studied psychological phenotypes, has been estimated to be as high as 0.80, while the heritability identified through GWAS is 0.07 (Schizophrenia Working Group of the Psychiatric Genomics Consortium, 2014). This gap represents a problem in identifying all gene variants associated with a phenotype. Several ideas have been proposed and discussed to explain missing heritability (Eichler et al., 2010; Maher, 2008; Manolio et al., 2009).

One way in which missing heritability could be explained is through overestimation of heritability estimates from twin and family studies. It may be the case that overestimation of heritability estimates leads to the appearance of a larger problem of missing heritability than actually exists. Overestimation can result from violations of the assumption of nonadditive genetic effects (Manolio et al., 2009; Zuk et al., 2012). Although this is a possible explanation for missing heritability, twin, family, and adoption studies use different samples, logic, and assumptions to estimate heritability. If all methods of heritability estimation overestimate the effect, they would have to do so by about the same amount. Twin and family studies generally agree quite closely on their estimates of heritability.

Table 12.2 Genome-wide association studies (GWAS) of psychopathology

	First author (year)	Sample size (N)	No. of independent genome-wide significant SNPs	Proportion of variance accounted for by significant SNPs	Total SNP heritability (SE)[a]
Major depression					
	Hyde et al. (2016)	75,607 cases; 231,747 controls	17	NR	0.19 (0.018)
Anxiety disorders					
	Otowa et al. (2016)	7,016 cases, 14,745 controls	1	NR	NR
Bipolar disorder					
	Muhleisen et al. (2014)	9,747 cases, 14,278 controls	5	NR	0.23 (0.010)
Schizophrenia					
	Ripke et al. (2014)	32,405 cases, 113,075 controls	108	0.034	0.19 (0.007)
ADHD[a]					
	Demontis et al. (2019)	20,183 cases, 35,191 controls	12	NR	0.22 (0.014)
Autism spectrum disorder					
	Autism Spectrum Disorders Working Group of the Psychiatric Genomics Consortium (2017)	16,539 cases, 157,234 controls	0	0	0.11 (0.017)
Eating disorders					
Anorexia nervosa					
	Boraska et al. (2014)	5,551 cases; 21,080 controls	0	0	NR
Substance use disorders					
Alcohol (dependence)					
	Gelernter et al. (2014)	7,677 cases, 7,102 controls	2	NR	NR
Tobacco (initiation)					
	Furberg et al. (2010)	143,023 total	8	NR	NR

[a] SNP heritabilities obtained from Lee et al. (2013).
NR = not reported.

Insufficient GWAS sample sizes may also contribute to missing heritability. If GWAS studies are underpowered they will not be capable of capturing a large proportion of the total heritability of a phenotype (Hong & Park, 2012). The statistical method known as Genome-wide complex trait analysis (GCTA)/Genome-based restricted maximum likelihood (GREML) can quantify the extent to which underpowered GWAS contribute to missing heritability (Yang et al., 2011a). This procedure uses GWAS findings to estimate the total heritability accounted for by all possible SNPs, often called the SNP heritability, without actually identifying the specific SNPs that contribute to phenotypic variance. The SNP heritability is typically much closer to heritability estimates from twin and family studies than is the total variance accounted for by variants identified in GWAS. For height, the estimate of the SNP heritability is 0.60 – much closer to the heritability estimate of 0.80 than the 0.20 accounted for by known height variants (Wood et al., 2014). For psychopathology, however, SNP heritability estimates typically diverge more from the estimates from twin and adoption studies than is the case for height. For example, the estimated SNP heritability of schizophrenia ranges between 0.19 and 0.24, which is still quite a bit less than the biometric heritability estimate of 60–80 percent (Lee et al., 2013). This implies that even with very large sample sizes much of the heritability of psychopathology will remain missing.

A third possible explanation of missing heritability is nonadditivity of genetic effects (Shao et al., 2008). Because each SNP variant is analyzed separately, GWAS only considers the additive effects of genetic variants, whereas nonadditive genetic effects can contribute to heritability estimates from twin studies. Nonadditive effects include dominance, which is due to the interactive effect of genetic variants at the same loci, and epistasis, which is due to interactive effects of genetic variants across loci. A comparison of MZ and DZ twin-pair correlations is often used to assess for the likelihood of genetic nonadditivity. If the MZ twin-pair correlation is approximately twice that of the DZ twin-pair correlation, this would suggest the existence of only additive genetic effects. Evidence for dominant or epistatic genetic effects occurs when the MZ twin-pair correlation is greater than twice the DZ twin-pair correlation. While nonadditive genetic effects would contribute to missing heritability, it is not likely to be a major factor due to the general lack of evidence for nonadditivity. That is, for most phenotypes the MZ correlation does not exceed twice the DZ correlation (Polderman et al., 2015).

Finally, missing heritability may be explained by the influence of genetic variants not assayed by standard genotyping arrays. GWAS have been designed to identify the contribution of common SNPs and are typically poorly suited to identify the contribution of rare and non-SNP variants even though these variants are likely to also contribute to disease risk (Hardy & Singleton, 2009; Manolio et al., 2009). Copy number variants (CNVs) provide an example of rare variants that are not captured in GWAS yet are known to contribute to risk for psychopathology (Addington & Rapoport, 2012; Malhotra & Sebat, 2012). Standard genetic theory holds that at each location in the genome we have two copies, one inherited from our mothers and the other inherited from our fathers. CNVs, however, are regions in the genome, spanning thousands up through millions of DNA bases, where individuals can have either duplicated or deleted genetic material. Any specific CNV is rare, typically occurring in less than 1 in 1,000 individuals (Sebat et al., 2004), yet the possession of CNVs is normative. As compared to the reference genome, the typical person has 170 CNVs, i.e., regions where they have a large number of contiguous DNA bases that are either duplicated or deleted (1000 Genomes Project Consortium et al., 2015).

Many CNVs have no discernable phenotypic effect, but some can have dramatic effects on the expression of a phenotype (Stranger, Stahl, & Raj, 2011). For example, a 593 kilobase de novo (meaning "new") deletion or duplication on chromosome 16p11.2 is associated with increased risk for autism (odds ratio (OR): 9.5–11.8), but accounts for only about 1 percent of autism cases (Weiss et al., 2008). Overall, large de novo CNVs, covering multiple genes, have been found to be associated with autism risk (OR = 5.6 for individuals carrying at least one de novo CNV encompassing more than one gene), with the total number of associated CNVs estimated to be between 130 and 234 (Sanders et al., 2011). Schizophrenia has also been associated with specific CNVs as well as with the overall CNV burden. The best known of these is the 22q11.2 deletion syndrome that accounts for approximately 1% of schizophrenia cases and increases relative risk by 20 to 25 times that of the lifetime general population risk (Bassett & Chow, 2008; Drew et al., 2011). Additionally, three de novo deletions on chromosomes 1q21.1, 15q11.2, and 15q13.3 have been shown to be significantly associated with increased schizophrenia risk (ORs 8.1, 2.1, 10.7, respectively; Stefansson et al., 2008; Stone et al., 2008). Each of these deletions, however, account for less than 1 percent of schizophrenia cases and so would account for little of the missing heritability. Several studies have also found high rates of de novo CNVs in cases of schizophrenia without a family history (Bin Xu et al., 2008; Kirov et al., 2012; Stone et al., 2008). Rare variant analysis, of which CNVs are a special case, can be exceedingly difficult to conduct, requiring very large sample sizes to detect. Additionally, while there is evidence that rare variants do contribute to risk for some types of psychopathology, identifying the specific rare variants has been very difficult to do (Gandal et al., 2016; Malhotra & Sebat, 2012).

Summary of Current Genetic Architecture of Psychopathology

Genetic architecture refers to the number, type, and effect size and frequency of risk variants associated with a phenotype (Manolio et al., 2009). The number of genetic variants contributing to any specific form of psychopathology risk is likely to be in the hundreds, possibly thousands. An estimate of the total number of SNPs associated with schizophrenia risk is currently about 8,300, accounting for approximately 32 percent of disease liability (Ripke et al., 2013). For several forms of psychopathology, we also know of multiple rare variants and CNVs that have a pronounced impact on psychopathology risk. Of note is that the existence of both rare variants with large phenotypic effect and common variants of small effect for any given form of psychopathology implicates etiological heterogeneity – there appear to be multiple pathways that can end in what is diagnosed as the same psychopathological condition (McClellan & King, 2010). Using autism as an example, rare mutations on chromosome 16p11.2 confer a marked increase in risk for the disorder but most individuals with autism do not carry this mutation.

At this point, we are not yet able to distinguish types of psychopathology based on genetic risk variants, meaning that diagnoses cannot yet be made based on the genetic variants an individual carries. Current methods have focused on common genetic variants, occurring in >1 percent of the population, that have very small phenotypic effects, as well as rare variants, occurring in <0.1 percent of the population, that have relatively large phenotypic effects. These methods are undoubtedly failing to capture all types of genetic variation. Common variants with exceedingly small effects will require even greater sample sizes to detect. While somewhat rare variants with moderate effects almost certainly exist, they have so far been elusive. The current increases in sample sizes in genetic studies will continue to spur progress in this area, though missing heritability is likely to remain an issue for the foreseeable future.

The need for very large samples sizes to detect genetic variants associated with risk prompts the question of whether this effort will pay off in the long run. The goal in genetic analyses is not necessarily to identify all genetic variants associated with risk, but rather to better understand the biological mechanisms underlying psychopathology. Using migraine as an example, a recent GWAS found 38 distinct loci significantly associated with risk. Several of the identified genes are implicated in vascular pathways, supporting the hypothesis that dysfunction in the vascular system contributes to migraine risk (Gormley et al., 2016). This opens avenues for future treatment of migraine that might not have been previously considered. For psychopathology, schizophrenia GWAS results also provide some promise in understanding the etiology of disease and implications for treatment. The most recent schizophrenia GWAS identified possible causal genes affecting proteins associated with the glutamatergic neurotransmission and synaptic plasticity, as well as genes known to encode for calcium channels (Schizophrenia Working Group of the Psychiatric Genomics Consortium, 2014). These insights might allow for better future therapeutic treatments of psychopathology.

Genetic analyses have also uncovered genetic structures that were not previously known. For example, schizophrenia and bipolar disorder share a substantial genetic correlation, implying the existence of multiple genetic variants that increase risk for both diseases (Cross-Disorder Group of the Psychiatric Genomics Consortium et al., 2013; Purcell et al., 2009). Additionally, schizophrenia and bipolar disorder share common genetic variants with creativity (Power et al., 2015), while variants associated with autism spectrum disorder (ASD) are linked with high cognitive functioning (Crespi, 2016). Findings such as these may help explain why diseases persist in a population despite their selective disadvantage; they may also help in elucidating the nosological boundaries among alternative forms of psychopathology. The causal genetic variants may be associated with both disease and normative, or even adaptive, behavior in a population.

Understanding an individual's genetic background may provide potential for advances in personalized medicine (Hamburg & Collins, 2010). The goal is to tailor medical intervention or preventative therapies based on an individual's genome. There is some evidence of success in terms of pharmacogenetic advances in treatment of schizophrenia. Patients with schizophrenia have shown variations in drug metabolism rates and drug side effects based on polymorphisms in dopamine and serotonin receptor genes, as well as drug metabolic polymorphisms in the cytochrome P450 genes (Arranz & de Leon, 2007). Additionally, genetic variation in glutamate decarboxylase-like 1 (GADL1) has been shown to affect a patient's response to lithium maintenance treatment for bipolar disorder (Chen et al., 2014). As progress is made in understanding the underlying biology of psychopathology, we will be better able to identify individuals at risk for disease and apply preventative interventions to reduce negative consequences of these diseases.

Genetics and the Environment

So far, we have only considered the effects of genetic and environmental factors separately. It could be the case, however, that genes and the environment interact in particular ways to influence psychopathology risk. The nature vs. nurture debate concluded in recognition of the importance of both genetics and the environment; the question is not whether but how genes and environment influence psychopathology risk (Anastasi, 1958). Here we will describe three of the main ways genes and the environment may interact: gene by environment correlation, gene by environment interaction, and epigenetics.

Genotype-Environment Correlation

Genotype-environment correlation (rGE) refers to the idea that an individual's genotype may be associated with their exposure to certain environments (Jaffee & Price, 2008; Plomin, DeFries, & Loehlin, 1977). There are three main

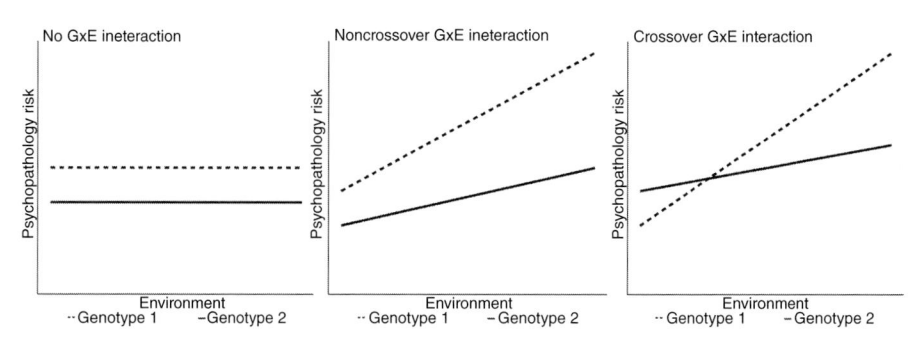

Figure 12.3 Examples of gene-by-environment interactions

types of rGE: active, passive, and evocative (reactive) genotype-environment correlations. Active rGE occurs when an individual's genotype influences the environment they seek out or select. For example, individuals with conduct problems may seek out environments in which rule-breaking or aggressive behavior is rewarded. Passive rGE occurs when parents contribute both genes and an environment that influence a trait. Parents of children with conduct disorder both pass on genes and help create a rearing environment that likely contributes to behavioral problems. Lastly, evocative rGE refers to genotype-environment correlations in which an individual's genotype influences how others respond to them. An example may be that a teacher applies a more negative teaching style to a student with externalizing behaviors. If conduct problems are heritable, which they are likely to be, then this will induce a correlation between the genes a child inherits and the child's experiences at school.

Genotype-environment correlations, while difficult to identify in some situations, have been examined most often using twin and adoption studies showing that environmental measures are heritable (Jaffee & Price, 2008; Kendler & Baker, 2007; Plomin & Bergeman, 1991). The existence of rGE has both methodological and substantive implications. The methodological implications derive primarily from concern that rGE can result in biased estimates of genetic and environmental influence, although twin and adoption studies can often identify and minimize these statistical biases (Plomin et al., 1977). The substantive implications derive from the recognition that active and reactive rGE imply that environmental factors may mediate some of the heritable effects on behavior. For example, perhaps one reason externalizing behaviors are heritable is that in our society individuals have some flexibility in constructing their experiences in a way that complements and reinforces their underlying genetically influenced behaviors.

Gene by Environment Interaction

Gene by environmental interaction (GxE) refers to the idea that the effect of the environment on a phenotype depends on the genetic makeup of the individual (Plomin et al., 1977). There are many different types of interaction that fall into two broad categories: noncrossover and crossover interaction effects. Examples of these categories, as well as the absence of GxE, are illustrated in Figure 12.3. Results with no GxE are shown in Figure 12.3a. Fan-shaped interactions occur when one genotype is always at higher risk than other genotypes but the magnitude of that elevated risk depends on the environment (Figure 12.3b). Crossover interactions, on the other hand, occur when one genotype is not always at highest risk but rather which genotype shows greatest risk depends on the level of environmental exposure (Figure 12.3c).

While the idea of GxE appears likely, there is much debate over the current results of GxE research. In general, most GxE results have not been successfully replicated, leading to speculation about whether the effects that have been reported are robust (Duncan & Keller, 2011). The most prominent example of this concerns the nature of association between the serotonin transporter (5-HTT) gene and risk for depression. In a landmark study, Caspi and colleagues (2003) found that the effect of a genetic variant in the promoter region of the gene (known as the 5-HTTLPR [serotonin-transporter-linked polymorphic region] variant with an "s" versus "l" allele) on depression risk was dependent on level of life stress. Among individuals who had not experienced childhood maltreatment or stress, risk of depression was roughly equal across 5-HTTLPR genotypes, but among individuals who experienced severe childhood maltreatment or stress, those who were homozygous s/s were at significantly increased risk of depression compared to those with a homozygous l/l genotype. The s/s genotype thus appeared to convey a vulnerability, but that vulnerability only resulted in elevated rates of depression if individuals with this genotype were exposed to trauma or high levels of stress.

Having been cited more than 8,000 times according to Google Scholar (October 31, 2018), Caspi and colleagues (2003) is one of the most highly cited studies in psychopathology research over the past 20 years. Nonetheless, despite considerable excitement over the initial finding and numerous attempts to replicate and extend the observed GxE effect, there continues to be uncertainty surrounding whether the effect is real. The most recent meta-analysis, which included 31 studies with over 38,000 participants, did not find evidence of an interaction between 5-HTTLPR and childhood maltreatment in the prediction of depression, although the researchers noted that they could not rule out the possibility of a GxE of modest size (Culverhouse et al., 2017). Failure to find consistent evidence for a GxE with 5-HTTLPR has raised concerns that the same sort of problems that have plagued candidate-gene studies of psychopathology (i.e., small samples, biased publication, failure to fully correct for multiple testing) may be similarly limiting GxE research.

A second example includes a possible GxE interaction for a link between cannabis use and psychotic symptoms moderated by the catechol-O-methyltransferase (COMT) gene. Within the COMT gene, there are "val" and "met" alleles corresponding to the different amino acids being coded. It was found that carriers of the "val" allele displayed higher rates of psychotic symptoms if they had used cannabis in adolescence, compared to those with the "met" allele (Caspi et al., 2005). Subsequent research has largely shown a failure to replicate these findings (Costas et al., 2011; van Winkel, 2011; Zammit et al., 2007). There have been some positive findings of GxE in psychopathology research, however, including a robust interaction between a variant in the aldehyde dehydrogenase 2 (ALDH2) gene and culture in the etiology of alcoholism. The ADLH2 gene contains two alleles: ALDH2-1 and ALDH2-2. Those who are heterozygous or homozygous for ALDH2-2 are lacking an enzyme produced by ALDH2 that helps to break down alcohol. When they consume alcohol, they experience facial flushing as well as other aversive physiological symptoms. This negative response to alcohol consumption creates a protective mechanism against alcohol abuse and dependence. Thus, association between ALDH2 and alcoholism depends on the environmental exposure of alcohol use (Dick & Kendler, 2012). GxE research is a growing field and should continue to be explored. Results from these studies, however, need robust replication in order to be generalizable.

Epigenetics

While the nucleus of every cell in the body contains the same DNA sequence, it is clear that there are a variety of different cell types. When these cells duplicate, they create another cell of their same type. For instance, liver cells produce liver cells and blood cells produce blood cells despite both types of cells containing the same DNA. The reason that liver cells do not produce blood cells, or vice versa, is because different genes are expressed in the different cell types due to epigenetic processes. Epigenetic modifications are changes that are applied on top of the DNA sequence that stably (i.e., over multiple cell divisions) alter gene expression without changing the DNA sequence. Gene expression can be driven up or down, causing a gene to be either expressed or repressed. These changes are passed on to all daughter cells, causing a liver cell to only create another liver cell.

There are several types of epigenetic modifications including DNA methylation, histone modification, and noncoding RNA, with DNA methylation being the most widely studied. This type of modification affects the transcription of genes. Upstream from the coding sequence of a gene is a promoter region. Within the promoter region, a cytosine (C) base is often followed by a guanine (G) base, connected by a phosphate (p) bond (known as CpG islands). Methyl groups can attach to these CpG islands (a process known as methylation) and methylation is associated with reduced transcription of the gene and is passed on to all duplicated cells.

The application of epigenetic research to behavior is a very new and exciting field. Epigenetic modification may explain why MZ twins are sometimes discordant for highly heritable traits. While there has been success in understanding how stress affects DNA methylation through animal studies (Weaver et al., 2004), less is known about the role of epigenetics in human psychopathology. The coming years will undoubtedly bring far more research on the effects of epigenetics on human behavior and psychopathology, and how environmental sources may effect epigenetic modifications.

Summary

Several conclusions can be drawn from behavior genetic research. It is clear from twin/adoption studies, as well as more recent molecular genetic research, that genetic factors influence risk for all types of psychopathology. Heritability estimates across the spectrum of psychopathology generally range between 0.3 and 0.6, indicating moderate genetic influences. After a general failure of candidate gene studies, genome-wide association studies (GWAS) have begun to find genetic variants associated with increased risk for psychopathology. It is now clear that instead of one, or a couple of, gene variants with a large effect influencing phenotypic risk, there are likely thousands of variants, each with a very small effect. GWAS will continue to identify additional gene variants of interest, though increasingly large sample sizes will be required. In addition, GWAS is only able to identify common genetic variants. Research incorporating rare variants, such as CNVs, is increasing and may help to address the problem of missing heritability. Studies comparing the simultaneous influence of gene variants on multiple types of psychopathology are continuing to help us understand the distinction and overlap between types of psychopathology.

Behavior genetic research has also highlighted the importance of environmental factors. Monozygotic twins are never completely concordant, indicating that the environment must have an effect on psychopathology. It appears that the nonshared environment is the overwhelming source of environmental variation contributing to psychopathology risk. Continuing work on gene-environmental correlation and interaction, as well as epigenetics, will help to address which environmental factors are important and how they influence psychopathology risk.

REFERENCES

1000 Genomes Project Consortium, Auton, A., Brooks, L. D., Durbin, R. M., Garrison, E. P., Kang, H. M., ... Abecasis, G. R. (2015). A Global Reference for Human Genetic Variation. *Nature, 526*(7571), 68–74.

Addington, A. M., & Rapoport, J. L. (2012). Annual Research Review: Impact of Advances in Genetics in Understanding Developmental Psychopathology. *Journal of Child Psychology and Psychiatry, 53*(5), 510–518.

Anastasi, A. (1958). Heredity, Environment, and the Question How? *Psychological Review, 65*(4), 197–208.

Andrew, T., Hart, D. J., Snieder, H., Lange, M. de, Spector, T. D., & MacGregor, A. J. (2001). Are Twins and Singletons Comparable? A Study of Disease-Related and Lifestyle Characteristics in Adult Women. *Twin Research and Human Genetics, 4*(6), 464–477.

Antonarakis, S. E., & Beckmann, J. S. (2006). Mendelian Disorders Deserve More Attention. *Nature Reviews Genetics, 7*(4), 277–282.

Arranz, M. J., & de Leon, J. (2007). Pharmacogenetics and Pharmacogenomics of Schizophrenia: A Review of Last Decade of Research. *Molecular Psychiatry, 12*(8), 707–747.

Autism Spectrum Disorders Working Group of the Psychiatric Genomics Consortium. (2017). Meta-analysis of GWAS of over 16,000 Individuals with Autism Spectrum Disorder Highlights a Novel Locus at 10q24.32 and a Significant Overlap with Schizophrenia. *Molecular Autism, 8*, 21.

Bassett, A. S., & Chow, E. W. C. (2008). Schizophrenia and 22q11.2 Deletion Syndrome. *Current Psychiatry Reports, 10*(2), 148–157.

Bin Xu, Roos, J. L., Levy, S., van Rensburg, E. J., Gogos, J. A., & Karayiorgou, M. (2008). Strong Association of De Novo Copy Number Mutations with Sporadic Schizophrenia. *Nature Genetics, 40*(7), 880–885.

Boomsma, D., Busjahn, A., & Peltonen, L. (2002). Classical Twin Studies and Beyond. *Nature Reviews Genetics, 3*(11), 872–882.

Boraska, V., Franklin, C. S., Floyd, J. a. B., Thornton, L. M., Huckins, L. M., Southam, L., ... Bulik, C. M. (2014). A Genome-Wide Association Study of Anorexia Nervosa. *Molecular Psychiatry, 19*(10), 1085–1094.

Buckholtz, J. W., & Meyer-Lindenberg, A. (2012). Psychopathology and the Human Connectome: Toward a Transdiagnostic Model of Risk for Mental Illness. *Neuron, 74*(6), 990–1004.

Bulik, C. M., Sullivan, P. F., Tozzi, F., Furberg, H., Lichtenstein, P., & Pedersen, N. L. (2006). Prevalence, Heritability, and Prospective Risk Factors for Anorexia Nervosa. *Archives of General Psychiatry, 63*(3), 305–312.

Burbridge, D. (2001). Francis Galton on Twins, Heredity and Social Class. *British Journal for the History of Science, 34*(3), 323–340.

Burt, S. A. (2009). Rethinking Environmental Contributions to Child and Adolescent Psychopathology: A Meta-Analysis of Shared Environmental Influences. *Psychological Bulletin, 135*(4), 608–637.

Bush, W. S., & Moore, J. H. (2012). Chapter 11: Genome-Wide Association Studies. *PLOS Computational Biology, 8*(12), e1002822.

Carlborg, Ö., Haley, C. S., & Carlborg, O. (2004). Epistasis: Too Often Neglected in Complex Trait Studies? *Nature Reviews Genetics, 5*(8), 618–625.

Caspi, A., Sugden, K., Moffitt, T. E., Taylor, A., Craig, I. W., Harrington, H., ... Poulton, R. (2003). Influence of Life Stress on Depression: Moderation by a Polymorphism in the 5-HTT Gene. *Science, 301*(5631), 386–389.

Caspi, A., Moffitt, T. E., Cannon, M., McClay, J., Murray, R., Harrington, H., ... Craig, I. W. (2005). Moderation of the Effect of Adolescent-Onset Cannabis Use on Adult Psychosis by a Functional Polymorphism in the Catechol-O-Methyltransferase Gene: Longitudinal Evidence of a Gene X Environment Interaction. *Biological Psychiatry, 57*(10), 1117–1127.

Chen, C.-H., Lee, C.-S., Lee, M.-T. M., Ouyang, W.-C., Chen, C.-C., Chong, M.-Y., ... Cheng, A. T.-A. (2014). Variant GADL1 and Response to Lithium Therapy in Bipolar I Disorder. *New England Journal of Medicine, 370*(2), 119–128.

Chial, H. (2008). Rare Genetic Disorders: Learning about Genetic Disease through Gene Mapping, SNPs, and Microarray Data. *Nature Education, 1*(1), 192.

Costas, J., Sanjuán, J., Ramos-Ríos, R., Paz, E., Agra, S., Tolosa, A., ... Arrojo, M. (2011). Interaction between COMT Haplotypes and Cannabis in Schizophrenia: A Case-Only Study in Two Samples from Spain. *Schizophrenia Research, 127*(1–3), 22–27.

Crespi, B. J. (2016). Autism as a Disorder of High Intelligence. *Frontiers in Neuroscience, 10*, 300.

Cross-Disorder Group of the Psychiatric Genomics Consortium, Lee, S. H., Ripke, S., Neale, B. M., Faraone, S. V., Purcell, S. M., ... International Inflammatory Bowel Disease Genetics Consortium (IIBDGC). (2013). Genetic Relationship between Five Psychiatric Disorders Estimated from Genome-Wide SNPs. *Nature Genetics, 45*(9), 984–994.

Culverhouse, R. C., Saccone, N. L., Horton, A. C., Ma, Y., Anstey, K. J., Banaschewski, T., ... Bierut, L. J. (2017). Collaborative Meta-Analysis Finds no Evidence of a Strong Interaction between Stress and 5-HTTLPR Genotype Contributing to the Development of Depression. *Molecular Psychiatry, 23*, 133–142.

Demontis, D., Walters, R. K., Martin, J., Mattheisen, M., Als, T. D., Agerbo, E., ... Neale, B. M. (2019). Discovery of the First Genome-Wide Significant Risk Loci for Attention Deficit/Hyperactivity Disorder. *Nature Genetics, 51*(1), 63–75.

Dick, D. M., & Kendler, K. S. (2012). The Impact of Gene-Environment Interaction on Alcohol Use Disorders. *Alcohol Research: Current Reviews, 34*(3), 318–324.

Drew, L. J., Crabtree, G. W., Markx, S., Stark, K. L., Chaverneff, F., Xu, B., ... Karayiorgou, M. (2011). The 22q11.2 Microdeletion: Fifteen Years of Insights into the Genetic and Neural Complexity of Psychiatric Disorders. *International Journal of Developmental Neuroscience, 29*(3), 259–281.

Duncan, L. E., & Keller, M. C. (2011). A Critical Review of the First 10 Years of Candidate Gene-by-Environment Interaction Research in Psychiatry. *American Journal of Psychiatry, 168*(10), 1041–1049.

Eichler, E. E., Flint, J., Gibson, G., Kong, A., Leal, S. M., Moore, J. H., & Nadeau, J. H. (2010). Missing Heritability and Strategies for Finding the Underlying Causes of Complex Disease. *Nature Reviews Genetics, 11*(6), 446–450.

Falconer, D. S. (1965). The Inheritance of Liability to Certain Diseases, Estimated from the Incidence among Relatives. *Annals of Human Genetics*, 29(1), 51–76.

Falconer, D. S., & Mackay, T. F. C. (1996). *Introduction to Quantitative Genetics*. Harlow: Pearson Education.

Faraone, S. V., Perlis, R. H., Doyle, A. E., Smoller, J. W., Goralnick, J. J., Holmgren, M. A., & Sklar, P. (2005). Molecular Genetics of Attention-Deficit/Hyperactivity Disorder. *Biological Psychiatry*, 57(11), 1313–1323.

Felson, J. (2014). What Can We Learn from Twin Studies? A Comprehensive Evaluation of the Equal Environments Assumption. *Social Science Research*, 43, 184–199.

Furberg, H., YunJung Kim, Dackor, J., Boerwinkle, E., Franceschini, N., Ardissino, D., ... Furberg, C. D. (2010). Genome-Wide Meta-Analyses Identify Multiple Loci Associated with Smoking Behavior. *Nature Genetics*, 42(5), 441–447.

Gandal, M. J., Leppa, V., Won, H., Parikshak, N. N., & Geschwind, D. H. (2016). The Road to Precision Psychiatry: Translating Genetics into Disease Mechanisms. *Nature Neuroscience*, 19(11), 1397–1407.

Gelernter, J., Kranzler, H. R., Sherva, R., Almasy, L., Koesterer, R., Smith, A. H., ... Farrer, L. A. (2014). Genome-Wide Association Study of Alcohol Dependence: Significant Findings in African- and European-Americans Including Novel Risk Loci. *Molecular Psychiatry*, 19(1), 41–49.

Gormley, P., Anttila, V., Winsvold, B. S., Palta, P., Esko, T., Pers, T. H., ... Palotie, A. (2016). Meta-Analysis of 375,000 Individuals Identifies 38 Susceptibility Loci for Migraine. *Nature Genetics*, 48(8), 856–866.

Hamburg, M. A., & Collins, F. S. (2010). The Path to Personalized Medicine. *New England Journal of Medicine*, 363(4), 301–304.

Hardy, J., & Singleton, A. (2009). Genomewide Association Studies and Human Disease. *New England Journal of Medicine*, 360(17), 1759–1768.

Hettema, J. M., Neale, M. C., & Kendler, K. S. (2001). A Review and Meta-Analysis of the Genetic Epidemiology of Anxiety Disorders. *American Journal of Psychiatry*, 158(10), 1568–1578.

Hill, W. G., Goddard, M. E., & Visscher, P. M. (2008). Data and Theory Point to Mainly Additive Genetic Variance for Complex Traits. *PLOS Genetics*, 4(2), e1000008.

Hirschhorn, J. N., & Daly, M. J. (2005). Genome-Wide Association Studies for Common Diseases and Complex Traits. *Nature Reviews Genetics*, 6(2), 95–108.

Hong, E. P., & Park, J. W. (2012). Sample Size and Statistical Power Calculation in Genetic Association Studies. *Genomics & Informatics*, 10(2), 117–122.

Hur, Y.-M., & Craig, J. M. (2013). Twin Registries Worldwide: An Important Resource for Scientific Research. *Twin Research and Human Genetics*, 16(1), 1–12.

Hyde, C. L., Nagle, M. W., Tian, C., Chen, X., Paciga, S. A., Wendland, J. R., ... Winslow, A. R. (2016). Identification of 15 Genetic Loci Associated with Risk of Major Depression in Individuals of European Descent. *Nature Genetics*, 48(9), 1031–1036.

Jaffee, S. R., & Price, T. S. (2008). Genotype-Environment Correlations: Implications for Determining the Relationship between Environmental Exposures and Psychiatric Illness. *Psychiatry*, 7(12), 496–499.

Johnson, R. C., Nelson, G. W., Troyer, J. L., Lautenberger, J. A., Kessing, B. D., Winkler, C. A., & O'Brien, S. J. (2010). Accounting for Multiple Comparisons in a Genome-Wide Association Study (GWAS). *BMC Genomics*, 11, 724.

Jorde, L. B. (2000). Linkage Disequilibrium and the Search for Complex Disease Genes. *Genome Research*, 10(10), 1435–1444.

Kendler, K. S., & Baker, J. H. (2007). Genetic Influences on Measures of the Environment: A Systematic Review. *Psychological Medicine*, 37(5), 615–626.

Kendler, K. S., Neale, M. C., Kessler, R. C., Heath, A. C., & Eaves, L. J. (1993). A Test of the Equal-Environment Assumption in Twin Studies of Psychiatric Illness. *Behavior Genetics*, 23(1), 21–27.

Kendler, K. S., Gatz, M., Gardner, C. O., & Pedersen, N. L. (2006). A Swedish National Twin Study of Lifetime Major Depression. *American Journal of Psychiatry*, 163(1), 109–114.

Kieseppä, T., Partonen, T., Haukka, J., Kaprio, J., & Lönnqvist, J. (2004). High Concordance of Bipolar I Disorder in a Nationwide Sample of Twins. *American Journal of Psychiatry*, 161(10), 1814–1821.

Kirov, G., Pocklington, A. J., Holmans, P., Ivanov, D., Ikeda, M., Ruderfer, D., ... Böttcher, Y. (2012). De Novo CNV Analysis Implicates Specific Abnormalities of Postsynaptic Signalling Complexes in the Pathogenesis of Schizophrenia. *Molecular Psychiatry*, 17(2), 142–153.

Kwon, J. M., & Goate, A. M. (2000). The Candidate Gene Approach. *Alcohol Research & Health*, 24(3), 164–168.

Lee, S. H., Ripke, S., Neale, B. M., Faraone, S. V., Purcell, S. M., Perlis, R. H., ... Asherson, P. (2013). Genetic Relationship between Five Psychiatric Disorders Estimated from Genome-Wide SNPs. *Nature Genetics*, 45(9), 984–994.

Li, M. D., Cheng, R., Ma, J. Z., & Swan, G. E. (2003). A Meta-Analysis of Estimated Genetic and Environmental Effects on Smoking Behavior in Male and Female Adult Twins. *Addiction*, 98(1), 23–31.

Maes, H. H. M., Neale, M. C., Kendler, K. S., Hewitt, J. K., Silberg, J. L., Foley, D. L., ... Eaves, L. J. (1998). Assortative Mating for Major Psychiatric Diagnoses in Two Population-Based Samples. *Psychological Medicine*, 28(6), 1389–1401.

Maher, B. (2008). Personal Genomes: The Case of the Missing Heritability. *Nature*, 456(7218), 18–21.

Malhotra, D., & Sebat, J. (2012). CNVs: Harbingers of a Rare Variant Revolution in Psychiatric Genetics. *Cell*, 148(6), 1223–1241.

Manolio, T. A., Collins, F. S., Cox, N. J., Goldstein, D. B., Hindorff, L. A., Hunter, D. J., ... Boehnke, M. (2009). Finding the Missing Heritability of Complex Diseases. *Nature*, 461(7265), 747–753.

Mazzeo, S. E., Mitchell, K. S., Bulik, C. M., Aggen, S. H., Kendler, K. S., & Neale, M. C. (2010). A Twin Study of Specific Bulimia Nervosa Symptoms. *Psychological Medicine*, 40(7), 1203–1213.

McClellan, J., & King, M.-C. (2010). Genetic Heterogeneity in Human Disease. *Cell*, 141(2), 210–217.

McGue, M., Keyes, M., Sharma, A., Elkins, I., Legrand, L., Johnson, W., & Iacono, W. (2007). The Environments of Adopted and Non-Adopted Youth: Evidence on Range Restriction from the Sibling Interaction and Behavior Study (SIBS). *Behavior Genetics*, 37(3), 449–462.

McGuffin, P., Rijsdijk, F., Andrew, M., Sham, P., Katz, R., & Cardno, A. (2003). The Heritability of Bipolar Affective Disorder and the Genetic Relationship to Unipolar Depression. *Archives of General Psychiatry*, 60(5), 497–502.

Mühleisen, T. W., Leber, M., Schulze, T. G., Strohmaier, J., Degenhardt, F., Treutlein, J., ... Cichon, S. (2014). Genome-Wide Association Study Reveals Two New Risk Loci for Bipolar Disorder. *Nature Communications*, 5, 3339.

Munafò, M. R. (2006). Candidate Gene Studies in the 21st Century: Meta-Analysis, Mediation, Moderation. *Genes, Brain & Behavior, 5*, 3–8.

Neale, M., & Cardon, L. (2013). *Methodology for Genetic Studies of Twins and Families*. New York: Springer Science & Business Media.

Nikolas, M. A., & Alexandra, S. (2010). Genetic and Environmental Influences on ADHD Symptom Dimensions of Inattention and Hyperactivity: A Meta-Analysis. *Journal of Abnormal Psychology, 119*(1), 1–17.

Otowa, T., Hek, K., Lee, M., Byrne, E. M., Mirza, S. S., Nivard, M. G., ... Hettema, J. M. (2016). Meta-Analysis of Genome-Wide Association Studies of Anxiety Disorders. *Molecular Psychiatry, 21*(10), 1391–1399.

Phillips, P. C. (2008). Epistasis – The Essential Role of Gene Interactions in the Structure and Evolution of Genetic Systems. *Nature Reviews Genetics, 9*(11), 855–867.

Plomin, R., & Bergeman, C. S. (1991). The Nature of Nurture: Genetic Influence on "Environmental" Measures. *Behavioral and Brain Sciences, 14*(3), 373–386.

Plomin, R., DeFries, J. C., & Loehlin, J. C. (1977). Genotype-Environment Interaction and Correlation in the Analysis of human behavior. *Psychological Bulletin, 84*(2), 309–322.

Plomin, R., Owen, M. J., & McGuffin, P. (1994). The Genetic Basis of Complex Human Behaviors. *Science, 264*(5166), 1733–1739.

Plomin, R., DeFries, J. C., Knopik, V. S., & Neiderheiser, J. (2013). *Behavioral Genetics* (6th edn.). New York: Worth Publishers.

Polderman, T. J. C., Benyamin, B., de Leeuw, C. A., Sullivan, P. F., van Bochoven, A., Visscher, P. M., & Posthuma, D. (2015). Meta-Analysis of the Heritability of Human Traits Based on Fifty Years of Twin Studies. *Nature Genetics, 47*(7), 702–709.

Power, R. A., Steinberg, S., Bjornsdottir, G., Rietveld, C. A., Abdellaoui, A., Nivard, M. M., ... Stefansson, K. (2015). Polygenic Risk Scores for Schizophrenia and Bipolar Disorder Predict Creativity. *Nature Neuroscience, 18*(7), 953–955.

Purcell, S. M., Wray, N. R., Stone, J. L., Visscher, P. M., O'Donovan, M. C., Sullivan, P. F., ... Moran, J. L. (2009). Common Polygenic Variation Contributes to Risk of Schizophrenia and Bipolar Disorder. *Nature, 460*(7256), 748–752.

Reich, D. E., & Lander, E. S. (2001). On the Allelic Spectrum of Human Disease. *Trends in Genetics, 17*(9), 502–510.

Ripke, S., O'Dushlaine, C., Chambert, K., Moran, J. L., Kähler, A. K., Akterin, S., ... Sullivan, P. F. (2014). Genome-Wide Association Analysis Identifies 13 New Risk Loci for Schizophrenia. *Nature Genetics, 45*(10), 1150–1159.

Robinson, M. R., Wray, N. R., & Visscher, P. M. (2014). Explaining Additional Genetic Variation in Complex Traits. *Trends in Genetics, 30*(4), 124–132.

Sanders, S. J., Ercan-Sencicek, A. G., Hus, V., Luo, R., Murtha, M. T., Moreno-De-Luca, D., ... State, M. W. (2011). Multiple Recurrent De Novo CNVs, Including Duplications of the 7q11.23 Williams Syndrome Region, Are Strongly Associated with Autism. *Neuron, 70*(5), 863–885.

Schizophrenia Psychiatric Genome-Wide Association Study (GWAS) Consortium. (2011). Genome-Wide Association Study Identifies Five New Schizophrenia Loci. *Nature Genetics, 43*(10), 969–976.

Schizophrenia Working Group of the Psychiatric Genomics Consortium. (2014). Biological Insights from 108 Schizophrenia-Associated Genetic Loci. *Nature, 511*(7510), 421–427.

Schwekendiek, D. (2009). Height and Weight Differences between North and South Korea. *Journal of Biosocial Science, 41*(1), 51–55.

Scriver, C. R. (2007). The PAH Gene, Phenylketonuria, and a Paradigm Shift. *Human Mutation, 28*(9), 831–845.

Sebat, J., Lakshmi, B., Troge, J., Alexander, J., Young, J., Lundin, P., ... Wigler, M. (2004). Large-Scale Copy Number Polymorphism in the Human Genome. *Science, 305*(5683), 525–528.

Shao, H., Burrage, L. C., Sinasac, D. S., Hill, A. E., Ernest, S. R., O'Brien, W., ... Nadeau, J. H. (2008). Genetic Architecture of Complex Traits: Large Phenotypic Effects and Pervasive Epistasis. *Proceedings of the National Academy of Sciences of the United States of America, 105*(50), 19910–19914.

Stefansson, H., Rujescu, D., Cichon, S., Pietiläinen, O. P. H., Ingason, A., Steinberg, S., ... Sigurdsson, A. (2008). Large Recurrent Microdeletions Associated with Schizophrenia. *Nature, 455*(7210), 232–236.

Stone, J. L., O'Donovan, M. C., Gurling, H., Kirov, G. K., Blackwood, D. H. R., Corvin, A., ... Macgregor, S. (2008). Rare Chromosomal Deletions and Duplications Increase Risk of Schizophrenia. *Nature, 455*(7210), 237–241.

Stoolmiller, M. (1999). Implications of the Restricted Range of Family Environments for Estimates of Heritability and Nonshared Environment in Behavior-Genetic Adoption Studies. *Psychological Bulletin, 125*(4), 392–409.

Stranger, B. E., Stahl, E. A., & Raj, T. (2011). Progress and Promise of Genome-Wide Association Studies for Human Complex Trait Genetics. *Genetics, 187*(2), 367–383.

Stulp, G., & Barrett, L. (2016). Evolutionary Perspectives on Human Height Variation. *Biological Reviews, 91*(1), 206–234.

Sullivan, P. F., Neale, M. C., & Kendler, K. S. (2000). Genetic Epidemiology of Major Depression: Review and Meta-Analysis. *American Journal of Psychiatry, 157*(10), 1552–1562.

Sullivan, P. F., Kendler, K. S., & Neale, M. C. (2003). Schizophrenia as a Complex Trait: Evidence from a Meta-Analysis of Twin Studies. *Archives of General Psychiatry, 60*(12), 1187–1192.

Taylor, S. (2011). Etiology of Obsessions and Compulsions: A Meta-Analysis and Narrative Review of Twin Studies. *Clinical Psychology Review, 31*(8), 1361–1372.

Tenesa, A., & Haley, C. S. (2013). The Heritability of Human Disease: Estimation, Uses and Abuses. *Nature Reviews Genetics, 14*(2), 139–149.

Tick, B., Bolton, P., Happe, F., Rutter, M., & Rijsdijk, F. (2016). Heritability of Autism Spectrum Disorders: A Meta-Analysis of Twin Studies. *Journal of Child Psychology and Psychiatry, 57*(5), 585–595.

Turkheimer, E. (2000). Three Laws of Behavior Genetics and What They Mean. *Current Directions in Psychological Science, 9*(5), 160–164.

Van Winkel, R. (2011). Family-Based Analysis of Genetic Variation Underlying Psychosis-Inducing Effects of Cannabis: Sibling Analysis and Proband Follow-Up. *Archives of General Psychiatry, 68*(2), 148–157.

Verhulst, B., Neale, M. C., & Kendler, K. S. (2015). The Heritability of Alcohol Use Disorders: A Meta-Analysis of Twin and Adoption Studies. *Psychological Medicine, 45*(5), 1061–1072.

Verweij, K. J. H., Zietsch, B. P., Lynskey, M. T., Medland, S. E., Neale, M. C., Martin, N. G., ... Vink, J. M. (2010). Genetic and Environmental Influences on Cannabis Use Initiation and Problematic Use: A Meta-Analysis of Twin Studies. *Addiction, 105*(3), 417–430.

Weaver, I. C. G., Cervoni, N., Champagne, F. A., D'Alessio, A. C., Sharma, S., Seckl, J. R., … Meaney, M. J. (2004). Epigenetic Programming by Maternal Behavior. *Nature Neuroscience, 7*(8), 847–854.

Weiss, L. A., Shen, Y., Korn, J. M., Arking, D. E., Miller, D. T., Fossdal, R., … Daly, M. J. (2008). Association between Microdeletion and Microduplication at 16p11.2 and Autism. *New England Journal of Medicine, 358*(7), 667–675.

Wilson, B. J., & Nicholls, S. G. (2015). The Human Genome Project, and Recent Advances in Personalized Genomics. *Risk Management and Healthcare Policy, 8*, 9–20.

Wood, A. R., Esko, T., Yang, J., Vedantam, S., Pers, T. H., Gustafsson, S., … Frayling, T. M. (2014). Defining the Role of Common Variation in the Genomic and Biological Architecture of Adult Human Height. *Nature Genetics, 46*(11), 1173–1186.

Yang, J., Lee, S. H., Goddard, M. E., & Visscher, P. M. (2011a). GCTA: A Tool for Genome-Wide Complex Trait Analysis. *The American Journal of Human Genetics, 88*(1), 76–82.

Yang, J., Weedon, M. N., Purcell, S., Lettre, G., Estrada, K., Willer, C. J., … Goddard, M. E. (2011b). Genomic Inflation Factors under Polygenic Inheritance. *European Journal of Human Genetics, 19*(7), 807–812.

Yengo, L., Sidorenko, J., Kemper, K. E., Zheng, Z., Wood, A. R., Weedon, M. N., … Consortium, G. (2018). Meta-Analysis of Genome-Wide Association Studies for Height and Body Mass Index in ~700,000 Individuals of European Ancestry. *BioRxiv*, 274654.

Zammit, S., Spurlock, G., Williams, H., Norton, N., Williams, N., O'Donovan, M. C., & Owen, M. J. (2007). Genotype Effects of CHRNA7, CNR1 and COMT in Schizophrenia: Interactions with Tobacco and Cannabis Use. *The British Journal of Psychiatry, 191*(5), 402–407.

Zondervan, K. T., & Cardon, L. R. (2004). The Complex Interplay among Factors that Influence Allelic Association. *Nature Reviews Genetics, 5*(2), 89–100.

Zuk, O., Hechter, E., Sunyaev, S. R., & Lander, E. S. (2012). The Mystery of Missing Heritability: Genetic Interactions Create Phantom Heritability. *Proceedings of the National Academy of Sciences of the United States of America, 109*(4), 1193–1198.

13 Concepts and Principles of Clinical Functional Magnetic Resonance Imaging

AVRAM J. HOLMES AND ANGUS W. MACDONALD III

The human brain is comprised of a dense web of interdigitated functional networks. Understanding how the brain's complex functions give rise to human cognitive abilities in both health and disease depends on unraveling the carefully coordinated interactions between networked brain regions and their responses to environmental change (Holmes & Patrick, 2018). Historically, substantial progress was made delineating this intricate architecture through postmortem dissections in humans and tract tracing and lesion studies in animals. Yet there remained many gaps in our understanding of how the brain influences behavior, particularly psychiatric illnesses. The limitations of these labor-intensive approaches have receded over the past 40 years with the advent of *in vivo* imaging approaches such as positron emission tomography (PET), electroencephalography (EEG), electrocorticography (ECoG), and magnetoencephalography (MEG; see Raichle, 2009, for a historical overview). The introduction of functional magnetic resonance imaging (fMRI), in particular, has sparked spectacular growth in psychiatry research.

Fueled by rapid methodological and analytic advances, fMRI has come to dominate the clinical literature, allowing us to study brain function in a rapid and noninvasive manner across ever larger samples. These technological developments have made it easy to become exceedingly optimistic about the future of clinical neuroscience. Although fMRI methods have evolved rapidly since the first brain scans in the 1980s, there remain core approaches and theoretical principles that can be used to understand the current state of the field and anticipate future innovations. Here we take a critical look at how fMRI measures can inform our understanding of brain functions in psychopathology. To help researchers select appropriate methods, we will cover fMRI study design, analysis, and interpretation and discuss some of the advantages and disadvantages of each design and analytic choice.

EXPERIMENTAL APPROACHES

A central goal of clinical cognitive neuroscience is to understand how common cognitive and neural systems may differ in people with, or at risk for, psychopathology. Among cognitive mechanisms we include perceptual processes, such as stimulus detection or facial recognition; salience-related processes, such as reward or threat detection; executive processes, such as cognitive control, emotion regulation or decision making; and motor processes, such as response initiation. Of course, there are many more, with names and descriptions that often overlap (Poldrack, 2010). The brain implements these cognitive processes in a manner that is evident at different levels of analysis, from molecular mechanisms (e.g., neurotransmission) to large-scale networks (e.g., spike-timing-dependent plasticity through which neurons assemble). Clinical cognitive neuroscience uses an array of methods to disentangle the complexities of psychopathology into finer, potentially discrete, deficits in specific aspects of brain biology. This approach is often most effective when it builds on established behavior-brain research that informs an understanding of individual differences. For example, a good deal of work on working memory impairments in psychosis patients extends upon foundational working memory studies in humans (Park, Holzman, & Goldman-Rakic, 1995) and nonhuman primates (e.g. Funahashi, Bruce, & Goldman-Rakic, 1989).

Measuring the Brain When Performing Tasks

It is intuitive to ask how the brains of people with mental illnesses differ while thinking. Pioneers such as Ingvar and Franzen (1974) used a forerunner of PET imaging to study resting cerebral blood flow (rCBF) in patients with schizophrenia during a cognitively demanding task. Although overall rCBF levels were similar to controls, in postcentral sulcus it was relatively higher in patients with schizophrenia, whereas in prefrontal cortex it was relatively lower. Thus began an era of function-based neuroimaging efforts that have increased in power and sophistication in the subsequent decades.

Table 13.1 delineates the four general approaches to task-evoked fMRI that are commonly used, and Figure 13.1

Table 13.1 Experimental approaches using functional magnetic resonance imaging

Method	Application	Strengths	Limitations	Key references
Block design (task evoked)	Contrasts conditions within an ongoing task or between an ongoing task and rest	Efficient data collection; maximizes potential activation differences	Difficult to attribute activation to a specific cognitive mechanism when contrasted tasks differ in several demands; difficult to exclude errors from analysis; requires task development	(Amaro & Barker, 2006)
Slow event-related design (task evoked)	Contrasts different trial types or cognitive demands sparsely timed	Different events may be intermixed in an unpredictable manner; relatively few assumptions about the nature of the hemodynamic response; few constraints on the interactions (or dependencies) between different cognitive demands; behavioral results often generalize to faster-paced versions; trials that fulfill a criterion (e.g., error trials) can be examined	Because hemodynamic response must return to baseline between trials, fewer trials can be collected; fewer trials for the evaluation of behavioral performance; boredom due to slow pace; requires task development	(Amaro & Barker, 2006)
Fast event-related design (task evoked)	Contrasts different trial types or cognitive demands more densely timed	As with slow event-related designs, different events may be intermixed; behavioral results are more robust because more trials are available; trials that fulfill a criterion (e.g., error trials) can be examined if sufficiently independent	Hemodynamic response function modeling required; events must be sufficiently independent; jittering time between events or using partial (catch) trials to make events independent may affect performance; requires task development	(Amaro & Barker, 2006; Ollinger, Shulman, & Corbetta, 2001)
Hybrid block/event-related designs (task evoked)	Nests an event-related design within a larger block design to allow multiple analyses	Allows for robust analyses present in block designs; additionally, trials that fulfill a criterion (e.g., error trials) can be examined separately	Hemodynamic response function modeling required; events only occur within the context the block; requires task development	(Braver, Reynolds, & Donaldson, 2003)
Intrinsic function ("resting state")	Examines on-going activity in the absence of specifically timed tasks or cognitive demands	Shorter development cycle; shorter training and fewer task demands facilitating data collection in special-needs populations; applicable to patients asleep or under sedation; easier to harmonize across sites and easier to combine data sets post hoc facilitating larger sample sizes	Brain activity cannot be related to specific cognitive events (but see Smith et al., 2009); group/individual differences findings may reflect differences in habitual thought patterns rather than ability to activate a region; connectivity metrics affected by subtle head motion	(Biswal et al., 1995; Smith et al., 2013)

illustrates how these approaches differ in stimulus presentation and associated models of the blood-oxygen-level-dependent (BOLD) response, which reflects the concentration of deoxygenated, relative to oxygenated hemoglobin. A ratio that is altered by local neural activity (Huettel, Song, & McCarthy, 2004; Logothetis et al., 2001). *Block designs* are far and away the most robust, and are therefore the most efficient, strategy for obtaining maps of where the BOLD response is occurring in the brain. A block design study consists of discrete "on" and "off" periods, each lasting from tens of seconds to minutes in duration. During the "on" times a stimulus is presented or a behavior is elicited. These blocks are contrasted with "off" periods that consist of rest, baseline, or alternate task

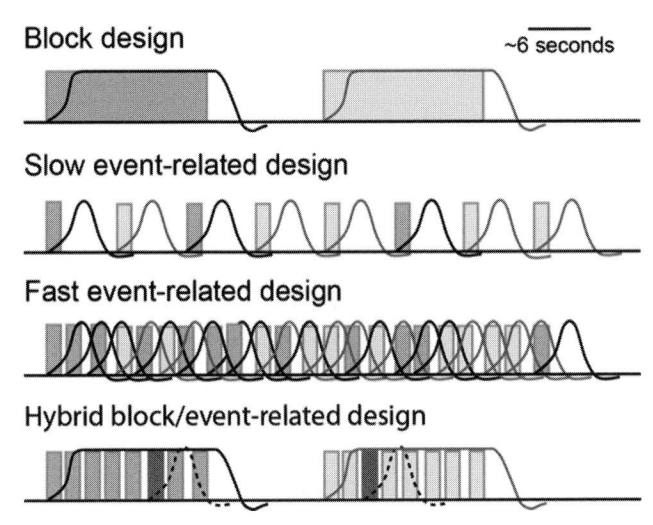

Figure 13.1 Different task conditions (gray boxes) and the corresponding convolution with a prototypical hemodynamic response function predict the rise and fall of the BOLD time course in an activated voxel across various experimental designs. Time proceeds from left to right. Note, the subtraction method of analysis (and derived brain maps) often compares the difference between the extent to which a voxel's time course resembles the black predicted time course relative to the lighter grey.

states. For example, the Human Connectome Project (HCP) collected seven tasks twice in block designs lasting 2–5 minutes (Barch et al., 2013). During the HCP working memory task, participants switched between performing a 2-back task (which required a running memory load of two items) and a 0-back attentional control task for 25 seconds each. All brain areas with metabolic demands when performing the task showed a rising BOLD response that reached a steady state within 5–6 seconds of the beginning of the block and then declined within 5–6 seconds when the cognitive load was removed. Because the metabolic demands associated with observing and responding to the stimuli were similar in the 2-back and 0-back conditions, a comparison between them would likely show little differential visual or motor activity. Instead, the biggest changes appeared in places where there were greater demands when maintaining a running load of two items. This strategy assumes that all items in a given condition are similarly difficult and that no aspect of the task (e.g., the occurrence of a repeated item) is of particular interest. The robustness of block designs can also be used to examine changes in activation across populations (e.g., case vs. comparison samples) or treatment conditions. Haut, Lim, and MacDonald (2010) compared 2-back to 0-back activity, this time in people with schizophrenia, to examine how cognitive training tasks changed activity more than controls in several regions of prefrontal cortex. Despite these advantages, block designs come with a number of potential interpretive problems for clinical research. For example, brain regions activated more in the 2-back than in the 0-back may be involved in many processes besides a higher working memory load, such as updating the stimuli after each trial, suppressing interference, monitoring conflict, expecting and preparing for another trial of the same kind, and experiencing frustration or even futility. For reasons we'll discuss further below in "Interpretation," performance differences between groups on the different blocks can also present a challenge, as the analysis of error trials is mixed together with the analysis of accurate trials.

In order to examine discrete trials and address constraints of block designs, many experimenters have employed event-related (later called "slow event-related") designs (Huettel, 2012) that leverage the hemodynamic time-course associated with local regional neural activity. This approach is characterized by large gaps in time between stimuli, allowing the BOLD response to rise and fall before presenting the next trial (which may have different tasks demands). Because this allows one to disentangle more components of task performance, many investigators gravitated toward this technique. For example, in a study of context processing-related deficits associated with the genetic liability to schizophrenia, MacDonald, Becker, and Carter (2006) differentiated between the task demand of maintaining a task representation, which involved dorsolateral prefrontal cortical (DLPFC) and was impaired in patients and first-degree relatives, and that of overcoming conflict, which evoked the anterior cingulate and was impaired in patients but not their relatives. What's more, these differences were observed on correct trials, thereby partially controlling for individuals' fluctuating task engagement. However, due to the slow nature of the hemodynamic response, fewer trials can be included using this approach, which provides reduced statistical power relative to a block design.

One way to introduce more trials is to group them closer together without allowing the hemodynamic response to resolve fully, which is commonly called the *fast event-related design*. This strategy owes its existence to several key contributions. As illustrated in Figure 13.1, convolution models for fMRI analysis (Friston et al., 1994) combined the time a stimulus occurred with an expectation about how the BOLD signal would respond. Subsequently, it was found that the BOLD response could be summated across successive trials even with short intertrial intervals (Dale & Buckner, 1997). This property of the BOLD signal allows task-relevant activations to be predicted based on the expected response to closely spaced stimuli and events, if the trials – or events within the trial – are sufficiently independent. This provides the opportunity to analyze closely spaced trials, or events within a task, drastically shortening the study collection times and the associated burden on research participants. For example, Poppe and colleagues (2016) took advantage of this design using a paradigm that required cue maintenance to control the response to a subsequent probe. In this case, if the cue was an A, then one would respond left to the probe, but if the cue was a B, then a right response to the probe was correct. To disentangle (to the degree possible) the relationships between the cues and probes, jitter (i.e., pseudorandom variation) was

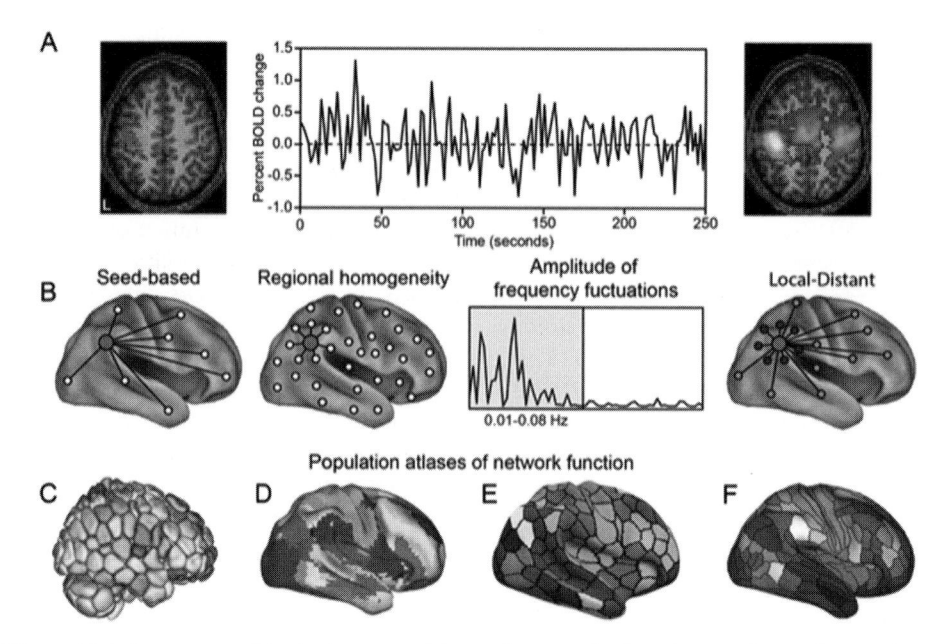

Figure 13.2 A summary of intrinsic connectivity analyses methods and cortical parcellations based on in-vivo brain imaging. (A) Intrinsic fluctuations in the fMRI BOLD signal exhibit patterns of covariation within functionally connected brain networks in the absence of overt task performance. Map of motor network from the seminal work by Biswal and colleagues (1995) as adapted by Vincent et al. (2006). (B) Selection of common analyses methods for intrinsic connectivity analyses (see Table 13.3). (C–E) Intrinsic fluctuations can be used to derive *in-vivo* brain parcellation. (C) Shen et al. (2013), (D) Power et al. (2011), and (E) Schaefer et al. (2018). (F) Multimodal parcellation using intrinsic connectivity, relative myelin mapping, cortical thickness and task-based fMRI (Glasser et al., 2016).

introduced into the interstimulus and intertrial intervals to facilitate modeling these independently.

Hybrid designs are used when the robustness of a block design is desirable, but where some questions can also be addressed using the more specific information that comes from examining individual trials. In such instances, one can blend block and event-related designs. This technique could be used to model individual trials within a block to identify and remove error trials, examine how different trials within a category interact with each other, or track different time courses across regions. The overarching point is that even as each design has various strengths and limitations, they need not be mutually exclusive. With planning, a given task might be conceptualized and analyzed from several perspectives, taking advantage of the associated strengths of each.

Measuring the Brain without Tasks: The "Resting State"

In addition to responding to stimuli, the brain also shows reliable patterns of activity in the absence of explicit task states or in so-called *resting state designs* (e.g., Biswal et al., 1995). As the brain is not particularly good at resting, we prefer the term *intrinsic function* or *intrinsic functional connectivity* for what is being measured when the mind is not directed to a particular task. The earliest clinical cognitive neuroscience studies examining patients' intrinsic brain functioning generally used PET to measure glucose metabolism. For example, individuals with schizophrenia display lower levels of rCBF in prefrontal and temporal regions (Farkas et al., 1984). This raised the question as to whether these differences reflected an inability to use those brain regions (a direct result of illness) or a disposition to use those brain regions less, perhaps because due to distraction, fatigue or some other factor (a downstream result of illness). This ambiguity led this approach to fall out of favor for a period of time, yet a number of advantages as well as promising empirical observations have reestablished resting-state, or intrinsic functioning, as a mainstay of neuroimaging.

Work in this domain by Biswal and colleagues (1995; Figure 13.2A) suggested an intrinsic organization to the brain that mirrored task-related functions. Their foundational study revealed that even when the brain was not engaged in a motor task, signal fluctuations in the motor cortex were highly correlated with neighboring voxels as well as spatially distinct regions associated with motor functioning. Functional connectivity between brain regions, such as reported by Biswal and colleagues, is usually analyzed in terms of correlation, signal coherence, or other temporal similarities in BOLD fluctuations. Providing converging evidence for intrinsic approaches to the study of brain functions, Koch, Norris, and Hud-Georgiadis (2002) combined diffusion-based and functional methods to reveal that intrinsic correlations between brain regions may depend on anatomic projections. This principle was expanded by Smith and colleagues (2009), who used a statistical algorithm called independent components analysis to identify a number of distributed functional networks (about 20 different such "components") and showed that the structure of these networks mapped quite closely to activation patterns from a large-scale meta-analysis of broad task categories. For example, meta-analysis indicated that regions of left prefrontal cortex and posterior parietal cortex frequently coactivated, along with a region of anterior cingulate cortex. The tasks most likely to coactivate these regions were working memory, explicit memory and language tasks. The observed locations closely resembled a network of voxels that coactivated at rest among a much smaller cohort of volunteers (a set of findings that has been replicated, e.g., Wisner et al., 2013), suggesting that intrinsic fluctuations may reflect coactivation among the regions with shared profiles of task-evoked function (Deco, Jirsa, & McIntosh, 2011).

Intrinsic approaches provide a new complement to task-based study designs, and the ease of collection and flexibility of intrinsic analyses has led to a rapid rise in their popularity. This is particularly true for clinical researchers, as the derived markers of intrinsic network function are more widely applicable than traditional measures of task-based fMRI. Since intrinsic network function can be assessed during sleep and under anesthesia, this functional mapping approach may be widely implemented in diverse populations including children, non-English-speaking participants, developmentally delayed patients, and patients who are under sedation. The promise of resting-state scanning has undergirded the accumulation of large-scale data sets that would be difficult if not impossible to obtain through more traditional task-based approaches. Open-access samples in the thousands are now widely available to the broader scientific community, such as the 1000 Functional Connectomes Project (Biswal et al., 2010), the Brain Genomics Superstruct Project (Holmes et al., 2015), the Human Connectome Project (Van Essen et al., 2013), and the UK Biobank (Ollier, Sprosen, & Peakman, 2005).

Despite the putative simplicity of intrinsic function studies, there remain a number of outstanding questions when acquiring and interpreting these data: what are the implications of collecting data with eyes open or closed (Van Dijk et al., 2010), with or without eye tracking, or if performing a low-level periodic response task (Krienen, Yeo, & Buckner, 2014)? How much data need to be collected to measure properties such as connectivity strength or network coherence reliably (Zuo et al., 2014)? This is a dynamic area of work with constantly emerging findings. Whatever the answers may be, intrinsic functional connectivity does have heritable characteristics (Ge et al., 2018) that may be informative about the nature of psychopathology (Baker et al., 2014).

Measurement and Neurometrics

Whether observing the brain during task performance or not, it is important to consider several general points as we move from asking "basic" questions regarding *how the brain works* to individual differences questions like *how do brains function differently*. Individual differences analyses evoke the 4Rs of measurement: robustness, repeatability, reliability, and replicability. Robustness is the likelihood a given analytic approach will provide a consistent answer. Repeatability is the likelihood that the same pattern of findings will occur if the same group is measured again. Reliability is the extent to which the participants are at the same point in a distribution when measured again, showing a consistent pattern of individual differences. Finally, replicability is the likelihood that the same pattern of findings will occur in a new sample.

The latter three, repeatability, reliability, and replicability, may be affected by various factors, such as caffeine (Laurienti et al., 2002), nicotine (Thiel & Fink, 2007), and ethanol (Seifritz et al., 2000) intake, but also by more subtle variables such as glucose levels (Anderson et al., 2006) and cardiac variability (Shmueli et al., 2007). Before becoming overwhelmed, the investigator should consider the extent to which these factors will affect *changes* in the pattern of evoked BOLD response or intrinsic connectivity for which they will be searching. Similarly, are these factors going to be a source of noise (reducing power) or biases (introducing confounds)?

ANALYSIS

Analytic approaches for fMRI data have rapidly increased in complexity since the initial discovery of the BOLD contrast and the early efforts to use it to map human mental operations (for review see Raichle, 2009). Nevertheless, there are several principles that can be used to understand the current state of the field, as well as guide our anticipation of what advances may wait on the horizon. In this section, we will introduce the methods most frequently used in event-related and intrinsic fMRI analyses, briefly discussing key advances that have shaped the field. For ease of interpretation, the analytic techniques, as well as their associated strengths and limitations, are presented in Tables 13.2 and 13.3. Critically, while these approaches can be applied individually, often two or more will be utilized within the same set of analyses.

Subtraction, Correlation, and Contrast Analyses

The evolution of block, fast-event-related, and hybrid fMRI designs was closely followed by the development of associated analytic methods. Initially, fMRI researchers leveraged approaches adapted from positron emission tomography where signal quality was greatly enhanced if participants were placed in a standard stereotaxic, or common, physical space (Fox et al., 1988). Once the individual participant data is registered to a common reference space the most straightforward and broadly applied method for obtaining results is to perform a simple

Table 13.2 Analytic techniques to examine task-evoked data sets

Method	Application	Strengths	Limitations	Key references
General linear model	Estimates the contribution of known predictors to BOLD signal fluctuations	Mathematically simple, relatively easy to interpret, available in standard analysis packages; can include multiple independent variables (e.g., scanner drifts, participant motion, etc.)	Relies on assumptions including a consistent hemodynamic response throughout the brain and the temporal stability of noise terms	(Friston et al., 1994)
Psychophysiological interaction	Examines the interaction between a task contrast of interest and the functional coupling between brain areas	Can reveal a task-specific change in correlation between areas that may not be evident through a shared effect of task	Can only examine a single source area; causal relations cannot be inferred	(Friston et al., 1997)
Structural equation model	Assesses the degree to which experimental manipulations influence the functional connectivity of brain regions	Can be used for both exploratory and confirmatory testing; based on prior knowledge of brain structure/function; can estimate causal relations and be used across multiple regions simultaneously	Can require a priori assumptions about causality, potentially obscuring other relations; lacks temporal information; assumes linearity	(McIntosh & Gonzalez-Lima, 1991)
Dynamic causal model	Uses biologically plausible neuronal models of the BOLD response to estimate the influence of experimental context on the functional coupling among brain regions	Uses hidden interactions at the neuronal level to study observable shifts in BOLD response; models bidirectional and modulatory interactions (for a comparison of SEM and DCM approaches see Penny et al., 2004)	Relies on pre-specified models and the inferences provided are only as valid as the priors used in the estimation procedure	(Friston, Harrison, & Penny, 2003)
Granger causality model	Assesses the degree to which one time series can predict another	Does not rely on a priori assumptions (e.g., regions of interest and associated connection)	Assumes (local) stationarity, incorrect inferences can result from measurement noise and/or hemodynamic response latencies across brain	(Kamiński et al., 2001)
Meta/mega-analysis	Assesses relations across multiple imaging data sets. Meta-analysis refers to the pooled analysis of published results; mega-analysis refers to the pooled analysis of raw data	Can increase power due to the large number of studies/participants available for analysis; other approaches (e.g., estimates of effective connectivity) can utilize meta/mega-analysis defined regions of interest	Experimental designs may not be uniform and/or adequately sample the full spectrum of behavior and function; meta-analyses often consider the distribution of activation peaks, rather than each study's/contrast's distributed pattern of activity; relies on traditional contrast analyses, this can serve as a confound if the process of interest is not successfully isolated	(Fox, Parsons, & Lancaster, 1998; Laird et al., 2005)

Table 13.3 Analytic techniques of intrinsic brain function

Method	Application	Strengths	Limitations	Key references
Seed-based correlations	Estimates the correlation between the BOLD signal in a predefined regions of interest with other regions, or rest of the brain	Mathematically simple and easy to interpret	Requires the a priori selection of regions; may provide illusory specificity	(Biswal et al., 1995)
Regional homogeneity	Uses Kendall's coefficient concordance to assess the similarity of the time series of a given voxel to those of its nearest neighbors	Mathematically simple and easy to interpret	Requires the a priori selection of regions; sensitive to spatial smoothing and the size of the region of interest	(Zang et al., 2004)
Local-distant	Takes into account local regional connections as well as remote or distant connections outside of a defined area	Allows for the analyses of relative weighting of local or distant connectivity in a region	Can conflate real cortical/anatomical distance with Euclidean distance	(Sepulcre et al., 2010)
Principal component analysis	Creates uncorrelated variables from best-fitting linear combinations of the variables in the raw data; reduces the dimensionality of complex data types	Can reveal hidden, simplified, features in high-dimensional data; does not require a priori task models or estimates of BOLD response	Based on a strong assumption of linearity and orthogonality in the resulting components; sensitive to noise and assumes a high signal-to-noise ratio in the data	(Friston et al., 1993; Viviani, Grön, & Spitzer, 2005)
Independent component analysis	An extension of principal component analysis that separates data into spatially or temporally independent patterns of activity	Few a priori assumptions; not restricted to deriving orthogonal components	Components are assumed to be statistically independent	(McKeown & Sejnowski, 1998; Calhoun et al., 2001)

subtraction across conditions of interest. Subtraction techniques, or more generally correlation analyses, are based on the expectation that the voxels/brain regions participating in a psychological or cognitive process should show dissociable functional responses during the completion of associated tasks. Rather than revealing absolute levels of cerebral blood flow or metabolism linked to a cognitive process, contrast analyses reveal relative changes in BOLD response across conditions. By averaging the time points acquired during an experimental condition and subtracting the average of all the time points associated with a control condition, differing in only one property, the brain regions associated with a cognitive process of interest can be identified.

While a growing proportion of fMRI studies go beyond subtraction logic to include parametric effects where the independent variable has a number of levels (e.g., task difficulty, stimulus intensity, monetary rewards), simple subtraction techniques are a powerful analytic approach. With an appropriate task design, they can be applied to preprocessed fMRI time courses using standard statistical techniques. Historically, subtraction analyses have provided foundational discoveries, characterizing the aspects of brain function that support key facets of cognition and behavior across health and disease. For example, consistent with a role in the modulation of affective functions, differential amygdala responses have been observed during the visual processing of emotional and neutral facial expressions in healthy populations (Breiter et al., 1996). Dysregulated amygdala response to emotional stimuli is hypothesized to underlie the onset and maintenance of affective illness (Mayberg, 1997). In line with these theories, in patient populations subtraction techniques have revealed abnormal amygdala responses in disorders marked by affective impairments (Price & Drevets, 2012) and in populations at increased genetic risk for onset (Smoller et al., 2014).

General Linear Models

The introduction of single-trial or event-related fMRI designs provided researchers the opportunity to separate mental operations into discrete moments in time, allowing for the differentiation of their associated fMRI signals (Huettel, 2012). The associated shift from representing

BOLD responses as static across blocks of time to considering moment-to-moment fluctuations allowed researchers to leverage dynamic analysis methods. In this area, general linear models (GLMs; introduced for fMRI analyses by Friston et al., 1994) are the primary analysis approach utilized in task-based research. The events in event-related designs often occur so rapidly that their associated BOLD responses overlap. GLM analyses assume that the observed BOLD signal is comprised of a linear combination of experimental factors (thereby allowing overlapping responses) and an uncorrelated noise term. A GLM analysis identifies voxels where the signal changes in response to experimental conditions, or events, calculating the significance/extent of effects based on how well the observed data fits the predicted model. These GLM-based approaches form the theoretical scaffolding that underlies most forms of fMRI data analysis, for instance the regression, prediction, and data exploration approaches detailed below. Importantly, GLM analyses are typically conducted in a mass univariate manner across each voxel, and there are several assumptions to keep in mind when utilizing GLM-based approaches that can constrain our interpretation of the results (see Monti, 2011). These include the use of a single model (design matrix) throughout the brain, that noise varies consistently across all time points (e.g., baseline relative to a contrast of interest), and the independence of associated statistical tests.

Multivariate Modeling and Predictive Approaches

An important limitation of traditional GLM-based analytic techniques is that they treat each voxel as independent, assessing if the signal within these discrete data points fluctuates in response to a task condition of interest. They do not account for the possible contribution of complex multivariate relations linking multiple voxels. As the field has developed beyond this mass univariate approach, an increased emphasis has been placed on computationally sophisticated approaches for identifying spatially distributed patterns of brain activity (e.g., multivoxel pattern analysis). For a more thorough treatment of the techniques from the field of machine learning and analytic approaches where specific mental states or task contexts are decoded from distributed activity patterns readers are referred to Chapter 34. In brief, these multivariate approaches are typically implemented in a two-step process. First, a classifier is trained to distinguish the occurrence of events for different conditions within a subset of the available data. Second, the trained model is then applied to an independent or held out sample where the classifier attempts to predict the events of interest. These approaches hold promise as a potential diagnostic tool for psychiatric illness, and their flexibility allows for their integration with other complementary processing and analysis techniques (Rosenberg, Casey, & Holmes, 2018).

For instance, machine-learning approaches have been used to discriminate male from female participants accurately (Chekroud et al., 2016), identify dissociable cognitive trajectories in Alzheimer's disease (Zhang et al., 2016) with gross morphometric estimates of brain anatomy, and predict individual participant attentional capacity, disease status (e.g., ADHD; Rosenberg et al., 2016), or symptomology through analyses of large-scale network function (e.g., presence of psychosis; Reinen et al., 2018).

Network Modeling

Recently, researchers have begun to shift their emphasis from the study of the specialization or segregation of brain functions in isolated regions toward an analytic framework that targets functional integration, working to characterize how signals covary across spatially distinct regions (for review see Sporns, 2014). These distributed processing models of brain function provide a powerful method to explain complex cognitive functions, individual variation, and the behavioral expression of psychiatric illness. Network models allow researchers to represent brain systems as distributed sets of neural elements and their associated interconnections. The generation of these network models requires partitioning or parcellation aspects of the brain into regions, or nodes, which share a consistent set of features. Broadly, brain networks reflect two different categories. Structural networks that describe the anatomical wiring properties of the brain, and functional networks reflect interactions among time series (e.g., correlations) across anatomical parcels or regions of interest. Unsurprisingly, network approaches encompass much of the current research on brain functions ranging from the biophysical modeling of task data through the estimation of the integrity of resting-state networks. Figure 13.2 displays a collection of population-intrinsic network parcellation schemes. Readers should note that the distinction between event-related (task-evoked) and resting-state (intrinsic) analytic techniques is in many ways arbitrary. These methods each probe specific features of brain function; with an appropriate study design they can be applied across data types.

Task-Evoked Functional Connectivity

These analyses can be broadly separated into two classes. The first examines functional connectivity, or the temporal correlation of observed BOLD responses between remote neural areas, similar to the intrinsic techniques detailed below. The second consists of model-based approaches that assess effective connectivity, or the putative influence one brain system or region may exert on another. Prototypical effective connectivity analyses include psychophysiological interaction, structural equation (McIntosh & Gonzalez-Lima, 1991), dynamic causal, and Granger models (Friston et al., 2003).

Psychophysiological interaction, for example, assesses whether connectivity varies between spatially distant brain regions in different psychological/task contexts (Friston et al., 1997). The presence of a psychophysiological interaction suggests that regional responses in the source area to an experimental or psychological factor are modulated by signals from a distal brain region. These approaches have revealed a host of key discoveries, for example, the aberrant development of amygdala-prefrontal connectivity following maternal deprivation, potentially reflecting an ontogenetic adaptation in response to early adversity (Gee et al., 2013).

Intrinsic Functional Connectivity

The convergence of new imaging technologies and increased computational resources has provided tools to map both local and distant connections in the brain (Holmes & Yeo, 2015). Recent work in this domain has established a strong correspondence between the structure of intrinsic (resting state) and extrinsic (coactivation) brain networks, suggesting that the brain's functional architecture at rest is closely linked to cognitive function (Smith et al., 2009; Tavor et al., 2016). Aberrant patterns of connectivity within these networks are evident across many major mental disorders, indicating that their breakdown can lead to diverse forms of psychological dysfunction (Buckholtz & Meyer-Lindenberg, 2012). For instance, impaired connectivity within the frontoparietal control network, which encompasses portions of the dorsolateral prefrontal, dorsomedial prefrontal, lateral parietal, and posterior temporal cortices, as well as corresponding aspects of the striatum and cerebellum (Yeo et al., 2011), is believed to underlie executive functioning deficits in psychotic illness (Baker et al., 2014; Reinen et al., 2018). A growing literature implicates frontoparietal network impairments as transdiagnostic markers of psychopathology (Cole, Repov, & Anticevic, 2014). A set of relationships may emerge through the generation of symptoms that are domain-specific (e.g., impaired executive function), but cut across many pathologies (Buckholtz & Meyer-Lindenberg, 2012).

There are myriad ways that network functions can be probed with intrinsic approaches (Table 13.3; Figure 13.2). From flexibility in the definition of networks of interest, including the use hypothesis derived "seed" regions defined through meta-analyses of task data (Yarkoni et al., 2011) and population atlases of network function (Schaefer et al., 2018; Yeo et al., 2011), through the use of complex dynamic (Hutchison et al., 2013; Reinen et al., 2018) and graph theoretical techniques (Sporns, 2014). Approaches that allow researchers to map functional network topography down within a single person, for example, are critical for clinical intervention and the study of individual differences (Kong et al., 2019; Wang et al., 2015). Research in this domain has led to the development of cortical parcellation methods to accurately map the brain's intrinsic functional organization at the individual level. Functional networks mapped by these techniques are highly reproducible among participants and effectively capture intersubject variability (Wang et al., 2015). Providing converging evidence for the use of intrinsic connectivity when defining participant-specific network topographies, these approaches have been validated by invasive cortical stimulation mapping in surgical patients, suggesting potential for use in clinical applications.

One key factor to consider across all fMRI analyses, but particularly those that examine functional connectivity, is the impact of participant motion. This is a concern for clinical researchers who frequently have to contend with study populations that differ markedly in terms of both disease status and data quality. In the area of intrinsic analyses, for instance, motion generates nonlinear effects on functional connectivity that can either artificially induce or obscure hypothesized results (Van Dijk, Sabuncu, & Buckner, 2012). While these effects cannot simply be regressed out, there are processing approaches that can limit the impact of motion on substantive findings (e.g., motion scrubbing; Ciric et al., 2018; Power et al., 2014). Additionally, given the availability of large-scale fMRI databases that measures of brain structure and function as well as multiple domains of cognition, behavior, and genetics (e.g., Holmes et al., 2015), some research groups have elected to carefully match patient and healthy comparison samples on the basis of data quality (e.g., Baker et al., 2014). The influence of data quality on connectivity analyses is a key point of consideration when interpreting case-control analyses, as patient populations often move more than healthy comparison samples. In the next section, we'll turn to several additional problems faced by clinical cognitive neuroscience.

INTERPRETATION

If brain functions are involved in mental illness, it would seem that methods akin to taking pictures of the living brain and then developing them would provide an objective, biological perspective on how that occurs. Of course, there are any number of reasons this simplistic optimism may not hold, but six criticisms of clinical cognitive neuroscience studies using neuroimaging stand out as particularly important to avoid. We hope the reader will note the challenge of satisfying all these constraints within a single study.

Mechanistic Specificity

The challenges of behavioral experimental psychopathology transfer quite directly to clinical cognitive neuroscience and neuroimaging in particular. A prominent challenge is the difficulty of demonstrating that a deficit in performance on a task is mechanistically relevant to the disorder and not an epiphenomenal, or secondary, effect associated with the presence of illness. Deficits observed in isolation are uninterpretable. For example, when patients perform worse on a facial affect recognition task and have reduced fusiform gyrus activation we are tempted to conclude that these two features are linked to the pathology. Yet the link may be tenuous. Rather than having a role in the symptoms of the disorder, the association between performance and brain activity may result from an earlier perceptual impairment, attention lapses, reduced effort on the task, or any of a number of other failures. More compelling would be to show that patients are worse on facial affect recognition *relative to another task demand measured with equivalent discriminating power*, or ability to distinguish between the groups being measured (see Salem, Kring, & Kerr, 1996, for this particular comparison). Such deficits have been called differential deficits, mechanism-specific or specific cognitive deficits (for review, see Macdonald, 2015). Experiments using one condition, without a second condition that has similar levels of discriminating power, are obviously not up to this standard of evidence. More subtly in experiments with multiple conditions, if the condition of interest is measured with more discriminatory power, then patients may perform worse on it without actually tapping a mechanism related to the disorder. That is, the difference between patients and controls may derive from a nonspecific raft of difficulties patients face when performing behavioral tasks.

Causality Confound

This confound refers to the concern that group differences in brain activity that occur when one group shows differences in performance may not be interpretable (Gur & Gur, 1995). In this case we wish to conclude that the difference in brain activation causes the observed performance differences, however we must also rule out the possibility that both the activation and performance differences reflect another, perhaps unmeasured, impairment. For example, lower motivation, compliance or visual acuity, misunderstanding the task, higher distress, or any number of other failures could also impair performance and reduce task-related brain engagement. This challenge has caused a great deal of aggravation in clinical research because it runs counter to the goal of demonstrating mechanistic specificity (discussed above), which alone is quite a daunting task.

Several approaches to this conundrum have been suggested, none of which fully addresses all of the potential concerns. The least satisfying approach has been to use tasks on which patients are unimpaired but which tap into a known deficiency (such as using a very easy working memory task). This is generally accomplished by taking advantage of a ceiling effect, rather than making the unmeasured impairment irrelevant. Three other approaches match performance in other ways. One way to match performance is to select patients and controls from their broader population distributions based on who performs at a comparable level. This solution falters because of the problem of generalization to the populations of interest. Another way to match performance is to train participants differently so that those who struggle more with the task receive more practice than those who naturally perform it better. This solution can be critiqued in so far as tasks that have become more automated often use different brain areas compared to more novel tasks. The third way to match performance is to titrate the difficulty of the task so all participants, and therefore groups, perform equally. This may be an ideal solution in many cases, however it means that group differences in activation reflect, in part, differences in the tasks they are performing. A final approach we have used is to examine only accurate trials using an event-related analysis, suggesting that the participant was engaged in the task during a given trial. One criticism of this approach is that to the extent that participants can respond accurately simply by chance then some proportion of those trials may still reflect an unmeasured spurious impairment. A second criticism is that it is overly conservative, insofar as part of the deficit of interest is the inability to respond accurately, and in this case there will be no group differences in brain regions that may be generally more difficult for patients to engage. These strategies and critiques are all an extension of concerns that come from using a quasi-experimental methodology, with both within- and between-subject effects. Whereas the quest for mechanistic specificity leads us to test within-subject effects, we are still hampered from strong causal claims by the challenges of differences in performance.

Diagnostic and Symptom Specificity

Diagnostic specificity refers to the extent to which an impairment is disorder-specific versus common across a number of disorders. This concern arises especially from testing patients sampled from a single categorical diagnosis. For example, early findings that patients with schizophrenia showed impairments in dorsolateral prefrontal cortical functioning were greeted with excitement (Ingvar & Franzen, 1974). Subsequent findings of DLPFC dysfunction in many other disorders, from depression (Goodwin, 1997) to substance abuse (Goldstein et al., 2004), may suggest that DLPFC dysfunction is less a cause of psychotic symptoms and perhaps more of a general psychopathology liability factor. Strikingly, a recent meta-analysis comparing task-related brain activation in people with a psychiatric illness and healthy controls reported few differences between the diagnostic constructs in terms of the distribution of case-control effects across the brain (537 studies, total $n = 21,427$; Sprooten et al., 2018). The challenge of diagnostic specificity is not limited to categorical discrimination, however. In the era of metastructural approaches to diagnosis (Holmes & Patrick, 2018), specificity refers to showing that an impairment relates more closely to a particular branch of psychopathology (e.g., thought disorder) than to other branches (e.g., externalizing or internalizing), or to general psychopathology (Lahey et al., 2012). This interpretive challenge can be addressed rigorously in a quasi-experimental design. Using a between-subjects design one can show that patients with an equal level of dysfunction with a different diagnosis show either performance or brain activation differences. Using a within-subject design, one can show that performance or brain activation differences correlate significantly more with one symptom factor relative to another using a Meng's Z or other appropriate test for correlated correlation coefficients. These complementary approaches allow researchers to demonstrate the presence of case/control differences in an aspect of brain biology and provide evidence that associated patterns of individual variability link with shifts in associated behaviors.

Forward and Reverse Inference

The vast majority of neuroimaging research uses an approach termed "forward inference" when probing the underlying biological architecture that supports cognitive functions (Henson, 2006). For example, when researchers manipulate stimuli to determine how the brain responds, forward inference proposes that a given experimental condition causes changes in local brain activity. Thus, dissociable BOLD responses can be used to distinguish between competing cognitive functions or theories. Critically, because forward inference is a correlational approach (see "Subtraction, Correlation, and Contrast Analyses" above), researchers cannot infer that the observed patterns of brain activity are either necessary or sufficient to support the associated cognitive process. However, as noted below, these shortcomings can be addressed through the integration of complementary methodology across levels, for instance the optogenetic modulation of neural activity within freely moving animals.

"Reverse inference" is a different inferential strategy utilized by much of the field, at least informally, and fraught with controversy. Here, researchers make a claim about the engagement of a specific cognitive process based on the activation of a given brain region (Poldrack, 2006). As an example, a researcher might observe that patients with schizophrenia exhibit heightened amygdala responses to images of scenes (e.g., mountains, plains, forests), leading them to erroneously conclude that scene viewing is associated with the experience of fear in psychotic illness. This sort of inference is common within the clinical literature where the core cognitive processes underlying psychiatric illnesses remain unknown. Reverse inference provides a useful deductive tool for expanding our understanding of the underlying brain mechanisms supporting behavior. However, this is a particularly weak standard of evidence, insofar as brain regions and networks generally activate in response to many different demands (Poldrack, 2006).

The issues pertaining to reverse inference are a widespread concern. Clearly, researchers should be cautious when making claims regarding their results, particularly when the functional properties of a given region have yet to be fully established, or in the absence of converging evidence from other methods. Even factors outside of a researcher's control can influence the accuracy of inferences, such as the number of voxels in a region of interest (ROI) or the selectivity of response in a given region of interest. Despite these limitations, reverse inferences can be exceptionally useful when applied judiciously, allowing researchers to relate cognitive processes across distinct theories and experimental contexts (Henson, 2006). Reverse inference can also be used to generate hypotheses, particularly when based on real data. Critically, both reverse and forward inferences can be formalized within a probabilistic framework. They can then be used for meta-/mega-analysis where they provide the opportunity for researchers to map links across diverse neural, cognitive, and disease states. These models provide the field with a powerful tool when coupled with the meta-analytic databases resulting from the recent development of text-mining and machine-learning techniques (Yarkoni et al., 2011).

Regional Differences in Sensitivity

Whenever we write that patients are impaired in one brain region, we imply that they are *not* impaired in the other brain regions examined. However, this implication is only true of other regions that we have measured at least as accurately, or sensitively, as the region where we found the group difference. The extent to which these other brain areas actually are measured as sensitively is largely ignored in the clinical imaging literature. Signal loss and susceptibility artifacts arise as a result of magnetic field inhomogeneities. In BOLD images, the decay in recoverable signal is exacerbated in regions where the brain is adjacent to air (e.g., sinus cavities). Clear spatial variation in voxel-level temporal SNR (the mean of the signal at each voxel over the BOLD run divided by the variance) is evident across the cortical mantle (Holmes et al., 2015). The associated problem is simple to illustrate: brain regions A and B are both impaired in patients, however brain region A (say the anterior cingulate) is measured with very good signal to noise and region B (say the orbitofrontal cortex, subject to susceptibility artifact) is measured with low signal to noise. In reporting our findings without acknowledging these differences in SNR, we end up implying that region B is unimpaired. We see the growing interest in neurometrics, the study of imaging measurement akin to psychometrics, as an important development in clinical cognitive neuroscience (Poppe et al., 2013). A ready-to-hand check on this assumption is to examine signal-to-noise maps across the brain to see that the areas implicated in group differences are not simply those with the highest signal-to-noise.

Cross-Modality Integration

fMRI provides a remarkably powerful technique for researchers to measure and map the functional networks in the human brain in both health and disease, albeit with the limitations inherent to all non-invasive approaches. For instance, recent fMRI work has demonstrated correspondence across the topographic structure of intrinsic and task-evoked functional networks of the human brain, suggesting that the features of the resting brain are closely linked to cognition (Crossley et al., 2013). Yet an integrated understanding of the complex neurobiological architecture of the human brain, from molecules through cells, circuits, and functional networks, will not be possible with a single method or approach. Rather, progress in clinical neuroscience will be made through the combined efforts of researchers working across levels of analyses and species (Holmes & Patrick, 2018). In this regard, work that can join the heterogeneous information provided through distinct analytic approaches, including genetics, brain metabolism, anatomy, electrophysiology, and behavior, has the potential to provide deep insights into the pathophysiology of psychiatric illness. The incorporation of methods that directly manipulate brain function, for instance lesion and optogenetic approaches in animal models or transcranial magnetic stimulation in humans, can allow researchers to test the causal relations between brain and behavior observed in fMRI (e.g., Deng, Yuan, & Dai, 2018). Coupling molecular and genetic approaches with fMRI, as another example, can nominate gene profiles that preferentially associate with functional brain networks (Anderson et al., 2018; Richiardi et al., 2015), revealing the molecular machinery of network communication.

SUMMARY AND FUTURE DIRECTIONS

This chapter sought to bridge between the basic experimental world of cognitive neuroscience and that of clinical research. We hope that basic researchers will find in it links to questions they want to resolve when entering the correlational science of individual differences and clinical problems. Clinical researchers, in turn, should find here the tools to inform a clinical cognitive neuroscience approach to their populations, or the ideas needed to be informed consumers of such research. But whether the reader is more at home with a basic or a clinical perspective, clinical cognitive neuroscience remains an uncanny domain in which the most important achievements seem to be just over the horizon.

On the one hand, advances in methodology and our understanding of brain functions seem to be advancing at an unprecedented speed. Within the last several years, there have been developments in spatial and temporal resolution of MRI equipment, larger samples allowing us to observe subtle effects, a growing number of algorithms to identify meaningful signals, and studies of the effects of genes on brain functioning, promising to remake the landscape of clinical cognitive neuroscience. At the same time, much of this excitement is familiar from previous episodes in which the field was enthusiastic about the potential of widespread noninvasive imaging (in the 1990s) and ever-increasing magnetic field strength (in the 2000s). While technological advances will continue to allow us to ask new questions, we should be sober about how these changes will affect our understanding of, and ultimately our ability to help, people with mental illness. The field is uncanny because that "just over the horizon" feeling drives us forward, but at the same time we need to gird ourselves for the likelihood that new insights may only fill in a few more pieces of a very large puzzle.

Ultimately, our understanding of psychopathology will not come from MRI, a high-throughput genetic chip, or a sophisticated data-mining algorithm. While these will continue to provide new and suggestive leads – and may even ultimately provide crucial elements for diagnosis and prognosis – such technologies cannot bridge the final gap between the biological measurement and the fundamental experience of distress, threat, or craving that make up the core of psychopathology. Researchers who are well-studied in these experiences, and those with firsthand knowledge, will need to work on both sides of the ledger – with these new sources of data, but also with the broad array of people's thoughts, feelings, and experiences, to assemble the final pieces of the puzzle of mental illness.

REFERENCE

Amaro, E., Jr., & Barker, G. J. (2006). Study Design in fMRI: Basic Principles. *Brain and Cognition, 60*(3), 220–232.

Anderson, A. W., Heptulla, R. A., Driesen, N., Flanagan, D., Goldberg, P. A., Jones, T. W., ... Gore, J. C. (2006). Effects of Hypoglycemia on Human Brain Activation Measured with fMRI. *Magnetic Resonance Imaging, 24*, 693–697.

Anderson, K. M., Krienen, F. M., Choi, E. Y., Reinen, J. M., Yeo, B. T. T., & Holmes, A. J. (2018). Gene Expression Links Functional Networks across Cortex and Striatum. *Nature Communications, 9*, 1428.

Baker, J. T., Holmes, A. J., Masters, G. A., Yeo, B. T. T., Krienen, F. M., Buckner, R. L., & Öngür, D. (2014). Disruption of Cortical Association Networks in Schizophrenia and Psychotic Bipolar Disorder. *JAMA Psychiatry, 71*(2), 109–118.

Barch, D. M., Burgess, G. C., Harms, M. P., Petersen, S. E., Schlaggar, B. L., Corbetta, M., ... Van Essen, D. C. (2013).

Function in the Human Connectome: Task-fMRI and Individual Differences in Behavior. *NeuroImage, 80*, 169–189.

Biswal, B. B., Mennes, M., Zuo, X.-N., Gohel, S., Kelly, C., Smith, S. M., ... Milham, M. P. (2010). Toward Discovery Science of Human Brain Function. *Proceedings of the National Academy of Sciences, 107*(10), 4734–4739.

Biswal, B. B., Yetkin, F. Z., Haughton, V. M., & Hyde, J. S. (1995). Functional Connectivity in the Motor Cortex of Resting Human Brain Using Echo-Planar MRI. *Magnetic Resonance in Medicine, 34*(4), 537–541.

Braver, T. S., Reynolds, J. R., & Donaldson, D. I. (2003). Neural Mechanisms of Transient and Sustained Cognitive Control during Task Switching. *Neuron, 39*(4), 713–726.

Breiter, H. C., Etcoff, N. L., Whalen, P. J., Kennedy, W. A., Rauch, S. L., Buckner, R. L., ... Rosen, B. R. (1996). Response and Habituation of the Human Amygdala during Visual Processing of Facial Expression. *Neuron, 17*(5), 875–887.

Buckholtz, J. W., & Meyer-Lindenberg, A. (2012). Psychopathology and the Human Connectome: Toward a Transdiagnostic Model of Risk for Mental Illness. *Neuron, 74*(6), 990–1004.

Calhoun, V. D., Adali, T., Pearlson, G. D., & Pekar, J. J. (2001). A Method for Making Group Inferences from Functional MRI Data Using Independent Component Analysis. *Human Brain Mapping, 14*, 140–151.

Chekroud, A. M., Ward, E. J., Rosenberg, M. D., & Holmes, A. J. (2016). Patterns in the Human Brain Mosaic Discriminate Males from Females. *Proceedings of the National Academy of Sciences, 113*(14), E1968.

Ciric, R., Wolf, D. H., Power, J. D., Roalf, D. R., Baum, G. L., Ruparel, K., ... Satterthwaite, T. D. (2018). Benchmarking of Participant-Level Confound Regression Strategies for the Control of Motion Artifact in Studies of Functional Connectivity. *NeuroImage, 154*, 174–187.

Cole, M. W., Repov, G., & Anticevic, A. (2014). The Frontoparietal Control System: A Central Role in Mental Health. *The Neuroscientist, 20*(6), 652–664.

Crossley, N. A., Mechelli, A., Vértes, P. E., Winton-Brown, T. T., Patel, A. X., Ginestet, C. E., ... Bullmore, E. T. (2013). Cognitive Relevance of the Community Structure of the Human Brain Functional Coactivation Network. *Proceedings of the National Academy of Sciences of the United States of America, 110*(28), 11583–11588.

Dale, A. M., & Buckner, R. L. (1997). Selective Averaging of Rapidly Presented Individual Trials Using fMRI. *Human Brain Mapping, 5*(5), 329–340.

Deco, G., Jirsa, V. K., & McIntosh, A. R. (2011). Emerging Concepts for the Dynamical Organization of Resting-State Activity in the Brain. *Nature Reviews Neuroscience, 12*(1), 43–56.

Deng, C., Yuan, H., & Dai, J. (2018). Behavioral Manipulation by Optogenetics in the Nonhuman Primate. *The Neuroscientist; 24*(5), 526–539.

Farkas, T., Wolf, A. P., Jaeger, J., Brodie, J. D., Christman, D. R., & Fowler, J. S. (1984). Regional Brain Glucose Metabolism in Chronic Schizophrenia: A Positron Emission Transaxial Tomographic Study. *Archives of General Psychiatry, 41*(3), 293–300.

Fox, P. T., Mintun, M. A., Reiman, E. M., & Raichle, M. E. (1988). Enhanced Detection of Focal Brain Responses Using Intersubject Averaging and Change-Distribution Analysis of Subtracted PET Images. *Journal of Cerebral Blood Flow and Metabolism, 8*(5), 642–653.

Fox, P. T., Parsons, L. M., & Lancaster, J. L. (1998). Beyond the Single Study: Function/Location Metanalysis in Cognitive Neuroimaging. *Current Opinion in Neurobiology, 8*(2), 178–187.

Friston, K.J., Frith, C. D., Liddle, P. F., & Frackowiak, R. S. J. (1993). Functional Connectivity: The Principal-Component Analysis of Large (PET) Data Sets. *Journal of Cerebral Blood Flow and Metabolism, 13*(1), 5–14.

Friston, K. J., Holmes, A. P., Worsley, K. J., Poline, J.-P., Frith, C. D., & Frackowiak R. S. J. (1994). Statistical Parametric Maps in Functional Imaging: A General Linear Approach. *Human Brain Mapping, 2*(4), 189–210.

Friston, K. J., Buechel, C., Fink, G. R., Morris, J., Rolls, E., & Dolan, R. J. (1997). Psychophysiological and Modulatory Interactions in Neuroimaging. *Neuroimage, 6*(3), 218–229.

Friston, K. J., Harrison, L., & Penny, W. (2003). Dynamic Causal Modelling. *Neuroimage, 19*(4), 1273–1302.

Funahashi, S., Bruce, C. J., & Goldman-Rakic, P. S. (1989). Mnemonic Coding of Visual Space in the Monkey's Dorsolateral Prefrontal Cortex. *Journal of Neurophysiology, 61*(2), 331–349.

Ge, T., Holmes, A. J., Buckner, R. L., Smoller, J. W., & Sabuncu, M. R. (2018). Heritability Analysis with Repeat Measurements and Its Application to Resting-State Functional Connectivity. *Proceedings of the National Academy of Sciences, 114*(21), 5521–5526.

Gee, D. G., Gabard-Durnam, L. J., Flannery, J., Goff, B., Humphreys, K. L., Telzer, E. H., ... Tottenham, N. (2013). Early Developmental Emergence of Human Amygdala-Prefrontal Connectivity after Maternal Deprivation. *Proceedings of the National Academy of Sciences of the United States of America, 110*(39), 15638–15643.

Glasser, M. F., Coalson, T. S., Robinson, E. C., Hacker, C. D., Harwell, J., Yacoub, E., ... Van Essen, D. C. (2016). A Multi-Modal Parcellation of Human Cerebral Cortex. *Nature, 536*(7615), 171–178.

Goldstein, R. Z., Leskovjan, A. C., Hoff, A. L., Hitzemann, R., Bashan, F., Khalsa, S. S., ... Volkow, N. D. (2004). Severity of Neuropsychological Impairment in Cocaine and Alcohol Addiction: Association with Metabolism in the Prefrontal Cortex. *Neuropsychologia, 42*(11), 1447–1458.

Goodwin, G. M. (1997). Neuropsychological and Neuroimaging Evidence for the Involvement of the Frontal Lobes in Depression. *Journal of Psychopharmacology, 11*(2), 115–122.

Gur, R. C., & Gur, R. E. (1995). Hypofrontality in Schizophrenia: RIP. *Lancet, 345*(8962), 1338–1340.

Haut, K. M., Lim, K. O., & MacDonald, A. (2010). Prefrontal Cortical Changes following Cognitive Training in Patients with Chronic Schizophrenia: Effects of Practice, Generalization, and Specificity. *Neuropsychopharmacology, 35*(9), 1850–1859.

Henson, R. (2006). Forward Inference Using Functional Neuroimaging: Dissociations versus Associations. *Trends in Cognitive Sciences, 10*(2), 64–69.

Holmes, A. J., & Patrick, L. M. (2018). The Myth of Optimality in Clinical Neuroscience. *Trends in Cognitive Sciences, 22*(3), 241–257.

Holmes, A. J., & Yeo, B. T. T. (2015). From Phenotypic Chaos to Neurobiological Order. *Nature Neuroscience, 18*(11), 1532–1534.

Holmes, A. J., Hollinshead, M. O., O'Keefe, T. M., Petrov, V. I., Fariello, G. R., Wald, L. L., ... Buckner, R. L. (2015). Brain Genomics Superstruct Project Initial Data Release with Structural, Functional, and Behavioral Measures. *Scientific Data, 2*, 150031.

Huettel, S. A. (2012). Event-Related fMRI in Cognition. *Neuroimage, 62*(2), 1152–1156.

Huettel, S. A., Song, A. W., & McCarthy, G. (2004). *Functional Magnetic Resonance Imaging* (Vol. 1). Sunderland, MA: Sinauer Associates Sunderland.

Hutchison, R. M., Womelsdorf, T., Allen, E. A., Bandettini, P. A., Calhoun, V. D., Corbetta, M., ... Chang, C. (2013). Dynamic Functional Connectivity: Promise, Issues, and Interpretations. *NeuroImage, 80*, 360–378.

Ingvar, D. H., & Franzen, G. (1974). Distribution of Cerebral Activity in Chronic Schizophrenia. *Lancet, 304*(7895), 1484–1486.

Kamiński, M., Ding, M., Truccolo, W. A., & Bressler, S. L. (2001). Evaluating Causal Relations in Neural Systems: Granger Causality, Directed Transfer Function and Statistical Assessment of Significance. *Biological Cybernetics, 85*(2), 145–157.

Koch, M. A., Norris, D. G., & Hund-Georgiadis, M. (2002). An Investigation of Functional and Anatomical Connectivity Using Magnetic Resonance Imaging. *Neuroimage, 16*(1), 241–250.

Kong, R., Li, J., Sun, N., Sabuncu, M. R., Schaefer, A., Scholz, M., ... Yeo, B. T. T. (2019). Spatial Topography of Individual-Specific Cortical Networks Predicts Human Cognition, Personality and Emotion. *Cerebral Cortex, 29*(6), 2533–2551.

Krienen, F. M., Yeo, B. T. T., & Buckner, R. L. (2014). Reconfigurable Task-Dependent Functional Coupling Modes Cluster around a Core Functional Architecture. *Philosophical Transactions of the Royal Society of London. Series B: Biological Sciences, 369*(1653), 20130526.

Lahey, B. B., Applegate, B., Hakes, J. K., Zald, D. H., Hariri, A. R., & Rathouz, P. J. (2012). Is There a General Factor of Prevalent Psychopathology during Adulthood? *Journal of Abnormal Psychology, 121*(4), 971–977.

Laird, A. R., Mickle Fox, P., Price, C. J., Glahn, D. C., Uecker, A. M., Lancaster, J. L., ... Fox, P. T. (2005). ALE Meta-Analysis: Controlling the False Discovery Rate and Performing Statistical Contrasts. *Human Brain Mapping, 25*(1), 155–164.

Laurienti, P. J., Field, A. S., Burdette, J. H., Maldjian, J. A., Yen, Y.-F., & Moody, D. M. (2002). Dietary Caffeine Consumption Modulates fMRI Measures. *Neuroimage, 17*, 751–757.

Logothetis, N. K., Pauls, J., Augath, M., Trinath, T., & Oeltermann, A. (2001). Neurophysiological Investigation of the Basis of the fMRI Signal. *Nature, 412*, 150–157.

MacDonald, A. W., Becker, T. M., & Carter, C. S. (2006). Functional Magnetic Resonance Imaging Study of Cognitive Control in the Healthy Relatives of Schizophrenia Patients. *Biological Psychiatry, 60*(11), 1241–1249.

MacDonald, A. W. (2015). Differential Deficit. In R. Cautin & S. Lilienfeld (Eds.), *The Encyclopedia of Clinical Psychology* (1st edn.). Hoboken, NJ: John Wiley.

Mayberg, H. S. (1997). Limbic-Cortical Dysregulation: A Proposed Model of Depression. *Journal of Neuropsychiatry and Clinical Neurosciences, 9*(3), 471–481.

McIntosh, A. R., & Gonzalez-Lima, F. (1991). Structural Modeling of Functional Neural Pathways Mapped with 2-Deoxyglucose: Effects of Acoustic Startle Habituation on the Auditory System. *Brain Research, 547*(2), 295–302.

McKeown, M. J., & Sejnowski, T. J. (1998). Independent Component Analysis of fMRI Data: Examining the Assumptions. *Human Brain Mapping, 6*, 368–372.

Monti, M. M. (2011). Statistical Analysis of fMRI Time-Series: A Critical Review of the GLM Approach. *Frontiers in Human Neuroscience, 5*(28), 1–13.

Ollier, W., Sprosen, T., & Peakman, T. (2005). UK Biobank: From Concept to Reality. *Pharmacogenomics, 6*(6), 639–646.

Ollinger, J. M., Shulman, G. L., & Corbetta, M. (2001). Separating Processes Within a Trial in Event-Related Functional MRI I: The Method. *NeuroImage, 13*(1), 210–217.

Park, S., Holzman, P. S., & Goldman-Rakic, P. S. (1995). Spatial Working Memory Deficits in the Relatives of Schizophrenic Patients. *Archives of General Psychiatry, 52*(10), 821–828.

Penny, W. D., Stephan, K. E., Mechelli, A., & Friston, K. J. (2004). Modelling Functional Integration: A Comparison of Structural Equation and Dynamic Causal Models. *Neuroimage, 23*(Suppl. 1), S264–274.

Poldrack, R. A. (2006). Can Cognitive Processes Be Inferred from Neuroimaging Data? *Trends in Cognitive Sciences, 10*(2), 59–63.

Poldrack, R. A. (2010). Mapping Mental Function to Brain Structure: How Can Cognitive Neuroimaging Succeed? *Perspectives on Psychological Science, 5*, 753–761.

Poppe, A. B., Wisner, K., Atluri, G., Lim, K. O., Kumar, V., & MacDonald, A. W. (2013). Toward a Neurometric Foundation for Probabilistic Independent Component Analysis of fMRI Data. *Cognitive, Affective, & Behavioral Neuroscience, 13*(3), 641–659.

Poppe, A. B., Barch, D. M., Carter, C. S., Gold, J. M., Ragland, J. D., Silverstein, S. M., & MacDonald, A. W. (2016). Reduced Frontoparietal Activity in Schizophrenia Is Linked to a Specific Deficit in Goal Maintenance: A Multisite Functional Imaging Study. *Schizophrenia Bulletin, 42*(5), 1149–1157.

Power, J. D., Cohen, A. L., Nelson, S. M., Wig, G. S., Barnes, K. A., Church, J. A., ... Petersen, S E. (2011). Functional Network Organization of the Human Brain. *Neuron, 72*(4), 665–678.

Power, J. D., Mitra, A., Laumann, T. O., Snyder, A. Z., Schlaggar, B. L., & Petersen, S. E. (2014). Methods to Detect, Characterize, and Remove Motion Artifact in Resting State fMRI. *NeuroImage, 84*, 320–341.

Price, J. L., & Drevets, W. C. (2012). Neural Circuits Underlying the Pathophysiology of Mood Disorders. *Trends in Cognitive Sciences, 16*(1), 61–71.

Raichle. M. E. (2009). A Brief History of Human Brain Mapping. *Trends in Neurosciences, 32*(2), 118–126.

Reinen, J. M., Chen, O. Y., Hutchison, R. M., Yeo, B. T. T., Anderson, K. M., Sabuncu, M. R., ... Holmes, A. J. (2018). The Human Cortex Possesses a Reconfigurable Dynamic Network Architecture That Is Disrupted in Psychosis. *Nature Communications, 9*, 1157.

Richiardi, J., Altmann, A., Milazzo, A.-C., Chang, C., Chakravarty, M. M., Banaschewski, T., ... Greicius, M. D. (2015). Correlated Gene Expression Supports Synchronous Activity in Brain Networks. *Science, 348*(6240), 1241–1244.

Rosenberg, M. D., Finn, E. S., Scheinost, D., Papademetris, X., Shen, X., Constable, R. T., & Chun, M. M. (2016). A Neuromarker of Sustained Attention from Whole-Brain Functional Connectivity. *Nature Neuroscience, 19*(1), 165–171.

Rosenberg, M. D., Casey, B. J., & Holmes, A. J. (2018). Prediction Complements Explanation in Understanding the Developing Brain. *Nature Communications, 9*, 589.

Salem, J. E., Kring, A. M., & Kerr, S. L. (1996). More Evidence for Generalized Poor Performance in Facial Emotion Expression in Schizophrenia. *Journal of Abnormal Psychology, 105*(3), 480–483.

Schaefer, A., Kong, R., Gordon, E. M., Laumann, T. O., Zuo, X.-N., Holmes, A., ... Yeo, B. T. (2018). Local-Global Parcellation of the Human Cerebral Cortex from Intrinsic Functional Connectivity MRI. *Cerebral Cortex, 28*, 3095–3114.

Seifritz, E., Bilecen, D., Hänggi, D., Haselhorst, R., Radü, E. W., Wetzel, S., ... Scheffler, K. (2000). Effect of Ethanol on BOLD Response to Acoustic Stimulation: Implications for Neuropharmacological fMRI. *Psychiatry Research Neuroimaging, 99*(1), 1–13.

Sepulcre, J., Liu, H., Talukdar, T., Martincorena, I., Yeo, B. T. T., & Buckner, R. L. (2010). The Organization of Local and Distant Functional Connectivity in the Human Brain. *PLoS Computational Biology, 6*(6), e1000808.

Shen, X., Tokoglu, F., Papademetris, X., & Constable, R. T. (2013). Groupwise Whole-Brain Parcellation from Resting-State fMRI Data for Network Node Identification. *Neuroimage, 82*, 403–415.

Shmueli, K., van Gelderen, P., de Zwart, J. A., Horovitz, S. G., Fukunaga, M., Jansma, J. M., & Duyn, J. H. (2007). Low-Frequency Fluctuations in the Cardiac Rate as a Source of Variance in the Resting-State fMRI BOLD Signal. *Neuroimage, 38*(2), 306–320.

Smith, S. M., Fox, P. T., Miller, K. L., Glahn, D. C., Mickle Fox, P., Mackay, C. E., ... Beckmann, C. F. (2009). Correspondence of the Brain's Functional Architecture during Activation and Rest. *Proceedings of the National Academy of Sciences, 106*(31), 13040–13045.

Smith, S. M., Vidaurre, D., Beckmann, C. F., Glasser, M. F., Jenkinson, M., Miller, K. L., ... Van Essen, D. C. (2013). Functional Connectomics from Resting-State fMRI. *Trends in Cognitive Sciences, 17*(12), 666–682.

Smoller, J. W., Gallagher, P. J., Duncan, L. E., McGrath, L. M., Haddad, S. A., Holmes, A. J., ... Cohen, B. M. (2014). The Human Ortholog of Acid-Sensing Ion Channel Gene ASIC1a Is Associated with Panic Disorder and Amygdala Structure and Function. *Biological Psychiatry, 76*(11), 902–910.

Sporns, O. (2014). Contributions and Challenges for Network Models in Cognitive Neuroscience. *Nature Neuroscience, 17*(5), 652–660.

Sprooten, E., Rasgon, A., Goodman, M., Carlin, A., Leibu, E., Lee, W. H., & Frangou, S. (2018). Addressing Reverse Inference in Psychiatric Neuroimaging: Meta-Analyses of Task-Related Brain Activation in Common Mental Disorders. *Human Brain Mapping, 38*(4), 1846–1864.

Tavor, I., Parker Jones, O., Mars, R. B., & Smith, S. M. (2016). Task-Free MRI Predicts Individual Differences in Brain Activity during Task Performance. *Science, 352*(6282), 216–220.

Thiel, C. M., & Fink, G. R. (2007). Visual and Auditory Alertness: Modality-Specific and Supramodal Neural Mechanisms and Their Modulation by Nicotine. *Journal of Neurophysiology, 97*(4), 2758–2768.

Van Dijk, K. R. A., Hedden, T., Venkataraman, A., Evans, K. C., Lazar, S. W., & Buckner, R. L. (2010). Intrinsic Functional Connectivity as a Tool for Human Connectomics: Theory, Properties, and Optimization. *Journal of Neurophysiology, 103*(1), 297–321.

Van Dijk, K. R. A., Sabuncu, M. R., & Buckner, R. L. (2012). The Influence of Head Motion on Intrinsic Functional Connectivity MRI. *Neuroimage, 59*(1), 431–438.

Van Essen, D. C., Smith, S. M., Barch, D. M., Behrens, T. E. J., Yacoub, E., & Ugurbil, K. (2013). The WU-Minn Human Connectome Project: An Overview. *Neuroimage, 80*, 62–79.

Vincent, J. L., Snyder, A. Z., Fox, M. D., Shannon, B. J., Andrews, J. R., Raichle, M. E., & Buckner, R. L. (2006). Coherent Spontaneous Activity Identifies a Hippocampal-Parietal Memory Network. *Journal of Neurophysiology, 96*(6), 3517–3531.

Viviani, R., Grön, G., & Spitzer, M., (2005). Functional Principal Component Analysis of fMRI Data. *Human Brain Mapping, 24*, 109–129.

Wang, D., Buckner, R. L., Fox, M. D., Holt, D. J., Holmes, A. J., Stoecklein, S., ... Liu, H. (2015). Parcellating Cortical Functional Networks in Individuals. *Nature Neuroscience, 18*(12), 1853–1860.

Wisner, K. M., Atluri, G., Lim, K. O., & MacDonald, A. W. (2013). Neurometrics of Intrinsic Connectivity Networks at Rest Using fMRI: Retest Reliability and Cross-Validation Using a Meta-Level Method. *Neuroimage, 76*, 236–251.

Yarkoni, T., Poldrack, R. A., Nichols, T. E., Van Essen, D. C., & Wager, T. D. (2011). Large-Scale Automated Synthesis of Human Functional Neuroimaging Data. *Nature Methods, 8*(8), 665–670.

Yeo, B. T. T., Krienen, F. M., Sepulcre, J., Sabuncu, M. R., Lashkari, D., Hollinshead, M., ... Buckner, R. L. (2011). The Organization of the Human Cerebral Cortex Estimated by Intrinsic Functional Connectivity. *Journal of Neurophysiology, 106*(3), 1125–1165.

Zang, Y., Jiang, T., Lu, Y., He, T., & Tian, L. (2004). Regional Homogeneity Approach to fMRI Data Analysis. *NeuroImage, 22*, 394–400.

Zhang, X., Mormino, E. C., Sun, N., Sperling, R. A., Sabuncu, M. R., & Yeo, B. T. T. (2016). Bayesian Model Reveals Latent Atrophy Factors with Dissociable Cognitive Trajectories in Alzheimer's Disease. *Proceedings of the National Academy of Sciences of the United States of America, 113*(42), E6535–E6544.

Zuo, X.-N., Anderson, J. S., Bellec, P., Birn, R. M., Biswal, B. B., Blautzik, J., ... Milham, M. P. (2014). An Open Science Resource for Establishing Reliability and Reproducibility in Functional Connectomics. *Scientific Data, 1*, 140049.

14 Clinical Computational Neuroscience

MICHAEL N. HALLQUIST* AND ALEXANDRE Y. DOMBROVSKI*

This chapter discusses mathematical models of learning in neural circuits with a focus on reinforcement learning. Formal models of learning provide insights into how we adapt to a complex, changing environment, and how this adaptation may break down in psychopathology. Their history can be traced to the turn of the twentieth century, when Edward Thorndike proposed that experience shaped neural connections between representations of stimuli and behavioral responses. Independently, Ivan Pavlov (1904) proposed that "conditioned reflexes," learnt associations between predictive stimuli and food, triggered digestive gland responses through a neural reflex arch. The twentieth century saw a remarkable development of these ideas, supported by a convergence between psychology and animal learning (Hull, Watson, Skinner, Estes, Bush and Mosteller, Rescorla and Wagner), as well as key contributions from economics (Simon, Kahneman and Tversky[1]), artificial intelligence (Bellman, Turing, Minsky, Samuel), and neuroscience (Marr, Montague, Dayan, Sejnowski).

The synergies among these fields supported the emergence of modern computational cognitive and behavioral neuroscience. Scholars have called this field decision neuroscience or neuroeconomics (for a useful introductory text, see Dreher & Tremblay, 2016). Recently, scientists have begun to use computational models of the brain to understand dysfunction and disease. Read Montague dubbed this area of research "computational psychiatry" (Montague et al., 2012), though in addition to theory-driven computational neuroscience approaches, computational psychiatry sometimes encompasses data-driven approaches to classification (see Wiecki et al., 2015). Here, we call this emerging interdisciplinary field "clinical computational neuroscience."[2] Broadly, this discipline describes in formal, mathematical terms how cognitive and specifically decision processes[3] are disrupted in psychopathology. As we discuss in the next section, mathematical models provide an explanatory bridge between behaviors and neural systems (Love, 2015).

This chapter reviews the basics of model specification, model inversion (parameter estimation), and model-based approaches to understanding individual differences in health and disease. We first touch on the epistemic foundations of computational neuroscience, the problem of levels, the definition of mechanisms and how models can describe them. We then review some early connectionist models to illustrate how learning could be implemented in neural networks. In the following practical sections, we illustrate how models can be specified based on theory and empirical observations, how they can be fitted to human behavior, and how we can decode model-predicted signals from neural recordings. We use a functional MRI (fMRI) study of social cooperation to illustrate the application of reinforcement learning (RL) to test hypotheses about neural underpinnings of human social behavior. Finally, we review some historically important models and their neural correlates.

We have made every effort to render this chapter accessible to the reader with a basic knowledge of probability and statistics. Thus, we have omitted derivations and other important mathematical details, particularly for model inversion and estimation. Books that fill these gaps are listed at the end of this chapter.

---

* Both authors contributed equally to this work.
[1] Kahneman was a cognitive psychologist and Tversky's background was in mathematical psychology, however their work had the greatest impact on the then-emerging field of behavioral economics.
[2] We wish to eschew the debate about boundaries between psychiatry and psychology. One of us (MH) is a psychologist and the other (AD) a psychiatrist, and both of us see the distinction as historical and institutional rather than substantive and natural. Perhaps the wall will come down one day. Mathematical psychology has a long and distinguished history, to which "computational psychiatry" does not seem to do justice.
[3] In this chapter, by "decision processes" we mean learning and choice processes.

LEVELS OF ANALYSIS

When playing chess, one assesses the position (stimuli) and makes a move (behavior), hoping to capture a piece or deliver checkmate (reinforcement). In a grocery store, one considers the available produce, its price, and one's needs and means; integrating these factors, one fills the basket. Often, we need to extract useful information from stimuli into *latent variables*, such as the quality of produce. Some latent variables are distilled across time and experiences, such as learning whether a store can be trusted to have fresh greens in its salad bar. To understand in biological detail how the brain produces adaptive behavior, we need to deduce what computations are performed on the inputs to produce a behavioral output.

Our neurons do not feel, think, or want. A neuron receives k inputs[4] on its soma and dendrites, processes them, and sends one output – the action potential – down its axon. In other words, based on k independent variables (inputs), a neuron *computes* a function whose values can be 0 (rest) or 1 (fire). At a much larger scale, the entire organism receives inputs from the environment in the form of perceptual stimuli and reinforcement and emits a behavioral output. The organism's inputs are not an objective, fixed reflection of the environment: they are shaped by goals and motivations (Minsky, 1967).

Thus, our primary goal is to describe how neural networks compute functions of their inputs to produce an output. Once we develop a candidate model, how can we be sure that it accurately describes the brain's computational architecture? First, given the experimental demands a human or animal is facing, we can test the model's output against observed behavior. Second, we can test whether model-predicted latent variables, such as the quality of produce in the store or the strength of a chess position, match neural activity. Unlike Pavlov, who relied on peripheral measures of salivary and gastric secretion in his studies of conditioning, we have many techniques for recording neural activity, covering almost the entire space of spatial and temporal resolutions (Sejnowski, Churchland, & Movshon, 2014). Thus, we can test the alignment of a mathematical model with neural data ranging from microcircuits (e.g., Moran et al., 2011) to large-scale brain networks. We also possess computing power and software to process "big data" and test computationally demanding models. Altogether, such technologies give us unprecedented opportunities to test detailed hypotheses about the brain's computational schemes.

To leverage these technologies to their greatest benefit, however, we need mathematically precise theories. Imagine alien scientists trying to understand how New York City works. Suppose they can record every conversation, message, and movement of people, and they can even conduct experiments such as closing streets or jamming certain radiofrequency bands. They may soon figure out traffic and communications patterns. However, they would need to develop theories of language, human behavior, and economy to understand the workings of an advertising agency, for example.

Another key challenge is to understand how to scale from learning processes in single neurons to learning about the environment in an organism. The brain is not a merely a large population of neurons. It is organized hierarchically: neurons form local ensembles, organized into larger-scale circuits. Local circuits in turn form structures such as nuclei or cortical areas. Finally, brain-mapping efforts of the early twenty-first century have revealed large-scale functional networks linking multiple cortical and subcortical structures (Laird et al., 2011). Where does one level end and the next level begin? The philosopher of science Carl Craver proposed a simple definition: "An entity (A) is at a lower level of mechanisms than an entity (B) when A is a component entity in the mechanism for B" (Craver, 2001). In other words, it is useful to think in terms of levels only when the workings of the lower level (e.g., a unit in a neural network) help explain the dynamics of the higher level (e.g., the entire network). Computational models embody mathematically precise hypotheses about the manner in which the elements of the lower level interact to produce the dynamics observed at the higher level. For example, the Hodgkin-Huxley model describes how the flow of ions through sodium and potassium channels gives rise to the action potential. Whereas Craver's definition of levels provides a useful viewpoint, we face a tension between the need for a biologically realistic account of computation at many lower levels (molecular, synaptic, microcircuit, and so on) and the ability of our model to integrate predictions across these levels to explain the organism's behavior.

Faced with the problem of describing the brain's computational architecture, David Marr (1982) preferred to consider levels of scientific analysis, rather than levels of structural organization. At Marr's highest, *computational*, level lies the problem to be solved. Marr asks: "What is the goal of the computation, why is it appropriate and what is the logic of the strategy by which it can be carried out?" The middle, *algorithmic*, level contains mathematical solutions to this problem – that is, what rules does the organism follow to solve the problem? Marr asks: "In particular, what is the representation of the input and output, and what is the algorithm for the transformation?" Finally, the question at the lower, *implementational*, level, is "How can the representation and algorithm be realized physically?" (Marr, 1982).

How do the distinctions among computational, algorithmic, and implementational levels apply to scientific questions in psychopathology? Clinical researchers are often interested in how internal states and traits control observed behaviors. To formulate clinical questions and hypotheses beginning with Marr's highest, computational

---

[4] k is in the thousands for a cortical neuron, and can be 16 times higher in the prefrontal cortex than in the primary visual cortex (Elston, 2003), giving us a sense that the amount of information to be integrated grows as problems become more abstract.

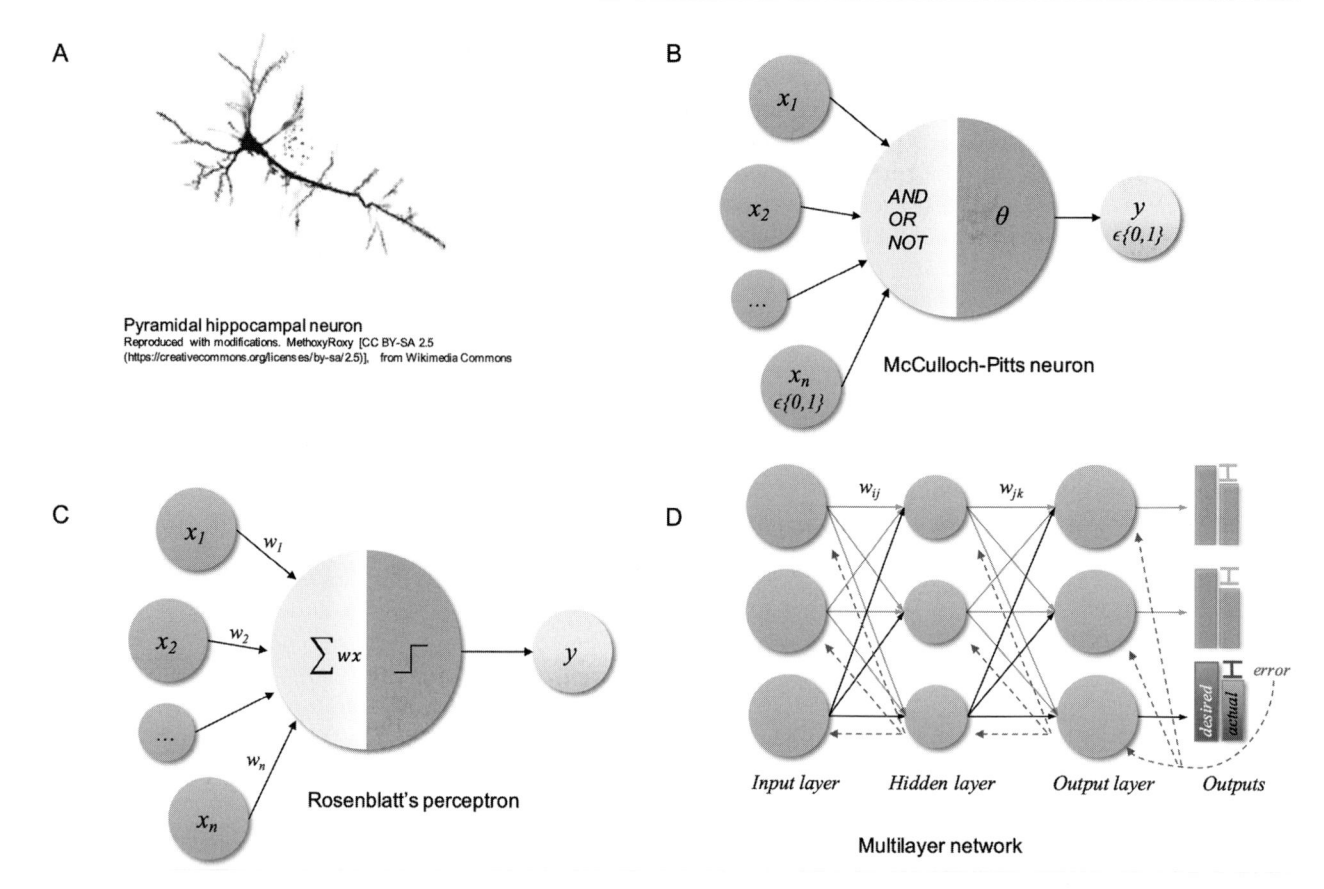

Figure 14.1 Neural network models. A. Pyramidal hippocampal neuron. Dendrites (7 and 2 o'clock) receive inputs, the axon (4 o'clock) delivers the output. B. The McCulloch-Pitts neuron performs logical operations on inputs. C. Rosenblatt's perceptron functions as a linear classifier of its inputs. D. Multilayer network with one hidden layer. The network learns by back-propagation wherein errors resulting from the discrepancy between the expected and the desired output (right-hand side) are used to update synaptic weights.

level, we need to specify what problem is being solved and how the solution may go wrong. Accordingly, at the algorithmic level we focus on models of behavioral adaptation, of which reinforcement learning is one successful exemplar. In general, such models seek to solve problems presented by the environment by adapting behavior to yield outcomes that are more aligned with the needs of the organism (e.g., approaching pleasure and avoiding pain). We view models of behavioral adaptation as useful tools for studying brain mechanisms of behavior (Kriegeskorte & Douglas, 2018), though George Box's (1979) maxim, "all models are wrong but some are useful," certainly applies.

COMPUTATION IN NEURAL NETWORKS: CONNECTIONIST MODELS OF LEARNING AND MEMORY

Thorndike and Pavlov intuited that experience was encoded in neuronal connections. If a neuron indeed computes an output as a function of inputs, can we define this function? The first mathematical model of a neuron (McCulloch & Pitts, 1943), performs logical operations (AND, OR) on binary inputs (Figure 14.1A). The McCulloch-Pitts neuron, however, cannot handle real-valued inputs (i.e., continuous stimulus dimensions). It also cannot weight inputs differentially, which is helpful if some features are more informative than others. These problems were solved in Rosenblatt's *perceptron* (1958), which has input weights and can be trained to perform as a classifier, making decisions about stimuli based on linear combinations of their features. A reader familiar with regression will appreciate that the perceptron's inputs, stimulus features, are analogous to predictors in logistic regression; the weights are regression coefficients; and the output, stimulus class (e.g., dog vs. cat), is the dependent variable (Figure 14.1B). The perceptron learns by finding the appropriate weights for a set of inputs that map onto desired outputs using a training set of data. These weights are computed using an offline learning delta-rule algorithm similar to the one we describe below for reinforcement learning. To continue with our logistic regression analogy, offline learning is akin to fitting regression coefficients to observed outcomes in order to obtain the best prediction.

Minsky and Papert (1969), however, pointed out the limitations of perceptrons, such as their inability to solve problems in which inputs are not linearly separable, such as the exclusive OR function (i.e., "one or the other, but not both"). This and other limitations are overcome by the addition of "hidden" layers between the input and the output layers (Figure 14.1C). Networks with many (often >10) hidden layers are termed "deep." Multilayer artificial neural networks (ANNs) can learn by back-propagation (Werbos, 1974), where the error signals are passed from the output layer through the hidden layers to the input layer and weights are repeatedly adjusted in small steps based on the derivative of the error function with respect to each weight. In the 1980s, McClelland and Rumelhart (1986), among others, developed neural network models of human cognition, preferring the term Parallel Distributed Processing (PDP) to "connectionism."

Neural network models provide a reasonably realistic account of how neuronal ensembles learn and likely approximate the implementational level in Marr's nomenclature. Furthermore, recent research has made substantial progress toward mapping neural network models onto empirical measurements of brain activity, although the exact nature of learning algorithms remains debated (Whittington & Bogacz, 2019). Indeed, one of the primary challenges of models with many hidden layers is that the latent dimensions that these neurons represent are difficult to decode using the fitted weights alone (see Chapter 34, by Coutanche and Hallion, for additional details).

REINFORCEMENT LEARNING (RL) MODELS OF BEHAVIORAL ADAPTATION

Overview of RL Models

RL models of behavior are unified by the common goal of predicting observable responses on the basis of rewarding and punishing experiences that influence hypothesized stimulus-outcome or action-outcome associations. Thus, RL models have a similar objective to neural network models in that they are motivated to understand how experiences (inputs) lead an individual to update their cognitive representation (hidden states) and to make choices that maximize subjective value (outputs). In contrast to neural networks, where the interpretability of the hidden layers is not of primary interest, in RL the scientist must specify what hidden/latent states are tracked and how these states are updated on the basis of learning.

Thus, the first step in specifying an RL model involves a thought experiment. What algorithm could learn to maximize gains and minimize losses in a given environment, without being explicitly instructed on what actions are "right" or "wrong"? In this way, RL models differ from supervised learning algorithms that are trained to predict the correct outputs using a set of inputs (e.g., the Perceptron or support vector machines; see Chapter 34, by Coutanche and Hallion). Supervised algorithms depend on desired outputs that are provided by an expert or objective knowledge. For example, given a set of digital photographs containing handwritten numbers, a human could read the numbers in each image (e.g., one image contains the digits "14091") and type these into a text file. A supervised learning algorithm could then be trained to provide the correct answer for each image (i.e., outputs) using the pixel values (i.e., inputs).

By contrast, in RL there is no single "truth" that can be learned. Rather, an agent can get better or worse at a given task by learning from rewarding or punishing feedback that is delivered sequentially. Therefore, being "good" at a particular task involves maximizing the value of one's choices – in units of reward or avoided punishment – based on ongoing feedback. Thus, an RL model tracks the *subjective*[5] (aka "learned") value of the actions, an approximation of their true values that reflects the specific reinforcement history. Applications of RL range from bug detection in software to the study of dopamine neurons, but here we focus specifically on RL models of human decision making.

Most RL models can be summarized by three components: a representation of the environment, a learning rule, and a choice rule. The representation is the most abstract piece of the model because it requires one to consider the thought problem of, "what aspects of the environment must an organism track in order to solve the task?" The elements that are tracked are viewed as *hidden states* that reflect latent decision processes that are maintained by the organism. The learning rule systematizes how hidden states should be updated on the basis of rewarding or punishing experiences. And the choice rule provides a mapping between the hidden states (typically, expected values) and the predicted choice probabilities on a given trial (i.e., "given what I know, what should I do?").

RL Model Specification

Consider a player seeking to earn the most money in a task where each of two choices results in a monetary outcome (aka a "two-armed bandit"). Imagine, for example, a cup and a bowl appearing on the screen. On each of 100 trials, the player chooses between them by pressing a corresponding button on the keyboard. Cup choices win 50¢ with 70 percent probability and 0¢ with 30 percent probability. Bowl choices win 80¢ or lose 10¢ with 50 percent probability. In the simplest economic account of expected value (for a more detailed treatment, see Hursh & Silberberg, 2008), the long-run expected return for a given choice, x (i.e., if it were sampled many times), is approximated by the product of the reinforcement probability and magnitude summed across all possible outcomes, i:

---

[5] This value is subjective in the sense that it reflects what is known to the subject, as opposed to the objective contingency set by the experimenter or the natural environment. It also reflects how fast a subject learns and how they value rewards/punishments, among other possible parameters.

$$E(x) = \sum_i p_i R_i$$

where p_i is the probability of a given outcome, and R_i amount won or lost. Thus, for the cup, the expected value would be $E(\text{cup}) = 0.7 * 50¢ + 0.3 * 0¢ = 35¢$; for the bowl it is $E(\text{bowl}) = .5 * 80¢ + .5 * -10¢ = 35¢$. The first step of building an RL model is to specify what information an agent needs to track in order to succeed. An agent on the cup versus bowl task might track separate estimates of expected value for choosing cup, $Q(\text{cup})$, or bowl, $Q(\text{bowl})$.[6] In addition, some agents might slowly forget expected values with time or a lack of sampling.

Learning Rule

Now that we have separate action values for cup or bowl choices, we need to formalize how these values are updated from one learning experience to the next. This formalism is called the *learning rule*, which is an equation that specifies how value representations are updated on the basis of experience. For example, perhaps an agent remembers the outcome of each previous experience of a given action, then takes the mean of these outcomes to approximate value. This agent would tack on the incremental outcome of trial t for choice i, $o_i(t)$, to a vector of values, $\mathbf{o}_i$. The agent would then compare the mean values of the two actions, $\bar{o}_\text{cup}$ versus $\bar{o}_\text{bowl}$, to decide which one to choose on the next trial, $t + 1$. This representation might work for a few trials, but memory limitations will soon make it impractical to remember every experience.

RL provides the solution to the seemingly intractable problem of needing to remember many individual experiences in order to estimate the value of an action.[7] Specifically, if one tracks a running value estimate (analogous to the mean) that represents the integration of all prior experience, then updates this estimate after each incremental experience, then the agent need not maintain a specific record of all individual experiences. Stated differently, decision problems are simplified greatly if the agent can decide among possible actions using only their current values at hand, rather than needing to recollect more remote experiences. This sequence of experiences and choices can be thought of as a *Markov decision process* (MDP). An MDP is a choice problem in which the agent is in a given "state" (e.g., the grocery store) and can choose among available actions (e.g., check out or search for apples). Based on the action selected, the agent then transitions to a new state from which new actions can be selected (e.g., choosing to get in the car after checking out). To be considered an MDP, the transition to the subsequent state can depend on the current state and action chosen, but the transition must be *conditionally independent* of all previous states and actions. Chess is a classic example of an MDP in that the choice of the best move to make next depends only on the current state of the board, not the history of the game to this point. Thus, MDPs are *memoryless*, since they retain no memory beyond the *state representation* (e.g., chess position).

Not all choice problems follow an MDP, and episodic memory plays an important role in learning (Wimmer et al., 2014). Nevertheless, MDP models including typical RL approaches are ubiquitous in the literature because they provide good solutions to many choice problems and their mathematics are straightforward. In RL, the Markov property means that the decision of what action to take on the next trial, $t + 1$, depends on the action values on the current trial, t, but not on more remote trials, $t - 1$, $t - 2$, and so on. That is, most RL models are *memoryless*.

Applying these principles to our cup versus bowl task, we could specify an MDP-compatible learning rule in which the agent updates the expected action value for cup and bowl based on each experience.

$$Q_\text{bowl}(t+1) = \begin{cases} Q_\text{bowl}(t) + \alpha[r(t) - Q_\text{bowl}(t)] & \text{if action}(t) = \text{bowl} \\ Q_\text{bowl}(t) & \text{if action}(t) = \text{cup} \end{cases}$$

$$Q_\text{cup}(t+1) = \begin{cases} Q_\text{cup}(t) + \alpha[r(t) - Q_\text{cup}(t)] & \text{if action}(t) = \text{cup} \\ Q_\text{cup}(t) & \text{if action}(t) = \text{bowl} \end{cases}$$

Here, t represents the current trial (aka "learning episode"), in which one chooses between cup or bowl. The reinforcement for the choice on this trial – here, money won or lost – is denoted $r(t)$. The degree to which one updates an action value estimate, Q_bowl or Q_cup, depends on the *learning rate* parameter, α (bounded between zero and one). This parameter scales the effect of the discrepancy between the reinforcement received, $r(t)$, and the expected reinforcement, $Q(t)$. This discrepancy, $r(t) - Q(t)$, is the *prediction error*, and reflects the magnitude and direction of surprise, or the degree to which the outcome was unexpected. As we review below, prediction errors appear to be signaled by the dopamine system (Glimcher, 2011) and in cortico-mesostriatal circuits supporting social and nonsocial decisions (Ruff & Fehr, 2014).

As one gains experience in a stable environment, prediction errors diminish as one's expectations approach the true expected values of the actions. Importantly, however, prediction errors diminish faster when the environment is more predictable. For example, if a given action returns 50 ± 3 cents, whereas another action returns 50 ± 90 cents, the latter, more volatile, option will generate larger prediction errors over a longer period. Furthermore, lower learning rates yield expected value representations that are updated more gradually and are therefore more stable (see Figure 14.2, top panel). Conversely, high learning rates make the agent more sensitive to recent reinforcement and promote local fluctuations around the true

---

[6] Here, we use the letter Q to denote the expected *action* value of choosing each option (i.e., what is the expected return on a given choice?). This notation aligns with Q-learning models in RL that focus on instrumental learning (Sutton & Barto, 2018).

[7] Sutton and Barto (2018) offer more context; they provide derivations and relate RL to dynamic programming and Monte Carlo methods.

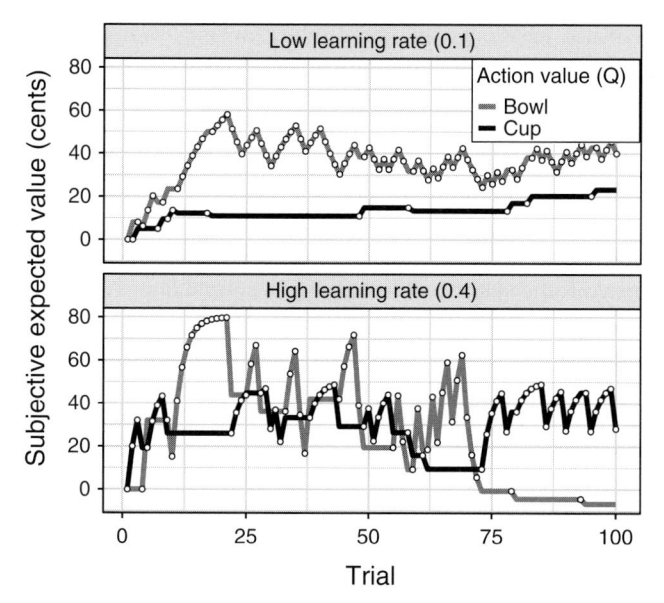

Figure 14.2 Learned action value estimates for cup and bowl according to a Q-learning reinforcement learning model. The top panel denotes an agent with a learning rate of $\alpha = 0.1$, whereas in the bottom panel $\alpha = 0.4$. Both agents use a softmax policy with a temperature of $\beta = 0.08$. The circles denote the chosen action on a given trial, after which its action value is updated according to the learning rule.

expected value, alternating between overestimation and underestimation (Figure 14.2, bottom panel).

Note, also, in our simple RL model, the action value, Q, is only updated for the chosen action, whereas the value of the unchosen action is unchanged by the learning rule. In this way, value estimates for unchosen alternatives are carried forward across trials, though RL models with memory constraints can decay them (Hallquist & Dombrovski, 2019).

As noted above, the long-run expected values for both cup and bowl are 35¢ per choice. But an RL agent is typically given no prior information about the contingencies. By chance, one might receive a string of wins on the bowl option, leading its *subjective* value to exceed the true value for a period of time, until the laws of probability catch up and depress the subjective value toward the true value (e.g., see Q_{cup} for trials 15–40 of Figure 14.2, bottom panel). These dynamics help describe, for example, why one is very excited upon being offered a pay rise (large positive prediction error), less so upon receiving one's first paycheck reflecting the raise (smaller prediction error) and almost not at all upon receiving the same paycheck a year later (asymptotically small prediction error).

Choice Rule

Given a state representation and a learning rule that specifies how states (here, learned values) are updated from experience, the next step in developing an RL model is to specify a *choice rule* or *policy*. This equation specifies how states are mapped onto behavioral outputs, thereby informing the decision of what to do next. For example,

if I moderately prefer sushi over hamburgers, should I choose to eat sushi whenever given the choice between these two foods? Or, in the long run, should I perhaps choose these foods in proportion to their relative values? As a brief introduction to policies, let us first consider ε-greedy and "softmax" approaches. These policies both address the *exploration/exploitation* dilemma (Kaelbling, Littman, & Moore, 1996) in which an agent must decide how often to select the best available action based on previous experience (exploitation) versus an action that has unknown or lower subjective value, but that could yield unexpectedly good outcomes (exploration). For example, in a familiar restaurant, should one order the steak, which has been delicious in the past, or try an unfamiliar daily special, shrimp jambalaya?

The ε-greedy policy for this problem would be to allocate a fixed proportion of choices to exploring (here, choosing shrimp jambalaya) and to exploit the best-known option the rest of the time (here, steak, which is the "greedy" choice). For example, if we set ε to 0.1, then we would choose jambalaya with a probability of 10 percent and steak with a probability of 90 percent. This reveals the fundamental point that most mathematical models of decision making view choice as a probabilistic process. That is, perhaps unsurprisingly, human (and generally animal) behavior is not a simple deterministic process in which we always choose actions with the highest value. If we did, we would likely converge prematurely on some actions and fail to explore new alternatives that could lead to higher total value in the long run.

A more sophisticated softmax policy weights the probability of choosing each action in proportion to its estimated value. This is achieved through a logistic transformation such that the probabilities sum to one. Moreover, a softmax policy includes a "temperature" parameter that controls the tendency to explore versus exploit. The softmax policy is specified as:

$$p(\text{action}(t) = i) = \frac{\exp(Q_i/\beta)}{\sum_{j=1}^{J} \exp(Q_j/\beta)}$$

In our cup–bowl problem, i would represent one of the possible actions (e.g., bowl) and J would represent all available actions: $J = \{\text{cup, bowl}\}$. The temperature is denoted by β. At higher temperatures, alternative options approach equiprobability (more exploratory). At lower temperatures, the probability of the best option approaches one, while the alternatives approach zero (more exploitative). The softmax will thus produce predicted probabilities of choosing each possible action on the current trial, t.

Does the Proposed Model Capture Human Behavior?

Developing a candidate RL model is an essential step in formalizing a hypothesis about decision processes. Whether humans likely implement the model, however, is a separate scientific question. With empirical data, we do not know the true model and must instead examine whether the model qualitatively fits observed behavior. Furthermore, as detailed below, we can evaluate the relative fit of a candidate model compared to reasonable alternatives.

To examine whether an RL model provides a useful description of behavior on a decision task, we can charge the model with predicting the choices of every participant. RL models generate predictions about observed decisions, and a good model should yield relatively few mispredictions. To examine how well a model captures learning in a given participant, we can apply *model inversion*[8] in which we search for parameter estimates that minimize the discrepancy between observed and model-predicted choices. This process is akin to simple linear regression, where the model adjudicates among slope estimates for a bivariate relationship on the basis of which value minimizes the sum of squared errors (i.e., the residuals). In the case of decision tasks, however, we are seeking to estimate RL model parameters, $\hat{\theta}$, that maximize the likelihood of the observed choice sequence, D, which consists state-action-reinforcement triplets:

$$\mathcal{L}(D|\hat{\theta}, \text{model}) = \prod_{t=1}^{T} p(a_t|s_t, r_{1,\ldots,t-1}, \hat{\theta})$$

This is the *individual likelihood* function, representing the model-predicted joint probability of observing an individual's choice sequence given a particular model with corresponding parameter estimates (e.g., learning rate and temperature). Returning to our cup-bowl decision task, the Q-learning model would be charged with trying to predict whether a participant chose cup or bowl on each of the 100 trials based on the learned values, Q_{cup} and Q_{bowl}, as well as the choice rule. The higher the model-predicted joint probability, the closer the model's predictions are to the reality of the participant's choices. Because the product of many trial probabilities becomes very small, we typically work with the log-likelihood function, which avoids precision constraints. Logarithmic transformation also simplifies the likelihood function to a sum of log probabilities, rather than a product.

To identify model parameter estimates that best fit a participant's choice sequence, one can employ standard numerical optimization methods (for a useful introduction, see Nocedal & Wright, 2006). These methods employ algorithms that calculate a measure of global fit (often called an objective function) for a candidate set of parameter estimates, then test whether different parameter estimates would improve the fit. Consider this analogy: if the space of all possible parameter values (e.g., learning rate and temperature) could be mapped on the physical surface of the Earth (longitude: learning rate, latitude: temperature), then any point on the globe would represent a unique combination of parameter values. In turn, every point on the globe has an elevation; in the analogy, the elevation represents the joint probability of observing the individual's choice sequence, D, at a given set of parameter values. Thus, numerical optimization algorithms can be likened to a zealous mountaineer who scales the highest peak. If the algorithm does its job, it should find parameter values that maximize the likelihood of observing the empirical data (i.e., it should reach Mount Everest).

Standard numerical optimization approaches search for parameter estimates that maximize the individual likelihood without any preference for particular parameter ranges. Importantly, however, fitting unique parameter estimates to each participant individually may overfit the peculiarities of a given choice sequence. We may also lack enough data to fit the model to a single subject reliably. Furthermore, there often are strong empirical or theoretical reasons to expect that some RL model parameters would fall in a certain range (e.g., Gershman, 2016). Bayesian approaches to RL parameter estimation can overcome some of these limitations by using formal prior distributions on model parameters (Wagenmakers, Morey, & Lee, 2016). Priors specify a statistical distribution (e.g., Gaussian with mean of 5 and variance of 2) that sets the expectation for parameter estimates in the absence of data. In Bayesian estimation, the likelihood function and the prior contribute to the predicted *posterior estimates* of parameters, representing the integration of prior expectation with observed data. In this way, priors can mitigate overfitting by *regularizing* parameter estimates (for a useful overview of Bayesian approaches, see Kruschke, 2014).

After we obtain parameter estimates that represent the best fit of a candidate model to a given subject, a more pressing question is whether the model's predictions align well with the sequence of choices, D. For example, if a participant chooses cup four times, then switches to bowl for ten trials, does the model predict this transition, and how many of these choices does the model predict accurately? Although model inversion tries to fit the model to the observed data, there is no guarantee that the result captures the qualitative dynamics of choice in a given subject. Thus, graphical examinations of model predictions, hidden states (e.g., learned action values), participant choices, and observed outcomes (e.g., wins versus losses) are essential to understand the quality of the fit (see Figure 14.3). We encourage researchers employing RL methods to examine plots of qualitative model fit for every subject in order to identify subjects whose choices display a peculiar or invalid pattern (e.g., always choosing one option) or whose behavior is not described well by the model. Often, by identifying subjects for whom the model misses a critical pattern of choices, one will feel encouraged to develop new model variants that could capture such patterns.

---

[8] That is, solving the inverse problem wherein we estimate from observed data the causal process that produced them. Think of hearing the music and deciding which instrument is playing.

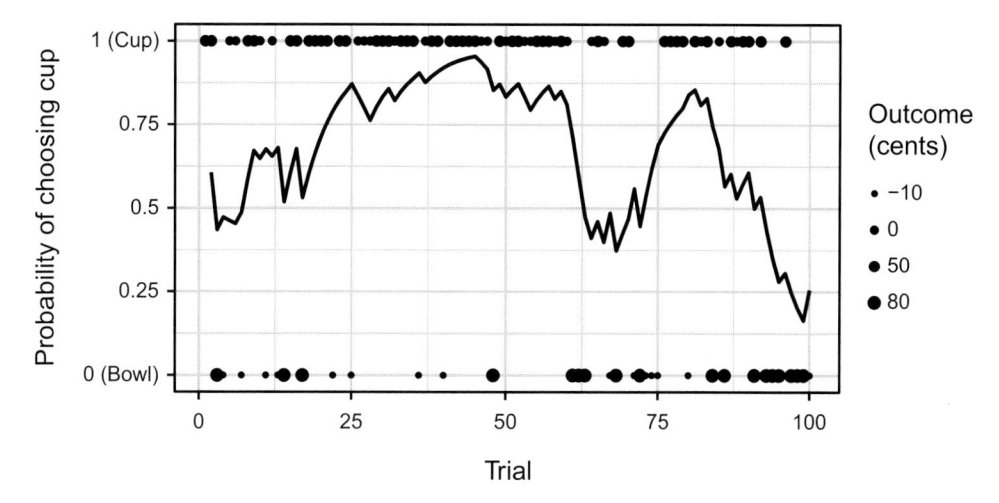

Figure 14.3 The model-predicted probability of choosing the cup (versus bowl) on each of 100 trials. The circles denote the participant's observed choices for each trial, and the size denotes the magnitude of the reinforcement. The black line denotes the model-predicted probability derived from a Q-learning model fitted to the subject's observed choices using maximum likelihood estimation. Note that the density of choices for cup is associated with greater model-predicted probabilities for this choice (e.g., trials 15–30), whereas bowl choices are associated with lower predicted probabilities for cup (e.g., trials 80–100).

Does the Proposed Model Explain Behavior Better than Plausible Alternatives?

Above, we described fitting a candidate model to subjects one at a time. This first step, however, does not address the question of whether a model adequately describes an entire sample. Simply put, some models should be better than others at describing human decisions. Recalling that RL models instantiate hypotheses about decision processes, we should test alternative hypotheses in part by testing alternative RL models.[9] How does one examine whether one model describes the sample better than another?

Model selection in statistics considers how to quantify the relative evidence of one model versus another (for an accessible introduction, see Burnham & Anderson, 2002). Importantly, more complex models (i.e., those with more parameters) are usually more capable of fitting a wider array of decision sequences, but at the expense of parsimony.[10] The fit-parsimony tradeoff is a key conceptual and mathematical consideration in model selection because we typically wish to identify the simplest model that nevertheless fits data sets reasonably well.

Although one can fit a model to a participant's choices by finding parameter estimates that maximize the individual likelihood, we must remember that likelihood is a *relative* measure of fit. Building on the idea of individual likelihood, there is also a *sample likelihood* function that quantifies the joint probability of decision sequences for *all* participants in the sample. The sample likelihood is the product of the individual likelihoods. Thus, the greater the sum of individual log-likelihoods, the more likely are the sample data given a model and corresponding parameter estimates. The simple log-likelihood summation approach, however, has a number of challenges. We discuss two.

First, models may fit one subject much better than another, which means that the distribution of individual likelihoods may be highly nonnormal. As a result, at the sample level, one model might have higher individual likelihoods for 95 percent of subjects compared to another, but its sample likelihood could be worse because the model fails rather badly (i.e., has very low individual likelihoods) for the remaining 5 percent. Second, while the sample likelihood quantifies how well a candidate model fits at the group level, it assumes that every subject used the same model to solve the task. There is the strong possibility, however, that subjects employ different models. Thus, we are often interested in quantifying the probability that one model is more likely than any of the others tested, called the *exceedance probability*. This reframes the problem of model selection in terms of estimating *model probabilities* given the sample data.

Altogether, an essential step in validating an RL model is to examine how well the model predicts decision sequences at the sample level compared to several theoretically plausible alternative models. For example, if we thought that subjects might update value estimates more from negative surprises than positive surprises, we could fit a model that has separate learning rate parameters for

---

[9] We are solving an inverse problem of a more general kind here: which model (rather than set of parameters of a given model) is more likely to have generated this behavior. Ideally, we would like to recover that model from the observations, but the space of possible models is so large that no algorithm can possibly search it. Thus, we need to formulate a set of hypotheses – specific models – and adjudicate among them. Competing models need to be specified on a theoretical basis rather than *ad hoc*, based on observed data alone.

[10] Why is parsimony preferred? For example, because an overly complex model may fit a given behavior with an implausible combination of parameters, which will not generalize to other instances (tasks or individuals). This is what we call overfitting the data.

negative and positive prediction errors. To adjudicate among models quantitatively at the group level, we encourage researchers to use a Bayesian model selection approach (for a useful treatment, see Stephan et al., 2009) that is robust to variation in individual model likelihoods and incorporates uncertainty about which model was employed by each subject (i.e., treating model as a random variable).

Can the Model Solve the Task?

Fitting an RL model to participant data (i.e., model inversion) can give us insights into latent decision processes and how they may be disrupted in psychopathology. Importantly, however, we will likely never know whether our model is close to the true decision processes or just a reasonable approximation. It is also common to encounter difficulties in model inversion such as parameter estimates that are highly correlated at the sample level, or poor qualitative fits in some subjects.[11] If we rely on empirical data alone, we will likely never know whether these challenges reflect a basic problem with the model itself or the peculiarities of the sample.

That is, at a basic level, we do not know if the model is good at its intended task. Stated differently, is the model *capable* of solving the task for which it was built? To understand the capabilities of the model, we should examine how well it performs in simulations, sometimes referred to as computational experiments. One intuitive way to frame this step is to imagine a robot player sitting at the same computer that will later be used by human subjects. Humans could be implementing any number of models, but with the robot, we know the model it is running to solve the task because we programmed it. Although we need not employ a physical robot, computational experiments largely focus on *forward modeling*, which refers to using the RL model to generate simulated behavior on the task.

To adjudicate whether model-simulated behavior supports or invalidates the capability of the model to solve the task, we must compare simulated choices to some criterion. Returning to the premise of reinforcement learning – that an agent must learn what to do by maximizing rewards and minimizing punishments – earnings are the most straightforward criterion by which to judge a model. For example, if we implemented an RL model to play chess, we could test its performance by simulating thousands of games and counting how often it won against a competing algorithm. In value-based decision-making tasks, one would typically quantify performance in terms of the relevant units of the task (e.g., number of points earned in a game or number of "likes" received in a chat room).[12] Thus, if a candidate RL model is capable of learning a given task environment, it should be able to maximize its earnings by choosing good actions.

However, what constitutes "good performance" in simulations? The answer is that both relative and absolute performance benchmarks are useful. Relative benchmarks compare the performance (in units of value/earnings) of one candidate model to another. Methods from standard experimental psychology can inform this judgment. For example, one can compare average earnings among models using a one-way analysis of variance (ANOVA). Likewise, if one is comparing many models across many experimental conditions, factorial ANOVA methods can provide resolution on which model does best in each experimental condition. Absolute benchmarks typically examine model earnings in relation to the best and worst possible performance on the task. For example, on a four-choice task in which reinforcement is provided in US dollars, how many dollars would an omniscient model earn if it always chose the best option? Likewise, what would be the total earned by a model that always chose by chance?

A few additional concerns must be addressed in computational studies. First, even if one has a preference for a given model, one should always compare performance across multiple models. Building on the view that every RL model embodies a representational hypothesis, testing alternative models is akin to comparing the evidence for alternative hypotheses. Second, one should test models across a range of experimental conditions to understand the limits of the model to solve tasks of a certain sort. For example, if a model can learn a stable contingency (e.g., 70 percent win, 30 percent loss), how quickly can the model update its value estimate if the contingency changes? Third, one should simulate model-based behavior over a large number of trials to provide a reasonable estimate of overall performance. If one only examined alternative models based on three learning trials, there would be too much sampling variability to consider this a reliable estimate of performance. This is analogous to sampling error in conventional statistics, where small samples are vulnerable to noisy estimates. Fourth, simulations should be repeated many times for each model and condition so that one can estimate the uncertainty associated with performance estimates, analogous to examining mean squared error in Monte Carlo studies.

---

[11] We note that problems in model inversion can also be a product of the parameter estimation method itself. For example, simple numerical optimization approaches to single-subject data may fail to converge if there are few constraints on the parameters. By contrast, hierarchical Bayesian approaches may have fewer difficulties because estimates of individual parameters borrow strength from the distribution in the sample and priors set on the parameters.

[12] If the task provides probabilistic feedback, it may be more informative to quantify model performance in terms of the expected value of its choices, rather than the reinforcements themselves. Thus, if a model chooses an action that has high average value, but receives a rare omission or punishment, focusing on expected value as a benchmark ensures that a good choice is not penalized for a rare bad outcome.

When comparing model performance in simulations, one important consideration is whether computational experiments should be allowed to search for parameter values that maximize earnings, as opposed to fixing parameters at rationally chosen values. If one uses numerical optimization to find parameter values that maximize earnings, we recommend two additional steps to aid in interpretability. First, compare the performance of alternative models at parameter values *optimized* in a different contingency. This ensures that performance differences are tested on out-of-sample predictions, reducing the likelihood that differences reflect peculiarities of the reinforcement history from the parameter optimization step. Second, simulate data from each model at its optimized parameter values and ensure that the qualitative pattern (e.g., using graphical diagnostics) suggests that the model is indeed learning the task across trials. Such qualitative checks can inform new alternative models – for example, if a candidate model learns well initially but soon stops improving or vice versa.

Are the Parameters of the Model Interpretable?

In addition to examining the performance of RL models in computational experiments, it is important to examine whether the parameters of a model are conceptually and numerically distinct. If the parameters are not distinct from each other or cannot be estimated accurately in simulations, then they may not be interpretable in empirical studies. This consideration relates to the concept of *identifiability* in statistics, which refers to whether one can obtain the true values of a model's parameters given enough data. If a model is identifiable, then parameter estimates of the model should converge on the population parameters as the number of observations increases. By contrast, if some model parameters are not identifiable, then parameter estimates may deviate from the population values either systematically, as in the case of biased estimates, or unsystematically, as in the case of widely varying, unstable estimates across samples. One of the most common sources of an *unidentified* model is when two parameters contribute similarly to learned representations or predicted outputs. For example, imagine that we believed that there are individual differences in sensitivity to learned values such that in some people, larger values influence choice more than in other people. We might amend the softmax choice rule as follows:

$$p(\text{action}(t) = i) = \frac{\exp(\eta Q_i / \beta)}{\sum_{j=1}^{J} \exp(\eta Q_j / \beta)}$$

Here, the η term that multiplies learned values, Q, could capture scaling differences in the influence of value on choice. However, a closer examination reveals an intractable problem with the model: the influence of value on the softmax choice rule now depends on the ratio of η and β, both of which are free parameters to be estimated. Consequently, there are an infinite number of parameter combinations for η and β that would yield the same model-predicted choices (i.e., outputs). For example, if $\eta = 10$ and $\beta = 2$, the model predictions would be the same as $\eta = 1000$ and $\beta = 200$ because both yield $\eta/\beta = 5$.

Further computational studies can examine whether all parameters of a model are uniquely identified. Instead of emphasizing model earnings, however, the goal of identifiability simulations is to test whether parameter estimates converge on known parameter values. That is, one should simulate model-predicted data sets across a range of parameter values (encompassing the *parameter space*). In each data set, the *groundtruth* parameters are necessarily known, in contrast to empirical data where we do not know the population values or even whether the organism implemented the proposed model to solve the task. After simulating data sets across the parameter space, one then fits each data set based on the behavior alone using the same model that generated the data.

This step, model inversion, is identical to fitting human behavior using a model, where one tries to identify parameter estimates that best fit the observed data. In the identifiability case, however, the goal is to test *parameter recovery* – do the parameter estimates converge on the groundtruth values? If estimates of all parameters align with known values (e.g., $R^2 > 0.9$), then the model is likely identified (at least across the range of parameter values tested). On the other hand, if there is not a tight correspondence between groundtruth and estimated values for some parameters, the offending parameters may not be identified. In this case, one should review whether every parameter is conceptually important and distinct from others. Likewise, one should test simpler models that combine correlated parameters in some way. For example, if one views decisions as a weighted sum of two signals, A and B, then a single weighting parameter, γ, will be identified, whereas two will probably be unidentified:

$$\text{Identified}: O = \gamma A + (1 - \gamma) B$$

$$\text{Not identified}: O = \alpha A + \nu B$$

Does the Model Yield Distinct Predictions and Capture Distinct Phenomena Relative to Other Models?

Having established that a model solves a particular task in simulations and that its parameters are identified, the last major step in computational experiments is to examine whether the model is distinguishable from alternatives. That is, does the model yield unique predictions about the dynamics of behavior that could not be captured equally well by another model? The model's uniqueness is closely related to identifiability, but instead of considering whether parameters within a model are confounded with each other, here we test whether predictions from two or more models could be *confused* with each other. If two models provide identical behavioral predictions in a given experimental context, they are considered to be *observationally equivalent*. If two RL models posit different cognitive mechanisms, but they are observationally equivalent in the task of interest, then one cannot adjudicate between the alternative mechanistic hypotheses no matter how many participants are collected. Tests of model confusion seek to examine the relative frequency and particular experimental conditions under which two or more models can be disambiguated from each other.

To examine model confusion, one first generates simulated data sets from each candidate model across a range of parameter values and relevant experimental conditions. Each model is then fit to every data set by *model inversion* – that is, we estimate parameters from each model that best describe each data set. Then, using model comparison methods (see above), one can compare the relative evidence that the behavior generated by a given model is also best described by that model, as opposed to alternatives. The posterior probabilities that a given model describes each data set can be plotted as a confusion matrix (also see Chapter 34, by Coutanche and Hallion) to quantify how well models can be discriminated. If a model is highly flexible in its ability to describe behavior, it may be able to fit many data sets, even those that it did not generate. On the other hand, if a model embodies a unique perspective on the processes underlying learning and decision making, its dynamics should not be easily described by alternative models. Thus, the overarching goal of model confusion tests is to examine whether a promising candidate model is not undermined by a lack of uniqueness or, in some situations, excessive flexibility.

Does the Model Capture Individual Differences?

Although there are indeed many steps involved in validating an RL model of decision making, these are essential to our ultimate goals as psychopathology researchers. More specifically, we are often interested in understanding *individual differences* in decision processes, such as whether people with depression fail to learn from rewards but not from punishments. To answer such questions, we must be confident that the model is capable of solving the decision task of interest, that we can estimate parameters in human data, and that the preferred model describes decision making better than plausible alternatives.

After surmounting these hurdles, we can test individual difference hypotheses at three levels: (1) which model is most likely, (2) variation in model parameter estimates, and (3) prediction of trial-level variables by model-based decision signals. As noted above, a Bayesian model selection approach provides posterior probabilities for each person that quantify the relative evidence that one model yielded the observed choices compared to another.[13] Thus, at the group level, there may be a mixture of k models, reflecting that people adopt meaningfully different approaches to the decision-making task. Using variants of multinomial regression (see Vermunt, 2010 for further details), we can test whether model tendency is associated with categorical (e.g., diagnostic group) or continuous (e.g., dimension of pathology) predictors. For example, if we hypothesize that psychotic participants generalize the subjective values among a set of unrelated stimuli due to aberrant salience processing (i.e., seeing patterns where none exist in the contingency), whereas healthy controls do not, we could fit models with and without a generalization process to all participants (e.g., by yoking action values in the learning rule). We could then test whether psychotic participants are more likely to have implemented the generalization model compared to controls, for example as quantified by an odds ratio of implementing the generalization model compared to the alternative given membership in the psychotic group. Altogether, if we believe that the RL model that best describes choices on a task varies as a function of individual differences, analyses of posterior model probabilities can be informative.

On the other hand, a single RL model may describe a heterogeneous sample well, with variation in model parameter estimates mapping onto individual difference variables. As detailed above, model inversion yields parameter estimates for every participant in the sample, akin to factor scores in a latent variable analysis (see Chapter 7, by Wright). Thus, individual differences in RL model parameter estimates can be tested using conventional between-subjects analyses such as linear regression and ANOVA. For example, if we believe that people with poorer working memory will have higher temperatures in the softmax choice rule (which scales choice randomness) in the cup-bowl task, we could correlate temperature parameter estimates (one per subject) with self-reported impulsivity. Likewise, in a case-control design, if we believe that a learning rate for punishments will be elevated in depressed participants compared to controls, we could use a t-test to examine whether the group means differ significantly.

Such between-subjects analyses can reveal intriguing individual differences in RL model parameters, but we must be sensitive to the challenge of uncertainty in our RL parameter estimates. That is, the reliability of our RL parameter estimates may differ from one subject to the next; subjects with very noisy or model-inconsistent data may have extreme or unreliable parameter estimates. However, propagating such parameter uncertainty to group analyses is difficult: typical analyses incorporate only the point parameter estimate per subject, but not its corresponding uncertainty (e.g., standard error). One promising solution to this problem is to conduct RL modeling within a hierarchical

---

[13] Importantly, these probabilities do not include the possibility that an untested model could describe the behavior even better than any model considered by the researcher.

Bayesian estimation framework (Katahira, 2016) that simultaneously fits the RL model to group data while also estimating individual differences in RL parameters (i.e., estimating both within- and between-subjects effects).

By their nature, RL models seek to capture decision processes that unfold trial by trial – indeed, they provide predictions about every choice in an experiment. Thus, the final major approach for making inferences about individual differences is to examine how trial-varying signals from the model are associated with or predict other outcomes. For example, Li and colleagues (2011) found that the learned associability of a predictive cue derived from a Rescorla-Wagner-type model (see below) was correlated with the magnitude of skin conductance responses to the cue. We recently found that decision times were slower during trials in which the difference between the best and the worst available option value was small, particularly in those who had previously attempted suicide compared to suicide ideators and nondepressed controls (Dombrovski et al., 2019). This suggests that suicide attempters struggled to distinguish between options that were close in value.

To leverage signals from RL models that vary trial by trial, we encourage researchers to use a multilevel modeling (MLM) framework, where trials are nested within participants. At a minimum, we will wish to estimate a random intercept per subject to capture residual mean differences in the dependent variable.[14] Typically, trial-level signals (aka "level 1") are derived from the hidden states, such as subjective value estimates, which are updated sequentially after every outcome. Returning to individual differences, we can include between-subject (aka "level 2") variables and introduce these either as predictors of mean tendency of the dependent variable (e.g., decision times), or as cross-level moderators that interact with within-subject predictors.

The appropriate dependent variable for MLMs depends on the task and hypothesis, but we will typically wish to examine how trial-level signals from the model are related to outcome variables other than observed choices. That is, in model inversion, the RL model is typically fit to the observed decision sequence. Thus, there is a risk of circularity if our trial-level MLMs essentially recapitulate the RL model. On the other hand, examining RL model-based prediction of independent signals such as peripheral physiological measures or trial-varying brain activity can inform us about the model's potential to provide a formal bridge between biology and behavior. For example, using fMRI, we have previously observed that depression is associated with disrupted corticostriatothalamic encoding of positive prediction errors (Dombrovski et al., 2013).

Finally, we wish to note that MLMs can also be a useful tool for describing influences on decisions prior to building an RL model. For example, in a two-choice task, one could use binary logistic multilevel models to examine trial-level and between-subject predictors of choices. Such analyses can inform the development of specific RL models that formalize these more descriptive "model-free" analyses. In general, if we are interested in prospective prediction, as opposed to contemporaneous effects, we must be careful to preserve the temporal sequence of the experiment in trial-level MLM. Specifically, we must avoid giving the regression model access to information before the participant would have had it. Extending the idea of a Markov decision process above, predictors that are lagged by one trial are often appropriate because they can enforce the constraint that prediction of the current choice depends only on information available during the previous trial.

Does the Model Link with Neurobiology?

In clinical computational neuroscience, we seek to link decision signals from a candidate RL model with neural activity. This relates to the interface between the algorithmic and implementation levels in Marr's framework. If an RL model is potentially implemented in the brain, we would anticipate a correspondence between neural activity and model predictions. For example, are there regions in the brain whose activity scales with the magnitude of a reward prediction error? Such questions are closely linked with the field of model-based cognitive neuroscience (for a useful overview, see Forstmann & Wagenmakers, 2015) in which predictions from computational models are compared against measures of neural activity.

In the case of RL models, the conventional approach is to extract a trial-varying signal putatively involved in decision making – for example, prediction errors or expected values – and test for a correlation between the magnitude of the RL-based signal and the magnitude of brain activity in a given region. In the case of fMRI data, we would typically align the trial-varying decision signal with the time grid of the experiment so that there is a temporal correspondence between model signals and brain activity. For example, reward prediction errors are thought to occur after feedback is received for a choice on a given trial. Thus, we could align prediction error estimates for every trial with the corresponding times during the fMRI scan at which feedback was delivered. Then we would typically convolve the time-aligned prediction error estimates with a canonical hemodynamic response function (HRF) in order to obtain a model-based prediction of neural activity measured by blood-oxygen-level-dependent (BOLD) signal. Importantly, in analyses of brain activity, we should typically include regressors that capture the occurrence of an event (e.g., feedback) alongside regressors that capture dimensional variation in a decision signal (e.g., prediction errors). Furthermore, prior to convolving a decision signal with an HRF, we should typically mean center the trial-varying estimates. By including regressors that are indicators of an event and mean-centered regressors of model-based signals, we effectively decompose brain

---

[14] In experiments with multiple runs of a given task – either with the same contingency or different contingencies – it is often appropriate to treat trials as nested within runs, which results in a nested random effects structure.

activity due to occurrence of an event from the linear association between the model-based and brain signals (see Chapter 13, by Holmes and MacDonald, for details about fMRI modeling). This approach has yielded an extensive literature on neural correlates of PE and value signals, discussed in the final sections of this chapter.

Altogether, the major difference between traditional and model-based cognitive neuroscience lies in the hypotheses that are being tested. Traditional approaches often compare the effects of two or more experimental conditions on brain activity. For example, if angry faces elicit more ventral striatal activity than happy faces in participants with borderline personality disorder, we might infer that this region is implicated in disrupted emotional processing. This analysis, however, does not yield a process account: how exactly does the ventral striatum contribute to normal emotions and what goes wrong in borderline personality? By contrast, in a model-based approach, we are testing whether there is a linear relationship between a parametrically varying decision signal (e.g., prediction errors) and activity in a region. If there is, this supports the inference that the region is involved in the decision process, although only neural manipulation such as brain stimulation can provide evidence of its causal involvement in this process. For example, if we observed a greater coupling between ventral striatum activity and prediction errors on a lottery task in patients with bipolar disorder compared to controls, we could infer that the ventral striatum supports exaggerated appetitive learning in bipolar disorder.

There are deeper extensions of a model-based cognitive neuroscience approach that seek to bridge brain and behavior. For example, one could test RL model variants in which activity from a brain region or network is entered into the model as a predictor of decisions (for an excellent example, see Cavanagh et al., 2013). If the inclusion of brain activity provides a better fit to behavior than alternatives that lack this input, this would support the importance of the region in decision making. One can also use approaches such as dynamic causal modeling to examine how task-related connectivity among brain regions predicts key task-related behaviors such as choices (Rigoux & Daunizeau, 2015). By using signals derived from an RL model as inputs that drive task-related connectivity, one can examine the role of a computational model in a functional network. Finally, there are even more advanced approaches that seek to model brain and behavior simultaneously according to a single model (for an overview, see Turner et al., 2017).

EXAMPLE APPLICATION: HOW DO WE LEARN FROM EXPERIENCE IN SOCIAL EXCHANGES?

Now that we have introduced the general formalism of RL, let us examine how it can help us answer a psychological question when applied to behavioral and neural data. This example is adapted from our recent study of social decision making in the trust game, and the reader can find additional background and methods detail in the original paper (Vanyukov et al., 2019). Below, we introduce the psychological question and review the steps we took to answer it: we converted alternative hypotheses to RL models, ensured that these models and their parameters were uniquely identified, and tested models against behavior and functional neuroimaging data.

Psychological Question

Imagine that you are in the market for a gently used Honda Accord and you come across an attractively priced top-trim car. The seller pressures you to buy, but your intuition tells you something is amiss and you decide to walk away. Later, you nevertheless decide to check the vehicle identification number and learn that the car had been flooded. This is, strictly speaking, bad news and you weren't going to buy the car anyway, but somehow you feel good about your discovery. This thought experiment informs the hypothesis that we can be gratified by negative information about others, as long as it reinforces our previous stance toward them. As articulated above, RL terms a choice strategy a *policy*, and we will thus refer to this account as the "policy hypothesis." Previous studies of human behavior and brain signals in economic exchanges, however, suggest that during our interactions with others we update their "social value," approximately corresponding to the reward rate that we expect from interacting with the other person. Under the social value hypothesis, we should encode all news of others' defection (e.g., failure to cooperate) as negative updates to social value.

These two hypotheses yield diverging predictions. Specifically, the policy hypothesis suggests that predicted defections are experienced as rewarding and would trigger positive (better than expected) neural prediction error signals, whereas the social value hypothesis suggests the opposite. More generally, the policy hypothesis predicts that the direction of prediction errors will depend on the individual's choice, whereas under the social value hypothesis, prediction errors would be insensitive to one's choices. In other words, the learning process is instrumental (action-outcome) under the policy hypothesis and Pavlovian (stimulus [identity of the counterpart]-outcome) under the social value hypothesis. We examined these hypotheses using functional neuroimaging during a modified trust game. More specifically, in three separate samples, we tested RL models that instantiated the policy and social value hypotheses.

Trust Game Experiment (Figure 14.4)

On each round, the first player, *the investor*, can entrust an investment (we'll call it "sharing" to be brief) to a second player, *the trustee*, or keep their money. If the money is invested, it grows threefold in the custody of the trustee, who can either cooperate and return half to the investor ("share") or defect and keep it all. Participants always played the role of the investor, whereas the trustee behavior was controlled by an algorithm that provided probabilistic

A.

Trustee Type: Presentation blocked and order counterbalanced across participants	Reinforcement Schedule: Order constant across participants			
	Block 1 (16 trials)	Block 2 (16 trials)	Block 3 (16 trials)	192 total trials
good	50%	88% (rich)	25% (poor)	48 trials
bad	50%	25% (poor)	88% (rich)	48 trials
neutral	50%	25% (poor)	88% (rich)	48 trials
computer	50%	88% (rich)	25% (poor)	48 trials

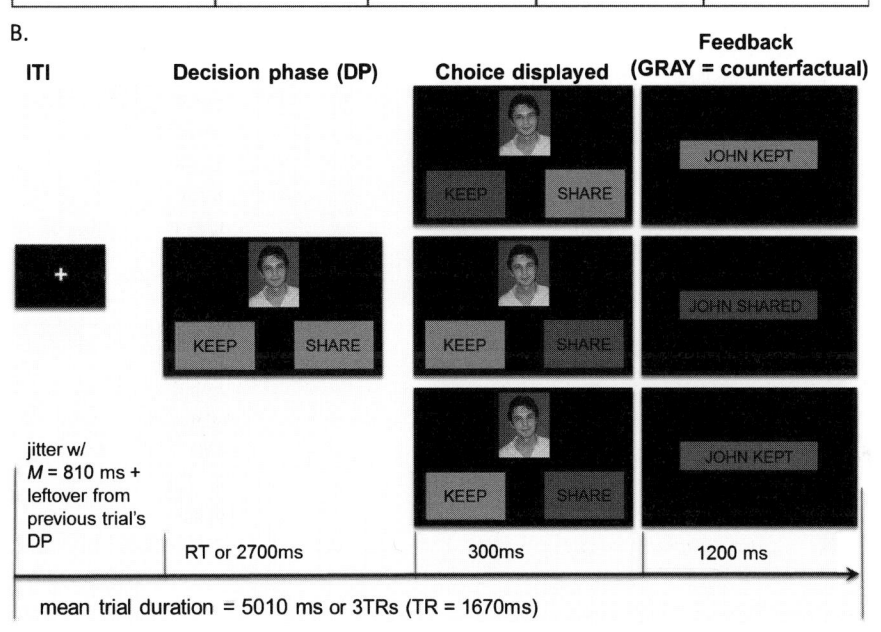

B.

Figure 14.4 Trust game experiment. A. Participants interacted with four trustees whose reputation (good, bad, neutral, computer) was manipulated orthogonally to trustee's rate of cooperation (50 percent, 88 percent, 25 percent) in a Latin square design. B. Trial design. Participants received counterfactual feedback after "KEEP" decisions (top panel).

feedback. Following earlier studies, we manipulated the reputation of the trustee (with good, bad, and neutral human trustees and a control computer trustee condition; Figure 14.4A) to ascertain whether the investor was sensitive to it (Delgado, Frank, & Phelps, 2005).[15] Critically, even when the participants decided to keep their money, we provided them with feedback about the trustee's would-be or *counterfactual* actions (Figure 14.4B). As with the flooded Accord, this counterfactual feedback allowed people to experience predicted defections that reinforced their decision to withhold the investment. Finally, our trustees first shared randomly (on 50 percent of trials) and then became first very trustworthy (88 percent) and then very untrustworthy (25 percent) or vice versa. In other words, we manipulated the trustees' reward rates to test whether they were reflected in investors' behavior and striatal learning signals.

RL Models Embodying Alternative Hypotheses

All models shared a standard delta rule for updating the expected value of sharing, Q(share), based on the last outcome:

---

[15] This was necessary to demonstrate that reputation effects were Pavlovian in nature and were additive to subsequent experiential instrumental learning from social exchanges. Replicating this effect also helped validate our experiment and maintain continuity with previous studies. The reader is referred to the original paper for details on reputation effects because they are peripheral to the RL model comparison.

Figure 14.5

	A. Payoff matrices		B. keep-keep vs. keep-return	C. Bayesian model comparison, exceedance probabilities			D. Striatal prediction error signals: $p_{voxelwise} < 0.001$, cluster size = 100, $p_{corr} < 0.05$
				Study # 1	2	3	$y = 11$
1. Actual reward	Trustee returns	Trustee keeps	0	0.0142	0.002	0	
Subject invests	1.5	0					
Subject keeps	1	1					
2. Regret	Trustee returns	Trustee keeps	0	0	0	0.0001	
Subject invests	1.5	-1.5					
Subject keeps	-0.5	-0.5					
3. Trustee-count.	Trustee returns	Trustee keeps	0	0	0	0	
Subject invests	1.5	-1.5					
Subject keeps	0	0					
4. Policy	Trustee returns	Trustee keeps	+1.5	0.9858	0.9998*	0.999*	
Subject invests	1.5	0					
Subject keeps	-0.5	1					

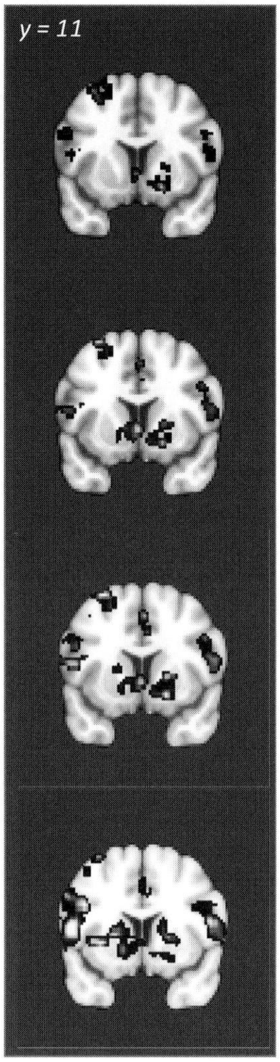

Figure 14.5 Comparison of RL models. A. Payoff matrices for each model. B. Reward difference between correctly anticipated trustee defection ("KEEP"/"KEEP") and a mispredicted defection ("KEEP"/"RETURN"). C. Bayesian model comparison, exceedance probabilities. D. Striatal responses to model-derived prediction errors ($p_{voxelwise} < 0.001$, cluster size = 100 voxels, $p_{corr} < 0.05$).

$$Q(\text{share})_t = Q(\text{share})_{t-1} + \theta * [\text{reward}_t - Q(\text{share})_t]$$

where θ is the learning rate. The choice policy was defined in terms of sharing versus keeping:

$$p(\text{share}) = \frac{1}{1 + e^{\kappa + \beta * Q(\text{share})}}$$

where β is the inverse temperature (value sensitivity parameter) and κ is the bias parameter, which is a sum of $\kappa_{subject}$, reflecting one's general tendency to cooperate and $\kappa_{trustee}$, reflecting one's tendency to cooperate with a particular (e.g., "good") trustee.

Models differed in how they represented outcomes (Figure 14.5A), referred to as *the payoff matrix* in game theory. The simplest social value model (#1) learns only from actual rewards. The policy model (#4) receives actual rewards on trials when the investor shared. On investor-keep trials the actual reward is always 0, but the policy model benefits from counterfactual feedback. Thus, its reward consists of the difference between the actual and the would-be reward associated with its own alternative action (sharing). This reward is positive for correctly predicted defections (i.e., when investor keeps and trustee keeps, then $r = 1$) and negative for unanticipated cooperation (i.e., when investor keeps and trustee shares, then $r = -0.5$). To rule out the possibility that the policy model was superior to the actual rewards model only by virtue of benefiting from counterfactual feedback, we also included a trustee-counterfactual model (#3), which learned from the difference between the outcomes associated with the trustee's actual and counterfactual actions. In other words, it learned with reference to the trustee's policy as opposed to its own. Finally, since human choices are often influenced by regret, we included a regret model (#2), which learned from the difference between the actual outcome and the best possible outcome. In summary, we wanted to rule out multiple alternative accounts embodied in models #1–3.

Model Comparison Using Behavioral Data

After reviewing individual learning curves and manipulation checks, we performed "model-free" analyses with multilevel logistic regression models predicting investor's choice as a function of trustee reputation and reinforcement. Briefly, investors were sensitive to trustee reputation and their cooperation rate. The effect of trustee cooperation increased over time whereas that of reputation remained unchanged, a fixed bias. This suggested that the two reflect separate processes as opposed to updates to a unitary social value estimate. Furthermore, investors learned from counterfactual feedback, sharing more after investor-keep/trustee-share trials than after investor-keep/trustee-keep trials. These results were consistent with the predictions of the policy model but did not provide unequivocal evidence of its superiority.

Next, we fitted the parameters of each of the four models to investors' choices and entered the model likelihoods into a Bayesian model comparison (see "Does the Proposed Model Better Explain Behavior than Plausible Alternatives?" above). The policy model dominated in each of the three samples (Figure 14.5C), as shown by an exceedance probability of >0.98. In two of three samples this finding withstood adjustment for the Bayesian omnibus error rate (BOR; Rigoux et al., 2014). These behavioral findings supported the policy model, but they did not necessarily demonstrate that predicted defections (investor-keep/trustee-keep) were encoded as rewarding. The functional imaging data can help to determine the direction of prediction error updates from predicted defections.

Simulations: Model and Parameter Identifiability

We generated the *model confusion* matrix as described above (*Does the model yield distinct predictions and capture distinct phenomena relative to other models?*). To ensure that the policy model yielded unique predictions about behavior, we verified that its actions could not be fitted by other models. Next, to ensure that the policy model did not benefit from greater flexibility than the comparators, we ascertained that it did not fit other models' behavior better than themselves. Finally, we verified that the policy model could uniquely identify its own parameters.

Model Comparison Using Imaging Data

Functional imaging is often criticized for its inability to resolve psychological questions (Uttal, 2001). Yet here the diverging predictions of alternative models about neural responses can be readily tested against fMRI data. The policy model predicts that when a trustee defection is correctly predicted (investor-keep/trustee-keep trials), people will encode a positive prediction error, reflecting an increase in the value of one's policy toward the trustee. All alternative models predict the opposite, reflecting a decrease in the estimated rate of reciprocation by the trustee (Figure 14.5B). As with behavioral data, it is best to conduct both model-based and model-free tests of this hypothesis. The analyses described below are informed by the larger literature on neural correlates of prediction errors discussed in a later section.

First, in a *model-based analysis* using whole-brain data, we regressed prediction error signals postulated by each model (along with covariates of no interest) against BOLD signal and found that the policy model predicts learning signals in the largest set of regions (Figure 14.5D). But perhaps some of the voxels indexed by the policy model represented a variable different from PE. To rule this out, we applied an independent mask from a meta-analysis of PE responses and tested whether activity within the regions that had responded to PEs in multiple previous studies (striatum, midbrain, anterior insula, and visual cortex; please see the section on neural signatures of prediction errors below) was best described by the policy model. Policy PEs explained activity in these regions better than the actual rewards model (t [312] = −3.80, $p = 0.001$); the regret model, (t [312] = 4.58, $p < 0.001$) and the trustee-counterfactual model (t [312] = 4.01, $p < 0.001$). For a *model-free* corroboration of these results, we simply contrasted responses to outcome across the four possible trial types in the striatum, the region most strongly implicated in PE signaling. Again, responses to predicted trustee defections were more positive than those to unpredicted defections (less cooperative than expected) and even unpredicted reciprocations (more cooperative than expected; data not shown).

SOME HISTORICALLY IMPORTANT MODELS OF ANIMAL LEARNING

Now that we have discussed the general formalism of RL, let us examine how models of animal learning emerged and evolved, shaped by empirical challenges. Some of the material in this and the following sections has been adapted from our meta-analysis of functional imaging studies employing RL (Chase et al., 2015).

Mathematical models of learning can be traced to the 1950s, when emerging experimental techniques for the study of conditioning demanded a new theoretical framework for understanding learning (Estes, 1950). Rejecting earlier attempts to find the ideal empirical learning curve, Skinner's student Estes pioneered a statistical approach to learning, *the sampling theory*, in which the dependent variables are classes of behavioral responses (R) with common quantitative properties, while the independent variables are statistical distributions of environmental stimuli (S). Estes proposed that all fundamental behavioral laws would be of the form $R = f(S)$, describing a probabilistic relationship between momentary changes of the stimuli and behavior (for a detailed review of Estes' work, see Bower, 1994). Yet by far the most influential of these early models was the analysis of instrumental learning experiments by Bush and Mosteller (1951). Analyzing the original Skinner box rat experiments, avoidance learning in dogs, and probabilistic reward learning in paradise fish, Bush and Mosteller came to two key conclusions. First, in agreement with Estes' sampling theory, the relationship between reinforcement and behavior had to be probabilistic or stochastic rather than deterministic. Second, the probability of response was updated on each trial following a learning rule (in the authors' original notation):

$$Qp = p + a(1-p) - bp$$

where Qp[16] is the response probability after reinforcement, p is the pre-reinforcement response probability, a is the parameter corresponding to reward, and b is the parameter corresponding to "punishment and the work required in making the response." One can see that the appetitive term $a(1-p)$ is larger early in learning when the reward is surprising to the animal and the response probability (p) is low, and the appetitive term is small when the reward is expected and the p is high. In other words, unpredicted rewards generate the learning signal. Thus, Bush and Mosteller introduced the notion of *prediction error* (without coining the term), intuiting that learning is driven by events which differ from expectations.

Unrealistically, the Bush–Mosteller model attempts to predict response probabilities directly, which reflect "performance" variables and not merely learning. This model of instrumental learning also cannot explain interactions among multiple Pavlovian stimuli presented at once, including the Kamin blocking effect, or diminished conditioned responding to stimulus X following AX → US pairing preceded by A → US (Kamin, 1968). It also fails to explain overshadowing, or diminished conditioned responding to stimulus X following AX → US pairing when both A and X were previously untrained. Both problems were solved in the influential model of Rescorla and Wagner (RW) (1972). Instead of response probabilities, the RW model updates the unobserved, but theoretically important, *associative strength* (V) corresponding to the expected *reward/punishment value* of a given stimulus. RW offered an elegant explanation of the Kamin blocking effect by combining the values of all stimuli present on a given trial in order to generate a prediction error. This means that an outcome is surprising only insofar as it is unpredicted by any of the stimuli. Here is how RW updates the values of the Pavlovian compound AX when it is followed by the US:

$$\Delta V_A = \alpha_A \beta_{US}(\lambda_{US} - V_{AX})$$

$$\Delta V_X = \alpha_X \beta_{US}(\lambda_{US} - V_{AX})$$

where α learning rate for each stimulus, β is the learning rate for the US, λ_{US} is the maximum value which this US will support, and $V_{AX} = V_A + V_X$ is the total predicted value. Here, after pre-training stimulus A to the asymptote, subsequent training with the AX compound yields no prediction error for X. Besides overshadowing and blocking, the RW model captures many, but not, all Pavlovian and instrumental phenomena (Miller, Barnet, & Grahame, 1995).

Temporal difference (TD) models learn from prediction errors like RW (Sutton & Barto, 2018), but do so in real time that is not artificially divided into experimental trials. TD predicts future rewards over the horizon, k, discounted over time:

$$R_t = r_{t+1} + \gamma r_{t+2} + \gamma^2 r_{t+3} + \cdots + \gamma^k r_{t+k+1}$$

where r is future expected reward and γ is the temporal discount factor, reflecting a preference for immediate over delayed rewards (since $0 < \gamma < 1$, raising it to successively higher powers discounts later rewards). While it cannot know the values of all future rewards in advance, TD estimates them online by iterating the following algorithm at each time step:

$$V_t \leftarrow V_t + \alpha(r_{t+1} + \gamma V_{t+1} - V_t)$$

where $\alpha(r_{t+1} + \gamma V_{t+1} - V_t)$ is the *temporal difference error*, and γV_{t+1} takes the place of the remaining terms $\gamma r_{t+2} + \gamma^2 r_{t+3} + \cdots + \gamma^k r_{t+k+1}$. To determine how much credit for a reward or punishment should be assigned to a preceding cue or action, TD introduces *eligibility traces*. The closer a cue or action is to the outcome, the more *eligible* is its value for the prediction error update. In summary, TD describes continuous evolution of value and temporal difference errors in real time, features that proved useful for electrophysiological studies discussed below.

NEURAL SIGNATURES OF PREDICTION ERRORS

Late-twentieth-century theories of midbrain dopaminergic function emphasized motor, attentional, and motivational processes. RL models, however, suggested a new hypothesis that phasic dopamine release in the mesostriatal pathway encoded prediction errors (Montague, Dayan, & Sejnowski, 1996). This hypothesis was soon corroborated in a monkey Pavlovian study employing a TD model (Schultz, Dayan, & Montague, 1997) and in subsequent blocking and conditioned inhibition experiments (Tobler, Dickinson, & Schultz, 2003; Waelti, Dickinson, & Schultz, 2001). Despite major empirical challenges (Fiorillo, Newsome, & Schultz, 2008) and refinements (Mohebi et al., 2018; Parker et al., 2016), the PE account of phasic dopamine signals remains influential.

---

[16] Bush and Mosteller introduce the response probability change operator Q. Qp thus reads: "Q operating on p."

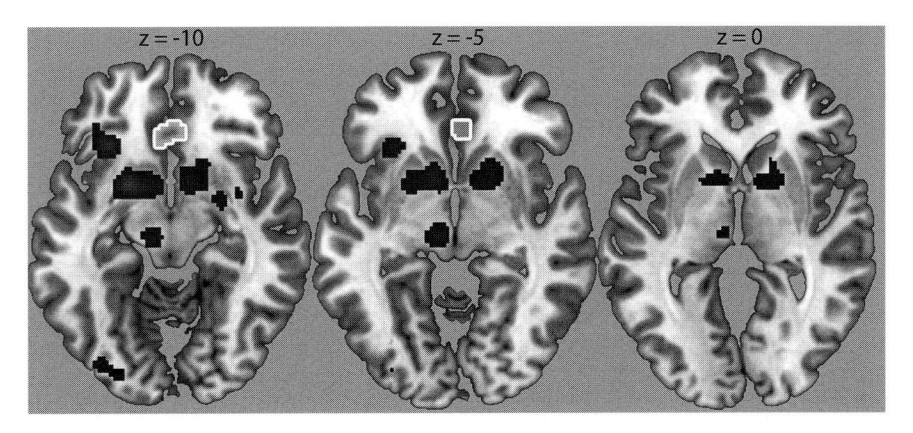

Figure 14.6 Meta-analytic maps of prediction error (dark) and learned value (light) signals from human fMRI studies (Chase et al., 2015). Prediction error signals are found in the striatum, thalamus/midbrain, and the frontal operculum/anterior insula as well as the visual cortex. Value signals are found in the ventromedial prefrontal cortex. Authors' original figure, previously unpublished, used here with the co-authors' permission.

Human electro- and magneto-encephalographic studies have detected prediction error signals (Cavanagh, 2015; Holroyd & Krigolson, 2007), but could not unequivocally localize responses to the midbrain or striatum. This task fell to fMRI experiments. Beginning with seminal studies (O'Doherty et al., 2003), researchers focused not on the midbrain itself, but on its monosynaptic target – the ventral striatum (VS) – because BOLD response reflects phasic inputs to a region (Logothetis & Pfeuffer, 2004) and also because the ventral tegmental area (VTA) containing reward-responsive midbrain dopaminergic neurons is small and obscured by arteries. Thus, VS BOLD responses likely reflect VTA firing. Subsequent fMRI studies, however, also detected prediction error-related activation in the VTA itself (D'Ardenne et al., 2008). A meta-analysis has found that fMRI studies consistently detect prediction errors in the VS and midbrain/thalamus (Chase et al., 2015). Whole-brain fMRI meta-analysis has also detected prediction error-like signals in the frontal operculum/anterior insula, particularly in studies of social rewards. In Pavlovian experiments, striatal prediction error signals extended into the amygdala, whereas instrumental experiments yielded caudate activations [(Chase et al., 2015; Figure 14.6). Do these widespread activations reflect signed (better or worse than expected) or unsigned PEs (any surprising outcome)? A more recent meta-analysis of fMRI studies found V-shaped responses, consistent with unsigned PE (high for large positive and negative PEs, low for no PE) in regions such as the anterior insular and mid-anterior cingulate cortices, as well as the dorsal striatum, although signed and unsigned PE responses overlapped in small areas of the same regions (Fouragnan, Retzler, & Philiastides, 2018).

LEARNED VALUE, ECONOMIC SUBJECTIVE VALUE, AND THEIR NEURAL CORRELATES

Soon after Estes (1950) and Bush and Mosteller (1951) developed statistical models of learning, Simon pointed out their implications for economic decision making (Simon, 1956). He highlighted these models' *subjective rationality*; that is, their behavior is optimal given limited knowledge of the environment and a simple representational structure. Simon thus emphasized the convergence between learned value and economic *expected utility* – the theoretical common currency used to compare disparate goods. However, whereas economic research is concerned with the way value-based choices integrate various attributes (e.g., a car's design, fuel consumption, dynamics, and safety), learned value generally integrates the history of rewards of a single type. On the other hand, in experimental and behavioral economics research, choices are typically description-based and hypothetical (e.g. what you will get, at what cost, with what probability and when), whereas animal learning is experience-based, producing the so-called description-experience gap in choice behavior (Hertwig & Erev, 2009). One empirical question, thus, is to what extent the neural substrates of economic and learned value overlap.

The evidence for neural value signals is considerably more complex than that of meso-striatal PE signals. Early lesion studies led the researchers to focus on the ventral prefrontal cortex (vPFC). The human vPFC encompasses the orbitofrontal cortex (OFC), the ventromedial prefrontal cortex (vmPFC), and ventrolateral prefrontal cortex (vlPFC). The vmPFC is an evolutionarily conserved part of human prefrontal cortex including agranular areas 25/24, 14c and dysgranular areas 32 and 14r; the phylogenetically newer granular areas 14m and 11m are also sometimes included. Orbital areas 14c, 14r, and 11m form the medial orbitofrontal cortex (mOFC) and areas 11l, 13, and 12, the lateral OFC (lOFC). Signals found in the mammalian central/lateral OFC possess various properties of economic value: sensitivity to delays and probability of reward, as well as to the presence of alternatives (Kennerley & Wallis, 2009; Kennerley et al., 2009; Kobayashi, Pinto de Carvalho, & Schulz, 2010; Padoa-Schioppa & Assad, 2006; Roesch & Olson, 2005; Tremblay & Schultz, 1999). Signals in the mOFC integrate internal states such as hunger (Bouret & Richmond, 2010;

Critchley & Rolls, 1996). Evolving value signals during learning have also been found in the OFC (Stalnaker, Cooch, & Schoenbaum, 2015). Human imaging studies, on the other hand, map both economic value and learned value to the more dorsal and posterior vmPFC (Bartra, McGuire, & Kable, 2013; Chase et al., 2015; Figure 14.6).

Animal electrophysiological studies mostly point to the lOFC, whereas human imaging studies consistently implicate the vmPFC. Nevertheless, the representations of economic and learned value spatially overlap in both cases, suggesting that Simon's parallel was correct. At the same time, value-like signals are found elsewhere in the brain, including the amygdala (Paton et al., 2006; Schoenbaum, Chiba, & Gallagher, 1999), striatum (Samejima et al., 2005), dopaminergic midbrain (Mohebi et al., 2018), anterior cingulate (Cai & Padoa-Schioppa, 2012; Jocham et al., 2009), and other prefrontal areas (reviewed in Wallis & Kennerley, 2010). Another complication is that the OFC is not required for many types of value-based decisions or for most instances of instrumental and Pavlovian learning (see Stalnaker et al., 2015, for a more nuanced discussion). Thus, in many cases value signals are likely broadcast to the OFC and vmPFC from the basal ganglia and amygdala rather than computed in situ. These empirical challenges have led to new hypotheses about OFC function, such as encoding cognitive maps of the state-space (Schuck et al., 2016) and credit assignment (Noonan et al., 2017; Walton et al., 2010). In summary, current RL models do not fully explain the nature of ventral prefrontal computations during learning and decision making, but they continue advancing the field by enabling researchers to generate and test falsifiable hypotheses.

CONCLUSIONS AND CAVEATS

Computational reinforcement learning models are invaluable tools to formalize falsifiable hypotheses about human behavior and its neural underpinnings. Relative to descriptive approaches that characterize clinical populations in terms of manifest behaviors such as reaction times or proportion correct, RL models specify the algorithmic processes by which individuals may solve a particular decision problem. Moreover, by formalizing our hypotheses about decision processes, we can further examine how these processes are disrupted or different in psychopathology. Supported by this computational machinery, we can also increase the precision of our inquiries into the neurobiological aspects of mental illness. More specifically, a model-based cognitive neuroscience approach to psychopathology allows us to test whether decision signals (e.g., prediction errors) correlate with neural activity, which would support the possibility that a given RL model is implemented by the brain. Furthermore, we can potentially develop better models of brain and behavior by integrating these data streams (Turner et al., 2017), as in the case of brain signals being included in RL model equations to enhance predictions of behavior (Cavanagh et al., 2013).

There are a few important caveats to this approach. First, model inversion and model comparison inform us about the *relative* fit of a proposed model compared to a null or alternative model. A superior fit does not guarantee that a model fits *well*, although posterior checks of model-predicted vs. actual responses in individual subjects can be very informative in this respect. It is harder to be convinced that the model captures the process by which our brain performs the task (Uttal, 2014), because different models can behave similarly, a phenomenon aptly termed "model mimicry" (Wagenmakers et al., 2004). Tests of model identifiability and analyses relating alternative models' internal variables to neural signals can reassure us about the specificity of the model's prediction, but the ultimate test remains whether the unique dynamics of a cognitive model can be reduced to lower levels of neural mechanisms. Second, any conclusions drawn from model-based studies are only as good as their experimental design. In our social trust example, it would be impossible to test alternative RL models of the trust game without showing participants the trustee decisions on trials where they did not invest (counterfactual feedback). Third, the complexity of cognitive modeling makes it hard to see what is under the hood, especially when model fits are less than perfect. A careful review of raw data and a model-free corroboration always help. These caveats notwithstanding, we view reinforcement learning approaches to clinical computational neuroscience as a fruitful new avenue for testing questions about decision making and its neurobiology in psychopathology.

FURTHER READING

For an in-depth treatment of reinforcement learning, we recommend Sutton and Barto's recently updated classic book, *Reinforcement Learning: An Introduction* (2018). The *Oxford Handbook of Computational and Mathematical Psychology* introduces the reader to cognitive modeling and contains Gureckis and Love's superb chapter on reinforcement learning (2015). Excellent computational neuroscience texts include Miller's *Introductory Course in Computational Neuroscience* (2018) and Dayan and Abbott's *Theoretical Neuroscience* (2005). Miller covers useful preliminary material, including mathematics, circuit physics and even computing and MATLAB (much of existing code for reinforcement learning modeling is written in MATLAB, but R and Python are becoming increasingly popular). Dayan and Abbot treat conditioning and reinforcement learning in greater detail. A more detailed treatment of model-based cognitive neuroscience can be found in *An Introduction to Model-Based Cognitive Neuroscience* (Forstmann & Wagenmakers, 2015).

REFERENCES

Bartra, O., McGuire, J. T., & Kable, J. W. (2013). The Valuation System: A Coordinate-Based Meta-Analysis of BOLD fMRI

Experiments Examining Neural Correlates of Subjective Value. *Neuroimage, 76*, 412–427.

Bouret, S., & Richmond, B. J. (2010). Ventromedial and Orbital Prefrontal Neurons Differentially Encode Internally and Externally Driven Motivational Values in Monkeys. *Journal of Neuroscience, 30*(25), 8591–8601.

Bower, G. H. (1994). A Turning Point in Mathematical Learning Theory. *Psychological Review, 101*, 290–300.

Box, G. E. P. (1979). Robustness in the Strategy of Scientific Model Building. In R. L. Launer & G. N. Wilkinson (Eds.), *Robustness in Statistics* (pp. 201–236). New York: Academic Press.

Burnham, K. P., & Anderson, D. R. (2002). *Model Selection and Multi-Model Inference: A Practical Information-Theoretic Approach* (2nd edn.). New York: Springer.

Bush, R. R., & Mosteller, F. (1951). A Mathematical Model for Simple Learning. *Psychological Review., 58*, 313–323.

Cai, X., & Padoa-Schioppa, C. (2012). Neuronal Encoding of Subjective Value in Dorsal and Ventral Anterior Cingulate Cortex. *Journal of Neuroscience, 32*, 3791–3808.

Cavanagh, J. F. (2015). Cortical Delta Activity Reflects Reward Prediction Error and Related Behavioral Adjustments, but at Different Times. *NeuroImage, 110*, 205–216.

Cavanagh, J. F., Eisenberg, I., Guitart-Masip, M., Huys, Q., & Frank, M. J. (2013). Frontal Theta Overrides Pavlovian Learning Biases. *Journal of Neuroscience, 33*, 8541–8548.

Chase, H. W., Kumar, P., Eickhoff, S. B., & Dombrovski, A. Y. (2015). Reinforcement Learning Models and Their Neural Correlates: An Activation Likelihood Estimation Meta-Analysis. *Cognitive, Affective, & Behavioral Neuroscience, 15*(2), 435–459.

Craver, C. F. (2001). Role Functions, Mechanisms, and Hierarchy. *Philosophy of Science, 68*, 53–74.

Critchley, H. D., & Rolls, E. T. (1996). Hunger and Satiety Modify the Responses of Olfactory and Visual Neurons in the Primate Orbitofrontal Cortex. *Journal of Neurophysiology, 75*(4), 1673–1686.

D'Ardenne, K., McClure, S. M., Nystrom, L. E., & Cohen, J. D. (2008). BOLD Responses Reflecting Dopaminergic Signals in the human Ventral Tegmental Area. *Science, 319*, 1264–1267.

Dayan, P., & Abbott, L. F. (2005). *Theoretical Neuroscience: Computational and Mathematical Modeling of Neural Systems* (Revised edn.). Cambridge, MA: MIT Press.

Delgado, M. R., Frank, R. H., & Phelps, E. A. (2005). Perceptions of Moral Character Modulate the Neural Systems of Reward during the Trust Game. *Nature Neuroscience, 8*, 1611–1618.

Dombrovski, A. Y., Szanto, K., Clark, L., Reynolds, C. F., & Siegle, G. J. (2013). Reward Signals, Attempted Suicide, and Impulsivity in Late-Life Depression. *JAMA Psychiatry, 70*, 1020–1030.

Dombrovski, A. Y., Hallquist, M. N., Brown, V. M., Wilson, J., & Szanto, K. (2019). Value-Based Choice, Contingency Learning, and Suicidal Behavior in Mid- and Late-Life Depression. *Biological Psychiatry, 85*(6), 506–516.

Dreher, J.-C., & Tremblay, L. (Eds.) (2016). *Decision Neuroscience: An Integrative Perspective* (1st edn.). Amsterdam: Academic Press.

Elston, G. N. (2003). Cortex, Cognition and the Cell: New Insights into the Pyramidal Neuron and Prefrontal Function. *Cerebral Cortex, 13*(11), 1124–1138.

Estes, W.K., 1950. Toward a Statistical Theory of Learning. *Psychological Review, 57*, 94–107.

Fiorillo, C. D., Newsome, W. T., & Schultz, W. (2008). The Temporal Precision of Reward Prediction in Dopamine Neurons. *Nature Neuroscience, 11*, 966–973.

Forstmann, B. U., & Wagenmakers, E.-J. (Eds.) (2015). *An Introduction to Model-Based Cognitive Neuroscience.* New York: Springer.

Fouragnan, E., Retzler, C., & Philiastides, M. G., 2018. Separate Neural Representations of Prediction Error Valence and Surprise: Evidence from an fMRI Meta-Analysis. *Human Brain Mapping, 39*, 2887–2906.

Gershman, S. J. (2016). Empirical Priors for Reinforcement Learning Models. *Journal of Mathematical Psychology, 71*, 1–6.

Glimcher, P. W. (2011). Understanding Dopamine and Reinforcement Learning: The Dopamine Reward Prediction Error Hypothesis. *Proceedings of the National Academy of Sciences of the United States of America, 108*(Suppl 3), 15647–15654.

Gureckis, T., & Love, B. (2015). Computational Reinforcement Learning. In *The Oxford Handbook of Computational and Mathematical Psychology* (pp. 99–117). Oxford: Oxford University Press

Hallquist, M. N., & Dombrovski, A. Y., 2019. Selective Maintenance of Value Information Helps Resolve the Exploration/Exploitation Dilemma. *Cognition, 183*, 226–243.

Hertwig, R., & Erev, I., 2009. The Description-Experience Gap in Risky Choice. *Trends in Cognitive Science, 13*, 517–523.

Holroyd, C. B., & Krigolson, O. E., 2007. Reward Prediction Error Signals Associated with a Modified Time Estimation Task. *Psychophysiology, 44*, 913–917.

Hursh, S. R., & Silberberg, A. (2008). Economic Demand and Essential Value. *Psychological Review, 115*, 186–198.

Jocham, G., Neumann, J., Klein, T. A., Danielmeier, C., & Ullsperger, M. (2009). Adaptive Coding of Action Values in the Human Rostral Cingulate Zone. *Journal of Neuroscience, 29*, 7489–7496.

Kaelbling, L. P., Littman, M. L., & Moore, A. W., 1996. Reinforcement Learning: A Survey. *Journal of Artificial Intelligence Research, 4*, 237–285.

Kamin, L. J. (1968). Predictability, Surprise, Attention, and Conditioning. In B. A. Campbell & R. M. Church (Eds.), *Punishment and Aversive Behavior* (pp. 279–296). New York: Appleton-Century-Crofts.

Katahira, K. (2016). How Hierarchical Models Improve Point Estimates of Model Parameters at the Individual Level. *Journal of Mathematical Psychology, 73*, 37–58.

Kennerley, S. W., & Wallis, J. D. (2009). Evaluating Choices by Single Neurons in the Frontal Lobe: Outcome Value Encoded across Multiple Decision Variables. *European Journal of Neuroscience, 29*, 2061–2073.

Kennerley, S. W., Dahmubed, A. F., Lara, A. H., & Wallis, J. D. (2009). Neurons in the Frontal Lobe Encode the Value of Multiple Decision Variables. *Journal of Cognitive Neuroscience, 21*, 1162–1178.

Kobayashi, S., Pinto de Carvalho, O., & Schultz, W. (2010). Adaptation of Reward Sensitivity in Orbitofrontal Neurons. *Journal of Neuroscience, 30*, 534–544.

Kriegeskorte, N., & Douglas, P. K. (2018). Cognitive Computational Neuroscience. *Nature Neuroscience, 21*, 1148–1160.

Kruschke, J. (2014). *Doing Bayesian Data Analysis: A Tutorial with R, JAGS, and Stan.* Cambridge, MA: Academic Press.

Laird, A. R., Fox, P. M., Eickhoff, S. B., Turner, J. A., Ray, K. L., McKay, D. R., ... Fox, P. T. (2011). Behavioral Interpretations of Intrinsic Connectivity Networks. *Journal of Cognitive Neuroscience, 23*, 4022–4037.

Li, J., Schiller, D., Schoenbaum, G., Phelps, E. A., & Daw, N. D. (2011). Differential Roles of Human Striatum and Amygdala in Associative Learning. *Nature Neuroscience, 14*, 1250–1252.

Logothetis, N.K., Pfeuffer, J., 2004. On the Nature of the BOLD fMRI Contrast Mechanism. *Magnetic Resonance Imaging, 22*, 1517–1531.

Love, B. C. (2015). The Algorithmic Level Is the Bridge between Computation and Brain. *Topics in Cognitive Science, 7*, 230–242.

Marr, D. (1982). Chapter 1: The Philosophy and the Approach. In *Vision: A Computational Investigation into the Human Representation and Processing Visual Information* (pp. 8–38). Cambridge, MA: MIT Press.

McClelland, J. L., & Rumelhart, D. E. (1986). *Parallel Distributed Processing, Explorations in the Microstructure of Cognition: Foundations.* Cambridge, MA: MIT Press.

McCulloch, W. S., & Pitts, W. (1943). A Logical Calculus of the Ideas Immanent in Nervous Activity. *Bulletin of Mathematical Biophysics, 5*, 115–133.

Miller, P. (2018). *An Introductory Course in Computational Neuroscience* (1st edn). Cambridge, MA: MIT Press.

Miller, R. R., Barnet, R. C., & Grahame, N. J. (1995). Assessment of the Rescorla-Wagner Model. *Psychological Bulletin, 117*, 363–386.

Minsky, M. L. (1967). *Computation: Finite and Infinite Machines*. Upper Saddle River, NJ: Prentice-Hall.

Minsky, M. L., & Papert, S. A. (1969). *Perceptrons: An Introduction to Computational Geometry*. Cambridge, MA: MIT Press.

Mohebi, A., Pettibone, J., Hamid, A., Wong, J.-M., Kennedy, R., & Berke, J. (2018). Forebrain Dopamine Value Signals Arise Independently from Midbrain Dopamine Cell Firing. *bioRxiv*, 334060.

Montague, P. R., Dayan, P., & Sejnowski, T. J. (1996). A Framework for Mesencephalic Dopamine Systems Based on Predictive Hebbian Learning. *Journal of Neuroscience, 16*, 1936–1947.

Montague, P. R., Dolan, R. J., Friston, K. J., & Dayan, P. (2012). Computational Psychiatry. *Trends in Cognitive Science, 16*, 72–80.

Moran, R. J., Symmonds, M., Stephan, K. E., Friston, K. J., & Dolan, R. J. (2011). An In Vivo Assay of Synaptic Function Mediating Human Cognition. *Current Biology, 21*, 1320–1325.

Nocedal, J., & Wright, S. (2006). *Numerical Optimization* (2nd edn.). New York: Springer.

Noonan, M. P., Chau, B. K. H., Rushworth, M. F. S., & Fellows, L. K. (2017). Contrasting Effects of Medial and Lateral Orbitofrontal Cortex Lesions on Credit Assignment and Decision-Making in Humans. *Journal of Neuroscience, 37*, 7023–7035.

O'Doherty, J. P., Dayan, P., Friston, K., Critchley, H., & Dolan, R. J. (2003). Temporal Difference Models and Reward-Related Learning in the Human Brain. *Neuron, 38*, 329–337.

Padoa-Schioppa, C., & Assad, J. A. (2006). Neurons in the Orbitofrontal Cortex Encode Economic Value. *Nature, 441*, 223–226.

Parker, N. F., Cameron, C. M., Taliaferro, J. P., Lee, J., Choi, J. Y., Davidson, T. J., ... Witten, I. B. (2016). Reward and Choice Encoding in Terminals of Midbrain Dopamine Neurons Depends on Striatal Target. *Nature Neuroscience, 19*, 845–854.

Paton, J. J., Belova, M. A., Morrison, S. E., & Salzman, C. D. (2006). The Primate Amygdala Represents the Positive and Negative Value of Visual Stimuli during Learning. *Nature, 439*, 865–870.

Pawlow, I. P. (1904). Nobel-Vortrag*. *Nordiskt Medicinskt Arkiv*, 37, 1–20.

Rescorla, R. A., & Wagner, A. R. (1972). A Theory of Pavlovian Conditioning: Variations in the Effectiveness of Reinforcement and Nonreinforcement. In: A. H. Black & W. F. Prokasy (Eds.), *Classical Conditioning II* (pp. 64–99). New York: Appleton-Century-Crofts.

Rigoux, L., & Daunizeau, J. (2015). Dynamic Causal Modelling of Brain-Behaviour Relationships. *NeuroImage, 117*, 202–221.

Rigoux, L., Stephan, K. E., Friston, K. J., & Daunizeau, J. (2014). Bayesian Model Selection for Group Studies – Revisited. *NeuroImage, 84*, 971–985.

Roesch, M. R., & Olson, C. R. (2005). Neuronal Activity in Primate Orbitofrontal Cortex Reflects the Value of Time. *Journal of Neurophysiology, 94*, 2457–2471.

Rosenblatt, F. (1958). The Perceptron: A Probabilistic Model for Information Storage and Organization in the Brain. *Psychological Review, 65*, 386–408.

Ruff, C. C., & Fehr, E. (2014). The Neurobiology of Rewards and Values in Social Decision Making. *Nature Reviews Neuroscience, 15*, 549–562.

Samejima, K., Ueda, Y., Doya, K., & Kimura, M. (2005). Representation of Action-Specific Reward Values in the Striatum. *Science, 310*, 1337–1340.

Schoenbaum, G., Chiba, A. A., & Gallagher, M. (1999). Neural Encoding in Orbitofrontal Cortex and Basolateral Amygdala during Olfactory Discrimination Learning. *Journal of Neuroscience, 19*, 1876–1884.

Schuck, N. W., Cai, M. B., Wilson, R. C., & Niv, Y. (2016). Human Orbitofrontal Cortex Represents a Cognitive Map of State Space. *Neuron, 91*, 1402–1412.

Schultz, W., Dayan, P., & Montague, P. R. (1997). A Neural Substrate of Prediction and Reward. *Science, 275*, 1593–1599.

Sejnowski, T. J., Churchland, P. S., & Movshon, J. A. (2014). Putting Big Data to Good Use in Neuroscience. *Nature Neuroscience, 17*, 1440–1441.

Simon, H. A. (1956). Rational Choice and the Structure of the Environment. *Psychological Review., 63*, 129–138.

Stalnaker, T. A., Cooch, N. K., & Schoenbaum, G. (2015). What the Orbitofrontal Cortex Does Not Do. *Nature Neuroscience, 18*, 620–627.

Stephan, K. E., Penny, W. D., Daunizeau, J., Moran, R. J., & Friston, K. J. (2009). Bayesian Model Selection for Group Studies. *NeuroImage, 46*, 1004–1017.

Sutton, R. S., & Barto, A. G. (2018). *Reinforcement Learning: An Introduction* (2nd edn.). Adaptive Computation and Machine Learning Series. Cambridge, MA: MIT Press.

Tobler, P. N., Dickinson, A., & Schultz, W. (2003). Coding of Predicted Reward Omission by Dopamine Neurons in a Conditioned Inhibition Paradigm. *Journal of Neuroscience, 23*, 10402–10410.

Tremblay, L., & Schultz, W. (1999). Relative Reward Preference in Primate Orbitofrontal Cortex. *Nature, 398*, 704–708.

Turner, B. M., Forstmann, B. U., Love, B. C., Palmeri, T. J., & Van Maanen, L. (2017). Approaches to Analysis in Model-Based Cognitive Neuroscience. *Journal of Mathematical Psychology, Model-Based Cognitive Neuroscience, 76*, 65–79.

Uttal, W. R. (2001). *The New Phrenology: The Limits of Localizing Cognitive Processes in the Brain, The New Phrenology: The Limits of Localizing Cognitive Processes in the Brain*. Cambridge, MA: MIT Press.

Uttal, W. R. (2014). *Psychomythics: Sources of Artifacts and Misconceptions in Scientific Psychology* (1st edn.). Hove: Psychology Press.

Vanyukov, P. M., Hallquist, M. N., Delgado, M. R., Szanto, K., & Dombrovski, A. Y. (2019). Neurocomputational Mechanisms of Adaptive Learning in Social Exchanges. *Cognitive, Affective, & Behavioral Neuroscience, 19*, 1–13.

Vermunt, J. K. (2010). Latent Class Modeling with Covariates: Two Improved Three-Step Approaches. *Political Analysis, 18*, 450–469.

Waelti, P., Dickinson, A., & Schultz, W. (2001). Dopamine Responses Comply with Basic Assumptions of Formal Learning Theory. *Nature, 412*, 43–48.

Wagenmakers, E.-J., Morey, R. D., & Lee, M. D. (2016). Bayesian Benefits for the Pragmatic Researcher. *Current Directions in Psychological Science, 25*, 169–176.

Wagenmakers, E.-J., Ratcliff, R., Gomez, P., & Iverson, G. J. (2004). Assessing Model Mimicry Using the Parametric Bootstrap. *Journal of Mathematical Psychology, 48*, 28–50.

Wallis, J. D., & Kennerley, S. W. (2010). Heterogeneous Reward Signals in Prefrontal Cortex. *Current Opinions in Neurobiology, 20*, 191–198.

Walton, M. E., Behrens, T. E., Buckley, M. J., Rudebeck, P. H., Rushworth, M. F. (2010). Separable Learning Systems in the Macaque Brain and the Role of Orbitofrontal Cortex in Contingent Learning. *Neuron, 65*, 927–939.

Werbos, P. (1974). *Beyond Regression: New Tools for Prediction and Analysis in the Behavioral Sciences.* Cambridge, MA: Harvard University Press.

Whittington, J. C. R., & Bogacz, R. (2019). Theories of Error Back-Propagation in the Brain. *Trends in Cognitive Science, 23*, 235–250.

Wiecki, T. V., Poland, J., & Frank, M. J. (2015). Model-Based Cognitive Neuroscience Approaches to Computational Psychiatry Clustering and Classification. *Clinical Psycholical Science, 3*, 378–399.

Wimmer, G. E., Braun, E. K., Daw, N. D., & Shohamy, D. (2014). Episodic Memory Encoding Interferes with Reward Learning and Decreases Striatal Prediction Errors. *Journal of Neuroscience, 34*, 14901–14912.

PART IV DEVELOPMENTAL PSYCHOPATHOLOGY AND LONGITUDINAL METHODS

15 Studying Psychopathology in Early Life

Foundations of Developmental Psychopathology

DANTE CICCHETTI

HISTORICAL ROOTS OF DEVELOPMENTAL PSYCHOPATHOLOGY

During the course of the past four decades, developmental psychopathology has emerged as a new science that integrates various disciplines, the efforts of which had been previously distinct and separate (Cicchetti, 1984, 1990, 2006; Masten, 2006). A number of eminent theorists from diverse disciplines have asserted that we can understand more about the normal functioning of an organism by investigating its pathology, and, likewise, more about pathology by examining its normal condition (Cicchetti, 1984). Because these early systematizers conceptualized psychopathology as an exaggeration or distortion of the normal condition, the investigation of pathological phenomena was thought to throw into sharper relief the comprehension of normal processes. Characterizing normal biological, psychological, and social processes is very helpful for understanding, preventing, and treating psychopathology. Moreover, understanding the deviations from and distortions of normal development that are seen in pathological processes indicate in exciting ways how normal development may be better investigated and understood (Sroufe, 2007, 2013). Indeed, for many theorists, the distinctiveness and uniqueness of a developmental psychopathology approach lie in its focus on both normal and abnormal, adaptive and maladaptive, developmental processes (Cicchetti, 1984, 1993; Rutter, 1986; Sroufe, 1990, 2007, 2013).

Because psychopathology unfolds over time in a developing organism, it is essential to adopt a developmental perspective in order to understand the processes underlying individual pathways to adaptive and maladaptive outcomes (Cicchetti, 2006; Sroufe, 2007). A developmental analysis presupposes change and novelty, highlights the critical role of timing in the organization of behavior, underscores multiple determinants, and cautions against expecting invariant relations between causes and outcomes. Furthermore, a developmental analysis is as applicable to the study of the gene or cell as it is to the investigation of the individual, family, or society (Cicchetti & Pogge-Hesse, 1982).

A developmental analysis is essential for tracing the roots, etiology, and nature of maladaptation so that interventions may be timed and guided as well as developmentally appropriate (Cicchetti & Toth, 2017; Toth et al., 2013). Moreover, a developmental analysis proves useful for discovering the compensatory mechanisms – biological, psychological, and social-contextual – that may promote resilient functioning despite the experience of significant adversity (Luthar & Cicchetti, 2000; Masten & Cicchetti, 2016).

Analysis strives to examine the prior sequences of adaptation or maladaptation that have contributed to a given outcome. Because developmental psychopathology assumes a lifespan view of developmental processes and aims to delineate how prior development influences later development, a major issue in the discipline is how to determine continuity in the quality of adaptation across developmental time. The same behaviors in different developmental periods may represent quite different levels of adaptation. For example, behaviors indicating competence within a developmental period may indicate incompetence within subsequent developmental periods. Normative behaviors manifested early in development may indicate maladaptation when exhibited at a later time. The manifestation of competence in different developmental periods is rarely indicated by isomorphism in behavior presentation.

THE ORGANIZATIONAL PERSPECTIVE

Although developmental psychopathology is not characterized by the acceptance of any unitary theoretical approach, the organizational perspective on development (Cicchetti & Schneider-Rosen, 1986; Cicchetti & Sroufe, 1978; Sroufe, 1979; Sroufe & Rutter, 1984) has offered a powerful framework for conceptualizing the intricacies of the lifespan perspective on risk and psychopathology, as well as on normality. Werner's (1948) orthogenetic principle specifies that the developing individual moves from a state of relatively diffuse, undifferentiated organization to states of greater articulation and complexity by differentiation and consolidation of the separate systems, followed by hierarchic integration within and between systems. Initially, separate systems within the infant are relatively undifferentiated; however, through development, the various systems increasingly become distinct or differentiated, and repeated hierarchic integration among these systems lead to increasingly complex levels of organization.

At each juncture of reorganization in development, the concept of hierarchic motility specifies that prior developmental structures are incorporated into later ones by means of hierarchic integration. In this way, early experience and its effects on the organization of the individual are carried forward (Cicchetti, 2017) within the individual's organization of systems rather than through reorganizations. The presence of prior structures within the current organization allows for possible future access by way of regressive activation of those previous structures in times of stress or crisis.

Organizational theorists believe that each stage of development confronts individuals with new challenges to which they must adapt. At each period of reorganization, successful adaptation or competence is signified by an adaptive integration within and among the behavioral and biological systems as the individual masters current developmental challenges.

To understand adaptive versus maladaptive development over time, it is important to consider the dialectic nature of two models employed to classify problem disorders – categorical approaches (e.g., *Diagnostic and Statistical Manual of Mental Disorders* 5th edition [*DSM-5*]; American Psychiatric Association [APA], 2013) and dimensional models (Research Domain Criteria [RDoC]; Beauchaine & Klein, 2017; Krueger et al., 2005). These two models reflect mutually enriching perspectives; however, each has limitations as well as benefits. From a developmental psychopathology perspective, we tend to be interested in continuities and discontinuities, both over time and with regard to typical and atypical developments. Developmental psychopathologists also are invested in understanding the entire continuum of functioning, from normal to abnormal. Developmental psychopathologists are committed to comprehending the origins of categorically definable conditions as well as to providing alternative ways of considering and describing psychopathology.

PRINCIPLES OF DEVELOPMENTAL PSYCHOPATHOLOGY

With the organizational perspective on development in mind, it is possible to examine some of the principles of developmental psychopathology. Inherent to a developmental psychopathology approach is a commitment to the importance of applying knowledge of normal development to the study of atypical populations. Even prior to the emergence of a mental disorder, certain pathways signify adaptational failures in normal development that probabilistically forebode subsequent psychopathology. Likewise, developmental psychopathologists are as interested in individuals who are at risk for the development of pathology but do not manifest it as they are in individuals who develop an actual disorder (Cicchetti & Garmezy, 1993; Masten & Cicchetti, 2016). Furthermore, developmental psychopathologists also are committed to understanding pathways to competent adaptation despite experiencing significant adversity (Luthar, Cicchetti, & Becker, 2000; Masten, 2014).

A related aspect of the developmental psychopathology perspective involves an interest in the mechanisms and processes that mediate or moderate the ultimate outcome of risk factors (Cicchetti, 1993; Kraemer et al., 1997; Rutter, 1988). The approach suggested by a developmental psychopathology framework requires a comprehensive assessment of functioning, including multidisciplinary, multidomain, multilevel, and multicontextual measurement strategies. Additionally, developmental psychopathology is a lifespan perspective, because it is only by examining a range of conditions from infancy through adulthood that developmental continuities and discontinuities can be elucidated fully.

Furthermore, because all periods of the life course usher in new biological and psychological challenges, strengths, and vulnerabilities, the process of development may embark on an unfortunate turn at any point in the lifespan (Rutter & Garmezy, 1983; Zigler & Glick, 1986). Whereas change in functioning remains possible at each transitional turning point in development, prior adaptation places constraints on subsequent adaptation. In particular, the longer an individual continues along a maladaptive pathway, the more difficult it is to reclaim a normal developmental trajectory (Cicchetti & Sroufe, 2000). Moreover, recovery of function to an adaptive level of developmental organization is more likely to occur following a period of pathology if the level of organization prior to the breakdown was a competent and adaptive one (Cicchetti, 1993, 2006; Sroufe, Egeland, & Kreutzer, 1990).

Developmental psychopathology also is committed to a multigenerational perspective that highlights an essential point: parents transmit their genes and provide a context for development. By studying gene-environment interactions over the course of development, the reciprocal interactions between constitutional and psychosocial factors can be better understood (Moffitt, 2005). Although genetic factors are not all expressed at birth, they play a prominent role at each phase of development. The age of onset of disorders is most likely affected by timed biological events (e.g., pruning of the connections among neurons; endocrine surges, etc.), as well as by the emergence of stage-salient issues of development (e.g., emotion regulation, secure attachment, effective peer relations, etc.; see Cicchetti, 1993; Sroufe, 1979; Waters & Sroufe, 1983).

Developmental psychopathologists view disorders of infancy, childhood, and adulthood from within the broader context of knowledge that has been accrued about biological, psychological, and social-contextual processes. Moreover, from the integrative perspective of developmental psychopathology it is critical to engage in a comprehensive evaluation of these factors and how they may influence the nature of individual differences, the continuity of adaptive or maladaptive behavioral patterns, and the different pathways by which the same developmental outcomes may be achieved (Cicchetti & Rogosch, 1996; Cicchetti & Schneider-Rosen, 1986).

This requires an understanding of an appreciation for the biological and psychological developmental transformations and reorganizations that occur over time; an analysis and appropriate weighing of the risk and protective factors and mechanisms operating in the individual and their environment throughout the life course; the investigation of how emergent functions, competencies, and developmental tasks modify the expression of risk conditions or disorders or lead to new symptoms and difficulties; and the recognition that a specific stress or underlying mechanism may, at difficult times in the developmental process and in varied contexts, lead to different behavioral difficulties (Cicchetti & Schneider-Rosen, 1986). Consequently, individuals may experience similar events differently, depending on their level of functioning across all levels of behavioral and biological development. Accordingly, various occurrences will have different meanings for an individual because of both the nature and the timing of their experience. The interpretation of the experience, in turn, will affect the adaptation or maladaptation that ensues.

Because of the interrelations involved in investigating normal and abnormal ontogenesis, developmental psychopathologists need to be aware of normal pathways, uncover deviations from these pathways, articulate the transformations that occur as individuals progress through these deviant developmental courses, and identify the factors and mechanisms that may divert an individual from a particular pathway and onto a more or less adaptive course (Masten & Cicchetti, 2016). A central tenet of developmental psychopathology is that persons may move between pathological and nonpathological forms of functioning. Additionally, developmental psychopathology underscores that, even in the midst of pathology, persons may display adaptive coping mechanisms. Moreover, depending on contextual constraints, the definition of normalcy may vary. For example, affective inhibition may be adaptive for a child with maltreating parents, but it may result in victimization by peers. This lifespan perspective on psychopathology enables developmental considerations to be brought into harmony with clinical concepts used to define the natural history of disorder, such as prodrome, onset, course, offsets, remission, and residual states.

ILLUSTRATIVE PRINCIPLES OF DEVELOPMENTAL PSYCHOPATHOLOGY

The Interface between Normality and Pathology

Embryologists, geneticists, ethologists, neuroscientists, psychiatrists, and psychologists have underscored that research on the normal and atypical must proceed hand in hand in order for an integrative theory of development that can account for normal as well as abnormal developmental processes.

Research conducted in embryology has made significant contributions to developmental theory (Fishbein, 1976; Gottlieb, 1976, 1983; Waddington, 1957; Weiss, 1969). From their empirical efforts to unravel the mysteries of normal embryological functioning through isolation, defect, and recombination experiments, early embryologists derived the principles of *differentiation*, of a *dynamically active organism*, and of a *hierarchically integrated system* (Waddington, 1966; Weiss, 1969). These three principles form the cornerstone beliefs of most contemporary developmental theorists (Cairns, 1983; Gottlieb, 1983). Unfortunately, despite the adherence of developmental psychopathologists to the belief that normal and abnormal ontogenetic processes should be investigated concurrently, most contemporary theory and research has focused on the contribution that normal development can make to advancing our knowledge of high-risk conditions and psychopathology.

The study of psychopathological phenomena and extreme risk conditions have made important contributions to theory development and refinement in other disciplines; however, it is curious that, until recent decades, there has been less recognition and acceptance that the investigation of high-risk and pathological conditions can affirm, augment, and challenge extant developmental theories. Often, the investigation of a system in its smoothly operating normal or healthy state does not afford the opportunity to comprehend the interrelations among its component subsystems. Because pathological conditions such as brain damage and developmental disorders, or growing up in malignant environments, enable us to isolate the components of the integrated system, their investigation sheds light on the normal structure of the system. If we choose simply to ignore or bypass the study of these atypical phenomena, then the eventual result is likely to be the construction of theories that are contradicted by the revelation of critical facts in psychopathology.

Extrapolating from abnormal populations with the goal of informing developmental theory, however, it is important that a range of populations and conditions be considered. Within the neurosciences, for example, progress in molecular cell biology has stemmed from developmental research on normal and abnormal variations in cells (research on development of cell specificity, cancer, immunity, etc.; Ciaranello et al., 1995). The study of a single psychopathological or risk process may result in spurious conclusions if generalizations are made based solely on that condition or disorder. However, if we view a given behavioral pattern in the light of an entire spectrum of diseased and disordered modifications, then we may be able to attain significant insight into the processes of development not generally achieved through sole reliance on studies of relatively more homogeneous nondisordered populations.

Pathways: Diversity in Process and Outcome

Diversity in process and outcome are among the hallmarks of the developmental psychopathology perspective (Kohlberg, LaCross, & Ricks, 1972; Sroufe, 1986, 1989). Developmental psychopathologists have articulated the expectation that there are multiple contributors to adaptive or maladaptive outcomes in any individual, and that there are myriad pathways to any particular manifestation of adaptive or disordered behavior (Cicchetti, 2006; Sroufe & Jacobvitz, 1989). Additionally, there is heterogeneity among individuals who develop a specific disorder with respect to the features of their disturbance, as well as among individuals who evidence maladaptation but who do not develop a disorder. In accord with this view, the principles of *equifinality* and *multifinality* derived from general systems theory (von Bertalanffy, 1968) are germane.

Equifinality refers to the observation that in any open system the same end state may be reached from a variety of different initial conditions and through different processes. In contrast, in a closed system the end state is inextricably linked to and determined by the initial conditions: if either the conditions change or the processes are modified, then the end state will also be modified.

Initial descriptions of equifinality originated from research in embryology. For example, research demonstrated that different initial sizes and different courses of growth can eventuate in the same ultimate size of an organism (von Bertalanffy, 1968; Waddington, 1957). Within the field of developmental psychopathology, equifinality has been invoked to explain why a variety of developmental pathways may eventuate in a given outcome rather than expecting a singular primary pathway to the adaptive or maladaptive outcome (Cicchetti & Rogosch, 1996). For example, there are a multiplicity of pathways to depression and other forms of psychopathology.

The principle of multifinality states that a particular adverse event, such as child maltreatment, should not necessarily be seen as leading the same psychopathological outcome.

Attending to the diversity of origins, processes, and outcomes in understanding developmental pathways does not suggest that prediction is futile because of the many potential individual patterns of adaptation (Sroufe, 1989). There are constraints on how much diversity is possible, and not all outcomes are equally likely (Cicchetti & Tucker, 1994; Sroufe et al., 1990). Nonetheless, the appreciation of equifinality and multifinality in development encourages theorists and researchers to entertain more complex and varied approaches to how they conceptualize and investigate development and psychopathology. Researchers should increasingly strive to demonstrate the multiplicity of processes and outcomes that may be articulated at the individual, person-oriented level within existing longitudinal data sets. Ultimately, future endeavors must conceptualize and design research at the outset with these differential pathways concepts as a foundation. In so doing, developmental psychopathologists will progress toward explaining the development of individual patterns of adaptation and maladaptation.

Multiple Levels of Analysis

A process-level understanding of mental disorder requires research designs and strategies that call for the simultaneous assessment of multiple domains of variables both within and outside of the developing person. Similarly, research in the area of resilience must adhere to these interdisciplinary multiple levels of analysis perspectives. Because each level of analysis both informs and constrains all other levels, psychopathology cannot be understood fully unless all levels are examined and integrated. Moreover, the influence of levels on one another is almost always bidirectional (Cicchetti & Cannon, 1999; Thelen & Smith, 1998). Because levels of organization and processes are reciprocally interactive, it is difficult, if not impossible, to impute ultimate causation to one level over another. It is the mutual relationship between at least two components of the developmental system that influences developmental organization or disorganization (Gottlieb, 1992).

Since levels of analysis constrain one another, as developmental psychopathologists learn more about multiple levels of analysis, researchers conducting their work at each level will need to develop theories that make testable predictions across all levels of inquiry. When disciplines function in isolation, they run the risk of creating theories that ultimately will be incorrect because vital information from other disciplines has either been ignored or is unknown. As is true in systems neuroscience, it is essential that psychopathology and resilience researchers utilize an integrative framework that incorporates all levels of analysis about relevant complex systems. Rather than adhering to a single domain or unitary disciplinary focus, striving for a multidomain, multilevel synthesis may impel researchers to broaden their visions and thereby lead to the formulation of integrative developmental theories that can elucidate both normal and abnormal forms of ontogenesis across developing systems.

Longitudinal Transactional Models of Development

Development is necessarily the result of the interdependence, coaction, and/or codetermination among multiple levels of influence over time (Gottlieb, 2001). In 1975, Sameroff and Chandler proposed a transactional model of development that recognized the importance of the transacting genetic, constitutional, neurobiological, biochemical, psychological, and social-contextual factors in the determination of behavior. Sameroff and Chandler's (1975) model stated that these factors change through their dynamic transaction and the processes by which compensatory, self-regulating tendencies are initiated whenever higher-level monitors detect deviances in a system (Sameroff, 2000; Waddington, 1966). An implication of the transactional model is that the continued manifestation of maladaptation depends on environmental support, whereas the child's reciprocal characteristics (e.g., temperament) partially determine the nature of the environment.

The call for longitudinal models of development has assumed a global profile, including two recent initiatives led by the National Institutes of Health. Founded in 2015, the Longitudinal Study of Adolescent Brain and Cognitive Development is recruiting 10,000 healthy children age 9–10 and is following them over 10 years into early adulthood, to measure brain maturation in the context of social, emotional, and cognitive development. Similarly, the Environmental Influences on Child Health Outcomes Program at the National Institutes of Health (ECHO) will support multiple longitudinal studies using existing study populations to investigate environmental exposures, including physical, chemical, biological, social, behavioral, natural, and built environments, on child health and development. These two national initiatives will yield rich longitudinal data sets that are ripe for the incorporation of longitudinal transactional models that examine the development of psychopathology across multiple developmental periods and use multilevel sources of data.

Longitudinal investigations offer great potential for recovery and resilience among children exposed to varied forms of adversity, regardless of whether the adversity stems from experiences in the families that children grow up in, their neighborhood environments, their peer networks, their genetic makeup, or a combination of these influences (Cicchetti, 2017). Longitudinal studies can provide evidence that children are not simply passive recipients of their environments, but, rather, they contribute to their own development in both passive and active ways, by eliciting behaviors from others and by selecting people and contexts within which to interact (Leve & Cicchetti, 2016; Scarr & McCartney, 1983).

The field has made significant strides over the last decade in developing methodological techniques and approaches to conduct strong tests of longitudinal transactional models of development (MacKinnon, 2008; Preacher & Hayes, 2008). This is important because stringent tests of longitudinal models of development and psychopathology should simultaneously consider continuity in behavior, mediational or indirect effects, reciprocal relationships, as well as covariates and potential moderating variables. Consequently, longitudinal transactional studies can present serious methodological challenges, particularly when the presentation or expression of psychopathology changes across development, for example, as in the transition from oppositional behavior, to conduct problems, to antisocial behavior.

Developmental Cascades

Developmental cascades refers to the cumulative consequences of the many interactions and transactions occurring in developing systems that result in spreading effects across levels, among domains at the same level, and across different systems or generations. Theoretically these effects may be direct and unidirectional, direct and bidirectional, or indirect through various pathways, but the consequences are not transient: developmental cascades alter the course of development. Such effects have gone by different names in the literature, including chain reactions, snowball, amplification, and spillover or progressive effects (Masten & Cicchetti, 2010).

Cascade effects could explain why some problems in childhood predict widespread difficulties in adulthood, whereas others do not. In their classic review of longitudinal data on adjustment, Kohlberg and colleagues (1972) observed decades ago that some indicators of childhood success or problems consistently forecast adult adjustment across multiple domains of outcome. In particular, they noted the robust and broad predictive validity of general cognitive competence (often indexed by academic achievement or intellectual ability) and socialized conduct versus antisocial behavior (often indexed by persistent rule-breaking behavior).

Given effects that spread over time for some kinds of psychopathology, well-timed and targeted interventions could interrupt negative or promote positive cascades. These efforts may work by counteracting negative cascades, by reducing problems in domains that often cascade to cause other problems, or by improving competence in domains that increase the probability of better function in other domains (Cicchetti & Curtis, 2006; Cicchetti & Gunnar, 2008; Masten, Burt, & Coatsworth, 2006; Masten et al., 2009). Moreover, if developmental cascades are common and often begin with adaptive behavior in early childhood, then that would explain why the evidence in prevention science indicates a high return on investment in early childhood interventions, such as high-quality preschool programs (Heckman, 2006; Reynolds & Temple, 2006).

Intervention designs that target mediating processes for change represent cascade models. The "theory of the intervention" represents a hypothesis that the intervention will change the mediating process that will in turn change the outcome of the person in key adaptive domains.

Cascade effects encompass a broad array of phenomena studied in developmental science within and across multiple levels of function, from the molecular to the macro level. Cascade models may account for the pathways by which gene-environment interplay unfolds over time in epigenesis to shape development, linking genes to brain function to behavior to social experience (Cicchetti, 2002; Cicchetti & Cannon, 1999; Gottlieb, 1998, 2007; Hanson & Gottesman, 2007). These cascades may flow downward across levels, upward, or across domains of function in a developing system and its interactions with other systems as development proceeds. The effects of pharmacological interventions on behavior can be viewed as intentional efforts to induce an upward cascade by means of neurochemical changes that influence neural function in the brain and subsequently adaptive behavior (Charney, 2004). The processes by which genetic disorders result in the development of behavioral anomalies have been described as cascades. Biological embedding of experience, as might happen when traumatic or negative experiences alter gene expression or the stress-response systems of a developing child, may begin as a downward cascade (experience alters functional systems in the child) and then subsequently these altered systems may have cascading upward consequences for brain development, stress reactivity, and symptoms of psychopathology through a complex sequence of processes (Cicchetti, 2002; Cicchetti & Cannon, 1999; Gunnar & Quevedo, 2007; Meaney, 2010; Shonkoff, Boyce, & McEwen, 2009).

Developmental cascades may be positive or negative in their consequences with respect to adaptive behavior. Developmental theories about competence (see Masten et al., 2006) invoke cascades in the fundamental assertion that effectiveness in one domain of competence in one period of life becomes the scaffold on which later competence in newly emerging domains develop: in other words, *competence begets competence*. Success in early developmental tasks of childhood are expected to foster competence in subsequent and later emerging developmental tasks (Cicchetti & Schneider-Rosen, 1986; Cicchetti & Tucker, 1994; Masten & Coatsworth, 1998; Masten & Wright, 2009; Sroufe, 1979).

Because earlier structures of the individual's organization are incorporated into later structures in the successive process of hierarchical integration, early competence tends to promote later competence. An individual who has adaptively met the developmental challenges of the particular stage will be better equipped to meet successive new challenges in development. This is not to say that early adaptation ensures successful later adaptation; major changes or stresses in the internal and external environment may tax subsequent adaptational capacities. However, early competence provides a more optimal organization of behavioral and biological systems, thus offering, in a probabilistic manner (Gottlieb, 1991), a greater likelihood that adaptive resources are available to encounter and cope with new developmental demands (Cicchetti, 1993; Erikson, 1950; Sroufe, 1979).

From a methodological standpoint, cascade models are challenging to test for several reasons. Testing cascade models requires longitudinal data that are often difficult and time-consuming to collect. Strong tests of cascade models also require repeated assessment of multiple domains or levels of function over time, and the application of complex statistical approaches. The most stringent tests of cascade models have similar requirements to those for testing mediating effects. It is important to control for the continuity or stability over time that one would expect within most domains of adaptive behavior. At the same time, it is important to control for the covariance of key measures assessed during an assessment window. If continuity and covariance are not controlled, then it is difficult to establish whether there is a unique and cumulative cascade effect from one domain to another over time and when this might be occurring. Without accounting for within-time covariance and across-time stability, an apparent cascade effect may reflect correlations that were already present at the beginning of the assessments or that represent an artifact of unmeasured outcome covariance.

A related challenge for research on developmental cascade occurs when a model to be tested included newly emerging domains of function or organization. In developmental task theory, for example, the domains of work and romantic competence emerge in adolescence in many developed nations, and subsequently become salient tasks of adulthood (Roisman et al., 2004). These emerging tasks are believed to have roots in academic and social competence earlier in childhood; success in earlier tasks builds the skills for success in these later tasks.

In summary, developmental cascade research has the potential to inform, test, and refine theories of change that are integral to understanding pathways of adaptation, maladaptation, psychopathology, and resilience. Such knowledge is central to the goal of designing preventive interventions with strategic timing and targets.

Resilience

Examinations of risk and psychopathology across the life course all too often portray the developmental process as somewhat deterministic, resulting in maladaptive and adverse outcomes. Investigations ranging from genetic and biological contributions to pathology, to assaults on development associated with inadequate caregiving and stress, graphically convey the multiplicity of risks that eventuate in maladaptation and/or psychopathology (Kraemer et al., 1997; Rutter & Garmezy, 1983). As the developmental perspective has assumed a more prominent role in psychopathology research, there has been a burgeoning interest in the study of resilience (Luthar, Cicchetti, & Becker, 2000; Masten, 2014; Masten & Cicchetti, 2016). From what individual, familial, or environmental societal risk mechanisms does the trajectory of risk for psychopathology stem, resulting in adaptive outcomes even in the presence of significant adversity? In addition, developmental psychopathologists emphasize the need to understand the functioning of individuals who, after having diverged onto deviant developmental pathways, "bounce back," resume normal functioning and achieve adequate adaptation (Cicchetti & Rogosch, 1997).

It is important that resilience not be conceptualized as a static or traitlike condition, but as being in dynamic transaction with intraorganismic and extraorganismic forces (Egeland, Carlson, & Sroufe, 1993). Research on the processes leading to resilient outcomes offers great promise as an avenue for facilitating the development of preventions and intervention strategies.

Prevention and Intervention

The preeminent objective of prevention science is to intervene in the course of development to reduce or eliminate the emergence of maladaptation and mental disorder. Prevention scientists also seek to foster the recovery of function and to promote resilient adaptation in individuals at high risk for psychopathology (Cicchetti & Hinshaw, 2002; Luthar & Cicchetti, 2000). To fulfill these laudable goals, it is important that prevention scientists possess a complex multilevel understanding of normality to formulate an in-depth representation of how deviations in normal developmental processes can eventuate in maladaptation and psychopathology (Cicchetti, 2017). The discipline of developmental psychopathology, with its major focus on the dialectic between normal and abnormal development, is uniquely poised to provide the theoretical foundation for prevention science (Institute of Medicine, 1994). Despite the logical links that exist between the provision of interventions to children, adolescents, and adults and developmental theory and research, far too few bridges have been forged between these realms of knowledge (Cicchetti, 2018; Cicchetti & Toth, 2017; Weisz & Kazdin, 2017). During the past several decades, there has been increased dialogue between guided prevention and intervention (Toth et al., 2013). Specifically, as the investigation of the relation between normal and abnormal development has burgeoned, so too has the application of findings conceptualized within this genre to prevention and intervention efforts.

Considerations such as when (i.e., the timing) and why a disorder occurs, how long it persists, and what precursors to the disordered functioning can be identified all require a developmental approach to ensure that prevention and intervention strategies are timed and guided. This lifespan perspective that characterizes a developmental psychopathology approach underscores that even adults and elderly persons continue to be confronted with challenges that present opportunities for new strengths or usher in new vulnerabilities. Thus, the developmental timing of a preventive intervention may be even more critical than its particular content. Minimally, we contend that the effect of an intervention will be enhanced or decreased in relation to its sensitivity to the temporal dimension of the developmental process.

Because periods of developmental transition and "sensitive periods" offer opportunities for reorganization and change, it might be especially important to target prevention and intervention efforts at these periods of intersystemic reorganization. For example, depending on when a pathology-inducing insult occurred or a deviant process was initiated, interventions might be more efficiently targeted. By developing more specific and circumscribed approaches, intervention effectiveness could increase and the need for more prolonged treatment might decrease. The knowledge of early developmental deviations and their link with subsequent psychopathology also could be used to prevent the emergence of full-blown psychopathology, thereby decreasing both time and money expended in the treatment of more entrenched and often severe clinical conditions (Institute of Medicine, 1994).

Likewise, the concepts of equifinality and multifinality should alert all clinicians to the importance of utilizing multiple strategies of treatment predicated on individuals' stage in the lifespan, their current level of functioning and developmental organization across psychological and biological/genetic domains, and any special characteristics of the group of individuals being treated. Moreover, because multiple risk factors, both intra- and extraorganismic, drawn from psychological, biological, and environmental forces, characterize the various pathways to maladaptive and disordered outcomes, interventions must address the broader causal matrix or be destined to produce ineffective and short-lived results (Cicchetti & Gunnar, 2008). Additionally, because the presence of many risk factors at multiple levels of the ecology and the individual are associated with more chronic and maladaptive outcomes, interventions for more serious pathological conditions (e.g., bipolar disorder, schizophrenia) must incorporate a longitudinal perspective and employ a series of interventions over time, in an effort to reduce the risk factors to a manageable minimum.

In summary, the prevention of maladaptation and mental disorder requires an in-depth knowledge of the dynamic relations among risk and protective factors and typical and atypical developmental processes. It is essential that prevention scientists conceptualize, design, and evaluate prevention trials in such a way that also enables their results to enhance our understanding of development and the pathways contributing to intervention efficacy at multiple levels of influence. From the integrative developmental psychopathology framework, randomized control trial interventions (RCTs) can be conceptualized as veridical experiments in modifying the course of development. Therefore, these RCTs can be viewed as tests of theory and causal mechanisms, thereby proffering insights into the etiology and pathogenesis of maladaptation and disordered outcome (Howe, Reiss, & Yuh, 2002). Determining the multiple levels at which change is engendered through randomized prevention trials will provide more insights into the mechanisms of change, the extent to which neural plasticity may be promoted, and the interrelations between biological and psychological processes in maladaptation, psychopathology, and resilience (Cicchetti & Curtis, 2006). Furthermore, preventive interventions with the most in-depth empirical support, based on integrative multilevel theories of normality, psychopathology, and resilience, can be implemented in effectiveness trials in community or real-world setting to reach the broadest number of people and prevent, or alleviate, suffering from mental disorders.

DEVELOPMENTAL NEUROBIOLOGY

Researchers concerned with developmental psychopathology have begun to focus more attention on brain development. There is increasing recognition that developmental disorders, particularly the severe disturbances such as infantile autism, arise from exogenous or endogenous disruptions in our knowledge of developmental neurobiology, that area of neuroscience that focuses on factors regulating the development of neurons, neural circuitry, and complex neuronal organization systems, including the brain. Furthermore, molecular genetics has made exciting contributions to understanding neurologic disease and mental disorder, which for the first time enable us to understand the genetic basis of certain disorders without requiring foreknowledge of the underlying biochemical abnormalities. Genetic disorders of behavior that are characterized by a variable age of onset and/or remissions and exacerbations should offer the most exciting possibilities for integrating biology, psychology, psychopathology, resilience, and development.

Investigations of the molecular neurobiology of developmental disorders can enhance our understanding of basic neurobiological processes (Thomas & Cicchetti, 2008). A lesson learned through developmental neurobiology research is that the normal development of the nervous system unfolds as a series of timed genetic events whose expression depends on properly timed and environmentally delivered experiences/events. We also know that the malleability of the nervous system – the ability of genetic expression to be influenced by environmental events – varies greatly across cell types, brain regions, and periods of time. Cellular phenotype is a good example: the maturational course of cells is "genetically determined," by which is usually meant that if cells are undisturbed then their developmental outcome is predictable. However, there is abundant evidence that cellular phenotype can be altered if cells are exposed to a different chemical milieu at a critical point in maturation. What, then, does the phrase genetically determined mean? We take it to mean that when a developing cell remains in its normal environment, the properties it assumes will be governed by its genetic programming. But if the environment is altered at a sensitive period in development, then gene expression, too, may be altered, and the cell's development may take a quite different course (Anacker, O'Donnell, & Meaney, 2014).

Although brain development is guided and controlled to some extent by genetic information (Rakic, 1988), a nontrivial portion of brain structure and function occurs through interaction with the environment (Eisenberg, 1995; Greenough, Black, & Wallace, 1987). Changes in the internal and external environment may lead to improvements in the ability of the individual to grapple with developmental challenges, including the experience of significant adversity.

Child maltreatment is a severe stressor that jeopardizes normative development and compromises adaptive functioning (Cicchetti & Toth, 2016). The growing neuroimaging literature on child maltreatment will play a critical role in understanding the mechanisms by which social experience affects brain development and functioning (Demers et al., 2018; Jedd et al., 2015) as well as subsequent psychopathology (DeBellis, 2001; McCrory, DeBrito & Viding, 2010).

Children endowed with normal brains may encounter a number of experiences (e.g., extreme poverty, community violence, physical abuse, sexual abuse, emotional maltreatment, and neglect) that can negatively impact upon developing brain structure, function, and organization and contribute to distorting their experiences of the world (Farah, 2017). Children may be especially vulnerable to the effects of pathological experiences during periods of rapid creation or modification of neuronal connections (Paus, Keshavan, & Giedd, 2008). Pathological experience may become part of a vicious cycle, as the resulting brain changes may distort the child's experience, with subsequent alterations in cognition or social interactions causing additional pathological experience and added brain pathology (Cicchetti & Tucker, 1994).

Finally, because some children develop in a resilient fashion even though they have experienced adversity, it is essential to keep in mind that such individuals may be likely to function adaptively in a wide array of environments. Children who function resiliently play an active role in seeking and receiving the experiences that are developmentally appropriate for them. Even though brain pathology, be it genetically driven or induced by experience, may set a maladaptive course for development it often makes efforts to overcome this (i.e., "self-righting" tendencies; Cicchetti & Rogosch, 1997; Sameroff & Chandler, 1975; Waddington, 1957). At one level, different parts of the brain may attempt to compensate, and at another, the organism may seek out new experiences in areas where it has strength. Because plasticity is a central feature of the mammalian brain (Kempermann, 2006; Kemperman, van Praag, & Gage, 2000), early brain anomalies or aberrant experiences should not be considered as determining the permanent fate of the organism (Cicchetti & Tucker, 1994). In addition to social and psychological

CONCLUSION

Research conducted within a developmental psychopathology framework may challenge assumptions about what constitutes health or pathology. It may also redefine the manner in which the mental health community operationalizes, assesses, classifies, communicates about, and treats the adjustment problems and functioning impairments of infants, children, adolescents, and adults. Through its principles and tenets, developmental psychopathology transcends disciplinary boundaries and provides fertile ground for moving beyond mere symptom description to a process-level understanding of normal and atypical developmental trajectories. The developmental psychopathology perspective provides a broad, integrative framework within which the contribution of separate disciplines can be fully realized in the broader context of understanding individual functioning and development.

The field of developmental psychopathology has contributed significantly to enhancing our understanding of risk factors, risk mechanisms, disorder, adaptation, and resilience across the life span. Its contributions to theory development and enhancements in research design and methodology also have been noteworthy. Much of the momentum of developmental psychopathology has stemmed from an openness to preexisting knowledge in combination with a willingness to question established theory, thereby continuing to promote growth of the field. Moreover, the integration of methods and concepts derived from areas that are too often isolated from each other has resulted in knowledge gains that might have been missed in the absence of cross-disciplinary dialogue. If its contributions to date presage future advances, then developmental psychopathology will continue to engender much enthusiasm and to foster significant advances in our knowledge of normal and atypical development. The key to further progress in the field is a continued emphasis on the concept of development.

REFERENCES

American Psychiatric Association (APA). (2013). *Diagnostic and Statistical Manual of Mental Disorders* (5th edn.). Arlington, VA: American Psychiatric Publishing.

Anacker, C., O'Donnell, K. J., & Meaney, M. J. (2014). Early Life Adversity and the Epigenetic Programming of Hypothalamic-Pituitary-Adrenal Function. *Dialogues of Clinical Neuroscience*, 16(3) 321–333.

Beauchaine, T. P. & Klein, D. N. (2017). Classifying Psychopathology: The DSM, Empirically Based Taxonomies, and the Research Domain Criteria. In T. P. Beauchaine & S. P. Hinshaw (Eds.), *Child and Adolescent Psychopathology* (3rd edn., pp. 33–67). Hoboken, NJ: John Wiley & Sons

Cairns, R.B. (1983). The Emergence of Developmental Psychology. In W. Kessen (Ed.), *Carmichael's Handbook of Child Psychology: Vol. I. History, Theory, and Methods* (4th edn., pp. 41–102). New York: Wiley.

Charney, D. (2004). Psychobiological Mechanisms of Resilience and Vulnerability: Implications for Successful Adaptation to Extreme Stress. *American Journal of Psychiatry*, 161, 195–216.

Ciaranello, R., Aimi, J., Dean, R. S., Morilak, D., Porteus, M. H., & Cicchetti, D. (1995). Fundamentals of Molecular Neurobiology. In D. Cicchetti & D. J. Cohen (Eds.), *Developmental Psychopathology: Theory and Method* (Vol. 1, pp. 109–160). New York: Wiley.

Cicchetti, D. (1984). The Emergence of Developmental Psychopathology. *Child Development*, 55(1), 1–7.

Cicchetti, D. (1990). A Historical Perspective on the Discipline of Developmental Psychopathology. In J. Rolf, A. Masten, D. Cicchetti, K. Nuechterlein, & S. Weintraub (Eds.), *Risk and Protective Factors in the Development of Psychopathology* (pp. 2–28). New York: Cambridge University Press.

Cicchetti, D. (1993). Developmental Psychopathology: Reactions, Reflections, Projections. *Developmental Review*, 13, 471–502.

Cicchetti, D. (2002). The Impact of Social Experience on Neurobiological Systems: Illustration from a Constructivist View of Child Maltreatment. *Cognitive Development*, 17, 1407–1428.

Cicchetti, D. (2006). Development and Psychopathology. In D. Cicchetti & D. J. Cohen (Eds.), *Developmental Psychopathology* (Vol. 1, 2nd edn., pp. 1–23). New York: Wiley.

Cicchetti, D. (2013). Annual Research Review: Resilient Functioning in Maltreated Children – Past, Present, and Future Perspectives. *Journal of Child Psychology and Psychiatry*, 54, 402–422.

Cicchetti, D. (Eds.) (2017). Biological and Behavioral Effects of Early Adversity on Multiple Levels Of Development. *Development and Psychopathology*, 29(5) [Special Issue], 1517–1986.

Cicchetti, D. (2018). A Multilevel Developmental Approach to the Prevention of Psychopathology in Children and Adolescents. In J. N. Butcher, J. Hooley, & P. D. Kendall (Eds.), *APA Handbook of Psychopathology* (pp. 37–53). Washington, DC: American Psychological Association Books.

Cicchetti, D., & Cannon, T. D. (1999). Neurodevelopmental Processes in the Ontogenesis and Epigenesis of Psychopathology. *Development and Psychopathology*, 11, 375–393.

Cicchetti, D., & Curtis, W. J. (2006). The Developing Brain and Neural Plasticity: Implications for Normality, Psychopathology, and Resilience. In D. Cicchetti & D. J. Cohen (Eds.), *Developmental Psychopathology: Developmental Neuroscience* (Vol. 2, 2nd edn., pp. 1–64). New York: Wiley.

Cicchetti, D., & Gunnar, M. R. (2008). Integrating Biological Processes into the Design and Evaluation of Preventive Interventions. *Development and Psychopathology*, 20(3), 737–743.

Cicchetti, D., & Garmezy, N. (1993). Prospects and Promises in the Study of Resilience. *Development and Psychopathology*, 5, 497–502.

Cicchetti, D., & Hinshaw, S. P. (Eds.) (2002). Editorial: Prevention and Intervention Science: Contributions to Developmental Theory. *Development and Psychopathology*, 14(4), 667–671.

Cicchetti, D., & Pogge-Hesse, P. (1982). Possible Contributions of the Study of Organically Retarded Persons to Developmental Theory. In E. Zigler & D. Balla (Eds.), *Mental Retardation: The Developmental-Difference Controversy* (pp. 277–318). Hillsdale, NJ: Lawrence Erlbaum Associates.

risk mechanisms, genetic and neurobiological protective and buffering mechanisms can serve as mediators or moderators of resilient adaptation (Cicchetti, 2013).

Cicchetti, D., & Rogosch, F. A. (1996). Equifinality and Multifinality in Developmental Psychopathology. *Development and Psychopathology, 8*, 597–600.

Cicchetti, D., & Rogosch, F. A. (1997). The Role of Self-Organization in the Promotion of Resilience in Maltreated Children. *Development and Psychopathology, 9*, 797–815.

Cicchetti, D., & Schneider-Rosen, K. (1986). An Organizational Approach to Childhood Depression. In M. Rutter, C. Izard, & P. Read (Eds.), *Depression in Young People, Clinical and Developmental Perspectives* (pp. 71–134). New York: Guilford.

Cicchetti, D., & Sroufe, L. A. (1978). An Organizational View of Affect: Illustration from the Study of Down's Syndrome Infants. In M. Lewis & L. Rosenblum (Eds.), *The Development of Affect* (pp. 309–350). New York: Plenum Press.

Cicchetti, D., & Sroufe, L. A. (2000). The Past as Prologue to the Future: The Times, They've Been a Changin'. *Development and Psychopathology, 12*, 255–264.

Cicchetti, D., & Toth, S. L. (2016). Child Maltreatment and Developmental Psychopathology: A Multilevel Perspective. In D. Cicchetti (Ed.), *Developmental Psychopathology* (3rd edn., Vol. 3, *Maladaptation and Psychopathology*, pp. 457–512). New York: Wiley.

Cicchetti, D., & Toth, S. L. (2017). Using the Science of Developmental Psychopathology to Inform Child and Adolescent Psychotherapy. In A. E. Kazdin & J. R. Weisz (Eds.), *Evidence-Based Psychotherapies for Children and Adolescents* (pp. 484–500). New York: The Guilford Press.

Cicchetti, D., & Tucker, D. (1994). Development and Self-Regulatory Structures of the Mind. *Development and Psychopathology, 6*, 533–549.

DeBellis, M. D. (2001). Developmental Traumatology: The Psychobiological Development of Maltreated Children and Its Implications for Research, Treatment, and Policy. *Development and Psychopathology, 13*, 539–564.

Demers, L., Jedd-McKenzie, K., Hunt, R. H., Cicchetti, D., Cowell, R. A., Rogosch, F. A., ... Thomas, K. M. (2018). Separable Effects of Maltreatment and Adult Functioning on Amygdala Connectivity during Emotion Processing. *Biological Psychiatry: Cognitive Neuroscience and Neuroimaging, 3*(2), 116–124.

Egeland, B., & Carlson, E., & Sroufe, L. A. (1993). Resilience as Process. *Development and Psychopathology, 5*(4), 517–528.

Eisenberg, L. (1995). The Social Construction of the Human Brain. *American Journal of Psychiatry, 152*, 1563–1575.

Erikson, E. H., (1950). *Childhood and Society*. New York: Norton.

Farah, M. J. (2017). The Neuroscience of Socioeconomic Status: Correlates, Causes, and Consequences. *Neuron, 96*(1), 56–71.

Fishbein, H. D. (1976). *Evolution, Development, and Children's Learning*. Pacific Palisades, CA: Goodyear Publishing.

Gottlieb, G. (1976). Conceptions of Prenatal Development: Behavioral Embryology. *Psychological Review, 83*(3), 214–234.

Gottlieb, G. (1983). Psychobiological Approach to Developmental Issues. In M. M. Haith & J Campos (Eds.), *Handbook of Child Psychology: Infancy and Biological Bases* (Vol. 2, pp. 1–26). New York: Wiley.

Gottlieb, G. (1991). Behavioral Pathway to Evolutionary Change. *Rivista Di Biologia, 84*(3), 385–409.

Gottlieb, G. (1992). *Individual Development and Evolution: The Genesis of Novel Behavior*. New York: Oxford University Press.

Gottlieb, G. (1998). The Significance of Biology for Human Development: A Developmental Psychobiological Systems View. In W. Damon & R. M. Lerner (Eds.), *Handbook of Child Psychology* (Vol. 1, 5th edn., pp. 233–273). New York: Wiley.

Gottlieb, G. (2001). The Relevance of Developmental-Psychobiological Metatheory to Developmental Neuropsychology. *Developmental Neuropsychology, 19*(1), 1–9.

Gottlieb, G. (2007). Probabilistic Epigenesis. *Developmental Science, 10*, 1–11.

Greenough, W., Black, J., & Wallace, C. (1987). Experience and Brain Development. *Child Development, 58*, 539–559.

Gunnar, M. R., & Quevedo, K. (2007). The Neurobiology of Stress and Development. *Annual Review of Psychology, 58*, 145–173.

Hanson, D., & Gottesman, I. I. (2007). Choreographing Genetic, Epigenetic, and Stochastic Steps in the Dances of Developmental Psychopathology. In A. S. Masten (Ed.), *Multilevel Dynamics in Developmental Psychopathology: Pathways to the Future: The Minnesota Symposia on Child Psychology* (Vol. 34, pp. 27–44). Mahwah, NJ: Lawrence Erlbaum Associates.

Heckman, J. J. (2006). Skill Formation and the Economics of Investing in Disadvantaged Children. *Science, 312*, 1900–1902.

Howe, G. W., Reiss, D., & Yuh, J. (2002). Can Prevention Trials Test Theories of Etiology? *Development and Psychopathology, 14*, 673–694.

Institute of Medicine. (1994). Reducing Risks for Mental Disorders: Frontiers for Preventive Intervention Research. In P. J. Mrazek & R. J. Haggerty (Eds.), *Committee on Prevention of Mental Disorders, Division of Biobehavorial Sciences and Mental Disorders*. Washington, DC: National Academies Press.

Jedd, K., Hunt, R. H., Cicchetti, D., Hunt, E., Rogosch, F., Toth, S., & Thomas, K. M. (2015). Long-term Consequences of Childhood Maltreatment: Altered Amygdala Functional Connectivity. *Development and Psychopathology, 27*(4pt2), 1577–1589.

Kempermann, G. (2006). *Adult Neurogenesis: Stem Cells and Neuronal Development in the Adult Brain*. New York: Oxford University Press.

Kemperman, G., van Praag, H., & Gage, F. H. (2000). Activity-Dependent Regulation of Neuronal Plasticity and Self-Repair. *Progress in Brain Research, 127*, 35–48.

Kohlberg, L., LaCrosse, J., & Ricks, D. (1972). The Predictability of Adult Mental Health from Childhood Behavior. In B. B. Wolman (Ed.), *Manual of Child Psychopathology* (pp. 1217–1284). New York: McGraw-Hill.

Kraemer, H. C., Kazdin, A. E., Offord, D. R., Kessler, R. C., Jensen, P. S., & Kupfer, D. J. (1997). Coming to Terms with the Terms of Risk. *Archives of General Psychiatry, 54*, 337–343.

Krueger, R. F., Markon, K. E., Patrick, C. J., & Iacono, W. G. (2005). Externalizing Psychopathology in Adulthood: A Dimensional-Spectrum Conceptualization and Its Implications for DSM-5. *Journal of Abnormal Psychology, 114*, 537–550.

Leve, L. D., & Cicchetti, D. (2016). Longitudinal Transactional Models of Development and Psychopathology. *Development and Psychopathology, 28*(3), 621–622.

Luthar, S. S., & Cicchetti, D. (2000). The Construct of Resilience: Implications for Intervention and Social Policy. *Development and Psychopathology, 12*, 857–885.

Luthar, S. S., Cicchetti, D., & Becker, B. (2000). The Construct of Resilience: A Critical Evaluation and Guidelines for Future Work. *Child Development, 71*, 543–562.

MacKinnon, D. P. (2008). *Introduction to Statistical Mediation Analysis*. Mahwah, NJ: Lawrence Erlbaum.

Masten, A.S. (2006). Developmental Psychopathology: Pathways to the Future. *International Journal of Behavior Disorders, 30*, 47–54.

Masten, A. S. (2014). *Ordinary Magic: Resilience in Development.* New York: Guilford.

Masten, A. S., Burt, K. B., & Coatsworth, J. D. (2006). Competence and Psychopathology in Development. In D. Cicchetti & D. J. Cohen (Eds.), *Developmental Psychopathology* (Vol. 3, 2nd edn., pp. 696–738). Hoboken, NJ: Wiley

Masten, A. S., & Cicchetti, D. (2010). Developmental Cascades. Developmental Cascades [Special Issue, Part 1], *Development and Psychopathology, 22*(3), 491–495.

Masten, A. S. & Cicchetti, D. (2016). Resilience in Development: Progress and Transformation. In D. Cicchetti (Ed.), *Developmental Psychopathology* (Vol. 4, *Risk, Resilience, and Intervention*, 3rd edn., pp. 271–333). New York: Wiley.

Masten, A. S., & Coatsworth, J.D. (1998). The Development of Competence in Favorable and Unfavorable Environments: Lessons from Research on Successful Children. *American Psychologist, 53*, 205–220.

Masten, A. S., Long, J. D., Kuo, S. I.-C., McCormick, C. M., & Desjardins, C. D. (2009). Developmental Models of Strategic Intervention. *European Journal of Developmental Science, 3*, 282–291.

Masten, A. S., & Wright, M. O. (2009). Resilience over the Lifespan: Developmental Perspectives on Resistance, Recovery, and Transformation. In J. W. Reich, A. J. Zautra, & J. S. Hall (Eds.), *Handbook of Adult Resilience* (pp. 213–237). New York: Guilford Press

McCrory, E., De Brito, S. A., & Viding, E. (2010). Research Review: The Neurobiology and Genetics of Maltreatment and Adversity. *Journal of Child Psychology and Psychiatry, 51*, 1079–1095.

Meaney, M. J. (2010). Epigenetics and the Biological Definition of Gene x Environment Interactions. *Child Development, 81*(1), 41–79.

Moffitt, T. E. (2005). The New Look of Behavioral Genetics in Developmental Psychopathology: Gene-Environment Interplay in Antisocial Behaviors. *Psychological Bulletin, 131*(4), 533–554.

Paus, T., Keshavan, M., Giedd, J. N. (2008). Why Do Many Psychiatric Disorders Emerge during Adolescence? *National Review of Neuroscience, 9*(12), 947–957.

Preacher, K. J., & Hayes, A. F. (2008). Asymptotic and Resampling Strategies for Assessing and Comparing Indirect Effects in Multiple Mediator Models. *Behavior Research Methods, 40*(3), 879–891.

Rakic, P. (1988). Specification of Cerebral Cortical Areas. *Science, 241*, 170–176.

Reynolds, A. J., & Temple, J. A. (2006). Economic Benefits of Investments in Preschool Education. In E. Zigler, W. Gilliam, & S. Jones (Eds.), *A Vision for Universal Prekindergarten* (pp. 37–68). New York: Cambridge University Press.

Roisman, G. I., Masten, A. S., Coatsworth, J. D., & Tellegen, A. (2004). Salient and Emerging Developmental Tasks in the Transition to Adulthood. *Child Development, 75*, 1–11.

Rutter, M. (1986). Child Psychiatry: The Interface between Clinical and Developmental Research. *Psychological Medicine, 16*, 151–160.

Rutter, M. (1988). Epidemiological Approaches to Developmental Psychopathology. *Archives of General Psychiatry, 45*(5), 486–495.

Rutter, M., & Garmezy, N. (1983). Developmental Psychopathology. In E. M. Hetherington (Ed.), *Mussen's Handbook of Child Psychology: Vol. 4 Socialization, Personality, and Social Development* (4th edn., pp. 775–911). New York: Wiley.

Sameroff, A. J. (2000). Developmental Systems and Psychopathology. *Development and Psychopathology, 12*, 297–312.

Sameroff, A. J., & Chandler, M. J. (1975). Reproductive Risk and the Continuum of Caretaking Causality. In F. D. Horowitz (Ed.), *Review of Child Development Research* (Vol. 4, pp. 187–244). Chicago: University of Chicago Press.

Scarr, S., & McCarney, K. (1983). How People Make Their Own Environments: A Theory of Genotype-Environment Effects. *Child Development, 54*, 424–435.

Shonkoff, J. P., Boyce, W. T., & McEwen, B. S. (2009). Neuroscience, Molecular Biology, and the Childhood Roots of Health Disparities: Building a New Framework for Health Promotion and Disease Prevention. *Journal of the American Medical Association, 301*, 2252–2259.

Sroufe, L. A. (1979). The Coherence of Individual Development: Early Care, Attachment, and Subsequent Developmental Issues. *American Psychologist, 34*, 834–841.

Sroufe, L. A. (1986). Bowlby's Contribution to Psychoanalytic Theory and Developmental Psychopathology. *Journal of Child Psychology and Psychiatry, 27*, 841–849.

Sroufe, L. A. (1989). Pathways to Adaptation and Maladaptation: Psychopathology as Developmental Deviation. In D. Cicchetti (Ed.), *Rochester Symposia on Developmental Psychopathology* (Vol. 1, pp. 13–40). Hillsdale, NJ: Erlbaum.

Sroufe, L. A. (1990). An Organizational Perspective on the Self. In D. Cicchetti & M. Beeghly (Eds.), *The Self in Transition: Infancy to Childhood* (pp. 281–307). Chicago, IL: University of Chicago Press.

Sroufe, L. A. (2007). The Place of Development in Developmental Psychopathology. In A. Masten (Ed.), *Multilevel Dynamics in Developmental Psychopathology: Pathways to the Future: The Minnesota Symposia on Child Psychology* (Vol. 34, pp. 285–299). Mahwah, NJ: Erlbaum.

Sroufe, L. A., (2013). The Promise of Developmental Psychopathology: Past and Present. *Development and Psychopathology, 25*, 1215–1224.

Sroufe, L. A., Egeland, B., & Kreutzer, T. (1990). The Fate of Early Experience Following Developmental Change: Longitudinal Approaches to Individual Adaptation in Childhood. *Child Development, 61*, 1363–1373.

Sroufe, L. A., & Jacobvitz, D. (1989). Diverging Pathways, Developmental Transformations, Multiple Etiologies and the Problem of Continuity in Development. *Human Development, 32*(3–4), 196–203.

Sroufe, L. A., & Rutter, M. (1984). The Domain of Developmental Psychopathology. *Child Development, 55*, 17–29.

Thelen, E., & Smith, L. B. (1998). Dynamic Systems Theories. In W. Damon & R. Lerner (Eds.), *Theoretical Models of Human Development: Vol. 1 Handbook of Child Psychology* (pp. 563–634). New York: Wiley.

Thomas, K., & Cicchetti, D. (Eds.) (2008). Imaging Brain Systems in Normality and Psychopathology [Special Issue]. *Development and Psychopathology, 20*(4), 1023–1349.

Toth, S. L., Gravener-Davis, J. A., Guild, D. J., & Cicchetti, D. (2013). Relational Interventions for Child Maltreatment: Past, Present, & Future Perspectives. *Development and Psychopathology, 25*(4pt2), 1601–1617.

Von Bertalanffy, L. (1968). *General System Theory: Foundations, Development, Applications.* New York: Braziller.

Waddington, C. H. (1957). *The Strategy of Genes.* London: Allen & Unwin.

Waddington, C. H. (1966). *Principles of Development and Differentiation*. New York: Macmillan.

Waters, E., & Sroufe, L. A. (1983). Competence as a Developmental Construct. *Developmental Review, 3*, 79–97.

Weiss, P. A. (1969). *Principles of Development: A Text in Experimental Embryology*. New York: Hafner Publishing.

Weisz, J. R. & Kazdin, A. E. (Eds.) (2017). *Evidence-Based Psychotherapies for Children and Adolescents*. New York: Guilford Press.

Werner, H. (1948). *Comparative Psychology of Mental Development*. New York: International Universities Press.

Zigler, E., & Glick, M. (1986). *A Developmental Approach to Adult Psychopathology*. New York: Wiley.

16 Adolescence and Puberty

Understanding the Emergence of Psychopathology

SARAH D. LYNNE, ALLISON S. METZ, AND JULIA A. GRABER

Adolescence is an important period in human development for the onset of behavioral and mental health disorders. In fact, the development of psychopathology has regularly been identified as one of the most studied areas of adolescent development (Graber & Sontag, 2009; Steinberg & Morris, 2001). For instance, rates of depressive disorders are relatively low during childhood (1–2 percent) and do not differ by gender (Egger & Angold, 2006). However, prevalence of depression increases during adolescence (see Table 16.1), with more marked increases among females (Center for Behavioral Health Statistics and Quality, 2017). This example, along with changes in prevalence of other disorders and substance use, have been the impetus for a plethora of research on developmental processes that address why rates change dramatically during adolescence and why they change differentially for specific subgroups of individuals (Caspi et al., 2014; Harris et al., 2006; Hottes et al., 2016; King et al., 2008). Moreover, changes in rates of disorders during adolescence are not uniform in age of onset or etiology. Some behaviors seem to emerge with age or social context, whereas others are linked with maturational processes such as puberty (Biro & Deardorff, 2013; Charalampopoulos et al., 2014; Ge & Natsuaki, 2009; Graber, 2013; Mendle & Ferrero, 2012; Negriff & Susman, 2011; Susman & Dorn, 2009; Ullsperger & Nikolas, 2017).

Understanding why psychopathology increases in this developmental period speaks to the timing and targets of preventive efforts. This chapter takes an ecological approach to evaluating the physiological, social, and contextual factors related to mental and behavioral health during adolescence. We begin by reviewing the physiological changes of puberty, highlighting differences by gender and race. We then review evidence of links between pubertal processes and mental health along with some of the more prominent theoretical frameworks for these associations. Next follows a review of the prevalence of mental and behavioral health disorders during adolescence with a focus on disparities within subgroups of the population. Finally, we close with a review of some of the methodological advances in the field of developmental psychopathology.

PHYSIOLOGICAL CHANGES DURING ADOLESCENCE

Puberty is best understood as a stage of development that culminates in reproductive capacity, rather than as a single life event. The onset of puberty is initially marked by changes in hormone levels, followed by physical changes such as growth spurt and changes in secondary sex characteristics, like the development of breasts, genitalia, and body hair (Byrne et al., 2017; Susman et al., 2010). The external physical signs of pubertal growth signal the early stages of the maturation process to outside observers.

There is high variability in both age of onset (i.e., pubertal timing) and duration of the pubertal transition (i.e., pubertal tempo). In general, the normative age of onset in the United States ranges from 8.5 to 13 years old for girls, while onset for boys' puberty begins about 1.5 years later on average, at 9.5–14 years old (Sun et al., 2002; Tanner & Whitehouse, 1976). Trained physician reports of pubertal status markers support the sex variation in onset, as well as variation by race/ethnicity (Herman-Giddens et al., 1997, 2012; Sun et al., 2002). Nationally representative estimates of the median age at the onset of pubic hair development are 9.4 years for non-Hispanic Black girls, 10.4 years for Mexican-American girls, and 10.6 years for non-Hispanic White girls. Median age at the onset of pubic hair development ranges from 11.2 years for non-Hispanic Black boys, 12.3 years for Mexican-American boys, and 12.0 years for non-Hispanic White boys. Overall, non-Hispanic Black girls and boys start puberty earlier than non-Hispanic White and Mexican-American children (Sun et al., 2002).

Menarche, the start of menstruation, is a significant milestone in female development, marking the beginning of reproductive capacity; however, this milestone occurs relatively late in the developmental process (Byrne et al., 2017; Susman et al., 2010). Duration, or tempo, of puberty

Table 16.1 Adolescent prevalence of major depressive episodes and substance use

Demographic characteristic	Past year (2016)				Past month
	MDE	Tobacco	Marijuana	Alcohol	Binge alcohol use
Total ages 12–17	12.8	10.5	12	21.6	4.9
Age group					
12–13	7.3	2.4	1.7	5	0.3
14–15	13.3	9.2	11.1	19.9	3.7
16–17	17.2	19.1	22.4	38.8	10.2
Gender					
Male	6.4	12.2	11.7	19.2	4.4
Female	19.4	8.7	12.3	24.1	5.4
Hispanic origin and race					
Not Hispanic or Latino	12.8	11.1	12.1	21.7	4.9
White	13.8	12.7	12.3	24	5.5
Black or African American	9.1	6.7	12.2	15.6	2.8
American Indian or Alaska Native	11.5	15.2	16.5	17	5.6
Asian	11.9	4.3	6.1	13.2	2.8
Two or more races	13.8	13.6	18	25.2	7.7
Hispanic or Latino	12.7	8.6	11.7	21.3	4.8

Notes: Major Depressive Episode (MDE) is defined as in the fourth edition of the *Diagnostic and Statistical Manual of Mental Disorders* (DSM-IV), which specifies a period of at least two weeks when a person experienced a depressed mood or loss of interest or pleasure in daily activities and had a majority of specified depression symptoms. Respondents with unknown past year MDE data were excluded. Binge Alcohol Use is defined as drinking five or more drinks (for males) or four or more drinks (for females) on the same occasion (i.e., at the same time or within a couple of hours of each other) on at least 1 day in the past 30 days.
Source: Center for Behavioral Health Statistics and Quality (2017)

refers to the rate of physical maturation from onset of puberty to full physical maturation. The typical progression through the pubertal stages spans approximately 3.5–6 years (Herman-Giddens et al., 2012; Mendle, 2014; Sun et al., 2002; Susman et al., 2010). However, pubertal tempo can vary dramatically at the individual level, with extremes ranging from one year to seven or more years within the range of normative development (Tanner & Whitehouse, 1976). It is also important to note that measurement of the duration of puberty is dependent on how full physical maturation is defined. For example, in some studies pubertal duration is defined for girls as the period from the age of onset of physical changes of puberty to the advent of menarche, resulting in duration averages of 1.96 ($SE = 0.06$) years (Marti-Henneberg & Vizmanos, 1997). As opposed to defining full maturation as the advent of menarche, full maturation may instead be defined as reaching adult-like physical development of the body (i.e., Tanner stage 5), which typically occurs about 1–2 years after menarche (Lee, Guo, & Kulin, 2001).

Recent developmental research has noted a long-term trend toward earlier onset of puberty in both males and females (Chavarro et al., 2017; Lee & Styne, 2013). The average age of menarche decreased from 17 to under 14 years in the US and parts of Western Europe between the mid-nineteenth and mid-twentieth centuries, with current median age of 12.5 years in the US (Chumlea et al., 2003; Lee et al., 2001; Parent et al., 2003).

As noted, however, the observable, physical changes of puberty in humans are preceded by hormonal changes, specifically in the adrenal androgen dehydroepiandrosterone (DHEA) and its sulfate (DHEA-S), as well as gonadal hormones such as testosterone and estradiol (the primary estrogen hormone measured in nonpregnant women). In fact, there are two distinct phases of pubertal development: adrenarche and gonadarche. Adrenarche, triggered by the maturation of the adrenal gland, begins well before the physical signs of pubertal development show, starting around 5–7 years of age. Gonadarche, which is associated with the maturation of the hypothalamic-pituitary-gonadal axis, triggers the release of gonadotropin-releasing hormone (GnRH) at around 10–11 years. In turn, GnRH triggers a rise in sex hormones, testosterone and estradiol, and thereby leads to the maturation of the primary and secondary sex characteristics and menarche. Gonadarche, then, is the phase of pubertal development that research on pubertal status generally refers to, as it relates to the observable, physical changes of pubertal development. In other species, changes in sexual bonding hormones oxytocin and vasopressin also undergo change during adolescence (Peper & Dahl, 2013); however, translational studies in humans of pubertal changes in oxytocin or vasopressin and links to subsequent behavioral health are limited (for relevant review see Sannino, Chini, & Grinevich, 2016).

Theories have been developed to understand potential hormone effects on brain development. Schulz, Molenda-Figueira, and Sisk (2009) provided evidence predominantly from animal models that gonadal steroid hormones have organizing effects on the brain, permanently changing its structure (e.g., neuronal number, myelination, dendritic branching), not just prenatally but also during puberty. For example, Ladouceur and colleagues (2012) review evidence of associations between hormonal changes during puberty and white matter volume in regions of the brain linked to risk for affective disorders. The research in this area is limited but findings highlight the need for future longitudinal research to determine if puberty-related changes in white matter represent a neurodevelopmental risk for affective disorders (Ladouceur et al., 2012). In addition, Sisk and Zehr (2005) hypothesize that there is a sensitive period for organizing effects of gonadal hormones on brain development, with sensitivity decreasing over the course of postnatal development, meaning brain plasticity decreases with age. In this model, individuals who experience the onset of puberty at younger ages have commensurate increases of gonadal steroids when their brain is still relatively more sensitive, or plastic. Thus, pubertal timing may have implications for the influence of hormones on brain development. Research by Lynne and colleagues has begun extending these models to humans and are finding initial evidence in support of the organizational hypothesis (Beltz et al., 2017).

Given both between-person and cohort variability in pubertal onset, it is currently standard procedure to examine for signs of these early physical changes during child health visits to the doctor. There is not a definitive age cutoff for determining sexual precocity; however, signs of pubertal development prior to age eight may warrant specialty consultation with a pediatric endocrinologist (Parent et al., 2003). Clinically diagnosed idiopathic central precocious puberty (CPP) is the appearance of secondary sexual characteristics prior to normative age of onset, in the absence of central nervous system (CNS) disorders or other disease (Styne & Grumbach, 2011). Prevalence of CPP is estimated to be 20–23 per 10,000 girls and <5 per 10,000 boys (Teilmann et al., 2005). Currently, pediatric endocrinologists treat only the most extreme cases of CPP to delay puberty with consideration given to the preservation of adult height potential and age-peer matching of psychosocial maturity.

In studies of normative pubertal development, variation in the age of onset of puberty is typically reported in terms of means/medians and standard deviations under the assumption of a normal distribution. However, the shape of the onset distribution varies by context given that extrinsic factors can be crucial for pubertal development. Variation in socioeconomic status, nutritional conditions, energy expenditure, physical activity, or stress can shift timing of puberty (Parent et al., 2003). For instance, asymmetrical distributions with increased variability toward late age of pubertal onset may be observed in underprivileged settings such as developing countries (Parent et al., 2003). Importantly, variability in these extrinsic factors within countries can also contribute to variability in pubertal timing. As such, it is critical to understand the role of extrinsic/contextual factors as they relate to puberty and future outcomes.

In studies of variability in pubertal timing in the general population, there are a variety of ways to measure pubertal development as well as a variety of ways to define early, on-time, and late maturation. Choices related to measurement should be based upon the feasibility and aims of the study, as different measurement techniques (e.g., physician assessment, self-report, parent report) result in different types of data and subsequently different methods for defining pubertal timing. For continuous measures of pubertal development, studies frequently regress pubertal status on chronological age to obtain sex-specific residuals, with negative residuals indicating later timing and positive residuals indicating earlier timing on a continuous scale (Ullsperger & Nikolas, 2017). Continuous definitions are useful when examining variation in timing across the normative range with a focus on earlier or later development in comparison with others in the same sample but without establishing that any specific child would be defined as early or late in development by pediatric standards. In addition, youth have been grouped categorically based on assessments of their pubertal status as early maturers (20 percent most developed), on-time maturers (60 percent), and late maturers (20 percent least developed), or by categorizing youth based on standard deviations above (early) or below (late) the mean (Graber, Petersen, & Brooks-Gunn, 1996; Lanza & Collins, 2002). Categorical definitions are useful when drawing comparisons to clinical subpopulations. The large disparity in the number of children considered early maturers in psychosocial studies of the general population (~20 percent) compared with those diagnosed with precocious puberty is made even more poignant when one considers the abundance of literature highlighting negative behavioral and psychological adjustment outcomes associated with early pubertal timing, particularly for girls.

Puberty and Psychopathology

Normative developmental variability in the timing and tempo of puberty inform an understanding of the biopsychosocial pathways to mental and behavioral health in adolescence, with timing of development especially salient. In girls, early pubertal timing is associated with higher rates of externalizing problems (e.g., substance use, delinquency) and internalizing problems (e.g., anxiety and depression; Ge & Natsuaki, 2009; Graber, 2013; Negriff & Susman, 2011). For boys, the evidence is mixed. For example, initial work suggested that late pubertal timing was associated with more mental health problems for boys, and that early development may have even been protective for boys (e.g., Mussen & Jones, 1957). More recent research, however, found early pubertal timing associated with internalizing symptoms for both boys and girls (e.g., Mendle & Ferrero, 2012; Susman & Dorn, 2009). Hence, early maturation is consistently and robustly associated with higher internalizing symptoms for girls while results are more inconsistent for boys (Dorn, Susman, & Ponirakis, 2003; Pomerantz et al., 2017).

A recent meta-analysis that included both clinical and community samples found that overall off-time maturation, especially early pubertal timing, is associated with higher levels of psychopathology (Ullsperger & Nikolas, 2017). Importantly, method of measurement (e.g., age of menarche; age of spermarche/oigarche; parent report, adolescent report, or physician report on pubertal status; parent or adolescent report on perceived pubertal timing) moderated this association while sex did not (Ullsperger & Nikolas, 2017). Larger effect sizes were observed when pubertal timing was assessed via Tanner stages or age at menarche and smaller but still significant effect sizes were observed for measures of adolescent or parent reports of perceived pubertal timing relative to the adolescents' peers.

However, it is important to note that the Ullsperger and Nikolas (2017) meta-analysis did not formally evaluate symptom severity as a moderator of pubertal timing effects. Sample composition (clinical versus community) was used as a proxy of symptom severity but more research is needed on pubertal timing and mental health diagnoses. In the few studies that have examined mental health diagnoses, consistent gender differences have been found, with early maturing girls having higher rates of disorder (see Graber, 2013, for a brief review). When examining variation in symptoms, gender differences are typically not observed. Overall, pubertal timing is consistently associated with psychopathology, and early timing in particular may be especially detrimental for girls compared with boys when examining pathology in the clinical range (Graber, 2013). Adrenarche, although less studied, appears to be associated with mental health symptoms and diagnoses as well. Specifically, both early timing of adrenarche and higher levels of adrenal hormones were associated with more mental health problems for girls. Patterns for boys seem to be similar, however less research exists examining boys' adrenal development (Byrne et al., 2017).

While associations between puberty and neurological, social, and behavioral outcomes may be small to moderate in their individual effects during adolescence, these associations may lead to trajectories of behavior that can have large effects over time. For example, robust evidence exists that early pubertal timing among girls is associated with modifiable health risk behaviors in adolescence that could increase the risk of mortality (e.g., sexual risk taking, tobacco, alcohol, and other substance use and abuse) as well as important public health and social problems (e.g., obesity, externalizing problems, early sexual debut, early pregnancy, internalizing symptoms and disorders including anxiety and depression, and lower academic achievement; Biro & Deardorff, 2013; Freedman et al., 2003; Ge & Natsuaki, 2009; Graber, 2013; Negriff & Susman, 2011; Walvoord, 2010). Yet early pubertal timing has been largely overlooked as an individual-level susceptibility factor.

Moreover, research on the timing of pubertal onset has revealed important associations with future health outcomes among women across the lifespan. A recent meta-analysis found lower mortality at higher ages of menarche, with a 3 percent decrease in all-cause mortality associated with each one-year increase in age at menarche (Charalampopoulos et al., 2014). This meta-analysis included studies from across the globe with cohorts ranging from 1959 to 2009, emphasizing the robustness of the inverse association between age at menarche and death from all causes. Possible mediating mechanisms of this robust connection include the well-established links between early pubertal timing and increased rates of modifiable risk behaviors as well as mental and behavioral health disorders. However, given challenges in conducting longitudinal studies that span decades, to date there has not been an empirical evaluation of mediating mechanisms between pubertal timing and mortality.

Understanding increases in mental and behavioral health disorders and the emergence of disparities in disorders during adolescence requires a multidisciplinary approach. Research on normative neurodevelopmental growth during adolescence has resulted in a number of cognitive theories aimed at explaining increased risk behaviors and mental/behavioral health disorders during adolescence (Crone & Dahl, 2012; Sisk & Zehr, 2005; Steinberg, 2010; Suleiman & Dahl, 2017). This area of inquiry has substantial potential for understanding underlying physiological mechanisms related to mental and behavioral health; however, it is important to emphasize that evidence from studies of neurodevelopment in adolescence is complex and there are significant gaps in our current understanding of these processes (Pfeifer & Allen, 2012).

The dual-systems model and the organizational hypothesis are two examples of theories that provide a structure for research on physiological mechanisms underlying risk behavior and the development of psychopathology in adolescence. The "dual systems" hypothesis has been widely disseminated and influential across disciplines with a variety of theories positing distinctions between two systems of thought (Epstein, 1994; Evans & Stanovich, 2013). Within the field of developmental neuroscience, the dual systems hypothesis provides an intuitive explanation for adolescent risk behavior. It posits that risk taking occurs due to the maturational mismatch between subcortical reward-seeking brain regions that develop early in concert with puberty, and the frontal cortical cognitive regions that develop in concert with age throughout adolescence and into adulthood (Steinberg, 2010). Empirical support for the hypothesis, however, has been mixed. In fact, a meta-analysis suggests that when neuroimaging studies have been conducted, results have not supported a simple model of a maturational mismatch (Crone & Dahl, 2012). There is some evidence of counterintuitive effects such that mature subcortical regions may lead to adaptive responses and mature frontal cognitive regions could confer risk (Pfeifer & Allen, 2012), particularly in the area of mental health. It is possible that variability in the findings could reflect a less automatic, more flexible cognitive control system in adolescence compared with childhood and adulthood. Adolescents may be more strongly influenced by the motivational salience of the context (e.g. presence of peers, affective appraisal or salience of performing the task, task instructions, and cognitive strategies) compared with other age groups (Crone & Dahl, 2012; Suleiman & Dahl, 2017). In addition, the changing context and greater potential opportunities to engage in risk behaviors in later adolescence may also explain higher rates of some risk behaviors in later adolescence and early adulthood. This possibility is not in conflict with recent reappraisals of the dual systems model which emphasizes that while the propensity for risk-taking behaviors may be highest in middle adolescence due to the maturational mismatch between the reward sensitivity and cognitive control neural processes, higher rates of some risk behaviors may still occur in later adolescence in response to context and opportunity (Shulman et al., 2016).

Understanding links between pubertal onset and neurodevelopment along with subsequent implications for mental and behavioral health is a critical area for continued research. Adolescence may mark a unique developmental sensitive period for learning and health. Suleiman and Dahl (2017) outline key domains for continued research in this area. For instance, more research is needed to understand how puberty affects neurodevelopment and specialized learning, particularly social and emotional responsivity. Advances in this domain require multidisciplinary collaborations between developmental experts, educators, clinicians, neuroscientists, and others.

Research from animal models has provided some of the first evidence dissociating puberty and age to understand the role that early gonadal hormone exposure may have on the developing brain. As indicated, Schulz and colleagues (2009) provided evidence predominantly from animal models that gonadal steroid hormones have organizing effects on the brain, not just in early development but also during puberty. Such organizing effects resulted in permanent changes to brain structure and behavior. Sisk and Zehr (2005) outlined the organizational hypothesis that there is a sensitive period for organizing effects of gonadal hormones on brain development, with sensitivity decreasing over the course of postnatal development. In this model, individuals who have earlier onset of puberty have commensurate increases of gonadal steroids when brain development is still relatively more sensitive to the organizing effects of these hormones; thus, timing of exposure may have permanent influences on brain structure and function. Translating this model to humans, pubertal timing should be inversely related to brain sensitivity to sex hormone surges associated with puberty. As such, it is plausible to hypothesize that individuals who experience earlier puberty may also demonstrate more sex-typed behaviors than those that mature later. A behavioral study provides partial support for the organizational hypothesis in human men where earlier maturation was associated with higher scores on a test of three-dimensional mental rotations, a cognitive ability that has shown large sex differences in prior research favoring males (Beltz & Berenbaum, 2013). While these findings are encouraging in terms of the translatability of animal research to humans, more work needs to be done to fully evaluate these complementary theories.

Measurement of Pubertal Timing
Despite variation in assessment, pubertal timing effects on mental and behavioral health outcomes are generally consistent (Ullsperger & Nikolas, 2017), especially regarding early maturation (Graber, 2013). This consistency of effects across forms of measurement speaks to the robustness of the role that puberty plays for the health and adjustment of youth during adolescence and into adulthood. It is also noteworthy that many studies aimed at understanding health and adjustment in adolescence do not include measures of puberty. Even if the evaluation of puberty is not a primary aim of a given research project, the inclusion of at least a brief measure of this construct in all studies of the development of mental and behavioral health during adolescence is recommended. The following section briefly reviews different measurement approaches to the assessment of puberty as well as recommendations for the field on when specific measures are most appropriate given research aims and constraints.

Measurement of pubertal timing requires two pieces of information. First, an assessment of pubertal status, or the level of physical development, is obtained. Then this information is placed in context, evaluating a child's status relative to their same-age, same-sex peers to provide information about an individual's pubertal development relative to others (e.g., early, on-time, or late). Historically, timing studies have often focused on girls and menarche. This is no longer true, however. Today, there are many studies with both female and male participants (Ullsperger & Nikolas, 2017).

For studies where understanding the role of puberty is a primary aim of the research, clinical relevance is a goal, or other physiological measures are being assessed (e.g., cortisol, brain development) researchers should consider assessing puberty using clinical-based methods. In clinical practice, physicians evaluate pubertal status by examining youth for physical signs of secondary sexual development and classifying the level of development using Tanner staging (Lee et al., 2001; Parent et al., 2003; Ullsperger & Nikolas, 2017). Information on Tanner stage and age can provide researchers with enough information to evaluate pubertal timing. In addition, changes in pubertal hormones such as DHEA, growth hormone (GH), luteinizing hormone (LH), follicle-stimulating hormone (FSH), gonadotropin releasing hormone, leptin, and sex hormones like estradiol and testosterone, can be used as measures of pubertal onset and can only be obtained through a blood draw in a medical setting. These forms of assessment are particularly important in studies of pubertal hormone influences on brain development. Medical record review of Tanner staging or menarche is less directly invasive but these data have other challenges including high rates of missingness and inconsistency between physician assessments. Clinical assessments of pubertal status and timing are challenging due to the sensitive nature of the information and invasiveness of some of the assessment methods (e.g., blood draws, physician assessment of physical development). In addition to being invasive, assessment of pubertal status by medical professionals can be costly and infeasible with larger samples of youth, which is when researchers should consider either medical record review or alternative forms of assessment.

Most research on pubertal timing has been conducted outside of medical settings. The most commonly used assessment tool to measure status is Petersen's Pubertal Development Scale (PDS; Petersen et al., 1988), which is a five-item parent/child report scale assessing perceived level of different indicators of physical development. These items are combined using a standardized protocol to create continuous measures of pubertal status as well as both categorical and continuous measures of pubertal timing compared to sample age norms. Menarche is assessed in the PDS and research has found that age at menarche can be reliably self-reported retrospectively (Natsuaki, Leve, & Mendle, 2011). While not included in the PDS, self-report of age at first ejaculation for males is sometimes used as a comparable pubertal milestone to menarche in females, but there is less information on the reliability of reporting this event compared with menarche (Dorn & Biro, 2011). The PDS is the least intrusive or explicit measure and is very short; hence its popularity for larger behavioral surveys or studies in which puberty is included but not a central focus.

Finally, for studies in which puberty may not be a primary aim, a direct assessment of perceived pubertal timing can be obtained from a valid single item where either the child or parent reports if they perceive the child's physical development to be earlier, about the same time as, or later than, same-sex, same-age peers (Petersen et al., 1988). This single-item perceptual measure of pubertal timing is correlated with more objective clinical measures but still represents a distinct aspect of the construct. Perceptions of pubertal timing more closely match objective measures when the objective measures are obtained during the pubertal transition and the perceptual measure is obtained after the pubertal transition has concluded (Dubas, Graber, & Petersen, 1991). Results of studies that use perceived pubertal timing as predictors of future psychopathology are consistent with other objective measures of pubertal timing, despite slightly smaller effect sizes (Ullsperger & Nikolas, 2017). As such, this single-item perceptual measure of timing is recommended but may be most accurate if used with older adolescents.

In addition to pubertal status and timing, there is also variability in pubertal tempo, or the time it takes for a child to progress fully through the pubertal transition. Data on pubertal tempo may be important for understanding associations between puberty and the development of behavioral and mental health during adolescence. Information on pubertal tempo is typically obtained via multiple assessments of pubertal status throughout the duration of the pubertal transition (Shirtcliff, Dahl, & Pollak, 2009). Tempo is most critical to assess in longitudinal studies where puberty is a primary aim of the research and can be estimated by either clinical or self/parent reports of pubertal status over time.

Recent technological advances hold promise not only for the measurement of puberty but in understanding the development of mental and behavioral health during adolescence. Technological advances in data collection, including web-based survey administration and use of smart devices, have the potential to allow measurement of pubertal status and timing at lower cost and with greater ease compared with medical assessments, while at the same time allowing for more objective measurement compared with traditional parent- and self-reported assessments. While challenges remain, the use of electronic medical records may have potential for large-scale surveys of links between puberty and adolescent mental, behavioral, and physical health. In addition, smartphones and other health-monitoring devices open opportunities for ecological momentary assessments (repeated and frequent data collection in real time), which maximize ecological validity and minimize recall bias. We will discuss specific examples of how these methods may enhance research by providing access to hard-to-reach populations (e.g., rural populations, underrepresented minorities) or more enriched assessments of developmental processes, symptoms, affect, and behaviors.

SOCIAL AND ENVIRONMENTAL INFLUENCES ON BEHAVIOR

In addition to neurodevelopmental pathways linking puberty to mental and behavioral health outcomes, there is evidence that social determinants of health play an important role in poor adjustment and future mortality and morbidity. Theories such as the stage termination model of pubertal timing effects or the deviance hypothesis outline ways in which social and contextual factors (e.g., peers) may explain links between pubertal timing and future psychopathology (Brooks-Gunn, Petersen, & Eichorn, 1985). Given that pubertal timing is itself determined by one's physical development (e.g., pubertal status) relative to a reference group, context and norms are important to understanding implications for adjustment. Specifically, the stage termination model hypothesizes that early maturing youth, particularly girls, experience more negative adjustment because they are perceived as older based on their physical appearance and are thus exposed to more adultlike situations at an earlier age through association with older/deviant peers (Brooks-Gunn et al., 1985). The deviance hypothesis predicts more negative adjustment for youth who deviate from their peers in timing of pubertal maturation, either early or late. Theories of biological sensitivity and differential susceptibility to the environment posit that individuals vary in neurobiological susceptibility to environmental influences (Ellis et al., 2011). These theories explicate the importance of context, both negative/stressful environments and positive/enriched environments, for adaptive outcomes in youth specifically highlighting interactions between these contexts and neurobiological susceptibility. For instance, early childhood exposure to stressful or enriching environments may influence somatic development in adolescence as well as future health (Ellis et al., 2011). Evaluations of these theories have found evidence supporting the role of social and contextual factors as moderators or mediators of pubertal timing effects on mental and behavioral health. Stressful life events in early childhood, inconsistent harsh parenting, and neighborhood disadvantage are associated, with both experiencing early puberty and subsequent negative adjustment outcomes (Lynne-Landsman, Graber, & Andrews, 2010; Negriff & Susman, 2011). Association with older, deviant peers is also a well-substantiated pathway to negative adjustment outcomes among early maturing adolescents (e.g., Lynne et al., 2007).

MENTAL AND BEHAVIORAL HEALTH IN ADOLESCENCE

A challenge for preventing mental and behavioral health disorders during adolescence is that multiple pathways/risk factors can lead to the same disorder (i.e., equifinality) and the same pathways/risk factors can lead to different disorders (i.e., multifinality). As such, there is a need to find appropriate methods for validly and reliably measuring both interindividual and intraindividual variability in factors related to mental and behavioral health. Technological advances in data collection have made frequent, within-person assessments of mood, affect, and other behaviors available and advances in statistical methodology are providing more sophisticated ways of evaluating these data (see Chapter 23, by Ferguson and colleagues). Yet obstacles remain regarding recruitment, study design, and measurement, particularly for sensitive topics like mental health disorders, substance use, deliberate self-harm, and suicide.

Prevalence across Adolescence

The Substance Use and Mental Health Services Administration (SAMHSA) conducts the National Survey on Drug Use and Health (NSDUH) annually, providing prevalence estimates for major depressive disorder and substance use by demographic characteristics such as age, gender, and race/ethnicity (see Table 16.1). Overall, the percentage of youth reporting a major depressive episode (MDE) or substance use over the past year increases across adolescence. In 2016, 19 percent of females reported an MDE in the past year compared to 6 percent of males. However, males and females reported roughly comparable levels of tobacco, alcohol, and marijuana use.

In addition to understanding common risk and protective factors related to the onset and progression of mental and behavioral disorders during adolescence, it is important to keep in mind that many of these disorders co-occur within individuals. In fact, comorbidity is particularly common for disorders during adolescence, especially when considering lifetime occurrence rates (Graber & Sontag, 2009; Lewinsohn et al., 1997, 2004). For example, rates of comorbidity of anxiety and depression from childhood, adolescence, and into adulthood are exceptionally high (over 70 percent in Lewinsohn et al., 1997), leading some researchers to question whether the disorders are distinct or can be meaningfully discriminated via empirical methods (Achenbach, 1993; Compas & Oppedisano, 2000). Even for larger classes of disorders such as externalizing versus internalizing disorders, symptoms have high correlations (e.g., >0.60) in studies of adolescence (e.g., Krueger et al., 1998). Caspi and colleagues (2014) provided evidence in support of a general psychopathology factor, the p-factor, which explained the co-occurrence, sequential comorbidity, dimensionality, and persistence of mental disorders from adolescence into midlife better than three separate factors of internalizing, externalizing, and thought disorders. While emerging evidence suggests a common underlying component to psychopathology that emerges over time, there is still a need to understand the developmental processes underlying the interplay among symptoms and behaviors (Caspi et al., 2014). Understanding common risk and protective factors across mental and behavioral health disorders, as well as potential direct effects between mental and behavioral health disorders during adolescence is critical to the development of effective prevention programming.

Disparities by Race/Ethnicity and the Role of Socioeconomic Status

Research on ethnic and racial disparities in mental health during adolescence are complex, with patterns of disparities differing by outcomes. Across a broad range of health indicators including physical and mental health, Asian and non-Hispanic White adolescents rank better overall compared with other ethnic/racial minority groups based on data from the National Longitudinal Study of Adolescent Health, while Native American and Black participants ranked lowest (Add Health; Harris et al., 2006). However, Black and Asian participants reported the lowest substance use and the least risk for depression and suicide. Another study specifically evaluated ethnic/racial disparities in mental health service use within both high and low poverty communities (Chow, Jaffee, & Snowden, 2003). Significantly more minority adolescents utilized public mental health services compared to their White counterparts; however, the disparity between White participants and racial/ethnic minorities in mental health service use was greater in low-poverty communities.

More research is needed to understand patterns of ethnic/racial disparities in mental health during adolescence as well as the underlying factors related to ethnic/racial disparities. Underlying mechanisms may include disparities in exposure to stress, socioeconomic inequality, and issues related to access to care and culturally compatible care providers. For example, the minority stress model provides a conceptual framework for understanding the heightened prevalence of disorder in terms of the prejudice and discrimination that create a hostile and stressful social environment for stigmatized minority populations (Meyer, 2003).

Socioeconomic status is one factor that has been evaluated as an explanation for ethnic/racial disparities in adolescent mental health outcomes. Higher levels of income and education, indicators of socioeconomic status, have been linked to better health outcomes overall (Harris et al., 2006; Williams et al., 2008). In addition, socioeconomically advantaged groups may be the most likely to experience benefits associated with advances in healthcare and technology, which may increase disparities between the most and the least socioeconomically advantaged (Williams et al., 2008). Research on ethnic/racial disparities in adolescent mental health outcomes have found that socioeconomic differences play a role but do not fully account for ethnic/racial disparities (Harris et al., 2006). Stress associated with racism can negatively affect health outcomes and it is possible that ethnic minorities with higher education or income could be exposed to higher levels of discrimination due to increased interaction with non-Hispanic White individuals (Braveman, Egerter, & Williams, 2011; Williams & Mohammed, 2009). Given complex patterns of disparities in mental and behavioral health outcomes by race/ethnicity as well as limited work with adolescents, future research should focus on changing patterns, risk factors, and protective factors within ethnic minority groups. Engaging underrepresented communities in this research will enhance the validity of the findings and maximize the effectiveness of prevention programs aimed at reducing disparities and improving health.

DISPARITIES BY SEXUAL MINORITY STATUS

Systematic reviews of the literature have found higher rates of mental health disorders, substance use, deliberate self-harm, and suicide among sexual minorities compared with heterosexual populations (Hottes et al., 2016; King et al., 2008; Meyer, 2003). Experiences of stressors related to stigmatization of sexual minorities, such as victimization, are theoretically linked to disparities in mental and behavioral health for sexual and gender minorities (Meyer, 2003; Rosario et al., 2002; Tebbe & Moradi, 2016). There is some evidence from cross-sectional retrospective studies of young adults that reaching sexual minority milestones (e.g., experiencing first same-gender attraction, age at first same-gender sexual behavior, age of first identifying as a sexual minority) earlier in development (early adolescence) is associated with greater depressive and anxious symptoms among lesbian women and gay men (Katz-Wise et al., 2017). This may be due to earlier exposure to minority stressors while neurocognition and emotion regulation strategies are less mature, similar to theories of early pubertal timing which assert that earlier maturation may expose adolescents to adult-like stresses or situations at younger ages, prior to developing the skills to cope with these stressors (Crone & Dahl, 2012; Steinberg, 2010; Suleiman & Dahl, 2017). Moreover, stress associated with the onset of puberty may confer risk for gender minority or transgender youth. In addition, younger adolescents may have less access to supportive resources, such as genders and sexualities alliance groups in schools, which could potentially serve a protective function for these youth. Additional longitudinal prospective research in this area is needed.

Importantly, research with sexual minority youth presents methodological challenges that have implications for validly estimating prevalence of mental and behavior health problems, identifying risk and protective factors, and developing effective prevention programs for these vulnerable youth. For instance, a meta-analysis found that methods for identifying and recruiting sexual minority adults accounted for 33 percent of the between-study variability in prevalence estimates of suicide attempts (Hottes et al., 2016). This study found that prevalence of suicide attempts was 20 percent among sexual minorities recruited through lesbian, gay, and bisexual (LGB) community venues but only 11 percent among sexual minorities recruited through population surveys and 4 percent among heterosexuals recruited through population surveys. Given variability in developmental trajectories for reaching sexual minority milestones during adolescence and young adulthood (Floyd & Stein, 2002) and potential information bias with underreporting of sexual minority identities, it is important to continue evaluating and improving upon the methods and measures used in research on sexual and gender minority mental and behavioral health.

Understanding the developmental roots of the emergence of disparities in mental and behavioral health outcomes is an important area of further inquiry. Disparities are complex because they are outcome- and group-specific. The interplay of individual, social, and contextual factors that combine to create disparities are different and vary not just between vulnerable populations and outcomes but also change over time in response to changes in society and technology. This requires continued population-wide monitoring of demographic trends in substance use and mental/behavioral health, as well as population-specific evaluations of risk and protective factors. The development of theoretical models specific to vulnerable populations provides a theoretical structure to begin to evaluate subgroup-specific risks for mental and behavioral health (e.g., minority-stress model; Meyer, 2003). Moreover, universal interventions to improve population health overall are important but it is critical to evaluate if innovations in mental and behavioral health interventions inadvertently exacerbate disparities due to differential access by the more socioeconomically affluent segments of the population (Williams et al., 2008).

METHODS IN DEVELOPMENTAL PSYCHOPATHOLOGY

There is a lot we have learned about the onset and progression of mental health and behavioral disorders during adolescence. In addition, evidence-based effective interventions have been developed to promote healthy adjustment during this period of life. Lists of interventions and evaluations of the strength of the evidence are available from a variety of sources (e.g., Substance Abuse and Mental Health Services Administration [SAMHSA]'s Evidence-Based Practices Resource Center, the National Institute on Drug Abuse [NIDA]'s research-based guide on Preventing Drug Use among Children and Adolescents, Blueprints for Healthy Youth Development). Numerous fields (e.g., epigenetics, developmental neuroscience, human development/developmental psychology, pediatrics) contribute theory and methods critical to understanding the development of psychopathology during adolescence. In addition, there has been a growth in quantity and quality of data as well as a commensurate growth in statistical methods. Of course, there continue to be gaps in our knowledge and this section reviews a few of the most promising areas for continued research regarding the development and prevention of mental and behavioral health disorders in adolescence.

Biological Basis of Mental and Behavioral Health

One domain in need of continued research is the biological basis of mental and behavioral health. Research on the interplay of pubertal hormones and brain development during adolescence is in its infancy, with the bulk of research in this area coming from animal models. Translating this research to human populations could provide critical insights into a biological basis of pubertal timing effects, particularly for girls. Noninvasive tools such as functional and structural magnetic resonance imaging and advances in data-analytic capabilities have the potential to advance our understanding of dynamic developmental processes. While there is evidence that changes in hormones during puberty may trigger a period of structural reorganization in the brain (Peper et al., 2011), the role of pubertal timing is unclear. Research is needed to understand how timing of neuroendocrine hormone exposure affects brain structure and function and how that may relate to cognitive, behavioral, and psychological outcomes among early maturing youth. This area of research requires interdisciplinary collaboration in the longitudinal assessment of pubertal development, neurodevelopment, physical health, mental health, behavioral health, and interpersonal dynamics.

In addition, a more nuanced understanding of how normative brain development facilitates adaptive growth could provide insight into creating more effective interventions to improve outcomes for youth. For instance, the dual systems theory provides an important heuristic structure for understanding why adolescence is a sensitive period for the development of risk behavior. However, it is critical to acknowledge that the differential growth of the cortical and limbic systems in the adolescent brain may also provide adaptive functions for youth and that changes in opportunity and context may lead to the emergence of risk behavior in later adolescence and emerging adulthood (Pfeifer & Allen, 2012; Shulman et al., 2016). Furthermore, a better understanding of how normative brain development contributes to adolescent differential preference and responsiveness to peers or emotional cues may also improve the effectiveness of interventions by facilitating the development of interventions that use peers and emotion to increase engagement.

Clinical Translation

Translating advances in technology and knowledge from research settings into primary healthcare requires partnerships with physicians and other health professionals to maximize applicability and sustainability of evidence-based research to the healthcare setting. For instance, recent advances in healthcare delivery have identified personalized medicine as a critical area of growth for enhancing treatment effectiveness and preventive medicine.

Personalized medicine refers to procedures that separate patients into different groups with medical decisions, practices, and interventions and/or products being tailored to the individual patient based on their predicted response or risk of disease (Insel, 2009). As such, current translational research aimed at promoting adolescent and young adult health through collaboration within the healthcare system requires a focus on individualized risks that can be identified by healthcare professionals during regular child health visits. It is standard procedure during child health visits to examine for signs of pubertal development. Only in extreme cases of early onset (prior to age six) are children typically referred out for specialty consultation to a pediatric endocrinologist to be evaluated for idiopathic central precocious puberty, the appearance of secondary sexual characteristics prior to normative age of onset, in the absence of CNS disorders or other disease (Styne & Grumbach, 2011).

The large disparity in the number of children considered early maturers in the psychological/behavioral sciences (~20% of youth) compared with those diagnosed with precocious puberty (20–23 per 10,000 girls and <5 per 10,000 boys; Teilmann et al., 2005) is even more poignant when one considers the abundance of literature highlighting negative behavioral and psychological adjustment outcomes associated with early pubertal timing. A disciplinary divide between medical research and psychological/behavioral research on early puberty has led to poor translation of early pubertal timing implications into the pediatric setting. Collaboration across disciplines is required to accelerate the translation of psychological and behavioral research into the clinical setting to improve the health and adjustment of adolescents.

Technological Advances

In addition, future research in understanding the role that technological advances (e.g., social media, wearable devices) play for mental and behavioral health in adolescence is needed. Individuals generate enormous amounts of data during the course of their daily lives given the widespread use of computers in almost every setting, including for personal use. For example, education and healthcare systems maintain electronic databases. There are administrative sources of data (e.g., data from local and state governments, healthcare organizations) that can provide unique information regarding trends in adolescent mental and behavioral health as well as information on risk and protective factors. Moreover, many adolescents engage with social media, use smartphones, or use health-monitoring devices such as Fitbits. These activities and devices are collecting unprecedented amounts of data on adolescent behaviors, social connections, and health. Unique opportunities for research and for reaching understudied and vulnerable populations are possible through these advances in technology. In addition, technological advances provide opportunities to develop innovative preventive interventions and advance both quantitative and qualitative research methods.

It is critical to continue and monitor demographic trends in mental health and substance use throughout adolescence, particularly given technological advances. For example, cyberbullying is a form of bullying that is a direct result of advances in social media technology, and electronic cigarettes represent a technological advance with implications for adolescent initiation of nicotine products. While there are certainly many benefits associated with technological advancement, it is important to monitor how advances in technology might affect the type of substances used by adolescents, the mode by which substance use is administered, or other iatrogenic effects of advances in technology on adolescent mental and behavioral health.

Law and Policy

Ecological systems theories of human development provide a theoretical foundation for understanding complex associations between individuals and the contexts in which they are born, grow, and live (Bronfenbrenner, 2009; Lerner, 2006). Laws and policies shape the social environment in which adolescents live and have implications for youth health and adjustment. As such it is critical to consider implementing and evaluating laws and policies as part of a comprehensive effort to improve mental and behavioral outcomes for youth as well as reduce disparities in outcomes. A substantial body of literature exists linking alcohol policies to youth alcohol use (Lynne-Landsman & Wagenaar, 2015). While there is a need for continued research on alcohol policy and youth health outcomes, there are also rapidly changing national, state, and local laws/policies with implications for youth mental and behavioral health. Research on marijuana policies are stimulating substantial interest given variability between states in the legality of this substance. In addition, legal rights and policies related to sexual and gender minorities also show jurisdictional variability. While it is beyond the scope of this chapter to review the current evidence base for the influence of these macrolevel policies on adolescent mental and behavioral health, these two examples are illustrative of the need for research utilizing diverse experimental designs in this area.

Many studies of policy effects on youth health outcomes use difference-in-differences designs which provide an evaluation of differences in change over time in an outcome of interest (e.g., adolescent marijuana use) between two groups: one that enacted the policy and another that did not (Dimick & Ryan, 2014; French & Heagerty, 2008). This analytic method assumes that the observed patterns of change over time between the two groups should be statistically similar had the policy change not occurred. Hence, differences in the pattern of change after the policy is enacted indicate an effect of the policy on the outcome of interest.

Most research on changing marijuana policies in the US has focused on medical marijuana laws (MMLs), with a number of studies finding no effects of MMLs on adolescent marijuana use (Choo et al., 2014; Harper, Strumpf, & Kaufman, 2012; Hasin et al., 2015; Lynne-Landsman, Livingston, & Wagenaar, 2013; Wall et al., 2016). Studies have begun expanding to evaluate developmental differences in the effects of MMLs and alternative outcomes beyond marijuana use (Cerdá et al., 2018; Wen, Hockenberry, & Cummings, 2015). Research on the influence of recreational marijuana legalization is just beginning to be assessed, with some evidence that this policy may increase frequency of marijuana use among youth already using (Rusby et al., 2018). Research findings on the effects of marijuana policies (e.g., medical marijuana, recreational marijuana) on adolescent outcomes are emerging; however, this is an area of research that needs continued and nuanced evaluation.

Similarly, there is a need for evaluation of how policies related to sexual and gender minorities may influence mental health outcomes for youth and disparities for sexual or gender minority youth. Using a difference-in-differences design, there is evidence that the enactment of same-sex marriage policies is associated with a reduction in high school student suicide attempts overall but with a concentrated reduction for sexual minority students (Raifman et al., 2017). In addition, there is evidence that youth from schools with genders and sexualities alliance groups (GSAs) report less truancy, smoking, drinking, suicide attempts, and sex with casual partners than youth from schools without GSAs, with positive effects stronger for sexual and gender minority youth (Poteat et al., 2012).

CONCLUSION

Adolescence is a sensitive period in human development for the onset and progression of mental and behavioral health disorders. The extent of physical, cognitive, and social changes that occur during adolescence are only surpassed by those of the prenatal and early childhood period. Many mental and behavioral health disorders manifest for the first time in adolescence. Moreover, it is also when a variety of disparities begin to emerge which persist into adulthood. This highlights the importance of research that focuses on within-person biological/physical growth and development, as well as research that studies interpersonal relationships, the role of the community, and culture taking an ecological systems perspective. Interdisciplinary teams will be critical to unpacking the complex associations between puberty, brain development and cognition, and implications for mental and behavioral health. This etiological research will enrich the foundation upon which preventive interventions are developed with the goal of reducing the prevalence of mental and behavioral health disorders overall as well as reducing disparities for vulnerable populations.

REFERENCES

Achenbach, T. M. (1993). Taxonomy and Comorbidity of Conduct Problems: Evidence from Empirically Based Approaches. *Development and Psychopathology*, 5(1–2), 51–64.

Biro, F. M., & Deardorff, J. (2013). Identifying Opportunities for Cancer Prevention during Preadolescence and Adolescence: Puberty as a Window of Susceptibility. *Journal of Adolescent Health*, 52(5), S15–S20.

Beltz, A. M., & Berenbaum, S. A. (2013). Cognitive Effects of Variations in Pubertal Timing: Is Puberty a Period of Brain Organization for Human Sex-Typed Cognition? *Hormones and Behavior*, 63(5), 823–828.

Beltz, A., Acharya, R., Graber, J. A., Nixon, S. J., & Lynne, S. D. (2017). *Puberty* Influences Reward-Related Neural Connectivity: Girls with Early versus Typical Maturation. Paper presented as part of a symposium at the Society for Research in Child Development Biennial Meeting, Austin, TX, April 6–8.

Braveman, P., Egerter, S., & Williams, D. R. (2011). The Social Determinants of Health: Coming of Age. *Annual Review of Public Health*, 32, 381–398.

Bronfenbrenner, U. (2009). *The Ecology of Human Development*. Cambridge, MA: Harvard University Press.

Brooks-Gunn, J., Petersen, A. C., & Eichorn, D. (1985). The Study of Maturational Timing Effects in Adolescence. *Journal of Youth and Adolescence*, 14(3), 149–161.

Byrne, M. L., Whittle, S., Vijayakumar, N., Dennison, M., Simmons, J. G., & Allen, N. B. (2017). A Systematic Review of Adrenarche as a Sensitive Period in Neurobiological Development and Mental Health. *Developmental Cognitive Neuroscience*, 25, 12–28.

Caspi, A., Houts, R. M., Belsky, D. W., Goldman-Mellor, S. J., Harrington, H., Israel, S., ... Moffitt, T. E. (2014). The P Factor: One General Psychopathology Factor in the Structure of Psychiatric Disorders? *Clinical Psychological Science*, 2(2), 119–137.

Center for Behavioral Health Statistics and Quality. (2017). 2016 National Survey on Drug Use and Health: Detailed Tables. Substance Abuse and Mental Health Services Administration, Rockville, MD.

Cerdá, M., Sarvet, A. L., Wall, M., Feng, T., Keyes, K. M., Galea, S., & Hasin, D. S. (2018). Medical Marijuana Laws and Adolescent Use of Marijuana and Other Substances: Alcohol, Cigarettes, Prescription Drugs, and Other Illicit Drugs. *Drug and Alcohol Dependence*, 183, 62–68.

Charalampopoulos, D., McLoughlin, A., Elks, C. E., & Ong, K. K. (2014). Age at Menarche and Risks of All-Cause and Cardiovascular Death: A Systematic Review and Meta-Analysis. *American Journal of Epidemiology*, 180(1), 29–40.

Chavarro, J. E., Watkins, D. J., Afeiche, M. C., Zhang, Z., Sánchez, B. N., Cantonwine, D., ... Peterson, K. E. (2017). Validity of

Self-Assessed Sexual Maturation against Physician Assessments and Hormone Levels. *Journal of Pediatrics, 186*, 172–178.

Choo, E. K., Benz, M., Zaller, N., Warren, O., Rising, K. L., & McConnell, K. J. (2014). The Impact of State Medical Marijuana Legislation on Adolescent Marijuana Use. *Journal of Adolescent Health, 55*, 160–166.

Chumlea, W. C., Schubert, C. M., Roche, A. F., Kulin, H. E., Lee, P. A., Himes, J. H., & Sun, S. S. (2003). Age at Menarche and Racial Comparisons in US Girls. *Pediatrics, 111*(1), 110–113.

Chow, J. C. C., Jaffee, K., & Snowden, L. (2003). Racial/Ethnic Disparities in the Use of Mental Health Services in Poverty Areas. *American Journal of Public Health, 93*(5), 792–797.

Compas, B. E., & Oppedisano, G. (2000). Mixed Anxiety/Depression in Childhood and Adolescence. In A. J. Sameroff, M. Lewis, & S. M. Miller (Eds.), *Handbook of Developmental Psychopathology* (2nd edn., pp. 531–548). New York: Plenum Press.

Crone, E. A., & Dahl, R. E. (2012). Understanding Adolescence as a Period of Social-Affective Engagement and Goal Flexibility. *Nature Reviews. Neuroscience, 13*(9), 636–650.

Dimick, J. B., & Ryan, A. M. (2014). Methods for Evaluating Changes in Health Care Policy: The Difference-in-Differences Approach. *JAMA Guide to Statistics and Methods, 312*(22), 2401–2402.

Dorn, L. D., Susman, E. J., & Ponirakis, A. (2003). Pubertal Timing and Adolescent Adjustment and Behavior: Conclusions Vary by Rater. *Journal of Youth and Adolescence, 32*(3), 157–167.

Dorn, L. D., & Biro, F. M. (2011). Puberty and Its Measurement: A Decade in Review. *Journal of Research on Adolescence, 21*(1), 180–195.

Dubas, J. S., Graber, J. A., & Petersen, A. C. (1991). A Longitudinal Investigation of Adolescents' Changing Perceptions of Pubertal Timing. *Developmental Psychology, 27*(4), 580–586.

Egger, H. L., & Angold, A. (2006). Common Emotional and Behavioral Disorders in Preschool Children: Presentation, Nosology, and Epidemiology. *Journal of Child Psychology and Psychiatry, 47*, 313–337.

Ellis, B. J., Boyce, W. T., Belsky, J., Bakermans-Kranenburg, M. J., & Van IJzendoorn, M. H. (2011). Differential Susceptibility to the Environment: An Evolutionary-Neurodevelopmental Theory. *Development and Psychopathology, 23*(1), 7–28.

Epstein, S. (1994). Integration of the Cognitive and the Psychodynamic Unconscious. *American Psychologist, 49*(8), 709–724.

Evans, J. S. B. T., & Stanovich, K. E. (2013). Dual-Process Theories of Higher Cognition: Advancing the Debate. *Perspectives on Psychological Science, 8*(3), 223–241.

Floyd, F. J., & Stein, T. S. (2002). Sexual Orientation Identity Formation among Gay, Lesbian, and Bisexual Youths: Multiple Patterns of Milestone Experiences. *Journal of Research on Adolescence, 12*(2), 167–191.

Freedman, D. S., Khan, L. K., Serdula, M. K., Dietz, W. H., Srinivasan, S. R., & Berenson, G. S. (2003). The Relation of Menarcheal Age to Obesity in Childhood and Adulthood: The Bogalusa Heart Study. *BMC Pediatrics, 3*(1), 3–11.

French, B., & Heagerty, P. J. (2008). Analysis of Longitudinal Data to Evaluate a Policy Change. *Statistics in Medicine, 27*, 5005–5025.

Ge, X., & Natsuaki, M. N. (2009). In Search of Explanations for Early Pubertal Timing Effects on Developmental Psychopathology. *Current Directions in Psychological Science, 18*(6), 327–331.

Graber, J. A. (2013). Pubertal Timing and the Development of Psychopathology in Adolescence and Beyond. *Hormones and Behavior, 64*(2), 262–269.

Graber, J. A., & Sontag, L. M. (2009). Internalizing Problems during Adolescence. In R. M. Lerner & L. Steinberg (Eds.), *Handbook of Adolescent Psychology: Individual Bases of Adolescent Development* (pp. 642–682). Hoboken, NJ: Wiley.

Graber, J. A., Petersen, A. C., & Brooks-Gunn, J. (1996). Pubertal Processes: Methods, Measures, and Models. In J. A. Graber, J. Brooks-Gunn, & A. C. Petersen (Eds.), *Transition through Adolescence: Interpersonal Domains and Context* (pp. 23–54). Mahwah, NJ: Erlbaum.

Harper, S., Strumpf, E. C., & Kaufman, J. S. (2012). Do Medical Marijuana Laws Increase Marijuana Use? Replication Study and Extension. *Annals of Epidemiology, 22*(3), 207–212.

Harris, K. M., Gordon-Larsen, P., Chantala, K., & Udry, J. R. (2006). Longitudinal Trends in Race/Ethnic Disparities in Leading Health Indicators from Adolescence to Young Adulthood. *Archives of Pediatrics & Adolescent Medicine, 160*(1), 74–81.

Hasin, D. S., Wall, M., Keyes, K. M., Cerdá, M., Schulenberg, J., O'Malley, P. M., ... Feng, T. (2015). Medical Marijuana Laws and Adolescent Marijuana Use in the USA from 1991 to 2014: Results from Annual, Repeated Cross-Sectional Surveys. *The Lancet, 2*(7), 601–608.

Herman-Giddens, M. E., Slora, E. J., Wasserman, R. C., Bourdony, C. J., Bhapkar, M. V., Koch, G. G., & Hasemeier, C. M. (1997). Secondary Sexual Characteristics and Menses in Young Girls Seen in Office Practice: A Study from the Pediatric Research in Office Settings Network. *Pediatrics, 99*(4), 505–512.

Herman-Giddens, M. E., Steffes, J., Harris, D., Slora, E., Hussey, M., Dowshen, S. A., ... Reiter, E. O. (2012). Secondary Sexual Characteristics in Boys: Data from the Pediatric Research in Office Settings Network. *Pediatrics, 130*(5), e1058–e1068.

Hottes, T. S., Bogaert, L., Rhodes, A. E., Brennan, D. J., & Gesink, D. (2016). Lifetime Prevalence of Suicide Attempts among Sexual Minority Adults by Study Sampling Strategies: A Systematic Review and Meta-Analysis. *American Journal of Public Health, 106*(5), e1–e12.

Insel, T. R. (2009). Translating Scientific Opportunity into Public Health Impact: A Strategic Plan for Research on Mental Illness. *Archives of General Psychiatry, 66*(2), 128–133.

Katz-Wise, S. L., Rosario, M., Calzo, J. P., Scherer, E. A., Sarda, V., & Austin, S. B. (2017). Endorsement and Timing of Sexual Orientation Developmental Milestones among Sexual Minority Young Adults in the Growing Up Today Study. *Journal of Sex Research, 54*(2), 172–185.

King, M., Semlyen, J., Tai, S. S., Killaspy, H., Osborn, D., Popelyuk, D., & Nazareth, I. (2008). A Systematic Review of Mental Disorder, Suicide, and Deliberate Self Harm in Lesbian, Gay and Bisexual People. *BMC Psychiatry, 8*(1), 70–87.

Krueger, R. F., Caspi, A., Moffitt, T. E., & Silva, P. A. (1998). The Structure and Stability of Common Mental Disorders (DSM-III-R): A Longitudinal-Epidemiological Study. *Journal of Abnormal Psychology, 107*, 216–227.

Ladouceur, C. D., Peper, J. S., Crone, E. A., & Dahl, R. E. (2012). White Matter Development in Adolescence: The Influence of Puberty and Implications for Affective Disorders. *Developmental Cognitive Neuroscience, 2*, 36–54.

Lanza, S. T., & Collins, L. M. (2002). Pubertal Timing and the Onset of Substance Use in Females during Early Adolescence. *Prevention Science, 3*(1), 69–82.

Lee, P. A., Guo, S. S., & Kulin, H. E. (2001). Age of Puberty: Data from the United States of America. *APMIS Journal of Pathology, Microbiology, and Immunology, 109*(S103), 81–88.

Lee, Y., & Styne, D. (2013). Influences on the Onset and Tempo of Puberty in Human Beings and Implications for Adolescent Psychological Development. *Hormones and Behavior, 64*(2), 250–261.

Lerner, R. M. (2006). Developmental Science, Developmental Systems, and Contemporary Theories. In R. M. Lerner (Ed.), *Handbook of Child Psychology: Vol. 1. Theoretical Models of Human Development* (pp. 1–17). Hoboken, NJ: Wiley.

Lewinsohn, P. M., Zinbarg, R., Seeley, J. R., Lewinsohn, M., & Sack, W. H. (1997). Lifetime Comorbidity among Anxiety Disorders and between Anxiety Disorders and Other Mental Disorders in Adolescents. *Journal of Anxiety Disorders, 11*(4), 377–394.

Lewinsohn, P. M., Shankman, S. A., Gau, J. M., & Klein, D. N. (2004). The Prevalence and Co-Morbidity of Subthreshold Psychiatric Conditions. *Psychological Medicine, 34*(4), 613–622.

Lynne, S. D., Graber, J. A., Nichols, T. R., Brooks-Gunn, J., & Botvin, G. J. (2007). Links between Pubertal Timing, Peer Influences, and Externalizing Behaviors among Urban Students followed through Middle School. *Journal of Adolescent Health, 40*(2), 181.e7–181.e13.

Lynne-Landsman, S. D. & Wagenaar, A. C. (2015). Alcohol Policy: Interventions to Prevent Youth Alcohol Use. In L. M. Scheier (Ed.), *Handbook of Adolescent Drug Use Prevention: Research, Intervention Strategies, and Practice* (pp. 329–341). Washington, DC: American Psychological Association.

Lynne-Landsman, S. D., Graber, J. A., & Andrews, J. A. (2010). Do Trajectories of Household Risk in Childhood Moderate Pubertal Timing Effects on Substance Initiation in Middle School? *Developmental Psychology, 46*(4), 853–868.

Lynne-Landsman, S. D., Livingston, M. D., & Wagenaar, A. C. (2013). Effects of State Medical Marijuana Laws on Adolescent Marijuana Use. *American Journal of Public Health, 103*(8), 1500–1506.

Martí-Henneberg, C., & Vizmanos, B. (1997). The Duration of Puberty in Girls Is Related to the Timing of Its Onset. *Journal of Pediatrics, 131*(4), 618–621.

Mendle, J. (2014). Beyond Pubertal Timing: New Directions for Studying Individual Differences in Development. *Current Directions in Psychological Science, 23*(3), 215–219.

Mendle, J., & Ferrero, J. (2012). Detrimental Psychological Outcomes Associated with Pubertal Timing in Adolescent Boys. *Developmental Review, 32*(1), 49–66.

Mussen, P. H., & Jones, M. C. (1957). Self-Conceptions, Motivations, and Interpersonal Attitudes of Late- and Early-Maturing Boys. *Child Development, 28*(2), 243–256.

Meyer, I. H. (2003). Prejudice, Social Stress, and Mental Health in Lesbian, Gay, and Bisexual Populations: Conceptual Issues and Research Evidence. *Psychological Bulletin, 129*(5), 674–697.

Natsuaki, M. N., Leve, L. D., & Mendle, J. (2011). Going Through the Rites of Passage: Timing and Transition of Menarche, Childhood Sexual Abuse, and Anxiety Symptoms in Girls. *Journal of Youth and Adolescence, 40*(10), 1357–1370.

Negriff, S., & Susman, E. J. (2011). Pubertal Timing, Depression, and Externalizing Problems: A Framework, Review, and Examination of Gender Differences. *Journal of Research on Adolescence, 21*(3), 717–746.

Parent, A. S., Teilmann, G., Juul, A., Skakkebaek, N. E., Toppari, J., & Bourguignon, J. P. (2003). The Timing of Normal Puberty and the Age Limits of Sexual Precocity: Variations around the World, Secular Trends, and Changes after Migration. *Endocrine Reviews, 24*(5), 668–693.

Peper, J. S., & Dahl, R. E. (2013). The Teenage Brain: Surging Hormones – Brain-Behavior Interactions during Puberty. *Current Directions in Psychological Science, 22*(2), 134–139.

Peper, J. S., Pol, H. E. H., Crone, E. A., & Van Honk, J. (2011). Sex Steroids and Brain Structure in Pubertal Boys and Girls: A Mini-Review of Neuroimaging Studies. *Neuroscience, 191*, 28–37.

Petersen, A. C., Crockett, L., Richards, M., & Boxer, A. (1988). A Self-Report Measure of Pubertal Status: Reliability, Validity, and Initial Norms. *Journal of Youth and Adolescence, 17*(2), 117–133.

Pfeifer, J. H., & Allen, N. B. (2012). Arrested Development? Reconsidering Dual-Systems Models of Brain Function in Adolescence and Disorders. *Trends in Cognitive Sciences, 16*(6), 322–329.

Pomerantz, H., Parent, J., Forehand, R., Breslend, N. L., & Winer, J. P. (2017). Pubertal Timing and Youth Internalizing Psychopathology: The Role of Relational Aggression. *Journal of Child and Family Studies, 26*(2), 416–423.

Poteat, V. P., Sinclair, K. O., DiGiovanni, C. D., Koenig, B. W., & Russell, S. T. (2012). Gay-Straight Alliances Are Associated with Student Health: A Multischool Comparison of LGBTQ and Heterosexual Youth. *Journal of Research on Adolescence, 23*(2), 319–330.

Raifman, J., Moscoe, E., Austin, S. B., & McConnell, M. (2017). Difference-in-Differences Analysis of the Association between State Same-Sex Marriage Policies and Adolescent Suicide Attempts. *JAMA Pediatrics, 171*(4), 350–356.

Rosario, M., Schrimshaw, E. W., Hunter, J., & Gwadz, M. (2002). Gay-Related Stress and Emotional Distress among Gay, Lesbian and Bisexual Youths: A Longitudinal Examination. *Journal of Consulting and Clinical Psychology, 70*(4), 967–975.

Rusby, J. C., Westling, E., Crowley, R., & Light, J. M. (2018). Legalization of Recreational Marijuana and Community Sales Policy in Oregon: Impact on Adolescent Willingness and Intent to Use, Parent Use, and Adolescent Use. *Psychology of Addictive Behaviors, 32*(1), 84–92.

Sannino, S., Chini, B., & Grinevich, V. (2016). Lifespan Oxytocin Signaling: Maturation, Flexibility, and Stability in Newborn, Adolescent, and Aged Brain. *Developmental Neurobiology, 77*, 158–168.

Schulz, K. M., Molenda-Figueira, H. A., & Sisk, C. L. (2009). Back to the Future: The Organizational-Activational Hypothesis Adapted to Puberty and Adolescence. *Hormones and Behavior, 55*(5), 597–604.

Shirtcliff, E. A., Dahl, R. E., & Pollak, S. D. (2009). Pubertal Development: Correspondence between Hormonal and Physical Development. *Child Development, 80*(2), 327–337.

Shulman, E. P., Smith, A. R., Silva, K., Icenogle, G., Duell, N., Chein, J., & Steinberg, L. (2016). The Dual Systems Model: Review, Reappraisal, and Reaffirmation. *Developmental Cognitive Neuroscience, 17*, 103–117.

Sisk, C. L., & Zehr, J. L. (2005). Pubertal Hormones Organize the Adolescent Brain and Behavior. *Frontiers in Neuroendocrinology, 26*(3–4), 163–174.

Steinberg, L. (2010). Commentary: A Behavioral Scientist Looks at the Science of Adolescent Brain Development. *Brain and Cognition, 72*(1), 160–164.

Steinberg, L., & Morris, A. S. (2001). Adolescent Development. *Annual Review of Psychology*, *52*(1), 83–110.

Styne, D. M. & Grumbach, M. M. (2011). Puberty, Ontogeny, Neuroendocrinology, Physiology, and Disorders. In S. Melmed, K. S. Polonsky, P. R. Larsen, & H. M. Kronenberg (Eds.), *Williams Textbook of Endocrinology* (12th edn., pp. 1054–1201). Philadelphia, PA: Saunders, Elsevier.

Suleiman, A. B., & Dahl, R. E. (2017). Leveraging Neuroscience to Inform Adolescent Health: The Need for an Innovative Transdisciplinary Developmental Science of Adolescence. *Journal of Adolescent Health*, *60*(3), 240–248.

Sun, S. S., Schubert, C. M., Chumlea, W. C., Roche, A. F., Kulin, H. E., Lee, P. A., ... Ryan, A. S. (2002). National Estimates of the Timing of Sexual Maturation and Racial Differences among US Children. *Pediatrics*, *110*(5), 911–919.

Susman, E. J., & Dorn, L. D. (2009). Puberty: Its Role in Development. In R. M. Lerner & L. Steinberg (Eds.), *Handbook of Adolescent Psychology: Individual Bases of Adolescent Development* (pp. 116–151). Hoboken, NJ: Wiley.

Susman, E. J., Houts, R. M., Steinberg, L., Belsky, J., Cauffman, E., DeHart, G., ... Halpern-Felsher, B. L. (2010). Longitudinal Development of Secondary Sexual Characteristics in Girls and Boys between Ages 9½ and 15½ Years. *Archives of Pediatrics & Adolescent Medicine*, *164*(2), 166–173.

Tanner, J. M., & Whitehouse, R. H. (1976). Clinical Longitudinal Standards for Height, Weight, Height Velocity, Weight Velocity, and Stages of Puberty. *Archives of Disease in Childhood*, *51*(3), 170–179.

Tebbe, E. A., & Moradi, B. (2016). Suicide Risk in Trans Populations: An Application of Minority Stress Theory. *Journal of Counseling Psychology*, *63*(5), 520–533.

Teilmann, G., Pedersen, C. B., Jensen, T. K., Skakkebæk, N. E., & Juul, A. (2005). Prevalence and Incidence of Precocious Pubertal Development in Denmark: An Epidemiologic Study Based on National Registries. *Pediatrics*, *116*(6), 1323–1328.

Ullsperger, J. M., & Nikolas, M. A. (2017). A Meta-Analytic Review of the Association between Pubertal Timing and Psychopathology in Adolescence: Are There Sex Differences in Risk?. *Psychological Bulletin*, *143*(9), 903–938.

Wall, M. M., Mauro, C., Hasin, D. S., Keyes, K. M., Cerdá, M., Martins, S. S., & Feng, T. (2016). Prevalence of Marijuana Use Does not Differentially Increase among Youth after States Pass Medical Marijuana Laws: Commentary on Stolzenberg et al. (2015) and Reanalysis of US National Survey on Drug Use in Households Data 2001–2011. *International Journal of Drug Policy*, *29*, 9–13.

Walvoord, E. C. (2010). The Timing of Puberty: Is It Changing? Does It Matter?. *Journal of Adolescent Health*, *47*(5), 433–439.

Wen, H., Hockenberry, J. M., & Cummings, J. R. (2015). The Effect of Medical Marijuana Laws on Adolescent and Adult Use of Marijuana, Alcohol, and Other Substances. *Journal of Health Economics*, *42*, 64–80.

Williams, D. R., & Mohammed, S. A. (2009). Discrimination and Racial Disparities in Health: Evidence and Needed Research. *Journal of Behavioral Medicine*, *32*(1), 20–47.

Williams, D. R., Costa, M. V., Odunlami, A. O., & Mohammed, S. A. (2008). Moving Upstream: How Interventions That Address the Social Determinants of Health Can Improve Health and Reduce Disparities. *Journal of Public Health Management and Practice*, *14*(Supplement), S8–S17.

17 Quantitative Genetic Research Strategies for Studying Gene-Environment Interplay in the Development of Child and Adolescent Psychopathology

ISABELL BRIKELL AND HENRIK LARSSON

During the past two decades there has been a steady increase in quantitative genetic research focused on the origins of childhood and adolescent psychopathology. This research relies on a range of genetically sensitive research designs which all use data from family samples to study the relative importance of (unmeasured) genes and environments for individual difference in psychopathology in the population. Based on over a thousand published research articles it is now clear that (1) genetic factors play an important role in all forms of psychopathology, and (2) no form of psychopathology is 100 percent heritable (Polderman et al., 2015), highlighting the importance of environmental factors in the development of psychopathology. Recent accumulation of longitudinal data and detailed assessments of environmental measures, coupled with advances in model-fitting techniques and multivariate analytic methods have now allowed quantitative genetics to address more complex questions about the role of gene-environment interplay for the development of psychopathology. The goal of this chapter is to provide an overview of quantitative genetic research methods, to summarize conclusions that can be drawn about the development of child and adolescent psychopathology from the past two decades of quantitative genetics research, and to highlight remaining questions that need to be addressed in future research.

A BRIEF INTRODUCTION TO QUANTITATIVE GENETIC METHODS

We begin by introducing the basic premise of univariate family, adoption, and twin designs, and the main methodological considerations and limitations of each of the study designs (Table 17.1). For a more detailed description of these methods see e.g., McGuffin et al, 2004, Plomin et al, 2013, and Rijsdijk & Sham, 2002.

The Family Design

Family studies investigate the extent to which disorders "run" in families by comparing the rate of a disorder among individuals with a relative affected by the same disorder to the rate in individuals without an affected relative. The familial risk is usually estimated via regression analyses and expressed as a relative risk (e.g., odds ratio or hazard ratio). An increased risk among the relatives of affected individuals suggests that factors shared within the family, either genetic or environmental (or more commonly, a combination of both) are important for the development of the disorder. Familial risk estimates are influenced and may be biased by numerous factors, such as whether the study population has lived through the age of risk, the choice of control group, and the population prevalence of the disorder (see Table 17.1). Family studies including multiple types of relatives (e.g., parent-offspring, full siblings, half siblings, and cousins) can also be used to infer whether the primary source of familial resemblance is genetic or environmental and to estimate heritability (Chen et al., 2017; Sandin et al., 2014).

The Adoption Design

Adoption provides a natural "experiment" for studying the relative importance of genetic and environmental factors for familial resemblance. Similarity between biological relatives adopted apart, who are genetically related but do not grow up together, provides direct evidence for the contribution of genetic factors to familial resemblance. Conversely, similarity between family members adopted together, who are genetically unrelated but share their family environment, can only be due to environmental factors and therefore provide a direct estimate for the importance of the shared family environment. There are two common types of adoption studies; (a) parent-offspring adoption studies, comparing the similarity between adoptees and their adoptive parents to the similarity between adoptees and their biological parents, and (b) sibling adoption studies, comparing the similarity between adoptive siblings with the similarity between biological siblings who were adopted apart. Whilst both types

Table 17.1 Methodological considerations and assumptions in quantitative genetic research (table adapted from Thapar and Stergiakouli, 2008)

Family design	**Control group:** The choice of control group will affect the relative risk estimate in family studies. Using a control group with another psychiatric disorder provides a stricter test, relative to using general population controls.
	Prevalence: Relative risk estimates depend on the prevalence of the disorder in the population.
	Age: Risk estimates will be biased if the study population has not passed through the age of risk, i.e., the age of typical onset for the disorder in question. Studies of later onset disorders such as schizophrenia therefore require an older study population than studies of early onset disorders like Autism Spectrum Disorder (ASD).
Adoption design	**Representativeness:** Families that partake in adoption may differ from families that do not, meaning results may not generalize to other populations.
	Assumes that biological mother–child resemblance is purely genetic: Higher similarity between biological mother and offspring can, in addition to genetic transmission, be due to early shared factors including the in-utero environment. If this is the case, genetic effects from adoption studies may be overestimated.
	Selective placement: If adoptive families are matched on certain characteristics (e.g., socioeconomic status) to the biological parents, estimation of the shared environment may be inflated.
	Open adoptions: Adoptions that allow for contact between the biological and the adoptive family are becoming more common, meaning the assumption that biological parents and adopted away children are only similar due to genetic factors may be violated.
Twin design	**Representativeness:** Twins are at higher risk for pre- and perinatal complications and lower birthweight, meaning that findings from twin populations may not generalize to non-twins. However, there is not strong evidence for different rates of psychopathology between twins and non-twins.
	Equal environment assumption: The twin design assumes that monozygotic and dizygotic twins share their environment to a very similar extent, which may not be true for every trait.
	Monozygotic twins are genetically identical: The twin design assumes monozygotic twins to be genetic clones, but there is some evidence for minor differences in noninherited genetic variations.
	No assortative mating: The twin design assumes no assortative mating, which refers to mate choice that is based on trait similarity. The extent to which this assumption holds will vary by trait, with evidence of assortative mating for traits like intelligence. If assortative mating occurs it will lead to an overestimation of the shared environment, as genetic and environmental similarity is increased between all types of relatives.
	No gene-gene or gene-environment interaction: Neither of these assumptions is likely to hold completely and should be considered, and if possible tested, when interpreting results from quantitative genetic designs. We discuss methods to investigating gene-environment interaction later in this chapter.

of designs allow for the estimation of genetic and environmental contributions to familial resemblance, parent-offspring and sibling adoption studies differ in several ways. Most notably, there are specific sources of the shared environment that can be tested, as the environment shared by siblings capture somewhat different processes than the environment provided by parents to their children. There are also different methodological considerations. For example, sibling-based adoption studies can be influenced by factors such as age differences between adoptive and biological siblings, whereas estimates from parent-offspring adoption studies will capture effects of the prenatal environment provided by the biological birth mother. Prenatal effects will be classified as genetic factors unless taken into account, thus resulting in potentially upward-biased estimates of genetic influences in parent-offspring adoption studies.

Adoption designs can also incorporate nonadoptive families, who share both genes and environment, and contrast their similarity to that of adopted-together family members who only share their environment (Plomin et al., 2013). All adoption designs, however, have several important limitations to consider (see Table 17.1), of which perhaps the most problematic is that families involved in adoption may not be representative of families who are not, thus limiting the generalizability of findings (McGuffin, Owen, & Gottesman, 2004).

The Classical Twin Design

The classical twin design is the most commonly used quantitative genetic approach. The design relies on the comparison of similarity between monozygotic (MZ) twins, who are (with very minor deviations) genetically identical, to the similarity between dizygotic (DZ) twins, who share on average 50 percent of their segregating genes, like any other full siblings. As both MZ and DZ twins are born at the same time and grow up in the same family, the twin design assumes that both types of twins share their environment to a very similar degree. To the extent that this assumption holds, a greater similarity within MZ twin pairs compared to DZ twin pairs can only be explained by their greater genetic similarity. Based on these assumptions of genetic and environmental sharing across MZ and DZ twins, the twin-pair correlations (or concordance rates for dichotomous outcomes) can be used to decompose the variance of a trait into four latent (or unmeasured; see chapter 7 for an in-depth discussion of latent variables) variance components; *additive genetics* (A), referring to the effects of alleles (alternative forms of a gene) "adding up," *dominance genetic effects* (D), referring to the interaction between alleles, *common or shared environment* (C), referring to all the factors that make family members similar beyond genetic resemblance, and *nonshared environments* (E), referring to any environmental factors that contribute to dissimilarity among family members. The E component also captures measurement error. In the classical twin design, including only twins who grow up together, C and D are confounded and cannot be estimated in the same model. This can be addressed by including other types of relatives (Rijsdijk & Sham, 2002).

What Is Heritability and How Is It Estimated?

Heritability refers to the proportion of the variance in a trait that can be attributed to genetic factors (A and D). It is important to note that heritability is a population statistic with no direct interpretation for a single individual, and that heritability can differ across populations, age, and with large shifts in the environment. Heritability, as well as the variance components C and E, can be calculated using Falconer's Formula (see Box 17.1) (Falconer, 1965). Because C and D are confounded and cannot be estimated in the same model, the estimation of D is missing in these derivations. Dominance (D) effects are implied when the DZ correlations are less than half the MZ correlations. Although Falconer's Formula can be used to gauge the relative importance of genetic and environmental factors, twin studies typically rely on structural equation modelling (SEM), as the method allows for comparison of alternative models, multivariate analyses, and incorporation of data from different types of relatives (Rijsdijk & Sham, 2002).

Box 17.1 Falconer's Formula for deriving A, C, and E from twin correlations

In the twin design, variance in a trait (V) is expressed as

$$(\mathbf{V}) = \mathbf{A} + \mathbf{C} + \mathbf{E} \qquad (1)$$

Resemblance within *MZ* twin pairs (r_{MZ}) is due to genes and shared environment, thus

$$r_{MZ} = \mathbf{A} + \mathbf{C} \qquad (2)$$

Resemblance within *DZ* pairs (r_{DZ}) is due to sharing half their segregating genes and all of their shared environment, such that

$$r_{DZ} = \frac{1}{2}\mathbf{A} + \mathbf{C} \qquad (3)$$

Based on equation (2) and (3), it is possible to estimate heritability (**A**) as twice the difference between the *MZ* and the *DZ* correlation

$$\mathbf{A} = 2(r_{MZ} - r_{DZ}) \qquad (4)$$

The shared environment (**C**) can be estimated as the difference between the *MZ* correlation and the heritability

$$\mathbf{C} = r_{MZ} - \mathbf{A} \qquad (5)$$

Nonshared environmental influences(**E**) are the only factors that make *MZ* twins differ from one another, and can therefore be calculated as the total trait variance, usually standardized to 1 for ease of interpretation, minus the *MZ* correlation.

$$\mathbf{E} = 1 - r_{MZ} \qquad (6)$$

KEY FINDINGS FROM UNIVARIATE QUANTITATIVE GENETIC RESEARCH ON CHILD AND ADOLESCENT PSYCHOPATHOLOGY

Genetic Factors Play an Important Role in Child and Adolescent Psychopathology

All types of psychopathology that typically emerge in childhood and adolescence show evidence of familial transmission and moderate-to-high heritability. Early family studies of Autism Spectrum Disorder (ASD) showed that 2–6 percent of siblings of children with ASD were themselves affected, which was about 100 times the population rate of diagnosed ASD at the time (Bailey, Phillips, & Rutter, 1996; Smalley, Asarnow, & Spence, 1988). Similarly, early family studies of Attention-Deficit/Hyperactivity Disorder (ADHD) reported a four- to fivefold risk increase among first-degree relatives of children with ADHD (Biederman et al., 1992; Faraone, Biederman, & Monuteaux, 2000). Twin studies have subsequently confirmed that ASD and ADHD are amongst the most heritable of all psychiatric disorders, with ASD heritability estimated between 60 and 90 percent, and ADHD heritability between 70 and 80 percent (Nikolas & Burt, 2010; Posthuma & Polderman, 2013; Tick et al., 2016). Adoption studies of ADHD also generally suggest that familial resemblance is largely genetic, with little influence from the shared family environment (Faraone & Larsson, 2018). There are to date few, if any, adoption studies on ASD.

Quantitative genetic studies of conduct problems show a somewhat different pattern, with a substantial contribution of the shared environment to trait variance. In a meta-analysis of twin and adoption studies on conduct problems, genetic effects accounted for 45 percent and shared environmental effects for 15–20 percent of variability in childhood and adolescent conduct problems (Rhee & Waldman, 2002). Family and twin studies on child and adolescent depression and anxiety suggest moderate heritability (<50 percent) for both disorders, and provide evidence for the role of both shared and nonshared environmental factors (Gregory & Eley, 2007; Rice, Harold, & Thapar, 2002). In contrast, adoption studies of childhood and adolescent depression and anxiety report little evidence for heritability and instead suggest substantial environmental transmission (Eley et al., 1998; Van den Oord, Boomsma, & Verhulst, 1994).

The Environment Matters as No Form of Child and Adolescent Psychopathology Is Entirely Heritable

Quantitative genetic studies have provided important insights on the role of the environment in the development of child and adolescent psychopathology. First, it is clear from family, adoption, and twin studies that environmental factors play a crucial role in all complex (non-Mendelian) forms of psychopathology, as no trait shows 100 percent heritability. Second, quantitative genetics also inform us about the way in which the environment acts to influence child and adolescent psychopathology (Plomin et al., 2016). In a meta-analysis of over 490 twin and adoption studies of childhood and adolescent psychopathology, shared familial factors (C) accounted for 10–19 percent of the variance in conduct disorder, oppositional defiant disorder, anxiety, depression, and broad dimensions of internalizing and externalizing problems. The only exception was ADHD, where familial resemblance appeared to be solely driven by genetic factors (Burt, 2009). Results also showed that the influence of the nonshared environment (E) was greater than C for all studied forms of child and adolescent psychopathology. This is not to say that the family environment doesn't matter; rather it suggests that it may affect children from the same family in different ways. Differences in how children respond to their environment may in turn be partly driven by genetic differences, a phenomenon referred to as gene-environment interplay.

The remainder of this chapter will focus on how multivariate and longitudinal quantitative genetic research designs can be used to go beyond merely estimating heritability, to address important questions regarding etiology across different levels of symptom severity, comorbidity, and development, and to study gene-environment interplay in child and adolescent psychopathology.

ETIOLOGIC FACTORS THAT ACCOUNT FOR VARIATION IN THE NORMAL RANGE OF BEHAVIOUR ARE SIMILAR TO THE ETIOLOGIC FACTORS THAT ACCOUNT FOR PSYCHOPATHOLOGY

The question of whether psychopathology should be classified and conceptualized in categorical or dimensional terms has been debated for some time. On the one hand, those who propose a categorical approach view psychopathology as qualitatively different from variation across the normal range of the trait expression in the population, and as having their own pattern of rather distinct causes, meaning that psychopathology differs from the trait expression in the general population in both degree and nature. On the other hand, those who consider psychopathology as an extreme expression of normal variation in the population emphasize continuity in the underlying causes, meaning that psychopathology and the trait expression in the general population differ only in degree, but not in nature.

A core issue in the categorical versus dimensional discussion is therefore the question of whether the etiological underpinnings are the same across different levels of severity of psychopathology. This question has to some extent been addressed in twin studies, using the so-called DeFries–Fulker extremes analysis (DeFries & Fulker, 1988). DeFries–Fulker extremes analysis explores genetic links between the extreme and the subthreshold levels of a trait or disorder by bringing together the dichotomized classification of psychopathology (e.g., yes/no diagnosis) and the underlying quantitative dimension of the psychopathological construct (e.g., the total score of a parent-rated symptom scale). Specifically, rather than assessing twin similarity in terms of individual differences on a quantitative trait or in terms of concordance for a certain cutoff, DeFries–Fulker extremes analysis assesses twin similarity as the extent to which the mean standardized quantitative trait score of the co-twins is as high as the mean standardized score of the selected probands (i.e., the disorder affected twin on whom the twin pair was selected into the study). This measure of twin similarity, called extreme group correlation, is calculated by dividing the quantitative traits scores of the co-twins by the probands mean, specific for each sex and zygosity group. Genetic influences are implied if extreme group correlations are greater for MZ than for DZ twin pairs. The extent to which genetic factors account for the mean differences between probands and the population is called group heritability.

Although DF extremes group heritability can be estimated by doubling the differences in MZ and DZ extreme group correlations, DF extremes analysis is more properly conducted using a regression model (DeFries & Fulker, 1988). Finding group heritability implies that there is a genetic link between the extreme of a quantitative dimension (e.g., clinical ADHD diagnosis) and normal variation of the same quantitative dimension (e.g. traits of hyperactivity, impulsivity and inattention) (Martin, Taylor, & Lichtenstein, 2017). This means that the same etiologic factors are involved across different levels of severity, including those diagnosed with a psychiatric disorder and those with subthreshold variations.

The available research using the DeFries–Fulker analytic method have demonstrated statistically significant group heritability for some forms of child and adolescent psychopathology. One study of 16,366 Swedish twins explored the genetic link between the extreme and the subthreshold variation of the *Diagnostic and Statistical Manual of Mental Disorders* (DSM)-IV ADHD symptoms. The estimated group heritability was around 0.60, depending on the level of severity for the imposed cutoff (Larsson et al., 2012). These study findings are consistent with one early study of 583 same-sexed twin pairs using DSM-III-R ADHD symptoms (Levy et al., 1997). Together, these studies suggest that ADHD is best viewed as the quantitative extreme of genetic and environmental factors operating dimensionally throughout the distribution of ADHD symptoms, indicating that the same etiological factors are involved in the full range of symptoms of inattention, hyperactivity, and impulsivity.

Two twin studies using the DeFries–Fulker analytic method also support significant group heritability using several different definitions of autistic-like traits (Lundström et al., 2011). Another study used a UK-based twin sample and applied a novel approach, referred to as the joint categorical-continuous twin model, to estimate the degree of etiological overlap between autistic-like traits and categorical diagnoses of ASD. This study reported a strong genetic correlation of 0.70 between autistic-like traits and categorical diagnoses of ASD (Colvert, Tick, & McEwen, 2015). Few, if any, twin studies have estimated the group heritability for other forms of child and adolescent psychopathology, such as depression, anxiety, and obsessive-compulsive disorder (Martin et al., 2017).

TRAIT CORRELATIONS BETWEEN DIFFERENT TYPES OF CHILD AND ADOLESCENT PSYCHOPATHOLOGY SHOW SUBSTANTIAL GENETIC MEDIATION

Comorbidity is the rule rather than the exception in psychopathology and multivariate quantitative genetic methods, which focus on investigating the covariance between two or more traits, can be used to study not merely *if* different forms of psychopathology co-occur, but also *why*.

Multivariate quantitative genetic analyses are commonly conducted using twin data, where covariance is measured via the cross-twin, cross-trait correlation (CTCT), referring to the correlation between trait one in one twin and trait two in the other twin. Similar to the univariate twin design, genetic influences on the covariance between traits is supported if the CTCT is greater in MZ than in DZ twins. Multivariate twin models provide estimates of A, C or D, and E for each of the included traits, as well as the correlation for each variance component across the traits. The genetic correlation (r_g) is a statistical measure of the extent to which genetic influences on one trait are correlated with the genetic influences on the other trait. r_g can vary from 0, meaning genetic influences are independent, to 1, suggesting a complete sharing of genetic influences across the traits. Shared and nonshared environmental correlations (r_a and r_e) across the traits are interpreted in a similar manner.

Multivariate quantitative genetic studies have greatly advanced our understanding of the underlying causes of comorbidity in child and adolescent psychopathology. ASD and ADHD provide a compelling example: these disorders could not be diagnosed together in the DSM-IV and the ICD-10. Yet, bivariate twin studies have reported a moderate to strong degree of overlap in genetic influences on ASD and ADHD (Martin et al., 2017). A UK study of over 6,000 twin pairs reported genetic correlations >0.50 between autistic-like traits and ADHD traits in the general population (Ronald et al., 2008). An even stronger genetic overlap was found in a study of parent-rated ASD and ADHD symptoms from 16,858 Swedish twins, where the genetic correlation was 0.80 (Lichtenstein et al., 2010). Evidence for shared genetic risk between clinically diagnosed ASD and ADHD have also been reported in a recent, large-scale family study ($N = 1,899,654$), where a highly elevated familial risk was observed in MZ twins (odds ratio [OR] = 17). Further, the familial risk across siblings was found to decrease along with decreasing genetic relatedness. Childhood and adolescent anxiety and depression provide another example where twin and family studies suggest that comorbidity is primarily driven by shared genetic factors, with limited evidence for overlapping environmental risk (Franić et al., 2010).

In recent years, a growing number of multivariate twin and family studies have gone beyond studying pairs of traits, to instead analyze shared etiology across broad dimensions of psychopathology. Two of the most common approaches to do so is the independent pathways model and the common pathway model (Rijsdijk & Sham, 2002), which can model genetic and environmental influences on covariance across multiple traits as well as estimate the extent to which etiological factors are trait-specific (see Rijsdijk & Sham, 2002, for an in-depth discussion of these models). Studies using variations of the independent and common pathway model have demonstrated that genetic influences tend to cluster into factors influencing an internalizing psychopathology dimension (i.e., depression and various forms of anxiety), and an externalizing psychopathology dimension (i.e., disruptive behaviors, aggression, and hyperactivity/impulsivity). These two sets of genetic risk are in turn correlated, suggesting the presence of a common genetic liability which increases the risk for virtually all forms of child and adolescent psychopathology (Caspi et al., 2014; Lahey et al., 2017). For example, Lahey and colleagues (2011) analyzed data on 11 internalizing and externalizing psychiatric traits from 1,571 pairs of 9-to-17-year-old twins. Results revealed two separate, but highly correlated (0.89), sets of genetic factors influencing the internalizing and externalizing traits. An alternative model which in addition included a general genetic factor (commonly referred to as a bifactor model, see chapter 7) that influenced covariance across all traits was found to provide a better fit to data. In contrast, environmental influences were mostly dimension-specific or trait-specific. Similar results have been reported for parent ratings of ASD, ADHD, tics, and learning difficulties in 6,595 twin pairs aged 9 or 12 years, where a single general genetic factor explained 31 percent of covariance across these childhood neurodevelopmental symptoms (Pettersson et al., 2013). Cosgrove and colleagues (2011) analyzed symptoms of depression, anxiety, ADHD, oppositional-defiant, and conduct problems in a sample of 1,162 twin pairs and 426 siblings. Results revealed that a latent internalizing factor and a latent externalizing factor provided the best fit to the data. Both factors were moderately heritable and correlated at 0.72, with 62 percent of the covariance between internalizing and externalizing problems explained by common genetic influences.

Together, multivariate quantitative genetic research suggests that virtually all trait correlations across child and adolescent psychopathology show genetic mediation, whereas environmental influences tend to be more trait-specific. Genetic risk is shared across internalizing disorders, externalizing disorders, and across a broader psychopathology dimension that influences risk across many forms of child and adolescent psychopathology. Future efforts in quantitative genetic research will hopefully generate a more in-depth understanding of the etiological (genetic and environmental) structure underlying not just psychopathology, but also a broader disease spectrum that includes somatic and neurologic conditions which commonly co-occur with psychopathology.

STABILITY AND CHANGE IN CHILD AND ADOLESCENT PSYCHOPATHOLOGY

A key question in child and adolescent psychopathology is how etiological factors contribute to the development of psychopathology over time. Longitudinal designs, in which the same trait has been measured in the same individuals on two or more occasions, allow researchers to study the relative importance of genetic and environmental factors at different ages, and to determine how etiological factors contribute to stability and change in psychopathology across ages.

One longitudinal, developmentally informative approach is the Cholesky decomposition (Neale & Cardon, 1992), which provides age-specific estimates of A, C or D, and E, as well as estimates of the extent to which the genetic and environmental factors that explain variance at one age continue to influence variance in the same trait at later ages, reflecting etiological stability. It also estimates the extent to which new genetic and environmental factors "come online" and influence change in the trait at each measured time point.

Another approach for analyzing longitudinal data is latent growth curve modelling (see chapter 19 for an in-depth discussion on Cholesky and latent growth models) These types of models estimate an intercept, which represents the mean baseline level of the trait, and a slope, which represents the mean systematic change in the trait over time (e.g., an overall linear decrease or increase in symptoms across time). In twin analysis, the intercept and the slope are treated as latent variables, with a variance and covariance that can be decomposed into genetic and environmental influences based on the pattern of twin correlations (Posthuma et al., 2003).

Several longitudinal twin studies have focused on the development of ADHD, reporting remarkably stable heritability of ADHD from childhood into early adulthood (Bergen, Gardner, & Kendler, 2007; Chang et al., 2013). The Cholesky decomposition approach was used in a study of 4,000 young twin pairs, where parent ratings of ADHD symptoms collected at ages two, three, four, seven, and eight showed high heritability at all ages (0.77–0.86), and moderate age-to-age phenotypic stability, which was mainly explained by stable genetic influences. In addition, there was evidence of new genetic influences, which were not shared with those acting earlier on, emerging at later ages. The contribution of environmental influences to the stability of the ADHD symptoms over time was negligible (Kuntsi et al., 2005). This pattern of results, with stable and high heritability of ADHD across ages and evidence of genetic stability, but also new genetic factors emerging at different developmental stages, has been replicated in several twin studies (Chang et al., 2013; Larsson, Lichtenstein, & Larsson, 2006; Rietveld et al., 2004). Latent growth curve modelling was also used in a recent twin study on individual differences in the developmental course of ADHD symptoms for ages 8–16 years in a sample of 8,395 UK twin pairs (Pingault et al., 2015). Hyperactivity/impulsivity symptoms were found to show a sharp linear decrease across ages, and high heritability both for baseline-level symptoms (90 percent) and for the change in symptoms over time (81 percent). However, only about 40 percent of the genetic factors influencing the developmental course of symptoms were shared with the genetic factors influencing the baseline hyperactivity/impulsivity.

Few studies have focused on etiological age-related changes in ASD, likely due to the developmentally more stable course of ASD in clinical populations. Nevertheless, two recent twin studies investigated developmental continuity in autistic-like traits in the general population. In a study of over 6,000 twin pairs followed from ages 8/9 to 12 years, autistic traits were found to be stable and show moderate to high heritability at each age. Phenotypic stability across ages was largely accounted for by genetic factors (Holmboe et al., 2014). Similar results were found in a sample of 2,500 twin pairs assessed at age 9 or 12 years and again at age 18. Although the stability of autistic-like traits from childhood to early adulthood were weaker than that reported in younger ages, 85 percent of the phenotypic correlation across ages was explained by stability in additive genetic factors (Taylor et al., 2017). Twin studies estimating the heritability of anxiety and depression at different ages have reported conflicting results, with some studies showing an increased heritability over time, and others reporting decreasing heritability from childhood to adolescence (Bergen et al., 2007; Nivard et al., 2014; Rice et al., 2002). Findings regarding the underlying sources of stability and change are somewhat more consistent, with two longitudinal twin studies reporting that genetic factors account for about 50 to 70 percent of phenotypic stability in anxious and depressive symptoms across childhood (Boomsma et al., 2008; Trzaskowski et al., 2012).

In summary, evidence from longitudinal quantitative genetic research has shown that stability of child and adolescent psychopathology across development is often largely explained by genetic stability, whereas environmental factors are often age-specific and responsible for change (Plomin et al., 2016). Age-dependent changes in heritability are evident for some forms of child and adolescent psychopathology, but not all, and have been suggested to reflect gene-environment interactions. Changes in raters and assessments across time can also influence results in longitudinal quantitative genetic studies, and remains an issue that requires careful consideration (Kan et al., 2014). Further research is needed to fully understand the origins of development of psychopathology across the lifespan, yet there is still a lack of prospectively collected data with sufficient follow-up time for many important conditions in child and adolescent psychopathology (e.g., ASD), and a paucity of research exploring the role of genetic and environmental factors for the development from childhood into older ages.

GENE-ENVIRONMENT INTERPLAY IN CHILD AND ADOLESCENT PSYCHOPATHOLOGY

Quantitative genetics includes two concepts of gene-environment interplay; gene-environment interaction (GxE) and gene-environment correlation (rGE). GxE refers to genetic sensitivity to environments, meaning that the effect of the environment on the expression of a given trait will vary depending on an individual's genetic makeup. rGE, on the other hand, refers to a situation in which genetic propensities are correlated with individual differences in environmental experiences. As the literature on GxE is still limited, in the following sections we will instead focus on the principles of rGE and review the evidence for rGE in child and adolescent psychopathology.

rGE Is Pervasive in Child and Adolescent Psychopathology

Measures of the environment widely used in studies of child and adolescent psychopathology, such as parenting styles and life events, can also be included in quantitative genetic analyses. An influential review of the available quantitative genetic studies of environmental measures found pervasive evidence for genetic influences. Statistically significant heritability estimates were reported for objective measures such as videotaped observations of parenting, as well as self-report measures of parenting styles and life events (Plomin & Bergeman, 2011). Since then, many quantitative genetic studies have found significant heritability estimates for many different environmental measures; a review of 55 quantitative genetic studies reported an average heritability estimate of 27 percent across 35 different environmental measures. This included not only measures of the family environment, but also measures of peer groups, classroom environments, neighborhood characteristics, and life events (Kendler & Baker, 2007).

What Processes Explain the Genetic Involvement in Putative Environmental Measures?

The issue of genetic influences on putative environmental measures is known as rGE, and can be defined as a correlation between an individual's genome and the environment that they inhabit. Three forms of rGE have been described in the quantitative genetic literature: passive, active, and evocative (Plomin et al., 2013). First, passive rGE describes the association between an individual's genotype and the environment in which the individual is raised, both of which are provided by the individual's biological parents. For example, a child may inherit genetic factors involved in conduct problems from their parent in whom the same genetic factors may be involved in harsh parental discipline. Second, active rGE involves the genetically influenced behavior of the individual seeking out an environment that "matches" their genotype. For example, a child's genetically influenced conduct problems may lead the child to actively seek confrontation with their parents. Third, evocative rGE involves the genetically influenced behavior of the individual seeking or evoking a particular response from the environment. For example, a child's genetically influenced conduct problems may evoke harsh discipline from their parents.

What Methods Can Be Used to Detect and Disentangle the Different Types of rGE?

Several methods are available to investigate rGE. One method for detecting rGE involves estimating correlations between biological parents' behavioral traits and the adoptive family's environment. This method is able to detect evocative rGE and active rGE, under the premise that the traits of the biological parents can be used as a proxy of the adopted child's genetic risk, and this genetic risk can be correlated with any measure of the adopted child's environment. Finding that biological parents' behavioral traits correlate with the environment of their adopted-away child suggests that the environmental measure in part reflects genetically influenced characteristics of the adopted child. This method has been used in a study (Ge et al., 1996) including data collected from biological and adoptive parents of children adopted at birth. The adopted children were classified as being at genetic risk or not at genetic risk for antisocial behavior based on diagnosed substance use or antisocial personality disorders in the biological parents. Antisocial behavior in the adopted children and the adoptive parents' harsh/inconsistent parenting were assessed via multiple sources of information (i.e., adoptive parents, the adoptee, and observers). The results showed that adoptive children at high genetic risk for antisocial behavior received more negative parenting from their adoptive parents compared to adopted children at low genetic risk. It was also demonstrated that the association between the child's genetic risk and negative parenting was partly mediated via the child's genetically influenced antisocial behavior (Ge et al., 1996). Another study, which linked 320 adopted children aged six years old to their adoptive and biological mothers, explored the role of evocative rGE using longitudinal assessments of maternal ADHD symptoms, parenting practices, child impulsivity/activation, and child ADHD symptoms (Harold et al., 2013). The results provided support for evocative rGE processes in which biologically related maternal ADHD symptoms were linked to adoptive maternal hostile parenting. This effect was mediated through early impulsivity/activation in the child. Maternal hostile parenting and impulsivity/activation in the child was in turn associated with later child ADHD symptoms. These results support evocative rGE as genetically influenced child ADHD traits evoke maternal hostility, which in turn links to later child ADHD symptoms.

Another method for detecting rGE involves comparisons of the correlation between an environmental measure and a behavioral trait in nonadoptive families with the corresponding correlation in adoptive families. Evidence for passive rGE would be suggested if the correlation between the environmental measure and the behavioral trait was stronger in nonadoptive families than in adoptive families. This method has been used to demonstrate that the association between broad measures of the home environment and the cognitive development in young children is higher in nonadoptive than in adoptive families (Plomin, Loehlin, & DeFries, 1985), which supports passive rGE. The approach has, unfortunately, not been used in relation to childhood and adolescent psychopathology.

Another approach for detecting rGE involves multivariate quantitative genetic analysis, described in previous sections, using data from an environmental measure and a measure of child and adolescent psychopathology. In this approach, any type of rGE (passive, evocative, or active) is indicated if genetic effects on the environmental measure overlap with genetic effects on the measured psychopathology. In one twin study, bivariate genetic analysis was used to test the contribution of genetic and environmental factors to the association between parent-child hostility and ADHD symptoms in a sample of 886 twin pairs, aged 11–17 years. The association between parent–child hostility and child ADHD symptoms was substantially attributed to genetic factors (Lifford, Harold, & Thapar, 2009). Another study using this approach indicated that a substantial part of the association between parental negativity and antisocial behavior and depressive symptoms in offspring could be attributed to genetic factors (Pike et al., 1996). Burt et al (2005) fitted a cross-lagged twin model to data on parent–child conflict and children's antisocial behavior assessed by maternal reports as well as child reports. Both parent–child conflict and antisocial behavior at age 11 were found to independently predict the other at age 14, providing support for a bidirectional effect model that allows for both parent-driven and child-driven processes. In addition, the result indicated that the parent-driven and the child-driven effects were a function of both genetic and environmental influences. Thus, this study suggests, on the one hand, that parent–child conflict contributes to childhood antisocial behavior via environmental mechanisms, but also that genetically influenced antisocial behavior evokes parent-child conflict. Similar findings of bidirectional effect for antisocial behavior have been observed in other samples and with other measures of parenting (Larsson et al., 2008), whereas genetically influenced child effects have been found to have a more pronounced effect for internalizing problems (Moberg et al., 2011). Although several quantitative genetic studies have explored the role of rGE for measures of ADHD, depression, and antisocial behaviors, little is known about such processes for the development of ASD.

Implications

One implication of rGE is that the correlation between a putative environmental measure and a behavioral trait may not necessarily reflect environmental causation. In other words, association between environmental risk factors and psychopathology outcomes can be confounded by genetic factors that are shared within the family. Disentangling whether association between environmental risk factors and outcomes in psychopathology reflect environmental causation, genetic and/or environmental confounding, or are indicative of other complex reciprocal processes is crucial to further our understanding of the etiology and course of disorders, and has important implications for prevention in psychopathology. One useful quantitative genetic approach to address such questions is the sibling-comparison design. Instead of comparing unrelated individuals who vary in their exposure to a risk factor, which is the typical study design, sibling-comparison studies explore differences among siblings who were differentially exposed to a specific risk factor (e.g., one sibling was exposed to smoking during pregnancy, but their sibling was not). These comparisons help account for certain types of confounding because siblings share familial factors that could confound the associations between the risk and outcome. For example, the comparison of siblings raised in the same family accounts for all environmental confounds that are shared by siblings, such as early life shared environmental factors. Furthermore, sibling comparisons help rule out some forms of genetic confounding (e.g., when genetic factors influence both risk of exposure and the outcome; D'Onofrio et al., 2013) The process of meiosis, the type of cell division that produces eggs and sperm, randomly distributes alleles from parents to each of their offspring. As such, this random process rules out systematic genetic confounding due to passive rGE. For example, the comparison of full siblings (siblings who share 50 percent of their genes), rules out the possibility that genetic factors passively passed down from both parents could account for the association between the exposure and the outcome. The design does not rule out genetic confounding arising from active or evocative rGE, which occurs when a child's genotype influences the exposure to the risk (and can vary between siblings). To completely rule out confounding by genetic factors researchers can compare MZ twins, for whom differences in exposure-outcome associations can only be explained by environmental factors as MZ twin share 100 percent of their genetic sequences (McGue, Osler, & Christensen, 2010)

CONCLUDING REMARKS

In this chapter, we have described how quantitative genetic methods can be used to move beyond merely estimating heritability. The key question in this line of research is not "if" genetic factors are important, but rather "how" genetic factors influence different forms of child and adolescent psychopathology across time, and in interaction with environmental factors. We conclude that multivariate and longitudinal quantitative genetic methods can be used to address fundamental questions about the etiology of child and adolescent psychopathology across different levels of symptom severity, comorbidity, and development, as well as the influence of gene-environment interplay in child and adolescent psychopathology.

We have also highlighted questions that need to be addressed in future quantitative genetic research. First, for many forms of psychopathology it is still unclear whether the etiological underpinnings are the same across different levels of severity of psychopathology. Second, the etiological (genetic and environmental) structure underlying not just psychopathology, but also a broader disease spectrum that includes somatic and neurologic conditions, is still unclear. Third, there is a lack of quantitative genetic studies with long-term follow-up for many important forms of child and adolescent psychopathology. Fourth, more quantitative genetic research is needed to clarify the role of rGE and also the impact of genetic confounding for associations between "environmental" risk factors and different forms of child and adolescent psychopathology.

REFERENCES

Bailey, A., Phillips, W., & Rutter, M. (1996). Autism: Towards an Integration of Clinical, Genetic, Neuropsychological, and Neurobiological Perspectives. *Journal of Child Psychology and Psychiatry and Allied Disciplines*, 37(1), 89–126.

Bergen, S. E., Gardner, C. O., & Kendler, K. S. (2007). Age-Related Changes in Heritability of Behavioral Phenotypes over Adolescence and Young Adulthood: A Meta-Analysis. *Twin Research and Human Genetics*, 10(3), 423–433.

Biederman, J., Faraone, S. V., Keenan, K., Benjamin, J., Krifcher, B., Moore, C., ... Steingard, R. (1992). Further Evidence for Family-Genetic Risk Factors in Attention Deficit Hyperactivity Disorder: Patterns of Comorbidity in Probands and Relatives in Psychiatrically and Pediatrically Referred Samples. *Archives of General Psychiatry*, 49(9), 728–738.

Boomsma, D. D., Van Beijsterveldt, C. E. M., Bartels, M., & Hudziak, J. J. (2008). Genetic and Environmental Influences on Anxious/Depression: A Longitudinal Study in 3- to 12-Year-Old Children. In J. J. Hudziak (Ed.), *Developmental Psychopathology and Wellness: Genetic and Environmental Influences*. Arlington, VA: American Psychiatric Publishing.

Burt, S. A. (2009). Rethinking Environmental Contributions to Child and Adolescent Psychopathology: A Meta-Analysis of Shared Environmental Influences. *Psychological Bulletin*, 135(4), 608–637.

Burt, S. A., McGue, M., Krueger, R. F., & Iacono, W. G. (2005). Sources of Covariation among the Child-Externalizing Disorders: Informant Effects and the Shared Environment. *Psychological Medicine*, 35(8), 1133–1144.

Caspi, A., Houts, R. M., Belsky, D. W., Goldman-Mellor, S. J., Harrington, H., Israel, S., ... Moffitt, T. E. (2014). The p Factor: One General Psychopathology Factor in the Structure of Psychiatric Disorders? *Clinical Psychological Science*, 2(2), 119–137.

Chang, Z., Lichtenstein, P., Asherson, P. J., & Larsson, H. (2013). Developmental Twin Study of Attention Problems: High Heritabilities throughout Development. *JAMA Psychiatry*, 70(3), 311–318.

Chen, Q., Brikell, I., Lichtenstein, P., Serlachius, E., Kuja-Halkola, R., Sandin, S., & Larsson, H. (2017). Familial Aggregation of Attention-Deficit/Hyperactivity Disorder. *Journal of Child Psychology and Psychiatry and Allied Disciplines*, 58(3), 231–239.

Colvert, E., Tick, B., & McEwen, F. (2015). Heritability of Autism Spectrum Disorder in a UK Population-Based Twin Sample. *JAMA Psychiatry*, 72(5), 415–423.

Cosgrove, V. E., Rhee, S. H., Gelhorn, H. L., Boeldt, D., Corley, R. C., Ehringer, M. A., ... Hewitt, J. K. (2011). Structure and Etiology of Co-occurring Internalizing and Externalizing Disorders in Adolescents. *Journal of Abnormal Child Psychology*, 39, 109–123.

D'Onofrio, B. M., Lahey, B. B., Turkheimer, E., & Lichtenstein, P. (2013). Critical Need for Family-Based, Quasi-Experimental Designs in Integrating Genetic and Social Science Research. *American Journal of Public Health*, 103(S1), S46–S55.

DeFries, J. C., & Fulker, D. W. (1988). Multiple Regression Analysis of Twin Data: Etiology of Deviant Scores versus Individual Differences. *Acta Geneticae Medicae et Gemellologiae*, 37(3–4), 205–216.

Eley, T. C., Deater-Deckard, K., Fombonne, E., Fulker, D. W., & Plomin, R. (1998). An Adoption Study of Depressive Symptoms in Middle Childhood. *The Journal of Child Psychology and Psychiatry and Allied Disciplines*, 39(3), 337–345.

Falconer, D. S. (1965). The Inheritance of Liability to Certain Diseases, Estimated from the Incidence among Relatives. *Annals of Human Genetics*, 29(1), 51–76.

Faraone, S., Biederman, J., & Monuteaux, M. (2000). Toward Guidelines for Pedigree Selection in Genetic Studies of Attention Deficit Hyperactivity Disorder. *Genetic Epidemiology*, 18(1), 1–16.

Faraone, S. V., & Larsson, H. (2018). Genetics of Attention Deficit Hyperactivity Disorder. *Molecular Psychiatry*, 24(4), 562–575.

Franić, S., Middeldorp, C. M., Dolan, C. V., Ligthart, L., & Boomsma, D. I. (2010). Childhood and Adolescent Anxiety and Depression: Beyond Heritability. *Journal of the American Academy of Child and Adolescent Psychiatry*, 49(8), 820–829.

Ge, X., Conger, R. D., Cadoret, R. J., Neiderhiser, J. M., Yates, W., Troughton, E., & Stewart, M. A. (1996). The Developmental Interface between Nature and Nurture: A Mutual Influence Model of Child Antisocial Behavior and Parent Behaviors. *Developmental Psychology*, 32(4), 574–589.

Ghirardi, L., Brikell, I., Kuja-Halkola, R., Freitag, C. M., Franke, B., Asherson, P., ... Larsson, H. (2017). The Familial Co-Aggregation of ASD and ADHD: A Register-Based Cohort Study. *Molecular Psychiatry*, 23(2), 257–262.

Gregory, A. M., & Eley, T. C. (2007). Genetic Influences on Anxiety in Children: What We've Learned and Where We're Heading. *Clinical Child and Family Psychology Review*, 10(3), 199–212.

Harold, G. T., Leve, L. D., Barrett, D., Elam, K., Neiderhiser, J. M., Natsuaki, M. N., ... Thapar, A. (2013). Biological and Rearing Mother Influences on Child ADHD Symptoms: Revisiting the Developmental Interface between Nature and Nurture. *Journal of Child Psychology and Psychiatry and Allied Disciplines*, 54(10), 1038–1046.

Holmboe, K., Rijsdijk, F. V., Hallett, V., Happe, F., Plomin, R., & Ronald, A. (2014). Strong Genetic Influences on the Stability of Autistic Traits in Childhood. *Journal of the American Academy of Child and Adolescent Psychiatry*, 53(2), 221–230.

Kan, K. J., van Beijsterveldt, C. E., Bartels, M., & Boomsma, D. I. (2014). Assessing Genetic Influences on Behavior: Informant and Context Dependency as Illustrated by the Analysis of Attention Problems. *Behavior Genetics*, 44(4), 326–336.

Kendler, K. S., & Baker, J. H. (2007). Genetic Influences on Measures of the Environment: A Systematic Review. *Psychological Medicine*, 37(5), 615–626.

Kuntsi, J., Rijsdijk, F., Ronald, A., Asherson, P., & Plomin, R. (2005). Genetic Influences on the Stability of Attention-Deficit/Hyperactivity Disorder Symptoms from Early to Middle Childhood. *Biological Psychiatry*, 57(6), 647–654.

Lahey, B. B., Van Hulle, C. A., Singh, A. L., Waldman, I. D., & Rathouz, P. J. (2011). Higher-Order Genetic and Environmental Structure of Prevalent Forms of Child and Adolescent Psychopathology. *Archives of General Psychiatry*, 68(2), 181–189.

Lahey, B. B., Krueger, R. F., Rathouz, P. J., Waldman, I. D., & Zald, D. H. (2017). A Hierarchical Causal Taxonomy of Psychopathology across the Life Span. *Psychological Bulletin*, 143(2), 142–186.

Larsson, H., Lichtenstein, P., & Larsson, J. O. (2006). Genetic Contributions to the Development of ADHD Subtypes from Childhood to Adolescence. *Journal of the American Academy of Child and Adolescent Psychiatry*, 45(8), 973–981.

Larsson, H., Viding, E., Rijsdijk, F. V., & Plomin, R. (2008). Relationships between Parental Negativity and Childhood Antisocial Behavior Over Time: A Bidirectional Effects Model in a Longitudinal Genetically Informative Design. *Journal of Abnormal Child Psychology*, 36(5), 633–645.

Larsson, H., Anckarsater, H., Rastam, M., Chang, Z., & Lichtenstein, P. (2012). Childhood Attention-Deficit Hyperactivity Disorder as an Extreme of a Continuous Trait: A Quantitative Genetic Study of 8,500 Twin Pairs. *Journal of Child Psychology and Psychiatry and Allied Disciplines*, 53(1), 73–80.

Levy, F., Hay, D., McStephen, M., Wood, C., & Waldman, I. (1997). Attention-Deficit Hyperactivity Disorder: A Category or a Continuum? Genetic Analysis of a Large-Scale Twin Study. *Journal of the American Academy of Child and Adolescent Psychiatry*, 36(6), 737–744.

Lichtenstein, P., Carlström, E., Råstam, M., Gillberg, C., & Anckarsäter, H. (2010). The Genetics of Autism Spectrum Disorders and Related Neuropsychiatric Disorders in Childhood. *American Journal of Psychiatry*, 167(11), 1357–1363.

Lifford, K. J., Harold, G. T., & Thapar, A. (2009). Parent-Child Hostility and Child ADHD Symptoms: A Genetically Sensitive and Longitudinal Analysis. *Journal of Child Psychology and Psychiatry and Allied Disciplines*, 50(12), 1468–1476.

Lundström, S., Chang, Z., Kerekes, N., Gumpert, C. H., Råstam, M., & Gillberg, C. (2011). Autistic-Like Traits and Their Association with Mental Health Problems in Two Nationwide Twin Cohorts of Children and Adults. *Psychological Medicine*, 41(11), 2423–2433.

Martin, J., Taylor, M. J., & Lichtenstein, P. (2017). Assessing the Evidence for Shared Genetic Risks across Psychiatric Disorders and Traits. *Psychological Medicine*, 48(11), 1–16.

McGue, M., Osler, M., & Christensen, K. (2010). Causal Inference and Observational Research: The Utility of Twins. *Perspectives on Psychological Science*, 5(5), 546–556.

McGuffin, P., Owen, M. J., & Gottesman, I. I. (2004). *Psychiatric Genetics and Genomics:* Oxford: Oxford University Press.

Moberg, T., Lichtenstein, P., Forsman, M., & Larsson, H. (2011). Internalizing Behavior in Adolescent Girls Affects Parental Emotional Overinvolvement: A Cross-Lagged Twin Study. *Behavior Genetics*, 41(2), 223–233.

Neale, M., & Cardon, L. (1992). *Methodology for Genetic Studies of Twins and Families*. Dordrecht: Kluwer Academic.

Nikolas, M. A., & Burt, S. A. (2010). Genetic and Environmental Influences on ADHD Symptom Dimensions of Inattention and Hyperactivity: A Meta-Analysis. *Journal of Abnormal Psychology*, 119(1), 1–17.

Nivard, M. G., Dolan, C. V., Kendler, K. S., Kan, K. J., Willemsen, G., van Beijsterveldt, C. E. M., ... Boomsma, D. I. (2014).

Stability in Symptoms of Anxiety and Depression as a Function of Genotype and Environment: A Longitudinal Twin Study from Ages 3 to 63 Years. *Psychological Medicine, 45*(5), 1039–1049.

Pettersson, E., Anckarsater, H., Gillberg, C., & Lichtenstein, P. (2013). Different Neurodevelopmental Symptoms Have a Common Genetic Etiology. *Journal of Child Psychology and Psychiatry and Allied Disciplines, 54*(12), 1356–1365.

Pike, A., McGuire, S., Hetherington, E. M., Reiss, D., & Plomin, R. (1996). Family Environment and Adolescent Depressive Symptoms and Antisocial Behavior: A Multivariate Genetic Analysis. *Developmental Psychology, 32*(4), 590–603.

Pingault, J., Viding, E., Galéra, C., Greven, C. U., Zheng, Y., Plomin, R., & Rijsdijk, F. (2015). Genetic and Environmental Influences on the Developmental Course of Attention-Deficit/Hyperactivity Disorder Symptoms from Childhood to Adolescence. *JAMA Psychiatry, 72*(7), 651–658.

Plomin, R., & Bergeman, C. S. (2011). The Nature of Nurture: Genetic Influence on "Environmental" Measures. *Behavioral and Brain Sciences, 14*(3), 373–386.

Plomin, R., Loehlin, J. C., & DeFries, J. (1985). Genetic and Environmental Components of "Environmental" Influences. *Developmental Psychology, 21*(3), 391–402.

Plomin, R., DeFries, J. C., Knopik, V. S., & Neiderhiser, J. (2013). *Behavioral Genetics*. New York: Worth Publishers.

Plomin, R., DeFries, J. C., Knopik, V. S., & Neiderhiser, J. M. (2016). Top 10 Replicated Findings from Behavioral Genetics. *Perspectives on Psychological Science: A Journal of the Association for Psychological Science, 11*(1), 3–23.

Polderman, T. J., Benyamin, B., de Leeuw, C. A., Sullivan, P. F., van Bochoven, A., Visscher, P. M., & Posthuma, D. (2015). Meta-Analysis of the Heritability of Human Traits Based on Fifty Years of Twin Studies. *Nature Genetics, 47*(7), 702–709.

Posthuma, D., & Polderman, T. J. (2013). What Have We Learned from Recent Twin Studies about the Etiology of Neurodevelopmental Disorders? *Current Opinion in Neurology, 26*(2), 111–121.

Posthuma, D., Beem, A. L., de Geus, E. J., van Baal, G. C., von Hjelmborg, J. B., Iachine, I., & Boomsma, D. I. (2003). *Theory and Practice in Quantitative Genetics. Twin Research, 6*(5), 361–376.

Rhee, S. H., & Waldman, I. D. (2002). Genetic and Environmental Influences on Antisocial Behavior: A Meta-Analysis of Twin and Adoption Studies. *Psychological Bulletin, 128*(3), 490–529.

Rice, F., Harold, G., & Thapar, A. (2002). The Genetic Aetiology of Childhood Depression: A Review. *Journal of Child Psychology and Psychiatry, 43*(1), 65–79.

Rietveld, M. J., Hudziak, J. J., Bartels, M., van Beijsterveldt, C. E., & Boomsma, D. I. (2004). Heritability of Attention Problems in Children: Longitudinal Results from a Study of Twins, Age 3 to 12. *Journal of Child Psychology and Psychiatry and Allied Disciplines, 45*(3), 577–588.

Rijsdijk, F. V., & Sham, P. C. (2002). Analytic Approaches to Twin Data Using Structural Equation Models. *Briefings in Bioinformatics, 3*(2), 119–133.

Ronald, A., Simonoff, E., Kuntsi, J., Asherson, P., & Plomin, R. (2008). Evidence for Overlapping Genetic Influences on Autistic and ADHD Behaviours in a Community Twin Sample. *Journal of Child Psychology and Psychiatry, and Allied Disciplines, 49*(5), 535–542.

Sandin, S., Lichtenstein, P., Kuja-Halkola, R., Hultman, C., Larsson, H., & Reichenberg, A. (2017). The Heritability of Autism Spectrum Disorder. *JAMA, 318*(12), 1182–1184. doi:10.1001/jama.2017.12141

Smalley, S. L., Asarnow, R. F., & Spence, M. A. (1988). Autism and Genetics: A Decade of Research. *Archives of General Psychiatry, 45*(10), 953–961.

Taylor, M. J., Gillberg, C., Lichtenstein, P., & Lundström, S. (2017). Etiological Influences on the Stability of Autistic Traits from Childhood to Early Adulthood: Evidence from a Twin Study. *Molecular Autism, 8*(1), 5.

Thapar, A., & Stergiakouli, E. (2008). Genetic Influences on the Development of Childhood Psychiatric Disorders. *Psychiatry, 7*(7), 277–281.

Tick, B., Bolton, P., Happe, F., Rutter, M., & Rijsdijk, F. (2016). Heritability of Autism Spectrum Disorders: A Meta-Analysis of Twin Studies. *Journal of Child Psychology and Psychiatry and Allied Disciplines, 57*(5), 585–595.

Trzaskowski, M., Zavos, H. M. S., Haworth, C. M. A., Plomin, R., & Eley, T. C. (2012). Stable Genetic Influence on Anxiety-Related Behaviours Across Middle Childhood. *Journal of Abnormal Child Psychology, 40*(1), 85–94.

Van den Oord, E. J., Boomsma, D. I., & Verhulst, F. C. (1994). A Study of Problem Behaviors in 10- to 15-Year-Old Biologically Related and Unrelated International Adoptees. *Behavior Genetics, 24*(3), 193–205.

18 Designing and Managing Longitudinal Studies

M. BRENT DONNELLAN AND DEBORAH A. KASHY

Longitudinal research is essential for understanding how psychopathology and psychological processes unfold over time – but it is an inherently risky endeavor. Longitudinal research requires considerable amounts of time and effort for an uncertain return. Many decisions are involved in designing longitudinal studies and these decisions, typically made early on in the research process, can have profound effects on the quality and utility of the data generated. Our goal is to provide a big-picture perspective and practical advice for designing and managing effective longitudinal studies to increase the chances that a longitudinal study is successful.[1]

The methodological literature contains many references for helping researchers analyze longitudinal data including many of the chapters in this volume (see also Collins, 2006; Ferrer & McArdle, 2003; Grimm, Davoudzadeh, & Ram, 2017; Little, 2013; Singer & Willett, 2003) but relatively fewer resources offering guidance about basic design and how to manage the study on a day-to-day basis (Donnellan & Conger, 2007; Taris, 2000; but see Faden et al., 2004; Graziotti et al., 2012; Morton et al., 2014; Stouthamer-Loeber, van Kammen, & Loeber, 1992; Wright & Markon, 2016). This asymmetry is unfortunate as a sophisticated analysis can rarely overcome deficits in the underlying data. If a study adopts poor measurement strategies, flawed sampling plans, questionable design choices, or if the study suffers high levels of attrition across time, there is little hope that a fancy state-of-the-art analysis will be able to turn lead into gold.

CRITICAL DESIGN QUESTIONS

The fundamental question to ask before designing a longitudinal study is whether such a study is necessary given the current literature. Just because a longitudinal design can be used does not mean it should be used. The inevitable lag between the start of a longitudinal study and the point when results are ready for publication means there should be solid reasons for conducting the study in the first place. Longitudinal studies should build on existing knowledge and hold the potential to fill important gaps. Researchers should strive to maximize the scientific pay-off for the time and resources invested in designing and executing a longitudinal study. Key questions to consider include: Is the literature sufficiently developed to warrant a longitudinal study? Is there a set of questions that need to be answered? Are there debates about stability and change? Are there debates about temporal precedence and causal dynamics? How exactly will this longitudinal study address such questions?

The four critical questions described in Table 18.1 are fundamental when designing a longitudinal study. We think it is valuable to engage in a thought experiment when considering these questions. That is, we suggest that researchers design the *ideal* longitudinal study when first considering the issues. Ignore concerns about feasibility and simply outline the optimal way to measure constructs and how often constructs will be measured. Select a sample size that will provide ideal power for key statistical tests and a high degree of precision in parameter estimation. The objective is to identify the kind of design that researchers *ought* to use to investigate a given topic. If this task proves too difficult, it is a good signal that more work is needed before one can meaningfully plan a longitudinal study.

The generation of an ideal design provides a point of contrast for evaluating the merits of specific design options that emerge when practical constraints are factored into the decision-making process. Time, money, and the availability of personnel inevitably place restrictions on the actual kinds of studies that can be conducted. The larger the gap between the ideal design and the actual design, the more worried researchers should be about the gamble of longitudinal research. In short, our first and most foundational piece of advice is to start with the ideal design in mind and work down to what is feasible. Make sure, however, that the feasible design is still able to address the motivating research questions.

---

[1] This is an update of a previous chapter by Donnellan and Conger (2007).

Table 18.1 Critical design decisions in longitudinal research

Question 1: What is the study about?

Key issues: What are the focal constructs? Who is the target population?

Basic advice: Make sure a longitudinal study is necessary. Study important constructs. Make sure foundational cross-sectional studies have been conducted.

Question 2: What is the basic design?

Key issues: How long will the study last? Across how many waves will data be collected? What is the frequency of assessment? What is the starting point of the study?

Basic advice. Strive for three or more waves of data collection. Make sure the timing of assessments is theoretically informed and tied to a model of stability and change. Select the starting point of the study for conceptual reasons rather than convenience. Anticipate statistical analyses.

Question 3: What is the measurement strategy?

Key issues: What measures and tests will be used? Will a multiple informant design be used?

Basic advice: Use measures that have evidence attesting to validity and reliability. Pilot test measures. Use the same core items at each wave. Strive for multiple informants. Have a plan for dealing with multiple measures of the same construct. Have a plan for dealing with multiple-informant data.

Question 4: What is the sample size?

Key issues: What level of statistical precision is desired? What is the smallest effect size of interest?

Basic advice: Bigger N is better (in the abstract). Effect sizes are more likely to be small than large. Plan for attrition. Strive to set beta at 0.05 (i.e., power set to 0.95), if at all possible.

the literature. These facts generate concerns about the generalizability of findings and make an honest consideration of external validity difficult. Hartmann (2005) suggests that researchers routinely identify their target populations so they can begin to consider issues of generalization.

Researchers should ideally draw a sample at random from the target population. However, this ideal approach is often impractical. Two sampling strategies that may be more externally valid relative to pure convenience sampling should be considered: deliberate sampling for heterogeneity and the impressionistic modal instance approach (Hartmann, 2005). Deliberate sampling for heterogeneity (or purposive sampling of heterogeneous instances; Shadish, Cook, & Campbell, 2002) attempts to maximize variation in the constructs of interest. The point is to find variation in the level of psychopathology or the mixture of risk and protective factors thought to be relevant for the etiology of the psychopathology. For example, if family environments are thought to be relevant for externalizing psychopathology, the point is to sample for a wide range of family environments. In contrast to this approach, the impressionistic modal instance model (or the purposive sampling of typical instances; Shadish et al., 2002) starts with the target(s) of generalization and then researchers sample cases that appear to be the closest match to the target(s) (i.e., the point is to find a sample that is "impressionistically" similar to the targets of generalization). Researchers might want to generate subsamples that match the prototype of highly advantaged family environments, average family environments, and highly disadvantaged environments. This focus on the importance of sampling reflects our perspective regarding the importance of understanding the ideal design when planning a longitudinal study.

Question One: What Is the Study About?

The ability to focus on important topics of enduring interest rather than research fads is indispensable when it comes to designing a longitudinal study. It is also helpful if there are competing theories about causal processes to help design an informative study. It can be helpful to consult colleagues and peers for advice in the design phases to help identify the myriad alternative explanations and competing theoretical claims that might be relevant.

After figuring out the general topic and approach, the next key issue involves the target population. Convenience samples are common in longitudinal research but this can be problematic because the biases introduced by convenience samples often go unrecognized and are nearly impossible to estimate (Hartmann, 2005). Sampling is arguably the Achilles' heel of psychology (Caspi, 1998) and WEIRD (Western, educated, industrialized, rich, and democratic; Henrich, Heine, & Norenzayan, 2010) samples dominate

Question Two: What Is the Basic Design?

A famous line in the literature about the analysis of longitudinal data is that "two waves of data are better than one, but maybe not much better" (Rogosa, Brandt, & Zimowski, 1982, p. 744). Two-wave studies are often a reasonable first longitudinal investigation but they provide limited information. First, participants can only increase, decrease, or remain constant across the two waves – there is no information about more complicated patterns of change. Second, two-wave studies do not provide an especially rigorous way to evaluate hypotheses about mediation (Cole & Maxwell, 2003; Mitchell & Maxwell, 2013; Selig & Preacher, 2009). Likewise, two-wave studies do not allow researchers to model how retest coefficients vary across multiple intervals, an approach that is useful for better understanding where constructs fall on the trait-state continuum (e.g., Fraley & Roberts, 2004; Kenny & Zautra, 1995).

Researchers who want to take advantage of the strengths of longitudinal designs will typically want to collect more than two waves of data. Such designs make it is possible to test the adequacy and appropriateness of a linear conception of change. Moreover, the precision of parameter estimates in individual growth models and the precision of estimates of the rate of change increases with additional waves (Rast & Hofer, 2014; Willett, Singer, and Martin, 1998). The increases are particularly pronounced when there are only a few waves of data. For example, adding a fourth wave to a three-wave study produces greater gains than adding the 19th wave to an 18-wave study (see Willett et al., 1998, figure 4).

A closely related issue to the number of waves of data collection is the onset and timing of the assessments (i.e., the time lag between measurements; see Dormann & Griffin, 2015 and Timmons & Preacher, 2015). These critical design issues are often decided by practical matters rather than a careful consideration of theoretical or conceptual factors. For example, it is common for researchers to write grant proposals in which they plan their longitudinal study to fit within a three-year grant budget period. An intervention study may propose that assessments occur across a six-month period (baseline, three months postintervention, and six months postintervention). With this schedule, and anticipating recruitment to be at a particular level each year, they can achieve their desired sample size across a two-year period. This leaves them the final year for data analysis and manuscript preparation. Such an approach – where timing is set to the grant schedule rather than the schedule of the psychological phenomena of interest – is likely to be a mistake and may render the design uninformative (Cole & Maxwell, 2003; Collins, 2006; Gollob & Reichardt, 1987; Mitchell & James, 2001). Psychological processes do not necessarily play out according to standard funding cycles.

The decision as to when to start a study in terms of the lives of participants is something that needs to be considered with care. It should be informed by relevant characteristics such as the age of participants and their first exposure to risk factors (e.g., a cancer diagnosis). Researchers may opt for heterogeneity (e.g., participants vary in age or exposure to risk factors) at baseline or uniformity at baseline (e.g., all participants are of similar ages or uniform with respect to exposure to risk such as being offspring of alcoholic parents). Different approaches will be better suited for different kinds of studies and we simply encourage researchers to consider their options. Make sure there is a strong conceptual rationale for the starting point of the study and complicated designs such as a cohort-sequential approach or accelerated longitudinal design (Anderson, 1993; Cattell, 1970; Schaie, 1965) might be necessary.

The timing of measurements is perhaps the most important and most challenging design consideration. If the interval is too short, data collection efforts are wasted as little change will have occurred. If the interval is too long, the design will be underpowered for detecting patterns of change because critical changes will already have occurred. It may be difficult to model trajectories with much fidelity. Likewise, the lag between assessments is important for studies designed to test the plausibility of causal hypotheses.

Cause-and-effect dynamics play out over time and the size of causal effects often varies over time (Gollob & Reichardt, 1987). The effects of aspirin on pain reduction from Gollob and Reichardt (1987) provides a clear example of timing concerns. Assume that there is no effect in the immediate few minutes after the drug is taken but that pain-reducing effects are noticeable 30 minutes after ingesting an aspirin. Further assume that the effects peak about three hours later and then decay. The effect vanishes after six hours. A researcher who uses a 10-hour interval will fail to detect a causal impact of aspirin whereas a researcher who uses a one-hour interval will find evidence for the substantial pain-relieving properties of aspirin. The same analogy works for psychological processes that play out at longer time scales than minutes and hours.

In short, insight into the suspected causal dynamics is critical for selecting intervals. In the absence of such insight, more frequent assessments with shorter time lags will likely lead to fewer Type II errors (see also Dormann & Griffin, 2015), unless of course the overall length of the study is too short to detect changes. Frequent assessments are also helpful during periods of rapid change such as childhood, school transitions, the initial phases of job loss, relationship dissolution, or entry into treatment. Figuring out the timing of assessments is one of the major challenges in crafting a useful longitudinal study.

Question Three: What Is the Measurement Strategy?

The gold standard is to use multiple sources of data and multiple measures to assess core constructs (e.g., Caspi, 1998). Assuming multiple measures are going to be used, a natural question becomes how many different types of measures should be included? We suggest that researchers strive for three methods as a heuristic, as this strikes a workable balance between construct coverage and resource constraints for use in latent variable models (Little, Lindenberger, & Nesselroade, 1999). Self-report surveys are an important source of data in many areas of psychology (e.g. Lucas & Baird, 2006) but they have limitations. Self-reports can be supplemented with informant reports or other methods of measuring core constructs. Informant reports may be a particularly important alternative source of data for studies of psychopathology. For example, there is evidence that informant reports of personality traits may have greater predictive validity relative to self-reports of personality traits (Connelly & Ones, 2010) and this might extend to psychopathology given the close overlap between personality traits and psychopathology (e.g., Kotov et al., 2010).

Observational data is another possibility that should be considered. For example, there is ongoing interest in how psychopathology is associated with relationship distress (e.g., Humbad et al., 2011) and observing interactions between romantic partners is a good way to gain insight into how relationships are functioning. Observed dyadic and family observations (see Conger, Lorenz, & Wickrama, 2004) might also prove useful for those who want to document changes over time in adjustment and interpersonal functioning. Other options to global self-reports such as aggregated momentary assessment techniques are a possibility as well. However, it should be noted that these techniques do not have a strong tradition of psychometric evaluation (see e.g., Ram et al., 2017; Wright & Zimmermann, 2018).

Regardless of the measurement source, it is important to select psychometrically strong instruments that are suited for detecting changes over time. Familiarity and pilot tests are invaluable. Willett and colleagues (1998, p. 411) noted that "the time for instrument modification is during pilot work, not data collection." The exact same instrument should be used at each wave (Willett et al., 1998) but this advice gets complicated when studies are designed to span long periods of the lifespan or if major contextual changes will likely happen over the course of the study.[2] Moreover administering the same instrument at each wave does not guarantee that the instrument will measure the latent construct in exactly the same way at each wave (see, e.g., Little, 2013; Schmitt & Kuljanin, 2008). It just provides a way to test this assumption, as we will explain.

When selecting measures for a longitudinal study, it is critical that researchers anticipate the types of data-analytic approaches they will need to use to address study questions. Many longitudinal statistical techniques make assumptions about the ways variables are measured. Standard growth models, for example, assume that variables are measured with interval-level scales where "a given value of the outcome on any occasion [represents] the 'same' amount of the outcome on every occasion" (Singer & Willett, 2003, p. 13). A commonly used stem in personality research is to have respondents rate how they compare with other people of their age and gender on a specific characteristic. However, because the reference group changes over time as people age, this stem renders the data largely inappropriate for standard growth models. It is likely that the same observed value on a scale (e.g., a score of 4.2) reflects different levels of the trait in question at different ages.

Whether the same observed score reflects the same value on the latent (unobserved) continuum is evaluated using tests of measurement invariance. These statistical procedures evaluate whether the same construct is being measured in the same way at each wave. Absent support for this claim, differences in scores between one wave and another "may be tantamount to comparing apples and spark plugs" (Vandenberg & Lance, 2000, p. 9). We therefore strongly recommend that researchers who use longitudinal designs familiarize themselves with the testing of invariance.

We also want to end by encouraging researchers who collect multiple informant data to have a coherent plan for how the different measures will be used. Latent variable models are a promising approach (e.g., Little et al., 1999) but researchers can sometimes be surprised by unexpectedly low levels of convergence (see Chapter 19, by Wood). This creates analytic challenges and interpretative difficulties. We recommend that researchers articulate a plan about how to deal with such issues in advance and to make sure substantive conclusions are not strongly bound to only one way of measuring core constructs. We return to this broader theme in the concluding section of this chapter.

Question four: What Is the Sample Size?

A final big question involves sample size planning. Small sample sizes are problematic because they typically have low statistical power and provide imprecise parameter estimates (e.g., Watson, 2004). Coming to the end of a multiyear longitudinal study only to discover that effect sizes were relatively small and the sample size was not sufficient to detect them is a bitter, but not uncommon outcome. Any discussion of the adequacy of a sample size has to grapple with two major issues – the underlying effect sizes in question and the desired level of accuracy in parameter estimation.

Effect sizes are often smaller than researchers would like to think for many areas of research involving multiply determined outcomes such as psychopathology (see, e.g., Ahadi & Diener, 1989; Hemphill, 2003). Average effect size estimates from many different literatures frequently hover around 0.20 in the r-metric (e.g., Bosco et al., 2015; Gignac, & Szodorai, 2016; Hill et al., 2008; Paterson et al., 2016; Richard, Bond, & Stokes-Zoota, 2003). Power analyses should therefore be rooted in the reality that effect sizes of interest are more likely to be "small" rather than "large." Moreover, published estimates are likely to be inflated due to factors like publication bias and the winner's curse (see Button et al., 2014). This again attests to the wisdom of anticipating small effect sizes and planning accordingly.

Traditional power analyses are generally predicated on the desired probabilities of making Type II errors. By tradition (following from Cohen, 1988), researchers tend

---

[2] Equating approaches drawing on latent variable and Item Response Theory methods (e.g., Reise, Ainsworth, & Haviland, 2005) could be used to identify a latent-metric from different measures of the same characteristic at different waves. Such approaches add complexity and analytic challenges and there is no guarantee these approaches will work. Thus, we stand by the advice to use the exact same measure at each wave or to at least preserve a core set of items across all waves.

to define an acceptable probability of a Type II error (i.e., beta) to be 0.20, whereas the acceptable probability of a Type I error (i.e., alpha) is conventionally set to 0.05.[3] Considering the large investment required for longitudinal research, conducting such a study knowing that it has only an 80 percent chance of finding an anticipated effect seems unacceptable. We recommend that researchers plan to power their studies to 0.95 such that the probability of the two types of errors is equal.

Researchers have also discussed setting alpha levels below 0.05 to levels such as 0.005 as a threshold for statistical significance for declaring new discoveries (Benjamin et al., 2018; but see Lakens et al., 2018). A blanket reduction in alpha for longitudinal studies is not necessarily unreasonable given that multiple hypothesis tests are typically conducted. Family-wise Type I error rates are hard to estimate but dropping alpha is a blunt-force option for gaining some control over false positives. The upshot of reducing alpha will also generate a pressure to increase sample sizes when considering statistical power.

The binary thinking associated with a classical inferential paradigm is not the only consideration when conducting research. The degree of precision in parameter estimation is also critical (see, e.g., Maxwell, Kelley, & Rausch, 2008). Wide confidence intervals around estimates cause consternation (Watson, 2004). The solution for more precise estimates is often to increase sample size. Accordingly, the simplest advice is to get as big a sample size as humanly possible.

When planning a study, attrition must be expected and drop-out is rarely a purely random phenomenon, a fact which may limit the applicability of many missing data estimation techniques (Schafer & Graham, 2002). What level of attrition is typical? Roberts and DelVecchio (2000) published an influential meta-analysis of stability coefficients and reported an average attrition rate of 42 percent. Roberts, Walton, and Viechtbauser (2006) reported a rate of 44 percent for longitudinal personality studies involving means. These are overall figures that can provide a ballpark estimate but attrition rates also vary by age, socioeconomic factors like education, and psychopathology (see, e.g., Claus, Kindleberger, & Dugan, 2002; Gustavson et al., 2012; Young, Powers, & Bell, 2006). Thus, researchers are advised to search for estimates related to their focal populations. It is an unpleasant reality that some participants will drop out of well-designed studies and this should factor into sample size planning.

In sum, we believe that addressing the four questions in Table 18.1 will help researchers design high-quality longitudinal studies. Astute readers will also notice that the questions in Table 18.1 are easiest to address when researchers have focused research questions and a considerable amount of background knowledge. Many actual longitudinal studies involve multiple research questions given the time and expense of the study. There is an inherent tension between the prospect of designing a strong study to address a focused set of research questions and making the study broad enough to justify the time and expense. We are sympathetic to such tensions but we ultimately think it is useful to prioritize specific research questions. This way, the study will be sensible for addressing at least some research questions. The alternative is to attempt to address too many questions simultaneously, which may impair the ability of the study to make any kind of contribution. This is a hard decision but we think there is increasing pressure toward specificity in psychological research, a topic we address in our concluding section. Now that we have covered foundational issues with designing longitudinal studies, the next section provides practical advice.

PRACTICAL ISSUES IN MANAGING LONGITUDINAL STUDIES

We suggest that researchers think about their longitudinal study as a factory that collects, processes, and analyzes data. Stouthamer-Loeber and colleagues (1992, p. 64) argued that "a large study has to be run like a production business whose function is to deliver high-quality data on time and within a predetermined budget." It is difficult to overstate the importance of thorough documentation and meticulous recording keeping. Block (1993, p. 17) noted that a "longitudinal study should make public and communicable just what was done" and Stouthamer-Loeber and colleagues (1992, p. 77) echoed this sentiment: "We cannot overemphasize the necessity of monitoring and *making visible* [italics in original] all aspects of a study's progress. Only when errors and actual or potential problems are made visible can action be taken."

One way to achieve this level of transparency is to expect that the data will be immediately archived for use by outside investigators who will demand thorough documentation about the procedures, measures, and design decisions. This expectation for outside use helps to instill a dedication toward documentation that will pay dividends when it comes to writing reports, responding to reviewer concerns, and designing future studies. It is also the case that memory fades and so a near-fanatical dedication to ongoing documentation has real value. This perspective also makes it easier to comply with many of the data-sharing requirements of certain funders and journals.

Given the importance of documentation and careful data collection, researchers should allocate plenty of time and energy to these tasks (Stouthamer-Loeber et al., 1992). Realistic timelines are important to make sure everything stays on track. Frequent monitoring of

---

[3] The rationale for selecting beta to 0.20 so power is set to 0.80 (i.e., power = 1 − beta) is described by Cohen (1988) on pages 55 to 56 of his classic text. It is valuable to consider his caveat: "This .80 desired convention is offered with the hope that it will be ignored whenever an investigator can find a basis in [their] substantive concerns in [their] research investigation to choose a value *ad hoc*." (p. 56).

progress and tracking mechanisms are highly useful. We now turn to some tips and insights based on published tutorials as well as personal insights from our own work.

Codebooks and Variable Labeling Conventions

A thorough codebook that details the names of specific variables, provides basic descriptive statistics, and details the scope of missing data is ideal. Including copies of the original assessments in appendices is especially helpful and should incorporate detailed notes about any deviations in the measures from original sources. The codebook should serve as the "bible" for data analyses and should be understandable to someone unfamiliar with the project. In fact, it should be able to serve as a guide to educate new project staff about the nature of the study.

We think establishing thoughtful variable naming conventions at the outset of the study is well worth the time and effort. A labeling system that ensures that each variable has intrinsic meaning will prevent errors and make data analysis easier to debug. In particular, it is useful for variable labels to reflect the name of the construct (or scale), wave of data collection, and source of the data. For example, consider a study in which depression is measured with a 20-item scale at three time points by mother report for a study of child depression. A possible scheme for wave 1 could be something like: W1Dep01M, W1Dep02M, up to W1Dep20M for the 20 items. In this case "W1" refers to Wave 1, "Dep01" to "Dep20" refers to the 1st to 20th items on the scale (and item content will ideally be the same for all other waves), and "M" refers to reports from mothers. The same 20 items for fathers could be labeled W1Dep01F to W1Dep20F. Parent reports at Wave 2 could be W2Dep01M to W2Dep20M for mothers and W2Dep01F to W2Dep20F for fathers. If a self-report was also solicited, then S could replace the M and F at the end of the variable name at each wave.

This kind of scheme facilitates the writing of code/statistical software syntax to do things like calculate scale reliabilities and create scale composites. The code or syntax is easier to debug because the labels have intrinsic meaning. Haphazard labels or labels that have no intuitive meaning (e.g., item1423) increase errors. We have unfortunately learned this lesson the hard way.

Open Communication and Defined Project Roles

Given the scope of a longitudinal study, it is helpful to have a research team with well-defined roles for project staff. The job is often too big to manage by a single person. Stouthamer-Loeber and colleagues (1992, p. 77) noted that "large longitudinal studies are clearly team efforts whose success depends largely on how well staff members work together." The leaders of projects should facilitate teamwork and foster open communication between project staff. (We use project staff inclusively to refer to personnel that can range from undergraduate volunteers to paid postdocs and project managers.) Well-run meetings that have recorded minutes are one way to develop group cohesion and to facilitate communication. The detailed notes from each meeting should be archived with other project documents. This record helps to create a history of the project. This is invaluable when looking backward to try to figure out the rationale for particular design decisions and processes. Meetings are especially important around major data collection periods.

Managing documents is another significant task in longitudinal studies. We have already discussed the importance of comprehensive codebooks, but other documents are relevant, such as Institutional Review Board records, consent forms, and recruitment mailings. Statistical analyses and manuscript tracking are other document-intensive tasks. Thus, a project librarian is often needed. This librarian can also maintain archives for computer programs and syntax to perform routine project tasks such as cleaning data, creating scales, and conducting basic descriptive analyses. A standard format for syntax along with a scrupulous system for maintaining version control is quite helpful.

The use of common syntax for routine analyses and data file construction saves individual analysts from having to duplicate prior efforts. It also likely reduces mistakes because the same syntax is used by multiple individuals. This should encourage multiple pairs of eyes to evaluate code. Accordingly, all analysts should be encouraged to check code for errors rather than assume everything is correct. Once detected, errors should be corrected, and all project staff members should be alerted. It is highly likely that a few errors will occur in the initial stages and a project philosophy that encourages collaboration and openness is helpful for making sure errors are corrected and that any damage created by such errors is minimized. Creating separate folders for specific papers that includes data files, key output, and code/statistical syntax is helpful when working on manuscripts and revisions.

Authorship issues are an occasional source of conflict among project staff. Thus, we recommend that formal procedures for identifying authorship and timelines are a part of all projects. A concept paper model is one way to handle these issues. This document includes a short writeup of the idea, a list of the target variables, a list of authors, a justification of author order, and timelines. This sort of document is especially useful for managing collaborations between faculty members and graduate students or postdocs (see also Fine & Kurdek, 1993). A standard policy for what happens when timelines are not followed is important to maximize productivity from the lab. If a lead author is no longer willing or able to complete a project described in a concept paper, then it might be reasonable to reconsider author order. This formal document can be submitted to the senior investigator(s) and stored in a common document folder. This approach prevents multiple people from pursuing similar papers and increases accountability.

Strategies to Manage Attrition

The loss of participants across time is a threat to the vitality and scientific rigor of a longitudinal study. Attrition limits the generalizability of studies (e.g., Barry, 2005; Ribisl et al., 1996; Shadish et al., 2002) and nonrandom attrition within experimental conditions also threatens internal validity (e.g., Barry, 2005; Shadish et al., 2002). Maintaining the sample is a major task when managing a longitudinal project. We recognize that statistical procedures are available to address issues related to missing data but we do not believe these are effective substitutes for sound project management. High retention rates are possible (e.g., Capaldi & Patterson, 1987).

Ribisl and colleagues (1996; see also Cotter et al., 2002; Davis, Broome, & Cox, 2002) offer a detailed list of strategies and this section of the chapter draws on that work. We also acknowledge that much of the literature about increasing retention is based on lore and experience rather than empirical evaluations of procedures (Fumagalli, Laurie, & Lynn, 2012). Additional empirical research is needed (Booker, Harding, & Benzeval, 2011). With this caveat in mind, three issues are especially critical: making sure the procedures of the study are pleasant for participants, creating investment in the project from participants, and developing a robust tracking system to maintain updated contact information such as email and postal addresses.

Compensation in the form of money or gift cards is an obvious way of making the participant experience pleasant (e.g., Ribisl et al., 1996; Stouthamer-Loeber et al., 1992). Incentives are an empirically supported method of reducing attrition (Booker et al., 2011). Participants should be compensated in a timely fashion. Moreover, if the study has a defined set of assessment occasions, a small bonus for completing all waves is a potential additional incentive. Increasing compensation gradually across waves can help with retention. One note of caution, however, is that different universities have different regulatory environments, so it is important to find a compensation scheme that works for the local Institutional Review Board (IRB). For example, there are limits to overall compensation levels that participants can receive before tax documents are required at our institution. Drawings for gift cards pose particular challenges and require input from general counsel due to concerns about gambling at our location. Although drawings can be cost-effective ways to gain informant data, alternatives are often preferred at our institution given the local hassles.

The length of the assessment protocol and the nature of the tasks are a major factor in the pleasantness of a study.[4] We noted the tendency for studies to measure a large number of constructs and this often increases the length of surveys and assessment protocols. Tedious surveys are not helpful for reducing attrition. The desire for exhaustive information has to be balanced with a recognition that participants have limited amounts of patience and attention. Adopt the perspective of the participant when considering assessment protocols. Issues like font size and visual appeal are useful to evaluate. Is the survey easy to read? How do the questions appear in print and on different electronic devices? Are researchers bored to tears when they complete the package themselves?

Conducting focus groups of individuals who complete planned assessment packages will often provide invaluable advice. These groups are in a position to tell researchers if surveys are an acceptable length and can help identify confusing or even potentially offensive questions. If researchers repeatedly receive negative comments about the length of a survey, something should be done. Be receptive to feedback. Reducing length by cutting assessments of ancillary constructs might be painful, but this is far less painful than the disaster of managing a longitudinal assessment that suffers from high levels of attrition. Make sure focus group members are drawn from the population of interest in the research. We have occasionally used graduate students or children of faculty members or graduate students as "test" participants. The feedback was generally positive and supportive. The children were focused and had high reading levels so they expressed no concerns about survey questions. However, these pilot runs were not always representative of the problems we encountered when conducting the study.

A second approach to increasing participation in subsequent waves is to find ways to encourage participant investment in the project. It is helpful to remind participants of the importance of the project and how they are helping to advance research in plain language. Another strategy to increase investment is to create a project identity (Ribisl et al., 1996). A fun acronym and an attractive study logo can help. This can serve as project branding for use in recruitment messages, surveys, and even a project website (Ribisl et al., 1996). Make sure the name of the project is something relatively innocuous and avoid anything that could stigmatize participants, convey hypotheses, or create reactance. Other ways to form connections between participants and the project are to send birthday greetings (Stouthamer-Loeber et al., 1992) and to generate project newsletters. An attractive website is also valuable. When sending emails or setting up a website, make sure to have an easy mechanism such as a link for participants to easily update contact information (Stouthamer-Loeber et al., 1992).[5]

The suggestion for having seamless mechanisms for updating contact information relates to the third major

---

[4] If short-forms are used to reduce participant burden, make sure the short forms have acceptable psychometric properties (see Smith, McCarthy, & Anderson, 2000). Do not simply shorten measures based on hunches and limited data!

[5] Make sure to verify that any emails sent by the project are not inadvertently trapped in spam filters. This has happened to us on occasion.

technique for reducing attrition – obtaining and maintaining extensive contact information (Ribisl et al., 1996). Names, addresses (local and permanent for students), email addresses (multiple if possible), as well as contact information for informants are all helpful to track. This latter kind of information can be very helpful for finding participants who have moved since the last wave of data collection. It is possible to contract with professional tracking services to find wayward participants but this adds costs.

Participation is something that can wax and wane across multiwave studies (Stouthamer-Loeber et al., 1992). Unless participants unequivocally tell investigators that they no longer wish to participate at all, we suggest that attempts are made to contact them in the future. People can have especially busy periods in their lives and skip participation at a particular wave. Individuals might not be as busy in the future and will again participate. It is also worthwhile to try to figure out why individuals are not participating. A quick follow-up might prove helpful, especially if participants identify issues that could be addressed such as rates of compensation that are too low or surveys that are too long and burdensome.

CONCLUDING COMMENTS: DESIGNING LONGITUDINAL STUDIES IN THE OPEN SCIENCE ERA

Different branches of psychology have wrestled with concerns over a crisis of confidence in the credibility of research findings in recent years due, in part, to a string of failures to replicate published findings, concerns about journal publication practices, and the prevalence of questionable research practices (e.g., Campbell, Loving, & Lebel, 2014; John, Loewenstein, & Prelec, 2012; Nelson, Simmons, & Simonsohn, 2018; Open Science Collaboration, 2015; Simmons, Nelson, & Simonsohn, 2011; for a review see Spellman, Gilbert, & Corker, 2018). Several recommendations for improving psychological science that have emerged from this discussion are relevant for designing and implementing longitudinal studies including the importance of transparency, increased sample sizes, and preregistration (see, e.g., Chapter 28, by Hopwood and Vazire; and Button et al., 2013; Donnellan et al., 2013; Fraley & Vazire, 2014; Kashy et al., 2009; Munafò et al., 2017; Nosek & Bar-Anan, 2012; Tackett et al., 2017). Many of these approaches to enhancing the rigor of research are fully consistent with the advice offered in this chapter.

For example, in terms of openness and transparency, we advocate the value of careful documentation and adopting the perspective that a study will be archived for outside users at the outset. These practices are highly beneficial for longitudinal researchers who need careful records to fully capitalize on their investments given the limits of human memory and the reality of project personnel turnover. As we have discussed, incorporating such practices into the research process makes it easier to comply with data-sharing policies in place at some journals and funding agencies (see Nosek et al., 2015). In terms of sample sizes, we advocate expecting small effects and setting beta low to protect against Type II errors given the investment of resources. We also noted the virtues of more precise parameter estimates. These considerations all push for larger sample sizes. Last, we emphasized the value of developing specific hypotheses to guide design decisions, advice that is consistent with the philosophy behind preregistration (Nosek et al., 2018). A concept paper can easily form the foundation for a formal preregistration.

At the same time, we acknowledge the complexity of applying advice about best practices developed in experimental contexts to longitudinal studies. It is probably easier and more feasible to preregister analyses and designs for a replication study or simple experiment than it would be for a large longitudinal study. Nonetheless, we doubt longitudinal researchers can afford to ignore the methodological advancements that have occurred in the last several years. Longitudinal studies involve many decisions and thus provide for a wide range of researcher degrees of freedom (Simmons et al., 2011). There are usually plenty of opportunities to p-hack (see Simonsohn, Nelson, & Simmons, 2014; Wicherts et al., 2016), often without full awareness, given the number of variables that are typically collected in a longitudinal study (see Donnellan et al., 2013).

Likewise, analyses from longitudinal studies can typically be conducted in multiple ways (e.g., models with different informants) and when using different control variables or selection criteria for inclusion. This again creates a situation where researchers can capitalize on chance and inadvertently cherry-pick results. Researchers, like all humans, are subject to confirmation and hindsight biases. Accordingly, we think researchers are well served when they find ways to test the sturdiness of their conclusions against alternative statistical models (see Steegen et al., 2016). We also think it is advisable to adopt a consistent set of practices regarding data cleaning, outlier detection, and the inclusion of a "standard" set of theoretically relevant covariates for all analyses from a study (e.g., SES and sex/gender). A form of preregistration for these procedures can provide a way to maintain quality control across the life of a project. Deviations are certainly possible, but they would require justification and help serve as a check on biases. More is being written about how open science practices can be applied to secondary data (Weston et al., 2018) and we believe these developments will prove useful for longitudinal researchers moving forward.

Any challenges to research rigor notwithstanding, we believe high-quality longitudinal studies are an indispensable tool for researchers interested in psychopathology. The insights one gains from such studies about the etiology of disorders and the time course of problems are an important complement to other kinds of designs. We hope the perspective and advice in this chapter proves useful to those who are committed to studying phenomena the long way – a phrase we borrowed from Block (1993). We acknowledged in the opening paragraph that longitudinal studies involve a gamble; however, we are convinced that the gamble is worth the risk when studies are well designed and effectively managed.

RECOMMENDED READING

Block, J. (1993). Studying Personality the Long Way. In D. C. Funder, R. D. Parke, C. Tomlinson Keasy, and K. F. Widaman (Eds.), *Studying Lives through Time: Personality and Development* (pp. 9–41). Washington, DC: American Psychological Association.

Block's chapter provides an insider perspective on longitudinal studies of personality and offers nine desiderata for studies.

Dormann, C., & Griffin, M. A. (2015). Optimal Time Lags in Panel Studies. *Psychological Methods, 20*, 489–505.

This paper summarizes the thorny issues involved in selecting time lags for longitudinal studies.

Stouthamer-Loeber, M., van Kammen, W., & Loeber, R. (1992). The Nuts and Bolts of Implementing Large-Scale Longitudinal Studies. *Violence and Victims, 7*, 63–78.

Although it is over 25 years old, this piece provides practical advice and tips for running a large-scale longitudinal study. The advice holds for many less expansive designs as well.

Ribisl, K. M., Walton, M. A., Mowbray, C. T., Luke, D. A., Davidson, W. S., & Bootsmiller, B. J. (1996). Minimizing Participant Attrition in Panel Studies through the Use of Effective Retention and Tracking Strategies: Review and Recommendations. *Evaluation and Program Planning, 19*, 1–25.

This article offers a wealth of advice about reducing attrition beyond what was covered in this chapter.

Weston, S. J., Ritchie, S. J., Rohrer, J. M., & Przybylski, A. (2018). Recommendations for Increasing the Transparency of Analysis of Pre-Existing Data Sets. *Advances in Methods and Practices in Psychological Science*, 2515245919848684.

Transparency and an awareness of researcher degrees of freedom are helpful for all kinds of psychological research. The guidelines in this article seem especially relevant when analyzing data from existing longitudinal studies.

Wicherts, J. M., Veldkamp, C. L., Augusteijn, H. E., Bakker, M., Van Aert, R., & Van Assen, M. A. (2016). Degrees of Freedom in Planning, Running, Analyzing, and Reporting Psychological Studies: A Checklist to Avoid p-Hacking. *Frontiers in Psychology, 7*, 1832.

This work provides an overview of the kinds of choices facing researchers and the checklist may increase awareness of p-hacking and improve the rigor of longitudinal analyses.

REFERENCES

Ahadi, S., & Diener, E. (1989). Multiple Determinants and Effect Size. *Journal of Personality and Social Psychology, 56*(3), 398–406.

Anderson, E. R. (1993). Analyzing Change in Short-Term Longitudinal Research Using Cohort-Sequential Designs. *Journal of Consulting and Clinical Psychology, 61*(6), 929–940.

Barry, A. E. (2005). How Attrition Impacts the Internal and External Validity of Longitudinal Research. *Journal of School Health, 75*, 267–270.

Benjamin, D. J., Berger, J. O., Johannesson, M., Nosek, B. A., Wagenmakers, E. J., Berk, R., ... & Cesarini, D. (2018). Redefine Statistical Significance. *Nature Human Behaviour, 2*(1), 6–10.

Block, J. (1993). Studying Personality the Long Way. In D. C. Funder, R. D. Parke, C. Tomlinson Keasy, & K. F. Widaman (Eds.). *Studying Lives through Time: Personality and Development* (pp. 9–41). Washington, DC: American Psychological Association.

Booker, C. L., Harding, S., & Benzeval, M. (2011). A Systematic Review of the Effect of Retention Methods in Population-Based Cohort Studies. *BMC Public Health, 11*(1), 249.

Bosco, F. A., Aguinis, H., Singh, K., Field, J. G., & Pierce, C. A. (2015). Correlational Effect Size Benchmarks. *Journal of Applied Psychology, 100*(2), 431–449.

Button, K. S., Ioannidis, J. P., Mokrysz, C., Nosek, B. A., Flint, J., Robinson, E. S., & Munafò, M. R. (2013). Power Failure: Why Small Sample Size Undermines the Reliability of Neuroscience. *Nature Reviews Neuroscience, 14*(5), 365–376.

Campbell, L., Loving, T. J., & Lebel, E. P. (2014). Enhancing Transparency of the Research Process to Increase Accuracy of Findings: A Guide for Relationship Researchers. *Personal Relationships, 21*(4), 531–545.

Capaldi, D., & Patterson, G. R. (1987). An Approach to the Problem of Recruitment and Retention Rates for Longitudinal Research. *Behavioral Assessment, 9*, 169–178.

Caspi, A. (1998). Personality Development across the Life Course. In W. Damon (Ed.). *Handbook of Child Psychology, Vol. 3: Social, Emotional, and Personality Development* (5th edn.) (pp. 311–388). New York: Wiley.

Cattell, R. B. (1970). Separating Endogenous, Exogenous, Epogenic, and Epogenic Component Curves in Developmental Data. *Developmental Psychology, 3*, 151–162.

Claus, R. E., Kindleberger, L. R., & Dugan, M. C. (2002). Predictors of Attrition in a Longitudinal Study of Substance Abusers. *Journal of Psychoactive Drugs, 34*(1), 69–74.

Cohen, J. (1988). *Statistical Power Analysis for the Behavioral Sciences* (2nd edn.). Hillsdale, NJ: Erlbaum.

Cole, D. A., & Maxwell, S. E. (2003). Testing Mediational Models with Longitudinal Data: Questions and Tips in the Use of Structural Equation Modeling. *Journal of Abnormal Psychology, 112*, 538–577.

Collins, L. M. (2006). Analysis of Longitudinal Data: The Integration of Theoretical Model, Temporal Design, and Statistical Model. *Annual Review of Psychology, 57*, 505–528.

Conger, R. D., Lorenz, F. O., & Wickrama, K. A. S. (2004). Studying Change in Family Relationships: The Findings and Their Implications. In R. D. Conger, F. O. Lorenz, & K. A. S. Wickrama (Eds.). *Continuity and Change in Family Relations: Theory, Methods, and Empirical Findings* (pp. 383–403). Mahwah, NJ: Lawrence Erlbaum.

Connelly, B. S., & Ones, D. S. (2010). An Other Perspective on Personality: Meta-Analytic Integration of Observers' Accuracy and Predictive Validity. *Psychological Bulletin, 136*(6), 1092–1122.

Cotter, R. B., Burke, J. D., Loeber, R., & Navratil, J. L. (2002). Innovative Retention Methods in Longitudinal Research: A Case Study of the Developmental Trends Study. *Journal of Child and Family Studies, 11*, 485–498.

Davis, L. L., Broome, M. E., & Cox, R. P. (2002). Maximizing Retention in Community-Based Clinical Trials. *Journal of Nursing Scholarship, 34*(1), 47–53.

Donnellan, M. B., & Conger, R. D. (2007). Designing and Implementing Longitudinal Studies. In R. W. Robins, R. Chris Fraley, & R. F. Krueger (Eds.). *Handbook of Research Methods in Personality Psychology* (pp. 21–36.). New York: Guilford Press.

Donnellan, M. B., Roisman, G. L., Fraley, R. C., & Lucas, R. E. (2013, Summer). Methodological Reform 101 for Developmental Researchers. *Developmental Psychologist Division 7 Newsletter, 34*–40.

Dormann, C., & Griffin, M. A. (2015). Optimal Time Lags in Panel Studies. *Psychological Methods, 20,* 489–505.

Faden, V. B., Day, N. L., Windle, M., Windle, R., Grube, J. W., Molina, B. S., ... & Sher, K. J. (2004). Collecting Longitudinal Data through Childhood, Adolescence, and Young Adulthood: Methodological Challenges. *Alcoholism: Clinical and Experimental Research, 28,* 330–340.

Ferrer, E., & McArdle, J. J. (2003). Alternative Structural Equation Models for Multivariate Longitudinal Data Analysis. *Structural Equation Modeling, 10,* 493–524.

Fine, M. A., & Kurdek, L. A. (1999). Reflections on Determining Authorship Credit and Authorship Order on Faculty-Student Collaborations. *American Psychologist, 48,* 1141–1147.

Fraley, R. C., & Roberts, B. W. (2005). Patterns of Continuity: A Dynamic Model for Conceptualizing the Stability of Individual Differences in Psychological Constructs across the Life Course. *Psychological Review, 112*(1), 60–74.

Fraley, R. C., & Vazire, S. (2014). The N-Pact Factor: Evaluating the Quality of Empirical Journals with Respect to Sample Size and Statistical Power. *PloS One, 9*(10), e109019.

Fumagalli, L., Laurie, H., & Lynn, P. (2013). Experiments with Methods to Reduce Attrition in Longitudinal Surveys. *Journal of the Royal Statistical Society: Series A (Statistics in Society), 176*(2), 499–519.

Gignac, G. E., & Szodorai, E. T. (2016). Effect Size Guidelines for Individual Differences Researchers. *Personality and Individual Differences, 102,* 74–78.

Gollob, H. F., & Reichardt, C. S. (1987). Taking Account of Time Lags in Causal Models. *Child Development, 58,* 80–92.

Graziotti, A. L., Hammond, J., Messinger, D. S., Bann, C. M., Miller-Loncar, C., Twomey, J. E., ... & Alexander, B. (2012). Maintaining Participation and Momentum in Longitudinal Research Involving High-Risk Families. *Journal of Nursing Scholarship, 44*(2), 120–126.

Grimm, K. J., Davoudzadeh, P., & Ram, N. (2017). IV. Developments in the Analysis of Longitudinal Data. *Monographs of the Society for Research in Child Development, 82*(2), 46–66.

Gustavson, K., von Soest, T., Karevold, E., & Røysamb, E. (2012). Attrition and Generalizability in Longitudinal Studies: Findings from a 15-Year Population-Based Study and a Monte Carlo Simulation Study. *BMC Public Health, 12*(1), 918.

Hartmann, D. P. (2005). Assessing Growth in Longitudinal Investigations: Selected Measurement and Design Issues. In D. M. Teti (Ed.), *Handbook of Research Methods in Developmental Science* (pp. 319–339). Malden, MA: Blackwell.

Hemphill, J. F. (2003). Interpreting the Magnitudes of Correlation Coefficients. *American Psychologist, 58,* 78–79.

Henrich, J., Heine, S. J., & Norenzayan, A. (2010). The Weirdest People in the World? *Behavioral and Brain Sciences, 33,* 62–135.

Hill, C. J., Bloom, H. S., Black, A. R., & Lipsey, M. W. (2008). Empirical Benchmarks for Interpreting Effect Sizes in Research. *Child Development Perspectives, 2*(3), 172–177.

Humbad, M. N., Donnellan, M. B., Klump, K. L., & Burt, S. A. (2011). Development of the Brief Romantic Relationship Interaction Coding Scheme (BRRICS). *Journal of Family Psychology, 25*(5), 759–769.

John, L. K., Loewenstein, G., & Prelec, D. (2012). Measuring the Prevalence of Questionable Research Practices with Incentives for Truth Telling. *Psychological Science, 23*(5), 524–532.

Kashy, D. A., Donnellan, M. B., Ackerman, R. A., & Russell, D. W. (2009). Reporting and Interpreting Research in PSPB: Practices, Principles, and Pragmatics. *Personality and Social Psychology Bulletin, 35*(9), 1131–1142.

Kenny, D. A., & Zautra, A. (1995). The Trait-State-Error Model for Multiwave Data. *Journal of Consulting and Clinical Psychology, 63,* 52–59.

Kotov, R., Gamez, W., Schmidt, F., & Watson, D. (2010). Linking "Big" Personality Traits to Anxiety, Depressive, and Substance Use Disorders: A Meta-Analysis. *Psychological Bulletin, 136*(5), 768–821.

Lakens, D., Adolfi, F. G., Albers, C. J., Anvari, F., Apps, M. A., Argamon, S. E., ... & Buchanan, E. M. (2018). Justify Your Alpha. *Nature Human Behaviour, 2*(3), 168–171.

Little, T. D. (2013). *Longitudinal Structural Equation Modeling.* New York: Guilford Press.

Little, T. D., Lindenberger, U., & Nesselroade, J. R. (1999). On Selecting Indicators for Multivariate Measurement and Modeling Latent Variables: When "Good" Indicators Are Bad and "Bad" Indicators Are Good. *Psychological Methods, 4,* 192–211.

Lucas, R. E., & Baird, B. M. (2006). Global Self-Assessment. In M. Eid & E. Diener (Eds.), *Handbook of Multimethod Measurement in Psychology* (pp. 29–42). Washington, DC: American Psychological Association.

Maxwell, S. E., Kelley, K., & Rausch, J. R. (2008). Sample Size Planning for Statistical Power and Accuracy in Parameter Estimation. *Annual Review of Psychology, 59,* 537–563.

Mitchell, T. R., & James, L. R. (2001). Building Better Theory: Time and the Specification of When Things Happen. *Academy of Management Review, 26*(4), 530–547.

Mitchell, M. A., & Maxwell, S. E. (2013). A Comparison of the Cross-Sectional and Sequential Designs When Assessing Longitudinal Mediation. *Multivariate Behavioral Research, 48*(3), 301–339.

Morton, S. M., Grant, C. C., Carr, P. E. A., Robinson, E. M., Kinloch, J. M., Fleming, C. J., ... & Liang, R. (2014). How Do You Recruit and Retain a Prebirth Cohort? Lessons Learnt from Growing Up in New Zealand. *Evaluation & the Health Professions, 37*(4), 411–433.

Munafò, M. R., Nosek, B. A., Bishop, D. V., Button, K. S., Chambers, C. D., du Sert, N. P., ... & Ioannidis, J. P. (2017). A Manifesto for Reproducible Science. *Nature Human Behaviour, 1*(1), 0021.

Nelson, L. D., Simmons, J., & Simonsohn, U. (2018). Psychology's Renaissance. *Annual Review of Psychology, 69,* 511–534.

Nosek, B. A., & Bar-Anan, Y. (2012). Scientific Utopia: I. Opening Scientific Communication. *Psychological Inquiry, 23*(3), 217–243.

Nosek, B. A., Alter, G., Banks, G. C., Borsboom, D., Bowman, S. D., Breckler, S. J., ... & Contestabile, M. (2015). Promoting an Open Research Culture. *Science, 348*(6242), 1422–1425.

Nosek, B. A., Ebersole, C. R., DeHaven, A. C., & Mellor, D. T. (2018). The Preregistration Revolution. *Proceedings of the National Academy of Sciences, 115*, 2600–2606.

Open Science Collaboration. (2015). Estimating the Reproducibility of Psychological Science. *Science, 349*(6251), 943–950.

Paterson, T. A., Harms, P. D., Steel, P., & Credé, M. (2016). An Assessment of the Magnitude of Effect Sizes: Evidence from 30 Years of Meta-Analysis in Management. *Journal of Leadership & Organizational Studies, 23*(1), 66–81.

Ram, N., Brinberg, M., Pincus, A. L., & Conroy, D. E. (2017). The Questionable Ecological Validity of Ecological Momentary Assessment: Considerations for Design and Analysis. *Research in Human Development, 14*(3), 253–270.

Rast, P., & Hofer, S. M. (2014). Longitudinal Design Considerations to Optimize Power to Detect Variances and Covariances among Rates of Change: Simulation Results Based on Actual Longitudinal Studies. *Psychological Methods, 19*, 133–154.

Reise, S. P., Ainsworth, A. T., & Havilan, M. G. (2005). Item Response Theory: Fundamentals, Applications, and Promise in Psychological Research. *Current Directions in Psychological Science, 14*, 95–101.

Ribisl, K. M., Walton, M. A., Mowbray, C. T., Luke, D. A., Davidson, W. S., & Bootsmiller, B. J. (1996). Minimizing Participant Attrition in Panel Studies through the Use of Effective Retention and Tracking Strategies: Review and Recommendations. *Evaluation and Program Planning, 19*, 1–25.

Richard, F. D., Bond Jr, C. F., & Stokes-Zoota, J. J. (2003). One Hundred Years of Social Psychology Quantitatively Described. *Review of General Psychology, 7*(4), 331–363.

Roberts, B. W., & DelVecchio, W. F. (2000). The Rank-Order Consistency of Personality Traits from Childhood to Old Age: A Quantitative Review of Longitudinal Studies. *Psychological Bulletin, 126*, 3–25.

Roberts, B. W., Walton, K. E., & Viechtbauer, W. (2006). Patterns of Mean-Level Change in Personality Traits across the Life Course: A Meta-Analysis of Longitudinal Studies. *Psychological Bulletin, 132*, 1–25.

Rogosa, D., Brandt, D., & Zimowski, M. (1982). A Growth Curve Approach to the Measurement of Change. *Psychological Bulletin, 92*, 726–748.

Schafer, J. L., & Graham, J. W. (2002). Missing Data: Our View of the State of the Art. *Psychological Methods, 7*, 147–177.

Schaie, K. W. (1965). A General Model for the Study of Developmental Problems. *Psychological Bulletin, 64*(2), 92–107.

Schmitt, N., & Kuljanin, G. (2008). Measurement Invariance: Review of Practice and Implications. *Human Resource Management Review, 18*(4), 210–222.

Selig, J. P., & Preacher, K. J. (2009). Mediation Models for Longitudinal Data in Developmental Research. *Research in Human Development, 6*(2–3), 144–164.

Shadish, W. R., Cook, T. D., & Campbell, D. T. (2002). *Experimental and Quasi-experimental Designs for Generalized Causal Inference.* New York: Houghton Mifflin.

Simmons, J. P., Nelson, L. D., & Simonsohn, U. (2011). False-Positive Psychology: Undisclosed Flexibility in Data Collection and Analysis Allows Presenting Anything as Significant. *Psychological Science, 22*(11), 1359–1366.

Simonsohn, U., Nelson, L. D., & Simmons, J. P. (2014). P-Curve: A Key to the File-Drawer. *Journal of Experimental Psychology: General, 143*(2), 534–547.

Singer, J. D., & Willett, J. B. (2003). *Applied Longitudinal Data Analysis: Modeling Change and Event Occurrence.* New York: Oxford University Press.

Smith, G. T., McCarthy, D. M., & Anderson, K. G. (2000). On the Sins of Short-Form Development. *Psychological Assessment, 12*, 102–111.

Spellman, B. A., Gilbert, E. A., & Corker, K. S. (2018). Open Science. In E. J. Wagenmakers & J. T. Wixted (Eds.), *Stevens' Handbook of Experimental Psychology and Cognitive Neuroscience Volume 5: Methodology* (4th edn., Vol. 5, pp. 1–47). Hoboken, NJ: Wiley.

Steegen, S., Tuerlinckx, F., Gelman, A., & Vanpaemel, W. (2016). Increasing Transparency through a Multiverse Analysis. *Perspectives on Psychological Science, 11*(5), 702–712.

Stouthamer-Loeber, M., van Kammen, W., & Loeber, R. (1992). The Nuts and Bolts of Implementing Large-Scale Longitudinal Studies. *Violence and Victims, 7*, 63–78.

Tackett, J. L., Lilienfeld, S. O., Patrick, C. J., Johnson, S. L., Krueger, R. F., Miller, J. D., ... & Shrout, P. E. (2017). It's Time to Broaden the Replicability Conversation: Thoughts for and from Clinical Psychological Science. *Perspectives on Psychological Science, 12*(5), 742–756.

Taris, T. W. (2000). *A Primer in Longitudinal Analysis.* Thousand Oaks, CA: Sage.

Timmons, A. C., & Preacher, K. J. (2015). The Importance of Temporal Design: How Do Measurement Intervals Affect the Accuracy and Efficiency of Parameter Estimates in Longitudinal Research? *Multivariate Behavioral Research, 50*, 41–55.

Vandenberg, R. J., & Lance, C. E. (2000). A Review and Synthesis of the Measurement Invariance Literature: Suggestions, Practices, and Recommendations for Organizational Research. *Organizational Research Methods, 3*, 4–69.

Watson, D. (2004). Stability versus Change, Dependability versus Error: Issues in the Assessment of Personality over Time. *Journal of Research in Personality, 38*, 319–350.

Weston, S. J., Ritchie, S. J., Rohrer, J. M., & Przybylski, A. (2018). Recommendations for Increasing the Transparency of Analysis of Pre-existing Datasets. *Advances in Methods and Practices in Psychological Science.* Retrieved from https://doi.org/10.1177%2F2515245919848684

Wicherts, J. M., Veldkamp, C. L., Augusteijn, H. E., Bakker, M., Van Aert, R., & Van Assen, M. A. (2016). Degrees of Freedom in Planning, Running, Analyzing, and Reporting Psychological Studies: A Checklist to Avoid p-Hacking. *Frontiers in Psychology, 7*, 1832

Willett, J. B., Singer, J. D., & Martin, N. C. (1998). The Design and Analysis of Longitudinal Studies of Development and Psychopathology in Context: Statistical Models and Methodological Recommendations. *Development and Psychopathology, 10*, 395–426.

Wright, A. G. C., & Markon, K. E. (2016). Longitudinal Designs. In J. C. Norcross, G. R. VandenBos, & D. K. Freedheim (Eds.), *American Psychological Association Handbook of Clinical Psychology, Vol. II: Theory and Research* (pp. 419–434). Washington, DC: American Psychological Association.

Wright, A. G., & Zimmermann, J. (2018). Applied Ambulatory Assessment: Integrating Idiographic and Nomothetic Principles of Measurement. *Psychological Assessment*, November 1. Retrieved from https://doi.org/10.31234/osf.io/6qc5x

Young, A. F., Powers, J. R., & Bell, S. L. (2006). Attrition in Longitudinal Studies: Who Do You Lose? *Australian and New Zealand Journal of Public Health, 30*(4), 353–361.

19 Measurement and Comorbidity Models for Longitudinal Data

PHILLIP K. WOOD

Many statistical models exist for longitudinal data involving psychopathology and related constructs. To name just a few, these include cross-lagged panel models, growth models (featuring linear, polynomial piecewise components), hierarchical linear models, or state-trait models (e.g., Jackson, Sher, & Wood, 2000; Jackson, O'Neill, & Sher, 2006; Park, Sher, & Krull, 2008; Sher et al., 1996b). The general goals of such analyses are to (a) compare relatively short-term patterns of behavioral persistence over time to more systematic long-term individual differences, (b) assess the relationship of level of initial severity to subsequent course, (c) identify periods of developmentally limited patterns of the problem behavior as distinct from more persistent patterns across the lifespan, and (d) assess the prospective effect of one problem behavior or psychopathology on some other behavior or psychopathology.

Although the sheer variety of longitudinal statistical models is exciting and growing by the year, the fact that so many alternatives exist can also lead to difficulty in adjudicating among them on the basis of substantive findings and statistical fit. Even if a researcher is so fortunate as to find that a chosen model fits well and contains parameter estimates which are both substantial and statistically significant, this model is often only the beginning of a dialogue about the reasonableness of the model and its estimated effects. For example, would a reasonable skeptic using a slightly different model reach substantially different conclusions? Have important longitudinal components of the behavior been overlooked or, alternatively, have some components been erroneously assumed? Are the psychometric measurement assumptions of the model met by the data? Based on these considerations, it seems worthwhile to consider whether exploration of a reasonably circumscribed subset of models could be used to identify plausible alternative explanations, provide insight into whether model assumptions are met, and, if influential observations are found, speak to the replicability of a proposed model's fit and parameter estimates.

Measurement Occasion Comorbidity, Prospective Comorbidity, Trait Comorbidity, and Comorbid Course

When specifying and evaluating structural models with longitudinal data, however, it is important to consider the various conceptual models of comorbidity implicit in many types of longitudinal models. In the interests of space, three qualitatively different models of comorbidity are possible. The first, and simplest, types of comorbidity deal with assessments of the degree of covariation between one type of problem behavior or psychopathology and another on a single or multiple measurement occasion. For example, we may be interested in the question of whether, across multiple occasions, some type of association exists between the construct of psychological distress and some measure of alcohol consumption. This type of comorbidity assessment can potentially address the question of whether some measurement occasions (such as the freshman year of college) show stronger associations between distress and alcohol consumption than other measurement occasions (such as post-college measurement). We may think of measurement occasion comorbidity as answering the question "Given that we observe a psychopathology or problem behavior at one measurement occasion to what extent is that associated with a different problem? Further: Is this degree of association similar across measurement occasions?

Prospective Comorbidity

Often, however, researchers are interested in determining whether one problem behavior or psychopathology is prospectively associated with some other problem behavior or psychopathology. Methodologically, these types of models fall under the general descriptor of Granger causality (see, e.g., Wiedermann & von Eye, 2016, for a discussion of Granger causality as it relates to several psychological constructs). In similar fashion to the previous paragraph, we might think of prospective comorbidity as answering the question "Given that we observe individuals' levels of one problem behavior or psychopathology at one measurement occasion, is there evidence for an association with some other problem behavior or psychopathology at a subsequent measurement occasion?" Often these longitudinal models, frequently termed cross-lagged panel models, adjust for the effects of measurement occasion comorbidity by specifying contemporaneous associations between the two constructs at each measurement occasion, either as a covariance or correlated error term.

Trait Comorbidity
Researchers may, however, be interested in the question of whether one psychopathology, expressed as a general trait, is associated with some other psychopathology, also expressed as a general trait. For example, a researcher may wish to examine whether individuals who are generally distressed (as assessed across all measurement occasions) also tend to consume alcohol (also assessed across all measurement occasions). Such models are generally thought of as assessing comorbidity via state-trait models (e.g., Jackson et al., 2003). State-trait models assess comorbidity as a stable interindividual difference across all measurement occasions via a covariance between traits, allow for the examination of additional measurement occasion comorbidity by reference to covariances between constructs at each measurement occasion and also permit specification of prospective comorbidity via cross-lagged associations between prospective measurement occasions across constructs. The partition of comorbidity in such models allows for a more refined understanding of the degree of general measurement occasion, and prospective associations between problem behaviors or psychopathologies.

Comorbid Course
Researchers, however, may be interested not only in the degree of association between psychopathologies or problem behaviors, within measurement occasions, and prospectively. They may be interested in assessing the degree to which the general course of one psychopathology/problem behavior is associated with some other psychopathology/problem behavior. These types of questions involve the specification of how, generally, multiple problem behaviors or psychopathologies increase or decrease in concert across measurement occasions. Such models are frequently described as "growth curve" models in which both a mean and variance are estimated in latent variables which span measurement occasions. If the researcher believes that a common course exists across the behaviors for all individuals and that individuals differ only as a matter of degree, the question of comorbid course can be expressed as growth curves in which covariances are estimated across the latent variables for each construct. As shown in the example below of alcohol use and psychological distress, growth curves may span all measurement occasions or only a subset, as in the case of a "developmentally limited" trend in the data.

The search for latent subgroups or typologies of comorbidity across such constructs is frequently approached extending the idea of a growth model using latent class or latent growth mixture models in which qualitatively different patterns of co-occurring course are thought to exist. Nelson, Van Ryzon, and Dishion (2015) summarize a variety of such models for longitudinal substance use course. Given space limitations these latent mixture models will not be explored here except to note that the right-sizing approach outlined below can be readily applied to mixture models and that evidence exists that inappropriate specification of the common growth model thought to underlie the mixtures likely leads to identification of artifactual subgroups (Sher, Jackson, & Steinley, 2011).

These three types of comorbidity can be modeled simultaneously when multiple problem behaviors or psychopathologies are assessed across several measurement occasions. Measurement occasion comorbidity conditional on general latent growth curve(s) can be modeled by correlating the error terms associated with a given measurement occasion; cross-lagged prospective effects can be modeled as paths from the measurement occasion for one problem behavior or psychopathology to some other behavior at the next measurement occasion, providing a test of Granger causality. Finally, covariances between latent variables associated with one behavior or psychopathology and some other behavior or psychopathology can also be specified. If these latent variables are modeled with an assumed mean of zero (and intercepts for all manifest variables are freely estimated), the result is a form of a state-trait model. If, however, factor means are freely estimated, the result is an investigation of comorbid course. Although the promise of being able to assess occasion-specific and cross-lagged prospective associations while taking into account general comorbid course is attractive, it becomes necessary to first address the question of whether a given construct has been successfully characterized in terms of course. For that reason, it becomes necessary to address whether alternative models should be considered and the ways that individuals frequently address such changes in structural modeling.

The Case for More than One Model

The thought of considering (let alone reporting) multiple candidate models rather than a single one is controversial for some. Preference for a single model stems from the belief that the model must be specified in advance of data analysis as part of a sound experimental method in order to avoid the problem of "p-hacking." The concern here is that "significant" effects in a model are actually spurious post-hoc effects cherry-picked from a universe of possible models. The argument presented in this paper, though, is that sound reportage of a data set should also involve consideration of multiple models. It may be that the complexity of an initial model has poor statistical power or, conversely, that an overly simple model obscures functional relationships discernible in a more complex model. As mentioned above, consideration of alternative critical models may operationalize counterarguments made by a reasonable skeptic. Models with alternative psychometric measurement assumptions can also constitute an assessment of whether a proposed effect is conditional on a particular psychometric assumption.

Questions such as these are increasingly being considered in the larger statistical literature, especially as they apply to the specification and replicability of claims regarding regression model fit and estimates. Simonsohn, Simmons, and Nelson (2015), for example, discuss these issues as they apply to the identification of potentially adversarial or alternative regression models and propose that "specification curves" can be used to assess the reasonableness of research claims across alternative models which vary in terms of their assumed functional forms, choice of covariates, and subsets of the data. Although this approach, as well as other penalized regression approaches such as the bridge or lasso (e.g., Fu, 1998) address the problem of models containing different exogenous variables, they are not taken up here but could be the object of extensions to the approach proposed here.

Single Parameter Modifications

Consideration of model alternatives in traditional structural equation modeling (SEM) usually assumes that the functional form and subset of relevant manifest variables are agreed upon. The question often arises as to whether a better model can be found by sequential modifications of the original model or whether a more adequate model exists which makes different psychometric assumptions of the data. Researchers using the sequential modification approach often consider additional parameters based on model modification indices of the original model under the proviso that such modifications should "appear reasonable on theoretical grounds." By contrast, however, poor model fit may also indicate that the chosen model is not "complex enough," and that addition of individual parameters will lead the researcher to consider simple but systematic changes in the basic measurement model for the constructs involved.

Improper Models

Alternatively, sometimes model fit is acceptable, but the estimated parameters represent a mathematically improper solution typified by, for example, negative estimates of variances or implied correlations which exceed unity in absolute value. Often, these offending parameter estimates are remediated by a variety of minor modifications of the original model. This can involve fixing an estimated negative variance component to zero in a final model or examining modification indices with an eye to adding correlations between measurement errors across constructs to decrease the magnitude of a latent variable correlation, which is estimated at a value larger than one or less than negative one. In recalcitrant or less clear-cut contexts, some researchers conduct a separate exploratory factor analysis with rotation in order to identify departures from simple structure which may be responsible for model misfit and then respecify the model with this revised structure in mind.

Alternatively, if the initial model is overly complex it may not be empirically identified and no unique solution exists (Savelei & Kolenikov, 2008). Researchers encountering such identification problems may incorrectly conclude that the convergence failure is a result of software limitations and that a more sophisticated minimization algorithm is needed or some change in program defaults for convergence are needed. While it is certainly possible that software and minimization algorithms sometimes fail, some reduction in the complexity of the model can also be entertained by refactoring the data or deleting selected variables from the model. In the cases of improper solutions and empirical nonidentification, the common problem appears to be that the initially chosen model is overly complex and a better model exists which is not as complex.

Even if a proposed structural model fits well, constitutes a proper solution, and shows significant parameter effects, it may be that the results are an artifact of the outsized effect of the inclusion of a small number of influential or atypical observations. For example, in my own work involving college students, much of the association between college grades and alcohol use appears due to the inclusion of a relatively small number of male, first-generation students involved in the campus Greek system. Identification of influential observations which unduly affect parameter estimates and model fit may be part of a counterargument proposed by a reasonable skeptic but also speaks to the replicability of a proposed effect, given that researchers may not be so fortunate as to ascertain such influential observations.

In summary, consideration of alternative models is a useful exercise regardless whether an initially chosen model fits well or identifies favored statistically significant parameters. If other well-fitting models exist which do not show the favored effect or if the model fit or parameter estimates are due to the inclusion of influential observations in the data, a reasonable skeptic should be made aware that these alternative explanations exist. Conversely, while it is often appreciated that ill-fitting, improper, or nonconvergent models require consideration of a different model, remedies which involve addition of a single parameter at a time are unlikely to inform the researcher of general changes to the assumed measurement model or dimensionality of the construct(s) under examination. Systematic consideration of dimensionality and psychometrics are especially important if the research question involves the comorbidity of multiple types of problem behaviors or psychopathologies.

Right-Sizing Growth Models

In practice, though, exploration of every possible prospective structural model is an impossible counsel of perfection given the logistics involved. It is, however, possible to investigate broad classes of alternatives in a systematic fashion. These classes can be broadly grouped into questions of the dimensionality, patterning, and mean-level effects for a longitudinal model. The three-step procedure for longitudinal data is analogous to that presented in Wood, Steinley, and Jackson (2015) but adapted to the particular analytic concerns involved in longitudinal psychopathology and comorbidity studies. In a nutshell, the steps outlined here represent an attempt to "right-size" a structural model by evaluating the number and patterning of the variance components, and the degree to which patterns of variance components so identified can be used to explain the developmental course of levels of functioning or behavior on the variables of interest across measurement occasions. After a promising candidate model has been identified, the identification of possibly influential observations can be used to bracket the magnitude of proposed effects as well as the decisions regarding the selection of a final model.

Number of Factors

As mentioned earlier, statistical models often vary in the number of latent variables thought to be present in prospective studies. One popular approach is to specify that covariation across all measurement occasions can be explained by reference to a common trait, for example. Regression paths between adjacent measurement occasions are often added to form a "state-trait" model. In many growth models, two factors are often assumed with one latent variable corresponding to a general random intercept factor and the other corresponding to a linear slope latent variable. (In some situations, additional polynomial latent variables are added as well.) Such growth models have been extended to model piecewise growth patterns in the data in which multiple slope variables are specified to represent linear growth patterns which mark the onset and ending of the piecewise growth patterns. Exploratory factor analysis of the construct across measurement occasions is one straightforward approach to assessing the number of factors present across measurement occasions.

Again, it must be remembered that such an approach is only suggestive of the dimensionality of the construct over time that may serve as the basis of an alternative model. It could be, for example, that the linear growth models that assume fixed values for loadings would fail to show a two-factor solution in an exploratory factor analysis due to the fact that linear growth models fix all values of factor loadings. Alternatively, if a developmentally limited period of change is present in the data, it would be necessary for that period to span three or more measurement occasions in order to be detected via exploratory factor analysis. Conversely, exploratory factor analyses which find evidence for more than one factor do not necessarily support the idea that a random intercept and slope factor exists, nor do they necessarily indicate the existence of a developmentally limited period. Additional patterns of covariation beyond a single factor could be due to autoregressive paths between measurement occasions, for example, and the two factors which are identified may not represent the random intercept and linear growth anticipated by the researcher. The point, however, of examining the dimensionality of the series is to explore whether a reasonable argument can be made for a different dimensionality than that initially assumed by the researcher.

Parsimonious Factor Patterns

It is well known that the latent variables in an exploratory factor analysis are rotationally indeterminate, and factor rotations can be applied to an initial matrix of loadings without loss of fit. Such rotations are useful in that they can be used to render the factor loadings more conceptually meaningful and are designed to identify a simple factor structure in the data in which each manifest variable loads on only one factor. In the case of growth data, it is often argued that the simple structure criteria are not appropriate, given the fact that it may be reasonable to think that the process under consideration may span all measurement occasions (Tucker, 1958, 1966; Wohlwill, 1973). Although several parsimonious confirmatory factor models can be fit to prospective data, as discussed in Wood and colleagues (2015), Figure 19.1 shows common factor patterns which have been discussed within the context of prospective psychopathology research.

The pattern shown in Figure 19.1A, labeled "Triangular/Cholesky," specifies a general factor spanning all measurement occasions and secures a mathematically identified subsequent factor by constraining one factor loading of a subsequent orthogonal factor to zero. The term "triangular" for this type of factor model gets its name from the fact that successive identification of factors produces a triangular pattern of loadings in the factor-loading matrix. Although this model permits specification of several factor loadings and is, therefore quite complex, one of the difficulties in specifying such models is that different factor loadings (and therefore different patterns of statistical significance) can arise depending on which measurement occasion is chosen to be fixed to zero in subsequent factors. In some situations, selection of the zero loading may not pose a problem, as prior research may suggest which loading is reasonable to fix to zero. For example, individuals may either mature out or show a delayed onset in the behavior under consideration. In cases of maturing out, the last measurement occasion may be fixed to zero and in the case of delayed onset, the first measurement

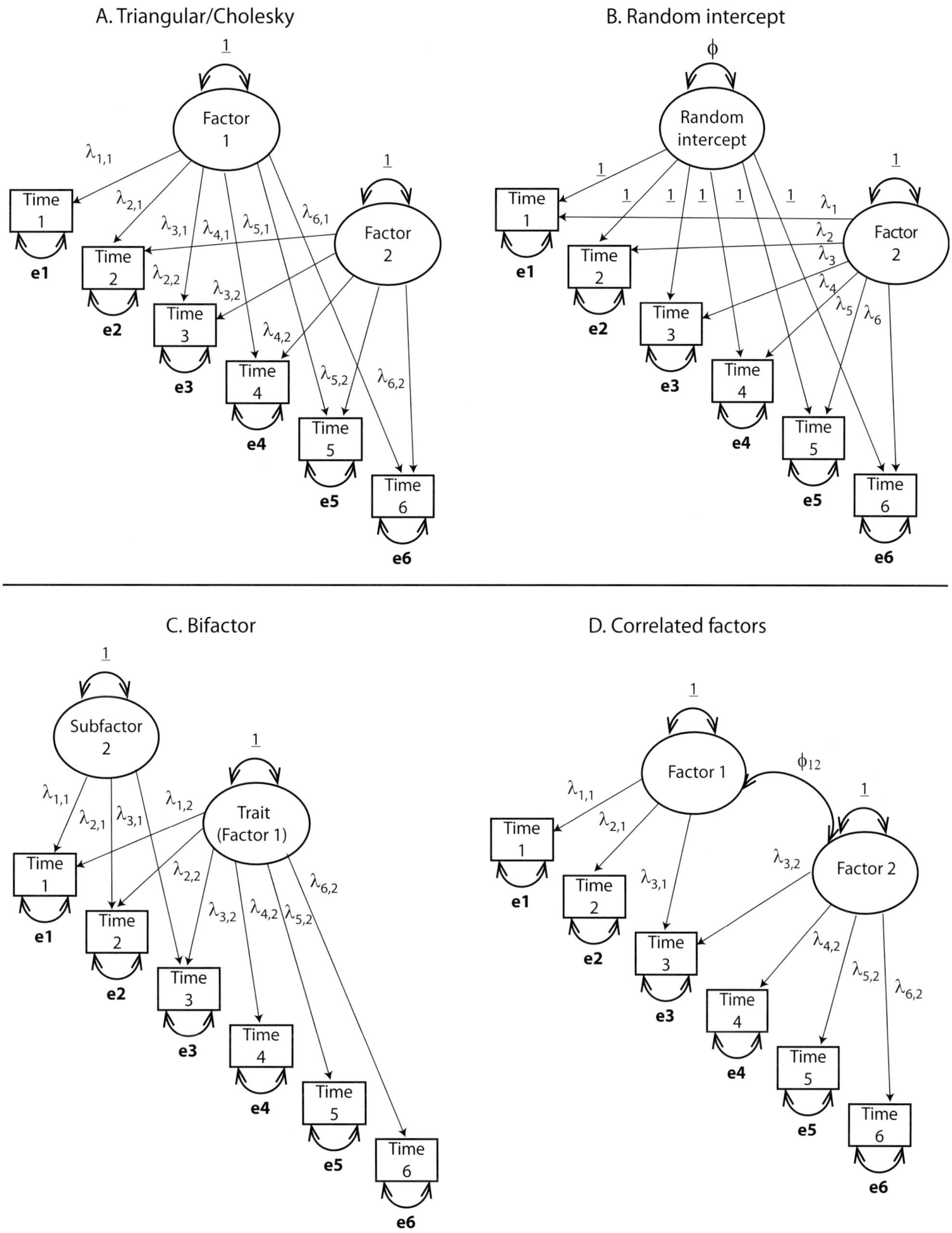

Figure 19.1 Parsimonious factor pattern models

occasion could be zeroed out. It is, of course, possible that a researcher could fit each of the possible two-factor models in turn and attempt to identify the "simplest structure" of the resulting factor by selecting that model for which the sum of the squared standardized loadings of the second factor is a minimum.

The random intercept model (Maydieu-Olivares & Coffman, 2006) shown in Figure 19.1B is another type of two-factor model in which the loadings of one factor are freely estimated while the loadings of the other, termed a random intercept factor, are set to unity and the factor variance freely estimated. Like the Triangular Decomposition model, the two latent variables are assumed to be orthogonal in this version, though an equally well-fitting model may also be specified by specifying one of the loadings of the freely estimated model to be zero. While Maydieu-Olivares and Coffman interpreted the factor loadings of the random intercept and freely estimated factors to represent method variance and the loadings of the "true" latent construct, respectively, in longitudinal data, a case can be made that both latent variables are components of the "real construct" as occurs when mean levels in the data are explained using the free curve slope intercept model discussed below. Wood and colleagues (2015) noted how the random intercept factor model is nested within the triangular factor model via an affine transformation, meaning that incremental fit indices can be used to assess whether the random intercept factor model is a plausible model or some other multifactor model demonstrates better fit.

Bifactor Model

It may be the case that the triangular model can be fit to the data, but that more than one loading can be fixed to zero in the second (and successive) factor(s). Such a model may be a parsimonious alternative to the triangular model if it is the case that a subset of measurement occasions demonstrates a relatively time-delimited influence on the behavior under consideration as shown in Figure 19.1C. For example, in studies of alcohol consumption across the years of young adulthood, it may be the case that alcohol use during the years a student lives in campus residence halls is different than subsequent years when living off campus. Alternatively, different covariances between alcohol use variables may obtain after the participant has reached legal drinking age than in measurement occasions prior to that. It is entirely possible that multiple such subfactors exist. It would be possible, for example, to also consider an additional subfactor which spanned the last three measurement occasions, for example, provided that three factors were indicated in the assessment of dimensionality.

Oblique Simple Structure Factors

Figure 19.1D shows a fourth possible two-factor structure in which two factors exist, but the factor loadings demonstrate the simple structure pattern often sought in exploratory factor models. For longitudinal data, such models suggest that there are two related distinct constructs under investigation but these two constructs are correlated to some degree.

Comparison of Competing Parsimonious Models

It is quite easy to see that a two-factor triangular decomposition contains the bifactor factor model as a special case, as additional loadings are set to zero in the bifactor model. Similarly, the correlated factor model can be seen as a special case of the bifactor model if the estimated factor loadings of the subfactor in the bifactor model are nearly proportional to those of the general factor. As noted in Wood and colleagues (2015), the random intercept factor model can be rotated to a triangular factor model via affine transformation. If comparison of multiple models yields relatively equivalent fit, these models may then be considered as candidates for explaining mean-level effects over time, where the predicted patterns of elevation are often quite different when latent variable means are included in the model.

Growth Curve Models

Growth curve models result when latent variable means are estimated in the model and a wide variety of growth models are often used in longitudinal research. A wide variety of parametric and psychometric growth curve models are presented in Wood and colleagues (2015) and, as with the previous discussion of parsimonious factor models, a subset of models which are presented in Figure 19.2 which represent growth curve models which appear reasonable for prospective research in psychopathology and problem behaviors. One of the simplest growth curve models for longitudinal data was proposed by McArdle and Epstein (1987), who christened the model the CURVE growth model shown in Figure 19.2A. One criticism of the CURVE model, however, is that it assumes that the repeated observed variables are measured in a common ratio level of assessment. Although this assumption may be thought of as similar to the physical measurements we might make of a variable such as height, different model fits can result if a constant is added to all observed variables. Consonant with the assumptions of traditional factor analysis, however, it may be that the observed variables associated with a construct are measured in a common interval-level scale of measurement.

The Factor Mean Shift (FM-Shift) model shown in Figure 19.2B has been proposed as a growth model which makes this less restrictive assumption of the data and involves the estimation of a common manifest variable intercept across measurement occasions (Wood et al., 2015). In practice, estimation of a shift parameter has been shown to greatly improve the fit of structural models that do not contain a random intercept factor model. Although the FM-Shift model was developed many years ago by McDonald (1967), the practical conceptual interpretation of the model has not been discussed outside of Wood and colleagues, so a brief explanation of the FM-Shift model as a measurement assumption is in order. If we were to assume that the manifest variable measured across measurement occasions was a ratio-level variable, the Factor Mean model would be appropriate. Under ratio measurement, we would expect a mean value of zero for

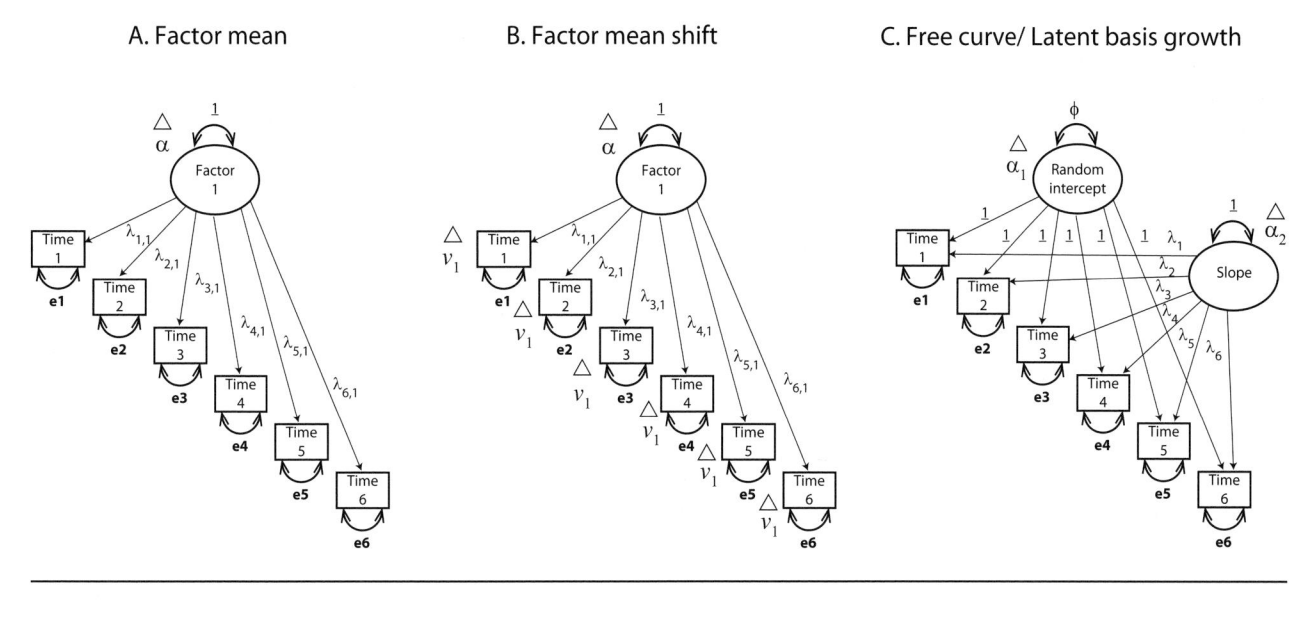

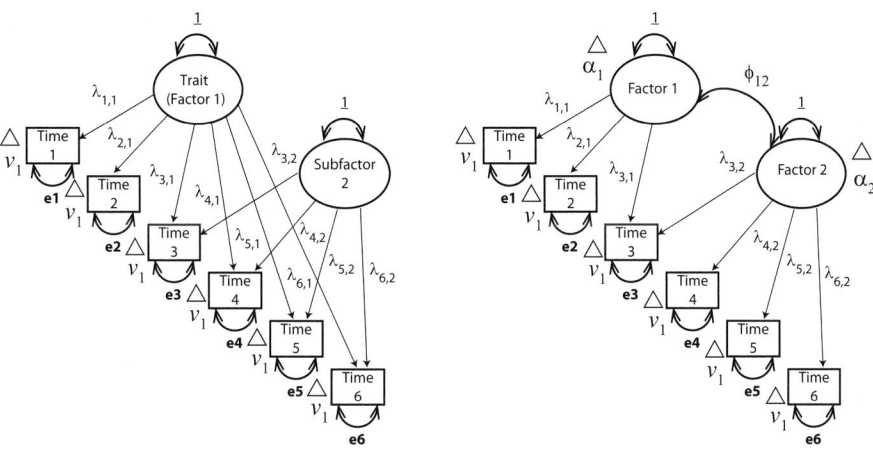

Figure 19.2 Commonly used growth curve models

the measurement occasion associated with having "none of" the latent variable (i.e., the starting point of growth). This may be a reasonable assumption for, say, stature (provided no shoes are worn). If, however, our measuring stick was always "broken" at some point on the scale (by, say, always measuring height beginning one foot off the ground, or by measuring the height of individuals who always wear three-inch-heeled shoes), we would not expect the starting point to represent a score of zero; we would adjust our expectations about height at the starting point by adding a foot to our score or subtracting three inches from our observed measurements. Analogously, given that scales often used in psychological measures consist of sums or averages or are based on Likert-format questions, the FM-Shift model adjusts observed scores to a common interval scale by adding or subtracting a constant across all measurement occasions. So, for example, if we administer a vocabulary test to children and a certain proportion of test items may be answered correctly by guessing, a shift operator adjusts the scores for this chance agreement.

The Free Curve Slope Intercept (or "Latent Basis") factor model shown in Figure 19.2C is the growth curve analogue of the random intercept factor model of Figure 19.1 (Meredith & Tisak, 1990). Given that the random intercept factor as factor loadings are set to unity, no separate shift effect may be estimated for such models. Wood and colleagues (2015) demonstrated how the Free Curve Slope Intercept model includes the more popular linear growth curve model as a special case when the estimated factor loadings of the slope factor fit a linear pattern across measurement occasions using nonlinear constraints. The Free Curve model also includes many other parametric growth curve models, such as the Gompertz or confined exponential growth curves as special cases via appropriate nonlinear constraints (Wood et al., 2015).

After the growth curve model is specified using factor means and, possibly, a shift operator, exploration of mean-level effects specific to one measurement occasion may be entertained. Although patterns of elevation which span more than one measurement occasion can be modeled using a latent variable spanning those measurement occasions, those specific to one measurement occasion cannot. It may be, for example, that young adults still living with their parents may not engage in behaviors they might otherwise engage in. Conversely, exposure to a new environment, such as the first year of college, may result in an overall elevation in behaviors such as alcohol consumption, psychological distress, or risky sex which are not typical of overall levels at other measurement occasions. Lagrange multipliers are one way to explore such questions. When, for example, a subset of individuals in a study goes on to college and such use induces additional patterns of covariation in the data, the parsimonious factor structure suggested by Figure 19.2D may be appropriate. When, by contrast, all of the participants after a certain measurement occasion change environments (such as college or changing residency location on campus), the correlated factor structure shown in Figure 19.2D may be more appropriate. Both of these latter models are shown specified in shifted form, as it may be appropriate to consider the manifest variables as representing interval level, as opposed to ratio-level measurement. Additional models which include a random intercept component are, of course, also possible.

Example Analyses

To illustrate how such right-sizing can operate within the context of multiple constructs an example will be presented of two constructs, general psychological distress and problematic alcohol consumption, assessed longitudinally in a sample of young adults drawn from a Midwestern US university. As details of participant ascertainment and the specific research questions involving reports of distress and other problem behaviors have been the subject of other published articles by our team, I will not repeat them in detail here (e.g., Rutledge & Sher, 2001). I note only that the data set considered here consists of an attempted assessment of all first-time college students entering the University of Missouri enrolling in the fall of 2002. The data are somewhat unusual in that an initial assessment of alcohol use was conducted in the summer prior to arrival on campus. The remaining eight assessments were conducted during the fall and spring semesters for the next four years. Two additional waves were collected in the fall of 2006, and the spring of 2009, representing assessments conducted approximately four and six years after the students first matriculated from the university, respectively.

These data have been the subject of several publications on the longitudinal course of problem behaviors such as alcohol consumption but often growth was examined using traditional growth curve models assuming a linear or polynomial growth curve with a random intercept variable that is usually scaled to the first measurement occasion. Our lab has also used hierarchical linear models for prospective data that assume a quadratic form over time to assess the comorbidity of alcohol and tobacco use, the relationship of alcohol-use disorders and psychological distress (Jackson & Sher, 2003; Sher, Wood, & Gotham, 1996c), the comorbidity of tobacco and alcohol-use disorders (Jackson et al., 2000; Sher et al., 1996a), the role of intoxication and role transition (Gotham, Sher, & Wood, 1997), and the role of personality and substance-dependence symptoms (Grekin, Sher, & Wood, 2006), just to name a few of the research questions undertaken. In many cases, getting these models to converge has been somewhat difficult, especially when multiple polynomials have been included in the model. Some of the identified prospective effects which were included based on modification indices in the models have been somewhat unanticipated, although not unreasonable in hindsight. Additionally, while the decision regarding the final model chosen from the specified candidates has been consistent across split half cross-validation, I have been somewhat concerned that some characteristics of the model, such as log-likelihood, have shown rather substantial variability in cross-validation. Finally, although the papers have been accepted for publication based on somewhat less than stellar model fits, I believe reviewers' acceptance of the approach is more informed by the generally lower model fits published in the area. The approach to assessing the general longitudinal comorbidity between psychological distress and problematic alcohol use proceeds by first assessing the psychometric models of two constructs, psychological distress and problematic alcohol use, and then joining these two latent variable models to produce a general model of comorbidity.

For each of the constructs, the three-step right-sizing procedure described will be conducted. The general psychological distress construct is slightly easier to construct because assessment at each measurement occasion consists of a single manifest variable. A similar factor model analogue for prospective data will be presented for problematic alcohol use where three different manifest variables were assessed at each measurement occasion. Finally, the production of a single model of comorbidity of these two constructs is presented as a multigroup investigation of the factorial invariance of the resulting right-sized model of comorbidity.

LONGITUDINAL ASSESSMENT OF PSYCHOLOGICAL DISTRESS

Entering college students often report elevated levels of psychological distress which appear to resolve for many during the course of college. For the study considered

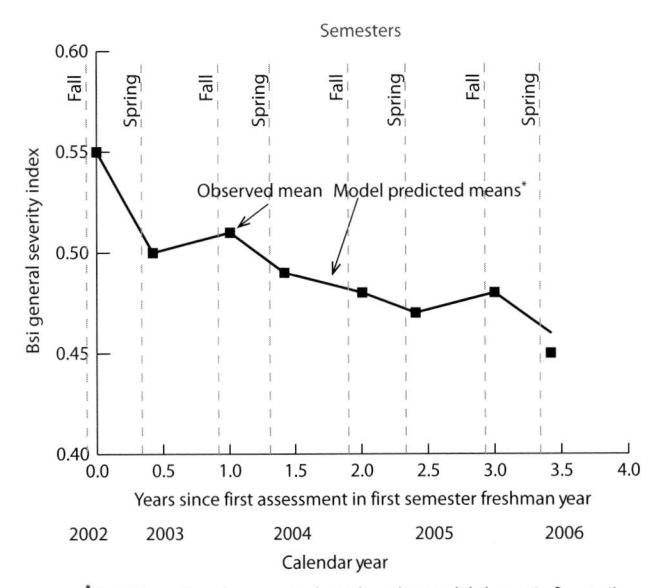

Figure 19.3 BSI General Severity Index (BSI-GSI) observed and model-predicted means

here, reported psychological distress was operationalized using a modified version of the 18-item version of the Brief Symptom Inventory (BSI; omitting the suicidal item) in which participants were asked to report their level of distress on average over the past week. The BSI was administered on eight occasions during the first four years of college. For each distress behavior (e.g., feeling faint, dizzy, or blue, feeling tense or nervous) participants were asked "how much that problem has distressed or bothered you during the past 7 days including today" (Derogatis, 1993, p. 4). Each item is scored on a five-point scale from 0 (not at all) to 4 (extremely). In this study, the BSI Global Severity Index (BSI-GSI) was used, which is calculated by dividing the sum of the response values by the total number of responses. Coefficient alphas for the BSI-GSI ranged between 0.91 to 0.93 across the eight waves of assessment. Figure 19.3 shows the longitudinal curve associated with the data (and the predicted means from the final model described below).

As can be seen from the figure, general distress appears to spike during the fall of the freshman year and shows a slight spike in the fall of the second year, but appears to decrease in subsequent assessments. The longitudinal means gathered over time, however, promote some questions regarding the nature of change over time. Do temporally elevated scores represent transient elevations in scores or are individuals who are more prone to reporting psychological distress also more likely to report an elevated level of distress at these particular measurement occasions? If elevations in scores are at this time largely transient, the strategies for possible intervention to reduce distress in the population are different than if they are indicative of a particularly stressful occasion in individuals likely to be distressed across all measurement occasions. Similarly, the curve plot is agnostic about the dimensionality associated with the psychological distress construct over time. As discussed below, it may be that systematic differences in reported psychological distress may be confined to a subset of measurement occasions or may typify a general response style across measurement occasions.

Earlier Analyses

In an earlier study of the role of familial risk of alcoholism in collegiate distress and alcohol, Sher and colleagues (1996c) used repeated measures analysis of variance and found that psychological distress decreased across the first four years of college and that individuals with a familial risk for alcoholism showed elevated levels of distress across the college years. As noted by Wood and colleagues (2015), however, the structural model of compound symmetry in repeated measures analysis of variance corresponds to a model of strictly parallel measurement across measurement occasions, an unlikely assumption given that some measurement occasions, such as the first semester of college, may represent occasions containing more measurement error due to the transient effects of moving to a new location, or may be less representative of trait distress assessed at other, more typical measurement occasions. As a result, the tau-equivalent factor model assumed by the compound symmetry assumption of repeated measures is likely violated due to the presence of increased measurement error at some occasions or the presence of different factor loadings across occasions (see Chapter 6, by Furr).

"Standard" (i.e., Polynomial) Growth Curve Analyses

As an alternative, a researcher may wish to explore polynomial growth models as a more plausible alternative. This may be particularly attractive, given that such growth models are popular in the research literature and due to the fact that, in many applications, any real-valued function can be successfully approximated by a polynomial expression of sufficient degree. The problems associated with using polynomials for longitudinal data such as that considered here is that the number of measurement occasions is rather large, meaning that the degree of the required polynomial is rather high, making interpretation of the resulting effects difficult. Moreover, addition of successive polynomial effects often results in rank-deficient parameter estimates, meaning that no single family of parameter estimates exists for the data, but rather that numerous solutions exist, due to the fact that some parameter estimates of the model are severely correlated and dependent on the estimated value of other parameter estimates. Further, the modification indices associated with polynomial solutions can provide rather counterintuitive modifications to the model, as was found in this example. Specifically, for these data, χ^2 statistics for the linear, quadratic, and cubic models across the eight measurement occasions were $\chi^2_{31} = 155.12$, $\chi^2_{27} = 84.99$, and $\chi^2_{22} = 49.63$, respectively. χ^2 difference statistics between these models favored the more complex model consistently as well (p's < 0.001). Under the assumption of a quadratic growth curve, the two largest modification indices suggest freeing the covariances between the spring semesters of 2004 and 2005 and between the fall of 2004 and spring of 2006. Although perhaps some case can be made for freeing covariances between similarly timed semesters, there seems no clear conceptual rationale for the second modification.

Conceptually, some could argue there is no reason to believe that the course of psychological distress follows a polynomial curve across measurement occasions. A more fine-grained prospective analysis could involve more than one latent variable but the relevant latent variables need not necessarily span all measurement occasions. For example, first-time college students at this particular Midwestern university, the University of Missouri, are required to live in university-operated residences during their first two years unless exempted and this may result in a time-delimited pattern of distress associated with those years. Alternatively, of course, the challenges associated with residence off campus may constitute additional strains on the student. Delimited patterns of distress may also occur due to the onset of problem behaviors such as alcohol consumption, which becomes legal for students who are 21 years old. Alternatively, as some students come closer to finishing their degrees, concern about the challenges associated with the next chapter of their lives may result in the presence of a second latent variable of distress associated with later measurement occasions.

Conversely, systematic changes in general level of distress may also be confined to a single measurement occasion. Arrival on campus and the changes associated with college courses may result in an initially higher level of distress limited to the fall of the freshman year, a phenomenon taken up in college student development and academic attrition research dealing with the freshman year (e.g., Martinez et al., 2009). Environmental effects, such as the general pattern that freshmen are more likely to be assigned early morning classes, may to some extent also temporarily elevate reports of distress (Wood, Sher, & Rutledge, 2007).

The identification of a parsimonious growth curve model for general psychological distress in these data will now be explored using the three-step approach outlined by Wood and colleagues (2015) in which dimensionality, parsimony, and the examination of mean level are considered in sequence.

Dimensionality

Exploratory factor analysis of longitudinal BSI-GSI scores using Mplus clearly support the existence of two factors in the data. Although χ^2 difference tests associated with the number of factors support the existence of a four-factor solution, no loadings in either the three- or four-factor solutions were statistically significant, and the statistical fit of the two-factor solution appeared quite adequate even though the χ^2 remained statistically significant ($\chi^2 = 61.18$, $p < 0.001$, CFI = 0.99, TLI = 0.99, RMSEA = 0.03, 95%CI = 0.03–0.04).

Rotation to Parsimonious Structure

Although there is no theory to suggest that BSI-GSI reports should necessarily show a two-factor simple structure in which each manifest variable has substantial loadings on only one latent variable, I explored whether such a rotation could be found using a variety of oblique rotations. Under all rotations, though, measurement occasions associated with the third, fourth, and fifth assessments cross-loaded on both factors. The Bi-CF-Quartimax rotation, which is designed to recover bifactor loading patterns, recovered a general factor and an additional factor limited to the first two years (the first four measurement occasions) was found. A corresponding Cholesky factor model was then fit in which the last occasion was fixed to zero. As the factor loadings associated with the fifth through seventh measurement occasions were both small and insignificant, they were constrained to zero. The resulting factor-pattern matrix loadings appeared similar to the Bi-CF-Quartimax rotated solution, and this bifactor model for the data in which a general factor is estimated as well as a subfactor consisting of the first four measurement occasions. This same pattern was recovered when the model was rerun after eliminating influential observations.

Random Intercept Factor (RI) Model

As described above, the random intercept model is an alternate plausible two-factor solution in many situations. In this case, for example, it may be that self-reports of the severity of psychological distress may contain both a random intercept (or tau-equivalent) factor in addition to a congeneric factor (in which factor loadings are different across measurement occasions). For these data, overall fits of the RI model were similar to the reduced Cholesky factor model. BIC values for the Cholesky model were lower (23,384 versus 23,400). This discrepancy persisted after 25 of the most influential observations were deleted from the data. Although some researchers believe that a BIC difference as large as 16 between competing models provides evidence that one model (in this case the Cholesky factor model) is better, Wood (2019) found, using simulations, that the power to discriminate between the (true) random intercept factor model and a two-factor alternative was poor until sample sizes became quite large (i.e., 3,000 or larger). The bifactor model was retained as preferable although the Cholesky model may be a close competitor to the bifactor model.

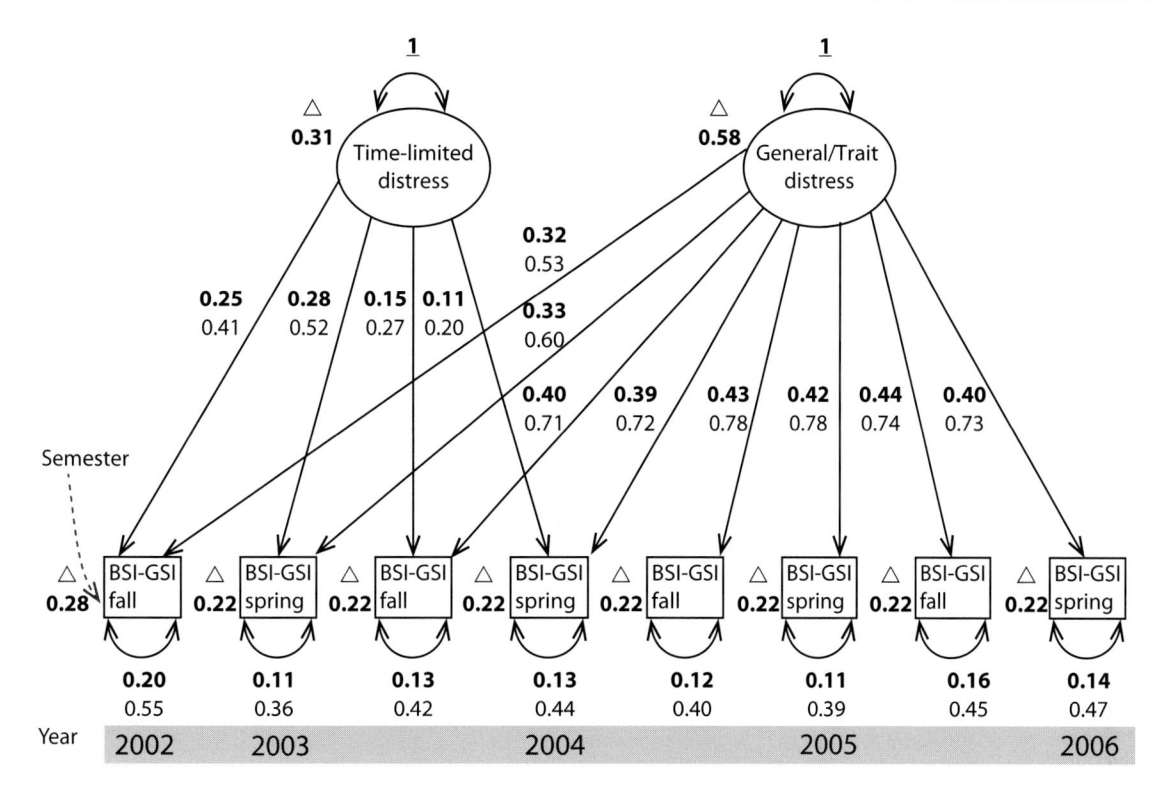

Figure 19.4 Growth curve model for BSI General Severity Index (BSI-GSI)

Underlined Font parameters = Fixed parameters
Bold Font parameters = Unstandardized values
Regular font parameters = Standardized values
△ = Factor mean or manifest variable intercept

Candidate Growth Curve Model

In the third step, the parsimonious factor model is then fit with estimated factor means and shift to produce a growth curve model. For these data, such a Factor Mean-Shift growth curve fit to the bifactor model appeared to do a good job of summarizing patterns of covariation in the data as well as overall level (χ_{21}^2 = 90.84, $p < 0.001$, CFI = 0.99, TLI = 0.99, RMSEA = 0.03, 95%CI = 0.03–0.04), and inclusion of the estimated shift operator is a statistically significant improvement over a model which estimates simple factor means.

In the final step, Lagrange multipliers associated with the manifest variable intercepts were examined to detect occasion-specific elevations for the model revealed that the assumption of intercept equality across measurement occasions should be removed for the first measurement occasion. Doing so resulted in the same fit statistics as before, but a large reduction in the χ_{21}^2 = 74.78 and a lower BIC value (23,384.63 versus 23,392.59 for the original model).

Conceptually, freeing the intercept associated with the first measurement occasion corresponds to the view that, on average, first-time college freshmen may experience a slightly elevated level of psychological distress which is unrelated to subsequent measurement occasions.

The final growth curve model is shown in Figure 19.4. Squaring and summing the standardized loadings across factors provides an estimate of the internal consistency of measurement. These are shown in the column labeled "BSIGSI" in Table 19.1. The lower internal consistency associated with the first semester of college may reflect psychological distress specific to the first semester of college. Similarly, the lower internal consistency at waves 7 and 8 may also reflect the fact that different environmental or interpersonal stressors are present at these measurement occasions which reflect, for most, that they have left the university. In terms of the psychometrics of the manifest variables, we may interpret this model as indicating that the manifest variable measurements are made at a common interval level except for the first semester of college, during which general severity scores appear to be temporarily elevated. The data support the existence of two latent variables, with a parsimonious model consisting of a general psychological distress factor and a delimited factor consisting of the first two academic years.

Influential Observations

Even though the proposed model appears reasonable and has acceptable fit, it is worth examining whether some

Table 19.1 Multigroup factor loadings, means, and manifest variable intercepts

	BSI		Alcohol		Method factors			Manifest variable intercepts			
	General	Limited	General	Limited	Drunk	High	Plus5	BSI	Drunk	High	Plus5
Wave	**Factor loadings**										
Precollege			0.52**	0.47**	0.04*	0.15**	0.1**		−0.01	0.19*	−0.03
Fall 2002	0.33**		0.49**	0.46**	0.08**	0.16**	0.11**	0.25	0.10**	0.40**	0.08*
Spring 2003	0.33**	0.24**	0.55**	0.44**	0.08**	0.22**	0.16**	0.18	0.05*	0.34**	0.03
Fall 2003	0.41**	0.30**	0.59**	0.36**	0.13**	0.20**	0.17**	0.18	0.05*	0.34**	0.03
Spring 2004	0.40**	0.15**	0.58**	0.27**	0.11**	0.24**	0.13**	0.18	0.05*	0.34**	0.03
Fall 2004	0.43**	0.10**	0.60**		0.10**	0.23**	0.22**	0.18	0.05*	0.34**	0.03
Spring 2005	0.42**		0.64**		0.16**	0.21**	0.24**	0.18	0.05*	0.34**	0.03
Fall 2005	0.44**		0.63**		0.14**	0.22**	0.22**	0.18	0.05*	0.34**	0.03
Spring 2006	0.41**		0.64**		0.15**	0.20**	0.23**	0.18	0.05*	0.34**	0.03
Wave 9			0.47**		0.10**	0.18**	0.26**		0.05*	0.34**	0.03
Wave 10			0.29**		0.09**	0.15**	0.20**		0.05*	0.34**	0.03
Means	Female	0.71*	0.39**	0.84**	0.17**	0.02	−0.65	−0.12			
	Male	0.64	0.20	1.68**	0.00	−0.98	−1.16**	0.21			
Variance	Female	<u>1</u>	<u>1</u>	<u>1</u>	<u>1</u>	<u>1</u>	<u>1</u>	<u>1</u>			
	Male	0.96**	0.78**	1.58**	2.29**	2.6**	1.44**	1.12**			
Standardized	Female	0.71*	0.39**	0.84**	0.17**	0.02	−0.65	−0.12			
Means	Male	0.65	0.23	1.11**	0.00	−0.61	−0.97**	0.20			

Underlined values denoted fixed parameters.

individuals show dramatically atypical patterns of response which may influence model fit or the estimated parameters. Figure 19.5 shows a scatterplot of the log-likelihood against the leverage for this model. Both of these indices provide an assessment of the degree of misfit of the model to individual observations in the data. The leverage differs from the log-likelihood in that the leverage documents the degree to which a particular observation may influence the estimated parameters and significance of a proposed model while the log-likelihood provides a general assessment of the likelihood of a particular observation given the model. In general, the magnitude of the influence values observed in the plot are somewhat unusual. General guidelines for determining a "large enough" leverage generally advise that values between 0.2 and 0.5 indicate moderate leverage, for example (Kutner et al., 2005, p. 399), and by that criterion a significant proportion of the sample would be dropped and the similar ranking on the basis of log-likelihoods suggests that these observations are not well fitted by the model. Closer inspection of only those observations in excess of 0.6 reveals that these individuals are more likely to be women and their BSI-GSI reports across measurement occasions are typified by a single high-distress report, with other reports being closer to more typical values. In this case, however, recalculating the model without these potentially influential observations did not produce substantial changes to the overall model fit and resulted in a modest reduction in measurement error values. Differences between the models considered remained, however, yielding the same decision regarding the final model. In sum, although some signal stressful events such as parental divorce, death of a close relative, or failure of a romantic relationship may be normative during young adulthood, these events are not well modeled by the general growth model and individuals who have experienced such events may be the object of other research questions.

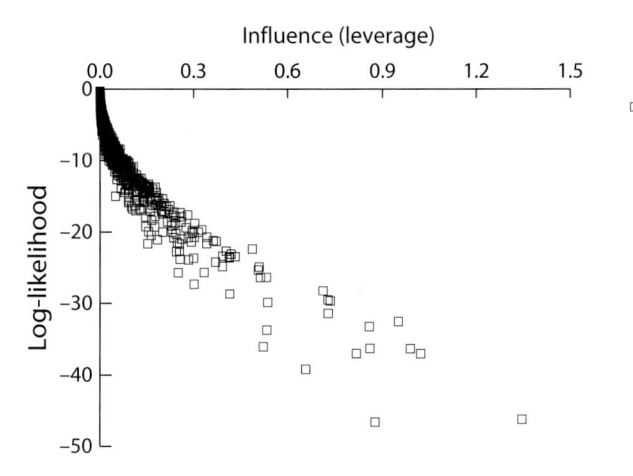

Figure 19.5 Scatterplot of log-likelihood and leverage for general psychological distress model

PROBLEMATIC ALCOHOL CONSUMPTION

We now turn to the right-sizing of another problem behavior, problematic alcohol consumption, to illustrate how the right-sizing approach may be applied in situations where multiple assessments exist within each measurement occasion. Three variables were administered to measure risky drinking in the assessments at precollege and in the first year of college: "In the past 30 days how many times a week have you had five or more drinks in a single sitting?" (Plus 5), "How many times a week during the past 30 days were you a little 'high' or lightheaded?" (High) and "How many times a week during the past 30 days were you drunk?" (Drunk). Participants responded to each item based on eight-point scales with response options of 0, 0.2, 0.6, 1.5, 3.5, 5.5, 6, and 7 times a week. The frequency of having five or more drinks in a single sitting has been widely used for assessing high-risk drinking (e.g., national surveys including the Monitoring the Future Survey and the CORE Alcohol and Drug Survey; also see Carey, 2001).

The general pattern of means across the measurement occasions is shown in Figure 19.6. As can be seen, the general pattern of endorsement of the alcohol behavior items is consistent across measurement occasions, with reports of being "high" or "lightheaded" being most endorsed, followed by rates of drinking five or more drinks in a sitting. The number of times that the individual reports being drunk are the least endorsed of the three. Across measurement occasions, a steadily increasing trend in alcohol consumption reports occurs up to spring

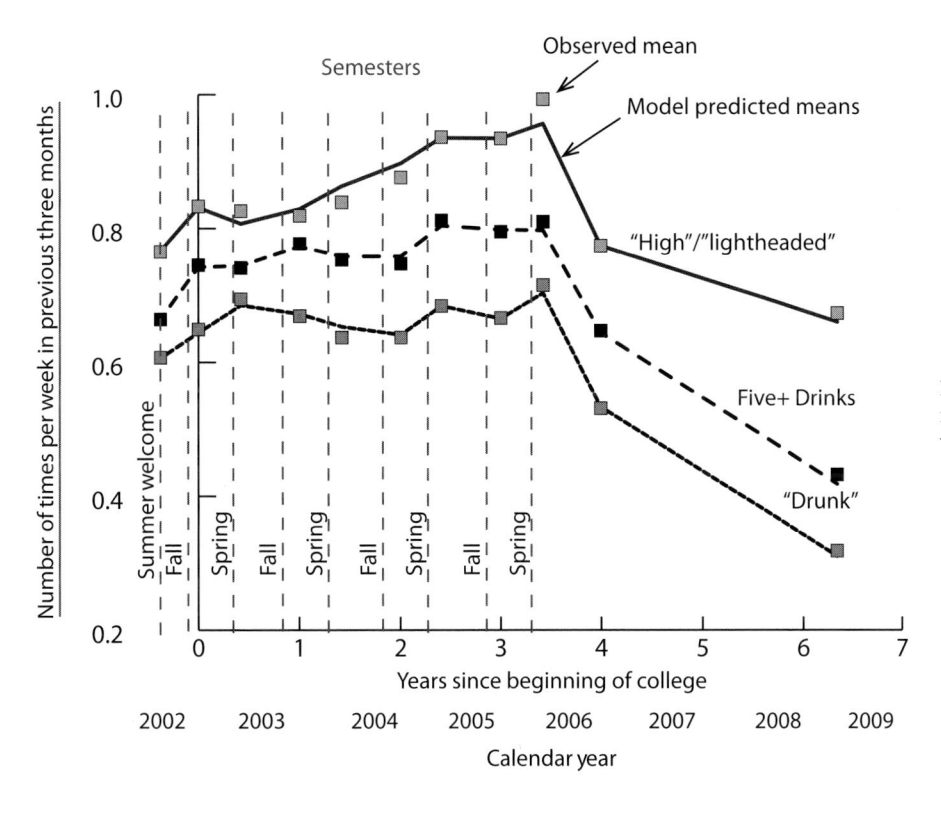

Figure 19.6 Observed and predicted means for alcohol consumption variables

2006 and the measurement occasions show a decline to levels which are below those of the first assessment.

In order to right-size each of these three variables over time, three separate exploratory factor analyses were conducted for these three variables. Consistently, two factors were found for each of the three measures across measurement occasions. As with the psychological distress construct, bifactor rotations of the measures yielded a general factor and a subfactor for each variable in which loadings occurred on the precollege through spring 2004 assessments, meaning there was evidence for a time-delimited factor corresponding to the first two years of college.

Although this factor may correspond to age-delimited patterns of alcohol use prior to age 21, it may also correspond to the period in college when students are first allowed to live in off-campus housing.

When including mean-level effects in the data, statistically significant shift operators were found for all three measures and, upon examining modification indices associated with intercept effects, large Lagrange multipliers were found for intercepts at the precollege and fall 2002 assessments. Consistently, the intercept was lower for the precollege assessment and higher for the first college semester than for the remaining measurement occasions. Accordingly, these parameters were freed as a result and taken to represent the phenomena that high school student alcohol consumption is somewhat hindered by their living environment or parental supervision and that the alcohol consumption during the first semester of college generally increased relative to subsequent measurement occasions.

Figures 19.7, 19.8, and 19.9 show the final growth model for the self-reports of the Number of Times High/Lightheaded report. By far the largest portion of changes in reports of being lightheaded across measurement occasions is due to the general factor, with a second pattern of mean trend associated with the time-delimited factor associated with the first five measurement occasions. Across all three constructs, factor loadings on the general factor are more pronounced during the later fall 2004 through spring 2006 occasions, where a majority of students attain the legal age for consumption of alcohol and students are allowed to live off campus. Factor loadings associated with early use decrease across the precollege and early college assessments and account for a proportionately smaller role in the observed mean level of endorsement of being lightheaded. When growth curve models are fit with estimating shifts (i.e., common intercepts across measurement occasions for each respective variable) the resulting models had a better fit than a corresponding standard factor mean model. Examination of Lagrange multipliers for the individual measurement occasions found lower means for the precollege and slightly higher means for the fall 2002 measurement occasions across all three variables.

When the leverage values of each of these growth models were examined, a relatively small number of influential observations were again found, as with the BSI-GSI analyses above. Although the influential BSI-GSI observations were largely women, influential observations for the alcohol variables were mainly men, who reported an extremely high rate of alcohol use at only one measurement occasion or, less frequently, a consistent but dramatic pattern of drinking more than five drinks during the early college years, followed by decreased reports on subsequent occasions with elevated reports during the postcollege assessments of waves 9 and 10. As before, these individuals did not significantly influence the estimated parameters of the growth model but the fit of the growth curve model was slightly improved when these individuals were removed from the data. As with the BSI-GSI plots, these individuals may constitute a subpopulation of individuals worthy of study in their own right, as they appear very atypical of the rest of the sample.

Summary of Alcohol Use Growth Curves

In summary, an identical dimensionality and similar factor structure was found across all three alcohol consumption items and supported the existence of two latent variables. These two latent variables, however, did not consist of a single random intercept factor and slope as found in many growth curve studies nor did traditional factor rotation techniques produce a simple structure solution. Rather, a bifactor solution consisted of a single general factor and a more delimited factor spanning the precollege and the first four semesters of college. For the general factors those assessment periods associated with the fall 2004 through spring 2006 assessments provided the most accurate measurements of the general factor. Factor loadings associated with the time-delimited factors decreased over the measurement occasions.

The consistency of the factor structure and measurement model associated with both the alcohol consumption and distress measures permits the specification of a general model of the comorbidity of alcohol report as well as explorations of the comorbidity of general and time-delimited aspects of general distress with alcohol consumption. Additionally, the resulting multifactor model can be extended as a multigroup model to characterize sex differences in these constructs as well. For this general model, the measurement model assumed by the factor mean shift measurement models identified in the single construct analyses above was used and the estimated shift operators for the manifest variables appeared to change little from the single construct models. Figure 19.10 presents the basic components of such a structural model in which the three alcohol consumption measures are assumed to be manifest variable indicators of an alcohol use factor at each measurement occasion (The full model is presented in Tables 19.1 through 19.3). These alcohol use latent variables at each measurement occasion in turn are taken to be indicators of higher-order factors representing general and limited alcohol use factors.

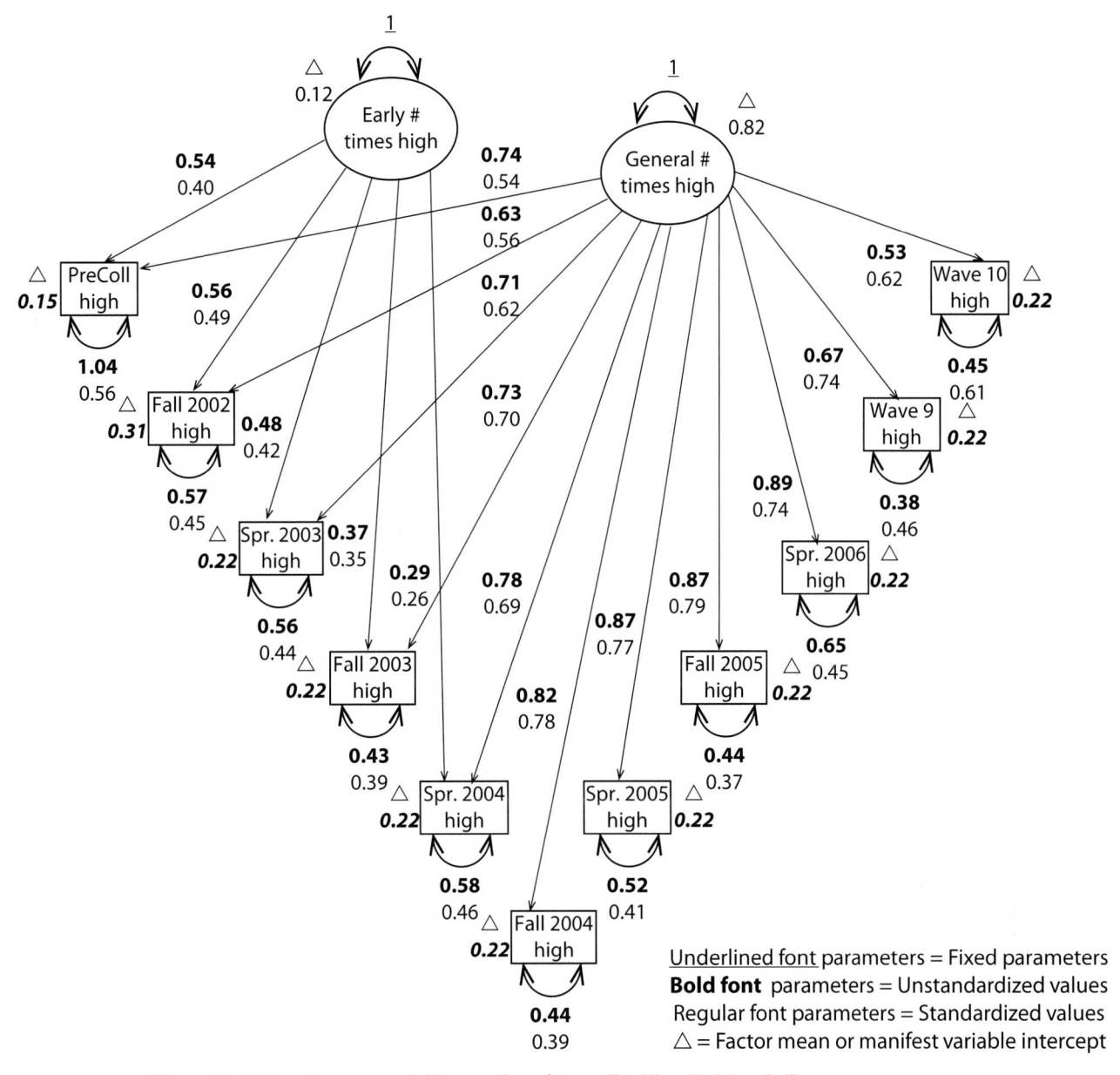

Figure 19.7 Growth curve model for number of times "high" or "lightheaded"

The general association between general distress and alcohol use is then assessed as a covariance between the general and limited factors. Sex differences at the level of the higher-order factors are then modeled by taking one of the groups (in this case females) as a reference scale with factor variance of zero and freely estimating the variance of these factors in the other group (in this case the males). Note that it is not necessary to constrain the factor means for the women to zero as is done in conventional multigroup structural models, as the factor mean in this group is identified.

Factor Mean and Variance Differences by Sex for Alcohol
Several characteristics of alcohol and general distress are evident from Figure 10.10 and Table 19.1. Unsurprisingly, the estimated mean of men on the general alcohol use factor is higher than for women (1.68 for men versus 0.84 for women in the figure and in the corresponding row labeled "Means" in Table 19.1) and the variability of general alcohol use is more pronounced for men than for women (factor variance of 2.29 for men relative to 1 for women in Figure 19.10 and in the corresponding row labeled "Variance" in Table 19.1). Although significant variability in the early alcohol factor was found for both men and women, it is somewhat surprising to see that the estimated factor mean for women on this factor is positive, while the estimated mean for men was close to zero. This suggests that the residential, social, or interpersonal environmental effects of the first two years has an effect on relatively time-delimited increases in alcohol use for women, while constituting more of a simple variance component for men. The general pattern of elevated alcohol

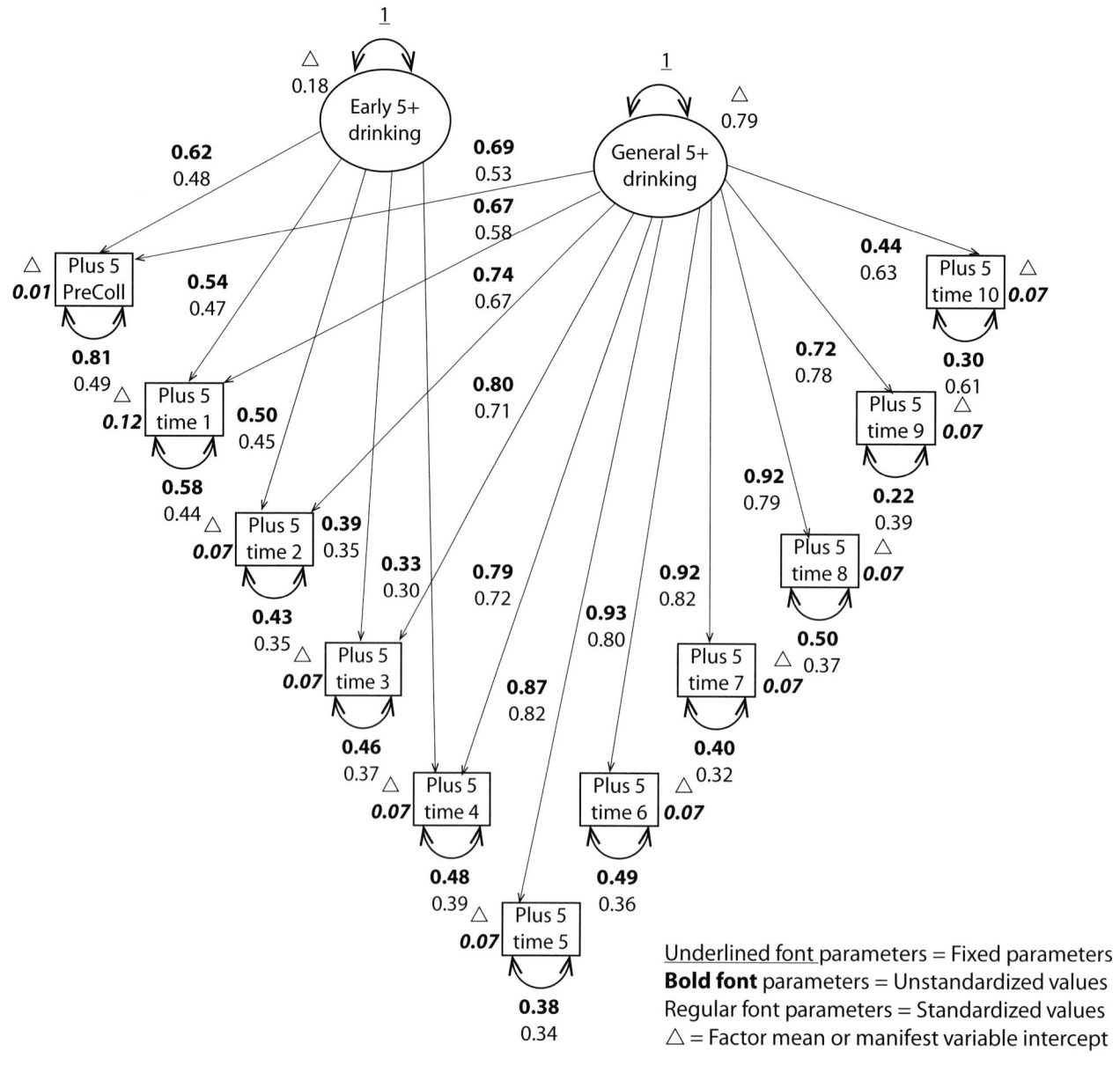

Figure 19.8 Growth curve model for five or more drinks in a sitting (plus 5)

consumption in men appears to be part of a larger process which spans all of the measurement occasions of the study.

Factor Mean and Variance Differences by Sex for General Distress

Women appear to endorse general distress more than men on both the general and early distress factors, and there appears to be slightly more variability in distress reports across women than men, as can be seen in Figure 19.10 and the corresponding mean and variance entries in Table 19.1. The conditional variability in alcohol use specific to each measurement occasion also appears to be markedly larger for men than for women across all measurement occasions (for example, 0.42 for women and 0.85 for mean at the precollege assessment).

Item-Specific General Method Factors

The formats of each alcohol use variable may result in a systematic method effect due to the item. This is accomplished by specifying a factor which has, as indicators, the individual manifest variables at each measurement occasion. Factor loadings associated with the method factors for the reports of number of times drunk, high, or drinking five or more drinks in a sitting is shown in the Method Factors column in Table 19.1. As for the other factors in the multigroup model, factor variances for women were fixed to unity and the factor variances for men were free to vary. The relative magnitude of these method effects is modest, with men systematically less likely to report being "high" or "lightheaded" than women. The fact that the estimated factor means for both men and women were negative supports the view that the high/

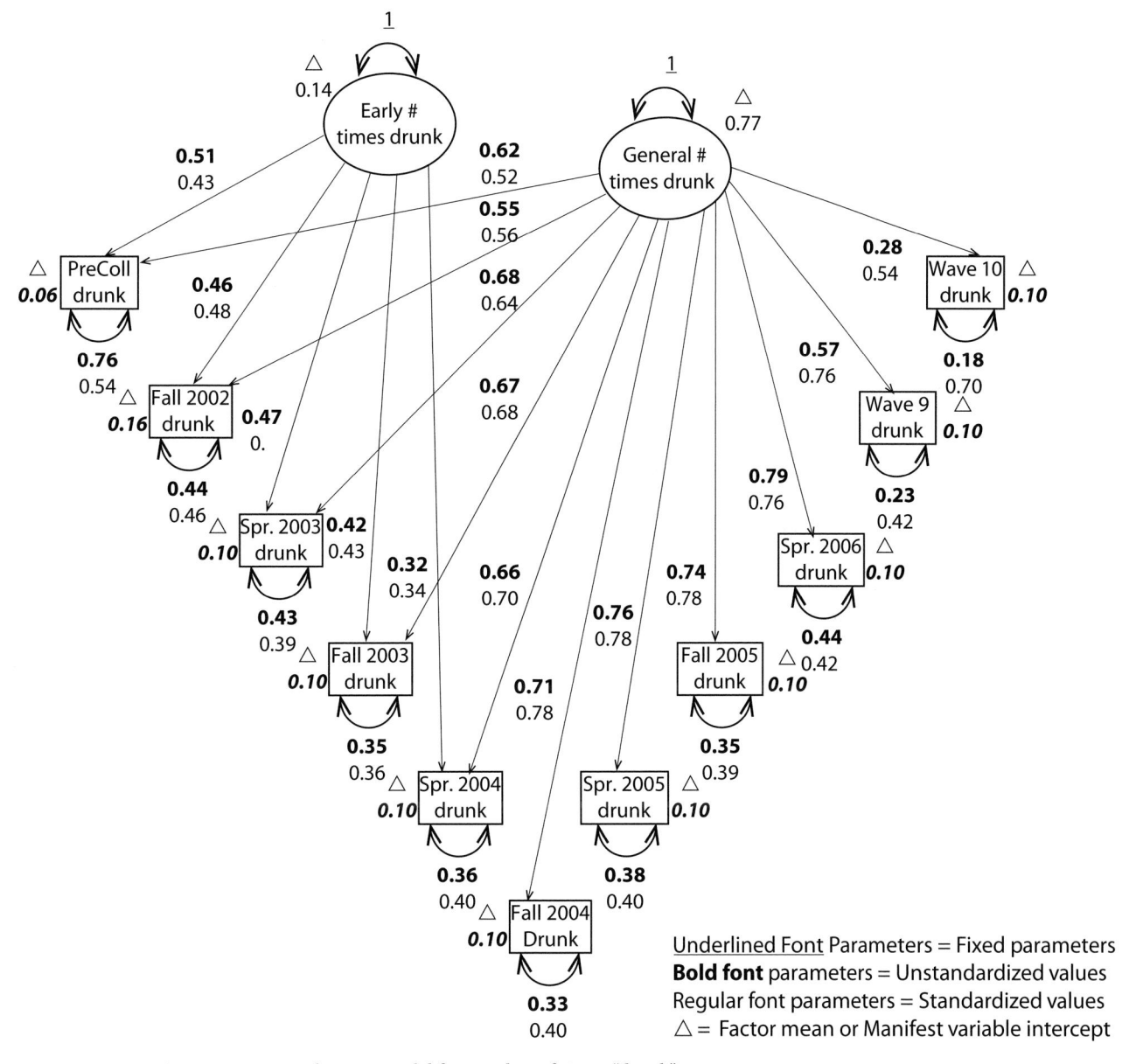

Figure 19.9 Growth curve model for number of times "drunk"

lightheaded item is assessing a lower level of intoxication than the reports of being drunk or consuming five or more drinks.

Patterns of Covariation between Alcohol and General Distress Factors

As can be seen in Table 19.2, covariation between the higher-order Alcohol factors and the general distress factors is modest, being around 0.20 and a small association between the General Distress factor and the Limited Alcohol factor is present for men. Small negative correlations are observed for the Plus 5 method factor and both distress factors, suggesting that those who report higher levels of distress may slightly underreport the number of times that five or more drinks are consumed.

Manifest Alcohol Variable Loadings on Occasion Subfactors

Factor loadings of the three alcohol variables on the alcohol occasion factors are shown in the top half of Table 19.3. As can be seen, several of the loadings appeared to be largely invariant across the measurement occasions, especially for the four later college semester assessments. At each measurement occasion, however, men consistently demonstrated more variability in their alcohol consumption behaviors which were unaccounted for by the higher-order alcohol factors.

Covariances between Constructs at Each Measurement Occasion

It is also possible to assess whether distress is associated with alcohol use at each individual measurement

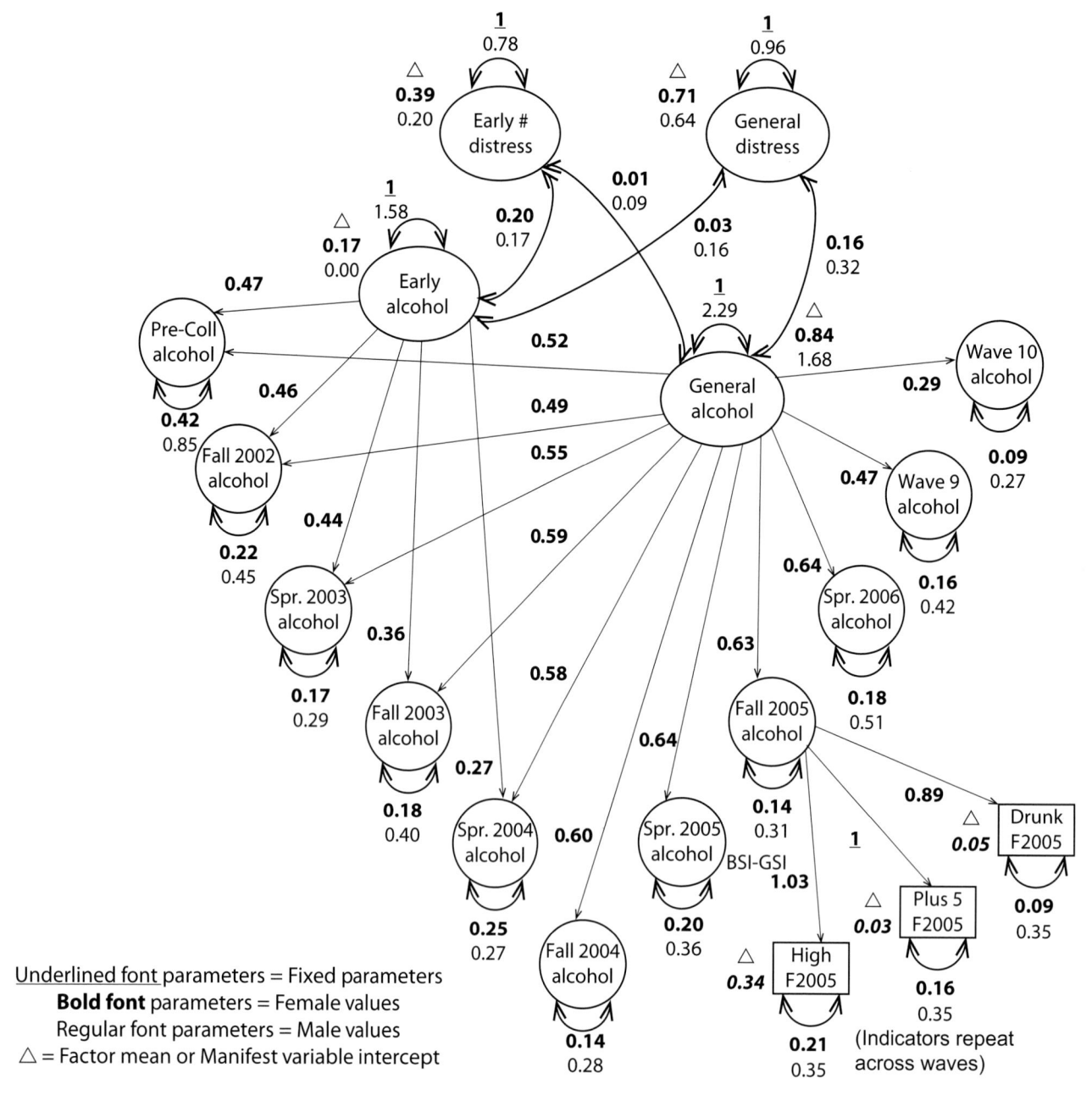

Figure 19.10 Comorbidity subdiagram by sex

occasion. These are shown in the bottom half of Table 19.3, and appear uniformly modest, with the slight exception of modest correlations between distress and alcohol use for the last two college assessments.

Variability of Errors of Measurement in Manifest Variables

Though not of general substantive importance, it is worth noting that errors of measurement associated with general distress were freely estimated across occasion and sex and are shown in Table 19.4. Although estimated errors of measurement in general distress at each measurement occasion appear largely similar across men and women, as shown in the first column of Table 19.4, variability in errors of measurement associated with the alcohol manifest variables appear consistently larger for men than for women, as shown in the columns labeling these error variances in Table 19.4. In general, errors of measurement in the alcohol variables during precollege and college assessments appear larger than errors of measurement during waves 9 and 10.

Although factor loadings were free to vary across the model, comparison of the respective factor loadings across measurement occasion reveal that they appear largely consistent, as seen in the rows of Table 19.3. Such a pattern suggests that these variables are largely invariant over time; this constraint was not imposed on the present solution, but could be imposed if desired.

Table 19.2 Factor correlations/covariances by sex

		Correlations (estimated covariance matrix)						
		BSI		Alcohol		Method		
	Female	General	Limited	General	Limited	Drunk	High	Plus5
BSI	General	<u>1</u>	<u>0</u>	0.16**	−0.04	−0.12	−0.01	−0.18
	Limited		1	0.01	0.20**	−0.25	−0.07	−0.02
Alcohol	General			1	0	0	0	0
	Limited				1	0	0	0
Method	Drunk					1	0	0
	High						1	0
	Pls5							1

Correlations

		BSI		Alcohol		Method		
	Male	General	Limited	General	Limited	Drunk	High	Plus5
BSI	General	1	0(F)	0.22**	0.13**	0.08	0.01	−0.32*
	Limited		1	0.07	0.08	−0.42	−0.1	−0.45
Alcohol	General			1	0	0	0	0
	Limited				1	0	0	0
Method	Drunk					1	0	0
	High						1	0
	Pls5							1

Estimated covariance matrix

		BSI		Alcohol		Method		
	Male	General	Limited	General	Limited	Drunk	High	Plus5
BSI	General	0.96**	0(F)	0.32**	0.16**	0.13	0.01	−0.33*
	Limited		0.78**	0.09	0.20**	−0.59	−0.11	−0.42
Alcohol	General			1.58**	0	0	0	0
	Limited				2.29**	0	0	0
Method	Drunk					2.6**	0	0
	High						1.44**	0
	Pls5							1.12**

Underlined values denote fixed parameters. Note that, for females, predicted covariance matrix is the same as the correlation matrix.

In terms of general fit, the overall model demonstrates a very good fit to the data (TLI = 0.97, CFI = 9.97, RMSEA = 0.03, 95%CI = 0.03–0.03, $\chi^2_{1,549}$ = 3,935.75, $p < 0.0001$).

The predicted mean curves of the BSI-GSI and the alcohol superfactors based on the general model are shown in Figure 19.11. It should first be noted that differences in elevation between the BSI and alcohol curves are not conceptually meaningful, given the conceptual status of the constructs. From this figure, however, the general pattern of increasing problematic alcohol consumption over the first four years of college is easily seen. The lower pattern of problematic alcohol use for women appears flatter, by contrast, due to the mean effect of the early alcohol factor for women. The evident larger amount of desistence in problematic alcohol consumption after the first four years of college appears attributable to the structure of the general alcohol factor and not to any differential mean change for women. As for the BSI-GSI curve, a consistent pattern of elevated scores for women is found relative to men, which appears to decrease across the four years of college.

Table 19.3 Alcohol subfactor loadings and conditional variances (disturbances)

	Alcohol loadings at each wave			Disturbances at each wave	
Wave	Drunk	High	Pls5	Female	Male
Precollege	0.95**	1.06**	<u>1</u>	0.42**	0.85**
Fall 2002	0.91**	0.96**	<u>1</u>	0.22**	0.45**
Spring 2003	0.99**	0.96**	<u>1</u>	0.17**	0.29**
Fall 2003	0.88**	0.92**	<u>1</u>	0.18**	0.40**
Spring 2004	0.88**	1.02**	<u>1</u>	0.25**	0.27**
Fall 2004	0.93**	1.03**	<u>1</u>	0.14**	0.28**
Spring 2005	0.90**	1.03**	<u>1</u>	0.20**	0.36**
Fall 2005	0.89**	1.03**	<u>1</u>	0.14**	0.31**
Spring 2006	0.91**	1.07**	<u>1</u>	0.18**	0.51**
Wave 9	0.90**	1.03**	<u>1</u>	0.16**	0.42**
Wave 10	0.79**	1.23**	<u>1</u>	0.09**	0.27**

Covariances and correlations between alcohol and BSI at each wave

	Cov(BSI,Wave)		Corr(BSI,Wave)	
Wave	Female	Male	Female	Male
Fall 2002	0.00	0.06**	−0.01	0.19**
Spring 2003	0.00	0	−0.01	−0.02
Fall 2003	0.02**	0.03**	0.11**	0.14**
Spring 2004	0.01	0.02*	0.06	0.10*
Fall 2004	0.01	0.01	0.05	0.02
Spring 2005	−0.01	0.02*	−0.07	0.11*
Fall 2005	0.01**	0.06**	0.11**	0.22**
Spring 2006	0.00	0.06**	0.02	0.23**

Underlined values denote fixed parameters.

Table 19.4 Manifest variable error variances

	Error$_{BSI}$		Error$_{Drunk}$		Error$_{High}$		Error$_{Pls5}$	
Wave	Female	Male	Female	Male	Female	Male	Female	Male
Precollege			0.13**	0.24**	0.26**	0.40**	0.25**	0.20**
Fall 2002	0.18**	0.25**	0.16**	0.21**	0.30**	0.36**	0.19**	0.42**
Spring 2003	0.11**	0.09**	0.23**	0.13**	0.34**	0.42**	0.19**	0.22**
Fall 2003	0.13**	0.14**	0.14**	0.25**	0.16**	0.32**	0.14**	0.21**
Spring 2004	0.13**	0.13**	0.15**	0.23**	0.22**	0.35**	0.14**	0.31**
Fall 2004	0.12**	0.14**	0.11**	0.19**	0.16**	0.36**	0.11**	0.36**
Spring 2005	0.10**	0.13**	0.10**	0.26**	0.19**	0.36**	0.19**	0.35**
Fall 2005	0.13**	0.21**	0.09**	0.35**	0.21**	0.35**	0.16**	0.35**
Spring 2006	0.14**	0.14**	0.15**	0.28**	0.20**	0.45**	0.16**	0.3**
Wave 9			0.05**	0.12**	0.13**	0.27**	0.05**	0.18**
Wave 10			0.06**	0.10**	0.16**	0.41**	0.06**	0.21**

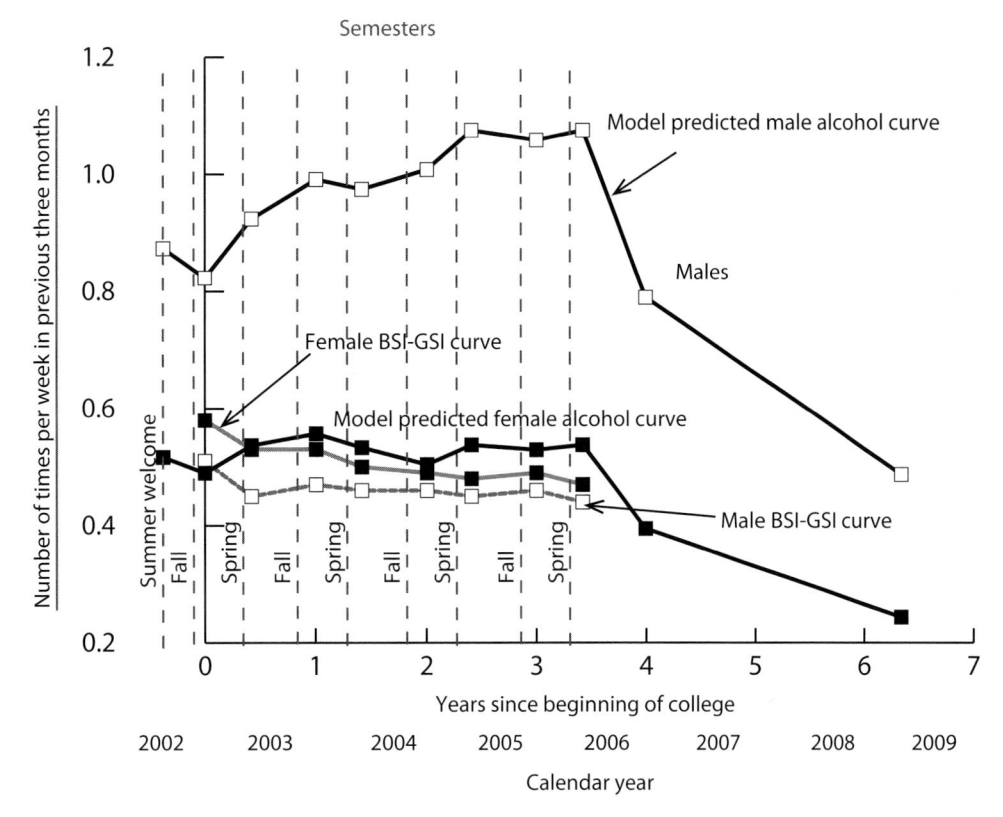

Figure 19.11 Model-predicted curves from Figure 19.10 for BSI-GSI and general alcohol use by sex

DISCUSSION

Although the final model presented in this paper is somewhat complex, the general form of the growth curve models produced seem similar across the constructs and model fit improved substantially if a shift operator was included in the measurement model. Final exploration of whether this shift was different for some measurement occasions also revealed support for the idea that levels of a problem behavior may be temporarily depressed or elevated due to living with one's parents or during the first semester of college. Although the similar growth components across constructs makes descriptions of comorbidity easier to summarize, this is not taken to mean that the three-step procedure has correctly identified the "true" model. Instead, it highlights that these common models "suffice" to explain covariation and mean level given the data. For small sample sizes or less reliably measured data, such right-sizing may involve cruder measurement models or even a reduced dimensionality while a more sophisticated measurement model or increased dimensionality may be appropriate for larger samples or more reliably measured phenomena. In other data sets there may be no "Goldilocks" statistical model which is not overcomplex or oversimple.

The fit of these models is, however, very different than those usually found in longitudinal assessments, which use polynomial growth curves. When the alcohol and distress variables for these data are fitted using multiple quadratic growth curves, unacceptable fit values are found and it is sometimes difficult for the models to even converge. Using program defaults for a multigroup quadratic growth curve model for these four constructs yields a poorly fitting model (TLI = 0.31, CFI = 0.30, RMSEA = 0.13 (95%CI = 0.13–0.13), χ^2_{1660} = 51882.96) and attempts to fit a quadratic analogue to the model presented in Figure 19.10 did not converge.

Including a shift operator in the growth models appeared to improve model fit substantially. Although the Factor Mean-Shift model was actually proposed by McDonald (1967), and has not been used in the study of psychopathology, it appears to be a reasonable measurement model and seems to fit well for other constructs in the data set. Such shifted models are also more generally applicable whenever the specific numeric values assigned to the manifest variables are essentially arbitrary in scale, such as Likert-format questions which may be coded 1–5 or just as easily as –2 to 2. Additionally, no support was found for the existence of more than two latent factors in any of the individual series, as all rotations involving more than two factors revealed the presence of singleton indicator solutions or were inestimable.

By the same token, some factor and growth models seem very unlikely for these data. Inclusion of a random intercept factor, for example, did not appear to be warranted for the alcohol variables and was only a close competitor to the distress measure. As such, hierarchical linear factor models which assume a random intercept component may be incorrect. Use of a latent basis growth curve model, often considered a very general model of growth, seems similarly unwarranted.

The longitudinal factor structure for both psychological distress and alcohol found no evidence for a simple factor structure even when factors were allowed to correlate. This pattern, however, is consistent with the fact that researchers think of these constructs as more trait-like than consisting of systematically defined epochs which span a well-delimited age range. That said, however, the presence of factors associated with the precollege and early college measurement occasions suggests that developmentally limited components to these behaviors also exist. Although, as noted above, it seems particularly reasonable in that this population experiences a great change in their life situation after the second year of college. To some extent some of the variability in alcohol consumption in the early assessments may have to do with the fact that alcohol consumption is not yet a legal activity, the fact that students during this time experience a significantly different social, residential, and interpersonal environment in the last six measurement occasions may account for the disruption of some of the factors represented by the early alcohol and distress factors.

In summary, although the proposed approach to fitting comorbidity appears to work well at least for data sets such as this which involve several measurement occasions, for smaller data sets involving only a few hundred observations the effect of influential observations may be more dramatic than found in this case. As with any study of comorbidity, the approach does not solve fundamental problems of experimental design and any study purporting to show comorbidity across constructs must carefully examine the possibility of third variable explanations for any associations found. The question of whether the estimated factor loadings actually reflect a linear association of the latent variable on the manifest variable is another possible limitation of the model, but to some extent this could be examined by Loess plots of factor scores used as predictors of the manifest variables or by exploring whether response options of the manifest variable could be modeled as ordered categorical, as opposed to continuous variables.

It is quite possible to extend the models presented here to other growth models as well. They could be extended to categorical manifest variables or count data (Newsom, 2015), though care must be taken in the case of count data for substance use, due to the fact that consumption patterns frequently do not conform to the Poisson process assumed. Individuals, for example, do not limit their consumption to a single cigarette or drink once they have decided to consume. Additionally, it has been observed that many problem behaviors are perhaps better modeled as zero-inflated, meaning that the population under consideration actually consists of one subpopulation that does not engage in the behavior (such as those who choose not to use alcohol or engage in risky sexual behavior) and another subpopulation that engages in the behavior to varying degrees (Olsen & Schafer, 2001). The extension of the three-step procedure outlined here can be extended when expressed as mixture models.

REFERENCES

Derogatis, F. R. (1975). *The Brief Symptom Inventory*. Baltimore, MD: Clinical Psychometric Research.

Fu, W. (1998). Penalized Regressions: The Bridge versus the Lasso. *Journal of Computational and Graphical Statistics, 7*(3), 397–416.

Gotham, H., Sher, K. J., & Wood, P. K. (1997). Predicting Stability and Change in Frequency of Intoxication from the College Years to Beyond: Individual-Difference and Role Transition Variables. *Journal of Abnormal Psychology, 106*, 619–629.

Grekin, E. R., Sher, K. J., & Wood, P. K. (2006). Personality and Substance Dependence Symptoms: Modeling Substance-Specific Traits. *Psychology of Addictive Behavior, 20*, 415–424.

Jackson, K. M., O'Neill, S. E., & Sher, K. J. (2006). Characterizing Alcohol Dependence: Transitions during Young and Middle Adulthood. *Experimental Clinical Psychopharmacology, 14*(2), 228–244.

Jackson, K. M., & Sher, K. J. (2003). Alcohol Use Disorders and Psychological Distress: A Prospective State-Trait Analysis. *Journal of Abnormal Psychology, 112*, 599–613.

Jackson, K. M., & Sher, K. J. (2003). Alcohol Use Disorders and Psychological Distress: A Prospective State-Trait Analysis. *Journal of Abnormal Psychology, 112*(4), 599-613.

Jackson, K. M., Sher, K. J., & Wood, P. K. (2000). Prospective Analyses of Comorbidity: Tobacco and Alcohol Use Disorders. *Journal of Abnormal Psychology, 109*, 679–694.

Kutner, M. H., Nachtsheim, C., Neter, J., & Li, W. (2005). *Applied Linear Statistical Models*. New York: McGraw-Hill Irwin.

McArdle, J. J., & Epstein, D. (1987). Latent Growth Curves within Developmental Structural Equation Models. *Child Development, 58*(1), 110–133.

McDonald, R. P. (1967). *Nonlinear Factor Analysis*. Psychometric Monograph, 15. Richmond, VA: Byrd Press.

Martinez, J. A., Sher, K. J., Krull, J. L., & Wood, P. K. (2009). Blue-Collar Scholars? Mediators and Moderators of University Attrition in First-Generation College Students. *Journal of College Student Development, 50*(1), 87–103.

Maydeu-Olivares, A., & Coffman, D.L. (2006). Random Intercept Item Factor Analysis. *Psychological Methods, 11*, 344–362.

Meredith, W., & Tisak, J. (1990). Latent Curve Analysis. *Psychometrika, 55*, 107–122.

Nelson, S., Van Ryzin, M., & Dishion, T. (2015). Alcohol, Marijuana, and Tobacco Use Trajectories from Age 12 to 24 Years: Demographic Correlates and Young Adult Substance Use Problems. *Development and Psychopathology, 27*(1), 253–277.

Newsom, J. T. (2015). *Longitudinal Structural Equation Modeling: A Comprehensive Introduction*. New York: Routledge.

Olsen, M. K., & Schafer, J. L. (2001). A Two-Part Random-Effects Model for Semicontinuous Longitudinal Data. *Journal of the American Statistical Association, 96*, 730–745.

Park, A., Sher, K. J., & Krull, J. (2008). Risky Drinking in College Changes as Fraternity/Sorority Affiliation Changes: A Person-Environment Perspective. *Psychology of Addictive Behaviors, 22*, 219–229.

Rutledge, P. C., & Sher, K. J. (2001). Heavy Drinking from the Freshman Year into Early Young Adulthood: The Roles of

Stress, Tension-Reduction Motives, Sex, and Personality. *Journal of Studies on Alcohol, 62*, 457–466.

Savalei, V., & Kolenikov, S. (2008). Constrained versus Unconstrained Estimation in Structural Equation Modeling. *Psychological Methods, 13*(2), 150–170.

Sher, K. J., Gotham, H., Erickson, D., & Wood, P. K. (1996a). A Prospective, High-Risk Study of the Relation between Tobacco Dependence and Alcohol Use Disorders. *Alcoholism: Clinical and Experimental Research, 20*, 485–492.

Sher, K. J., Wood, M. D., Wood, P. K., & Raskin, G. (1996b). Alcohol Outcome Expectancies and Alcohol Use: A Latent Variable Cross-Lagged Panel Study. *Journal of Abnormal Psychology, 105*, 561–574.

Sher, K. J., Wood, P. K., & Gotham, H. (1996c). The Course of Psychological Distress in College: A Prospective High-Risk Study. *Journal of College Student Development, 37*, 42–51.

Sher, K. J., Jackson, K. M., & Steinley, D. (2011). Alcohol Use Trajectories and the Ubiquitous Cat's Cradle: Cause for Concern? *Journal of Abnormal Psychology, 120*(2), 322–335.

Simonsohn, U., Simmons, J. P., & Nelson, L. D. (2015). Specification Curve: Descriptive and Inferential Statistics on All Reasonable Specifications. Retrieved from SSRN: https://ssrn.com/abstract=2694998 or http://dx.doi.org/10.2139/ssrn.2694998

Tucker, L. R. (1958). Determination of Parameters of a Functional Relation by Factor Analysis. *Psychometrika, 23*, 19–23.

Tucker, L. R. (1966). Learning Theory and Multivariate Experiment: Illustration of Determination of Generalized Learning Curves. In R. B. Cattell (Ed.), *Handbook of Multivariate Experimental Psychology* (pp. 476–501). New York: Rand McNally.

Wiedermann, W., & von Eye, A. (2016). *Statistics and Causality: Methods for Applied Empirical Research*. Hoboken, NJ: Wiley.

Wohlwill, J. (1973). *The Study of Behavioral Development*. New York: Academic Press.

Wood, P. K. (2019). *Approaches to Understanding Structural Models: Models of Relationships between Variables, Occasions, and People*. Seattle, WA: Amazon.

Wood, P. K., Sher, K. J., & Rutledge, P. C. (2007). College Student Alcohol Consumption, Day of the Week, and Class Schedule. *Alcoholism: Clinical and Experimental Research, 31*, 1195–1207.

Wood, P. K., Steinley, D., & Jackson, K. M. (2015). Right-Sizing Statistical Models for Longitudinal Data. *Psychological Methods, 20*(4), 470–488.

PART V INTERVENTION APPROACHES

20 The Multiphase Optimization Strategy for Developing and Evaluating Behavioral Interventions

KATE GUASTAFERRO, CHAD E. SHENK, AND LINDA M. COLLINS

Intervention science faces a number of challenges to moving forward. For the purpose of this chapter, we examine one field, clinical psychology, with a focus on behavioral interventions[1] for children who have experienced trauma, such as child maltreatment. This example illustrates universal challenges in modern intervention science in four ways. First, almost all interventions applied with children who have been maltreated are multicomponent behavioral interventions. It is usually unknown whether each of the components included in a particular intervention actively contributes to a desired treatment effect. Second, new interventions developed for traumatized children often recycle components from existing interventions. A way is needed to rapidly develop and evaluate both established and new components for inclusion in novel interventions that enhances treatment effects beyond existing interventions. Third, funding priorities at the National Institutes of Health now encourage the identification and targeting of causal processes over discrete psychiatric outcomes or disorders. A method is needed for evaluating the impact of specific components on unique proximal mediators explaining the relationship between child maltreatment and subsequent psychopathology. Finally, children and families experiencing maltreatment cannot benefit from services they do not receive. Approaches for improving the dissemination and implementation of interventions so that they address the needs of patients within an acceptable cost range and at the desired scale are needed.

Like other fields in which interventions are developed and evaluated, trauma treatment for children with a history of maltreatment and, more generally, the field of clinical psychology, finds itself at a crossroads: continue developing interventions as usual or seek innovative solutions that provide a way past existing challenges like those described above. In this chapter, we describe how the multiphase optimization strategy (MOST) offers methodological solutions to these universal challenges. We provide an overview of the MOST framework and demonstrate its application in a hypothetical example of the optimization of a clinical intervention targeting posttraumatic stress disorder (PTSD) symptoms among children who have experienced maltreatment. We offer a discussion of considerations applicable to clinical psychology researchers and make suggestions to encourage the application of MOST. As one introductory chapter cannot adequately discuss all aspects of MOST, interested readers are directed to Collins (2018), which provides a comprehensive introduction to MOST, and Collins and Kugler (2018), which reviews a number of related advanced topics.

CONTEXT: CHILD MALTREATMENT

Addressing PTSD in children who have experienced maltreatment is only one area of clinical psychology, but one with significant magnitude. Child maltreatment affects over 683,000 children each year in the United States (US Department of Health et al., 2017), and it is one of the strongest predictors of PTSD and PTSD symptom severity in the pediatric population (Copeland et al., 2007). Indeed, between 20 and 37 percent of youth with a history of maltreatment will meet diagnostic criteria for PTSD (Kahana et al., 2006; Widom, 1999) or experience clinically significant PTSD symptoms (Berkowitz, Stover, & Marans, 2011). Fortunately, there are several well-established or putatively efficacious behavioral interventions to treat PTSD following exposure to child maltreatment (Cohen, Mannarino, & Deblinger, 2006; Dorsey et al., 2017; Silverman et al., 2008). The magnitude of the prevalence of PTSD among children who have experienced maltreatment and the number of interventions that have been developed to treat PTSD in this population motivated the use of this public health concern as an example here.

---

[1] MOST is applicable for behavioral, biobehavioral, or biomedical interventions. For clarity, and given the audience of this chapter, we are restricting our example to only behavioral interventions. In this context, the term "behavioral" broadly refers to any nonpharmacological intervention, such as behavioral, cognitive, or cognitive-behavioral interventions.

THE POTENTIAL BENEFITS OF MOST FOR CLINICAL PSYCHOLOGY

In the classical approach to development and evaluation of an intervention, a researcher identifies treatment components relevant for the disorder or clinical population of interest, packages those components together into an intervention, and then evaluates the effect of this new intervention through a randomized controlled trial (RCT). However, approaching development of a packaged intervention in this way makes it nearly impossible to identify which components are actively contributing to therapeutic change in the outcome of interest. It is also unclear whether one component is affected, either positively or negatively, by another component (Collins, MacKinnon, & Reeve, 2013) or if all components are necessary to achieve the desired outcome or effect (Collins, 2014; Collins et al., 2014c).

An example in the field of clinical psychology is cognitive therapy (CT) for major depressive disorder (MDD). Based on existing theory and research on the causes of MDD, Beck and colleagues (1979) assembled a packaged intervention made up of multiple components and then evaluated this packaged intervention in extensive RCT research over the course of several decades. Without question, the CT package is one of the most effective and widely disseminated behavioral interventions that exists today. However, this classical approach to intervention development left unanswered questions regarding which CT components contribute to the overall treatment effect and whether all are needed to achieve the desired outcome. In a seminal study by Jacobson and colleagues (1996), major CT components were evaluated as standalone interventions for MDD and then compared to the overall efficacy of the complete CT package (i.e., component analysis). This study demonstrated that the behavioral activation component of CT alone was as efficacious as the entire CT package. This led to the development of Behavioral Activation as a standalone treatment (Jacobson, Martell, & Dimidjian, 2001), which upon replication and extension demonstrated superiority to CT and had efficacy comparable to that of antidepressant medication in the treatment of moderate to severe MDD (Dimidjian et al., 2006). The MOST framework, through its emphasis on the identification and evaluation of components prior to evaluation of packaged interventions in RCTs, can address these challenges at the beginning of intervention development to arrive at optimal interventions more rapidly and effectively.

MOST: AN OVERVIEW

MOST is an engineering-inspired framework for optimizing multicomponent behavioral interventions that offers solutions to some of the methodological and conceptual challenges arising from the classical approach to intervention development. MOST inserts a phase of research between the identification of components and the evaluation of the multicomponent intervention in the RCT. In this phase, called optimization, the intervention scientist empirically examines the mechanics of intervention components to understand the way in which each component contributes to the outcome of interest and if there are interactions between components that affect that outcome. Using MOST there is the potential to focus on important attributes of an intervention other than effectiveness (i.e., efficiency, economy, and scalability) and thereby achieve steady, systematic progress in intervention science. For more than a decade, MOST has been applied across a variety of behavioral contexts including: smoking cessation (Baker et al., 2015; Piper et al., 2016); adolescent substance use and risky sexual behavior (Caldwell et al., 2012); obesity (Kugler et al., 2016; Pellegrini et al., 2014); and antiretroviral therapy adherence (Gwadz et al., 2017). The goal of this chapter is to introduce MOST to the clinical psychology audience.

MOST emphasizes the *optimization* of an intervention in addition to its overall effectiveness (Wyrick et al., 2014). Optimization is a process to identify a multicomponent intervention that provides the best expected outcome obtainable within identified constraints (Collins, 2014). This is accomplished in the phase added between the identification of the set of components and the RCT (Figure 20.1). In this section, we will describe the three phases of MOST (preparation, optimization, and evaluation) using a hypothetical example of an intervention designed to reduce PTSD symptoms among maltreated youth as measured by the UCLA PTSD Reaction Index (Steinberg et al., 2004, 2013). Specifically, this hypothetical example emphasizes the preparation and optimization phases of MOST to demonstrate its unique ability to address methodological challenges early in intervention development and immediately prior to testing in an RCT.

Preparation

The first phase of MOST, the preparation phase, lays the foundation for the subsequent optimization and evaluation phases. There are two primary goals of the preparation phase: (1) to develop a conceptual model for the intervention in which a set of candidate components is identified and (2) to specify the optimization criterion. Many activities of the preparation phase, such as pilot testing and developing a conceptual model, will be familiar to those trained in the classical approach. Others, such as specification of an optimization criterion, may be unfamiliar.

MOST is a framework for multicomponent interventions. A *component* is anything that may be separated out for study. There are several types of components including: (1) intervention *content* (e.g., group or individual sessions, psychoeducation, or a specific activity, such as cognitive processing or relaxation); (2) features of the

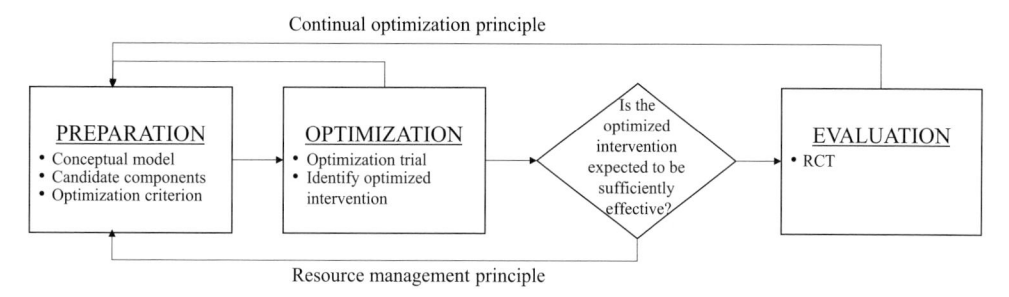

Figure 20.1 The multiphase optimization strategy (MOST) process

program that promote *engagement* (e.g., text message reminders or homework assignment); or (3) procedures aimed at improving *fidelity* of a program (e.g., enhanced training or supervision of program implementation staff). The idea of a component is not novel or unique to MOST; in fact the field of clinical child psychology has long embraced the concept of *common elements* shared by a number of psychotherapeutic interventions (Barth et al., 2012; Becker et al., 2015; Chorpita et al., 2007; Lindsey et al., 2014), which has conceptual overlap with *components*. A literature analysis procedure, the common elements approach identifies practice techniques or strategies common across treatments and the contexts in which these techniques or strategies have been successful (Boustani et al., 2017). These elements may be used to inform intervention decisions by providers, though this approach was not designed to inform treatment decisions. Rather, the common elements approach is an analytic tool applied to the literature that may be useful in identifying components to be included in an intervention. For example, researchers identified the most commonly used components in the treatment of traumatic stress in children across a number of interventions (Chorpita & Daleiden, 2009; Chorpita et al., 2011). Even with common elements identified, it is clear neither how these components may affect the desired outcome nor how these components affect each other. The common elements approach can identify potential intervention components, but MOST extends the concept by calling for the optimization of an intervention based on careful experimental manipulation and evaluation of components, existing and new.

Conceptual Model

The researcher using MOST applies the theoretical basis and specifies a mechanism of change with regard to the relationship between intervention components and the intended outcome via a conceptual model. The conceptual model is a model of the intervention and the process being intervened upon. Driven by theory and informed by empirical literature, the conceptual model clearly articulates how the selected candidate components affect the causal pathway of the intervention, or the way in which the intervention will directly intervene on the causal process. Specifying a conceptual model requires explicitly specifying the expected effect of the intervention components on proximal mediators that lead to the desired distal outcome. A proximal mediator is a variable on the causal pathway, explaining the relationship between the independent variable, in this case the candidate intervention component, and the dependent variable, or the distal outcome (MacKinnon, Fairchild, & Fritz, 2007). The a priori specification of the causal process in this detailed way, accompanied by the subsequent empirical test in the optimization phase, provides a "powerful method to investigate processes of health behavior" (Collins et al., 2013, p. 587).

The conceptual model should be explicit and highly detailed; however, the figure representing the conceptual model may not necessarily include every detail in the model. For example, a figure representing a conceptual model is usually most useful when kept as simple as reasonably possible, and accompanied by some text explaining the model in full (Kugler et al., 2018). Figure 20.2 depicts the conceptual model of our hypothetical intervention targeting PTSD symptoms in children with a history of maltreatment. On the left of the figure, five candidate intervention components are listed and the proximal mediator that the component is designed to affect is specified in the middle column. The right side of the figure depicts the way in which each mediator will affect the desired outcome of the intervention. The following sections describe each of the aspects of the conceptual model in detail. Note that Figure 20.2 does not represent a structural equation model; it does not contain every expected nonzero regression path. Instead, its purpose is to depict the thinking behind the development of the intervention.

Selection of Intervention Components

Candidate components, as discussed above, may come from a number of sources, including existing research and theory. In our hypothetical intervention targeting PTSD symptom severity among children with a history of maltreatment, we identified five candidate components that can independently target each identified mediator (Figure 20.2, far left). One, *relaxation training*, including techniques such as progressive muscle relaxation and diaphragmatic breathing, has an extensive literature demonstrating efficacy in the improvement of short- and long-term stress regulation for those who have experienced

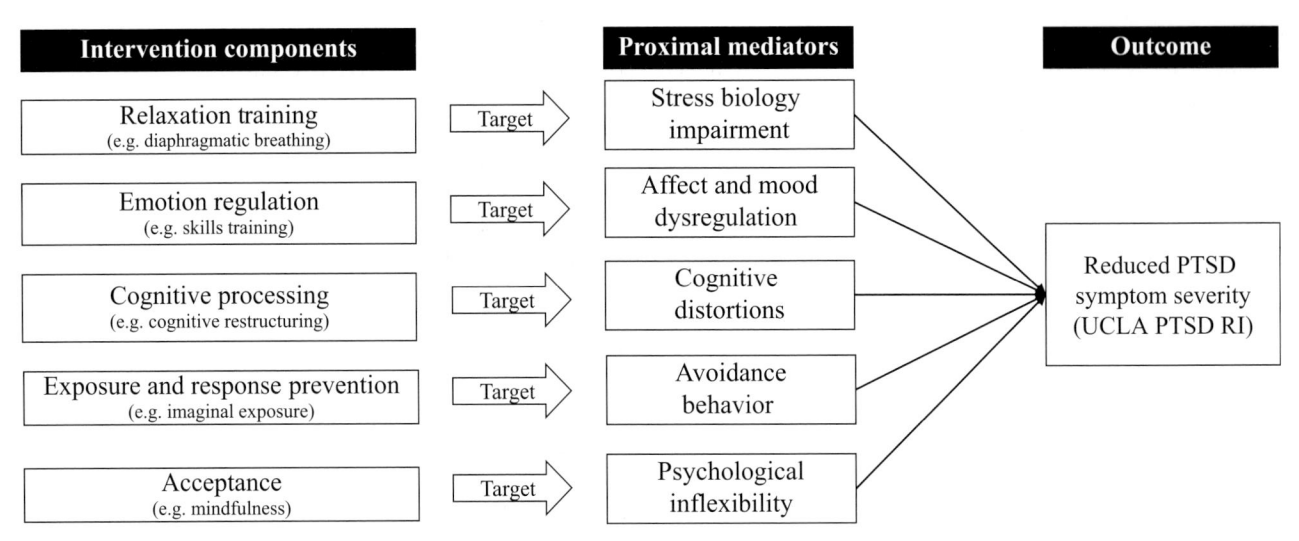

Figure 20.2 Conceptual model for the hypothetical intervention targeting PTSD symptoms in children

maltreatment. Two, skills training involving the acquisition of new *emotion regulation* techniques, such as emotion identification and labeling as well as daily engagement in pleasurable activities, can have a direct impact on daily fluctuations in affective states and promote stable elevations in affect and mood. Three, *cognitive processing* can involve traditional cognitive restructuring techniques, such as tracking the frequency and content of distorted cognitions, gathering evidence to refute the thought, and disputation of distorted thoughts, as well as techniques specific to traumatized populations, such as generating an accurate account of the trauma and promoting healthy expressions of anger or blame. Four, *exposure and response prevention* is perhaps one of the more valuable tools a clinical psychologist has in treating PTSD symptoms following child maltreatment. The use of gradual, imaginal, and in vivo exercises to expose patients to conditioned stimuli, prevent engagement in avoidance behaviors, and promote acquisition of new approach behaviors that foster emotional processing can have a significant impact on resulting PTSD symptoms. Finally, promoting *acceptance* of painful private events, such as aversive thoughts, memories, or physiological states, through the use of mindfulness or validating communication with caregivers, can promote resilience and psychological flexibility following exposure to child maltreatment.

Our hypothetical intervention is comprised of components common to many existing interventions for traumatized children such as relaxation training, emotion regulation, cognitive processing, and exposure and response prevention (Dorsey et al., 2017). Although initial work on the effectiveness of some of these established components has been done (Deblinger et al., 2011), it remains unknown which of these components actively contribute to desired change in proximal mediators and ultimately reductions in PTSD symptoms. The hypothetical intervention also introduces a novel component, acceptance. The inclusion of this component was driven by theory and empirical literature. In this case, it is also unknown whether the novel component achieves a desired level of effectiveness beyond existing, established components. The advantage of simultaneously testing existing and new components is to identify those components actively contributing to the desired treatment effect for ultimate inclusion in a new, packaged intervention that provides added therapeutic value to enhance overall treatment effects for traumatized children. Such results can also inform existing psychological theories on the etiology and maintenance of certain clinical conditions as well as the potential refinement of conceptual models for further improvement in treatment effects.

Proximal Mediators

Although sometimes it is necessary for a component to target more than one mediator, we recommend that each intervention component target one proximal mediator whenever possible. The figure representing the conceptual model should explicitly link each component to a proximal mediator. Extensive basic science research demonstrates that exposure to child maltreatment results in significant changes in multilevel processes (e.g., biological and behavioral) increasing the probability of subsequent PTSD symptoms and greater symptom severity. Exposure to a traumatic event can result in prolonged activation of biological stress-mediating systems, particularly the neuroendocrine and autonomic systems, that results in globalized hypervigilance, difficulty sleeping, and irritability (Carrion et al., 2002; Trickett et al., 2010). The relaxation training component targets the stress biology impairment mediator.

Similarly, child maltreatment can engender frequent reexperiencing of the event, including nightmares and painful memories, which can overwhelm a child's ability to regulate subsequent affect, leading to prolonged mood disturbance (Shenk, Griffin, & O'Donnell, 2015). As depicted in the conceptual model, the emotion-regulation component targets the affect and mood dysregulation mediator.

Children who have been maltreated often experience cognitive distortions, such as "It was my fault for not doing what I was told" or "I can never trust anyone ever again," which can change individual goals and views about the world, increase a feeling of distance from others, and delay recovery during treatment (Hayes et al., 2017). The cognitive processing component directly targets the cognitive distortion proximal mediator.

Like exposure to other traumatic events, child maltreatment increases the avoidance of people, places, and situations that remind children of the trauma, which can reinforce behaviors that delay recovery and maintain PTSD symptoms (Foa et al., 2013). The exposure and response prevention component targets the proximal mediator of avoidance behavior.

Finally, child maltreatment can result in psychological inflexibility, that is, reduced engagement in goal-directed, value-based behaviors resulting from frequent attempts to suppress or change painful private experiences, leading to greater PTSD symptom severity over time (Shenk et al., 2014). The acceptance component targets the proximal mediator of psychological inflexibility.

The proximal mediators (Figure 20.2, middle) then have a direct effect on the distal outcome (Figure 20.2, far right), as specified in the conceptual model. Indeed, the hypothesized causal processes depicted in our conceptual model map on closely to the PTSD symptom clusters in the current diagnostic conceptualization of PTSD in the Diagnostic and Statistical Manual of Mental Disorders (DSM-5). This view is also consistent with the National Institute of Mental Health's strategic focus on experimental therapeutics, where causal processes of psychiatric conditions are identified across multiple levels of analysis and then targeted through the use of active intervention components (Insel & Gogtay, 2014). Thus, if the proximal mediators are on the causal pathway, and the selected intervention components have a unique effect on each of these mediators, then a reduction in overall PTSD symptom severity should be achieved.

Optimization Criterion

MOST requires the specification of an *optimization criterion*, or the operational definition of the optimized intervention, which will form part of the basis for decisions regarding the components to be included in the optimized intervention. The optimization criterion considers constraints, that is, anything that may affect the implementation of an intervention such as provider time to deliver the intervention, participant burden, or cost of implementation. The optimization criterion can be quite simple. For example, a researcher could select the *all active components* optimization criterion to achieve the best expected outcome on reducing PTSD symptoms irrespective of cost. At the end of the optimization phase, the effect of each component is evaluated and those components with a net positive effect on the desired outcome will be included and those with a negative effect will not. Alternatively, the optimization criterion could specify constraints, or implementation parameters, such as cost per participant (e.g., "the best expected outcome on reducing PTSD symptoms for less than $400 per participant") or time to deliver (e.g., "the best expected outcome on reducing PTSD symptoms in less than 15 weeks"). These parameters may be specified by payers of or stakeholders in an intervention. Selecting an optimization criterion that considers constraints is likely to increase the probability of achieving immediate scalability at the conclusion of evaluation. For illustrative purposes, let us assume we have selected an optimization criterion specifying an upper limit of $400 per-participant cost for our hypothetical example.

Optimization

The optimization phase sets MOST apart from the classical approach. Informed by the conceptual model, in the optimization phase the intervention scientist will select a highly efficient experimental design to examine the effect of each intervention component. The experiment designed to collect this information is called an *optimization trial*. In the preceding preparation phase, the researcher has identified candidate components, specified the conceptual model, and selected an optimization criterion. Now, in the optimization phase, the researcher must select component levels to examine and the appropriate experimental design. A *component level* represents a variation of the candidate component to be examined in the optimization trial. Generally two levels, such as "on vs. off" or "high vs. low," are selected for each component. Overall, the purpose of the optimization phase is to build an optimized intervention by experimentally examining components and component levels and then selecting the set that provides the best expected outcome within constraints. In contrast to the typical RCT in which a multicomponent intervention is evaluated en bloc, the results of the optimization trial provide empirical rationale for what components, at which levels and in which combination, produce the greatest effect while considering the prespecified constraints.

Selecting an Experimental Design

The selection of the experimental design for the optimization trial is a critical decision and must be guided by the *resource management principle*. This principle states that the best experimental design for the optimization trial is one that "provides the greatest scientific benefit, without exceeding available resources" (Collins et al., 2014a, p. 503). In other words, the resource management principle holds that an investigator must strive for the most efficient use of available resources when obtaining scientific information (Collins, 2018; Collins, Dziak, & Li,

2009; Dziak, Nahum-Shani, & Collins, 2012). Money is a likely resource to consider, but time, personnel, equipment, and experimental subjects as well as myriad intervention-specific resources must also be considered. Additionally, the type of intervention being developed (fixed or adaptive) will inform which type of experimental design is selected (Almirall et al., 2014; Collins, Nahum-Shani, & Almirall, 2014b). The most common experimental design for an optimization trial is some variation on the factorial design due to the efficiency and economy of this approach, but any of a wide variety of experimental designs may be used.

The factorial experiment is made up of factors corresponding to the candidate components. Each factor will be experimentally manipulated to take on each of the levels of its corresponding component (e.g., off, on; low, high). A complete factorial design includes all combinations of factor levels. A factorial experiment enables the estimation of individual main effects of the candidate components and the examination of component interactions (Collins et al., 2009, 2014c). Thus, the factorial experiment answers different research questions than an RCT. The length of this chapter precludes much detail about the factorial design; readers are referred to Collins and colleagues (2014c), which provides a brief introduction to factorial experiments for those trained primarily in the RCT. A detailed treatment of experimental designs for the optimization trial can be found in Collins (2018). However, to head off one common misconception (Collins et al., 2014a), we want to point out that factorial experiments are powered differently than RCTs. Factorial designs actually require fewer experimental subjects than alternative designs, such as the comparative treatment experiment, a component analysis, or conducting an individual experiment for each component under consideration, while maintaining comparable power (Collins et al., 2009, 2011). This is because a factorial design does not compare the mean of one condition to another head to head, but instead compares means based on combinations of conditions. In fact, factorial experiments can be adequately powered even with relatively small per-condition sample sizes, as long as the overall sample size is sufficient. Analysis of data collected in a factorial experiment requires a standard ANOVA to estimate main effects of factors and interactions between factors.

Returning to our example, we have five candidate components (which are called factors in the factorial experiment) each with two levels (e.g., the subject gets the component [yes] or does not [no]) in our factorial experiment. This translates to a 2 × 2 × 2 × 2 × 2, or 2^5, factorial experiment with 32 experimental conditions (Table 20.1).

With our design identified, we must power our factorial experiment. Once again, space precludes a full discussion of power calculations for a factorial experiment; readers are referred to Collins and colleagues (2009) and Dziak and colleagues (2012). The Methodology Center at the Pennsylvania State University has a SAS macro available for download, called *FactorialPowerPlan*, that can assist with these estimates (Dziak, Collins, & Wagner, 2013; https://methodology.psu.edu/downloads/factorialpowerplan).

When data are not clustered, investigators typically select a Type I error rate, say $\alpha = 0.05$, specify an effect size, and determine what sample size is needed to achieve satisfactory statistical power. (When data are clustered, the design effect is an additional consideration.) When conducting an optimization trial, one option for the specified effect size is the smallest main effect that a component needs to demonstrate in the experiment to be eligible for inclusion in the optimized intervention, in the investigator's judgment. For example, an investigator might decide that a component must demonstrate a main effect that is at least small using Cohen's rule of thumb (Cohen, 1988).

Suppose during our hypothetical preparation phase we selected the following statistical attributes for the optimization trial based on recommended literature and standard rules of thumb: $\alpha = 0.05$, an effect size of $d = 0.20$, and statistical power of at least 0.80 ($\beta = 0.20$). Also affecting our power calculation is the pretest value for our outcome of interest (symptom level on the UCLA PTSD Reaction Index). The correlation of the pretest and any other covariates with the posttest must be taken into account in the power calculation. We estimate a correlation of 0.6 for this example. These values are entered into the *FactorialPowerPlan* macro to determine sample size. Our 2^5 factorial experiment would require 128 participants, 4 participants in each experimental condition (see Table 20.1). Note that even with only 4 participants per condition, this experiment achieves power = 0.80.

Evaluation

Assuming in the optimization trial enough components have detectable effects to comprise a sufficiently promising intervention, the optimized intervention moves to the final phase of MOST, the evaluation phase. The RCT remains the gold standard in determining the effectiveness of an intervention. In contrast to the classical approach, the RCT conducted in the evaluation phase of MOST compares the effect of an optimized intervention, for which we know how the components work implicitly, to a suitable control or comparison intervention. Although the RCT remains resource-intensive, the comparison of an optimized intervention to a suitable control is, in the view of the current authors, a more responsible use of those resources than the evaluation of an intervention package made up of components that have at best been pilot tested. If the RCT confirms that the optimized intervention has a statistically significant and clinically meaningful effect, the optimized intervention can then be released and scaled up.

It is important to note, MOST is *not* in opposition to nor is it a replacement for the RCT. We are suggesting that, alone, the RCT (even a three-arm design) does not provide

Table 20.1 Experimental conditions and per-condition n in 2^5 factorial design ($N = 128$)

Exp. condition	RELAX	EMOTION	PROCESS	EXPOSE	ACCEPT	n
1	No	No	No	No	No	4
2	No	No	No	No	Yes	4
3	No	No	No	Yes	No	4
4	No	No	No	Yes	Yes	4
5	No	No	Yes	No	No	4
6	No	No	Yes	No	Yes	4
7	No	No	Yes	Yes	No	4
8	No	No	Yes	Yes	Yes	4
9	No	Yes	No	No	No	4
10	No	Yes	No	No	Yes	4
11	No	Yes	No	Yes	No	4
12	No	Yes	No	Yes	Yes	4
13	No	Yes	Yes	No	No	4
14	No	Yes	Yes	No	Yes	4
15	No	Yes	Yes	Yes	No	4
16	No	Yes	Yes	Yes	Yes	4
17	Yes	No	No	No	No	4
18	Yes	No	No	No	Yes	4
19	Yes	No	No	Yes	No	4
20	Yes	No	No	Yes	Yes	4
21	Yes	No	Yes	No	No	4
22	Yes	No	Yes	No	Yes	4
23	Yes	No	Yes	Yes	No	4
24	Yes	No	Yes	Yes	Yes	4
25	Yes	Yes	No	No	No	4
26	Yes	Yes	No	No	Yes	4
27	Yes	Yes	No	Yes	No	4
28	Yes	Yes	No	Yes	Yes	4
29	Yes	Yes	Yes	No	No	4
30	Yes	Yes	Yes	No	Yes	4
31	Yes	Yes	Yes	Yes	No	4
32	Yes	Yes	Yes	Yes	Yes	4

the opportunity to examine components and component levels that produce the best outcome. There is still a time and a place for the RCT. In fact, the RCT is included within the MOST framework: the optimized intervention is compared to a nonoptimized intervention or a standard of care through a classical RCT. MOST offers a potential solution for the need to know which components, at what levels, and in what combination, produce the best-designed outcome.

Returning to the hypothetical example, let us imagine that, at the end of the optimization trial, we have identified the combination of components and component levels that produces the greatest expected reduction in PTSD symptom severity among children with a history of trauma resulting from child maltreatment, that can be obtained for a per-participant cost of less than $400, meeting our optimization criterion. This is the optimized intervention. Suppose it comprises three components:

relaxation training, emotion regulation, and acceptance. This optimized intervention would be tested in an RCT and compared to a standard of care. If the optimized intervention had more favorable outcomes than the standard of care, it would be disseminated widely.

The Continual Optimization Principle

The three phases of MOST are interdependent and, though conducted sequentially, do not always follow a lockstep process. As depicted in Figure 20.1, the optimization phase is informed by the preparation phase; however, results of an optimization trial may suggest a return to the preparation phase. That is, if the product of the optimization trial is not expected to be sufficiently effective, rather than progressing to the evaluation phase, the researcher would return to the preparation phase to refine the conceptual model and begin the process anew.

Further, at the conclusion of the evaluation phase, the optimization process is not inherently complete, because there is always an opportunity to improve the intervention or update it in a subsequent study. We refer to this as the *continual optimization principle*. Not only might there be new constraints or literature to consider but a multicomponent behavioral intervention can always be better – more effective, more efficient – through further rounds of optimization. The continual optimization of an intervention provides a limitless opportunity to advance intervention science and improve the reach and impact of interventions.

At the conclusion of the evaluation phase, our hypothetical optimized intervention was comprised of three components (relaxation training, emotion regulation, and acceptance) and cost $400 or less per participant. According to the continual optimization principle, we would return to the preparation phase to consider new literature that might suggest the viability of a novel component. Additionally, we could seek to reduce the per-participant cost while maintaining effectiveness or increase effectiveness while maintaining the per-participant cost steady. Through one cycle of MOST we have an optimized intervention that can be disseminated on a wide scale; however, in the spirit of the MOST framework, to advance intervention science we can subject this optimized intervention to the cycle again so as to continually improve its public health impact.

ADDRESSING CONCERNS IN APPLYING MOST TO THE FIELD OF CLINICAL PSYCHOLOGY

What follows is a discussion of common considerations relevant to those in the clinical psychology field who are thinking of applying MOST in their research. The intention here is to answer or offer solutions to some of the most prominent concerns, but by no means is the list exhaustive.

"My Intervention Is Gestalt"

Applying the MOST framework requires the acknowledgment that at the end of the process, the intervention may be different from how it was initially conceived. Indeed, the resulting optimized intervention may not include all of the original components. This may prompt a researcher to state that their intervention must be retained as a whole, and therefore MOST is irrelevant. Intervention scientists who conceive of their intervention as a gestalt may still find utility in the MOST framework. Rather than focusing on the content of the intervention, as we did in our hypothetical example, the intervention scientist could consider engagement or fidelity components in their optimization trial. An engagement component might be one that improves a participant's adherence to the intervention. A fidelity component might be one that improves the delivery of the intervention. Let us use our hypothetical example to explore this notion.

Suppose our hypothetical intervention is gestalt; that is, as developers, we were unwilling to disaggregate the intervention. Rather than walk away from MOST, we might consider optimizing the intervention with components focused on improving the engagement in our intervention. Perhaps instead of designing the intervention to reduce PTSD symptom severity, we optimize the intervention to improve its completion rate. In the preparation phase, we review the literature and identify three candidate engagement strategies that we wish to consider: clinician check-ins, duration of treatment, and goal setting. We select an all-active-components optimization criterion; that is, only components with an empirically detectable positive effect will be included. The two levels of our components are set to on and off. Our engagement-focused optimization trial would use a factorial design to examine three components each with two levels: check-ins from the clinician between sessions (e.g., phone call or text message); varied duration of treatment sessions (e.g., shorter sessions delivered more frequently or longer sessions delivered less frequently); and the use of goal setting at the end of sessions (e.g., yes or no). The results of this 2^3 factorial experiment would produce strategies to ensure completion of the intervention. The intervention stays intact, but has now been optimized to improve engagement.

Management of a High Number of Experimental Conditions in the Field

A factorial experiment inherently means more than two conditions to manage in the field. By definition, the smallest number of conditions to be included in a factorial experiment is four (a factorial experiment requires at least two factors with two levels each), but the number of conditions increases exponentially as more factors are added. It is natural to feel some concern about potential challenges associated with managing the conditions in a large factorial experiment conducted in a field setting, but there is growing evidence it can be done. For interventions delivered online, a factorial experiment can be relatively easily managed and implemented. For interventions delivered in person, a strategy used increasingly by those adopting the MOST framework is to use Research Electronic Data Capture (REDCap) to manage random assignment, data collection, and maintenance of the database (Harris et al., 2009). This can be tremendously helpful, because a factorial optimization trial has many more conditions than an investigator may have previously dealt with. For example, investigators using MOST to optimize a behavioral intervention designed to increase engagement with the HIV care continuum for vulnerable persons living with HIV are using REDCap to track participants' assignments to the randomization scheme of the 16 experimental conditions in their optimization trial (Gwadz et al., 2017). These investigators are also using REDCap to cue interventionist action steps. That is, an interventionist can easily look up and track which component a participant is or is not to receive. REDCap provides an interface for data entry, management, and export to common statistical packages. It should be noted that other investigators have been successful in keeping meticulous records and managing up to 80 experimental conditions without the use of REDCap software (Cook et al., 2016; Piper et al., 2016, 2018; Schlam et al., 2016).

Commonly, optimization trials using a traditional factorial design in MOST are 2^4 (i.e., 4 components, with 2 levels each resulting in 16 experimental conditions), 2^5 (32 experimental conditions), or 2^6 (64 experimental conditions). It is recommended to select the number of candidate components that makes sense conceptually. The number of experimental conditions required is related to the number of components being examined. If the resulting design requires implementation of more conditions than a team can manage, there is an option to use a fractional factorial design (Collins et al., 2009). A full discussion of this design is precluded by space but, briefly, a fractional factorial is a variation of the factorial design in which a smaller set of conditions selected from the complete factorial are retained. If managing the 32 conditions in our hypothetical 2^5 factorial experiment was unrealistic, then we might consider a half-fraction fractional factorial design, 2^{5-1}, in which there are only 16 conditions to manage. A fractional factorial may be more manageable in the field than a complete factorial, but there are important tradeoffs that must be considered before selecting a fractional factorial design (Dziak et al., 2012). In general, for researchers conducting an optimization trial to examine a relatively small number of components, or for whom the costs associated with implementing experimental conditions are relatively small (e.g., internet-delivered interventions), fractional factorial designs are not recommended. On the other hand, researchers who are conducting an optimization trial to examine a large number of components, and for whom the costs associated with implementing experimental conditions are high, may wish to consider a fractional factorial.

How Is MOST Used in the Real World?

At this point, there is a substantial literature using the MOST framework, demonstrating that MOST can readily be used in the real world. In particular, increasingly interventions to be delivered online are using the MOST framework (Kugler et al., 2018). Managing a high number of experimental conditions in a complete factorial design for an online intervention is relatively easy, but MOST is not only for online interventions. An important concern of many researchers is the assumption that there must be one psychologist, or provider, per condition. Though having 32 unique providers for each experimental condition might be the cleanest way to conduct the optimization trial, it is not inherently required. It is feasible to train fewer providers to implement all of the experimental conditions. Suppose there were only 3 available clinicians to implement the 32 conditions in our hypothetical 2^5 factorial experiment. These clinicians would be trained to deliver all components (i.e., relaxation training, emotion regulation, cognitive processing, exposure and response prevention, and acceptance) and would use REDCap (or some other approach) to help them to ensure a participant got the correct components corresponding to the experimental condition to which they were randomized. Though there is an increased probability of error (e.g., delivering a component to a person in a condition that does not include that component) when clinicians are trained in all components, using software such as REDCap and doing fidelity checks are ways to minimize this error. Sessions could be recorded and adherence to the protocol could be monitored by other staff; protocol deviations would be reported and dealt with accordingly.

What Happens if Something Goes Wrong?

It is always possible for something, somewhere along the way to go wrong with applied research. Usually all is not lost, as long as the project begins with a detailed protocol and adherence to that protocol is monitored continuously. This is true for MOST as well. Even with a complex optimization trial, it is possible to recover from situations in which something goes wrong. For example, in an early application of MOST by Strecher and colleagues (2009) optimizing an online smoking cessation intervention, the original design called for a 32-condition fractional factorial. However, due a programming error, 1 of the 32 conditions was not filled. As a solution, 1 factor was collapsed over and the design was "folded" into 16 experimental conditions. Though not ideal, this solution salvaged the optimization trial. Other situations from which it may be more difficult to recover include nonignorable dropout, contamination across conditions, protocol deviations, poor implementation, or poor adherence. Of course, any of these problems may occur in any field research, whether it is an optimization trial or an RCT.

Dealing with Clustering

There are any number of situations in clinical psychology settings in which participants may be clustered (e.g., clinics, group sessions). In the situations where there is clustering (i.e., different clinics), but cluster randomization is not required, the best approach is to replicate the factorial experiment within each cluster (i.e., within-cluster randomization). In this way, the effect of clustering is minimized. When cluster randomization is necessary (e.g., between-cluster randomization), it is still possible to use a factorial design in the optimization phase (Caldwell et al., 2012; Schlam et al., 2016), but different power calculation is required (Dziak et al., 2012; Nahum-Shani & Dziak, 2018; Nahum-Shani, Dziak, & Collins, 2017). All else being equal, between-cluster randomization typically requires a greater overall sample size to maintain power compared to within-cluster randomization. It is also necessary to have enough clusters to assign at least one to each experimental condition. The *FactorialPowerPlan* macro mentioned above can assist in the calculation of sample size, effect size, and power of factorial designs with and without clustering (Dziak et al., 2013).

CONCLUSION

The purpose of this chapter has been to present a brief introduction to the multiphase optimization strategy (MOST) to the field of clinical psychology. We described the hypothetical application of MOST to build an optimized intervention from the ground up based on relevant basic science research supporting multilevel changes in hypothesized causal processes leading to PTSD symptoms following child maltreatment. This intervention reflects components included in current treatment approaches for youth experiencing PTSD symptoms following a traumatic event, such as Trauma-focused Cognitive-Behavioral Therapy and Prolonged Exposure. The addition of a detailed conceptual model and the targeting of hypothesized mediators, where observation of change during treatment delivery could inform mechanisms of change research, advances theoretical work in this area about which process or processes are most affected by individual components and most relevant for creating change in the outcome of interest. Alternatively, an existing intervention may also be optimized using MOST as long as the intervention scientist is willing to disaggregate the parts of the whole and accept that all components may not end up in the optimized intervention. Our optimized intervention includes only components and component levels that positively affect the outcome of interest, reduced PTSD symptoms. Assuming that the optimized intervention demonstrates effectiveness in an RCT when assessed against a comparable intervention, the optimized intervention would be disseminated widely and we would return to the preparation phase with the goal of further optimizing our intervention.

MOST is an innovative framework across disciplines. There are unique future directions particularly relevant to clinical psychology. In our example, we included only one outcome: a reduction in PTSD symptoms. It is an open field of inquiry to apply MOST to a situation in which there are multiple outcomes of interest. This is particularly relevant for psychological conditions that are highly comorbid, such as PTSD and MDD. Comorbid diagnoses are often the result of an incomplete or faulty conceptual model or nosology where distinct psychiatric outcomes have shared and not unique underlying causal mechanisms (Sanislow et al., 2010). The identification and engagement of mechanisms affected by child maltreatment and leading to multiple outcomes, called transdiagnostic mechanisms, can also advance efforts at optimizing behavioral interventions in terms of efficiency and effectiveness (Nolen-Hoeksema & Watkins, 2011).

REFERENCES

Almirall, D., Nahum-Shani, I., Sherwood, N. E., & Murphy, S. A. (2014). Introduction to SMART Designs for the Development of Adaptive Interventions: With Application to Weight Loss Research. *Translational Behavioral Medicine*, 4(3), 260–274.

Baker, T. B., Collins, L. M., Mermelstein, R., Piper, M. E., Schlam, T. R., Cook, J. W., ... Fiore, M. C. (2015). Enhancing the Effectiveness of Smoking Treatment Research: Conceptual Bases and Progress. *Addiction*, 111, 107–116.

Barth, R. P., Lee, B. R., Lindsey, M. A., Collins, K. S., Strieder, F., Chorpita, B. F., ... Sparks, J. A. (2012). Evidence-Based Practice at a Crossroads: The Timely Emergence of Common Elements and Common Factors. *Research on Social Work Practice*, 22(1), 108–119.

Beck, A. T., Rush, A. J., Shaw, B. F., & Emery, G. (1979). *Cognitive Therapy of Depression*. New York: Guilford.

Becker, K. D., Lee, B. R., Daleiden, E. L., Lindsey, M., Brandt, N. E., & Chorpita, B. F. (2015). The Common Elements of Engagement in Children's Mental Health Services: Which Elements for Which Outcomes? *Journal of Clinical Child & Adolescent Psychology*, 44(1), 30–43.

Berkowitz, S. J., Stover, C. S., & Marans, S. R. (2011). The Child and Family Traumatic Stress Intervention: Secondary Prevention for Youth at Risk of Developing PTSD. *Journal of Child Psychology and Psychiatry and Allied Disciplines*, 52(6), 676–685.

Boustani, M. M., Gellatly, R., Westman, J. G., & Chorpita, B. F. (2017). Advances in Cognitive Behavioral Treatment Design: Time for a Glossary. *The Behavior Therapist*, 40(6), 199–208.

Caldwell, L. L., Smith, E. a., Collins, L. M., Graham, J. W., Lai, M., Wegner, L., ... Jacobs, J. (2012). Translational Research in South Africa: Evaluating Implementation Quality Using a Factorial Design. *Child & Youth Care Forum*, 41(2), 119–136.

Carrion, V. G., Weems, C. F., Ray, R. D., Glaser, B., Hessl, D., & Reiss, A. L. (2002). Diurnal Salivary Cortisol in Pediatric Posttraumatic Stress Disorder. *Biological Psychiatry*, 51(7), 575–582.

Chorpita, B. F., & Daleiden, E. L. (2009). Mapping Evidence-Based Treatments for Children and Adolescents: Application of the Distillation and Matching Model to 615 Treatments from 322 Randomized Trials. *Journal of Consulting and Clinical Psychology*, 77(3), 566–579.

Chorpita, B. F., Becker, K. D., Daleiden, E. L., & Hamilton, J. D. (2007). Understanding the Common Elements of Evidence-Based Practice. *Journal of the American Academy of Child & Adolescent Psychiatry*, 46(5), 647–652.

Chorpita, B. F., Daleiden, E. L., Ebesutani, C. K., Young, J., Becker, K. D., Nakamura, B. J., ... Starace, N. (2011). Evidence-Based Treatments for Children and Adolescents: An Updated Review of Indicators of Efficacy and Effectiveness. *Clinical Psychology: Science and Practice*, 18(2), 154–172.

Cohen, J. (1988). *Statistical Power Analysis for the Behavioral Sciences* (2nd edn.). London: Routledge.

Cohen, J. A., Mannarino, A. P., & Deblinger, E. (2006). *Treating Trauma and Traumatic Grief in Children and Adolescents*. New York: Guilford Press.

Collins, L. M. (2014). Optimizing Family Intervention Programs: The Multiphase Optimization Strategy (MOST). In S. M. McHale, P. Amato, & A. Booth (Eds.), *Emerging Methods in Family Research* (pp. 19–37). Basel: Springer International.

Collins, L. M. (2018). *Optimization of Behavioral, Biobehavioral, and Biomedical Interventions: The Multiphase Optimization Strategy (MOST)*. New York: Springer.

Collins, L. M., & Kugler, K. C. (Eds.). (2018). *Optimization of Multicomponent Behavioral, Biobehavioral, and Biomedical Interventions: Advanced Topics*. New York: Springer.

Collins, L. M., Dziak, J. J., & Li, R. (2009). Design of Experiments with Multiple Independent Variables: A Resource Management Perspective on Complete and Reduced Factorial Designs. *Psychological Methods*, 14(3), 202–224.

Collins, L. M., Baker, T. B., Mermelstein, R. J., Piper, M. E., Jorenby, D. E., Smith, S. S., ... Fiore, M. C. (2011). The Multiphase Optimization Strategy for Engineering Effective Tobacco Use Interventions. *Annals of Behavioral Medicine*, 41(2), 208–226.

Collins, L. M., MacKinnon, D. P., & Reeve, B. B. (2013). Some Methodological Considerations in Theory-Based Health Behavior Research. *Health Psychology*, 32(5), 586–591.

Collins, L. M., Dziak, J. J., Kugler, K. C., & Trail, J. B. (2014a). Factorial Experiments: Efficient Tools for Evaluation of Intervention Components. *American Journal of Preventive Medicine*, 47(4), 498–504.

Collins, L. M., Nahum-Shani, I., & Almirall, D. (2014b). Optimization of Behavioral Dynamic Treatment Regimens Based on the Sequential, Multiple Assignment, Randomized Trial (SMART). *Clinical Trials*, 11(4), 426–434.

Collins, L. M., Trail, J. B., Kugler, K. C., Baker, T. B., Piper, M. E., & Mermelstein, R. J. (2014c). Evaluating Individual Intervention Components: Making Decisions Based on the Results of a Factorial Screening Experiment. *Translational Behavioral Medicine*, 4(3), 238–251.

Cook, J. W., Collins, L. M., Fiore, M. C., Smith, S. S., Fraser, D., Bolt, D. M., ... Mermelstein, R. (2016). Comparative Effectiveness of Motivation Phase Intervention Components for Use with Smokers Unwilling to Quit: A Factorial Screening Experiment. *Addiction*, 111(1), 117–128.

Copeland, W. E., Keeler, G., Angold, A., & Costello, E. J. (2007). Traumatic Events and Posttraumatic Stress in Childhood. *Archives of General Psychiatry*, 64(5), 577–584.

Deblinger, E., Mannarino, A. P., Cohen, J. A., Runyon, M. K., & Steer, R. A. (2011). Trauma-Focused Cognitive Behavioral Therapy for Children: Impact of the Trauma Narrative and Treatment Length. *Depression & Anxiety*, 28(1), 67–75.

Dimidjian, S., Hollon, S., Dobson, K., Schmaling, K., Kohlenberg, R., Addis, M., ... Jacobson, N. (2006). Randomized Trial of Behavioral Activation, Cognitive Therapy, and Antidepressant Medication in the Acute Treatment of Adults with Major Depression. *Journal of Consulting and Clinical Psychology*, 74(4), 658–670.

Dorsey, S., Mclaughlin, K. A., Kerns, S. E. U., Harrison, J. P., Lambert, H. K., Briggs, E. C., ... Amaya-Jackson, L. (2017). Evidence Base Update for Psychosocial Treatments for Children and Adolescents Exposed to Traumatic Events. *Journal of Clinical Child & Adolescent Psychology*, 46(3), 303–330.

Dziak, J. J., Nahum-Shani, I., & Collins, L. M. (2012). Multilevel Factorial Experiments for Developing Behavioral Interventions: Power, Sample Size, and Resource Considerations. *Psychological Methods*, 17(2), 153–175.

Dziak, J. J., Collins, L. M., & Wagner, A. T. (2013). *FactorialPowerPlan SAS Macro Suite Users' Guide (No. 1.0)*. University Park: Methodology Center, Pennsylvania State University.

Foa, E. B., McLean, C. P., Capaldi, S., & Rosenfield, D. (2013). Prolonged Exposure vs. Supportive Counseling for Sexual Abuse-Related PTSD in Adolescent Girls: A Randomized Clinical Trial. *JAMA*, 310(24), 2650–2657.

Gwadz, M. V., Collins, L. M., Cleland, C. M., Leonard, N. R., Wilton, L., Gandhi, M., ... Scott Braithwaite, R. (2017). Using the Multiphase Optimization Strategy (MOST) to Optimize an HIV Care Continuum Intervention for Vulnerable Populations: A Study Protocol. *BMC Public Health*, 17, 1–20.

Harris, P. A., Taylor, R., Thielke, R., Payne, J., Gonzalez, N., & Conde, J. G. (2009). Research Electronic Data Capture (REDCap) – A Metadata-Driven Methodology and Workflow Process for Providing Translational Research Informatics Support. *Journal of Biomedical Informatics*, 42(2), 377–381.

Hayes, A. M., Yasinski, C., Grasso, D., Ready, C. B., Alpert, E., McCauley, T., ... Deblinger, E. (2017). Constructive and Unproductive Processing of Traumatic Experiences in Trauma-Focused Cognitive-Behavioral Therapy for Youth. *Behavior Therapy*, 48(2), 166–181.

Insel, T. R., & Gogtay, N. (2014). National Institute of Mental Health Clinical Trials: New Opportunities, New Expectations. *JAMA Psychiatry*, 71(7), 745–746.

Jacobson, N., Dobson, K., Truax, P., Addis, M., Koerner, K., Gollan, J., ... Prince, S. (1996). A Component Analysis of Cognitive-Behavioral Treatment for Depression. *Consulting and Clinical Psychology*, 64(2), 295–304.

Jacobson, N., Martell, C. R., & Dimidjian, S. (2001). Behavioral Activation Treatment for Depression: Returning to Contextual Roots. *Clinical Psychology: Science and Practice*, 8(3), 255–270.

Kahana, S. Y., Feeny, N. C., Youngstrom, E., & Drotar, D. (2006). Posttraumatic Stress in Youth Experiencing Illnesses and Injuries: An Exploratory Meta-Analysis. *Traumatology*, 12(2), 148–161.

Kugler, K. C., Balantekin, K. N., Birch, L. L., & Savage, J. S. (2016). Application of the Multiphase Optimization Strategy to a Pilot Study: An Empirical Example Targeting Obesity among Children of Low-Income Mothers. *BMC Public Health*, 16(1), 1181–1191.

Kugler, K. C., Wyrick, D. L., Tanner, A. E., Milroy, J. J., Chambers, B. D., Ma, A., ... Collins, L. M. (2018). Using the Multiphase

Optimization Strategy (MOST) to Develop an Optimized Online STI Preventive Intervention Aimed at College Students: Description of Conceptual Model and Iterative Approach to Optimization. In L. M. Collins & K. C. Kugler (Eds.), *Optimization of Multicomponent Behavioral, Biobehavioral, and Biomedical Interventions: Advanced Topics* (pp. 1–21). New York: Springer.

Lindsey, M. A., Brandt, N. E., Becker, K. D., Lee, B. R., Barth, R. P., Daleiden, E. L., & Chorpita, B. F. (2014). Identifying the Common Elements of Treatment Engagement Interventions in Children's Mental Health Services. *Clinical Child and Family Psychology Review*, *17*, 283–298.

MacKinnon, D. P., Fairchild, A. J., & Fritz, M. S. (2007). Mediation Analysis. *Annual Review of Psychology*, *58*, 593–614.

Nahum-Shani, I., & Dziak, J. J. (2018). Multilevel Factorial Designs in Intervention Development. In L. M. Collins & K. C. Kugler (Eds.), *Optimization of Multicomponent Behavioral, Biobehavioral, and Biomedical Interventions: Advanced Topics* (pp. 47–87). New York: Springer.

Nahum-Shani, I., Dziak, J. J., & Collins, L. M. (2017). Multilevel Factorial Designs with Experiment-Induced Clustering. *Psychological Methods*, *22*(3), 458–479.

Nolen-Hoeksema, S., & Watkins, E. R. (2011). A Heuristic for Developing Transdiagnostic Models of Psychopathology: Explaining Multifinality and Divergent Trajectories. *Perspectives on Psychological Science*, *6*(6), 589–609.

Pellegrini, C. A., Hoffman, S. A., Collins, L. M., & Spring, B. (2014). Optimization of Remotely Delivered Intensive Lifestyle Treatment for Obesity Using the Multiphase Optimization Strategy: Opt-IN Study Protocol. *Contemporary Clinical Trials*, *38*(2), 251–259.

Piper, M. E., Fiore, M. C., Smith, S. S., Fraser, D., Bolt, D. M., Collins, L. M., … Baker, T. B. (2016). Identifying Effective Intervention Components for Smoking Cessation: A Factorial Screening Experiment. *Addiction*, *111*(1), 129–141.

Piper, M. E., Schlam, T. R., Fraser, D., Oguss, M., & Cook, J. W. (2018). Implementing Factorial Experiments in Real-World Settings: Lessons Learned While Engineering an Optimized Smoking Cessation Treatment. In L. M. Collins & K. C. Kugler (Eds.), *Optimization of Multicomponent Behavioral, Biobehavioral, and Biomedical Interventions: Advanced Topics* (pp. 23–45). New York: Springer.

Sanislow, C. A., Pine, D. S., Quinn, K. J., Kozak, M. J., Garvey, M. A., Heinssen, R. K., … Cuthbert, B. N. (2010). Developing Constructs for Psychopathology Research: Research Domain Criteria. *Journal of Abnormal Psychology*, *119*(4), 631–639.

Schlam, T. R., Fiore, M. C., Smith, S. S., Fraser, D., Bolt, D. M., Collins, L. M., … Baker, T. B. (2016). Comparative Effectiveness of Intervention Components for Producing Long-Term Abstinence from Smoking: A Factorial Screening Experiment. *Addiction*, *111*(1), 142–155.

Shenk, C. E., Putnam, F. W., Rausch, J. R., Peugh, J. L., & Noll, J. G. (2014). A Longitudinal Study of Several Potential Mediators of the Relationship between Child Maltreatment and Posttraumatic Stress Disorder Symptoms. *Development and Psychopathology*, *26*(1), 81–91.

Shenk, C. E., Griffin, A. M., & O'Donnell, K. J. (2015). Symptoms of Major Depressive Disorder Subsequent to Child Maltreatment: Examining Change across Multiple Levels of Analysis to Identify Transdiagnostic Risk Pathways. *Development and Psychopathology*, *27*(4 pt 2), 1503–1514.

Silverman, W. K., Ortiz, C. D., Viswesvaran, C., Burns, B. J., Kolko, D. J., Putnam, F. W., & Amaya-Jackson, L. (2008). Evidence-Based Psychosocial Treatments for Children and Adolescents Exposed to Traumatic Events. *Journal of Clinical Child & Adolescent Psychology*, *37*(1), 156–183.

Steinberg, A. M., Brymer, M. J., Decker, K. B., & Pynoos, R. S. (2004). The University of California at Los Angeles Posttraumatic Stress Disorder Reaction Index. *Current Psychiatry Reports*, *6*(2), 96–100.

Steinberg, A. M., Brymer, M. J., Kim, S., Briggs, E. C., Ippen, C. G., Ostrowski, S. A., … Pynoos, R. S. (2013). Psychometric Properties of the UCLA PTSD Reaction Index: Part I. *Journal of Traumatic Stress*, *26*, 1–9.

Strecher, V. J., Mcclure, J. B., Alexander, G. L., Nair, V. N., Konkel, J. M., Greene, S. M., … Pomerleau, O. F. (2009). Web-Based Smoking-Cessation Program: Results of a Randomized Trial. *American Journal of Preventive Medicine*, *34*(5), 373–381.

Trickett, P. K., Noll, J. G., Susman, E. J., Shenk, C. E., & Putnam, F. W. (2010). Attenuation of Cortisol across Development for Victims of Sexual Abuse. *Development and Psychopathology*, *22*(1), 165–175.

US Department of Health & Human Services, Administration for Children and Families, Administration on Children Youth and Families, & Children's Bureau. (2017). *Child Maltreatment 2015*. Retrieved from www.acf.hhs.gov/programs/cb/research-data-technology/statistics-research/child-maltreatment

Widom, C. S. (1999). Posttraumatic Stress Disorder in Abused and Neglected Children Grown Up. *American Journal of Psychiatry*, *156*(8), 1223–1229.

Wyrick, D. L., Rulison, K. L., Fearnow-Kenney, M., Milroy, J. J., & Collins, L. M. (2014). Moving beyond the Treatment Package Approach to Developing Behavioral Interventions: Addressing Questions That Arose during an Application of the Multiphase Optimization Strategy (MOST). *Translational Behavioral Medicine*, *4*(3), 252–259.

21 Future Directions in Developing and Evaluating Psychological Interventions

SHANNON SAUER-ZAVALA

Over the past 40 years, the field of clinical psychology has seen promising advances in development of efficacious treatments for a range of mental health conditions (see Nathan & Gorman, 2007). Specifically, the emergence of a classification system largely focused on differentiating psychopathology into thinly sliced but reliably identified categories, an approach exemplified by the *Diagnostic and Statistical Manual* (DSM; American Psychiatric Association [APA], 1980, 2000, 2013), allowed for enormous progress in treatment outcome research. Splitting broad, nonempirical categories (e.g., neurosis, psychosis) into discrete diagnoses (e.g., panic disorder, social anxiety, major depressive disorder) with objective-symptom-criteria-initiated meaningful research on outcomes of pharmacological and individual psychological treatments (e.g., Barlow et al., 1984). These psychotherapeutic treatments mirrored the DSM approach by targeting specific forms of psychopathology (e.g., Mastery of Your Anxiety and Panic for panic disorder; Barlow & Craske, 2007) and were found to be efficacious for symptom reduction in a variety of formats, uses, and settings (Barlow, 1996, 2004, 2008; Barlow et al., 2000; Heimberg, Liebowitz, & Hope, 1998; Nathan & Gorman, 2007).

There is, however, more progress to be made. Specifically, only a third of treatment-seeking individuals receive evidence-based care (Thornicroft, 2007; Wang et al., 2007) and remission rates are far from optimal even among those who are exposed to empirically supported interventions (Barlow, 2008; Hollon, Thase, & Markowitz, 2002). Thus, it is critical that future treatment development efforts address the fact that patients do not respond uniformly to our treatments and widespread dissemination is startlingly slow.

It is possible that the same advances that moved the field forward in past decades are currently holding it back. Specifically, emerging research suggests that the "splitting" approach to diagnosis, while enhancing reliability, may come at the expense of validity (see Barlow et al., 2014). In other words, DSM disorders may not represent unique constructs, but are, rather, relatively trivial variations in a common underlying syndrome (Brown & Barlow, 2009). The most obvious manifestation of this issue is the high rate of co-occurrence amongst related disorders. For example, data suggest that 55 percent of individuals with a principal anxiety disorder will meet criteria for an additional anxiety, depressive, or related disorder, and these estimates rise to 75 percent when lifetime comorbidity is considered (Brown et al., 2001). Practically, this means that treatment protocols designed to target a single DSM diagnosis are unlikely to match patient presentations in community practice, impeding their efficacy; indeed, therapists' efforts to address co-occurring conditions have been shown to interfere with treatment of the primary problem (Craske et al., 2007; Gibbons & DeRubeis, 2008). Furthermore, interventions focused on a single disorder represent a large training burden for therapists who are required to gain competence in a wide array of protocols in order to provide evidence-based care to their patients; the time and monetary costs associated with training, coupled with the fact that most treatments are not equipped to handle comorbidity, likely dampens enthusiasm for clinicians to use these interventions – limiting dissemination efforts.

In contrast, focusing on core processes that are relevant for the development and maintenance of a range of common conditions may be a more robust and efficient approach to treatment. This view is consistent with the National Institute of Mental Health's Research Domain Criteria (RDoC; Insel et al., 2010), challenging researchers to identify putative psychopathological mechanisms that cut across diagnoses and can become the focus of treatment (rather than focusing on individual disorder symptoms).

The overarching goal of this chapter is to delineate various approaches to treatment development and outcome work that are suited to address the suboptimal response rates and limited dissemination stemming from the "splitting approach" to mental health classification and intervention. Specifically, this chapter will emphasize the importance of identifying core, transdiagnostic processes that can become the focus of treatment and may lead to simultaneous symptom reduction across diagnoses. Additionally, techniques to confirm that components included in our treatment packages each engage these core processes will be highlighted with the goal of ensuring that interventions are comprised of only active ingredients. Next, considerations for conducting randomized-controlled trials specifically with transdiagnostic interventions will be addressed. Finally, this chapter will describe the advantages of a personalized approach to psychological medicine, taking the field away from the nomothetic, one-size-fits-all treatment protocols that work for many, but not all, individuals.

IDENTIFICATION OF TRANSDIAGNOSTIC PSYCHOPATHOLOGICAL TARGETS

An important first step in improving treatment outcomes for a wider range of individuals with mental health conditions is to clarify exactly what our interventions should target. As noted above, treatments developed in recent decades have focused on addressing symptoms associated with a discrete DSM disorder. In the past several years, however, the field has shifted toward the exploration of transdiagnostic processes that are implicated in the development and maintenance of a range of conditions. Although the term "transdiagnostic" has been increasingly used to describe psychological constructs, there is some variability in its application (see Sauer-Zavala et al., 2017b). For example, Mansell and colleagues (2009) consider a process to be transdiagnostic if it has been assessed in both clinical and nonclinical samples and is present in a minimum of four disorders. Harvey and colleagues (2011) reserve the term "descriptively transdiagnostic" for processes that are simply present in a range of diagnoses, consistent with Mansell and colleagues' (2009) criteria. In contrast, processes that reflect causal mechanisms for co-occurrence across disorders are designated as "mechanistically transdiagnostic."

Processes that are causally related to a range of psychopathology (i.e., mechanistically transdiagnostic) confer greater promise as candidate treatment targets given that interventions can be developed to focus on these core deficits rather than on addressing disorder correlates (i.e. descriptively transdiagnostic constructs). For example, overevaluation of shape and weight has been described as a core vulnerability relevant for the development and maintenance of anorexia, bulimia, and eating disorders not otherwise specified (e.g., Fairburn et al., 1993; Wilson et al., 2002). Directly targeting this construct in treatment has been in associated with decreases in symptoms across the full range of eating disorders (Fairburn et al., 2009). In contrast, although DSM panic attacks can occur in the context of any disorder (descriptively transdiagnostic), there is little evidence that panic attacks are functionally related to the onset or maintenance of schizophrenia, for example, and that targeting them in treatment would lead to substantial clinical improvement.

Continued identification of mechanistically transdiagnostic processes may lead to more efficient treatments as targeting underlying mechanisms has been shown to lead to clinical improvement across comorbid conditions (e.g., Farchione et al., 2012). The RDoC matrix, initiated by the National Institute of Mental Health (see Insel et al., 2010) provides guidance into potential research domains in which transdiagnostic constructs may be identified: the negative valence system, the positive valence system, the cognitive system, the arousal/regulatory system, and the system for social processes.

CLASSIFICATION OF MENTAL DISORDERS: IMPLICATIONS OF FOCUSING ON CORE MECHANISMS

Given the value in identifying core processes that can become the focus of treatment, there is also a need for reliable assessment of these constructs and, more broadly, their integration into the classification of mental illness. It is important to note that the use of underlying vulnerabilities as the basis for a nosological system would not represent a return to the nonempirical system of classification that existed prior to DSM-III (APA, 1980). Rather, advances in empirical methods and analyses make it possible to explore constructs that underlie the full range of disorders without the constraints of artificial categories (see Chapter 9, by Lenzenweger).

One result of these advancements has been the proposal of dimensional classification systems that convey more information about the nature of an individual's difficulties beyond crossing an arbitrary threshold to receive a categorical diagnosis. The most notable example of such a system is the alterative model for personality disorders in DSM-5 (APA, 2013). Briefly, as a more detailed account of this approach to classification can be found in Chapter 3 (by Venables and Patrick), the alternative model was created in response to phenotypic overlap and high rates of co-occurrence within personality disorders, great heterogeneity within specific diagnoses, and overreliance on the unspecified personality disorder category (Widiger & Trull, 2007). These issues suggest that receipt of a given personality disorder diagnosis is not particularly informative with regard to the core underlying processes that should be targeted in treatment. As such, the DSM-5 alternative model includes severity ratings on a range of traits, allowing for more precision in communicating the deficits that drive symptoms (Hopwood et al., 2012). Instead of, for example, relying on a borderline personality disorder diagnosis and applying a one-size-fits-all treatment, this new system communicates salient traits within an individual that may be traditionally associated with different personality disorders (e.g., hostility, grandiosity, intimacy avoidance).

Another example of dimensional classification has been proposed for the broad range of anxiety, depressive, and related disorders (Brown & Barlow, 2009) – referred to as emotional disorders (Barlow, 1991). This model was formulated in response to the mounting evidence that all considerable covariance in the expression of these DSM constructs could be accounted for by higher-level dimensions of temperament, most notably the tendency to experience negative affect (Brown & Barlow 2009; Brown, Chorpita, & Barlow, 1998). Importantly, low positive affect is also a relevant temperamental dimension that is specifically associated with depressive disorders, social anxiety disorder, and agoraphobia (Rosellini et al., 2010). Several other key features are found across the emotional disorders that, in general, pertain to the specific stimuli that produce interfering negative affect. These can include somatic symptoms, autonomic surges, intrusive cognitions, social evaluation, and past traumatic experiences. Similar to the ratings of traits in the alternative model for personality disorders (APA, 2013), evaluation of these dimensions provides more information to clinicians than a categorical diagnosis; a new assessment tool has recently been developed to capture these salient features relevant to the range of emotional disorders (Rosellini et al., 2015).

While many agree that classification based on core, underlying difficulties has advantages over the current categorical system, there have been challenges to the widespread adoption of dimensional nosology. First, even in the area of personality disorders, where the notion of a dimensional approach is widely accepted (Widiger & Crego, 2013), disagreement remains with regard to precisely which lower-order traits should be included in the model (e.g., Widiger & Krueger, 2013). Additionally, the "user friendliness" for clinicians who are accustomed to categorical labels (e.g., "she is depressed") is also a barrier that that needs to be addressed. Nevertheless, assessment of transdiagnostic processes is an important component in the goal of shifting the field away from emphasizing categorical differences and continued efforts must be dedicated to this pursuit.

TRANSDIAGNOSTIC, MECHANISM-FOCUSED TREATMENT DEVELOPMENT

Reliable assessment of core underlying processes has the potential to ignite treatment development efforts similar to the manner in which the publication of DSM-III (APA, 1980) prompted the creation of protocols focused on improving symptoms associated with each diagnosis. Instead of focusing on symptoms, however, contemporary interventions should address transdiagnostic targets implicated in the development and maintenance of a range of disorders. Indeed, this transition has already begun.

For example, Cognitive Behavioral Therapy-Enhanced (CBT-E; Fairburn, 2008), which was created to address the full range of eating disorder diagnoses, represents one of the first attempts to develop an intervention targeting shared mechanisms. This treatment is based on the theoretical model of eating disorders (described above) that implicates overevaluation of shape and weight as a core maintaining factor for anxiety, bulimia, and eating disorder not otherwise specified (Fairburn, Cooper, & Shafran, 2003). The core process is addressed via several intermediary targets, including decreasing body checking, decreasing rigid rules around food, implementing regular eating, and challenging distorted thinking. Additionally, care is taken to enhance other areas of an individual's life beyond shape and weight (e.g., friends, family, school/work) to reduce the disproportionate importance placed there.

Perhaps the best-known example of a transdiagnostic intervention is Barlow and colleagues' Unified Protocol (UP) for the Transdiagnostic Treatment of Emotional Disorders (Barlow et al., 2018). The UP is a cognitive-behavioral intervention that was explicitly designed to target the core mechanisms that maintain symptoms across emotional disorders. Specifically, the UP encourages a more willing, accepting attitude toward emotional experiences that, in turn, reduces reliance on the avoidant coping strategies that actually increase negative affectivity in the long term (Purdon, 1999; Wenzlaff & Wegner, 2000). There is ample evidence to support the UP's efficacy in treating emotional disorders, including the range of anxiety disorders (Barlow et al., 2017; Ellard et al., 2010; Farchione et al., 2012), depression (Boswell, Anderson, & Barlow, 2014), and borderline personality disorder (Sauer-Zavala, Bentley, & Wilner, 2016). Additionally, several studies have demonstrated that the UP is indeed associated with reductions in its putative target – aversive reactions to emotional experiences (Boswell et al., 2013; Sauer-Zavala et al., 2012).

IDENTIFICATION OF TREATMENT COMPONENTS THAT ADDRESS CORE PROCESSES

Historically, multicomponent treatment packages have been evaluated in their entirety. With a primary focus on symptom reduction, this approach makes sense as exposure to one element from an intervention (e.g., two sessions of cognitive appraisal) may be unlikely to result in diagnostic remission. However, as transdiagnostic mechanisms increasingly become the target of care, there may be an opportunity to explore whether each skill included in a treatment package indeed engages the core processes thought to maintain symptoms. To be considered a psychopathological mechanism, change on candidate processes must precede and predict change in symptoms (Kazdin, 2007); thus, if a given skill does not contribute to mechanism engagement, it is unlikely to result in downstream symptom reduction and should be removed from the package. Consistent with this mechanism-based understanding of improvement in treatment, the National Institute of Mental Health has been prioritizing experimental therapeutics studies aimed at elucidating the impact of discrete treatment components on the engagement of psychopathological targets (Insel & Gogtay, 2014).

Single-case experimental design (SCED) represents one low-cost, efficient option for exploring core process engagement in response to specific treatment components (Barlow, Nock, & Hersen, 2008). SCEDs are within-subject studies in which each participant serves as their own control and levels of an intervention are varied within each individual to explore their effect on a given target. Alternating treatment designs (ATDs) represent one commonly used type of SCED; here, participants are repeatedly exposed to two intervention strategies, both presented as credible, in a randomly assigned order. For example, in a laboratory-based ATD aimed at determining whether opposite action from Dialectical Behavior Therapy (DBT; Linehan, 1993) reduces emotional intensity, participants ($N = 16$) attended six study sessions and were instructed to act opposite to an induced mood in half the sessions and to act consistent to the mood in the other half (Sauer-Zavala et al., 2019). Results suggested that, in response to sadness and guilt, opposite-action sessions resulted in significantly greater reductions in emotional intensity than did act consistent sessions; however, this pattern of results was not found in response to anxiety and anger.

In a follow-up SCED, this time using a multiple-baseline design, Sauer-Zavala and colleagues (manuscript in preparation) explored the effect of a brief opposite action intervention on the frequency of negative emotions across time. In a multiple-baseline study, participants complete an assessment-only baseline phase prior to the introduction of treatment and the randomly assigned length of this phase varies (e.g., two weeks, four weeks). A significant treatment effect occurs when, regardless of baseline length, change in variables of interest occurs only after the introduction of the intervention; in the context of the opposite-action study ($N = 8$), results suggest that negative emotions were significantly less frequent during the treatment phase than the baseline phase. Emotional vulnerability has been described as an important risk factor for the development of borderline personality disorder (Linehan, 1993) and, taken together, these SCEDs confirm that opposite action, a commonly used treatment element for this population, indeed engages this target mechanism.

SCEDs have also been used to isolate the unique effects of each UP module on target mechanisms using SCEDs (e.g., Brake et al., 2016; Sauer-Zavala et al., 2017a). First, Brake and colleagues (2016) utilized an alternating treatment design to determine whether the mindful emotion exposure module of the UP indeed enhances patients' willingness to tolerate strong emotions. Each participant completed six exposure sessions with varied (alternating) instructions for how to engage with the emotions produced by the study procedures; in half the sessions, participants were instructed to mindfully approach emotions, while attempting to avoid them in the remaining sessions.

Results suggest that the mindful sessions, consistent with how exposures are generally conducted in the UP, led greater willingness to approach emotions than the "seek to avoid" sessions. Additionally, Sauer-Zavala and colleagues (2017a) conducted another SCED, a multiple baseline design in this case, in which each UP skill module was presented in isolation following an assessment-only baseline phase. The putative psychopathological mechanism targeted by the UP, aversive reactions to strong emotions, remained stable during the baseline period and decreased significantly only after the introduction of each UP module. Across these two studies, this pattern of results suggests that each UP module indeed engages the core process thought to maintain symptoms across emotional disorders.

SCEDs are advantageous at early stages in the treatment development process given that their rigorous within-individual methods require a small number of participants to demonstrate an effect. Once evidence suggesting that a treatment component engages a core mechanistic process has been established, larger nomothetic investigations of the effects of discrete treatment strategies can be undertaken. Although dismantling efficacious treatments may seem counterproductive given that they have already been proven to improve core psychopathological processes and reduce symptoms, there is great value in understanding exactly what each individual skill contributes. The initiative to confirm that all treatment components engage target mechanisms allows the developers of psychological interventions to craft together protocols that contain only active ingredients. Treatments containing skills that are confirmed to engage shared mechanisms important across diverse diagnoses are likely to be more robust and efficient, though trials comparing their effects against the gold-standard symptom-focused approaches are necessary.

CONSIDERATIONS FOR TESTING TRANSDIAGNOSTIC INTERVENTIONS

A number of considerations are necessary when conducting clinical trials testing transdiagnostic interventions that were developed explicitly to address psychopathological mechanisms. First, care should be taken to ensure that measurement of core processes and symptoms are collected frequently enough to draw the conclusion that change on mechanisms precedes and predicts change on diagnostic outcomes (refer to Chapters 30 and 31 for more information on this topic). Additionally, for decades, a hallmark of clinical trials for psychological interventions was rigorous control with regard to inclusion and exclusion criteria, usually restricting the sample to one diagnostic category (e.g., Barlow et al., 1989). Testing transdiagnostic treatments calls for diagnostically diverse samples that include broader classes of DSM categories maintained by similar functional processes (e.g., Barlow et al., in press). Alternatively, and perhaps more consistent with the above recommended departure from categorical classification, trial inclusion criteria could be explicitly based on threshold levels of dimensional processes, eschewing diagnoses altogether (e.g., Rapee et al., 2005).

Further, it is also important to consider what might serve as an appropriate comparison condition for trials evaluating transdiagnostic interventions. After initial efficacy is established, equivalence designs powered to confirm that two treatments do not differ from each other may represent a useful option (Greene et al., 2008). An equivalence trial addresses the most pressing question facing treatment development researchers – not whether transdiagnostic interventions are efficacious, but whether targeting transdiagnostic mechanisms is just as good at addressing the primary condition as the targeted protocol designed explicitly for that disorder. Barlow and colleagues (manuscript) recently conducted the first equivalence trial comparing a transdiagnostic treatment (the UP) to four single-disorder protocols for principal anxiety disorders: *Managing Social Anxiety* (MSA-II; Hope, Heimberg, & Turk, 2006); *Mastery of Anxiety and Panic* (MAP-IV; Barlow & Craske, 2007); *Mastery of Anxiety and Worry* (MAW-II; Craske & Barlow, 2006); and *Treating Your OCD with Exposure and Response (Ritual) Prevention Therapy* (Yadin, Foa, & Lichner, 2012). Findings suggest that the UP's transdiagnostic approach results in symptom reduction for principal anxiety disorders equivalent to these gold-standard diagnosis-specific protocols. Additionally, at posttreatment, 62 percent of individuals in the UP condition no longer met diagnostic criteria for any emotional disorder, compared to 47 percent in the single-disorder protocol condition.

The results of Barlow and colleagues' (manuscript) UP equivalence trial underscores the advantages of transdiagnostic interventions that can lead to simultaneous symptom reduction across a range of conditions by targeting shared mechanisms. This efficient approach to addressing comorbid conditions also has the potential to significantly reduce therapists' burden as providers need only learn one protocol in order to provide evidence-based care to a wide swath of individuals. Future research should continue to prioritize testing mechanistically transdiagnostic interventions over treatments that more narrowly apply to a single condition.

PERSONALIZED PSYCHOLOGICAL INTERVENTIONS

Finally, shifting away from nomothetic, one-size-fits-all interventions in favor of personalized delivery of relevant skills is another important consideration for future treatment development. Modular interventions in which patients only receive treatment components that best fit with their presentations represent one approach to personalized medicine (Chorpita, Daleiden, & Weisz, 2005). These treatments, most prominent in the child literature (e.g., Chorpita et al., 2004) and in low- and middle-income countries (Murray et al., 2013), have demonstrated steeper trajectories of improvement compared to traditional manualized care (Weisz et al., 2012). Of course, decisions regarding which modules to administer have traditionally been based on DSM symptom presentations (e.g., Chorpita & Weisz, 2005).

However, given that transdiagnostic interventions often consist of multiple components designed to target the same core vulnerability (as described above), it is possible that personalized, modular delivery may be applicable; some skills may be more or less robust at target engagement for a particular patient. Despite the fact that personalization of psychological interventions has been delineated as an NIMH strategic priority (Insel et al., 2010), the question remains with regard to how best to select or order modules from any treatment package, transdiagnostic or otherwise. Several potential strategies have been articulated. For example, baseline evaluation of relative strengths and deficits in skills could be used to individualize their sequence (Cheavens et al., 2012). A capitalization model refers to prioritizing the presentation of skills according to a patient's relative strengths, while a compensation model involves prioritizing areas of greatest weakness. Most of the evidence cited to support these personalization strategies has come from post-hoc examination of pretreatment characteristics that predict differential response to interventions in randomized trials and the literature is mixed with regard to the most advantageous approach (see Simon & Perlis, 2010). Recently, though, Cheavens and colleagues (2012) randomly assigned depressed individuals to receive treatment strategies based on prioritizing strengths or weaknesses and observed steeper trajectories of change on depressive symptoms for patients in the capitalization condition.

Another approach to personalizing the delivery of psychological interventions is planning treatment through shared decision making (SDM). SDM is a collaborative process through which therapists and patients discuss how care should progress based on the available research evidence and the patient's values and preferences (Elwyn, Edwards, & Thompson, 2016). SDM is theoretically grounded in the judgment and decision-making literature (Fagerlin et al., 2012; Weber & Johnson, 2009), and has recently been adapted for use in healthcare settings (Charles, Gafni, & Whelan, 1997). Use of SDM is consistent with policy imperatives that promote patient autonomy and person-centered care (Patient Protection and Affordable Care Act, 42 U.S.C. § 18001 [2010]), and it is especially useful in situations in which there is limited evidence regarding which skills to prioritize.

Finally, advances in technological and statistical techniques (e.g., machine learning; Kessler et al., 2016) may also provide useful information with which to personalize treatment. Continued understanding of individual predictors of treatment outcomes should allow for adaptive interventions that may even utilize computer programming to far more efficiently and accurately individualize treatments for specific patients. For example, Kelly and colleagues (2012) describe using machine learning to create intelligent real-time therapy in which treatment packages are tailored for maximum attenuation of pathological pathways. In addition, see Chapter 34 for more information this topic.

SUMMARY AND CONCLUSIONS

Over the past 40 years, treatments for mental health diagnoses have come a long way, in large part due to the publication of DSM-III (APA, 1980), which allowed researchers to move from broad nonempirical diagnoses to objective, reliably assessed symptom sets. This advancement spurred a proliferation of treatment development focused on each discrete DSM diagnosis, along with clinical trials demonstrating that associated symptoms decrease in response to these interventions. Despite this progress, there have been some challenges. Specifically, dissemination of evidence-based protocols for mental health disorders developed in academic settings has been quite slow; in fact, estimates suggest that only 30 percent of treatment seekers actually receive empirically based interventions (Thornicroft, 2007; Wang et al., 2007). Additionally, even when recommended care is provided, all patients do not respond satisfactorily (Barlow, 2008; Hollon et al., 2002).

Emerging research suggests that although categorical classification, exemplified by the DSM approach, enhanced reliability of mental health diagnoses, it may have come at the expense of validity (see Barlow et al., 2014). Instead of representing true, unique entities, studies examining the latent structure of related disorders suggest that they can be better accounted for by higher-order processes (e.g., Brown & Barlow, 2009). Despite this evidence, the field has continued to largely subscribe to a symptom-focused approach, with different treatment protocols developed to address each discrete DSM diagnosis (McHugh, Murray, & Barlow, 2009). High rates of comorbidity (Kessler et al., 1996, 1998), likely due to shared underlying processes, complicate care and gold-standard treatment protocols for a given diagnosis are often not equipped to handle co-occurring conditions (McHugh et al., 2009).

The next generation of treatment development for mental health conditions must shift away from its focus on discrete DSM diagnoses. Instead, transdiagnostic treatments that address shared mechanisms important in the development and maintenance of a range of disorders represent a more efficient, and ultimately disseminable approach. Continued identification of core transdiagnostic processes that can become the focus of treatment is an important step toward this goal. Next, the ability to reliably assess these mechanisms is necessary to ensure that the treatments developed to target them indeed lead to improvements in specified processes. It is also important to confirm that individual treatment components, in addition to entire intervention protocols, engage these transdiagnostic targets with the goal of including only active ingredients in the package. Once treatments addressing core processes are developed, it is essential to consider clinical trial methodology appropriate for testing their efficacy, including the use of repeated measurement to ensure mechanisms indeed precede symptom improvement, inclusion criteria based on psychopathological processes (instead of symptoms), and equivalence trials demonstrate that transdiagnostic approaches are just as effective as gold-standard single-disorder protocols. Finally, in recent decades, research has focused on the development of one-size-fits-all treatment protocols; although the resulting interventions have helped many, not all individuals respond. Future research should emphasize not only the identification of efficacious treatment components, but also on strategies to personalize their delivery so that a wider range of patients benefits.

REFERENCES

American Psychiatric Association (APA). (1980). *Diagnostic and Statistical Manual of Mental Disorders* (3rd edn.). Washington, DC: APA.

American Psychiatric Association (APA). (2000). *Diagnostic and Statistical Manual of Mental Disorders* (4th edn., text rev.). Washington, DC: APA.

American Psychiatric Association (APA). (2013). *Diagnostic and Statistical Manual of Mental Disorders* (5th edn.). Washington, DC: APA.

Barlow, D. H. (1991). Disorders of Emotion. *Psychological Inquiry, 2*, 58–71.

Barlow, D. H. (1996). The Effectiveness of Psychotherapy: Science and Policy. *Clinical Psychology: Science and Practice, 3*, 236–240.

Barlow, D. H. (2004). Psychological Treatments. *American Psychologist, 59*, 869–878.

Barlow, D. H. (2008). *Clinical Handbook of Psychological Disorders: A Step-by-Step Treatment Manual* (4th edn.). London: Oxford University Press.

Barlow, D. H., & Craske, M. G. (2007). *Mastery of Your Anxiety and Panic* (4th edn.) New York: Oxford University Press.

Barlow, D. H., Cohen, A. S., Waddell, M. T., Vermilyea, B. B., Klosko, J. S., Blanchard, E. B., & DiNardo, P. A. (1984). Panic and Generalized Anxiety Disorders: Nature and Treatment. *Behavior Therapy, 15*, 431–449.

Barlow, D. H., Craske, M. G., Cerny, J. A., & Klosko, J. S. (1989). Behavioral Treatment of Panic Disorder. *Behavior Therapy, 20* (2), 261–282.

Barlow, D. H., Gorman, J. M., Shear, M. K., & Woods, S. W. (2000). Cognitive-Behavioral Therapy, Imipramine, or Their Combination for Panic Disorder: A Randomized Controlled Trial. *Journal of the American Medical Association, 283*, 2529–2536.

Barlow, D. H., Nock, M., & Hersen, M. (2008). *Single Case Research Designs: Strategies for Studying Behavior Change*. New York: Allyn & Bacon.

Barlow, D. H., Sauer-Zavala, S., Carl, J. R., Bullis, J. R., & Ellard, K. E. (2014). The Nature, Diagnosis, and Treatment of Neuroticism: Back to the Future. *Clinical Psychological Science, 2*(3), 344–365.

Barlow, D. H., Farchione, T. J., Bullis, J. R., Gallagher, M. W., Murray-Latin, H., Sauer-Zavala, S., ... Cassiello-Robbins, C. (2017). The Unified Protocol for Transdiagnostic Treatment of Emotional Disorders Compared with Diagnosis-Specific Protocols for Anxiety Disorders: A Randomized Clinical Trial. *JAMA Psychiatry, 74*(9), 875–884.

Barlow, D. H., Sauer-Zavala, S., Farchione, T. J., Latin, H., Ellard, K. K., Bullis, J. R., ... Cassiello-Robbins, C. (2018). *Unified Protocol for Transdiagnostic Treatment of Emotional Disorders: Patient Workbook*. New York: Oxford University Press.

Boswell, J. F., Farchione, T. J., Sauer-Zavala, S., Murray, H. W., Fortune, M. R., & Barlow, D. H. (2013). Anxiety Sensitivity and Interoceptive Exposure: A Transdiagnostic Construct and Change Strategy. *Behavior Therapy, 44*(3), 417–431.

Boswell, J. F., Anderson, L. M., & Barlow, D. H. (2014). An Idiographic Analysis of Change Processes in the Unified Transdiagnostic Treatment of Depression. *Journal of Consulting and Clinical Psychology, 82*(6), 1060–1071.

Brake, C. A., Sauer-Zavala, S., Boswell, J. F., Gallagher, M. W., Farchione, T. J., & Barlow, D. H. (2016). Mindfulness-Based Exposure Strategies as a Transdiagnostic Mechanism of Change: An Exploratory Alternating Treatment Design. *Behavior Therapy, 47*(2), 225–238.

Brown, T. A. & Barlow, D. H. (2009). A Proposal for a Dimensional Classification System Based on the Shared Features of the DSM-IV Anxiety and Mood Disorders: Implications for Assessment and Treatment. *Psychological Assessment, 21*(3), 256–271.

Brown, T. A., Chorpita, B. F. & Barlow, D. H. (1998). Structural Relationships among Dimensions of the DSM-IV Anxiety and Mood Disorders and Dimensions of Negative Affect, Positive Affect, and Autonomic Arousal. *Journal of Abnormal Psychology, 107*, 179–192.

Brown, T. A., Campbell, L. A., Lehman, C. L., Grisham, J. R., & Mancill, R. B. (2001). Current and Lifetime Comorbidity of the DSM-IV Anxiety and Mood Disorders in a Large Clinical Sample. *Journal of Abnormal Psychology, 110*, 49–58.

Charles, C., Gafni, A., & Whelan, T. (1997). Shared Decision-Making in the Medical Encounter: What Does It Mean? (Or It Takes at Least Two to Tango). *Social Science & Medicine (1982), 44*(5), 681–692.

Cheavens, J. S., Strunk, D. R., Lazarus, S. A., & Goldstein, L. A. (2012). The Compensation and Capitalization Models: A Test of Two Approaches to Individualizing the Treatment of Depression. *Behaviour Research and Therapy, 50*(11), 699–706.

Chorpita, B. F., & Weisz, J. R. (2005). *Modular Approach to Therapy for Children with Anxiety, Depression, or Conduct Problems*. Honolulu: University of Hawaii at Manoa; Boston, MA: Harvard Medical School.

Chorpita, B. F., Taylor, A. A., Francis, S. E., Moffitt, C. E., & Austin, A. A. (2004). Efficacy of Modular Cognitive Behavior Therapy for Childhood Anxiety Disorders. *Behavior Therapy, 35*, 263–287.

Chorpita, B. F., Daleiden, E., & Weisz, J. (2005). Identifying and Selecting Common Elements of Evidence-Based Treatment: A Distillation and Matching Model. *Mental Health Services Research, 7*, 5–20.

Craske, M., & Barlow, D. H. (2006). *Mastery of Your Anxiety and Worry* (2nd edn.). New York: Oxford University Press.

Craske, M. G., Farchione, T. J., Allen, L. B., Barrios, V., Stoyanova, M., & Rose, R. (2007). Cognitive Behavioral Therapy for Panic Disorder and Comorbidity: More of the Same or Less of More? *Behaviour Research and Therapy, 45*(6), 1095–1109.

Ellard, K. K., Fairholme, C. P., Boisseau, C. L., Farchione, T. J., & Barlow, D. H. (2010). Unified Protocol for the Transdiagnostic Treatment of Emotional Disorders: Protocol Development and Initial Outcome Data. *Cognitive and Behavioral Practice, 17*(1), 88–101.

Elwyn, G., Edwards, A., & Thompson, R. (Eds.). (2016). *Shared Decision-Making in Health Care: Achieving Evidence-Based Patient Choice* (3rd edn.). Oxford: Oxford University Press.

Fagerlin, A., Pignone, M., Abhyankar, P., Col, N., Feldman-Stewart, D., Gavaruzzi, T., ... Witteman, H. O. (2013). Clarifying Values: An Updated Review. *BMC Medical Informatics and Decision Making, 13*(2), 2–8.

Fairburn, C. (2008). *Cognitive Behavioral Therapy and Eating Disorders*. New York: Guilford Press.

Fairburn, C. G., Peveler, R. C., Jones, R., Hope, R. A., & Doll, H. A. (1993). Predictors of 12 Month Outcome in Bulimia Nervosa and the Influence of Attitudes to Shape and Weight. *Journal of Consulting and Clinical Psychology, 61*(4), 696–698.

Fairburn, C., Cooper, Z., & Shafran, R. (2003). Cognitive Behaviour Therapy for Eating Disorders: A "Transdiagnostic" Theory and Treatment. *Behaviour Research and Therapy, 41*(5), 509–528.

Fairburn, C.G., Cooper, Z. Doll, H. A., O'Connor, M. E., Bohn, K., Hawker, D. M., ... Palmer, R. L. (2009). Transdiagnostic Cognitive-Behavioral Therapy for Patients with Eating Disorders: A Two-Site Trial with 60-Week Follow-Up. *American Journal of Psychiatry, 166*, 311–319.

Farchione, T. J., Fairholme, C. P., Ellard, K. K., Boisseau, C. L., Thompson-Hollands, J., Carl, J. R., ... Barlow, D. H. (2012). Unified Protocol for Transdiagnostic Treatment of Emotional Disorders: A Randomized Controlled Trial. *Behavior Therapy, 43*(3), 666–678.

Foa, E. B., Yadin, E., & Lichner, T. K. (2008). *Obsessive-Compulsive Disorder: A Cognitive Behavioral Therapy Approach*. New York: Oxford University Press.

Gibbons, C. J., & DeRubeis, R. J. (2008). Anxiety Symptom Focus in Sessions of Cognitive Therapy for Depression. *Behavior Therapy, 39*, 117–125.

Greene, C. J., Morland, L. A., Durkalski, V. L., & Frueh, B. C. (2008). Noninferiority and Equivalence Designs: Issues and Implications for Mental Health Research. *Journal of Traumatic Stress, 21*(5), 433–439.

Harvey, A., Murray, G., Chandler, R., & Soehner, A. (2011). Sleep Disturbance as Transdiagnostic: Consideration of Neurobiological Mechanisms. *Clinical Psychology Review, 31*, 225–235.

Heimberg, R. G., Liebowitz, M. R., & Hope, D. A. (1998). Cognitive Behavioral Group Therapy versus Phenelzine Therapy for Social Phobia: 12-Week Outcome. *Archives of General Psychiatry, 55*, 1133–1141.

Hollon, S. D., Thase, M. E., & Markowitz, J. C. (2002). Treatment and Prevention of Depression. *Psychological Science in the Public Interest, 3*, 39–77.

Hope, D. A., Heimberg, R. G., & Turk, C. L. (2006). *Managing Social Anxiety: A Cognitive-Behavioral Therapy Approach: Therapist Guide*. San Antonio, TX: Oxford University Press.

Hopwood, C. J., Thomas, K. M., Markon, K. E., Wright, A. G., & Krueger, R. F. (2012). DSM-5 Personality Traits and DSM-IV Personality Disorders. *Journal of Abnormal Psychology, 121*(2), 424–432.

Insel, T. R., & Gogtay, N. (2014). National Institute of Mental Health Clinical Trials: New Opportunities, New Expectations. *JAMA Psychiatry, 71*(7), 745–746.

Insel, T., Cuthbert, B., Garvey, M., Heinssen, R., Pine, D. S., Quinn, K., ... & Wang, P. (2010). Research Domain Criteria (RDoC): Toward a New Classification Framework for Research on Mental Disorders. *American Journal of Psychiatry, 167*(7), 748–751.

Kazdin, A. E. (2007). Mediators and Mechanisms of Change in Psychotherapy Research. *Annual Review of Clinical Psychology, 3*(1), 1–27.

Kelly, J., Gooding, P., Pratt, D., Ainsworth, J., Welford, M., & Tarrier, N. (2012). Intelligent Real-Time Therapy: Harnessing the Power of Machine Learning to Optimise the Delivery of Momentary Cognitive-Behavioural Interventions. *Journal of Mental Health, 21*(4), 404–414.

Kessler, R. C., Nelson, C. B., McGonagle, K. A., Liu, J., Swartz, M., & Blazer, D. G. (1996). Comorbidity of DSM-III-R Major Depressive Disorder in the General Population: Results from the US National Comorbidity Survey. *British Journal of Psychiatry. Supplement,* (30), 17–30.

Kessler, R. C., Stang, P. E., Wittchen, H. U., Ustun, T. B., Roy-Burne, P. P., & Walters, E. E. (1998). Lifetime Panic-Depression Comorbidity in the National Comorbidity Survey. *Archives of General Psychiatry, 55*(9), 801–808.

Kessler, R. C., van Loo, H. M., Wardenaar, K. J., Bossarte, R. M., Brenner, L. A., Cai, T., ... & Nierenberg, A. A. (2016). Testing a Machine-Learning Algorithm to Predict the Persistence and Severity of Major Depressive Disorder from Baseline Self-Reports. *Molecular Psychiatry, 21*, 1366–1371.

Linehan, M. (1993). *Cognitive-Behavioral Treatment of Borderline Personality Disorder*. New York: Guilford Press.

Mansell, W., Harvey, A., Watkins, E., & Shafran, R. (2009). Conceptual Foundations of the Transdiagnostic Approach to CBT. *Journal of Cognitive Psychotherapy, 23*(1), 6–19.

McHugh, R. K., Murray, H. W., & Barlow, D. H. (2009). Balancing Fidelity and Adaptation in the Dissemination of Empirically-Supported Treatments: The Promise of Transdiagnostic Interventions. *Behaviour Research and Therapy, 47*(11), 946–953.

Murray, L. K., Dorsey, S., Haroz, E., Lee, C., Alsiary, M. M., Haydary, A., ... & Bolton, P. (2013). A Common Elements Treatment Approach for Adult Mental Health Problems in Low-and Middle-Income Countries. *Cognitive and Behavioral Practice, 21*(2), 111–123.

Nathan, P. E., & Gorman, J. M. (Eds.). (2007). *A Guide to Treatments That Work* (3rd edn.). London: Oxford University Press.

Purdon, C. (1999). Thought Suppression and Psychopathology. *Behaviour Research and Therapy, 37*(11), 1029–1054.

Rapee, R. M., Kennedy, S., Ingram, M., Edwards, S., & Sweeney, L. (2005). Prevention and Early Intervention of Anxiety Disorders in Inhibited Preschool Children. *Journal of Consulting and Clinical Psychology, 73*(3), 488–497.

Rosellini, A. J., Boettcher, H., Brown, T. A., & Barlow, D. H. (2015) A Transdiagnostic Temperament-Phenotype Profile Approach to Emotional Disorder Classification: An Update. *Psychopathological Review, 2,* 110–128.

Rosellini, A. J., Lawrence, A., Meyer, J., & Brown, T. A. (2010). The Effects of Extraverted Temperament on Agoraphobia in Panic Disorder. *Journal of Abnormal Psychology, 119,* 420–426.

Sauer-Zavala, S., Boswell, J. F., Gallagher, M. W., Bentley, K. H., Ametaj, A., & Barlow, D. H. (2012). The Role of Negative Affectivity and Negative Reactivity to Emotions in Predicting Outcomes in the Unified Protocol for the Transdiagnostic Treatment of Emotional Disorders. *Behaviour Research and Therapy, 50*(9), 551–557.

Sauer-Zavala, S., Bentley, K. H., & Wilner, J. G. (2016). Transdiagnostic Treatment of Borderline Personality Disorder and Comorbid Disorders: A Clinical Replication Series. *Journal of Personality Disorders, 30*(1), 35–51.

Sauer-Zavala, S., Cassiello-Robbins, C., Conklin, L., Bullis, J. R., Thompson-Hollands, J., & Kennedy, K. (2017a). Isolating the Unique Effects of the Unified Protocol Treatment Modules Using Single-Case Experimental Design. *Behavior Modification, 40,* 286–307.

Sauer-Zavala, S., Gutner, C., Farchione, T. J., Boettcher, H. T., Bullis, J. R., & Barlow, D. H. (2017b). Current Definitions of "Transdiagnostic" in Treatment Development: A Search for Consensus. *Behavior Therapy, 48*(1), 128–138.

Sauer-Zavala, S., Wilner, J. G., Cassiello-Robbins, C., Saraff, P., & Pagan, D. S. (2019). Opposite Action in Borderline Personality Disorder: A Laboratory-based Alternating Treatment Design. *Behaviour Research and Therapy, 117,* 79–86.

Simon, G. E., & Perlis, R. H. (2010). Personalized Medicine for Depression: Can We Match Patients with Treatments? *American Journal of Psychiatry, 167*(12), 1445–1455.

Thornicroft, G. (2007). Most People with Mental Illness Are not Treated. *The Lancet, 370*(9590), 807–808.

Wang, P. S., Aguilar-Gaxiola, S., Alonso, J., Angermeyer, M. C., Borges, G., Bromet, E. J., ... & Haro, J. M. (2007). Use of Mental Health Services for Anxiety, Mood, and Substance Disorders in 17 Countries in the WHO World Mental Health Surveys. *The Lancet, 370*(9590), 841–850.

Wenzlaff, R. M., & Wegner, D. M. (2000). Thought Suppression. *Annual Review of Psychology, 51*(1), 59–91.

Widiger, T. A., & Crego, C. (2013). Diagnosis and Classification. In I. B. Weiner, G. Stricker, & T. A. Widiger (Eds.), *Handbook of Psychology: Clinical Psychology* (2nd edn., pp. 3–18). Hoboken, NJ: Wiley.

Widiger, T. A., & Krueger, R. F. (2013). Personality Disorders in the DSM-5: Current Status, Lessons Learned, and Future Challenges. *Personality Disorders: Theory, Research, & Treatment, 4,* 341.

Widiger, T. A., & Trull, T. J. (2007). Plate Tectonics in the Classification of Personality Disorder: Shifting to a Dimensional Model. *American Psychologist, 62*(2), 71–83.

Wilson, G. T., Fairburn, C. C., Agras, W. S., Walsh, B. T., & Kraemer, H. (2002). Cognitive-Behavioral Therapy for Bulimia Nervosa: Time Course and Mechanisms of Change. *Journal of Consulting and Clinical Psychology, 70*(2), 267–274.

Weisz, J. R. (2012). Testing Standard and Modular Designs for Psychotherapy Treating Depression, Anxiety, and Conduct Problems in Youth: A Randomized Effectiveness Trial. *Archives of General Psychiatry, 69*(3), 274–282.

Weber, E. U., & Johnson, E. J. (2009). Mindful Judgment and Decision Making. *Annual Review of Psychology, 60*(1), 53–85.

Yadin, E., Foa, E., & Lichner, T. (2012). *Treating Your OCD with Exposure and Response (Ritual) Prevention* (2nd edn.). New York: Oxford University Press.

Zinbarg, R., Craske, M., & Barlow, D. H. (2006). *Therapist's Guide for the Mastery of Your Anxiety and Worry Program*. New York: Oxford University Press.

22 Health Psychology and Behavioral Medicine

Methodological Issues in the Study of Psychosocial Influences on Disease

TIMOTHY W. SMITH

The intellectual and scientific roots of behavioral medicine and health psychology run deep into the history of biomedical and behavioral science (Friedman & Adler, 2011), but their formal histories are relatively brief. Behavioral medicine, defined as the "interdisciplinary field concerned with the development and integration of behavioral, psychological, and biomedical science knowledge and techniques relevant to the understanding of health and illness, and the application of this knowledge and these techniques to prevention, diagnosis, treatment, and rehabilitation" (Schwartz & Weiss, 1978, p. 250), was established in the late 1970s. Though defined similarly (Matarazzo, 1980) and emerging during the same period, the single-discipline emphasis places health psychology within behavioral medicine (Larkin & Klonoff, 2014). The specialty of clinical health psychology emerged more recently, with a specific focus on applied research and professional services at the interface of psychology and medicine (France et al., 2008).

The overlapping scientific focus of behavioral medicine and health psychology reflects their emergence from the mid-twentieth-century sentiment that the traditional biomedical model guiding research and practice lacked adequate attention to behavior and other psychosocial processes (Engel, 1977). Strong multi- and interdisciplinary research methods and a critical scientific approach have been central throughout the history of health psychology and behavioral medicine. This has meshed well with the emergence of the evidence-based model in medicine (Sackett et al., 1996), an approach explicitly adopted in both health psychology and behavioral medicine (Davidson et al., 2004; Larkin & Klonoff, 2014). Both fields have endorsed statements in biomedical science regarding reporting standards for randomized clinical trials for non-pharmacological interventions (Boutron et al., 2008), observational studies (Von Elm et al., 2007), and systematic reviews, meta-analyses, and metasyntheses (Moher et al., 2009). The maturing of the evidence base has placed increased importance on high-quality systematic reviews and meta-analyses of accumulated research (Siddaway, Wood, & Hedges, 2019).

This chapter provides an overview of methodological issues in behavioral medicine and health psychology, with an emphasis on psychosocial influences on physical health and the psychobiologic mechanisms underlying such effects. Given the interdisciplinary nature of the field and the breadth of the basic and applied research agenda, a more complete review is not possible. However, a basic framework for understanding the field's structure, conceptual foundations, and examples of key methodological challenges are provided as a guide to the application of issues discussed elsewhere in this volume to the context of physical health.

FOUNDATIONS AND SCOPE

The Biopsychosocial Model: Integration across Levels of Analysis

The biopsychosocial (BPS) model guides both research and healthcare services (Engel, 1977, 1980), and the framework is central in behavioral medicine and health psychology. The BPS model (see Figure 22.1) is based in systems theory (Von Bertalanffy, 1968), and holds that health and disease reflect reciprocal influences among biological, psychosocial, and social-cultural processes. Hierarchically arranged levels of analysis range from cells and tissues to whole persons, their personal relationships, groups, and cultures.

There are conceptual perspectives and research methods associated with each level of analysis, and the BPS framework requires their integration, as each level is seen as influencing and influenced by adjacent levels. For example, current research examines how the individual's physiological stress responses are reciprocally related to underlying brain processes (Gianaros et al., 2017), and how an individual's emotional adjustment is reciprocally related to the quality of intimate relationships, as closely interrelated influences on physical health (Smith & Baucom, 2017). At higher levels of the BPS model, individuals' health behaviors (e.g., physical activity levels, diet) and

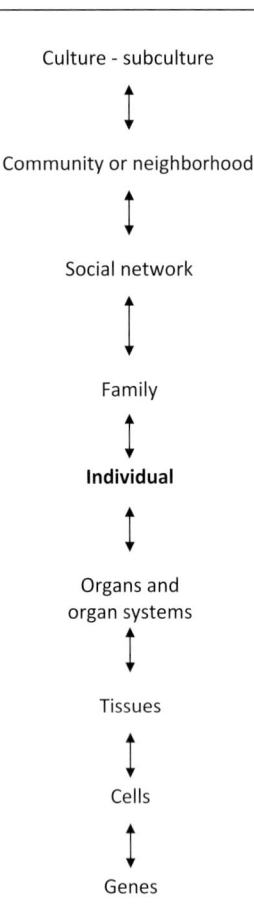

Culture - subculture
↕
Community or neighborhood
↕
Social network
↕
Family
↕
Individual
↕
Organs and organ systems
↕
Tissues
↕
Cells
↕
Genes

Figure 22.1 Systems hierarchy in the biopsychosocial model (based on Engel, 1977)

physiological stress responses are influenced by characteristics of the places where they live and work (Diez Roux & Mair, 2010; Karmeniemi et al., 2018). Hence, the concepts, methods, and findings of a wide range of scientific disciplines, up and down the levels of analysis, are relevant to researchers in health psychology and behavioral medicine, requiring considerable methodological breadth for individual researchers and interdisciplinary teams.

Evolving Views of Health: Endpoints and Outcomes

Earlier in the history of health psychology and behavioral medicine, methods for measuring health varied widely and were often uncritically interpreted as reflecting underlying disease processes. A common example was the use of self-report physical symptom checklists as an indication of actual physical health, an attractive measurement approach given its low cost. Certainly, symptom reports contain substantial variance reflecting underlying disease, and other inexpensive self-report subjective health measures (e.g., self-rated health, subjective age) predict objective outcomes such as longevity even when controlling multiple biological markers of health (Idler & Benyamini, 1997; Stephan, Sutin, & Terracciano, 2018b). However, symptom reports and other measures of illness behaviors (i.e., things that people do that typically reflect illness) also contain systematic variance reflecting other constructs, as exemplified by the excessive somatic complaints that characterize hypochondriasis and related aspects of personality and emotional adjustment (Smith, 2011a). That is, illness behavior does not necessarily reflect illness itself. Further, statistical associations of psychosocial risk factors (e.g., personality traits, emotional adjustment, aspects of personal relationships) most commonly measured through self-reports with health outcomes also measured through self-reports could be inflated by common method variance (Podsakoff, MacKenzie, & Podsakoff, 2012), rather than reflecting the associations with underlying health versus disease.

These ambiguities meant that many early studies often had limited impact, and drove the fields' increasing use of more compelling health outcomes, such as longevity, mortality, and objectively assessed presence and severity of specific diseases. Robust behavioral and psychosocial predictors of higher rates of mortality from all causes (i.e., reduced longevity) have obvious importance and occupy a central place in the evidence base. However, associations with all-cause mortality and longevity also raise questions about specific causes of death and the mechanisms through which psychosocial factors might have influenced some subset of health conditions captured in all-cause mortality and similar outcomes. Often, researchers are interested in a specific disease, and questions about mechanisms linking psychosocial factors with specific diseases are highly relevant to the biological plausibility of statistical associations.

More routine use of objective measures of health and disease has increased the impact and standing of behavioral medicine and health psychology. However, physical health outcomes are not fully captured by the presence and severity of objectively documented disease. Health-related quality of life (Passhier & Busschbach, 2015; Ryff & Singer, 1998) goes beyond the presence of specific medical diagnoses to include aspects of experience (e.g., pain, other subjective symptoms) and functioning (e.g., mobility, activity, employment, recreation). Psychosocial factors can predict – and related interventions can enhance – these broader aspects of health. Health-related quality of life is especially important for medical conditions where subjective symptoms and levels of functioning are a primary concern, such as in chronic pain (Dixon, 2016). Thus, after an initial period of disillusionment with subjective health outcomes and other illness behaviors, growing interest in broader conceptual models of health has encouraged their more sophisticated incorporation.

Given challenges posed by rising healthcare expenditures, behavioral medicine and health psychology increasingly examine costs associated with healthcare utilization and intervention outcomes relative to the costs required to produce them, as a guide to healthcare policy decisions (Kaplan & Groessl, 2002). For example, providing mental health treatment to patients with co-occurring chronic medical illness and major depression is associated with substantially lower healthcare costs (Mausbach, Bos, & Irwin, 2018; Simon et al., 2007).

Three Interrelated Domains

Behavioral medicine and health psychology include three major, interrelated topics. The first examines *health behavior and risk reduction*, specifically the identification of modifiable behaviors that predict the development and course of disease, the determinants of those behaviors, and the efficacy, effectiveness, and health benefits of related behavior change interventions. Smoking, diet, physical activity, sleep, and various preventive actions (e.g., sunscreen and seatbelt use, participation in medical screening exams, adherence to prescribed medications) are widely studied examples, given their robust associations with common and expensive sources of morbidity and mortality, as well as the benefits of modifying these behaviors in the prevention and management of disease (Fisher et al., 2011).

The second topic involves *psychosocial aspects of medical illness and medical care*. Research in this area examines the impact of various medical conditions and related treatment on functioning (e.g., emotional adjustment, social relationships, functional activity vs. disability, pain, and other subjective symptoms), as well as the benefits of adjunctive psychosocial interventions as additions to routine medical and surgical care (Helgeson & Zajdel, 2017). This topic overlaps with the first, as health behavior change (e.g., smoking cessation, exercise, dietary change) is a key component in the medical management of many conditions.

The final and perhaps oldest topic – *psychosocial and psychobiological influences on disease* – examines the effect of psychological stress and related aspects of personality, emotional adjustment, and social context on the development and course of disease, through the general mechanism of cumulative effects of physiological stress responses in the pathophysiology of disease. This topic is also a central focus of the older and highly active field of psychosomatic medicine. Psychosocial epidemiology is the portion of this topic that identifies associations of aspects of personality, emotional adjustment, and social relationships and contexts with the onset and course of disease. Psychophysiological research examines associations of those risk factors with neuroendocrine, cardiovascular, and immune system physiology related to the onset and progression of disease. This topic overlaps with the previous two, given that stress can influence health behavior, producing a less direct effect of psychosocial risk factors on physical illness, and stress reduction interventions are often useful in the treatment for chronic medical illness.

The remainder of this chapter emphasizes methodological issues in this third central topic. A more complete discussion of methodological issues across all three major topics is beyond our limited scope. However, many of the issues relevant to research on psychosocial and psychobiological influences on disease are central in the other areas as well.

THE PSYCHOBIOLOGY OF STRESS AND DISEASE

Psychosocial Epidemiology

Low socioeconomic status is among the best established of the psychosocial predictors of reduced longevity and heightened risk of morbidity and mortality from specific causes (Adler, 2009). The presence and quality of close personal relationships are similarly important, as social isolation, loneliness, low social support, and interpersonal conflict all predict the onset and course of major diseases and reduced longevity (Holt-Lunstad, 2018; Robles et al., 2014). Chronic work stress also predicts serious illness, especially cardiovascular disease (Eller et al., 2009; Kivimaki et al., 2013). Among aspects of personality and emotional adjustment, anger and hostility (Chida & Steptoe, 2009), depression (de Miranda Azevedo et al., 2016), anxiety (Celano et al., 2015; Roest et al., 2010), and posttraumatic stress disorder (Edmondson et al., 2013a) predict the development and course of cardiovascular disease, and many of these psychosocial characteristics similarly predict the development and/or course of diabetes (Chida & Hamer, 2008b; Kelly & Ismail, 2015; Knol et al., 2006), cancer (Chida et al., 2008; Lutgendorf & Andersen, 2015), and HIV disease (Chida & Vedhara, 2009). Conversely, several positive psychosocial characteristics such as optimism, positive affectivity, and meaning in life predict all-cause mortality, longevity, and the development and course of major diseases (Pressman, Jenkins, & Moskowitz, 2019).

In typical psychosocial epidemiological studies, the association of a given risk or protective factor with the health endpoint is tested in logistic regression models or survival analyses, controlling for demographic variables (e.g., age, sex, race/ethnicity, education income) and biomedical risk variables (e.g., obesity, blood lipids) in an effort to establish risk independent of these potential third variables. Often, health behaviors (e.g., smoking, physical activity) are added to the predictive model, to determine whether or not these established behavioral risk variables account for effects of psychosocial factors. If not, then it is presumed that psychobiological mechanisms discussed below are implicated in the association. Importantly, it is possible that health behaviors and biological risk factors could be seen as mechanisms accounting for associations of emotional, personality, or social relationship characteristics with subsequent health, rather than confounds to be controlled (e.g., Turiano et al., 2015).

Despite robust evidence, psychosocial epidemiological research on initially healthy samples (i.e., studies of disease onset) and individuals with preexisting disease (i.e., clinical epidemiological studies of disease course) faces challenges in conceptualization, measurement, and analysis of these risk factors. A major issue involves the treatment of correlated aspects of psychosocial risk, both within and across categories of such factors. Within the category of individual-level characteristics such as personality traits and aspects of emotional adjustment, risk factors are often considered separately, despite instances of well-established associations. For example, symptoms of anxiety and depression are often discussed and analyzed as if they are fully distinct constructs, despite their well-established, close association as personality traits (i.e., facets of neuroticism), symptoms of emotional distress, and comorbid diagnoses (Smith, 2010). Thus, it is highly likely that rather than reflecting two fully separate psychosocial influences on disease, anxiety and depression could have overlapping effects. The same is true for many of the positive (i.e., protective) and negative psychosocial factors described above.

Similarly, trait anger, hostility, depressive symptoms, PTSD, optimism and other aspects of individual functioning are typically conceptualized and analyzed as a distinct class of risk factors, without consideration of their robust association with the quality close relationships and other aspects of social functioning known to predict important health outcomes (Smith & Baucom, 2017; Smith et al., 2013). That is, within and across the individual level and social or interpersonal level of psychosocial risk, the modal approach to describing, measuring, analyzing, interpreting, and combining characteristics for systematic review and meta-analysis has been a default piecemeal model. Natural aggregations within and across individual and social levels of risk have typically been ignored.

When these aggregations or correlated risks are not simply ignored, their possibly overlapping effects of multiple risk factors are often addressed in the traditional epidemiological approach of evaluating independent risk through statistical control, typically treating one risk factor as primary and testing its independent effects through statistical control of correlated or "confounding" risks, similar to demographic, biomedical, or health behavior variables as described above. This approach has well-established limitations (Christenfeld et al., 2004; Phillips & Davey Smith, 1991). Most obviously, incomplete lists of possible confounds and imperfectly reliable and valid measures of these third variables result in their undercorrection and residual confounding in multivariate analyses, and control of closely related risk factors can produce unstable statistical results.

Less widely recognized is the possibility that statistical control of related variables can create ambiguity in the meaning of the variable of primary interest. The psychometric characteristics (i.e., reliability and construct validity) of that risk factor measure after partialing the overlapping variance with the measure of the confounding factor may differ from the original, unpartialed version (Lynam, Hoyle, & Newman, 2006; Sleep et al., 2017). Yet psychometric properties of the partialed versions of predictors are rarely examined. For example, considered separately, anxiety predicts increased risk for cardiovascular disease. However, one relevant study found that when depressive symptoms were controlled the expected positive association with disease was reversed; uncorrected anxiety symptoms predicted increased risk but anxiety symptoms that were independent of depression (via statistical control) predicted significantly *reduced* risk of disease (Grossardt et al., 2009). Apparently, variance in self-reported anxiety that was independent of depressive symptoms assessed a protective factor (e.g., caution, adaptive worry), compared to the full version of the scale.

Given the known and quite understandable association of trait anger and hostility with marital strain and disruption (Smith & Baucom, 2017), it is not clear what measures of anger and hostility reflect when their shared variance with measures of interpersonal functioning is held constant; they represent effects of such traits if somehow disagreeable individuals enjoyed the same relationship quality as their more warm and even-tempered counterparts, a scenario well known to be false. Thus, the common practice of evaluating independent risk through statistical control can create ambiguous measures and counterfactual conditions, complicating interpretations of many psychosocial risk studies.

One obvious implication of such limitations of statistical control is the importance of interpretive caution. However, it also suggests the value of exploring directly the overlapping and correlated risks as an alternative to forcing their independence. Rather than a nuisance to be managed, the overlapping variance may contain much of the relevant influence on subsequent health. For example, the common core among seemingly distinct negative affective characteristics and among positive protective factors may be the essential ingredient conferring risk or protection. Similarly, processes that account for or emerge from the reciprocal associations between characteristics of individuals and the levels of relationship quality they enjoy or endure could account for their associations with health.

Newer structural models of emotional adjustment and personality reviewed elsewhere in this volume (see Chapter 4, by Williams and Simms, and Chapter 7, by Wright) can provide invaluable conceptual guides to aggregation within the individual-level risk factors, but are not yet widely used in psychosocial epidemiology. Recently developed statistical models can provide direct translations of such hierarchical structures into empirical tests (e.g., de Miranda Azevedo et al., 2016). As an alternative to the more commonly used variable-based approaches to aggregating correlated risk factors (e.g., factor analysis), case-based analytic methods (e.g., cluster analysis, latent class analysis, latent profile analysis) can be used to model the co-occurrence of multiple risks, such as aspects of personality, emotional adjustment, and social functioning (e.g., Gallo & Smith, 1999). Here, conceptual models of the dynamic associations of personality characteristics and aspects of emotional adjustment with interpersonal functioning and social relationships (Pincus & Ansell, 2013; Wilson, Stroud, & Durbin, 2017) can provide a useful framework for overlapping risks (Smith & Baucom, 2017). The traditional focus on evaluation of independent risk is often useful in identifying which elements of correlated risks are most predictive, but this practice can be usefully supplemented by direct examination of naturally overlapping psychosocial risks. The most important influences on disease – and opportunities for risk-reducing intervention – may reside in the overlap. For aggregations of individual-level risk factors, the overlapping variance may provide a more precise description of the characteristic that confers risk or protection. For cross-level aggregations of individual characteristics with relationship or social risk factors, the overlapping variance may describe similarly important interpersonal processes.

Basic issues in measurement of psychosocial risk factors also pose important challenges. The most common approach is self-report scales, which is understandable given the need to minimize respondent burden and protocol complexity in large epidemiological or clinical studies. However, as in research on other important life outcomes (Connelly & Ones, 2010), informant reports and behavioral ratings of personality and emotional adjustment have been found to have better utility in predicting objective health outcomes compared to parallel self-reports (Newman et al., 2011; Smith et al., 2008), perhaps because respondents are often less than fully willing or capable in providing accurate self-description when psychosocial characteristics have high (e.g., optimism, positive affectivity) or low (e.g., depression, anger) social desirability. Similarly, observational assessments of couple interaction patterns have been found to predict objective disease indicators better than self-reports of relationship quality (Smith et al., 2011).

Self-reports are sometimes used to assess psychosocial risk and resilience factors that reflect cognitive abilities. For example, the self-regulatory processes often grouped as the cognitive executive functions are implicated in a wide variety of health contexts (Williams & Thayer, 2009). Although well-established behavioral measures are available, self-report scales of attentional control, behavioral inhibition, and other aspects of self-control are often utilized given their relative ease and low cost in administration. However, self-reports and the more direct behavioral measures of these self-regulatory abilities are minimally correlated (Williams et al., 2017), and such behavioral assessments have been found to have better utility in predicting health outcomes (Emery & Levine, 2017).

Emerging assessment methods have considerable promise in this context. For example, traditional retrospective global self-reports of trait anxiety and anger predict only a small portion of the stable variance in these affective risk factors as assessed via ecological momentary assessments (EMA) (Edmondson et al., 2013b), suggesting that this more intensive approach might capture individual differences in affective risk factors more effectively. Affective reactivity to daily stress assessed through similar daily experience sampling predicts mortality in patients with chronic disease (Chiang et al., 2018). Daily experience assessments of the quality of personal relationships also predict objective measures of disease (e.g., Joseph et al., 2014). If EMA and other daily experience methods, informant reports, and behavioral assessments provide more valid measures of psychosocial risk factors, it is possible that much of the psychosocial epidemiology literature to date provides an underestimate of these influences on health, given the common reliance on more limited self-reports.

Many studies of psychosocial risk utilize measures of personality, emotional adjustment, and interpersonal processes with limited evidence of construct validity. Scales are often interpreted as measuring the construct of interest largely on the basis of scale labels and item content. This creates the potential for what Block (1995) described as the "jingle" and "jangle" fallacies. Scales that "sound alike" (i.e., jingle) are interpreted as measuring the same construct, even in the absence of evidence of convergent validity. This can produce a misleading inconsistency in the overall pattern of statistical associations with health outcomes for a given psychosocial risk factor. The inconsistent results could reflect variations in the validity of measures across studies of supposedly identical constructs, rather than a weak or variable association of the construct with health outcomes. In the jangle fallacy, measures with distinct-sounding labels are interpreted as assessing distinct constructs, even in the absence of evidence of discriminant validity. This error has the potential to lead to an unnecessary and unwieldy proliferation of psychosocial risk factors and a lack of integration in the field, where previously studied risk factors are unwittingly rediscovered as a distinct predictor of health.

The jingle and jangle problems can be largely eliminated by consistent use of a construct validity approach (Strauss & Smith, 2009), including tests of convergent and discriminant validity and the regular use of well-established conceptual and measurement frameworks to identify similarities and differences across risk-factor measures and constructs. The five-factor model can serve this function in the integration of personality risk factors (Smith & Williams, 1992), and the interpersonal circumplex can similarly serve as a point of integration for individual differences and social relationship risk factors (Smith et al., 2010).

Mechanisms

It is quite possible that some of the association of psychosocial risk factors with subsequent health outcomes reflects the intervening effect of health behaviors (e.g., Sutin et al., 2016). Although most of the well-established psychosocial risk factors have effects even when health behaviors are measured and statistically controlled, as noted previously health behavior could still play a role. In most conceptual models underlying psychosocial epidemiological studies, the effects of personality, emotional adjustment, and features of social relationships and contexts are believed to alter physiological stress responses, which in turn influence the development and course of disease. This raises two broad types of psychobiological research – the associations of psychosocial factors with stress responses, and the association of those stress responses with subsequent disease. It is important to note that tests of the full mediational model in which psychosocial risk and protective factors influence disease through the mediational effects of physiological stress responses are very rare. They are certainly possible, but the complexity and scope of such studies needed to produce definitive results makes them far less common than studies of the components of the general model.

A wide variety of physiological stress responses have been implicated in the development and course of physical illnesses, with heightened sympathetic nervous system activation and its neuroendocrine (i.e., release of catecholamines) and cardiovascular consequences (i.e., increases in heart rate and blood pressure), activation of the hypothalamic-pituitary-adrenocortical (HPA) axis (i.e., release of cortisol), changes in the immune system (i.e., suppression of some aspects of cell proliferation and cytotoxicity, increases in inflammatory processes), and decreases in parasympathetic processes (e.g., reduced vagally mediated heart rate variability) among the most widely studied (Lovallo, 2015). There are well-described and empirically grounded procedures for evoking and measuring these components of the stress response (Luecken & Gallo, 2008), and care in following such procedures is essential in obtaining reliable and valid measures of these processes. The general process of fluctuating physiological adaptations to stressors has been labeled *allostasis*, and the cumulative total physiological burden associated with these physiological adaptations is often referred to as *allostatic load* (McEwen, 1998).

Research supports a role for these responses in the development and course of cardiovascular disease (Chida & Steptoe, 2010; Steptoe & Kivimaki, 2013), cancer (Lutgendorf & Anderson, 2015), Type 2 diabetes (Kelly & Ismail, 2015), the progression of HIV disease (Cole, 2008), and other sources of premature morbidity and mortality (Lovallo, 2015). Psychosocial risk factors have been linked to many of the specific components of the stress response (Cacioppo et al., 2015; Chida & Hamer, 2008a; Uchino & Way, 2017).

Methodological issues in the conceptualization and measurement of psychosocial risk factors are also highly relevant in research on their associations with physiological stress responses. Other issues in the conceptualization and measurement of elements of these stress mechanisms are relevant as well. For example, most research has examined the psychosocial correlates and health consequences of the magnitude of various aspects or types of physiological *reactivity* to stressors (i.e., changes from a resting state to levels during stress exposure); a smaller literature has examined the degree of physiological *recovery* from such events (Chida & Hamer, 2008a; Chida & Steptoe, 2010). Reactivity and recovery are most commonly studied in laboratory settings, where standardized stressors are presented in a tightly controlled manner and responses are measured before, during, and after the stressor.

As elements of overall physiological activation or burden, reactivity and recovery are obviously important. However, such laboratory-based studies might actually underestimate the effects of personality, emotional adjustment, the quality of close relationships, and other psychosocial risk factors on allostatic load, because this paradigm relies on what Wachtel (1973) labeled the *implacable experimenter*. Psychosocial risk factors might influence physiological burden not only by moderating the magnitude of reactivity during and recovery following a given stressor; they might also directly reflect or be an influence on the frequency, magnitude, and duration of *exposure* to such stressors. Traditional laboratory reactivity paradigms intentionally limit variations in the experimenter's behavior and other aspects of stress exposure, because the moderating effects of the psychosocial factor on responses to identical stimuli are the specific focus of the research.

But a key reflection or consequence of personality, emotional adjustment, and other types of psychosocial risk factors are effects on the social environment and other recurring contexts (e.g., work demands), rendering them more or less stressful. These risk factors are related to the types of situations individuals experience, the situations they elect to enter or avoid, and their intended or unintended impact on the responses of others. Psychosocial characteristics that confer risk are generally associated with greater stress exposure through such processes; those that confer protection are associated with less. Figure 22.2 depicts overall physiological stress response across these three sources of activation, as well as a more recently identified stress mechanism involving the degree of physiological *restoration* during sleep (Hall, Brindle, & Buysse, 2018; Irwin, 2015). Psychosocial factors including personality, emotional distress, and the quality of personal relationships are all related to sleep duration and/or quality (Kent de Grey et al., 2018; Stephan et al., 2018b), and these aspects of sleep influence the extent to which physiological processes return to resting states and even fall below waking levels (e.g., nocturnal blood pressure dipping).

The three panels of Figure 22.2 represent varying degrees of standing on a given psychosocial risk versus resilience factor (e.g., optimism vs pessimism; agreeableness vs antagonism; high vs low marital quality). The vertical axis represents the extent of physiological

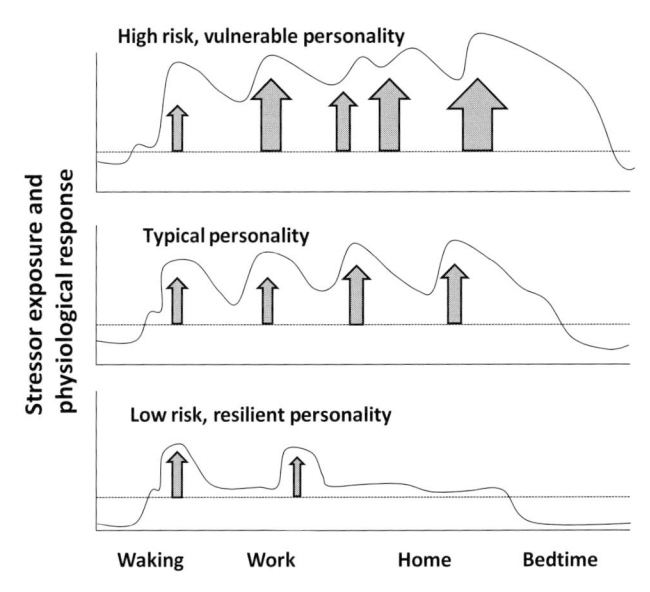

Figure 22.2 Effects of stress exposure, reactivity, recovery, and restoration on physiological response, across levels of psychosocial risk and resilience

activation or response, such as rising and falling levels of heart rate or circulating catecholamine concentrations; the horizontal axis reflects exposure to stressors over the course of a day. In response to an identical stressor experienced by each individual, the higher-risk individual shows both greater reactivity and more limited recovery, whereas the resilient, lower-risk individual shows lower reactivity and better recovery. These sources in overall physiological burden (i.e., area under the curve) are compounded across the day through differences in the frequency (number of stressor "bars"), magnitude (height of bars), and duration (width of bars) of stressors. Importantly, these variations in stress exposure can reflect objective external events (e.g., arguments with co-workers or family members), differing appraisals of similar events, and the mental rehearsal of future and past events, as in the case of worry and rumination, respectively, cognitive processes that reliably evoke physiological activation (Ottaviani et al., 2016). The overall differences in waking physiological activation resulting from exposure, reactivity, and recovery are compounded further by additional differences in physiological restoration during sleep (Williams et al., 2011). The somewhat inconsistent literature on associations between psychosocial risk factors and measures of allostatic load (Wiley, Bei, & Bower, 2017) may reflect the fact that few studies capture both a broad range of physiological mechanisms and the full set of four stress mechanisms that contribute to overall burden (i.e., area under the curve).

These physiological stress mechanisms have also been examined in ambulatory studies in which the occurrence of perceived stressors and related experiences (e.g., negative affect) is measured through EMA or similar daily experience methods while recording some physiological response (e.g., ambulatory blood pressure; Kamarck et al., 2005). This approach permits assessment of exposure, reactivity, recovery, and nighttime restoration. The precision and power of causal inference inherent in experimental manipulation of stress exposure is necessarily traded for reliance on observational design and self-report assessment of stressors, but with the important increase in ecological validity and more complete assessment of the various processes influencing the overall physiological burden. Hence, the laboratory and ambulatory/daily experience approaches are especially valuable as complementary or converging paradigms.

The Value of Adjunctive Psychosocial Interventions

Answers to age-old questions regarding effects of stress, emotion, and social circumstances on disease are inherently valuable. However, the potential for this research to inform efforts to prevent or manage serious illness is a central goal. In the cardiovascular literature, an early major randomized clinical trial showed that group therapy could reduce Type A behavior in heart patients, as well as reduce the rate of recurrent coronary events (Friedman et al., 1986). Subsequent randomized clinical trials (RCTs) of stress management for individuals with cardiovascular disease have replicated these effects (e.g., Gulliksson et al., 2011). RCTs of cognitive behavioral treatments of depression and anxiety in these populations have demonstrated clear benefits for emotional adjustment, but have generally not demonstrated significant effects on cardiac outcomes (Reavell et al., 2018). Randomized trials of interventions for cancer patients also suggest beneficial effects on emotional adjustment, symptoms (e.g., fatigue), and functional activity levels (Lutgendorf & Andersen, 2015; Stanton, Rowland, & Ganz, 2015). The effects on cancer progression and underlying psychobiological mechanisms have been more variable, with positive and null results emerging from well-controlled trials (Lutgendorf & Andersen, 2015).

Many landmark studies of stress reduction interventions on physical health predate recent reporting guidelines that have generally strengthened clinical trials, and few studies attempt to assess and test hypothesized mediating mechanisms involving stress physiology. The choice of comparison groups in such studies is complicated for practical, ethical, and scientific reasons (Freedland et al., 2011). The field has a tendency to interpret the outcomes of such trials as reflecting effects of the specific content of the treatment delivered, when the construct actually evaluated consists of all of the differences between treatment and comparison arms.

EMERGING ISSUES AND FUTURE DIRECTIONS

The Guiding Role of Conceptual Models

Across multiple types of research described here, a recurring theme is the need for more explicit and effective use of theory as a guide. This is essential in the selection, refinement, and interpretation of measures of psychosocial influences on health and the health outcomes themselves, as discussed above. Clear conceptual models are essential as a guide to appropriate statistical analysis as well. For example, tests of the association of the Type D "distress prone" personality, a widely studied psychosocial risk model that posits synergistic effects of two distinct individual differences (i.e., negative affectivity, social inhibition), with subsequent cardiovascular health (Grande, Romppel, & Barth, 2012) often fail to utilize the statistical test most directly related to this interactive construct (Smith, 2011b). Elsewhere, physiological reactivity is sometimes described as a mediating variable accounting for effects of psychosocial risk factors on health, and sometimes as a moderating variable in which individual differences in characteristic physiological responsiveness alter effects of other risk factors (Smith & Gerin, 1998; see also Chapter 30, by Rockwood and Hayes). These distinct conceptualizations can be translated quite precisely, as when blood pressure reactivity is tested as a moderator of the effect of stress exposure on atherosclerosis (Kamarck et al., 2018). And as noted previously, clear conceptual models are essential when testing effects of correlated psychosocial risks.

Emerging Developmental Perspectives

A rapidly growing literature suggests that psychosocial risks for serious illness emerge early in life, as low socioeconomic status (SES) in one's family of origin and exposure to other adverse childhood experiences (ACEs) predict disease in adulthood (Jakubowski, Cundiff, & Matthews, 2018). This work has potentially invaluable implications for understanding psychosocial risk and the development of risk-reducing interventions. However, current research is largely retrospective, and there are substantial challenges in measuring early adversity (Appleton et al., 2017), quantifying cumulative risk (Evans, Li, & Whipple, 2013), examining possible mechanisms of action (Bush, Lane, & McLaughlin, 2016; Winning et al., 2018), and testing the causality in this type of developmental model (Gage, Munafo, & Davey Smith, 2016).

For example, novel evidence that exposure to early family conflict predicts adult atherosclerosis through the mediating effects of the quality of social interactions in adulthood (John-Henderson et al., 2016) suggests potentially important developmental continuities in psychosocial risks. Understanding these risk trajectories over the life course (Smith, Baron, & Grove, 2014) will require not only increased use of concepts, methods, and findings from developmental behavioral science, but also recent approaches to explicating genetic and environmental mechanisms (Manuck & McCaffery, 2014) and confounds (Klahr & Burt, 2014; McAdams et al., 2014) in associations between parent and child characteristics, including the role of epigenetic processes (Jones, Moore, & Kobor, 2018).

Between-Person versus Within-Person Approaches

Health psychology and behavioral medicine provide opportunities for pursuing the recent methodological/conceptual paradigm shift involving between- versus=within person associations (Molenaar & Campbell, 2009; see also Chapter 25, by Molenaar and Beltz, and Chapter 32, by Bauer et al.). Between-person associations of psychosocial factors with health outcomes or physiological responses (or the lack of those associations) are often assumed to reflect similar associations within individuals over time, but that assumption has only recently been examined. For example, in between-person analyses, the association between the magnitude of change in the subjective experience of anger and blood pressure reactivity is often nonsignificant, but is highly significant when tested as a within-person effect (Zawadzki et al., 2017). Further, in within-person analyses, the association between daily stress and exercise is bidirectional, but there are significant individual differences in these associations (Burg et al., 2017). Finally, expected associations between intentions to eat well and actual dietary intake are evident in within-person analyses, even when absent in between-person analyses (Inauen et al., 2016).

Health Disparities and Diversity

Disparities related to SES, race, and ethnicity are evident across most major health outcomes and are a central concern in behavioral medicine and health psychology (Whitfield et al., 2002). Race, ethnicity, and SES are interrelated predictors of health, complicating efforts to describe their independent effects, and each is associated with complexities in their quantification or representation in research (Nuru-Jeter et al., 2018). These disparities certainly could reflect the operation of the psychosocial and psychobiological processes described previously (Chen & Miller, 2013), as well as the additional effects of discrimination, which involve other conceptual and measurement issues (Lewis, Cogburn, & Williams, 2015). However, health disparities also reflect health behavior mechanisms and variability in access to and quality of healthcare services (Barrera et al., 2013; Fiscella & Sanders, 2016; Stuart-Shor et al., 2012).

Methodological Education in Health Psychology and Behavioral Medicine

As is obvious from this review, a broad range of research methods and approaches – from neuroimaging and epigenetics to public health and healthcare economics – are relevant in health psychology and behavioral medicine. Individual researchers are likely to develop considerable depth in some of the issues and approaches discussed here, and virtually all of the topics covered in this handbook could prove useful. But it is also the case that sufficient expertise in approaches outside the usual set in behavioral science are valuable by facilitating the interdisciplinary research that is common in the field. In a survey of faculty and students in health psychology training programs, the top training needs identified included interdisciplinary teamwork, advanced statistics and methods, and theory-driven research, emphasizing several of the themes discussed in this chapter (Goldstein et al., 2017). Further, given the health disparities related to SES, race, and ethnicity, broadly defined cultural competence is also an important element of research training in the field, as is the development of relevant diversity in research teams (Alcaraz et al., 2017; Huey et al., 2014). Broad training in theory-driven measurement, design, and analysis within one's home discipline, combined with exposure to the methods in related fields, will maximize the quality and impact of biopsychosocial research.

REFERENCES

Adler, N. E. (2009). Health Disparities through a Psychological Lens. *American Psychologist, 64*, 663–673.

Alcaraz, K. I., Sly, J., Ashing, K., Fleisher, L., Gil-Rivas, V., Ford, S., ... Gwende, C. (2017). The ConNECT Framework: A Model for Advancing Behavioral Medicine Science and Practice to Foster Health Equity. *Journal of Behavioral Medicine, 40*, 23–38.

Appleton, A., Holdsworth, E., Ryan, M., & Tracy, M. (2017). Measuring Childhood Adversity in Life Course Cardiovascular Research: A Systematic Review. *Psychosomatic Medicine, 79*, 434–440.

Barrera Jr., M., Castro, F. G., Stryker, L. A., & Tolbert, D. J. (2013). Cultural Adaptations of Behavioral Health Interventions: A Progress Report. *Journal of Consulting and Clinical Psychology, 81*, 196–205.

Block, J. (1995). A Contrarian View of the Five-Factor Approach to Personality Description. *Psychological Bulletin, 117*, 187–215.

Boutron, I., Mosher, D., Altman, D., Schultz, K., & Ravaud, P. (2008). Extending the CONSORT Statement to Randomized Trials of Nonpharmacologic Treatments: Explanation and Elaboration. *Annals of Internal Medicine, 148*, 295–309.

Burg, M. M., Schwartz, J., Kronish, I., Diaz, K., & Alcantara, C. (2017). Does Stress Result in You Exercising Less? Or Does Exercising Result in You Being Less Stressed? Or Is It Both? Testing the Bi-Directional Stress-Exercise Association at the Group and Person (N of 1) Level. *Annals of Behavioral Medicine, 51*, 799–809.

Bush, N. R., Lane, R. D., & McLaughlin, K. A. (2016). Mechanisms Underlying the Association between Early-Life Adversity and Physical Health: Charting a Course for the Future. *Psychosomatic Medicine, 78*, 114–119.

Cacioppo, J. T., Cacioppo, S., Capitanio, J., & Cole, S. (2015). The Neuroendocrinology of Social Isolation. *Annual Review of Psychology, 66*, 733–767.

Celano, C., Millstein, R., Bedoya, C., Healey, B., Roest, A., & Huffman, J. (2015). Association between Anxiety and Mortality in Patients with Coronary Artery Disease: A Meta-Analysis. *American Heart Journal, 170*, 1105–1115.

Chen, E., & Miller, G. (2013). Socioeconomic Status and Health: Mediating and Moderating Factors. *Annual Review of Clinical Psychology, 9*, 723–749.

Chiang, J., Turiano, N., Mroczek, D., & Miller, G. (2018). Affective Reactivity to Daily Stress and 20-Year Mortality Risk in Adults with Chronic Disease: Findings from the National Study of Daily Experience. *Health Psychology, 37*, 170–178.

Chida, Y., & Hamer, M. (2008a). Chronic Psychosocial Factors and Acute Physiological Responses to Laboratory-Induced Stress in Healthy Populations: A Quantitative Review of 30 Years of Investigations. *Psychological Bulletin, 134*, 829–884.

Chida, Y., & Hamer, M. (2008b). An Association of Adverse Psychosocial Factors with Diabetes Mellitus: A Meta-Analytic Review of Longitudinal Cohort Studies. *Diabetologia, 51*, 2168–2178.

Chida, Y., & Steptoe, A. (2009). The Association of Anger and Hostility with Future Coronary Heart Disease: A Meta-Analytic Review of Prospective Evidence. *Journal of the American College of Cardiology, 53*, 774–778.

Chida, Y., & Steptoe, A. (2010). Greater Cardiovascular Responses to Laboratory Mental Stress Are Associated with Poor Subsequent Cardiovascular Risk Status: A Meta-Analysis of Prospective Evidence. *Hypertension, 55*, 1026–1032.

Chida, Y., & Vedhara, K. (2009). Adverse Psychosocial Factors Predict Poorer Prognosis in HIV Disease: A Meta-Analytic Review of Prospective Investigations. *Brain, Behavior, and Immunity, 23*, 434–445.

Chida, Y., Hamer, M., Wardle, J., & Steptoe, A. (2008). Do Stress-Related Psychosocial Factors Contribute to Cancer Incidence and Survival? *Nature Clinical Practice: Oncology, 5*, 466–475.

Christenfeld, N. J. S., Sloan, R. P., Carroll, D., & Greenland, S. (2004). Risk Factors, Confounding, and the Illusion of Statistical Control. *Psychosomatic Medicine, 66*, 868–875.

Cole, S. W. (2008). Psychosocial Influences on HIV-1 Disease Progression: Neural, Endocrine, and Virologic Mechanisms. *Psychosomatic Medicine, 70*, 562–568.

Connelly, B. S., & Ones, D. S. (2010). An Other Perspective on Personality: Meta-Analytic Integration of Observers' Accuracy and Predictive Validity. *Psychological Bulletin, 136*, 1092–1122.

Davidson, K. W., Trudeau, K. J., Ockene, J. K., Orleans, C. T., & Kaplan, R. M. (2004). A Primer on Current Evidence-Based Review Systems and Their Implications for Behavioral Medicine. *Annals of Behavioral Medicine, 28*, 226–238.

De Miranda Azevedo, R., Roest, A., Carney, R., Denollet, J., Freedland, K., Grace, S., ... De Jonge, P. (2016). A Bifactor Model of the Beck Depression Inventory and Its Association with Medical Prognosis after Myocardial Infarction. *Health Psychology, 35*, 614–624.

Diez Roux, A. V., & Mair, C. (2010). Neighborhoods and Health. *Annals of the New York Academy of Sciences, 1186*, 125–145.

Dixon, K. E. (2016). Pain and Pain Behavior. In Y. Benyamini, M., Johnston, & E. Karademas (Eds.), *Assessment in Health Psychology* (pp. 147–159). Gottingen: Hogrefe.

Edmondson, D., Kronish, I., Shaffer, J., Falzon, L., & Burg, M. (2013a). Posttraumatic Stress Disorder and Risk for Coronary Heart Disease: A Meta-Analytic Review. *American Heart Journal, 166*, 806–814.

Edmondson, D., Shaffer, J., Chaplin, W., Burg, M., Stone, A., & Schwartz, J. (2013b). Trait Anxiety and Trait Anger Measured by Ecological Momentary Assessment and Their

Correspondence with Traditional Trait Questionnaires. *Journal of Research in Personality, 47*, 843–852.

Eller, N. H., Netterstrom, B., Gyntelberg, F., Kristensen, T., Nielsen, F., Steptoe, A., & Theorell, T. (2009). Work-Related Psychosocial Factors and the Development of Ischemic Heart Disease: A Systematic Review. *Cardiology Reviews, 17*, 83–97.

Emery, R. L., & Levine, M. D. (2017). Questionnaire and Behavioral Task Measures of Impulsivity Are Differentially Associated with Body Mass Index: A Comprehensive Meta-Analysis. *Psychological Bulletin, 143*, 868–902.

Engel, G. L. (1977). The Need for a New Medical Model: A Challenge for Biomedicine. *Science, 196*, 129–136.

Engel, G. L. (1980). The Clinical Application of the Biopsychosocial Model. *American Journal of Psychiatry, 137*, 535–544.

Evans, G. W., Li, D., & Whipple, S. (2013). Cumulative Risk and Child Development. *Psychological Bulletin, 139*, 1342–1396.

Fiscella, K., & Sanders, M. (2016). Racial and Ethnic Disparities in the Quality of Health Care. *Annual Review of Public Health, 37*, 375–394.

Fisher, E. B., Fitzgibbons, M. L., Glasgow, R. E., Haire-Joshu, D., Hayman, L. L., Kaplan, R. M., ... Okene, J. K. (2011). Behavior Matters. *American Journal of Preventive Medicine, 40*(5), e15–e30.

France, C. R., Masters, K. S., Belar, C. D., Kerns, R. D., Klonoff, E. A., Larkin, K. T., ... Thorn, B. E. (2008). Application of the Competency Model to Clinical Health Psychology. *Professional Psychology: Research and Practice, 39*, 573–580.

Freedland, K. E., Mohr, D. C., Davidson, K. W., & Schwartz, J. E. (2011). Usual and Unusual Care: Existing Practice Control Groups in Randomized Controlled Trials of Behavioral Interventions. *Psychosomatic Medicine, 73*, 323–325.

Friedman, H. S., & Adler, N. E. (2011). The Intellectual Roots of Health Psychology. In H. S. Friedman (Ed.), *The Oxford Handbook of Health Psychology* (pp. 3–14). New York: Oxford University Press.

Friedman, M., Thoreson, C. E., Gill, J. J., Ulmer, D., Powell, L. H., Price, V. ... Dixon, T. (1986). Alteration of Type A Behavior and Its Effects on Cardiac Recurrences in Post Myocardial Infarction Patients: Summary of Results of the Recurrent Coronary Prevention Project. *American Heart Journal, 112*, 653–665.

Gage, S. H., Munafo, M., & Davey Smith, G. (2016). Causal Inference in Developmental Origins of Health and Disease (DOHaD) Research. *Annual Review of Psychology, 67*, 567–585.

Gallo, L. C., & Smith, T. W. (1999). Patterns of Hostility and Social Support: Conceptualizing Psychosocial Risk Factors as Characteristics of the Person and the Environment. *Journal of Research in Personality, 33*, 281–310.

Gianaros, P. J., Sheu, L., Uyar, F., Koushik, J., Jennings, J. R., ... Vertynen, T. (2017). A Brain Phenotype for Stressor-Related Blood Pressure Reactivity. *Journal of the American Heart Association, 6*, e006053.

Goldstein, C. M., Minges, K., Schoffman, D., & Cases, M. (2017). Preparing Tomorrow's Behavioral Medicine Scientists and Practitioners: A Survey of Future Directions for Education and Training. *Journal of Behavioral Medicine, 40*, 214–226.

Grande, G., Romppel, M., & Barth, J. (2012). Association between Type D Personality and Prognosis in Patients with Cardiovascular Diseases: A Systematic Review and Meta-Analysis. *Annals of Behavioral Medicine, 43*, 299–310.

Grossardt, B. R., Bower, J. H., Geda, Y. F., Colligan, R. C., & Rocca, W. A. (2009). Pessimistic, Anxious, and Depressive Personality Traits Predict All-Cause Mortality: The Mayo Clinic Cohort Study of Personality and Aging. *Psychosomatic Medicine, 71*, 491–500.

Gulliksson, M., Burrell, G., Vessby, B., Lundin, L., Toss, H., & Svardsudd, K. (2011). Randomized Controlled Trial of Cognitive Behavior Therapy vs. Standard Treatment to Prevent Recurrent Cardiovascular Events in Patients with Coronary Heart Disease: Secondary Prevention in Uppsala Primary Care Project (SUPRIM). *Archives of Internal Medicine, 171*, 134–140.

Hall, M. H., Brindle, R. C., & Buysse, D. J. (2018). Sleep and Cardiovascular Disease: Emerging Opportunities for Psychology. *American Psychologist, 73*, 994–1006.

Helgeson, V. S., & Zajdel, M. (2017). Adjusting to Chronic Health Conditions. *Annual Review of Psychology, 68*, 545–571.

Holt-Lunstad, J. (2018). Why Social Relationships Are Important for Physical Health: A Systems Approach to Understanding and Modifying Risk and Protection. *Annual Review of Psychology, 69*, 437–458.

Huey, S., Tilley, J., Jones, E., & Smith, C. (2014). The Contribution of Cultural Competence to Evidence-Based Care for Ethnically Diverse Populations. *Annual Review of Clinical Psychology, 10*, 305–338.

Idler, E. L., & Benyamini, Y. (1997). Self-Rated Health and Mortality: A Review of Twenty-Seven Community Studies. *Journal of Health and Social Behavior, 38*, 21–37.

Inauen, J., Shrout, P., Bolger, N., Stadler, G., & Scholz, U. (2016). Mind the Gap? An Intensive Longitudinal Study of Between-Person and Within-Person Intention-Behavior Relations. *Annals of Behavioral Medicine, 50*, 516–522.

Irwin, M. R. (2015). Why Sleep Is Important for Health: A Psychoneuroimmunology Perspective. *Annual Review of Psychology, 66*, 143–172.

Jakubowski, K. P., Cundiff, J. M., & Matthews, K. A. (2018). Cumulative Childhood Adversity and Adult Cardiometabolic Disease: A Meta-Analysis. *Health Psychology, 37*, 701–715.

John-Henderson, N., Kamarck, T., Muldoon, M., & Manuck, S. (2016). Early Life Family Conflict, Social Interactions, and Carotid Artery Intima-Media Thickness in Adulthood. *Psychosomatic Medicine, 78*, 319–326.

Jones, M. J., Moore, S., & Kobor, M. (2018). Principles and Challenges of Applying Epigenetic Epidemiology to Psychology. *Annual Review of Psychology, 69*, 459–485.

Joseph, N. T., Kamarck, T. W., Muldoon, M. F., & Manuck, S. B. (2014). Daily Marital Interaction Quality and Carotid Artery Intima-Medial Thickness in Healthy Middle-Aged Adults. *Psychosomatic Medicine, 76*, 347–354.

Kamarck, T. W., Schwartz, J., Shiffman, S., Muldoon, M., Sutten-Tyrrell, K., & Janicki, D. (2005). Psychosocial Stress and Cardiovascular Risk: What Is the Role of Daily Experience? *Journal of Personality, 73*, 1749–1774.

Kamarck, T. W., Li, X., Wright, A. G. C., Muldoon, M., & Manuck, S. (2018). Ambulatory Blood Pressure Reactivity as a Moderator in the Association between Daily Life Psychosocial Stress and Carotid Artery Atherosclerosis. *Psychosomatic Medicine, 80*, 774–782.

Kaplan, R. M., & Groessl, E. J. (2002). Applications of Cost-Effectiveness Methodologies in Behavioral Medicine. *Journal of Consulting and Clinical Psychology, 70*, 482–493.

Karmeniemi, M., Lankila, T., Ikaheimo, T., Koivumaa-Honkanen, H., & Korpelainen, R. (2018). The Built Environment as a Determinant of Physical Activity: A Systematic Review of Longitudinal Studies and Natural Experiments. *Annals of Behavioral Medicine, 52*, 239–251.

Kelly, S., & Ismail, M. (2015). Stress and Type 2 Diabetes: A Review of How Stress Contributes to the Development of Type 2 Diabetes. *Annual Review of Public Health, 36*, 441–462.

Kent de Grey, R., Uchino, B. N., Trettvik, R., Cronan, S., & Hogan, J. (2018). Social Support and Sleep: A Meta-Analysis. *Health Psychology, 37*, 787–798.

Kivimaki, M., Nyberg, S. T., Batty, G. D., Fransson, E. I., Heikkla, K., Alfredsson, L., ... Theorell, T. (2013). Job Strain as a Risk Factor for Coronary Heart Disease: A Collaborative Meta-Analysis of Individual Participant Data. *Lancet, 380*, 1491–1497.

Klahr, A., & Burt, S. A. (2014). Elucidating the Etiology of Individual Differences in Parenting: A Meta-Analysis of Behavioral Genetics Research. *Psychological Bulletin, 140*, 544–586.

Knol, M. J., Twisk, J. W. R., Beekman, A. T. F., Heine, R. J., Snoek, F. J., & Fouwer, F. (2006). Depression as a Risk Factor for the Onset of Type 2 Diabetes Mellitus. A Meta-Analysis: *Diabetologia, 49*, 837–845.

Larkin, K. T., & Klonoff, E. A. (2014). *Specialty Competencies in Clinical Health Psychology*. New York: Oxford University Press.

Lewis, T. T., Cogburn, C., & Williams, D. (2015). Self-Reported Experiences of Discrimination and Health: Scientific Advances, Ongoing Controversies, and Emerging Issues. *Annual Review of Clinical Psychology, 11*, 407–440.

Lovallo, W. R. (2015). *Stress and Health: Biological and Psychological Interactions* (3rd edn.). Thousand Oaks, CA: Sage.

Luechen, L. J., & Gallo, L. C. (Eds.). (2008). *Handbook of Physiological Research Methods in Health Psychology*. Los Angeles: Sage.

Lutgendorf, S., & Andersen, B. (2015). Biobehavioral Approaches to Cancer Progression and Survival: Mechanisms and Interventions. *American Psychologist, 70*, 186–197.

Lynam, D. R., Hoyle, R. H., & Newman, J. P. (2006). The Perils of Partialling: Cautionary Tales from Aggression and Psychopathy. *Assessment, 12*, 328–341.

Manuck, S. B., & McCaffery, J. M. (2014). Gene-Environment Interaction. *Annual Review of Psychology, 65*, 41–70.

Matarazzo, J. D. (1980). Behavioral Health and Behavioral Medicine: Frontiers for a New Health Psychology. *American Psychologist, 35*, 807–817.

Mausbach, B. T., Bos, T., & Irwin, S. (2018). Mental Health Treatment Dose and Annual Healthcare Costs in Patients with Cancer and Major Depressive Disorder. *Health Psychology, 37*, 1035–1040.

McAdams, T., Neiderhiser, J., Rijsdijk, F., Narusyte, J., Lichtenstein, P., & Eley, T. (2014). Accounting for Genetic and Environmental Confounds in Associations between Parent and Child Characteristics: A Systematic Review of Children of Twins Studies. *Psychological Bulletin, 140*, 1138–1173.

McEwen, B. S. (1998). Stress, Adaptation, and Disease. *Allostasis and Allostatic Load: Annals of the New York Academy of Sciences, 840*, 33–44.

Moher, D., Liberati, A., Tetzlaff, J., & Altman, D. (2009). Preferred Reporting Items for Systematic Reviews and Meta-Analyses: The PRISMA Statement. *British Medical Journal, 339*, b2535.

Molenaar, P., & Campbell, C. (2009). The New Person-Specific Paradigm in Psychology. *Current Directions in Psychological Science, 18*, 112–117.

Newman, J. D., Davidson, K. W., Shaffer, J. A., Schwartz, J. E., Chaplin, W., Kirkland, S., & Shimbo, D. (2011). Observed Hostility and the Risk of Incident Ischemic Heart Disease: A Prospective Population Study from the 1995 Canadian Nova Scotia Health Survey. *Journal of the American College of Cardiology, 58*, 1222–1228.

Nuru-Jeter, A. M., Michaels, E., Thomas, M., Reeves, A., Thorpe, R., & LaVeist, T. (2018). Relative Roles of Race versus Socioeconomic Position in Studies of Health Inequalities: A Matter of Interpretation. *Annual Review of Public Health, 39*, 169–188.

Ottaviani, C., Thayer, J. F., Verkuil, B., Lonigro, A., Medea, B., Couyoumdjian, A., & Brosschot, J. F. (2016). Physiological Concomitants of Perseverative Cognition: A Systematic Review and Meta-Analysis. *Psychological Bulletin, 142*, 231–259.

Passhier, J., & Busschbach, J. (2015). Quality of Life. In F. Andrasik, J. L. Goodie, and A. L. Peterson (Eds.), *Biopsychosocial Assessment in Clinical Health Psychology* (pp. 182–194). New York: Guilford Press.

Phillips, A. N., & Davey Smith, G. (1991). Bias in Relative Odds Estimation Owing to Imprecise Measurement of Correlated Exposures. *Statistics in Medicine, 11*, 953–961.

Pincus, A. L., & Ansell, E. B. (2013). Interpersonal Theory of Personality. In T. Millon & M. J. Lerner (Eds.), *Handbook of Psychology (Vol. 5): Personality and Social Psychology* (2nd edn., pp. 141–159). New York: John Wiley.

Podsakoff, P. M., MacKenzie, S. B., & Podsakoff, N. (2012). Sources of Method Bias in Social Science Research and Recommendations on How to Control It. *Annual Review of Psychology, 63*, 539–569.

Pressman, S. D., Kenkins, B., & Moskowitz, J. (2019). Positive Affect and Health: What Do We Know and Where Should We Go? *Annual Review of Psychology, 70*, 627–650.

Reavell, J., Hopkinson, M., Clarkesmith, D., & Lane, D. (2018). Effectiveness of Cognitive Behavioral Therapy for Depression and Anxiety in Patients with Cardiovascular Disease: A Systematic Review and Meta-Analysis. *Psychosomatic Medicine, 80*, 742–753.

Robles, T. F., Slatcher, R. B., Trombello, J. M., & McGinn, M. M. (2014). Marital Quality and Health: A Meta-Analytic Review. *Psychological Bulletin, 140*, 140–187.

Roest, A., Martens, E., de Jong, P., & Denollet, J. (2010). Anxiety and Risk of Incident Coronary Heart Disease: A Meta-Analysis. *Journal of the American College of Cardiology, 56*, 38–46.

Ryff, C. D., & Singer, B. (1998). The Contours of Positive Human Health. *Psychological Inquiry, 9*, 1–28.

Sackett, D., Rosenberg, W., Muir Gray, J., Haynes, R., & Richardson, W. (1996). Evidence Based Medicine: What It Is and What It Isn't. *British Medical Journal, 312*, 71–72.

Schwartz, G. E., & Weiss, S. M. (1978). Behavioral Medicine Revisited: An Amended Definition. *Journal of Behavioral Medicine, 1*, 249–251.

Siddaway, A. P., Wood, A., & Hedges, L. (2019). How to Do a Systematic Review: A Best Practice Guide for Conducting and Reporting Narrative Reviews, Meta-Analyses, and Meta-Syntheses. *Annual Review of Psychology, 70*, 747–770.

Simon, G. E., Katon, W. J., Lin, E. H., Rutter, C., Manning, W. G., Von Korff, M. ... Young, B. A. (2007). Cost-Effectiveness of Systematic Depression Treatment among People with Diabetes Mellitus. *Archives of General Psychiatry, 64*, 65–72.

Sleep, C. E., Lynam, D. R., Hyatt, C. S., & Miller, J. D. (2017). Perils of Partialing Redux: The Case of the Dark Triad. *Journal of Abnormal Psychology, 126*, 939–950.

Smith, T. W. (2010). Conceptualization, Measurement, and Analysis of Negative Affective Risk Factors. In A. Steptoe (Ed.), *Handbook of Behavioral Medicine Research: Methods and Applications* (pp. 155–168). New York: Springer.

Smith, T. W. (2011a). Measurement in Health Psychology Research. In H. S. Friedman (Ed.), *Oxford Handbook of Health Psychology* (pp. 42–72). New York: Oxford University Press.

Smith, T. W. (2011b). Toward a More Systematic, Cumulative, and Applicable Science of Personality and Health: Lessons from Type D. *Psychosomatic Medicine, 73,* 528–532.

Smith, T. W., & Baucom, B. R. W. (2017). Intimate Relationships, Individual Adjustment, and Coronary Heart Disease: Implications of Overlapping Associations in Psychosocial Risk. *American Psychologist, 72,* 578–589.

Smith, T. W., & Gerin, W. (1998). The Social Psychophysiology of Cardiovascular Response: An Introduction to the Special Issue. *Annals of Behavioral Medicine, 20,* 243–246.

Smith, T. W., & Williams, P. G. (1992). Personality and Health: Advantages and Limitations of the Five Factor Model. *Journal of Personality, 60,* 395–423.

Smith, T. W., Uchino, B. N., Berg, C. A., Florsheim, P., Pearce, G., Hawkins, M., ... Yoon, H. C. (2008). Associations of Self-Reports versus Spouse Ratings of Negative Affectivity, Dominance and Affiliation in Coronary Artery Disease: Where Should We Look and Who Should We Ask when Studying Personality and Health? *Health Psychology, 27,* 676–684.

Smith, T. W., Traupman, E., Uchino, B. N., & Berg, C. (2010). Interpersonal Circumplex Descriptions of Psychosocial Risk Factors for Physical Illness: Application to Hostility, Neuroticism, and Marital Adjustment. *Journal of Personality, 78,* 1011–1036.

Smith, T. W., Uchino, B. N., Florsheim, P., Berg, C. A., Butner, J., Hawkins, M., ... Yoon, H. C. (2011). Affiliation and Control during Marital Disagreement, History of Divorce, and Asymptomatic Coronary Artery Calcification in Older Couples. *Psychosomatic Medicine, 73,* 350–357.

Smith, T. W., Ruiz, J. M., Cundiff, J. M., Baron, K. G., & Nealey-Moore, J. B. (2013). Optimism and Pessimism in Social Context: An Interpersonal Perspective on Resilience and Risk. *Journal of Research in Personality, 47,* 553–562.

Smith, T. W., Baron, C., & Grove, J. (2014). Personality, Emotional Adjustment, and Cardiovascular Risk: Marriage as a Mechanism. *Journal of Personality, 82,* 502–514.

Stanton, A. L., Rowland, J. H., & Ganz, P. A. (2015). Life after Diagnosis and Treatment of Cancer in Adulthood: Contributions from Psychosocial Oncology Research. *American Psychologist, 70,* 159–174.

Stephan, Y., Sutin, A., Bayard, S., Krizan, Z., & Terracciano, A. (2018a). Personality and Sleep Quality: Evidence from Four Prospective Studies. *Health Psychology, 37,* 271–281.

Stephan, Y., Sutin, A., & Terracciano, A. (2018b). Subjective Age and Mortality in Three Longitudinal Samples. *Psychosomatic Medicine, 80,* 659–664.

Steptoe, A., & Kivimaki, M. (2013). Stress and Cardiovascular Disease: An Update on Current Knowledge. *Annual Review of Public Health, 34,* 337–354.

Stuart-Shor, J., Berra, K. A., Kamau, M. W., & Kumanyika, S. K. (2012). Behavioral Strategies for Cardiovascular Risk Reduction in Diverse and Underserved Racial/Ethnic Groups. *Circulation, 125,* 171–184.

Strauss, M. E., & Smith, G. T. (2009). Construct Validity: Advances in Theory and Methodology. *Annual Review of Clinical Psychology, 5,* 1–25.

Sutin, A. R., Stephan, Y., Luchetti, M., Artese, A., Oshio, A., & Terracciano, A. (2016). The Five-Factor Model of Personality and Physical Inactivity: A Meta-Analysis of 16 Samples. *Journal of Research in Personality, 63,* 22–28.

Turiano, N., Chapman, B., Gruenwald, T., & Mroczek, D. (2015). Personality and the Leading Behavioral Contributors of Mortality. *Health Psychology, 34,* 51–60.

Uchino, B. N., & Way, B. (2017). Integrative Pathways Linking Close Family Ties to Health: A Neurochemical Perspective. *American Psychologist, 72,* 590–600.

Von Bertalanffy, L. (1968). *General Systems Theory*. New York: Braziller.

Von Elm, E., Altman, D., Egger, M., Popcock, S., Getzche, P., & Vanderbroucke, J. (2007). STrengthening the Reporting of OBservational Studies in Epidemiology (STROBE) Statement: Guidelines for Reporting Observational Studies. *British Medical Journal, 335,* 806–808.

Wachtel, P. L. (1973). Psychodynamics, Behavior Therapy, and the Implacable Experimenter: An Inquiry in to the Consistency of Personality. *Journal of Abnormal Psychology, 82,* 324–334.

Whitfield, K. E., Weidner, G., Clark, R., & Anderson, N. B. (2002). Sociodemographic Diversity and Behavioral Medicine. *Journal of Consulting and Clinical Psychology, 70,* 463–481.

Wiley, J., Bei, B., & Bower, J. (2017). Relationship of Psychosocial Resources with Allostatic Load: A Systematic Review. *Psychosomatic Medicine, 79,* 283–292.

Williams, P. G., & Thayer, J. F. (2009). Executive Functioning and Health: Introduction to the Special Series. *Annals of Behavioral Medicine, 37,* 101–105.

Williams, P. G., Smith, T. W., Gunn, H. E., & Uchino, B. N. (2011). Personality and Stress: Individual Differences in Exposure, Reactivity, Recovery, and Restoration. In R. Contrada & A. Baum (Eds.), *Handbook of Stress Science: Biology, Psychology, and Health* (pp. 231–245). New York: Springer.

Williams, P. G., Rao, H., Suchy, Y., Thorgusen, S., & Smith, T. W. (2017). On the Validity of Self-Report Assessment of Cognitive Abilities: Attentional Control Scale Associations with Cognitive Performance, Emotional Adjustment, and Personality. *Psychological Assessment, 29,* 519–530.

Wilson, S., Stroud, C., & Durbin, C. E. (2017). Interpersonal Dysfunction in Personality Disorders: A Meta-Analytic Review. *Psychological Bulletin, 143,* 677–734.

Winning, A., McCormick, M., Glymour, M., Gilsanz, P., & Kubzansky, L. (2018). Childhood Psychological Distress and Health Cardiovascular Lifestyle 17–35 Years Later: The Potential Role of Mental Health in Primordial Prevention. *Annals of Behavioral Medicine, 52,* 621–632.

Zawadzki, M. J., Smyth, J. M., Sliwinski, M., Ruiz, J. M., & Gerin, W. (2017). Revisiting the Lack of Association between Affect and Physiology: Contrasting Between-Person and Within-Person Analyses. *Health Psychology, 36,* 811–818.

PART VI INTENSIVE LONGITUDINAL DESIGNS

23 Ambulatory Assessment

STUART G. FERGUSON, TINA JAHNEL, KATHERINE ELLISTON, AND SAUL SHIFFMAN

In this chapter we present a conceptual and methodological discussion of the relevance of ambulatory assessment (AA) methods for the study of psychopathology. Here we cover the unique strengths of AA methods for the study of psychopathology and discuss the conceptual and methodological challenges involved with using these techniques to further our understanding of psychopathology. We also touch on potential uses of AA methods and data for improving treatments for psychological disorders, focusing on the use of AA methods for monitoring the onset, and/or the exacerbation of symptoms, with an eye toward intervening earlier in the time course of symptom management than is possible using conventional strategies.

AA – or, as it is often referred to in the psychological literature, ecological momentary assessment (EMA; Stone & Shiffman, 1994) – refers to the repeated collection of real-time – or near real-time – data from an individual as they go about their day-to-day life. (Hereafter we will predominantly use the term AA, however the terms EMA and AA can be used interchangeably.) AA methods have been used extensively in research for decades now and the methods and techniques used to gather such real-time data have been discussed in detail in earlier publications (see Shiffman, 2014; Shiffman, Stone, & Hufford, 2008).

Briefly, AA methods involve monitoring, recording, and assessing behaviors (e.g., binge eating, self-harm), cognitions (e.g., obsessive or self-deprecating thoughts), emotions of interest (e.g., sadness, stress, and anger), physiology (e.g., heart rate), social (e.g., being with others), and situational (e.g., environmental) factors that are believed to influence behaviors of interest. Contemporary AA studies generally involve the use of handheld electronic technologies such as mobile phones or tablets to collect data and monitor individuals. Whilst AA studies vary in terms of their design, they all capture momentary changes in thoughts and behaviors, involve repeated assessments over time, and are completed within individuals' day-to-day environments.

Perhaps the most dominant characteristic of AA methods is that they attempt to capture data in real time, or at least close to the moment that it is experienced. Assessments are focused on individuals' current state or situation in order to obviate the need to retrieve responses from individuals' memory, which can significantly bias the information reported (Hammersley, 1994). Thus, one specific advantage of using AA methods is that – compared to other methods involving self-reported cognitions and/or behaviors – the potential for biased recall is limited (Shiffman et al., 2008; Spook et al., 2013). Collecting information at the time the event of interest occurs is the "momentary" aspect of AA.

To overcome the artificial nature of laboratory studies, where situational and environmental contexts that affect individuals' experiences and behaviors are either absent or contrived, data is collected in individuals' natural environments. This is the "ecological" aspect of AA methods. By collecting data in real-world settings, researchers are able to capture naturally occurring variations in individuals' experiences and the contexts they are exposed to. This allows for generalization to individuals' real lives and to explore the links between behavior, the environment, and internal states that individuals experience (i.e., improving ecological validity).

Moments in an individual's day can be sampled using a mixture of event- or time-based sampling techniques (Wheeler & Reis, 1991; Table 23.1). Event-based sampling involves asking individuals to actively report[1] when an event of interest – e.g., hallucinations (Delespaul, deVries, & van Os, 2002), binge eating (Smyth et al., 2009), obsessive thoughts (Tilley & Rees, 2014), etc. – occurs. On the other hand, time-based sampling involves assessing an individual's state and other relevant variables (e.g.,

---

[1] Traditionally, participants in AA studies have been required to actively report the occurrence of an event of interest (e.g., by tapping a button to initiate an assessment on an electronic diary). As is discussed later in the chapter, more recently researchers have begun to explore using external sensors (e.g., the motion sensors built into smartphones) to automatically detect when events of interest are occurring. While such techniques are still in their infancy, such monitoring procedures promise to improve the accuracy of reporting, and to aid with the delivery of active support.

Table 23.1 Types of sampling used in ambulatory assessment (AA) studies

Assessment type	Characteristics	Key advantages	Key disadvantages
Event-based	• Event of interest triggers the recording and assessment (e.g., smoking a cigarette)	• Allows a focus directly on the specific event of interest • Enables data collection on the context and situation around the time the event occurs • Allows for documenting behaviors and events that are occurring and their sequencing	• Difficult to estimate the true rate of events and thus difficult to determine non- – or biased – compliance with reporting • Potential for reactivity; the act of reporting means that an event has occurred which might influence future behavior
Time-based	• Scheduling assessments to sample participants' states over time (e.g., level of anxiety present throughout the day)		
• Randomly scheduled[a]	• Randomly scheduled moments for assessment	• Easy test of compliance; calculate the number of assessments answered • Approximates representative and unbiased sample of moments within participants daily lives	• The unpredictable nature of when participants receive random assessments might be burdensome
• Fixed scheduled	• Fixed intervals (e.g., occurring every four hours) • Fixed periods (e.g., assessments in the morning and evening)	• Useful for cyclical temporal phenomena • Less intrusive for participants due to the predictable timing of assessments	• Crucial information at other times might be missed • Participants can anticipate assessments which might affect responses

[a] Randomly scheduled assessments may not necessarily occur at entirely "random" points throughout the day, instead they are timed to occur at "non-event"-based times.

location, activities) at various time points over the course of day-to-day life, irrespective of what the individual is currently doing. With time-based sampling, assessment is not triggered by the initiation of an event, but rather uses a predefined time-based algorithm to initiate data collection (Conner & Lehman, 2012; Shiffman, 2009). As such, time-based sampling can be used to gather information about conditions that are not marked by specific events; that is, conditions that are not episodic in nature (e.g., anxiety levels in generalized anxiety disorder).

A key strength of AA studies that utilize both time- and event-based sampling schemes is that they can be used to monitor how different contextual factors (social and situational factors), and internal events or states (affects, symptoms, and thoughts) may change in the days, hours, or even minutes, leading up to an event of interest, and how such antecedents alter the probability that the behavior of interest will occur (Maclure & Mittleman, 2000). Through such sampling, AA methods can capture the natural flow of human behavior over time and across settings (Myin-Germeys et al., 2009); a strategy that, as we will discuss below, is particularly useful for understanding and treating psychopathology.

ADOPTING AA METHODS FOR THE STUDY OF PSYCHOPATHOLOGY

Given the characteristics of psychopathology – discussed in detail in earlier chapters – it is optimal to study individuals in the field, as they go about their daily lives. As discussed previously, the experience of psychopathology is linked to contextual factors, for example, certain environments and/or situations may drive the occurrence and exacerbation of psychopathological symptoms. With the emphasis on recording information in real-time and in real-world settings, AA methods can be used to assess the dynamic interplay between individuals and the environmental influences on their behaviors. AA methods also make it possible to examine within-subject variability of symptoms independent from specific events, for example to provide a measurement of mood lability (Ebner-Priemer et al., 2007). Thus, AA methods are particularly well suited to study characteristics of psychopathology, especially those that vary between (e.g., age, gender, diagnostic status, personality) and within individuals (i.e., mood, environment, or social context). To understand why, it is useful to explore the types of questions that AA methods are particularly well suited to address.

Ecologically Valid Descriptions

Traditional clinical assessments focus on symptoms; according to the *Diagnostic and Statistical Manual*, symptoms define a disorder. This serves formal diagnoses but may give clinicians very little sense of the client's daily life and how the disorder and its symptoms manifest. For example, traditional clinical assessment does not tell you whether the client's symptoms are pervasive and nearly always present, or whether the symptoms are episodic, with periods of symptom-free function. Symptoms aside, what is the client's life like – do they have frequent contact with others, or are they isolated? How do they spend their time? Do they experience dysphoric moods that are not part of the typical phenotypic expression of the defined problem or diagnosis? Assessments collected from individual's real lives – such as through AA studies – can give clinicians a broader, 360-degree view of their client's experience than would be obtained from traditional assessment methods.

Laboratory studies are often used to gather data on the experiences, triggers and severity of psychopathology (e.g., assessing inattention and impulsivity among children with Attention-Deficit Hyperactivity Disorder; Barley, 1991). However, laboratory-based studies can lack ecological validity; when examining behaviors in this way it is unclear which social or physical environments are important for the behavior of interest to occur, and individuals themselves may not have insight into this. Furthermore, psychopathology is hard to simulate in the laboratory because of its complex interactions with environmental contexts. In some anxiety disorders, for example, symptoms may occur seemingly unpredictably (e.g., panic attacks in panic disorder; Alpers, 2009), or may only occur in specific situations which are difficult to reproduce – especially with any fidelity – in the laboratory. Furthermore, laboratory stimuli (e.g., imagining a difficult discussion with a spouse) may be very abstract or weak, especially in their evocative potency, compared to real-world stimuli (an actual argument with one's spouse). These examples demonstrate the need to assess behaviors of interest in real-word settings.

In addition, intraindividual fluctuation in mood and behavior is better represented in AA studies than it is in laboratory environments, as the time and context in which individuals are asked questions can influence their responses. For example, an individual in the midst of a manic episode is likely to report their overall mood differently than if they were experiencing an episode of depression. Having information from numerous time points can provide indication of mood and behavior variability, as well as the antecedents leading to events of interest (Moskowitz & Young, 2006). In sum, laboratory studies are unable to replicate the contextual cues that are important in driving specific behaviors of interest.

Gaudino and colleagues (2015) describe how electronic diaries could be used to better understand contextual factors such as the occurrence of stressful experiences, medication side effects and how experiencing negative affect may contribute to the expression of medication nonadherence and psychotic symptoms for conditions such as schizophrenia. Understanding what drives medication nonadherence and exacerbates symptoms can assist health professionals to identify and work with patients to better target risk factors and ultimately improve interventions for psychological conditions. Due to their portable nature, electronic diaries allow for real-time contextual information to be gathered in a way that is not possible with traditional methods investigating medication nonadherence, e.g., self-reported recall during appointments with specialists.

Information gathered through AA methods can provide answers to key questions relating to psychopathology in a way that traditional clinical assessment methods cannot ascertain. By having individuals report on their symptoms throughout the day, contextual information is gathered, thus, allowing for a deeper understanding of the drivers underlying pathological symptoms and behaviors. This information can enable mental health professionals to better address the cues surrounding psychopathological symptoms and create more personalized treatments that focus on preventing symptoms before they occur.

Temporal Sequences

While AA data could, in principle, collect just a few assessment points over a short period of time, AA studies almost always collect long streams of data with closely spaced observations: what has come to be referred to as intensive longitudinal data. By having individuals complete intensive repeated assessments over a few days or weeks, AA methods allow for detailed information to be collected on within-person fluctuations in behaviors and situations of interest. This means that AA can be used to obtain information about the antecedents and consequences of events and experiences, and to analyze and describe cascades of events and their interactions between events that shape behaviors over a period of time. Within these types of analyses, it is crucial to consider the order of events. Many theories of psychopathology focus on microprocesses underlying behaviors, that is, the interplay of cognitive, affective, or behavioral variables over a short period of time (e.g., a negative social comment leading to anxiety or obsession with social acceptance). Thus, AA methods are particularly well suited to study these theories.

Psychopathology and its functional impairments fluctuate over time (e.g., panic attacks) and across different contexts (e.g., social vs. solitary situations; Myin-Germeys et al., 2009). Understanding how psychologically relevant factors such as mood, stress, self-esteem, and other phenomena that fluctuate throughout the day or across days can be used to better understand the daily experiences of individuals living with conditions such as anxiety disorders (Walz, Nauta, & Aan Het Rot, 2014). However, clinical diagnosis of psychopathology typically results in assessments being made based on the patient's or others' general description of behavior, implying an unrealistic uniformity of behavior. This assessment approach neglects to consider both the fluctuating nature of psychological symptoms and the inconsistencies between individuals' and others' reports on their behavior (Achenbach, 2006). Understanding the lived experience of psychological conditions and appreciating the varying nature of symptom severity may lead to establishing better preventive measures and more targeted and personalized treatments.

In summary, AA methods enable real-time monitoring of contextual information driving pathological symptoms and behaviors, thereby increasing our understanding of the lived experience of psychological disorders. Overall, AA is particularly well suited to achieve a greater understanding on the trajectory of symptoms and factors that influence the day-to-day fluctuations of these symptoms than is possible with traditional clinical assessments (Armey et al., 2015). As will be discussed later in the chapter, the ability to ascertain temporal sequencing of events affords the possibility of novel real-time treatments to manage psychological disorders.

Individual Differences

Certain events or experiences related to psychopathology (e.g., experiencing anxiety) are common across the entire population (Martin, 2003). What marks someone with a diagnosis of an anxiety disorder, however, is the frequency of extreme anxiety-related symptoms, their occurrence within inappropriate contexts, and their interference with normal functioning. For example, anxiety when in social situations may be indicative of social phobia if the anxiety is extreme, persists across all social situations and if it impairs normal functioning. Experiencing anxiety when giving a presentation, for example, would not in itself be sufficient to warrant a diagnosis of an anxiety disorder. The frequency and functional impairment relating to extreme symptoms and the context in which they occur is best measured using AA methods. AA may therefore be a useful tool in assisting with the diagnosis of psychological disorders.

Compared to other methods such as laboratory, case-control studies, or epidemiological survey methods, AA methods have the advantage of examining within-person variation, by obtaining multiple assessments. To characterize individual differences, AA data is aggregated to attain an average measure that is collapsed across multiple assessments over time. When a variable of interest fluctuates over time or in response to environmental or situational factors, an aggregate of multiple assessments is more likely to be accurate than is a single measure at a specific point in time. For example, Ben-Zeev and colleagues (2012a) used AA to study the impact of self-stigma on the recovery of individuals with schizophrenia and found that the experience of self-stigma changes based on individuals' internal states and external circumstances.

Beyond capturing individuals' average characteristics, AA data can capture dynamic within-person processes that are themselves an important source of individual differences. For example, the volatility of affect among individuals with borderline personality is best captured through within-person data collection over time and across relevant contexts (Trull et al., 2008).

In order to understand individual differences in psychological symptoms, both situational and internal cues driving symptoms and behaviors of interest are crucial for the diagnosis and treatment of psychopathology. Cross-sectional instruments are unable to assess the variations and determinants of individual characteristics of psychopathology. However, AA methods are able to overcome these issues through the collection of repeated assessments within individuals over time. AA methods are therefore useful in identifying when psychological symptoms interfere with individuals' day-to-day lives and may also be developed to identify psychopathology.

ISSUES IN AA MONITORING

Compliance

Recording events as they occur can place a substantial burden on individuals, especially if the behavior or state of interest occurs frequently. Missing assessments – specifically, nonrandom missing assessments – have the potential to bias the obtained data. When individuals selectively choose to complete time-based assessments or record events, they may, for example, only complete assessments when they are not stressed, thus biasing the reported results. Therefore, it is important that researchers using AA methods focus their attention to individuals' compliance with the assessment protocol.

Timely compliance refers to the reporting of information at the time events are occurring, or in close proximity to when random assessments are initiated. Timely compliance varies according to the method of monitoring and can also affect the investigator's ability to understand and monitor compliance. Paper diaries, for example, carry the potential risk of backfilling, that is when individuals' complete assessments, maybe even en masse, at wrong times, whilst seeming compliant (Stone et al., 2003b). Without objective measures of timeliness, the researcher has no certainty of the participants' compliance rate. A common method to detect compliance with the protocol is using electronic diaries to automatically timestamp each report as it is completed, thus allowing for the verification of the time each data entry was made. While timely compliance with prompts issued by the AA device can be objectively determined, this is not the case for event-based recording. There is usually no way of determining whether events occurred but were not recorded. Issues with compliance are particularly important when it comes to incorporating AA monitoring into treatment, as will be discussed in detail below.

To complicate the problem of compliance with AA methods, individuals suffering from psychopathology may not have the structure in their daily lives to accommodate the intensive monitoring that AA methods require (Boivin, 2000). For example, individuals suffering from bipolar disorder experience extreme fluctuations in mood, impairing their daily functioning, and rendering a protocol of repeated assessments throughout the day difficult to complete. However, with new developments in technology it is increasingly possible to automatically monitor individuals, as will be discussed below. Using technology to automatically record events reduces the burden of recording and may improve understanding of symptom management and treatment.

Reactivity

Reactivity refers to the possibility that the research methods themselves influence the state or behavior of interest and thus bias the findings. As assessments in AA occur in close proximity to behaviors of interest, AA methods have the potential to be vulnerable to reactivity. Additionally, the reactivity induced through self-monitoring is dependent on the time of event recording. Prospective recording may maximize reactivity by making the participant aware of the behavior. However, there is currently little evidence to suggest AA methods induce reactivity (Hufford & Shiffman, 2002; Hufford et al., 2002; Stone et al., 2003a). Nevertheless, researchers applying AA methods should be alert to the potential of reactivity. This is particularly true in treatment contexts, as evidence suggests that self-monitoring is most reactive when the person is trying to change their behavior (McFall, 1977). From a treatment (rather than assessment) perspective, change following self-monitoring can be a positive outcome, and, indeed, self-monitoring is often incorporated into behavioral treatments for psychological conditions.

BEYOND SELF-REPORT

As discussed above, self-reporting and monitoring can be burdensome, specifically for individuals with psychopathological relevant symptoms. With the development of new technologies, such as sensors that synchronize with AA measurement tools, research on the triggers and fluctuations of mental health symptoms can be improved while potentially lowering participant burden. Sensors embedded in electronic diaries or in everyday devices, such as smartphones and wearable devices, enable continuous and passive measurement of physical properties such as heart rate and sleep cycles.

The use of such sensor data has five potential advantages: (1) by relieving subjects of the responsibility of reporting, they reduce subject burden; (2) they enable collection of data more continuously than is realistically done by self-report (e.g., actigraphy can report physical movement every second, rather than every hour or two, as we might ask a subject to do; Tahmasian et al., 2013); (3) by circumventing self-report, sensing provides more objective insight into the moment, without relying on what the subject notices (e.g., there might be food cues present that the subject is not consciously aware of or is unwilling to report; Elliston et al., 2017); (4) sensing can track processes that are simply not available to consciousness and self-report (e.g., heart rate, skin conductance; Picard et al., 2016); (5) sensors may record data at a more basic and "raw" form, allowing for the researcher to impose various constructs on the data (e.g., actigraphy can be processed to track spurts of intense activity rather than overall movement; recorded conversations can be coded for strain, on the one hand, and support, on the other, without having the subject classify or rate these dimensions; Mehl, 2017).

By passively collecting objective data, participant engagement and burden can be minimized and self-reported information verified. Additionally, novel information sensors can track everyday physical activity, location, social interaction, movement, communication, light, sound, and more. As sensor technology has advanced over the past decade, with sensors becoming lighter, smaller, and increasingly accurate, they have become ubiquitous and embedded into networks (Mohr, Zang, & Schueller, 2017). Therefore, sensors can provide vast amounts of data measuring behavior both instantaneously and continuously. Together, sensors can provide rich information surrounding individual's behavior, emotions, and ultimately their mental health status.

The utilization of ubiquitous sensing data to estimate and study behavior is often referred to as "personal sensing" (Klasnja et al., 2009, Mohr et al., 2017). The theory and the challenge of applying personal sensing to mental health sees large amounts of raw sensor data captured and converted into meaningful information related to behavior, emotions, or clinical states. To date, most research applying personal sensing for mental health has used sensors embedded in mobile phones, or other wearable technologies such as smartwatches and accelerometers. Sensors such as global positioning system (GPS) receivers built into electronic diaries can passively track individuals' movements and locations whenever they carry the device and can provide information not only on individuals' geographical location but can also be used to derive information about location-specific characteristics. Combined with diary and survey information, GPS information provides a rich data set to answer location-based questions. For example, Kirchner and colleagues (2013) showed that the proximity of tobacco retail outlets was related to risk of relapse among smokers trying to quit. More recently, Saeb and colleagues (2015) examined the relationship between GPS location and depression; demonstrating that the variability in time spent in different locations was related to depression.

Similarly, sensors built into smartphones or wearables can measure the amount and type of physical activity without active engagement from the user. A study in which accelerometer-based wearables tracked the intensity of physical activity identified that lower odds of depression were associated with increasing levels of physical activity (Vallance et al., 2011). In addition, other commercial wearables, such as smartwatches or fitness trackers, contain sensors that measure heart rate or skin conductance (Picard et al., 2016).

Other forms of mobile technology open the possibility to obtain a variety of data and insight into individuals' daily lives and may help gain a greater understanding of psychopathology. For example, AA studies can record sound or take photographs to provide information about social and physical environments. The Electronically Activated Recorder (EAR; Mehl, 2017) is an audio recording device which participants wear attached to their clothes during their waking hours. It periodically samples brief audio snippets of people's daily lives, enabling researchers to unobtrusively collect authentic real-life data about the sonic environment, and about conversations throughout the day. In a study by Tomko and colleagues (2014), EAR was used to assess interpersonal behavior and affect in patients with borderline personality disorder and patients with depressive disorder and showed that when experiencing anger, individuals with depression were less likely to spend time with others compared to individuals with borderline personality disorder.

Electronic diaries can be used in combination with other forms of ambulatory monitoring that enable real-time, objective, and passive data collection on physiological measures continuously or near-continuously. In a study assessing emotional response systems to better understand daily stress and the relationship with health and performance, Wilhelm, Pfaltz, and Grossman (2006) used various forms of continuous cardiovascular and respiratory monitoring in combination with an electronic diary. Other physiological measures can be feasibly recorded, such as hypothalamic-pituitary-adrenal axis activity by collecting salivary cortisol during daily life. It is believed that those with emotion regulation problems show enhanced cortisol reactivity to stressful events (Havermans et al., 2011).

Regardless of the benefits personal sensing may add to traditional AA methods, some challenges should be considered. Although there is a growing number of studies collecting self-reported information in combination with phone sensor data, findings are usually not generalizable as studies that appear to address the same behavior often apply different methods of measuring the behavior, different sensors, and varying research designs (e.g., in some studies participants use their own smartphone, whereas in other studies participants are provided with study phones). In addition, data quality varies depending on manufacturer, different environments, and characteristics of participants. Sensors in smartphones, for example, vary from manufacturer to manufacturer. Where participants carry the smartphone affects the sensor data (e.g., in a pocket vs. in a backpack); and participant characteristics (e.g., age, gender) might influence usage patterns. Furthermore, passive collection of digital data raises the issue of security (Mohr et al., 2017). An important aspect here is trust, which refers to the notion that data will be used and handled appropriately in accordance with ethical considerations such as anonymity. Compared to traditional collection of personal data, digital data collection tools may pose a greater risk of identification. For example, only four spatiotemporal points are needed to identify 95 percent of individuals (de Montjoye et al., 2013). Better standards and best-practice guidelines are needed to ensure participant's privacy and the secure storage of their data.

Despite the challenges, employing sensor technology in AA methods has considerable potential to continuously and objectively identify behavior and recognize people who are at risk or in need of treatment. Ultimately, personal sensing can be used to develop a new generation of intervention technologies that can reduce participants' burden through passive data collection while producing highly personalized, contextualized, and dynamic mobile health (mhealth) tools that learn and interact with the user.

Therapeutic Uses of AA

The ultimate goal of using technology to better understand psychopathology is to improve the treatment options available to patients. Providing support to patients – for example, via booklets or helplines – has been the cornerstone of interventions for decades, and numerous methods of delivering education, advice, and interventions to patients have been trialed. One way to characterize these different delivery modalities is in terms of the intensity of the interaction with a patient they provide. On the passive end of the spectrum, education and advice has been provided to patients via self-help books and tapes (both audio and visual), with little or no human interaction. While on the active end of the spectrum, clinicians have used repeated one-on-one counseling, which necessitates high levels of interaction and discussion with the patient by trained counselors. Other delivery options (e.g., internet-based self-help programs; Gainsbury & Blaszczynski, 2011, telephone counseling, etc.) fall at various points along this spectrum. As one might expect, greater interaction with patients generally results in better treatment outcomes (Carroll, 2016).

As AA methods allow for examining direct environmental and situational influences on behavior, such data can be used to improve the delivery of more traditional forms of treatment. For example, real-time data can be fed back to a clinician to allow them to monitor the progress of their clients, potentially allowing them to adjust treatment schedules and/or content based on between-session experiences. Such data could also be used to identify triggers which could then be used to guide in-person counseling sessions. Oorschot and colleagues (2012), for example, demonstrate how mobile devices could be used in a therapeutic context to elucidate idiosyncratic symptom patterns and dynamic changes within individuals with schizophrenia to improve personalized psychoeducation.

The potential for AA-based interventions extends beyond simply updating clinicians on their clients' day-to-day experiences. While traditional clinical approaches aim to change the client's behavior by working with them in a setting removed from their challenges (i.e., the consulting room), AA-based interventions recognize that problematic behavior (as well as nonproblematic behavior) can be driven by cues within the immediate environment and vary on a moment-to-moment basis. Small interventions, delivered at the relevant moment, are likely to have as much as or even more of an effect than large interventions delivered at a distance from the key moments influencing behavior.

Over the past decade, researchers have experimented with the provision of education and advice materials via mobile phone. Such programs are particularly common in the area of behavior change but have also been developed for the monitoring and treatment of psychopathology. So-called mobile health – or mHealth – solutions were first delivered in the form of text messages (Whittaker et al., 2016). More recently researchers have trialed standalone behavior change applications administered via mobile phones (e.g., via "apps"; Zhao, Freeman, & Li, 2016).

The provision of content via mobile phones is more than just a "clever use of technology." Individual mHealth applications have demonstrated success with assisting and promoting treatment for psychotic disorders (Firth et al., 2016), reducing depression symptoms and negative cognitions (Birney et al., 2016), reducing stress levels and/or anxiety (Firth et al., 2017; Przeworski & Newman, 2004; Thornorarinsdottir, Kessing, & Faurholt-Jepsen, 2017), promoting self-awareness (Runyan et al., 2013), and quitting smoking (Businelle et al., 2016a; Rodgers et al., 2005; Ubhi et al., 2015).

One of the insights provided by AA data is to emphasize the degree to which behavior, conditions, and emotions – dysfunctional and otherwise – vary over time and are influenced by the immediate context. This, in turn, suggests the potential importance of intervening at the right time and place. In contrast to traditional therapeutic interventions, which attempt to use interactions in an artificial context removed from the source of the problem (i.e., in the consulting room) to implement change, just-in-time interventions attempt to intervene at the very moment that dysfunction is triggered by context. For example, a smoker who is quitting might get an intervention as their craving and temptation for a cigarette is escalating, and a person with borderline personality disorder might receive an intervention when their anger and resentment are rising. Just-in-time interventions require underlying monitoring of the person's state and an algorithm to determine when an intervention is needed, and what intervention is appropriate in the context. The incremental effectiveness of just-in-time interventions has not yet been established with clinical evidence and evaluating such interventions may require new research paradigms (Klasnja et al., 2015), but the concept holds great promise for bringing clinical interventions into peoples' daily lives.

Establishing how to best to capitalize on the potential benefits of using mobile phones to monitor symptoms and/or deliver interventions is the focus of considerable research effort. The key premise of mobile interventions is the potential to improve outcomes by harvesting and interpreting information about a user's life and using this information to provide personalized intervention content and delivery (Businelle et al., 2016b). Recent technological advances, as well as the near-universal availability of handheld devices such as mobile phones, have significantly improved the viability of assessment and intervention strategies utilizing such technologies (Kaplan & Stone, 2013). One example is the Mobylize! system (Burns et al., 2011). Mobylize! is a context-aware intervention system for depression that uses a combination of AA-reported mood and context variables with sensors embedded within a smartphone (e.g., light detection, phone usage). Machine-learning methods are then used to predict when participants are experiencing low mood and to trigger an intervention aimed at improving mood or supporting coping strategies. Though continued research on this kind of preemptive approach to behavioral intervention is warranted, these examples highlight how AA methods enable the development of personalized treatment by identifying and delivering interventions at critical times.

The delivery of the inventions via apps has several appealing characteristics. Firstly, app interventions are relatively easy to scale up to serve a dispersed population and can be distributed widely without incurring significant additional costs, increasing their potential net public health benefit. This is particularly relevant in countries with relatively low population density and rural and remote population centers (because of the challenges this brings to providing the whole population adequate access to medical resources), and in less-developed countries (where the cost of delivering an intervention is a particularly important barrier).

Achieving personalized interventions through the use of mobile technology requires careful examination of real-time data collected in AA studies. Using AA to assess behaviors and symptoms on a daily or multiday basis allows researchers and clinicians to examine how interventions affect a client's symptoms and experience of psychopathology. This is particularly relevant for assessing micro-randomized interventions delivered via mobile technology that randomize treatments within participants – that is, compare days on which participants were delivered the intervention content with days on which participants were delivered control content (Klasnja et al., 2015).

UTILIZING REAL-TIME DATA IN INTERVENTIONS: PROMISE, PROBLEMS, AND FUTURE DIRECTIONS

As noted earlier, the key benefit of AA-based interventions is that they have the potential to collect real-time information and use this to deliver personalized interventions. Achieving this requires not only developing and evaluating methods of gathering data from users, but also determining how best to use this information to inform treatment content and delivery. Both challenges can be informed by learning from real-time AA data collection studies.

One promise of interventions enriched with real-time data is that such interventions will be able to recognize environmental antecedents of psychopathological events and prospectively intervene. Put another way, researchers hope they will one day be able to successfully predict when a user is likely to encounter a stressor that could trigger a psychopathological event, or exacerbate an underlying condition, and intervene accordingly. Perhaps the best current example of such an intervention comes from the smoking cessation literature: by tracking individuals' movements via GPS and implementing geofencing[2] strategies, the Q Sense app (Naughton et al., 2016) can deliver smoking cessation education and advice to users as they near an area known to be associated with their smoking. A challenge with the current application of this technology is that it relies on users to initially "tag" locations where they believe they will be most at risk of relapse. This is somewhat problematic as it relies on users to recognize their personal risk factors, a task that research suggests that many smokers cannot successfully accomplish. Future advances in geofencing might allow for the automatic selection of high-risk locations. For example, research is currently underway (Ferguson et al., 2015; Schüz, Bower, & Ferguson, 2015; Shiffman et al., 2014; Shiffman, Dunbar, & Ferguson, 2015) looking at whether it is feasible to use pre-quit monitoring to determine personalized environmental triggers of behaviors and use this information to automatically implement geofences around high-risk locations – even, potentially, high-risk locations that a patient has never physically visited before (e.g., by tagging all bars for alcoholics, not just those that the patient has previously identified as being their regular drinking locations).

AA-based methods of data collection can be time-demanding for participants to complete (Moskowitz & Young, 2006). One of the key challenges facing researchers is how to design monitoring programs and interventions so that enough data is collected, but users do not feel burdened with responding to the apps (Wright et al., 2016). Burke and colleagues (2017) conducted an AA-based eating study where participants reported their triggers of lapses and relapses of unplanned eating following weight loss over a 12-month period. They designed their study to reduce participant burden as much as possible. Their app was programmed to utilize skip patterns among assessments so that questions could be filtered according to situational relevance. Users could prespecify when both the beginning-of-day and end-of-day reports were issued. If the participants were busy, the random assessments could be delayed for up to 20 minutes at a time. Finally, users could silence alarms in situations where random assessments would be disruptive. Despite the intensity and the length of this study, there was a high participant compliance rate (88 percent with random assessments and 90 percent of time-contingent assessments were completed), and an 87 percent retention rate at 12 months, demonstrating that with appropriate measures in place, the burden of AA monitoring can be overcome.

App developers and researchers need to guard against inadvertently introducing additional barriers to treatment. While data collection might be useful for informing intervention delivery, this needs to be weighed against user burden and the risk of users abandoning it if the data-collection burden becomes too heavy. Nevertheless, AA methods have been shown to be feasible, acceptable, and valid methods of data collection for individuals with severe schizophrenic symptoms (Ben-Zeev et al, 2011), depression and anxiety (Ben-Zeev et al., 2012b), acute daily stress (Kimhy et al., 2010), or bipolar disorder (Schwartz et al., 2016). Improved modeling of behaviors – such as models that require fewer data points to reach meaningful conclusions – and/or incidental data-collection techniques that gather data passively, will undoubtedly assist in reducing the risks associated with patient burden. In addition, it is important to provide the right type and amount of support at the right time to promote intervention engagement and to prevent intervention fatigue (Gunlicks-Stoessel et al., 2016; Nahum-Shani et al., 2016). Just as an intervention delivered at the right time can help, interventions delivered at the wrong time, or too many times, can cause harm. Similarly, designing mHealth apps to drive engagement – for

---

[2] Geofencing involves using spatial information to create a virtual boundary, or fence, around a given location. Software running on a handheld device can then be programmed to respond when a patient approaches, enters, or leaves this virtual location.

example by incorporating gaming characteristics (Bindoff et al., 2016) and/or prompting users throughout the day (Ubhi et al., 2015) – may guard against patients abandoning monitoring and treatment.

Briefly, consideration also needs to be given to the ethical and legal issues that may arise as these technologies enable clinicians and other stakeholders to monitor patients in real time (Labrique et al., 2013). The potential involvement of other stakeholders such as family members, social media platforms and their owners, insurance companies, governments and employers, and the vast amount of data that can be potentially collected about individuals have clinical, policy, and legislative implications (Kumar et al., 2013). Appropriate data storage allowing for privacy of personal information is a key concern surrounding the use of mHealth services. However, a study by McClure, Hartzler, and Catz (2016) found that providers, not users, were the ones most concerned about the privacy of mobile phone apps in tracking behavior change. Nevertheless, these are issues that researchers and developers will need to continue to monitor as the mHealth field matures.

CONCLUDING REMARKS

Using AA concepts to develop systems to monitor psychopathological conditions and symptoms may be a cost-effective and accessible way to facilitate treatment. Real-time behavioral monitoring may facilitate individuals' self-awareness of the cues triggering symptoms thereby encouraging self-driven behavioral change (Thornorarinsdottir et al., 2017) and ultimately enhance the treatment of mental health conditions (Donker et al., 2013). AA-based monitoring and intervention programs have the potential to effectively reduce a range of mental health concerns including depression, anxiety, stress, and substance use (Donker et al., 2013).

While, to date, AA technologies have largely been employed within a research setting, there is growing interest in using such techniques to improve clinical treatments and interventions. The ability of AA technologies to provide tailored interventions that respond to the individual's real-time behaviors, moods, social contexts, and geographical locations has the potential to revolutionize the treatment of psychopathology. Considerable work remains to ensure that AA-based interventions are appropriate, theoretically derived, and appropriately ethical in their effects on privacy and confidentiality.

REFERENCES

Achenbach, T. (2006). As Others See Us: Clinical and Research Implications of Cross-Informant Correlations of Psychopathology. *Current Directions in Psychological Science, 15*(2), 94–98

Alpers, G. W. (2009). Ambulatory Assessment in Panic Disorder and Specific Phobia. *Psychological Assessment, 21*(4), 476–485.

Armey, M. F., Schatten, H. T., Haradhvala, N., & Miller, I. W. (2015). Ecological Momentary Assessment (EMA) of Depression-Related Phenomena. *Current Opinion in Psychology, 4*, 21–25.

Barley, R. (1991). The Ecological Validity of Laboratory and Analogue Assessment Methods of ADHD Symptoms. *Journal of Abnormal Child Psychology, 19*(2), 149–178.

Ben-Zeev, D., Ellington, K., Swendsen, J., & Granholm, E. (2011). Examining a Cognitive Model of Persecutory Ideation in the Daily Life of People with Schizophrenia: A Computerized Experience Sampling Study. *Schizophrophrenia Bulletin, 37*(6), 1248–1256.

Ben-Zeev, D., Frounfelker, R., Morris, S. B., & Corrigan, P. W. (2012a). Predictors of Self-Stigma in Schizophrenia: New Insights Using Mobile Technologies. *Journal Dual Diagnosis, 8* (4), 305–314.

Ben-Zeev, D., Morris, S., Swendsen, J., & Granholm, E. (2012b). Predicting the Occurrence, Conviction, Distress, and Disruption of Different Delusional Experiences in the Daily Life of People with Schizophrenia. *Schizophrophrenia Bulletin, 38*(4), 826–837.

Bindoff, I., de Salas, K., Peterson, G., Ling, T., Lewis, I., Wells, L., ... Ferguson, S. G. (2016). Quittr: The Design of a Video Game to Support Smoking Cessation. *Journal of Medical Internet Research Serious Games, 4*(2), e19.

Birney, A. J., Gunn, R., Russell, J. K., & Ary, D. V. (2016). Mood-Hacker Mobile Web App With Email for Adults to Self-Manage Mild-to-Moderate Depression: Randomized Controlled Trial. *Journal of Medical Internet Research Mhealth Uhealth, 4*(1), e8.

Boivin, D. (2000). Influence of Sleep-Wake and Circadian Rhythm Disturbances in Psychiatric Disorders. *Journal of Psychiatry & Neuroscience, 25*(5), 446–458.

Burke, L. E., Shiffman, S., Music, E., Styn, M. A., Kriska, A., Smailagic, A., ... Rathbun, S. L. (2017). Ecological Momentary Assessment in Behavioral Research: Addressing Technological and Human Participant Challenges. *Journal of Medical Internet Research, 19*(3), e77.

Burns, M. N., Begale, M., Duffecy, J., Gergle, D., Karr, C. J., Giangrande, E., & Mohr, D. C. (2011). Harnessing Context Sensing to Develop a Mobile Intervention for Depression. *Journal of Medical Internet Research, 13*(3), e55.

Businelle, M. S., Ma, P., Kendzor, D. E., Frank, S. G., Vidrine, D. J., & Wetter, D. W. (2016a). An Ecological Momentary Intervention for Smoking Cessation: Evaluation of Feasibility and Effectiveness. *Journal of Medical Internet Research, 18*(12), e321.

Businelle, M. S., Ma, P., Kendzor, D. E., Frank, S. G., Wetter, D. W., & Vidrine, D. J. (2016b). Using Intensive Longitudinal Data Collected via Mobile Phone to Detect Imminent Lapse in Smokers Undergoing a Scheduled Quit Attempt. *Journal of Medical Internet Research, 18*(10), e275.

Carroll, J. F. X. (2016). What Constitutes "Success" in Addiction Treatment and How Do We Determine What Works Best? *Alcoholism Treatment Quarterly, 34*(2), 252–260.

Conner, T., & Lehman B. (2012). Getting Started: Launching a Study in Daily Life. In M. Mehl & T. Conner (Eds.), *Handbook of Research Methods for Studying Daily Life* (pp. 89–107). New York: The Guilford Press.

Delespaul, P., deVries, M., & van Os, J. (2002). Determinants of Occurrence and Recovery from Hallucinations in Daily Life. *Social Psychiatry and Psychiatric Epidemiology, 37*(7), 97–104.

De Montjoye, A., Hidalgo, C. A., Verleysen, M., & Blondel, V. D. (2013). Unique in the Crowd: The Privacy Bounds of Human Mobility. *Scientific Reports*, 3(1376), 1–5.

Donker, T., Petrie, K., Proudfoot, J., Clarke, J., Birch, M. R., & Christensen, H. (2013). Smartphones for Smarter Delivery of Mental Health Programs: A Systematic Review. *Journal of Medical Internet Research*, 15(11), e247.

Ebner-Priemer, U. W., Welch, S. S., Grossman, P., Reisch, T., Linehan, M. M., & Bohus, M. (2007). Psychophysiological Ambulatory Assessment of Affective Dysregulation in Borderline Personality Disorder. *Psychiatry Research*, 150(3), 265–275.

Elliston, K. G., Ferguson, S. G., Schuz, N., & Schuz, B. (2017). Situational Cues and Momentary Food Environment Predict Everyday Eating Behavior in Adults with Overweight and Obesity. *Health Psychology*, 36(4), 337–345.

Ferguson, S. G., Frandsen, M., Dunbar, M. S., & Shiffman, S. (2015). Gender and Stimulus Control of Smoking Behavior. *Nicotine & Tobacco Research*, 17(4), 431–437.

Firth, J., Cotter, J., Torous, J., Bucci, S., Firth, J. A., & Yung, A. R. (2016). Mobile Phone Ownership and Endorsement of "mHealth" Among People with Psychosis: A Meta-Analysis of Cross-Sectional Studies. *Schizophrophrenia Bulletin*, 42(2), 448–455.

Firth, J., Torous, J., Nicholas, J., Carney, R., Rosenbaum, S., & Sarris, J. (2017). Can Smartphone Mental Health Interventions Reduce Symptoms of Anxiety? A Meta-Analysis of Randomized Controlled Trials. *Journal of Affective Disorders*, 218, 15–22.

Gainsbury, S., & Blaszczynski, A. (2011). A Systematic Review of Internet-Based Therapy for the Treatment of Addictions. *Clinical Psychology Review*, 31(3), 490–498.

Gaudiano, B. A., Moitra, E., Ellenberg, S., & Armey, M. F. (2015). The Promises and Challenges of Ecological Momentary Assessment in Schizophrenia: Development of an Initial Experimental Protocol. *Healthcare (Basel)*, 3(3), 556–573.

Gunlicks-Stoessel, M., Mufson, L., Westervelt, A., Almirall, D., & Murphy, S. (2016). A Pilot SMART for Developing an Adaptive Treatment Strategy for Adolescent Depression. *Journal of Clinical Child & Adolescent Psychology*, 45(4), 480–494.

Hammersley, R. (1994). A Digest of Memory Phenomena for Addiction Research. *Addiction*, 89(3), 283–293.

Havermans, R., Nicolson, N. A., Berkhof, J., & deVries, M. W. (2011). Patterns of Salivary Cortisol Secretion and Responses to Daily Events in Patients with Remitted Bipolar Disorder. *Psychoneuroendocrinology*, 36(2), 258–265.

Hufford, M. R., Shields, A. L., Shiffman, S., Paty, J., & Balabanis, M. (2002). Reactivity to Ecological Momentary Assessment: An Example Using Undergraduate Problem Drinkers. *Psychology of Addictive Behaviors*, 16(3), 205–211.

Hufford, M. R., & Shiffman, S. (2002). Methodological Issues Affecting the Value of Patient-Reported Outcomes Data. *Expert Review on Pharmacoeconomic Outcomes Research*, 2(2), 119–128.

Kaplan, R. M., & Stone, A. A. (2013). Bringing the Laboratory and Clinic to the Community: Mobile Technologies for Health Promotion and Disease Prevention. *Annual Reviews Psychology*, 64, 471–498.

Kimhy, D., Delespaul, P., Ahn, H., Cai, S., Shikhman, M., Lieberman, J. A., ... Sloan, R. P. (2010). Concurrent Measurement of "Real-World" Stress and Arousal in Individuals with Psychosis: Assessing the Feasibility and Validity of a Novel Methodology. *Schizophrenia Bulletin*, 36(6), 1131–1139.

Kirchner, T. R., Cantrell, J., Anesetti-Rothermel, A., Ganz, O., Vallone, D. M., & Abrams, D. B. (2013). Geospatial Exposure to Point-of-Sale Tobacco: Real-Time Craving and Smoking-Cessation Outcomes. *American Journal of Preventive Medicine*, 45(4), 379–385.

Klasnja, P., Hekler, E. B., Shiffman, S., Boruvka, A., Almirall, D., Tewari, A., & Murphy, S. A. (2015). Microrandomized Trials: An Experimental Design for Developing Just-in-Time Adaptive Interventions. *Health Psychology*, 34S, 1220–1228.

Klasnja, P., Consolvo, S., McDonald, D. W., Landay, J. A., & Pratt, W. (2009). Using Mobile & Personal Sensing Technologies to Support Health Behavior Change in Everyday Life: Lessons Learned. *American Medical Informatics Association Annual Symposium*, 2009, 338–342.

Kumar, S., Nilsen, W. J., Abernethy, A., Atienza, A., Patrick, K., Pavel, M., ... Swendeman, D. (2013). Mobile Health Technology Evaluation: The mHealth Evidence Workshop. *American Journal of Preventive Medicine*, 45(2), 228–236.

Labrique, A. B., Vasudevan, L., Kochi, E., Fabricant, R., & Mehl, G. (2013). mHealth Innovations as Health System Strengthening Tools: 12 Common Applications and a Visual Framework. *Global Health Science and Practice*, 1(2), 160–171.

Maclure, M., & Mittleman, M. A. (2000). Should We Use a Case-Crossover Design? *Annual Review of Public Health*, 21, 193–221.

Martin, P. (2003). The Epidemiology of Anxiety Disorders: A Review. *Dialogues in Clinical Neuroscience*, 5(3), 281–298.

McFall, R. M. (1977). Parameters of Self-Monitoring. In R. B. Stuart (Ed.), *Behavioral Self-Management: Strategies, Techniques, and Outcome* (pp. 196–214). New York: Brunner/Mazel.

McClure, J. B., Hartzler, A. L., & Catz, S. L. (2016). Design Considerations for Smoking Cessation Apps: Feedback From Nicotine Dependence Treatment Providers and Smokers. *Journal of Medical Internet Research Mhealth Uhealth*, 4(1), e17.

Mehl, M. R. (2017). The Electronically Activated Recorder (EAR): A Method for the Naturalistic Observation of Daily Social Behavior. *Current Directions in Psychological Science*, 26(2), 184–190.

Mohr, D. C., Zhang, M., & Schueller, S. M. (2017). Personal Sensing: Understanding Mental Health Using Ubiquitous Sensors and Machine Learning. *Annual Review of Clinical Psychology*, 13, 23–47.

Moskowitz, D. S., & Young, S. N. (2006). Ecological Momentary Assessment: What It Is and Why It Is a Method of the Future in Clinical Psychopharmacology. *Journal of Psychiatry and Neuroscience*, 31(1), 13–20.

Myin-Germeys, I., Oorschot, M., Collip, D., Lataster, J., Delespaul, P., & van Os, J. (2009). Experience Sampling Research in Psychopathology: Opening the Black Box of Daily Life. *Psychology & Medicine*, 39(9), 1533–1547.

Nahum-Shani, I., Smith, S. N., Spring, B. J., Collins, L. M., Witkiewitz, K., Tewari, A., & Murphy, S. A. (2016). Just-in-Time Adaptive Interventions (JITAIs) in Mobile Health: Key Components and Design Principles for Ongoing Health Behavior Support. *Annals of Behavioral Medicine*, 52(6), 446–462.

Naughton, F., Hopewell, S., Lathia, N., Schalbroeck, R., Brown, C., Mascolo, C., ... Sutton, S. (2016). A Context-Sensing Mobile Phone App (Q Sense) for Smoking Cessation: A Mixed-Methods Study. *Journal of Medical Internet Research Mhealth Uhealth*, 4(3), e106.

Oorschot, M., Lataster, T., Thewissen, V., Wichers, M., & Myin-Germeys, I. (2012). Mobile Assessment in Schizophrenia:

A Data-Driven Momentary Approach. *Schizophrenia Bulletin, 38*(3), 405–413.

Picard, R. W., Fedor, S., & Ayzenberg, Y. (2016). Multiple Arousal Theory and Daily-Life Electrodermal Activity Asymmetry. *Emotion Review, 8*(1), 62–75.

Przeworski, A., & Newman, M. G. (2004). Palmtop Computer-Assisted Group Therapy for Social Phobia. *Journal of Clinical Psychology, 60*(2), 179–188.

Rodgers, A., Corbett, T., Bramley, D., Riddell, T., Wills, M., Lin, R. B., & Jones, M. (2005). Do U Smoke After Txt? Results of a Randomised Trial of Smoking Cessation Using Mobile Phone Text Messaging. *Tobacco Control, 14*(4), 255–261.

Runyan, J. D., Steenbergh, T. A., Bainbridge, C., Daugherty, D. A., Oke, L., & Fry, B. N. (2013). A Smartphone Ecological Momentary Assessment/Intervention "App" for Collecting Real-Time Data and Promoting Self-Awareness. *PLOS ONE, 8*(8), e71325.

Saeb, S., Zhang, M., Karr, C. J., Schueller, S. M., Corden, M. E., Kording, K. P., Mohr, D.C. (2015). Mobile Phone Sensor Correlates of Depressive Symptom Severity in Daily-Life Behavior: An Exploratory Study. *Journal of Medical Internet Research Mhealth Uhealth, 17*, e175.

Schüz, B., Bower, J., & Ferguson, S. G. (2015). Stimulus Control and Affect in Dietary Behaviours: An Intensive Longitudinal Study. *Appetite, 87*, 310–317.

Schwartz, S., Schultz, S., Reider, A., & Saunders, E. F. (2016). Daily Mood Monitoring of Symptoms Using Smartphones in Bipolar Disorder: A Pilot Study Assessing the Feasibility of Ecological Momentary Assessment. *Journal of Affective Disorders, 191*, 88–93.

Shiffman, S. (2009). Ecological Momentary Assessment (EMA) in Studies of Substance Use. *Psychological Assessment, 21*(4), 486–497.

Shiffman, S. (2014). Ecological Momentary Assessment. In K. J. Sher (Ed.), *The Oxford Handbook of Substance Use Disorders* (Vol. 2). Oxford: Oxford University Press.

Shiffman, S., Stone, A. A., & Hufford, M. R. (2008). Ecological Momentary Assessment. *Annual Review of Clinical Psychology, 4*, 1–32.

Shiffman, S., Dunbar, M. S., Li, X., Scholl, S. M., Tindle, H. A., Anderson, S. J., & Ferguson, S. G. (2014). Smoking Patterns and Stimulus Control in Intermittent and Daily Smokers. *PLOS ONE, 9*(3), e89911.

Shiffman, S., Dunbar, M. S., & Ferguson, S. G. (2015). Stimulus Control in Intermittent and Daily Smokers. *Psychology of Addictive Behaviors, 29*(4), 847–855.

Smyth, J. M., Wonderlich, S. A., Sliwinski, J. M., Crosby, D. R., Engel, S. G., Mitchell, J. E., & Calogero, R. M. (2009). Ecological Momentary Assessment of Affect, Stress, and Binge-Purge Behaviors: Day of Week and Time of Day Effects in the Natural Environment. *International Journal of Eating Disorders, 42*(5), 429–436.

Spook, J. E., Paulussen, T., Kok, G., & Van Empelen, P. (2013). Monitoring Dietary Intake and Physical Activity Electronically: Feasibility, Usability, and Ecological Validity of a Mobile-Based Ecological Momentary Assessment tool. *Journal of Medical Internet Research, 15*(9), e214.

Stone, A. A., & Shiffman, S. (1994). Ecological Momentary Assessment (EMA) in Behavioral Medicine. *Annals of Behavioral Medicine, 16*(3), 199–202.

Stone, A. A., Broderick, J. E., Schwartz, J. E., Shiffman, S., Litcher-Kelly, L., & Calvanese, P. (2003a). Intensive Momentary Reporting of Pain with an Electronic Diary: Reactivity, Compliance, and Patient Satisfaction. *Pain, 104*(1), 343–351.

Stone, A. A., Shiffman, S., Schwartz, J. E., Broderick, J. E., & Hufford, M. R. (2003b). Patient Compliance with Paper and Electronic Diaries. *Controlled Clinical Trials, 24*(2), 182–199.

Tahmasian, M., Khazaie, H., Golshani, S., & Avis, K. T. (2013). Clinical Application of Actigraphy in Psychotic Disorders: A Systematic Review. *Current Psychiatry Reports, 15*(6), 359.

Thornorarinsdottir, H., Kessing, L. V., & Faurholt-Jepsen, M. (2017). Smartphone-Based Self-Assessment of Stress in Healthy Adult Individuals: A Systematic Review. *Journal of Medical Internet Research, 19*(2), e41.

Tilley, P. J., & Rees, C. S. (2014). A Clinical Case Study of the Use of Ecological Momentary Assessment in Obsessive Compulsive Disorder. *Frontiers in Psychology, 5*, 339.

Tomko, R. L., Brown, W. C., Tragesser, S. L., Wood, P. K., Mehl, M. R., & Trull, T. J. (2014). Social Context of Anger in Borderline Personality Disorder and Depressive Disorders: Findings from a Naturalistic Observation Study. *Journal of Personality Disorders, 28*(3), 434–448.

Trull, T. J., Solhan, M. B., Tragesser, S. L., Jahng, S., Wood, P. K., Piasecki, T. M., & Watson, D. (2008). Affective Instability: Measuring a Core Feature of Borderline Personality Disorder with Ecological Momentary Assessment. *Journal of Abnormal Psychology, 117*(3), 647–661.

Ubhi, H. K., Michie, S., Kotz, D., Wong, W. C., & West, R. (2015). A Mobile App to Aid Smoking Cessation: Preliminary Evaluation of SmokeFree28. *Journal of Medical Internet Research, 17*(1), e17.

Vallance, J. K., Winkler, E. A., Gardiner, P. A., Healy, G. N., Lynch, B. M., & Owen, N. (2011). Associations of Objectively-Assessed Physical Activity and Sedentary Time with Depression: NHANES (2005–2006). *Preventive Medicine, 53*(4–5), 284–288.

Walz, L. C., Nauta, M. H., & Aan Het Rot, M. (2014). Experience Sampling and Ecological Momentary Assessment for Studying the Daily Lives of Patients with Anxiety Disorders: A Systematic Review. *Journal of Anxiety Disorders, 28*(8), 925–937.

Wheeler, L., & Reis, H. T. (1991). Self-Recording of Everyday Life Events: Origins, Types, and Uses. *Journal of Personality, 59*(3), 339–354.

Whittaker, R., McRobbie, H., Bullen, C., Rodgers, A., & Gu, Y. (2016). Mobile Phone-Based Interventions for Smoking Cessation. *Cochrane Database of Systematic Reviews, 4*, CD006611.

Wilhelm, F. H., Pfaltz, M. C., & Grossman, P. (2006). Continuous Electronic Data Capture of Physiology, Behavior and Experience in Real Life: Towards Ecological Momentary Assessment of Emotion. *Interacting with Computers, 18*(2), 171–186.

Wright, C. J., Dietze, P. M., Crockett, B., & Lim, M. S. (2016). Participatory Development of MIDY (Mobile Intervention for Drinking in Young People). *BMC Public Health, 16*, 184.

Zhao, J., Freeman, B., & Li, M. (2016). Can Mobile Phone Apps Influence People's Health Behavior Change? An Evidence Review. *Journal of Medical Internet Research, 18*(1), e287.

24 Modeling Intensive Longitudinal Data

MARLIES HOUBEN, EVA CEULEMANS, AND PETER KUPPENS

Research in clinical psychology is largely concerned with the investigation of the presence and severity of a range of different maladaptive feeling states, cognitions and behaviors, which we will refer to as psychopathological symptoms. Typical ways that have been used to assess such symptoms are clinical interviews or retrospective self-reported questionnaires, inquiring how people typically felt and behaved in a certain period of time or in a certain situation. For example, according to the Structured Clinical Interview for DSM disorders (First et al., 2015) and the Center for Epidemiologic Studies Depression Scale (Radloff, 1977), respectively, depressed mood is typically assessed using the following items: "In the last month, has there been a period in which you felt down or depressed for the largest part of the day," and "During the past week, I felt depressed."

However, research increasingly finds that the way these symptoms behave over time – not their mere presence or absence – is a crucial feature that needs to be studied for a better understanding of the phenomenology and mechanisms underlying psychopathology (Myin-Germeys et al., 2009). Indeed, many psychopathological symptoms are not stable or constant over time; rather, they fluctuate and change at a moment-to-moment level, often in response to changes in the environment. Moreover, several symptoms or diagnostic criteria are inherently dynamic, such as emotional instability in borderline personality disorder (BPD; American Psychiatric Association [APA], 2013; Ebner-Priemer et al., 2015; Houben et al., 2016), or persistent depressed mood for a depressive episode (APA, 2013), implying the importance of studying psychopathology from a dynamic viewpoint.

In order to examine the ups and downs of symptoms over time, a data collection method that allows for repeated moment-to-moment assessments of these symptoms in daily life or in the lab is required. In daily life, ambulatory assessment methods (AA) are a very promising approach in which self-reported symptoms, behaviors, or physiological processes are assessed in real time using computerized devices (i.e., smartphones, ambulatory blood pressure devices, etc.), while participants undertake their normal daily activities (Trull & Ebner-Priemer, 2013) (see Chapter 23 by Ferguson, Jahnel, Elliston and Shiffman). Similarly, repeated moment-to-moment assessments can also be collected in the lab, for example, in response to standardized stimuli that are presented to participants on a computer screen (Koval et al., 2013b), or during conversations with social interaction partners (Hollenstein & Lewis, 2006; Kuppens, Allen, & Sheeber, 2010). Repeated assessments can involve self-reported data, observational data, physiological parameters, and so forth.

These data collection approaches typically result in intensive longitudinal data, consisting of many repeated measurements from single participants, that are typically collected over many different time points during the day and potentially also over several consecutive days. These long strings of data contain rich dynamic information, as they allow us to explore the frequency and duration of symptoms over time, as well as dynamic properties of symptoms that are obscured by more traditional data collection methods. Relatedly, they allow us to examine the relationship among symptoms within a person. Rather than focusing on how symptoms co-occur on a person level (e.g., people with higher levels of depressed mood also tend to experience fatigue), intensive longitudinal data allow us to examine how symptoms co-occur or predict one another in the moment within an individual person (e.g., at times where a participant experiences more depressed mood, this co-occurs with higher levels of fatigue), and how these within-person relations differ between people, as a function of person-level features such as diagnostic status or gender.

In this chapter, we present an overview of some major dynamic properties of single or multiple symptoms (and their interrelations) that can be studied with intensive longitudinal data, and how they can be calculated or modeled statistically. We focus primarily on properties that characterize time series as a whole, thereby implicitly assuming that these dynamic features remain unchanged throughout the time series (i.e., the models we present here assume stationary processes). However, it is important to note that these properties can change abruptly or more gradually over time under certain conditions, such as when symptoms become less unstable over time due to treatment progress, or become more or less correlated with one another after the occurrence of a significant event. For simplicity, the majority of methods and models discussed in this chapter do not address such changes in dynamics, and assume stationarity. However, in the discussion section, some models and techniques will be briefly discussed that can be used to investigate such changes in dynamics.

We will first describe methods and models to examine the dynamic properties of a single symptom. Next, we will present methods and models to explore the dynamic relationships between two or more symptoms. For each approach, we will provide information on how to calculate simple indices on a more descriptive level, as well as how to model the dynamic features using more complex models. See Table 24.1 for an overview of the different dynamic properties that can be examined, involving one symptom (i.e., univariate approaches), and two or more symptoms (multivariate approaches), which indices or models to use, and what they exactly reflect. Note that this is not an exhaustive list, and that this chapter addresses frequently used methods and approaches. Last, in the discussion section, we will end with some additional considerations that should be taken into account when dealing with intensive longitudinal data.

UNIVARIATE APPROACHES: THE DYNAMICS OF ONE SYMPTOM OVER TIME

Using intensive longitudinal data, we can explore how a single symptom, for example, depressed mood, behaves over time. In doing so, one can focus on features related to the central tendency of the symptom, the spread and the dynamics over time.

Central Tendency

On a descriptive and most simple level, several indices can be calculated to summarize the central tendency of the time series of one symptom, such as depressed mood. The average or median level of depressed mood of an individual reflects the typical or average moment-to-moment level of depressed mood for that person. Similarly, one can examine the frequency with which a certain symptom, such as depressed mood or the occurrence of nonsuicidal self-injurious behavior, occurs throughout the time series. For symptoms that are assessed on a binary scale (such as nonsuicidal self-injury) or after first dichotomizing symptoms that are assessed on a continuous scale at a clinically relevant cutoff score, one can calculate the proportion of time points in which a symptom was present for each person. These indices provide information about overall or average intensity and frequency of occurrence of a symptom across a certain period of time.

When data from multiple individuals are available, one can examine between-person patterns in average symptom intensity and frequency using multilevel regression models (Hox, Moerbeek, & Schoot, 2010; Nezlek, 2008; Nezlek & Plesko, 2001). If time series data of an individual are short, these models often provide more accurate estimates of the parameters of interest, because data from multiple people are combined in one model (Bulteel et al., 2018). In case of repeated assessments nested within persons, the average or overall intensity of a symptom, such as depressed mood, can be modeled with a two-level linear regression model that includes an intercept that is allowed to differ between people (i.e., a so-called random intercept), and no other predictors (unless predictors are entered person-mean centered; see also Chapter 32 by Bauer, McNeish, Baldwin and Curran). These intercepts reflect the average depressed mood across all time points for each person. One can next investigate how these average levels of depressed mood differ as a function of certain person-specific variables, such as diagnostic status, by adding this variable as a predictor to the person level of the model. For example, Trull and colleagues (2008) used such an approach to examine differences between persons with a diagnosis of borderline personality disorder and major depressive disorder, and found that both groups tend to experience similar levels of overall momentary positive and negative affect.

In a similar way, the frequency or likelihood of occurrence of symptoms can be modeled with a multilevel logistic regression model, in which the log-odds for a symptom to occur at the moment-to-moment level is modeled with a random intercept, again allowing for individual differences between people, that can again be modeled in function of person-specific features. For instance, Houben and colleagues (2016) used this approach to study the occurrence of emotional switches between positive and negative emotional states in individuals who were diagnosed with borderline personality disorder versus healthy controls. However, no differences were found, which means that healthy participants and persons with a borderline personality disorder were equally likely to jump from a positive emotional state to a negative emotional state (or vice versa) on consecutive time points.

Spread Measures

Moving beyond summaries of central tendency using means, medians, and frequencies, the examination of the dispersion of these repeated assessments of symptoms can provide additional interesting insights. Dispersion scores of repeated intensity assessments in an individual reflect how much an individual tends to deviate from their own overall or typical intensity of a symptom over time, reflecting the degree of variability in the intensity of a symptom over time. On a descriptive level, the within-person standard deviation of the repeated scores per person provides the simplest index of variability. This measure reflects the

Table 24.1 Overview of some major dynamic properties of single or multiple symptoms (and their interrelations) and how they can be calculated or modeled

Univariate approaches: examining one variable

Feature		How to calculate/model	Interpretation
Central tendency			
	Descriptive	Mean or median per person	Typical moment-to-moment intensity level of a symptom
		Count or proportion of time points a symptom was present per person	Frequency of a symptom
	Model	(Multilevel) linear regression model with (random) intercept	Average symptom intensity across all time points
		(Multilevel) Logistic regression model	Likelihood of occurrence of a symptom
Spread			
	Descriptive	Within-person standard deviation or variance, within-person interquartile range	Variability; degree of overall deviation or dispersion around typical intensity levels of a symptom
	Model	Error variance at the moment-to-moment level in (multilevel) regression models	Degree of variability of a symptom within-person over time
Dynamics			
Instability	Descriptive	Mean square successive difference and related metrics per person	Average magnitude of changes in intensity of a symptom from one time point to the next
	Model	Squared or absolute successive difference modeled using a (random) intercept in (multilevel) regression models	Average magnitude of changes in intensity of a symptom from one time point to the next
Duration	Descriptive	Count of (successive) time (points) a symptom is present	Duration of a symptom
	Model	Survival analyses	Expected duration until the termination of a symptom
Autocorrelation	Descriptive	Within-person autocorrelation	Inertia, self-predictability of a symptom over time
	Model	(Multilevel) AR models	Inertia, self-predictability of a symptom over time
Time course	Descriptive	Visual plots	Visual inspection of overall time trends in a symptom per person
	Model	(Multilevel) regression models with time as a predictor	Testing time effects in the intensity of symptoms over time
		Growth curve models	Modeling of growth or change of time-related symptoms
		K-spectral centroid clustering method	Match the time course of a symptom with reference intensity profiles, that differ in shape and amplitude

degree of fluctuation or variability around the mean intensity level of a symptom. Similarly, and in combination with the median as a summary measure, one can compute the interquartile range for each person, indicating the range of the middle 50 percent of scores for each person.

In a regression context and for data of multiple individuals, variability of a symptom over time can also be modeled using multilevel models. In case of repeated assessments nested within persons, the error variance at the moment-to-moment level would capture the variability

Table 24.1 (cont.)

Multivariate approaches: examining the relation between two (or more) variables

Feature		How to calculate/model	interpretation
Concurrent relationships			
	Descriptive	Within-person correlation	Strength and direction of the association between two symptoms assessed at a moment-to-moment level for each person
		Intraclass correlation per person	consistency between a group of related variables across different time points for each person
	Model	(Multilevel) regression models with one or more predictors at the moment-to-moment level	Strength and direction of the association between two symptoms on a moment-to-moment level
		PCA and (multilevel) simultaneous component analysis	Summarizes a set of symptoms by extracting one or more components that explain as much of the variance across the time points as possible
Lagged relationships			
	Descriptive	Partial lagged correlation coefficient per person	Strength and direction of the association between two symptoms measured at consecutive time points for each person
	Model	(Multilevel) VAR-models with one lagged predictor	Strength and direction of the association between two symptoms measured at consecutive moments in time; symptom reactivity to context at the moment-to-moment level
Advanced techniques			
	Descriptive	State-space analysis	visual inspection of how two categorical variables co-evolve over time; descriptive indices regarding the dynamics of two categorical variables at the moment-to-moment level simultaneously
	Model	(Multilevel) VAR models with multiple predictors in a network	Network models showing unique and direct effects between a set of symptoms
		Group Iterative Multiple Model Estimation (GIMME)	Approach to examine several lagged and contemporaneous relationships within-person over time
		Bayesian Ornstein-Uhlenbeck Model (BOUM)	Estimation of three different dynamic parameters of two separate continuous variables, simultaneously: home base, the variability around this home base, and the attractor strength.

of a symptom within a person over time. For instance, such an approach was used by Peeters and colleagues (2006), in which variance at both moment-to-moment and day levels was compared between participants with major depressive disorder and healthy controls. Results showed larger moment-to-moment variability in negative affect in the depressed group. However, note that models are needed that allow for heterogeneity of variance across levels of a covariate (e.g., diagnostic status, thereby allowing variance estimates to differ between diagnostic groups), or across subjects (i.e., within-person variance estimates that are allowed to vary between people). This is not a standard assumption of multilevel models, as usually this within-person variance is assumed to be the same (i.e., homogeneous) across subjects. Hedeker, Mermelstein, and Demirtas (2012) describe such a model in which both the mean and within-person variance of repeated assessments are allowed to differ between subjects.

Additionally, one can also examine the proportion of variance in the dependent variable at each level. Estimating empty multilevel models with only intercepts, the estimated variance at the moment-to-moment level and the person level will typically be reported in the output. Next, one can compute the proportion of variance in the dependent variable at each level by dividing the variance at each level of the model by the total variance. In this context, an intraclass correlation coefficient (ICC) is a frequently used metric, calculated as the ratio of between-person variance and total variance. This reflects the proportion of variance located at the between-person level. The proportion of within-person variance is 1-ICC. This provides insight into where variability in the intensity of a symptom is located: is the observed variance in symptom scores mainly driven by differences between people, or by differences from moment to moment within individuals?

Although these measures of spread provide some interesting information, one disadvantage is that they do not take into account *when* the changes in the intensity of a symptom occurred. Two participants could exhibit high levels of variability in the intensity of their depressed mood despite different temporal patterns. For example, one person could experience an increase in depressed mood followed by a decrease, following an inverted U-shape, while a second person could experience frequent ups and downs. Nevertheless, across the full time series, both participants may deviate from their mean levels to a similar extent, resulting in a similar overall variance. To capture the time course of symptoms, the dimension of time has to be taken into account.

Dynamics

When we are interested in how emotions behave over time, we can go beyond the dispersion measures and focus on different aspects of the dynamics of a symptom over time. As such, a researcher can examine how unstable or abrupt symptoms change over time, what the duration of symptoms is, how self-predictive symptoms are over time, or we can explore the full time-course of a symptom.

Instability

A first interesting temporal aspect is the degree to which symptoms abruptly go up and down over time, i.e., the degree of instability. One way of examining the ups and downs of symptoms, while taking into account the temporal dependency of the changes, is to calculate the mean squared successive difference (MSSD) or mean absolute successive difference (MASD) between consecutive assessments of a symptom. With this approach, we calculate the successive difference in the intensity of a symptom. Next, we square these differences (implying that more weight is given to larger changes). Or we take the absolute value of these differences, which is less influenced by extreme changes. This is done to remove signs, because we want to capture the magnitude of changes, irrespective of the direction of the change. Next, the average is taken of these successive (squared or absolute) differences, yielding an index of instability. The square root of the MSSD (RMSSD) can also be computed, to obtain an instability index that is in the original unit of the symptom scores. Because the MSSD (and related indices) quantify the average magnitude of change within a smaller time interval, rather than across the entire time series as is the case for dispersion measures, these measures are better able to capture frequent ups and downs in the intensity of symptoms over time. Indeed, large abrupt changes that occur within short time intervals will yield larger MSSD values, while dispersion measures (such as the within-person standard deviation) can result in similar values both if changes occur in an orderly fashion (i.e., the intensity slowly increases and subsequently decreases), or more abruptly (i.e., the intensity of a symptom rapidly shifts up and down). To illustrate, Figure 24.1 shows time series data for two fictive subjects. Both time series consist of exactly the same data points, therefore yielding the same mean and standard deviation across the time series data for both subjects. For subject 1, data points are ordered in such a way that intensity gradually increases and decreases, resulting in a low MSSD index. On the other hand, for subject 2, the intensity rapidly shifts up and down from time point to time point, resulting in a high MSSD index. This illustrates that, in contrast to measures of spread (reflecting overall variability), the MSSD indices capture time-ordered instability in symptoms. This approach has, for instance, been used by Thompson and colleagues (2012), who found significantly higher levels of emotional instability in negative affect but not positive affect in persons with a diagnosis of major depressive disorder compared to healthy controls.

Next to computing these metrics for each person on a descriptive level, we can again model MSSD and related metrics with multilevel models for multiple people at the same time. In case of repeated assessments of depressed mood, nested within persons, we can estimate two-level linear regression models in which the squared successive difference or the absolute successive difference of a symptom between any two time points is used as the outcome, and modeled with a random intercept at the moment-to-moment level. This random intercept will again allow for individual differences between people in the average estimate, which can again be modeled as a function of person-specific characteristics, such as diagnostic status at the person level of the model. Such an approach was used by

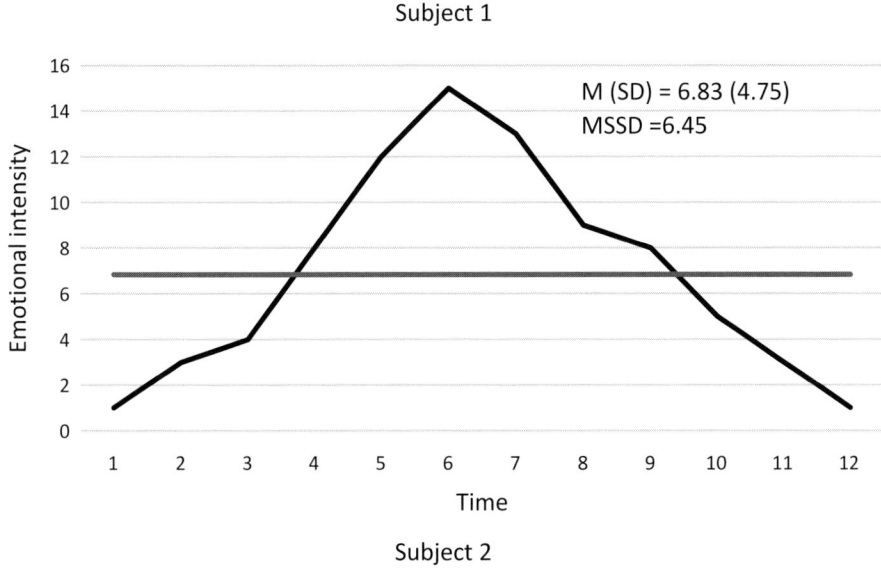

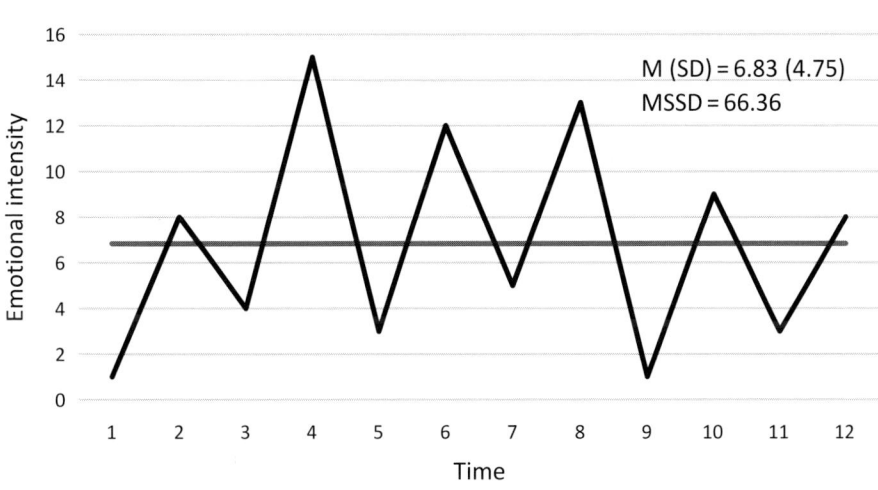

Figure 24.1 Graphs showing simulated time series data for two fictive subjects. Both time series data strings consist of the same data points. Therefore, they have the same mean and standard deviation, showing the same degree of deviation or fluctuation from the average intensity level. However, for subject 2, changes in intensity occur more frequently and more abruptly (reflecting greater instability), which is captured by a higher MSSD index. In contrast, in the time series data of subject 1, changes are more gradual, resulting in a low MSSD index.

Koval and colleagues (2013a), for instance, when they showed that emotional instability in positive affect, as measured with the MSSD of emotional states over time, was related to lower heartrate variability, a physiological indicator of emotion regulation capacity. More detailed information about the MSSD measure and how it can be modeled can also be found in Jahng, Wood, & Trull (2008).

Of note with measures of instability is that a correction for differences in time intervals between consecutive measurements is sometimes necessary. In case the length of time intervals differs largely between consecutive measurements, one can adjust each successive change for the actual time that elapsed, as was done by Thompson and colleagues (2012) and Trull and colleagues (2008). However, even if time intervals between consecutive measurements (within a day) are relatively equal, one should consider removing or correcting for overnight changes, as changes in a symptom between the evening and the next morning might not be meaningful. Additionally, note that instability measures can be highly influenced and driven by average intensity levels. More information can be found in the discussion section of this chapter.

Duration

Duration is another dynamic aspect of symptoms over time. One might be interested to know how long a symptom, such as depressed mood, persisted at or above a clinically meaningful severity level. To obtain this information on a descriptive level, a count of the number of time points, or a sum of the continuous time, that a symptom was consecutively rated or observed at or above a prespecified intensity level would provide this information.

Survival analyses can be used to model the expected duration until a certain event happens, such as the termination of an ongoing process. Hence, they can be used to model the duration of symptoms, such as the time until depression levels decrease below a clinical cutoff score. An example of the application of the use of discrete-time survival analyses to model duration can be found in Verduyn and colleagues (2009), where they used these models to estimate the duration of an emotional episode. Such an analysis provides several statistics related to symptom duration, such as the hazard rate, which reflects the conditional probability that a process such as an ongoing depressive episode that has not yet ended at the beginning of a certain time interval will end during that interval. This hazard rate can next be modeled as a function of person-level predictors. More information can also be found in a study by Sbarra (2006).

Autocorrelation

Intensive longitudinal data allow for the examination of autocorrelations or autoregressive effects of symptoms over time. The autocorrelation captures how self-predictive a symptom is over time. A high autocorrelation reflects high self-predictability, meaning that the current intensity of a symptom is highly predictive of the intensity of the symptom at a following time point, suggesting a strong carry-over effect from one moment to the next. A high autocorrelation thus implies that a symptom is lingering, spiraling off and getting stuck, showing relatively little homeostatic recovery to normative states. It reflects strong resistance to change, or resistance to outside influences: independent of what is happening in the environment, symptoms are self-perpetuating. In the clinical and non-clinical literature, high levels of self-predictability are often referred to as "inertia," named after the concept of inertia in physics, which refers to the slowness and resistance of an object to changing its current state of motion. To illustrate, depression has been consistently linked to high levels of emotional inertia, especially for negative affect (Houben, Van Den Noortgate, & Kuppens, 2015), showing that depression is characterized by patterns of negative emotions that have become stuck, and resistant to change. The negative emotions have spiraled off, without homeostatic return to baseline. Moreover, research has consistently shown that persons with a diagnosis of borderline personality disorder exhibit large within-person variance in their emotional experiences. Additionally, some indications were found for stronger autoregressive effects in their emotional states (Ebner-Priemer et al., 2015; Houben et al., 2015). These findings imply that persons with borderline personality disorder tend to deviate more strongly from their emotional baseline levels, and subsequently tend to linger, and show slower return to baseline, indicating lower emotional recovery.

On a descriptive level, an autocorrelation measure for each person can be calculated by taking the bivariate correlation of a symptom, with a time-lagged version of itself (i.e., the variable is shifted down with one or more time points), reflecting how strongly the intensity of a symptom is related to the intensity of the same symptom at the previous time point. Using a more advanced modeling approach for data from multiple individuals, we can use multilevel autoregressive models (multilevel AR model), where a symptom is predicted by a random intercept, and a time-lagged version of itself at the moment-to-moment level. Making this autoregressive slope a random effect facilitates capturing individual differences between people in the strength of this autoregressive effect, which can further be modeled as a function of person-specific variables. Similar approaches have been used in many studies, in which inertia of, for example mood, was related to low levels of wellbeing and higher levels of psychopathology (e.g., Kuppens et al., 2010a; Thompson et al., 2012). More information regarding the modelling of inertia can also be found in Jahng et al. (2008).

Often, a lag of 1 is used in such AR models (i.e., first-order AR models). However, note that depending on the time scale on which a process of interest is likely to occur and the sample frequency that was used in a study, a different number of lags might be required. A way to determine the correct order of the AR model (i.e., the number of lags needed) is to check for the absence of serial dependencies in the residuals. Different methods are available to do so (for the AR model see, e.g., Box & Jenkins, 1970; for the VAR model see e.g. Brandt & Williams, 2007).

Commonly expected values for such autoregressive effects in psychological data typically range between 0 and 1 (Rovine & Walls, 2006), with 0 reflecting no self-predictive effects, and 1 very high self-predictive effects. However, sometimes negative autoregressive effects can also be found. For example, processes following a sine curve, in which low intensity at one point in time predicts high intensity at another time point, can result in negative autocorrelations, when captured at the correct time scale. Examples of such processes might be diurnal hormone levels, mood patterns in those with rapid cycling bipolar disorder, or food intake.

Note that measures of spread (i.e., within-person variance), instability (i.e., MSSD), and autocorrelation are not independent, but are statistically related. More information about these relations can be found in Jahng and colleagues (2008). Relatedly, a recent study by Dejonckheere and colleagues (2019) also pointed out that different emotion dynamic measures are highly interrelated.

Time Course

Last, for several research questions, it can be interesting to examine the overall course of a symptom over time. For example, how does the intensity of someone's depressed mood progress during the day? Such a question would involve the investigation of diurnal patterns of symptoms. However, the course of a symptom might also be examined over a longer period of time, such as a year. On a descriptive level, the simplest way to examine the time course of a symptom is to plot repeated measurements of a symptom over time, and visually inspect the time course. Such a visual inspection can provide insight into possible time trends in the intensity of a symptom over time. For example, does depressed mood increase or decrease in intensity across the course of a typical day, and how steep is the increase or decrease? On a more complex level, time courses can also be modeled in different ways. First, if data from multiple individuals are available, one can examine how symptom severity changes as a function of time, by using multilevel models, and adding time as a predictor of symptom severity at the moment-to-moment level. As such, we can examine whether symptom severity such as depressed mood increases or decreases linearly with time, or whether a quadratic relationship with time can be observed, which reflects a U-shaped or inverted U-shaped time course of symptom intensity over time. For example, this approach was used by Trull and colleagues (2008), where time trends in positive and negative affect over days and within days in persons with borderline personality disorder and depression were examined and compared, and by Peeters and colleagues (2006), who examined linear and quadratic diurnal patterns in affect, in persons with and without depression.

More broadly, growth curves are a highly flexible class of models that can be used to model growth and change in time-related phenomena and symptoms, and can accommodate a variety of shapes and structures of change. More detailed information can be found in McArdle and Nesselroade (2003).

The time course of symptoms can take many different shapes, however, not always following linear or quadratic functions. As such, some people can experience depressed mood very intensely at the start of the day, and then experience a slow decrease, while other could well experience the reverse pattern or an intense onset at the start of the day, with a second occurrence later that day. Therefore, more advanced models have been developed that allow the modeling of more complex features of the time course of the intensity of symptoms, or intensity profiles. For example, the K-spectral centroid clustering method (Heylen et al., 2016; Yang & Leskovec, 2011) was developed to model different shapes (such as different steepness of onset and skewness) and different amplitudes (height of the profile) of such intensity profiles over time. With this method, first, based on all available data, reference or example profiles are determined, each of which captures a distinct profile shape that can be observed in the data based on shape and amplitude. An example of a typical shape is the early blooming shape, with a steep onset and peak in the intensity of a symptom at the beginning, followed by a slow return to baseline. More examples can be found in Heylen and colleagues (2015, 2016). Next, each observed intensity profile is assigned to one of the determined reference profiles, and receives an amplitude coefficient which indicates the extent to which the reference profile needs to be inflated or deflated to optimally approximate the observed intensity profile. As such, for each person it can be determined which of several possible intensity profile shapes best describes the course of their symptoms over time. Next, assignment of profile shape, but also duration and amplitude, can be predicted by person-level variables, such as diagnostic status. This approach is illustrated in a study by Heylen and colleagues (2015), where they explored different shapes of intensity profiles of episodes of experienced anger. They also examined how these different shapes were related to emotion regulation use. First, they identified two types of anger episodes in the data: early-blooming episodes of anger, during which experienced anger lasts a relatively short time and reaches a peak in intensity soon after the start of the episode, and late-blooming episodes of anger, which last longer and reach their peak relatively later in the episode. Next, they found that early-blooming episodes of anger were associated with adaptive emotion regulation strategies, such as cognitive reappraisal, while late-blooming episodes of anger were linked to maladaptive strategies, such as rumination. Moreover, emotion regulation strategy use was related to the amplitude (reflecting overall intensity) and duration of anger episodes, with adaptive strategies being linked to shorter duration and lower amplitudes. In case several intensity profiles per person are collected, for example one for each day or profiles of multiple symptoms (meaning that intensity profiles are nested within person), a hierarchical (i.e., multilevel) extension of this model can be used to examine differences within the same person regarding the shape of intensity profiles they experience. Next, these differences can be related to other variables, such as overall wellbeing that day (Heylen et al., 2016).

MULTIVARIATE APPROACHES: THE DYNAMICS OF MULTIPLE SYMPTOMS

The first part of the chapter focuses on how to model several dynamic features of a single symptom. Researchers or clinicians are often interested in how several symptoms behave over time and, more importantly, how they mutually co-occur and influence each other. Several options exist to explore such dynamic properties of multiple symptoms. We first propose some ways to examine concurrent relationships (i.e., relations between variables, measured at the same time) between two or more symptoms, and next discuss some options to investigate possible prospective relationships further. We will end with some more advanced techniques that allow for simultaneously modeling of several of these properties and relations.

Concurrent Relationships

Some research questions involve the examination of how two or more symptoms are related or co-occur in the moment. For example, does depressed mood typically co-occur with fatigue on a moment-to-moment level, and is the association positive or negative? On a descriptive level, such questions can be answered by computing indices of co-occurrence for each individual. First, a within-person correlation per person quantifies the strength and the direction of the moment-to-moment relationship between two symptoms for each individual. In case of data from multiple people, one can similarly center the variable within-person (i.e., center each person's time series on their mean) and next compute a normal correlation coefficient for all data points pooled across individuals. In the case of more than two variables, the intraclass correlation coefficient can be used. Although different types of ICCs exist, the most often used ICC in this research context quantifies consistency between a group of related variables across different time points (Shrout & Fleiss, 1979), which is equivalent to a Cronbach's alpha. A high ICC indicates a high correlation between several variables across different measurements points. For example, the ICC has been extensively used in research on emotion differentiation, which reflects the degree to which people label their emotions in a differentiated and specific way, rather than in global ways. In this field, ICC is used as a measure of emotion differentiation, where a high ICC is assumed to reflect low emotion differentiation (i.e., emotions are rated as rising and falling together), as emotions covary strongly with each other across the different measurement points, and therefore people do not strongly distinguish between these emotions (Erbas et al., 2014; Tomko et al., 2015).

Using a modeling approach for data from multiple individuals, the relationship between two or more symptoms can be examined using multilevel regression models, in which the intensity of one symptom is predicted by one or more other symptoms at the moment-to-moment level. Such analyses can provide insight into how strongly and in which direction (positively or negatively) one symptom, such as depressed mood, is related to one or more other symptoms, such as fatigue and concentration problems, in the moment. Using random slopes, individual differences in these relationships can be modeled as a function of other person-specific variables, such as diagnostic status. An example of such models can be found in Hepp and colleagues (2016), who examined how several core features of borderline personality disorder, such as negative affectivity, impulsivity, and interpersonal problems, are related to close social contact on a momentary level, and how personality traits further impact these relationships.

Another approach to examine concurrent relationships between several symptoms across time is to apply principal component analysis (PCA) to the data of each individual. This dimension-reduction approach summarizes the variables (i.e., symptoms) to one or more components that explain as much of the variance across the time points as possible (Cattell, 1952; Jones & Nesselroade, 1990) and are a linear combination of the symptoms. As in standard principal component analysis or factor analysis of cross-sectional data, the interpretation of the components is based on the loadings of the variables on the components. For example, if depressed mood and other mood-related symptoms would load highly on the same component, that component would likely reflect affect-related disturbances. Symptoms that load strongly on the same component typically covary significantly over time. Each time point has a score on each of the components, which can be plotted against time to visualize the fluctuations over time. Extensions of this approach have been proposed that allow symptom data from multiple individuals to be analyzed simultaneously, such as (multilevel) simultaneous component analysis (Ceulemans et al., 2016; Timmerman, 2006; Timmerman & Kiers, 2003). These models allow one set of components to be extracted for all persons (i.e., the loadings are fixed across persons), and component scores to be derived for each time point of each individual. These component scores allow examination of differences between persons in within-person variance (e.g., one person might show more variability in affect-related disturbances over time than others). Moreover, one can inspect differences between people in within-person component correlations: the extracted components might co-occur more strongly over time for some persons and be almost independent for others.

Lagged or Prospective Relationships

One major strength of intensive longitudinal data is that it allows us to look beyond concurrent relationships between two or more symptoms, enabling researchers to examine prospective relationships between symptoms. That is, detailed analyses can reveal how symptom intensity, such as momentary fatigue, is related to other symptoms, such as depressed mood, at the next point in time, providing information about the temporal directionality of the relationship.

On a descriptive level, an index for such prospective associations is the partial lagged correlation coefficient. This is computed by taking the correlation of a variable, such as depressed mood, with a second variable, such as fatigue, after being lagged one time point. Such a correlation quantifies the strength and the direction of the association between current depressed mood, and preceding fatigue. However, to preclude that the obtained association is driven by concurrent rather than prospective relationships, the partial correlation is preferred, which is additionally controlled for the autocorrelation effect (i.e., the lagged version) of the first variable (i.e., depressed mood in our example), and possibly also the concurrent version of the second variable (i.e., fatigue in our example).

On a modeling level, using data from multiple persons, prospective associations can be modeled with multilevel vector-autoregressive (VAR) models. Such models are extensions of the AR models described earlier in which, next to the lagged version of the dependent variable (i.e., autoregressive effects), additional predictors measured at the previous time point are also added (i.e., cross-regressive effects). Using the same example, examining how fatigue is associated with depressed mood at the following time point, multilevel VAR models can be used where depressed mood is predicted by fatigue at the previous time point (i.e., lagged with one time point), also adding depressed mood at the previous time point as a predictor. Possibly current fatigue can also be added as a predictor on the moment-to-moment level. This allows us to examine to what degree current depressed mood is predicted by or associated with fatigue at the previous time point, above and beyond depressed mood at the previous time point, and current fatigue. Using random slopes, individual differences in these prospective associations can be modeled, also as a function of person-level features, such as diagnostic status. For example, Houben and colleagues (2017) used VAR models to examine the effect of the occurrence of nonsuicidal self-injury on the intensity of subsequent positive and negative affect, correcting for preceding positive and negative affect, and found that the occurrence of NSSI is related to high levels of negative affect and low levels of positive affect at the following time point.

Note that next to the associations between symptoms, context information can also be included in such models, allowing for the investigation of symptom reactivity to contextual variables. As such, estimating models in which depressed mood is predicted by depressed mood at the previous time point, and a context variable, such as the occurrence of a negative event or a social encounter, allows us to examine depressed mood reactivity on event occurrence. Such models have been used in research examining mood reactivity in psychopathology (e.g. Bylsma, Taylor-Clift, & Rottenberg, 2011).

Advanced Techniques

Last, next to examining concurrent or prospective relationships between two or more variables, more advanced techniques exist that allow for the modeling of several types of relationships or more complex relations between two or more symptoms simultaneously. While we will not provide an exhaustive list of different models that exist, we will give a short overview of some possibilities.

Descriptive Techniques

On a descriptive level, the state-space approach (extensively used by Hollenstein and colleagues; e.g., Hollenstein & Lewis, 2006; Lougheed & Hollenstein, 2016) provides a method to visually inspect how two categorical variables coevolve over time and to derive descriptive indices regarding the dynamics of two variables simultaneously. Typically using specialized software, *GridWare 1.1* (Lamey et al., 2004), these methods have been used to inspect and describe the trajectory of dyadic emotions (i.e., combinations of emotional states from two individuals) of parent-child pairs during social interactions. These different dyadic emotional states are typically depicted in a grid, with the emotional state of one person on the x-axis, and the emotional state of the second person on the y-axis. Each combination represents a possible dyadic emotional state. Next, the trajectory of these dyadic states can be inspected, and indices can be calculated to quantify the dynamics of these states. As such, a dispersion measure can be computed that reflects the range of different dyadic states that are endorsed across the grid. This dispersion measure then reflects whether all behavior occurred in one cell versus whether behavior was equally distributed across the grid. Second, a measure of transitions can be computed that captures the number of changes or transitions between different cells on the grid. In the context of dyadic emotion states, both measures are assumed to reflect greater flexibility, and have been shown to be associated with better psychosocial adjustment (Hollenstein & Lewis, 2006; Lougheed & Hollenstein, 2016).

Modeling Approaches

Using a modeling approach, different options exist to examine the dynamics of two or more symptoms and complex interrelations between the symptoms. (Multilevel) vector-autoregressive (VAR) models can be used, in which, next to the lagged version of the dependent variable, a range of different predictors measured at the previous time point are also added. As such, a VAR model can be used to determine to what degree a symptom is predicted by itself, and by a range of other additional symptoms, measured at the previous time point. Next, these VAR models can form the basis for network models (Bringmann et al., 2016), in which the unique and direct effects between a set of symptoms are visualized. Coefficients derived from such VAR models, in which each time a different symptom serves as outcome, are used to determine the strength of the connections or "edges" between the different symptoms or "nodes" in such a network model (for some cautionary notes, however, see Bulteel et al., 2016). Additionally, several features can be derived to describe the dynamical interplay between the symptoms or nodes in the network model, such as the density of the overall network, which indicates how strongly the symptoms in the network are interconnected (Pe et al., 2015), or node centrality, referring the importance of a single node in the network (Bringmann et al., 2013). For some additional cautionary notes, see Bringmann and colleagues (2019).

To examine the interrelationships between several symptoms within persons, Group Iterative Multiple Model Estimation (GIMME) is also a promising approach in which, next to several lagged relationships, as is the case for VAR models, contemporaneous relationships within a person over time can also be examined. It is a structural equation method (SEM)-based method for identifying group-, subgroup-, and individual-level relations within time series data of several variables. Unlike the multilevel VAR model, GIMME estimates person-specific models for each individual, as well as searching for common features across individuals. This takes a different approach to dealing with heterogeneity of effects across individuals. This method can be applied, using the GIMME R package. Initially developed for functional magnetic resonance imaging (fMRI) data, it has recently been proven to work well with intensive repeated behavioral measures (e.g., daily diary data; Lane et al., 2019; Wright et al., 2019; more information can be found in Lane & Gates, 2017).

Another approach to simultaneously model several dynamic features of two separate symptoms is the Bayesian (and Hierarchical) Ornstein-Uhlenbeck Model (BOUM; Kuppens, Oravecz, & Tuerlinckx, 2010b; Oravecz, Tuerlinckx, & Vandekerckhove, 2011). It is based on the Ornstein-Uhlenbeck (OU) model, with a hierarchical (i.e., multilevel) extension. This implies that it can take into account individual differences between people. Moreover, it is estimated in a Bayesian framework. Using the BOUM model, three different dynamic parameters of two separate variables or dimensions can be estimated simultaneously: (1) it estimated a home base for each dimension, which reflects the average or normative state for a variable; (2) it estimates the amount of variability around this home base over time; and (3) it estimates the attractor strength, which reflects regulatory forces that pull a variable back toward the home base. Because these processes are estimated for two variables at the same time, covariation between the two dimensions can be considered. As the model also allows for individual differences in the estimated parameters, these parameters can be regressed on covariates to explain these individual differences. Moreover, this model takes time into account as a continuous variable, meaning that it takes into account the actual time that elapsed between consecutive ratings. Therefore, it is ideal for measurements taken at possibly irregular time points.

The BOUM model has been used in several studies (Ebner-Priemer et al., 2015; Santangelo et al., 2016) on affective dysregulation in borderline personality disorder, as defined by Linehan (1993). According to Linehan (1993), this dysregulation is reflected in heightened sensitivity to emotional stimuli, strong emotional reactivity, and a slow return to baseline. These three parameters can be mapped, one-on-one, onto the home base, variability, and attractor strength parameters that are estimated in BOUM. Therefore, the BOUM model offered a way to estimate all three parameters underlying emotional dysregulation in BPD simultaneously, and to examine possible differences between persons with a diagnosis of BPD and healthy controls in these parameters, by adding diagnostic status as a covariate in the model. For detailed statistical information about the BOUM model and custom-made statistical software to run the analyses, see Oravecz et al. (2011) and Oravecz, Tuerlinckx, & Vandekerckhove (2012).

DISCUSSION

The behaviors, thoughts, and feelings that indicate psychopathology are often not of a static nature, but rather change and fluctuate over time in response to changes in the environment and daily life situations. Therefore, clinical psychology research can benefit from focusing on how psychopathological features behave over time, as it can provide new perspectives and insights concerning the phenomenology and mechanisms underlying psychopathology.

The collection of intensive longitudinal data, consisting of many repeated measurements from single participants, allows for the investigation of several dynamic properties of single or multiple symptoms (and their interrelations). In this chapter, we presented an overview of some major dynamic properties that can be studied with intensive longitudinal data, and how they can be calculated or modeled statistically. First, we focused on several univariate approaches, allowing the examination of one single symptom or feature over time. Next, we discussed some methods and models to further examine the dynamic relationships between two or more symptoms. The diagram shown in Figure 24.2 provides a further guide on which method or model to use for which kind of data and/or research questions.

To end, some general issues should be taken into account when dealing with intensive longitudinal data. First, in this chapter we described several methods and models to examine dynamic features of symptoms and the interrelations between symptoms. However, as noted in the introduction, most methods and models discussed implicitly assume that these dynamic features of symptoms or the interrelations between symptoms remained unchanged throughout the time series. This is of course not always the case, as properties or relations between symptoms can change or shift abruptly, for example after the occurrence of a significant event, or gradually, for example during and after treatment. If it is known where the change occurred, because a participant initiated a treatment, say, single case study designs can be used to further explore these changes (for more information about such models, see, e.g., Manolov & Onghena, 2018). If it is not known where and how many changes occurred, changes can be detected using change point analyses (for a comparison of several techniques, see, e.g., Cabrieto et al., 2017). Such a technique was also used in Wichers and colleagues (2016) to detect a sudden change in depressive symptoms in data from one participant who had a history of multiple episodes of major depression (for a further analysis of this data, see Cabrieto et al., 2018). When changes in dynamic features occur more gradually, one can consider detrending the data first (Jahng et al., 2008), or applying methods that can deal with gradual changes in dynamic features such as time-varying AR and VAR models (Bringmann et al., 2017, 2018).

Second, when examining dynamic features of symptoms time plays an essential role. However, note that for some models and indices (such as the MSSD, AR, and VAR models) time is considered as a discrete variable, meaning that time is seen as a categorical variable consisting of mere consecutive time points. On the other hand, time

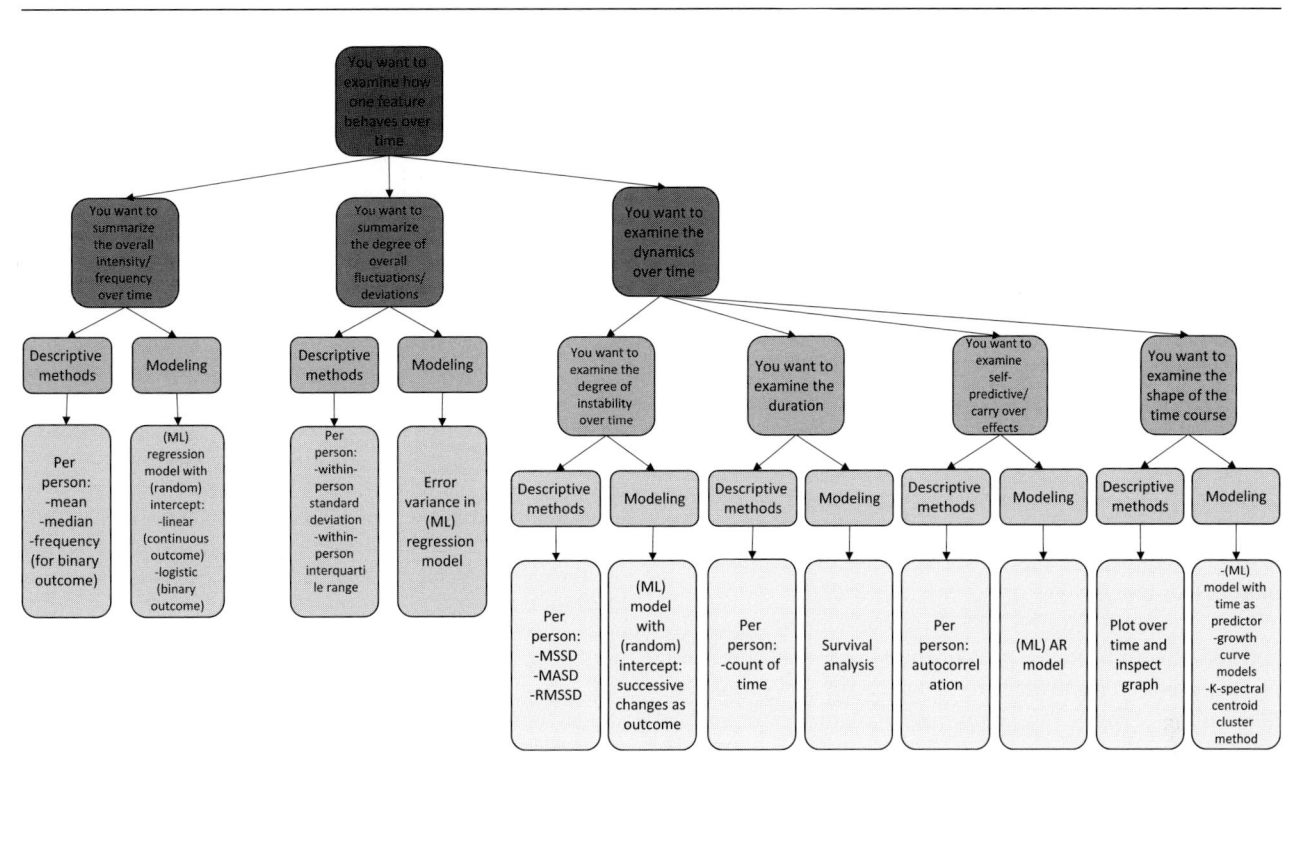

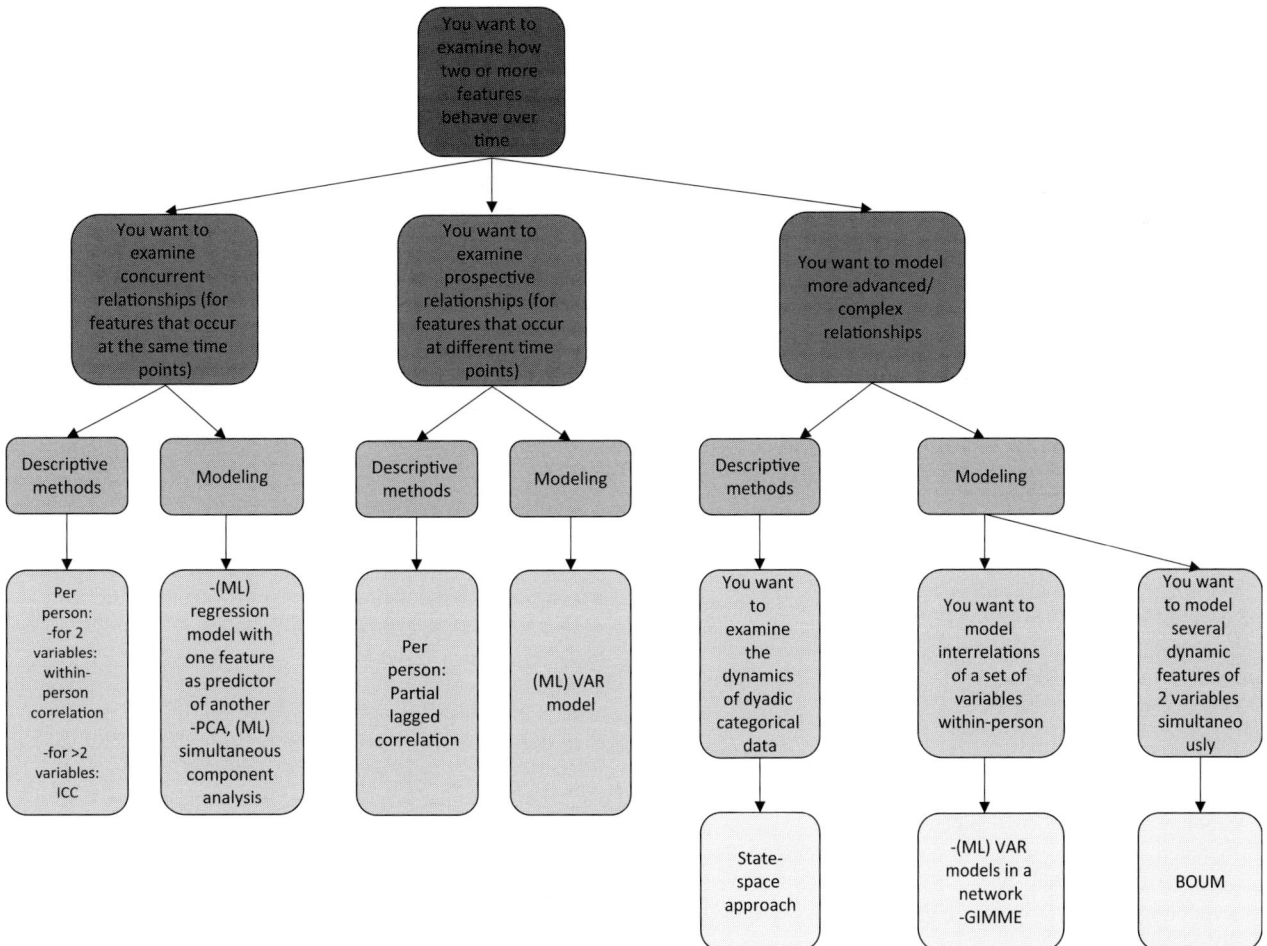

Figure 24.2 Diagram showing which models to use for several types of research questions, related to one variable (above) or multiple variables (below).

can also be considered as a continuous variable, where the actual elapsed time is taken into account. Examples of such models that are suited for time series data are the BOUM model described above (Oravecz et al., 2011), and the time-varying structural equation modeling (that comes with R package ctsem; Driver, Oud, & Voelkle, 2017). Such models would be more optimal when there is large variability in the time intervals between consecutive measurements.

Third, the more advanced models might offer elegant ways of modeling complex dynamic features and complex interrelations between symptoms. Still note that very complex models are not always the best choice, and more simple indices of dynamic features and interrelations between symptoms are often to be preferred in several situations, for example if the length of the time series data is limited (Bulteel et al., 2018; Krone, Albers, & Timmerman, 2016; Liu, 2017).

Last, dynamic features of symptoms can be typically correlated with average levels of the symptom. Therefore, often correction for average levels of a symptom should be taken into account when examining dynamic features (for example, see a method proposed by Mestdagh et al., 2018). To illustrate, Dejonckheere and colleagues (2019) indicated that the predictive value of different emotion dynamic measures for psychological wellbeing is limited after they corrected for overlap with average affect levels.

To conclude, intensive longitudinal data allow us to examine how clinical symptoms and features change over time and dynamically influence each other. As a consequence, it offers us a new approach for studying psychopathology: it enables us to look beyond static traits, and model processes that change over time. It allows us to look beyond group effects, and focus on person-specific effects that unfold within individuals over time. As such, it is a promising approach to further explore phenomenology and mechanisms underlying psychopathology, and model change in response to treatment.

REFERENCES

American Psychiatric Association (APA) (Ed.). (2013). *Diagnostic and Statistical Manual of Mental Disorders: DSM-5* (5th edn.). Washington, DC: American Psychiatric Association.

Box, G. E. P., & Jenkins, G. M. (1970). *Time Series Analysis: Forecasting and Control*. San Francisco, CA: Holden-Day.

Brandt, P. T., & Williams, J. T. (2007). *Multiple Time Series Models*. Thousand Oaks, CA: Sage.

Bringmann, L. F., Vissers, N., Wichers, M., Geschwind, N., Kuppens, P., Peeters, F., … Tuerlinckx, F. (2013). A Network Approach to Psychopathology: New Insights into Clinical Longitudinal Data. *PLOS ONE, 8*(4), e60188.

Bringmann, L. F., Pe, M. L., Vissers, N., Ceulemans, E., Borsboom, D., Vanpaemel, W., … Kuppens, P. (2016). Assessing Temporal Emotion Dynamics Using Networks. *Assessment, 23*(4), 425–435.

Bringmann, L. F., Hamaker, E. L., Vigo, D. E., Aubert, A., Borsboom, D., & Tuerlinckx, F. (2017). Changing Dynamics: Time-Varying Autoregressive Models Using Generalized Additive Modeling. *Psychological Methods, 22*(3), 409–425.

Bringmann, L. F., Ferrer, E., Hamaker, E. L., Borsboom, D., & Tuerlinckx, F. (2018). Modeling Nonstationary Emotion Dynamics in Dyads using a Time-Varying Vector-Autoregressive Model. *Multivariate Behavioral Research, 53*, 293–314.

Bringmann, L. F., Elmer, T., Epskamp, S., Krause, R. W., Schoch, D., Wichers, M., … Snippe, E. (2019). What Do Centrality Measures Measure in Psychological Networks? *Journal of Abnormal Psychology*. Retrieved from https://doi.org/10.1037/abn0000446

Bulteel, K., Tuerlinckx, F., Brose, A., & Ceulemans, E. (2016). Using Raw VAR Regression Coefficients to Build Networks can be Misleading. *Multivariate Behavioral Research, 51*(2–3), 330–344.

Bulteel, K., Mestdagh, M., Tuerlinckx, F., & Ceulemans, E. (2018). VAR(1) Based Models Do not Always Outpredict AR(1) Models in Typical Psychological Applications. *Psychological Methods, 23*, 740–756.

Bylsma, L. M., Taylor-Clift, A., & Rottenberg, J. (2011). Emotional Reactivity to Daily Events in Major and Minor Depression. *Journal of Abnormal Psychology, 120*(1), 155–167.

Cabrieto, J., Tuerlinckx, F., Kuppens, P., Grassmann, M., & Ceulemans, E. (2017). Detecting Correlation Changes in Multivariate Time Series: A Comparison of Four Non-Parametric Change Point Detection Methods. *Behavior Research Methods, 49*(3), 988–1005.

Cabrieto, J., Tuerlinckx, F., Kuppens, P., Hunyadi, B., & Ceulemans, E. (2018). Testing for the Presence of Correlation Changes in a Multivariate Time Series: A Permutation Based Approach. *Scientific Reports, 8*(1), 769.

Cattell, R. B. (1952). The Three Basic Factor-Analytic Research Designs: Their Interrelations and Derivatives. *Psychological Bulletin, 49*, 499–520.

Ceulemans, E., Wilderjans, T. F., Kiers, H. A. L., & Timmerman, M. E. (2016). MultiLevel Simultaneous Component Analysis: A Computational Shortcut and Software Package. *Behavior Research Methods, 48*, 1008–1020.

Dejonckheere, E., Mestdagh, M., Houben, M., Rutten, I., Sels, L., Kuppens, P., & Tuerlinckx, F. (2019). Complex Affect Dynamics Add Limited Information to the Prediction of Psychological Well-Being. *Nature Human Behaviour, 3*, 478–491.

Driver, C. C., Oud, J. H. L., & Voelkle, M. C. (2017). Continuous Time Structural Equation Modeling with R Package ctsem. *Journal of Statistical Software, 77*(5), 1–35.

Ebner-Priemer, U. W., Houben, M., Santangelo, P., Kleindienst, N., Tuerlinckx, F., Oravecz, Z., … Kuppens, P. (2015). Unraveling Affective Dysregulation in Borderline Personality Disorder: A Theoretical Model and Empirical Evidence. *Journal of Abnormal Psychology, 124*(1), 186–198.

Erbas, Y., Ceulemans, E., Pe, M. L., Koval, P., & Kuppens, P. (2014). Negative Emotion Differentiation: Its Personality and Well-Being Correlates and a Comparison of Different Assessment Methods. *Cognition and Emotion, 28*(7), 1196–1213.

First, M. B., Williams, J. B. W., Karg, R. S., & Spitzer, R. L. (2015). *Structured Clinical Interview for DSM-5 – Research Version (SCID-5 for DSM-5, Research Version; SCID-5-RV)*. Arlington, VA: American Psychiatric Association.

Hedeker, D., Mermelstein, R. J., & Demirtas, H. (2012). Modeling Between-Subject and Within-Subject Variances in Ecological

Momentary Assessment Data Using Mixed-Effects Location Scale Models. *Statistics in Medicine, 31*(27), 3328–3336.

Hepp, J., Carpenter, R. W., Lane, S. P., & Trull, T. J. (2016). Momentary Symptoms of Borderline Personality Disorder as a Product of Trait Personality and Social Context. *Personality Disorders, 7*(4), 384–393.

Heylen, J., Verduyn, P., Van Mechelen, I., & Ceulemans, E. (2015). Variability in Anger Intensity Profiles: Structure and Predictive Basis. *Cognition and Emotion, 29*(1), 168–177.

Heylen, J., Van Mechelen, I., Verduyn, P., & Ceulemans, E. (2016). KSC-N: Clustering of Hierarchical Time Profile Data. *Psychometrika, 81*(2), 411–433.

Hollenstein, T., & Lewis, M. D. (2006). A State Space Analysis of Emotion and Flexibility in Parent-Child Interactions. *Emotion, 6*(4), 656–662.

Houben, M., Van Den Noortgate, W., & Kuppens, P. (2015). The Relation between Short-Term Emotion Dynamics and Psychological Well-Being: A Meta-Analysis. *Psychological Bulletin, 141*(4), 901–930.

Houben, M., Vansteelandt, K., Claes, L., Sienaert, P., Berens, A., Sleuwaegen, E., & Kuppens, P. (2016). Emotional Switching in Borderline Personality Disorder: A Daily Life Study. *Personality Disorders: Theory, Research, and Treatment, 7*(1), 50–60.

Houben, M., Claes, L., Vansteelandt, K., Berens, A., Sleuwaegen, E., & Kuppens, P. (2017). The Emotion Regulation Function of Nonsuicidal Self-Injury: A Momentary Assessment Study in Inpatients with Borderline Personality Disorder Features. *Journal of Abnormal Psychology, 126*(1), 89–95.

Hox, J. J., Moerbeek, M., & van de Schoot, R. (2010). *Multilevel Analysis: Techniques and Applications* (2nd edn.). New York: Routledge.

Jahng, S., Wood, P. K., & Trull, T. J. (2008). Analysis of Affective Instability in Ecological Momentary Assessment: Indices Using Successive Difference and Group Comparison via Multilevel Modeling. *Psychological Methods, 13*(4), 354–375.

Jones, C. J., & Nesselroade, J. R. (1990). Multivariate, Replicated, Single-Subject, Repeated Measures Designs and P-Technique Factor Analysis: A Review of Intraindividual Change Studies. *Experimental Aging Research, 16*, 171–183.

Koval, P., Ogrinz, B., Kuppens, P., den Bergh, O. V., Tuerlinckx, F., & Sütterlin, S. (2013a). Affective Instability in Daily Life Is Predicted by Resting Heart Rate Variability. *PLOS ONE, 8*(11), e81536.

Koval, P., Pe, M. L., Meers, K., & Kuppens, P. (2013b). Affect Dynamics in Relation to Depressive Symptoms: Variable, Unstable or Inert? *Emotion, 13*(6), 1132–1141.

Krone, T., Albers, C. J., & Timmerman, M. E. (2016). Comparison of Estimation Procedures for Multilevel AR(1) Models. *Frontiers in Psychology, 7*, 486.

Kuppens, P., Allen, N. B., & Sheeber, L. B. (2010a). Emotional Inertia and Psychological Maladjustment. *Psychological Science, 21*(7), 984–991.

Kuppens, P., Oravecz, Z., & Tuerlinckx, F. (2010b). Feelings Change: Accounting for Individual Differences in the Temporal Dynamics of Affect. *Journal of Personality and Social Psychology, 99*(6), 1042–1060.

Lamey, A., Hollenstein, T., Lewis, M. D., & Granic, I. (2004). GridWare (Version 1.1). [Computer software]. Retrieved from http://statespacegrids.org

Lane, S. T., & Gates, K. M. (2017). Automated Selection of Robust Individual-Level Structural Equation Models for Time Series Data. *Structural Equation Modeling: A Multidisciplinary Journal, 24*(5), 1–15.

Lane, S. T., Gates, K. M., Pike, H. K., Beltz, A. M., & Wright, A. G. C. (2019). Uncovering General, Shared, and Unique Temporal Patterns in Ambulatory Assessment Data. *Psychological Methods, 24*, 54–69.

Linehan, M. (1993). *Cognitive-Behavioral Treatment of Borderline Personality Disorder*. New York: Guilford Press.

Liu, S. (2017). Person-Specific versus Multilevel Autoregressive Models: Accuracy in Parameter Estimates at the Population and Individual Levels. *British Journal of Mathematical and Statistical Psychology, 70*, 480–498.

Lougheed, J. P., & Hollenstein, T. (2016). Socioemotional Flexibility in Mother-Daughter Dyads: Riding the Emotional Rollercoaster across Positive and Negative Contexts. *Emotion, 16*(5), 620–633.

Manolov, R., & Onghena, P. (2018). Analyzing Data from Single-Case Alternating Treatments Designs. *Psychological Methods, 23*, 480–504.

McArdle, J. J., & Nesselroade, J. R. (2003). Growth Curve Analysis in Contemporary Psychological Research. In J. A. Schinka & W. F. Velicer (Eds.), *Handbook of Psychology: Research Methods in Psychology* (Vol. 2, pp. 447–480). New York: Wiley.

Mestdagh, M., Pe, M. L., Pestman, W., Verdonck, S., Kuppens, P., & Tuerlinckx, F. (2018). Sidelining the Mean: The Relative Variability Index as a Generic Mean-Corrected Variability Measure for Bounded Variables. *Psychological Methods, 23*, 690–707.

Myin-Germeys, I., Oorschot, M., Collip, D., Lataster, J., Delespaul, P., & van Os, J. (2009). Experience Sampling Research in Psychopathology: Opening the Black Box of Daily Life. *Psychological Medicine, 39*(9), 1533–1547.

Nezlek, J. B. (2008). An Introduction to Multilevel Modeling for Social and Personality Psychology. *Social and Personality Psychology Compass, 2*(2), 842–860.

Nezlek, J. B., & Plesko, R. M. (2001). Day-to-Day Relationships among Self-Concept Clarity, Self-Esteem, Daily Events, and Mood. *Personality and Social Psychology Bulletin, 27*(2), 201–211.

Oravecz, Z., Tuerlinckx, F., & Vandekerckhove, J. (2011). A Hierarchical Latent Stochastic Differential Equation Model for Affective Dynamics. *Psychological Methods, 16*(4), 468–490.

Oravecz, Z., Tuerlinckx, F., & Vandekerckhove, J. (2012). BHOUM: Bayesian Hierarchical Ornstein-Uhlenbeck Modeling [Computer Software]. Retrieved from www.zitaoravecz.net/

Pe, M. L., Kircanski, K., Thompson, R. J., Bringmann, L. F., Tuerlinckx, F., Mestdagh, M., ... Gotlib, I. H. (2015). Emotion-Network Density in Major Depressive Disorder. *Clinical Psychological Science, 3*(2), 292–300.

Peeters, F., Berkhof, J., Delespaul, P., Rottenberg, J., & Nicolson, N. A. (2006). Diurnal Mood Variation in Major Depressive Disorder. *Emotion, 6*(3), 383–391.

Radloff, L. S. (1977). The CES-D Scale: A Self-Report Depression Scale for Research in the General Population. *Applied Psychological Measurement, 1*(3), 385–401.

Rovine, M., & Walls, T. (2006). A Multilevel Autoregressive Model to Describe Interindividual Differences in the Stability of a Process. In J. Schafer and T. Walls (Eds.), *Models for Intensive Longitudinal Data* (pp. 124–147). New York: Oxford University Press.

Santangelo, P. S., Limberger, M. F., Stiglmayr, C., Houben, M., Coosemans, J., Verleysen, G., ... Ebner-Priemer, U. W. (2016). Analyzing Subcomponents of Affective Dysregulation in Borderline Personality Disorder in Comparison to Other Clinical Groups Using Multiple e-Diary Datasets. *Borderline Personality Disorder and Emotion Dysregulation, 3*(1), 5.

Sbarra, D. A. (2006). Predicting the Onset of Emotional Recovery Following Nonmarital Relationship Dissolution: Survival Analyses of Sadness and Anger. *Personality & Social Psychology Bulletin, 32*(3), 298–312.

Shrout, P. E., & Fleiss, J. L. (1979). Intraclass Correlations: Uses in Assessing Rater Reliability. *Psychological Bulletin, 86*, 420–428.

Thompson, R. J., Mata, J., Jaeggi, S. M., Buschkuehl, M., Jonides, J., & Gotlib, I. H. (2012). The Everyday Emotional Experience of Adults with Major Depressive Disorder: Examining Emotional Instability, Inertia, and Reactivity. *Journal of Abnormal Psychology, 121*(4), 819–829.

Timmerman, M. E. (2006). Multilevel Component Analysis. *British Journal of Mathematical and Statistical Psychology, 59*(2), 301–320.

Timmerman, M. E., & Kiers, H. A. L. (2003). Four Simultaneous Component Models for the Analysis of Multivariate Time Series from More than One Subject to Model Intraindividual and Interindividual Differences. *Psychometrika, 68*(1), 105–121.

Tomko, R. L., Lane, S. P., Pronove, L. M., Treloar, H. R., Brown, W. C., Solhan, M. B., ... Trull, T. J. (2015). Undifferentiated Negative Affect and Impulsivity in Borderline Personality and Depressive Disorders: A Momentary Perspective. *Journal of Abnormal Psychology, 124*(3), 740–753.

Trull, T. J., & Ebner-Priemer, U. (2013). Ambulatory Assessment. *Annual Review of Clinical Psychology, 9*, 151–176.

Trull, T. J., Solhan, M. B., Tragesser, S. L., Jahng, S., Wood, P. K., Piasecki, T. M., & Watson, D. (2008). Affective Instability: Measuring a Core Feature of Borderline Personality Disorder with Ecological Momentary Assessment. *Journal of Abnormal Psychology, 117*, 647–661.

Verduyn, P., Delvaux, E., Van Coillie, H., Tuerlinckx, F., & Van Mechelen, I. (2009). Predicting the Duration of Emotional Experience: Two Experience Sampling Studies. *Emotion, 9*(1), 83–91.

Wichers, M., Groot, P. C., & Psychosystems, ESM Group, EWS Group. (2016). Critical Slowing Down as a Personalized Early Warning Signal for Depression. *Psychotherapy and Psychosomatics, 85*(2), 114–116.

Wright, A. G. C., Gates, K. M., Arizmendi, C., Lane, S. T., Woods, W. C., & Edershile, E. A. (2019). Focusing Personality Assessment on the Person: Modeling General, Shared, and Person Specific Processes in Personality and Psychopathology. *Psychological Assessment, 31*, 502–515.

Yang, J., & Leskovec, J. (2011). Patterns of Temporal Variation in Online Media. In *Proceedings of the Fourth ACM International Conference on Web Search and Data Mining* (pp. 177–186). New York: ACM. Retrieved from http://dl.acm.org/citation.cfm?id=1935863

25 Modeling the Individual

Bridging Nomothetic and Idiographic Levels of Analysis

PETER C. M. MOLENAAR AND ADRIENE M. BELTZ

Intensive longitudinal and ambulatory assessment data are increasingly used in clinical psychology; they include ecological momentary assessment, daily diary, and physiological data from wearable devices (see Chapter 23, by Ferguson et al.; Schiffman, Stone, & Hufford, 2008). They generally consist of the repeated collection of data on subjects' behavior and experience in their natural environments. The data thus obtained are replicated time series data, given that multiple subjects are participating in the study. These data can in principle be analyzed in two distinct ways: either by pooling across the replications (subjects) or by analyzing the data of each subject (replication) individually. The first approach yields nomothetic results that are valid at the level of the population from which the subjects (replications) have been sampled, while the second approach yields idiographic results that are valid at the level of each individual subject.

Suppose there are N subjects who each have been repeatedly observed on a p-variate variable at T equidistant time points. The observations pertain to the psychological states characterizing each subject at a particular time point (or aggregation of time points, e.g., across a day). Then at each time point a standard (cross-sectional) factor analysis can be carried out by pooling across the N subjects. Also N distinct factor analyses of the p-variate time series associated with each subject can be carried out by pooling across time points (see below for further description). The important question then arises whether the factor solutions obtained under both scenarios are the same. It was shown in Molenaar (2004) that this is in general not the case: factor solutions obtained by pooling across subjects do not correspond to solutions obtained by pooling across time points within each subject (i.e., subject-specific data analysis). Only if a psychological process obeys strict criteria (to be specified shortly) will the two types of factor analysis yield the same solution. Because most psychological processes are expected to violate these criteria – even clinically relevant processes, such as personality, that were once thought to be traits (Molenaar & Campbell, 2009) – these processes need to be analyzed in subject-specific ways in order to obtain results which apply at the individual level.

In what follows, the rationale of subject-specific data analysis is explained and compared with person-oriented approaches (Bergman, Magnusson, & El-Khouri, 2003). It is based on Molenaar (2015). Person-oriented approaches in which the person is taken as the unit of analysis have some prominence in clinical psychology (e.g., Mäkikangas & Kinnunen, 2016) and are related to subject-specific data analysis in a way to be explained shortly.

SUBJECT-SPECIFIC DATA ANALYSIS

The origin of subject-specific data analysis considered in this chapter is mathematical-statistical theory, and a concise presentation of the relevant part of this theory is provided here. First, interindividual and intraindividual variation are explained; they are essential to understand the necessity of subject-specific data analysis. Second, the relation between results obtained with data analysis of interindividual variation and intraindividual variation is discussed, making clear that generally no substantial relation exists, even if the same variables are measured using identical instruments. It will be explained that this lack of relation has fundamental consequences for the analysis of intensive longitudinal and ambulatory assessment data, necessitating the use of subject-specific data analysis. For extensive background material on subject-specific methods, readers are referred to Molenaar and Newell (2010), Molenaar, Lerner, and Newell (2014), and Valsiner and colleagues (2009).

DEFINITION OF INTER- AND INTRAINDIVIDUAL VARIATION

A heuristic is introduced that has proven to be very useful during several decades of psychological research to make the concepts of inter- and intraindividual variation more concrete – the data box (Cattell, 1952). It is especially relevant to clinical time series data. The basic data box is

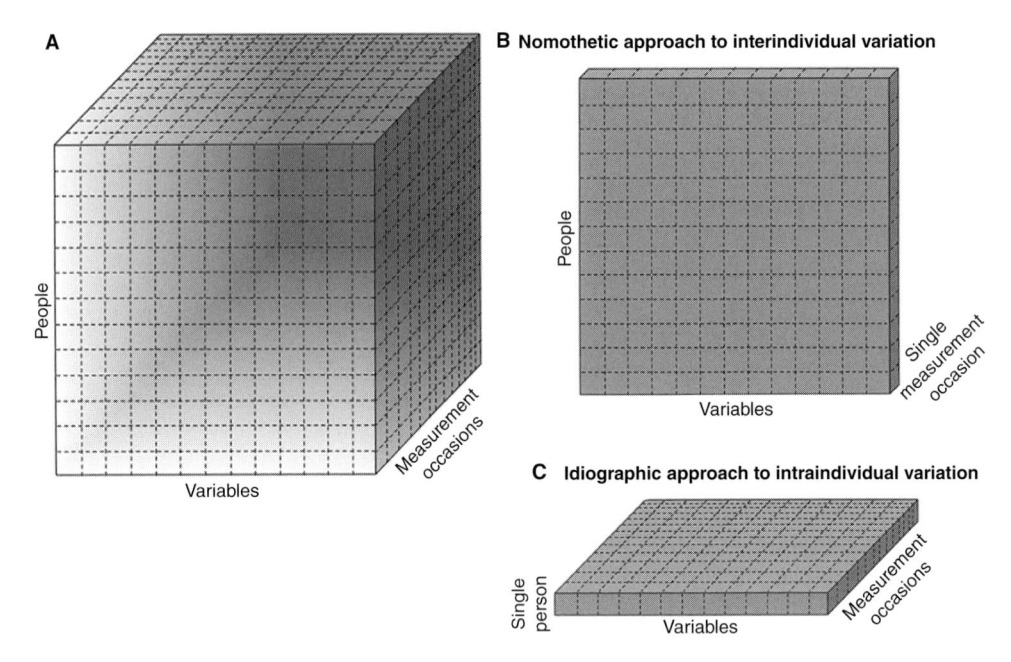

Figure 25.1 Cattell's (1952) data box (A) with two orthogonal slices corresponding to interindividual variation (B) and intraindividual variation (C).
Permission to reproduce from Beltz et al. (2016) by SAGE Publications.

a cube defined by an axis for persons, an axis for variables, and an axis for occasions of measurement. Each element in the data box is a datum representing an intersection of axes and is thus a single score for a given person on a given variable at a given occasion of measurement. In Figure 25.1, a data box is depicted organized by person, variable, and occasion of measurement. Two orthogonal slices taken from this data box correspond, respectively, to inter- and intraindividual data. In its purest form the interindividual data slice involves the data of a sample of persons assessed at a single occasion on a given set of variables. Again, in its purest form the intraindividual data slice involves the data of a single person assessed at a sampled stretch of consecutive occasions on the same given set of variables.

Given that inter- and intraindividual data slices taken from the same data box are orthogonal to each other, the question arises how the structures of these obviously distinct types of variation can be related. For instance, using the same set of variables, does factor analysis of interindividual variation yield the same results as factor analysis of intraindividual variation? This question is of fundamental importance to clinical science insofar as its results are motivated by the goal of understanding individual subjects. If there is no relation between results obtained in statistical analyses of interindividual and intraindividual variation, then results obtained by analyzing interindividual variation cannot be validly applied at the level of individual assessment. That is, such results would not pertain to individuals and therefore might be relevant in nonpsychological settings (e.g., epidemiology, sociology) but would not be of direct importance to psychology understood as the science of individual human functioning. Likewise, if results from interindividual analyses do not apply to individuals, this could undermine applications to clinical work that focus on understanding the etiology, presentation, progression, and ultimately treatment of a disorder in an individual.

RELATION BETWEEN INTER- AND INTRAINDIVIDUAL VARIATION

What follows in this and the next section is a summary of parts of Molenaar and Nesselroade (2015), to which the reader is referred for a complete treatment. The standard approach to statistical analysis in clinical science is to draw a random sample of subjects from a presumably homogeneous population of subjects, analyze the structure of interindividual variation in this sample, and then generalize the results thus obtained to the population. Such analyses of interindividual variation underlie all standard statistical techniques in psychology, including analysis of variance, regression analysis, factor analysis, multilevel (e.g., latent growth curve) modeling, cluster analysis, mixture modeling, etc. Consequently, the standard approach to psychological data analysis aims to describe the state of affairs at the population level, not at the level of individual subjects. Accordingly, the individuality of each of the subjects in the sample and population is considered to be of no consequence: the subjects are considered to be replications (i.e., exchangeable random draws describing the same phenomenon). This is expressed by the assumption that subjects are homogeneous in all respects relevant to the analysis. This homogeneity assumption allows for the averaging (pooling) of the scores of the sampled subjects in the estimation of statistics (means, variances, correlations, etc.) to be generalized to the population. Pooling across subjects is the hallmark of analyses of interindividual variation.

In contrast, the natural mathematical-statistical model for intraindividual variation is a dynamic systems model of the time-dependent changes of an individual's behavior. In its standard form a dynamic systems model explains time-dependent changes in the dependent variable as a function of the current value of that dependent variable as well as external input. Given that the standard statistical approach to the analysis of psychological processes unfolding in time is based on longitudinal factor analysis of interindividual variation, not intraindividual variation, the fundamental question arises whether such a psychometric approach yields results which are valid at the individual level. In other words, do results from dynamic models of person-specific variation captured by, for example, ecological momentary assessments, map onto factor-analytic results collapsed across people in a sample? This question has been addressed before, for instance in Wohlwill's (1973) monograph on developmental processes and Lamiell's (2003) work in personality psychology. In this chapter a definitive negative answer is presented.

The standard statistical approach to dynamic systems modeling of psychological processes based on analysis of interindividual variation yields results that cannot be validly applied at the individual level if these processes do not obey stringent conditions (Molenaar, 2004). The proof is based on classical ergodic theory, a set of theorems of extreme generality which apply to all measurable processes (see Choe, 2005, for a modern proof of the so-called individual ergodic theorem of Birkhoff, 1931). To understand the implications of these theorems, the elementary methodological situation in psychological measurement is explained. Instead of postulating an abstract population of subjects, we consider an ensemble of human subjects as they actually exist, whose measurable psychological processes are functions of time. The resulting basic scientific representation of each human subject in psychology therefore requires a high-dimensional dynamic system (i.e., having many concurrent measurements, as do intensive longitudinal data) which generates a set of time-dependent processes. The system includes important functional subsystems such as the perceptual, emotional, cognitive, and physiological subsystems, as well as their dynamic interrelations. The complete set of measurable time-dependent variables characterizing the system's behavior can be represented as the coordinates of a high-dimensional space which will be referred to as the *behavior space*. The behavior space contains all the scientifically relevant information about a human subject (see De Groot, 1954).

A useful dictionary definition of variation is: "The degree to which something differs, for example, from a former state or value, from others of the same type, or from a standard" (Molenaar, 2004). To simplify the following discussion, variation will be understood to be quantified in terms of covariance matrices, although the gist of what follows also applies to alternative operationalizations of the dictionary definition. Within the behavior space, interindividual variation is defined as follows:

(i) select a fixed subset of variables;
(ii) select one or a few fixed time points as measurement occasions;
(iii) draw a random sample of N subjects
(iv) determine the variation of the scores (i.e., covariance matrices) on the selected variables at the selected fixed time points by pooling across the sampled subjects.

Analysis of interindividual variation thus defined has been called R-technique by Cattell (1952). In contrast, intraindividual variation is defined as follows:

(i) select a fixed subset of variables;
(ii) select one or more fixed subject(s);
(iii) determine the variation of the scores (i.e., covariance matrices) of each single subject on the selected variables by pooling across a sampled time interval.

Analysis of intraindividual variation thus defined has been called (replicated) P-technique by Cattell (1952).

With these preliminary specifications in place, the following heuristic description of the content of Birkhoff's (1931) individual ergodic theorem details the conditions that must be met in order to generalize results obtained in analyses of interindividual variation to results obtained in analyses of intraindividual variation, and vice versa. A process is *non-ergodic* if the results of analyses of interindividual variation do not generalize to the level of intraindividual change over time, and vice versa. In what follows, only Gaussian (normally distributed) processes are considered. The criteria that a Gaussian process must meet in order to be *ergodic* is twofold (see Hannan, 1970).

1. The process has to be *homogeneous in time*, having constant mean levels, no cycles, and sequential dependencies which only depend upon relative time differences (lags). Such a process is called "weakly stationary."
2. The process has to be *homogeneous across different subjects* in the population. That is, each subject in the population (ensemble) has to obey exactly the same dynamic model.

In the context of longitudinal factor analysis, for instance, the first criterion implies that all model parameters (factor loadings, etc.) have to be constant in time while the latter criterion implies that each subject has to obey the same factor model in which the number of factors, the factor loadings, the measurement error variances, and the factor score intercorrelations are invariant across subjects. Invariance should be considered theoretically (e.g., are parameters expected to align across subjects) and statistically (in which case inferences will depend upon features of the analyses and data, such as sample size and variability).

If a Gaussian process is either nonstationary (violating the homogeneity in time criterion) or heterogeneous across subjects (violating the homogeneity across subjects criterion), or both, then it is non-ergodic. This means that there does not exist any lawful relation between the structure of interindividual variation at the population level and the structures of intraindividual variation at the level of individual subjects belonging to the population. Put another way, if the conditions of ergodicity are violated, no lawful relations exist between results obtained in an analysis of interindividual variation (R-technique) and results obtained in an analogous analysis of intraindividual variation (P-technique). Consider personality as an example (Molenaar & Campbell, 2009). Analyses of interindividual variation suggest there are five factors; they represent the relations among the complete set of cognitions and behaviors *that can be experienced in a population*. But, individuals do not similarly engage in all possible cognitions and behaviors. In fact, one set of analyses of intraindividual variation suggested no single person had a five factor model of personality that aligned with the validated population model; some people had as few as two factors (Molenaar & Campbell, 2009)! These individual-level personality dimensions represent the relations among cognitions and behaviors *actually engaged in by an individual over time*. For instance, "irritable" aligns with other cognitions and behaviors to indicate the neuroticism factor in the population, but for an individual, it may align with "sociable" and "silent," perhaps reflecting a person-specific social anxiety factor.

The consequences of the classical ergodic theorems affect all psychological statistical methodology (Borsboom, 2005; Molenaar, Huizenga, & Nesselroade, 2003). Because a wide range of central psychological processes like learning, information processing, habituation, disorder progression or remission, development, and adaptation generally imply that some kind of growth or decline occurs, these processes are almost always nonstationary (violating the homogeneity in time criterion for ergodicity) and are, therefore, non-ergodic. This implies that their analysis should be based on intraindividual variation so that valid information at the level of individual subjects is obtained. Importantly, starting with analyses of intraindividual variation does not preclude valid generalization across subjects. But such valid generalization cannot proceed in the standard way of pooling across subjects, as in analysis of interindividual variation techniques.

THE PRINCIPLES OF PERSON-ORIENTED THEORY

There is an excellent general exposition of the holistic-interactionistic perspective of person-oriented theory (e.g., Bergman et al., 2003), but the focus in this chapter is on the six principles or tenets underlying person-oriented theory as presented in Sterba and Bauer (2010a) in the context of developmental psychopathology research. These are listed with brief descriptions in the first two columns of Table 25.1: (1) Individual Specificity, (2) Complex Interactions, (3) Interindividual Differences in Intraindividual Change, (4) Pattern Summary, (5) Holism, (6) Pattern Parsimony.

Sterba and Bauer (2010a) then consider four types of methods to test these six principles, which are listed in columns 3–6 in Table 25.1. These four types are: (a) Less Restrictive Variable Oriented (latent growth curve modeling), (b) Classification (latent class growth analysis and latent Markov modeling), (c) Hybrid Classification (growth mixture modeling and mixed latent Markov modeling), and (d) Single Subject (dynamic factor analysis). Sterba and Bauer (2010a) cross-list these four types of methods with the six principles of person-oriented theory in terms of how well they allow for testing these principles. Their conclusions are summarized with check marks in Table 25.1, with larger check marks denoting greater testability. In what follows the focus is on the Single Subject type of method (highlighted in gray in Table 25.1).

Sterba and Bauer (2010a) conclude that the Single Subject type of method enables testing of principles 1 (Individual Specificity), 4 (Pattern Summary), 5 (Holism) and 6 (Pattern Parsimony). In contrast, they consider principle 2 (Complex Interactions) untestable with Single Subject methods, while principle 3 (Interindividual Differences in Intraindividual Change) is considered to have limited testability.

Dynamic factor analysis in its current form, however, is much more general than assumed by Sterba and Bauer (2010a), and therefore, the Single Subject type of method enables testing of person-oriented principles 2 (Complex Interactions) and 3 (Interindividual Differences in Intraindividual Change) (Molenaar, 2010a); this is denoted by plus signs in Table 25.1. In particular, the following generalizations of dynamic factor analysis are important for the present discussion. First, the assumption of weak stationarity (criterion 1 for ergodicity), on which initial versions of the dynamic factor model in psychometrics have been based (Molenaar, 1985), can be dropped (Molenaar et al., 2009, 2016). This yields dynamic factor models in which arbitrary subsets of parameters (factor loadings, auto- and cross-lagged regression parameters, mean vectors) can be time-varying. Second, the dynamic factor model can be extended to include time-varying covariates (called measured input in engineering). (For an interesting application of this extension involving subject-specific optimal control of diabetes type 1 patents in real time, see Wang et al., 2014). Third, the dynamic factor model can be applied to replicated (e.g., multiple subjects) intensive longitudinal and ambulatory assessment data. In fact, this can be done in various ways, as

Table 25.1 Person-oriented principles and methods adapted from Sterba and Bauer (2010a) and updated by Molenaar (2010a)

Principle	Brief description	Less Restrictive Variable Oriented	Classification	Hybrid Classification	Single Subject
1. Individual Specificity	Psychological processes are at least in part unique to the individual	✔		✔	✔
2. Complex Interactions	Psychological processes involve interactions at multiple levels	✔	✔	✔	✚
3. Interindividual Differences in Intraindividual Change	Psychological processes show between-person structure in within-person variation	✔		✔	✚
4. Pattern Summary	Psychological processes show lawful patterns of the involved factors			✔	✔
5. Holism	Factors derive their meanings from their mutual interactions	✔	✔	✔	✔
6. Pattern Parsimony	At a sufficiently macro level psychological processes show a finite number of patterns		✔	✔	✔

Note. Adapted from Tables 1 and 2 in Sterba and Bauer (2010a); ✔ denotes testability according to Sterba and Bauer (2010a), with larger sizes representing greater testability; ✚ denotes testability according to Molenaar (2010a).

will be considered below. Given this, the label Single Subject for this type of data analysis is unfortunate because it incorrectly suggests that application to multiple subjects is impossible.

Thus, the current generalized form of the dynamic factor model enables testing of all six principles characterizing person-oriented theory (Molenaar, 2010a) – and Sterba and Bauer (2010b) now agree. They claim, however, that time-varying factor loadings in the context of dynamic factor models violate the criteria for measurement invariance, and therefore that the meanings of the latent dynamic factors continuously change in time (Sterba & Bauer, 2010b). This requires further comment.

GENERALIZED MEASUREMENT INVARIANCE IN DYNAMIC FACTOR MODELS

The operationalization of measurement invariance in terms of invariant factor loadings is well established and almost universally accepted by psychometricians (see Millsap, 2011) and is related to edgodicity. Using this operationalization, it is immediately clear that the presence of time-varying factor loadings implies a continuous violation of measurement invariance and, again given this operationalization, leads to the conclusion that the meanings of the latent factor series also change continuously. However, recently it has been suggested that the standard operationalization of measurement invariance in terms of invariant factor loadings is too limited for the purpose of analysis of intraindividual variation (Molenaar, 2014). If a factor loading in a dynamic factor model is changing continuously then this may not always be due to a change in the meaning of the latent factor series. One salient (though not psychological) example concerns the detection of the direction of arrival of N airplanes (Nesselroade & Molenaar, 2016). Given a linear array of radar stations, a (complex-valued) N-factor model (in the frequency domain) is used to solve this, in which each airplane constitutes a factor and its reflected radar signals on each station constitute the factor loadings. The resulting factor model has time-varying factor loadings because the planes are constantly moving (i.e., the solid angle of each moving plane with each radar station is changing continuously) and hence continuously violate the criterion of invariant factor loadings for measurement invariance. But it would be nonsensical to conclude that the meanings of the N factors, that is, the identity of the N airplanes, change in time-dependent ways (see Nesselroade & Molenaar, 2016, for further details and mathematical specifications).

Other examples could be given (Molenaar, 2014) and extensions to clinical science are warranted. In the meantime, however, focus on the theoretical innovations in measurement equivalence testing in the context of dynamic models (e.g., factor and state space) provide some useful insight. They are described in Box 25.1.

Box 25.1 Recent innovations in measurement invariance testing in dynamic models

For a given linear dynamic q-factor model the q-variate latent factor series is identified up to rotation, similar to the situation in standard factor analysis of interindividual variation. That implies that two different factor models are equivalent if the matrix of factor loadings of one can be adequately aligned by Procrustes rotation. If that is indeed possible then the two models are measurement invariant. Hence the set of measurement invariant linear (dynamic) factor models constitutes a group with (q,q)-dimensional matrix rotation as group action.

The theoretical question now can be raised of how to generalize this algebraic perspective on measurement invariance in terms of sets of measurement invariant models in a way that accommodates time-varying factor loadings. Which alternative group with different appropriate group action could be considered? A key observation in this context is the following. Suppose attention is restricted to linear state space models with time-varying factor loadings, and autoregressive and/or cross-lagged regression coefficients. To fit such a state space model with a q-variate latent state process to the data it is transformed by adding all M free model parameters to the latent state process (latent factor series; Molenaar et al., 2009, 2016). This extension to a $(q + M)$-dimensional state process transforms the initial linear state space model into a nonlinear equivalent one. This transformation of linear state space models with time-varying parameters to nonlinear equivalent models suggests that the question stated above can be answered by considering equivalence transformations in nonlinear state space models. Hence what is the analog for nonlinear state spacemodels of factor rotation in linear state space models?

The following theorem in Isidori (1985) provides the answer. Let $\mathbf{z} = \boldsymbol{\Gamma}(\mathbf{x})$ be a nonlinear invertible smooth state transformation (global diffeomorphism). Let

$$(\blacklozenge) \quad \mathbf{y}(t) = \mathbf{h}[\mathbf{x}(t),t] + \mathbf{v}(t)$$
$$\mathbf{x}(t+1) = \mathbf{f}[\mathbf{x}(t),t] + \mathbf{w}(t+1)$$

be a nonlinear state space model where $\mathbf{y}(t)$ is a p-variate observed series, $\mathbf{x}(t)$ is a q-variate latent state process, $\mathbf{h}[.]$ and $\mathbf{f}[.]$ are, respectively, p- and q-variate smooth nonlinear functions, and $\mathbf{v}(t)$ and $\mathbf{w}(t)$ are, respectively p-variate measurement error and q-variate process noise, which each lack any sequential dependencies. Then applying the global diffeomorphism $\mathbf{z} = \boldsymbol{\Gamma}(\mathbf{x})$ it follows that $(\blacklozenge)$ is equivalent to (Isidori, 1985):

$$(\blacklozenge\blacklozenge) \quad \mathbf{y}(t) = \mathbf{h}^*[\mathbf{z}(t),t] + \mathbf{v}(t)$$
$$\mathbf{z}(t+1) = \mathbf{f}^*[\mathbf{z}(t),t] + \mathbf{g}^*[\mathbf{z}(t),t]\mathbf{w}(t+1)$$

where $h^*[\mathbf{z},t] = \mathbf{h}[\mathbf{x},t]$ at the point $\mathbf{x} = \boldsymbol{\Gamma}^{-1}(\mathbf{z})$, while $f^*[\mathbf{z},t] = [\frac{\partial \Gamma}{\partial x} f[\mathbf{x},t]]$, and $g^*[\mathbf{z},t] = [\frac{\partial \Gamma}{\partial x}]$, taking the $\mathbf{x}$-value in the derivatives at the right-hand sides of these equations again at the point $\mathbf{x} = \boldsymbol{\Gamma}^{-1}(\mathbf{z})$. It is noted that if $\boldsymbol{\Gamma}(x)$ is linear, $\mathbf{z} = \boldsymbol{\Gamma}\mathbf{x}$ with $\boldsymbol{\Gamma}$ a (q,q)-dimensional nonsingular matrix, then this diffeomorphic transformation reduces to standard rotation in the linear state space model.

The conjecture is that this diffeomorphic equivalence transformation for smooth nonlinear state space models opens up the possibility to substantially generalize the operationalization and testing of measurement invariance. In particular, it allows for a principled generalization of measurement equivalence in linear state space models with time-varying parameters. Further implementation and testing of this generalized measurement equivalence is a work in progress – it was presented for the first time before a psychometric audience only recently (Molenaar, 2014, 2015).

OPTIMAL CONTROL OF PSYCHOLOGICAL PROCESSES

A final topic concerns the possibility of optimal control of dynamic psychological processes (also discussed in Molenaar, 2010a); it extends an engineering subfield to behavioral science. This has huge implications for clinical psychology, especially for personalized medicine. Consider behaviors assessed in a series of psychotherapy sessions as the observed time series (reflecting a single subject's dynamical system), and the words and actions of the therapist as external input (i.e., time-varying covariate) intended to alter that system (into a more optimal or stable functional state).

If a dynamic factor model yields a satisfactory fit to an observed p-variate time series $\mathbf{y}(t)$, and this model involves a dynamic regression of the q-variate latent factor series $\boldsymbol{\eta}(t)$ on a measured s-variate time-varying covariate process $\mathbf{u}(t)$, then optimal control of $\boldsymbol{\eta}(t)$ is possible by judicious choice of $\mathbf{u}(t)$ (Molenaar, 2015). The measured covariate process $\mathbf{u}(t)$ is referred to as the input.

To apply optimal control, nothing else is required but a dynamic factor model with time-varying external input as described above. (Computational control theory in the psychological context has been previously described; Molenaar, 2010b.) Suppose that the required dynamic factor model is available at time t. Then at this time t the expected value of $\boldsymbol{\eta}(t + 1)$ is determined and the (partly) manipulable components of $\mathbf{u}(t)$ are computed such that a so-called cost function is minimized. This cost function usually consists of the squared deviation of $\boldsymbol{\eta}(t)$ from its desired value $\boldsymbol{\eta}^*(t)$ in combination with the costs of applying control actions. The latter cost usually is quantified by the squared deviation of $\mathbf{u}(t)$ from its desired value $\mathbf{u}^*(t)$. It is noted that this computational scheme can be applied recursively in real time at $t, t + 1, t + 2$, etc., as the observed series $y(t)$ becomes available sequentially.

The determination of the desired values of the state $\eta^*(t)$ and manipulated input $u^*(t)$ are up to the controller. For instance, if $\eta(t)$ relates to disease symptoms then an obvious choice for $\eta^*(t)$ is zero, although more sophisticated choices are possible in which the development of particular patterns of values for $\eta(t)$ (syndromes) is especially penalized. Together the determination of the details of the cost function and the mathematical techniques used to determine optimal input are called the design parameters (see Molenaar, 2010b, for example parameters).

Actual application of mathematical optimal guidance/control to psychological processes is still rare, although consideration of feedback and homeostasis at the theoretical level appears to be quite popular. An illuminating example of a successful application to the optimization of an individual psychotherapeutic process is, however, available (Molenaar, 1987). A 26-year-old client suffering from physical tension, anxiety, lack of self-confidence, and social poise is followed during 61 therapy sessions (twice per week). At the start of each session the client is assessed with respect to his emotional tone, self-esteem, social activities, and physical fitness during the period since the last session. At the close of each session the therapist is assessed with respect to the time spent counseling/modeling, positive and negative reinforcement, and their style. A dynamic systems model with time-varying parameters is fitted to these data and optimal control is computed for each session. That is, the amount of counseling/modeling, reinforcement, and interactive style are determined in such a way that the discrepancy between the predicted client behavior at the next session and the desired values are minimized. It is found that the feedback functions with which these optimal therapist actions are determined are substantially time-varying across the sequence of 61 sessions.

Wang and colleagues (2014) presented a recent application to the optimal control of individual diabetes type 1 patients in real time under normal living conditions that permitted evaluation of model performance. Through recursively fitting a dynamic factor model with time-varying parameters to each individual patient, it was possible to predict each patient's blood glucose level 30 minutes ahead with more than 90 percent fidelity. This time interval of 30 minutes sufficed for fast-acting insulin to have noticeable effects. It was necessary to fit the model to each patient individually because the time-dependent changes in blood glucose and insulin effects vary substantially in subject-specific ways both within and between patients.

BROADER TENETS OF SUBJECT-SPECIFIC DATA ANALYSIS: BRIDGING NOMOTHETIC AND IDIOGRAPHIC LEVELS OF ANALYSIS

The practical implications of the classical ergodic theorems are straightforward. To obtain results about non-ergodic psychological processes that are valid at the level of individual subjects, the analysis must be based on intraindividual, not interindividual, variation. Because most psychological processes violate the two necessary criteria for ergodicity, subject-specific analysis of intraindividual variation as described above should become the new norm in psychometrics. This, however, may raise the specter of a completely fragmented psychological science – in the words of one anonymous reviewer, "a different psychological theory for each individual subject." On closer scrutiny, this appears not to be the case, however.

Consider the application to optimal control of diabetes type 1 patients discussed previously. This application has to be entirely subject-specific in order to obtain high-fidelity predictions and control. Yet the same mathematical-statistical dynamic model was applied to each individual patient. Only the estimated time-varying model parameters were subject-specific. This allows not only for *a posteriori* comparison of parameter estimates, but also opens up possibilities to carry out standard analyses of interindividual variation of these parameter estimates (MANOVA [multivariate analysis of variance], cluster analysis, etc.). This is a popular data analysis plan in cognitive neuroscience: in the first phase, obtain parameter estimates based on single-subject analyses of intraindividual variation (EEG [electroencephalography], MEG [magnetoencephalography], fMRI [functional magnetic resonance imaging]), and then in a second phase, carry out analysis of interindividual variation of these estimated parameter values to arrive at conclusions that can be generalized to the population level (Friston et al., 2007). Accordingly, nomothetic knowledge about idiographic processes can be obtained.

GIMME

If the number of replications in a time series design (N) is large and if heterogeneity also affects model structure (that is, different subjects obey different dynamic models), then a new approach to arrive at nomothetic knowledge even in this more fundamentally heterogenic situation has been developed (Gates & Molenaar, 2012). While this so-called GIMME (group iterative multiple model estimation) approach has been mainly applied in the context of fMRI data analysis, it can also be applied much more generally to the analysis of any psychological process assessed in a replicated time series design (see Beltz et al., 2013, for an application to social interaction processes, and see Beltz et al., 2016, for an exemplar application to personality pathology and discussion of clinical implications). GIMME consists of a data-driven automatic two-step approach using standard likelihood criteria. In the first step, a common (group) dynamic model structure is determined for all N replications. While this common structure is shared by all N replications, the beta weights associated with each directed link are allowed to vary across replications. In the second step, subject-specific directed links are added to the common structure until the goodness-of-fit no longer can be significantly improved. GIMME has been validated with truly large-scale simulated data (see Gates & Molenaar, 2012) and shown to be extremely successful in recovering the true state of affairs. GIMME can be freely accessed at http://CRAN.R-project.org/package=gimme. An extension of GIMME to determine a finite set of multiple common submodels is described in Gates and colleagues (2014). This set of common submodels is determined in a purely data-driven way using a modularity approach for community detection.

Clearly, the common (group) models detected by means of GIMME with possibly heterogeneous replicated time series data again constitute nomothetic knowledge about idiographic processes. Thus, the broader tenets of subject-specific data analysis are commensurable with standard analyses of interindividual variation; the goal is still to arrive at nomothetic knowledge that can be generalized to the population level. But instead of assuming ad hoc (as in analyses of interindividual variation) that psychological processes are ergodic, subject-specific analyses recognize that ergodicity is a rare feature of such processes. The remaining sections address related approaches to bridge the gap between nomothetic and idiographic analysis.

A Unified Framework for the Study of Inter- and Intraindividual Variation

In a noteworthy publication by Voelkle and colleagues (2014), a unified framework for the study of inter- and intraindividual variation using replicated Gaussian (normally distributed) time series data was presented. Because of its general relevance to studying psychological processes, the key features are summarized here. Let $\mathbf{y}(i,t)$ be the observed p-variate time series of subject $i = 1,2,\ldots,N$ across times $t = 1,2,\ldots,T$. Then a standard interindividual factor model was fit to the data at each time point t, yielding T such models. Also, an intraindividual factor model was fit to the data of each person, yielding N such models. For each of the $N*T$ cells in the data box (refer to Figure 25.1) a log-likelihood was created under the following restrictive assumption: the parameters in the T interindividual factor models and the N intraindividual factor models are all equal to each other (representing ergodicity). The log-likelihood of each interindividual factor model is additive over the persons, and the intraindividual factor model was fit by means of MKFM6 (Dolan, 2010), which is software based on a recursive estimator yielding additive components of the log-likelihood over the times. In an application to simulated data obeying strict ergodicity, it was shown that the log-likelihood surface across the $N*T$ cells of the data box obeys the nominal type I error rate.

The next step in providing a bridge between inter- and intraindividual structures was to consider possible violations of strict ergodicity due to the presence of serial dependencies, time-varying mean trends, and group differences. Again, data were simulated according to these scenarios and the test of strict ergodicity described above was carried out. This yielded log-likelihood surfaces with distinctive patterns of significant results, indicating violations of ergodicity. These violations were handled by considering log-likelihoods of conditional structural equivalence among inter- and intraindividual structures, conditioning on adding serial dependency to the intraindividual structures, partialing out (detrending) the mean trends in data preprocessing (for the handling of time-varying mean trends), and/or splitting the total number of N persons into adequate subgroups. The resulting conditional log-likelihoods again obeyed the nominal type I error rates.

Conclusions on the alignment of findings regarding inter- and intraindividual variation from simulated data seem to generalize to empirical data. In a clinically relevant application to the time series data of 101 subjects which were assessed each day during 100 consecutive days, the focus was on 3 variables constituting indicators of the factor *attentiveness* of the Positive and Negative Affect Schedule (PANAS; see Watson, Clark, & Tellegen, 1988). Likelihoods were computed conditional on within-person means, serial dependency, and grouping of men and women for all 10,100 cells in the data box. Thus, it was found that equality of inter- and intraindividual structures was rejected in about 50 percent of the cases. This conditional result shows that for each individual subject making up the sample, there is a possible lack of ergodicity, even if this occurs only a few times during the 100 days.

Clustering Vector Autoregressive Models

Bulteel and colleagues (2016) present a cluster analysis of standard VAR(1) modeling of multivariate time series data, specifically daily assessments of depressive symptoms, from young female subjects. Conditional on the choice of K clusters, an alternating least squares (ALS) algorithm was applied to determine the allocation of each subject to a cluster yielding minimal Euclidean distance, where subjects within a cluster all have the same VAR(1). The clustering algorithm was the well-known Ward's algorithm (Ward, 1963). The number of clusters was determined by varying K between a minimum and maximum value and choosing the solution for which Chull (a statistic used to determine the elbow in a scree plot of squared prediction errors as function of K; see Ceulemans & Kiers, 2006) is maximal. Results yielded a two-cluster solution ($K = 2$) where the main distinction between the clusters was the degree of sequential dependencies characterizing predictions, with one cluster having low degree (showing flexible dynamics) and the second cluster a higher degree (showing more rigidity in their dynamics).

Mixed VAR Modeling

In Epskamp and colleagues (2017), VAR modeling for multiple subjects was discussed, based on a combination of the standard VAR(1) and mixed modeling. Consider the following representation for N subjects of the VAR(1) for a p-variate observed time series $\mathbf{y}(t)$:

$$\mathbf{y}^k(t+1) = \boldsymbol{\mu}^k + \boldsymbol{\Phi}^k \mathbf{y}^k(t) + \boldsymbol{\varepsilon}^k(t+1), k = 1, 2, \ldots, N;$$
$$t = 1, 2, \ldots, T$$

where $\boldsymbol{\mu}$ is a p-variate vector of intercepts, $\boldsymbol{\varepsilon}(t)$ denotes a so-called white noise innovation series which lacks any sequential dependencies and $\boldsymbol{\Phi}$ is a (p,p)-dimensional matrix of autoregressive (along the diagonal) and cross-lagged regression (off-diagonal) coefficients. It is assumed that $\boldsymbol{\mu}^k$ and $\boldsymbol{\Phi}^k$ vary across the N subjects according to a multivariate normal distribution.

Application of the mixed VAR model yields three kinds of networks: N networks of the lagged connections among the p component series in $\mathbf{y}(t)$ (reflecting intraindividual variation), N network of contemporaneous connections associated with the covariance matrix of the innovation series $\boldsymbol{\varepsilon}(t)$ (reflecting intraindividual variation), and a network of the interindividual variation of the mean (fixed effects) parameters associated with $\boldsymbol{\mu}^k$ and $\boldsymbol{\Phi}^k$. The mixed VAR model can be fitted to the data by means of the R package mlVAR (available at https://CRAN.R-project.org/web/packages/mlVAR/index.html). It is beginning to be utilized in work aimed at capturing the temporal relations in clinical ambulatory assessment data (e.g., examining how symptoms are related to, predict, and are predicted by context variables for individuals and groups).

CONCLUSION

The relation between subject-specific and person-oriented approaches is still quite indirect. The necessity to use subject-specific approaches has its origin in mathematical-statistical theory, while person-oriented approaches originate from holistic-interactionistic perspectives within psychological theory. The important paper by Sterba and Bauer (2010a), however, shows that subject-specific data analysis also enables testing of most principles underlying person-oriented theory. Molenaar (2010a) argued that subject-specific data analysis enables testing of all six principles of person-oriented theory. In this chapter, Molenaar's (2010a) argument has been extended in a number of ways, thus further elaborating the importance of subject-specific data analysis for the testing of person-oriented theory. In view of the spectacular increase in intensive repeated measurement designs in psychology in general and clinical science in particular, it is expected that subject-specific data analysis also will become the new norm for applications of person-oriented theory.

REFERENCES

Beltz, A. M., Beekman, C., Molenaar, P. C. M., & Buss, K. A. (2013). Mapping Temporal Dynamics in Social Interactions with Unified Structural Equation Modeling: A Description and Demonstration Revealing Time-Dependent Sex Differences in Play Behavior. *Applied Developmental Psychology, 17*, 152–168.

Beltz, A. M., Wright, A. G. C., Sprague, B. N., & Molenaar, P. C. M. (2016). Bridging the Nomothetic and Idiographic Approaches to the Analysis of Clinical Data. *Assessment, 23*, 447–458.

Bergman, L. R., Magnusson, D., & El-Khouri, B. (2003). *Studying Individual Development in an Interactive Context: A Person-Oriented Approach*. Mahwah, NJ: Erlbaum.

Birkhoff, G. (1931). Proof of the Ergodic Theorem. *Proceedings of the National Academy of Science of the USA, 17*, 656–660.

Borsboom, D. (2005). *Measuring the Mind: Conceptual Issues in Contemporary Psychometrics*. Cambridge: Cambridge University Press.

Bulteel, K., Tuerlinckx, F., Brose, A., & Ceulemans, E. (2016). Clustering Vector Autoregressive Models: Capturing Qualitative Differences in Within-Person Dynamics. *Frontiers in Psychology, 7*, 1540.

Cattell, R. B. (1952). The Three Basic Factor-Analytic Designs – Their Interrelations and Derivatives. *Psychological Bulletin, 49*, 499–520.

Ceulemans, E., & Kiers, H. A. L. (2006). Selecting among Three-Mode Principal Component Models of Different Types and Complexities: A Numerical Convex Hull Based Method. *British Journal of Mathematical and Statistical Psychology, 59*, 133–150.

Choe, G. H. (2005). *Computational Ergodic Theory*. Berlin: Springer-Verlag.

De Groot, A. D. (1954). Scientific Personality Diagnosis. *Acta Psychologica, 10*, 220–241.

Dolan, C. V. (2010). *MKFM6: Multi-Group, Multi-Subject Stationary Time Series Modeling Based on the Kalman Filter (Computer Software)*. Amsterdam: University of Amsterdam.

Epskamp, S., Waldorp, L. J., Mottus, R., & Borsboom, D. (2017). Discovering Psychological Dynamics: The Gaussian Graphical Model in Cross-Sectional and Time-Series Data. *arXiv:1609.04156v5* [stat.ME], 15 September.

Friston, K. J., Ashburner, J. T., Kiebel, S. J., Nichols, T. E., & Penny, W. D. (Eds.). (2007). *Statistical Parametric Mapping: The Analysis of Functional Brain Images*. New York: Academic Press.

Gates, K. M., & Molenaar, P. C. M. (2012). Group Search Algorithm Recovers Effective Connectivity Maps for Individuals in Homogeneous and Heterogeneous Samples. *NeuroImage, 63*, 310–319.

Gates, K. M., Molenaar, P. C. M., Iyer, S. P., Nigg, J. T., & Fair, D. A. (2014). Organizing Heterogeneous Samples Using Community Detection of GIMME-Derived Resting State Functional Networks. *PLoS-ONE, 9*, e91322.

Hannan, E. J. (1970). *Multiple Time Series*. New York: Wiley.

Isidori, A. (1985). *Nonlinear Control Systems*. London: Springer-Verlag.

Lamiell, J. T. (2003). *Beyond Individual and Group Differences: Human Individuality, Scientific Psychology, and William Stern's Critical Personalism*. Thousand Oaks, CA: Sage.

Mäkikangas, A., & Kinnunen, U. (2016). The Person-Oriented Approach to Burnout: A Systematic Review. *Burnout Research, 3*, 11–23.

Millsap, R. E. (2011). *Statistical Approaches to Measurement Invariance*. New York: Routledge.

Molenaar, P. C. M. (1987). Dynamic Assessment and Adaptive Optimisation of the Therapeutic Process. *Behavioral Assessment, 9*, 389–416.

Molenaar, P. C. M. (2004). A Manifesto on Psychology as Idiographic Science: Bringing the Person back into Scientific Psychology, This Time Forever. *Measurement: Interdisciplinary Research and Perspectives, 2*, 201–218.

Molenaar, P. C. M. (2010a). Testing All Six Person-Oriented Principles in Dynamic Factor Analysis. *Development and Psychopathology, 22*, 255–259.

Molenaar, P. C. M. (2010b). Note on Optimization of Psychotherapeutic Processes. *Journal of Mathematical Psychology, 54*, 208–213.

Molenaar, P. C. M. (2014). *Equivalent Dynamic Models*. Sells Award Address. 54th Annual Society of Multivariate Experimental Psychology Meeting, Nashville, TN, October 8–11.

Molenaar, P. C. M. (2015). On the Relation between Person-Oriented and Subject-Specific Approaches. *Journal for Person-Oriented Research, 1*, 34–41.

Molenaar, P. C. M., & Campbell, C. G. (2009). The New Person-Specific Paradigm in Psychology. *Current Directions in Psychological Science, 18*, 112–117.

Molenaar, P. C. M., & Nesselroade, J. R. (2015). Systems Methods for Developmental Research. In W. F. Overton & P. C. M. Molenaar (Eds.), *Handbook of Child Psychology and Developmental Science* (7th edn., Vol. 1, pp. 652–682). Hoboken, NJ: Wiley.

Molenaar, P. C. M., & Newell, K. M. (Eds.). (2010). *Individual Pathways of Change: Statistical Models for Analyzing Learning and Development*. Washington, DC: American Psychological Association.

Molenaar, P. C. M., Huizenga, H. M., & Nesselroade, J. R. (2003). The Relationship between the Structure of Interindividual and Intraindividual Variability: A Theoretical and Empirical Vindication of Developmental Systems Theory. In U. M. Staudinger & U. Lindenberger (Eds.), *Understanding Human Development: Dialogues with Lifespan Psychology* (pp. 339–360). Dordrecht: Kluwer Academic Publishers.

Molenaar, P. C. M., Sinclair, K. O., Rovine, M. J., Ram, N., & Corneal, S. E. (2009). Analyzing Developmental Processes on an Individual Level Using Non-Stationary Time Series Modeling. *Developmental Psychology, 45*, 260–271.

Molenaar, P. C. M., Lerner, R. M., & Newell, K. M. (Eds.). (2014). *Handbook of Developmental Systems Theory & Methodology*. New York: Guilford Press.

Molenaar, P. C. M., Beltz, A. M., Gates, K. M., & Wilson, S. J. (2016). State Space Modeling of Time-Varying Contemporaneous and Lagged Relations in Connectivity Maps. *NeuroImage, 125*, 791–802.

Nesselroade, J. R., & Molenaar, P. C. M. (2016). Some Behavioral Measurement Concerns and Proposals. *Multivariate Behavioral Research, 51*, 396–412.

Schiffman, S., Stone, A. A., & Hufford, M. R. (2008). Ecological Momentary Assessment. *Annual Review of Clinical Psychology, 4*, 1–31.

Sterba, S. K., & Bauer, D. J. (2010a). Matching Method with Theory in Person-Oriented Developmental Psychopathology Research. *Development and Psychopathology, 22*, 239–254.

Sterba, S. K., & Bauer, D. J. (2010b). Statistically Evaluating Person-Oriented Methods Revisited. *Development and Psychopathology, 22*, 287–294.

Valsiner, J., Molenaar, P. C. M., Lyra, M. C. D. P., & Chaudary, N. (Eds.). (2009). *Dynamic Process Methodology in the Social and Developmental Sciences*. Berlin: Springer-Verlag.

Voelkle, M. C., Brose, A., Schmiedek, F., & Lindenberger, U. (2014). Toward a Unified Framework for the Study of Between-Person and Within-Person Structures: Building a Bridge between Two Research Paradigms. *Multivariate Behavioral Research, 49*, 193–213.

Wang, Q., Molenaar, P. C. M., Harsh, S., Freeman, K., Xie, J., Zhou, J., … Rovine, M. J. (2014). Personalized State-Space Modeling of Glucose Dynamics for Type 1 Diabetes Using Continuously Monitored Glucose, Insulin Dose and Meal Intake: An Extended Kalman Filter Approach. *Journal of Diabetes Technology and Science, 8*, 331–345.

Ward, J. H. Jr. (1963). Hierarchical Grouping to Optimize an Objective Function. *Journal of the American Statistical Association, 58*, 236–244.

Watson, D., Clark, L. A., & Tellegen, A. (1988). Development and Validation of Brief Measures of Positive and Negative Affect: The PANAS Scales. *Journal of Personality and Social Psychology, 54*, 1063–1070.

Wohlwill, J. F. (1973). *The Study of Behavioral Development*. New York: Academic Press.

26 Social Processes and Dyadic Designs

ROBERT D. VLISIDES-HENRY, SHEILA E. CROWELL, ERIN A. KAUFMAN, AND BETTY LIN

Humans are an interdependent and inherently social species. Some have argued that, like other eusocial animals, human survival *depends* on relationships (e.g., via multigenerational and communal caregiving, shared resource allocation, risk distribution, and load sharing; Joiner et al., 2016). Others have hypothesized humans are so uniquely adapted to communal living that homeostasis and adaptive functioning are most likely to occur in the context of social connection (social baseline theory; Coan, 2008, 2010). Indeed, a wealth of research has demonstrated that healthy parent–child relationships (i.e., attachment bonds), adaptive peer groups, and intimate pair bonds can promote resilience, extend the lifespan, and support physical and mental health (e.g., Yang et al., 2016). By contrast, low status, isolation, loneliness, and social rejection are associated with biosocial dysregulation and poor health (e.g., Hawkley et al., 2010).

Research designs examining the influence of social processes are continually evolving. Most studies focus on the individual level of analysis in spite of well-established associations between social relationships, health, and wellbeing. Those examining dynamic interactions among parents, peers, partners, families, and other groups are sorely needed to effectively advance our understanding of health and psychopathology and to develop more targeted treatments. Despite obvious benefits, dyadic research comes with unique challenges related to study design, participant recruitment/retention, data collection, scoring, processing, and statistical analyses/interpretation. The goals of this chapter are to highlight current research using dyadic methods, identify innovative approaches to this type of study design, and describe challenges associated with this work. We first discuss several key theories and concepts for understanding social processes via dyadic designs.

KEY THEORIES AND CONCEPTS: ATTACHMENT THEORY

Attachment is understood as the way individuals create and maintain social relationships (Hofer, 2006). Importantly, attachment relationships emerge across the lifespan, with early interactions informing those occurring later in development. Early experiences with caregivers influence children's beliefs, attitudes, expectations, and behaviors in relationships with others, and manifest as attachment styles that can be characterized as secure or insecure. For example, anxiously attached adults often experienced unreliable care as children and may intensely seek connection and intimacy from romantic partners (e.g., Collins & Feeney, 2004). Children may also develop an avoidant attachment style when childhood caregiving was stable yet low in warmth and support, which has been linked to reluctance to seek help and support in adulthood (e.g., Feeney & Noller, 1990; Gillath et al., 2006). Thus, lifespan attachment provides one important context for understanding dyadic social process.

Child-Caregiver Attachment

Attachment theory provides a foundation for much of the research on dyadic relationships and an overarching framework for conceptualizing the centrality of human relationships to survival and wellbeing (see e.g., Bretherton, 1985; Cassidy & Shaver, 1999). Research on human attachment burgeoned following John Bowlby's (1951) observation that mothers play a central role in shaping early child development (both adaptive and maladaptive). Mary Ainsworth expanded this work by classifying infant-mother relationships as positive, ambivalent, and indifferent or hostile. Ainsworth reasoned that a stable attachment figure provides a "secure base," from which the child can comfortably detach in order to explore the world (e.g., Bowlby et al., 1956). Shortly after, Bowlby (1960) discussed how theories of the time did not explain adequately why children experience intense anxiety upon separation from caregivers. This underscored the utility of attachment theory for clinical purposes and spurred Bowlby's attempts to characterize children's dynamic behaviors in response to *separation* from mothers (Bowlby, 1982). Although Bowlby's (1951, 1960) early work relied on case studies and individual observation, Ainsworth (1969) argued that attachment theory required researchers to study infants in their family context across time. Innovative attachment researchers heeded Ainsworth's suggestions, and the theory became a popular framework for understanding infant-caregiver interactions.

Animal studies have also drawn insight from attachment theory, and have contributed importantly to understanding the nature of social processes in humans. In parallel to the separation anxiety observed in human infants, rat pups display protest-despair responses upon separation (see Hofer, 1994). Hofer (1994, 2006) hypothesized that these responses emerge because pups lose sight of the secure bases with which they synchronize. Upon separation, the pups' physiological systems abruptly lose their "guiding system," similar to a musician suddenly being unable to hear a metronome. Presumably, when a parent is present, the pup's (or child's) physiological activity coordinates naturally with the caregiver's and each member of the dyad dynamically influences the other. This coregulation of physiological, social, and emotional systems undergirds an infant's development of self-regulatory processes and, potentially, lifetime risk for emotion dysregulation (see, e.g., Cole, Michel, & Teti, 1994; Ostlund et al., 2017). As such, it is unsurprising that coregulation (in the context of attachment) is an essential tenet of all types of dyadic research.

Mutually influential dyadic processes go by many names (e.g., synchrony, concordance, attunement, coregulation; Butler & Randall, 2013). Although these terms have nuanced definitions, researchers often use them interchangeably when referring to dyadic behavioral, emotional, or physiological coordination, particularly among infant-caregiver dyads (e.g., Tronick et al., 1978). Consistent with Butler and Randall (2013), we operationalize synchrony, concordance, and attunement as coordination of a single construct within dyads. However, we operationalize coregulation as *dynamic coordination of two or more constructs within dyads across time* (Butler & Randall, 2013). Thus, bona fide coregulation can only be considered when researchers measure coordination with multiple measures as they contribute to dyadic stability. For example, when two members of an attachment relationship experience a simple linkage of physiological activity (e.g., their heart rate responses synchronize), this would be described as physiological attunement, synchrony, or concordance (Butler & Randall, 2013). Covariation *and* dynamic coupling of positive and negative affect would be considered coregulation, as researchers use multiple measures and examine how they contribute to affective stability among couples (Butner, Diamond, & Hicks, 2007). In this chapter, we argue that including assessments of some form of synchrony is foundational to dyadic research, and that this research should ideally also include assessments of coregulatory processes when possible. As Ainsworth (1969) noted, when examining any attachment relationship, researchers should utilize designs that assess responses from both members of the dyad over time.

Many argue that infant-caregiver synchrony is the mechanism by which parent-child dyads build secure attachments (Feldman, 2012). This "bottom-up" hypothesis claims that, through proximity, infants and caregivers begin to coordinate physiological activity. In turn, this physiological synchrony serves as a foundation for the development of a secure base. Empirical work supports this theory. For instance, there is a curvilinear (i.e., inverted-U) relationship between physiological synchrony and attachment, such that dyads with mean levels of synchrony have the most secure attachment (Feldman, 2007). Leclére and colleagues (2014) considered associations linking mother-infant physiological coordination and attachment security across multiple studies, and found that greater dyadic coordination was consistently associated with infants' secure attachment styles. In sum, available evidence underscores the utility of attachment theory as a guiding framework when studying infant-caregiver and child-caregiver dyads. Attachment theory also provides an important framework for understanding subsequent relationships, such as relationships with peers and romantic partners.

Peer-Peer Attachment

Although peers are also known to be formative contributors to development, peer relationships are studied far less often than child-caregiver dyads and romantic relationships (Ravitz et al., 2010). Historically, effects of attachment on peer interactions have been less clear (e.g., Dalton, Frick-Horbury, & Kitzmann, 2006). Recent theories suggest that aspects of infant-caregiver attachment relationships may be replicated among peers later in development. These effects may be mediated by psychophysiological changes. For example, researchers have established that negative parenting practices predict heightened stress responsivity in the context of later social interactions (Diamond, 2015). Early-life experiences may thus set a child on a path toward emotion dysregulation, potentially leading to peer rejection and continued insecure attachments with others (e.g., Crowell, Puzia, & Yaptangco, 2015).

Coan's (2008, 2010) social baseline theory focuses primarily on adult attachment relationships and suggests that proximity to others helps people self-regulate. Coan argues that mammalian brains are wired for attachment relationships, naturally preferring (1) distribution of risk within groups, and (2) load sharing, both of which are adaptive for survival. There are numerous survival advantages when mammals live in groups. For example, individual herbivores are more likely to become prey when alone as compared to when in a group. By maintaining proximity to others, mammals *distribute risk* amongst the group and increase the likelihood of every member's survival.

Load sharing is another phenomenon that ultimately increases all group members' wellbeing. When mammals form groups, trust between individuals builds over time, which increases the likelihood that resources will be shared. As a result, important tasks can also be shared (e.g., caring for young). Interdependence becomes most prominent among long-term groups. As trust builds, humans are able to mutually disclose, or "share the load," of their emotional distress. For instance, individuals expend less effort to regulate themselves in the presence of an attachment figure (Coan, 2010). By sharing the burden of stressful experiences, organisms ultimately conserve valuable energy resources and are more equipped for survival and adaptive functioning.

Butler (2015) has also proposed a model of *interpersonal emotional regulation* that shares core elements with social baseline theory. This model identifies three interdependent sequential processes that affect dyadic relationship dynamics. Ideally, there is first convergence between each individual's emotional responses to external stimuli. Second, the individuals then react emotionally to each other. Finally, this results in what Butler describes as interpersonal emotion regulation, whereby dyads coregulate emotional experiences. Butler's model shares features of the load-sharing process described in social baseline theory (which can also be conceptualized as a form of interpersonal emotion regulation; e.g., Beckes & Coan, 2011). These theoretical perspectives allow us to apply attachment theory to understand dynamic, interpersonal relationships between peers. Ultimately, attachment may affect peer dyads in two ways. First, like romantic couples, individuals bring their own working model of attachment into any relationship, affecting the interpersonal dynamics described by Butler (2015; see also Pietromonaco, Uchino, & Dunkel Schetter, 2014). Second, social baseline theory holds that the attachment relationship between peers may moderate individual and even interpersonal regulatory effort (Coan, 2010).

Positive attachment relationships are characterized by a sense of safety (Gillath et al., 2006). Members of a secure relationship tend to distribute risk, load share, and facilitate emotion regulation. When relationships serve these functions, individuals are more likely on a trajectory toward healthy development and homeostasis, and are at reduced risk for psychopathology (Feldman, 2007). Without functional regulatory relationships, individuals are at far greater risk for poor mental health. Insecure attachments are characterized by a lack of coregulation and synchrony and, ultimately, less effective independent self-regulatory skills (e.g., Hughes et al., 2012). Emotion dysregulation is a primary contributing factor to the emergence of psychopathology (e.g., Beauchaine, 2015; Beauchaine & Crowell, in press) and attachment theory is a common framework for understanding how emotion dysregulation develops. Further research is needed to characterize attachment processes during the transitions from parent–child, to peer–peer, to adult romantic relationships.

Partner–Partner Attachment

Researchers have also established the utility of attachment theory for understanding romantic dyads[1] (e.g., Collins & Read, 1990; Ravitz et al., 2010; Shaver & Mikulincer, 2010). Interestingly, there appears to be a discrepancy between self-reported stress and indicators of physiological stress reactivity among insecurely attached adults (Diamond & Fagundes, 2010). That is, adults with avoidant or dismissive (i.e., insecure) attachment styles tend to report low stress yet demonstrate heightened physiological stress responsivity in the laboratory. Avoidant individuals may fail to either experience or report their distress. When individuals with insecure attachments are part of a couple, they are at higher risk for negative outcomes as well. In one early study, Hazan and Shaver (1987) used a large-scale self-report design and found that anxiously attached individuals demonstrated more emotional lability in their relationships compared with securely attached individuals. Those with anxious attachment styles also report greater need for their partners and greater relationship insecurity (Joel, MacDonald, & Shimotomai, 2011). Similarly, Gallo and Smith (2001) also used a self-report design and found that anxiously attached married couples described their relationships as having greater conflict and lower support. The authors concluded that anxiously attached individuals experience greater internal conflict with regard to their romantic relationship. Yet these three studies utilized self-report alone, which is not suited ideally to assessing attachment (Diamond & Fagundes, 2010).

Researchers have extended these findings to physical functioning. Pietromonaco and colleagues (2014) created a theoretical framework for understanding how dyadic processes shape relationship behaviors and health. Within their model, individuals' attachment styles predict their own relationship behaviors (e.g., support-seeking) in addition to other relationship mediators and outcomes (e.g., commitment). The individuals' behaviors, mediators, and outcomes then interact with their partners' behaviors, mediators, and outcomes, creating complex dyadic processes that ultimately influence physiological responses and medical outcomes (see, e.g., Uchino, Smith, & Berg, 2014). Diamond and Hicks (2005) found that lower vagal tone (indexed by respiratory sinus arrhythmia [RSA]) was associated with insecure attachment styles. Moreover, secure attachment predicted a quicker return to a baseline emotional state following laboratory anger induction, though RSA mediated this relation. RSA (a measure of heart rate variability) is a hallmark of overall health (see Billman, 2011), suggesting that insecure attachment has implications for lifespan mental *and* physical health. Overall, these findings emphasize the importance of

---

[1] A detailed review of the effect of adult attachment style on romantic relationships is beyond the scope of this chapter. Interested readers are referred to Pietromonaco and Beck (2015).

considering attachment when researching romantic dyads. We next discuss several methodologies used for dyadic research design, including the benefits and drawbacks to each.

COMMON MULTIPERSON DESIGNS AND METHODOLOGIES

Child-Caregiver Designs

As discussed previously, dyad members are mutually influential. Researchers can tap into dynamic, real-world social processes by using creative approaches designed to examine synchrony and coregulation. One common laboratory methodology used among infants and their caregivers is the still-face paradigm (Tronick et al., 1978). Within this design, caregivers first play with their infant as they typically would, followed by a period where the caregiver stares at the infant with a neutral, unresponsive expression. Finally, there is a reunion play phase in which mother and child play normally again. This paradigm is based on the theory that caregivers drive physiological synchrony during early childhood. In other words, infant physiology tends to follow the parent's rather than the other way around. The still-face paradigm is a particularly effective method for assessing infant-caregiver dyads because the task evokes stress in both caregiver and child over the span of several minutes. Thus, this task provides a controlled opportunity to assess aspects of physiological, emotional, and behavioral synchrony between dyad members (e.g., Ostlund et al., 2017). Researchers have also found that the still-face paradigm can be used to examine how maternal sensitivity mediates infant physiological recovery (e.g., Conradt & Ablow, 2010; Moore et al., 2009).

Another common paradigm for assessing childcaregiver dyads is the strange situation task (Ainsworth et al., 1978). Here, researchers assess a child's reactions to a series of separations and reunions from their caregiver, which can be used to classify dyad attachment style. Researchers often pair the strange situation with other tasks to assess maternal responsiveness or sensitivity (Ainsworth et al., 1978). As one example of this type of design, Raby and colleagues (2012) used a strange situation task among infants at ages 12 and 18 months (though they first assessed how responsive mothers were to infants at six months), and also examined how attachment style interacted with the infants' serotonin-transporter-linked polymorphic region (5-HTTLPR) genotype. The authors found infants with the short allele for 5-HTTLPR experienced greater distress during the strange situation, whereas greater maternal responsiveness predicted infant attachment security. No gene-environment interactions emerged, suggesting that further dyadic research is needed on biology-environment interactions and infant attachment outcomes. This study, which examined multiple characteristics of dyad members (i.e., responsiveness, attachment, and the 5-HTTLPR genotype), serves as one example of how dyadic research can benefit from using multiple measures across time.

Much of the research on child-caregiver dyads has involved young children – yet dyadic processes continue to be important with older children. Adolescent-caregiver pairs are particularly unique from other dyads, as the two members are closer in age than when the child is young but the parent still holds more relationship power. Our work (Crowell et al., 2008, 2013, 2017a) has employed tasks that elicit conflict between emotionally dysregulated adolescents (e.g., depressed and/or engaging in self-injurious behavior) and their mothers in order to examine important social processes. We have found that more dysregulated adolescents display greater negative affect, less positive affect, and more defiance during interpersonal conflict (Crowell et al., 2008). We have also found that self-injuring adolescents and their mothers tended to match or escalate their dyad partner's aversive behaviors – only deescalating once the highest levels of aversiveness were reached (thereby negatively reinforcing conflict escalation). In contrast, mothers in healthy control dyads were more likely to deescalate conflict, regardless of the adolescent's behavior (Crowell et al., 2013). The use of dyadic design allowed us to demonstrate that conflict behavior may be one mechanism by which emotion dysregulation is shaped within family environments of self-injuring adolescents.

As another example, we have used a dynamic systems approach to examine the extent to which adolescent and maternal behavior drive their dyad partner's behavior and physiological responding during conflict (Crowell et al., 2017). We found that both depressed and self-injuring adolescents are more *behaviorally* reactive to maternal behavior than control adolescents. Moreover, self-injuring adolescents were also more *physiologically* reactive to maternal behavior than either depressed or typical control teens. In contrast, adolescents did not drive maternal behavior or physiology during conflict across any of the three groups. This work replicates infant findings reviewed above and demonstrates that emotion dysregulated adolescents may also be more reactive to parent behavior than vice versa. This pattern of reactivity may be especially true for adolescents who are more sensitive. Understanding teenagers and teen-caregiver relationships is critically important for elucidating trajectories to healthy development and/or risk for psychopathology (e.g., Cicchetti & Rogosch, 2002). Further work is needed to assess coregulation, attachment, and dynamic social processes during adolescence and later developmental stages.

Peer–Peer Designs

Peer relations throughout life are critical for health (see Bukowski, Laursen, & Rubin, 2018). For example, positive peer interactions during adolescence can place youth on a developmental course toward health and buffer against risk. In contrast, adverse peer experiences may place adolescents on a path toward maladaptive behavior and psychopathology (Prinstein & Giletta, 2016). Researchers have found that adolescent girls' self-reported negative peer relations predict reassurance seeking among peers, which, in turn, predicts poor friendship quality and stability as assessed 11 months later (Prinstein et al., 2005). While Prinstein and colleagues' (2005) results are bolstered by the measurement of multiple constructs across time, the researchers relied on survey methodology alone, ultimately limiting the scope of their findings. Across samples, researchers tend to find that adolescent behaviors and attitudes mirror those of their peers, though again much of this work has been survey-based, rarely utilizing dyadic designs (e.g., Brechwald & Prinstein, 2011; Lopes, Gabbard, & Rodrigues, 2013).

The importance of the peer group to adolescents cannot be overstated (see Cicchetti & Rogosch, 2002). One of the few experimental studies on adolescents and young adults found that the presence of a randomly assigned peer increased adolescent risk-taking (Gardner & Steinberg, 2005). In another study, researchers randomly assigned adolescents to a confederate with high or low perceived popularity. Adolescents were more likely to engage in prosocial behaviors when their responses were public and the confederate was perceived as high-ranking (Choukas-Bradley et al., 2015). Though these studies provide us with important information on adolescents and their peers, unfortunately, neither of these studies assessed key constructs across time in both members of the dyad. Adolescence is a critical developmental stage with implications for the emergence of psychopathology. Thus, more research is needed examining how adolescent dyads (and larger peer groups) influence one another dynamically over the course of an interaction.

Even less dyadic work has been conducted with adult peers, though researchers often examine the effects of a stranger's presence on participant responding while completing study tasks (e.g., Butler, Wilhelm, & Gross, 2006). Such work highlights the importance of researching adult dyadic responses at the pair level, as most researchers to date have focused on individual responses. Furthermore, if adult peers are strangers, it becomes difficult to interpret results in the context of extant theories. In general, further theoretically driven work (e.g., social baseline, attachment, coregulation) is needed with adult peer–peer dyads.

Romantic Partner Designs

Romantic relationship functioning has critical implications for mental and physical health (see Robles et al., 2014). In one study of romantic couples, Butner and colleagues (2007) used a daily-diary method in which both members of a couple reported on positive and negative emotions over several days. Reported affect covaried between partners, and this effect was amplified when partners spent more time together. Though the researchers did not assess response data from both partners simultaneously (i.e., in the lab together), the study did benefit from data collected from both members over time. In a well-known study of heterosexual married couples, Coan, Schaefer, and Davidson (2006) examined wives' fMRI responses to a shock threat while they were holding their husband's hand, a stranger's hand, or no hand (controls). Neural threat responses were attenuated (e.g., reduced anterior cingulate cortex activity) when holding any hand compared to when in the control condition. Yet the effect was stronger for spousal handholding than when participants held the stranger's hand, and this effect was moderated by marriage quality. Results align well with theories of adult attachment in that having a secure attachment figure in close physical proximity helped women regulate their own stress response. This study is strengthened by having both dyad members present, though similar research could also benefit from simultaneous measurements from both dyad members, as this study cannot address questions about dyadic dynamics (Ainsworth, 1969; Butler & Randall, 2013). For example, researchers may be interested in understanding dyadic synchrony (i.e., dynamic co-occurrence across a single measure) or coregulation (i.e., dynamic co-occurrence across multiple measures).

Self-report-based methodologies have helped inform the field (e.g., Feeney & Noller, 1990; Joel et al., 2011), though experimental designs allow for a more objective examination of couples' relationship dynamics. For instance, Gallo and Smith (2001) asked couples to sit at a table separated by a partition. Couples were randomly assigned to debate the same or opposite side of an issue in either a high- or low-stakes situation (i.e., those assigned to the high-stakes condition were told their speech would be evaluated for quality). This design allowed the researchers to tap into relationship dynamics via real-life discussions in a controlled environment. Similarly, other researchers have used experimental designs to assess psychophysiology in response to emotion induction and partner/friend presence (e.g., Diamond & Hicks, 2005).

In a recent innovative study, Baucom and colleagues (2018) examined ambulatory cardiovascular activity in couples during planned and spontaneous conflicts both in a laboratory setting and in participants' homes. Heart rate reactivity was significantly greater at home as compared with the laboratory setting, particularly during spontaneous conflicts. Interestingly, physiological responses at home were also more highly correlated with psychological predictors (e.g., positive and negative affect, emotion regulation, relationship quality) relative to laboratory responses. By measuring physiological responses in both members across multiple settings *and* time, the researchers have a rich data set with which to examine dyadic synchrony with attachment dynamics in mind. In sum, several methodologies exist for examining dynamics of a romantic couple, though most experimental designs involve a laboratory-based discussion or argument. In sum, synchronic and coregulatory processes are most comprehensively investigated with innovative experimental and longitudinal designs (Ainsworth, 1969; Butler & Randall, 2013).

KEY DYADIC MEASURES

As many studies have shown, attachment, coregulation, and dyadic relationships in general may be best examined through a multiple-levels-of-analysis approach (see Beauchaine, 2015), including measures of peripheral physiology (e.g., heart rate, RSA), self-report and informant report (e.g., questionnaires, interviews), and behavioral coding (e.g., microfacial coding, global dyad-level coding). Here we briefly discuss each of these levels and their uses in dyadic designs.

Central and Peripheral Psychophysiology

Physiological indices are an indirect means of indexing dyad-level constructs. For example, RSA is a popular tool for assessing dyadic coregulation in adults, and refers to the upregulation and downregulation of one's heart rate across the respiratory cycle. RSA is an index of vagal influences on heart rate (see Beauchaine & Thayer, 2015). Under appropriate stimulus conditions, RSA is a useful predictor of emotional responsiveness, emotion dysregulation, and regulatory effort in both individuals and dyads (e.g., Butler et al., 2006; Crowell et al., 2017b). Thus, RSA is a commonly selected peripheral psychophysiological marker for dyadic coregulation.

Researchers have also used measures of neural activity to index mutual dyadic influence – most commonly through functional magnetic resonance imaging, or fMRI. fMRI offers the unique benefit of real-time neuroanatomical scanning in response to various stimuli. For example, individuals can be presented with images of other people, including their romantic partner, when in the scanner (e.g., Cheng et al., 2010), and can also be subjected to various external conditions, such as the threat of a shock (e.g., Coan et al., 2006). By measuring baseline neural activity prior to stimulus presentation, researchers can examine relevant brain activity changes in response to experimental conditions. Recent innovations in two-person fMRI assessments may allow for more sophisticated tests of key theories (e.g., Chauvigné, Belyk, & Brown, 2018). If both members of a dyad can be scanned simultaneously, researchers could assess their dynamic neurological activity across time – allowing for the possibility of assessing dyadic synchrony or even coregulation.

Questionnaires and Interviews

Although an exhaustive review of self-report instruments used within dyadic designs is beyond the scope of this chapter, we discuss briefly several measures that are both useful and commonly selected. Whereas attachment is typically assessed through observation in children, self-report is the most widely used method for examining this construct among adults. The Adult Attachment Interview (Hesse, 1999) is a semistructured interview commonly seen within the developmental psychology literature (see, e.g., Roisman et al., 2007). Researchers use this instrument to determine a current working model of attachment based on recalled childhood experiences. The Experiences in Close Relationships-Relationships Structure is another valid and widely used measure of adult attachment (see Fraley, Waller, & Brennan, 2000; Fraley et al., 2011). Facets of coregulation are rarely assessed via self-report, and may be better understood with physiological indices and behavior coding. Nonetheless, emotional synchrony can be examined partially through correlations between parent reports of child emotional reactivity (e.g., Carter & Briggs-Gowan, 2006) and parent self-reports of their own emotion dysregulation (e.g., Gratz & Roemer, 2004).

Observational Coding

Finally, observational (or behavioral) coding schemes are a valuable tool for dyad-level assessment, as they are less vulnerable to bias compared with self-report instruments and allow researchers to assess both the content and the structure of interpersonal interactions (Aspland & Gardner, 2003). Here, we discuss observational coding for dyadic data broadly. We also provide examples of different types of coding systems, including those at global and microanalytic levels.[2]

Global systems are useful for assessing constructs at the dyad level and require relatively little coder training. However, it may be more difficult to establish acceptable reliability with global coding schemes, and many provide little information on individual behaviors (Kerig & Baucom, 2004). One system useful for quantifying aspects of parent–child dyads is the System for Coding Interactions in Family Functioning (SCIFF; Lindahl & Malik, 2000). The SCIFF allows coders to rate the dyads on various qualities, such as rejection, invalidation, and coercion. The Parent–Child Interaction System (PARCHISY) is another example of a widely used global coding scheme (Deater-Deckard, Pylas, & Petrill, 1997). The PARCHISY is designed to assess child and maternal affect and behavior during an interaction. Codes capture reciprocity, positive and negative affect, control, and verbalization. Fogel's (1994) relational coding system can be used to assess both child–parent synchrony and more qualitative features of the relationship. Dyadic synchrony is assessed by examining the amount of time dyads spend focused jointly on the

---

[2] Interested readers are referred to Aspland and Gardner (2003); Baucom et al. (2017); Chorney et al. (2015); and Kerig and Baucom (2004) for more detailed reviews of specific observational coding systems.

same object, action, or behavior (see e.g., Evans & Porter, 2009). For more on child–parent coding systems, see Lotzin and colleagues (2015).

Older dyads have traditionally been assessed with microanalytic systems, or coding schemes designed to quantify moment-to-moment behaviors. Such systems tend to be highly reliable and specific, though they can require intensive coder training (Kerig & Baucom, 2004). For instance, the Specific Affect Coding System (SPAFF; Coan & Gottman, 2007) is a well-known system in dyadic research on emotion regulation. The SPAFF is a large, complex microanalytic coding system that allows researchers to dynamically assess individual and dyadic affect based on microlevel facial movements. To code verbal "behaviors" at the microanalytic level, researchers may use the Family and Peer Process Code (FPPC; Stubbs et al., 1998). Trained coders use the FPPC to quantify individual verbal affect during a dyadic interaction. Of course, dyadic researchers may choose to utilize both microanalytic and global coding systems to more thoroughly capture constructs of interest (see, e.g., Crowell et al., 2013).

Recently, researchers have begun to apply machine-learning algorithms to dyadic interactions (see Chapter 34, by Coutanche and Hallion). For instance, researchers have begun to use Automatic Speech Recognition (ASR) to quantify empathy in patient-provider psychotherapeutic interactions (Xiao et al., 2015). ASR appears to be a robust tool that may help improve patient-therapist dyadic interactions through greater empathy concordance (Xiao et al., 2015). Other researchers have discussed the implications of these machine-learning models, as they can easily begin to incorporate human observations across multiple modalities (e.g., speech, vocal pitch, eye movements; see Baucom et al., 2017). These approaches have only begun to emerge in the dyadic designs literature, and require further examination.

INNOVATIVE APPROACHES AND DESIGNS

Although dyadic research is complex, newly emerging paradigms are continually improving our ability to capture intricate social processes in increasingly sophisticated ways. As discussed above, dyadic studies can take place in the context of spontaneous, semistructured, or structured interactions and may occur in a controlled laboratory setting or in more naturalistic environments. Moreover, researchers investigating dyadic social processes frequently examine associations between members' self-reported and/or objective ratings of emotion and behavior.

Ratings of emotions and behaviors can be used to capture individual-level behaviors (e.g., negative affect, maternal sensitivity) and dyad-level behaviors (e.g., dyadic reciprocity) in the context of an interaction. Researchers may analyze transactional associations between individual members' behaviors using advanced statistical modeling techniques, such as cross-lagged panel models (e.g., Kenny, 2005). Extant studies have typically relied on teams of trained coders to identify key behaviors while maintaining standards of interrater reliability (i.e., agreement between two coders; Kerig & Baucom, 2004). Yet researchers are continually seeking to reduce bias associated with subjective evaluation through incorporating sophisticated methods like artificial intelligence as well as individual and dyadic eye-tracking (e.g., Black et al., 2013; Campbell et al., 2014).

Some research groups combine behavioral coding and physiological methods for studying dyadic synchrony and coregulation. However, few have measured such indices among young children. In one innovative study with a sample of mother-toddler dyads, Lunkenheimer and colleagues (2015) found patterns of both physiological concordance (i.e., child-mother RSA both moving in the same direction) and discordance (i.e., opposite direction). Discordance was associated uniquely with toddlers' externalizing behavior problems.

Another novel approach to studying emotional reactivity involves assessing patterns of vocal arousal among partners. Fundamental frequency (FF) is an index of vibrational patterns that occur while speaking and represents a relatively new area of behavioral coding. FF is associated with involuntary expressions of internal emotional arousal (e.g., through pitch; Baucom, 2010). Black and colleagues (2013) utilized machine-learning programming to analyze acoustic speech features of married couples' conversations. The researchers found that the programs were almost entirely better than chance (accuracy ranging from 49.3 to 86 percent) in detecting problem-solving behaviors in the conversations (though still well below human-level accuracy). In another study, Baucom and colleagues (2011) examined the correspondence between vocal demand/withdrawal behavior and vocal arousal among heterosexual couples. They found that higher levels of demand/withdrawal behavior were associated with higher levels of emotional arousal overall. Finally, Fischer and colleagues (2015) examined vocal synchrony among individuals with psychological disorders and their romantic partners. Individuals with anorexia nervosa exhibited concordance with their partners, suggesting that these participants and their partners were highly reactive to one another's levels of arousal. In contrast, vocal arousal among participants with obsessive-compulsive disorder was unrelated to their partners' – though partners' vocal arousal *was* related to patients'.

Innovations in dyadic research have also emerged at levels of data quantification and analysis. For example, recent modeling techniques borrow from mathematical and computational fields such as dynamic systems analysis and coupled linear oscillator models (see Chapter 27, by Estrada, Sbarra, and Ferrer). These techniques offer groundbreaking approaches for analyzing data captured during dyadic tasks. Dynamic systems models are a series

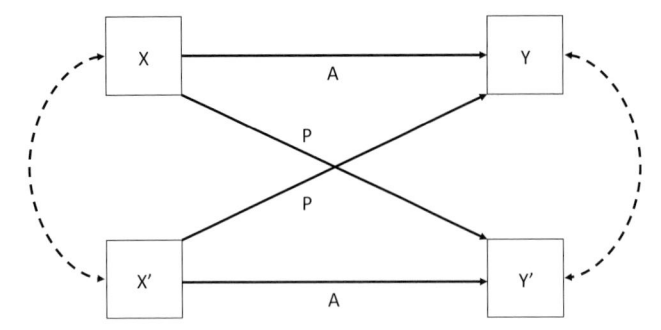

Figure 26.1 Schematic of the actor-partner interdependence model. X = Person 1 data at time point 1, and Y = Person 1 data at time point 2. X' = Person 2 data at time point 1, and Y' = Person 2 data at time point 2. Solid lines with single arrowheads represent predictive paths, whereas dashed lines with two arrowheads represent correlations. Paths labeled "A" are actor effects, whereas paths labeled "P" represent partner effects. For more on this model, see also Cook and Kenny (2005).

of equations that can be used to model dyadic change across time (Butner, Behrends, & Baucom, 2017). Dynamic systems models represent a unique class of modeling techniques with both conceptual and mathematical considerations and may provide one of the best analytic techniques for assessing dyadic synchrony and coregulation. Using these models, one can examine how changes in one member of a dyad dynamically influence the other. Furthermore, points at which changes occur can be readily identified because dynamic systems models are focused on emergence of dyadic patterns over time. If one assumes some patterns are more likely than others, it becomes possible to characterize different types of dyads by their emerging and reemerging patterns. Characterizing the pattern is mathematical, yet identifying and predicting common patterns is conceptually driven. This innovation in dyadic data analysis is only just beginning to burgeon in the social sciences (see Butner et al., 2017).

Similar to dynamic systems models, dyadic social science researchers also use actor-partner interdependence models (APIM). APIMs allow researchers to examine more traditional regression pathways (i.e., for one dyad member) while also addressing how individuals affect one another (Cook & Kenny, 2005; see Figure 26.1). There are numerous empirical studies that utilize APIMs for dyadic studies and researchers are now using theory and other models to improve its utility. Perry and colleagues (2017) discuss how repeated measures APIMs coupled with vector field diagrams can help researchers better interpret and disseminate the results of standard APIM studies. These modern, cutting-edge analytic techniques allow researchers to utilize more scientifically rigorous designs. However, there are still several unique challenges associated with dyadic designs.

UNIQUE CHALLENGES WITH DYADIC DESIGNS

Dyadic designs offer exceptional benefits for understanding dynamic actor-partner effects across multiple levels of analysis. Through such studies, researchers can examine concordance across psychophysiological and observational measures as they occur in the context of important social relationships (e.g., Crowell et al., 2013). Even so, executing dyadic designs presents a unique set of challenges across several phases of the research process.

Effective and efficient recruitment typically proves difficult, even when sampling typical community participants. The complexity of recruitment is magnified for studies using dyadic designs – particularly for research involving specialized groups (e.g., clinical populations, marginalized groups). Furthermore, participant engagement can be more challenging than in individual-level designs, as the research team must generate sufficient interest and motivation for both members of the dyad to participate. For example, fathers are chronically underrepresented in pediatric research (Davison et al., 2017; Parent et al., 2017). We have also found recruiting fathers to be exceptionally difficult, despite oversampling efforts.

Participants typically benefit from individualized recruitment strategies whereby recruiters adopt a demeanor sensitive to each person's background, values, and developmental status (Davison et al., 2017; Gyure et al., 2014). Effectively building rapport and commitment from multiple parties is simply more complex, effortful, and time-consuming, with double the opportunities for problems to occur. Researchers must also pay special attention to whether both parties are electing to participate voluntarily, particularly when working with child-caregiver pairs, or any other dyad in which there is likely to be a significant power discrepancy (American Psychological Association [APA], 2017).

Coordinating two (or more) persons' schedules with the research team can also result in delays between recruitment and active participation. For example, family members often have competing demands, and work/school schedules often conflict. Thus, research teams may find that days, weeks, or even months pass between participant screening and study visits. Unfortunately, time elapsed between participant recruitment and appointment may significantly increase preinclusion attrition rates (Walsh & Brinker, 2015). Implementing child/adolescent peer research can be particularly challenging to organize. Parental consent must be obtained from two sources prior to study enrollment (APA, 2017), and participants often travel to the lab independently – presenting challenges related to locating the lab and parking for both parties. Finally, vulnerable and/or stressed populations often encounter barriers that influence their ability to schedule, attend, and/or participate fully in study appointments (e.g., transportation, work schedules). Such challenges are not unique to dyadic research, but can be compounded in such designs, as disruptions for one member of the pair often preclude participation for both. As another example, mother-infant dyads may be affected by disrupted sleep or frequent medical appointments, and are thus a particularly challenging population to sample. Repeated contact with the research team (needed to manage screening and participation for multiple individuals) may also place excess burdens on potential participants.

Dyadic designs also typically involve more time and effortful coordination during study visits. Participants may require additional time and staff to complete procedures, including screening, consent/assent, debriefing, electrode placement, etc. Maintaining confidentiality and/or deception can also be more challenging when compared with individual-level research, as participants may interact with one another during or between study procedures (see Wittenborn, Dolbin-MacNab, & Keiley, 2013). In such cases, special care is needed to inform and remind participants that they should keep their answers and experiences private, unless otherwise instructed. The order and relative length of study tasks is also important to consider for designs where each dyad member completes multiple procedures. Research teams should pilot and refine their method for rotating pairs through various study components in order to ensure both members can complete their respective procedures efficiently and without lengthy waiting periods between tasks. Without doing so, participants may be unnecessarily burdened or receive compensation that is incongruent with the time spent.

Perhaps one of the most potent benefits to the dyadic researcher is the ability to examine real-life social processes carried out in a laboratory space. Thus, researchers are strongly advised to identify dyadic tasks that will adequately evoke meaningful social interaction. Laboratory studies can elicit spontaneous emotional experiences that closely resemble those occurring in everyday life (e.g., Roberts, Tsai, & Coan, 2007). However, genuine social processes can also be difficult to capture in artificial settings (e.g., Baucom et al., 2018). For example, there are many considerations that arise when designing a discussion-based task. Topics should be well suited to the participants' relationship, poignant enough to elicit sufficient emotional arousal, and discussed long enough for participants to habituate to their surroundings. For example, in our work on mother-daughter conflict we often observe an inflection point after approximately five minutes where mothers begin interacting more naturalistically and clinical participants begin to diverge from controls (Crowell et al., 2011). The still-face paradigm, for example, stimulates a social situation for infant-mother dyads in which the mother becomes entirely unresponsive to her infant. Parent–child discussions are often based on contentious topics that are relevant to day-to-day life such as chores and curfew (see, e.g., Crowell et al., 2008, 2013, 2017a; Granic et al., 2007). Thus, both time and topic salience are important for helping pairs settle into their interaction and overcome limitations associated with a laboratory environment.

A final challenge involves arranging data sets for appropriate dyadic data analyses. Multilevel models (see Chapter 32) can account for nesting individuals within dyads. They can determine (1) the extent to which dyad member's affect/behaviors/physiological responses covary with one another and (2) differences in the magnitude of those associations across different types of dyads (i.e., clinical vs. control). When constructing data sets, it is often helpful to create a dyad-level code as well as individual codes that differentiate each dyad member.

SUMMARY AND FUTURE DIRECTIONS

From conception onward, humans are interconnected and communal. Our physiological, emotional, cognitive, and behavioral systems are hardwired to respond to social inputs and to signal distress in the context of abandonment, rejection, or isolation. From this perspective, it is not surprising that loneliness is a leading cause of morbidity and mortality (Patterson & Veenstra, 2010). Similarly, suicide and suicidal behaviors nearly always occur in the context of interpersonal distress, including conflict, status loss, or a perceived lack of belongingness (Joiner et al., 2016; Trout, 1980). From a developmental perspective, early relationships function as a template for later social interactions and expectancies (Hughes et al., 2012). By adolescence or adulthood, interpersonal behaviors and expectations about others' responses begin to crystalize. In some cases, this may contribute to the emergence of one or more diagnosable psychological conditions characterized by mistrust, interpersonal conflict, anger, or interpersonal instability such as depression, anxiety, conduct disorder, or personality disorders (e.g., Crowell, 2016).

In this chapter, we have briefly reviewed attachment theory across the lifespan, innovative methods for measuring dyads, and challenges associated with dyadic research. Dyadic research necessitates a multiple-levels-of-analysis approach. For instance, emotion dysregulation is often assessed with self-report measures, interaction tasks, and RSA (e.g., Beauchaine & Thayer, 2015; Gratz & Roemer, 2004). However, tasks, methods, and measures vary depending upon the research question and developmental stage of the target participant. Nonetheless, nearly all dyadic stressor tasks involve withdrawal of support, warmth, or physical/emotional presence of a loved one, whereas support tasks involve physical or emotional connection, such as validation or handholding. A clear theme of our review is that researchers assessing dyadic social processes must proactively consider methodological issues. This work involves multiple participants and often requires complex longitudinal designs to answer developmentally relevant questions across multiple levels of analysis.

There are myriad future directions for research on social processes and dyads. First, researchers should extend dyadic methods to novel contexts, such as therapy sessions. For example, recent research finds that patient-clinician synchrony in emotion states during clinical crisis encounters is associated with stronger emotional bonds (Bryan et al., 2018). In turn, emotional bonding is associated with mutual down-regulation of arousal in both clinicians and patients, consistent with coregulation theories. Second, dyadic research should continue to move outside of the laboratory by capitalizing on wearable technology and the "internet of things" to benefit participants and potentially improve interactions in real time (Stankovic, 2014). Third, there is a clear need for innovations beyond dyads to triads (e.g., child and two caregivers), families, therapy groups, classrooms, and other multiperson groups. Methodologically, many of these future directions will require careful consent and attention to privacy, as well as techniques for processing, reducing, storing, analyzing, and interpreting vast amounts of data. Novel statistical approaches, such as machine learning and dynamic systems, will need to become more commonplace, which will require sophisticated statistical courses in graduate training programs. However, in spite of this complexity, dyadic and group designs hold tremendous promise. Indeed, if approached thoughtfully, dyadic research has potential to advance understanding of social interactions, development, psychopathology, health, and wellbeing.

REFERENCES

Ainsworth, M. D. S. (1969). Object Relations, Dependency, and Attachment: A Theoretical Review of the Infant-Mother Relationship. *Child Development, 40*, 969–1025.

Ainsworth, M. D. S., Blehar, M. C., Waters, E., & Wall, S. (1978). *Patterns of Attachment: A Psychological Study of the Strange Situation.* Hillsdale, NJ: Erlbaum.

American Psychological Association (APA). (2017). *Ethical Principles of Psychologists and Code of Conduct* (2002, Amended June 1, 2010 and January 1, 2017). Retrieved from http://www.apa.org/ethics/code/index.aspx

Aspland, H., & Gardner, F. (2003). Observational Measures of Parent-Child Interaction: An Introductory Review. *Child and Adolescent Mental Health, 8*, 136–143.

Baucom, B. (2010). Power and Arousal: New Methods for Assessing Couples. In K. Hahlweg, M. Grawe-Gerber, & D. H. Baucom (Eds.), *Enhancing Couples: The Shape of Couple Therapy to Come* (pp. 171–184). Cambridge, MA: Hogrefe.

Baucom, B. R., Atkins, D. C., Eldridge, K., McFarland, P., Sevier, M., & Christensen, A. (2011). The Language of Demand/Withdraw: Verbal and Vocal Expression in Dyadic Interactions. *Journal of Family Psychology, 25*, 570–580.

Baucom, B. R. W., Leo, K., Adamo, C., Georgiou, P., & Baucom, K. J. W. (2017). Conceptual and Statistical Issues in Couples Observational Research: Rationale and Methods for Design Decisions. *Journal of Family Psychology, 31*, 972–982.

Baucom, B. R. W., Baucom, K. J. W., Hogan, J. N., Crenshaw, A. O., Bourne, S. V., Crowell, S. E., ... Goodwin, M. S. (2018). Cardiovascular Reactivity during Marital Conflict in Laboratory and Naturalistic Settings: Differential Associations with Relationship and Individual Functioning across Contexts. *Family Process, 57*, 662–678.

Beauchaine, T. P. (2015). Future Directions in Emotion Dysregulation and Youth Psychopathology. *Journal of Clinical Child and Adolescent Psychology, 44*, 875–896.

Beauchaine, T. P., & Crowell, S. E. (Eds.). (In press). *Oxford Handbook of Emotion Dysregulation.* New York: Oxford University Press.

Beauchaine, T. P., & Thayer, J. F. (2015). Heart Rate Variability as a Transdiagnostic Biomarker of Psychopathology. *International Journal of Psychophysiology, 98*, 338–350.

Beckes, L., & Coan, J. A. (2011). Our Social Baseline: The Role of Social Proximity in Economy of Action. *Social and Personality Psychology Compass, 12*, 89–104.

Billman, G. E. (2011). Heart Rate Variability – A Historical Perspective. *Frontiers in Physiology, 2*, 1–13.

Black, M. P., Katsamanis, A., Baucom, B. R., Lee, C. C., Lammert, A. C., Christensen, A., ... Narayanan, S. S. (2013). Toward Automating a Human Behavioral Coding System for Married Couples' Interactions Using Speech Acoustic Features. *Speech Communication, 55*, 1–21.

Bowlby, J. (1951). *Maternal Care and Mental Health* (Vol. 2). Geneva: World Health Organization.

Bowlby, J. (1960). Separation Anxiety. *International Journal of Psychoanalysis, 41*, 89–114.

Bowlby, J. (1982). Attachment and Loss: Retrospect and Prospect. *American Journal of Orthopsychiatry, 52*, 664–678.

Bowlby, J., Ainsworth, M., Boston, M., & Rosenbluth, D. (1956). The Effects of Mother-Child Separation: A Follow-Up Study. *British Journal of Medical Psychology, 29*, 211–247.

Brechwald, W. A., & Prinstein, M. J. (2011). Beyond Homophily: A Decade of Advances in Understanding Peer Influence Processes. *Journal of Research on Adolescence, 21*, 166–179.

Bretherton, I. (1985). Attachment Theory: Retrospect and Prospect. *Monographs of the Society for Research in Child Development, 50*, 3–35.

Bryan, C. J., Baucom, B. R., Crenshaw, A. O., Imel, Z., Atkins, D. C., Clemans, T. A., ... Rudd, M. D. (2018). Associations of Patient-Rated Emotional Bond and Vocally Encoded Emotional Arousal among Clinicians and Acutely Suicidal Military Personnel. *Journal of Consulting and Clinical Psychology, 86*, 372–383.

Bukowski, W. M., Laursen, B., & Rubin, K. H. (Eds.). (2018). *Handbook of Peer Interactions, Relationships, and Groups* (2nd edn.). New York: Guilford Press.

Butler, E. A. (2015). Interpersonal Affect Dynamics: It Takes Two (and Time) to Tango. *Emotion Review, 7*, 336–341.

Butler, E. A., & Randall, A. K. (2013). Emotional Coregulation in Close Relationships. *Emotion Review, 5*, 202–210.

Butler, E. A., Wilhelm, F. H., & Gross, J. J. (2006). Respiratory Sinus Arrhythmia, Emotion, and Emotion Regulation during Social Interaction. *Psychophysiology, 43*, 612–622.

Butner, J., Behrends, A. A., & Baucom, B. R. (2017). Mapping Co-Regulation in Social Relations through Exploratory Topology Analysis. In R. R. Vallacher, S. J. Read, & A. Nowak (Eds.), *Computational Social Psychology* (pp. 144–177). New York: Routledge.

Butner, J., Diamond, L. M., & Hicks, A. M. (2007). Attachment Style And Two Forms of Affect Coregulation between Romantic Partners. *Personal Relationships, 14*, 431–455.

Campbell, D. J., Shic, F., Macari, S., & Chawarska, K. (2014). Gaze Response to Dyadic Bids at 2 Years Related to Outcomes at 3 Years in Autism Spectrum Disorders: A Subtyping Analysis. *Journal of Autism and Developmental Disorders, 44*, 431–442.

Carter, A. S., & Briggs-Gowan, M. J. (2006). *ITSEA: Infant-Toddler Social and Emotional Assessment Examiner's Manual.* San Antonio, TX: PsychCorp.

Cassidy, J., & Shaver, P. R. (1999). *Handbook of Attachment: Theory, Research, and Clinical Applications.* New York: Guilford Press.

Cheng, Y., Chen, C., Lin, C. P., Chou, K. H., & Decety, J. (2010). Love Hurts: An fMRI Study. *Neuroimage, 51,* 923–929.

Chauvigné, L. A. S., Belyk, M., Brown, S. (2018). Taking Two to Tango: fMRI Analysis of Improvised Joint Action with Physical Contact. *PLoS ONE, 13,* e0191098.

Chorney, J. M., McMurtry, C. M., Chambers, C. T., & Bakeman, R. (2015). Developing and Modifying Behavioral Coding Schemes in Pediatric Psychology: A Practical Guide. *Journal of Pediatric Psychology, 40,* 154–164.

Choukas-Bradley, S., Giletta, M., Cohen, G. L., & Prinstein, M. J. (2015). Peer Influence, Peer Status, and Prosocial Behavior: An Experimental Investigation of Peer Socialization of Adolescents' Intentions to Volunteer. *Journal of Youth and Adolescence, 44,* 2197–2210.

Cicchetti, D., & Rogosch, F. A. (2002). A Developmental Psychopathology Perspective on Adolescence. *Journal of Consulting and Clinical Psychology, 70,* 6–20.

Coan, J. A. (2008). Toward a Neuroscience of Attachment. In J. Cassidy & P. R. Shaver (Eds.), *Handbook of Attachment: Theory, Research, and Clinical Applications* (2nd edn., pp. 241–265). New York: Guilford Press.

Coan, J. A. (2010). Adult Attachment and the Brain. *Journal of Social and Personal Relationships, 27,* 210–217.

Coan, J. A., & Gottman, J. M. (2007). The Specific Affect Coding System (SPAFF). In J. A. Coan & J. J. B. Allen (Eds.), *Handbook of Emotion Elicitation and Assessment: Series in Affective Science* (pp. 267–285). New York: Oxford University Press.

Coan, J. A., Schaefer, H. S., & Davidson, R. J. (2006). Lending a Hand: Social Regulation of the Neural Response to Threat. *Psychological Science, 17,* 1032–1039.

Cole, P. M., Michel, M. K., & Teti, L. O. D. (1994). The Development of Emotion Regulation and Dysregulation: A Clinical Perspective. *Monographs of the Society for Research in Child Development, 59,* 73–100.

Collins, N. L., & Feeney, B. C. (2004). Working Models of Attachment Shape Perceptions of Social Support: Evidence from Experimental and Observational Studies. *Journal of Personality and Social Psychology, 87,* 363–383.

Collins, N. L., & Read, S. J. (1990). Adult Attachment, Working Models, and Relationship Quality in Dating Couples. *Journal of Personality and Social Psychology, 58,* 644–663.

Conradt, E., & Ablow, J. (2010). Infant Physiological Response to the Still-Face Paradigm: Contributions of Maternal Sensitivity and Infants' Early Regulatory Behavior. *Infant Behavior and Development, 33,* 251–265.

Cook, W. L., & Kenny, D. A. (2005). The Actor-Partner Interdependence Model: A Model of Bidirectional Effects in Developmental Studies. *International Journal of Behavioral Development, 29,* 101–109.

Crowell, S. E. (2016). Biting the Hand That Feeds: Current Opinion on the Interpersonal Causes, Correlates, and Consequences of Borderline Personality Disorder. *F1000Research, 5.*

Crowell, S. E., Beauchaine, T. P., McCauley, E., Smith, C. J., Vasilev, C. A., & Stevens, A. L. (2008). Parent-Child Interactions, Peripheral Serotonin, and Self-Inflicted Injury in Adolescents. *Journal of Consulting and Clinical Psychology, 76,* 15–21.

Crowell, S. E., Beauchaine, T. P., Bride, D. L., Hsiao, R. C., & McCauley, E. (2011). *Dyadic Psychophysiology during Conflict: Family Process and Emotion Dysregulation in Self-Injuring and Depressed Adolescents.* Paper presented at the Society for Research in Child Development, Montreal, Quebec, Canada. April.

Crowell, S. E., Baucom, B. R., McCauley, E., Potapova, N. V., Fitelson, M., Barth, H., . . . Beauchaine, T. P. (2013). Mechanisms of Contextual Risk for Adolescent Self-Injury: Invalidation and Conflict Escalation in Mother-Child Interactions. *Journal of Clinical Child and Adolescent Psychology, 42,* 467–480.

Crowell, S. E., Butner, J. E., Wiltshire, T. J., Munion, A. K., Yaptangco, M., & Beauchaine, T. P. (2017a). Evaluating Emotional and Biological Sensitivity to Maternal Behavior among Self-Injuring and Depressed Adolescent Girls Using Nonlinear Dynamics. *Clinical Psychological Science, 5,* 272–285.

Crowell, S. E., Price, C. J., Puzia, M. E., Yaptangco, M., & Cheng, S. C. (2017b). Emotion Dysregulation and Autonomic Responses to Film, Rumination, and Body Awareness: Extending Psychophysiological Research to a Naturalistic Clinical Setting and a Chemically Dependent Female Sample. *Psychophysiology, 54,* 713–723.

Crowell, S. E., Puzia, M. E., & Yaptangco, M. (2015). The Ontogeny of Chronic Distress: Emotion Dysregulation across the Life Span and Its Implications for Psychological and Physical Health. *Current Opinion in Psychology, 3,* 91–99.

Dalton, W. T., Frick-Horbury, D., & Kitzmann, K. M. (2006). Young Adults' Retrospective Reports of Parenting by Mothers and Fathers: Associations with Current Relationship Quality. *Journal of General Psychology, 133,* 5–18.

Davison, K. K., Charles, J. N., Khandpur, N., & Nelson, T. J. (2017). Fathers' Perceived Reasons for Their Underrepresentation in Child Health Research and Strategies to Increase Their Involvement. *Maternal and Child Health Journal, 21,* 267–274.

Deater-Deckard, K., Pylas, M., & Petrill, S. A. (1997). *The Parent-Child Interaction System (PARCHISY).* London: Institute of Psychiatry.

Diamond L. M. (2015) The Biobehavioral Legacy of Early Attachment Relationships for Adult Emotional and Interpersonal Functioning. In V. Zayas & C. Hazan (Eds.), *Bases of Adult Attachment* (pp. 79–105). New York: Springer.

Diamond, L. M., & Fagundes, C. P. (2010). Psychobiological Research on Attachment. *Journal of Social and Personal Relationships, 27,* 218–225.

Diamond, L. M., & Hicks, A. M. (2005). Attachment Style, Current Relationship Security, and Negative Emotions: The Mediating Role of Physiological Regulation. *Journal of Social and Personal Relationships, 22,* 499–518.

Evans, C. A., & Porter, C. L. (2009). The Emergence of Mother-Infant Co-Regulation during the First Year: Links to Infants' Developmental Status and Attachment. *Infant Behavior and Development, 32,* 147–158.

Feeney, J. A., & Noller, P. (1990). Attachment Style as a Predictor of Adult Romantic Relationships. *Journal of Personality and Social Psychology, 58,* 281–291.

Feldman, R. (2007). Parent-Infant Synchrony and the Construction of Shared Timing; Physiological Precursors, Developmental Outcomes, and Risk Conditions. *Journal of Child Psychology and Psychiatry, 48,* 329–354.

Feldman, R. (2012). Parent-Infant Synchrony: A Biobehavioral Model of Mutual Influences in the Formation of Affiliative Bonds. *Monographs of the Society for Research in Child Development, 77,* 42–51.

Fischer, M. S., Baucom, D. H., Kirby, J. S., & Bulik, C. M. (2015). Partner Distress in the Context of Adult Anorexia Nervosa: The Role of Patients' Perceived Negative Consequences of AN and Partner Behaviors. *International Journal of Eating Disorders*, 48, 67–71.

Fraley, R. C., Waller, N. G., & Brennan, K. A. (2000). An Item Response Theory Analysis of Self-Report Measures of Adult Attachment. *Journal of Personality and Social Psychology*, 78, 350–365.

Fraley, R. C., Heffernan, M. E., Vicary, A. M., & Brumbaugh, C. C. (2011). The Experiences in Close Relationships – Relationship Structures Questionnaire: A Method for Assessing Attachment Orientations across Relationships. *Psychological Assessment*, 23, 615–625.

Fogel, A. (1994). *Co-Regulation Coding System*. Unpublished manual, University of Utah.

Gallo, L. C., & Smith, T. W. (2001). Attachment Style in Marriage: Adjustment and Responses to Interaction. *Journal of Social and Personal Relationships*, 18, 263–289.

Gardner, M., & Steinberg, L. (2005). Peer Influence on Risk Taking, Risk Preference, and Risky Decision Making in Adolescence and Adulthood: An Experimental Study. *Developmental Psychology*, 41, 625–635.

Gillath, O., Mikulincer, M., Fitzsimons, G. M., Shaver, P. R., Schachner, D. A., & Bargh, J. A. (2006). Automatic Activation of Attachment-Related Goals. *Personality and Social Psychology Bulletin*, 32, 1375–1388.

Granic, I., O'Hara, A., Pepler, D., & Lewis, M. D. (2007). A Dynamic Systems Analysis of Parent-Child Changes Associated with Successful "Real-World" Interventions for Aggressive Children. *Journal of Abnormal Child Psychology*, 35, 845–857.

Gratz, K. L., & Roemer, L. (2004). Multidimensional Assessment of Emotion Regulation: Development, Factor Structure, and Initial Validation of the Difficulties in Emotion Regulation Scale. *Journal of Psychopathology and Behavioral Assessment*, 26, 41–54.

Gyure, M. E., Quillin, J. M., Rodríguez, V. M., Markowitz, M. S., Corona, R. A., Borzelleca, J. F., ... Bodurtha, J. N. (2014). Practical Considerations for Implementing Research Recruitment Etiquette. *IRB: Ethics and Human Research*, 36, 7–12.

Hawkley, L. C., Thisted, R. A., Masi, C. M., & Cacioppo, J. T. (2010). Loneliness Predicts Increased Blood Pressure: 5-Year Cross-Lagged Analyses in Middle-Aged and Older Adults. *Psychology and Aging*, 25, 132–141.

Hazan, C., & Shaver, P. (1987). Romantic Love Conceptualized as an Attachment Process. *Journal of Personality and Social Psychology*, 52, 511–524.

Hesse, E. (1999). The Adult Attachment Interview: Historical and Current Perspectives. In J. Cassidy & P. R. Shaver (Eds.), *Handbook of Attachment: Theory, Research, and Clinical Applications* (pp. 395–433). New York: Guilford Press.

Hofer, M. A. (1994). Hidden Regulators in Attachment, Separation, and Loss. *Monographs of the Society for Research in Child Development*, 59, 192–207.

Hofer, M. A. (2006). Psychobiological Roots of Early Attachment. *Current Directions in Psychological Science*, 15, 84–88.

Hughes, A. E., Crowell, S. E., Uyeji, L., & Coan, J. A. (2012). A Developmental Neuroscience of Borderline Pathology: Emotion Dysregulation and Social Baseline Theory. *Journal of Abnormal Child Psychology*, 40, 21–33.

Joel, S., MacDonald, G., & Shimotomai, A. (2011). Conflicting Pressures on Romantic Relationship Commitment for Anxiously Attached Individuals. *Journal of Personality*, 79, 51–74.

Joiner, T. E., Hom, M. A., Hagan, C. R., & Silva, C. (2016). Suicide as a Derangement of the Self-Sacrificial Aspect of Eusociality. *Psychological Review*, 123, 235–254.

Kenny, D. A. (2005). Cross-Lagged Panel Design. In B. S. Everitt & D. Howell (Eds.), *Encyclopedia of Statistics in Behavioral Science* (pp. 240–451). Hoboken, NJ: Wiley.

Kerig, P. K., & Baucom, D. H. (Eds.). (2004). *Couple Observational Coding Systems*. Mahwah, NJ: Erlbaum.

Leclère, C., Viaux, S., Avril, M., Achard, C., Chetouani, M., Missonnier, S., & Cohen, D. (2014). Why Synchrony Matters during Mother-Child Interactions: A Systematic Review. *PloS One*, 9, e113571.

Lindahl, K. M., & Malik, N. M. (2000). *System for Coding Interactions and Family Functioning*. Miami, FL: University of Miami.

Lopes, V. P., Gabbard, C., & Rodrigues, L. P. (2013). Physical Activity in Adolescents: Examining Influence of the Best Friend Dyad. *Journal of Adolescent Health*, 52, 752–756.

Lotzin, A., Lu, X., Kriston, L., Schiborr, J., Musal, T., Romer, G., & Ramsauer, B. (2015). Observational Tools for Measuring Parent-Infant Interaction: A Systematic Review. *Clinical Child and Family Psychology Review*, 18, 99–132.

Lunkenheimer, E., Tiberio, S. S., Buss, K. A., Lucas-Thompson, R. G., Boker, S. M., & Timpe, Z. C. (2015). Coregulation of Respiratory Sinus Arrhythmia between Parents and Preschoolers: Differences by Children's Externalizing Problems. *Developmental Psychobiology*, 57, 994–1003.

Moore, G. A., Hill-Soderlund, A. L., Propper, C. B., Calkins, S. D., Mills-Koonce, W. R., & Cox, M. J. (2009). Mother-Infant Vagal Regulation in the Face-to-Face Still-Face Paradigm Is Moderated by Maternal Sensitivity. *Child Development*, 80, 209–223.

Ostlund, B. D., Measelle, J. R., Laurent, H. K., Conradt, E., & Ablow, J. C. (2017). Shaping Emotion Regulation: Attunement, Symptomatology, and Stress Recovery within Mother-Infant Dyads. *Developmental Psychobiology*, 59, 15–25.

Parent, J., Forehand, R., Pomerantz, H., Peisch, V., & Seehuus, M. (2017). Father Participation in Child Psychopathology Research. *Journal of Abnormal Child Psychology*, 45, 1259–1270.

Patterson, A. C., & Veenstra, G. (2010). Loneliness and Risk of Mortality: A Longitudinal Investigation in Alameda County, California. *Social Science and Medicine*, 71, 181–186.

Perry, N. S., Baucom, K. J., Bourne, S., Butner, J., Crenshaw, A. O., Hogan, J. N., ... Baucom, B. R. (2017). Graphic Methods for Interpreting Longitudinal Dyadic Patterns from Repeated-Measures Actor-Partner Interdependence Models. *Journal of Family Psychology*, 31, 592–603.

Pietromonaco, P. R., & Beck, L. A. (2015). Attachment Processes in Adult Romantic Relationships. In M. Mikulincer, P. R. Shaver, J. A. Simpson, & J. F. Dovidio (Eds.), *APA Handbooks in Psychology. APA Handbook of Personality and Social Psychology, Vol. 3: Interpersonal Relations* (pp. 33–64). Washington, DC: American Psychological Association.

Pietromonaco, P. R., Uchino, B., & Dunkel Schetter, C. (2014). Close Relationship Processes and Health: Implications of Attachment Theory for Health and Disease. *Health Psychology*, 32, 499–513.

Prinstein, M. J., & Giletta, M. (2016). Peer Relations and Developmental Psychopathology. In D. Cicchetti (Ed.), *Developmental Psychopathology* (3rd edn., pp. 529–579). Hoboken, NJ: Wiley

Prinstein, M. J., Borelli, J. L., Cheah, C. S., Simon, V. A., & Aikins, J. W. (2005). Adolescent Girls' Interpersonal Vulnerability to Depressive Symptoms: A Longitudinal Examination of Reassurance-Seeking and Peer Relationships. *Journal of Abnormal Psychology, 114*, 676–688.

Raby, K. L., Cicchetti, D., Carlson, E. A., Cutuli, J. J., Englund, M. M., & Egeland, B. (2012). Genetic and Caregiving-Based Contributions to Infant Attachment: Unique Associations with Distress Reactivity and Attachment Security. *Psychological Science, 23*, 1016–1023.

Ravitz, P., Maunder, R., Hunter, J., Sthankiya, B., & Lancee, W. (2010). Adult Attachment Measures: A 25-Year Review. *Journal of Psychosomatic Research, 69*, 419–432.

Roberts, N. A., Tsai, J. L., & Coan, J. A. (2007). Emotion Elicitation Using Dyadic Interaction Tasks. In J. A. Coan & J. B. Allen (Eds.), *Handbook of Emotion Elicitation and Assessment* (pp. 106–123). New York: Oxford University Press.

Robles, T. F., Slatcher, R. B., Trombello, J. M., & McGinn, M. M. (2014). Marital Quality and Health: A Meta-Analytic Review. *Psychological Bulletin, 140*, 140–187.

Roisman, G. I., Holland, A., Fortuna, K., Fraley, R. C., Clausell, E., & Clarke, A. (2007). The Adult Attachment Interview and Self-Reports of Attachment Style: An Empirical Rapprochement. *Journal of Personality and Social Psychology, 92*, 678–697.

Shaver, P. R., & Mikulincer, M. (2010). New Directions in Attachment Theory and Research. *Journal of Social and Personal Relationships, 27*, 163–172.

Stankovic, J. A. (2014). Research Directions for the Internet of Things. *IEEE Internet of Things Journal, 1*, 3–9.

Stubbs, J., Crosby, L., Forgatch, M. S., & Capaldi, D. M. (1998). *Family and Peer Process Code: A Synthesis of Three Oregon Social Learning Center Behavior Codes.* Eugene: Oregon Social Learning Center.

Tronick, E. Z., Als, H., Adamson, L., Wise, S., & Brazelton, T. B. (1978). The Infant's Response to Entrapment between Contradictory Messages in Face-to-Face Interaction. *Journal of American Academy of Child Psychiatry, 17*, 1–13.

Trout, D. L. (1980). The Role of Social Isolation in Suicide. *Suicide and Life-Threatening Behavior, 10*, 10–23.

Uchino, B. N., Smith, T. W., & Berg, C. A. (2014). Spousal Relationship Quality and Cardiovascular Risk: Dyadic Perceptions of Relationship Ambivalence Are Associated with Coronary-Artery Calcification. *Psychological Science, 25*, 1037–1042.

Walsh, E. I., & Brinker, J. K. (2015). Delay between Recruitment and Participation Impacts on Preinclusion Attrition. *Quarterly Journal of Experimental Psychology, 68*, 635–640.

Wittenborn, A. K., Dolbin-MacNab, M. L., & Keiley, M. K. (2013). Dyadic Research in Marriage and Family Therapy: Methodological Considerations. *Journal of Marital and Family Therapy, 39*, 5–16.

Xiao, B., Imel, Z. E., Georgiou, P. G., Atkins, D. C., & Narayanan, S. S. (2015). "Rate My Therapist": Automated Detection of Empathy in Drug and Alcohol Counseling via Speech and Language Processing. *PLoS One, 10*, e0143055.

Yang, Y. C., Boen, C., Gerken, K., Li, T., Schorpp, K., & Harris, K. M. (2016). Social Relationships and Physiological Determinants of Longevity across the Human Life Span. *Proceedings of the National Academy of Sciences, 113*, 578–583.

27 Models for Dyadic Data

EDUARDO ESTRADA, DAVID A. SBARRA, AND EMILIO FERRER

Humans are social animals, and most human experience unfolds in a social context. Indeed, many of the problems addressed in clinical psychology can only be understood by taking into consideration the interdependence of two (or more) people within their social system. For example, social anxiety, depression, violence and aggression, psychopathic and narcissistic behaviors, eating disorders, and marital satisfaction, are manifest in and shaped by the social context in which they exists (Girard et al., 2017; Sbarra & Wishman, 2013). Clinical interventions for childhood and developmental problems typically require working with the child's social environment, including teachers, parents, and other people who shape the daily lives of children.

The most basic social system is that of two individuals, or a dyad. Despite its simplicity, many forms of complex social interactions take place in the context of an interactive dyadic system – for example, the bond between a child and its mother, a romantic relationship between two partners, or the relationship between a patient and therapist. In this chapter, we describe models for analyzing data from such dyadic systems. We begin by defining key characteristics of dyadic systems and then identify clinical research questions related to dyadic systems and processes that unfold over time. We use these questions to select models and illustrate data analytic techniques.

WHY DO WE NEED MODELS FOR DYADIC DATA?

The key statistical concept underpinning dyadic research is *nonindependence*. This term means that the members of a dyad are associated in such a way that the assumption of independence in standard analyses does not hold (Kenny, 1996; Kenny & Judd, 1986; Kenny, Kashy, & Cook, 2006). Other terms such as interdependence (Gonzalez & Griffin, 1997; Thorson, West, & Mendes, 2017) or linked scores (Kenny et al., 2006) are used to refer to the same broad idea. Nonindependence can be understood from both a conceptual and a statistical standpoint. Conceptually, nonindependence means that the behavior of one unit in a dyadic system is best explained by also taking into consideration the behavior of the other unit. For example, the meaning of a six-month-old baby's crying or gazing at his caregiver is not fully grasped unless also observing the concurrent behavior of his caregiver (e.g., smiling back to the child, feeding it, soothing it). Likewise, the caregiver's behavior is better explained – and might be understood only – by also observing the child's behavior. Similarly, an individual's adherence to, and progress in, therapy can only be understood in the context of their interactions with the therapist.

From a statistical standpoint, nonindependence manifests in a nonzero correlation between the scores of the dyad members. The requirement for independent observations is one of the key assumptions for ANOVAs, t-tests, linear regression, and other models based in the general linear model; in the case of correlated data from two members of a dyad, the independence assumption is violated. Consequently, these models cannot be used for analyzing dyadic data because the corresponding inferential statistics (T, F or regression coefficients) have biased standard errors and decreased degrees of freedom, leading to either too conservative or too liberal significance tests. This, in turn, leads to higher rates of Type 1 and Type 2 errors (Kenny et al., 2006).

In our description of models, we refer to specifications that pertain to distinguishable dyads. A dyad is *distinguishable* or *nonexchangeable* when there is at least one variable that allow differences to be established between the members (Gonzalez & Griffin, 1997; Kenny et al., 2006). For example, heterosexual couples are potentially distinguishable because one can measure differences between men and women. Different age siblings or therapist-client dyads are also distinguishable (Bolger & Laurenceau, 2013). In contrast, same-sex couples or twins are indistinguishable if there is no clear factor establishing a difference between the members of the dyad. Although this distinction is not necessary, some researchers have used structural equation models for distinguishable dyads, and multilevel models for indistinguishable dyads (Nestler, Grimm, & Schönbrodt, 2015). However, this is not always the case (e.g., Baucom et al., 2015a). Methods for evaluating distinguishability are beyond the scope of this chapter and are available in the literature (Gonzalez & Griffin, 1997; Kenny et al., 1998; Ledermann & Macho, 2014).

KEY RESEARCH QUESTIONS IN CLINICAL PSYCHOLOGY RELATED TO DYADIC SYSTEMS: OUTCOMES VERSUS PROCESS RESEARCH

Understanding modeling approaches for dyadic data, or any other data, requires a context in which research questions can be stated and models selected accordingly. In clinical psychology, one possible framework is to outline questions about dyadic systems that distinguish between *outcomes* of a process and the *process* itself. One example of clinical research in dyads focused on *outcomes* would be a study of the characteristics of a counselor that are associated with client's perceived working alliance (Kivlighan et al., 1993). One example of research focused on *processes* would be a study examining how the counselor's behavior and the client's perception of working alliance change over time. Here, the study would focus on understanding how the two trajectories can be described, whether or not they are related, and what factors can explain different features of the change.

In the next section, we illustrate a number of clinical research questions that can be considered as related to outcomes, and describe a statistical model typically used to address them. In the following section, we illustrate a variety of questions focused on processes, and describe statistical models that can be used to address them. Table 27.1 provides a list of the models covered in this chapter, examples of research questions addressed by each them, and data features suited to such models.

Importantly, the models we describe below may be used flexibly to study dyadic outcomes and processes ranging from laboratory-based psychophysiological investigations to prospective epidemiological studies. Moreover, such processes – and potential changes in the processes – can operate at multiple levels, in multiple dimensions, and multiple time metrics (e.g., minutes, days, years).

MODELS FOR STUDYING OUTCOMES

Studies focusing on outcomes typically have either cross-sectional or pre-post designs. In the first case, all the variables are measured at a single occasion. In the second case, dyadic change in the relevant variables (e.g., symptoms) is measured before and after an intervention. Ideally, these two measures are taken as part of a randomized control trial, so the difference in the change of control and treated group can be attributed to the intervention. Regardless of the specific research design and question, these pre-post studies also take an outcomes approach, largely because they focus on whether the intervention is associated with differences in the total amount of change in the outcome variable.

In both cases, the research takes an *outcomes approach* because the goal is to explain individual differences in the outcome variable of interest by exploring its associations with one or more covariates and predictors, without characterizing the processes that led to that outcome or investigating how those associations unfolded over time. Some other examples of questions focusing on outcomes in cross-sectional studies are include: What type of spouse support is associated with higher marital satisfaction in patients with cancer (Hagedoorn et al., 2000)? What features of the family's environment and expressed emotion help explaining inpatients' eating disorders (Medina-Pradas et al., 2011)? Is expressed emotion of parent and child related to child obsessive-compulsive disorder symptoms (Przeworski et al., 2012)? All of these studies are dyadic in nature because the outcome variable (marital satisfaction, therapy working alliance, eating disorders, etc.) is part of a system of interconnected constructs, and evaluating the association between those constructs requires gathering information about both members in the dyad.

Some examples of questions from pre-post studies are: In couple therapy, what styles of couple alliance predict couple outcomes (Anker et al., 2010)? What coping strategies protect infertile couples from depression over a failed insemination attempt (Berghuis & Stanton, 2002)? What characteristics of counselor and client are associated with psychotherapy working alliance (Kivlighan, Marmarosh, & Hilsenroth, 2014)? When treating children with social anxiety problems, under what conditions is it beneficial to involve their parents in therapy (Garcia-Lopez et al., 2014)? Most of the clinical questions in dyadic cross-sectional or pre-post studies are and can be addressed with the actor-partner interdependence model, or APIM (Kenny et al., 2006), which is described below.

The Actor-Partner Interdependence Model

When studying dyadic systems, researchers focusing on outcomes often use the APIM (Kenny, 1996; Kenny et al., 2006). For example, in couples with one member affected by cancer, Hagedoorn and collaborators (2000) applied the model to study the extent to which marital satisfaction was explained by the adoption by the nonaffected partner of a supportive style termed *active engagement* (this style entails engaging the patient in discussions, inquiring how the patient feels, asking about help and information, and using other constructive problem-solving methods). Here, we use this topic to illustrate how the APIM can address questions about dyadic outcomes – although their study used a more complex version of the model involving more, and slightly different, variables.

Figure 27.1 depicts the most basic form of an APIM for this research question. X represents the active engagement of the caregiver, as perceived by the caregiver (Xc) and the patient (Xp). Yc and Yp represent marital satisfaction of

Table 27.1 Overview of dyadic models

Model	Type of data	Example of clinical research question
Focus on outcomes		
Actor-partner interdependence model (APIM)	Cross-sectional. At least two variables measured on each dyad member	Couples with one partner affected by cancer, "Perception of active engagement of the nonaffected partner" and "marital satisfaction" are measured on both partners. Does engagement explain satisfaction? Are the interpersonal and intrapersonal effects equally important? Are the influences reciprocal or is one partner more influential than the other?
	Pre-post. At least one variable measured pre and post for each dyad member (same or different variables measured at each occasion)	"Attachment style" of therapists and clients are measured at the beginning of therapy sessions, and "working alliance" is reported by both individuals after three weeks. Does attachment predict alliance? Are the interpersonal and intrapersonal effects equally important? Are the influences reciprocal or is one partner more influential than the other?
Focus on processes		
Large sample size, few measurement occasions per case ($\uparrow N$, $\downarrow T$)	At least one variable measured on each dyad member; >2 occasions	Standard goal: To characterize dyadic change and interrelations at the group level
Latent growth curve model (LGC)		Couples anticipating an upcoming stressful event for one member. "Anxiety" is being measured on the target member and "relational dissatisfaction" is measured on the partner. What are the trajectories for anxiety and dissatisfaction, as the event approaches? What is the relation between the trajectories in both partners?
Latent change score model (LCS)		"Depressive symptoms" are measured in a sample of mothers, while "anxiety" is measured on their daughters. What is the trajectory of both variables over time? What is the relation between both trajectories? Does mother's depression predict subsequent increases in the daughter's anxiety? What about the other way around?
Many measurement occasions per case ($\uparrow T$)	Stationarity is typically assumed	Standard goal: To characterize change for each dyad member and interrelations among them
Autoregressive model (AR) (equivalent to a repeated measures [RM-]APIM)	At least one variable measured on each dyad member; multiple occasions	Daily measures of "anxiety" in couples for four months. Are both partners equally influenced by their own previous anxiety? Are partner effects equally important? How many previous measures are needed to explain the state of the system at any given time (i.e., what is the time dependency of the system)?

Table 27.1 (cont.)

Model	Type of data	Example of clinical research question
Dynamic factor analysis (DFA)	Multiple variables measured on each dyad member; multiple occasions. These variables measure one or more common latent factors on each partner. Measurement invariance is assumed	Daily reports of negative mood reported by both dyad members. A latent factor representing "negative affect" is extracted from the observed variables. What is the structure of negative affect? What is the time dependency of negative affect? Are there differences between both partners? Regarding the latent processes of each partner, are they influenced by their own previous states or by those of the partner? How long until such influences dissipate?
Differential equations model (DEM)	One of multiple variables measured on each dyad member; multiple occasions. Time is treated as continuous	Daily measures of negative affect in couples for two months. What are the dynamics of affect for each person in the dyad? What are the dynamics between both individuals? Does each partner self-regulate affect autonomously? Do they coregulate each other's affect? Is the system stable and progress toward an equilibrium point?

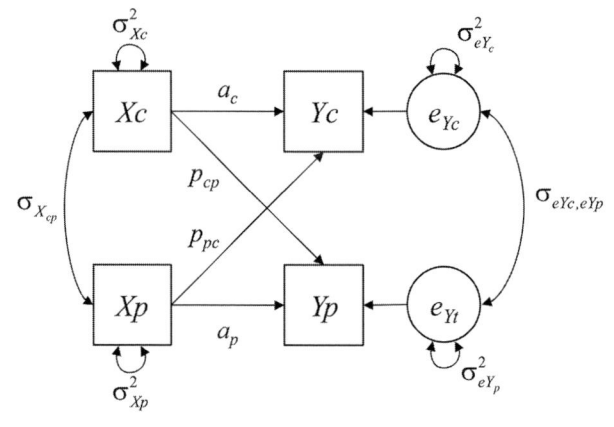

Figure 27.1 Path diagram of an actor-partner independence model (APIM)

each member of the couple. The effect of each individual's perception on their own satisfaction is called *actor effect* (*a*), whereas the effect on the partner's satisfaction is called *partner effect* (*p*): a_c and a_p are the actor effects of the caregiver's and patient's perceptions on their own marital satisfaction, whereas p_{cp} and p_{pc} are the influences of each member of the dyad on their counterpart's satisfaction. Two error terms, e_{Yc} and e_{Yp}, are typically included to account for the variance of the outcome variables that are not explained by the actor or partner effects.

This model typically includes a covariance between Xc and Xp, and another covariance between e_{Yc} and e_{Yp}. The former accounts for the association between the initial states of both elements of the dyad, often called *compositional effects* (Kenny et al., 2006). The latter accounts for the relation between the final states that are not explained by the model.

The APIM allows quantifying and comparing the interpersonal and intrapersonal effects for both members of the dyads, and testing hypotheses about their statistical significance: Is the patient's satisfaction explained by the perception of the partner's active engagement? Is it explained by the partner's self-perception of active engagement? Which of these two has a greater influence? One approach to evaluate hypotheses such as these is to constrain some of the model parameters. This implies restricting the values that one or several parameters can take, and evaluating whether the model fit decreases as a consequence of such restriction. For example, to test a hypothesis positing a symmetrical relation, one could constrain the two partner effects (*p*) to have the same value. If one of the partners exerts a stronger influence than the other, the model fit will be significantly degraded by this constraint, and the hypothesis of symmetrical relation will be formally rejected. Similarly, if we hypothesize that the relationship is nonreciprocal, we may constrain one of the partner effects to zero, while freely estimating the other, and evaluate the increase of misfit through nested models comparison. Note that a cross-sectional design without experimentation (i.e., an observational study) can only show relations among constructs but provides no information about potential causal links between them.

The APIM can also be applied in pre-post designs. For example, Marmarosh and colleagues studied the relation between the attachment style of therapists and clients at the beginning of psychotherapy (predictive variable X) and the working alliance reported by each one after three weeks (criterion variable Y; Marmarosh et al., 2014). The interpretation of the actor and partner effects is analog to the cross-sectional example detailed above, and similar hypotheses could be tested formally. It is also possible to include more than one predictive variable at time 1, so several actor and partner effects are evaluated. For example, Marmarosh and colleagues (2014) included two different attachment styles (avoidance and anxiety) for both clients and therapists, plus the interaction between the styles of clients and therapists, resulting in a more complex APIM. In their study, they found a negative interaction between therapists' and clients' initial attachment styles, which significantly predicted later client-perceived alliances: the effect of therapists' anxiety on client-perceived alliances was different depending on whether the clients themselves had high or low levels of anxiety. Applying a pre-post design ensures that the predictive variable antecedes the outcome. However, a proper experimental design – with random assignment to different conditions – would provide better evidence for attributing causality.

If the same variables are measured at both occasions, the APIM described here is equivalent to an autoregressive cross-lagged model with two time points (Nestler et al., 2015). The stability and cross-lagged coefficients are equivalent to the actor and partner effects, respectively. In the context of psychotherapy, this model is sometimes termed a repeated measures (RM) APIM (see Baucom et al., 2015a; Crowell et al., 2014; Perry et al., 2017). Extending the model to further include additional occasions allows for testing whether the actor and partner effects can explain the longitudinal interrelations – i.e., whether the processes are interdependent over time (Laurenceau & Bolger, 2005). This leads to an autoregressive model with more than two occasions. We describe these models later in the chapter.

APIMs can be used to address a number of different questions in clinical psychology. Some examples involve health behavior in couples (Butterfield & Lewis, 2002; Franks et al., 2004), attachment styles and relationship dependence on behaviors under stress (Campbell et al., 2001), outcomes in couple's therapy (Cook & Snyder, 2005), relations among aggression, victimization, anxiety and depression in couples (Lawrence et al., 2009), and health-related quality of life in children with cystic fibrosis and their caregivers (Driscoll et al., 2012), among others. Lederman, Macho, and Kenny (2011) provide a discussion on how the APIM can be extended to account for mediation effects.

MODELS FOR STUDYING PROCESSES

Dynamic Processes in Dyads

Whereas some studies focus on outcomes, other types of studies focus on psychological processes as they unfold over time. We refer to this as *process approach* or *process research* (see Goldfried, Greenberg, & Marmar, 1990; Kopta et al., 1999; Laurenceau, Hayes, & Feldman, 2007). In this approach, questions relate to change and its underlying mechanisms. For example, Laurenceau et al. (2007) defined three sets of questions relevant to change in psychotherapy. The first question pertains to the course – and shape – of change, including: (a) comparing the observed trajectory with the theoretically predicted pathway; (b) evaluating whether the individual trajectories are well represented by the group trajectory; (c) identifying sections of the trajectory with higher rates of change and studying what factors explain them (e.g., some researchers have found periods of rapid improvement during depression treatment, interspersed with periods of symptom stability or exacerbation. What variables explain these periods?); and (d) finding differences between responders and nonresponders (e.g., did the nonresponders show an improvement, but then return to the initial level? How can this pattern be explained?).

The second type of question that can be answered using a process approach relates to moderators of change. Moderators may change the effect of the causal variables in the study. In the context of psychotherapy, for example, a client's age and sex may differentially predict treatment development. Process research often focuses on identifying moderators that are associated with changes of interest (e.g., Is the course of change the same for younger and older couples?). The third question focuses on the mediators of change. A mediator is a variable that is changed by the intervention and causes later changes in the relevant outcome (e.g., Is a reduction in OCD behavioral symptoms preceded by a reduction of caregivers' overprotective behaviors, which follow from the intervention itself?).

A process approach is, in principle, more involved, with efforts to reveal mechanisms underlying changes in the dyads. Such processes typically involve complex interrelations among multiple variables. Any attempt to identify interrelations will require a number of factors, including: (a) rich data that represent the multivariate nature of the process as well as their time dependency, and (b) models that can uncover such dynamics at the appropriate level of analysis and timescale resolution (e.g., minutes, days, weeks) (Nesselroade & Boker, 1994).

Any system of human interrelations is dynamic, inasmuch as it unfolds over time. Hence, a dynamic approach is required for studying and understanding many clinical problems. A dynamic model is a set of equations that expresses the time dependency of the system. For example, a dynamic model can include a function describing the changes in a system as a function of its previous state (Hamaker, Zhang, & Maas, 2009). A dynamic model applied to dyadic systems includes at least two processes – one for each member of the dyad – and must account for the existence of both between- and within-person influences (Ferrer & Steele, 2012, 2014; Levenson & Gottman, 1983). For instance, if one is interested in emotional valence in a system composed of two romantic partners, the valence for one of them at the present moment is likely related to their own previous valence (intraindividual influence), as well as to the previous state of the partner (interindividual influence). A model to describe a dyadic system must account for both self and partner effects (Felmlee & Greenberg, 1999). Dynamic models allow not only describing the present state as a function of previous states, but also present rate and direction of change.

In clinical psychology research, as in any other field, matching theory to methods is vital. To study dyadic processes we need models able to characterize and differentiate processes within and between people, and within and between dyads. In this respect, to answer specific hypotheses, model selection must be guided by one's theory about the phenomenon. In the remainder of this chapter we focus on models for dyadic data that are suited to capture features of dynamic processes in dyads.

Latent Growth Curve Models

Latent growth curve (LGC) modeling is a technique used to characterize changes in a process over time (Ferrer & McArdle, 2003; McArdle & Epstein, 1987; Meredith & Tisak, 1990; Rao, 1958). Here, we focus on a bivariate LGC specification and illustrate how it can be used to address clinical research questions involving dyads. Suppose a researcher would like to study the effect of an upcoming stressful event on a person's emotions and those of their partner. This was the topic addressed by Thompson and Bolger (1999) in their study of daily stress among candidates for the Bar Examination to obtain a lawyer license and their partners. In a scenario like this, several questions would be pertinent that involve both dyad members. For example, focusing on the emotional effects of the examination, one could be interested in: (a) the trajectory of the examinee's anxiety as the date of the exam approaches, (b) the trajectory of the partner's dissatisfaction with the relationship, and (c) the potential relation between these two processes. Given a number of measurement occasions on both variables, a bivariate LGC model could be applied to address these questions. Figure 27.2 depicts a path diagram of such a model (top panel) and trajectories in variables X and Y from a sample of 100 couples (bottom panel).

In the LGC model of Figure 27.2, X represents the examinee's self-reported anxiety and Y represents the partner's dissatisfaction with the relationship. Both variables are measured on four occasions. At each occasion, the observed scores for each process ($Y_{[t]}$ and $X_{[t]}$) are a function of two latent variables: an intercept (y_0 and x_0) and a slope (y_s and x_s) of the growth trajectory, for each variable. The latent intercept has loadings of the same value – usually 1 – on all occasions. The latent slope has loadings that define the shape of the curve over time. For example, if the trajectory is expected to increase linearly and the observations are equally spaced in time, the loadings could be fixed to 0, 1, 2, 3. Other values for the slope's loadings allow different specifications for the trajectory; for example, it is possible to fix only the first and last loading to 0 and 1, and freely estimate the rest, so the model can capture any shape of change (i.e., latent basis growth model; McArdle & Epstein, 1987; Meredith & Tisak, 1990). If the first slope loading is fixed to zero, the latent intercept represents the value of the process at the initial occasion.

This model allows estimation of the means of the two latent variables (μ_0 and μ_s), which capture the mean intercept and slope for the whole sample. If the slope loadings are specified with increasing values, a positive slope represents an increase in the average trajectory, and vice versa. Because the model also yields estimates of the variances in the latent variables, it is possible to capture between-individual differences in both initial level and rate of change. The covariance between the intercept and slope (σ_{0s}) captures the relation between the initial state and the growth for the sample. In Figure 27.2, the covariances between the two latent intercepts (σ_{00}) and between the two latent slopes (σ_{ss}) are also estimated. The first parameter represents the relation between the initial values in the examinees' anxiety and their partners' dissatisfaction. The second parameter captures the relation between the rates of growth of both processes. To accommodate further associations between the partners' processes, other parameters can be included in the model. For example, the covariance between one person's intercept and the partner's slope would represent the association between, say, the examinee's anxiety at the initial assessment and increases (or decreases, depending on the slope valence) in the partner's dissatisfaction with the relationship. To account for the nonindependence between both partners' scores, the unique variance of the scores for each individual are typically allowed to covary. This covariance accounts for any interdependence between the processes that are not related to the bivariate growth.

Although we describe standard uses of the LGC models, alternative specifications can be used to capture other forms of associations between both dyad members. For example, Ledermann and Macho (2014) described a LGC model for *common fate effects*, defined as variables affecting both members of the dyad equally (Kenny et al., 2006; Ledermann & Kenny, 2012). In Ledermann and Macho's model, the scores from the two members of a couple were regressed on the same latent variable or process at each time point. Therefore, it was assumed that they are different expressions of the same unobserved construct. In their study, the scores represented negative sentiments about the relationship expressed by each member of the couple, whereas the common latent process was the negative affect in the dyad. Then, a single LGC model was specified to describe the trajectory of the latent process for the dyad. Because this model specifies a measurement structure for the latent process, *measurement invariance* tests are required to ensure that the construct is the same across measurement occasions. Measurement invariance, also called *factorial invariance*, is a statistical procedure for formally testing whether the

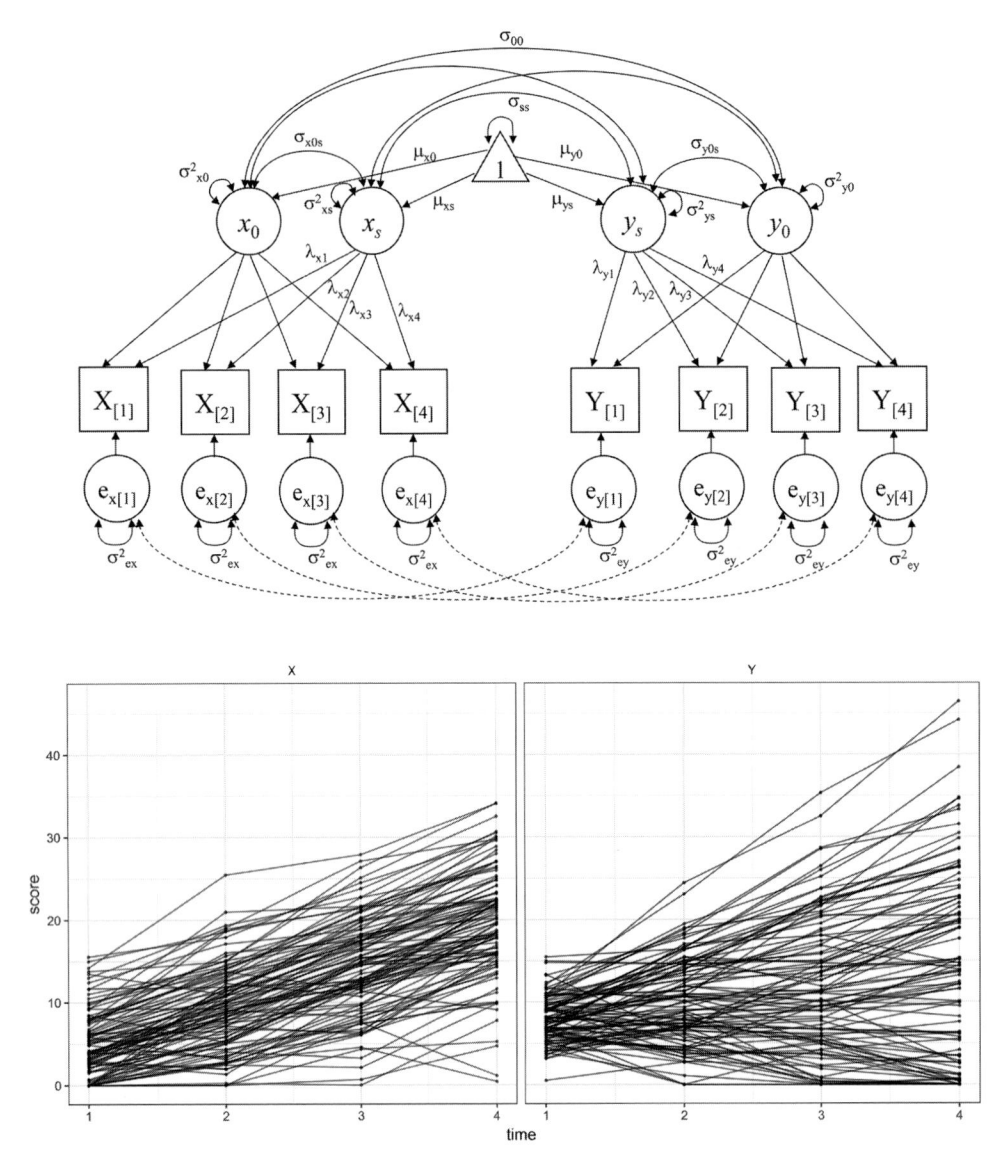

Figure 27.2 Bivariate (dyadic) latent growth curve (LGC) model. Path diagram (top panel) and trajectories of a sample of $n = 100$ (bottom panel)

relations between a set of observed and latent variables are constant over time, or across different groups (see Bollen & Curran, 2006; Ferrer, Balluerka, & Widaman, 2008; Ledermann & Macho, 2014; Meredith, 1993). Other specifications of multivariate processes are also possible in the context of LGC models (McArdle, 1988).

One general limitation of most LGC models for dyadic interactions is that all variability not related to the within- or between-person influences goes to the residuals, without further partitioning. One relatively recent technique suited to overcome this criticism about residuals is the mixed-effects location scale model (Hedeker, Mermelstein, & Demirtas, 2008; Rast, Hofer, & Sparks, 2012), an extension of the mixed-effects model that allows partitioning the residuals. In the context of dyadic data, this model was used to characterize daily emotional ups and downs of couples as a function of individual and partner effects. In addition, the within-person variance over time was explained by factors outside the dyad (e.g., weather), which permeated the daily ups and downs in affect (Ferrer & Rast, 2017).

Bivariate LGC models for dyadic data allow the identification of growth in each variable as well as the relationship between the changes in both variables. But such a relation, expressed by the covariance between slopes, is not time-dependent, thus overlooking possible interrelations between the variables over time. It is also possible to specify a LGC with direct cross-lagged associations between the observed scores of both individuals (Bollen & Curran, 2006; Curran & Bollen, 2001). These parameters would capture time-lagged influences between the two processes, beyond those modeled by the growth process. Whereas the latent slope in this model captures smooth (typically linear) change over time, the addition of time-lagged parameters allows time-specific features of change to be described. These parameters can be constant over time, or time-specific, representing discrete periods in the data. Typically, however, a dyadic LGC is specified to model associations between the changes in dyadic processes through the covariances between latent intercepts and slopes. If a main interest is in examining time-lagged influences and sequences between the processes, other models exist that focus more specifically on such relations (Ferrer & McArdle, 2003; McArdle, 2009). We describe these models in the next sections.

Latent Change Score Models

Latent change score (LCS) models (McArdle, 2001, 2009; McArdle & Hamagami, 2001) are a general framework to model the change in a process as well as its time-related sequences. As such, they subsume LGC models as one particular specification. However, whereas the standard LGC model includes parameters capturing information about the whole trajectories, LCS models typically combine those parameters with additional time-lagged effects, which capture time-specific sequences of influences. Here we provide a conceptual overview of this family of models through one clinical example, and we explain how to interpret the main results. More extended descriptions and applications are available elsewhere (e.g., Ferrer & McArdle, 2003, 2010; Ferrer et al., 2007; McArdle & Hamagami, 2001).

There is a large literature suggesting that child and adolescent psychopathology is associated with features of parents and family environment. For example, there is a strong association between depressive symptoms in mothers and anxiety in their adolescent daughters (Garcia-Lopez et al., 2014). Suppose we want to describe the longitudinal trajectory of both processes and to study the dynamic interrelations between these two constructs in a sample of adolescents. We could hypothesize that higher levels of mothers' depressive symptoms would lead to subsequent increases in their adolescent daughters' anxiety, but high anxiety in daughters would not be associated with increases in mother's depressive symptoms. A LGC model would capture whether the initial state and the growth of each process are related, but would not provide information about the order in which these processes influence each other. A potential model to address this question is a dyadic LCS model (Ferrer & McArdle, 2003, 2010; McArdle, 2001, 2009; McArdle & Hamagami, 2001).

Suppose we recruit a sample of mother–daughter dyads and measure mothers' depression and daughters' anxiety every two weeks during four months, leading to nine assessments in total. A potential bivariate LCS specification to examine questions from these data is depicted in Figure 27.3 (path diagram in top panel, trajectories for 100 dyads in bottom panel). Here, X and Y represent mothers' depressive symptoms and daughters' anxiety, respectively. In the standard LCS model, a latent variable representing changes is specified for each process at each repeated occasion. This latent variable captures the changes in the latent scores of a process between one occasion and the next.

At each occasion, the latent process is a function of the initial unobserved level, y_0 and x_0, plus the accumulation of changes up to that occasion. Various specifications for the unobserved changes Δ_y and Δ_x are possible, depending on one's hypothesis of change. In Figure 27.3, these changes are a function of three elements: (a) an additive component (y_s and x_s), (b) the latent levels of the same process at the previous occasion (i.e., self-feedback, β_y and β_x), and (c) the latent levels of the other process at the previous occasion (i.e., coupling, γ_y and γ_x). All three parameters capture a relevant feature of the trajectory, and therefore they must be interpreted together.

Let us assume the hypothesis that mothers' depressive symptoms (X) lead to changes in daughters' anxiety (Y), but not the other way around. Support for this hypothesis data would be represented by the γ_y coupling parameter being different from zero, but not γ_x. Additionally, the self-feedback effects (β_y and β_x) would indicate the time-lagged influence of each process on itself, whereas the mean of each of the additive components (i.e., the means of y_s and x_s) would indicate the presence of a constant influence in the changes at each occasion, and the magnitude of such influence. As in a dyadic LGC model, the LCS model captures the variability in the latent initial states and the additive components of the processes (quantified by the variances of y_0, x_0, y_s, and x_s) as well as the linear relations among them (quantified by the covariances between those four latent variables). Dyadic LCS models have been applied, to study depressive symptoms in spouses (Kouros & Cummings, 2010), relation between marital satisfaction and self-rated health (Proulx & Snyder-Rivas, 2013), effects of intimate safety, acceptance, and activation on marital satisfaction (Hawrilenko, Gray, & Córdova, 2016) and the effect of self-directed interventions on couples' communication (Bodenmann et al., 2014), among others.

Finally, the dyadic LCS model could be extended to include additional processes. Suppose that, in addition to examining mothers' depressive symptoms and daughters' anxiety, we are interested in mothers' expressed emotion and its interplay with the other two processes. This new variable could be included in a model where different patterns of lagged influences among the three processes could be evaluated. Because each additional process can lead to a large number of additional parameters, the model becomes more complex. Hence, we recommend relying on theory for model specification and selection of paths to be examined. For examples of such multivariate systems, see Ferrer and McArdle (2004) and McArdle et al. (2000).

Time Series Analysis

The models described above (APIM, LGC, and LCS) are typically applied in scenarios with relatively large sample sizes. The goal is often to characterize the relevant aspects of change at the group level. In contrast, the term "time

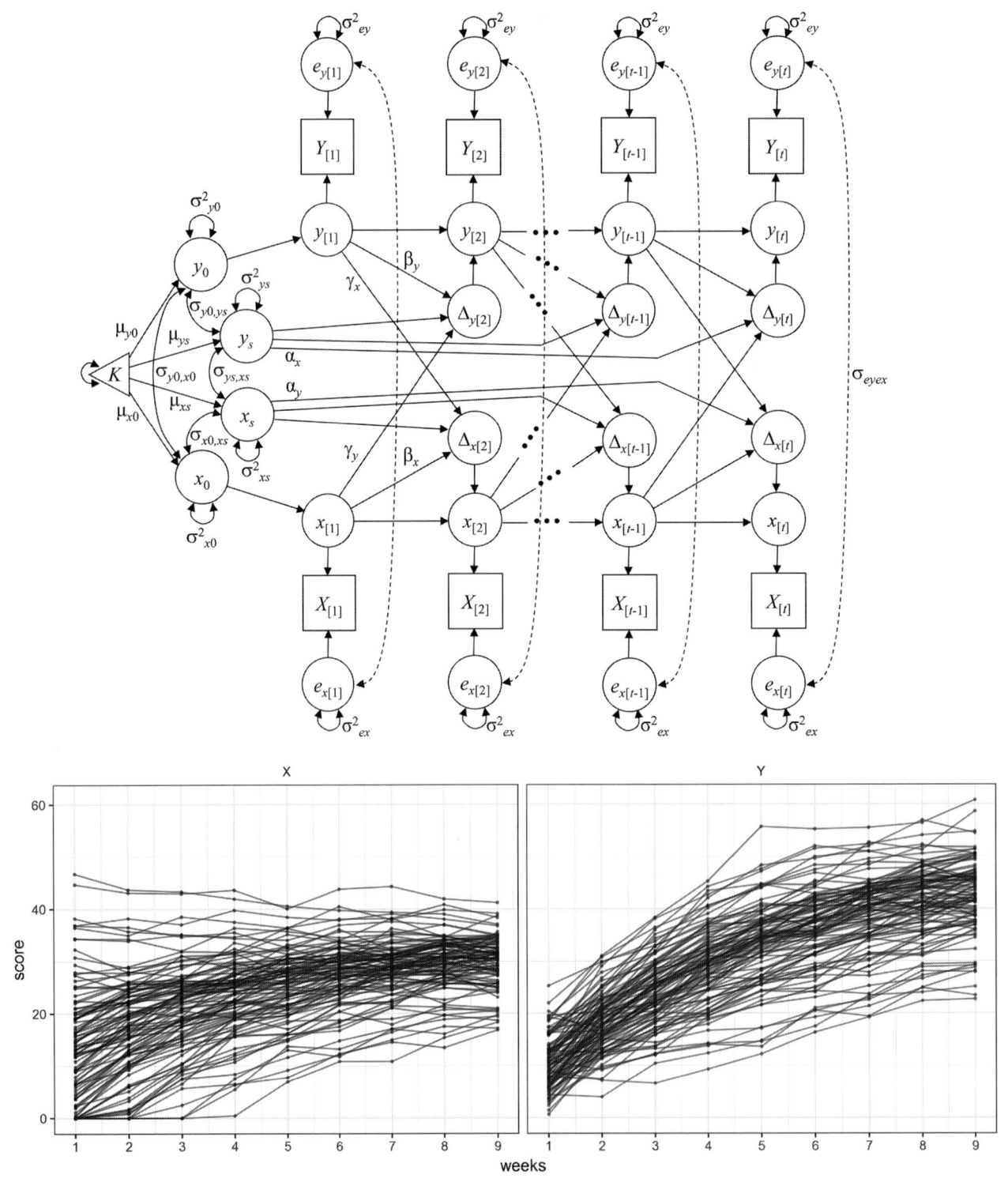

Figure 27.3 Bivariate (dyadic) latent change score (LCS) model. Path diagram (top panel) and trajectories of a sample of $n = 100$ (bottom panel)

series analysis" is commonly used in situations with fewer dyads – one is enough – and a relatively long string of data points of the same process per case. Researchers using these methods are typically interested in characterizing the idiosyncratic features of change in each of the dyads under study. Some examples of time series include the hourly stock market value of a given company or the daily temperature of a specific location. In psychological research on dyads, time series are primarily applied to the study of emotion, affect, and intensively sampled psychophysiological data (Ferrer & Zhang, 2009; Gottman & Notarius, 2000; Levenson & Gottman, 1983). A critical feature of time series is the self-dependency of the data points. This dependency indicates the degree to which the state of the system at any given moment is correlated with itself at previous times. Time self-dependency can be quantified through the autocorrelation (ACF) and partial autocorrelation (PACF) functions. The former refers to the correlation between two given scores separated for a particular time lag k. For example, an autocorrelation of order one represents the correlation between adjacent scores (Ferrer & Zhang, 2009). The *partial autocorrelation*, on the other hand, represents the dependence between the scores at times t and $t + k$ after accounting for all the intermediate occasions.

Another key concept in time series processes is that of stationarity. Conceptually, stationarity indicates that the statistical characteristics of the process do not change over time. Therefore, any particular window of time is equally representative of the whole process. For most time series models, weak stationarity is required (Ferrer & Zhang, 2009), implying that: (a) the mean levels of the variables in the process do not change (i.e., there is no increase or decrease over time), (b) the variances and covariances between the different variables are constant at any particular measurement occasion, and (c) the pattern of lagged dependency only changes as a function of interval length. When a series is not stationary due to trends or seasonal components, it is common to preprocess the data to remove these components before applying any model. A number of methods have been proposed for detecting and analyzing nonstationarity (Bringmann et al., 2017; Cabrieto et al., 2017; for more information about general properties of time series, see Bisgaard & Kulahci, 2011; Shumway & Stoffer, 2011).

Autoregressive Models

The autoregressive (AR) models described here are mathematically equivalent to a repeated measures APIM, and also to a LCS model without the latent additive components (x_s and y_s). For example, the classic cross-lagged panel model is an application of AR models applied to repeated measures panel data. In general, however, AR models are a class of models typically used in time series where a long string of data is available. In the context of dyadic research, an important goal of AR models is to describe the idiosyncratic features of each dyad. Figure 27.4 depicts a path diagram of an AR model (top panel), and the observed scores over 95 days for three different couples (bottom panel). This figure represents a bivariate autoregressive model for the two members of a couple. The variables Y and X could denote, say, the anxiety of the female and male, respectively, in each dyad. At any time t, the female's anxiety in the relationship Y_t is a function of both her own anxiety at the previous time Y_{t-1} as well as that of her partner at the previous occasion X_{t-1}. The first influence is typically termed the autoregressive coefficient, stability or inertia (Gottman, Swanson, & Murray, 1999; Hamaker et al., 2009) of the process, and is captured by the parameters β. The second influence is the cross-lagged regression (CL), or partner's effect, captured by the parameter γ_x or γ_y, for the female and male, respectively. Together, these parameters define an autoregressive model of order one (AR1). The residuals in the processes (d_y and d_x) are typically termed innovations or shock variables and account for the variance in the process not explained by previous states. Such shocks can also be specified so they predict the system at future occasions. In this case, the model becomes an autoregressive moving-average model, ARMA.

For simplicity and parsimony, the values of the AR and CL parameters are often constrained to be equal across occasions. But such a constraint can be relaxed in order to account for potential influences of lags greater than one or to test for differences in the parameters across various segments in the data. As noted previously, if an APIM is applied to more than two measures, and the actor and partner influences are constrained to be equal over time, the resulting RM-APIM is equivalent to the autoregressive model described here (Nestler et al., 2015; Perry et al., 2017).

AR models allow answering questions pertinent to dyadic interactions. Examples of such questions are: Are both individuals in the dyad equally influenced by their own previous states? Are the partner effects the same? Does the model lose its predictive value if one of the partner effects is constrained to be zero – e.g., from male to female? By adding AR and CL parameters of higher order (e.g., lag 2, lag 3) it is possible to evaluate the time dependency of the model: How many previous measures are needed to explain the state of the system at any given time?

The basic AR model described here can be extended in different ways. For example, Gottman and collaborators (2002) proposed a set of dynamic functions – linear and nonlinear – that allows the autoregressive components in the model to capture different patterns of partners' influence. They used these equations to describe dynamic affect changes in couples' interactions and to predict later marriage success. Recent developments of this framework (Hamaker et al., 2009) allow comparing competing models and evaluating which one is better for each particular couple (Madhyastha, Hamaker, & Gottman, 2011).

Dynamic Factor Analysis

When studying many psychological phenomena, researchers must account for the fact that the observed scores are not perfect measures of the processes under study – i.e., there is error of measurement. To deal with this issue, multiple observed variables are often used to measure a latent factor representing the unobserved *process*. One model that can accommodate the resulting multivariate time series is the dynamic factor analysis (DFA; Engle & Watson, 1981; McArdle, 1982; Molenaar, 1985). DFA integrates features of factor analysis and time series analysis into a single model that captures the measurement structure of a latent variable as well as its time dependency.

The basic specification of a DFA model allows for addressing questions such as: Which of the observed indicators is contributing more strongly to the latent process? How do the latent processes in the dyad influence each other and themselves over time? Are they influenced by

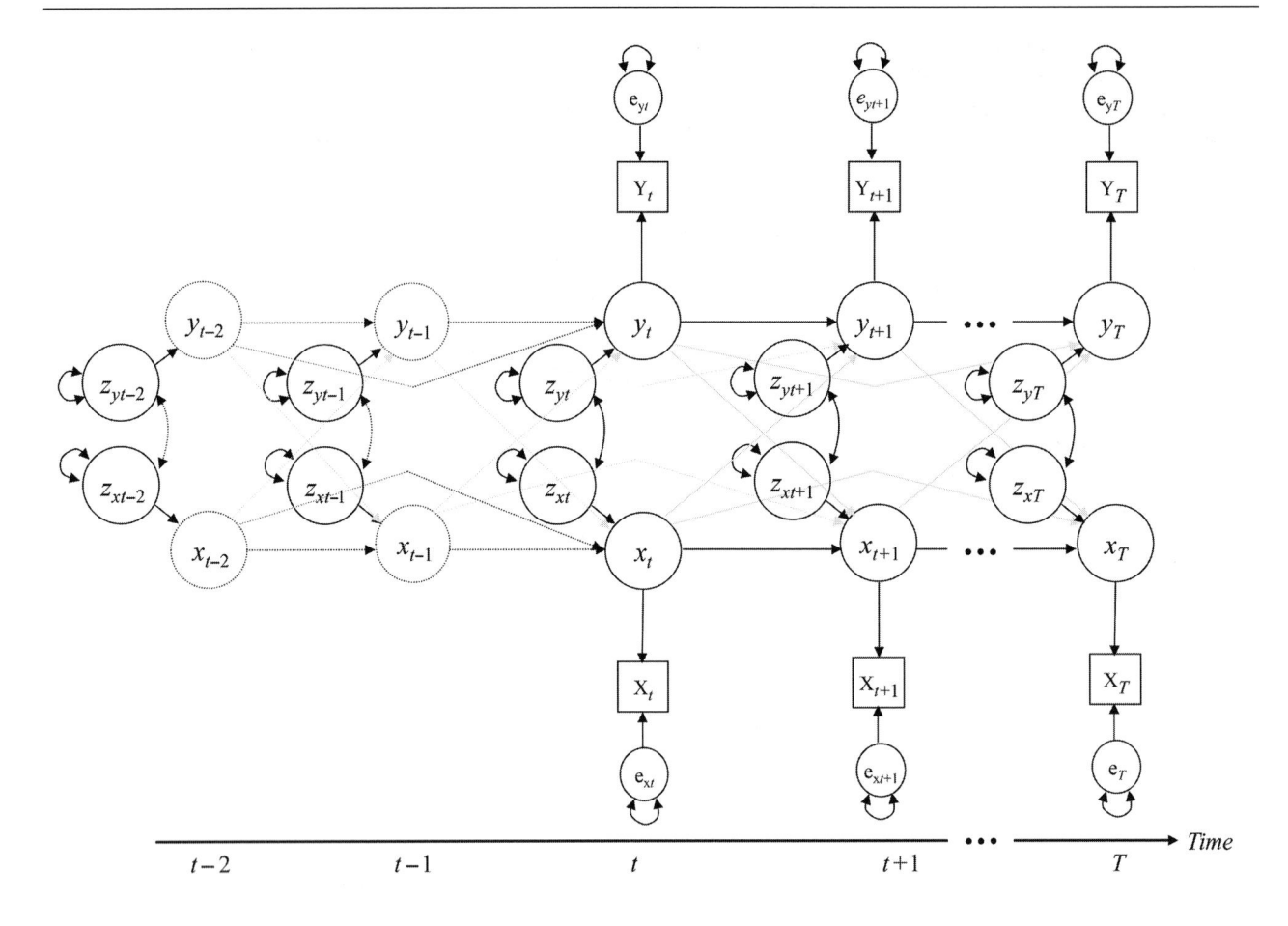

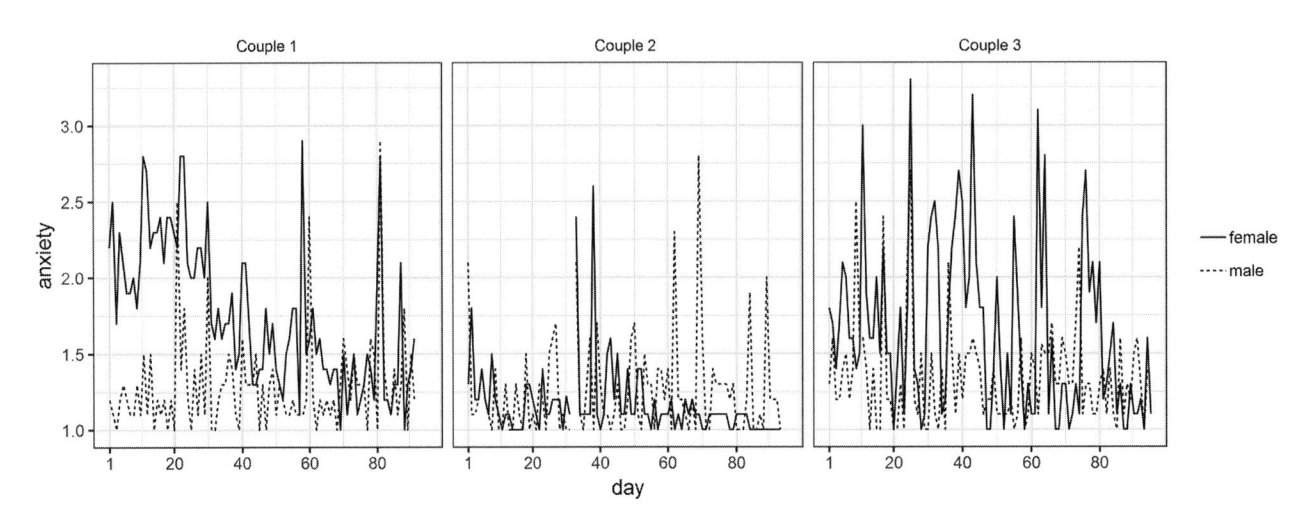

Figure 27.4 Bivariate (dyadic) autoregressive cross-lagged model. Path diagram (top panel) and trajectories of three different couples (bottom panel)

their own previous states or by those of the partner? For how long can such influences be detected? Or, put differently, how long until such influences dissipate? DFA has been primarily used to study fluctuations of emotions over time. For example, it has been applied to examine the change in emotional states of patients with Parkinson's disease (Chow et al., 2004; Shifren et al., 1997), the different patterns of emotional evolution after a romantic breakup (Sbarra & Ferrer, 2006), or the changes in anxiety in patients following psychotherapy (Fisher, Newman, & Molenaar, 2011).

In dyadic data, DFA has been predominantly used to study the interrelations between the emotional ups and downs of both dyad members. For example, Ferrer and Nesselroade (2003) used it to examine the structure and dynamics of affect in a husband and wife couple. One interesting finding from that study is that the husband's negative affect on a given day influenced the wife's negative affect one day later, while at the same time inhibiting her positive mood. Ferrer and Widaman (2008) applied DFA to a large sample of dyads to investigate similar questions of structure and dynamics. Some findings of particular relevance included: (a) large differences between results from a model fitted to the entire sample and a model fitted to each individual dyad, and (b) very large variability in both affect structure and dynamics across dyads, without support for a model that represents all couples – "the average couple." In another study, Castro-Schilo and Ferrer (2013) employed DFA to extract information about daily affective dynamics in couples and compared the predictive validity of such information against standard measures, such as mean and standard deviation, of the time series. The DFA parameters representing interrelations between the two individuals in the dyad over time were predictive of relationship quality – but not breakup – one and two years later after the daily data.

Figure 27.5 depicts a path diagram of a DFA (top panel) and the manifest and latent scores of one couple over 95 days (bottom panel). In this figure, the latent variables y_t and x_t represent negative affect over time for a female and a male in the couple. Suppose we want to address the question: Is the negative affect of each dyad member equally important for predicting the negative affect of the other member? In other words, are the time-lagged partner effects symmetrical? The latent factors representing negative affect are measured by three manifest variables (Y_1–Y_3 and X_1–X_3, for the female and male, respectively). Different specifications are possible for such a multivariate dyadic system. Here, we describe a *process* DFA model in which time dependencies are modeled through relations among the factors (Browne & Nesselroade, 2005; Browne & Zhang, 2007; Ferrer & Zhang, 2009).

In a standard specification of a process DFA, at any given occasion, there is a measurement structure representing how the different observed variables relate to the latent factor. Such a measurement structure is typically held invariant across occasions, so relations among the factors over time can be examined. For example, according to Figure 27.5, the female's and male's negative affect factors are being measured by three variables. Each of these variables relates to its latent factor and also has its unique variance. Thus, at each measurement occasion, the model describes the measurement structure of negative affect for each person. The associations among the latent factors are specified to examine time dependencies. In the path diagram of Figure 27.5, the latent negative affect factor for each person is influenced by their own affect, and the partner's affect at the previous occasion (lag-1 effects), as well as their own affect at the occasion before the previous (lag-2 effects). These auto- and cross-lagged influences are analogous to those described for the AR model, but relate latent factors that take into consideration their measurement structure. Two random shock variables, z_y and z_x, represent the variance in each factor that is not accounted for by the system.

In a standard specification, stationarity of the system is assumed, and these parameters are constrained to be invariant over time, indicating that the effects are the same throughout the entire time series. This constraint, however, can be relaxed to test hypotheses of temporal sequences (Ferrer & Nesselroade, 2003), examine possible time-varying effects in the parameters (Chow et al., 2011b), or investigate trends in the data (Molenaar & Nesselroade, 2001; Molenaar, Gooijer, & Schmitz, 1992). Similarly, the processes can be specified to be influenced not only by the adjacent occasion but also by other occasions before, by including additional regression coefficients from $t - 2, t - 3 \ldots$ up to $t - k$, leading to a process of kth *order*. If the members of the dyad are influenced by each other in a symmetrical manner, the two cross-lagged influences will have similar values. If the influence is asymmetrical, one of them will have a lower value.

Differential Equations Models for Dyads

Differential equation models (DEMs) are a class of models used to examine the dynamics of a process in continuous time. DEMs are particularly useful for examining dyadic interactions, as these models explicitly consider the two members of a dyad as an interdependent system.

DEMs have been primarily used as a theoretical framework to investigate the predictive behavior of a dyadic system under formal mathematical assumptions (Felmlee, 2006; Felmlee & Greenberg, 1999). In psychology, they have been used to examine a wide array of topics, including intimacy and disclosure in marriage (Boker & Laurenceau, 2006), attachment and affect coregulation (Butner, Diamond, & Hicks, 2007), dynamics in families (Chow, Mattson, & Messinger, 2014; Ram et al., 2014), daily emotional ups and downs in couples (Chow, Ferrer, & Nesselroade, 2007; Steele & Ferrer, 2011), self-organization and nonstationarity in dyadic interactions (Chow et al., 2017), temporal evolution of emotions in romantic partners (Butler, 2017; Reed, Barnard, & Butler, 2015), and the dynamics of physiological signals between partners in romantic couples (Ferrer & Helm, 2013; Helm, Sbarra, & Ferrer, 2012, 2014).

In general terms, DEMs directly model the changes in the system – and each of its units – as a continuous process. As such, they are a general framework to study dynamics. A desirable feature of these models is that such generality allows researchers to specify a particular model that best represents the potential mechanisms underlying such dynamics. For dyads, the model would involve two differential equations, one for each person in the dyad. One such model developed in the context of dyadic interactions was developed by Felmlee and colleagues

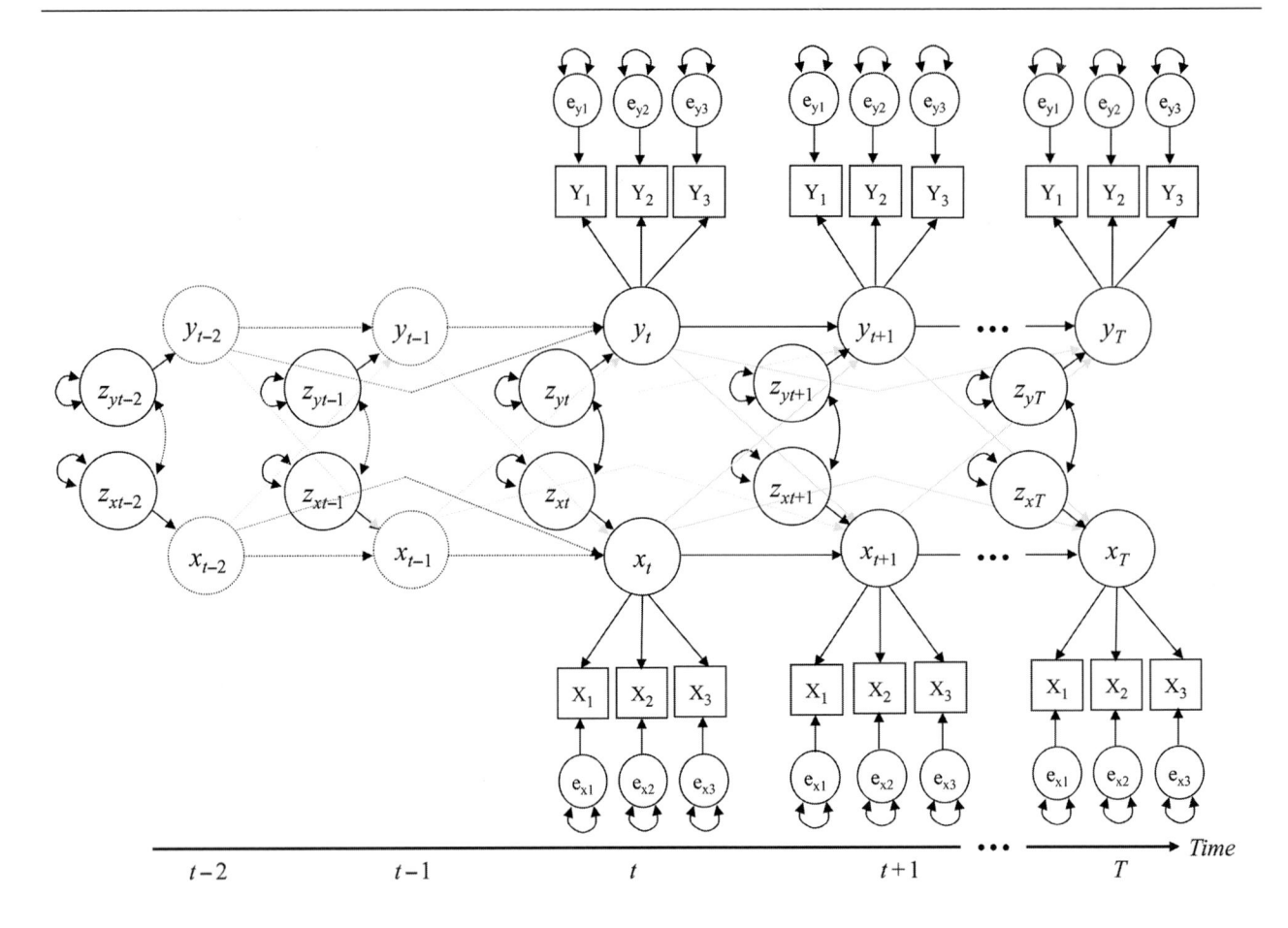

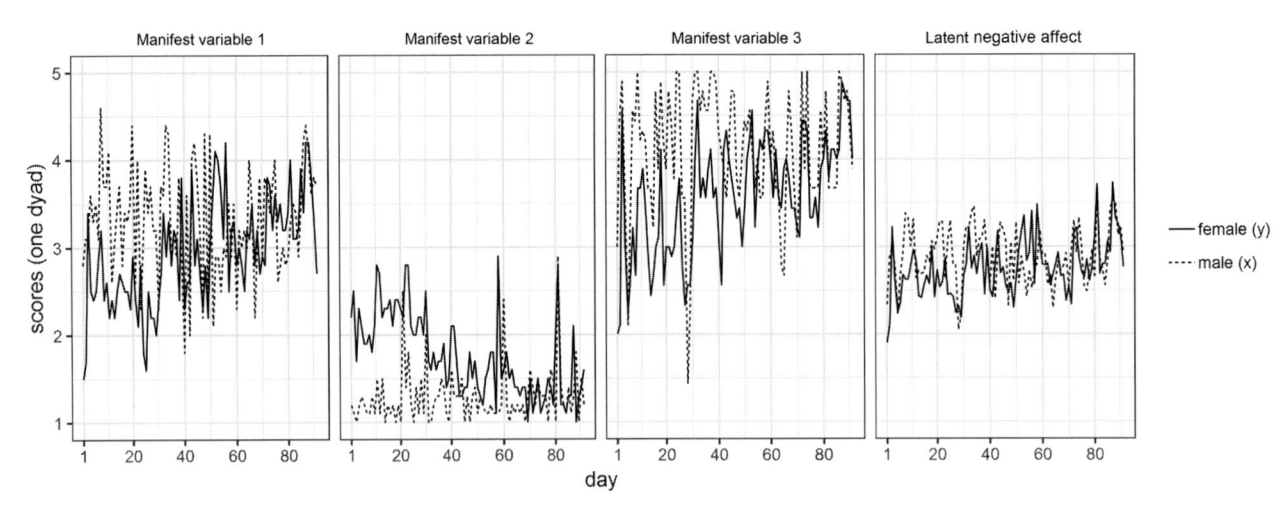

Figure 27.5 Bivariate (dyadic) dynamic factor analysis (DFA) model. Path diagram (top panel) and trajectories of one couple in the manifest and latent variables (bottom panel)

(Felmlee, 2006; Felmlee & Greenberg, 1999) to study individual goals and dyadic influences.

In the context of, say, daily emotions of two individuals in a married couple, this model would specify the rate of change in the daily emotions for each individual as a function of two parameters: one representing the extent to which each person changes as a function of the partner's current emotion. Therefore, the first component can be understood as representing self-regulation, whereas the second term would represent coregulation, or the degree of emotional interrelation between both individuals. Different values and restrictions for these four parameters lead to different behaviors of the system over time (Felmlee, 2006; Ferrer & Steele, 2012, 2013; Steele, Gonzales, & Ferrer, 2018). The top panel in Figure 27.6 depicts four combinations of parameters, leading to different

Both cooperative

$$dx/dt = a_1 \cdot (x^* - x) + a_2 \cdot (y - x)$$
$$dy/dt = b_1 \cdot (y^* - y) + b_2 \cdot (x - y)$$

Both independent

$$dx/dt = a_1(x^* - x_t)$$
$$dx/dt = a_2(y^* - y_t)$$

One uncooperative (x) and one dependent (y)

$$dx/dt = a_1 \cdot (x^* - x) - a_2 \cdot (y - x)$$
$$dy/dt = b_2 \cdot (x - y)$$

Both reactionaries

$$dx/dt = -a_2 \cdot (y - x)$$
$$dy/dt = -b_2 \cdot (x - y)$$

(Adapted from Felmlee, 2006; Felmlee & Greenberg, 1999)

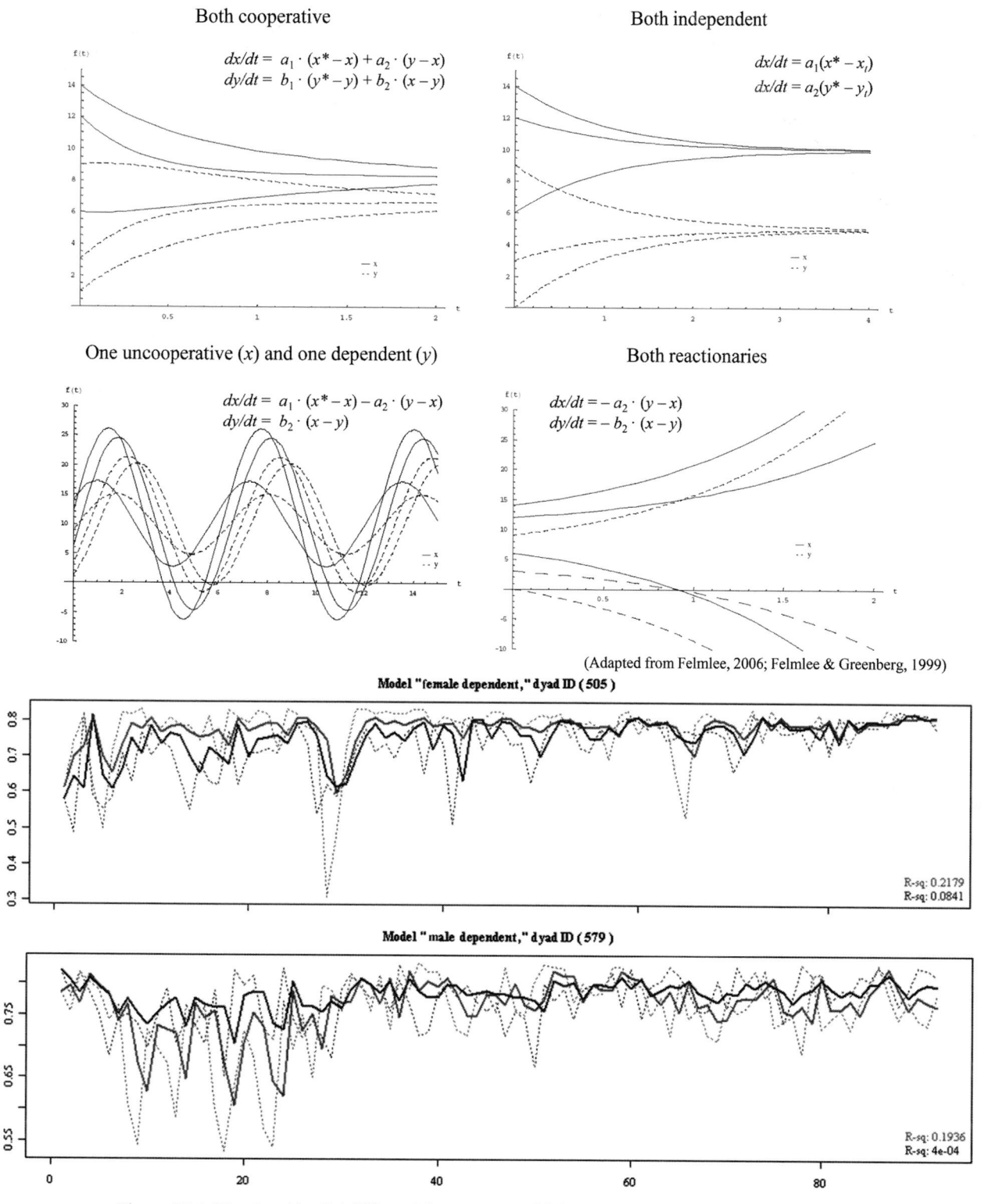

Figure 27.6 Bivariate (dyadic) differential equation model (DEM). Examples of different trajectories expressed by model parameters (top panel) and observed and predicted trajectories of two couples (bottom panel)

regulatory patterns. The bottom panel in Figure 27.6 depicts two examples of predicted and observed trajectories.

If the parameters representing self-regulation are positive for both individuals, they are expected to change toward their own emotional equilibrium. If they are negative, their affect tends to drift away from their corresponding equilibrium points – i.e., the within-partner influence does not regulate the system, because the farther they are from the equilibrium, the faster they will move away. If the parameters representing coregulation are positive for both individuals, the partners engage in a cooperative process in which they tend to converge over time. If they are negative, both individuals tend to drift away from the partner's emotional state.

EXPLORATORY AND COMPUTATIONAL METHODS FOR STUDYING DYADIC DATA

In addition to the models described thus far, several newer exploratory methods are now available for capturing relevant features of the dyad without relying on statistical assumptions such as normality, homoscedasticity, or stationarity (Chow, Ferrer, & Hsieh, 2011a). The application of some of these methods to dyadic interactions is still in its nascent form but they will become increasingly common in the near future. Some examples of this methodology include computational algorithms to identify periods of similarity in couples' daily emotional fluctuations (Ferrer et al., 2010, p. 20; Ferrer, Steele, & Hsieh, 2012; Hsieh et al., 2010), or network analysis for investigating the structure and dynamics of affect in couples (Bringmann et al., 2013, 2016; Ferrer et al., 2010). Ferrer (2016) provides a brief overview of these innovative approaches.

A NOTE ON COMPUTER PROGRAMS FOR DYADIC DATA ANALYSIS

All the models described in this chapter can be specified within different modeling frameworks. The APIM, LGC, LCS and AR models are often specified through structural equation modeling (SEM). Some of the most popular computer software for estimating SEMs are standalone programs such as Mplus (Muthén & Muthén, 1998), and the packages lavaan (Rosseel, 2012) and OpenMx (Boker et al., 2011; Neale et al., 2016) for the R programming language (R Core Team, 2017). Kline (2016) provides a comprehensive review of computer tools for SEM. Most of these models can also be specified in the multilevel framework (hierarchical linear models, mixed-effects models), using standard multilevel modeling software such as SAS, HLM, or, within the R environment, the packages lme4 (Bates et al., 2015) and MCMCglmm (Hadfield, 2010).

Regarding the dynamic models described here (i.e., AR, DFA, DEM), multiple software options are available for fitting them. Some recently developed packages are flexible enough for specifying and fitting any of these dynamic systems models: OpenMx (Boker et al., 2011; Neale et al., 2016), dynr (Ou, Hunter, & Chow, 2017) and ctsem (Driver, Oud, & Voelkle, 2017) are powerful tools for fitting any of these dynamic models to dyadic data. Other programs are also available that were developed specifically for time series (DyFA; Browne & Zhang, 2003; Zhang & Browne, 2008).

CONCLUSION

Research on dyads in clinical research – as in any other social and behavioral science –fundamentally involves studying systems over time. This premise relies on the interdependence between the two people in a dyad, which leads to many interesting questions about processes and dynamics. Understanding the emotional underpinnings of romantic couples, for example, requires knowledge about how each person's emotions, thoughts, and behaviors influence those of the other person. Knowing how the working alliance between a therapist and a client develops over time depends on insights about the interactions between both individuals, as they exchange information and emotion. Finally, fully understanding child psychopathology requires knowledge about their parents' psychological wellbeing as well. To answer these questions, analytic techniques are needed that can capture key aspects of the interdependence between the members of a dyad.

Many authors emphasize that research in clinical psychology must focus on the process of change and the factors that moderate and mediate this change (Granic & Patterson, 2006; Hayes & Strauss, 1998; Hollon et al., 2002; Laurenceau et al., 2007). This implies transitioning from asking questions about the outcomes (i.e., Did the therapy work?) to questions about the processes and mechanisms (i.e., How did it work?). Accordingly, clinical researchers need models that can describe the temporal dynamics of these processes. Recent advances in quantitative methodology have yielded multiple techniques that allow researchers to examine such questions about processes and dynamics. In this chapter we reviewed a number of models developed to capture different aspects of dyadic interactions. We emphasized models that focus on change and dynamics because they are the best suited to answer questions related to processes in dyads. We hope the information provided here is useful as psychological scientists deepen their focus on the study of dyadic processes.

REFERENCES

Anker, M. G., Owen, J., Duncan, B. L., & Sparks, J. A. (2010). The Alliance in Couple Therapy: Partner Influence, Early Change, and Alliance Patterns in a Naturalistic Sample. *Journal of Consulting and Clinical Psychology, 78*(5), 635–645.

Bates, D., Mächler, M., Bolker, B., & Walker, S. (2015). Fitting Linear Mixed-Effects Models Using lme4. *Journal of Statistical Software, 67*(1). Retrieved from www.jstatsoft.org/article/view/v067i01

Baucom, B. R., Dickenson, J. A., Atkins, D. C., Baucom, D. H., Fischer, M. S., Weusthoff, S., ... Zimmermann, T. (2015a). The Interpersonal Process Model of Demand/Withdraw Behavior. *Journal of Family Psychology, 29*(1), 80–90.

Baucom, B. R., Sheng, E., Christensen, A., Georgiou, P. G., Narayanan, S. S., & Atkins, D. C. (2015b). Behaviorally-Based Couple Therapies Reduce Emotional Arousal during Couple Conflict. *Behaviour Research and Therapy, 72*, 49–55.

Berghuis, J. P., & Stanton, A. L. (2002). Adjustment to a Dyadic Stressor: A Longitudinal Study of Coping and Depressive Symptoms in Infertile Couples over an Insemination Attempt. *Journal of Consulting and Clinical Psychology, 70*(2), 433–438.

Bisgaard, S., & Kulahci, M. (2011). *Time Series Analysis and Forecasting by Example*. Hoboken, NJ: Wiley.

Bodenmann, G., Hilpert, P., Nussbeck, F. W., & Bradbury, T. N. (2014). Enhancement of Couples' Communication and Dyadic Coping by a Self-Directed Approach: A Randomized Controlled Trial. *Journal of Consulting and Clinical Psychology, 82*(4), 580–591.

Boker, S. M., & Laurenceau, J.-P. (2006). Dynamical Systems Modeling: An Application to the Regulation of Intimacy and Disclosure in Marriage. In T. A. Walls & J. L. Schafer (Eds.), *Models for Intensive Longitudinal Data* (pp. 195–218). New York: Oxford University Press.

Boker, S. M., Neale, M., Maes, H., Wilde, M., Spiegel, M., Brick, T., ... Bates, T. (2011). OpenMx: An Open Source Extended Structural Equation Modeling Framework. *Psychometrika, 76*(2), 306–317.

Bolger, N., & Laurenceau, J.-P. (2013). Design and Analysis of Intensive Longitudinal Studies of Distinguishable Dyads. In *Intensive Longitudinal Methods: An Introduction to Diary and Experience Sampling Research* (pp. 143–175). New York: Guilford Press.

Bollen, K. A., & Curran, P. J. (2006). *Latent Curve Models: A Structural Equation Perspective*. Hoboken, N.J: Wiley-Interscience.

Bringmann, L. F., Hamaker, E. L., Vigo, D. E., Aubert, A., Borsboom, D., & Tuerlinckx, F. (2017). Changing Dynamics: Time-Varying Autoregressive Models Using Generalized Additive Modeling. *Psychological Methods, 22*(3), 409–425.

Bringmann, L. F., Pe, M. L., Vissers, N., Ceulemans, E., Borsboom, D., Vanpaemel, W., ... Kuppens, P. (2016). Assessing Temporal Emotion Dynamics Using Networks. *Assessment, 23*(4), 425–435.

Bringmann, L. F., Vissers, N., Wichers, M., Geschwind, N., Kuppens, P., Peeters, F., ... Tuerlinckx, F. (2013). A Network Approach to Psychopathology: New Insights into Clinical Longitudinal Data. *PLOS ONE, 8*(4), e60188.

Browne, M. W., & Nesselroade, J. R. (2005). Representing Psychological Processes with Dynamic Factor Models: Some Promising Uses and Extensions of Arma Time Series Models. In A. Maydeu-Olivares & J. J. McArdle (Eds.), *Advances in Psychometrics: A Festschrift for Roderick P. McDonald* (pp. 415–452). Mahwah, NJ: Lawrence Erlbaum Associates.

Browne, M. W., & Zhang, G. (2003). DyFA 2.03 User Guide. Retrieved from https://psychology.osu.edu/people/browne.4

Browne, M. W., & Zhang, G. (2007). Developments in the Factor Analysis of Individual Time Series. In R. Cudeck & R. C. MacCallum (Eds.), *Factor Analysis at 100: Time Series in Psychology. Historical Developments and Future Directions* (pp. 265–291). Mahwah, NJ: Lawrence Erlbaum Associates.

Butler, E. A. (2017). Emotions Are Temporal Interpersonal Systems. *Current Opinion in Psychology*, 17(Supplement C), 129–134.

Butner, J., Diamond, L. M., & Hicks, A. M. (2007). Attachment Style and Two Forms of Affect Coregulation between Romantic Partners. *Personal Relationships, 14*(3), 431–455.

Butterfield, R. M., & Lewis, M. A. (2002). Health-Related Social Influence: A Social Ecological Perspective on Tactic Use. *Journal of Social and Personal Relationships, 19*(4), 505–526.

Cabrieto, J., Tuerlinckx, F., Kuppens, P., Grassmann, M., & Ceulemans, E. (2017). Detecting Correlation Changes in Multivariate Time Series: A Comparison of Four Non-Parametric Change Point Detection Methods. *Behavior Research Methods, 49*(3), 988–1005.

Campbell, L., Simpson, J. A., Kashy, D. A., & Rholes, W. S. (2001). Attachment Orientations, Dependence, and Behavior in a Stressful Situation: An Application of the Actor-Partner Interdependence Model. *Journal of Social and Personal Relationships, 18*(6), 821–843.

Castro-Schilo, L., & Ferrer, E. (2013). Comparison of Nomothetic Versus Idiographic-Oriented Methods for Making Predictions about Distal Outcomes from Time Series Data. *Multivariate Behavioral Research, 48*(2), 175–207.

Chow, S.-M., Ferrer, E., & Hsieh, F. (2011a). *Statistical Methods for Modeling Human Dynamics: An Interdisciplinary Dialogue*. New York: Taylor & Francis.

Chow, S.-M., Ferrer, E., & Nesselroade, J. R. (2007). An Unscented Kalman Filter Approach to the Estimation of Nonlinear Dynamical Systems Models. *Multivariate Behavioral Research, 42*(2), 283–321.

Chow, S.-M., Mattson, W. I., & Messinger, D. S. (2014). *Representing Trends and Moment-to-Moment Variability in Dyadic and Family Processes Using State-Space Modeling Techniques*. In *Emerging Methods in Family Research* (pp. 39–55). Cham: Springer.

Chow, S.-M., Nesselroade, J. R., Shifren, K., & McArdle, J. J. (2004). Dynamic Structure of Emotions among Individuals with Parkinson's Disease. *Structural Equation Modeling: A Multidisciplinary Journal, 11*(4), 560–582.

Chow, S.-M., Ou, O., Cohn, J. F., & Messinger, D. S. (2017). Representing Self-Organization and Non-Stationarities in Dyadic Interaction Processes Using Dynamic Systems Modeling Techniques. In A. Von Davier, P. C. Kyllonen, & M. Zhu (Eds.), *Innovative Assessment of Collaboration*. New York: Springer.

Chow, S.-M., Zu, J., Shifren, K., & Zhang, G. (2011b). Dynamic Factor Analysis Models with Time-Varying Parameters. *Multivariate Behavioral Research, 46*(2), 303–339.

Cook, W. L., & Snyder, D. K. (2005). Analyzing Nonindependent Outcomes in Couple Therapy Using the Actor-Partner Interdependence Model. *Journal of Family Psychology, 19*(1), 133–141.

Crowell, S. E., Baucom, B. R., Yaptangco, M., Bride, D., Hsiao, R., McCauley, E., & Beauchaine, T. P. (2014). Emotion Dysregulation and Dyadic Conflict in Depressed and Typical Adolescents: Evaluating Concordance across Psychophysiological and Observational Measures. *Biological Psychology, 98*(Supplement C), 50–58.

Curran, P. J., & Bollen, K. A. (2001). The Best of Both Worlds: Combining Autoregressive and Latent Curve Models. In L. M. Collins & A. G. Sayer (Eds.), *New Methods for the Analysis of Change* (pp. 107–135). Washington, DC: American Psychological Association.

Driscoll, K. A., Schatschneider, C., McGinnity, K., & Modi, A. C. (2012). Application of Dyadic Data Analysis in Pediatric Psychology: Cystic Fibrosis Health-Related Quality of Life and Anxiety in Child-Caregiver Dyads. *Journal of Pediatric Psychology, 37*(6), 605–611.

Driver, C. C., Oud, J. H. L., & Voelkle, M. C. (2017). Continuous Time Structural Equation Modeling with R Package ctsem. *Journal of Statistical Software, 77*(5). Retrieved from www.jstatsoft.org/article/view/v077i05

Engle, R., & Watson, M. (1981). A One-Factor Multivariate Time Series Model of Metropolitan Wage Rates. *Journal of the American Statistical Association, 76*(376), 774–781.

Felmlee, D. H. (2006). Application of Dynamic Systems Analysis to Dyadic Interactions. In D. Ong & M. van Dulmen (Eds.), *Oxford Handbook of Methods in Positive Psychology* (pp. 409–422). New York: Oxford University Press.

Felmlee, D. H., & Greenberg, D. F. (1999). A Dynamic Systems Model of Dyadic Interaction. *Journal of Mathematical Sociology, 23*(3), 155–180.

Ferrer, E. (2016). Exploratory Approaches for Studying Social Interactions, Dynamics, and Multivariate Processes in Psychological Science. *Multivariate Behavioral Research, 51*(2–3), 240–256.

Ferrer, E., & Helm, J. L. (2013). Dynamical Systems Modeling of Physiological Coregulation in Dyadic Interactions. *International Journal of Psychophysiology, 88*(3), 296–308.

Ferrer, E., & McArdle, J. J. (2003). Alternative Structural Models for Multivariate Longitudinal Data Analysis. *Structural Equation Modeling: A Multidisciplinary Journal, 10*(4), 493–524.

Ferrer, E., & McArdle, J. J. (2004). An Experimental Analysis of Dynamic Hypotheses about Cognitive Abilities and Achievement from Childhood to Early Adulthood. *Developmental Psychology, 40*(6), 935–952.

Ferrer, E., & McArdle, J. J. (2010). Longitudinal Modeling of Developmental Changes in Psychological Research. *Current Directions in Psychological Science, 19*(3), 149–154.

Ferrer, E., & Nesselroade, J. R. (2003). Modeling Affective Processes in Dyadic Relations via Dynamic Factor Analysis. *Emotion, 3*(4), 344–360.

Ferrer, E., & Rast, P. (2017). Partitioning the Variability of Daily Emotion Dynamics in Dyadic Interactions with a Mixed-Effects Location Scale Model. *Current Opinion in Behavioral Sciences, 15*(Supplement C), 10–15.

Ferrer, E., & Steele, J. (2013). Differential Equations for Evaluating Theoretical Models of Dyadic Interactions. In P. C. M. Molenaar, R. M. Lerner, & K. M. Newell (Eds.), *Handbook of Developmental Systems Theory and Methodology* (pp. 345–368). New York: Guilford.

Ferrer, E., & Steele, J. (2014). *Differential Equations for Evaluating Theoretical Models of Dyadic Interactions: Handbook of Developmental Systems Theory and Methodology*. New York: Guilford.

Ferrer, E., & Steele, J. S. (2012). Dynamic Systems Analysis of Affective Processes in Dyadic Interactions Using Differential Equations. In G. Hancock & J. Harrings (Eds.), *Advances in Longitudinal Methods in the Social and Behavioral Sciences* (pp. 111–134). Charlotte, NC: Information Age Publishing.

Ferrer, E., & Widaman, K. F. (2008). Dynamic Factor Analysis of Dyadic Affective Processes with Inter-Group Differences. In N. A. Card, J. P. Selig, & T. D. Little (Eds.), *Modeling Dyadic and Interdependent Data in the Developmental and Behavioral Sciences* (pp. 107–137). New York: Routledge.

Ferrer, E., & Zhang, G. (2009). Time Series Models for Examining Psychological Processes: Applications and New Developments. In R. E. Millsap & A. Maydeu-Olivares (Eds.), *Modeling Dyadic and Interdependent Data in the Developmental and Behavioral Sciences* (pp. 107–137). London: Sage Publications.

Ferrer, E., Balluerka, N., & Widaman, K. F. (2008). Factorial Invariance and the Specification of Second-Order Latent Growth Models. *Methodology: European Journal of Research Methods for the Behavioral and Social Sciences, 4*(1), 22–36.

Ferrer, E., Chen, S., Chow, S.-M., & Hsieh, F. (2010). Exploring Intra-Individual, Inter-Individual and Inter-Variable Dynamics in Dyadic Interactions. In S.-M. Chow, E. Ferrer, & F. Hsieh (Eds.), *Statistical Methods for Modeling Human Dynamics: An Interdisciplinary Dialogue* (pp. 381–411). New York: Taylor and Francis.

Ferrer, E., McArdle, J. J., Shaywitz, B. A., Holahan, J. M., Marchione, K., & Shaywitz, S. E. (2007). Longitudinal Models of Developmental Dynamics between Reading and Cognition from Childhood to Adolescence. *Developmental Psychology, 43*(6), 1460–1473.

Ferrer, E., Steele, J. S., & Hsieh, F. (2012). Analyzing the Dynamics of Affective Dyadic Interactions Using Patterns of Intra- and Interindividual Variability. *Multivariate Behavioral Research, 47*(1), 136–171.

Fisher, A. J., Newman, M. G., & Molenaar, P. C. M. (2011). A Quantitative Method for the Analysis of Nomothetic Relationships between Idiographic Structures: Dynamic Patterns Create Attractor States for Sustained Posttreatment Change. *Journal of Consulting and Clinical Psychology, 79*(4), 552–563.

Franks, M. M., Wendorf, C. A., Gonzalez, R., & Ketterer, M. (2004). Aid and Influence: Health-Promoting Exchanges of Older Married Partners. *Journal of Social and Personal Relationships, 21*(4), 431–445.

Garcia-Lopez, L. J., Díaz-Castela, M. del M., Muela-Martinez, J. A., & Espinosa-Fernandez, L. (2014). Can Parent Training for Parents with High Levels of Expressed Emotion Have a Positive Effect on Their Child's Social Anxiety Improvement? *Journal of Anxiety Disorders, 28*(8), 812–822.

Girard, J., Wright, A., Beeney, J., Lazarus, S., Scott, L., Stepp, S., & Pilkonis, P. (2017). Interpersonal Problems across Levels of the Psychopathology Hierarchy. *Comprehensive Psychiatry, 79*, 53–69.

Goldfried, M. R., Greenberg, L. S., & Marmar, C. (1990). Individual Psychotherapy: Process and Outcome. *Annual Review of Psychology, 41*(1), 659–688.

Gonzalez, R., & Griffin, D. (1997). On the Statistics of Interdependence: Treating Dyadic Data with Respect. In S. Duck (Ed.), *Handbook of Personal Relationships: Theory, Research and Interventions* (pp. 271–302). Hoboken, NJ: John Wiley.

Gottman, J. M., & Notarius, C. I. (2000). Decade Review: Observing Marital Interaction. *Journal of Marriage and Family, 62*(4), 927–947.

Gottman, J. M., Murray, J. D., Swanson, C. C., Tyson, R., & Swanson, K. R. (2002). *The Mathematics of Marriage: Dynamic Nonlinear Models*. Cambridge, MA: MIT Press.

Gottman, J. M., Swanson, C., & Murray, J. (1999). The Mathematics of Marital Conflict: Dynamic Mathematical Nonlinear Modeling of Newlywed Marital Interaction. *Journal of Family Psychology, 13*(1), 3–19.

Granic, I., & Patterson, G. R. (2006). Toward a Comprehensive Model of Antisocial Development: A Dynamic Systems Approach. *Psychological Review, 113*(1), 101–131.

Hadfield, J. D. (2010). MCMC Methods for Multi-Response Generalized Linear Mixed Models: The MCMCglmm R Package. *Journal of Statistical Software, 33*(2). Retrieved from www.jstatsoft.org/article/view/v033i02

Hagedoorn, M., Kuijer, R. G., Buunk, B. P., DeJong, G. M., Wobbes, T., & Sanderman, R. (2000). Marital Satisfaction in Patients with Cancer: Does Support from Intimate Partners Benefit Those Who Need It Most? *Health Psychology, 19*(3), 274–282.

Hamaker, E. L., Zhang, Z., & van der Maas, H. L. J. (2009). Using Threshold Autoregressive Models to Study Dyadic Interactions. *Psychometrika, 74*(4), 727–745.

Hawrilenko, M., Gray, T. D., & Córdova, J. V. (2016). The Heart of Change: Acceptance and Intimacy Mediate Treatment Response in a Brief Couples Intervention. *Journal of Family Psychology, 30*(1), 93–103.

Hayes, A. M., & Strauss, J. L. (1998). Dynamic Systems Theory as a Paradigm for the Study of Change in Psychotherapy: An Application to Cognitive Therapy for Depression. *Journal of Consulting and Clinical Psychology, 66*(6), 939–947.

Hedeker, D., Mermelstein, R. J., & Demirtas, H. (2008). An Application of a Mixed-Effects Location Scale Model for Analysis of Ecological Momentary Assessment (EMA) Data. *Biometrics, 64*(2), 627–634.

Helm, J. L., Sbarra, D., & Ferrer, E. (2012). Assessing Cross-Partner Associations in Physiological Responses via Coupled Oscillator Models. *Emotion, 12*(4), 748–762.

Helm, J. L., Sbarra, D. A., & Ferrer, E. (2014). Coregulation of Respiratory Sinus Arrhythmia in Adult Romantic Partners. *Emotion, 14*(3), 522–531.

Hollon, S. D., Muñoz, R. F., Barlow, D. H., Beardslee, W. R., Bell, C. C., Bernal, G., … Sommers, D. (2002). Psychosocial Intervention Development for the Prevention and Treatment of Depression: Promoting Innovation and Increasing Access. *Biological Psychiatry, 52*(6), 610–630.

Hsieh, F., Ferrer, E., Chen, S.-C., & Chow, S.-M. (2010). Exploring the Dynamics of Dyadic Interactions via Hierarchical Segmentation. *Psychometrika, 75*(2), 351–372.

Kenny, D. A. (1996). Models of Non-Independence in Dyadic Research. *Journal of Social and Personal Relationships, 13*(2), 279–294.

Kenny, D. A., & Judd, C. M. (1986). Consequences of Violating the Independence Assumption in Analysis of Variance. *Psychological Bulletin, 99*(3), 422–431.

Kenny, D. A., Kashy, D. A., & Bolger, N. (1998). Data Analysis in Social Psychology. In D. Gilbert, S. Fiske, & G. Lindzey (Eds.), *Handbook of Social Psychology* (4th edn., Vol. 1, pp. 233–265). Boston: McGraw-Hill.

Kenny, D. A., Kashy, D. A., & Cook, W. L. (2006). *Dyadic Data Analysis*. New York: Guilford Press.

Kivlighan, D. M., Clements, L., Blake, C., Arnzen, A., & Brady, L. (1993). Counselor Sex Role Orientation, Flexibility, and Working Alliance Formation. *Journal of Counseling & Development, 72*(1), 95–100.

Kivlighan, D. M., Marmarosh, C. L., & Hilsenroth, M. J. (2014). Client and Therapist Therapeutic Alliance, Session Evaluation, and Client Reliable Change: A Moderated Actor-Partner Interdependence Model. *Journal of Counseling Psychology, 61*(1), 15–23.

Kline, R. B. (2016). Computer Tools. In *Principles and Practice of Structural Equation Modeling* (4th edn., pp. 97–113). New York: Guilford Press.

Kopta, S. M., Lueger, R. J., Saunders, S. M., & Howard, K. I. (1999). Individual Psychotherapy Outcome and Process Research: Challenges Leading to Greater Turmoil or a Positive Transition? *Annual Review of Psychology, 50*(1), 441–469.

Kouros, C. D., & Cummings, E. M. (2010). Longitudinal Associations Between Husbands' and Wives' Depressive Symptoms. *Journal of Marriage and Family, 72*(1), 135–147.

Laurenceau, J.-P., & Bolger, N. (2005). Using Diary Methods to Study Marital and Family Processes. *Journal of Family Psychology, 19*(1), 86–97.

Laurenceau, J.-P., Hayes, A. M., & Feldman, G. C. (2007). Some Methodological and Statistical Issues in the Study of Change Processes in Psychotherapy. *Clinical Psychology Review, 27*(6), 682–695.

Lawrence, E., Yoon, J., Langer, A., & Ro, E. (2009). Is Psychological Aggression as Detrimental as Physical Aggression? The Independent Effects of Psychological Aggression on Depression and Anxiety Symptoms. *Violence and Victims, 24*(1), 20–35.

Ledermann, T., & Kenny, D. A. (2012). The Common Fate Model for Dyadic Data: Variations of a Theoretically Important but Underutilized Model. *Journal of Family Psychology, 26*(1), 140–148.

Ledermann, T., & Macho, S. (2014). Analyzing Change at the Dyadic Level: The Common Fate Growth Model. *Journal of Family Psychology, 28*(2), 204–213.

Ledermann, T., Macho, S., & Kenny, D. A. (2011). Assessing Mediation in Dyadic Data Using the Actor-Partner Interdependence Model. *Structural Equation Modeling: A Multidisciplinary Journal, 18*(4), 595–612.

Levenson, R. W., & Gottman, J. M. (1983). Marital Interaction: Physiological Linkage and Affective Exchange. *Journal of Personality and Social Psychology, 45*(3), 587–597.

Madhyastha, T. M., Hamaker, E. L., & Gottman, J. M. (2011). Investigating Spousal Influence Using Moment-to-Moment Affect Data from Marital Conflict. *Journal of Family Psychology, 25*(2), 292–300.

Marmarosh, C. L., Kivlighan, D. M., Bieri, K., LaFauci Schutt, J. M., Barone, C., & Choi, J. (2014). The Insecure Psychotherapy Base: Using Client and Therapist Attachment Styles to Understand the Early Alliance. *Psychotherapy, 51*(3), 404–412.

McArdle, J. J. (1982). *Structural Equation Modeling of an Individual System: Preliminary Results from "A Case Study in Episodic Alcoholism."* Unpublished Manuscript, Department of Psychology, University of Denver.

McArdle, J. J. (1988). Dynamic but Structural Equation Modeling of Repeated Measures Data. In J. R. Nesselroade & R. B. Cattell (Eds.), *Handbook of Multivariate Experimental Psychology* (pp. 561–614). Boston, MA: Springer.

McArdle, J. J. (2001). A Latent Difference Score Approach to Longitudinal Dynamic Structural Analysis. In R. Cudeck, S. du Toit, & D. Sörbom (Eds.), *Structural Equation Modeling, Present and Future: A Festschrift in Honor of Karl Jöreskog* (pp. 7–46). Lincolnwood, IL: Scientific Software International.

McArdle, J. J. (2009). Latent Variable Modeling of Differences and Changes with Longitudinal Data. *Annual Review of Psychology, 60*(1), 577–605.

McArdle, J. J., & Epstein, D. (1987). Latent Growth Curves within Developmental Structural Equation Models. *Child Development, 58*(1), 110–133.

McArdle, J. J., & Hamagami, F. (2001). Latent Difference Score Structural Models for Linear Dynamic Analyses with Incomplete Longitudinal Data. In L. M. Collins & A. G. Sayer (Eds.), *New Methods for the Analysis of Change* (pp. 139–175). Washington, DC: American Psychological Association.

McArdle, J. J., Hamagami, F., Meredith, W., & Bradway, K. P. (2000). Modeling the Dynamic Hypotheses of Gf–Gc Theory Using Longitudinal Life-Span Data. *Learning and Individual Differences, 12*(1), 53–79.

Medina-Pradas, C., Navarro, J. B., López, S. R., Grau, A., & Obiols, J. E. (2011). Dyadic View of Expressed Emotion, Stress, and Eating Disorder Psychopathology. *Appetite, 57*(3), 743–748.

Meredith, W. (1993). Measurement Invariance, Factor Analysis And Factorial Invariance. *Psychometrika, 58*(4), 525–543.

Meredith, W., & Tisak, J. (1990). Latent Curve Analysis. *Psychometrika, 55*(1), 107–122.

Molenaar, P. C. M. (1985). A Dynamic Factor Model for the Analysis of Multivariate Time Series. *Psychometrika, 50*(2), 181–202.

Molenaar, P. C. M., & Nesselroade, J. R. (2001). Rotation in the Dynamic Factor Modeling of Multivariate Stationary Time Series. *Psychometrika, 66*(1), 99–107.

Molenaar, P. C. M., Gooijer, J. G. D., & Schmitz, B. (1992). Dynamic Factor Analysis of Nonstationary Multivariate Time Series. *Psychometrika, 57*(3), 333–349.

Muthén, L. K., & Muthén, B. O. (1998). *Mplus User's Guide* (6th edn.). Los Angeles, CA: Muthén & Muthén.

Neale, M. C., Hunter, M. D., Pritikin, J. N., Zahery, M., Brick, T. R., Kirkpatrick, R. M., ... Boker, S. M. (2016). OpenMx 2.0: Extended Structural Equation and Statistical Modeling. *Psychometrika, 81*(2), 535–549.

Nesselroade, J. R., & Boker, S. M. (1994). Assessing Constancy and Change. In T. Heatherton & J. Weinberg (Eds.), *Can Personality Change?* (pp. 121–147). Washington, DC: American Psychological Association.

Nestler, S., Grimm, K. J., & Schönbrodt, F. D. (2015). The Social Consequences and Mechanisms of Personality: How to Analyse Longitudinal Data from Individual, Dyadic, Round-Robin and Network Designs. *European Journal of Personality, 29*(2), 272–295.

Ou, L., Hunter, M. D., & Chow, S.-M. (2017). *What's for dynr. A Package for Linear and Nonlinear DYNamic Modeling in R*. Retrieved from https://quantdev.ssri.psu.edu/sites/qdev/files/OuHunterChow_Dynr.pdf

Perry, N. S., Baucom, K. J. W., Bourne, S., Butner, J., Crenshaw, A. O., Hogan, J. N., ... Baucom, B. R. W. (2017). Graphic Methods for Interpreting Longitudinal Dyadic Patterns from Repeated-Measures Actor-Partner Interdependence models. *Journal of Family Psychology, 31*(5), 592–603.

Proulx, C. M., & Snyder-Rivas, L. A. (2013). The Longitudinal Associations between Marital Happiness, Problems, and Self-Rated Health. *Journal of Family Psychology, 27*(2), 194–202.

Przeworski, A., Zoellner, L. A., Franklin, M. E., Garcia, A., Freeman, J., March, J. S., & Foa, E. B. (2012). Maternal and Child Expressed Emotion as Predictors of Treatment Response in Pediatric Obsessive-Compulsive Disorder. *Child Psychiatry & Human Development, 43*(3), 337–353.

R Core Team. (2017). *R: A Language and Environment for Statistical Computing*. Vienna: R Foundation for Statistical Computing.

Ram, N., Shiyko, M., Lunkenheimer, E. S., Doerksen, S., & Conroy, D. (2014). *Families as Coordinated Symbiotic Systems: Making Use of Nonlinear Dynamic Models*. In *Emerging Methods in Family Research* (pp. 19–37). Cham: Springer.

Rao, C. R. (1958). Some Statistical Methods for Comparison of Growth Curves. *Biometrics, 14*(1), 1–17.

Rast, P., Hofer, S. M., & Sparks, C. (2012). Modeling Individual Differences in Within-Person Variation of Negative and Positive Affect in a Mixed Effects Location Scale Model Using BUGS/JAGS. *Multivariate Behavioral Research, 47*(2), 177–200.

Reed, R. G., Barnard, K., & Butler, E. A. (2015). Distinguishing Emotional Coregulation from Codysregulation: An Investigation of Emotional Dynamics and Body Weight in Romantic Couples. *Emotion, 15*(1), 45–60.

Rosseel, Y. (2012). lavaan: An R Package for Structural Equation Modeling. *Journal of Statistical Software, 48*(2). Retrieved from www.jstatsoft.org/article/view/v048i02

Sbarra, D. A., & Ferrer, E. (2006). The Structure and Process of Emotional Experience following Nonmarital Relationship Dissolution: Dynamic Factor Analyses of Love, Anger, and Sadness. *Emotion, 6*(2), 224–238.

Sbarra, D. A., & Wishman, M. A. (2013). Marital and Relational Discord. In L. Castonguay & T. C. Oltmans (Eds.), *Psychopathology: Bridging the Gap between Basic Empirical Findings and Clinical Practice* (pp. 393–418). New York: Guilford Press.

Shifren, K., Hooker, K., Wood, P., & Nesselroade, J. R. (1997). Structure and Variation of Mood in Individuals with Parkinson's Disease: A Dynamic Factor Analysis. *Psychology and Aging, 12*(2), 328–339.

Shumway, R. H., & Stoffer, D. S. (2011). *Time Series Analysis and Its Applications: With R Examples* (3rd edn.). New York: Springer.

Steele, J. S., & Ferrer, E. (2011). Latent Differential Equation Modeling of Self-Regulatory and Coregulatory Affective Processes. *Multivariate Behavioral Research, 46*(6), 956–984.

Steele, J. S., Gonzales, J. E., & Ferrer, E. (2018). Uses and Limitations of Continuous Time Models to Examine Dyadic Interactions. In K. van Montfort, J. Oud, & M. C. Voelkle (Eds.), *Continuous Time Modeling in the Behavioral and Related Sciences* (pp. 135–162). Cham: Springer International.

Thompson, A., & Bolger, N. (1999). Emotional Transmission in Couples under Stress. *Journal of Marriage and the Family, 61*(1), 38–48.

Thorson, K., West, T., & Mendes, W. (2017). *Measuring Physiological Influence in Dyads: A Guide to Designing, Implementing, and Analyzing Dyadic Physiological Studies*. PsyArXiv. Retrieved from https://doi.org/10.17605/OSF.IO/9NDKF

Zhang, G., & Browne, M. W. (2008). *DyFA Bootstrap: Dynamic Factor Analysis of Lagged Correlation Matrices with Bootstrap Standard Errors and Goodness of Fit Test. Version: Beta 1*. Retrieved from https://psychology.osu.edu/people/browne.4

PART VII GENERAL ANALYTIC CONSIDERATIONS

28 Reproducibility in Clinical Psychology

CHRISTOPHER J. HOPWOOD AND SIMINE VAZIRE

There is a long history of invalid ideas in clinical psychology, many of which have had profound negative effects on public health and individual lives. "Refrigerator mothers" were once blamed for autism and schizophrenia (Greydnaus & Toledo-Pereya, 2012). Leeching (De Young, 2015), animal magnetism (Ellenberger, 1970), and the orgone box (Isaacs, 1999) were once proposed to treat mental illness. In each of these cases and many others, invalid ideas were corrected by scientific research, to the benefit of the public interest. The ability to identify and correct bad ideas about etiology and intervention is the primary virtue of a scientific approach to clinical psychology. Although the causes of psychopathology and the best way to prevent and treat problems in living remain poorly understood, there has been clear progress in the field that can be attributed directly to the scientific method.

Invalid and harmful theories are not isolated to the distant past. In 1998, Wakefield and colleagues published a paper in a prestigious journal ostensibly linking the measles, mumps, and rubella (MMR) vaccine to autism. Although it was based on a small sample of only 12 children, the finding was highly publicized. Even though a number of epidemiological studies found no association between the MMR vaccine and autism, rates of MMR vaccinations decreased and rates of measles, mumps, and rubella increased in the United Kingdom following the publication of the study (McIntyre & Leask, 2008). After journalists uncovered multiple conflicts of interest, questionable research practices, and several ethical violations, the journal retracted the paper. Wakefield had apparently reported fraudulent data after having been paid to "find" a link between the MMR vaccine and autism (Godlee, 2011). He was found guilty of malpractice and lost his medical license.

Since the time of the Wakefield and colleagues study there has been a crescendo of high-profile false positives in the psychology literature (Klein et al., 2014). The issue has been widely discussed in the scientific literature (Baker, 2016; Nosek, 2012), the blogosphere (e.g., Gelman, 2016; Srivastava, 2016; Vazire, 2016b), and the popular press (Aschwanden, 2015; Belluz, 2015; Engber, 2016; Yong, 2016). This is a serious problem. Faulty science in clinical psychology negatively affects patients and the public because unhelpful or harmful practices are disseminated to ill effect and because persistent reports of invalid findings in the popular press can erode public trust in the scientific method (Vazire, 2017).

But it is also a fortuitous opportunity to improve clinical psychological research (Munafo et al., 2017; Tackett et al., 2017). An upshot of recent discoveries of invalid findings has been a movement dedicated to teaching and disseminating more rigorous scientific methods (e.g., Open Science Collaboration, 2015). A central focus of this movement has involved investigating the *reproducibility* of reported effects. Genuine effects should be reproducible, and effects that cannot be reproduced should generally not be taken to be true. The purposes of this chapter are to describe the importance of reproducibility in the scientific method, review recent issues in the social sciences that contributed to the recognition of reproducibility problems, and describe best practices for conducting reproducible research in clinical psychology.

FOUNDATIONS OF THE SCIENTIFIC METHOD

The scientific method is one way of explaining phenomena. It can be contrasted with other methods, such as explanation via tradition or metaphysics, by a few foundational principles. In this section, we briefly review the distinguishing principles of the scientific method, with a focus on reproducibility.

Observations and Explanations

An explanation must be able to predict observations to be convincing in a scientific sense (Hempel & Oppenheim, 1948). Thus, science is fundamentally about the link between observations in nature and explanations about why those observations occurred. When explanations predict the same observations more than once they are increasingly convincing, and when they can predict similar kinds of observations across different contexts and situations, they become increasingly general.

Unlike in most other approaches to knowing, science rests on the principle of falsification (Popper, 1950), or the idea that scientists try to prove their explanations wrong rather than trying to prove them right. It follows from the idea of falsification that observations that are consistent with an explanation add confidence for that explanation but do not prove that it is true, whereas observations that are inconsistent with an explanation indicate that the explanation is at least partly inaccurate. This makes science really difficult, because our interest generally lies in proving something to be true rather than incrementally increasing our confidence that our explanation probably is not wrong. And as a general rule, it is easy to fool ourselves, particularly when we are motivated to see things a certain way (Feynman & Leighton, 1988).

Replication

This is why replication is so important in science. Replication means making the same explanation-relevant observations more than once in order to test the validity of the explanation. There are different levels of replication, ranging from completely direct to completely conceptual or constructive (Lykken, 1968). In the most direct replication, an effect would be observed twice by the same person under maximally similar conditions. A replication becomes less direct when the same observations are made under highly similar conditions but by different people. This increases confidence in the observation.

Conceptual replications push the boundaries of the explanation by changing the conditions of the experiment. If you knew that mixing vinegar (an acid) and baking soda (a base) together creates a reaction that produces carbon dioxide, and you knew that carbon dioxide is less dense than regular air, it would follow that mixing vinegar and baking soda together in a plastic bottle would expand the bottle until it burst. Other things would follow, too. For instance, this should also work in closed spaces other than a plastic bottle, and it should also work with other acid-base combinations. The underlying explanation for why the bottle blew up when vinegar was added to baking soda provides a basis for setting up conceptual replications that could test that explanation further and evaluate the boundary conditions of the effect.

To use a more clinical example, it is probable that an astute physician observed many years ago that psychotic patients had an unusually high incidence of family members with mental health problems. Having shared this observation with colleagues, others may have made similar observations. Eventually, early psychiatrists tested the concordance rates of family members in a controlled manner and confirmed an association, even across multiple ways of assessing psychotic conditions and evaluating family concordance (e.g., Jelliffe, 1911). This important set of observations could be used to support multiple possible explanations, including those related to heritability, the problematic "refrigerator mother" hypothesis, and others. These explanations were eventually tested as well, leading to the contemporary understanding that the increased incidence of psychotic symptoms among family members is mostly genetically mediated (Gottesman & Shields, 1973). While the etiology of psychotic phenomena remains poorly understood, replicated research on familial concordance that can be explained by nature to a greater degree than nurture advances our understanding significantly. From a scientific perspective, this kind of replicated evidence is always more convincing than any well-crafted argument untethered to replicated observations.

Both direct and conceptual replication are important, for different reasons. Conceptual replications are a riskier test of a theory, extrapolating from past findings to make new predictions (e.g., predicting the effect in a new setting, population, or using a different measure). However, the meaning of a failed conceptual replication is ambiguous. It could mean that the original theory was wrong, and even the early successful studies should be discounted as evidence for the theory, or it could mean that the researcher extrapolated one step too far, but the earlier successful studies should still be taken as strong evidence in support of the theory. Direct replications are useful because they eliminate this ambiguity. If all attempts to directly replicate a finding fail, this suggests that the original finding should not be taken as evidence for the theory, and that the theory is likely wrong. Thus, from a falsification perspective, direct replications are stronger tests because they expose the researcher to the risk of being forced to abandon a theory. Conceptual replications, on the other hand, help extend a theory when they are successful, but rarely put the researcher at risk of having to discard the entire theory.

Transparency

One of the hallmarks of science is that the basis for scientists' claims are verifiable by others (Lupia & Elman, 2014). This is one of the key features that makes science different from other ways of knowing (e.g., intuition, faith, authority). Unlike other ways of knowing that are only accessible to some people or some types of people, science is fundamentally democratic because, in principle, anyone can do it, and anyone who can show via observation that they have identified a novel explanation for something interesting and important should get credit for having done so. From a scientific perspective, we should always be skeptical of people who claim they have made observations that we may not be able to make, or who will not share their methods or results.

As it turns out, transparency was a key issue in the recent reproducibility problems reported in the scientific and popular literature. Economic models can be used to show that a lack of transparency (i.e., "information asymmetry" between the researcher and the consumer of research) will lead to an erosion of trust in the scientific "market," because consumers will eventually realize that they cannot tell rigorous scientific products from flimsy ones, and so will eventually refuse to put much stock in any scientific findings (Vazire, 2017). The growing concerns about the replicability of scientific findings is one sign of this erosion of trust.

ORIGINS OF THE REPRODUCIBILITY PROBLEM[1]

Methodologists have been emphasizing the issues that led to reproducibility problems for many decades. Meehl (1967) expressed concerns about the standard approach to determining whether an observation fits a hypothesis or not, which is generally referred to as null hypothesis significance testing (NHST). The essence of the problem with NHST is that it is probabilistic but not intuitive. A finding is "statistically significant" in an NHST framework when the likelihood of observing a particular effect is less than some predetermined value, usually 5 percent, if the effect does not actually exist in nature. However, it is possible to find apparently positive evidence for an effect that does not exist, and in fact all things being equal this would be expected about 5 percent of the time when nonexistent effects are tested. That means that if a researcher set their significance (alpha) level to 0.05 and ran 20 experiments, they would expect one to be "significant" even if there was no effect in the population.

However, one common misinterpretation of this 5 percent threshold is that this means that only 5 percent of significant findings are false positives. This is incorrect. Even under ideal circumstances (i.e., when researchers follow all the rules of NHST), the 5 percent threshold would guarantee that when the null hypothesis is true, 5 percent of results would be false positives. But this is not the same thing as guaranteeing that only 5 percent of significant results are false positives. Moreover, a number of common practices inflate the false positive rate (and violate the conditions under which NHST is supposed to be carried out). First, researchers can "peek" at the data as they are being collected, and decide when to stop data collection when the p-value crosses below the 0.05 threshold. After data collection, researchers can run many analyses (e.g., for many different subgroups, or using many different measures) and only report significant effects, claiming that those were the effects they predicted from the beginning (i.e., hypothesizing after the results are known, or HARKing; Kerr, 1998). Even in the absence of HARKing, running more tests increases the likelihood of spuriously finding false positives, because whereas 1/20 false positive findings would be expected at a 5 percent significance threshold, that percentage increases with every new test, even under perfectly unbiased conditions. In addition, researchers can tinker with their analyses by, for example, removing outliers in an outcome-dependent fashion (i.e., based on whether the key result becomes significant), transforming variables in such a way that the results come out significant, or adding or removing covariates until the results become significant. Finally, if none of these efforts lead to a significant result, researchers can put the "failed" study in a "file drawer" and start over, and then report only the studies that "worked" (i.e., produced a significant result) in their submitted manuscript. All of these practices, collectively referred to as "p-hacking" or "questionable research practices," increase the chances of reporting a significant effect that is actually a false positive. Many of them are quite common (John, Loewenstein, & Prelec, 2012).

The fact that journals have a strong bias to publish findings that are statistically significant compounds this problem (Rosenthal, 1979). Pressures to publish, complete degrees, get jobs, obtain tenure, and enhance one's status motivate researchers to find and report significant effects, and the foibles of human reasoning make it possible even for the most well-intentioned to justify these research practices (Tversky & Kahneman, 1971). During the last few decades, it has also become increasingly common to report novel findings in the popular press. The popular press is biased toward surprising, "sexy" findings. However, by definition surprising and counterintuitive findings are less likely to be observed, otherwise they would not surprise us. In the end, some researchers have built up large bodies of work seeming to support a particular explanation for what turn out to be false positive observations, by cleverly (and often unwittingly) violating some of the rules of NHST.

Finally, while the most common concern with small samples involves failing to find a true effect because of a lack of statistical power (false negative), the common practice of running studies with too-small samples can also increase the rate of false positive findings in the literature (Fraley & Vazire, 2014). This occurs for two reasons. First, studies with samples that are too small to reliably detect the true effect produce inconsistent and imprecise results. Second, journals are biased toward publishing significant results, as described above. Because small samples can produce both overly large and overly small effects, but only the overly large effects are likely to get published, these small sample studies lead to a literature that is biased toward exaggerated effects and false positives.

IMPACTS OF THE REPRODUCIBILITY PROBLEM

The discovery of substantial reproducibility issues in psychological research has caused significant upheaval in the

---

[1] This section borrows extensively from a blog post by Andrew Gelman (2016).

field. A major transition is occurring from an old way of doing things to a new, and hopefully better, way. This transition has revealed both bad news and good news for psychological research.

The Bad News

The most obvious bad news is that false positive effects have been disseminated in the scientific literature. Ideally, when an article claims to have found something, it is true. Counterinstances should be rare. When the majority of published research findings are true, consumers can have confidence in the results of empirical reviews or meta-analyses. However, when there is a bias for positive findings in the literature, the proportion of false positives is sure to increase (Smaldino & McElreath, 2016). With a literature that is untrustworthy, even meta-analyses with many samples will overestimate effects (Simonsohn, Nelson, & Simmons, 2014). In such cases, what appear to be valid explanations for observations are invalid.

This is potentially even more damaging than situations in which the received explanations for phenomena do not have empirical support, such as mother-blaming for psychotic conditions, because explanations that appear to have the imprimatur of science can and should get extra weight in evidence-based mental health. Evidence-based clinical practice depends upon the publication of valid explanations that connect observations in laboratories with observations seen in the consulting room (Anderson, 2006; Drake et al., 2001). When those explanations are inaccurate, even the most up-to-date and conscientiousness practitioner can use invalid techniques that do not help or cause harm to their patients.

For instance, a 2011 study in a high-profile journal reported the results of a large-scale, well-funded trial on treatments for chronic fatigue syndrome (White et al., 2011). The study authors concluded that cognitive-behavior therapy and exercise were moderately effective to improve functioning in individuals who suffer from chronic fatigue, and that adverse consequences were very rare. This result was covered widely in the popular press and recommended as an approved treatment by trusted institutions such as the United States Center for Disease Control and Mayo Clinic. However, patients with chronic fatigue questioned the results, and after a prolonged court battle the authors of the original study were compelled to release their data (Rehmeyer, 2016). Reanalyses by independent observers showed that therapy and exercise had minimal benefits for individuals with chronic fatigue (McGrath, 2015; Tuller, 2015).

The retraction of findings that have been previously disseminated can have a negative impact on the public perception of science as well (Carlson, 2006). Public understanding of mental health issues is generally poor and stigmas persist regarding mental health problems and treatments (Schomerus et al., 2012). Constant popular press reports about how explanations and effective interventions for mental health problems turn out to be untrue cannot help the field's image.

Finally, reproducibility problems can have a personal cost for individual scientists and create justice challenges for the field. Researchers who have followed the old rules to produce novel but nonreproducible effects have been rewarded with degrees, jobs, tenure, and esteem. In contrast, other researchers who have tried to replicate or build upon false effects that they believed have had difficulties publishing their research, with attendant impacts on their careers and status in the field.

The Good News

On the positive side, there is currently a robust movement, in the sciences generally and psychology in particular, focused on "meta-science," or issues related to how science is conducted (Munafo et al., 2017). Reproducibility is perhaps the core value driving this movement. As described above, there are a number of blogs whose content often involves issues related to reproducibility. There are also other social media and internet platforms focused on the issue. For instance, www.retractionwatch.com is a website that publishes content related to research integrity, including recent retractions in the scientific literature. Another site, www.pubpeer.com, is an online "journal club" in which articles can be searched and on which there are blogs and discussions related to metascience. The Open Science Framework (www.osf.io) is an online data repository sponsored by the Center for Open Science, which provides a variety of resources for researchers interested in doing transparent and reproducible work. The Center for Open Science also sponsored research aimed at replicating existing findings across many independent labs (e.g., Open Science Collaboration, 2015), as well as the *Society for the Improvement of Psychological Science* (http://improvingpsych.org/), an organization of researchers interested in issues related to reproducibility. Major journals have changed their editorial policies in order to encourage reproducible work (e.g., Cooper, 2016; Lindsay, 2015; Vazire, 2016a), and many individual researchers have dedicated resources to improving their methods and replicating previous studies.

Collectively these efforts suggest that the old way of doing things will eventually give way to a more reproducible science. However, this movement has differentially affected the subdisciplines of psychology, and it has not yet made as much progress in clinical psychology as some other branches (Tackett et al., 2017). The remainder of this chapter focuses on best practices for reproducible clinical psychological research.

BEST PRACTICES IN REPRODUCIBLE CLINICAL RESEARCH

Best practices can ensure that the scientific literature rests on a solid and reproducible foundation. Ideally, these practices should be adopted for intrinsic reasons, but the extrinsic pressures are growing. Below we provide guidelines for how to conduct reproducible research (see also: Funder et al., 2014; Munafo et al., 2017; Nosek et al., 2015).

Be Transparent

A virtue of scientific epistemology is that arguments rest on their merits. Cases are made empirically, rather than through appeals to authority or intuition. Making an empirical argument requires the collection of data that would provide a critical test of a hypothesis. In order to convince critical observers, the hypothesis, methods used to test the hypothesis, and results need to be available for scrutiny. Observers can then draw their own conclusions about the quality of the test, the validity of the reported results, and the degree to which the results provide a convincing argument regarding the hypothesis. When information about the test is unavailable, observers cannot draw reasonable conclusions, and the scientific method – and trust in it – breaks down (Vazire, 2017).

The lack of transparency about why and how scientists conducted their research was a major enabling factor in the replication problems reviewed above. Rather than clearly stating hypotheses prior to data collection, researchers often described their results as if they were confirming hypotheses that were actually generated after the results were known (Kerr, 1998). Instead of publishing all of the variables collected in the study, people reported only the variables that could be used to find significant results. Rules for when data collection would stop, which participants would be included, and which analyses would be conducted were hidden or were simply not decided on ahead of time. And the data researchers used to draw their conclusions were not made available to other scientists.

These kinds of practices led to results that could not be reproduced by other researchers, often because the methods themselves could not be reproduced. This is a clear violation of the basic premises of the scientific approach to knowing. Researchers can take several specific steps to ensure that their own work is transparent and reproducible, including preregistering their studies, reporting everything, and distinguishing between exploratory and confirmatory hypotheses.

Preregister

In the basketball game HORSE, the goal is to make shots that others cannot make. A common strategy in this game is to add a twist that makes a certain shot more difficult, such as bouncing the ball off the backboard before it goes in, or shooting with your opponent's nondominant hand. In order for this game to work, you have to call your shots before shooting. If you make a shot that happens to hit the backboard before going in, your opponent does not know if this was intentional or not, and cannot fairly be expected to make the same kind of shot. On the other hand, if you "call your shot" before shooting, they have to follow the same technique.

Good science works the same way (Nosek & Lakens, 2014). Statistically significant effects can typically be found in any data set if enough tests are conducted, just as very difficult basketball shots can occur by chance even when they were not intended to be difficult. Therefore, if you believe in a hypothesis and have carefully thought through the best way you can think of to test it, your argument will be most convincing if your data work out the way you predicted beforehand.

A number of resources are currently available to preregister studies, including individual research websites, the Open Science Framework (OSF) website referenced above (www.osf.io) or other repositories (see www.nature.com/sdata/policies/repositories#social; https://aspredicted.org/). The OSF site allows researchers to post hypotheses, data, methods, analytic plans, results, research reports, and other aspects of the research process with time stamps so that independent observers can evaluate scientific arguments. This information does not need to be made fully public right away, in case there are concerns about "scooping," data theft, or other worries related to research propriety. However, because documentation is timestamped, preregistering through such sites allows researchers to demonstrate that they made their decisions about how to collect and analyze their data before collecting the data. They also motivate good scientific practice in general by compelling researchers to fully think through their designs and analyses (i.e., "call their shots"). Another advantage of using resources like OSF is that it provides a free resource for storing research materials and methods. Obviously, some materials in clinical research are confidential or otherwise sensitive, and care needs to be taken to ensure that posting materials or data to sites like OSF is done ethically (Tackett et al., 2017).

When preregistering specific hypothesis plans is not possible, a researcher can still "pre-register" that a study will include only exploratory tests (which essentially commits them to honesty and prevents them from later claiming that something was predicted). In addition, researchers can preregister "Standard Operating Procedures" (SOP) for their lab, such as how they typically transform variables, deal with outliers, or select covariates (Lin & Green, 2016).

Report Everything

Reproducible science depends upon the full reporting of everything relevant to a given study (Funder et al., 2014), because independent researchers cannot evaluate or replicate studies without all of the information relevant to those studies. This includes information pertinent to different steps in the process, such as the instrumentation used in the study, the nature of the sample, data acquisition and management, the analyses used to test hypotheses, the results of those analyses, and the actual data used in the study. With respect to instrumentation, it is important to report exactly which instruments or methods were used, including which version of the instrument in cases where there are multiple versions. It is also critical to report estimates of the reliability of study instruments within the study in which inferences are being made, because reliability is a characteristic of an assessment tool within a particular sample, not a characteristic of a tool in the abstract.

Sample information includes the population from which the sample was drawn, relevant demographic characteristics, and data acquisition and management procedures. It is often of interest to examine and report differences between subsamples (e.g., women and men) on study variables, and in some cases to establish measurement equivalence between subsamples on those variables. At times there are reasons to remove individuals from the sample prior to analysis, because of evidence that the participant did not produce reliable data or due to extreme scores on study variables. At other times an effect can only be observed with dubious data points included. Thus, either the inclusion or exclusion of outliers can potentially be used to "find" statistically significant effects in a subsample that is not present in the full sample. This approach raises concerns about which came first – the principled removal of certain outliers based on predetermined criteria or the discovery of significant effects after certain participants have been removed. We therefore recommend establishing outlier removal rules prior to data collection and analysis (e.g., by posting your group's Standard Operating Procedures for data cleaning; Lin & Green, 2016), reporting results with and without outliers, using statistics that account for the impact of outliers, and making the data available so that others can access the full sample.

There are many ways to analyze data to test a certain hypothesis. Decisions such as which variable to use as the primary outcome, whether or not to use covariates, which covariates to use, how to score variables (e.g., dimensional measures or extreme groups), and which specific analytic technique to use will impact the results. It is common for different techniques to give different answers, particularly when effects are small, as they often are in clinical psychology. This creates significant opportunities for "researcher degrees of freedom" (Simmons, Nelson, & Simohnson, 2011). In other words, the availability of alternative approaches to analyzing data means that researchers motivated to find and publish a significant effect can analyze the same data multiple ways and only report significant results (i.e., "*p*-hack"). This is a probable pathway to false-positive findings that will not replicate. As an alternative, we recommend determining analytic approaches to test hypotheses before conducting the analyses based on the research question, reporting the results of all analyses conducted, and making the data available to others who might want to analyze the data differently. We would also suggest that, all things being equal, the simplest analysis that provides an adequate test of the focal hypothesis is generally best.

Researchers sometimes only report whether or not an effect fell below a statistical significance value, without providing information about the magnitude of the effect or the actual test statistic value. There are at least four problems with this practice. First, results without information about effect sizes and confidence intervals do not provide enough information for readers to interpret the practical significance of the finding. Second, insufficient reporting makes it impossible for future reviewers and meta-analysts to collate the results with those from other samples. Third, techniques designed to evaluate replicability (e.g., *p*-curve analysis) rely on exact *p*-values and other detailed information about statistical tests. Studies that do not report this information cannot easily be examined for replicability. Finally, we all make mistakes in calculations, syntax, and other aspects of testing research hypotheses. When all of the information about a hypothesis test is reported, such mistakes can more readily be identified and corrected than when information is limited to the statement that the test statistic fell below a certain threshold. We recommend reporting effect sizes, confidence intervals, and exact *p*-values (or analogous results in other, e.g., Bayesian, frameworks) for all statistical tests in a study.

In an ideal case, researchers would make the data and methods used in their study available for other researchers to use. As described above, the OSF and other sites make this possible to do with relative ease, and it safeguards against scooping and other concerns. There are situations in which sharing data is still not desirable. For instance, there are large-scale collaborative projects in which funded researchers are actively working on a project. Having put significant resources into obtaining funding and collecting data, those researchers should perhaps have first rights to analyze and write up the data. In that case, it might be reasonable to post only the variables used in a given manuscript rather than the entire data set, and to have a plan for posting the full data after they have been fully collected and analyses related to primary aims are completed and published. There may also be situations where ethical issues such as reidentification risk would prevent full posting of data (Tackett et al., 2017). These issues aside, we recommend that researchers make data and other study features available whenever they can. We hope that this eventually becomes the default practice in psychological research.

Distinguish Exploratory and Confirmatory Hypotheses

It is possible to misinterpret recommendations to preregister studies and report everything as strictures against exploratory research. Objections might follow that exploratory research is an integral part of the creative scientific process (Goldin-Meadow, 2016). The best ideas sometimes occur by accident, because people find unexpected effects in their data. To lose the possibility of principled scientific exploration would slow progress in an already slow-moving endeavor.

We agree that there needs to be room for exploration in psychological research. This includes the principled "fishing" for significant effects, meaning trying to find meaningful relationships in data that were not hypothesized. The critical issue is distinguishing between confirmatory and exploratory research. Confirmatory research involves testing hypotheses that were predicted before the data were collected, using preregistered methods. Exploratory research involves examining data for potentially meaningful results that could form the basis for future confirmatory research.

Both confirmatory and exploratory research are critical aspects of science, but it is also critical to distinguish them. Obviously, successful confirmatory research should inspire more confidence than positive exploratory research. This distinction gets blurry when results that are actually exploratory are written up as if they were confirmatory (Kerr, 1998). From this perspective, preregistration and full reporting enable a clearer distinction between confirmatory and exploratory research. We would hope that this kind of clear distinction does not reduce opportunities to publish results, but instead allows for calibration of conclusions to the strength of the evidence.

Exploratory and Confirmatory Hypotheses in Multivariate Modeling

Many of the issues described above have most often been discussed in the case of experimental, bivariate research (e.g., where experimental and control groups are compared on some outcome variable). However, all of the suggestions for best practice extend to multivariate research in which models are tested based on the covariance structure of some set of variables (e.g., confirmatory factor analyses, structural equation models; see Hoyle, 2012 or Kline, 2016 for general reviews). In these kinds of models, the relationships among a number of variables are conceptualized in terms of some underlying theory or conception about how they can be represented in nature.

For example, a researcher might hypothesize that the 10 items on a depression measure correlate because they are all indicators of the same latent construct, depression. In a slightly more complex model, there might be two constructs such as depression and anxiety, and a model might be fit that posits that 10 items load on depression and only depression, 10 other items load on anxiety and only anxiety, and any covariance between the depression and anxiety items can be accounted for by the correlation between the latent depression and anxiety constructs. The adequacy of the model is judged by indicators of how well the data statistically "fit" the theoretical model. These indicators essentially answer the question: if nature really worked the way the model proposes, would the variables correlate this way in these data?

This kind of data analysis can risk the creation of models that are unlikely to generalize in order to achieve acceptable statistical fit between model and data within a certain sample (i.e., overfitting). There are many ways to arrange variables in a covariance model, particularly as the model becomes more complex. There are also a variety of ways to alter models to achieve "good" fit. For instance, you can allow the error terms associated with two specific items (i.e., the variance that is not explained by a higher-level factor) to covary, even though there is not necessarily a good theoretical reason to do this. Or parameters can be fixed, such as when you specify that there should be no variation in a certain variable. As a general rule, the more models are altered, the less likely they will generalize to new samples (Preacher, 2006). Thus, post hoc model alterations should be done with caution. Problems associated with altering models to achieve good fit are compounded when these alterations are presented as if they were predicted all along. As with experimental studies, we would view preregistration as a necessary condition for calling a hypothesis test (i.e., model result) "confirmatory."

In summary, the most reproducible models are not altered to achieve acceptable fit in a particular sample, and are preregistered and cross-validated in multiple samples. When models are modified, we recommend that researchers are transparent about what those alterations were and on what basis they were made.

Shift Focus from Statistical Significance to Effect Sizes and Confidence Intervals

As discussed above, philosophers of science have been calling for a shift in emphasis from NHST to effect estimation for decades (e.g., Cohen, 1994; Fraley & Marks, 2007; Meehl, 1978). Unlike statistical significance tests that do not provide direct information about the effect of interest and can only be interpreted in the context of sample size and other factors related to statistical power, effect sizes directly convey information about the direction and magnitude of a hypothesized effect. There are a variety of effect size coefficients, but the two most common standardized metrics for effect sizes are the correlation coefficient and Cohen's d-coefficient. When squared, correlations indicate the amount of variability in one variable that can be explained by variability in another variable. Cohen's d indicates how different two group means are in the metric of the pooled standard deviation across groups. Effect sizes can also be expressed in unstandardized units or estimated for multivariate models. The important point here is that these values indicate what we are generally most interested in when testing univariate hypotheses: the degree to which variables are related or groups differ. However, effect sizes should never be interpreted in the absence of their corresponding confidence intervals. Confidence intervals convey the precision with which we measured the effect, and smaller confidence intervals are better (i.e., indicate more precise estimates). With small samples, it is relatively easy to get extremely large effects by chance, but the confidence intervals around these point estimates of the effect size will also be very large. Thus, the point estimates alone can be very misleading. Conversely, very small effects can nevertheless be practically meaningful (e.g., if we have identified a predictor of longevity), but only if they are estimated very precisely.

The importance of effect sizes can be illustrated by the following example. If a clinician were selecting from between two treatment approaches for a patient diagnosed with schizophrenia, and all they knew was that one treatment was more effective than the other, the responsible decision would be to choose the more effective treatment. However, if they also knew that the more effective treatment required twice as many resources and came with a higher risk of adverse effects, the decision would become more complicated. In this situation, the effect size would be extremely valuable information. If the effect size difference between the two treatments were small, the clinician would be more inclined to choose the treatment that was slightly less effective but also less costly and which came with fewer risks. If the effect size difference were large, that treatment might still be desirable despite these other factors.

Conduct Adequately Powered Studies

Statistical power means the probability that a true effect in nature would be statistically significant in a given experiment. From a NHST perspective, power is the probability of detecting a real effect (i.e., rejecting the null hypothesis of no effect) if a real effect exists. Studies should generally be designed that have sufficient power to detect the effects being tested. A typical convention for power is 0.80, which means that there is a 20 percent chance that a researcher would fail to detect an effect even though there is a real effect in nature (Cohen, 1988). However, in order to determine what sample size is needed to achieve 80 percent power (or any other level of power), we must make an assumption about the likely size of the effect. It is very easy to go wrong at this step, especially because if we knew the size of the effect we were looking for, we would not need to conduct research on it. Thus, some researchers have called for alternative ways to determine sample size, such as planning for precision (Cumming, 2014) or sequential analysis (Lakens, 2014).

There are several ways to enhance the power to find effects. The most obvious is to increase one's sample size, and there have been many calls for researchers to take this recommendation very seriously (e.g., Fraley & Vazire, 2014). An advantage of large samples, in addition to power, is that they lend to more precise estimates of effect sizes (i.e., confidence intervals around effect sizes decrease as sample sizes increase). Another approach to increasing power is to use instruments that are as reliable as possible, because measurement error lowers the sensitivity of instrumentation to detect effects. In group comparison designs, increasing the homogeneity of groups can have a similar effect, because within-group heterogeneity can be associated with statistical error.

Power is related to reproducibility because many findings that have failed to replicate were originally found in very small samples. This is intuitive when you consider that most published effect sizes in psychology are around $r = 0.20$ (Fraley & Marks, 2007; Richard, Bond, & Stokes-Zoota, 2003), and this is likely an overestimate because publication bias inflates effect sizes: 194 participants are required to have 0.80 power to observe statistical significance when a 0.20 correlation is anticipated. Thus, any time a sample in a correlational study is less than 200 and the effect seems likely to be in the average range of effects found in psychology, it is reasonable to be skeptical about a significant finding.

Power can be difficult to achieve, because psychological measurement is imperfect and there are practical issues associated with getting large samples. Indeed, small samples can be defended in certain cases, such as when there are ethical issues related to sampling (e.g., in research with nonhuman animals or vulnerable human populations or research that uses resource-intensive methods). However, even in these cases, it is not clear that running underpowered studies with small samples is more ethical than running well-powered studies, because even though the latter accrue more costs, they also produce more benefits (i.e., more reliable findings). In other words, if it is ethically questionable to kill 50 mice to produce a robust finding, it does not follow that it is more justifiable to kill only six mice to produce a flimsy finding that is not likely to be true. Moreover, the fact that large samples are difficult or impossible in such research does not protect the results from issues related to low power. There are solutions to underpowered studies, including methodological adjustments (e.g., within-person designs), the explicit reporting of confidence intervals accompanied by an acknowledgement regarding statistical power, or Bayesian data-analytic frameworks in which decreased confidence in the effect based on power is an explicit aspect of the model.

Bayesian Inference

Controversies between a focus on statistical significance and effect size actually go back to the earliest days of inferential statistics. Eighteenth-century statistician Sir Ronald Fisher and others (e.g., Neyman and Pearson) advocated a frequentist hypothesis testing framework that led to modern NHST and effect estimation approaches. Thomas Bayes, who lived a century before Fisher, is credited with an alternative approach that focused on confidence in observed effects rather than the probability of inaccurate conclusions (Kruschke, 2015). In a Bayesian approach, a prior (prediction) is made before data are collected about the effect that will be observed in an experiment (for an accessible introduction to Bayesian approaches, see Dienes, 2008). This prior can be based on previous evidence regarding the effect or a researcher's educated guess. In either case it is quantified as a formal aspect of evaluating the validity of a hypothesis when the Bayesian researcher derives a posterior estimate that updates the prior belief based on the new (observed) information. If the prior is extreme (i.e., strongly favors one expected result), a result that does not match that prior may not have much of an effect on the posterior. On the other hand, if the result was based on strong methods (e.g., precise instrumentation in a well-controlled study with a large sample) and the prior was weak (e.g., gave many different results equal probability of being true), then the posterior will be significantly different than the prior following the observation.

It has been argued that Bayesian approaches are more intuitive than frequentist approaches like NHST. That is, normally scientists have expectations about what they will find but are open to adjusting those expectations based on the results of their research (Morey & Rouder, 2011). In contrast, frequentist approaches are somewhat counterintuitive, because scientists, like other people, do not usually think in terms of the likelihood that they would observe something if it were not actually true.

Bayesian approaches to data analysis have been offered as one solution to reproducibility problems. In general, any approach which requires researchers to make formal predictions prior to data collection and which focuses on effect sizes and precision of effect estimates could help yield more reproducible science. However, there is nothing inherent in the Bayesian approach that requires formal predictions to be made ahead of time (i.e., preregistration), or prevents tinkering with analyses to get a desired result, any more than the rules of NHST technically prohibit these practices.

Another touted advantage of Bayesian approaches is that they build upon previous research by definition (Gershman, 2016). There would seem to be more value in previous results which can inform prior estimates than would be the case in frequentist research, in which every study tests hypotheses anew. However, we caution that Bayesian approaches to data analysis are not a panacea to reproducibility problems. In principle, researchers could still change their priors after the fact, exercise researcher degrees of freedom in selecting their participants, variables, or analyses, or selectively report their results. Recent trends suggest that the field may move in the direction of greater balance between Bayesian and frequentist approaches to data analysis. All things being equal that is probably beneficial, if for no other reason than this balance provides researchers with more tools with which to engage in reproducible science (though it may also provide more opportunities to obfuscate).

Replicate

Replication is the *sine qua non* of the scientific method. Without replication, there is no way to know how reliable a particular observation is, and thus no way to know how well an observation fits with a given explanation for some natural phenomenon. As described above, replications range from direct to conceptual. Direct replications increase confidence in the particulars of a given study or experiment, whereas conceptual replications increase generalizability and test the boundaries of a given effect.

Historically, replications, and particularly direct replications, have been thought of as relatively unexciting because they do not add new information to the literature. However, it is important to remember that this view assumes that the original effects were valid, which has often not been the case in psychological research. As a result of this view, replications have been difficult to publish, with some journals explicitly stating that they are only interested in novel research. There has been a related distaste for null results, leading to the "file drawer effect" (Rosenthal, 1979) in which there is a systematic bias in the literature in favor of positive results. This includes false positive results, which skew empirical or meta-analytic reviews of the literature.

One positive outcome of the metascience movement is that a number of journals now explicitly state their interest in replication reports. Whereas replications used to be very difficult to publish, there are now outlets for this kind of work. We encourage researchers to replicate their own work and the work of others. In general, we would suggest a progression from direct to more conceptual replication, such that studies in which novel effects are reported are initially followed by direct replications to ensure the reliability of those effects, and then eventually followed by conceptual replications to examine the boundaries of the effect. In clinical treatment research, for example, a randomized controlled effectiveness trial suggesting the superiority of a specific treatment approach over treatment as usual might first be followed by a replication by different researchers. Having established a reliable effect, further research might be conducted in the form of efficacy trials, which test the validity of the treatment in the community, or dismantling trials, in which specific elements of the treatment are evaluated for their specific effect on the clinical outcome.

In some instances, replication can be stressful for personal or political reasons. If researcher A publishes a study suggesting that the treatment they developed is superior to other treatments, researcher B might be loath to publish data suggesting that the treatment is similar in effectiveness to other approaches. Adversarial collaborations are useful in such situations. In an adversarial collaboration, researcher A and B would work together to design and conduct a study, so that both parties would presumably agree regarding the value of that study and accept the results, whatever they may be.

When there are power differences, such as when researcher A is a senior scholar with an editorial position at a major journal whereas researcher B is a graduate student, the stress associated with replicating others' work can be amplified. It is also the case that replication studies, despite their importance, may never be valued as much as novel research. For this reason, young researchers whose careers depend on producing impactful research may be punished for focusing on replication studies. We hope that these views will change, and that all scholars will be rewarded fairly for contributing to science, including via conducting replications. However, in the current state of the field we would suggest that there is a particular responsibility for more senior, tenured psychologists to welcome replications of their own work and conduct replication research themselves.

SUMMARY

There is now clearly a wrong but common way to do psychology research. First, you plan a study (with or without a prediction in mind), and begin collecting data. You collect data until you get the effect you expected, aiming for the smallest sample that will allow you to detect the effect. You use several tools at your disposal to try to uncover the effect you're looking for, including eliminating outliers, swapping variables, adding covariates, or analyzing the data in different ways to see if you can find the effect. If you still do not get the effect you expected, you change your hypotheses but write up the results as if the modified hypothesis was predicted all along. If you get no effect at all, you sweep the study under the rug (or into the file drawer) and try again, or move onto a different question, without publishing the results of your study. All replications are conceptual extensions of the original effect. Journals, employers, funders, and the press reward novel and interesting findings and do not require preregistration, transparency, or robustness checks. The literature is littered with false positives to the point that it is unclear how to interpret any effects, leading to a crisis of confidence in the entire endeavor.

We have described a way to do psychological research that is more likely to lead to reproducible findings. In this alternative approach, scientists think through and preregister studies. They make all methods and data open to the public to the extent possible. Studies have adequate power and use maximally precise measures. Null effects are valued as much as positive effects, and the question shifts from "is this effect real?" to "how substantial is the effect and how precisely can we estimate it?" Effects are routinely replicated. Journals and universities adjust their incentive structures to reward reproducible science, and the popular press and media follow suit.

It seems clear that researchers from all areas of psychology have engaged in poor practices for a long time, and many were probably unaware they were doing so. This has led to a situation in which much of the published research is of questionable value. In clinical psychology, false positive effects can have a negative effect on public health and individual lives, as illustrated by some of the examples above. The upside of this situation is that a lot can be learned from the discoveries of the last few decades. One hopes that the hard-earned lessons of the reproducibility problem will lead to a permanent correction and ultimately to more reproducible effects that can contribute to a better understanding of psychopathology, health, prevention, and intervention in clinical psychology.

REFERENCES

Anderson, N. B. (2006). Evidence-Based Practice in Psychology. *American Psychologist, 61*, 271–285.

Aschwanden, C. (2015). *Science Isn't Broken*. Retrieved from http://fivethirtyeight.com/features/science-isnt-broken/

Baker, M. (2016). Is There a Reproducibility Crisis? *Nature, 533*, 452–455.

Belluz, J. (2015). *Scientists Often Fail When They Try to Reproduce Studies: This Scientist Explains Why*. Retrieved from www.vox.com/2015/8/27/9212161/psychology-replication

Carlson, E. A. (2006). *Times of Triumph, Times of Doubt: Science and the Battle for Public Trust*. Cold Spring Harbor, NY: Cold Spring Harbor Press.

Cohen, J. (1988). *Statistical Power Analysis for the Behavioral Sciences* (2nd edn.). Hillsdale, NJ: Lawrence Erlbaum.

Cohen, J. (1994). The Earth is Round ($p < .05$). *American Psychologist, 49*, 997–1003.

Cooper, L. (2016). Editorial. *Journal of Personality and Social Psychology, 110*, 431–434.

Cumming, G. (2014). The New Statistics: Why and How. *Psychological Science, 25*, 7–29.

De Young, M. (2015). *Encyclopedia of Asylum Therapeutics, 1750–1950*. Jefferson, NC: McFarland.

Dienes, Z. (2008). *Understanding Psychology as a Science: An Introduction to Scientific and Statistical Inference*. Basingstoke: Palgrave Macmillan.

Drake, R. E., Goldman, H. H., Leff, H. S., Lehman, A. F., Dixon, L., Mueser, K. T., & Torrey, W. C. (2001). Implementing Evidence-Based Practices in Routine Mental Service Settings. *Psychiatric Services, 52*, 179–182.

Ellenberger, H. (1970). *The Discovery of the Unconscious*. New York: Basic Books.

Engber, D. (2016). Cancer Research Is Broken. Retrieved from www.slate.com/articles/health_and_science/future_tense/2016/04/biomedicine_facing_a_worse_replication_crisis_than_the_one_plaguing_psychology.html

Feynman, R. P., & Leighton, R. (1988). *Surely You're Joking, Mr. Feynman*. New York: W. W. Norton.

Fraley, R. C., & Marks, M. J. (2007). The Null Hypothesis Significance Testing Debate and Its Implications for Personality Research. In R. W. Robins, R. C. Fraley, & R. F. Krueger (Eds.), *Handbook of Research Methods in Personality Psychology* (pp. 149–169). New York: Guilford Press.

Fraley, R. C., & Vazire, S. (2014). The N-Pact Factor: Evaluating the Quality of Empirical Journals with Respect to Sample Size and Statistical Power. *PLOS One, 9*, e109019.

Funder, D. C., Levine, J. M., Mackie, D. M., Morf, C. C., Sansone, C., Vazire, S., & West, S. G. (2014). Improving the Dependability of Research in Personality and Social Psychology: Recommendations for Research and Educational Practice. *Personality and Social Psychology Review, 18*, 3–12.

Gelman, A. (2016). What Has Happened Down Here Is the Winds Have Changed. *Statistical Modeling, Causal Inference, and Social Science*. Retrieved from http://andrewgelman.com/2016/09/21/what-has-happened-down-here-is-the-winds-have-changed/

Gershman, S. J. (2016). Empirical Priors for Reinforcement Learning Models. *Journal of Mathematical Psychology, 71*, 1–6.

Godlee, F. (2011). The Fraud Behind the MMR Scare. *British Medical Journal, 342*, d22.

Goldin-Meadow, S. (2016). Why Pre-Registration Makes Me Nervous. *Observer, 7*. Retrieved from www.psychologicalscience.org/observer/why-preregistration-makes-me-nervous#.WJzFz7YrLUo

Gottesman, I. I., & Shields, J. (1973). Genetic Theorizing and Schizophrenia. *British Journal of Psychiatry, 122*, 15–30.

Greydnaus, D. E., & Toledo-Pereya, L. H. (2012). Historical Perspectives on Autism: Its Past Record of Discovery and Its Present State of Solipsism, Skepticism, and Sorrowful Suspicion. *Pediatric Clinics of North America, 59*, 1–11.

Hempel, C., & Oppenheim, P. (1948). Studies in the Logic of Explanation. *Philosophy of Science, 15*, 135–175.

Hoyle, R. H. (2012). *Handbook of Structural Equation Modeling*. New York: Guilford.

Isaacs, K. (1999). Searching for Science in Psychoanalysis. *Journal of Contemporary Psychotherapy, 29*, 235–252.

Jelliffe, S. E. (1911). Predementia Praecox: The Hereditary and Constitutional Features of the Dementia Praecox Makeup. *Journal of Nervous and Mental Disease, 38*, 1–26.

John, L. K., Loewenstein, G., & Prelec, D. (2012). Measuring the Prevalence of Questionable Research Practices with Incentives for Truth Telling. *Psychological Science, 23*, 524–532.

Kerr, N. L. (1998). HARKing: Hypothesizing after the Results Are Known. *Personality and Social Psychology Review, 2*, 196–217.

Klein, R. A., Ratliff, K. A., Vianello, M., Adams, R. B., Bahnik, S., Bernstein, M. J., ... Nosek, B. A. (2014). Investigating Variation in Replicability: A "Many Labs" Replication Project. *Social Psychology, 45*, 142–152.

Kline, R. B. (2016). *Principles and Practice of Structural Equation Modeling* (4th edn.). New York: Guilford.

Kruschke, J. K. (2015). *Doing Bayesian Data Analysis (2nd ed): A Tutorial with R, JAGS, and Stan*. Cambridge, MA: Academic Press.

Lakens, D. (2014). Performing High-Powered Studies Efficiently with Sequential Analysis. *European Journal of Social Psychology, 44*, 701–710.

Lin, W., & Green, D. P. (2016). Standard Operating Procedures: A Safety Net for Pre-Analysis Plans. *PS: Political Science & Politics, 49*, 495–500.

Lindsay, D. S. (2015). Replication in Psychological Science. *Psychological Science, 26*, 1827–1832.

Lykken, D. T. (1968). Statistical Significance in Psychological Research. *Psychological Bulletin, 70*, 151–159.

Lupia, A., & Elman, C. (2014). Openness in Political Science: Data Access and Research Transparency. *PS: Political Science and Politics, 47*, 19–42.

McGrath, S. (2015). Omission of Data Weakens the Case for Causal Mediation in the PACE Trial. *The Lancet, 2*, e7–e8.

McIntyre, P., & Leask, J. (2008). Improving Uptake of MMR Vaccine. *British Medical Journal, 336*, 729–30.

Meehl, P. E. (1967). Theory-Testing in Psychology and Physics: A Methodological Paradox. *Philosophy of Science, 34*, 103–115.

Meehl, P. E. (1978). Theoretical Risks and Tabular Asterisks: Sir Karl, Sir Ronald, and the Slow Progress of Soft Psychology. *Journal of Consulting and Clinical Psychology, 46*, 806–834.

Morey, R. D., & Rouder, J. N. (2011). Bayes Factor Approaches for Testing Interval Null Hypotheses. *Psychological Methods, 16*, 406–419.

Munafo, M. R., Nosek, B. A., Bishop, D. V. M., Button, K. S., Chambers, C. D., Percie Du Sert, N., ... Ioannidis, J. P. A. (2017). A Manifesto for Reproducible Science. *Nature: Human Behavior, 1*(1), 0021.

Nosek, B. A. (2012). An Open, Large-Scale, Collaborative Effort to Estimate the Reproducibility of Psychological Science. *Perspectives on Psychological Science, 7*, 657–660.

Nosek, B. A., & Lakens, D. (2014). Registered Reports: A Method to Increase the Credibility of Published Results. *Social Psychology, 45*, 137–141.

Nosek, B. A., Alter, G., Banks, G., Boorsboom, D., Bowman, S., Breckler, S. J., ... DeHaven, A. C. (2015). Promoting an Open Research Culture: The TOP Guidelines for Journals. Retrieved from https://osf.io/vj54c/

Open Science Collaboration. (2015). Estimating the Reproducibility of Psychological Science. *Science, 349*, aac4716-1–aac4716-8.

Popper, K. (1950). *The Logic of Scientific Discovery*. New York: Routledge.

Preacher, K. J. (2006). Quantifying Parsimony in Structural Equation Modeling. *Multivariate Behavioral Research, 41*, 227–259.

Rehmeyer, J. (2016). Bad Science Misled Millions with Chronic Fatigue Syndrome. Here's How We Fought Back. Retrieved from www.statnews.com/2016/09/21/chronic-fatigue-syndrome-pace-trial/

Richard, F. D., Bond, C. F., & Stokes-Zoota, J. J. (2003). One Hundred Years of Social Psychology Quantitatively Described. *Review of General Psychology, 7*, 331–363.

Rosenthal, R. (1979). The File Drawer Problem and Tolerance for Null Results. *Psychological Bulletin, 86*, 638–641.

Schomerus, G., Schwahn, C., Holzinger, A., Corrigan, P. W., Grabe, H. J., Carta, M. G., & Angermeyer, M. C. (2012). Evolution of Public Attitudes about Mental Illness: A Systematic Review and Meta-Analysis. *Acta Pyschiatrica Scandinavica, 125*, 440–452.

Simmons, J. P., Nelson, L. D., & Simonsohn, U. (2011). False-Positive Psychology: Undisclosed Flexibility in Data Collection and Analysis Allows Presenting Anything as Significant. *Psychological Science, 22*(11), 1359–1366.

Simonsohn, U., Nelson, L. D., & Simmons, J. P. (2014). P-Curve and Effect Size: Correcting for Publication Bias Using Only Significant Results. *Perspectives on Psychological Science, 9*, 666–681.

Smaldino, P. E., & McElreath, R. (2016). The Natural Selection of Bad Science. *Royal Society Open Science*. Retrieved from https://royalsocietypublishing.org/doi/full/10.1098/rsos.160384

Srivastava, S. (2016). Everything Is Fucked: The Syllabus. *The Hardest Science*. Retrieved from https://hardsci.wordpress.com/2016/08/11/everything-is-fucked-the-syllabus/

Tackett, J. L., Lilienfeld, S. O., Patrick, C. J., Johnson, S. L., Krueger, R. F., Miller, J. D., ... Shrout, P. E. (2017). It's Time to Broaden the Replicability Conversation: Thoughts for and from Clinical Psychological Science. *Perspectives on Psychological Science, 12*, 742–756.

Tuller, D. (2015). Trial by Error: The Troubling Case of the PACE Chronic Fatigue Syndrome Study. Virology Blog. Retrieved from www.virology.ws/2015/10/21/trial-by-error-i/

Tversky, A., & Kahneman, D. (1971). Belief in the Law of Small Numbers. *Psychological Bulletin, 1971*, 105–110.

Vazire, S. (2016a). Editorial. *Social Psychological and Personality Science, 7*, 3–7.

Vazire, S. (2016b). I Have Found the Solution and It Is Us. *Sometimes I'm Wrong*. Retrieved from http://sometimesimwrong.typepad.com/wrong/2016/08/solution-is-us.html

Vazire, S. (2017). Quality Uncertainty Erodes Trust in Science. *Collabra: Psychology, 3*, 1.

Wakefield, A. J., Murch, S. H., Anthony, A., Linnell, J., Clsson, D. M., Malik, M., ... Walker-Smith, J. A. (1998). Ileal-Lymphoid-Nodular Hyperplasia, Non-Specific Colitis, and Pervasive Developmental Disorder in Children. *The Lancet, 351*, 637–641.

White, P. D., Goldsmith, K. A., Johnson, A. L., Potts, L., Walwyn, R., DeCesare, J. C., ... PACE Trial Management Group (2011). Comparison of Adaptive Pacing Therapy, Cognitive Behavior Therapy, Graded Exercise Therapy, and Specialist Medical Are for Chronic Fatigue Syndrome (PACE): A Randomized Trial. *The Lancet, 377*, 823–836.

Yong, E. (2016). Psychology's Replication Crisis Can't Be Wished Away. Retrieved from www.theatlantic.com/science/archive/2016/03/psychologys-replication-crisis-cant-be-wished-away/472272/

29 Meta-Analysis

Integration of Empirical Findings through Quantitative Modeling

KRISTIAN E. MARKON

Replicability of results has become a central concern in a number of scientific fields, with empirical studies suggesting that only a fraction of initial scientific results are likely to be confirmed in subsequent replications (Munafo et al., 2017). Lack of replicability poses fundamental challenges to the scientific process, in that it is difficult to evaluate or develop theories when the empirical evidence pertaining to those theories is itself dubious.

Although the public crisis surrounding replicability is in many ways recent, inquiry into replicability and generalizability is much older, having been established with meta-analysis: statistical analysis that combines or integrates in some way the results of multiple scientific studies. Many issues considered fundamental to quantitative understanding of scientific process, such as generalizability of conclusions, publication bias and *p*-hacking, the role of power in reproducibility, and so forth, have origins in meta-analysis or became salient in scientific literature through meta-analytic findings. Meta-analysis has its origins in the early twentieth century, but began rising in popularity in the late twentieth century to become a dominant contemporary presence (see, e.g., Shadish, 2015, for historical discussions of meta-analysis).

Concerns that catalyzed interest in meta-analysis at that time, such as questions about the efficacy of psychotherapy, remain illustrative of meta-analysis as a method (Glass, 2015; Shadish, 2015). By the mid-1970s numerous studies of psychotherapy efficacy had been conducted. These studies were challenging to interpret, however, as they provided different estimates of treatment effects, possibly due to random variation across samples, differences in method, sample size, and other factors. Relatedly, reviews at the time were controversial, due to the idiosyncratic approaches of reviews and lack of a statistically rigorous framework for summarizing the literature. Meta-analysis addressed these concerns, providing a rigorous, formal, quantitative framework for integrating the results of studies, rather than relying on idiosyncratic and subjective appraisals of the literature. It allowed for pooling of different effect estimates based on smaller samples into a single estimate based on studies in aggregate, quantifying the efficacy of psychotherapy. It also allowed for formal statistical testing pertaining to more focused questions, such as questions about the relative efficacy of different treatment approaches.

This early application of meta-analysis – integration of results from randomized treatment studies – remains an important, prototypical use of meta-analysis today, but it also finds many other uses. Meta-analysis can be used to develop more accurate estimates of population trends by aggregating over individual studies; to develop complex statistical models of a phenomenon that would not be possible based on the data from any individual existing study; and to address basic questions about the scientific process, such as how different features of study quality relate to outcomes, or how representative published reports are of studies in general. It is arguably not exaggerating to say that meta-analysis has become critical to any area of study where a body of scientific results has developed, and there are questions about that body of results as a whole.

FORMS OF META-ANALYSIS AND THEIR RATIONALES

With ubiquity of shared information in numerous forms – from open data sets to various types of partial or summarized data – the difference between meta-analysis and other forms of quantitative analysis can be unclear, as can be the rationale for using one or another. Fundamentally, meta-analysis is a form of statistical analysis that quantitatively integrates results of multiple studies. It is related to, but distinct from, types of analysis such as integrative data analysis, where one integrates raw data sets of multiple studies. Intermediate between the two are analyses of summary statistics that may be nearly equivalent to the raw data, informationally speaking – as when multiple group structural equation modeling is used with covariance matrices from different studies (e.g., Krueger & Markon, 2006). Generally speaking, decisions between these options are based on availability of raw data:

integration of raw data is generally preferred, but doing so is often infeasible, and investigations suggest that meta-analysis can provide equivalent results under certain conditions (Lin & Zeng, 2010).

Why choose meta-analysis over nonquantitative alternatives? Relative to something like a systematic review, meta-analysis offers quantitative precision and formality. If a hypothesis can be stated, it often can be stated quantitatively, even if very vaguely ("less than" or "more often than"), and formalizing those quantitative statements via meta-analysis offers advantages in this regard. Formulating hypotheses meta-analytically helps clarify what constitutes sufficient evidence in favor of or against the hypotheses, and precision about what, exactly, the evidence suggests overall.

There are as many aims of meta-analysis as there are meta-analyses, but in general meta-analyses tend to have one or more of three types of goals: summative modeling, higher-order modeling, or metascientific inference. Of course, a meta-analysis can and often does pursue several of these goals simultaneously, and the goals are not entirely distinct, but they provide a rubric for understanding the aims of many meta-analyses in practice.

One goal is summative modeling, where the meta-analysis summarizes estimates from multiple studies, or to combine estimates from multiple studies to make better inferences about a population value they are all approximating. This is in many ways the canonical use of meta-analysis. Examples include estimating the magnitude of a population treatment effect from individual sample estimates, or the correlation between two variables using sample correlation estimates from multiple studies. With this type of goal, the focus of a meta-analysis is often in central tendencies of estimates cross-studies, such as mean or median effect size estimate.

A second goal is in higher-order modeling, where effect sizes are reparameterized or decomposed in terms of higher-order population parameters. In this case, the interest is not in summarizing estimates across studies, but in using those estimates to make inferences about some other model parameters of interest. Examples include meta-analytic structural equation modeling, where correlation or covariance estimates from multiple studies are used to estimate parameters of a structural equation model (SEM) or path model (e.g., Viswesvaran & Ones, 1995); or meta-analytic diffusion modeling, where signal detection theory statistics from multiple studies are reparameterized and used to estimate diffusion model parameters (Huang-Pollock et al., 2012). Many times, in these types of analyses, results from numerous studies are used to estimate models that could not be estimated within each study individually, because each study only provides partial information about the model parameters. In other cases, higher-order meta-analytic modeling is used to reframe or reconstruct information in existing estimates to provide a novel account of empirical phenomena.

A third goal of meta-analysis is in making metascientific inferences, namely, inferences about the scientific process itself, or the processes involved in conducting the original studies. A common example of this is in quantifying publication bias, or estimating the probability of replication. In such cases, the focus is on explaining the scientific process, such as in determining what factors contribute to eventual publication, how study design increases or decreases estimates of effect size, or how study characteristics are associated with more or less heterogeneity in estimates. Often meta-analyses with this type of emphasis are focused on variability or higher-order distributional characteristics of effect size estimates, such as the variance of effect size estimates, or their skew.

META-ANALYSIS AS MODELING

A meta-analysis can be thought of as a model, of the distribution of estimates (e.g., of correlations, odds ratios, or other effect sizes) across studies, or of the processes by which the studies arrived at those estimates. A meta-analytic model might be relatively flexible in its assumptions and more nonparametric (requiring estimation of secondary features of the data such as distributional shape along with primary features of interest, such as group means), or stricter and more parametric (requiring assumptions about data features rather than estimating them). Similarly, a meta-analytic model might be relatively minimalistic, concentrated on the distribution of estimates per se, or it might be broader and incorporate some causal or structural theory of the phenomena of interest. Regardless, the meta-analysis comprises a quantitative representation of studies' results, in a form that can be compared and evaluated against alternative representations.

General Features of Meta-Analytic Models

In many ways, meta-analytic models are similar to models of raw data, but with important differences. One important distinction is that meta-analytic observations – that is, study estimates – are generally not independently and identically distributed, but instead comprise a mixture of distributions, due to differences in study designs. Correlation estimates from studies differing in sample size will not be distributed the same, for example, nor will treatment effect estimates from studies using measures that differ in reliability. For this reason, meta-analytic models are generally implicitly or explicitly formulated as a mixed effect, mixture, or latent variable models (Berkey et al., 1995; Cheung, 2015; Stram, 1996; Van Houwelingen, Zwinderman, & Stijnen, 1993). In general, a typical meta-analytic model has a form resembling

$$Y_i \sim f(\alpha_i, X_i\beta, \varepsilon_i, \gamma) \qquad (29.1)$$

where Y_i is the estimate from study i (e.g., the study estimate of treatment effect), which is distributed as a function of various parameters. α_i reflects the population value of the statistic for study i not attributable to covariates – for example, the population value of Cohen's d, correlation, eta squared, or whatever the case might be, from which the value for study i is sampled. $X_i\beta$ reflects the effects of study characteristics on estimates – for example, study-level covariates X_i with covariate effects β, such as measures used, gender composition of sample, and so forth. ε_i reflects random dispersion of estimates across studies (i.e., sampling variance), and γ reflects other miscellaneous parameters in the model, such as skew due to publication bias, or other parameters constituting some theory of the phenomena.

Fixed versus Random Effects Models

A classic normal-distribution *fixed-effect* meta-analytic model assumes that the studies all estimate the same population effect size (that is, α_i all equal a single α), with no study covariates, so that

$$Y_i = \alpha + \varepsilon_i \quad (29.2)$$

and Y_i distributed normally, so $\sim N(\alpha, \sigma_{\varepsilon i}^2)$. If the studies are all estimating some mean – for example, if the purpose of the meta-analysis is to obtain norms for a cognitive test, and the studies all estimate a common mean score – then $\sigma_{\varepsilon i}^2 = \sigma_i^2/N$, the squared standard error of the mean estimate in study i. If a different effect size was being estimated, such as a correlation, then $\sigma_{\varepsilon i}^2$ would be defined differently. In a fixed effect model, the sample estimates vary due to stochastic, chance factors; aside from sample size, two studies are assumed to differ in the same way as if one of the studies were perfectly replicated over and over again, producing estimates with variance $\sigma_{\varepsilon i}^2$ across replications. Two studies are seen as replicates of some underlying population effect, and $\sigma_{\varepsilon i}^2$ is a representation of the variance associated with replication alone.

In a *random effects* meta-analysis, each study can reflect its own population effect. The population values of α_i differ, with variance σ_α^2, so that the sample estimates have total variance equal to the sum of the error variance and the variance between population values, or $\sigma_{\varepsilon i}^2 + \sigma_\alpha^2$. In a random effects model, the sample estimates randomly vary due to random variation in populations of persons, protocols, etc., as well as chance factors; two different studies are assumed to differ because of meaningful differences between them as well as because of stochastic factors. There are many possible reasons for this, such as when different studies are done in different but unspecified sociocultural groups, or use different methodologies in a way that is not captured by any study-level covariates or predictors. In random effects meta-analysis, sometimes the overall mean effect – that is, the mean of the α_i – might be of interest; in other cases, focusing on an overall mean of the α_i might obscure important heterogeneity of effects,

captured by σ_α^2 (Higgins, Thompson, & Spiegelhalter, 2009).

Estimation

Models such as those in the second equation could be estimated using maximum likelihood methods, with hypotheses regarding effect sizes tested using likelihood ratio tests, likelihood-based model fit statistics, and so forth (Hardy & Thompson, 1996; Lee et al., 2016). However, for certain cases, estimation can be simplified. For example, it can be shown that using the fixed-effect meta-analytic model in the second equation is equivalent to using a weighted mean effect size, where the effect sizes are weighted inversely to their sampling variance (Cochran, 1954; Lee et al., 2016). That is,

$$\hat{\alpha} = \frac{\sum w_i Y_i}{\sum w_i} \quad (29.3)$$

where $w_i = 1/\sigma_{\varepsilon i}^2$. A similar formula can be developed for random effects meta-analysis, with $w_i = 1/(\sigma_{\varepsilon i}^2 + \sigma_\alpha^2)$, but in that case there are advantages to using a likelihood-based approach pertaining to estimation accuracy and optimization (Hardy & Thompson, 1996).

Reparameterized Models

Note that in reparameterized or higher-order models, a meta-analytic model becomes reparameterized in terms of theoretical parameters of interest rather than the population values of the observed statistics per se. For example, the population value of a correlation between two variables itself might not be of interest; rather, what is of interest are other parameters that correlation can be used to estimate, such as direct and indirect effects of a theoretical mediation model, or parameters of a measurement model. The population correlation itself is not directly estimated; the correlation is estimated indirectly as a function of the parameters of interest.

In this sort of higher-order or reparameterized model, Equation 29.1 might be rewritten something like

$$Y_i \sim f(\alpha_i(\theta_i), X_i\beta, \varepsilon_i, \gamma) \quad (29.4)$$

where θ represents theoretical parameters in the model of interest, and α becomes a function of θ. In a meta-analytic structural equation model, for example, α might be a correlation between scales, which could be parameterized in terms of path parameters such as loadings or regression coefficients. Then, for example, in a one-factor meta-analytic SEM, an $\alpha(\theta)$ might be equal to $r_{12}(\lambda_1, \lambda_2)$, where λ_1, λ_2 are the loadings of variables 1 and 2 – the correlation between two variables r_{12} is reformulated as a function of the loadings λ_1 and λ_2. When estimating a meta-analytic model of sample correlations, the population correlations would be estimated indirectly through the loading parameters. To fit such a model, there would likely be multiple correlations, and therefore multiple α, but they would be a function of the SEM parameters. In this type of meta-analysis, the observed estimates are reformulated in terms of higher-order parameters of some theoretical model of interest.

Accounting for Estimation Error: Variance and Bias

In order to understand the role of study characteristics in overall estimation error, it is helpful to understand the bias-variance decomposition of estimation error. In general, the mean squared error of an estimate – any estimate – can be decomposed as

$$MSE = \sum(\hat{a} - a)^2 = (\bar{a} - a)^2 + \sum(\hat{a} - \bar{a})^2 \quad (29.5)$$

where the sum is taken over all the estimates $\hat{a}$ and $\bar{a}$ is the mean estimate of a (see, for example, Kass, Eden, & Brown, 2014, for an in-depth discussion). Put differently,

mean squared error = squared bias + random variance

One way to conceptualize this in the current context is that, if any given study were to be repeated over and over, the mean squared difference of an estimate from the true value can be decomposed into difference due to systematic bias, and difference due to random variability around its mean value. In a high-bias, low-variance scenario, for example, a literature might consist of reports from reasonably large studies that nevertheless are biased due to *p*-hacking or other practices. In this case, there is significant error in the reported estimates, but it is driven by systematic biases in reporting rather than random noise due to sampling – the studies' estimates may in fact appear relatively consistent with one another, but be biased away from their true value. In a low-bias, high-variance scenario, in contrast, one might have a literature characterized by reports from extremely small studies that are nevertheless conducted impartially and openly. In that case, the significant error in the estimates would be driven by random sampling variation – even without any bias, the small sample sizes of studies would lead to significant scatter around the true values.

The bias-variance decomposition is significant in meta-analyses because it provides a framework for addressing important elements of study accuracy. Within a meta-analysis, accuracy can be treated primarily in terms of the two components of bias and variance; these two components are the two major contributors to variability that should be considered when modeling or conceptualizing variability of study estimates around population values. The ways these two components contribute to meta-analytic estimates differ somewhat, however: the conditions under which bias and variance decrease with sample size differs, as do the corrections that are available for both. Depending on the scenario, for example, the impact of variance might decrease with larger studies or more studies in a meta-analysis even though bias might not.

Often in meta-analyses, error components of individual studies are modeled in terms of known relationships, such as the w_i in Equation 29.3 above, which are the inverse of the sampling variances, and thereby linked directly to the variance in Equation 29.5. In some meta-analyses, however, error components for individual studies have been formulated not in terms of variables with a known relationship with mean squared estimation error (e.g., sample size, the population value of the statistic), but in terms of more general, vague constructs, such as overall study quality. In such cases, study quality is typically coded by researchers, possibly as a sum of structured criteria, or using a rating scale (e.g., based on factors such as sample representativeness, quality of blinding, reproducibility of protocol, and so forth). The rationale behind this approach is that accuracy of a study's estimates might include factors that are unrecognized by the theory of the sampling distribution of the statistic itself, such as study design characteristics or risk of bias due to experimenter decisions.

The challenge with this approach is in finding a formal rationale for explicitly relating coded study quality to the accuracy of estimates. Studies have suggested that different protocols for coding study quality can lead to significantly different conclusions in a meta-analysis (Jüni et al., 1999), raising questions about the reliability or generalizability of study quality coding decisions. Others have noted that when the relationship between overall error and study quality is unknown, it is possible to overcorrect for either bias or variance, which might lead to increased error overall. Weighting by study quality might also correct either bias or variance at the expense of the other, such as when variance is decreased but bias is increased (Greenland & O'Rourke, 2001).

One approach is to model the sensitivity of conclusions to various specific study quality variables – this allows for examination of variables related to estimation accuracy other than those captured by sampling variance, without making implicit assumptions regarding the form of relationship between a quality variable and estimation precision. Greenland and O'Rourke (2001), for example, suggested regressing study estimates on variables hypothesized to relate to study quality, to delineate whether and how the former changes as a function of the latter. If study estimates are related to a study quality variable (for example, if double-blinded trials produced smaller effects than unblinded trials) on average, it suggests possible estimation error effects associated with that study quality variable.

Variance

As is reflected in the discussion of Equation 29.2, a study's estimation error is generally modeled through an error variance term $\sigma_{\varepsilon i}^2$, which generally depends on the study's sample size and other variables related to the statistic being modeled. The relationship between variables contributing to estimation error and the estimates themselves underlies the rationale for funnel plots, an example of which is presented for a hypothetical meta-

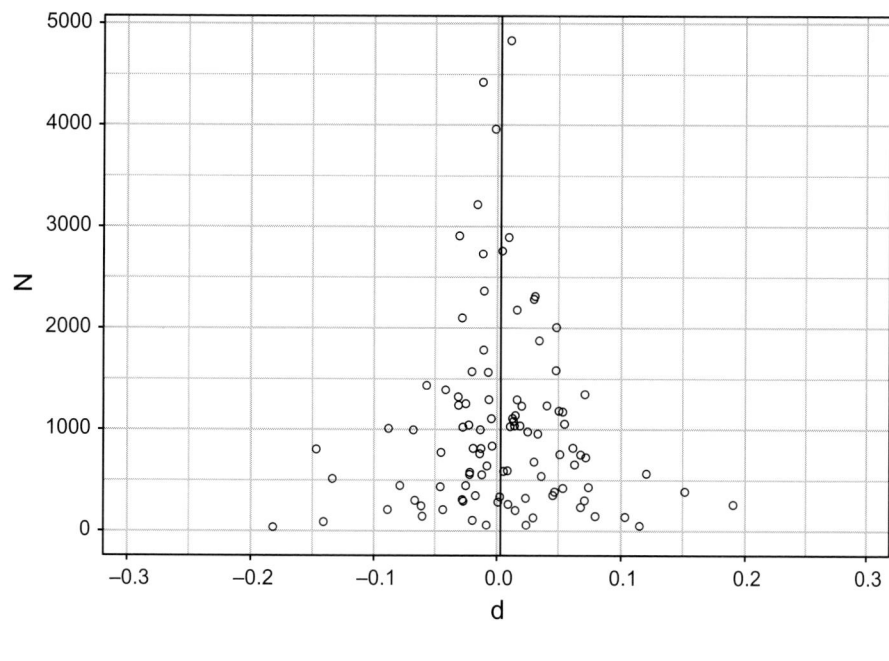

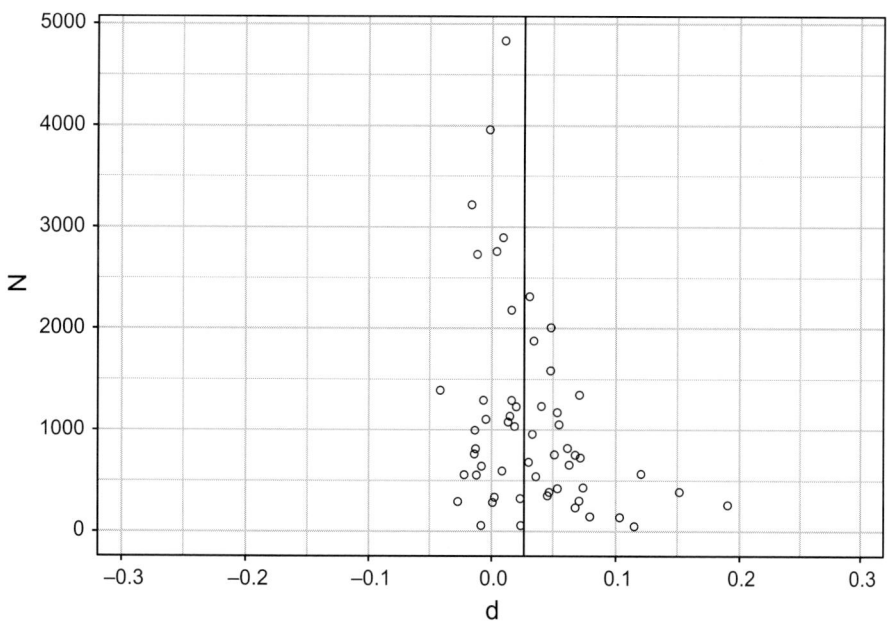

Figure 29.1 Funnel plots for fixed-effect meta-analyses of 100 hypothetical studies reporting Cohen's d values. The top part illustrates an unbiased case; the bottom part illustrates a case of bias, such as publication bias. In both cases, sample size (N) is being plotted against d, where each circle represents a study estimate. The vertical line in each figure represents the meta-analytic estimate of population value (note that the population value is 0 in both cases).

analysis at the top of Figure 29.1. The funnel plot represents a hypothetical meta-analysis of d values, as might be the case in a meta-analysis of treatment effects or group comparisons, with a population d value of 0. Without any boundaries on possible values of the statistic, its sample values are symmetrically distributed around an inferred population value. Moreover, the variance decreases around the population value as the sample size increases.

Heterogeneity When a random effects model is appropriate, some of the variability is not due to error (σ_{ei}^2), but rather variance due to between-subpopulation differences (σ_{ei}^2), such as subpopulations of persons (e.g., different sociodemographic groups) or subpopulations of study methods (e.g., different measures or experimental manipulations). One way to conceptualize this is that in a fixed effect model, in a funnel plot such as Figure 29.1, as the sample size increases, the sample estimates will theoretically get closer and closer to the population value. In a random effects model, however, they will approach some constant level of scatter around their mean, reflecting the between-subpopulation differences. This scatter does not reflect estimation error, but rather true differences among subpopulation values of the statistic. As such, it becomes important to distinguish between random error variance and true subpopulation variance in conceptualizing and inferring estimation error.

Numerous methods have been developed for estimating components of variance and decomposing variance into random error variance (σ_{ei}^2) and subpopulation variance (σ_a^2). Veroniki and colleagues (2016) provided a thorough review of many different approaches and their advantages and disadvantages, including generalized method of moment (GMM) estimators, normal-distribution maximum likelihood (ML) estimators, and Bayesian estimators. Their review highlighted areas of uncertainty in the literature, as well as the tradeoffs involved in estimation accuracy, such as balancing increases in bias against improvements in overall estimation accuracy. Normal-distribution ML estimators of variance, for example, achieve the smallest level of estimation error among the estimators they examined, and so might be preferable in that regard, but they can also be biased downward, which complicates decisions about which estimators to use. Many patterns of bias, variance, and overall estimation error had complex relationships with sample size, number of studies in the meta-analysis, and other variables, which further complicates interpretation.

Tests of Heterogeneity. Tests for heterogeneity in effect size (e.g., of the null hypothesis H_0: $\sigma_a^2 = 0$) have also received substantial attention in the literature. Most of this literature focuses on the Q statistic, which is closely related to GMM estimators of variance, and a derived statistic, I^2. Q is a weighted sum of squared differences between the sample estimates and a fixed-effect estimate of the population value, with weights equal to the expected sampling variance of each study. That is,

$$Q = \sum w_i (a_i - a_w)^2 \qquad (29.6)$$

where $w_i = 1/\sigma_{ei}^2$, the fixed effect weight, and a_w is the fixed effect estimate using those weights. When Q is large relative to what would be expected with no subpopulation variance in effect sizes, one can infer significant heterogeneity in effect size between subpopulations. I^2 can be interpreted in different ways, either as the proportion of "typical" variance due to between-subpopulation variance, or as a Q statistic scaled to the number of studies:

$$I^2 = \frac{\sigma_a^2}{\sigma_{ei}^2 + \sigma_a^2} = \frac{Q - k + 1}{Q} \qquad (29.7)$$

where k is the number of studies (e.g., Li et al., 2015).

Various lines of research have suggested that Q and I^2 are both influenced by the amount of information in a meta-analysis. As a meta-analysis increases in the number of studies and the number of participants, Q and I^2 will increase as well (e.g., Li et al., 2015), leading to false findings of heterogeneity (that is, Type I errors) as numbers of studies and participants increase. Conversely, to complicate interpretation further, in meta-analyses with a small number of small studies, Q and I^2 are often underpowered (Ioannidis, 2008), leading to failures to detect heterogeneity when it is actually present (that is, Type II errors). This raises questions about their utility in practice, if they are underpowered in small analyses and counterintuitively tend toward Type I errors as the informativeness of the meta-analysis increases beyond some ideal point. Moreover, in meta-analyses of behavioral phenomena, unless replications across studies are exact, in similar participant populations it seems theoretically reasonable to usually assume some level of heterogeneity, suggesting that tests of heterogeneity often might be unnecessary. It is arguably more important to obtain accurate estimates of the error and subpopulation variances, as these have implications for accurately modeling estimation error and making inferences about population values of statistics.

Bias

The bottom portion of Figure 29.1 illustrates a funnel plot in the presence of hypothetical publication bias. The distribution of estimates is asymmetric and positively skewed, with larger estimates being associated with numerous smaller samples, which have greater sampling variance, and fewer larger samples remaining close to the population value. This pattern represents a prototypic pattern of bias assumed to result from selective reporting of results (e.g., *p*-hacking, file drawer problems). Selective reporting is often assumed to be more feasible with smaller studies, which are presumably less expensive and more numerous than larger studies; larger studies are presumed to be more costly, and to attract more attention regardless of findings. Moreover, because of the tendency to publish statistically significant findings, small studies, which require large effects for significance, are selectively published (see Chapter 28, by Hopwood and Vazire). Overall, bias and variance are assumed to be positively correlated.

One common approach to testing for publication bias is found in regression tests (for a review, see Jin, Zhou, & He, 2015), which follow the aforementioned rationale of regressing sample estimates on specific study quality variables. In this case, the study quality variable being modeled – sample size or inverse sampling variance – is not itself the source of bias, but is rather correlated with the source. In the absence of bias, the estimates are expected to be equally distributed around a mean value regardless of sample size (Figure 29.1); in the presence of bias, estimates are expected to be larger for smaller samples or those with greater expected sampling variance. Because of this, publication bias is inferred if regression slopes or correlations relating estimates to the inverse of the sampling variance are significantly negative. In the data presented in the bottom half of Figure 29.1, for example, Kendall's tau correlation between sample size and estimate is significantly negative (−0.332, $p < 0.001$).

Various approaches to estimating and correcting for bias assume some sort of pattern in either sample estimates or their associated statistics when unbiased, and then adjust or reweight the observed data or effect size estimate to approximate this unbiased ideal pattern. In the trim-and-fill approach (Duval & Tweedie, 2000), for example, a symmetric funnel plot distribution is assumed in the unbiased ideal case; the observed, possibly asymmetric distribution of sample estimates is "filled in" with imputed values until symmetry is obtained around a trimmed estimate of the population effect size. Similarly, in p-methods approaches (van Aert, Wicherts, & van Assen, 2016), given a certain value of true effect size, significant p-values are assumed to be equally (that is, uniformly) distributed. As such, the empirical distribution of significant p-values given effect size estimates is examined, and the unbiased estimate of true effect size is taken to be that which produces as close as possible to a uniform distribution of significant p-values for the effect size estimates that were observed.

Other, more formal models of bias have also been suggested. In selection methods, for example, models in which the estimate distributions are assumed to be truncated or censored in some way due to publication bias are statistically compared to those in which the estimate distributions are uncensored without publication bias (Copas, 1999, 2013; Hedges, 1984; Iyengar & Greenhouse, 1988). Selection methods generally involve the creation of formal models of the distribution of estimates under various bias conditions, and can be estimated via likelihood-based methods, including Bayesian methods. Different selection methods differ in what is being modeled, and in the assumptions being made about the censoring process. Recent reviews have generally concluded that selection methods are preferable to other methods for addressing publication bias, and have outlined how selection methods relate to other methods for examining bias (Jin et al., 2015; McShane, Böckenholt, & Hansen, 2016).

Although helpful, concerns have also been raised about the process of modeling publication bias. Many of these concerns ultimately stem from the unknowable nature of bias: the population effect is the target of the meta-analysis, so it is difficult to know how different it is from what might be inferred from observed study results. Truncation models of population bias, for example, require assumptions about the nature of the effects being examined, so the models must be interpreted with reference to those assumptions and how plausible they might be. Various confounds can mimic the presence of publication bias in many tests and estimates of bias, which can lead to false conclusions about the existence or magnitude of bias. Certain forms of heterogeneity, for example, can lead to erroneous conclusions about publication bias, as can certain types of substantive effect patterns (Ioannidis, 2005).

From this perspective, the most defensible approach to bias might be to outline the risk of bias due to different sources, and to examine estimated levels of bias under different scenarios of varying plausibility, much as in power analyses. The Cochrane Collaboration, for example, has published a bias risk assessment guide (Higgins et al., 2011), which outlines different sources of bias that could occur at each phase of a study, from sample recruitment (e.g., various sampling biases) and assignment (e.g., in randomization procedures), to manipulation (e.g., blinding of condition), and measurement (e.g., blinded assessments), to biases in patterns of missingness (e.g., attrition bias) and reporting (e.g., selective publication of results). The group recommends systematically evaluating risk of bias due to each of these possible contributing factors, and to scale each study in the meta-analysis relative to the magnitude of each source of bias risk. Selection methods for addressing bias could be used to formalize this approach further, by way of sensitivity analyses delineating how much each source of bias could influence overall meta-analytic estimates.

ELEMENTS OF META-ANALYSIS: DECISIONS IN CONDUCTING A META-ANALYSIS

Most meta-analyses share certain elements, regardless of domain; these lead to commonly faced issues and decisions in meta-analysis across areas of inquiry. In general, these decisions pertain to how the data being analyzed is structured – what data is included or excluded, the scale of the estimates, and the reliability of coded variables. Other decisions pertain to how the data are approached in terms of analytic strategy, especially whether the data are treated as univariate or multivariate in nature.

Consensus Schemas

Although it is important to approach each meta-analysis as unique, to rigorously evaluate and address issues at hand in any specific meta-analysis, consensus schemas regarding typical expected meta-analytic elements and decisions have been developed to improve the quality and transparency of meta-analyses in general. These consensus schemas provide guidance in conducting and reporting meta-analyses, and also help users of meta-analyses point to potential biases or omissions in the literature.

The Preferred Reporting Items for Systematic Reviews and Meta-Analyses (PRISMA; www.prisma-statement.org) statement, for example, comprises a flowchart and checklist of different elements of a meta-analysis, which can be used to structure a meta-analysis as well as communicate decisions that were made. The flowchart conveys information about the sampling and selection of studies included in a meta-analysis; the checklist covers a large range of elements, from sections of a meta-analytic report, to elements of the meta-analytic methodology, such as whether or not the meta-analysis was preregistered, how sources and studies were sampled and searched, the risk of bias in sources and estimates, and how estimates are summarized or synthesized. Various extensions to the PRISMA schema have been developed for particular types of meta-analyses, such as network meta-analyses.

Even when formal consensus schemas such as PRISMA are not explicitly reported – although they often are – they still provide helpful tools for organizing, conducting, and interpreting meta-analyses, and point to issues that are typically encountered. In many ways, consensus meta-analytic schemas such as PRISMA represent basic starting points for meta-analyses, in that a typical meta-analysis will involve many more design, analytic, and communication decisions than are reflected in such resources. However, given this, it is arguably all the more important to have structured schemas to build from.

Selection and Sampling of Studies

One of the first – and in many ways, one of the most important – sets of decisions facing a meta-analyst is how to select and sample studies. This actually comprises a series of decisions in which study eligibility criteria are determined, along with methods for identifying and sampling eligible studies. Roughly, meta-analysis proceeds through a series of steps, possibly iterative at points, in which a preliminary pool of potential studies are sampled, those potential studies are screened with regard to inclusion and exclusion criteria, and a final set of studies are obtained.

Various challenges can arise in identifying an initial pool of studies, especially challenges related to maintaining representativeness of the literature. One major challenge – and a major motivation for registered replication studies – is to identify studies in such a way as to address publication biases in the literature, such as those due to file drawer problems or p-hacking, where null results are underrepresented in published outlets. Often this is addressed by searching gray literature such as dissertations, conference papers, and other works appearing outside published journals and books. Authors and researchers can also be contacted directly, to inquire about results that are referenced but unreported, or, if appropriate, for leads in identifying unpublished results.

Although often the goal in identifying studies is comprehensiveness, to include all the available literature in a meta-analysis, at times this becomes infeasible due to the size of the literature or other considerations. In such cases, eligibility criteria or the pool might be shifted, to focus on a subset of representative studies, such as studies focused on particularly prominent measures or protocols, or studies otherwise having design features that are representative of the broader literature. Another possibility is to randomly sample from identified studies, much as one treats samples of persons in primary studies as representative of a broader population.

Scaling of Estimates

As discussed, an advantage of traditional meta-analysis is in affording inferences about the generalizability of phenomena across different populations of measures, designs, and persons. The variability underlying these generalizability inferences often creates challenges in scaling different estimates to a common metric, however. Differences in measures, for example, might alter expected effect sizes due to issues related to reliability or constraints on the types of statistics that are appropriate. The choices of statistics being reported in studies themselves can represent design choices to generalize inferences across, in the sense that different statistics might have different sampling properties (e.g., variance across replications) or might represent slightly different operationalizations that might not be completely interchangeable (e.g., different forms of correlations).

Numerous sources have provided formulae for converting one statistic into another (e.g., Fritz, Morris, & Richler, 2012; Rosenberg, 2010; Ruscio, 2008), depending on the domain. These conversions sometimes require ancillary information not contained in the original statistic itself; if this information is unavailable elsewhere in reports, either directly or indirectly through inferences, it may create even greater challenges in comparing studies. For this reason, effect size statistics (e.g., r equivalent; Rosenthal & Rubin, 2003) have been proposed that are relatively general and rely only on commonly available information such as sample size. These general-application statistics are useful in providing a common metric across studies, but have also been criticized for losing information about design features of original studies that might important (e.g., in affecting power; Kraemer, 2005). In this way, the tension between preserving specific features of studies and generalizing across them remains.

At times, rescalings of estimates can have unintended consequences. Formulae for effect size corrections for unreliability (Schmidt & Hunter, 2015), for example, are based on true values of reliabilities, and often fail to recognize that different reliability estimates have different levels of bias. Coefficient a, for example, tends to be a lower-bound estimate of reliability and is biased downward (e.g., Revelle & Zinbarg, 2009). Applying formulae for unreliability that do not recognize this bias can lead to overcorrections of estimates, sometimes beyond plausible values (e.g., overcorrected correlations greater than 1).

Univariate versus Multivariate Analysis

One major analytic decision is whether to approach analyses as univariate or multivariate in nature – that is, whether to model each effect size separately, or to model multiple effect sizes simultaneously. When a single effect size is of interest, it must be treated as univariate, but when multiple effect sizes are of interest it is often possible to approach them either way.

Adopting a multivariate approach generally provides statistical and modeling advantages, in that one is able to leverage information from one set of sample estimates in modeling others. In network treatment effect meta-analyses, for example, effects of each intervention among a set are simultaneously estimated from a large set of pairwise comparisons between interventions (Lumley, 2002). That is, rather than estimating the effect of each intervention relative to a control, one at a time, the estimates of the effect of each intervention relative to one another are modeled simultaneously. This has the advantage of leveraging all the information from all the possible comparisons simultaneously, and also integrates information from multiple studies that would otherwise be treated separately. Similar advantages to multivariate analysis have been demonstrated in other areas as well, such as with meta-analytic SEM (Furlow & Beretvas, 2005); in certain cases, where a fully parameterized model is being meta-analytically estimated, a multivariate approach allows one to fit a model in the presence of missing information (e.g., Sharma, Markon, & Clark, 2014).

Although multivariate meta-analytic models often afford numerous advantages, they also can be more complex to implement. This is especially true with regard to inferences about the meta-analytic estimates, in that the uncertainty intervals around the estimates often involve calculating a sampling variance–covariance matrix, rather than a vector of variances per se. For example, univariate meta-analysis of five correlations involves understanding the sampling variance of each of the five correlations independently; in a multivariate meta-analysis, however, the joint sampling properties of the correlations might be considered, which means understanding the five sampling variances, as well as the 10 sampling covariances (the covariances of estimates across samples, or how an estimate of one correlation covaries with an estimate of a different correlation). Sometimes these variance–covariance matrices require additional information than is typically reported in any individual study (e.g., correlations between measures involved), which then in turn can require collection of additional data. Multivariate meta-analytic methods are becoming more accessible, but their advantages will have to be weighed against their feasibility in any given application.

Coding and Reliability of Meta-Analytic Variables

Another important consideration is which potential moderators (e.g., sample composition) and other variables to code as part of a meta-analysis, and how. Which variables should be coded depends heavily on the meta-analysis, but some common variables to consider coding are study-identifying (e.g., authors, year, journal) and sociodemographic variables (e.g., average age, variance in age, gender composition, ethnic composition, language), other sample characteristics (e.g., clinical versus community), measure characteristics, and statistic type. Potential bias-associated variables (such as preregistration status or data availability; see Center for Open Science, 2018, for examples) or variance-inducing variables (design quality variables not related to bias per se) are also important to consider. It is often beneficial to be overly broad in the set of variables to be coded, in that the process of coding can be expensive either financially or in time or effort; returning to a set of papers to recode additional variables is akin to collecting follow-up data from a sample.

Although it might seem obvious, it is important to consider what constitutes a record. For example, many studies report results for multiple samples, suggesting that it might be most efficient to code based on sample or estimate, rather than study, as many studies have multiple samples, and many samples involve multiple estimates. Decisions regarding the nature of a record will necessarily involve other decisions, such as whether to conduct analyses as univariate or multivariate in nature.

Although some variables are reasonably straightforward to code (e.g., gender composition of sample), other variables may be less so (e.g., sample type). For this reason, it is worth considering reliability analyses to demonstrate the accuracy and generalizability of the coded variables. Even when variables seem straightforward, double-coding allows for identification of possible errors.

The Nature of Replications in a Meta-Analysis: Control versus Generalizability

Although meta-analyses traditionally have often been conducted on results from studies that existed prior to the initiation of the meta-analysis, this is increasingly not always the case. In a registered replication study – where replication attempts are publicly registered a priori, by registering a description of the intended study and analysis plans, along with predictions, before the study is ever conducted – a meta-analysis is generally conceived of as part of the study's design. In this scenario, there is an a priori stated attempt to replicate a finding, often across multiple samples by multiple research groups, and meta-analysis is often used to integrate results. Registered replication affords more control over replications than in traditional meta-analysis of preexisting studies, and because the attempts are known a priori, it better avoids difficulties such as the file-drawer problem or p-hacking, where study results are withheld from publication based on their significance, contradiction of a priori hypotheses, or for other reasons.

Although registered replication studies have benefits, it is not always feasible to organize them. Even when it is, the resources available for a registered replication initiative might not match those that arise from unanticipated opportunities, such as previously unrecognized data sets or funding opportunities. Moreover, although they provide insight into the replicability of findings in a stochastic sense, on their own they do not necessarily address concerns about processes "in the wild," in terms of understanding how, outside of a registered replication context, published research results vary and why. Presumably, the processes involved in designing, implementing, and communicating novel studies, or, more broadly, studies conducted outside the framework of a replicated pre-registration paradigm, are different from those conducted within a registered replication framework. It might even be argued that registered replication studies are subject to their own biases, like all research. Regardless, the registered replication framework is arguably different from many scientific processes; understanding the latter likely requires traditional, post-hoc meta-analysis at least in part.

These differences parallel a broader debate in the metascientific literature about direct versus conceptual replications (e.g., Zwan et al., 2018), with direct replications being focused on replicating effects under conditions that are as similar as possible, and conceptual replications focused on replicating effects under conditions that are deemed theoretically central. Direct replications prevent changes in analysis or design that afford capitalization on chance; conceptual replications afford opportunities to examine how estimates generalize or not across variations in design or other variables.

Although theoretically useful, the distinction between direct and conceptual replications might be moot, in that both are important to understanding a phenomenon. The combination of these types of replications allow one to better parse variability in study results into components due to pure randomness and chance, and those due to systematic moderators that vary across studies. This is a central question in meta-analysis, as it helps us understand the boundaries of a phenomenon, and one that is best addressed using a combination of both approaches to replication.

At some level, the distinctions between registered and unregistered, and direct and conceptual replication, reflect different paradigms valuing control and generalizability, respectively. There is no reason why a conceptual replication couldn't be registered, or a direct replication unregistered, but the different positions on these issues do generally reflect different perspectives on the advantages and disadvantages of homogeneity and flexibility during the process of replication. In this regard, advantages and disadvantages of registered replications and traditional meta-analysis parallel those of experimental and observational studies, with the former approaches offering more control, and the latter offering more generalizability.

An interesting question moving forward is how to integrate findings from meta-analyses of registered versus unregistered replications, and whether differential results between the two approaches can be used to improve meta-analytic methodology in general. For example, an initial meta-analysis of the ego-depletion effect – which posits that self-control is a limited resource that can be depleted, leading to subsequent self-control failures – found evidence for a moderate effect (Hagger et al., 2010). However, when publication bias was modeled in a later meta-analysis, the effect was found to be nonsignificant (Carter et al., 2015). In part to reconcile these conclusions, a later registered replication study was conducted, which again found a nonsignificant effect (Hagger et al., 2016). Is it possible to use cases such as these to calibrate meta-analytic methodology, such as tests of publication bias? What is the best way to approach registered and unregistered replications in this context?

AREAS FOR DEVELOPMENT IN META-ANALYSIS

Meta-analytic methodology is well-established, having been developed and refined greatly over the last few decades. Nevertheless, there are important areas for further development, some of which pertain to lingering challenges in the field, and some of which pertain to changes in the landscape of science and statistical applications.

Nonparametric Approaches to Meta-Analysis

An implicit assumption underlying most meta-analytic approaches is that given a reported statistic – a correlation, eta squared, odds ratio, or whatever the case might be – the sampling distribution of that statistic is known and can be used to account for the statistic's bias and variance in deriving an overall meta-analytic estimate. This is a reasonable assumption, given that the theoretical sampling distributions of commonly reported statistics are generally known.

Although reasonable under most circumstances, the assumption that the sampling distribution is known can be questionable in many meta-analytic scenarios. In some cases, the estimated sampling distribution of a statistic might not reduce to an analytic formula, or might otherwise be infeasible to calculate. In other cases, the sampling distribution is known, but depends on information that is not often reported. The sampling variance of kappa, for example, is known, but depends on cell counts that are frequently unreported in practice (Fleiss, Cohen, & Everitt, 1969).

In still other cases, it seems reasonable to question whether the theoretical sampling distribution of a statistic under idealized scenarios reflects the actual sampling distribution under actual conditions as they occur. Substantial evidence indicates that actual empirical sampling distributions often deviate from theoretical expectations due to various biases in study design and implementation; it seems reasonable to argue that sampling distributions might deviate from theoretical expectations in other ways as well, such as in increases or decreases of variability due to unrecognized dependencies or other factors.

Considerations such as this suggest a need for greater focus on nonparametric approaches to meta-analysis, which would avoid assumptions about the distribution of statistics and their sampling distributions. Work in this area has tended to focus on nonparametric effect size metrics (e.g., Kraemer & Andrews, 1982; Ruscio, 2008) or nonparametric approaches to modeling the distribution or moderation of random effects (Switzer, Paese, & Drasgow, 1992; Van Houwelingen et al., 1993), with less work on nonparametric approaches to stochastic error and overall model misspecification. As observations in their own right, primary study results could be treated within many nonparametric frameworks in the same way as any other observations. Presumably, some loss of statistical power would follow from not incorporating information about sampling distributions into a model – studies with lower sampling error wouldn't be weighted more than studies with greater sampling error – but assuming lack of bias (which is often also the case in parametric meta-analysis) such an approach would still have greater power than any one of the primary studies alone.

An interesting question is whether or not sampling distributions can be estimated from observed studies nonparametrically – that is, rather than assuming the statistics have a certain distribution (for example, a normal distribution), or that the variability of the statistics relates to sample size in a certain way, could the relationship between a statistic's distribution and sample size be estimated? Many variables related to sampling variance, such as sample size, are generally reported, raising the possibility that the sampling distribution might be modeled directly as part of the meta-analysis. Questions might be raised about possible unrepresentativeness of studies due to bias and other factors, but it may be that these problems can be addressed analogously to how they are addressed in parametric meta-analyses, possibly by drawing on what is known about the sampling distribution of estimates (e.g., boundary values). Moreover, even if the sampling distribution itself cannot be estimated, it may be that lower or upper bounds could be formulated (Park, Serpedin, & Qaraqe, 2013; Stein & Nossek, 2017), leading to robust meta-analytic inferences.

Meta-Analysis of Bayesian Estimates

With advances in computational statistics, adoption of Bayesian inference has become increasingly common. As the adoption of Bayesian inference grows, there will be increasing need for the development of methods for integrating Bayesian estimates meta-analytically. Bayesian meta-analytic methods have been developed substantially (Bodnar et al., 2017; Smith, Spiegelhalter, & Thomas, 1995), and have seen substantial application, but important questions and issues remain. Many Bayesian treatments implicitly or explicitly assume frequentist or point estimates from primary studies, rather than Bayesian estimates, or at least assume that the two types of estimates are interchangeable; fewer treatments have focused on estimates that are Bayesian themselves.

For instance, in primary studies, full posterior distributions are often not explicitly reported; more common is to report the maximum, expected, or median posterior values, possibly together with a posterior variance. This raises the question of how to integrate Bayesian estimates when the full posterior distributions are missing, and whether or when there is any meaningful disadvantage to relying on point estimates or partial posterior information in meta-analytic integration of estimates, relative to using the full posteriors. In some cases, posterior densities are analytically known or could be derivable, even if they are not reported, but this is not always the case. Moreover, a Bayesian version of the central limit theorem (Clarke, 1999) suggests using normal distributions as approximations when the posteriors are unknown, but in general, how best to meta-analytically integrate, within a Bayesian framework, partial information on estimates is not entirely clear (Xu & Raginsky, 2017, provide a mathematical treatment of related issues).

Another set of related questions pertains to how to meta-analytically model the influence of prior distributions. Unlike frequentist estimates, which depend largely on the design, data, and model, Bayesian estimates are also influenced by prior distributions (Bickel & Blackwell, 1967; Blackwell & Girshick, 1954; Noorbaloochi & Meeden, 1983). As such, it seems reasonable to quantify the extent to which differences in estimates among studies are due to different prior assumptions. Two different analyses, using the same data and same model, might produce different estimates because of differing priors. Methods for quantifying the contribution of priors to posterior variance have been developed (Gustafson & Clarke, 2004); these could be applied to meta-analysis to improve models of sample estimates and decompose sources of heterogeneity in Bayesian estimates.

An example of some of the issues involved can be found in a series of articles about application of Bayesian meta-analysis to the study of social priming. Scheibehenne, Jamil, and Wagenmakers (2016) applied a relatively standard Bayesian rationale for integration of information, in which posterior estimates from one study are sequentially used as priors in the next analysis. In doing so, they concluded that the literature provides strong support for the priming effect examined. In a commentary, however, Carlsson and colleagues (2017) approached the same set of data and, using a Bayesian version of traditional random-effects meta-analysis, argued that evidence in support of the effect was inconclusive and sensitive to choice of priors. Further research is needed to resolve these types of issues, and to more fully characterize meta-analytic estimation and inference in a Bayesian context.

SUMMARY

Meta-analysis has become essential to modern scientific discourse, allowing for rigorous, quantitative summaries of empirical findings, and metascientific examination of the scientific process itself. Meta-analysis is broad in scope, and ultimately comprises an analytic paradigm or focus, rather than a specific method per se: there are myriad approaches to meta-analysis depending on the scenario, such as what types of results are reported in a domain, how they report the heterogeneity in those findings, the types of meta-analytic or metascientific questions at hand, and the statistical philosophy of the meta-analyst. The methods introduced here hopefully provide a springboard from which many of these issues can be approached in a particular question domain.

REFERENCES

Berkey, C. S., Hoaglin, D. C., Mosteller, F., & Colditz, G. A. (1995). A Random-Effects Regression Model for Meta-Analysis. *Statistics in Medicine*, *14*(4), 395–411.

Bickel, P. J., & Blackwell, D. (1967). A Note on Bayes Estimates. *Annals of Mathematical Statistics*, *38*, 1907–1911.

Blackwell, D., & Girshick, M. A. (1954). *Theory of Games and Statistical Decisions*. New York: John Wiley.

Bodnar, O., Link, A., Arendacká, B., Possolo, A., & Elster, C. (2017). Bayesian Estimation in Random Effects Meta-Analysis Using a Non-informative Prior. *Statistics in Medicine*, *36*(2), 378–399.

Carlsson, R., Schimmack, U., Williams, D. R., & Bürkner, P.-C. (2017). Bayes Factors from Pooled Data Are no Substitute for Bayesian Meta-Analysis: Commentary on Scheibehenne, Jamil, and Wagenmakers (2016). *Psychological Science*, *28*(11), 1694–1697.

Carter, E. C., Kofler, L. M., Forster, D. E., & McCullough, M. E. (2015). A Series of Meta-Analytic Tests of the Depletion Effect: Self-Control Does Not Seem to Rely on a Limited Resource. *Journal of Experimental Psychology: General*, *144*, 796–815.

Center for Open Science. (2018). *Transparency and Openness Promotion Guidelines*. Retrieved from https://cos.io/our-services/top-guidelines/

Cheung, M. (2015). *Meta-Analysis: A Structural Equation Modeling Approach*. Chichester: Wiley.

Clarke, B. S. (1999). Asymptotic Normality of the Posterior in Relative Entropy. *IEEE Transactions on Information Theory*, *45*(1), 165–176.

Cochran, W. G. (1954). The Combination of Estimates from Different Experiments. *Biometrics*, *10*(1), 101–129.

Copas, J. B. (1999). What Works? Selectivity Models and Meta-Analysis. *Journal of the Royal Statistical Society: Series A (Statistics in Society)*, *162*, 95–109.

Copas, J. B. (2013). A Likelihood-Based Sensitivity Analysis for Publication Bias in Meta-Analysis. *Journal of the Royal Statistical Society: Series C (Applied Statistics)*, *62*, 47–66.

Duval, S., & Tweedie, R. (2000). Trim and Fill: A Simple Funnel-Plot-Based Method of Testing and Adjusting for Publication Bias in Meta-Analysis. *Biometrics*, *56*, 455–463.

Fleiss, J. L., Cohen, J., & Everitt, B. S. (1969). Large Sample Standard Errors of Kappa and Weighted Kappa. *Psychological Bulletin*, *72*(5), 323–327.

Fritz, C. O., Morris, P. E., & Richler, J. J. (2012). Effect Size Estimates: Current Use, Calculations, and Interpretation. *Journal of Experimental Psychology: General*, *141*(1), 2–18.

Furlow, C. F., & Beretvas, S. N. (2005). Meta-Analytic Methods of Pooling Correlation Matrices for Structural Equation Modeling under Different Patterns of Missing Data. *Psychological Methods*, *10*(2), 227–254.

Glass, G. V. (2015). Meta-Analysis at Middle Age: A Personal History. *Research Synthesis Methods*, *6*(3), 221–231.

Greenland, S., & O'Rourke, K. (2001). On the Bias Produced by Quality Scores in Meta-Analysis, and a Hierarchical View of Proposed Solutions. *Biostatistics*, *2*(4), 463–471.

Gustafson, P., & Clarke, B. (2004). Decomposing Posterior Variance. *Journal of Statistical Planning and Inference*, *119*(2), 311–327.

Hagger, M. S., Chatzisarantis, N. L. D., Alberts, H., Anggono, C. O., Batailler, C., Birt, A. R., ... Zwienenberg, M. (2016). A Multilab Preregistered Replication of the ego-Depletion Effect. *Perspectives on Psychological Science*, *11*(4), 546–573.

Hagger, M. S., Wood, C., Stiff, C., & Chatzisarantis, N. L. D. (2010). Ego Depletion and the Strength Model of Self-Control: A Meta-Analysis. *Psychological Bulletin*, *136*, 495–525.

Hardy, R. J., & Thompson, S. G. (1996). A Likelihood Approach to Meta-Analysis with Random Effects. *Statistics in Medicine*, *15*(6), 619–629.

Hedges, L. V. (1984). Estimation of Effect Size under Nonrandom Sampling: The Effects of Censoring Studies Yielding Statistically Insignificant Mean Differences. *Journal of Educational and Behavioral Statistics*, *9*, 61–85.

Higgins, J. P. T., Altman, D. G., Gøtzsche, P. C., Jüni, P., Moher, D., Oxman, A. D., ... Cochrane Statistical Methods Group (2011). The Cochrane Collaboration's Tool for Assessing Risk of Bias in Randomized Trials. *British Medical Journal*, *343*, d5928.

Higgins, J. P. T., Thompson, S. G., & Spiegelhalter, D. J. (2009). A Re-evaluation of Random-Effects Meta-Analysis. *Journal of the Royal Statistical Society: Series A (Statistics in Society)*, *172*(1), 137–159.

Huang-Pollock, C. L., Karalunas, S. L., Tam, H., & Moore, A. N. (2012). Evaluating Vigilance Deficits in ADHD: A Meta-Analysis of CPT Performance. *Journal of Abnormal Psychology*, *121*(2), 360–371.

Ioannidis, J. P. A. (2005). Differentiating Biases from Genuine Heterogeneity: Distinguishing Artifactual from Substantive Effects. In H. Rothstein, A. J. Sutton, & M. Borestein (Eds.), *Publication Bias in Meta-Analysis: Prevention, Assessment and Adjustments* (pp. 287–302). Chichester: Wiley.

Ioannidis, J. P. A. (2008). Interpretation of Tests of Heterogeneity and Bias in Meta-Analysis. *Journal of Evaluation in Clinical Practice*, *14*(5), 951–957.

Iyengar, S., & Greenhouse, J. B. (1988). Selection Models and the File Drawer Problem. *Statistical Science*, *3*(1), 109–117.

Jin, Z.-C., Zhou, X.-H., & He, J. (2015). Statistical Methods for Dealing with Publication Bias in Meta-Analysis. *Statistics in Medicine*, *34*(2), 343–360.

Jüni, P., Witschi, A., Bloch, R., & Egger, M. (1999). The Hazards of Scoring the Quality of Clinical Trials for Meta-Analysis. *Journal of the American Medical Association*, *282*(11), 1054–1060.

Kass, R. E., Eden, U. T., & Brown, E. N. (2014). *Analysis of Neural Data*. New York: Springer.

Kraemer, H. C. (2005). A Simple Effect Size Indicator for Two-Group Comparisons? A Comment on Requivalent. *Psychological Methods*, *10*(4), 413–419.

Kraemer, H. C., & Andrews, G. (1982). A Nonparametric Technique for Meta-Analysis Effect Size Calculation. *Psychological Bulletin*, *91*(2), 404–412.

Krueger, R. F., & Markon, K. E. (2006). Reinterpreting Comorbidity: A Model-Based Approach to Understanding and Classifying Psychopathology. *Annual Review of Clinical Psychology, 2*(1), 111–133.

Lee, C. H., Cook, S., Lee, J. S., & Han, B. (2016). Comparison of Two Meta-Analysis Methods: Inverse-Variance-Weighted Average and Weighted Sum of z-Scores. *Genomics & Informatics, 14*(4), 173–180.

Li, S., Jiang, H., Yang, H., Chen, W., Peng, J., Sun, M., ... Zeng, J. (2015). The Dilemma of Heterogeneity Tests in Meta-Analysis: A Challenge from a Simulation Study. *PLOS ONE, 10*(5), e0127538.

Lin, D. Y., & Zeng, D. (2010). On the Relative Efficiency of Using Summary Statistics versus Individual-Level Data in Meta-Analysis. *Biometrika, 97*(2), 321–332.

Lumley, T. (2002). Network Meta-Analysis for Indirect Treatment Comparisons. *Statistics in Medicine, 21*(16), 2313–2324.

McShane, B. B., Böckenholt, U., & Hansen, K. T. (2016). Adjusting for Publication Bias in Meta-Analysis: An Evaluation of Selection Methods and Some Cautionary Notes. *Perspectives on Psychological Science, 11*(5), 730–749.

Munafo, M. R., Nosek, B. A., Bishop, D. V. M., Button, K. S., Chambers, C. D., Percie du Sert, N., ... Ioannidis, J. P. A. (2017). A Manifesto for Reproducible Science. *Nature Human Behaviour, 1*(1), 0021.

Noorbaloochi, S., & Meeden, G. (1983). Unbiasedness as the Dual of Being Bayes. *Journal of the American Statistical Association, 78*, 619–623.

Park, S., Serpedin, E., & Qaraqe, K. (2013). Gaussian Assumption: The Least Favorable but the Most Useful. *IEEE Signal Processing Magazine, 30*(3), 183–186.

Revelle, W., & Zinbarg, R. E. (2009). Coefficients Alpha, Beta, Omega, and the GLB: Comments on Sijtsma. *Psychometrika, 74*(1), 145–154.

Rosenberg, M. S. (2010). A Generalized Formula for Converting Chi-Square Tests to Effect Sizes for Meta-Analysis. *PLoS ONE, 5*(4), e10059.

Rosenthal, R., & Rubin, D. B. (2003). Requivalent: A Simple Effect Size Indicator. *Psychological Methods, 8*(4), 492–496.

Ruscio, J. (2008). A Probability-Based Measure of Effect Size: Robustness to Base Rates and Other Factors. *Psychological Methods, 13*(1), 19–30.

Scheibehenne, B., Jamil, T., & Wagenmakers, E.-J. (2016). Bayesian Evidence Synthesis Can Reconcile Seemingly Inconsistent Results: The Case of Hotel Towel Reuse. *Psychological Science, 27*(7), 1043–1046.

Schmidt, F. L., & Hunter, J. E. (2015). *Methods of Meta-Analysis: Correcting Error and Bias in Research Findings*. Thousand Oaks, CA: Sage.

Shadish, W. R. (2015). Introduction to the Special Issue on the Origins of Modern Meta-Analysis: Introduction to Special Issue. *Research Synthesis Methods, 6*(3), 219–220.

Sharma, L., Markon, K. E., & Clark, L. A. (2014). Toward a Theory of Distinct Types of "Impulsive" Behaviors: A Meta-Analysis of Self-Report and Behavioral Measures. *Psychological Bulletin, 140*(2), 374–408.

Smith, T. C., Spiegelhalter, D. J., & Thomas, A. (1995). Bayesian Approaches to Random-Effects Meta-Analysis: A Comparative Study. *Statistics in Medicine, 14*(24), 2685–2699.

Stein, M. S., & Nossek, J. A. (2017). A Pessimistic Approximation for the Fisher Information Measure. *IEEE Transactions on Signal Processing, 65*(2), 386–396.

Stram, D. O. (1996). Meta-Analysis of Published Data Using a Linear Mixed-Effects Model. *Biometrics, 52*, 536–544.

Switzer, F. S., Paese, P. W., & Drasgow, F. (1992). Bootstrap Estimates of Standard Errors in Validity Generalization. *Journal of Applied Psychology, 77*(2), 123–129.

Van Aert, R. C. M., Wicherts, J. M., & van Assen, M. A. L. M. (2016). Conducting Meta-Analyses Based on p Values: Reservations and Recommendations for Applying p-Uniform and p-Curve. *Perspectives on Psychological Science, 11*(5), 713–729.

Van Houwelingen, H. C., Zwinderman, K. H., & Stijnen, T. (1993). A Bivariate Approach to Meta-Analysis. *Statistics in Medicine, 12*(24), 2273–2284.

Veroniki, A. A., Jackson, D., Viechtbauer, W., Bender, R., Bowden, J., Knapp, G., ... Salanti, G. (2016). Methods to Estimate the Between-Study Variance and Its Uncertainty in Meta-Analysis. *Research Synthesis Methods, 7*(1), 55–79.

Viswesvaran, C., & Ones, D. S. (1995). Theory Testing: Combining Psychometric Meta-Analysis and Structural Equations Modeling. *Personnel Psychology, 48*(4), 865–885.

Xu, A., & Raginsky, M. (2017). Information-Theoretic Lower Bounds on Bayes Risk in Decentralized Estimation. *IEEE Transactions on Information Theory, 63*(3), 1580–1600.

Zwaan, R., Etz, A., Lucas, R., & Donnellan, M. (2018). Making Replication Mainstream. *Behavioral and Brain Sciences, 41*, e120.

30 Mediation, Moderation, and Conditional Process Analysis

Regression-Based Approaches for Clinical Research

NICHOLAS J. ROCKWOOD AND ANDREW F. HAYES

Establishing causal effects is the focus of much clinical psychology research, such as identifying if a new substance abuse treatment program is more effective than treatment as usual, or if the experience of certain psychological symptoms results in certain negative physical health outcomes. But merely identifying the existence of such effects is not sufficient. Knowing that, for instance, some treatment is more effective than another – a question of *whether* or *if* – raises the question of *how* or *why* this effect exists or operates. For example, perhaps a treatment for alcohol abuse reduces substance cravings which translate to lower use, or maybe it encourages patients to seek support from others, who in turn discourage use or help a person find alternative means of dealing with their anxieties, depression, or other factors that give rise to substance use.

Several clinical researchers have observed the importance and advantages of understanding how causal effects operate. In the context of treatments and interventions, these advantages include the ability to target key components of an intervention while placing less emphasis on ineffective components to enhance the treatment as a whole, as well as better understanding of the underlying psychological disorders addressed by an effective treatment (Frazier, Tix, & Barron, 2004; Kazdin, 2007; Kraemer et al., 2002; Magill, 2011; Onken et al., 2014; Schmidt & Schimmelmann, 2015; Windgassen et al., 2016). Furthermore, understanding the nature of effects is important for theory building and testing competing theories of psychological phenomena and experiences.

Seemingly related questions, but conceptually distinct from how an effect operates, are *for whom* or *when* does the effect operate or exist and for whom or when does it not. Is the treatment effective for everyone? Most likely not. Maybe it is effective for people with addictions to certain classes of drugs, but not for people addicted to other kinds of drugs. Or perhaps the effectiveness of the treatment depends on the amount of emotional support that the patient receives. Individual and clinician characteristics are also likely to influence the effectiveness of a particular treatment. Understanding these boundary conditions, enhancers, and inhibitors of an effect allows treatments to be tailored to different people, or groups of people, in an effort to maximize treatment effectiveness and reduce wasted resources (Finney, 1995; Magill, 2011; Norcross & Wampold, 2011). These questions address the role of individual differences and situational contexts in their impact on causal effects of interest, which can also further help test and build theory.

The questions of *how* and *why* refer to identifying or testing the mechanism, or process, that underlies an effect, whereas the questions of *for whom* and *when* refer to identifying the contingencies of or conditions on that effect. When combined with strong theory, *statistical mediation analysis* and *statistical moderation analysis* can be used to explore or test hypotheses about the mechanisms and contingencies of effects, respectively.

Exemplars of mediation and moderation analysis are in abundance in published clinical research, yet many researchers still struggle with the distinction between them, both conceptually and statistically. The purpose of this chapter is to provide an overview of mediation and moderation analysis, mapping the concepts to statistical models. Further, we describe the integration of mediation and moderation into a unifying framework known as *conditional process analysis* (Hayes, 2018). After detailing the framework, we present an example of a conditional process analysis, with the hope that it will provide researchers insight into how to construct and test a conditional process model related to their own research questions and hypotheses of interest.

This is a vast topic, and the methodology literature in this area is massive, so it is necessary for us to focus our discussion, lest it be too vague to be useful. Throughout the chapter, we focus on the analysis of continuous outcomes that can be estimated with linear regression analysis using the least squares criterion for optimizing fit of a model. Most clinical researchers are familiar with least squares regression analysis, and it is built into most statistical software and is a staple of graduate training in the social and health sciences. Linear models are most commonly used when answering questions about mediation and moderation, and understanding the concepts and statistical methods in this framework provides a nice foundation to build upon. We also limit our discussion to scenarios involving dichotomous (e.g., a single treatment versus a control condition, as in a simple clinical trial) or continuous independent and moderator variables. Readers interested in mediation or moderation models involving multicategorical independent variables can consult Hayes and Preacher (2014), Hayes and Montoya (2017), and Hayes (2018). Although our discussion is limited to studies with these data properties, many of the concepts introduced here can be extended to more complex types of data and research designs with alterations, sometimes minor, sometimes more significant.

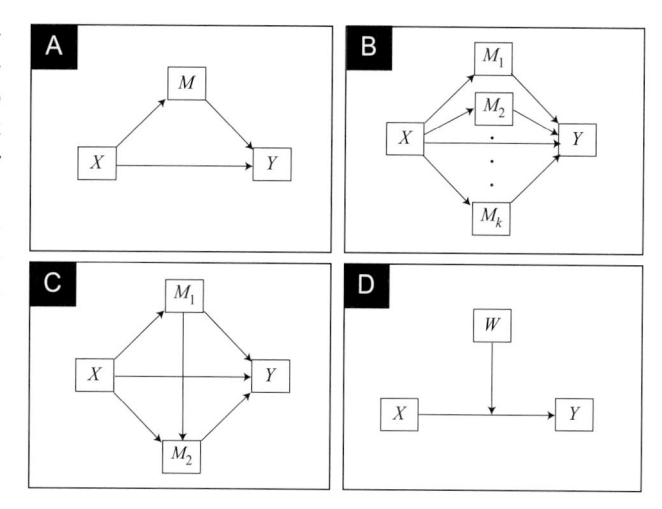

Figure 30.1 A simple mediation model (panel A), parallel multiple mediator model (panel B), serial multiple mediator model (panel C) and a conceptual representation of moderation (panel D)

MEDIATION

Statistical mediation analysis is frequently used to test hypotheses about the process by which one variable X causally transmits its effect on another variable Y through one or more intervening or *mediator* variables M. A mediation process takes the form of a train of causal events of the form "X influences M, which, in turn, influences Y." The purpose of mediation analysis is to identify and explain *how* the effect of X on Y operates. Several clinical researchers have implemented mediation analysis to assess the mechanism by which an effect operates. For example, Mason and colleagues (2015) found that relative to participants assigned to a control condition, participants in a text-based smoking-cessation intervention program had enhanced feelings of readiness to stop smoking, which was in turn related to fewer cigarettes smoked during a 30-day period. And Duchesne and colleagues (2017) identified self-esteem as a mediator of the effect of body dissatisfaction on depression and anxiety in adolescents. Higher body dissatisfaction was associated with lower self-esteem, which resulted in higher depression and anxiety.

THE SIMPLE MEDIATION MODEL

When M and Y are continuous, most researchers interested in mediation begin an analysis by estimating the following equation:

$$Y = i_Y + cX + e_Y \quad (30.1)$$

where Y is the ultimate outcome of interest (e.g., symptoms experienced, frequency of substance use, etc.) and X is the presumed cause (a new treatment versus therapy as usual, or some continuous individual difference) of differences between people on Y. This is a simple linear regression model, where i_Y is the intercept, e_Y is the error term, and c represents the association between X and Y. Specifically, c estimates how much two cases that differ by one unit on X differ on Y. Whether X is continuous or dichotomous, this interpretation does not change. But when X is dichotomous and the two categories are coded one unit apart (e.g., therapy as usual = 0, new treatment = 1), c is the mean difference in Y between the two groups. In the parlance of mediation analysis, c is referred to as the *total effect* of X.

Although Equation 30.1 is often of substantive interest to researchers, in mediation analysis it plays only a peripheral role in interpretation. The purpose of mediation analysis is to better understand the causal process *by which*, or *how*, X affects Y through a sequence of causal events in which X influences at least one mediator variable M which, in turn, causally influences Y. Some simple rules of path analysis algebra tell us that the total effect of X, c in Equation 30.1, can be partitioned into a *direct effect* and an *indirect effect* of X, estimated using a set of two regression models

$$M = i_M + aX + e_M \quad (30.2)$$

$$Y = i_Y + c'X + bM + e_Y \quad (30.3)$$

Figure 30.1, panel A, displays a path diagram corresponding to this model. The direct effect, represented by c' in Equation 30.3, quantifies the expected difference in Y between two units that differ by one unit in X, *but have the same value for M*. Put another way, it is the effect of X on Y while statistically controlling for, or holding constant, M. Thus, it is X's effect on Y that is independent of M. Referring back to the dichotomous X example (therapy as usual = 0, new treatment = 1), the direct effect is an *adjusted mean difference* on Y between the groups. It is an estimate of the difference between the groups on Y, on average, if they were equal on the proposed mediator.

The coefficient a in Equation 30.2 quantifies the effect of X on M, and the coefficient b in Equation 30.3 quantifies the effect of M on Y statistically controlling for X. The indirect effect of X on Y is the product of these effects. Specifically, it is the product of a and b, ab, and it represents the expected difference in Y for two cases that differ by one unit on X due to X's effect on M, which in turn affects Y. Thus, ab is a quantification of the mediation process. When Equations 30.1–30.3 are estimated using least squares regression analysis, the sum of the indirect and direct effects is equal to the total effect. That is, $ab + c' = c$. Because the indirect effect is the product of a and b, a positive indirect effect can correspond to a and b values that are both positive or both negative, and a negative indirect effect corresponds to a and b values that are opposite in sign. Therefore, the signs of a and b are important to interpretation of the sign of the indirect effect.

Inference about the Indirect Effect

Since the indirect effect quantifies the mediation process, an inferential test of the indirect effect should be undertaken before concluding whether a pattern of relationships is consistent with a mediation process. Yet because the indirect effect is the product of the coefficients a and b, inference about the indirect effect is not as straightforward as it is for such statistics as a difference between two means or a correlation coefficient. Simple, though now outdated, methods for making inferences about the indirect effect include those based on the casual steps criteria and its variants (Baron & Kenny, 1986; Kraemer et al., 2008) and the Sobel test (Sobel, 1982). These methods are limited by their reliance on a series of significance tests based on the individual a and b coefficients (rather than ab, which is what actually quantifies what matters in a mediation analysis) or an assumption that the sampling distribution of ab is normal or symmetrical, which typically is not the case (Aroian, 1947). Today, many methodologists recommend using *bootstrap confidence intervals* for inference about the indirect effect (Hayes, 2018; Hayes & Scharkow, 2013; MacKinnon, Lockwood, & Williams, 2004; Preacher & Selig, 2012; Shrout & Bolger, 2002) as bootstrapping does not make an assumption about the shape of the sampling distribution of the indirect effect.

Bootstrapping is a resampling procedure used to generate an empirical approximation of the sampling distribution of a statistic (Efron & Tibshirani, 1994) that can be used for inference. In the context of mediation analysis, this statistic is the indirect effect, ab. To construct a bootstrap confidence interval for an indirect effect, the indirect effect is estimated in many *bootstrap samples* of the data. A bootstrap sample is constructed by taking a simple random sample of n cases (or rows) from the existing data, where n is the original sample size, and cases are sampled *with replacement*. The resulting bootstrap sample is then analyzed. Using the bootstrap sample, the indirect effect is estimated using Equations 30.2 and 30.3 and then saved. This process is repeated over and over, for a total of k times (where k is at least 1,000; more is better), producing a dataset containing k bootstrap estimates of the indirect effect. This bootstrap distribution of ab values serves as an empirical estimate of the sampling distribution of the indirect effect, and a 95 percent percentile-based bootstrap confidence interval can be computed as 2.5th and 97.5th percentiles of the distribution of k bootstrap estimates. Simulation studies have demonstrated that compared to alternative methods, inference using a confidence interval for the indirect effect constructed using this *percentile-based* bootstrap approach provides a good balance of power, coverage, and Type I error rate (e.g., Hayes & Scharkow, 2013).

Establishing a Total Effect

Historically, mediation analysis has been used only after one has established evidence of an association between X and Y. This is predicated on the widespread belief that without association between X and Y, there presumably is no effect to explain, and therefore no reason to try to establish or test the process by which that (non)effect operates. That is, how can an effect that seems not to exist be mediated? But it is now widely accepted among methodologists that a statistically significant total effect (path c in the mediation model in Figure 30.1, panel A, estimated with Equation 30.1) is not a requirement of mediation or a prerequisite of mediation analysis (Cerin & MacKinnon, 2009; Hayes, 2009; Hayes & Rockwood, 2017; O'Rourke & MacKinnon, 2018; Rucker et al., 2011; Shrout & Bolger, 2002). As Bollen (1989, p. 52) aptly stated "a lack of correlation does not disprove causation."

This is counterintuitive. An example will make it less so. But first, recall that although c, the total effect of X, quantifies differences in Y between cases that differ on X, the total effect is the sum of the direct and indirect effects of X, and there is no mathematical restriction that these effects be in the same direction. So the indirect effect of X on Y may be positive (or negative) while the direct effect is opposite in sign. If these effects are similar in strength, their sum will be close to zero. Given this background, now consider the effect of aerobic activity (X) on body weight (Y). Common sense and conventional wisdom would predict a negative relationship between aerobic activity and body weight. All else being equal, increasing aerobic activity should result in a reduction in body weight. Yet an increase in aerobic activity may result in consumption of more calories (calories consumed, M) due to an increase in metabolism, which could *increase* body weight, all other things being held fixed. Thus, aerobic activity could *positively* influence body weight through calorie consumption indirectly, but when holding calorie consumption constant, more aerobic activity should be negatively correlated with body weight. If the increase in weight resulting from aerobic activity operating indirectly through calorie consumption is as strong as the weight reduction from aerobic activity operating independent of calorie consumption, the result is no total effect of aerobic activity on weight. Without identifying the mediating role of calorie consumption in the (apparent lack of) relationship between X and Y, you could mistakenly conclude that aerobic activity is not an effective method of reducing weight. But knowing the direct and indirect effects are opposite in sign, this information could be used to develop an intervention that targets and reduces the added calorie consumption that occurs as a result of aerobic activity, thereby allowing the negative effect of aerobic activity on weight to manifest itself. Some people call this phenomenon *suppression*, though labeling it as such does not facilitate and is not the same as understanding.

Differences in sign of the direct and indirect effects are not a requirement for finding an indirect effect absent a statistically significant total effect. Remember that $c = c' + ab$. Suppose the direct effect c' equals zero but this is estimated with considerably more uncertainty than the indirect effect ab (i.e., the standard error of c' is larger than the standard error of ab). As a result, the total effect c is estimated with more uncertainty than is the indirect effect ab, and so c is harder to detect than ab. Indeed, Kenny and Judd (2014) show that equivalent total and indirect effects (i.e., when the direct effect is zero) are tested with inequivalent power. More specifically, tests on the indirect effect (such as a bootstrap confidence interval) tend to be conducted with higher power than tests of the total effect. Since the indirect effect is what carries information about the mediation of the effect of X on Y, rejection of the null hypothesis of no total effect is not a sensible prerequisite of mediation analysis.

Causal Inference

Mediation is a causal process. One cannot talk about mediation without using causal language. Yet statistical methods are agnostic to cause-effect and say nothing about whether an observed association is causal. An inference is a product of the mind of the researcher interpreting an analysis rather than the mathematics of the analysis and what is found in a statistical output (see Hayes, 2018). The statistical tools presented throughout this chapter can be used to piece together an argument, but causal claims must rely on more than statistically significant indirect effects. Within the mediation literature, there has been growing attention on explicitly stating the assumptions that must hold for valid causal inference of indirect (and direct) effects. Although these assumptions are often difficult or impossible to test, it is important to be aware of them for research design and statistical analysis purposes.

One of the more important assumptions is that there are no variables that confound the observed associations between the variables in the putative causal system. An association between two variables is confounded by a third if that third variable causally influences those two variables. In mediation, this assumption can be met, but only partly, by using an experimental design in which participants are randomly assigned to X. In that case, assuming other good experimental design practices are followed (such as eliminating contamination of manipulation by other variables and differential attrition or dropout in one condition), the $X \rightarrow M$ (path a) and $X \rightarrow Y$ (paths c and c') relationships can be inferred as causal because random assignment would render the correlation between X and any potential confounds zero. However, random assignment to X does *not* ensure that M causes Y, as other variables that are correlated with and influence M and Y could be responsible for the observed association between M and Y even when X is controlled. To deal with this, potential confounding variables can be included as covariates in the equations for M and Y (see Hayes, 2018, for a discussion of the path analysis algebra). This can also help when X is merely observed rather than manipulated. Unfortunately, one can never know for certain if the true confounds, if any, have been controlled.

Cause-effect claims also presume proper causal ordering, with the cause preceding the effect. In the case of mediation, we assume X temporally precedes M, which in turn precedes Y. Random assignment to X helps to meet this assumption for the $X \rightarrow M$ and $X \rightarrow Y$ components of a mediation model. But it does *not* ensure that M precedes Y. It could be that Y precedes M, and so Y is actually a mediator of the effect of X on M. Temporal precedence is hard to establish in observational (i.e., purely correlational) studies. For this reason, some have argued that mediation analysis should not be conducted with cross-sectional data (Maxwell & Cole, 2007; Maxwell, Cole, & Mitchell, 2011), with longitudinal or experimental data collection being preferred. While we agree that longitudinal data can help strengthen a cause-effect claim, the perspective that mediation cannot or should not be undertaken with correlational data is overly cynical for our taste. We believe that mediation analysis can be a useful tool with cross-sectional data, though the researcher will need to advance a strong theoretical justification for the presumed temporal ordering in an effort to mitigate the limitation with the research design. Further, the possibility of other causal orderings should be understood and addressed in discussion or analysis.

The use of ab and c' in Equations 30.2 and 30.3 as estimates of the indirect and direct effects of X assumes that the effect of M on Y does not depend on X, which also implies that X's effect on Y does not depend on M. Recently, there has been research focused on identifying the conditions in which this assumption can be relaxed. Discussions of modeling the interaction between X and M in a mediation model can be found in Hayes (2015, 2018), Imai, Keele, and Tingley (2010), Preacher, Rucker, and Hayes (2007), and Valeri and VanderWeele (2013).

More than One Mediator

X can and often does transmit its effect on Y through multiple mechanisms simultaneously. The simple mediation model just described doesn't capture the complexity of many real-world processes, but mediation analysis is easily extended to include additional mediators. Multiple mediator models are important for clinical research because they allow researchers to test competing theories while also facilitating a comparison between the sizes or magnitudes of more than one mechanism that may be responsible for the effect of X on Y. When X affects Y through multiple processes simultaneously, a failure to accurately model such complexities can result in misstatements, misunderstandings, or, at best, an oversimplified characterization of how X affects Y.

In a model involving more than one mediator, there may or may not be a causal relationship among the mediators. When there is no hypothesized causal pathway between the mediators, the model is termed a *parallel mediator model*. Figure 30.1, panel B, displays an example of a parallel mediation model with k mediators. For example, Zhang and colleagues (2013) tested a mediation model in which wellbeing and coping were hypothesized as mediators of the relationships between a culturally sensitive intervention program for suicidal African-American women and depressive symptoms and suicide ideation. They found that existential wellbeing mediated each of these relationships, while religious wellbeing, adaptive coping, and maladaptive coping did not.

A mediation model in which one or more mediators exerts its effect on another mediator is known as a *serial mediation model*. Figure 30.1, panel C, displays an example of a serial mediation model with two mediators. These models have also been popular in clinical psychology research. For instance, Stanley, Joiner, and Bryan (2017) found that anger and depression mediated the relationship between mild traumatic brain injury and suicide risk for a clinical sample of deployed military personnel. In their model, anger was estimated as causally preceding depression. According to the theory they were testing, brain injury causes anger, which in turn causes depression, which subsequently leads to an elevated risk of suicide.

Models that combine parallel and serial models can also be theorized and estimated. Space precludes an exhaustive discussion of the mathematics of complex multiple mediator models, so we focus on the parallel multiple mediator model here, as doing so will help to inform our discussion of the example conditional process analysis introduced later. For a discussion of the serial and combined parallel and serial model, see Hayes (2018).

The direct and indirect effects of X on Y in a parallel multiple mediator model, as in Figure 30.1, panel B, can be estimated with a set of $k + 1$ regression equations, k for the mediators (one for each mediator), and one for Y:

$$M_j = i_{M_j} + a_j X + e_{M_j} \text{ for } j = 1 \text{ to } k \qquad (30.4)$$

$$Y = i_Y + c'X + \sum_{j=1}^{k} b_j M_j + e_Y \qquad (30.5)$$

The direct effect of X is c' in Equation 30.5 and quantifies the effect of X on Y operating independent of all k mediators. Now there are multiple indirect effects, one through each of the k mediators, with the indirect effect through mediator j quantified as the product of a_j from Equation 30.4 and b_j from Equation 30.5. Thus, as in the simple mediation model, indirect effects are products of the constituent effects linking X to Y. In models involving multiple mediators, the indirect effect through a given mediator is termed a *specific indirect effect*. The sum of the specific indirect effects across all k mediators is the *total indirect effect* of X. For instance, in a three-mediator model ($k = 3$), the total indirect effect is $a_1 b_1 + a_2 b_2 + a_3 b_3$. The total indirect effect is interpreted as the effect of X on Y operating through all mediators in aggregate, and a specific indirect effect is the component of the total indirect effect that is unique to a particular mediator, independent of the indirect effect through all other mediators in the model. The total indirect effect and the direct effect of X add up to the total effect c, estimated with Equation 30.1.

There are a few important properties of a parallel mediation model worth acknowledging. First, the indirect effect of X through a given mediator is expected to change with the inclusion of additional mediators if the additional mediators are correlated with the mediator in question. The b path for mediator j in Equation 30.5 (i.e., the estimated effect of M_j on Y) will vary depending on the other mediators in a model, just as regression coefficients in a regression analysis are dependent on what other variables are in the model. Second, the specific indirect effects are not influenced by the scale of the mediators. That is, each specific indirect effect is scaled only in terms of the scales of X and Y. This makes specific indirect effects through different mediators directly comparable even when the measurement scales of the mediators are different (Preacher & Hayes, 2008). As a consequence, you can compare specific indirect effects to test whether they are the same or different. Two different theories may postulate different mediators of the effect of a therapeutic intervention on some kind of treatment outcome. If both theories' mediators are in the model simultaneously, you can test whether one theory's specific indirect effect is statistically different from another theory's. This is a more refined test of competing theories than asking which indirect effect is different from zero, or whether one or both is. Difference in significance does not imply statistical difference, and equivalence in significance does not equate with statistical equivalence. For a more detailed discussion of mechanics of comparing specific indirect effects, see Hayes (2018) or Preacher and Hayes (2008).

MODERATION

Moderation is used to explore and test the conditional nature of effects. X's effect on Y is *moderated* if the strength or direction of the causal effect of X on Y depends on the value or level of a third variable, with that third variable called a *moderator*. Moderators can be either categorical or continuous, and they may moderate the effect of a categorical or continuous variable. For example, Shanmugam and Davies (2015) were interested in the relationship between self-critical perfectionism and eating pathology in athletes. They found that gender moderated this relationship, where the effect was strong and positive for females, though largely nonexistent for males.

The utility of moderation analysis in clinical research is obvious if you are interested in the tailoring of treatments and therapies to individuals. Researchers can detect and better understand how the effectiveness of a given treatment varies across individuals, where a particularly effective treatment for one group may be ineffective, or even harmful, for another. An additional benefit of moderation analysis in a clinical setting is the ability to identify treatment inhibitors and enhancers. As an example of such phenomena, Singewald and colleagues (2015) reviewed previous research demonstrating that some pharmacological treatments can enhance the effectiveness of psychotherapeutic treatments of anxiety, fear, and trauma-related disorders, while other medications, such as anxiolytics, can inhibit these psychotherapeutic treatments' effectiveness. Moderation can be used in such scenarios to quantify and test the differential treatment effects when combined with various medications.

This section will provide a brief overview of moderation analysis. Topics include conceptualizing the model, making inferences about moderation, and interpreting the regression coefficients.

The Linear Moderation Model

The effect of X on Y is moderated by W if the effect of X on Y varies as a function of W. Moderation is represented in conceptual form in Figure 30.1, panel D. Mathematically, such a model can be represented as

$$Y = i_Y + f(W)X + b_2W + e_Y \qquad (30.6)$$

where $f(W)$ is some function of W. The linear function $b_1 + b_3W$ is most popular, which, when substituted into Equation 30.6, yields

$$Y = i_Y + (b_1 + b_3W)X + b_2W + e_Y \qquad (30.7)$$

In this model, X's effect on Y is of most interest and so X is sometimes called the *focal predictor*. Its effect is a linear function of W and so depends on W, making W the *moderator* of the effect of X on Y. The *conditional* effect of X on Y, which we can denote as $\theta_{X \to Y}$, is the linear function. In Equation 30.7, $\theta_{X \to Y} = b_1 + b_3W$. Observe that the dependency between the effect of X on Y and W is carried in b_3. If b_3 is zero, the relationship between X and Y is constrained to b_1 for all values of W and therefore is linearly independent of W. Therefore, an inference about whether b_3 is different from zero is used as a test of *linear* moderation of the effect of X on Y by W. If it is, it is sometimes said that X and W *interact*. So, interaction and moderation are synonyms, at least in our discussion. Not everyone uses these terms interchangeably.

To estimate the model, Equation 30.7 is typically expanded by distributing X across the linear function, resulting in

$$Y = i_Y + b_1X + b_2W + b_3XW + e_Y \qquad (30.8)$$

where XW is the product of X and W. Thus, all of the regression coefficients can be estimated by including X, W, and XW as predictors of Y in a linear regression model. The *p*-value for the weight for XW or a confidence interval, provided by standard regression analysis software, is used for inference as to whether X's effect on Y is linearly moderated by W, or that X and W interact in their influence on Y.

It is a widely believed myth that testing for interaction between X and W requires building a regression model using hierarchical entry of predictors, where X and W (and any covariates, if desired) are entered into the first model, followed by the XW product. An *F*-rest for the change in R^2 when XW is added is used as a test of interaction. Although there is nothing wrong with this procedure – the validity of this procedure is not a myth – it is not necessary to use hierarchical entry in this fashion. The *p*-value for the test of the null hypothesis that the weight for XW in Equation 30.7 is zero is the same as the *p*-value for the *F*-test comparing the fit of the model with and without the product, so the same conclusion about interaction will be reached either way.

Interpretation of Regression Coefficients

The regression coefficient b_3 quantifies the difference in the effect of X on Y for each one-unit difference in W. When the moderator W is dichotomous (e.g., in a clinical trial, a variable coding whether a participant is assigned to therapy as usual or a new experimental therapy) and the two values are coded one unit apart (e.g., 0 and 1), then b_3 corresponds to the difference between the effect of X on Y in the two groups.

The proper interpretation of b_1 and b_2 is often misunderstood. One of the common misinterpretations of b_1 and b_2 is that they represent "main effects" of X and W, respectively. This misinterpretation is a result of overgeneralizing concepts from the factorial ANOVA literature, textbooks, and statistics classrooms to *all* linear models. ANOVA is just a special form of a linear model, but not all ANOVA concepts generalize to any linear model. In a factorial ANOVA, the main effect of X is defined as the effect of X averaged across the groups that define W, and the main effect of W is the effect of W averaged across the groups that define X. However, as discussed below, b_1 and b_2 are actually closer to what are known in ANOVA-speak as *simple effects*. Coefficients b_1 and b_2 are main effects only when X and W are categorical variables and the categories are coded a specific way. See Darlington and Hayes (2017) and Hayes (2018) for a discussion of the difference between a *simple effects* and a *main effects parameterization* of factorial ANOVA in a regression context.

To better see the interpretation of b_1 and b_2, suppose XW were excluded from Equation 30.8. In that case, b_1 is the effect of X on Y holding W fixed *at any value*, and b_2 is the effect of W on Y holding X fixed *at any value*. These are *unconditional* effects or *partial* effects. But when XW is included, as in Equation 30.8, this changes the interpretation of b_1 and b_2. In Equation 30.8, the effect of X on Y depends on the value of W, and the effect of W on Y depends on the value of X and so become conditional effects. More specifically, b_1 is the estimated effect of X on Y when $W = 0$. Remember that X's effect is a function $f(W) = b_1 + b_3W$, which equals b_1 when $W = 0$. Similarly, b_2 is the estimated effect of W on Y conditional on $X = 0$. So b_1 and/or b_2 are meaningful only if $X = 0$ and/or $W = 0$ are meaningful. If $X = 0$ is not meaningful (e.g., if it is impossible value on a measurement scale), then b_2 is not meaningful, and if $W = 0$ is not meaningful, then b_1 is not meaningful.

One way to ensure that b_1 and b_2 are meaningful is to mean-center X and W before the creation of XW. A variable is mean-centered by subtracting the sample mean from each value in the data. When this is done, b_1 quantifies the effect of X on Y when $W = \bar{W}$ and b_2 quantifies the effect of W on Y when $X = \bar{X}$. However, this does not turn b_1 and b_2 into "main effects" as the term is used in analysis of variance, except in certain circumstances. Several books and journal articles state that mean-centering of X and W in a moderation analysis *must* be done to reduce multicollinearity, as XW is usually highly correlated with X and W. Although mean-centering does reduce multicollinearity, doing so has no effect on the test of moderation, as b_3, its standard error, t- and p-values, and confidence interval will be the same regardless. Mean-centering X and W is a personal choice one makes. It is not a requirement. See Hayes (2018) for further discussion of this topic as well as other myths of moderation analysis.

Understanding the correct interpretation of b_1 and b_2 also sheds light on why the lower order terms (i.e., X and W) should be included as predictors in the model in addition to the interaction term XW. Removing X (or W) from the model would be equivalent to constraining b_1 (or b_2) to zero. It is rare that we would have such prior information on these conditional effects and so the constraints, if inaccurate, can have negative consequences on our estimate of the interaction effect and its corresponding inferential test.

Plotting and Probing a Moderation Effect

A substantively interesting moderation effect can correspond to several different trends in the data. To better understand a moderation effect and what it is telling you about a phenomenon of interest, you should generate a visual depiction of the model that allows you to see how X's effect on Y varies with the moderator. This involves choosing combinations of values of X and W and then plugging them into the regression equation, which generates estimates of Y for those combinations. When W or X is dichotomous, use the values in the data coding the two groups. For continuous X or W, the most commonly used values are the sample mean and one standard deviation above and below the mean, although these are mere conventions and entirely arbitrary. To ensure the values chosen are not outside the range of data (as interpretation of a model in a region of space where one has no data is ill-advised), the 25th, 50th, and 75th percentile may be used instead. Hayes (2018) recommends the 16th, 50th, and 84th percentiles. But other strategies for making the choice could be used. Regardless, once the plot of the model is generated, it becomes easier to see the pattern in how variation in X is related to variation in Y differently for different values of W. These are the conditional effects of Y. Recognizing a pattern is the first step in substantively interpreting it. For more detail on visualizing an interaction in a regression model, see Hayes (2018).

In addition to plotting a model to visualize the conditional nature of the relationship between X and Y, it is useful to use inferential methods to test for the presence of a conditional effect of X on Y at values of W chosen. A popular method is the *pick-a-point approach* to *probing an interaction* (Bauer & Curran, 2005; Hayes, 2018), also called an *analysis of simple slopes*. The conditional effect of X on Y is easy to generate for any value of W you choose. It is $b_1 + b_3W$. However, the standard error, needed for inference for a given value of W, can be tricky to calculate. Several textbooks provide formulas to calculate the standard error of a linear function of regression coefficients, but the *regression-centering approach* makes inference about a conditional effect in a regression model easy. This method is best understood by recalling that the coefficient b_1 represents the conditional effect of X on Y when $W = 0$. By centering W around the value of interest w (i.e., subtracting w from every value of W in the data, even if W is just two arbitrary codes for two groups) prior to constructing XW and estimating the model, then b_1 estimates the conditional effect of X on Y when $W = w$, and the standard error of b_1 is a valid estimate of the standard error of this conditional effect. Thus, the pick-a-point approach can be conducted by repeating the analysis multiple times with W centered around each of the values of interest.

It is important to keep in mind that the conclusions drawn from the inferential tests may change depending on the values of the moderator at which the moderation effect is probed. This is one of the drawbacks of the pick-a-point approach. For this reason, some recommend using the Johnson-Neyman technique, which does not require you to choose values of W but, rather, analytically derives values of W that help you find where in the distribution of W that X is significantly related to Y and where it is not (for details, see Bauer & Curran, 2005; Hayes, 2018; Hayes & Matthes, 2009). Regardless, don't be lured into thinking that if X's effect on Y is significantly different from zero at some value of W but not significantly different from zero at another value, this implies moderation. Differences in the significance of two conditional effects does not mean they differ from each other. It is the test on b_3 in Equation 30.8 that provides formal support for a moderation hypothesis, which is about how X's effect on Y varies with W. And recall that the test of b_3 does not depend on the choice of centering, so the method used to probe the interaction should only influence our understanding of the nature of the interaction, not the existence of it.

CONDITIONAL PROCESS ANALYSIS

To this point we have discussed how statistical mediation and moderation analysis can be used to help understand the process by which an effect operates and the contingencies of an effect. Although these methods are useful as distinct analytic tools with different purposes, they can be combined to explore and test the conditional nature of the process by which one variable influences another. Several clinical researchers have stressed the importance of combining mediation and moderation analyses to better understand psychological phenomena and the effectiveness of treatments in various populations. For example, in an overview of methods for improving psychotherapy and psychological intervention research, Emmelkamp and colleagues (2014) ask why and for whom cognitive behavior therapy methods are more effective than other therapeutic methods. They note that "relatively little is known about why a certain type of psychotherapy is effective for one child but fails to produce a positive effect in another child" (Emmelkamp et al., 2014, p. 76). Collins, Murphy, and Bierman (2004) describe how variables identified as mediators and moderators in treatment research can be used to develop adaptive interventions, where repeated measurements of these variables over time are used to adaptively tailor the treatment to the individual. When used in combination, mediation and moderation analysis can be used to address such questions.

Conditional process analysis is the integration of mediation and moderation analyses in a unified statistical model. Thus, mediation and moderation models are the building blocks of a *conditional process model*. When conducting a conditional process analysis, interest lies in not only identifying the causal mechanisms linking X to Y, but also understanding the boundary conditions or conditional nature of these mechanisms. Conditional process models subsume models that test for *moderated mediation*, where the strength or direction of an indirect effect depends on the value of a moderator. They can take many forms and are built by combining features from the mediation and moderation models discussed in the previous sections. A conditional process model can include one or more mediators while allowing the strength or direction of one or more of the paths linking X to Y (i.e., X's effect on M, M's effect on Y, or both) to depend on a moderator W. Some examples can be found in Figure 30.2, each of which

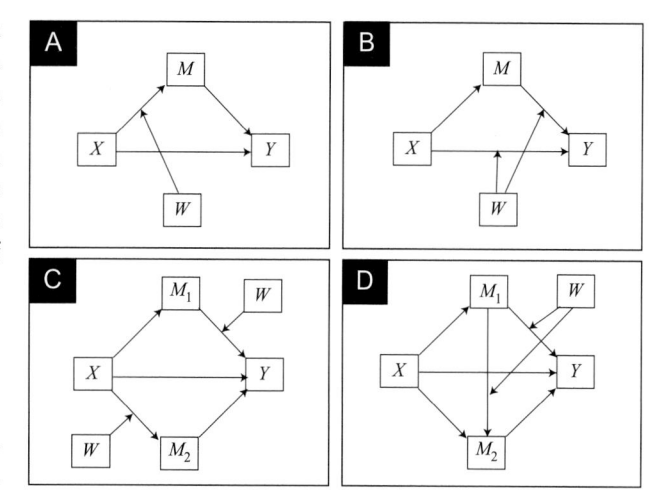

Figure 30.2 Four examples of conditional process models with a single moderator W.

contains an indirect effect of X on Y that is specified as moderated by W. The examples in panels A and B contain only a single mediator, but they differ with respect to whether W moderates the effect of X on M, the effect of M on Y, or the direct effect of X on Y. The models in panels C and D of Figure 30.2 are multiple mediator models (in parallel form in panel C, and in serial form in panel D), with W moderating one or more of the paths in the indirect effect of X on Y through at least one mediator. These are only a few of the numerous possibilities.

A growing number of researchers in clinical psychology have used conditional process analysis to explore the boundary conditions of a causal mechanism. For instance, Mason and colleagues (2015, discussed earlier) were interested not only in the effectiveness of text-based interventions to reduce smoking brought about by increasing readiness to stop (the mediator), but also the effect of friends' smoking behaviors on the strength of this mechanism. They found that the size of the indirect effect from treatment (X) to reduction of cigarettes (Y) through readiness to stop smoking (M) depended on the number of friends that smoke (W). That is, this mechanism was in operation more among those who had fewer friends that smoke. Gaume and colleagues (2016) tested a conditional process model where motivational interviewing (MI) experience (W) moderated the indirect effect of MI-consistent behaviors (X) on drinking at three-month follow-up (Y) through client change talk (M). The indirect effect was stronger for therapists that had more MI experience. And using a Latino sample, Torres and Taknint (2015) found that ethnic identity and self-efficacy (W) moderated the effect of ethnic microaggressions (X) on traumatic stress symptoms (M), which, in turn, affected depression (Y).

A Simple Conditional Process Model

Consider a conditional process model that allows the indirect effect of X on Y through M to be moderated by

W while the direct effect of X is fixed to be constant. If the moderation of the indirect effect operates through the moderation of the $X \rightarrow M$ path only, the model appears as in Figure 30.2, panel A. Using the principles of moderation and mediation analysis described earlier, and under the assumption that all effects are linear and that W is linearly related to the size of the effect of X on M, this model can be specified with a set of two regression equations:

$$M = i_M + a_1 X + a_2 W + a_3 XW + e_M \tag{30.9}$$

$$Y = i_Y + c'X + bM + e_Y \tag{30.10}$$

Note the only difference between this conditional process model and the simple mediation model using Equations 30.2 and 30.3 is the equation for M, which now includes W and XW as predictors. Including XW as a predictor allows the effect of X on M to depend linearly on W.

By rearranging Equation 30.9 as $M = i_M + (a_1 + a_3 W) X + a_2 W + e_M$, it is clear that X's effect on M is a linear function of W: $\theta_{X \rightarrow M} = a_1 + a_3 W$. But the effect of M on Y (controlling for X) is fixed to be independent of W. It is simply b, just as in a simple mediation model. And the direct effect of X on Y remains c'. It is fixed to be independent of W in this model.

Recall that the indirect effect in a simple mediation model is the product of the effect of X on M and the effect of M on Y when statistically controlling for X. In a conditional process model, an indirect effect is still defined as a product of its constituent effects. So in this model, the indirect effect of X on Y through M is

$$\text{Indirect effect of } X = (a_1 + a_3 W) b = a_1 b + a_3 bW \tag{30.11}$$

(see, e.g., Edwards & Lambert, 2007; Hayes, 2018; Preacher et al., 2007). Notice that Equation 30.11 is a linear function of W. As the value of the indirect effect of X will depend on the value of W (unless $a_3 = 0$ or $b = 0$), the output of this function is called a *conditional indirect effect*. If W is dichotomous, then there are two conditional indirect effects of X, one for each value of W. But if W is a numerical continuum, then there are many, or even an infinite number of conditional indirect effects, one for each possible value of W.

Index of Moderated Mediation

In Equation 30.11, $a_1 b$ is the indirect effect of X on Y when $W = 0$. If $W = 0$ is not meaningful, neither is $a_1 b$, and it estimates nothing interpretable. The term $a_3 b$ in Equation 30.11 is analogous to the regression weight for the product of X and W in a moderation model, in that it quantifies the relationship between the effect of X and the moderator. Except here it is the product of two regression coefficients, and it quantifies the change in the *indirect effect* of X as W changes by one unit. Hayes (2015) calls $a_3 b$ the *index of moderated mediation* for this model. An inference about moderation of the indirect effect – moderated mediation – is undertaken by testing whether the index of moderated mediation is different from zero. A confidence interval for the index of moderated mediation can be constructed using the bootstrapping method described earlier. If the confidence interval does not include zero, this is evidence of moderated mediation, but if the confidence interval includes zero, you cannot definitively claim that the indirect effect depends on the moderator. Note that with this approach, a claim of moderated mediation is not reliant on a series of significance tests of the individual components, a_3 and b. Inference about the index of moderated mediation is all that is needed when determining if the indirect effect is moderated by W (see Hayes, 2015, 2018, for details).

Probing the Moderation of Mediation

After establishing that the indirect effect of X is moderated by W, this moderation effect can be visualized and probed, just as in the simple moderation model. Now, however, the interest is on the conditional indirect effect of X on Y through M at different values of W. Equation 30.11 is a linear function of W. A line graph depicting the relationship between W and the size of X's indirect effect can be generated by plotting the output of Equation 30.11 as a function of many values of W. The resulting plot will take the form of a straight line, with the steepness of the slope determined by $a_3 b$. When the output of the function results in a value less than zero for a given value of W, then X is negatively related to Y through M when W is that value, whereas when output of the function is greater than zero, X is positively related to Y through M at that value of W. The moderation of the indirect effect can also be probed using an analogue of the pick-a-point approach discussed earlier, choosing various values of W and plugging them into Equation 30.11 after the regression coefficients are estimated. However, this produces only point estimates of the conditional indirect effect at those values of W. Additional work is required for inference. Because a conditional indirect effect is a product of regression coefficients and so its sampling distribution will be nonnormal, a bootstrap confidence interval for the conditional indirect effect can be used for inference. Special software is needed, such as a good structural equation modeling program or a special tool such as the PROCESS macro used in the following section (see Hayes, 2018, for a discussion).

A Worked Example

Now that the fundamentals of the framework for integrating mediation and moderation analyses have been discussed, we walk through an example loosely based on a study conducted by Barnett and colleagues (2010), who implemented a brief intervention targeting alcohol use in emergency department patients. The participants consisted of patients admitted to a hospital emergency department who had been under the influence of alcohol at the time of admission. Based on various diagnostic assessments, participants were classified as severe ($W = 1$) or not severe ($W = 0$) abusers of alcohol. The participants were then randomly assigned to a treatment that consisted of a motivational interview with a counselor followed by personalized feedback ($X = 1$) or were provided only personalized feedback without any interaction with a counselor ($X = 0$). This is the independent variable in the analysis below. Two potential mediators of the effect of the intervention, posttreatment perceived risk and benefits of alcohol use (M_1) and degree of treatment seeking (M_2), were measured. Higher scores on these represent greater perceived risks and more treatment seeking. The outcome variable (Y) is a composite measure of alcohol use (frequency and amount) at 12-month follow-up, with higher scores reflecting more use. In this example, the moderator W is a dichotomous variable. However, the procedure we discuss here applies whether W is dichotomous or a numerical continuum.

Note that the analysis we describe below is not based on the actual data from this study but instead uses simulated data constructed for the purpose of this example. Furthermore, although Barnett and colleagues (2010) were interested in both moderators and mediators of the effect of motivational interviewing, they did not estimate a conditional process model as we do here. We offer no guarantee that the results of the analysis we present here match what would be found using real data (theirs or anyone else's), and readers of this chapter should not use our analysis to argue for the existence of any empirical support in Barnett and colleagues (2010) or elsewhere for the findings we report.

Research Questions

Suppose it is hypothesized that perceived risks of alcohol use and degree of treatment seeking, M_1 and M_2, are mediators of the relationship between the motivational interviewing treatment (X) and alcohol use (Y) at 12-month follow-up. Furthermore, perhaps there is reason to believe that the indirect effects through perceived risks and treatment seeking will differ in size depending on severity of the patient's abuse of alcohol (W). Specifically, perhaps the indirect effect of motivational interviewing on alcohol use through perceived risks is proposed to be larger than the indirect effect through treatment seeking for patients with less severe alcohol abuse, while the reverse is expected for patients with severe alcohol use. This expectation is the result of the prediction that the difference in effects between these two types of abusers would stem from differences between them in the effect of the treatment on the mediators (i.e., $X \rightarrow M_1$ and $X \rightarrow M_2$). That is, W is proposed as moderating the relationships between X and M_1 and X and M_2.

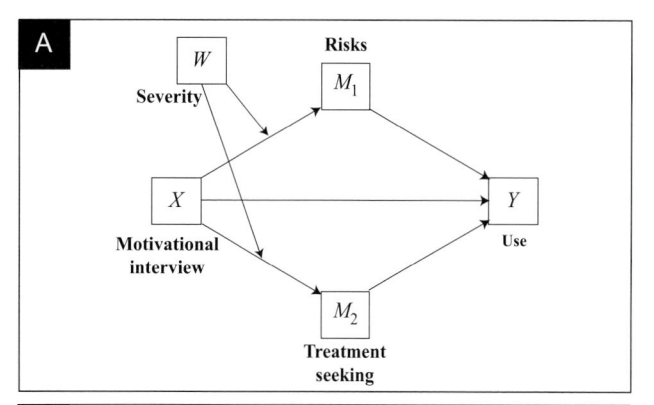

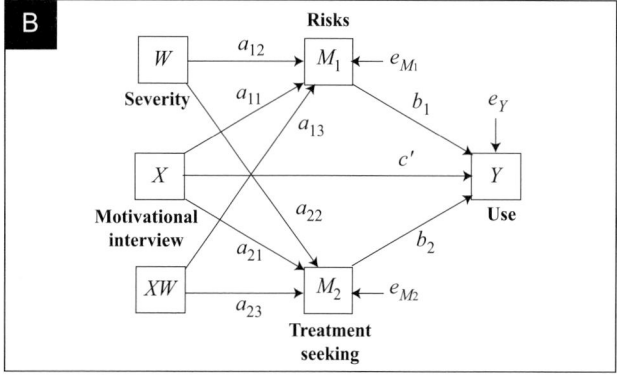

Figure 30.3 A first stage moderated parallel multiple mediator model in conceptual form (panel A) and in the form of a path diagram (panel B)

The Model in Visual and Equation Form

A visual representation of such a conditional process model can be found in Figure 30.3, panel A. It depicts a *first-stage moderated parallel multiple mediator model*, as it contains two mediators, and only the effect of X on the mediators (the first stage of the mediation process) is proposed as moderated. The diagram can be represented mathematically in the form of a set of three regression equations, one for each of the mediators and one for Y, using the principles described in the prior sections. These equations are

$$M_1 = i_{M2} + a_{11}X + a_{12}W + a_{13}XW + e_{M1} \quad (30.12)$$

$$M_2 = i_{M2} + a_{21}X + a_{22}W + a_{23}XW + e_{M2} \quad (30.13)$$

$$Y = i_Y + c'X + b_1M_1 + b_2M_2 + e_Y \quad (30.14)$$

and represented in the form of a path diagram in Figure 30.3, panel B. When specified in this form, the assumption is that all effects are linear. Furthermore, the XW products in Equations 30.12 and 30.13 specify X's effect on M_1 and M_2 as linear functions of W.

Model Estimation

The regression coefficients in Equations 30.12–30.14 can be estimated using any statistical software that can conduct regression analysis. However, many of the details we discussed throughout this paper, such as probing interactions and constructing bootstrap confidence intervals

Table 30.1 Regression coefficients, standard errors, p-values, and confidence intervals from the motivational interviewing conditional process analysis

Outcome	Predictor	Parameter	Estimate	SE	p	95% confidence interval
Risks (M_1)	Intercept	i_{M1}	3.228	0.114	<0.001	(3.004, 3.453)
	Interview (X)	a_{11}	0.453	0.165	0.006	(0.129, 0.777)
	Severity (W)	a_{12}	0.189	0.162	0.244	(−0.130, 0.508)
	X × W	a_{13}	−0.516	0.236	0.030	(−0.981, −0.051)
Treatment seeking (M_2)	Intercept	i_{M2}	3.534	0.115	<0.001	(3.308, 3.761)
	Interview (X)	a_{21}	−0.127	0.166	0.444	(−0.453, 0.199)
	Severity (W)	a_{22}	−0.308	0.163	0.060	(−0.629, 0.012)
	X × W	a_{23}	0.867	0.238	<0.001	(0.398, 1.335)
Use (Y)	Intercept	i_Y	6.227	0.277	0.006	(0.227, 1.318)
	Interview (X)	c'	−0.416	0.117	<0.001	(−0.646, −0.187)
	Risks (M_1)	b_1	−0.513	0.056	<0.001	(−0.624, −0.403)
	Treat seeking (M_2)	b_2	−0.276	0.055	<0.001	(−0.384, −0.168)

Table entries come from the PROCESS output in the Appendix.

for products or functions of regression coefficients, can be quite tedious to implement without the assistance of special tools designed for these purposes. One such tool is the PROCESS macro for SPSS and SAS (Hayes, 2018). PROCESS was designed specifically to simplify mediation, moderation, and conditional process analysis, as the estimation of indirect effects, the index of moderated mediation, and the probing of moderated mediation and estimation of conditional (indirect) effects is automated with minimal syntax or an easy-to-use point-and-click interface (for SPSS users). Output for models containing moderators also includes code that can be used to plot the conditional effects and conduct various additional analyses. We used the PROCESS macro for SPSS to conduct the analysis in this example. PROCESS is freely available and can be downloaded from www.processmacro.org. Guidance on its use can be found in Hayes (2018).

Mediation, moderation, and conditional process analysis can also be conducted using structural equation modeling (SEM) software. An advantage of using SEM is that doing so allows for the estimation of latent variable models, which can help reduce the effects of measurement error on the estimation of effects when a structural model is combined with a good measurement model of the latent variables. Although any SEM program can be used, some programs provide features that make conditional process modeling easier than do others. For example, Mplus (Muthén & Muthén, 2015) and lavaan (Rosseel, 2012) can construct bootstrap confidence intervals for functions of model coefficients, allowing the user to make inferences about indirect effects, the index of moderated mediation, and conditional indirect effects. SEM programs offer more flexibility than some of the other options for conditional process modeling, but they also often require substantially more programming effort. And as Hayes, Montoya, and Rockwood (2017) discuss, SEM and a regression-based tool like PROCESS generally produce equivalent results for the kinds of observed variable models (i.e., nothing latent) discussed in this chapter, so the extra programming skills and effort required to use SEM software may not be the most efficient use of time and resources, although we encourage all clinical researchers to work toward developing such programming skills at some point in their careers.

Model Components

When working with a complex conditional process model, it is often useful to begin by looking at smaller components of the bigger model. Understanding the smaller components is accomplished by looking at and interpreting the regression coefficients from the individual equations before bringing those results together as needed to calculate statistics and test hypotheses that require integration across the equations. The estimated regression coefficients in Equations 30.12–30.14, along with standard errors, p-values, and confidence intervals can be found in Table 30.1. These come from the PROCESS output in Appendix 30.A, sections 1, 2, and 3. In the PROCESS output, Y, X, M_1, M_2, and W are named "use," "intview," "riskbene," "trtseek," and "severity," respectively.

Table 30.2 Conditional indirect effects of motivational interviewing on alcohol use and the index of moderated mediation, with bootstrap confidence intervals, from the conditional process analysis

Mediator	Group (W)	Parameter	Estimate	95% bootstrap confidence interval
Risks (M_1)	Not severe (0)	$a_{11}b_1$	−0.233	(−0.408, −0.072)
	Severe (1)	$(a_{11} + a_{13})b_1$	−0.032	(−0.149, 0.222)
	Index of MM	$a_{13}b_1$	0.265	(0.020, 0.533)
Treatment seeking (M_2)	Not severe (0)	$a_{21}b_2$	0.035	(−0.055, 0.132)
	Severe (1)	$(a_{21} + a_{23})b_2$	−0.204	(−0.340, −0.095)
	Index of MM	$a_{23}b_2$	−0.239	(−0.413, −0.098)

Estimates and confidence intervals come from the PROCESS output in the Appendix.

The effect of the motivational interview, X, on each of the mediators depends on abuse severity, W, and each of the M equations take the form of a simple moderation model as discussed in the moderation section above. For the risk/benefits measure, M_1, the effect of X is $\theta_{X \to M_1} = a_{11} + a_{13}W = 0.453 - 0.516W$. This conditional effect of X for severe abusers and less severe abusers can be found by plugging in $W = 0$ and $W = 1$ into this function. For less severe abusers, the effect is $a_{11} + a_{13}(0) = 0.453$, whereas for severe abusers, the effect is $a_{11} + a_{13}(1) = -0.063$. Because the treatment indicator X is dichotomous with codes in the data that differ by one unit (0 and 1), these effects can be interpreted as mean differences. Specifically, less severe abusers who received the motivational interviewing treatment scored 0.453 units higher on the risk and benefits measure, on average, than less severe abusers who received only personalized feedback. And severe abusers who received the motivational interview treatment scored slightly lower (−0.063), on average, than severe abusers who received feedback only. The difference in the effect of X on M_1 between the two types of patients is $a_{13} = -0.516$, the coefficient for XW. This difference is statistically significant ($p = 0.030$), indicating that the treatment effect on the risk/benefit measure changes depending on the patient type.

The equation for the effect of the motivational interview on treatment seeking, the second mediator M_2, is constructed similarly. The conditional effect of X on M_2 is $\theta_{X \to M_2} = a_{21} + a_{23}W = -0.127 + 0.867W$, and the difference between the treatment effect for the two severity groups is $a_{23} = 0.867$, $p < 0.001$. Here, the effect of the motivational interview on treatment seeking is larger for severe abusers, $-0.127 + 0.867(1) = 0.740$, than for less severe abusers, $-0.127 + 0.867(0) = 0.127$.

The equation for drinking at 12-month follow-up, Y, is $6.227 - 0.416X - 0.513M_1 - 0.276M_2$. As set up in this model, none of the effects of the independent variable or mediators are expressed as functions of a moderator, so these effects can be interpreted as are any partial regression coefficients. The direct effect of the motivational interview on later use is $c' = -0.416$, $p < 0.001$, meaning that, controlling for perceived risks and treatment seeking, those who received the motivational interview used alcohol less at 12-month follow-up. The effects of each of the mediators on later use are both statistically significant ($b_1 = -0.513$, $p < 0.001$; $b_2 = -0.276$, $p < 0.001$). Those who perceived alcohol as riskier used alcohol less later, as did those who engaged in more treatment seeking.

Conditional Indirect Effects

Having interpreted the components of the larger model, we can now integrate this information across the components to calculate the indirect effects of the motivational interview on later alcohol use through perceived risks (M_1) and treatment seeking (M_2). The indirect effect through perceived risks is the product of the effect of X on M_1, $\theta_{X \to M_1} = a_{11} + a_{13}W$, and the effect of M_1 on Y, b_1. Multiplying these effects results in $(\theta_{X \to M_1})b_1 = (a_{11} + a_{13}W)b_1 = a_{11}b_1 + a_{13}b_1W = -0.233 + 0.265W$. So, the indirect effect through perceived risks depends on alcohol abuse severity, W. As we did with the conditional effect of X on M_1, we can calculate the conditional indirect effect of X on Y through M_1 for the two abuse severity groups by plugging in values of W into this equation. The indirect effect through perceived risk is $a_{11}b_1 + a_{13}b_1W = -0.233 + 0.265(0) = -0.233$ for less severe abusers ($W = 0$). This is the product of the conditional effect of X on M_1 when $W = 0$ (0.453) and the effect of M_1 on Y (−0.513). But for severe abusers of alcohol, the indirect effect through perceived risk is $a_{11}b_1 + a_{13}b_1W = -0.233 + 0.265(1) = 0.032$. This is the product of the conditional effect of X on M_1 when $W = 1$ (−0.063) and the effect of M_1 on Y (−0.513).

Table 30.2 includes each of these conditional indirect effects as well as 95 percent bootstrap confidence intervals based on 5,000 bootstrap samples. These can be found in section 4 of the PROCESS output. The 95 percent bootstrap confidence interval for less severe abusers is −0.408 to −0.072 and does not contain zero, but the confidence interval of −0.149 to 0.222 for severe abusers does. Yet difference in significance does not necessarily mean that these conditional indirect effects differ from each other (see Hayes, 2015, 2018). The difference between these two conditional indirect effects is the index of moderated mediation, which here is $a_{13}b_1 = 0.265$. A 95 percent bootstrap confidence interval for this index is 0.020–0.533, which is entirely above zero. So, we can say that these two indirect effects are significantly different from each other. For less severe abusers, motivational interviewing decreases alcohol use by increasing the perceived risks of use, which reduces use. Such a process does not operate among severe abusers.

The same procedure can be used to quantify the conditional indirect effect of motivational interviewing on alcohol use through treatment seeking, M_2, which is $(\theta_{X \to M_2}) b_2 = (a_{21} + a_{23}W)b_2 = a_{21}b_2 + a_{23}b_2W = 0.035 - 0.239W$. For less severe abusers ($W = 0$), the conditional indirect effect is $0.035 - 0.239(0) = 0.035$, which is the product of the conditional effect of X on M_2 when $W = 0$ (–0.127) and the effect of M_2 on Y (–0.276). For severe abusers ($W = 1$), the conditional indirect effect is $0.035 - 0.239(1) = -0.204$, which is the product of the conditional effect of X on M_2 when $W = 1$ (0.740) and the effect of M_2 on Y (–0.276). The difference between these two conditional indirect effects is $a_{23}b_2 = -0.293$, the index of moderated mediation for this indirect effect, with a 95 percent bootstrap confidence interval from –0.413 to –0.098. So the difference between these conditional indirect effects is statistically significant. Probing this moderation of the indirect effect shows that the conditional indirect effect of motivational interviewing on use through treatment seeking is not definitively different from zero among less severe abusers (95 percent bootstrap confidence interval = –0.055 to 0.132), but among severe abusers, this indirect effect is significantly negative (95 percent bootstrap confidence interval = –0.340 to –0.095). So, among severe abusers, motivational interviewing results in an increase in treatment seeking, which in turn is related to a reduction in alcohol use. No such process is in operation among less severe abusers.

Substantive Conclusion

To summarize, and using a data set fabricated for the purpose of this example, we have used a combination of a parallel mediator model and the simple moderation model to better understand the contingencies of the processes by which motivational interviewing affects alcohol use. Perceived risk of alcohol use and treatment seeking were the mediators of interest in this example. The effect of the motivational interview on each of the mediators was dependent on severity of the patient's alcohol abuse, and each of the mediators was negatively correlated with later alcohol use. Furthermore, the indirect effect through each mediator was conditional on alcohol use severity. The indirect effect through perceived risk was larger for less severe abusers than for severe abusers, but the indirect effect through treatment seeking was larger for severe abusers than it was for less severe abusers.

Summary

Throughout this chapter, we've provided an overview of mediation, moderation, and conditional process analysis, with an emphasis on mapping particular research questions of interest to their respective statistical models. We've described how statistical mediation analysis can be used to help identify, quantify, and understand a causal sequence of events in which one variable influences another through one or more intermediary variables, while statistical moderation analysis can be used to test the boundary conditions or contingencies of an effect. These methods have proved useful for clinical research, both separately and when integrated within a unifying conditional process model. After detailing the framework of mediation, moderation, and conditional process analysis, with a focus on the substantive meaning of model parameters and how they relate to hypotheses of interest, we presented an example analysis that synthesized many of the concepts discussed throughout this chapter. We believe this exposition should provide clinical researchers with a good understanding of the conceptual and statistical foundations of mediation and moderation analysis and also aid them in the development of their own conditional process models for testing and exploring their own hypotheses and substantive interests.

REFERENCES

Aroian, L. A. (1947). The Probability Function of the Product of Two Normally Distributed Variables. *Annals of Mathematical Statistics*, 265–271.

Barnett, N. P., Apodaca, T. R., Magill, M., Colby, S. M., Gwaltney, C., Rohsenow, D. J., & Monti, P. M. (2010). Moderators and Mediators of Two Brief Interventions for Alcohol in the Emergency Department. *Addiction*, *105*(3), 452–465.

Baron, R. M., & Kenny, D. A. (1986). The Moderator-Mediator Variable Distinction in Social Psychological Research: Conceptual, Strategic, and Statistical Considerations. *Journal of Personality and Social Psychology*, *51*(6), 1173–1182.

Bauer, D. J., & Curran, P. J. (2005). Probing Interactions in Fixed and Multilevel Regression: Inferential and Graphical Techniques. *Multivariate Behavioral Research*, *40*, 373–400.

Bollen, K. A. (1989). *Structural Equation Modeling with Latent Variables*. New York: Wiley.

Cerin, E., & MacKinnon, D. P. (2009). A Commentary on Current Practice in Mediating Variable Analyses in Behavioural Nutrition and Physical Activity. *Public Health Nutrition*, *12*(8), 1182–1188.

Collins, L. M., Murphy, S. A., & Bierman, K. L. (2004). A Conceptual Framework for Adaptive Preventive Interventions. *Prevention Science*, *5*(3), 185–196.

Darlington, R. B., & Hayes, A. F. (2017). *Regression Analysis and Linear Models: Concepts, Applications, and Implementation*. New York: Guilford Press.

Duchesne, A.-P., Dion, J., Lalande, D., Bégin, C., Émond, C., Lalande, G., & McDuff, P. (2017). Body Dissatisfaction and Psychological Distress in Adolescents: Is Self-Esteem a Mediator? *Journal of Health Psychology*, 22(12), 1563–1569.

Edwards, J. R., & Lambert, A. L. (2007). Methods for Integrating Moderation and Mediation: A General Analytical Framework Using Moderated Path Analysis. *Psychological Methods, 12*, 1–22.

Efron, B., & Tibshirani, R. J. (1994). *An Introduction to the Bootstrap*. Boca Raton, FL: Chapman and Hall.

Emmelkamp, P. M., David, D., Beckers, T., Muris, P., Cuijpers, P., Lutz, W., ... Berking, M. (2014). Advancing Psychotherapy and Evidence-Based Psychological Interventions. *International Journal of Methods in Psychiatric Research, 23*(S1), 58–91.

Finney, J. W. (1995). Enhancing Substance Abuse Treatment Evaluations: Examining Mediators and Moderators of Treatment Effects. *Journal of Substance Abuse, 7*(1), 135–150.

Frazier, P. A., Tix, A. P., & Barron, K. E. (2004). Testing Moderator and Mediator Effects in Counseling Psychology Research. *Journal of Counseling Psychology, 51*(1), 115–134.

Gaume, J., Longabaugh, R., Magill, M., Bertholet, N., Gmel, G., & Daeppen, J.-B. (2016). Under What Conditions? Therapist and Client Characteristics Moderate the Role of Change Talk in Brief Motivational Intervention. *Journal of Consulting and Clinical Psychology, 84*(3), 211–220.

Hayes, A. F. (2009). Beyond Baron and Kenny: Statistical Mediation Analysis in the New Millennium. *Communication Monographs, 76*, 408–420.

Hayes, A. F. (2015). An Index and Test of Linear Moderated Mediation. *Multivariate Behavioral Research, 50*(1), 1–22.

Hayes, A. F. (2018). *Introduction to Mediation, Moderation, and Conditional Process Analysis: A Regression-Based Approach* (2nd edn.). New York: Guilford Press.

Hayes, A. F., & Matthes, J. (2009). Computational Procedures for Probing Interactions in OLS and Logistic Regression: SPSS and SAS Implementations. *Behavior Research Methods, 41*, 924–936.

Hayes, A. F., & Montoya, A. K. (2017). A Tutorial on Testing, Visualizing, and Probing an Interaction Involving a Multicategorical Variable in Linear Regression Analysis. *Communication Methods and Measures, 11*(1), 1–30.

Hayes, A. F., & Preacher, K. J. (2014). Statistical Mediation Analysis with a Multicategorical Independent Variable. *British Journal of Mathematical and Statistical Psychology, 67*(3), 451–470.

Hayes, A. F., & Rockwood, N. J. (2017). Regression-Based Statistical Mediation and Moderation Analysis in Clinical Research: Observations, Recommendations, and Implementation. *Behaviour Research and Therapy, 98*, 39–57.

Hayes, A. F., & Scharkow, M. (2013). The Relative Trustworthiness of Inferential Tests of the Indirect Effect in Statistical Mediation Analysis Does Method Really Matter? *Psychological Science, 24*(10), 1918–1927.

Hayes, A. F., Montoya, A. K., & Rockwood, N. J. (2017). The Analysis of Mechanisms and Their Contingencies: PROCESS versus Structural Equation Modeling. *Australasian Marketing Journal, 25*, 76–81.

Imai, K., Keele, L., & Tingley, D. (2010). A General Approach to Causal Mediation Analysis. *Psychological Methods, 15*(4), 309–334.

Kazdin, A. E. (2007). Mediators and Mechanisms of Change in Psychotherapy Research. *Annual Review of Clinical Psychology, 3*, 1–27.

Kenny, D. A., & Judd, C. M. (2014). Power Anomalies in Testing Mediation. *Psychological Science, 25*, 334–339.

Kraemer, H. C., Kiernan, M., Essex, M., & Kupfer, D. J. (2008). How and Why Criteria Defining Moderators and Mediators Differ between the Baron & Kenny and MacArthur Approaches. *Health Psychology, 27*, S101–S108.

Kraemer, H. C., Wilson, G. T., Fairburn, C. G., & Agras, W. S. (2002). Mediators and Moderators of Treatment Effects in Randomized Clinical Trials. *Archives of General Psychiatry, 59*(10), 877–883.

MacKinnon, D. P., Lockwood, C. M., & Williams, J. (2004). Confidence Limits for the Indirect Effect: Distribution of the Product and Resampling Methods. *Multivariate Behavioral Research, 39*(1), 99–128.

Magill, M. (2011). Moderators and Mediators in Social Work Research: Toward a More Ecologically Valid Evidence Base for Practice. *Journal of Social Work, 11*(4), 387–401.

Mason, M., Mennis, J., Way, T., & Campbell, L. F. (2015). Real-Time Readiness to Quit and Peer Smoking within a Text Message Intervention for Adolescent Smokers: Modeling Mechanisms of Change. *Journal of Substance Abuse Treatment, 59*, 67–73.

Maxwell, S. E., & Cole, D. A. (2007). Bias in Cross-Sectional Analyses of Longitudinal Mediation. *Psychological Methods, 12*(1), 23–44.

Maxwell, S. E., Cole, D. A., & Mitchell, M. A. (2011). Bias in Cross-Sectional Analyses of Longitudinal Mediation: Partial and Complete Mediation under an Autoregressive Model. *Multivariate Behavioral Research, 46*(5), 816–841.

Muthén, L. K., & Muthén, B. (2015). *Mplus Version 7.4* [Computer software manual]. Los Angeles: Muthén & Muthén.

Norcross, J. C., & Wampold, B. E. (2011). What Works for Whom: Tailoring Psychotherapy to the Person. *Journal of Clinical Psychology, 67*(2), 127–132.

Onken, L. S., Carroll, K. M., Shoham, V., Cuthbert, B. N., & Riddle, M. (2014). Re-envisioning Clinical Science: Unifying the Discipline to Improve the Public Health. *Clinical Psychological Science, 2*(1), 22–34.

O'Rourke, H. P., & MacKinnon, D. P. (2018). Reasons for Testing Mediation in the Absence of an Intervention Effect: A Research Imperative in Prevention and Intervention Research. *Journal of Studies on Alcohol and Drugs, 79*, 171–181.

Preacher, K. J., & Hayes, A. F. (2008). Asymptotic and Resampling Strategies for Assessing and Comparing Indirect Effects in Multiple Mediator Models. *Behavior Research Methods, 40*(3), 879–891.

Preacher, K. J., & Selig, J. P. (2012). Advantages of Monte Carlo Confidence Intervals for Indirect Effects. *Communication Methods and Measures, 6*(2), 77–98.

Preacher, K. J., Rucker, D. D., & Hayes, A. F. (2007). Addressing Moderated Mediation Hypotheses: Theory, Methods, and Prescriptions. *Multivariate Behavioral Research, 42*(1), 185–227.

Rosseel, Y. (2012). lavaan: An R Package for Structural Equation Modeling. *Journal of Statistical Software, 48*(2), 1–36.

Rucker, D. D., Preacher, K. J., Tormala, Z. L., & Petty, R. E. (2011). Mediation Analysis in Social Psychology: Current Practices and New Recommendations. *Social and Personality Psychology Compass, 5*(6), 359–371.

Schmidt, S. J., & Schimmelmann, B. G. (2015). Mechanisms of Change in Psychotherapy for Children and Adolescents: Current State, Clinical Implications, and Methodological and Conceptual Recommendations for Mediation Analysis. *European Child and Adolescent Psychiatry, 24*, 249–253.

Shanmugam, V., & Davies, B. (2015). Clinical Perfectionism and Eating Psychopathology in Athletes: The Role of Gender. *Personality and Individual Differences, 74*, 99–105.

Shrout, P. E., & Bolger, N. (2002). Mediation in Experimental and Nonexperimental Studies: New Procedures and Recommendations. *Psychological Methods, 7*(4), 422–445.

Singewald, N., Schmuckermair, C., Whittle, N., Holmes, A., & Ressler, K. J. (2015). Pharmacology of Cognitive Enhancers for Exposure-Based Therapy of Fear, Anxiety and Trauma-Related Disorders. *Pharmacology & Therapeutics, 149*, 150–190.

Sobel, M. E. (1982). Asymptotic Confidence Intervals for Indirect Effects in Structural Equation Models. *Sociological Methodology, 13*(1982), 290–312.

Stanley, I. H., Joiner, T. E., & Bryan, C. J. (2017). Mild Traumatic Brain Injury and Suicide Risk among a Clinical Sample of Deployed Military Personnel: Evidence for a Serial Mediation Model of Anger and Depression. *Journal of Psychiatric Research, 84*, 161–168.

Torres, L., & Taknint, J. T. (2015). Ethnic Microaggressions, Traumatic Stress Symptoms, and Latino Depression: A Moderated Mediational Model. *Journal of Counseling Psychology, 62*(3), 393–401.

Valeri, L., & VanderWeele, T. J. (2013). Mediation Analysis Allowing for Exposure-Mediator Interactions and Causal Interpretation: Theoretical Assumptions and Implementation with SAS and SPSS Macros. *Psychological Methods, 18*(2), 137–150.

Windgassen, S., Goldsmith, K., Moss-Morris, R., & Chalder, T. (2016). Establishing How Psychological Therapies Work: The Importance of Mediation Analysis. *Journal of Mental Health, 25*(2), 93 99.

Zhang, H., Neelarambam, K., Schwenke, T. J., Rhodes, M. N., Pittman, D. M., & Kaslow, N. J. (2013). Mediators of a Culturally-Sensitive Intervention for Suicidal African-American Women. *Journal of Clinical Psychology in Medical Settings, 20*(4), 401–414.

Appendix 30.A

Output from the PROCESS Macro for SPSS

```
***************PROCESS Procedure for SPSS Release 3.1**********


          Written by Andrew F.Hayes,Ph.D.   www.afhayes.com
   Documentation available in Hayes(2018).  www.guilford.com/p/hayes3

**************************************************************

Model:7
   Y :use
   X :intview
   M1:riskbene
   M2:trtseek
   W :severity
Sample
Size:300
**************************************************************
```

1 OUTCOME VARIABLE:
riskbene

Model Summary

R	R-sq	MSE	F	df1	df2	p
.1617	.0261	1.0422	2.6478	3.0000	296.0000	.0492

Model

	coeff	se	t	p	LLCI	ULCI
constant	3.2281	.1141	28.2826	.0000	3.0035	3.4528
intview	.4530	.1647	2.7509	.0063	.1289	.7770
severity	.1892	.1619	1.1685	.2435	-.1295	.5079
Int_1	-.5161	.2363	-2.1837	.0298	-.9812	-.0510

Product terms key:
  Int_1   :   intview   x   severity

```
Test(s) of highest order unconditional interaction(s):
         R2-chng          F       df1          df2          p
X*W        .0157     4.7687    1.0000     296.0000      .0298
----------
        Focal predict: intview (X)
          Mod var: severity (W)

Conditional effects of the focal predictor at values of the moderator(s):

   severity    Effect       se        t        p      LLCI     ULCI
     .0000     .4530     .1647   2.7509    .0063     .1289    .7770
    1.0000    -.0632     .1696   -.3725    .7098    -.3968    .2705

Data for visualizing the conditional effect of the focal predictor:
Paste text below into a SPSS syntax window and execute to produce plot.


DATA LIST FREE/
    intview     severity    riskbene   .
BEGIN DATA.
     .0000       .0000      3.2281
    1.0000       .0000      3.6811
     .0000      1.0000      3.4173
    1.0000      1.0000      3.3542
END DATA.
GRAPH/SCATTERPLOT=
   intview WITH   riskbene  BY    severity.

************************************************************
```

2 OUTCOME VARIABLE:
   trtseek

```
Model Summary
           R       R-sq       MSE         F        df1         df2         p
        .2513     .0631    1.0564    6.6484     3.0000    296.0000     .0002

Model
              coeff        se         t        p        LLCI      ULCI
constant     3.5344     .1149   30.7575    .0000      3.3082    3.7605
intview      -.1272     .1658    -.7674    .4435      -.4534     .1990
severity     -.3084     .1630   -1.8919    .0595      -.6293     .0124
Int_1         .8665     .2379    3.6415    .0003       .3982    1.3348

Product terms key:
 Int_1   :    intview    x    severity

Test(s) of highest order unconditional interaction(s):
         R2-chng          F       df1          df2          p
X*W        .0420    13.2608    1.0000     296.0000      .0003
----------
```

```
           Focal predict: intview  (X)
             Mod var: severity(W)

Conditional effects of the focal predictor at values of the moderator(s):

     severity     Effect         se          t          p       LLCI       ULCI
       .0000     -.1272      .1658     -.7674      .4435     -.4534      .1990
      1.0000      .7393      .1707     4.3309      .0000      .4033     1.0752

Data for visualizing the conditional effect of the focal predictor:
Paste text below into a SPSS syntax window and execute to produce plot.


DATA LIST FREE/
    intview     severity     trtseek   .
BEGIN DATA.
       .0000        .0000     3.5344
      1.0000        .0000     3.4072
       .0000       1.0000     3.2259
      1.0000       1.0000     3.9652
END DATA.
GRAPH/SCATTERPLOT=
   intview WITH   trtseek   BY      severity.

****************************************************************
```

3 OUTCOME VARIABLE:
  use

Model Summary
          R   R-sq   MSE   F   df1   df2   p
      .5591   .3126   .9842   44.8607   3.0000   296.0000   .0000

Model
                coeff   se   t   p   LLCI   ULCI
constant   6.2274   .2772   22.4618   .0000   5.6818   6.7730
intview   -.4163   .1165   -3.5747   .0004   -.6455   -.1871
riskbene   -.5134   .0560   -9.1626   .0000   -.6237   -.4031
trtseek   -.2759   .0548   -5.0316   .0000   -.3839   -.1680

****************** DIRECT AND INDIRECT EFFECTS OF X ON Y ***********

4 Direct effect of X on Y
       Effect   se   t   p   LLCI   ULCI
       -.4163   .1165   -3.5747   .0004   -.6455   -.1871

Conditional indirect effects of X on Y:

```
INDIRECT EFFECT:
intview   ->   riskbene   ->   use

       severity      Effect      BootSE     BootLLCI    BootULCI
         .0000       -.2325       .0860      -.4080      -.0717
        1.0000        .0324       .0934      -.1490       .2215

Index of moderated mediation (difference between conditional in direct effects):
             Index      BootSE     BootLLCI    BootULCI
severity     .2650       .1297       .0201       .5331
---

INDIRECT EFFECT:
intview   ->   trtseek   ->   use

       severity      Effect      BootSE     BootLLCI    BootULCI
         .0000        .0351       .0471      -.0554       .1317
        1.0000       -.2040       .0630      -.3398      -.0947

Index of moderated mediation (difference between conditional in direct effects):
             Index      BootSE     BootLLCI    BootULCI
severity    -.2391       .0817      -.4131      -.0975
---

************************ANALYSIS NOTES AND ERRORS***************


Level of confidence for all confidence intervals in output:
   95.0000

Number of bootstrap samples for percentile bootstrap confidence intervals:
   10000
```

31 Statistical Inference for Causal Effects in Clinical Psychology

Fundamental Concepts and Analytical Approaches

REAGAN MOZER, DONALD B. RUBIN, AND JOSE ZUBIZARRETA

A central problem with studies in clinical psychology is how to draw inferences about the causal effects of treatments, interventions, or exposures. For example, can new medications for treating a psychological condition reduce the occurrence of negative physical health outcomes? Does chronic substance abuse cause mental impairment? Does a new treatment for depression reduce suicide rates?

This chapter provides a structure for thinking about and drawing inferences about such causal effects in psychopathology research using the potential outcomes framework, now commonly referred to as the "Rubin Causal Model" (RCM; Holland, 1986). A number of alternative mathematical frameworks for causal inference exist (e.g., Pearl, 2009; Robins, 2001); however, our focus here is on the insights provided by the potential outcomes formulation, rather than a comparison with other approaches.

The chapter begins with a description of the first part of the RCM, which uses potential outcomes to define causal effects in all situations, whether from a randomized experiment or observational study. The second part of the RCM concerns the assignment mechanism, a probabilistic model for how some units receive treatment and others receive control, and is described in the third section of this chapter. Three approaches to the analysis of randomized experiments, are then presented, including two randomization-based modes of inference due to Neyman (1923) and Fisher (1925) and a third, model-based, or Bayesian, mode of inference due to Rubin (1978). Some complications that commonly arise in randomized experiments and extensions of the RCM for inference in these settings are then discussed. A further section considers causal inference in observational studies and reviews analytical approaches for drawing inferences about causal effects when data are nonrandomized. Finally, the use of sensitivity analyses for investigating the robustness of causal inference in any setting is assessed.

CAUSAL INFERENCE PRIMITIVES

There are three fundamental concepts required for defining causal effects: units, treatments, and potential outcomes. A "unit" is a physical object, for example a person, at a particular point in time. A "treatment" is an action that can be applied or withheld from a unit. We focus here on settings with two treatments, namely, one active treatment and one control treatment, but the principles and methods described in this chapter extend to settings with more than two treatments.

The objective of causal inference is to learn about the effect of administering the active treatment compared to the control version on an outcome variable Y for a given unit, sample of units, or population of units. Causal estimands are comparisons of two "potential outcomes," which represent the value that Y would take at a future point in time if the active treatment were applied, and the value of Y at the same future point in time if instead the control treatment were applied.

The formal mathematical notation for defining causal effects in randomized experiments as comparisons of potential outcomes dates back to Neyman (1923). However, these ideas were not generally used for causal inference in practice until Rubin (1974) extended the framework to encompass nonrandomized treatment assignment and other complications that arise in clinical practice, such as unintended missing data and noncompliance to assigned treatment.

First suppose that we are interested in inference on a single unit. For example, the unit could be a person "now" with a headache, the active treatment could be taking aspirin for the headache, and the control could be not taking aspirin. The outcome Y could be the intensity of the headache pain in two hours, with the two potential outcomes being the headache intensity with and without

---

This chapter is a modification and condensation of material from Imbens and Rubin (2015), as well as analogous chapters in other texts, including the second chapter in volume 27 of the *Handbook of Statistics* edited by C. R. Rao, J. P. Miller, and D. C. Rao (Rubin, 2007b) and a chapter in the second edition of *The New Palgrave Dictionary of Economics* edited by S. N. Durlauf and L. Blume (Imbens and Rubin, 2008).

aspirin. The causal effect of aspirin is the comparison of the person's headache intensity with and without the aspirin.

Let W be an indicator for which treatment is actually received, with $W = 1$ indicating receipt of the active treatment, and $W = 0$ indicating receipt of the control treatment. Also let $Y(1)$ be the value of the potential outcome if the unit received the active treatment, and $Y(0)$ be the value of the potential outcome if the unit received the control. The causal effect of the active treatment relative to the control is the comparison of $Y(1)$ and $Y(0)$. This is typically defined by the difference, $Y(1) - Y(0)$, but may be defined by any comparison of potential outcomes, such as the ratio, $Y(1)/Y(0)$, or difference in logs, $\log(Y(1)) - \log(Y(0))$.

Covariates have values that are unaffected by the treatments, such as age or sex of the unit in the headache example, and are denoted by X. Even when X represents a lagged value of the outcome, such as intensity of the headache two hours ago, $Y(1) - X$ is not the causal effect of aspirin unless $Y(0) = X$, but rather a change of headache intensity across time.

The "fundamental problem facing causal inference" (Rubin, 1978) is that we cannot observe both potential outcomes, $Y(0)$ and $Y(1)$. Rather, we get to observe only the value of the potential outcome under the treatment actually received, $Y^{obs} = WY(0) + (1 - W)Y(1)$. In particular, if we assign aspirin to treat the headache, then we observe $Y(1)$ but do not observe $Y(0)$, the outcome we would have seen if aspirin were not assigned. The outcomes under the treatment not received can be therefore be regarded as missing, and the problem of causal inference is one of drawing inferences about these missing potential outcomes using the observed data (Rubin, 1974).

Suppose now that there are N units, indexed by $i = 1, \ldots, N$, where each unit i has covariates X_i, treatment assignment W_i, and potential outcomes $(Y_i(0), Y_i(1))$. A causal estimand involves a comparison of $Y_i(0)$ and $Y_i(1)$ across all N units, or on a common subset of units; for example, the average causal effect across all units that are female as indicated by X_i. The array of values of X, $Y(0)$ and $Y(1)$ therefore contains all values about which we want to learn and is referred to as the "Science." Tables 31.1 and 31.2 illustrate the difference between the data that is actually observed in a study and the "Science," which contains the "truth" underlying the observed data.

When the definition of causal effects through potential outcomes is applied to a set of units, a possible complication arises in that the outcomes for a particular unit may depend on the treatments and outcomes for other units in the set. For example, treating some patients with an antidepression medication may affect the outcomes of other patients with whom they regularly interact. In these settings, the number of potential outcomes defined for each unit increases exponentially with the total number of units in the study. This problem is avoided with an assumption

Table 31.1 An example of the observed data for N units, where * denotes observed data and ? denotes missing data

Unit (i)	Covariates (X_i)	Treatment (W_i)	Potential outcomes $Y_i(0)$	$Y_i(1)$	Observed outcome (Y_i^{obs})
1	X_1	1	?	*	$Y_1(1)$
2	X_2	1	?	*	$Y_2(2)$
⋮	⋮	⋮	⋮	⋮	⋮
N–1	X_{N-1}	0	*	?	$Y_{N-1}(0)$
N	X_N	0	*	?	$Y_N(0)$

Table 31.2 An example of the "Science" for N units under SUTVA

Unit (i)	Covariates (X_i)	Potential outcomes $Y_i(0)$	$Y_i(1)$	Unit-level causal effect
1	X_1	$Y_1(0)$	$Y_1(1)$	$Y_1(1) - Y_1(0)$
2	X_2	$Y_2(0)$	$Y_2(1)$	$Y_2(1) - Y_2(0)$
⋮	⋮	⋮	⋮	⋮
N–1	X_{N-1}	$Y_{N-1}(0)$	$Y_{N-1}(1)$	$Y_{N-1}(1) - Y_{N-1}(0)$
N	X_N	$Y_N(0)$	$Y_N(1)$	$Y_N(1) - Y_N(0)$

called the Stable Unit Treatment Value Assumption (SUTVA; Rubin, 1980), which asserts that there is "no interference between different units" (Cox, 1958, p. 19) and also that there are no hidden versions of the treatments for the units (Neyman & Iwaszkiewicz, 1935; Rubin, 1990a). Under SUTVA, the potential outcomes for the ith unit depend only on the treatment assignment of that unit. Clearly, this is a strong assumption; however, without some such restrictions, which limit the number of potential outcomes that must be considered when defining causal effects, causal inference is impossible.

THE ASSIGNMENT MECHANISM AND ASSIGNMENT-BASED CAUSAL INFERENCE

The second part of the RCM framework is the specification of an "assignment mechanism," which defines a probabilistic model for how some units received the active treatment and how other units received the control.

To illustrate the need to posit an assignment mechanism, consider the situation in which each of four patients undergoing treatment for a psychological condition is assigned to one of two possible medications, medication A or medication B. Suppose we are interested in

Table 31.3 Artificial example of a confounded assignment mechanism

	The science			The observed data	
Patient	$Y_i(A)$	$Y_i(B)$	$Y_i(A) - Y_i(B)$	W_i	$Y_i(A)$	$Y_i(B)$
1	6	1	5	A	6	?
2	12	3	9	A	12	?
3	8	9	−1	B	?	9
4	10	11	−1	B	?	11
Mean	9	6	3		9	10

evaluating the causal effect of medication A compared to medication B on, say, the number of psychotic episodes that occur within one year of treatment. An artificial example of data that might be observed in this situation is depicted in Table 31.3.

Here, W is a vector of assignments actually received, with W = A for patients assigned and receiving medication A, and W = B for patient assigned and receiving medication B. The values Y(A) and Y(B) represent the potential outcomes under each medication, and Y(A) − Y(B) is a vector of individual causal effects, defined as the difference in number of psychotic episodes within one year of treatment under medications A and B. Implicit within this representation is the SUTVA assumption. Examining the last two columns in Table 31.3, we see that each patient is assigned to the treatment that leads to the maximum individual benefit for that patient. But what general conclusions do the observed data suggest? The average observed number of psychotic episodes for those given medication A is 1 less than for those given medication B, so the obvious, but incorrect, conclusion is that medication A is superior for the pool of patients for which these four patients are representative. This conclusion is wrong because the average causal effect (i.e., the average of the individual causal effects in Table 31.3, column 4) clearly favors medication B for these patients, giving an average reduction of three psychotic episodes. The discrepancy arises here because the comparison of observed results assumes that treatments were randomly assigned, rather than as they were, to provide the maximal benefit to each individual patient.

This simple example shows why consideration of the assignment mechanism is crucial for valid inference about causal effects. Without a model for how treatments are assigned to units, formal causal inference is impossible. This means that probabilistic statements about causal effects (such as confidence intervals or claims that estimates are unbiased) cannot be made without knowledge of, or assumptions about, the nature of the mechanism.

The assignment mechanism specifies the conditional probability of the vector of assignments $W = (W_1, \ldots, W_N)^T$ given the matrix of all covariates and potential outcomes:

$$P(W|X, Y(0), Y(1)). \quad (31.1)$$

An assignment mechanism is called "unconfounded" (Rubin, 1990a) if the probability of assignment to active treatment versus control does not depend on either Y(0) or Y(1) and is governed by the observed covariates:

$$P(W|X, Y(0), Y(1)) = P(W|X). \quad (31.2)$$

Assignment mechanisms are "probabilistic" if each unit has a positive probability of receiving either treatment:

$$0 < P(W_i = 1|X, Y(0), Y(1)) < 1, \text{ for all } i = 1, \ldots, N. \quad (31.3)$$

"Strongly ignorable" assignment mechanisms (Rosenbaum & Rubin, 1983b) satisfy Equations 31.2 and 31.3 and thus have unit level probabilities, or "propensity scores," $P(W_i = 1|X_i)$, that are strictly between 0 and 1, and are free of all potential outcomes. These assignment mechanisms typically allow straightforward estimation of causal effects. Randomized experiments are special in that they have known strongly ignorable assignment mechanisms by design. Strongly ignorable assignment mechanisms are also the basic template for the analysis of observational, nonrandomized studies, because two units with the same propensity score but different treatments can essentially be viewed as if they were randomized into the two treatment conditions. Therefore, in observational studies, it is possible to incorporate a "design phase" of the analysis, which restores an assumed underlying experimental design, after which inference is straightforward based only on the assignment mechanism. This is presented in greater detail in the section on Design and Analysis of Observational Studies.

APPROACHES TO INFERENCE IN RANDOMIZED STUDIES

Statistical analysis of observed data to draw inferences about causal effects can be carried out with only the first two parts of the RCM using randomization-based approaches for inference, which are based only on a model for the assignment mechanism, $P(W|X, Y(0), Y(1))$. There are two distinct forms of randomization-based inference, one due to Neyman (1923) and the other due to Fisher (1925). Both of these approaches view the potential outcomes as fixed quantities and regard the treatment assignment as the sole source of randomness. Alternatively, the third and optional part of the RCM implements a model-based, or Bayesian, approach for inference that allows specification of a distribution on the science, $P(X, Y(0), Y(1))$, and views the potential outcomes as random variables. Rubin (1990b) provides a general discussion and comparison of these modes of inference for causal effects, and a full-length text that presents each of these perspectives in greater detail is Imbens and Rubin (2015, chapters 5, 6, and 8). There are also alternative approaches to inference in the context of randomized experiments, such as those that also consider stochastic potential outcomes (VanderWeele & Robins, 2012). However, we focus here on the three major modes for inference commonly used within the RCM.

Fisherian Randomization-Based Inference

Given data from a randomized experiment, Fisher's mode of inference focuses on assessing the "sharp null hypothesis," which is usually $Y_i(0) = Y_i(1)$ for all units and asserts that receipt of active treatment versus control has no effect on the potential outcomes. Under this null hypothesis, all potential outcomes are known from the observed values of the potential outcomes, Y^{obs}, because $Y(1) = Y(0) = Y^{obs}$. Thus, using Fisher's approach, the missing potential outcomes, Y_i^{mis}, are implicitly imputed as $Y_i^{mis} = Y_i^{obs}$ for all $i=1, \ldots, N$.

Consider a statistic, S, defined as a function of the assignment vector, W, the observed outcomes, Y^{obs}, and possibly covariates, X (e.g., the difference of the observed averages for units in the study exposed to active treatment and units exposed to control treatment, denoted $\bar{y}_1 - \bar{y}_0$. Under the sharp null hypothesis, the value of any such statistic is known, not only for the observed assignment, but for all possible assignments W. Further, the assignment mechanism allows us to calculate the probability of each possible assignment. For example, in a completely randomized experiment with $N = 2n$ units, n units are randomly assigned to the active treatment ($W_i = 1$) and the remaining n are assigned to the control treatment ($W_i = 0$). Therefore, any assignment W that has n 1's and n 0's has probability $\binom{N}{n}^{-1}$ and all other assignments have probability 0.

Under the null hypothesis and given the assignment mechanism, it follows that we can generate the "randomization distribution" of S, which is the distribution of the values of S that would be observed under all possible randomizations of units into active treatment and control groups. Using this distribution, we can compare the value of the statistic actually observed, S^{obs}, against the randomization distribution of S. If S^{obs} is "unusual," given the distribution of S induced by the null hypothesis, it is taken as evidence against the null hypothesis. "Unusual" is measured by the probability that a value as extreme or more extreme (in terms of magnitude) would have been observed if the null hypothesis were true, and is defined a priori by a p-value or significance level. Thus, unless the data suggest that the sharp null hypothesis of no unit-level treatment effects is false (for an appropriate choice of statistic, S), it is not easy to claim evidence for differing efficacies of the treatments.

In summary, Fisher's approach requires three steps: (1) specifying a sharp null hypothesis, (2) choosing a statistic, and (3) calculating the probability of obtaining a test statistic at least as extreme as the observed one under the sharp null hypothesis. The first two choices should be made based on the scientific nature of the problem. Although it is common to test the null hypothesis of zero treatment effect, Fisher's approach can be extended to other sharp null hypotheses, that is, a null hypothesis such that from knowledge of Y^{obs}, $Y(1)$ and $Y(0)$ are known. For example, we might consider an additive null, which asserts that for each unit, $Y(1) - Y(0)$ is a specified constant, perhaps $Y_i(1) - Y_i(0) = 3$ for all $i = 1, \ldots, N$. The collection of such null hypotheses that do not lead to an extreme p-value can be used to create interval estimates of the causal effect, for example by assuming additivity. In particular, by repeatedly evaluating an additive null, $Y_i(1) = Y_i(0) + \tau$ over a range of values τ, we can identify a set of values that have p-values larger than 0.05, which provides a 95 percent "Fisher interval" for the treatment effect.

This form of inference is intuitive and straightforward to implement, but is limited in the sense that the sharp null hypothesis is very restrictive, the choice of statistic is somewhat arbitrary, and a small p-value does not necessarily imply that deviations from the null hypothesis are substantively important. Simply being confident that there is some nonzero treatment effect for some units is often not sufficient to inform decisions in clinical practice. Also notice that Fisher's perspective provides no ability to generalize beyond the particular set of units in the experiment. These limitations are not present in Neyman's approach, which is presented below.

Neymanian Randomized-Based Inference

In many situations, researchers are interested in making inferences about the average treatment effect defined across a population of units without being concerned about the unit-level effects. Neyman (1923) developed an approach for inference in randomized experiments that applies in these settings.

Unlike the other two approaches presented here, which draw inferences about causal effects through implicit or explicit imputation of the missing potential outcomes, Neyman's form of randomization-based inference relies on properties of statistical procedures under repeated sampling. In particular, inferences are made by evaluating the expectations of statistics over the distribution induced by the assignment mechanism. Typically, an unbiased estimator of the causal estimand is created, and an unbiased, or upwardly biased, estimator of the sampling variance of that unbiased estimator is found (bias and sampling variance are both defined with respect to the randomization distribution). Then, an appeal is made to the central limit theorem for the normality of the estimator over its randomization distribution, which allows for the construction of a large sample confidence interval for the causal estimand.

The causal estimand typically evaluated under Neyman's approach is the average treatment effect $E[Y(1) - Y(0)]$, where the expectation is over all units in the population being studied, rather than just those units in the experiment. The traditional estimator for this estimand is the difference in observed averages between the treatment and control groups, $\bar{y}(1) - \bar{y}(0)$, which can be shown to be unbiased for $E[Y(1) - Y(0)]$ in randomized experiments. The common choice for estimating the variance of $\bar{y}(1) - \bar{y}(0)$ over its randomization distribution, in a randomized experiment with $N = n_1 + n_2$ units, is $se^2 = \frac{s_1^2}{n_1} + \frac{s_0^2}{n_0}$, where s_1^2, s_0^2, n_1, and n_0 are the observed sample variances and sample sizes in the two treatment groups, respectively. Neyman (1923) showed that this estimator overestimates the actual sampling variance of $\bar{y}(1) - \bar{y}(0)$, unless additivity holds (i.e., unless all individual causal effects are the same), in which case se^2 is unbiased for the variance of $\bar{y}(1) - \bar{y}(0)$. The standard 95 percent confidence interval for $Y(1) - Y(0)$ is $\bar{y}(1) - \bar{y}(0) \pm 1.96 se$. Note that unlike the Fisher interval, which makes a precise probabilistic statement about the possible values of the causal estimand given the observed data, a 95 percent confidence interval constructs a range of plausible values for the causal estimand based on the observed data such that, in large enough samples, this range will include the "true" value of the estimand in at least 95 percent of the possible random assignments. That is, if we were to repeat the sampling procedure a large number of times and calculate a confidence interval for each sample, then 95 percent of the resulting intervals would contain the "true" value of the causal estimand.

Neyman's mode of randomization-based inference forms the theoretical foundation for much of what is done in important areas of application, including in clinical psychology. However, like Fisher's approach, there are many situations where we are interested in questions that cannot be adequately addressed using this approach. For example, we may want to estimate average treatment effects while conditioning on the distributions of covariates within each of the treatment and control groups. Such settings can often be more flexibly addressed using a Bayesian, or model-based approach, discussed below.

Posterior Predictive, or Model-Based, Casual Inference

Bayesian inference for causal effects directly and explicitly confronts the missing potential outcomes, Y^{mis}, by using the specification for the assignment mechanism, $p(W | X, Y(0), Y(1))$, and the specification for the science, $p(X, Y(0), Y(1))$, to derive the posterior predictive distribution of Y^{mis}:

$$P(Y^{mis} | X, Y^{obs}, W). \quad (31.4)$$

This distribution is posterior because it conditions on all observed values (X, Y^{obs}, W) and is predictive because it predicts (probabilistically) the missing potential outcomes. Using this distribution and all of the observed data (including the observed potential outcomes, Y^{obs}, the observed assignments, W, and observed covariates, X), the posterior distribution of any causal effect can, in principle, be calculated. Specifically, the posterior predictive distribution specifies how to take a random draw of Y^{mis}. Once a value of Y^{mis} is drawn, any causal effect can be directly calculated by comparing the drawn value of Y^{mis} to the observed values, Y^{obs}, possibly while conditioning on the observed values of X. Repeatedly drawing values of Y^{mis} in this way and calculating the causal effect for each draw generates the posterior distribution of the causal estimand of interest. Performing inference in this way regards causal inference entirely as a missing data problem, where we multiply-impute (Rubin, 1987; 2004; Rubin & Schenker, 1987) the missing potential outcomes in order to generate a posterior distribution for causal effects.

For example, in the setting of a classical randomized experiment, units assigned active treatment have $Y_i(1)$ observed and $Y_i(0)$ missing. Under strong ignorability, which is true by design in randomized experiments, the conditional distribution of potential outcomes under control among treated units, $p(Y_i(0) | X_i, W_i = 1)$ can be shown to be the same as the conditional distribution among control units, $p(Y_i(0) | X_i, W_i = 0)$, where the latter distribution can be inferred using the observed data. Thus, the RCM tells us to build a realistic model of $Y_i(0)$ given X_i among control subjects, and use this model to (multiply) impute the missing $Y_i(0)$ among the treated from their X_i values, while being wary of issues of extrapolation beyond the observed range of X_i's among the controls. Analogously, we can build a model of $Y_i(1)$ given X_i among the treated, and use it to (multiply) impute the missing $Y_i(1)$ among controls. Repeating this process generates a distribution of treatment effects, which can then be used to evaluate causal estimands defined over multiple units, e.g., the sample average treatment effect. The general structure is outlined in Rubin (1978), and is developed in detail in Imbens and Rubin (2015); a chapter-length summary appears in Rubin (2007b).

The Bayesian, model-based approach to inference is sometimes regarded as more flexible than randomization-based approaches in the sense that it is intuitive to implement, it is not limited to inference about sharp null hypotheses, and it allows covariates to be easily incorporated into the analyses. However, model-based approaches achieve these benefits by specifying a distribution for the data, $p(Y(0), Y(1) | X)$, which randomization-based approaches avoid. Models must be carefully chosen to reflect the key features of the problem and should be checked for important violations of key assumptions and modified where necessary.

In summary, one should be willing to use the best features of all three of these inferential approaches. In simple randomized experiments with data that are normally or close to normally distributed, the three approaches lead to similar results, but they diverge in more complex settings. In these situations, each perspective provides different strengths. Ideally, in any practical setting, results obtained using each of the perspectives will not be in conflict with each other, at least with appropriate models, proper conditioning, and large samples.

The Role for Covariates in Randomized Experiments

As stated earlier, covariates are variables whose values are not affected by the treatment assignment: for example, variables that are recorded before units are randomized into treatment and control groups (e.g., year of birth, prior medical history in a clinical evaluation). Covariates are typically critical in pyschopathology, especially in observational studies. But it is important to clarify first the role of covariates in randomized settings. In classical randomized experiments, if a covariate is used in the assignment mechanism, as with a blocking variable in an experiment with a randomized block design, that covariate must be considered in the analysis because it affects the randomization distribution induced by the assignment mechanism. Further, even in experiments in which covariates are not used in the assignment mechanism, they can still be used to increase efficiency of estimation, regardless of the inferential approach.

The point about efficiency gains can be seen in the context of a randomized experiment where interest is on evaluating the efficacy of a treatment for depression, where X is baseline level of self-reported depression on a scale of one to ten and Y is self-reported depression on the same scale after treatment. From either the Fisherian or Neymanian perspectives, we can use covariates to define a new statistic to estimate the causal estimands of interest. For example, to estimate $Y(1) - Y(0)$, rather than the difference in average outcomes, $\bar{y}(1) - \bar{y}(0)$, one can use the difference in average observed gain scores, $(\bar{y}(1) - \bar{x}(1)) - (\bar{y}(0) - \bar{x}(0))$, where $\bar{x}(1)$ and $\bar{x}(0)$ are the average observed baseline measurements for the treatment and control groups, respectively, Suppose X and Y are correlated, which is to be expected in many clinical settings. From the Neymanian perspective, the sampling variance of the difference in average gain scores will be less than the sampling variance of the difference in average outcomes, which translates into smaller estimated sampling variances and therefore shorter confidence intervals for this estimator. From the Fisherian perspective, the reduced sampling variance of the difference in average gain scores translates into more sensitive tests of the sharp null hypothesis; that is, smaller deviations from the null hypothesis are more likely to lead to more significant p-values.

Under the Bayesian approach, baseline level of self-reported depression is likely predictive of missing posttreatment outcomes. Therefore, using baseline measurements will lead to more precise predictions of Y^{mis}, which will reduce the variance of the imputed missing potential outcomes. For instance, when using a linear regression model for the distribution of $(Y(0), Y(1))$ given X, the residual variance of posttreatment outcomes after adjusting for baseline measurements is less than the variance of the raw posttreatment outcomes.

COMPLICATIONS IN RANDOMIZED STUDIES

Although randomized experiments have been used for nearly a century, for decades they were only used with unconscious units, such as plants, animals, or industrial objects, none of which could be influenced by the knowledge that they were objects of experimentation or could act intentionally on this knowledge. Historically, it has been recognized that humans are different and can be influenced by the knowledge that they are part of an active experiment. In some cases, that knowledge alone has been shown to influence participants behavior, as with the well-known "Hawthorne effect" (Landsberger, 1958), where awareness of participation in a study influences outcomes. In other examples, experimentation with human subjects has led to complications such as noncompliance to assigned treatment, dropout (Angrist, J. D., Imbens, G. W., & Rubin, D. B., 1996), and censoring due to death (Frangakis, C. & Rubin, D. B., 2002). These and other such issues commonly occur in real-world studies, especially in clinical settings, and can pose a number of challenges for inference on causal effects (e.g., *see* Rubin, D. B., 2019). In the subsections below, we describe a number of these complications that may be particularly relevant for practitioners in clinical psychology and discuss how the underlying issues can be conceptualized using the RCM.

Noncompliance

Analysis of a randomized study is complicated when units do not comply with their assigned treatment. For example, some of the units assigned to take the active treatment may take the control treatment instead, and some assigned to take the control may manage to take the active treatment. An intention-to-treat (ITT) analysis (Sheiner & Rubin, 1995) compares the outcomes of units between treatment and control groups, ignoring the compliance information. The ITT effect measures the effect of being assigned to receive treatment rather than the effect of actually taking the active treatment. The ITT estimator is protected from selection bias by randomized treatment assignment, but in general is a biased measure of the "true" treatment effect. An alternative to intention-to-treat analysis is "as-treated" analysis, which compares the outcomes between units that actually took the active treatment versus those that actually took the control treatment, regardless of their original assignments. The problem then is that randomization is violated, and confounding factors associated with switching from the assigned treatment may corrupt the causal interpretation of treatment effects.

The definition of causal effects through potential outcomes provides a useful basis for understanding noncompliance problems and assumptions implied by various estimation strategies. There is a relatively large and growing literature on this topic (e.g., Hirano et al., 2000; Imbens & Rubin, 1997a, 1997b; Mealli & Mattei, 2012; Mealli & Rubin, 2002). For example, using the principal stratification framework of Angrist, Imbens, and Rubin (1996), units are classified as one of four types in the context of a particular randomized experiment: compliers, defiers, never-takers, and always-takers. Compliers are individuals who would take whatever treatment they are assigned, never-takers are those who would take the control treatment regardless of what assignment they receive, always-takers would take the active treatment regardless of what assignment they receive, and defiers are those who would take the opposite treatment to their assignment. Under this formulation, separate causal estimands can be defined within each principal stratum (e.g., the average causal effect among compliers). In practice, we can observe compliance only for the assigned treatment, so the full compliance status of each unit must be inferred using missing data techniques. The general topic of causal inference when there is noncompliance or other posttreatment complications is currently an active area of research (Feller, Mealli, & Miratrix, 2017; Forastiere, Mealli, & Miratrix, 2017; Mealli & Pacini, 2013).

Placebo Effects

Often in clinical settings, the knowledge that some individuals will receive an active drug with a particular anticipated effect creates the expectation among experimental units that this anticipated effect will be achieved among all participants, a version of so-called expectancy effects (Rosenthal & Fode, 1963; Rosenthal & Jacobson, 1966). This can lead to large "placebo effects" (i.e., large observed outcomes within the control group), which complicate the interpretation and implication of the results related to the efficacy of the active treatment. Discussion of this issue in various clinical settings can be found in Kessels, Mozer, and Bloemers (2017). The problem is formulated in the RCM framework in Mozer, Kessels, and Rubin (2018), which uses principal stratification to disentangle active treatment and placebo effects in RCTs using Bayesian inference. Related work based on latent class modeling and factor analysis in psychological studies is described in Loken (2004), and a comparison of Bayesian and frequentist approaches to inference in these settings is presented in Lu, Chow, and Loken (2017).

Unintended Missing Data

Missing data is a pervasive problem in clinical studies. Statistical analyses when there are missing values are often confined to somewhat ad hoc fixes, such as simply discarding the incomplete measured units. These analyses make strong assumptions about the nature of the missing data mechanism and often yield biased inferences about causal effects. A more attractive strategy in these settings is to use all of the information from both the complete and incomplete units in a manner that yields inferences that account for the loss of information arising from the fact that some values are missing. A number of statistical tools have been developed for this purpose, including methods that multiply-impute the missing data (Raghunathan et al., 2001; Schafer, 1999; Van Buuren, 2007) and likelihood-based methods such as the Expectation Maximization (EM) algorithm (Dempster, Laird, & Rubin, 1977) and its relatives, including data augmentation (Tanner & Wong, 1987) and the Gibbs sampler (Geman & Geman, 1984). A review of these techniques is presented in Graham (2009) and a text on this subject is Little and Rubin (2020).

Until recently, most methods for handling missing data have assumed the missing data are missing at random (MAR), meaning that the missingness does not depend on the missing values after conditioning on observed values (Rubin, 1976). This assumption is realistic in some settings, but is often violated in clinical studies where missing data arises because a unit may either drop out or "die" before its final outcome can be measured. For example, in longitudinal substance abuse trials, dropout is non-MAR when it is associated with treatment outcome, as is the case if participants fail to show for a clinic visit because they have died from substance-abuse-related issues or have relapsed and do not want to admit it. In these cases, outcomes can be regarded as "censored due to death" (Zhang & Rubin, 2003). On the other hand, participants may also fail to show because they have ceased all drug use and are not motivated to participate further in the study. Other forms of dropout, such as that arising from units relocating, may be plausibly unrelated to the outcome and hence potentially MAR. When data are not MAR, a model for the missing-data mechanism needs to be incorporated into the analysis to yield appropriate inferences. Approaches to inference in these settings have only recently been formulated (e.g., Ding et al., 2011; McConnell, Stuart, & Devaney, 2008; Rubin, 2006; Zhang & Rubin, 2004) and applied in practice (Frumento et al., 2012; Zhang, Rubin, & Mealli, 2008).

CAUSAL INFERENCE IN OBSERVATIONAL STUDIES

In observational studies, the researcher has no control over the assignment of treatment to units. This lack of control makes such studies inherently more controversial than evaluations based on randomized assignment, because units may select their own treatments, or their environments may impose treatments upon them, in a manner that leads to systematic differences between treatment and control groups. For example, in an observational study evaluating the effect of smoking on longevity, individuals who choose to smoke may be more likely than nonsmokers to engage in other risky behaviors that decrease life expectancy, so comparisons of outcomes between these groups may reflect these differences rather than effects of smoking itself. Thus, removing biases based on observed covariates and addressing uncertainty about other potentially hidden biases are central issues for causal inference in observational studies.

Design and Analysis of Observational Studies

Carefully designed randomized experiments are generally considered the gold standard for causal inference

(Cochran & Chambers, 1965). When designing an experiment, we do not have any outcome data, but we plan the collection, organization, and analysis of the data to improve our chances of obtaining precise causal answers. The same principle can and should be applied when approaching inference in observational studies (Rosenbaum, 2010; Rubin, 2007a, 2008).

When considering how to estimate causal effects in a nonrandomized study, a key first step is to conceptualize the hypothetical randomized experiment that could have led to the observed data set (Rubin, 2008; Bind, MAC., and Rubin, D. B., 2017). In this process, observational studies can be designed to approximate randomized experiments, specifically, without examining any outcome data. Analyses can then be planned in a manner such that the inferences obtained will be as close as possible to the inferences that would have been obtained in an analogous randomized experiment.

The objective of the design phase is to assemble a subsample of the original data such that, within this sample, the covariate distribution for those in the control group is approximately the same as in the treatment group. Thus, researchers can select a sample so as to approximate the balance between treatment and control groups that is achieved, in expectation, by randomization. An important advantage of this process is that the investigator may discover that there is essentially no overlap in the distributions of the covariates in the treated and control groups. In that case, there is no hope of drawing valid causal inferences from the data without making strong external assumptions involving extrapolation (Rubin, 1977), so it is important to allow that some data sets cannot support strong causal inference.

When there is sufficient overlap in the covariate distributions to allow for credible causal inference, the analysis phase of the study can then be carried out to draw inferences about causal effects. For example, one might use covariate data to create matched samples of treated and control units, then analyze the outcome data for these samples as if they had arisen from a randomized experiment with a matched pairs design. Alternatively, one might identify subgroups of units with similar covariate distributions, then compare outcomes between treatment and control groups within each subgroup and average across subgroups to estimate the average treatment effect. There are a number of strategies for assessing the degree of balance in the covariate distributions between treatment and control groups (see Imbens & Rubin, 2015, chapter 14), and identifying balanced samples of treated and control units is the subject of a large and rapidly expanding body of work on methods for matching (e.g., Rubin, D. B., 2006b) and subclassification (e.g., Rosenbaum, 2012; Zubizarreta et al., 2013; Zubizarreta, Paredes, & Rosenbaum, 2014).

A Critical Assumption

When designing and analyzing observational studies where the assignment mechanism is unknown, an important first step is to posit a particular form for the assignment mechanism. A crucial yet unverifiable assumption that is often made here is that the treatment assignment is unconfounded (Rubin, D. B., 1980) or strongly ignorable (Rosenbaum, P. R., & Rubin, D. B., 1983b). That is, we assume that all covariates that affect both the treatment assignment and the potential outcomes are observed in the study, such that it is plausible that the unknown assignment mechanism is unconfounded given these covariates. It is this assumption that allows us to interpret the assignment mechanism for an observational study as if, within populations of units with the same value for the covariates, a randomized experiment was conducted. Thus, we can analyze data from each subsample of units that have the same covariate values as if they came from a randomized experiment.

Obviously, the assumption of an unconfounded treatment assignment is a strong one and may not be plausible in all settings, especially in studies with few observed covariates. Other approaches offer alternatives to the strong ignorability assumption and still allow for essentially unbiased estimation of some causal estimands. For example, the Instrumental Variables (IV) approach (Angrist, J. D., Imbens, G. W., & Rubin, D. B., 2006) relies on the presence of additional information in the data made available through an "instrument," which is a variable known a priori to have a causal effect on the treatment assignment, but no direct effect on the outcome of interest. This instrument is used to uncover the causal effect of treatment. For example, continuing the earlier example on studying the effect of smoking on longevity, one might use the tax rate on cigarettes as an instrument, because this variable may be related to longevity only through its influence on the development of smoking habits. These methods have been widely applied in practice (Angrist et al., 1996; Imbens & Rubin, 1997b) and are described in detail in Angrist and Krueger (2001). Other methods for causal inference with observational data that relax the assumption of strong ignorability or rely on alternative assumptions have been the subject of much recent work (e.g., regression discontinuity designs; Lee & Lemieuxa, 2010; Li, Mattei, & Mealli, 2015; Mattei & Mealli, 2016).

Covariate Adjustment

When the assignment mechanism of an observational study can be assumed to be unconfounded i.e., after conditioning on observed covariates valid causal inferences can be obtained by adjusting for the covariates. These adjustments are often incorporated in the design phase of an observational study to improve balance in covariate distributions, for example using matching or subclassification, or by "trimming" the sample to discard treated units with covariate values for which there are no suitable counterparts in the control group. Covariate adjustments can also be incorporated in the analysis phase of the study to directly adjust the estimated treatment effects. For example, when the covariates are expected to be linearly related to the outcome, one common approach to covariate adjustment uses regression of the outcome on the indicator for treatment assignment and on the covariates. Regression-based adjustments are straightforward to implement, but can be unreliable when the treatment and control groups differ substantially in their covariate distributions, because the extrapolation implied by the regression model relies strongly on assumptions such as linearity (Cochran & Rubin, 1973; Rubin, 1973).

Many methods for covariate adjustment, in both the design and analysis phases of an observational study, adjust for an estimate of the propensity score. When propensity scores are unknown (which is usually the case in observational studies), they must first be estimated using one of a number of methods (e.g., discriminant analysis, logistic regression, generalized boosted models). Adjustment methods based on the propensity score include matching or subclassification of treated units and controls based on the estimated propensity score, adding the estimated propensity score as a covariate in a regression model with limited covariates, or weighting units by estimates of the propensity score. For example, using matching, treated-control pairs can be formed with similar values of the propensity score (and perhaps the covariates); outcomes can then be compared across the resulting matched samples to estimate the causal effect of treatment. Alternatively, through subclassification, subjects might be cross-classified by treatment group and by quintiles of the estimated propensity score; separate estimates of treatment effects are then made within each propensity score subclass and then combined across the five subclasses. These techniques are described in (Imbens & Rubin, 2015, chapter 12), and in (Rosenbaum, 2010, part II).

Recent methods that directly and flexibly balance covariate distributions include Zubizarreta (2012, 2015). These methods are implemented in the packages "designmatch" and "sbw" in R (Zubizarreta & Kilcioglu, 2016). Other recent work has focused on extending methods from machine learning to this setting, either to estimate the outcome surface or propensity scores, or to combine them using a "doubly-robust estimator" (Hirano & Imbens, 2001; Robins, Rotnitzky, & Zhao, 1995). See Van der Laan and Rose (2011) or Athey and Imbens (2015) for an introduction to these methods and a discussion of their utility for causal inference in both observational studies and randomized experiments.

SENSITIVITY ANALYSES

Typically, in observational studies in clinical psychology we do not know that we have available a set of covariates that is adequate to support the claim of an unconfounded assignment mechanism. Thus, even if we have successfully adjusted for all covariates at hand, we cannot be sure that there is not some hidden unmeasured covariate that, if included, would alter the results. A sensitivity analysis posits the existence of such a covariate and how it relates to both treatment assignment and outcome, and it examines how the results change if this variable were used for adjustment. A classic example is the analysis by Cornfield et al. (1959; reprinted in 2009) of the relationship between lung cancer and smoking. Other approaches include methods based on maximum likelihood (Rosenbaum & Rubin, 1983a), methods based on the randomization distribution (Rosenbaum, 2002), and bounding methods (Horowitz & Manski, 2000; Manski & Nagin, 1998).

Sensitivity analyses are also important in randomized experiments subject to complications, such as those described above, because methods that attempt to address these complications are typically sensitive to model specification and other design decisions required by the researcher. For example, the approach developed in Mozer and colleagues (2018) for disentangling active treatment and placebo effects in randomized experiments posits a parametric model on the joint distribution of potential outcomes under the active treatment and under the control. Here, the choice of different models may lead to different results, so researchers are advised to explore whether results are robust to this specification by performing sensitivity analyses. Rosenbaum (2010, chapter 3) provides a general framework for the use of sensitivity analyses in observational studies, and their use in both observational studies and randomized experiments is discussed in Ding and VanderWeele (2016).

CONCLUSION

In this chapter, we have attempted to convey the power of the potential outcomes formulation of causal effects in a variety of settings and provide an introduction to a number of methodologies that may be relevant for psychopathology research. We by no means provide a comprehensive review of the vast literature on the analysis of causal effects, which cannot be covered even in a single book. Current research in causal inference is exceedingly lively, involving extensive interactions between psychologists, behavioral scientists, computer scientists, economists, epidemiologists, philosophers, statisticians, and others. Armed with simple but powerful ideas such as the potential outcomes definition of causal effects, we look forward to continued methodological developments in this crucial and fascinating area of empirical research, of critical importance to clinical psychology.

REFERENCES

Angrist, J., & Krueger, A. B. (2001). Instrumental Variables and the Search for Identification: From Supply and Demand to Natural Experiments. *Journal of Economic Perspectives, 15*(4), 69–85.

Angrist, J. D., Imbens, G. W., & Rubin, D. B. (1996). Identification of Causal Effects Using Instrumental Variables. *Journal of the American Statistical Association, 91*(434), 444–455.

Athey, S., & Imbens, G. W. (2015). Machine Learning Methods for Estimating Heterogeneous Causal Effects. *Stat, 1050*(5), 1–26.

Bind, MAC., & Rubin, D. B. (2017). Bridging observational studies and randomized experiments by embedding the former in the latter. *Statistical Methods in Medical Research 28*(7), 1958–1978.

Cochran, W. G., & Chambers, S. P. (1965). The Planning of Observational Studies of Human Populations. *Journal of the Royal Statistical Society. Series A (General), 128*(2), 234–266.

Cochran, W. G., & Rubin, D. B. (1973). Controlling Bias in Observational Studies: A Review. *Sankhyā: The Indian Journal of Statistics, Series A, 35*(4), 417–446.

Cornfield, J., Haenszel, W., Hammond, E. C., Lilienfeld, A. M., Shimkin, M. B., & Wynder, E. L. (1959). Smoking and Lung Cancer: Recent Evidence and a Discussion of Some Questions. *Journal of the National Cancer Institute, 22*(1), 173–203.

Cornfield, J., Haenszel, W., Hammond, E. C., Lilienfeld, A. M., Shimkin, M. B., & Wynder, E. L. (2009). Smoking and Lung Cancer: Recent Evidence and a Discussion of Some Questions. *International Journal of Epidemiology, 38*(5), 1175–1191.

Cox, D. R. (1958). *Planning of Experiments*. New York: Wiley.

Dempster, A. P., Laird, N. M., & Rubin, D. B. (1977). Maximum Likelihood from Incomplete Data via the EM Algorithm. *Journal of the Royal Statistical Society. Series B (Methodological), 39*(1), 1–38.

Ding, P., & VanderWeele, T. J. (2016). Sensitivity Analysis without Assumptions. *Epidemiology (Cambridge, Mass.), 27*(3), 368–377.

Ding, P., Geng, Z., Yan, W., & Zhou, X.-H. (2011). Identifiability and Estimation of Causal Effects by Principal Stratification with Outcomes Truncated by Death. *Journal of the American Statistical Association, 106*(496), 1578–1591.

Feller, A., Mealli, F., & Miratrix, L. (2017). Principal Score Methods: Assumptions, Extensions, and Practical Considerations. *Journal of Educational and Behavioral Statistics, 42*(6), 726–758.

Fisher, R. A. (1925). *Statistical Methods for Research Workers*. Edinburgh: Oliver and Boyd.

Forastiere, L., Mealli, F., Miratrix, L. (2017). Posterior Predictive p-Values with Fisher Randomization Tests in Noncompliance Settings: Test Statistics vs Discrepancy Measures. *Bayesian Analysis, 13*(3), 681–701.

Frangakis, C., & Rubin, D. B. (2002). Principal Stratification in Causal Inference. *Biometrics, 58*(1), 21–29.

Frumento, P., Mealli, F., Pacini, B., & Rubin, D. B. (2012). Evaluating the Effect of Training on Wages in the Presence of Noncompliance, Nonemployment, and Missing Outcome Data. *Journal of the American Statistical Association, 107*(498), 450–466.

Geman, S., & Geman, D. (1984). Stochastic Relaxation, Gibbs Distributions, and the Bayesian Restoration of Images. *IEEE Transactions on Pattern Analysis and Machine Intelligence, 6*, 721–741.

Graham, J. W. (2009). Missing Data Analysis: Making It Work in the Real World. *Annual Review of Psychology, 60*, 549–576.

Hirano, K., & Imbens, G. W. (2001). Estimation of Causal Effects Using Propensity Score Weighting: An Application to Data on Right Heart Catheterization. *Health Services and Outcomes Research Methodology, 2*(3), 259–278.

Hirano, K., Imbens, G. W., Rubin, D. B., & Zhou, X.-H. (2000). Assessing the Effect of an Influenza Vaccine in an Encouragement Design. *Biostatistics, 1*(1), 69–88.

Holland, P. W. (1986). Statistics and Causal Inference. *Journal of the American Statistical Association, 81*(396), 945–960.

Horowitz, J. L., & Manski, C. F. (2000). Nonparametric Analysis of Randomized Experiments with Missing Covariate and Outcome Data. *Journal of the American Statistical Association, 95*(449), 77–84.

Imbens, G. W., & Rubin, D. B. (1997a). Bayesian Inference for Causal Effects in Randomized Experiments with Noncompliance. *The Annals of Statistics, 25*(1), 305–327.

Imbens, G. W., & Rubin, D. B. (1997b). Estimating Outcome Distributions for Compliers in Instrumental Variables Models. *Review of Economic Studies, 64*(4), 555–574.

Imbens, G. W., & Rubin, D. B. (2008). Rubin Causal Model. *The New Palgrave Dictionary of Economics, 7*, 255–262.

Imbens, G. W., & Rubin, D. B. (2015). *Causal Inference in Statistics, Social, and Biomedical Sciences*. New York: Cambridge University Press.

Kessels, R., Mozer, R., & Bloemers, J. (2017). Methods for Assessing and Controlling Placebo Effects. *Statistical Methods in Medical Research, 28*(4), 1141–1156.

Landsberger, H. A. (1958). *Hawthorne Revisited: Management and the Worker, Its Critics, and Developments in Human Relations in Industry*. Ithaca, NY: Cornell University Press.

Lee, D. S., & Lemieuxa, T. (2010). Regression Discontinuity Designs in Economics. *Journal of Economic Literature, 48*(2), 281–355.

Li, F., Mattei, A., & Mealli, F. (2015). Evaluating the Causal Effect of University Grants on Student Dropout: Evidence from a Regression Discontinuity Design Using Principal Stratification. *Annals of Applied Statistics, 9*(4), 1906–1931.

Little, R. J., & Rubin, D. B. (2020). *Statistical Analysis with Missing Data*, 3rd edition. New York: John Wiley.

Loken, E. (2004). Using Latent Class Analysis to Model Temperament Types. *Multivariate Behavioral Research, 39*(4), 625–652.

Lu, Z.-H., Chow, S.-M., & Loken, E. (2017). A Comparison of Bayesian and Frequentist Model Selection Methods for Factor Analysis Models. *Psychological Methods, 22*(2), 361–381.

Manski, C. F., & Nagin, D. S. (1998). 3. Bounding Disagreements about Treatment Effects: A Case Study of Sentencing and Recidivism. *Sociological Methodology, 28*(1), 99–137.

Mattei, A., & Mealli, F. (2016). Regression Discontinuity Designs as Local Randomized Experiments. *Observational Studies, 2*, 156–173.

McConnell, S., Stuart, E. A., & Devaney, B. (2008). The Truncation-by-Death Problem: What to Do in an Experimental Evaluation When the Outcome Is Not Always Defined. *Evaluation Review, 32*(2), 157–186.

Mealli, F., & Mattei, A. (2012). A Refreshing Account of Principal Stratification. *International Journal of Biostatistics, 8*(1), 1–19.

Mealli, F., & Pacini, B. (2013). Using Secondary Outcomes to Sharpen Inference in Randomized Experiments with Noncompliance. *Journal of the American Statistical Association, 108*(503), 1120–1131.

Mealli, F., & Rubin, D. B. (2002). Assumptions When Analyzing Randomized Experiments with Noncompliance and Missing Outcomes. *Health Services and Outcomes Research Methodology, 3*(3), 225–232.

Mozer, R., Kessels, R., & Rubin, D. B. (2018). Disentangling Treatment and Placebo Effects in Randomized Experiments Using Principal Stratification – An Introduction. In M. Wiberg, S. Culpepper, R. Janssen, J. González, & D. Molenaar (Eds.), *Quantitative Psychology* (pp. 11–23). Cham: Springer.

Neyman, J. (1923). On the Application of Probability Theory to Agricultural Experiments: Essay on Principles, Section 9

(Translated) in *Statistical Science* (1990), 5(4), 465–480). *Annals of Agricultural Sciences*, *10*, 1–51.

Neyman, J., & Iwaszkiewicz, K. (1935). Statistical Problems in Agricultural Experimentation. *Supplement to the Journal of the Royal Statistical Society*, *2*(2), 107–180.

Pearl, J. (2009). Causal Inference in Statistics: An Overview. *Statistics Surveys*, *3*, 96–146.

Raghunathan, T. E., Lepkowski, J. M., Van Hoewyk, J., & Solenberger, P. (2001). A Multivariate Technique for Multiply Imputing Missing Values Using a Sequence of Regression Models. *Survey Methodology*, *27*(1), 85–96.

Robins, J. M. (2001). Data, Design, and Background Knowledge in Etiologic Inference. *Epidemiology*, *12*(3), 313–320.

Robins, J. M., Rotnitzky, A., & Zhao, L. P. (1995). Analysis of Semiparametric Regression Models for Repeated Outcomes in the Presence of Missing Data. *Journal of the American Statistical Association*, *90*(429), 106–121.

Rosenbaum, P. R. (2002). Observational Studies. In *Observational Studies* (pp. 1–17). New York: Springer.

Rosenbaum, P. R. (2010). *Design of Observational Studies*. New York: Springer.

Rosenbaum, P. R. (2012). Optimal Matching of an Optimally Chosen Subset in Observational Studies. *Journal of Computational and Graphical Statistics*, *21*(1), 57–71.

Rosenbaum, P. R., & Rubin, D. B. (1983a). Assessing Sensitivity to an Unobserved Binary Covariate in an Observational Study with Binary Outcome. *Journal of the Royal Statistical Society. Series B (Methodological)*, *45*(2), 212–218.

Rosenbaum, P. R., & Rubin, D. B. (1983b). The Central Role of the Propensity Score in Observational Studies for Causal Effects. *Biometrika*, *70*(1), 41–55.

Rosenthal, R., & Fode, K. L. (1963). The Effect of Experimenter Bias on the Performance of the Albino Rat. *Systems Research and Behavioral Science*, *8*(3), 183–189.

Rosenthal, R., & Jacobson, L. (1966). Teachers' Expectancies: Determinants of Pupils' IQ Gains. *Psychological Reports*, *19*(1), 115–118.

Rubin, D. B. (1973). The Use of Matched Sampling and Regression Adjustment to Remove Bias in Observational Studies. *Biometrics*, *29*(1), 185–203.

Rubin, D. B. (1974). Estimating Causal Effects of Treatments in Randomized and Nonrandomized Studies. *Journal of Educational Psychology*, *66*(5), 688–701.

Rubin, D. B. (1976). Inference and Missing Data. *Biometrika*, *63*(3), 581–592.

Rubin, D. B. (1977). Assignment to Treatment Group on the Basis of a Covariate. *Journal of Educational Statistics*, *2*(1), 1–26.

Rubin, D. B. (1978). Bayesian Inference for Causal Effects: The Role of Randomization. *Annals of Statistics*, *6*(1), 34–58.

Rubin, D. B. (1980). Randomization Analysis of Experimental Data: The Fisher Randomization Test Comment. *Journal of the American Statistical Association*, *75*(371), 591–593.

Rubin, D. B. (1987, 2004). *Multiple Imputation for Nonresponse in Surveys*. New York: Wiley.

Rubin, D. B. (1990a). Comment: Neyman (1923) and Causal Inference in Experiments and Observational Studies. *Statistical Science*, *5*(4), 472–480.

Rubin, D. B. (1990b). Formal Mode of Statistical Inference for Causal Effects. *Journal of Statistical Planning and Inference*, *25*(3), 279–292.

Rubin, D. B. (2006a). Causal Inference through Potential Outcomes and Principal Stratification: Application to Studies with "Censoring" Due to Death. *Statistical Science*, *21*(3), 299–309.

Rubin, D. B. (2006b). Matched Sampling for Casual Inference. New York: Wiley.

Rubin, D. B. (2007a). The Design versus the Analysis of Observational Studies for Causal Effects: Parallels with the Design of Randomized Trials. *Statistics in Medicine*, *26*(1), 20–36.

Rubin, D. B. (2007b). Statistical Inference for Causal Effects, with Emphasis on Applications in Epidemiology and Medical Statistics. *Handbook of Statistics*, *27*, 28–63.

Rubin, D. B. (2008). For Objective Causal Inference, Design Trumps Analysis. *Annals of Applied Statistics*, *2*(3), 808–840.

Rubin, D. B. (2019).

Rubin, D. B., & Schenker, N. (1987). Interval Estimation from Multiply-Imputed Data: A Case Study Using Census Agriculture Industry Codes. *Journal of Official Statistics*, *3*(4), 375–387.

Schafer, J. L. (1999). Multiple Imputation: A Primer. *Statistical Methods in Medical Research*, *8*(1), 3–15.

Sheiner, L. B., & Rubin, D. B. (1995). Intention-to-Treat Analysis and the Goals of Clinical Trials. *Clinical Pharmacology & Therapeutics*, *57*(1), 6–15.

Tanner, M. A., & Wong, W. H. (1987). The Calculation of Posterior Distributions by Data Augmentation. *Journal of the American Statistical Association*, *82*(398), 528–540.

Van Buuren, S. (2007). Multiple Imputation of Discrete and Continuous Data by Fully Conditional Specification. *Statistical Methods in Medical Research*, *16*(3), 219–242.

Van der Laan, M. J., & Rose, S. (2011). *Targeted Learning: Causal Inference for Observational and Experimental Data*. New York: Springer.

VanderWeele, T. J., & Robins, J. M. (2012). Stochastic Counterfactuals and Stochastic Sufficient Causes. *Statistica Sinica*, *22*(1), 379–392.

Zhang, J. L., & Rubin, D. B. (2003). Estimation of Causal Effects via Principal Stratification When Some Outcomes Are Truncated by Death. *Journal of Educational and Behavioral Statistics*, *28*(4), 353–368.

Zhang, J. L., & Rubin, D. (2004). Censoring Due to Death via Principal Stratification. *Journal of Educational and Behavioral Statistics*, *28*(4), 353–368.

Zhang, J. L., Rubin, D. B., & Mealli, F. (2008). Evaluating the Effects of Job Training Programs on Wages through Principal Stratification. *Modelling and Evaluating Treatment Effects in Econometrics*, *21*, 117–145.

Zubizarreta, J. R. (2012). Using Mixed Integer Programming for Matching in an Observational Study of Kidney Failure after Surgery. *Journal of the American Statistical Association*, *107*(500), 1360–1371.

Zubizarreta, J. R. (2015). Stable Weights That Balance Covariates for Estimation with Incomplete Outcome Data. *Journal of the American Statistical Association*, *110*(511), 910–922.

Zubizarreta, J. R., & Kilcioglu, C. (2016). Designmatch: Construction of Optimally Matched Samples for Randomized Experiments and Observational Studies that are Balanced by Design. [Computer software manual]. Retrieved from http://CRAN.R-project.org/package=designmatch (R package version 0.1.1).

Zubizarreta, J. R., Paredes, R. D., & Rosenbaum, P. R. (2014). Matching for Balance, Pairing for Heterogeneity in an Observational Study of the Effectiveness of For-Profit and Not-For-Profit High Schools in Chile. *Annals of Applied Statistics*, *8*(1), 204–231.

Zubizarreta, J. R., Small, D. S., Goyal, N. K., Lorch, S., & Rosenbaum, P. R. (2013). Stronger Instruments via Integer Programming in an Observational Study of Late Preterm Birth Outcomes. *Annals of Applied Statistics*, *7*(1), 25–50.

32 Analyzing Nested Data

Multilevel Modeling and Alternative Approaches

DANIEL J. BAUER, DANIEL M. MCNEISH, SCOTT A. BALDWIN, AND PATRICK J. CURRAN

Many of the phenomena studied in clinical psychology involve inherently nested data structures, such as clients nested within therapists, siblings nested within families, or repeated measures nested within person over time. It is thus critical for clinical researchers to have a general understanding of the unique conceptual and statistical issues that nested data present. The goal of this chapter is to provide an introductory overview of the analysis of nested data.

As a general definition, nested data exists whenever multiple observations are sampled from each of multiple units. The data then consist of two levels of sampling, with observations sampled at Level 1 nested (or clustered) within units sampled at Level 2.[1] In this chapter, we will discuss two principal types of nested data. In hierarchically nested data, multiple individuals are sampled from each of many groups, such as siblings within families. In contrast, in longitudinally nested data, individuals are observed on multiple occasions over time, such as in a treatment evaluation study. As we describe in this chapter, researchers must be carefully attuned to the conceptual and statistical issues that arise when working with nested data structures like these, both to fully capitalize on the available information within the data and to avoid drawing false conclusions about the population under study.

We begin by describing scenarios in which nested data commonly arise in clinical psychology research, drawing on examples in the research literature. We then describe two key issues that complicate the analysis of nested data. The first is dependence of observations: that is, the tendency for observations within a unit to be more similar to one another than observations from different units. This data feature violates the assumption of independence made in many statistical models, and calls for the application of models specifically designed for use with nested data. The second critical issue is the need to identify the level of the data structure at which predictors exert their effects. For instance, in hierarchical data, a predictor might operate at the individual level, the group level, or both.

We show how to address these key analytic issues through the application of appropriate statistical methods. Our primary focus will be on the application of multilevel models with random effects, but we also follow this with a brief survey of other approaches for analyzing nested data, including an assessment of relative strengths and weaknesses. We close by considering a number of special situations that often occur in clinical psychology research, including partial nesting (where only a subset of the data has a nested structure), small sample sizes at the upper level, and pre/posttreatment designs. Overall, our hope is that this chapter will provide researchers with both a general orientation to the issues that arise with nested data and a useful entry point to the broader methodological literature on this topic.

NESTED DATA IN CLINICAL PSYCHOLOGY RESEARCH

Nested data structures are common in many areas of clinical psychology, including intervention trials, psychotherapy process research, and observational psychopathology studies. Both hierarchical and longitudinal structures commonly arise in clinical psychology research, sometimes in combination. For example, in psychotherapy trials patients are nearly always nested within therapists, meaning that each patient sees only one therapist but each therapist treats several patients. Thus, patients are *clustered* within therapist (Crits-Christoph & Mintz, 1991), resulting in a hierarchical structure. Likewise, in a trial involving a group-administered treatment, patients are nested within groups – each patient is part of only one group, but each group has multiple patients (Baldwin, Murray, & Shadish, 2005). Patients could also be clustered within a couple in marriage therapy or within a family in family therapy (Kenny, 1995). Hierarchical clustering can also occur in other ways, such as when conducting a multisite trial, a school-based study, or when assessing peers within friendship groups.

---

[1] Three or more levels of sampling are also possible.

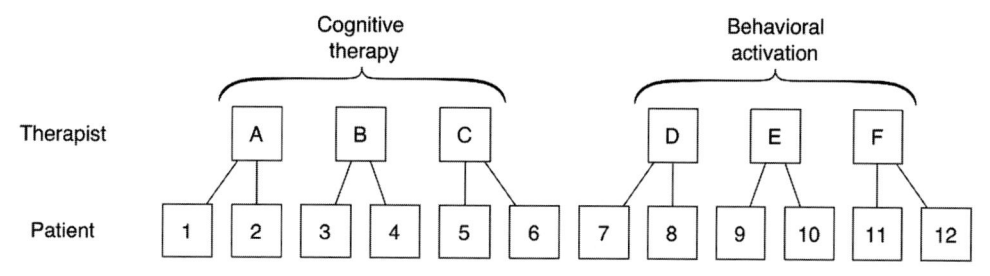

Figure 32.1 Schematic representation of the therapy arms of Dimidjian et al. (2006). We limited the drawing to two patients per therapist to enhance the readability of the figure

As one example, consider the psychotherapy trial by Dimidjian and colleagues (2006) that compared cognitive therapy, behavioral activation, antidepressant medication, and placebo in the treatment of depression. We focus just on the cognitive therapy and behavioral activation arms of this study. Eighty-eight patients were randomly assigned to cognitive therapy (n = 45) and behavioral activation (n = 43). Three therapists provided cognitive therapy and three therapists provided behavioral activation (Dimidjian et al., 2006, p. 660). On average, therapists in the cognitive therapy condition saw 15 patients and therapists in the behavioral activation saw 14.3 patients. Figure 32.1 shows an overview of the nature of the design, which indicates the nesting of patients within therapist and the assignment of therapists to treatment (note that it is also possible for therapist to be crossed with treatment, where each therapist provides all treatments; for example, see the design in Taylor et al., 2003).

This schematic also allows us to draw an important distinction between random versus fixed factors in our models. Random factors involve units that are sampled from a broader population of potential units. For instance, the investigators might have recruited different patients into the study or might have employed different therapists. Both patients and therapists are thus samples from populations of potential patients and therapists. These factors are referred to as *random* based on the idea that the particular units in the sample were randomly drawn from a population. This idea applies even if neither therapists nor patients were actually chosen randomly but still represent samples from broader populations. In analyzing the data, the goal is to make inferences that pertain to these broader populations (i.e., any patient seeing any therapist). In contrast, assignment to cognitive therapy or behavioral activation is a *fixed* factor. There are only two treatments of interest and these were not sampled from a population of potential treatments. One would not seek to make inferences about the universe of all possible treatments, but would instead restrict inferences to just these two. When analyzing nested data, fixed factors are treated as nominal predictors (e.g., treatment assignment is just a characteristic of the therapist), as in analysis of variance (ANOVA) or linear regression. The primary concern is how to account properly for the random factors.

Longitudinal data structures are also common in clinical research. For instance, developmental psychopathology research relies on longitudinal designs to evaluate age-related within-person changes in symptomatology. Likewise, modern psychotherapy trials are longitudinal, with two or more time points to evaluate both short- and long-term outcomes. One example of longitudinally nested data comes from a study by Lenzenweger, Johnson, and Willet (2004), who assessed the stability of personality disorder symptoms in 258 participants over their first, second, and fourth years of college using the International Personality Disorders Examination. Repeated measurements of personality disorder symptoms are grouped within person in this example, just as patients were grouped within therapists in the hierarchical example. In the longitudinal setting, both the repeated observations and the individuals are viewed as random factors, reflecting the sampling of observations from a set of individuals. The goal is to make inferences about individual differences in change over time within the population as a whole.

It is also common for a data set to be both hierarchical and longitudinal. For example, in the Dimidjian (2006) trial introduced earlier, patients were measured at intake, eight weeks post-intake, and 16 weeks post-intake. Thus, the data consisted of three levels of sampling: repeated measures within persons within therapists. Since treatment is a fixed factor to which therapists are assigned, this would typically be viewed as a therapist-level predictor rather than a level of the data.

THE PROBLEM OF DEPENDENCE IN NESTED DATA

As we have just seen, many of the phenomena studied in clinical psychology involve inherently nested data structures. Unfortunately, many familiar statistical models, like ANOVA, regression, and structural equation modeling, are ill-suited to the analysis of nested data because they assume independence of observations (for more extensive discussion, see Hox, 2010; Raudenbush & Bryk, 2002; Snijders & Bosker, 2012). To be more precise, these models assume that the observations are *conditionally* independent, or unrelated after accounting for the shared effects of any predictors that may be in the model. To clarify this assumption, consider the standard linear regression model. Cases with similar values on the predictors are expected to show similar outcomes; that is, their predicted values will be similar to one another. The residuals relative to these predicted values are assumed, however, to be unrelated across cases. Knowing any one person's residual should tell us nothing about the residuals of any other individuals.[2] In more technical parlance, the observations should be independent after conditioning on the predictors. In many research settings, this assumption makes sense. If we were to draw a simple random sample of college students and predict their alcohol use on the basis of their anxiety, we might expect that similarly anxious individuals would have closer alcohol use values than individuals with dissimilar anxiety. After conditioning on anxiety, however, we would have no reason to expect the residuals for alcohol use to be related from one case to another. The observations would be conditionally independent.

With nested data, however, the assumption of conditional independence is rarely met. For example, suppose that instead of taking a simple random sample of college students, we recruited our sample from fraternities and sororities on campus. Now we have individuals nested within Greek letter organizations. Such organizations might vary in their promotion versus prohibition of alcohol use. As a consequence, we might expect that, independent of anxiety, the alcohol use of two members of the same fraternity or sorority would be more similar than that of two individuals from different fraternities or sororities. That is, even after conditioning on anxiety, we would expect alcohol use to be related for individuals in the same organization. This dependence violates a key assumption of the linear regression model (and many other statistical models) which is required to make accurate inferences about the effects of our predictors. Failing to account for dependence often results in standard errors that are too small, in turn producing confidence intervals that are too narrow, p-values that are too small, and an elevated Type I error (false positive) rate (Hox, 2010, chapter 1; Kutner et al., 2004, pp. 481–482; Kenny, Kashy, & Cook, 2006, pp. 43–46). Using the standard linear regression model in this instance would thus likely lead us to overstate the precision of the estimated effect of anxiety on alcohol use, perhaps even leading us to falsely identify an effect where none is actually present.

Dependence arises in both hierarchical and longitudinal data for a variety of reasons. In hierarchical data, dependence can arise through processes of selection and socialization, through the accrual of shared experiences, or even through biological relatedness (e.g., for siblings within families). For instance, consider again the case of patients nested within therapists within the Dimidjian and colleagues (2006) study. The patients of more skilled therapists may show less depression even after accounting for any other known predictors, such as whether the patients were assigned to cognitive therapy or behavioral activation. The outcome values observed for patients who share a therapist would then be dependent. In general, psychologists know a great deal about the kinds of processes that give rise to dependence in hierarchical data, yet many know less about how to analyze such data appropriately. It is still common to see published research reports in which dependence in hierarchically nested data has been ignored, calling into question the validity of the inferences made and the conclusions drawn.

In contrast, psychologists more fully understand the need to address dependence in the analysis of longitudinal data. Any researcher who has examined the correlations between repeated measures over time has observed dependence. Repeated measures tend to be positively correlated, indicating that, from one point in time to the next, individual differences are maintained to some degree (that is, the rank ordering of individuals has some stability). For instance, a child who is well above the mean in externalizing behaviors at age five is likely to still be above the mean at age six and maybe also at age seven. This relative consistency in individual differences over time results in positively correlated repeated measures and hence dependent observations. Often, these correlations decay in magnitude with longer time intervals, reflecting that this consistency is not inviolate – people change over time in different ways. One child might start low but increase in externalizing whereas another might start high and decrease, leading them to switch relative positions over time.

In the past, dependence in longitudinal data was often addressed by employing repeated measures ANOVA or multivariate analysis of variance (MANOVA). Although these techniques remain useful in some settings, they share a number of limitations that make them less widely applicable for contemporary research in clinical psychology. First, repeated measures ANOVA and MANOVA are both fit using least squares, which requires complete data. Using these techniques with longitudinal data results in listwise deletion of any individuals with missing assessments, reducing power and introducing potential bias if the missing data is not missing completely at random.[3] Second, these models assume that all individuals are assessed on the same schedule. They cannot accommodate individually varying assessments, such as occur in experience sampling designs where data is collected at random intervals. Third, both repeated measures ANOVA and MANOVA treat time as a categorical variable and thus are not well-suited to modeling smooth trajectories of change over time. Fourth, both of these models treat

---

[2] One possible point of confusion here is that the sample-estimated residuals obtained from fitting a linear regression model are actually correlated to a small degree even when observations are sampled independently (Kutner, Nachtsheim, & Neter, 2004, pp. 102–103). This correlation, however, does not hold at the population level and it is at this level that we assume residuals to be unrelated.

[3] Multiple imputation techniques can be implemented to overcome this limitation (see Schafer, 2003, for a nice overview).

dependence as a nuisance rather than as a source of information upon which to draw insights about individual differences in change. Fortunately, more modern methods of analyzing nested data can overcome these limitations. Before introducing these models, however, we must next consider a second critical issue that arises in the context of analyzing nested data, namely the need to identify the levels within the nested data structure at which predictors exert their effects.

LEVELS OF EFFECTS

When working with nested data another major concern is to properly identify the levels of the data at which effects operate. For example, suppose we wish to examine racial disparities in the prescription of psychostimulants to children with attention deficit hyperactivity disorder (ADHD) using medical record data from children nested within clinics throughout North Carolina. In analyzing these data, we might observe that, overall, psychostimulants are prescribed less often for black than white children with ADHD. One possible explanation for such a disparity would be explicit or implicit prejudices on the part of the providers. If so, we would expect that, within any given clinic, providers would be less likely to prescribe psychostimulants to black patients. But such disparities could also reflect a kind of institutional racism. Clinics that predominantly serve black communities might be less likely to prescribe psychostimulants in general, regardless of the particular race of the individual patients seen. Either or both of these effects could be producing the overall racial disparity in prescription rates. Clearly, isolating the level at which such effects operate is critical for drawing valid conclusions, informing practice and public policy, and directing intervention efforts.

More generally, with two-level data we can distinguish between three types of effects for predictors observed at the lower level of the nesting structure. There is an overall effect for all people pooling across all clusters (the *total* effect), an effect at the level of individual people nested within a given cluster (the *within-group* effect), and an aggregate effect at the level of the clusters (the *between-group* effect). Depending on how the model is specified (which we describe in more detail in the next section), the estimated relation between a lower-level predictor and an outcome can represent different combinations of these effects.[4] Indeed, without taking care in the model specification and interpretation, a researcher might be estimating one type of effect (e.g., the overall effect) but interpreting it as another (e.g., the within-group effect).

Logical fallacies of this kind (specific versions of which are known as Simpson's Paradox and the Ecological Fallacy) are particularly problematic when effects differ in magnitude or even direction across levels of the data (see Diez-Roux, 1998; Kievit et al., 2013). It is thus critically important to be aware of which effects are being estimated and their proper interpretation.

To better understand these different types of effects, we will consider a study by Baldwin, Wampold, and Imel (2007), which investigated how the therapeutic alliance between therapists and patients is related to patient outcomes. Very broadly, the essence of the therapeutic alliance is that both the therapist and patient are cooperatively working toward a shared goal. Hypothetical data similar to the Baldwin and colleagues (2007) study are shown in Figure 32.1. These data are intended to represent the relation between the Working Alliance Inventory (higher values indicate strong patient-therapist alliances) and a questionnaire assessing patient outcomes (where higher scores indicate better social functioning[5]). Each therapist saw multiple patients, so patients are nested within therapists

If the hierarchical structure of the data is ignored and a standard regression model is fit to all patients in the sample, this yields the *total effect*. In Figure 32.2, the total effect is positive such that, on average, strong alliances tend to result in improved patient outcomes. However, what we do not yet know is whether this overall effect is due to a within-therapist effect or a between-therapist effect.

To better examine these effects, Figure 32.3 includes information about the nested structure of the data by grouping the patients by therapist. These data contain 10 therapists where each black oval approximately captures the patients belonging to each therapist and the points within the oval represent a patient's observed data. We can now see that the *within-therapist effect* (sometimes called the *conditional effect* because it is conditioned on therapist affiliation) is essentially null – within any particular psychotherapist, there are no discernable differences in patient outcomes for patients with different patient-therapist alliance strengths.

But if the within-group effect is functionally null, why is the total effect clearly positive? The answer is that we must simultaneously consider the *between-group effect*. The between-group effect is conceptually a function of the therapist-specific means of the predictor and the outcome. That is, we can imagine aggregating the data on patients within each therapist to obtain 10 cluster means for social functioning and alliance strength.[6] The relation

---

[4] Note that this is not a concern for upper-level predictors. In a two-level structure, upper-level predictors (i.e., characteristics of the cluster) can only have between-group effects, since they vary in value only between and not within clusters. Thus the need to determine levels of effects is exclusively a concern for lower-level predictors (i.e., characteristics of individuals within clusters).

[5] Note that in the original study, *lower* outcome scores were associated with better patient outcomes. For this hypothetical example, *higher* outcome scores indicate better patient outcomes, so the trend is reversed to be consistent with the original findings.

[6] When using a multilevel model this effect is not actually estimated by calculating the means of the dependent and independent variables and regressing one upon the other, but this is a conceptually useful way to think about what the between-group effect represents.

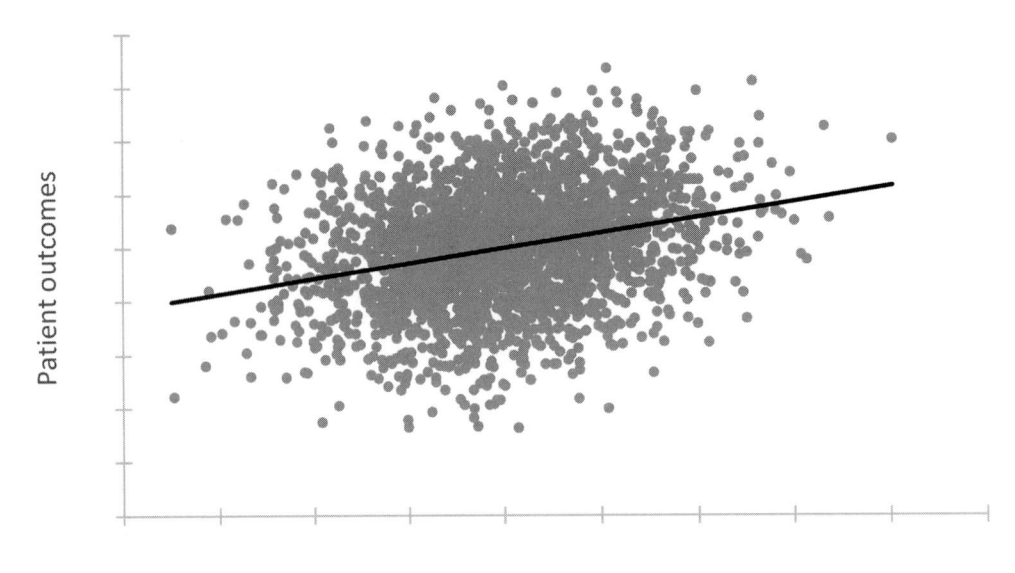

Figure 32.2 Plot of hypothetical data showing the relation between patient outcomes and patient-therapist alliance strength. Each point represents a patient and the solid black line shows the trendline if the nesting of patients within psychotherapists is ignored

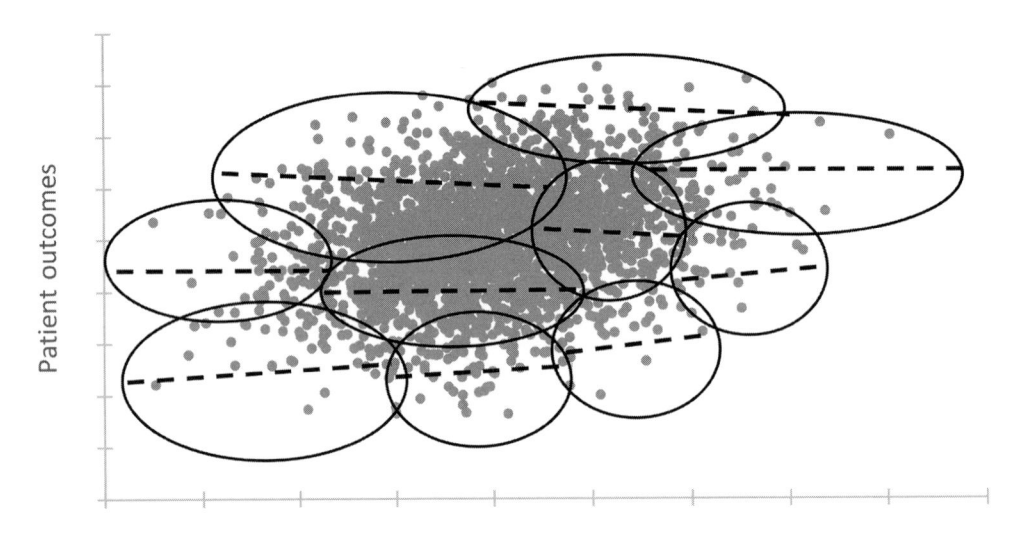

Figure 32.3 Same data as Figure 32.2 with nested structure shown. Ovals represent psychotherapists, data points represent patients. Dashed lines show the within effects for each psychotherapist. There are 10 psychotherapists, so there are 10 unique within effects

between the therapist means for patient outcome and alliance strength is then assessed, as shown in Figure 32.4. The relation, similar to the total effect, is demonstrably positive such that therapists who have stronger alliances with their patients also, on average, have patients who have better outcomes. Sometimes the difference between the within- and between-group effects is referred to as the *contextual effect*, particularly when the group-level effect exceeds the individual-level effect (as it does in this example). When such a difference exists, it indicates that the predictor operates differently as a contextual variable than as an individual variable and points to the importance of separating the within- and between-group effects.

Figure 32.5 shows all three effects together in one plot. As can be seen, the total effect that we initially observed (the bold black line) is actually a weighted combination of the within and between effects (the solid and dashed gray lines, respectively). Splitting out these effects, we can see that the relationship in this example arises solely due to the positive between-therapist effect.

We would interpret this as showing that therapists who form stronger alliances, on average, tend to have patients

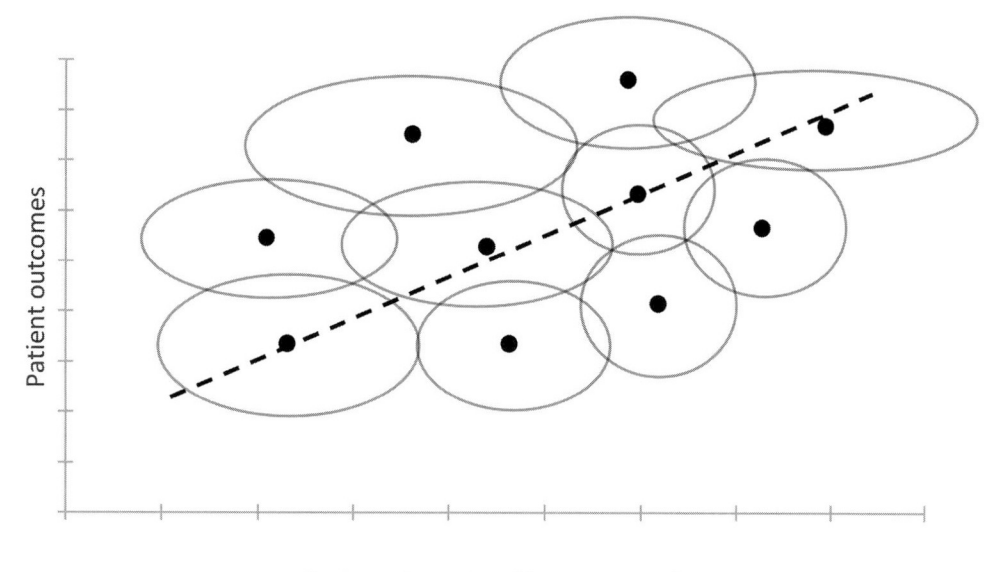

Figure 32.4 Plot of the psychotherapist means (black dots) and the effect of patient-therapist alliance strength on patient outcomes between psychotherapists

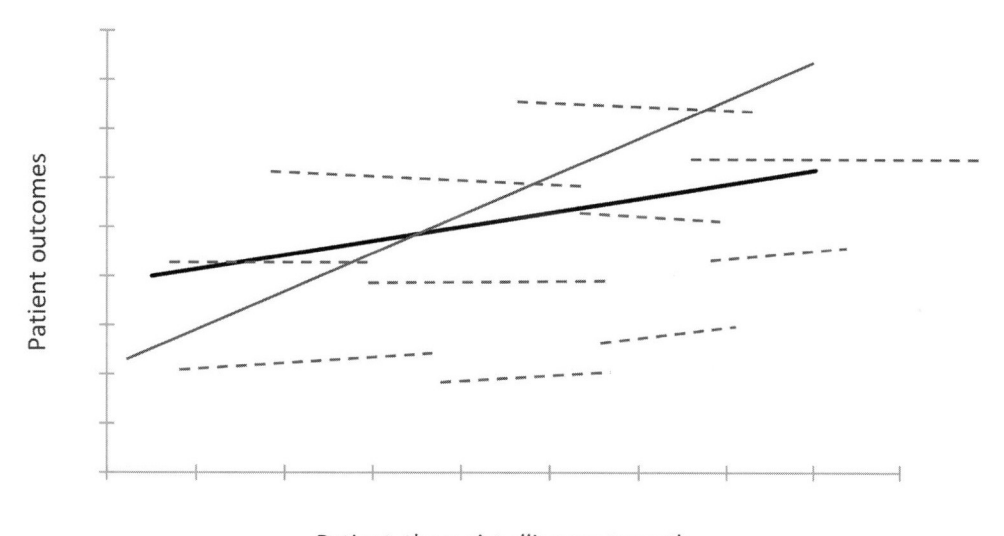

Figure 32.5 Depiction of the total effect (solid black line), the between effect (solid gray line), and the within effects (dashed gray lines). The total effect is equal to a weighted sum of the within effects and the between effect

with more favorable outcomes. For any given therapist, however, we do not see any differences among patient outcomes based on alliance strength. That is, patients who see the same therapist but differ in alliance strength do not appear to differ in their outcomes.

Though nuanced, this distinction is important for properly interpreting the relationships within the data. In our hypothetical example, it is the between-group effect that matters most. Perhaps therapists who develop stronger working alliances with their patients tend help their patients achieve better outcomes. An alternative explanation, however, is that it is purely a selection effect. That is, some therapists might work in a practice that tends to see patients whose issues are less severe and these patients might in turn form stronger alliances in general. That is, it may not be the strength of the alliance itself that matters; "alliance strength" may simply be a proxy for other variables that are driving the effect. Thus, it would not be appropriate to conclude that increasing the strength of patient-therapist alliance for all psychotherapists would lead to improved outcomes for all patients without conducting additional research to isolate the precise causal mechanisms behind the effect. Decomposing the effect into its within- and between-group components helps greatly, however, in understanding how to target this follow-up research.

MULTILEVEL MODELS

In the prior sections we described two key issues that must be considered when analyzing nested data, namely the dependence that exists between observations within a cluster and the need to distinguish between different levels when considering the effects of predictors. Multilevel models offer a number of important advantages for addressing these issues. In this section, we briefly describe the specification and interpretation of these models. To help illustrate their generality, we will consider two hypothetical examples. In the first example, we evaluate how children's anxiety varies both within and across families as a function of perceptions of family conflict. In the second example, we examine age-related changes in negative affect from childhood to adolescence in a longitudinal study. We wish to determine whether trajectories of negative affect differ between boys and girls or as a function of parental alcoholism.

Earlier we made a distinction between fixed and random factors, with the key difference being that a random factor arises if units are drawn from a broader population to which we wish to make inferences. Each of our examples includes two random factors. In the first example, we sampled families and then siblings within each family.[7] In the second example, we have time points nested within individuals, each of which also constitutes a random factor. As we will see, when specifying a multilevel model, one of the primary differences from a standard regression model is how we treat the random factors that represent our levels of sampling.

Example 32.1 Family conflict and anxiety among children nested within families

We begin by considering how to specify a model for the anxiety of children nested within families. Specifying a multilevel model often proceeds by writing an equation for each level of the data, with Level 1 referring to the lower level of sampling (here children) and Level 2 referring to the upper level of sampling (here families). Given the nested structure of the data, we use these two equations to specify a model that accounts for variation both within and across families, respectively. We can start by simply parsing these two sources of variance in the outcome variable, anxiety, without yet considering our predictor, perceived family conflict. This initial model is sometimes called a *random-effects ANOVA model*, since it partitions the variance in the outcome into within- and between-groups components similar to a standard ANOVA but with the difference that the group is a random rather than fixed factor. It may also be referred to as an *empty model* because no predictors are included.

We begin by expressing the sources of within-family variation in Anxiety at Level 1. Using the notation of Raudenbush and Bryk (2002), our Level 1 equation is

$$\text{Level 1 (Within-Family)}: \text{Anxiety}_{ij} = \beta_{0j} + r_{ij} \qquad (32.1)$$

Here i indexes the child and j indexes the family, our two levels of sampling (and our two random factors). The term β_{0j} is an intercept for the family. Since there are no predictors in the model, the intercept is interpretable as the mean level of anxiety for children for the jth family. The term r_{ij} is the residual for the child, representing the discrepancy between the observed level of anxiety for that child and the mean of their family. This residual captures within-family variation in anxiety.

At Level 2, we express the sources of between-family differences in anxiety. Our Level 2 equation is

$$\text{Level 2 (Between-Family)}: \beta_{0j} = \gamma_{00} + u_{0j} \qquad (32.2)$$

All we have done is decompose the family mean levels of anxiety, as represented by β_{0j}, into an overall across-family average, γ_{00}, and a residual u_{0j} that represents the deviation between family j's average anxiety and the overall across-family average.

We can substitute the right-hand side of Equation 32.2 into Equation 32.1 (replacing β_{0j}) to arrive at a single equation for the model, variously referred to as the combined, reduced-form, or mixed-model equation:

$$\text{Combined}: \text{Anxiety}_{ij} = \gamma_{00} + u_{0j} + r_{ij} \qquad (32.3)$$

Here we see that the model decomposes anxiety scores into an overall average, γ_{00}, between-family differences, u_{0j}, and within-family differences, r_{ij}. Equation 32.3 looks much like a standard regression model, with the difference being that there are distinct residual terms for each random factor in the model. That is, we have a residual for

---

[7] The siblings obtained for a given family may be more a census than sample (i.e., we would likely assess all siblings) but they would still usually be treated as a random factor in the model specification. The inferences we seek to make are not about the specific children within the families in the study but to a broader population of children similar to those included in the study. Thus, treating the siblings as a random factor is appropriate.

Example 32.1 (*cont.*)

families, u_{0j}, as well as siblings within families, r_{ij}. Because the units i and j are associated with random factors, these residuals represent *random effects*. Just as i and j are used to indicate specific children from specific families within the population, r_{ij} and u_{0j} represent specific draws from the population distributions of possible within- and between-family discrepancies.

To make inferences to the broader population of families we need to estimate the parameters of the population distributions for these within- and between-family random effects. This involves making parametric assumptions about how within- and between-family differences are distributed. Customarily, we assume these to be normally distributed and independent of one another within the population. Using the shorthand of statistical notation, these assumptions would be expressed as $r_{ij} \sim N(0, \sigma^2)$ and $u_{0j} \sim N(0, \tau_{00})$, indicating that each random effect is normally distributed with a mean of zero and a nonzero variance. Specifically, σ^2 is the within-family variance and τ_{00} is the between-family variance in anxiety. In fitting the model, we estimate these variances, known as the *variance components* of the model, and not the individual values of the residuals/random effects themselves. That is, we do not estimate r_{ij} and u_{0j} directly for our sampled units, but rather we estimate the variances σ^2 and τ_{00} that characterize the distributions of the random effects in the population as a whole.[8] This focus on the variance components reflects the inferential goals of the study. We want to make inferences to the broader population of children and families. Thus, variance components that capture variability in the outcome in the population are more relevant than residuals for specific families or children in the sample.

Although the notions of random effects and variance components might feel foreign at first, it may help to realize that we also use these concepts within the familiar single-level regression model. Within a typical regression model, we have a single residual for the unexplained discrepancies among the observations, and we commonly assume that residual to be normally distributed with a mean of zero and a nonzero variance. In fitting the model, we obtain an estimate of this variance, typically labeled "mean squared error" in the regression output (reflecting the use of ordinary least-squares estimation). Thus, a standard regression model includes one variance component for a single level of sampling. By extension, the multilevel model includes multiple variance components for multiple levels of sampling.

In addition to random effects, a multilevel model will also include *fixed effects*. For this model, the term γ_{00} is the sole fixed effect. Not to be confused with a fixed factor, a fixed effect is simply an effect within the model that does not vary in magnitude over levels of a random factor (and therefore does not carry an i or j subscript). The fixed effect in this model, γ_{00}, represents the overall average level of anxiety in the population, and is the same (fixed in its value) regardless of which child and which family we might happen to consider. Unlike random effects, we do obtain direct estimates of the fixed effects, and we can interpret these similarly to the coefficients from a regular regression analysis.

For the very simple model in Equations 12.1 through 12.3 we would thus obtain three estimates: an estimate of the fixed effect representing the overall average level of anxiety and estimates of the two variance components, capturing within- and between-family variance in anxiety. Of these, our primary focus would likely be on the relative size of the variance component estimates. The *intraclass correlation* (ICC) expresses the degree of between-family relative to within-family variance in an intuitive metric. The ICC is defined as

$$ICC = \frac{\text{Variance Between}}{\text{Total Variance}} = \frac{\tau_{00}}{\tau_{00} + \sigma^2} \qquad (32.4)$$

The ICC ranges from 0 to 1 and indicates the proportion of variance in the outcome (anxiety) that is due to between-cluster differences (differences between families).[9] The proportion of within-cluster variance is whatever is left, or 1 − ICC. Thus, the ICC provides a standardized measure of between- versus within-cluster variance. An ICC of 0 indicates that all variation occurs within clusters – anxiety scores for children from the same family are no more related than those of children from different families. At the other extreme, an ICC of 1 indicates that all of the variation occurs between clusters – all children within a given family have identical levels of anxiety, with differences only between families. For hierarchically nested data in psychology and allied fields, ICC values tend to

---

[8] After fitting the model, we can obtain estimates of the random effects and residuals for our sample units, which are known as *empirical Bayes' estimates*, but these are not model parameters that we estimate when fitting the model itself (see Raudenbush & Bryk, 2002, for greater detail).

[9] It is worth noting that alternative model structures can allow for negative ICCs with a range from −1 to 1 (Kenny et al., 2002) but it is not especially common to encounter negatively correlated values within clusters.

Example 32.1 (*cont.*)

fall roughly between 0.05 and 0.50 (Bliese, 1998; Hedges & Hedberg, 2007), with higher values indicating greater between-cluster variation.

Another way of interpreting the ICC helps to clarify how the random effects in the model address the problem of dependence. Notice that in Equation 32.3 the random effect of family, u_{0j}, is shared by siblings within a given family. Suppose u_{0j} is positive, reflecting increased familial risk for anxiety, whether of biological or environmental origin. All siblings within family j share this risk, which in turn leads us to expect higher than average levels of anxiety for these children. Because we expect their scores to be elevated above the overall average, γ_{00}, the presence of u_{0j} in the model implies the anxiety scores of siblings within the family are positively correlated. Conversely, if u_{0j} is negative, then all of the children within family j would be expected to have lower than average anxiety (below γ_{00}), again resulting in a positive within-family correlation. The ICC expresses this correlation of anxiety scores within a family. Thus, if we obtained an ICC of 0.3, this would indicate both that 30 percent of the variance in anxiety reflects between-family differences and that anxiety scores are correlated 0.30 within families. This dual interpretation of the ICC clarifies that between-cluster differences and within-cluster similarities are really flip sides of the same coin, each captured through the introduction of random effects to the model.

In addition to using the variance component estimates from this model to calculate the ICC, we can also use them as benchmarks for evaluating the effects of predictors included in subsequent models. As in a single-level regression model, when predictors are added to the model the unexplained variance in the outcome will diminish, with greater decreases indicating larger predictive effects. What is different from single-level regression, however, is that this decrease in unexplained variance may occur at either or both levels of the model, depending on whether a predictor has a within-family or between-family effect or both. The empty model, including no predictors, provides estimates of the total variance in the outcome that resides at each level of the model. As we add predictors, we will use these initial estimates to see how much variance the predictors explain at each level of the model.

To see this, we can extend the model to include our predictor, perceived family conflict. Each sibling within a family has their own perception of the degree of family conflict, making this a child-level predictor. We include this predictor in the Level 1 equation as follows

$$\text{Level 1 (Within-Family)}: \text{Anxiety}_{ij} = \beta_{0j} + \beta_{1j} \text{Family Conflict}_{ij} + r_{ij} \qquad (32.5)$$

Notice that we now have two coefficients within the Level 1 model, representing the intercept and slope of the linear regression of anxiety on family conflict for family j. For each Level 1 coefficient, we must then write a Level 2 equation:

$$\text{Level 2 (Between-Family)}: \begin{array}{l} \beta_{0j} = \gamma_{00} + u_{0j} \\ \beta_{1j} = \gamma_{10} \end{array} \qquad (32.6)$$

Here we have included a family-level random effect for the intercept, u_{0j}, to account for any across-family variation in levels of anxiety that remain after controlling for perceptions of family conflict. The effect of family conflict is, however, represented only by a fixed effect and is thus assumed to be constant across families. We will discuss random effects for predictors (i.e., random slopes) shortly, when we turn to our longitudinal example.

Substituting Equation 32.6 into Equation 32.5 we obtain the combined model

$$\text{Combined}: \text{Anxiety}_{ij} = \gamma_{00} + \gamma_{10} \text{Family Conflict}_{ij} + u_{0j} + r_{ij} \qquad (32.7)$$

The model includes two fixed effects, γ_{00} and γ_{10}, which are interpreted much like the coefficients of a standard regression model: γ_{00} is the expected level of anxiety for a child who rates family conflict as zero, and γ_{10} is the expected increase in anxiety associated with a one-unit increase in family conflict ratings. The model also includes two random terms, u_{0j} and r_{ij}, to account for between- and within-family differences in anxiety that persist even after accounting for family conflict. The variance components associated with these random effects now capture residual variance in anxiety net the effect of family conflict. We can thus compare the variance component estimates obtained from this model to those obtained from the previous null model to see how much variance in anxiety is explained by family conflict at each level of the data.

Adding family conflict to our model in this way seems reasonable, but it fails to address our second issue with nested data, the need to separate the effects of our predictors across levels of the model. Perhaps there is a within-family effect of conflict on anxiety, such that children who perceive greater conflict within the home also express greater anxiety than their siblings. Alternatively, there might be a between-family effect, such that families with higher average ratings of family conflict also tend to be those with higher average anxiety ratings. More likely, both effects are operating simultaneously, although they may differ in magnitude. Currently, our model allows for only

Example 32.1 (*cont.*)

one effect of family conflict on anxiety, γ_{10}, and thus does not allow for the separation of within- versus between-family effects.

The most common way to obtain distinct estimates of within- and between-group effects for lower-level predictors is to use group-mean centering (see Enders & Tofighi, 2007, for an overview). With this approach, we calculate two variables for our predictor that segregate between-family variability from within-family variability. First, we compute the mean level of perceived family conflict within each family. These family means capture the between-family variability in family conflict. Next, we center the individual family conflict scores for each child relative to the mean of their family. That is, we subtract the family mean from the child's own perceived family conflict score to create the new group-mean-centered (GMC) version of the predictor:

$$GMC\ Family\ Conflict_{ij} = Family\ Conflict_{ij} - Mean\ Family\ Conflict_j \qquad (32.8)$$

The resulting group-mean-centered family conflict variable contains only within-family variability.

Once this data management has been accomplished, we enter the group-mean-centered version of the predictor at Level 1 to capture the within-group effect and we enter the group means at Level 2 to capture the between-group effect:

$$\text{Level 1 (Within - Family)}: \ Anxiety_{ij} = \beta_{0j} + \beta_{1j} GMC\ Family\ Conflict_{ij} + r_{ij} \qquad (32.9)$$

$$\text{Level 2 (Between-Family)}: \ \begin{matrix} \beta_{0j} = \gamma_{00} + \gamma_{01} Mean\ Family\ Conflict_j + u_{0j} \\ \beta_{1j} = \gamma_{10} \end{matrix} \qquad (32.10)$$

$$\text{Combined}: \ Anxiety_{ij} = \gamma_{00} + \gamma_{01} Mean\ Family\ Conflict_j + \gamma_{10} GMC\ Family\ Conflict_{ij} + u_{0j} + r_{ij} \qquad (32.11)$$

We now have two effects of family conflict: γ_{01}, which represents the between-family effect, and γ_{10}, which represents the within-family effect. If we were to compare two families with a one-unit difference in average family conflict ratings we would expect to observe a γ_{01}-unit difference in the average anxiety ratings of the children within these families. This is the between-family relation between perceived family conflict and anxiety. Aggregated ratings of perceived family conflict, perhaps more reliably capturing the actual conflict within the home and the shared experiences of the siblings, are predictive of the anxiety of all of the children living within that home. In contrast, if we were to select any given family and compare two siblings who differed by one unit in their individual perceptions of family conflict, we would expect to observe a γ_{10}-unit difference in their anxiety scores. This is the within-family relation between perceived family conflict and anxiety. One possible interpretation of this effect is that children who are more sensitive to conflict within the family are also likely to be more anxious than their less perceptive siblings.

Although straightforward, group-mean centering is not without problems when assessing within- and between-group effects. Lüdtke and colleagues (2008) noted that the group means we calculate contain sampling variability, yet we treat these as if they were error free. Interestingly, Lüdtke and colleagues show that between-group effect estimates are affected by this problem but within-group effect estimates are not. The degree of bias in the between-group effect estimate depends both on cluster size (more bias with smaller clusters) and on the ICC for the *predictor* (more bias with less between-group variability in the predictor). To obtain better estimates for between-group effects, Lüdtke and colleagues recommend moving to a latent variable modeling approach (see also Marsh et al., 2009), however this requires a fairly large number of clusters (i.e., 100 or more) to be most effective (see Gottfresdon, 2019, for a simpler alternative that may work better with fewer clusters). Further, Lüdtke and colleagues note an important "finite-population" caveat to these results: The simple group-mean centering approach will often show little to no bias even when few observations are sampled per cluster if these observations constitute most or all of the available units for the clusters. For instance, if we sampled all of the children within each family then our estimates of the within- and between-family effects of family conflict on anxiety will be unbiased, even if we have only one to four siblings per family.

In sum, multilevel modeling provides a natural way to model hierarchically nested data. It captures and capitalizes on the dependence within the data, separating within- and between-group variability into distinct variance components. Further, it can be used to localize the effects of predictors to specific levels of the data hierarchy through the explicit estimation of within- and between-group effects for lower-level predictors. The simple multilevel analysis illustrated here could be extended in a variety of ways. Before describing possible extensions, however, we first describe a longitudinal application of the multilevel model.

Example 32.2 Modeling change in negative affect over time

In our second example, we examine trajectories of negative affect from childhood to adolescence. Suppose we obtained annual measures of negative affect on each child from 6 to 14 years of age (although fewer time points or different age spans for different children could also be accommodated). Again, we would write a model for each level of the data. At Level 1 we have repeated measures within person and we are modeling intraindividual variability. Assuming change to be linear, we might write our Level 1 model as

$$\text{Level 1 (Intraindividual)}: \textit{Negative Affect}_{ti} = \beta_{0i} + \beta_{1i} Age_{ti} + r_{ti} \quad (32.12)$$

such that negative affect for child i at time t follows a straight-line trajectory characterized by an intercept β_{0i} and slope β_{1i} and with time-specific residuals r_{ti}. As in simple regression, the intercept reflects the expected level of negative affect at age zero and the slope represents the rate of change in negative affect per year of age, here specific to child i. To enhance interpretation, the time variable (Age) is often scored to have a value of zero at the first time point (e.g., computing Age = age in years − 6, since individuals were six years old at the first assessment). This makes the intercept interpretable as "initial status" (e.g., expected negative affect at six years of age; see Biesanz et al., 2004, for a thorough discussion of this and other time scoring options). The left panel of Figure 32.6 shows what this model might imply for a single child. The underlying trajectory, which is decreasing over time, is determined by a child-specific intercept and slope. The scatter around the child's trajectory is captured by the residuals. Of course, this is but one trajectory, and we ultimately want to make inferences to the broader population of all children. We will thus now turn to the Level 2 equations to express how children differ from one another in their trajectories of change over time.

In the Level 2 equations, we reexpress the growth coefficients in terms of fixed and random effects as

$$\text{Level 2 (Interindividual)}: \begin{matrix} \beta_{0i} = \gamma_{00} + u_{0i} \\ \beta_{1i} = \gamma_{10} + u_{1i} \end{matrix} \quad (32.13)$$

This is sometimes referred to as an *unconditional growth model* because we have not yet included any predictors of the growth coefficients. For this model, the fixed effects, γ_{00} and γ_{10}, can be interpreted as the average intercept (initial status, assuming age is set to zero at the earliest assessment) and average slope (rate of change) within the population. Together, these fixed effects describe the average trajectory that a person in this population follows. The person-level random effects, u_{0i} and u_{1i}, represent individual-specific deviations from these averages and allow each child within the population to have a unique intercept and slope. That is, they capture individual differences in trajectories of change over time. The right panel of Figure 32.6 shows the average trajectory defined by the fixed

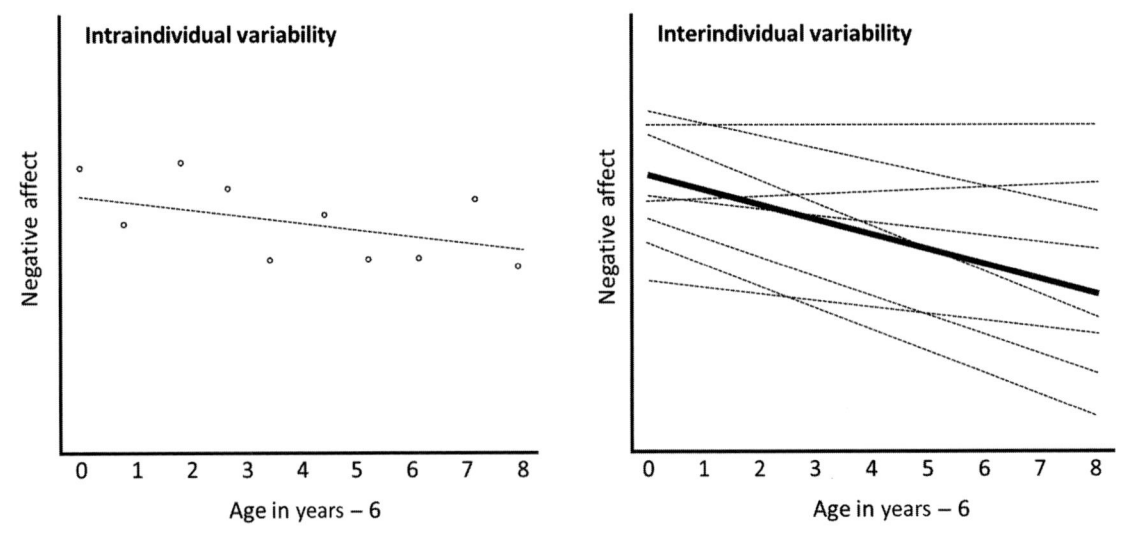

Figure 32.6 Visual depiction of two levels of variability represented with a multilevel growth model with random intercepts and slopes. The left panel shows the observed repeated measures and model-implied linear trajectory for a single child, providing a visual depiction of the Level 1 model for intraindividual variability. The right panel shows a collection of individual trajectories (light, dashed lines) as well as their mean (bold, solid line), providing a depiction of the Level 2 model for interindividual differences

Example 32.2 (*cont.*)

effects (bold) as well as a handful of individual trajectories (dashed) that vary around the average in both initial level and rate of change due to the contributions of the random effects. Notice that the inclusion of the random slope, new to this example, allows the effect of the predictor (Age) to vary across upper-level units (individuals).

Together, the two panels of Figure 32.6 convey the two levels of variability captured by the time-specific residuals at Level 1 and the person-level random effects at Level 2. These can also be seen in the combined equation

$$\text{Combined}: \textit{Negative Affect}_{ti} = \gamma_{00} + \gamma_{10}Age_{ti} + u_{0i} + u_{1i}Age_{ti} + r_{ti} \tag{32.14}$$

For each of the random terms in the model, u_{0i}, u_{1i}, and r_{ti}, there exists a distribution of possible values in the population. Again, we customarily assume these distributions to be normal, with each random effect/residual characterized by a mean of zero and a nonzero variance. In addition, because the person-level random intercept and slope reside at the same level of the data structure, we conventionally include a covariance parameter to capture any relationship that may exist between them (e.g., a possible association between initial levels of negative affect and the rate of change in negative affect over time). Thus, we have the following distributions for the random terms of the model:

$$\begin{pmatrix} u_{0i} \\ u_{1i} \end{pmatrix} \sim N\left[\begin{pmatrix} 0 \\ 0 \end{pmatrix}, \begin{pmatrix} \tau_{00} & \\ \tau_{10} & \tau_{11} \end{pmatrix} \right] \tag{32.15}$$

and

$$r_{ti} \sim N(0, \sigma^2)$$

Equation 32.14 specifies a bivariate normal distribution for the person-level random effects with means of zero, intercept and slope variances of τ_{00} and τ_{11}, respectively, and an intercept-slope covariance of τ_{10}. When fitting an unconditional growth model, these variance–covariance parameter estimates are often of great interest, as we use these to make inferences about the magnitude of individual differences in trajectories of change over time. In this application, the variance of the intercepts (τ_{00}) indicates the extent of initial differences in negative affect at six years of age. Likewise, the variance of the slopes (τ_{11}) quantifies individual differences in the rate of change of negative affect from 6 to 15 years of age. Finally, the covariance (τ_{10}) tells us whether there is any relationship between early levels of negative affect and later changes.

Ultimately, however, we would likely also wish to include predictors of these individual differences in our model. Here we extend our Level 2 model to include a measure of parental alcoholism (*PAlc*) and the sex of the child (*Sex*):

$$\text{Level 2 (Interindividual)}: \begin{aligned} \beta_{0i} &= \gamma_{00} + \gamma_{01}PAlc_i + \gamma_{02}Sex_i + u_{0i} \\ \beta_{1i} &= \gamma_{10} + \gamma_{11}PAlc_i + \gamma_{12}Sex_i + u_{1i} \end{aligned} \tag{32.16}$$

Such a model is often referred to as a *conditional growth model* because we can now examine whether initial levels and rates of change in negative affect are conditional on (related to) parental alcoholism and the child's sex. Substituting for the terms in Equation 32.12 we obtain the combined model equation:

$$\text{Combined}: \textit{Negative Affect}_{ti} = \gamma_{00} + \gamma_{01}PAlc_i + \gamma_{02}Sex_i + \gamma_{10}Age_{ti} + \gamma_{11}PAlc_iAge_{ti} + \gamma_{12}Sex_iAge_{ti} + u_{0i} + u_{1i}Age_{ti} + r_{ti} \tag{32.17}$$

Comparing Equations 32.17 and 32.18, we can see that predictions of the trajectory intercepts in the Level 2 equations (γ_{01} and γ_{02}) manifest as main effects in the combined model equation. Likewise, predictions of the slopes in the Level 2 equations (γ_{11} and γ_{12}) manifest as *cross-level interactions* in the combined model equation. In a conditional growth model, these fixed effect estimates tend to be of greatest interest. To facilitate their interpretation, we can make use of many of the same tools used to probe and plot interactions in single-level regression models, including conditional effect plots and the Johnson-Neyman technique (Bauer & Curran, 2005). For instance, we might find that girls and boys have similar initial levels of negative affect, but that boys show decreases in negative affect from childhood to adolescence whereas girls do not. We might also find that children of alcoholics have both higher initial levels of negative affect and less steep decreases over time. These are substantively powerful inferences about individual development.

> **Example 32.2** (*cont.*)
>
> Aside from the fixed effects, we also obtain estimates of the variances and covariances of the random effects. These now capture residual differences in the intercepts and slopes of the individual negative affect trajectories after accounting for parental alcoholism and sex. Comparing the magnitude of these variance estimates to those obtained from the unconditional growth model shows how much of the variance in intercepts and slopes is explained by the joint effects of parental alcoholism and sex.
>
> Why did we not attempt to separate the effects of our predictors across the two levels of the model, as we did in our first example? First, parental alcoholism and child sex are Level 2 predictors. By definition, they contain only between-person variability (i.e., parental alcoholism and sex are time-invariant, having the same values across all time points) and thus they can exert only between-person effects. The need to separate within- and between-person effects only presents itself for Level 1 predictors. In this example, the only Level 1 predictor in the model was age. Age is often roughly balanced (i.e., all individuals are often assessed at the same points in time, barring missing data), so that there is typically very little between-person variability with which to assess between-person effects. Even when age is not balanced, this is usually by design (e.g., use of an accelerated longitudinal design) rather than something informative about the individuals being studied (i.e., person mean age is not a useful predictor). It is thus very uncommon to see the decomposition of within- and between-person effects for age (or any other time variable) within a multilevel growth model. Had we included some other time-varying covariate in the Level 1 equation (such as the level of depression of the mother when child negative affect was assessed), then we might have felt a greater need to separate within- from between-person effects. However, time-varying covariates often themselves evince time trends that can complicate the assessment of within- and between-person effects in longitudinal models (Curran & Bauer, 2011; Hoffman & Stawski, 2009; Wang & Maxwell, 2015). Thus, part of why we did not illustrate estimating within- versus between-person effects in the second example is because it's complicated and our mandate was to write a short chapter (a mandate with which we have already taken some liberty).

Model Extensions

Overall, our primary goals in presenting the hierarchical and longitudinal examples above were to convey a sense of the generality of multilevel models and to explicate how these models are specified in practice. Both examples, however, involved quite simple models. More realistically, a researcher would likely wish to include multiple predictors at each level of the model. Key decisions when specifying more complex models are (1) whether to decompose within- and between-cluster effects for Level 1 predictors; and (2) which predictors at Level 1 should have random slopes. These can be difficult decisions and may be motivated as much from the constraints of the data at hand as by substantive theory. For instance, when working with a relatively small number of clusters (20 or fewer), including between-group effects for all Level 1 predictors can quickly exhaust the degrees of freedom at Level 2. Prioritizing some effects over others may be necessary. Likewise, including random slopes for all Level 1 predictors may at first seem logical, since it is often possible to imagine the effect of a predictor could vary in magnitude across clusters. Moreover, not including random effects in the model that actually are present in the population can lead to problems, including bias in other variance component estimates and incorrect inferential tests of fixed effects. Yet models with many random effects can be difficult to estimate and often fail to converge. One approach is to determine which random effects to include based on the magnitude of the variance estimates.[10] Typically, random intercepts would only be omitted from a model if their inclusion resulted in a zero (or negative) variance estimate or caused estimation problems. In contrast, sometimes even a small nonzero variance component estimate will be used to justify trimming a random slope from the model, under the logic that between-cluster heterogeneity in the effect of the predictor is too insignificant to warrant the added model complexity. Thus, just like any statistical model, researchers must take care in how they specify multilevel models and be transparent about how specification decisions were made.

NOT THE ONLY GAME IN TOWN: OTHER APPROACHES FOR ANALYZING NESTED DATA

Multilevel modeling is the overwhelming favorite for analyzing nested data in psychology (Bauer & Sterba, 2011; Hoffman & Rovine, 2007; West et al., 2011). This preference likely stems from the flexibility of the model and the diverse set of research questions that it can be used to address. Concomitant with this flexibility, however, comes

---

[10] Variance components can be inferentially tested so that *p*-values can be used to help make this decision. These tests are not always straightforward, however, because there are unique difficulties involved with testing variance parameters that are not present in other parameters (e.g., variance cannot be negative, the null distribution is not symmetric).

the complexity of separating between- and within-cluster effects, proper specification of random effects, interpretation of variance components, and meeting sometimes strong assumptions. Though multilevel models rightfully deserve their position as the first-choice method for analyzing nested data, they are not the only method by which researchers can analyze nested data. In some contexts, the added complexity of multilevel models is not necessary and alternative methods may more simply but effectively address the research questions of interest. These methods include computation of cluster-robust standard errors, generalized estimating equations, and fixed-effects models. We briefly survey these alternative methods here.

Cluster-Robust Standard Errors

Multilevel models accommodate nested data structures by explicitly modeling the different levels of the data. However, in some cases the research questions may not require information at each level to be considered in isolation; that is, the total effect may be of prevailing interest. In such cases, the nested structure may be viewed as a nuisance, perhaps arising only as a byproduct of the data collection design (e.g., subjects are recruited from schools for convenience rather than due to any scientific interest in school effects). If we ignore the nesting and fit a standard regression model, we will still obtain accurate estimates of the total effects of our predictors. Only the standard errors of these effects are compromised by failing to account for the dependence in the data. The multilevel model addresses this dependence through the introduction of random effects, but an alternative approach is to retain the standard regression model and implement *cluster-robust standard errors* (Huang, 2016; McNeish, Stapleton, & Silverman, 2017).[11] That is, after fitting the model under the usual assumption of conditionally independent observations, we can apply a statistical correction to the standard errors of the estimates to take into account the dependence that is actually present in the data. This adjustment usually results in the standard errors being larger than their "naïve" counterparts that do not account for clustering, with the correction serving to maintain the nominal Type I error rate.

An advantage of this approach is that predictors from either level of the data structure can be included in the model with effects specified and interpreted just like in single-level regression. Further, because the outcome variance is pooled rather than partitioned across levels of the data, we obtain the usual R^2 measure of overall explained variance (whereas R^2-type measures for multilevel models are less straightforward; Rights & Sterba, 2019, for review). The usual standardized effect measures are also available (e.g., partial or semipartial correlations, standardized regression coefficients, etc.), the computations of which are mostly unchanged when implementing cluster-robust standard errors (whereas how to compute standardized effects in multilevel models remains a topic of discussion). But there are also disadvantages of using cluster-robust standard errors. First, researchers usually only estimate total effects for their Level 1 predictors and may be tempted to misinterpret these as within- or between-cluster effects. Second, the cluster-robust correction tends to be less effective for large ICCs (roughly above 0.30; Zeger, Liang, & Albert, 1988) or with small or moderate samples (e.g., more than 50 upper-level units are typically required; Cameron, Gelbach, & Miller, 2008; McNeish & Harring, 2017), making it less suitable for longitudinally nested data in particular.

Generalized Estimating Equations

Using generalized estimating equations (GEE) is another option for fitting linear and generalized linear models with nested data (Diggle et al., 2002; Hardin & Hilbe, 2003; Liang & Zeger, 1986). GEE is similar to the correction approach described above in avoiding the use of random effects and in implementing cluster-robust standard errors. It differs, however, by explicitly taking account of the dependence of observations within clusters in the estimation phase as well. That is, rather than assume independence, GEE involves the specification of a *working correlation matrix* for the within-cluster residuals. So, for instance, if we obtained four repeated measures per person, we would specify a 4 × 4 working correlation matrix for the residuals of these measures over time. Assuming all of the correlations to be zero would imply independent residuals and make the GEE equivalent to the cluster-robust standard error approach described above. With GEE, however, other types of structures may be specified, including exchangeable (a.k.a., compound symmetric; all correlations are equal but nonzero), autoregressive (correlations decay as time points become further separated), or unstructured (each correlation is uniquely estimated), among other possibilities. Postestimation, robust standard errors are still computed to address any possible misspecification of the working correlation matrix structure. Inferences will thus be valid even if the working correlation matrix is misspecified (although more accurate working correlation structures lead to more efficient estimates with smaller standard errors).

A key advantage of GEE is again that the model is specified and interpreted much like a single-level model with the primary exception being the need to indicate the structure of the working correlation matrix. Additionally, the estimates and inferences provided by GEE are robust even when the working correlation matrix is incorrectly

---

[11] Note that there are a variety of robust standard errors available to address different problems, including heteroscedasticity, nonnormality, and clustering. Unfortunately, these are often all named similarly. When implementing cluster-robust standard errors in general software like SAS, SPSS, Stata, or Mplus, it is thus wise to double-check that the appropriate type of robust error has been selected, and that this is correcting for possible dependencies among the residuals within clusters.

specified. In contrast, the multilevel model is less robust to incorrect specifications of the random effects. The disadvantage of GEE, however, is that because there are no random effects in the model and no corresponding variance component estimates, there is no way to examine between-cluster heterogeneity in the model coefficients (e.g., between-person differences in the intercepts and slopes of a linear growth process) other than those differences associated with predictors included in the model.

Fixed Effects Models

The fixed-effect modeling approach addresses nesting using a quite different strategy than the other approaches we have considered so far. It relies on the standard linear (or generalized linear) regression model and maintains the usual assumption of conditional independence of the observations. To make this assumption plausible, clustering effects are included directly within the model itself. That is, to account for all possible between-cluster differences that might give rise to within-cluster dependence, we simply include *Cluster* as another predictor in our model. Cluster membership is treated like any other nominal predictor (i.e., like a fixed factor, as discussed previously), using coding variables to capture between-cluster mean differences. The regression coefficients for these coding variables are fixed effects, so the name for this approach reflects the fact that we are using fixed effects to account for the dependence in the data rather than random effects, as in the multilevel model. Once the between-cluster differences are absorbed into the fixed effects, the residuals are left uncorrelated after conditioning on the predictors (which includes cluster affiliation), permitting us to conduct all of the usual inferential tests without modification.

A key advantage of the fixed effects approach is that it avoids making assumptions at the level of the clusters. Clusters need not be selected randomly and there are no random effects about which to assume normality. We also need not worry about omitted variables at the cluster level, since all cluster-level variability has been removed by controlling for cluster. For the same reason, all effects within the model can be unambiguously interpreted as within-cluster effects. Nothing comes without costs, however. Because clusters are not treated as a level of sampling, no inferences can be made to a broader population of clusters. Additionally, because all between-cluster variability is already absorbed by the cluster coding variables, we can estimate no additional between-cluster effects. Thus, we can neither examine the between-cluster effects of lower-level predictors nor the effects of upper-level predictors.

Summary of Approaches

To summarize, although the most common approach to analyzing nested data in psychological research is to use a multilevel model with random effects, other options can also be useful depending on the research questions under investigation. Using a standard single-level model with cluster-robust standard errors can yield accurate inferences about total effects. For greater accuracy, especially when there is greater within-cluster dependence (e.g., in longitudinal applications), the GEE approach may be preferable. In contrast, the fixed-effects modeling approach may be useful when interest centers on within-cluster effects and there is neither a desire to make inferences to a broader population of clusters nor to explain between-cluster differences. The reasons why multilevel models with random effects remain so popular, however, is that they allow inferences about the broader population and provide insights into both within- and between-cluster differences, including the prediction of this variation at each level of the model.

WHEN DATA DEFY THE TEXTBOOK

In this section, we briefly describe several situations that often arise within the context of clinical psychology research but fall outside of standard "textbook" examples.

Partially Nested Data

The designs discussed up to this point are fully nested designs. For example, in the Dimidjian and colleagues (2006) trial, patients were nested within therapists in both the cognitive therapy and behavioral activation conditions. Also common are partially nested designs, where nesting is present for only a subset of observations (Sterba, 2017). For example, Stice and colleagues (2006) compared four interventions for the prevention of eating disorders. In two conditions, the dissonance condition and the healthy weight condition, participants were nested within groups. In the other two conditions, the expressive writing condition and assessment-only condition, there was no nesting. To appropriately model partially nested data like these, the multilevel model needs to be adapted to produce different variance decompositions for different conditions (Bauer, Sterba, & Hallfors, 2008; Roberts & Roberts, 2005). For partially nested data we recommend (a) only adding a random effect for cluster in conditions with clustering, (b) adjusting degrees of freedom using either the Satterthwaite or Kenward–Roger method, and (c) allowing for heteroscedasticity in the residual variance (Baldwin et al., 2011). In some cases it may also be useful to consider implementing Bayesian methods to fit the model (Baldwin & Fellingham, 2013).

Small Number of Groups

Hierarchically nested data in clinical research often involves relatively few clusters. Fitting multilevel models to such data is contraindicated by commonly cited guidelines that recommend at least 30 clusters in order to use multilevel models (Kreft, 1996). Fortunately, methods now exist that facilitate using multilevel modeling with fewer clusters. With few clusters, the primary statistical concerns are (1) biased estimates of variance components and (2) underestimated standard errors for fixed-effect estimates, the consequence of which is an inflated Type I error rate. The first problem, bias in the variance component estimates, can be mitigated by estimating the model with restricted maximum likelihood instead of full maximum likelihood (Bell et al., 2014). The second problem, biased standard errors for the fixed effects, can be addressed by implementing the Kenward–Roger method for testing fixed effects. This method has been shown to yield reasonable estimates and Type I error rates with about 10 groups (McNeish & Stapleton, 2016a). For fewer than 10 groups, it may be necessary to use Bayesian methods to circumvent estimation issues (Hox, van de Schoot, & Matthijsse, 2012). However, when using Bayesian methods with a small number of groups, it is vital to carefully consider prior distributions; otherwise estimates can be quite biased (McNeish, 2016; van de Schoot et al., 2015). When the number of groups is particularly small, it may ultimately be necessary to turn to a fixed-effects approach. For additional details on analyzing nested data with few groups, readers may wish to consult a recent review by McNeish and Stapleton (2016b).

Pre-Post Design

Pre-post designs are one of the mainstays of treatment evaluation studies. Since two repeated measures are observed per person, it is natural to assume that methods for analyzing longitudinally nested data should be applied to data obtained from pre-post designs. Often, however, it is preferable to treat the pre-test measure as a covariate in the model, as in analysis of covariance (Rausch, Maxwell, & Kelley, 2003). That is, the model is specified so that only the post-test measure is the outcome, pre-test is a covariate, and treatment is a fixed factor. The pre-test measure is included principally to reduce the mean-squared error of the model, thereby enhancing power to detect a treatment effect. Since there is only one post-test measure per person, this model formulation obviates the need to account for dependent observations. Of course, when multiple pre-tests, post-tests, and/or follow-ups are available, then the need to address longitudinal nesting once again presents itself.

Both Longitudinally and Hierarchically Nested Data

Although we considered hierarchical and longitudinal data structures separately, we also noted that they often occur together. For instance, we might evaluate trajectories of negative affect for children nested within families. Similarly, in a multi-site clinical trial we might evaluate change as a function of weeks on treatment. In each of these cases, repeated measures are nested within person, which in turn are nested within groups, producing a three-level data structure. Fortunately, multilevel models with random effects readily expand to include higher levels of nesting. The model is extended to include random effects at both Level 2, to account for dependence within person over time, and at Level 3, to account for dependence of persons within clusters. The substantive interpretation of these random effects is that they capture within- and between-cluster differences in individual trajectories of change over time.

Discrete Dependent Variables

Many outcomes of interest in clinical psychology are discrete rather than continuous in nature, such as the presence or absence of a disorder or the number of marijuana cigarettes smoked in a week. Yet the models presented in this chapter assumed conditional normal distributions for the outcomes. These models also assumed linear effects, which generally won't hold for discrete outcomes with limited ranges, such as binary, ordinal, or count outcomes. Fortunately, the multilevel generalized linear model offers a straightforward way to accommodate these kinds of outcomes. Model specification includes two additional steps: selecting an appropriate conditional response distribution for the outcome and selecting a nonlinear link function to keep predictions for the outcome within the valid range. For instance, suppose that we obtained weekly repeated measures on binge drinking episodes for college students with an interest in modeling whether binge drinking changed over the course of the academic year and in response to specific sporting events. For each week, students indicated that they had or had not engaged in binge drinking, resulting in a binary outcome measure. For binary outcomes, an appropriate conditional distribution is the Bernoulli distribution (equivalent to the binomial distribution with one "trial"). Specifying the Bernoulli distribution implies that, for any given set of values for the predictors and random effects, there is some probability of binge drinking and some complementary probability of not binge drinking (rather than a continuous normal curve of infinitely many possible values of binge drinking). Probabilities have a range from zero to one, so we employ a logistic link function to model differences in probabilities of binge drinking due to our predictors and random effects (in contrast to a linear model which might imply probabilities less than zero or greater than one). These generalized models offer a powerful way to extend the use of multilevel models to discrete outcomes; however, the estimation and interpretation of these models is also more challenging and nuanced (for greater detail see Bauer & Sterba, 2011; McCulloch, Searle, & Neuhaus, 2008; Raudenbush & Bryk, 2002, chapter 10).

Other Special Data Structures: Dyadic Data and Intensive Longitudinal Data

Multilevel models are also important tools for analyzing dyadic data, such as data on romantic partners or best friends (Kenny et al., 2006), and for analyzing intensive longitudinal data, such as data arising from daily diary studies, experience sampling, ecological momentary assessment, or ambulatory measures of psychophysiology (Schwartz & Stone, 1998). Chapters 24 and 27 of this handbook provide additional details on the analysis of these types of data.

CONCLUSIONS

In sum, multilevel models offer a powerful and flexible framework for analyzing nested data that provides us with the opportunity to test complex theories of human behavior more fully. We can use these models to accommodate dependence in nested data, to locate and quantify variability across levels of the data structure, and to separate levels of effects for our predictors. As we hope we have conveyed, these are important advantages, whether we are interested in capturing group effects (e.g., schools, families, dyads, providers), change over time (e.g., age-related changes, treatment effects, short-term fluctuations), or both. Of course, entire textbooks have been written on the analysis of nested data (e.g., Hox, 2010; Raudenbush & Bryk, 2002; Snijders & Bosker, 2012) and this chapter is rather cursory by comparison. Nevertheless, we hope it has provided a useful orientation to the principal issues involved in analyzing nested data and may serve as a point of entry for researchers seeking to explore the broader methodological literature on this topic.

REFERENCES

Baldwin, S. A., Bauer, D. J., Stice, E., & Rohde, P. (2011). Evaluating Models for Partially Clustered Designs. *Psychological Methods*, *16*, 149–165.

Baldwin, S. A., & Fellingham, G. W. (2013). Bayesian Methods for the Analysis of Small Sample Multilevel Data with a Complex Variance Structure. *Psychological Methods*, *18*, 151–164.

Baldwin, S. A., Murray, D. M., & Shadish, W. R. (2005). Empirically Supported Treatments or Type I Errors? Problems with the Analysis of Data from Group-Administered Treatments. *Journal of Consulting and Clinical Psychology*, *73*, 924–935.

Baldwin, S. A., Wampold, B. E., & Imel, Z. E. (2007). Untangling the Alliance-Outcome Correlation: Exploring the Relative Importance of Therapist and Patient Variability in the Alliance. *Journal of Consulting and Clinical Psychology*, *75*, 842–852.

Bauer, D. J., & Curran, P. J. (2005). Probing Interactions in Fixed and Multilevel Regression: Inferential and Graphical Techniques. *Multivariate Behavioral Research*, *40*, 373–400.

Bauer, D. J., & Sterba, S. K. (2011). Fitting Multilevel Models with Ordinal Outcomes: Performance of Alternative Specifications and Methods of Estimation. *Psychological Methods*, *16*, 373–390.

Bauer, D. J., Sterba, S. K., & Hallfors, D. D. (2008). Evaluating Group-Based Interventions when Control Participants Are Ungrouped. *Multivariate Behavioral Research*, *43*, 210–236.

Bell, B. A., Morgan, G. B., Schoeneberger, J. A., Kromrey, J. D., & Ferron, J. M. (2014). How Low Can You Go? *Methodology*, *10*, 1–10.

Biesanz, J. C., Deeb-Sossa, N., Aubrecht, A. M., Bollen, K. A., & Curran, P. J. (2004). The Role of Coding Time in Estimating and Interpreting Growth Curve Models. *Psychological Methods*, *9*, 30–52.

Bliese, P. D. (1998). Group Size, ICC Values, and Group-Level Correlations: A Simulation. *Organizational Research Methods*, *1*, 355–373.

Cameron, A. C., Gelbach, J. B., & Miller, D. L. (2008). Bootstrap-Based Improvements for Inference with Clustered Errors. *Review of Economics and Statistics*, *90*, 414–427.

Crits-Christoph, P., & Mintz, J. (1991). Implications of Therapist Effects for the Design and Analysis of Comparative Studies of Psychotherapies. *Journal of Consulting and Clinical Psychology*, *59*, 20–26.

Curran, P. J., & Bauer, D. J. (2011). The Disaggregation of Within-Person and Between-Person Effects in Longitudinal Models of Change. *Annual Review of Psychology*, *62*, 583–619.

Diez-Roux, A. V. (1998). Bringing Context Back into Epidemiology: Variables and Fallacies in Multilevel Analysis. *American Journal of Public Health*, *88*, 216–222.

Diggle, P. J., Heagerty, P., Liang, K. Y., & Zeger, S. L. (2002). *Analysis of Longitudinal Data*. Oxford: Oxford Statistical Science Series.

Dimidjian, S., Hollon, S. D., Dobson, K. S., Schmaling, K. B., Kohlenberg, R. J., Addis, M. E., … Jacobson, N. S. (2006). Randomized Trial of Behavioral Activation, Cognitive Therapy, and Antidepressant Medication in the Acute Treatment of Adults with Major Depression. *Journal of Consulting and Clinical Psychology*, *74*, 658–670.

Enders, C. K., & Tofighi, D. (2007). Centering Predictor Variables in Cross-Sectional Multilevel Models: A New Look at an Old Issue. *Psychological Methods*, *12*, 121–138.

Gottfredson, N. C. (2019). A Straightforward Approach for Coping with Unreliability of Person Means When Parsing Within-Person and Between-Person Effects in Longitudinal Studies. *Addictive Behaviors*, *94*, 156–161.

Hardin, J., & Hilbe, J. (2003). *Generalized Estimating Equations*. London: Chapman and Hall/CRC.

Hedges, L. V., & Hedberg, E. C. (2007). Intraclass Correlation Values for Planning Group-Randomized Trials in Education. *Educational Evaluation and Policy Analysis*, *29*, 60–87.

Hoffman, L., & Rovine, M. J. (2007). Multilevel Models for the Experimental Psychologist: Foundations and Illustrative Examples. *Behavior Research Methods*, *39*, 101–117.

Hoffman, L., & Stawski, R.S. (2009). Persons as Contexts: Evaluating Between-Person and Within-Person Effects in Longitudinal Analysis. *Research in Human Development*, *6*, 97–120.

Hox, J. J. (2010). *Multilevel Analysis: Techniques and Applications*. New York: Routledge.

Hox, J. J., van de Schoot, R., & Matthijsse, S. (2012, July). How Few Countries Will Do? Comparative Survey Analysis from a Bayesian Perspective. *Survey Research Methods*, *6*, 87–93.

Huang, F. L. (2016). Alternatives to Multilevel Modeling for the Analysis of Clustered Data. *Journal of Experimental Education*, *84*, 175–196.

Kenny, D. (1995). The Effect of Nonindependence on Significance Testing in Dyadic Research. *Personal Relationships*, *2*, 67–75.

Kenny, D. A., Kashy, D. A., & Cook, W. L. (2006). *Dyadic Data Analysis*. New York: Guilford.

Kenny, D. A., Mannetti, L., Pierro, A., Livi, S., & Kashy, D. A. (2002). The Statistical Analysis of Data from Small Groups. *Journal of Personality and Social Psychology*, *83*, 126–137.

Kievit, R., Frankenhuis, W. E., Waldorp, L., & Borsboom, D. (2013). Simpson's Paradox in Psychological Science: A Practical Guide. *Frontiers in Psychology, 4,* 1–14.

Kutner, M. H., Nachtsheim, C. J., & Neter, J. (2004). *Applied Linear Regression Models.* Boston, MA: McGraw-Hill

Kreft, I. G. G. (1996). Are Multilevel Techniques Necessary? An Overview, Including Simulation Studies. California State University at Los Angeles.

Lenzenweger, M. F., Johnson, M. D., & Willett, J. B. (2004). Individual Growth Curve Analysis Illuminates Stability and Change in Personality Disorder Features: The Longitudinal Study of Personality Disorders. *Archives of General Psychiatry, 61,* 1015–1024.

Liang, K. Y., & Zeger, S. L. (1986). Longitudinal Data Analysis Using Generalized Linear Models. *Biometrika, 73,* 13–22.

Lüdtke, O., Marsh, H. W., Robitzsch, A., Trautwein, U., Asparouhov, T., & Muthén, B. (2008). The Multilevel Latent Covariate Model: A New, More Reliable Approach to Group-Level Effects in Contextual Studies. *Psychological Methods, 13,* 203–229.

Marsh, H. W., Lüdtke, O, Robitzsch, A., Trautwein, U., Asparouhov, T., Muthén, B., & Nagengast, B. (2009). Doubly-Latent Models of School Contextual Effects: Integrating Multilevel and Structural Equation Approaches to Control Measurement and Sampling Error. *Multivariate Behavioral Research, 44,* 764–802.

McCulloch, C. E., Searle, S. R., & Neuhaus, J. M. (2008). *Generalized, Linear, and Mixed Models.* New York: Wiley.

McNeish, D. (2016). On Using Bayesian Methods to Address Small Sample Problems. *Structural Equation Modeling, 23,* 750–773.

McNeish, D., & Harring, J. R. (2017). Clustered Data with Small Sample Sizes: Comparing the Performance of Model-Based and Design-Based Approaches. *Communications in Statistics: Simulation and Computation, 46,* 855–869.

McNeish, D. M., & Stapleton, L. M. (2016a). The Effect of Small Sample Size on Two-Level Model Estimates: A Review and Illustration. *Educational Psychology Review, 28,* 295–314.

McNeish, D., & Stapleton, L. M. (2016b). Modeling Clustered Data with Very Few Clusters. *Multivariate Behavioral Research, 51,* 495–518.

McNeish, D., Stapleton, L. M., & Silverman, R. D. (2017). On the Unnecessary Ubiquity of Hierarchical Linear Modeling. *Psychological Methods, 22,* 114–140.

Raudenbush, S. W., & Bryk, A. S. (2002). *Hierarchical Linear Models.* Newbury Park, CA: Sage.

Rausch, J. R., Maxwell, S. E., & Kelley, K. (2003). Analytic Methods for Questions Pertaining to a Randomized Pretest, Posttest, Follow-Up Design. *Journal of Clinical and Consulting Psychology, 32,* 467–486.

Rights, J. D., & Sterba, S. K. (2019). Quantifying Explained Variance in Multilevel Models: An Integrative Framework for Defining R-Squared Measures. *Psychological Methods, 24,* 309–338.

Roberts, C., & Roberts, S. A. (2005). Design and Analysis of Clinical Trials with Clustering Effects Due to Treatment. *Clinical Trials, 2,* 152–162.

Schafer, J. L. (2003). Multiple Imputation: A Primer. *Statistical Methods in Medical Research, 8,* 3–15.

Schwartz, J. E., & Stone, A. A. (1998). Strategies for Analyzing Ecological Momentary Assessment Data. *Health Psychology, 17,* 6–16.

Snijders, T., & Bosker, R. (2012). *Multilevel Analysis: An Introduction to Basic and Advanced Multilevel Modeling* (2nd edn.). London: Sage.

Stice, E., Shaw, H., Burton, E., & Wade, E. (2006). Dissonance and Healthy Weight Eating Disorder Prevention Programs: A Randomized Efficacy Trial. *Journal of Consulting and Clinical Psychology, 74,* 263–275.

Sterba, S. K. (2017). Partially Nested Designs in Psychotherapy Trials: A Review of Modeling Developments. *Psychotherapy Research, 27,* 425–436.

Taylor, S., Thordarson, D. S., Maxfield, L., Fedoroff, I. C., Lovell, K., & Ogrodniczuk, J. (2003). Comparative Efficacy, Speed, and Adverse Effects of Three PTSD Treatments: Exposure Therapy, EMDR, and Relaxation Training. *Journal of Consulting and Clinical Psychology, 71,* 330–338.

Wang, L., & Maxwell, S. E. (2015). On Disaggregating Between-Person and Within-Person Effects with Longitudinal Data Using Multilevel Models. *Psychological Methods, 20,* 63–83.

West, S. G., Ryu, E., Kwok, O. M., & Cham, H. (2011). Multilevel Modeling: Current and Future Applications in Personality Research. *Journal of Personality, 79,* 2–50.

van de Schoot, R., Broere, J. J., Perryck, K. H., Zondervan-Zwijnenburg, M., & Van Loey, N. E. (2015). Analyzing Small Data Sets Using Bayesian Estimation: The Case of Posttraumatic Stress Symptoms Following Mechanical Ventilation in Burn Survivors. *European Journal of Psychotraumatology, 6*(1), 25216.

Zeger, S. L., Liang, K. Y., & Albert, P. S. (1988). Models for Longitudinal Data: A Generalized Estimating Equation Approach. *Biometrics, 44,* 1049–1060.

33 Missing Data Analyses

AMANDA N. BARALDI AND CRAIG K. ENDERS

Over the last century, methodologists have made a great deal of progress developing and improving methods for handling analyses with missing data (Anderson, 1957; Beale & Little, 1975; Dempster & Laird, 1977; Orchard & Woodbury, 1972; Rubin, 1978). This chapter specifically addresses two of these methods: multiple imputation and maximum likelihood estimation. Although the methodological literature has long touted the importance of using these tools with missing data (e.g., Kenward & Molenberghs, 1998; Little & Rubin, 2002; Schafer & Graham, 2002; Shafer, 1997), researchers in psychology and related fields have been slow to adopt them (e.g., Jeličić, Phelps, & Lerner, 2009; Peugh & Enders, 2004; Sullivan et al., 2017). Many researchers still rely on older missing data approaches such as listwise and pairwise deletion. At best, these older methods produce unbiased estimates with a loss of statistical power under strict assumptions. More insidiously, these methods have the potential to introduce bias. In contrast, multiple imputation and maximum likelihood estimation enable researchers to include all available data in the analysis, resulting in increased statistical power. Furthermore, these methods produce unbiased parameter estimates under less strict assumptions. Altogether, multiple imputation and maximum likelihood estimation represent best practices for handling missing data and, consequently, are becoming the standard for psychological research.

This chapter serves as an introductory primer on missing data analyses in clinical psychology research. First, we introduce a motivating example from pediatric clinical psychology that we will use to illustrate the concepts presented in the chapter. Next, we introduce important missing data terminology, including a description of Rubin's missing data mechanisms (Rubin, 1976); these mechanisms can be thought of as assumptions underlying the performance of a particular analytic technique when data are missing. After reviewing pertinent concepts, we describe multiple imputation and maximum likelihood estimation. For both methods, we present analyses using the motivating example for illustration and provide sample syntax. Finally, we describe practical issues in using these methods. Specifically, we discuss choosing between multiple imputation and maximum likelihood estimation and provide examples of using multiple imputation for more complex models such as models including mixtures of categorical and continuous variables, models with composite variables, nonlinear models, and multilevel models.

MOTIVATING EXAMPLE

In order to facilitate the explanation of multiple imputation and maximum likelihood estimation with missing data, we draw on a motivating analysis throughout the chapter. The example is loosely based on data from a study of predictors of depressive symptoms in children with juvenile rheumatic diseases published by Chaney and colleagues (2016). We used Monte Carlo computer simulation to create an artificial data set with 100 observations mimicking a variety of parent- and child-level variables. Readers who want to replicate analyses in this chapter are encouraged to download the artificial data from the *Handbook*'s companion website (www.cambridge.org/ 9781107189843). Table 33.1 provides an overview of the study variables along with descriptive statistics, and Table 33.2 presents the correlations among the variables. We used maximum likelihood estimation (described later) to estimate both the descriptive statistics (Table 33.1) and correlations (Table 33.2), thus incorporating the observed data from the full sample. To illustrate estimating descriptive statistics and correlations when data are missing, Appendix 33.A provides the syntax for M*plus* (Muthén & Muthén, 1998–2017).

To illustrate the application of missing data-handling techniques, we use a regression model where child depressive symptoms is regressed on age, parent distress, caregiver demand, and parent perceptions of uncertainty:

$$DEPRESS_i = b_0 + b_1(AGE_i) + b_2(DISTRESS_i) \\ + b_3(DEMAND_i) + b_4(PARUNCERT_i) + e_i. \tag{33.1}$$

Table 33.1 Summary of variables in motivating example

Variable	Description	% complete	M	SD
DEPRESS	Child depressive symptoms	81	7.40	5.39
AGE	Child age	100	13.35	2.64
DISTRESS	Parent distress	91	0.40	0.32
DEMAND	Caregiver (parent) demand	100	46.87	13.34
PARUNCERT	Parent perceptions of uncertainty	90	79.25	16.35
CHIUNCERT	Child perceptions of uncertainty	100	81.59	12.88
ACTIVITY	Doctor rated disease activity	100	3.75	1.72
COPING	Doctor rated patient coping	100	5.54	0.99
SEVERITY	Doctor rated disease severity	100	3.63	2.30

Note. Estimates calculated using maximum likelihood estimation with full sample ($N = 100$); See Appendix 33.A for sample syntax. Only 70 cases have complete data on all variables.

Table 33.2 Correlations among study variables

Variable	1	2	3	4	5	6	7	8
1. AGE	1.000							
2. DISTRESS	−0.036	1.000						
3. DEMAND	0.003	**0.469**	1.000					
4. DEPRESS	0.033	**0.397**	**0.366**	1.000				
5. PARUNCERT	0.084	**0.300**	**0.315**	0.242	1.000			
6. CHIUNCERT	0.115	0.000	**−0.208**	**−0.473**	**−0.194**	1.000		
7. ACTIVITY	−0.072	−0.036	**0.217**	0.134	**0.349**	−0.156	1.000	
8. COPING	−0.169	**0.246**	0.068	**0.279**	−0.023	−0.176	0.015	1.000
9. SEVERITY	−0.130	−0.122	0.030	0.110	**0.452**	**−0.242**	**0.658**	0.035

Note. Correlations based on maximum likelihood estimation with full sample; see Appendix 33.A for sample syntax. Bolded values denote correlations where $p < 0.05$.

Figure 33.1 provides a path diagram of the regression model in Equation 33.1. As shown in Table 33.1, three of the analysis variables have missing observations: depressive symptoms (*DEPRESS*), parent distress (*DISTRESS*), and parent perceptions of uncertainty (*PARUNCERT*). Although nine variables are available, the analysis model in Equation 33.1 only utilizes five. As we will elaborate on later, the missing data-handling procedures can incorporate additional variables included in the data set as *auxiliary variables* in order to improve the quality and precision of the estimates.

MISSING DATA MECHANISMS AND TERMINOLOGY

To convey the assumptions underlying a missing data analysis, we briefly describe the missing data mechanisms proposed by Rubin and colleagues (Little & Rubin, 2002; Rubin, 1976). These mechanisms are a classification system used to describe the relationships between analysis variables and the probability of missing data. From a more applied perspective, these missing data mechanisms can be thought of as assumptions that dictate the performance of various missing data analysis techniques. Before delving into the missing data mechanisms, we note that a missing data *mechanism* is not the same as a missing data *pattern*. Missing data *patterns* are a characteristic of the data set that describes the configuration of observed and missing values, whereas missing data mechanisms describe possible relations between the data (observed or missing) and the propensity for missingness.

The most restrictive missing data mechanism is missing completely at random (MCAR), a condition which some argue may rarely be met in practice (Muthén, Kaplan, &

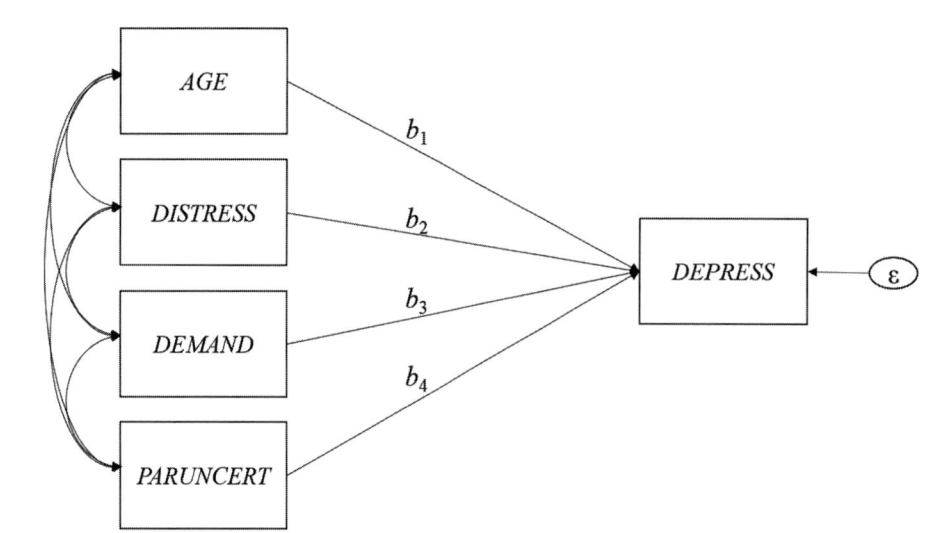

Figure 33.1 A path diagram for the multiple regression model in Equation 33.1. Single-headed straight lines denote regression coefficients, double-headed curved arrows denote correlations, rectangles denote manifest variables, and the ellipse denotes a latent variable. The predictor outcome variables are manifest variables and the residual term is a latent variable that captures the influences on the outcome variable not captured by the model.

Hollis, 1987; Raghunathan, 2004). The MCAR mechanism applies when missingness is completely *unsystematic* in the sense that it is unrelated to the data, observed or missing. In the context of the model from Equation 33.1, the analysis would satisfy the MCAR mechanism if the probability of missing *DEPRESS*, *DISTRESS*, and *PARUNCERT* is unrelated to one's scores (both observed and hypothetical unobserved) on these variables *and* is unrelated to the observed values of the complete analysis variables, *AGE* and *DEMAND*. In other words, the probability of missingness is unrelated to the data, and every participant has the same potential for missing values. Importantly, MCAR is the only mechanism with readily testable propositions (Dixon, 1988; Kim & Bentler, 2002; Little, 1988; Raykov, 2011). Specifically, each incomplete variable can be recoded into a corresponding missing data indicator, R, such that complete and incomplete values are assigned codes of zero and one, respectively. If MCAR holds for a particular variable, its missing data indicator should be unrelated to other variables. Conversely, the MCAR mechanism is implausible if an indicator is related to an analysis variable. Importantly, the MCAR assumption is typically required when excluding incomplete cases (i.e., listwise or pairwise deletion). Moreover, this practice is generally not robust to violations of MCAR – that is, when missingness is associated with any variable(s) in the analysis.

The missing at random (MAR) mechanism describes situations where missingness on some variable Y is related to other observed variables in the analysis but not to the values of Y itself. Although the name of this mechanism is potentially confusing because the word "random" implies a haphazard process, MAR actually describes a specific type of systematic missingness. Returning to the motivating example, the dependent variable, *DEPRESS*, is incomplete. The MAR mechanism stipulates that one's level of depression is unrelated to nonresponse; however, other variables in the analysis may predict missingness, either directly or indirectly via mutual associations with variables not in the analysis. Notice that we cannot empirically evaluate the MAR assumption because this mechanism involves a proposition about the *unobserved* data (e.g., the likelihood of missing the outcome unrelated to one's level of depression). From a practical perspective, MAR requires that we fully condition on variables that simultaneously correlate with an incomplete variable and its missing data indicator. This requirement is often approximately satisfied by introducing additional *auxiliary variables* into the analysis (Collins, Schafer, & Kam, 2001; Graham, 2003) or imputation model. As elucidated in the next section, variables that are not part of the research question, auxiliary variables, may be incorporated to support missing data analyses; these variables can be used to help satisfy the MAR mechanism without changing the substantive interpretation of the results. Maximum likelihood estimation, multiple imputation, and Bayesian estimation (not discussed here; Gelman et al., 2014, chapter 18) are MAR-based methods that thus require less stringent assumptions than deletion. From a practical perspective, this means that multiple imputation and maximum likelihood will yield accurate estimates in situations that deletion will not.

Finally, an analysis is consistent with the not missing at random (NMAR) mechanism if the probability of missing data on a particular variable is related to the unobserved values of that variable. Returning to the motivating example, an NMAR mechanism would result if depressed children were more likely to refuse the child depression inventory, *DEPRESS*, even after conditioning on other variables. Here again, note that we cannot empirically evaluate the NMAR assumption because this mechanism also involves a proposition about the unobserved data (nor can we distinguish it from an MAR mechanism for the same reason). Analyses that satisfy the NMAR mechanism tend to be the most challenging from an analytical standpoint because the analysis model needs to incorporate additional parameters that link the missing data indicators to the analysis variables. Longitudinal versions of such models are readily available and summarized in a variety of resources (Enders, 2010, 2011; Muthén et al., 2011). Because of the complexities of NMAR modeling, we chose to focus this chapter on the MAR-based methods.

AUXILIARY VARIABLES

Auxiliary variables are variables that are not part of the intended research model, but can be included to support the missing data analysis. Auxiliary variables have the capability to improve power and reduce nonresponse bias. Importantly, we can include these auxiliary variables without changing the substantive interpretation of the results. To include auxiliary variables in multiple imputation, we include these variables in the process that fills in the data, but not in the analysis phase of the procedure. In maximum likelihood, the auxiliary variables must be incorporated into the estimation routine. Although inclusion of auxiliary variables in the missing data analysis can be somewhat straightforward (and elaborated on later in this chapter), researchers often struggle with choosing appropriate auxiliary variables. The best auxiliary variables meet two criteria: (1) correlate with model variables that have missingness, and (2) correlate with the missing data indicators previously described (i.e., recoded variables where complete and incomplete values are assigned codes of zero and one). Auxiliary variables that meet these two criteria can potentially improve statistical power through the relationships with model variables with missingness *and* reduce nonresponse bias through the support of the MAR mechanism. Variables that meet both criteria should be the first type of variables that researchers seek as potential auxiliary variables. However, variables that only correlate with model variables that have missingness (criterion 1) may also serve as useful auxiliary variables. Although these auxiliary variables do nothing to reduce nonresponse bias, these variables can potentially increase power if the magnitude of the correlation between the variables is sufficiently large.

MULTIPLE IMPUTATION

Multiple imputation consists of three phases. In the imputation phase, the researcher creates several copies of the original data set (e.g., 20 or more; Graham, Olchowski, & Gilreath, 2007), each of which contains different filled in values or imputations. The imputation step uses regression equations to define distributions of plausible replacement values, with each regression predicting an incomplete variable from complete and previously imputed variables. Next, in the analysis phase, the researcher performs the statistical analyses of interest on each of the imputed data sets. Finally, in the pooling phase, the estimates and standard errors from the analysis phase are combined to provide a single set of results similar to those obtained from a complete data analysis.

Imputation Phase

The imputation phase creates multiple copies of the observed data set with a unique imputation in each. Joint model imputation (Schafer, 1997) and fully conditional specification (Raghunathan et al., 2001; van Buuren, 2012; van Buuren et al., 2006) are the most common imputation frameworks implemented in software packages. (Fully conditional specification is also referred to as chained equations and sequential regression imputation in the literature.) The joint model generates imputations from a multivariate regression with the set of incomplete variables regressed on the complete variables. In contrast, fully conditional specification imputes variables one at a time from a series of univariate models featuring complete and previously imputed variables as predictors. These two algorithmic approaches generally yield equivalent estimates when applied to normally distributed variables, but fully conditional specification is arguably more flexible for mixtures of categorical and continuous variables because each imputation model can be tailored to match the incomplete variable's metric (Raghunathan et al., 2001; van Buuren, 2007). However, some joint model imputation schemes can accommodate categorical variables by specifying an underlying normal latent variable for each discrete response (Asparouhov & Muthén, 2010b; Goldstein et al., 2009). For a demonstration of an imputation procedure with the analysis example, see Box 33.1. In the interest of space, we limit the discussion in Box 33.1 to fully conditional specification imputation because this approach is widely available in software packages.

After imputing the missing depression scores, the MCMC algorithm treats depressive symptoms as a complete variable and applies the same estimation and imputation sequence to parental distress and uncertainty. To generate the desired number of imputations (20 or more is a good rule of thumb; Graham et al., 2007), the MCMC algorithm cycles for hundreds and possibly thousands of iterations, saving a data set at specified intervals (e.g., after every 500th iteration; details on choosing this interval are discussed in the Multiple Imputation analysis example).

Analysis and Pooling Phases

After generating imputations, the researcher performs the statistical analyses of interest on each of the imputed data sets. Although this sounds tedious, most software packages automate the process of repeatedly analyzing the filled-in data. The analysis phase yields parameter estimates and standard errors for each data set, and the pooling phase combines these quantities into a single set

Box 33.1 Fully conditional specification for the analysis example

Fully conditional specification employs an iterative Markov chain Monte Carlo (MCMC) algorithm that repeatedly estimates a regression model and uses the resulting parameter values to define a distribution of replacement scores for each case. Applying fully conditional specification to the analysis in Equation 33.1 requires three regression models (as mentioned above, there are no missing observations for AGE or DEMAND).

$$DEPRESS_i = \gamma_0^{(1)} + \gamma_1^{(1)}(AGE_i) + \gamma_2^{(1)}(DEMAND_i) \\ + \gamma_3^{(1)}(DISTRESS_i) + \gamma_4^{(1)}(PARUNCERT_i) + \varepsilon_i^{(1)} \quad (33.2)$$

$$DISTRESS_i = \gamma_0^{(2)} + \gamma_1^{(2)}(AGE_i) + \gamma_2^{(2)}(DEMAND_i) \\ + \gamma_3^{(2)}(PARUNCERT_i) + \gamma_4^{(2)}(DEPRESS_i) + \varepsilon_i^{(2)} \quad (33.3)$$

$$PARUNCERT_i = \gamma_0^{(3)} + \gamma_1^{(3)}(AGE_i) + \gamma_2^{(3)}(DEMAND_i) \\ + \gamma_3^{(3)}(DEPRESS_i) + \gamma_4^{(3)}(DISTRESS_i) + \varepsilon_i^{(3)} \quad (33.4)$$

To emphasize that the imputation regressions are distinct from the analysis model in Equation 33.1, we now use γ's and ε to represent coefficients and the residual, respectively, and the numeric superscripts on these quantities indicate that each regression requires unique estimates (e.g., $\gamma_0^{(1)}$ and $\gamma_0^{(2)}$ take on different numeric values).

The composition of the imputation regression models warrants a brief discussion. First, notice that the outcome in one equation (an incomplete variable) functions as a predictor in all other equations. From an algorithmic perspective, fully conditional specification imputes an incomplete variable (the outcome in each equation) then subsequently treats the filled-in variable as though it is complete, moving it to the right side of all other regression equations. Second, the imputation equations (33.2–33.4) include only variables from the analysis model, but we would typically introduce additional auxiliary variables in order to help satisfy the MAR assumption and reduce nonresponse bias. Ideal auxiliary variables simultaneously correlate with one or more analysis variables and their missing data indicators. Third, obtaining accurate estimates from multiple imputation is only possible when the imputation regression equations preserve important features of the analysis model. The model in Equation 33.1 posits straightforward linear relations among the variables, and so Equations 33.2–33.4 will yield appropriate imputations. However, we often need to invoke specialized imputation routines to address more complex scenarios (e.g., categorical variables, interactive effects, and multilevel data structures). We address a few of these scenarios later in the chapter.

As noted previously, fully conditional specification uses an iterative MCMC algorithm to implement the estimation and imputation steps. The computational machinery underpinning the imputation procedure is rooted in methodology from Bayesian statistics. In contrast to the "traditional" (i.e., frequentist paradigm), Bayesian methodology considers each parameter has having a distribution (as opposed to being fixed). Although we ultimately bring our analyses back to the frequentist framework, for each incomplete variable, the algorithm uses Bayesian estimation methods to obtain parameter values, after which it uses the estimates to define a distribution of plausible replacement values for each missing observation. To illustrate the imputation step, consider the imputation model for depressive symptoms in Equation 33.2. At the beginning of iteration t, the MCMC algorithm first obtains Bayesian estimates of the coefficients and residual variance. To emphasize that these estimates vary across iterations, we henceforth omit the numeric superscript and instead attach an iteration superscript to the parameters (e.g., $\hat{\gamma}_0^{(t)}, \ldots, \hat{\gamma}_4^{(t)}, \hat{\sigma}_\varepsilon^{2(t)}$ denote the parameter estimates from iteration t).

Collectively, the regression model parameter estimates define a normal distribution of plausible replacement values for each case, with a predicted value and residual variance defining the center and spread, respectively. More formally, we can write the distribution of missing values as

$$DEPRESS_{mis}^{(t)} \sim N\left(DEPRESS_i^{*(t)}, \hat{\sigma}_\varepsilon^{2(t)}\right) \quad (33.5)$$

where $DEPRESS_{mis}^{(t)}$ is an imputed value, $\sim N$ denotes a normal distribution, $DEPRESS_i^{*(t)}$ is the predicted depression score for observation i

$$DEPRESS_i^{*(t)} = \hat{\gamma}_0^{(t)} + \hat{\gamma}_1^{(t)}(AGE_i) + \hat{\gamma}_2^{(t)}(DEMAND_i) + \hat{\gamma}_3^{(t)}\left(DISTRESS_i^{(t-1)}\right) + \hat{\gamma}_4^{(t)}\left(PARUNCERT_i^{(t-1)}\right) \quad (33.6)$$

and $\hat{\gamma}_0^{(t)}, \ldots, \hat{\gamma}_4^{(t)}, \hat{\sigma}_\varepsilon^{2(t)}$ are the parameter values from the preceding Bayesian estimation sequence. For variables with missing values (i.e., DISTRESS and PARUNCERT), the predicted value of depression for each participant is calculated using imputed values from the previous estimation sequence $(t-1)$. To impute the missing values, the algorithm uses Monte Carlo computer simulation to randomly "select" synthetic scores from the normal distribution in Equation 33.5.

of results. Pooled parameter estimates are derived by taking the arithmetic mean of each parameter estimate from each data set (Rubin, 1987):

$$\bar{\theta} = \frac{\sum \hat{\theta}_m}{M} \qquad (33.7)$$

where $\hat{\theta}_m$ represents a parameter estimate from data set m, and M is the total number of data sets. Returning to the previous analysis example, the pooling equation could be applied to regression coefficients, the residual variance, R^2 statistic, and descriptive quantities such as means, standard deviations, and correlations.

Pooling standard errors is not quite as straightforward as pooling estimates, as the combining rules take into account both complete data sampling error (called *within-imputation variance*) and the additional variation that results from imputing missing data (called *between-imputation variance*). Within-imputation variance estimates the squared standard error that would have been obtained had the data been complete. This component is computed as the arithmetic mean of the squared standard errors (i.e., sampling variances) from each imputed data set, as follows.

$$V_W = \frac{\sum SE_m^2}{M} \qquad (33.8)$$

The additional uncertainty due to missing data is quantified by the variation in the estimates across the imputed data sets. This between-imputation variance is obtained by applying the familiar formula for the sample variance to the estimates themselves, as follows.

$$V_B = \frac{\sum (\hat{\theta}_m - \bar{\theta})^2}{M - 1} \qquad (33.9)$$

Between-imputation variance increases as the amount of missing data increases, but the associations among the analysis variables moderate this relation, with stronger correlations reducing missing data uncertainty. Finally, the pooled standard error combines within- and between-imputation variance, as follows.

$$SE = \sqrt{V_W + V_B + \frac{V_B}{M}} \qquad (33.10)$$

The equation shows that the between-variance terms inflate the standard error to compensate for uncertainty about the imputations. Single-parameter significance tests can be performed by referencing the ratio of the pooled estimate to its standard error to either a t or z distribution (Asparouhov & Muthén, 2010a; Barnard & Rubin, 1999; Li, Raghunathan, & Rubin, 1991; Rubin, 1987), and multivariate test statistics (e.g., Wald or likelihood ratio tests) are also widely available in software (Li et al., 1991; Meng & Rubin, 1992).

MULTIPLE IMPUTATION ANALYSIS EXAMPLE

Having provided an overview of multiple imputation, we now use the Blimp software program (Enders, Keller, & Levy, 2017b; Enders, Du, & Keller, in press; Keller & Enders, 2019) to apply multiple imputation to the regression analysis from Equation 33.1. Blimp is available as a free download for macOS, Windows, and Linux at www.appliedmissingdata.com/multilevel-imputation. Although many popular software packages offer multiple imputation routines, we use Blimp because it handles mixtures of categorical and continuous variables in models with up to three levels, and it also includes newly developed features for interactive effects that are not yet widely available in other packages (Enders, Du, & Keller, 2017a). We describe some of these capabilities later in the chapter.

Prior to generating imputations for analysis, it is important to examine diagnostics to verify that the imputation algorithm has converged (i.e., is selecting imputations from distributions with stable center and spread) and to determine how many iterations are required between saved imputations. The size of the interval between saved imputations (called the thinning or between-imputation interval) depends on a variety of factors, including the sample size, number of variables, missing data rates, and correlations among the variables. Imputation software packages typically provide numeric or graphical diagnostics that can guide the specification of this interval. Blimp provides a numeric diagnostic for these purposes called the potential scale reduction factor [PSR]; Gelman & Rubin, 1992; Gelman et al., 2014). The PSR is scaled such that values close to the theoretical minimum of 1.0 reflect convergence, with values less than 1.05–1.10 commonly regarded as acceptable; higher scores generally suggest that more iterations are required (Gelman & Shirley, 2011). Appendix 33.B gives the Blimp script for a preliminary diagnostic run. The PSR values from this analysis suggested that only 50 iterations were required to achieve convergence. We generated 20 imputations (conservatively based on Graham et al., 2007), saving the first data set after the 100th iteration and saving subsequent data sets at every 100th iteration thereafter (i.e., based on PSR values that suggested 50 iterations were required, we specified a 100-iteration burn-in interval and a 100-iteration thinning interval). Appendix 33.C gives the Blimp script for the final MCMC algorithm, and imputation can also be performed from the application's graphical interface.

To illustrate the use of auxiliary variables, the Blimp scripts include all analysis variables as well as child perceptions of uncertainty and disease severity (*CHIUNCERT* and *SEVERITY*, respectively). The procedure for choosing the auxiliary variables, *CHIUNCERT* and *SEVERITY*, is typically done prior to beginning the analysis. As stated previously, the best auxiliary variables correlate with both variables that have missingness and missing data indicators. Including *CHIUNCERT* is important because this variable simultaneously correlates with the outcome variable *DEPRESS* and its missing data indicator (recall that MAR can only be satisfied if we condition on such

Table 33.3 Pooled estimates and standard errors from multiple imputation analysis example

Variable	Est.	SE	t	df	p
INTERCEPT (b_0)	−1.19	4.21	−0.28	56.28	0.78
AGE (b_1)	0.05	0.21	0.24	71.49	0.82
DISTRESS (b_2)	4.59	2.04	2.25	59.42	0.03
DEMAND (b_3)	0.09	0.05	1.75	62.22	0.09
PARUNCERT (b_4)	0.03	0.04	0.69	49.22	0.49

variables). Although disease severity does not predict missingness, including this variable in the imputation models may improve precision because it correlates strongly with the incomplete *PAUNCERT* variable (see Table 33.2). Because the auxiliary variables are complete (they need not be), they enter the imputation regression models as covariates, exactly like *AGE* and *DEMAND* in Equations 33.2–33.4 in Box 33.1.

We use M*plus* and *R* to illustrate the analysis and pooling phase, with the respective syntax files given in Appendices 33.D and 33.E. Analysis and pooling scripts for SAS, SPSS, and Stata can be downloaded from the chapter's companion website. Three points are worth noting about the analysis scripts. First, most software packages require a single file with imputed data sets stacked one on top of each other. The *R* script in Appendix 33.E follows this format, for example. Although Blimp defaults to this setting, the imputation script in Appendix 33.C generates imputations as separate data files (specified by listing the SEPARATE keyword on the OUTPUT line), a format consistent with M*plus*. Blimp also generates a text file containing the file paths to the imputed data sets, which the M*plus* script in Appendix 33.D subsequently uses as input data. Second, any auxiliary variables used during the imputation phase can be omitted when analyzing the data because the filled-in values naturally condition on these additional variables. Consequently, the M*plus* and *R* scripts do not use the two auxiliary variables from the imputation phase in the regression analysis. Finally, software packages differ in their level of automation. The MITML package (Grund, Robitzsch, & Lüdke, 2016) that we use in the *R* script pools unstandardized estimates and provides a number of useful multivariate significance testing options, but it does not pool standardized coefficients. M*plus*, on the other hand, automatically pools every quantity of interest, making it an attractive option for analyzing multiply imputed data sets.

Table 33.3 provides the pooled estimates and standard errors from the regression analysis. For the sake of brevity, we leave it to the reader to derive interpretations from the results. However, it is important to note that the substantive interpretation of the estimates using multiple imputation is no different from that of a complete data analysis.

For example, we would expect a 4.59 increase in depressive symptoms for every one-unit increase in *DISTRESS*, controlling for other predictors. Although not shown in the table, the *R* and M*plus* scripts use a Wald test to evaluate the joint significance of the four regression slopes; MITML uses an approximate *F* statistic, $F(4, 1333.74) = 4.58$, $p < 0.01$, whereas M*plus* uses a chi-square statistic, $\chi^2(4) = 19.29$, $p < 0.001$. Consistent with omnibus tests from a complete data regression analysis, these Wald statistics suggest that the set of explanatory variables reliably explains some portion of the variance in depressive symptoms. Note that SAS and Stata also provide these multivariate test statistics, but SPSS does not.

MAXIMUM LIKELIHOOD ESTIMATION

Maximum likelihood estimation is the second MAR-based missing data handling method that enjoys widespread use in the psychology literature. Importantly, maximum likelihood estimation is a common complete data estimator that readily extends to situations with incomplete data. At the broad level, maximum likelihood estimation identifies the population parameter values that most likely produced the sample data, given some assumed distribution for the population data; for the purposes of illustration, we assume multivariate normality, which is consistent with the treatment of incomplete variables in multiple imputation. Multivariate normality generalizes the univariate normal distribution to multiple variables simultaneously. The fit of the data to the parameter values is quantified by the natural log of the density function for the multivariate normal distribution (i.e., log of the formula that defines the shape of the multivariate normal curve). Critically, to identify well-fitting parameters, ML estimation depends on computing the Mahalanobis distance (MD) for each individual *i*, which quantifies the distance of the score vector, $\mathbf{Y}_i$, from the center of the multivariate normal distribution. See Box 33.2 for details about ML estimation.

Maximum Likelihood with Incomplete Data

Maximum likelihood estimation readily extends to missing data analysis with a slight alteration to the log-likelihood equation; see Equation 33.13, Box 33.2. More specifically, the estimate of an individual's MD – and therefore their contribution to the overall fit of the model – is estimated using all available data. To illustrate the missing data log-likelihood, we return to the regression model in Equation 33.1. For simplicity, we continue to use the mean vector and covariance matrix in our notation, reminding readers that these quantities are functions of the regression model parameters (i.e., μ_i and Σ_i are model-implied or model-predicted moments). The set of analysis variables is characterized by seven missing data patterns: a group of complete cases, three patterns with a single missing value, two patterns with a pair of missing

Box 33.2 Details on maximum likelihood estimation with multivariate normality

The natural log of the density function for a multivariate normal distribution is used to quantify the relative probability (expressed on a logarithmic scale) that the sample data originate from a multivariate normal population with a particular mean vector, μ, and covariance matrix, Σ.

$$\log L = \sum_{i=1}^{N} \left[-\frac{k}{2} \log(2\pi) - \frac{1}{2} \log |\Sigma| \right.$$
$$\left. -\frac{1}{2} (\mathbf{Y}_i - \mu)^T \Sigma^{-1} (\mathbf{Y}_i - \mu) \right] \quad (33.11)$$

The term $\mathbf{Y}_i$ denotes a set of k observed scores for an individual i, and the terms in brackets quantify the log-likelihood (i.e., "fit" to the parameters) for that participant; geometrically, the log-likelihood for a single individual corresponds to the height of the multivariate normal distribution at that individual's data. Finally, summing the individual log-likelihood values summarizes fit for the entire sample and is called the sample log-likelihood.

Maximum likelihood estimation identifies the values of μ and Σ that maximize the sample log-likelihood, as these are the population parameters that most likely produced the sample data. The critical aspect of this process is a sum of squared standardized deviation scores.

$$\text{MD} = (\mathbf{Y}_i - \mu)^T \Sigma^{-1} (\mathbf{Y}_i - \mu) \quad (33.12)$$

This expression, known as Mahalanobis distance (MD), quantifies the distance of the score vector, $\mathbf{Y}_i$, from the center of the multivariate normal distribution (i.e., the mean vector, μ). Numerically, MD can be interpreted as a squared z-score or χ^2 value. To maximize the sample log-likelihood, the estimation procedure must identify the parameter values that minimize the sum of the squared standardized deviation scores across the entire sample. Although some simple models do not require iterative optimization (e.g., a regression model with complete data), software packages typically use iterative algorithms to determine the parameters that maximize the sample log-likelihood (or minimize the sum of the MD values). Conceptually, the software auditions different parameter values until it finds those that are most likely to have produced the sample data.

Although we have discussed maximum likelihood estimation in terms of a mean vector and covariance matrix, the procedure accommodates a variety of common linear models (e.g., regression models, structural equation models), in which case μ and Σ

Box 33.2 (cont.)

are model-implied or model-predicted moments (i.e., the elements in μ and Σ are functions of model parameters such as regression slopes, factor loadings, etc.). For example, reconsider the regression model in Equation 33.1 or Figure 33.1. In a maximum likelihood regression analysis, predicted values (i.e., conditional means) define the mean vector and variances, covariances, and regression coefficients that combine to define the elements in Σ (e.g., the total variance of the outcome variable is the sum of the explained variance from the regression model and the residual variance). Importantly, defining μ and Σ as functions of other model parameters does not change the interpretation of the MD values in Equation 33.12, nor does it change the underlying logic of estimation (i.e., identify the model parameters that minimize the sum of the MD values).

Box 33.3 Extension of maximum likelihood estimation to incomplete data

Maximum likelihood estimation readily extends to missing data analysis with a slight alteration to the log-likelihood equation from Equation 33.11 in Box 33.2.

$$\log L = \sum_{i=1}^{N} \left[-\frac{k_i}{2} \log(2\pi) - \frac{1}{2} \log |\Sigma_i| - \frac{1}{2} (\mathbf{Y}_i - \mu_i)^T \right.$$
$$\left. \Sigma_i^{-1} (\mathbf{Y}_i - \mu) \right] \quad (33.13)$$

The distinctive difference between Equations 33.11 and 33.13 is the subscript i attached to the number of scores, k_i, the mean vector, μ_i, and covariance matrix, Σ_i. This subscript i indicates that the size and contents of the matrices in each participant's log-likelihood expression depend on the observed data pattern for that person. Because the data vector $\mathbf{Y}_i$ has missing values, the deviation scores in $(\mathbf{Y}_i - \mu_i)$ can be computed using only the mean vector elements that correspond to the observed variables (i.e., the data available for this quantity). Similarly, the size of the covariance matrix used to standardize the deviation scores must adjust, such that the elements in Σ_i also correspond to the observed variables.

values, and one pattern with three variables missing. The data vector $\mathbf{Y}_i$ is distinct for each of these patterns, potentially differing in size (e.g., three elements versus five) and contents (e.g., the two patterns missing a pair of scores have different variables in $\mathbf{Y}_i$). Similarly, the size and contents of $\boldsymbol{\mu}_i$ and $\boldsymbol{\Sigma}_i$ adjust to accommodate the observed data for each pattern. To illustrate, consider an individual with *AGE* and *DEMAND* scores but missing values on all other variables. The deviation scores for this case are computed using two of the five means, and $\boldsymbol{\Sigma}_i$ is a 2 × 2 matrix containing the variances and covariance of *AGE* and *DEMAND*, as follows.

$$\text{MD} = (\mathbf{Y}_i - \boldsymbol{\mu}_i)^T \boldsymbol{\Sigma}_i^{-1}(\mathbf{Y}_i - \boldsymbol{\mu}_i) = \left(\begin{bmatrix} AGE_i \\ DEMAND_i \end{bmatrix} - \begin{bmatrix} \mu_{AGE} \\ \mu_{DEMAND} \end{bmatrix} \right)^T \begin{bmatrix} \sigma^2_{AGE} & \sigma_{AGE,DEMAND} \\ \sigma_{AGE,DEMAND} & \sigma^2_{DEMAND} \end{bmatrix} ^{-1} \left(\begin{bmatrix} AGE_i \\ DEMAND_i \end{bmatrix} - \begin{bmatrix} \mu_{AGE} \\ \mu_{DEMAND} \end{bmatrix} \right)$$

(33.14)

The MD expression for the remaining patterns follows a similar logic (e.g., for the complete cases, $\mathbf{Y}_i$ and $\boldsymbol{\mu}_i$ each have five elements, and $\boldsymbol{\Sigma}_i$ is a 5 × 5 matrix; for the two patterns missing a pair of scores, $\mathbf{Y}_i$ and $\boldsymbol{\mu}_i$ each have three elements, and $\boldsymbol{\Sigma}_i$ is a 3 × 3 matrix).

Importantly, adjusting each individual's log-likelihood to accommodate the missing data pattern implies that no observations are discarded during estimation. Rather, individuals with at least one observed score can be included in estimation. The goal of estimation is still to identify the model parameters that minimize the sum of the MD values, but each squared z-score now reflects a different configuration and amount of observed data. It isn't immediately obvious from the previous log-likelihood expressions, but including the partial data records can improve both the accuracy and precision of the resulting parameter estimates. Although maximum likelihood estimation does not literally fill in the missing values during iterative optimization, methodologists have described the procedure as "implicit imputation" (Widaman, 2006) because the multivariate normality assumption implies plausible values for the incomplete scores during estimation (e.g., for a particular *AGE* value, the normal distribution implies that some *DEPRESS* scores are more likely than others, and leveraging this knowledge under multivariate distributional assumptions produces an adjustment to the estimates). For example, under the assumption that depression scores and age are normally distributed and associated with each other, estimating the covariance between depression scores and age from all available cases can help provide better estimates of the parameters that define the multivariate distribution. Conceptually, each iteration of the optimization procedure attempts to identify the parameter values that are most consistent with the observed data and the "implicit" imputations.

MAXIMUM LIKELIHOOD ESTIMATION ANALYSIS EXAMPLE

Now that we have provided an overview of maximum likelihood estimation for missing data, we use M*plus* (Muthén & Muthén, 1998–2017) and the lavaan package (Rosseel, 2012) in *R* to estimate the regression model from Equation 33.1. Structural equation modeling (SEM) programs such as M*plus* and lavaan are particularly flexible for maximum likelihood estimation because they offer researchers myriad options for modeling multivariate normal data (e.g., regression analyses, factor and latent variable models, longitudinal analyses, etc.). We chose these programs because they enjoy widespread use in behavioral science research, but almost any SEM software package offers similar functionality.

Returning to the analysis example, readers may recall from their regression training that predictor scores are treated as fixed values, meaning that they require no distributional assumptions; rather, normality applies only to the outcome variable's residuals. However, distributional assumptions are required for missing data handling, as incomplete variables must appear in the $\mathbf{Y}_i$ vector of Equation 33.13. The SEM framework naturally applies normality to all manifest variables, thus providing a mechanism for handling missing data. To give readers some intuition about the underlying model setup, Figure 33.2 shows the parameterization of a regression model that assumes multivariate normality. Three features of the path diagram are worth highlighting. First, all manifest (observed) variables have arrows feeding into them, implying that they function as outcome variables in $\mathbf{Y}_i$. Second, each exogenous variable is linked to a corresponding pseudo-latent variable via a unit regression weight, and its residual variance is fixed at zero. The former constraint places the pseudo-latent variables on the same metric as their manifest variable counterparts, and the latter constraint transfers all variance to the latent variables, effectively defining them as carbon copies of the manifest variables. Third, and perhaps most importantly, the SEM parameterization has no impact on the substantive interpretation of the results, as the estimates have the same interpretation as those from ordinary least squares. More simply, each observed predictor is recast as a latent variable with a loading of one and a residual of zero, thus creating identical predictor variables that have distributional assumptions and appear in the $\mathbf{Y}_i$ vector of Equation 33.13.

Recall that the multiple imputation procedure above included two auxiliary variables, child perceptions of uncertainty and disease severity (*CHIUNCERT* and *SEVERITY*, respectively). Conditioning on *CHIUNCERT* is important because this variable simultaneously correlates with the outcome variable *DEPRESS* and its missing data indicator, and MAR can only be satisfied if we condition on such variables. Although disease severity does not predict missingness, conditioning on this variable can improve precision because it correlates strongly with the incomplete *PAUNCERT* variable (see Table 33.2). Including auxiliary variables is relatively straight-forward in multiple imputation, but incorporating these additional variables in maximum likelihood estimation is slightly more complex. The literature describes two ways to

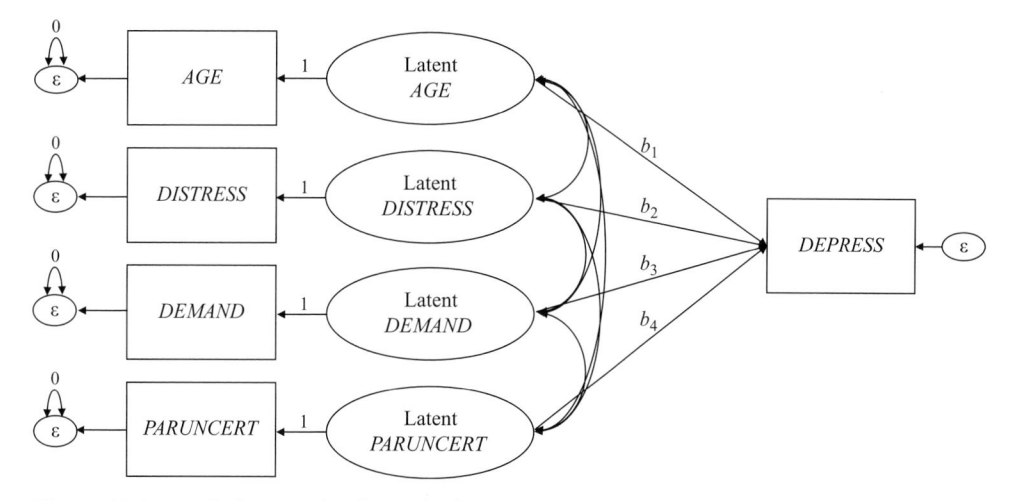

Figure 33.2 A path diagram for the multiple regression model in Equation 33.1 recast as a latent variable model where the latent variables have a single manifest indicator with a loading of zero and a residual variance of one. Single-headed straight lines denote regression coefficients, double-headed curved arrows denote correlations, rectangles denote manifest variables, and ellipses denote latent variable

include auxiliary variables in a maximum likelihood estimation analysis: the saturated correlates approach and the extra dependent variable approach (Graham, 2003). Both approaches work well in single-level analyses with normally distributed variables, but the latter approach can be applied to a wide range of models (e.g., multilevel models).

In the saturated correlates approach, auxiliary variables are incorporated via a series of correlations with the analysis variables and their residual terms (Enders, 2010; Graham, 2003). A path diagram of the saturated correlates model is given in Figure 33.3; because the pseudo-latent variables in Figure 33.2 are not typically shown in path diagrams (nor do they need to be specified by the user), we revert to usual path diagramming conventions. The figure shows that auxiliary variables correlate with manifest predictor variables, other auxiliary variables, and the residual terms of the outcome variable. Appendix 33.F provides M*plus* syntax for the saturated correlates approach. Conveniently, the M*plus* AUXILIARY command automates the saturated correlates model setup, whereas the researcher must explicitly specify the model in other packages. Importantly, listing all analysis variables in the MODEL part of the program declares these variables as multivariate normal, and it is not necessary to specify the pseudo-latent variables from Figure 33.2 manually. The extra dependent variable model incorporates auxiliary variables as additional outcomes, and it links all outcomes via correlated residuals (Graham, 2003; Graham et al., 1997). A path diagram of the analysis model using this specification is shown in Figure 33.4. Software packages do not automate the extra dependent variable approach, so the model must be specified manually. Appendix 33.G implements this approach using the lavaan package in R (Rosseel, 2012).

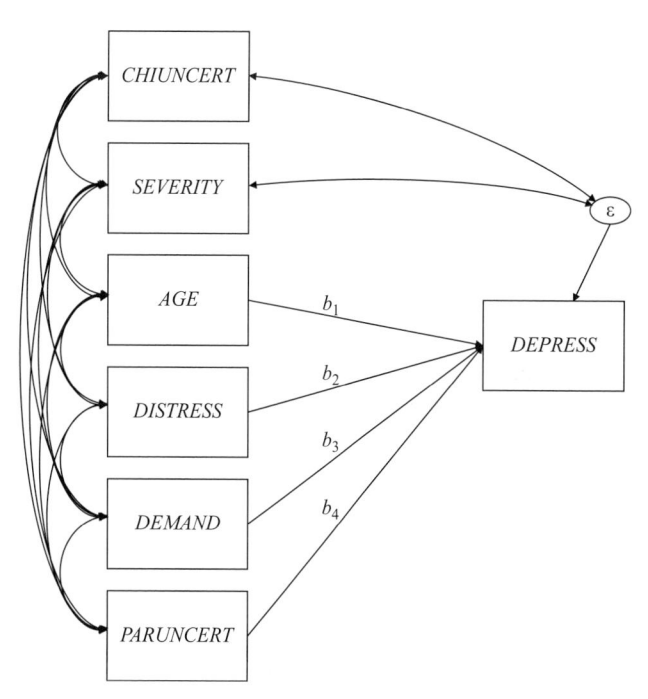

Figure 33.3 A path diagram demonstrating use of the saturated correlates approach to include auxiliary variables in the analysis model; because the pseudo-latent variables in Figure 33.2 are not typically shown in path diagrams (nor do they need to be specified by the user), we revert to usual path diagramming conventions. The path diagram shows that auxiliary variables correlate with manifest predictor variables, other auxiliary variables, and the residual terms of the outcome variable

Table 33.4 gives the maximum likelihood estimates from the analysis example. The table omits the additional auxiliary variable correlations because these estimates are not of substantive interest though they do typically appear in the software output. Further, we display a single set of results because the saturated correlates and extra dependent variable models are equivalent and thus produced identical estimates, within rounding error. Two points are important to highlight. First, the maximum likelihood estimates are remarkably similar to those from multiple imputation, which is to be expected with normally distributed variables. Second, as in the multiple imputation case, despite the presence of additional auxiliary variables that are not part of the original analysis, the substantive interpretation of the regression coefficients is no different from that of a complete data analysis. For example, for every one-unit increase in DISTRESS, the predicted value of DEPRESS is expected to increase by 4.48 points, controlling for AGE, DEMAND, and PARUNCERT. Although not shown in the table, the analysis scripts use a Wald chi-square test to evaluate the joint significance of the four regression slopes, $\chi^2(4) = 19.80, p \leq 0.001$. Consistent with omnibus tests from complete data regression analyses, the statistically significant Wald test suggests that the set of predictor variables explain a nonzero proportion of variation in child depressive symptoms.

Table 33.4 Estimates and standard errors from maximum likelihood analysis example

Variable	Est.	SE	z	p
INTERCEPT (b_0)	−1.21	3.86	−0.31	0.76
AGE (b_1)	0.04	0.21	0.22	0.83
DISTRESS (b_2)	4.48	1.99	2.26	0.02
DEMAND (b_3)	0.08	0.05	1.75	0.08
PARUNCERT (b_4)	0.03	0.04	0.80	0.43

PRACTICAL ISSUES: IMPUTATION VERSUS DIRECT ESTIMATION

The results in Tables 33.3 and 33.4 showed negligible differences between multiple imputation and maximum likelihood estimation. Given their apparent similarity, why would we choose one method over the other? When applied to single-level linear models with a common set of normally distributed variables and a sufficiently large sample size, personal preference is the sole determining factor because the two methods are equivalent (Collins et al., 2001; Gelman et al., 2014; Meng, 1994; Schafer, 2003). Under these conditions, maximum likelihood estimation is probably preferable on the basis of simplicity alone because the researcher need only translate the desired analysis to a capable software package, rather than the multiple step procedure involved in multiple imputation. SEM programs provide a flexible platform for estimation because they can accommodate a variety of common analyses with different configurations of missing data (e.g., incomplete predictors, as in our analysis example). Commercial programs have offered maximum likelihood missing data handling for over 20 years (Arbuckle, 1996), and powerful free programs such as lavaan offer most of the same capabilities (Rosseel, 2012).

Of course, behavioral science research often involves complexities that go well beyond linear models with normally distributed variables. Published studies in the behavioral sciences often feature mixtures of categorical and continuous variables, composite scores with missing components, nonlinear or interactive effects, and possibly multilevel data structures. These features are currently difficult to deal with in the likelihood framework, but multiple imputation can readily handle any of these problems. For this reason, multiple imputation is often a better

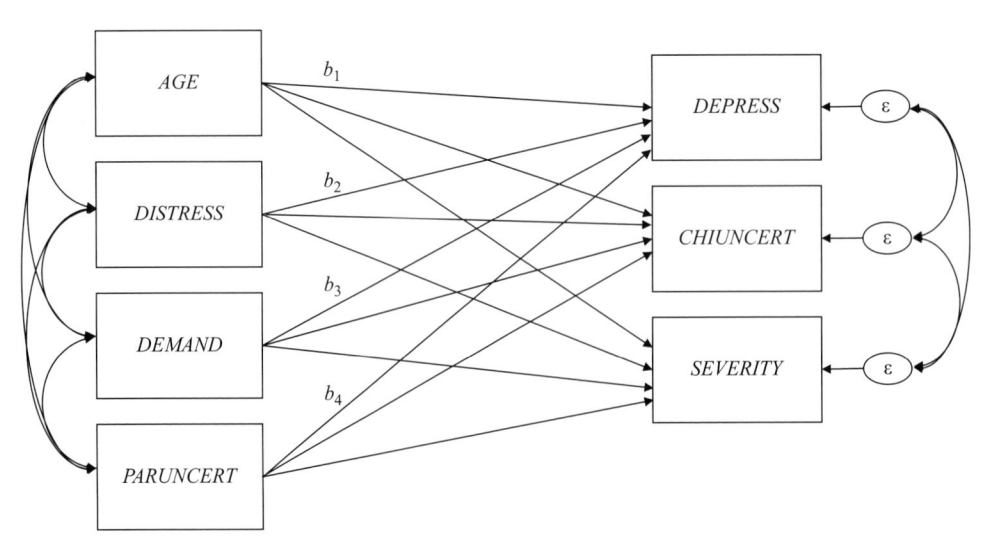

Figure 33.4 A path diagram demonstrating use of the extra dependent variable approach to include auxiliary variables in the analysis model; because the pseudo-latent variables in Figure 33.2 are not typically shown in path diagrams (nor do they need to be specified by the user), we revert to usual path diagramming conventions. The extra dependent variable model incorporates auxiliary variables as additional outcomes, and it links all outcomes via correlated residuals.

tool for behavioral research because it gives researchers the flexibility to tailor the missing data-handling procedure to match specific features of the variables in a particular analysis model. For the remainder of the chapter, we discuss the use of imputation to address the abovementioned complexities.

To begin, consider analyses with mixtures of categorical and continuous variables. As an example, suppose we expanded the regression analysis from Equation 33.1 to include an incomplete nominal covariate (e.g., race, diagnostic category, gender). As noted previously, maximum likelihood estimation in the SEM framework requires distributional assumptions for all variables, and software packages would typically force the user to apply a normal distribution assumption to the incomplete categorical variable. This assumption is particularly problematic for nominal variables represented by a set of dummy codes, as SEM programs would treat the code variables as multivariate normal. In contrast, categorical imputation routines are widely available in software packages, and researchers have a wealth of options for imputing these variables. The Blimp package illustrated earlier in the chapter can impute nominal or ordinal variables, and categorical imputation is triggered by simply listing variables on the NOMINAL or ORDINAL command lines, respectively. The Blimp User Guide (Keller & Enders, 2019) includes a number of analysis examples that illustrate this functionality.

Composite variables are a second situation where imputation offers distinct advantages to maximum likelihood estimation. For example, it is exceedingly common for researchers to compute scale scores by summing or averaging a set of Likert-type responses, a task made difficult when the component questionnaire items are incomplete. Returning to the regression model in Equation 33.1, depressive symptoms (*DEPRESS*) is a scale score computed by summing a set of ordinal questionnaire items, as is parent distress (*DISTRESS*) and parent perceptions of uncertainty (*PARUNCERT*). Perhaps the most common way of dealing with item-level missing data is to compute scale scores by averaging the available items (e.g., if a respondent answered 7 out of 10 items, the scale is the mean of the complete responses). Although exceedingly common in published studies, these so-called prorated scale scores require the restrictive MCAR mechanism, and they further assume that the scale items fit a unidimensional factor model with equal loadings and measurement intercepts (Mazza, Enders, & Ruehlman, 2015; Schafer & Graham, 2002).

Because it does not fill in the data, maximum likelihood strongly encourages the user to treat the scale as missing when one or more of its component items is missing. This practice produces a rather dramatic decrease in power that is only mitigated by introducing a subset of the questionnaire items as auxiliary variables (Eekhout et al., 2015; Gottschall, West, & Enders, 2012; Mazza et al., 2015). In contrast, multiple imputation provides a simple and ideal strategy for incomplete scale scores: apply a categorical imputation routine to the incomplete item responses, then compute scale scores from the filled-in data. This procedure maximizes power because it leverages typically strong correlations among items measuring the same construct. Further, common imputation routines such as the MCMC algorithm in Blimp invoke a saturated imputation model that makes no assumptions about the internal structure of the scale.

A third situation where multiple imputation currently offers a dramatic advantage over maximum likelihood estimation is with analysis models involving nonlinear or interactive (moderation) effects. Returning to the previous example, suppose we hypothesize that the influence of parental distress on child depressive symptoms is stronger for parents with greater caregiver demands (i.e., caregiver demands moderate the relation between parent distress and child depressive symptoms). The moderated regression analysis model is as follows.

$$DEPRESS_i = b_0 + b_1(AGE_i) + b_2(DISTRESS_i) + b_3(DEMAND_i)$$
$$+ b_4(PARUNCERT_i) + b_5(DISTRESS_i)(DEMAND_i) + e_i$$
(33.15)

Importantly, the product term is incomplete because one of its components, parental distress, is incomplete. Until very recently, researchers were forced to compute the product term and treat it like any other incomplete variable in the missing data-handling procedure. For example, fully conditional specification would regress the product variable on all other variables, then draw imputations from a normal curve. Similarly, maximum likelihood estimation would treat the product as a normally distributed variable that is not constrained to equal the product of its constituents (Enders, Baraldi, & Cham, 2014). This so-called just-another-variable approach to interactions is problematic because it assumes an MCAR mechanism and is prone to substantial biases, even when applying maximum likelihood estimation and conventional multiple imputation routines (Bartlett et al., 2014; Carpenter & Kenward, 2013; Enders et al., 2014; Yuan & Savalei, 2014).

The Blimp application offers a recently developed imputation routine termed substantive model-compatible imputation that yields accurate interaction effects under an MAR mechanism (Bartlett et al., 2014; Enders et al., 2017a). Whereas the standard fully conditional specification (FCS) imputation procedure described previously would impute the product term from a normal distribution, the substantive model-compatible approach does not directly impute the product term at all, thus avoiding bias-inducing model-specification problems. Rather, the procedure imputes the product term's component variables, using a special algorithm to select imputations that are consistent with the hypothesized interaction effect. Currently, the Blimp application can accommodate polynomial and interactive effects in models with up to three levels, and simulation results suggest that the procedure virtually eliminates the bias endemic to standard missing data handling schemes (Enders et al., 2017a).

To illustrate the approach, Appendix 33.H shows the Blimp script that applies substantive model-compatible imputation to the analysis in Equation 33.16. For brevity, we do not present the analysis results here; see Appendix 33.H for implementation details and www.appliedmissingdata.com/multilevel-imputation for a variety of analysis scripts.

Multilevel data structures are a final example of a common analytic problem where multiple imputation tends to be superior to maximum likelihood estimation. These data structures are exceedingly common throughout behavioral science and include cases where students are nested within classrooms or schools, repeated measures are nested within individuals, and individuals are nested within dyads or families, to name a few. Maximum likelihood estimation for multilevel models with incomplete data is often limited because software packages tend to restrict missing data handling to outcome variables, and programs that can handle incomplete predictors typically limit this functionality to random intercept models with normally distributed variables (Asparouhov & Muthén, 2010b; Shin & Raudenbush, 2007, 2010). In contrast, the Blimp application handles a wide variety of multilevel analysis problems, including models with random slopes, categorical variables, and data structures with up to three levels (Enders et al., 2017b). The program also applies the substantive model-compatible approach described previously to multilevel moderation effects such as random coefficients and cross-level interactions (Enders et al., 2017a). For the sake of brevity, we refer interested readers to the Blimp User Guide for examples of multilevel imputations, and analysis-phase scripts for a variety of popular software programs are available on the program's website.

CONCLUSION

The methodological literature offers a relatively clear and unambiguous recommendation: multiple imputation or maximum likelihood estimation should generally be the standard for published research. Relative to older methods such as deletion, these approaches are preferable because they rely on a less stringent assumption about missingness (i.e., an MAR mechanism where missingness is related to observed variables). In this chapter, we provided an overview of these two approaches, and we illustrated their use in popular software packages. When applied to normally distributed variables, the choice between multiple imputation and maximum likelihood estimation is largely one of personal preference, as the methods are asymptotically equivalent and tend to yield similar results (as they did in our examples). However, multiple imputation is arguably more flexible for many common analytic problems. For example, clinical psychology researchers often work with mixtures of categorical and continuous variables, composite scores from multiple-item scales, interactive effects, and multilevel data. Currently, multiple imputation can readily accommodate these analysis features, whereas maximum likelihood estimation is generally less adept.

REFERENCES

Anderson, T. W. (1957). Maximum Likelihood Estimates for a Multivariate Normal Distribution When Some Observations Are Missing. *Journal of the American Statistical Association*, 52, 200–203.

Arbuckle, J. L. (1996). Full Information Estimation in the Presence of Incomplete Data. In G. A. Marcoulides & R. E. Schumacker (Eds.), *Advanced Structural Equation Modeling* (pp. 243–277). Mahwah, NJ: Lawrence Erlbaum.

Asparouhov, T., & Muthén, B. (2010a). Chi-Square Statistics with Multiple Imputation. Retrieved from www.statmodel.com/download/MI7.pdf

Asparouhov, T., & Muthén, B. (2010b). Multiple Imputation with Mplus. Retrieved from www.statmodel.com/download/Imputations7.pdf

Barnard, J., & Rubin, D. B. (1999). Small-Sample Degrees of Freedom with Multiple Imputation. *Biometrika*, 86, 948–955.

Bartlett, J. W., Seaman, S. R., White, I. R., & Carpenter, J. R. (2014). Multiple Imputation of Covariates by Fully Conditional Specification: Accommodating the Substantive Model. *Statistical Methods in Medical Research*, 24, 462–487.

Beale, E. M. L., & Little, R. J. A. (1975). Missing Values in Multivariate Analysis. *Journal of the Royal Statistical Society, Series B (Methodological)*, 37, 129–145.

Carpenter, J. R., & Kenward, M. G. (2013). *Multiple Imputation and Its Application*. Chichester: Wiley.

Chaney, J. M., Gamwell, K. L., Baraldi, A. N., Ramsey, R. R., Cushing, C. C., Mullins, A. J., ... & Mullins, L. L. (2016). Parent Perceptions of Illness Uncertainty and Child Depressive Symptoms in Juvenile Rheumatic Diseases: Examining Caregiver Demand and Parent Distress as Mediators. *Journal of Pediatric Psychology*, 41, 941–951.

Collins, L. M., Schafer, J. L., & Kam, C.-M. (2001). A Comparison of Inclusive and Restrictive Strategies in Modern Missing Data Procedures. *Psychological Methods*, 6, 330–351.

Dempster, A. P., & Laird, N. M. (1977). Maximum Likelihood from Incomplete Data via the EM Algorithm. *Journal of the Royal Statistical Society, Series B (Methodological)*, 39, 1–38.

Dixon, W. J. (1988). *BMDP Statistical Software*. Los Angeles: University of California Press.

Eekhout, I., Enders, C. K., Twisk, J. W. R., de Boer, M. R., de Vet, H. C. W., & Heymans, M. W. (2015). Analyzing Incomplete Item Scores in Longitudinal Data by Including Item Score Information as Auxiliary Variables. *Structural Equation Modeling: A Multidisciplinary Journal*, 22, 1–15.

Enders, C. K. (2010). *Applied Missing Data Analysis*. New York: Guilford Press.

Enders, C. K. (2011). Missing Not at Random Models for Latent Growth Curve Analyses. *Psychological Methods*, 16, 1–16.

Enders, C. K., Baraldi, A. N., & Cham, H. (2014). Estimating Interaction Effects with Incomplete Predictor Variables. *Psychological Methods*, 19, 39–55.

Enders, C. K., Du, H., & Keller, B. T. (2017a). A Fully Bayesian Imputation Procedure for Random Coefficient Models (and Other Pesky Product Terms). Paper presented at the Society for Multivariate Experimental Psychology, Minneapolis, MN, October 4–7.

Enders, C. K., Du, H., & Keller, B. T. (In press). A Model-Based Imputation Procedure for Multilevel Regression Models with Random Coefficients, Interaction Effects, and Other Nonlinear Terms. *Psychological Methods.* Retrieved from doi:10.1037/met0000148

Enders, C. K., Keller, B. T., & Levy, R. (2017b). A Fully Conditional Specification Approach to Multilevel Imputation of Categorical and Continuous Variables. *Psychological Methods, 23* (2), 298–317.

Gelman, A., & Rubin, D. B. (1992). Inference from Iterative Simulation Using Multiple Sequences. *Statistical Science, 7,* 457–472.

Gelman, A., & Shirley, K. (2011). Inference from Simulations and Monitoring Convergence. In S. Brooks, A. Gelman, G. Jones, & X. L. Meng (Eds.), *Handbook of Markov Chain Monte Carlo.* Boca Raton, FL: CRC Press.

Gelman, A., Carlin, J. B., Stern, H. S., Dunson, D. B., Vehtari, A., & Rubin, D. B. (2014). *Bayesian Data Analysis* (3rd edn.). Boca Raton, FL: CRC Press.

Goldstein, H., Carpenter, J., Kenward, M. G., & Levin, K. A. (2009). Multilevel Models with Multivariate Mixed Response Types. *Statistical Modelling, 9,* 173–197.

Gottschall, A. C., West, S. G., & Enders, C. K. (2012). A Comparison of Item-Level and Scale-Level Multiple Imputation for Questionnaire Batteries. *Multivariate Behavioral Research, 47,* 1–25.

Graham, J. W. (2003). Adding Missing-Data Relevant Variables to FIML-Based Structural Equation Models. *Structural Equation Modeling: A Multidisciplinary Journal, 10,* 80–100.

Graham, J. W., Hofer, S. M., Donaldson, S. I., MacKinnon, D. P., & Schafer, J. L. (1997). Analysis with Missing Data in Prevention Research. In K. J. Bryant, M. Windle, & S. G. West (Eds.), *The Science of Prevention: Methodological Advances from Alcohol and Substance Abuse Research* (pp. 325–366). Washington, DC: American Psychological Association.

Graham, J. W., Olchowski, A. E., & Gilreath, T. D. (2007). How Many Imputations Are Really Needed? Some Practical Clarifications of Multiple Imputation Theory. *Prevention Science, 8,* 206–213.

Grund, S., Robitzsch, A., & Lüdke, O. (2016). mitml: Tools for Multiple Imputation in Multilevel Modeling. Retrieved from https://cran.r-project.org/web/packages/mitml/

Jeličić, H., Phelps, E., & Lerner, R. M. (2009). Use of Missing Data Methods in Longitudinal Studies: The Persistence of Bad Practices in Developmental Psychology. *Developmental Psychology, 45,* 1195–1199.

Keller, B. T., & Enders, C. K. (2019). *Blimp User's Manual* (Version 2.0). Los Angeles, CA.

Kenward, M. G., & Molenberghs, G. (1998). Likelihood Based Frequentist Inference when Data Are Missing at Random. *Statistical Science, 13,* 236–247.

Kim, K. H., & Bentler, P. M. (2002). Tests of Homogeneity of Means and Covariance Matrices for Multivariate Incomplete Data. *Psychometrika, 67,* 609–624.

Li, K. H., Raghunathan, T. E., & Rubin, D. B. (1991). Large-Sample Significance Levels from Multiply Imputed Data Using Moment-Based Statistics and an F Reference Distribution. *Journal of the American Statistical Association, 86,* 1065–1073.

Little, R. J. A. (1988). A Test of Missing Completely at Random for Multivariate Data with Missing Values. *Journal of the American Statistical Association, 83,* 1198–1202.

Little, R. J. A., & Rubin, D. B. (2002). *Statistical Analysis with Missing Data* (2nd edn.). Hoboken, NJ: Wiley

Mazza, G. L., Enders, C. K., & Ruehlman, L. S. (2015). Addressing Item-Level Missing Data: A Comparison of Proration and Full Information Maximum Likelihood Estimation. *Multivariate Behavioral Research, 50,* 504–519.

Meng, X.-L. (1994). Multiple-Imputation Inferences with Uncongenial Sources of Input. *Statistical Science, 9,* 538–558.

Meng, X.-L., & Rubin, D. B. (1992). Performing Likelihood Ratio Tests with Multiply-Imputed Data Sets. *Biometrika, 79,* 103–111.

Muthén, L. K., & Muthén, B. O. (1998–2017). *Mplus User's Guide* (8th edn.). Los Angeles, CA: Muthén & Muthén

Muthén, B., Asparouhov, T., Hunter, A. M., & Leuchter, A. F. (2011). Growth Modeling with Nonignorable Dropout: Alternative Analyses of the STAR*D Antidepressant Trial. *Psychological Methods, 16,* 17–33.

Muthén, B., Kaplan, D., & Hollis, M. (1987). On Structural Equation Modeling with Data that Are Not Missing Completely at Random. *Psychometrika, 52,* 431–462.

Orchard, T., & Woodbury, M. A. (1972). A Missing Information Principle: Theory and Applications. In *Proceedings of the Sixth Berkeley Symposium on Mathematical Statistics and Probability* (Vol. 1, pp. 697–715). Berkeley: University of California Press.

Peugh, J. L., & Enders, C. K. (2004). Missing Data in Educational Research: A Review of Reporting Practices and Suggestions for Improvement. *Review of Educational Research, 74,* 525–556.

Raghunathan, T. E. (2004). What Do We Do with Missing Data? Some Options for Analysis of Incomplete Data. *Annual Review of Public Health, 25,* 99–117.

Raghunathan, T. E., Lepkowski, J. M., Van Hoewyk, J., & Solenberger, P. (2001). A Multivariate Technique for Multiply Imputing Missing Values Using a Sequence of Regression Models. *Survey Methodology, 27,* 85–95.

Raykov, T. (2011). On Testability of Missing Data Mechanisms in Incomplete Data Sets. *Structural Equation Modeling: A Multidisciplinary Journal, 18,* 419–429.

Rosseel, Y. (2012). lavaan: An R Package for Structural Equation Modeling. *Journal of Statistical Software, 48*(2), 1–36.

Rubin, D. B. (1976). Inference and Missing Data. *Biometrika, 63,* 581–592.

Rubin, D. B. (1978). Multiple Imputations in Sample Surveys – A Phenomenological Bayesian Approach to Nonresponse. In *Proceedings of the Survey Research Methods Section of the American Statistical Association* (pp. 30–34).

Rubin, D. B. (1987). *Multiple Imputation for Nonresponse in Surveys.* Hoboken, NJ: Wiley.

Schafer, J. L. (1997). *Analysis of Incomplete Multivariate Data.* New York: Chapman & Hall.

Schafer, J. L. (2003). Multiple Imputation in Multivariate Problems When the Imputation and Analysis Models Differ. *Statistica Neerlandica, 57,* 19–35.

Schafer, J. L., & Graham, J. W. (2002). Missing Data: Our View of the State of the Art. *Psychological Methods, 7,* 147–177.

Shin, Y., & Raudenbush, S. W. (2007). Just-Identified versus Overidentified Two-Level Hierarchical Linear Models with Missing Data. *Biometrics, 63,* 1262–1268.

Shin, Y., & Raudenbush, S. W. (2010). A Latent Cluster-Mean Approach to the Contextual Effects Model with Missing Data. *Journal of Educational and Behavioral Statistics, 35*, 26–53.

Sullivan, T. R., Yelland, L. N., Lee, K. J., Ryan, P., & Salter, A. B. (2017). Treatment of Missing Data in Follow-Up Studies of Randomised Controlled Trials: A Systematic Review of the Literature. *Clinical Trials, 14*, 387–395.

Van Buuren, S. (2007). Multiple Imputation of Discrete and Continuous Data by Fully Conditional Specification. *Statistical Methods in Medical Research, 16*, 219–242.

Van Buuren, S. (2012). *Flexible Imputation of Missing Data*. New York: Chapman & Hall.

Van Buuren, S., Brand, J. P. L., Groothuis-Oudshoorn, C. G. M., & Rubin, D. B. (2006). Fully Conditional Specification in Multivariate Imputation. *Journal of Statistical Computation and Simulation, 76*, 1049–1064.

Widaman, K. F. (2006). Missing Data: What to Do with or without Them. *Monographs of the Society for Research in Child Development, 71*, 42–64.

Yuan, K.-H., & Savalei, V. (2014). Consistency, Bias and Efficiency of the Normal-Distribution-Based MLE: The Role of Auxiliary Variables. *Journal of Multivariate Analysis, 124*, 353–370.

Appendix 33.A

M*plus* Script for Correlations and Descriptive Statistics

```
DATA:
file = example.dat;
VARIABLE:
names = id age distress demand depress paruncert chiuncert
activity coping severity;
usevariables = age - severity;
missing = all(-99);
ANALYSIS:
estimator = ml;
MODEL:
age - severity;
age - severity with age - severity;
  [age - severity];
OUTPUT:
standardized (stdyx);
```

Appendix 33.B

Blimp Script for Diagnosing Convergence of the MCMC Algorithm

```
DATA: ~/desktop/examples/example.dat;
VARIABLES: id age distress demand depress paruncert chiuncert
    activity coping severity;
MISSING: -99;
MODEL: ~ age distress demand depress paruncert
    chiuncert severity;
NIMPS: 2;
BURN: 3000;
THIN: 1;
SEED: 90291;
OUTFILE: ~/desktop/examples/imps.csv;
OPTIONS: psr;
CHAINS: 2 processors 2;
```

Appendix 33.C

Blimp Script for Multiple Imputation

```
DATA: ~/desktop/examples/example.dat;
VARIABLES: id age distress demand depress paruncert chiuncert
   activity coping severity;
MISSING: -99;
MODEL: ~ age distress demand depress paruncert
   chiuncert severity;
NIMPS: 20;
BURN: 100;
THIN: 100;
SEED: 90291;
OUTFILE: ~/desktop/examples/imp*.csv;
OPTIONS: separate;
CHAINS: 2 processors 2;
```

Appendix 33.D

M*plus* Script for Analysis and Pooling Phases

```
DATA:
file = implist.csv;
type = imputation;
VARIABLE:
names =  id age distress demand depress paruncert chiuncert
   activity coping severity;
usevariables = depress age distress demand paruncert;
MODEL:
! regression model with parameter labels;
depress on age (b1)
    distress (b2)
    demand (b3)
    paruncert (b4);
MODEL TEST:
! Wald test;
b1 = 0; b2 = 0; b3 = 0; b4 = 0;
OUTPUT:
standardized(stdyx);
```

Appendix 33.E

mitml Script for Analysis and Pooling Phases

```r
# load mitml
library(mitml)
# import stacked data
filepath <- "~/desktop/examples/imps.csv"
impdata <- read.csv(filepath, header = F)
names(impdata) <- c("imputation", "id", "age", "distress",
    "demand", "depress", "paruncert", "chiuncert",
    "activity","coping", "severity")
# analyze data and pool estimates
implist <- as.mitml.list(split(impdata, impdata$imputation))
model <- with(implist,
              lm(depress ~ age + distress + demand + paruncert))
estimates <- testEstimates(model, var.comp = T, df.com = (95))
estimates
# specify the Wald test
emptymodel <- with(implist, lm(depress ~ 1))
testModels(model, emptymodel, method = "D1")
```

Appendix 33.F

M*plus* Script for the Saturated Correlates Regression Model

```
DATA:
file = example.dat;
VARIABLE:
names = id age distress demand depress paruncert chiuncert
 activity coping severity;
usevariables = depress age distress demand paruncert;
auxiliary = (m) chiuncert severity;
missing = all(-99);
ANALYSIS:
estimator = ml;
MODEL:
! declare variables as multivariate normal;
depress age distress demand paruncert;
! regression model with parameter labels;
depress on age (b1)
      distress (b2)
      demand (b3)
      paruncert (b4);
MODEL TEST:
! Wald test;
b1 = 0; b2 = 0; b3 = 0; b4 = 0;
OUTPUT:
standardized (stdyx) patterns;
```

Appendix 33.G

lavaan Script for the Extra Dependent Variable Regression Model

```
# load lavaan
library(lavaan)
# import text file and recode missing values
rawdat <- read.csv("~/desktop/examples/example.dat", header = F)
names(rawdat) <- c("id", "age", "distress", "demand", "depress",
    "paruncert", "chiuncert", "activity", "coping", "severity")
rawdat[ rawdat == -99] <- NA
# laavan regression model
model <- '
depress ~ b1 * age + b2 * distress
    + b3 * demand + b4 * paruncert
chiuncert ~ age + distress + demand + paruncert
severity ~ age + distress + demand + paruncert
depress ~~ chiuncert + severity
chiuncert ~~ severity
'
# fit model and get estimates
fit <- sem(model, rawdat, missing = 'fiml', fixed.x = F)
summary(fit, fit.measures = T, rsquare = T, standardize = T)
# use parameter labels to specify the Wald test
lavTestWald(fit, 'b1 == 0; b2 == 0; b3 == 0; b4 == 0')
```

Appendix 33.H

Blimp Script and Documentation for Substantive Model-Compatible Imputation for Interactive Effects

```
DATA: ~/desktop/example.dat;
VARIABLES: age distress demand depress paruncert chiuncert
    activity coping severity;
MISSING: -99;
MODEL: ~ age distress demand depress paruncert chiuncert
    severity distress*demand;
OUTCOME: depress;
NIMPS: 20;
BURN: 100;
THIN: 100;
SEED: 90291;
OUTFILE: ~/desktop/imp*.csv;
OPTIONS: separate;
CHAINS: 2 processors 2;
```

Note. Three points are noteworthy about this Blimp script. First, because the imputation algorithm must choose imputations that are consistent with the hypothesized interaction effect, it is necessary to specify the dependent variable from the analysis model on the OUTCOME line. Second, the imputation routine does not impute the interaction directly and thus does not require a product variable in the data set. Rather, the interaction is specified by joining its constituent lower-order effects by an asterisk on the MODEL command. Third, because Blimp does not impute the interaction directly, the product term does not appear in the output data sets and must be computed prior to analysis. This feature provides a great deal of flexibility because researchers can center the lower-order predictors in each data set and compute the product by multiplying centered variables, just as they would with complete data (Aiken & West, 1991).

34 Machine Learning for Clinical Psychology and Clinical Neuroscience

MARC N. COUTANCHE AND LAUREN S. HALLION

In all its complexity, the question towards which all outcome research should ultimately be directed is the following: *What treatment, by whom, is most effective for this individual with that specific problem, and under which set of circumstances?* (Paul, 1967, p. 111)

This question, posed by Gordon Paul in 1967 (more recently distilled to its essence as, "what works for whom, and under what circumstances?"), has remained one of the foundational and as yet unresolved problems of modern clinical psychology. Although the original quotation refers to questions of treatment selection, the spirit is broadly applicable as we seek to ask: what are the personal characteristics, sociocultural contexts, and situation-specific factors that result in the onset, development, maintenance, and remission of psychopathology? More recently, the question has also been extended to the level of the individual, with the goal of identifying the factors that are most relevant for understanding *this individual's* clinical presentation and treatment needs.

These types of question form the core of much basic and applied research in the field of clinical psychology. Although many studies have made significant and meaningful progress toward answering one or two facets of these questions in isolation, questions of "what, for whom, under what circumstances, and why" remain largely unaddressed at an integrative level. Current challenges include small sample sizes, large and unwieldy bodies of research that have developed in relative isolation, and – most importantly for the purposes of this chapter – limited access to the statistical tools that would be best-suited to addressing these problems. Unfortunately, even the best efforts to implement **precision medicine** in a clinical psychology context have been constrained by a mismatch between the inherent complexity of these questions (e.g., Cohen & DeRubeis, 2018) and the relatively narrow "statistical toolbelt" that comes standard-issue in most clinical psychology training programs. In our experience, some of the most significant barriers to clinical psychology graduate students being able to use these particular advanced statistical methods include: (1) a limited availability of formal training opportunities that are geared to noncomputer scientists; (2) informed decisions to deprioritize methods that are not directly applicable to current research projects; and (3) textbooks that were written primarily with computer scientists or statisticians in mind, such that they assume a level of background knowledge that extends well beyond what most clinical psychology students have obtained in prior coursework.

The goal of this chapter is to help address some of the most significant barriers to entry with machine learning, with an eye toward questions of "what, for whom, under what circumstances, and why?" Our goal is to present a clear and concise introduction and how-to guide that covers basic principles and example applications of machine learning to questions in psychology.

WHAT IS MACHINE LEARNING?

The term "machine learning" refers to a range of mathematical techniques that leverage the computational power made available by modern computers to identify meaningful signals within large and often complex data sets. A common characteristic of these techniques is that they all identify patterns or relationships that are present in a given data set and then extend those patterns to independent data sets to validate the model (Yarkoni & Westfall, 2017). The first phase, in which an algorithm is applied to identify patterns and relationships in one set of data, is described as the **learning** or **training** phase. The second phase, in which the resulting model is applied to a new data set, is called the **testing** phase.

A common theme across techniques is a strong emphasis placed on prediction, or the extent to which a model that is trained on one set of data can successfully predict patterns and relationships in a new (untrained) data set. As in traditional statistical approaches (e.g., linear regression), machine learning models are based on the patterns and relationships in data. However, whereas a traditional linear regression is evaluated on the basis of how well it accounts for patterns in an original data set, a trained machine learning model is evaluated on how accurately it can predict patterns and relationships in new data. The ability of a model that was trained in one data set to predict patterns in another is described as the model's ability to **generalize**.

Another unique and central feature of machine learning is its focus on multidimensionality. Although there are certainly good reasons to ask whether a given predictor is independently associated with an outcome absent the effects of all other constructs, this artificial scenario does little to capture the complex interplay that exists in reality. Instead, it may be more meaningful to ask how an outcome is predicted in the context of many predictors. This multidimensionality allows a model to identify subtle patterns in data sets, which have often been likened to **fingerprints**. Indeed, it has been suggested that in areas of psychology in which simple mechanistically understood models fail to successfully predict any behavior beyond a particular data set, it is time to switch to prediction (rather than explanation alone) as a goal, even at the cost of interpretability (Yarkoni & Westfall, 2017).

BEYOND TRADITIONAL STATISTICS: WHAT DOES MACHINE LEARNING ADD?

The question of what "counts" as machine learning can be somewhat murky at times. This is because, at their core, some machine learning techniques rely on the same basic mathematical principles that are applied in traditional statistical analyses (e.g., logistic regression and the general linear model). In the definition of machine learning above, we suggested two key aspects that define and distinguish this set of methods, prediction and multidimensionality, which we expand on below.

Prediction and Generalization

In traditional statistical methods, a result (such as an *F* or *t* statistic) reflects the extent to which the model provides a strong fit to the current data set. However, this value does not reflect the extent to which this *particular* model will apply beyond the data on which it was tested (Yarkoni & Westfall, 2017). For example, the specific pattern of beta coefficients from a linear regression analysis may not adequately describe a different set of patients or research participants. This lack of external validation has been identified as one major reason for recent failures to replicate particular biomarkers of psychiatric treatment outcome (Gillan & Whelan, 2017). In line with these observations, the recent push toward open science and replicability has led to an increased emphasis on the importance of replication as a criterion to evaluate the robustness of a given finding (Open Science Collaboration, 2015). However, the decision to conduct a replication study is nontrivial. Even a straightforward replication can require significant time, effort, and expense – none of which most independent scientists (let alone most graduate students) have to spare. Further, a failure to replicate can arise for a number of reasons, and as such can sometimes be difficult to interpret.

A machine learning approach offers a researcher several distinct advantages here, particularly the ability to quantify the generalizability of a "learned" pattern or algorithm to new data. Indeed, success is typically operationalized as the concordance between the values predicted from a previously trained model and actual values observed in different data. We discuss the best ways to determine which data points are used for training versus testing below, but one common and intuitive approach is to **hold out** anywhere from a single participant at a time (repeated iteratively for each participant) to up to half the sample (through **cross-validation**). These held-out participants are considered the "test" data set and are excluded from any analyses that are performed in the training data set. This includes any imputation, transformation, or other data cleaning or preprocessing procedures that might use the entire data set.

Alternatively, a researcher may approach the question of generalization from a more conservative or theoretically driven perspective by asking whether a model that is trained on a data set with certain characteristics (e.g., undergraduate students) can provide accurate predictions in a data set with different sample characteristics (e.g., treatment-seeking adults) or involving the administration of a different treatment. The more "steps removed" between the training and testing data sets (that is, the more the training and testing data sets differ in research context, study participants, and so on), the stronger the test of generalization (see Gillan & Whalen, 2017 for a detailed consideration of this issue). From a theoretical perspective, discovering that a model performs better for some data sets than others can be informative, particularly if there are potentially meaningful differences between data sets. For example, Chekroud and colleagues (2016) applied machine learning with cross-validation to clinical trial data of response to a 12-week course of citalopram (a selective serotonin reuptake inhibitor; SSRI) in patients with depression (in the Sequenced Treatment Alternatives to Relieve Depression trial; STAR*D). Patient-reported variables were used to predict symptom remission with 64.6 percent accuracy in the STAR*D cohort. Reported clinical features such as insomnia and somatic complaints contributed to the model's performance. The authors also examined the extent to which the model generalized to an independent clinical trial of response to escitalopram in a similar sample (Combining Medications to Enhance Depression Outcomes; COMED). The STAR*D model predicted escitalopram (versus placebo) response in this separate (COMED) trial at a level significantly above chance (59.6 percent). Notably, prediction accuracy dropped to near-chance (51.4 percent) when predicting response to a serotonin-norepinephrine reuptake inhibitor (SNRI; venlafaxine-mirtazapine) in the COMED trial: evidence that the model was specific to predicting response to SSRIs.

Multidimensionality

A further advantage of machine learning lies in its ability to account for the multidimensionality of psychological data. With a linear regression analysis, we typically ask if a particular variable (represented by a coefficient) is a significant predictor. For most machine learning techniques, the focus is less on "does this one variable predict the outcome?" and is closer to "does this suite of variables predict the outcome?" Successful solutions can include scenarios where individual variables only have marginal predictive ability but (when pooled together) form a highly predictive *set*. This point – of using overall prediction accuracy, rather than the coefficient for any one or more variables – is especially important because a given individual variable could contribute to successful predictions in a way that is not clearly revealed by inspecting beta weights (i.e., as in traditional linear regression). Instead, we consider unique predictor profiles, patterns, or fingerprints, without necessarily emphasizing specific details of that pattern.

As an illustrative example, one study found that individual symptoms did not differentiate unipolar from bipolar depression, but that the *pattern* of symptoms did (Perlis et al., 2006). This concept has also played a significant role in the rapid development and application of machine learning to functional magnetic resonance imaging (fMRI) data. The discovery that patterns of brain activity can contain information about perceptual, cognitive, and affective processes, when signals in individual voxels (or region averages) do not, continues to advance our understanding of brain systems with a degree of specificity that was not previously possible (Coutanche, 2013; Haxby et al., 2001; Tong & Pratte, 2012).

Considering multidimensional "profiles" of variables is not always intuitive, but it has nevertheless shown promise. For example, Astle and colleagues (2019) examined a heterogeneous sample of 530 children who were referred due to problems in attention, memory, language, or poor school performance. An unsupervised machine learning analysis (discussed further below) was applied to performance on seven cognitive tasks. The authors employed Self Organizing Maps (SOMs; Kohonen, 1989), where an artificial neural network projects a multidimensional input space into two dimensions. The relevant output, a two-dimensional grid, represents the degree of similarity in the input data – here, task performance in individual children. This was combined with a **k-means clustering** approach (described in more detail below), which allocated the children to four empirically derived clusters based on their cognitive profiles. These clusters represented profiles corresponding to: (1) broad cognitive difficulties and significant reading, spelling and math problems; (2) age-typical cognitive and learning abilities; (3) working memory problems; or (4) phonological difficulties. Interestingly, the cognitive profiles of the children were not predicted by diagnostic status or reason for their referral. The clusters did, however, correspond to differences in white matter brain networks measured through diffusion tensor MRI. These brain data had not been used to construct the SOMs or clusters, bolstering the validity of the detected clusters. This example identified data-driven neurocognitive dimensions underlying learning-related difficulties. Moreover, it highlights an advantage of recruiting and analyzing samples with varied, rather than "pure," deficits, which can often ignore people with comorbidity (a substantial portion of a population) and overemphasize similarity within groups. A focus on sampling along continuous dimensions in study design can provide new targets for approaches to early detection and intervention, as well as help to identify etiological mechanisms.

As we expand on below, machine learning techniques are also able to use a larger number of variables than standard statistical approaches. However, these approaches remain constrained by statistical power and are subject to the broad signal-to-noise and sample size considerations that also affect more traditional methods.

HOW CAN MACHINE LEARNING INFORM CLINICAL THEORY?

The relevance and utility of machine learning techniques for addressing questions related to diagnosis, treatment selection, and prediction of critical behaviors is, in our opinion, fairly uncontroversial. A model that can predict how well an individual will respond to a treatment, or the likelihood that a given individual will die by suicide in a given 12-month period, would represent a significant and important advance with major clinical implications. However, one limitation of these techniques is that the full set of relationships between a model's predictors and its predictions are not always transparent. This limitation has resulted in machine learning approaches at times earning the (somewhat unflattering) "black box" descriptor. One response to this critique is that even the most opaque of black boxes can have its place. If an algorithm enhances prediction of important clinical information, it has value, irrespective of its transparency. However, the opacity of machine learning techniques does raise challenges when approached from a theoretical perspective. In this section, we consider whether, how, and to what extent machine learning can be leveraged to inform theoretical models of psychopathology and its treatment.

Although theory testing is not traditionally considered to be a strength of machine learning, there are nevertheless a number of ways in which these approaches can and have been creatively leveraged to advance theoretical understanding. For example, we have used a **feature ablation** (lesion) approach (described below) in our own work (Hallion et al., 2019) to test theoretical predictions about the distinctions between different classes of perseverative thought (e.g., worry versus rumination). In that study, we identified features of thoughts that have been proposed to characterize and distinguish between different classes of perseverative thought, such as temporal orientation (worry is traditionally defined as future oriented, whereas rumination is defined as present or past oriented), valence, form, intrusiveness, and others. First, we asked participants to rate a sampling of their own thoughts according to each of those features. We then tested whether a model trained on those features could successfully predict scores on conventional self-report measures of worry, rumination, and other kinds of perseverative thought. Next, we examined the impact of ablating (removing) each feature that has been proposed as theoretically important for defining and distinguishing between the different types of perseverative thought. Within an ablation framework, a significant reduction in accuracy after removing a variable indicates that the variable contributed unique predictive power (i.e., offered predictive power that could not be obtained from the other variables in the model). Conducting this ablation process for each predictor in turn gives values for the relative influence of each variable. The analyses lent support to some theoretical distinctions between types of perseverative thought and raised challenges for others. For example, self-referential processing has been proposed to be a central feature of depressive rumination (Nolen-Hoeksema, Wisco, & Lyubomirsky, 2008). Consistent with that account, ablating items related to self-referential processing significantly reduced the model's accuracy for predicting self-reported rumination, but did not impact accuracy for predicting other kinds of thought. However, although temporal orientation has previously been identified as the defining (and only reliable) difference between worry and rumination (Watkins, 2008), ablating items related to temporal orientation did not significantly or differentially impact the model's accuracy for predicting or discriminating between scores on traditional measures of worry and rumination. Altogether, these findings (along with a broader set of results not described here) pointed away from a categorical "subtype" model of perseverative thought and instead toward a fully dimensional model in which thoughts are most accurately described in terms of underlying features, rather than in terms of classes characterized a priori by the presence or absence of certain features.

This ablation approach could easily be adapted to examine the impact of including versus excluding any number of theory-derived predictors. Example questions that might be examined include:

- What is the impact of removing theoretically important (versus theoretically peripheral) predictors on predicting a symptom or syndrome of interest?
- Does including or removing a proposed risk factor change the model's ability to predict the development of psychopathology prospectively?
- Which baseline clinical features have the largest influence on the model's accuracy for predicting symptom trajectory or treatment response? (Inferred by an especially large drop in prediction when those features are removed.)
- Does the addition or removal of certain clinical characteristics improve the model's ability to discriminate between two closely related disorders (e.g., social anxiety disorder versus avoidant personality disorder) or constructs (e.g., fear versus anxiety)?

Generalization, or the validation of a model's success through prediction, can also be leveraged to test certain theoretical questions. Accuracy for predicting observations in a new data set can both validate the original model and quantify the ability of this model to explain and predict observations from a set of data that differs in theoretically interesting ways. We discussed an example of generalization testing above in Chekroud and colleagues (2016), where a model trained on treatment response to citalopram was more accurate when making treatment predictions for escitalopram (which has a similar neural mechanism) versus an SNRI combination. Comparisons of generalizability can be extended to test theoretical ideas related to mechanism (do cognitive task outcomes improve prediction?), sample characteristics (does prediction vary by gender?), nosology (does the model distinguish between disorders?), and so on. By emphasizing the performance of a model in predicting outcomes in a new data set, machine learning approaches are well suited to testing specific, concrete, and testable predictions about the generality versus specificity of a given finding.

HOW CAN MACHINE LEARNING INFORM CLINICAL NEUROSCIENCE?

Machine learning techniques have a long history of application within a neuroimaging context. More recently, these techniques have been applied specifically to data from clinical samples and clinically relevant paradigms to enhance understanding of a wide variety of symptoms and disorders. These methods are more widely explored within those literatures so we do not provide a thorough treatment here (we refer the interested reader to Woo et al., 2017, for a review). We will instead touch briefly on some of the ways in which machine learning can be applied to neural data to make predictions about risk assessment, early detection, differential diagnosis, treatment response, and biological mechanisms.

Clinically predictive neural fingerprints can include a wide variety of neural signatures (see Sundermann et al., 2014, for a review), such as patterns of activity across voxels (Coutanche et al., 2011), resting-state connectivity (Du et al., 2012), task-based connectivity (Deshpande et al., 2013), and volumetric differences (Nouretdinov et al., 2011), among others. New developments in integrating machine learning with other brain measures, such as connectivity (Anzellotti & Coutanche, 2018), continue to present new targets for clinical investigations.

The availability of large data sets online, and prediction competitions (such as the attention deficit-hyperactivity disorder [ADHD]-200 Global Competition; Bellec et al., 2017), have rapidly increased the number of classification studies that are focused on discriminating among patient groups or identifying transdiagnostic dimensional features. A prediction competition typically involves sharing a large data set (including assigned labels, such as diagnostic status) with contestants who then develop (i.e., train) classifier models. A test data set is then shared (without any labels). Contestants typically use their trained model to generate label predictions for the test data set (e.g., diagnosis), which they submit to the contest organizers. The contest organizers can then grade each model based on the accuracy of its predictions. A welcome component of many recent contests is holding the test data set in escrow, reducing the chance that final performance metrics might be inflated by accidental influence of test data.

In parallel, over the last 10 years, the number of multisite neuroimaging studies has risen significantly. These studies have several advantages to a researcher seeking to employ machine learning. Sample sizes in such studies can often number in the thousands, and having data from multiple sites allows models to be trained that are capable of generalizing to new sites and scanners (Sundermann et al., 2014).

An important consideration for the neuroimaging investigator is that a model can learn to use any **features** that are predictive of the relevant classification. On the one hand, this is why such models are so powerful, but on the other, this warrants caution, particularly because neuroimaging models can be influenced by clinically relevant factors that do not reflect the neural signals of interest, such as in-scanner head motion for predicting ADHD (Eloyan et al., 2012) or a strong effect of eye-blinks on predicting an autism spectrum disorder diagnosis (Eldridge et al., 2014). Such features are valid predictors for each respective disorder, but do not give insights into their neural underpinnings.

The application of machine learning techniques to understand the neural basis of psychopathology has yielded important insights. However, the utility of these techniques for matching individuals to treatments (which some might describe as the "holy grail" of clinical neuroscience) is as yet unclear. In their consideration of some of these issues, Gillan and Whelan (2017) suggest that efforts to develop a comprehensive computational neural model of specific clinical symptoms could result in a "dead end," not because of methodological limitations, but because symptoms are rarely (if ever) pathognomonic. Just as a headache could result from any number of underlying pathologies, so too can many clinical symptoms. We do not believe that this equifinality renders futile any efforts to develop neural models of these symptoms, but it does raise theoretical, methodological, and interpretational challenges. Similarly, a classifier may be trained to differentiate between different disorders or syndromes, but if the a priori distinctions are artifactual or otherwise spurious, a classifier with perfect accuracy will nevertheless still fail to truly "carve nature at its joints."

DOWN TO BRASS TACKS

A wide array of methods is available to the researcher wishing to apply machine learning to address clinical problems. These techniques can be broadly divided into **supervised learning** or **unsupervised learning** methods. Here, supervision (or lack thereof) refers to whether the training data set includes labels or other information that corresponds to the criterion (or "correct answer"). An example of supervised learning, which we discuss further below, would be a **machine learning classifier**. In a classifier approach, the analyst assigns a label (e.g., diagnostic status) to each observation, which reflects the "true" class. This label could come from sources such as an expert's diagnosis or a genetic analysis, but (independent of the scientific truth) this class membership is treated as "true" by the classifier. The classifier model is trained to successfully separate observations into the classes using the values of the set of predictors (or features).

In contrast, unsupervised models take a more data-driven approach, which does not assume that a certain criterion (such as an observation's class, or even the number of classes) is known. Instead, an unsupervised model is trained to provide the best fit to the underlying structure of a data set. This can be useful when a priori classification schemes (e.g., a diagnostic taxonomy) may not accurately reflect scientific reality. An experimenter

---

Box 34.1 An example application

Suppose we are broadly interested in understanding the cognitive and neural mechanisms that contribute to maintaining anxiety-related psychopathology. Let's further suppose that we have access to a large, multimodal data set that includes an array of predictors and outcomes of interest. Our participants (cases) include over 1,000 individuals who responded to an advertisement for a research study on "worry and anxiety." Our variables include performance on a range of traditional neuropsychological and other cognitive tasks, resting state data from fMRI, self-report measures for a variety of anxiety symptoms, and *Diagnostic and Statistical Manual of Mental Disorders* (DSM-5) diagnostic status as determined in a clinician-administered, semistructured diagnostic interview. There is a multitude of potentially promising questions that we could ask using this data set. As the chapter progresses, feel free to refer to these boxes to see example applications to this hypothetical data set for the techniques we discuss.

> **Box 34.2** Applying a supervised learning classifier
>
> To continue our example, we might use a supervised learning model to predict diagnostic status from resting-state fMRI data and cognitive testing data. Our diagnostic interview data has labels for each participant (case): a Yes/No (binary) label for each diagnosis of interest (e.g., Generalized Anxiety Disorder [GAD]; panic disorder; social anxiety disorder; specific phobia). A training data set is first created – with fMRI data and cognitive testing data as predictors, and diagnostic status (Yes/No for each of the four diagnoses of interest) as outcomes. Using cross-validation (discussed further below), we could test how well a trained model can classify new cases (i.e., cases that were not included in the training data set) with respect to their clinician-assigned diagnosis. One classification approach would be to ask a separate (binary; two-way) question for each diagnosis (i.e., GAD – Yes/No; panic disorder – Yes/No; social anxiety disorder – Yes/No; specific phobia – Yes/No). In this case, we train and test separate models for each of the four diagnoses/outcome variables. Chance performance would be 50 percent accuracy for predicting each outcome.

> **Box 34.3** Applying an unsupervised learning model
>
> In our ongoing example, we might seek to detect distinct profiles of cognitive and neural functioning within our sample. We could then examine how subjects with distinct profiles differ in their diagnoses or symptoms, and ask what this means for a particular diagnosis or symptom. In this case, the examined data would include our cognitive and neural variables, but without information about diagnostic status or symptoms (i.e., without labels). An unsupervised algorithm is used to detect robust patterns (such as clusters of cognitive and neural predictors) in the training set. The testing set can then be used to evaluate the reliability of these profiles for new cases. Alternatively, we could ask if the naturally emerging profiles of neurocognitive functioning can predict differences in diagnoses or symptoms in independent cases.

might dictate certain parameters (e.g., how many clusters to form), but the clusters themselves are derived without labels corresponding to one specific criterion or outcome of interest. This is not to say that unsupervised analyses are always preferable. The class labels provided in supervised approaches are often meaningful and can provide important information to guide the identification of a solution.

Study Design

Here, we provide a brief overview of study designs that are well suited to a machine learning analytic approach. These often vary in whether they are between- or within-subject designs, and whether we use categorical labels (classes) or continuous values.

Between- or Within-Subject Designs

Classifying between (or across) subjects involves training and testing a model using data from different sets of subjects, where a positive result reflects the model making successful predictions about new (untrained) participants.

An underappreciated alternative is to apply machine learning to within-subject data (e.g., trial-by-trial performance in a cognitive task). Within-subject study designs that result in many observations per person (e.g., ecological momentary assessment, eye-tracking, computerized cognitive tasks, and neuroimaging), can be submitted to machine learning approaches to obtain a measure of the extent to which trial types (e.g., a certain class of stimuli) are distinguishable within a given participant. This metric can in turn be related to an individual difference of interest.

For example, in an fMRI study of face perception in adolescents with autism spectrum disorder, we used patterns of activity in the fusiform gyrus to classify whether each participant was viewing a face or a house on a given trial (Coutanche, Thompson-Schill, & Schultz, 2011). We then related participants' "face-versus-house" classifier accuracies to their social scores on the Autism Diagnostic Observation Schedule (ADOS), a clinician-administered measure of symptom severity. Here, instead of attempting to classify participants, we were interested in relating an individual difference in symptom severity to the distinctiveness of their neural codes for faces. We found that neural activity patterns for faces and nonfaces were more confusable in adolescents with more severe social impairments.

Predicting Group Membership

If we are classifying cases, how many (and what) classes should we include? When extending the principles of machine learning to classify three or more classes, there are several considerations to bear in mind. For instance, with multiway (i.e., >two groups) classifications, although the overall accuracy speaks to discriminability among classes, a significant accuracy value does *not* mean that all the classes can be successfully predicted. This is analogous to the case of an omnibus ANOVA, where a significant F-test indicates that at least two groups differ, but does not tell us which ones (instead, post hoc tests are used to answer this).

Predicting Continuous and Discrete Outcomes

Classifiers often seem to get the machine learning headlines. The value of continuous models, however, is

> **Box 34.4** Applying class predictions
>
> Previously, we asked whether our model could predict the presence or absence of each of four diagnoses (GAD; panic disorder; social anxiety disorder; specific phobia). Instead, we might ask if a model can use predictors to discriminate *between* individuals with different diagnoses (i.e., four-way classification). For illustrative purposes, imagine that each of our 1,000 participants meets criteria for one, and only one, disorder (unlikely, but such is the power of hypotheticals). In this case, the trained model would assign each case to one of four classes (GAD *or* panic disorder *or* social anxiety *or* specific phobia). Here, chance performance is 25 percent when the cases are distributed evenly across diagnoses (i.e., there is a one in four chance of classifying correctly on the basis of chance alone).

> **Box 34.5** Applying caution to interpretations of accuracy
>
> When conducting multiway classifications (i.e., discriminating more than two groups), it is important to be cognizant that some predictors could be highly specific to one (and only one) outcome class. For example, suppose our model used "frequent uncued panic attacks" as a predictor. This predictor would have high accuracy for classifying cases as "panic disorder" or "not panic disorder," but classification could be at chance for the other comparisons. In this example, panic disorder would be classified with (close to) 100 percent accuracy, while GAD, social anxiety disorder, and specific phobia would each be classified with 33.3 percent accuracy (because they are confusable with each other, but not with panic disorder). This particular scenario would give an *overall* classification performance of 50 percent, twice the level of chance (25 percent) and typically significant (see "Outputs and Interpretation" below for a discussion of testing statistical significance). Yet it would not be entirely accurate to conclude that this model successfully discriminates between the four anxiety disorders. Instead, we have actually trained an excellent discriminator of one disorder. This example shows the importance of moving beyond classifier accuracy alone. Here, a **confusion matrix** (discussed further below) would give us insights into the structure of our multiway classifier.

> **Box 34.6** Applying models to predict continuous outcomes
>
> As we learn more about the underlying structure of psychopathology, transdiagnostic approaches become increasingly important. Rather than attempting to classify individuals into discrete groups, we may be interested in developing a model to predict severity of a certain type of symptom (e.g., autonomic arousal; interpersonal difficulties). Self-report measures are generally well suited to these questions. To continue our example, we might be interested in whether a model trained on neural and cognitive data can accurately predict how much (or how little) control an individual believes they have over their worry. Alternatively, we might be interested in whether our model is more (or less) effective for predicting the severity of cognitive symptoms (e.g., unwanted thought; difficulty concentrating) versus physical symptoms (e.g., muscle tension; autonomic arousal). Combined with a feature-ablation approach (discussed below), we might develop insights about mechanisms that underlie each symptom.

increasingly being recognized (e.g., Dosenbach et al., 2010). Just like classifiers, these models are trained on a subset of the data and tested on a held-out set. In a continuous case, however, model success is determined from the difference (or prediction loss) between the true (continuous) values and the model's predictions. This is a form of regression analysis (e.g., support vector regression; SVR).

Data Analysis

Once you have decided on your design and data set, how do you analyze your data?

Selecting Features

A key advantage of machine learning models is their robustness to using a large number of "features" or predictors, relative to the number of observations. A standard linear regression model is quickly overwhelmed by a large number of predictors, but machine learning models are robust to a high predictor-to-observation ratio, including where predictors outnumber observations (as can happen in, for example, fMRI studies with hundreds or even thousands of voxels predicting 100 observations for each condition). As the number of features rise, so does the chance of **overfitting**. Overfitting occurs when a large number of features lead a model to closely fit the unique idiosyncrasies of the training data, at the cost of limited generalizability to new (test) data. It is for this reason that machine learning techniques are rarely applied to the whole brain's set of voxels (which can easily number 60,000) without an intermediate feature selection step.

As the name suggests, "feature selection" aims to reduce the number of predictors in the final model, while optimizing the ability to detect a signal (i.e., to remove features with minimal impact on the quality of prediction). Entire research programs are dedicated to analytically selecting a subset of available features (Guyon & Elisseeff, 2003) but it is valuable for a psychologist applying machine learning techniques to be familiar with a number of basic approaches.

One vital rule closely governs the choice and application of feature selection techniques: training and testing sets must always be independent. This is especially important during feature selection because an accidental violation of this rule can artificially inflate a model's (apparent) performance, even using data that is just noise. Consider a data set of completely randomly generated values (i.e., noise), with 10,000 observations and 1,000 predictors. We could randomly split these observations into two classes that we then attempt to classify. Ordinarily, we would expect to obtain chance performance from this pure-noise data set. But now imagine that before classification, we quantify how well each variable separates the observations. Even by chance, some variables will have higher values in one condition than another. If we were to then choose the top 10 most "discriminating" variables, and classify our observations using these predictors, we would obtain above-chance performance. Clearly, this is not right – the data set is pure noise – so how did we erroneously reach this conclusion? Our main error was to choose predictors (i.e., select features) based on the full data set, including those observations that our model will test on. It is not surprising that our model could separate the classes because we picked those features that differed between them (by chance). Here, our model's success is closer to a redescription of how we picked the variables ("top 10 most discriminative variables") than reflecting successful prediction. Fortunately, we can still select features without this error.

One possible approach to selecting features is to use a difference among groups/classes that is unrelated to the one we are classifying. For example, suppose we are classifying participants into one of two psychological disorders based on their responses to a number of surveys. The surveys to be ultimately selected as features might be chosen based on unrelated dimensions, such as reliability metrics, from prior work. These dimensions might also be extractable from the data itself, though great care is needed to ensure the dimensions are truly orthogonal (Kriegeskorte et al., 2009).

Some analytical methods allow one to reduce the dimensionality of a data set in a way that preserves most of the variance. For example, a principal component analysis (PCA) can be used to reduce an original set of predictors to a smaller set of components that still explain a large part of the variance. This is a popular approach to feature selection when classifying fMRI data, as often 90 percent of the variance in the signal of hundreds of voxels can be captured in 10–30 new dimensions. Using these components reduces the number of predictors (thus reducing the risk of overfitting) while keeping much of the variance intact.

Another approach (not mutually exclusive with the first) is using cross-validation (described further below), which by design separates training from testing. The concern in the above example of predicting noise was that our predictors were selected based on their ability to discriminate the classes using *all* the data – including the testing set that was ultimately used to evaluate the model. A cross-validation structure allows us to avoid this situation because for each cross-validation iteration, we can select features based on training data only. Even if the contrast used to select features is identical to the ultimate contrast that we care about, the part of the data set used to evaluate its predictive potential is held out (independent), so that above-chance classification is no longer inevitable. Instead, success requires the trained model to generalize to a subset of the data that was not involved in feature selection. An important consideration with using cross-validation in this manner is that the features selected can (and often will) differ across cross-validation iterations. This might not be a concern if optimizing accuracy is a primary goal, but it could present complications when asking which features are contributing to the final mean accuracy. One way to evaluate predictors in this case is to quantify (and possibly rank) how often a feature is selected across the full set of iterations.

Another form of selecting features is inherent to certain classification algorithms. Perhaps one of the simplest examples is **regularization**. As a model is fit during training, regularization can make certain solutions less likely by penalizing certain ways in which weights are assigned to features. For example, a least absolute shrinkage and selection operator (lasso) regression classifier tends to allocate weight values more sparsely by fixing some coefficients to zero, leading to fewer predictors contributing to predictions. Alternatively, a ridge regression classifier penalizes solutions that have many very large positive or negative weights. This can be helpful for ensuring that a model does not arbitrarily rely on one of several highly correlated predictors. When using a classifier that requires a user to specify a parameter (such as a penalty value), the value must be selected using independent data (such as piloting or prior work) or from the training (but never testing) data.

Cross-Validation

A fundamental principle underlying machine learning is training a model on a subset of data, and then making predictions on an independent set. The manner in which we separate our training and testing sets is thus more than a question of logistics – it is fundamental to ensuring the validity of a model's performance. How should one divide data to best ensure this? A simple approach is to divide the data set into halves randomly, train on the first half, and

> **Box 34.7** Applying cross-validation
>
> To conduct cross-validation in our ongoing example, we might choose to randomly assign each of our 1,000 cases to one of 10 separate data sets ("folds"), each containing 100 cases. In the first training iteration, the model would be trained on folds 2–10 (corresponding to cases 101–1,000) and then tested on fold 1 (cases 1–100). The second iteration would train on folds 1 and 3–10 (observations 1–100 and 201–1,000) and test on fold 2 (observations 101–200), and so on. By proceeding through 10 iterations, every case is included in the test set once, but always remains independent from the training set.

> **Box 34.8** Applying caution to unequal groups
>
> To illustrate the difficulties with having imbalanced classes, suppose that in our example data set, 700 participants meet criteria for GAD, while only 100 each meet criteria for panic disorder, social anxiety disorder, and specific phobia. Here, a model that always predicts that a given case should be classified as "GAD" would be accurate 70 percent of the time. This level of accuracy might seem impressive if the researcher is not accounting for the imbalanced data set, but in reality it does not reflect anything meaningful about each class (beyond frequency in this data set).

then test on the second. Although this split-half approach clearly separates the data, it is not the best approach from a statistical power perspective because it does not maximize the amount of training data. Cross-validation fills this role. Through cross-validation, a full sample is split into k subsamples (or **folds**). A classifier is trained on data in all but one ($k - 1$) folds and tested on the remaining (held-out) fold. This is repeated across k iterations with the data in each fold acting as the test set in turn.

In addition to maintaining training/testing independence, cross-validation helps maximize the amount of data used to train a model. This is particularly valuable because there is much to gain by increasing the percentage of data for training, with relatively little cost. This is most evident when considering how any single prediction is impacted by adding training versus testing data. The addition of new observations to a training set results in a more accurate model that is capable of making better predictions of new observations. In contrast, adding or removing a testing observation typically does not affect other predictions (an exception here is if a prior dependency is created, such as through z-scoring a test set).

Beyond maximizing the number of folds (and therefore the percentage of training data), what factors should we consider when deciding how to assign observations to folds? The most straightforward and intuitive approach would be to randomly assign cases to folds or use a simple leave-one-subject-out approach. In certain circumstances, however, a different organizational structure might be preferred. For example, the organization of folds determines which data points are included in training and testing for each iteration, where data points in the same fold are kept together (e.g., all contributing to training or testing) and data points in different folds will at some point be split across training and testing. It is therefore important to ensure that the data to be split are independent. For example, z-scoring might be conducted across observations within a fold, but doing the same with observations that cross folds could create dependencies between training and testing (as the values of the test set would then be influenced by values in the training set). If the data is collected from multiple sites, a "leave-one-site-out" approach can be a valuable structure. Successful predictions for held-out sites (e.g., predicting treatment outcomes for first-episode psychosis with approximately 70 percent accuracy in Koutsouleris et al., 2016) indicate that models are sufficiently robust to differences across sites, such as sample characteristics, geography, and others.

Another consideration relates to the balance (or ratio) of different classes, such as groups (in a between-subjects design) or trial-types (in a within-subjects design). Unbalanced data sets are common in clinical research, in large part because some groups of participants are easier to recruit than others. Missing data issues and flagged participants are another source of imbalance. Imbalanced samples can lead to difficulties in interpreting classifier performance. As a result, it is good practice to strive for equal representation of the classified groups, to the extent that is possible within the constraints of the research context.

Sometimes unbalanced designs are unavoidable, however, and when this is the case, solutions can be applied at the level of assessment, selection, and refinement of classifier models. In most cases, the most elegant solution is to balance the classes in each fold. One strategy is to randomly subsample observations in the more prevalent class to reach an equal number of data points in every class. To ensure a classification performance that is representative of the full sample, this can be repeated by randomly sampling the observations that are included from the larger class each time.

Classifier and Regression Models

With dozens of classifier variants available, how should we choose which type of model to use? A key distinction among alternatives is whether a model is linear or nonlinear. Linear models can learn to distinguish observations of classes that are "linearly separable." With just

> **Box 34.9** Applying linear models
>
> Suppose we are interested in predicting whether an individual was diagnosed with panic disorder during a diagnostic interview. Two likely relevant dimensions would be the frequency of uncued panic attacks in the past month (dimension 1), and the extent to which the individual fears or avoids future panic attacks (dimension 2). A higher value on each dimension is associated with a greater likelihood of diagnosis, so a straight diagonal line across the dimensions is one possible way to make this prediction (as such a line can separate higher from lower values in both dimensions). This is a linear solution.

one dimension, a linear separation would be equivalent to using a simple cut-off value. For two dimensions, a linear method of separation is a straight line.

As the number of dimensions increases beyond three, it can be difficult to visualize, but whereas a straight line can separate two dimensions, a straight plane is used for three, and a straight (multidimensional) hyperplane is used for more than three dimensions. However, so far these models have assumed a linear separation. If the line/plane/hyperplane must be curved to separate observations (such as when very high and very low – but not intermediate – predictor values are associated with an outcome), a linear model cannot learn the class distinction. Nonlinear models allow classes to be distinguished in ways beyond linear models, but also have a significant cost in interpretability: successful solutions can be difficult to visualize or understand. Interestingly, relatively few neuroimaging investigations have reported an advantage from using nonlinear, compared to linear, models (e.g., Kamitani & Tong, 2005), so a linear classifier such as a support vector machine (SVM) is a common choice.

In deciding on the particular model to use, the characteristics of the predictors and outcome variable are often important. If there are no unique properties of the employed predictors, a powerful linear classifier that is robust to overfitting, such as a SVM, might be recommended. This particular classifier can be trained relatively quickly and copes well with a large number of features relative to observations. If there is reason to believe that the predicting features explain very similar parts of the variance, a ridge regression classifier (discussed above) that is penalized for particularly extreme weights might be preferred. Some classifiers (such as SVMs) are binary (two-way only) though they can be assembled for multiway classifications (Bishop, 2006), whereas others (such as Gaussian Naïve Bayes) are intrinsically multiclass. Perhaps the most significant property of a predicted variable is whether it has a categorical or continuous structure. Although classifiers of categorical outcomes are the most frequently employed form of machine learning, alternative (continuous) models, such as SVR, can predict variables where value differences are meaningful (e.g., symptom severity, rather than categorical diagnoses). In this case, accuracy is not determined from correctly predicting a class, but from the distance between predicted and actual values.

Outputs and Interpretation

The first output typically examined by an analyst is the accuracy of their model's predictions of the testing set. For a classifier, the model predicts the testing set's observation labels – whether an observation belongs to a particular class. Classification accuracy simply reflects the percentage of classifier predictions that are correct. Studies using cross-validation typically report the mean accuracy from all iterations, so that the overall accuracy reflects predictions made for every observation (as each observation would be held out in the test set once). As with accuracy measured from a human behavioral experiment, a model's accuracy values can be submitted to various signal detection analyses. Measures such as d', area under the curve (AUC), and others, are all applicable depending on the precise question being asked about a model's performance (for a primer on signal detection theory, see McNicol, 2005). For a continuous model, such as SVR, predictions are continuous and in the same scale as the outcome variable. Here, accuracy could be quantified in several different ways, including mean squared error, loss, or a correlation between the true and predicted values.

For machine learning analyses conducted within subjects (e.g., to predict trial types), an accuracy value can be generated for each participant. The group's values can then be tested against chance (such as through an ANOVA). An example of this would be predicting the emotional expression of different observed faces, based on an observer's trial-by-trial eye-tracking. The ability to classify observed faces might differ based on the observer's diagnostic status – an ANOVA that compares groups of individuals could be used to test this possibility.

When subjects are being predicted (i.e., across-subject classification), we typically have a single overall accuracy value from the analysis. One can test the significance of this value through resampling techniques. In **permutation testing**, the labels of the observations are scrambled, then the same machine learning analysis (such as classification) is reconducted using the misaligned labels. This is repeated thousands of times with a new label order each time. The resulting set of accuracy values can act as a null distribution of the outcome variable (i.e., a set of values that would be expected if the labels are actually uninformative). The accuracy obtained from the true analysis (with nonscrambled labels) is compared to this distribution to calculate a p-value (e.g., by asking if the true value falls within the top or bottom 2.5 percent of the

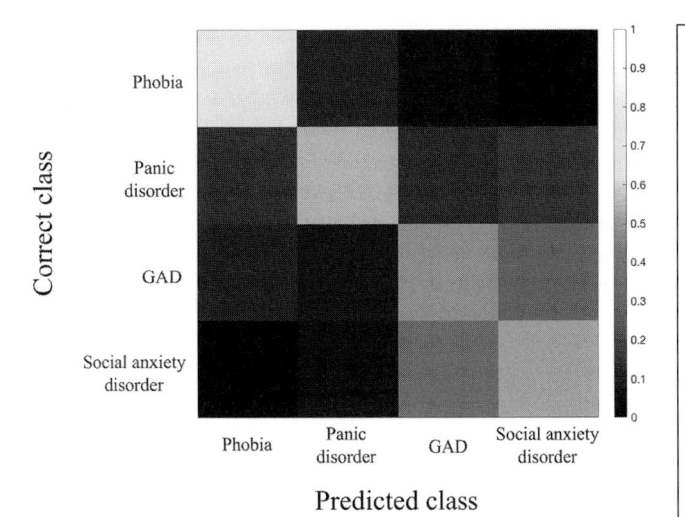

Figure 34.1 An example confusion matrix. The color scale reflects proportions of class predictions, organized by the actual (correct) class. This hypothetical example is a classification of participants into one of four diagnoses (see Box 34.10 for discussion).

Box 34.10 Applying a confusion matrix to multiway classification results

Imagine that we conduct a four-way classification of cases into one of four diagnoses – GAD, panic disorder, specific phobia, or social anxiety disorder – based on validated self-report measures of perfectionism, depressive symptom severity, interpersonal concerns, and cued and uncued panic attacks. We obtain impressive classification accuracy (where chance is 25 percent) but wish to know how well or poorly our model discriminates different diagnoses from each other. A confusion matrix (Figure 34.1) shows the proportion of guesses made for each class. In our example, GAD and social anxiety disorder are more confusable by the model, suggesting greater similarity in their predictor profiles.

---

distribution for a two-tailed test at $p < 0.05$), as well as confidence intervals. There are several considerations in deciding how to permute labels, such as ensuring that resampled values are "exchangeable" and that a sufficient number of permutations is run. Bootstrapping is a related technique that is commonly used to establish confidence intervals (discussed in Efron & Tibshirani, 1993).

The above output is typically only a first step. Multiway classifications in particular can give above-chance performance due to a variety of possible combinations of discriminability and confusion between classes. One way to better understand a trained model's solution is to examine the confusion matrix. A confusion matrix breaks down a model's overall accuracy to reveal the kinds of errors being made. The confusion matrix organizes the predictions into their (true) classes, and displays the proportions of guesses (Figure 34.1). This allows the reader to examine which classes are more likely to be confused (hence "confusion matrix"). The proportion of guesses made in each class is often represented by a color scheme, which quickly communicates the pattern of easily confused classes (e.g., GAD and social anxiety disorder in Figure 34.1). Because the diagonal represents correct guesses (e.g., when the observation is in the first class, correctly guessing "class one"), the diagonal of the matrix will often have higher values than off-diagonal parts, if a classifier is successful.

Understanding which classes are more or less confusable can be invaluable in understanding how a classifier is performing, and how classes relate to one another. The confusion matrix from a classifier can itself be statistically compared to a behavioral confusion matrix. For example, a researcher might be interested in how their model's successes and failures correspond to rates of misdiagnosis among clinicians. A confusion matrix of misdiagnosis can be created using relevant rates and then statistically compared (e.g., through a correlation) to test for similarity in how the disorders are distinguished or confused.

As well as examining overall performance, we might be interested in how different features are contributing to a model's predictive power. A trained model's set of weights (or an "importance map") reflects how the model is using each predictor, which can give insights into the nature of the model. Caution is needed, however, in how the set of weights is interpreted. First, the weight given to a feature reflects its contribution as part of a pattern; a set of variables might each have a small weight, but still have substantial predictive power as a set. Second, the model's weights reflect how this particular model was formed, but it is possible that similar prediction accuracies could be obtained with different sets of weights (i.e., the allocated weights are sufficient, but might not be necessary). Third, in cross-validation, a number (k) of models are trained (one per iteration), giving one set of weights for each iteration. A common way to deal with this is to average the weight maps across iterations, or to train a model on the entire data set purely for the purpose of inspecting the associated weights. It is important to note, however, that we cannot then strictly say that weights from a model fit to the entire data set yielded the generated classification accuracy, since this accuracy did not come from a model trained on all the data. Fourth, when a nonlinear model is used, the weight matrix cannot be easily understood.

In addition to visualizing the weights, it is possible to visualize the product of the weights with the corresponding inputs (e.g., Polyn et al., 2005), which can be easier to relate to the sign of the input data. For example, with a weight matrix alone, a negative weight can seem easy to interpret (lower values are more predictive of a class) but when the input data is itself negative for a particular class, this can become difficult (e.g., a negative predicting variable with a negative weight is actually a positive contribution of that variable). The weight-input product reflects both the input and the weight, removing this confusion. If a researcher is interested in the necessity of a given variable for successful prediction, or wishes to compare the relative predictive powers of individual predictors, one elegant approach is ablating a predictor from the data set, repeating the training and testing procedure, and comparing performance using this reduced model to performance with the full model. The prior consideration – that individual variables might not be particularly predictive, but can still contribute to an informative set – is still an important factor when interpreting the results of the ablation approach. Gotsopoulos and colleagues (2018) discuss these and other approaches for the classification of fMRI data.

Identifying Clusters on the Basis of Latent Structure

Unsupervised learning approaches aim to discover latent structures in a data set, without explicit guidance such as a set of labels. Instead, clusters or factors are discovered based on the predicting variables, in a training portion of the data set. This structure can then be validated in a testing set, allowing a researcher to ask whether the discovered structure is robust, as well as giving the opportunity to test whether the structure conforms to one or more hypotheses.

Two of the most popular clustering techniques are K-mean clustering, and hierarchical clustering. In K-mean clustering, data points are assigned to one of K clusters based on distance in multidimensional space. Multiple solutions (differing numbers of clusters, K) can often accurately reflect a data set, where latent structures are present at differing degrees of coarseness/specificity. The number of clusters (K) can be specified according to an existing theory, or from the data. One popular data-driven way of selecting K is through the "elbow method," where the percentage of variance explained is plotted against the number of clusters. Adding a cluster has diminishing returns on the variance explained, so the analyst can look for the point at which the explained variance decreases significantly less than it has before (i.e., the plot resembles an elbow-shape) to select K.

Hierarchical clustering is another approach. This technique allocates each data point to its own cluster, before merging nearby clusters until they form a hierarchical branching tree structure. This approach can reveal multiple degrees of specificity for a given clustered data set (at different levels of the hierarchy), allowing one to observe multiple structures within a data set. An example of this being applied is in Chekroud and colleagues (2017b), who used hierarchical clustering to group baseline symptoms in patients with depression using data from the Quick Inventory of Depressive Symptomatology (QIDS-SR) and clinician-rated Hamilton Depression (HAM-D) rating scale. Three clusters ("Core emotional," "sleep (insomnia)," "atypical") were identified, which were differentially responsive to antidepressant treatments. In another study, Drysdale and colleagues (2017) applied hierarchical clustering to resting-state fMRI connectivity data in a large multisite sample of patients with depression. The researchers identified four subtypes based on resting-state connectivity values. The dissociated subtypes had differing clinical symptom profiles and varied in their responsiveness to repetitive transcranial magnetic stimulation therapy.

Common Pitfalls

There are a number of important considerations for psychology researchers seeking to apply machine learning techniques to their data. First, although the prediction framework gives confidence in a model's validity in new data, we are still always constrained by the variability (or lack thereof) that is present in the data. Characteristics such as the demographics of our sample, the noise structure of data from a particular device (e.g., MRI scanner), particular nuances of how a clinical team verifies diagnoses, and more, will often affect data quality. It frequently requires an explicit effort, resources, and often collaborations, to apply trained models to applicable data sets that have been collected by different teams, with different samples. For example, "leave-one-site-out" cross-validation often produces lower accuracies than leave-one-subject-out, indicative of the effect that such characteristics have on model performance. One way to increase the ability of a model to generalize is to expose it to observations with broad variance (i.e., recruiting a sample that is diverse across relevant characteristics). A model trained with such variance will be more robust at predicting the signal in new data (for a similar principle in neuroimaging, see Coutanche & Thompson-Schill, 2012).

With computational power improving every year, and increasing access to computing clusters, analyses that once took hours can now take minutes or even seconds. Nonetheless, a researcher can still find that a planned analysis requires days to complete. In addition to "housekeeping" to make a code more efficient, it can be important to consider the influence that certain analysis choices (such as leave-one-subject-out classification) have on the time required. Combining a 1,000-fold classification (for 1,000 subjects) with techniques that measure the impact of excluding features (feature ablation, discussed above), or with nested cross-validation (in which another cross-validation is run inside the training data, usually to select optimal classification parameters), can quickly compound the required computation time. In such cases, changing the fold structure to, for example, a leave-10-subjects-out organization is a quick way to reduce computation time (as it reduces the number of models to be trained by a factor of 10), while preserving independence between training and testing.

LOOKING AHEAD

A number of recent analyses have focused on relating treatment outcomes to underlying dimensions rather than diagnostic status alone. For example, Webb and colleagues (2019) examined endophenotypes in Major Depressive Disorder to predict responsiveness to sertraline versus placebo. Better outcomes were predicted by greater depression severity, higher neuroticism, older age, less impairment in cognitive control, and employment. Successful prediction in this case serves the dual purpose of identifying how patients might be assigned to treatments and giving insights into dimensions that underlie depression and treatment-induced change. Relatedly, efforts to recruit for a diverse array of symptoms (e.g., Astle et al., 2019) show promise for developing models that apply to the real heterogeneity present within the population, and avoid artificially inflating reliability within examined groups.

Another promising direction is to combine information across modalities in the same model to improve predictive power. The diagnostic signal present in behavioral self-report, clinicians' reports, fMRI data, genotypes, and more, are very likely to explain distinct portions of individual variance. Bringing these diverse predictors into the same model presents both a challenge for interpretation and an opportunity for detecting meaningful differences across individuals.

The translation of findings from the laboratory to medical practice has its own set of challenges. As models are developed, the question of practicality (in terms of relative cost, efficiency, etc.) becomes important (Cohen & DeRubeis, 2018; Gillan & Whelan, 2017). The precise role taken by machine-informed decisions on treatment or diagnosis has yet to be determined. Rather than replacing clinician judgment altogether, any generalizable and successful model is more likely to act as a supplement to clinical intuitions and experience (Linden, 2012). The degree of influence of any such model is likely to be greater for particular questions or decisions that are frequently difficult for today's clinicians to determine alone, such as the identification of Unipolar versus Bipolar Depression (e.g., Perlis et al., 2006). If particular models are employed in clinical practice, it will also be important to maintain a suitably open line of communication between commercial entities and independent scientists regarding the precise basis for any treatment recommendations. For example, patents have been filed in relation to predicting treatment response in depression using machine learning techniques (e.g., Chekroud et al., 2017a; Patent WO2017210502A1). Likewise, improving treatment predictions depends on an effective relationship between data scientists and commercial entities involved in dissemination.

SUMMARY

The recent application of machine learning techniques in clinical psychology has presented exciting opportunities and a daunting training challenge for psychology researchers. Fortunately, the number of free learning resources available to new machine learning investigators grows every day (see Further Resources below for some jumping-off points). We hope that the research and approaches reviewed above illustrate that – more than simply providing an increase in sensitivity – machine learning approaches provide an opportunity to select treatments, refine diagnoses, understand cognitive and neural bases for symptoms and clinical dimensions, and more. In this sense, the most powerful tool in the investigator's box remains their creativity and willingness to apply these techniques to problems in new ways. As more clinical investigators master the approaches we have discussed, new opportunities will arise to combine their content expertise with machine learning, leading to research studies and applications that are truly unique and groundbreaking. We look forward to enjoying your work.

FURTHER RESOURCES

The modern machine learning researcher has a plethora of resources to choose from. You should choose a resource that bests suits your own learning preferences, while operating at the right level of theory and practical application. There is no substitute for getting your hands on data, so we recommend applying what you learn, even with a sample data set, whenever possible. Below we have listed some compilations, along with some of our personal favorites

Compilations of Resources and Online Guides

- *Machine Learning for Humans* A series and free e-book that provides simple, plain-English explanations accompanied by math, code, and real-world examples: https://medium.com/machine-learning-for-humans/why-machine-learning-matters-6164faf1df12
- *My Curated List of AI and Machine Learning Resources from Around the Web* An enormous list of noteworthy machine learning blogs, video courses, code, forum discussions and more: https://medium.com/machine-learning-in-practice/my-curated-list-of-ai-and-machine-learning-resources-from-around-the-web-9a97823b8524

Books

- James, G., Witten, D., Hastie, T., & Tibshirani, R. (2013). *An Introduction to Statistical Learning: With Applications in R* (1st edn.). New York: Springer.
- Flach, P. (2012). *Machine Learning: The Art and Science of Algorithms That Make Sense of Data* (1st edn.). Cambridge: Cambridge University Press.

Software and Packages

- R – www.r-project.org/
- Python – www.python.org/
- The Python scikit learn machine learning toolbox – https://scikit-learn.org/stable/
- MATLAB – www.mathworks.com/products/matlab.html
- Octave – www.gnu.org/software/octave/

Neuroimaging-Specific Resources
- The Princeton MVPA MATLAB toolbox – https://github.com/PrincetonUniversity/princeton-mvpa-toolbox
- PyMVPA – www.pymvpa.org/
- Nilearn – https://nilearn.github.io/
- Mur, M., Bandettini, P. A., & Kriegeskorte, N. (2009). Revealing Representational Content with Pattern-Information fMRI – An Introductory Guide. *Social Cognitive and Affective Neuroscience*, 4(1), 101–109.
- Pereira, F., Mitchell, T., & Botvinick, M. (2009). Machine Learning Classifiers and fMRI: A Tutorial Overview. *NeuroImage*, 45 (1 Suppl), S199–209.

REFERENCES

Anzellotti, S., & Coutanche, M. N. (2018). Beyond Functional Connectivity: Investigating Networks of Multivariate Representations. *Trends in Cognitive Sciences*, 22(3), 258–269.

Astle, D. E., Bathelt, J., CALM Team, & Holmes, J. (2019). Remapping the Cognitive and Neural Profiles of Children Who Struggle at School. *Developmental Science*, 22(1), 1–17.

Bellec, P., Chu, C., Chouinard-Decorte, F., Benhajali, Y., Margulies, D. S., & Craddock, R. C. (2017). The Neuro Bureau ADHD-200 Preprocessed Repository. *NeuroImage*, 144(Pt B), 275–286.

Bishop, C. M. (2006). *Pattern Recognition and Machine Learning*. New York: Springer.

Chekroud, A. M., Zotti, R. J., Shehzad, Z., Gueorguieva, R., Johnson, M. K., Trivedi, M. H., ... Corlett, P. R. (2016). Cross-Trial Prediction of Treatment Outcome in Depression: A Machine Learning Approach. *The Lancet. Psychiatry*, 3(3), 243–250.

Chekroud, A., Krystal, J. H., Gueorguiva, R., & Chandra, A. (2017a). *WO2017210502A1*. World Intellectual Property Organization. Retrieved from https://patents.google.com/patent/WO2017210502A1/en

Chekroud, A. M., Gueorguieva, R., Krumholz, H. M., Trivedi, M. H., Krystal, J. H., & McCarthy, G. (2017b). Reevaluating the Efficacy and Predictability of Antidepressant Treatments: A Symptom Clustering Approach. *JAMA Psychiatry*, 74(4), 370–378.

Cohen, Z. D., & DeRubeis, R. J. (2018). Treatment Selection in Depression. *Annual Review of Clinical Psychology*, 14, 209–236.

Coutanche, M. N. (2013). Distinguishing Multi-Voxel Patterns and Mean Activation: Why, How, and What Does It Tell Us? *Cognitive, Affective & Behavioral Neuroscience*, 13(3), 667–673.

Coutanche, M. N., & Thompson-Schill, S. L. (2012). The Advantage of Brief fMRI Acquisition Runs for Multi-Voxel Pattern Detection across Runs. *NeuroImage*, 61(4), 1113–1119.

Coutanche, M. N., Thompson-Schill, S. L., & Schultz, R. T. (2011). Multi-Voxel Pattern Analysis of fMRI Data Predicts Clinical Symptom Severity. *NeuroImage*, 57(1), 113–123.

Deshpande, G., Libero, L. E., Sreenivasan, K. R., Deshpande, H. D., & Kana, R. K. (2013). Identification of Neural Connectivity Signatures of Autism Using Machine Learning. *Frontiers in Human Neuroscience*, 7, 1–15.

Dosenbach, N. U. F., Nardos, B., Cohen, A. L., Fair, D. A., Power, J. D., Church, J. A., ... Schlaggar, B. L. (2010). Prediction of Individual Brain Maturity Using fMRI. *Science*, 329(5997), 1358–1361.

Drysdale, A. T., Grosenick, L., Downar, J., Dunlop, K., Mansouri, F., Meng, Y., ... Liston, C. (2017). Resting-State Connectivity Biomarkers Define Neurophysiological Subtypes of Depression. *Nature Medicine*, 23(1), 28–38.

Du, W., Calhoun, V. D., Li, H., Ma, S., Eichele, T., Kiehl, K. A., ... Adali, T. (2012). High Classification Accuracy for Schizophrenia with Rest and Task FMRI Data. *Frontiers in Human Neuroscience*, 6, 1–12.

Efron, B., & Tibshirani, R. J. (1993). *An Introduction to the Bootstrap* (1st edn.). New York: Chapman and Hall/CRC.

Eldridge, J., Lane, A. E., Belkin, M., & Dennis, S. (2014). Robust Features for the Automatic Identification of Autism Spectrum Disorder in Children. *Journal of Neurodevelopmental Disorders*, 6(1), 1–12.

Eloyan, A., Muschelli, J., Nebel, M. B., Liu, H., Han, F., Zhao, T., ... Caffo, B. (2012). Automated Diagnoses of Attention Deficit Hyperactive Disorder Using Magnetic Resonance Imaging. *Frontiers in Systems Neuroscience*, 6, 1–9.

Gillan, C. M., & Whelan, R. (2017). What Big Data Can Do for Treatment in Psychiatry. *Current Opinion in Behavioral Sciences*, 18, 34–42.

Gotsopoulos, A., Saarimäki, H., Glerean, E., Jääskeläinen, I. P., Sams, M., Nummenmaa, L., & Lampinen, J. (2018). Reproducibility of Importance Extraction Methods in Neural Network Based fMRI Classification. *NeuroImage*, 181, 44–54.

Guyon, I., & Elisseeff, A. (2003). An Introduction to Variable and Feature Selection. *Journal of Machine Learning Research*, 3 (March), 1157–1182.

Hallion, L. S., Wright, A. G. C., Coutanche, M. N., Joormann, J., and Kusmierski, S. N. (2019). Toward an Empirically-Derived Taxonomy of Perseverative Thought: Support for a Multidimensional Approach to Classification. Retrieved from https://psyarxiv.com/gu7xp/

Haxby, J. V., Gobbini, M. I., Furey, M. L., Ishai, A., Schouten, J. L., & Pietrini, P. (2001). Distributed and Overlapping Representations of Faces and Objects in Ventral Temporal Cortex. *Science*, 293(5539), 2425–2430.

Kamitani, Y., & Tong, F. (2005). Decoding the Visual and Subjective Contents of the Human Brain. *Nature Neuroscience*, 8(5), 679–685.

Kohonen, T. (1989). *Self-Organization and Associative Memory* (3rd edn.). Berlin: Springer-Verlag.

Koutsouleris, N., Kahn, R. S., Chekroud, A. M., Leucht, S., Falkai, P., Wobrock, T., ... Hasan, A. (2016). Multisite Prediction of 4-Week and 52-Week Treatment Outcomes in Patients with First-Episode Psychosis: A Machine Learning Approach. *The Lancet Psychiatry*, 3(10), 935–946.

Kriegeskorte, N., Simmons, W. K., Bellgowan, P. S. F., & Baker, C. I. (2009). Circular Analysis in Systems Neuroscience: The Dangers of Double Dipping. *Nature Neuroscience*, 12(5), 535–540.

Linden, D. E. J. (2012). The Challenges and Promise of Neuroimaging in Psychiatry. *Neuron*, 73(1), 8–22.

McNicol, D. (2005). *A Primer of Signal Detection Theory*. Hove: Psychology Press.

Nolen-Hoeksema, S., Wisco, B. E., & Lyubomirsky, S. (2008). Rethinking Rumination. *Perspectives on Psychological Science*, 3(5), 400–424.

Nouretdinov, I., Costafreda, S. G., Gammerman, A., Chervonenkis, A., Vovk, V., Vapnik, V., & Fu, C. H. Y. (2011). Machine Learning Classification with Confidence: Application of Transductive Conformal Predictors to MRI-Based Diagnostic and Prognostic Markers in Depression. *NeuroImage*, 56(2), 809–813.

Open Science Collaboration (2015). Estimating the Reproducibility of Psychological Science. *Science, 349*(6251), 943.

Paul, G. L. (1967). Strategy of Outcome Research in Psychotherapy. *Journal of Consulting Psychology, 31*(2), 109–118.

Perlis, R. H., Brown, E., Baker, R. W., & Nierenberg, A. A. (2006). Clinical Features of Bipolar Depression versus Major Depressive Disorder in Large Multicenter Trials. *American Journal of Psychiatry, 163*(2), 225–231.

Polyn, S. M., Natu, V. S., Cohen, J. D., & Norman, K. A. (2005). Category-Specific Cortical Activity Precedes Retrieval during Memory Search. *Science, 310*(5756), 1963–1966.

Sundermann, B., Herr, D., Schwindt, W., & Pfleiderer, B. (2014). Multivariate Classification of Blood Oxygen Level-Dependent FMRI Data with Diagnostic Intention: A Clinical Perspective. *AJNR: American Journal of Neuroradiology, 35*(5), 848–855.

Tong, F., & Pratte, M. S. (2012). Decoding Patterns of Human Brain Activity. *Annual Review of Psychology, 63*, 483–509.

Watkins, E. R. (2008). Constructive and Unconstructive Repetitive Thought. *Psychological Bulletin, 134*(2), 163–206.

Webb, C. A., Trivedi, M. H., Cohen, Z. D., Dillon, D. G., Fournier, J. C., Goer, F., ... Pizzagalli, D. A. (2019). Personalized Prediction of Antidepressant v. Placebo Response: Evidence from the EMBARC Study. *Psychological Medicine, 49*(7), 1118–1127.

Woo, C.-W., Chang, L. J., Lindquist, M. A., & Wager, T. D. (2017). Building Better Biomarkers: Brain Models in Translational Neuroimaging. *Nature Neuroscience, 20*(3), 365–377.

Yarkoni, T., & Westfall, J. (2017). Choosing Prediction over Explanation in Psychology: Lessons from Machine Learning. *Perspectives on Psychological Science, 12*(6), 1100–1122.

GLOSSARY

Confusion matrix: A matrix that presents a classifier's guesses, organized by each (true) class. This allows us to examine the classes that a classifier is more or less likely to misclassify (i.e., confuse). Often visualized with colors reflecting the number of guesses given for each class.

Cross-validation: A process by which a model is trained on all but one ($k - 1$) subsamples of a data set ("folds") and then tested on the subsample held out from training. Iterations of training and testing are conducted to hold out each fold in turn (see: *holding-out*; *fold*).

Feature: A predictor that is used as part of a machine learning model. Each observation will typically have values on a large set of features.

Feature ablation: A method for quantifying the unique predictive contribution of individual or subsets of predictors ("features") to a model's success by systematically removing feature(s) and measuring the subsequent drop in model accuracy (see: *feature*).

Fingerprint: The weighted pattern of predictors that is obtained during the training phase for a given model.

Fold: A subsample of a data set that is held out for testing during cross-validation (see: *cross-validation*).

Generalization: The accuracy with which a machine learning model that is trained on one data set can predict patterns and relationships in a new (untrained) data set.

Holding out: Excluding one or more cases from a training set for testing a model.

k-means clustering: A machine learning approach that groups data into groups (clusters) based on similarity in the values of their features. The number of clusters (k) is determined by the analyst (see: *unsupervised learning*).

Learning: A process by which an algorithm identifies patterns or relationships that are present in a given data set (see: *training*).

Machine learning classifier: A computational model that is trained to separate data (such as observations) based on labels that are typically assigned by the analyst (such as diagnostic status). The model then gives class (label) predictions for separate data based on the patterns it was able to extract from the trained data.

Overfitting:	The act of a model being trained so precisely to the training data set that it becomes less able (or even unable) to generalize to independent test data (which will almost never have exactly the same values as a training set). Often caused by having too many features relative to observations (see *feature*; *regularization*).
Permutation testing:	A statistical tool allowing one to compare a result such as classification accuracy to a null distribution generated by repeatedly running an analysis with shuffled labels each time. Comparing the true accuracy to this distribution can give a *p*-value, confidence intervals, and other useful properties.
Precision medicine:	The broad effort to match patients to treatments on the basis of personal or contextual factors.
Regularization:	A technique for reducing the chance of overfitting a model to the training data by constraining the types of solutions that are learned (through how weights are allocated to features).
Supervised learning:	A type of machine learning model for which a label (or other information; i.e., "correct" answer) is assigned to each class or observation in a training set.
Testing:	The process by which a model developed on one dataset (or part of a data set) is applied to a new (untrained) data set in order to assess the validity or generalizability of the model (see: *cross-validation*).
Training:	The process by which a model is exposed to a data set and identifies patterns or relationships within this data (see: *learning*)
Unsupervised learning:	A process in which a model is trained to detect structure in data without using labels (i.e., the analyst does not teach the model the "right answer").

Index

AA. *See* ambulatory assessment
active engagement, in dyadic data, 351
active rGE, 146–147, 225–226
Actor-Partner Interdependence Models (APIMs), 344
  dyadic data outcomes and, 351–354
  repeated measures in, 354
additive genetics, 221
ADHD. *See* attention-deficit hyperactivity disorder
adjunctive psychosocial interventions, 293
adolescence, psychopathology during. *See also* puberty; quantitative genetic research
  behavioral health in, 212
    biological basis of, 213
    ethnic disparities in, 212
    prevalence estimates of, 211–212
    racial disparities in, 212
    socioeconomic status factors for, 212
  biological sensitivity theories for, 211
  in child-caregiver designs, 340
    adolescent-caregiver pairs, 340
  depressive disorders, 205–206
  environmental influences on, 211
  for ethnic minorities, 212
  for LGBT community, 212–213
  mental health in, 212
    biological basis of, 213
    ethnic disparities in, 212
    prevalence estimates of, 211–212
    racial disparities in, 212
    sexual minority status and, disparities as result of, 212–213
    socioeconomic status factors for, 212
  methodological applications, 213–215
    evidence-based interventions, 213
    through law, 214–215
    personalized medicine procedures, 214
    through public policy, 214–215
    with technological data sources, 214
  MMLs and, 215
  in peer–peer designs, 341
  physiological changes, 205–210
  social influences on, 211
  substance abuse and, 206
  theoretical approach to, 205
adolescent-caregiver pairs, 340

adoption studies, 139
  biometric modeling in, 136–137
  clinical findings from, 139, 143
    individual analyses from, 140–141
    meta-analyses from, 140–141
  epigenetics and, 148
    modifications in, 148
  genetic risk factor determination in, 136
  GWAS and, 142–144
    CDCV model, 142–143
    clinical findings from, 143
    missing heritability and, 145
    for schizophrenia, 143
    SNPs, 142–143
  heritability factors in, 136–137
    definition of, 137
    MFT model, 137–138
    misinterpretation of, 137
  overview of, 136
  quantitative genetic research in, 219–220
adrenarche, 206–208
adverse childhood experiences, 294
age of onset, for puberty, 207
  gender differences in, 205
  trends in, earlier onset, 206
alcohol consumption. *See* problematic alcohol consumption
allostatic models, 119
alternating treatment designs (ATDs), 282
ambulatory assessment (AA)
  applications of, 309
    in interventions, 308–309
  data collection in, 303, 305–306
    with EAR, 306
    through mobile technology, 306
    through personal sensing, 305–306
    by sensors, 305
    time demands of, 308
  definition of, 301
  disadvantages of, 308–309
  future directions of, 308–309
  geofencing, 308
  individual differences in, 304
  methodology of, 301–304
    data collection. *See* data collection
    ecological aspects of, 301, 303
    laboratory studies, 303

    participants in, 301
    sampling techniques in, 301–302
    temporal sequences in, 303–304
  monitoring issues with, 304–305
    compliance factors, 304–305
    reactivity in, 305
  scope of, 301
  therapeutic uses of, 306–308
    for client's behaviors, 307
    in mHealth applications, 307
analysis of variance (ANOVA), 428–429
animal studies
  attachment theories and, 338
  puberty and, psychopathology of, 209
ANNs. *See* artificial neural networks
ANOVA. *See* analysis of variance
ANS functioning. *See* autonomic nervous system functioning
anxious attachment styles, 339
APIMs. *See* Actor-Partner Interdependence Models
artificial neural networks (ANNs), 171
ASDs. *See* autism spectrum disorders
ASR. *See* Automatic Speech Recognition
assignment mechanism, 415–417
  strongly ignorable, 417
  unconfounded, 417
assignment-based causal inferences, 416
associative validity, of psychometrics, 61–62
ATDs. *See* alternating treatment designs
atheoretical analytic approach, to latent variable models, 70
attachment, definition of, 337
attachment theories, 337–340
  animal studies and, 338
  child-caregiver attachment, 337–338
    bottom-up approach to, 338
    dyadic processes, 338
  partner–partner attachment, 339–340
    anxious attachment styles, 339
    dyadic processes in, 339–340
  peer–peer attachment, 338–339
    emotional dysregulation in, 339
    insecure, 339
    interpersonal emotional regulation, 339
    positive relationships in, 339
    risk distribution in, 338–339

attachment theories (cont.)
    social baseline theory, 338–339
attention-deficit hyperactivity disorder (ADHD)
    cognitive studies for, 110–111
        control group selection for, 107
    nested data for, 429
    quantitative genetic research for
        genetic factors in, 221–222
        heritability estimations for, 221–222
authorship issues, in longitudinal studies, 235
autism spectrum disorders (ASDs)
    cognitive studies for, 105
    quantitative genetic research for
        etiologic factors in, 224–225
        genetic factors in, 221–222
autocorrelations, in intensive longitudinal data, 318–319
    for borderline personality disorder, 318
    for depression, 318
    time course, 318–319
Automatic Speech Recognition (ASR), 343
autonomic nervous system (ANS) functioning, 118–119
    enteric system in, 118
    PSNS in, 118
        respiratory sinus arrhythmia, 118
    SNS in, 118
        skin conductance responses, 118
auto-regressive models, in process research, 359–360
auxiliary variables, in missing data, 447, 453–454

BAS. *See* Behavioral Activation System
Bayes, Thomas, 378
Bayesian inferences
    meta-analysis through, 393
    in randomized studies, 419–420
    reproducibility and, 378–379
Bayesian Ornstein-Uhlenbeck Model (BOUM), 322
Beck Depression Inventory, 14
behavior space, 329
behavioral activation, in nested data, 427
Behavioral Activation System (BAS), 120
behavioral genetics
    in candidate-gene studies, 141–142
    environmental variation factors in, 136
    historical development of, 136
    risk factor determinations in, 136
    three laws of, 139
behavioral health, during adolescence, 212
    biological basis of, 213
    ethnic disparities in, 212
    prevalence estimates of, 211–212
    racial disparities in, 212
    socioeconomic status factors for, 212
Behavioral Inhibition System (BIS), 120
behavioral medicine
    adjunctive psychosocial interventions, 293
    adverse childhood experiences and, 294
    between-person approaches, 294
    biopsychosocial model, 287–288
        systems hierarchy in, 288
    conceptual models for, 293–294

definition of, 287
endpoints of, evolution of, 288–289
ethnicity and, health disparities and, 294
future directions for, 293–295
health behaviors and, risk reduction and, 289
medical illness and, psychosocial aspects of, 289
methodological education on, 294–295
outcomes for, evolution of, 288–289
psychobiological influence on disease and, 289–293
    assessment measures, 291
    epidemiology of, 289–291
    implacable experimenter, 292
    mechanisms of, 291–293
    risk factors, 289–291
quality of life issues, 288
race and, health disparities and, 294
RCTs in, 293
socioeconomic status, health disparities and, 294
theoretical approach to, 287
within-person approaches, 294
behavioral neuroscience
    algorithmic levels of, 169
    challenges in, 169–170
    computational levels of, 169
    implementational level of, 169
    latent variables in, 169
    neural networks, 169
    reinforcement learning approaches in, 168
Berkson's bias, 106
best practices, for reproducibility, 374–380
    pre-registering of studies, 375
    replication methods in, 379–380
    reporting guidelines, 375–376
    transparency approach, 375
between-person structural frameworks, 331
bias
    in estimation errors, 388–389
    predictive, in psychometrics, 62
    selection
        Berkson's bias, 106
        clinician's illusion bias, 106
bifactor comorbidity models, 246
bifactor models, 73–75
    conceptual issues with, 75
    estimation of, 73–74
    features and structures of, 73
    hierarchical, 74
big data, clinical psychology and, applications of, 3
biological models, in descriptive psychopathology, 37–38
    for BPD, 38
    RDoC in, 38
biological reductionism, 26
biological sensitivity theories, 211
biomarkers, endophenotypes and, 14
biometric modeling, in adoption and twin studies, 136–137
biopsychosocial model, 287–288
    systems hierarchy in, 288
BIS. *See* Behavioral Inhibition System

Blimp Scripts, 460–461, 466
block designs, for fMRI, 154–155
blood-oxygen-level dependent (BOLD) responses, 153–154, 159–160
bootstrap confidence intervals, 398
borderline personality disorder (BPD)
    in biological models, 38
    clinical theories of, 36
    hysteria and, 7
    intensive longitudinal data on, 318
    modern descriptive psychiatric approach to, 36–37
    quantitative modeling perspective on, 37
BOUM. *See* Bayesian Ornstein-Uhlenbeck Model
Bowlby, John, 337–338. *See also* attachment theories
Box, George, 170
boys. *See* males
BPD. *See* borderline personality disorder
brain
    during puberty, 207
    at resting state, 156–157
Bridgman, Percy, 5, 9
    on operationalism, 15
Brief Symptom Inventory (BSI), 248–249
    General Severity Index for, 249
    growth curve models for, 251
BSI. *See* Brief Symptom Inventory
Bush–Mosteller model, 184

candidate-gene studies, 141–142
capitalization model, 283
cardiovascular measures, in peripheral psychophysiology, 122–124
    biological mechanisms, 122
    EKG, 122–124
        Einthoven's Triangle Hypothesis, 122
        HR and, 123–124
        HRV and, 123–124
        respiratory sinus arrhythmia, 123–124
    history of, 122
    ICG, 123–124
        pre-ejection period, 124–125
    testing preparations for, 122
    vagotonia hypothesis, 122
caregivers. *See* child-caregiver attachment
Carnap, Rudolf, 7
cascade effects, 197–198
case-control studies, 83–85
    advantages of, 84
    implementation challenges of, 84–85
    source populations in, 83–84
categorical latent variable models, 75–77
causal effects, 396
causal inferences. *See also* randomized studies
    assignment-based, 416
    objectives of, 415
    in observational studies, 423
        analysis of, 422
        assumptions for identification in, 422
        covariate adjustments in, 423
        design of, 422
    primitives, 415–416
    in PROCESS output, 416
    RCM, 415
        assignment mechanism, 415–417

sensitivity analyses of, 423
  on single units, 415–416
  Stable Unit Treatment Value
      Assumption, 416
  in statistical mediation analysis, 399
    of indirect effects, 399
    ordering in, 399
  theoretical approach to, 415
causality, 102
  in longitudinal studies, 232
causality confound, 162
CBT. *See* cognitive behavioral therapy
CDCV model. *See* common disease/
    common-variant model
central precocious puberty (CPP), 207, 214
central psychophysiology, in dyadic
    relationships, 342
central tendencies, in intensive
    longitudinal data modeling, 313
CFA. *See* confirmatory factor analysis
Chang, Hasok, 6
chaos theory, 120
Chapman, Jean, 94
Chapman, Loren, 94
child abuse. *See* maltreatment
child maltreatment. *See* maltreatment
child psychopathology. *See* cognitive
    studies
child-caregiver attachment, 337–338
  bottom-up approach to, 338
  dyadic processes, 338
child-caregiver designs, for dyadic
    relationships, 340
  adolescent behavior factors, 340
  adolescent-caregiver pairs, 340
  maternal behavior factors, 340
  strange situation tasks in, 340
choice rules, in RL models, 173
Cholesky decomposition approach,
    224
  in classical twin studies, 224
Classical Test Theory, 49–50
classifiers, from machine learning, 471,
    475–476, 481
  supervised learning, 471–472
clinical psychology. *See also* open
    concepts; reproducibility; *specific
    topics*
  big data and, 3
  as evolving discipline, 3
  factor analysis in, 11
  invalid theories in, history of, 371
  MOST and, 268, 274–275
    in cognitive therapy, 268
    for MDD, 268, 276
  physics as influence on, 10–12
  scientific crisis of confidence with, 5
  theoretical approach to, 3–4
  theory building in, 12
clinical theory
  for descriptive psychopathology,
      35–36
  for BPD, 36
  machine learning as influence on,
      469–470
clinician's illusion bias, 106
clustering, in MOST, 276
CNVs. *See* copy number variants

codebooks, in longitudinal studies, 235
cognition, child psychopathology and,
    109–112
  with diffusion modeling, 111
  executive function, 110
  with RT variables, 110–111
cognitive behavioral therapy (CBT), in
    transdiagnostic processes, 281
cognitive neuroscience
  model-based, 180
  traditional approaches to, 180
cognitive paradigm ontology, 15
cognitive psychopathology, theoretical
    scope of, 105–106
cognitive studies, in child
    psychopathology. *See also* child-
    caregiver attachment
  for ADHD, 110–111
  control group selection for, 107
  assessment strategies for, 108–112
    of cognition, 109–112
    in cross-informant studies, 108
    for degree of psychopathology,
        108–109
    with planned missingness approach,
        109
    rating scales in, 108
    with TMM, 109
  for autism, 105
  for cognition
    assessment strategies for, 109–112
    with diffusion modeling, 111
    executive function, 110
    with RT variables, 110–111
  control group selection in, 107–108
    for ADHD, 107
    types of controls in, 108
  data collection for, pragmatic
      considerations in, 112
    for research assistant training, 112
  descriptive designs for, 105–106
  endophenotypes in, 105–106
  experimental designs for, 105
  RDoC processes in, 105–106
  target group selection in, 106–107
    Berkson's bias in, 106
    clinician's illusion bias in, 106
    sampling strategies in, 106–107
cognitive therapy, 268
  nested data in, 427
cohort studies, 80–82
  exposure definitions and measurements
      in, 81–82
  identification of outcomes in, 80–82
  planning strategies for, 80–81
  prospective, 80
  retrospective, 80
  target populations in, 81
common disease/common-variant (CDCV)
    model, 142–143
common factor variable models, 69
common fate affects, 355
common pathway model, in quantitative
    genetic research, 223–224
comorbid course models, 242
comorbidity models, for longitudinal data.
    See also psychological distress
  alternative, 243–244

analyses of, 248
bifactor, 246
comorbid course, 242
growth curve, 248
  commonly used, 246–247
  Factor Mean Shift model, 246–247
  Free Curve Slope Intercept factor
      model, 247
improper, 243–244
multiple, 242–243
oblique simple structure factors, 246
parsimonious factor pattern, 245–246
prospective, 241–242
purpose and goals of, 241
Random Intercept, 245–246
right-sizing growth, 244–246
  number of variables in, 244
single parameter modifications, 243
trait, 242
Triangular Decomposition, 246
compensation, for longitudinal studies,
    236
complex sampling designs, 88
compliance, in AA monitoring, 304–305
computational methods, for dyadic data,
    364
  computer programs for, 364
  SEM in, 364
conceptual replication
  in meta-analysis, 392
  in scientific method, 372
concurrent relationships, in intensive
    longitudinal data modeling,
    319–320
conditional indirect effects, 406–408
conditional process analysis, 403–408
  definition of, 403
conditional process model, 403
  simple, 403–404
  with single moderator, 403
confirmatory factor analysis (CFA)
  ESEM procedures, 57
  of externalizing proneness, 23–24
  of latent variable models, 68, 71–72,
      76–77
    in exploratory-confirmatory
        spectrum, 72–73
  of psychometrics, 56–57
confirmatory hypotheses, for
    reproducibility, 376–377
  exploratory hypotheses as distinct from,
      376–377
  in multivariate modeling, 377
confusion matrix, 473, 477, 481
connectionist models, RL models and,
    170–171
consequential validity, of psychometrics,
    62
content validity, of psychometrics,
    60–61
continual optimization principle, 274
contrast analysis, in fMRI, 157–159
control groups, selection of. *See also* target
    populations
  in cognitive studies, 107–108
    for ADHD, 107
    types of controls, 108
conventionalism, 6

copy number variants (CNVs), missing heritability and, 145
correlation analysis, in fMRI, 157–159
Coupled Linear Oscillator Models, 343–344
CPP. *See* central precocious puberty
Craver, Carl, 169
Cromwell, Rue, 94
cross-twin-cross-trait correlation (CTCT) measures, 223

DAGs. *See* directed acyclical graphs
data analysis. *See also* big data; missing data; nested data
　from machine learning, 476
　　cross-validation through, 475, 481
　　feature selection in, 474
　　overfitting, 473, 482
　　PCA in, 474
　　in regression models, 475–476
　　regularization and, 474, 482
　in NHST, 96–97
　　discovery context in, 97–98
　　heterogeneity factors in, 98–99
　　justification context in, 97–98
　　levels of, 99–100
　　post-hoc, 97–98
data boxes, in subject-specific data analysis, 328
data cleaning, in NHST, 97
data collection. *See also* data analysis; missing data; nested data
　in AA, 303, 305–306
　　with EAR, 306
　　through mobile technology, 306
　　through personal sensing, 305–306
　　by sensors, 305
　　time demands of, 308
　in cognitive studies, pragmatic considerations in, 112
　　for research assistant training, 112
　　for intensive longitudinal modeling, 312–313
　in longitudinal studies, 231–232
DeFries–Fulker extreme analytic method, 222–223
dehydroepiandrosterone (DHEA), 206–207, 210
dementia praecox, definition of, 8
dependence, in nested data, 427–429
　ANOVA and, 428–429
　in hierarchical data, 428
　in longitudinal data, 428
depression
　during adolescence, 205–206
　Beck Depression Inventory, 14
　Hamilton Rating Scale for Depression, 14
　intensive longitudinal data on, 318
　MMPI-2 depression scale, 14
　operational definition of, 5
descriptive designs, for cognitive studies, 105–106
　endophenotypes in, 105–106
descriptive psychiatry. *See* modern descriptive psychiatry
descriptive psychopathology, 39–40
　alternatives to, 39–40
　biological models in, 37–38

　　for BPD, 38
　　RDoC in, 38
　causation-based mechanisms in, 33
　clinical theory for, 35–36
　　for BPD, 36
　critiques of, 39
　future recommendations for, 40–41
　HiTOP, 39–40
　integration of, challenges to, 40
　modern descriptive psychiatry, 36–37
　　assessment approaches in, 36
　　BPD and, 36–37
　　defining features of, 36
　　theoretical approach to, 38–39
　process-based mechanisms in, 33
　quantitative models in, 37
　　assessment strategies in, 37
　　for BPD, 37
　　validity criterion in, 37
　RDoC, 39–40
　　in biological models, 38
　theoretical approaches to, 33–39. *See also specific approaches*
　　external validation in, 34
　　internal validation in, 34
　　scientific descriptive models in, 33–35
　　summary of, 34
　　theory formulation in, 34
　transdiagnostic clinical processes and, 39
developmental cascades, 197–198
　cascade effects, 197–198
　models of, 198
　theory of the intervention, 197
developmental neurobiology, developmental psychopathology and, 200–201
　maltreatment of children as influence on, 200
　resilience development, 200
developmental psychopathology. *See also* longitudinal transactional models of development
　developmental cascades, 197–198
　　cascade effects, 197–198
　　models of, 198
　　theory of the intervention, 197
　developmental neurobiology and, 200–201
　　maltreatment of children as influence on, 200
　　of resilience, 200
　diversity in outcome in, 196
　diversity in process in, 196
　embryology research and, 195
　　equifinality in, 196
　equifinality principle, 196, 199
　　in embryology research, 196
　hierarchic motility in, 194
　historical roots of, 193
　individual experiences in, 195
　levels of analysis in, 196
　life-span perspective on, 199
　longitudinal transactional models of development, 196–197
　　ECHO Program, 197
　measurement strategies, 194

　multigenerational perspective on, 194–195
　normality interface in, 195–196
　organizational perspective on, 193–194
　orthogenetic principle, 193–194
　pathology in, 195–196
　prevention science and, 199–200
　principles of, 194–200
　　differentiation, 195
　　for dynamically active organisms, 195
　　equifinality, 196, 199
　　for hierarchically integrated systems, 195
　　multifinality, 196, 199
　RCTs and, 199–200
　resilience and, 198–199
　　developmental neurobiology of, 200
　theoretical approach to, 193
DHEA. *See* dehydroepiandrosterone
Diagnostic and Statistical Manual of Mental Disorders (*DSM*)
　categorization in, 279
　classification strategies in, 9, 284
　core mechanisms of, 280–281
　diagnosis targets of, 279
　　RDoC matrix of, 280
　　transdiagnostic processes, 280
　NHST and, 101–102
　operational definitions in, 101
　transdiagnostic interventions in, 284
　　capitalization model in, 283
　　in CBT, 281
　　mechanism-focused treatment development and, 281
　　targets in, 280
　　testing of, 282–283
　　UP in, 281
differential equation models, 361–364
differentiation principle, 195
diffusion modeling, in cognitive studies, 111
dimensional latent variable models, 75–77
dimensionality
　of psychological distress, longitudinal assessment of, 250
　in psychometrics, 54–57
　　CFA evaluation of, 56–57
　　EFA evaluation of, 55–56
　　importance of, 55
　　meaning of, 54
　　statistical software for, 56–57
direct effect inferences, 400
direct replication
　in meta-analysis, 392
　in scientific method, 372
directed acyclical graphs (DAGs), 87
discovery context, in data analysis, 97–98
disease. *See* psychobiological influence on disease
dispersion, in intensive longitudinal data modeling, 313–314
distress. *See* psychological distress
diversity in outcome, in developmental psychopathology, 196
diversity in process, in developmental psychopathology, 196

INDEX 487

dizygotic twins, 137–138
　DeFries–Fulker extreme analysis approach, 222–223
DLPFC functioning. *See* dorsolateral prefrontal cortical functioning
dominance genetic effects, 221
dorsolateral prefrontal cortical (DLPFC) functioning, 162
DSM. *See* Diagnostic and Statistical Manual of Mental Disorders
DSM-III. *See* Diagnostic and Statistical Manual of Mental Disorders
DSM-5. *See* Diagnostic and Statistical Manual of Mental Disorders
DSMs. *See* Dynamical Systems Models
DST. *See* Developmental Systems Theory
dual systems model, for puberty, 209
dyadic data. *See also* process research
　computational methods for, 364
　　computer programs for, 364
　　SEM in, 364
　definition of, 350
　distinguish ability of, 350–351
　exploratory methods for, 364
　　computer programs for, 364
　　SEM in, 364
　models for, 350–351
　　for outcomes, 351–354
　　overview of, 351–352
　non-exchangeability of, 350–351
　non-independence of, 350
　outcomes of, 351–354
　　active engagement in, 351
　　APIM, 351–354
　　models for, 351–354
dyadic nested data, 441–442
dyadic relationships, research approaches to. *See also* attachment theories
　child-caregiver designs, 340
　　adolescent behavior factors, 340
　　adolescent-caregiver pairs, 340
　　maternal behavior factors, 340
　　strange situation tasks in, 340
　design challenges, 344–345
　　in laboratory studies, 345
　　recruitment strategies, 344–345
　future directions for, 345–346
　innovative approaches to, 343–344
　　APIMs, 344
　　Coupled Linear Oscillator Models, 343–344
　　DSMs, 343–344
　　rating of emotions and behaviors, 343
　　vocal arousal patterns, 343
　key measures for, 342–343
　　ASR, 343
　　central psychophysiology, 342
　　FPPC, 343
　　observational coding, 342–343
　　peripheral psychophysiology, 342
　　in questionnaires and interviews, 342
　　RSA software, 342
　　SCIFF, 342–343
　　SPAFF, 343
　peer–peer designs, 340–341
　　during adolescence, 341
　　with adult peers, 341
　　research approach to, 337

　romantic partner designs, 341–342
　　self-report methodologies, 341
　　theoretical approach to, 337
dynamic factor analysis, 330–332
　measurement invariance tests for, 331–335
　person-oriented theory and, 330–331
　in process research, for dyadic data, 362
Dynamical Systems Models (DSMs), 343–344

EAR. *See* Electronically Activated Recorder
ECG/EKG. *See* electrocardiogram
ECHO Program. *See* Environmental Influences on Child Health Outcomes Program
ECoG. *See* electrocorticography
ecological momentary assessment (EMA), 327, 329
EDA. *See* electrodermal activity
education. *See* methodological education
EEG. *See* electroencephalography
EF. *See* executive functions
EFA. *See* exploratory factor analysis
Einthoven's Triangle Hypothesis, 122
electrocardiogram (ECG/EKG), 122–124
　Einthoven's Triangle Hypothesis, 122
　HR and, 123–124
　HRV and, 123–124
　respiratory sinus arrhythmia, 123–124
electrocorticography (ECoG), 153
electrodermal activity (EDA), 125–127
　ambient temperature and humidity, 126
　biological mechanisms, 125
　　psychological sweating, 119–125
　clinical findings with, 127
　electrode placement with, 125–126
　endosomatic measures, 125
　exosomatic measures, 125
　history of, 125
　skin conductance quantification, 126
electroencephalography (EEG), 153
Electronically Activated Recorder (EAR), 306
electro-oculography (EOG) methods, 129
EMA. *See* ecological momentary sssessment
embryology research, 195
　equifinality in, 196
emergence, 102
emotional dysregulation, in peer–peer attachment, 339
emotional regulation
　in MOST, 269–270
　in peer–peer attachment, interpersonal elements of, 339
emotions, James–Lange theory, 122
endophenotypes, 12–14
　biomarkers and, 14
　definitional criteria for, 13
　in descriptive cognitive study designs, 105–106
　"gloomy prospects" in, 13
　in GWAS, 13
　for neuroticism, 14

　in psychopathology, 13
　as risk factor, 13
　validators and, 13
endosomatic measures, for EDA, 125
enteric system, 118
environmental influences, on adolescence, 211
Environmental Influences on Child Health Outcomes (ECHO) Program, 197
EOG methods. *See* electro-oculography methods
epidemiologic experiments, 82–83
　Opportunity experiment, 83–86
　RCTs, 82–83
epidemiology methods. *See* psychiatric epidemiology methods
epigenetics, 148
　modifications in, 148
equifinality principle, 99
　in developmental psychopathology, 196, 199
　embryology research and, 196
ergodic theorem, 329, 333
　consequences of, 330
　non-ergodic processes, 329
ESEM procedures. *See* Exploratory Structural Equation Modeling procedures
ESI. *See* Externalizing Spectrum Inventory
ESI-Disinhibition (ESI-DIS) scale, 21
estimate scaling, in meta-analysis, 390
estimation errors, in meta-analysis, 386–389
　bias in, 388–389
　heterogeneity in, 387–388
　variance in, 386–388
estradiol, 206–207, 210
ethnic minorities, adolescence for, psychopathology of, 212
ethnicity, behavioral medicine and, health disparities influenced by, 294
etiological heterogeneity, in NHST, 99
evocative rGE, 146–147, 225–226
exceedance probability, for RL models, 175
executive functions (EF)
　in cognitive studies, 110
　with externalizing proneness, 21–22
exosomatic measures, for EDA, 125
expectancy effects, in randomized studies, 421
experiences, through social exchange, 180
experimental designs, for cognitive studies, 105
experimental psychopathology. *See also* null hypothesis statistical testing
　causality in, 102
　clinical observation in, 95
　conceptual foundations of, 93–96, 102–103
　development of, 94
　correlates in, 95–96, 102
　　counting of, 96
　　ratings of, 95–96
　definitions of, 93
　determinants in, 95–96
　　counting of, 96
　　ratings of, 95–96

experimental psychopathology. (cont.)
  emergence in, 102
  genetic influences in, 102
  goals of, 95
  schizophrenia and, 93–94
explanations, in scientific method, 371–372
exploration/exploitation dilemmas, 173
exploratory factor analysis (EFAs)
  ESEM procedures, 57
  for externalizing proneness, 22–24
  of latent variable models, 70–71
    atheoretical analytic approach, 70
    in exploratory-confirmatory spectrum, 72–73
    function and purpose of, 71
    retention of factors in, 70
    rotation of factors in, 71
  of psychometrics, 55–56
exploratory hypotheses, for reproducibility, 376–377
  confirmatory hypotheses as distinct from, 376–377
  in multivariate modeling, 377
exploratory methods, for dyadic data, 364
  computer programs for, 364
  SEM in, 364
Exploratory Structural Equation Modeling (ESEM) procedures, 57
  in latent variable models, 72–73
    advantages of, 72
    estimation of, 72
    for personality disorder syndromes, 72–73
exploratory-confirmatory spectrum, 72–73
exposure and response prevention, in MOST, 269–270
externalizing proneness, 21–24
  CFA approach to, 23–24
  EF factors for, 21–22
  EFA approach to, 22–24
  inhibition–disinhibition models, 22–24
  multi-method multivariate model of, 22–24
externalizing psychopathology dimension, 223–224
Externalizing Spectrum Inventory (ESI)
  ESI-DIS scale, 21
  for neurobehavior traits, 21

facial measures, for peripheral psychophysiology, 127–131
  EOG methods, 129
  oculometry, 129–130
    biological mechanisms, 129
    early history of, 129
    methodological guidelines for, 129–130
    point of regard measures, 129–130
    smooth pursuit eye movements, 129–130
  POG methods, 129
  pupillometry, 131–135
    biological mechanisms, 130
    clinical findings for, 131
    early history of, 130
    methodological guidelines for, 130–131

  startle response, 127–129
    biological mechanisms, 127
    FPS, 127, 129
    identification of, 127
    methodological guidelines for, 127–128
    PAMR, 127–129
    startle-eye blink measures, 127–128
Factor Mean Shift model, 246–247, 261
factor mixture modeling, 75–77
  complexity of, 77
Falconer Formula, in twin studies, 221
Family and Peer Process Code (FPPC), 343
family studies, in quantitative genetic research, 219
fast event-related designs, for fMRI, 155–156
fear-potentiated startle (FPS), 127, 129
feature ablation approach, in machine learning, 469–470, 481
feedback, in longitudinal studies, 236
Feigl, Herbert, 14
FF. See fundamental frequency of vocal patterns
Fight or Flight system, 120
fingerprints, 468, 481
first-stage parallel multiple moderation model, 405
Fisher, Ronald A., 136, 378
Fisherian randomized-based inference, 418
Five Factor Model of Personality, 10–11
fixed effects meta-analytic models, 385, 387
fMRI. See functional magnetic resonance imaging
folds, machine learning and, 481
follicular-stimulating hormone (FSH), 210
formative latent variable models, 68
forward inference, of fMRI, 163
forward modeling, in RL models, 176
FPPC. See Family and Peer Process Code
FPS. See fear-potentiated startle
Free Curve Slope Intercept factor model, 247
FSH. See follicular-stimulating hormone
functional magnetic resonance imaging (fMRI)
  analytic approaches, 157–161. See also specific models
    contrast, 157–159
    correlation, 157–159
    of intrinsic functional connectivity, 159–161
    subtraction, 157–159
    for task-evoked datasets, 158
    of task-evoked functional connectivity, 160–161
  for BOLD responses, 153–154, 159–160
  for brain resting state, 156–157
  cross-modality integration of, 164
  of DLPFC dysfunction, 162
  evolution of, 153
  experimental approaches to, 153–157
    block designs, 154–155
    fast event-related designs, 155–156

    hybrid designs, 156
    resting state designs, 156–157
    slow event-related designs, 154–155
    for task performances, brain measurement with, 153–156
  future directions of, 164
  HCP and, 154–155
  interpretation of, 161–164
    of causality confound, 162
    of diagnostic specificity, 162–163
    of forward inference, 163
    of mechanistic specificity, 161–162
    of regional-based sensitivity, 163–164
    of reverse inference, 163
    of symptom specificity, 162–163
    of intrinsic functional connectivity, 156–157
  through models
    general linear, 159–160
    network modeling, 160
    neurometrics, 157
    PET as influence on, 157–159
    prediction errors through, 184–185
fundamental frequency (FF) of vocal patterns, 343

GAD Questionnaire. See Generalized Anxiety Disorder Questionnaire
Galton, Francis, 137
game theory, 182
GEE. See generalized estimating equations
gender. See also males; women
  age of onset of puberty by, 205
  problematic alcohol consumption by, longitudinal assessment of factor correlations, 259
    gender factors and, variance differences in, 255–256
    in growth curve models, 255–258
    psychological distress and, 256
gene by environmental interaction (GxE), 147–148
  in quantitative genetic research, 225
general linear models (GLM), in fMRI, 159–160
Generalized Anxiety Disorder (GAD) Questionnaire, 54–55
generalized estimating equations (GEE), 439–440
genetic variant identification, 139–141
genetics. See also behavioral genetics; molecular genetics
  additive, 221
  dominance genetic effects, 221
  experimental psychopathology influenced by, 102
  molecular
    historical development of, 136
    risk factor determinations in, 136
  quantitative research, for child and adolescent psychopathology, 221–222
    for ADHD, 221–222
    for ASDs, 221–222
  risk factors in, determination of, 136

INDEX

genome-wide association studies (GWAS)
   in adoption and twin studies, 142–144
      CDCV model, 142–143
      clinical findings from, 143
      missing heritability and, 145
      for schizophrenia, 143
      SNPs, 142–143
   endophenotypes in, 13
   twin studies and, 142–144
genotype-environment correlation (rGE), 146–147
   active, 146–147, 225–226
   evocative, 146–147, 225–226
   passive, 146–147, 225
   in quantitative genetic research, 225–227
      active rGE, 225–226
      detection of, 225–226
      evocative rGE, 225–226
      long-term implications for, 226–227
      passive rGE, 225
geofencing, 308
GH. *See* growth hormone
GIMME models. *See* Group Iterative Multiple Model Estimation models
GLM. *See* general linear models
"gloomy prospects," in endophenotypes, 13
GnRH. *See* gonadotropin releasing hormone
gonadarche, 206–207
gonadotropin releasing hormone (GnRH), 206–207, 210
Gray's Motivational Theory, 120
   interconnected neuropsychological systems, 120
groundtruth parameters, of RL models, 177
Group Iterative Multiple Model Estimation (GIMME) models, 321–322, 333–334
growth curve models, 248
   commonly used, 246–247
   Factor Mean Shift model, 246–247
   Free Curve Slope Intercept factor model, 247
   polynomial, 249–250
   for problematic alcohol consumption, 255–257, 261–262
      alcohol use in, summary of, 254–255
      gender factors in, 255–258
   for psychological distress, 251
      for BSI, 251
growth hormone (GH), 210
Guttman, Louis, 10–12
Guze, Samuel, 8
   poor prognosis syndrome, 8
GWAS. *See* genome-wide association studies
GxE. *See* gene by environmental interaction

Hamilton Rating Scale for Depression, 14
HARKing effects. *See* Hypothesizing After the Results are Known effects
Hawthorne effect, 420
HCP. *See* Human Connectome Project
health psychology

adjunctive psychosocial interventions, 293
adverse childhood experiences and, 294
between-person approaches, 294
biopsychosocial model, 287–288
   systems hierarchy in, 288
conceptual models for, 293–294
definition of, 287
endpoints of, evolution of, 288–289
ethnicity and, health disparities and, 294
future directions for, 293–295
health behaviors and, risk reduction and, 289
medical illness and, psychosocial aspects of, 289
methodological education on, 294–295
outcomes for, evolution of, 288–289
psychobiological influence on disease and, 289–293
   assessment measures, 291
   epidemiology of, 289–291
   implacable experimenter, 292
   mechanisms of, 291–293
   risk factors, 289–291
quality of life issues, 288
race and, health disparities and, 294
RCTs in, 293
socioeconomic status, health disparities and, 294
theoretical approach to, 287
within-person approaches, 294
heart rate (HR), 123–124
heart rate variability (HRV), 123–124
hemodynamic response functions (HRFs), 179–180
heritability
   in classical twin studies, 221
   in quantitative genetic research
      for ADHD, 221–222
      for ASDs, 221–222
      in classical twin studies, 221
      environmental factors in, 222
Hess, Eckhard, 130
heterogeneity
   in estimation errors, 387–388
   in NHST
      in data analysis, 98–99
      etiological, 99
      laboratory task performance, 99
heterostatic models, 119
hierarchic motility, 194
hierarchical data, nested, 427, 441
   dependence in, 428
hierarchical factor models, 74
Hierarchical Taxonomy of Psychopathology (HiTOP) model, 21, 39–40
hierarchically integrated systems, in developmental psychopathology, 195
higher-order meta-analytic models, 384
HiTOP model. *See* Hierarchical Taxonomy of Psychopathology model
holding-out, 468–482
Holzman, Philip S., 94
homeorhetic models, 119
homeostasis, peripheral psychophysiology and, 118–119

hormones, during puberty, 206–207. *See also specific hormones*
   brain development and, 207
HR. *See* heart rate
HRFs. *See* hemodynamic response functions
HRV. *See* heart rate variability
human behaviors, through RL models, 173–175
   for individual differences, 178–179
   individual likelihood function, 174
   model inversion of, 174–176
   parameter estimations for, 174
   prediction of, 177–178
Human Connectome Project (HCP), 154–155
human phenotype ontology, 15
Hume, David, 9
hybrid designs, for fMRI, 156
Hyman, Steven, 9
hypochondriasis, 6
Hypothesizing After the Results are Known (HARKing) effects, 373
hypothetical variable models, 66
hysteria
   borderline personality disorder and, 7
   epilepsy and, 6–7
   as open concept phenotype, 6–7
      explanatory hypotheses for, 6
      gender and, 6
      redefinition of, 6–7
      social stigma from, 6
      symptoms of, 6
   prodromal psychosis and, 6–7

Iavaan script for the Extra Dependent Variable Regression Model, 465
ICG. *See* impedance cardiography
idiographic levels, of subject-specific data analysis, 333–335
imaging modalities. *See specific imaging*
impedance cardiography (ICG), 123–124
   pre-ejection period, 124–125
implacable experimenter, 292
independent pathways model, in quantitative genetic research, 223–224
indirect effect inferences
   conditional, 406–408
   in statistical mediation analysis, 398, 400
      bootstrap confidence intervals, 398
      causal elements of, 399
inferences. *See also* causal inferences
   direct effect, 400
   indirect effect, 398, 400
      bootstrap confidence intervals, 398
      causal elements of, 399
      conditional, 406–408
inhibition–disinhibition models
   of externalizing proneness, 22–24
   liabilities of, 25–26
inhibitory control, in psychopathology, 21
insecure peer–peer attachment, 339
intensive longitudinal data, modeling of autocorrelations in, 318–319
   for borderline personality disorder, 318
   for depression, 318
   time course, 318–319

intensive longitudinal data, modeling of (cont.)
　collection techniques for, 312–313
　dynamics of, 316–318
　　duration, 317–318
　　instability, 316–317
　　for multiple symptoms, 313–314, 319–322
　　for single symptoms, 313–316
　MASD in, 316
　modeling approaches, 321–323
　　BOUM, 322
　　GIMME, 321–322
　　regression, 313
　　VAR, 320–321
　MSSD in, 316–317
　for multiple symptoms, dynamical properties of, 313–314, 319–322
　　advanced techniques for, 321–322
　　concurrent relationships, 319–320
　　descriptive techniques for, 321
　　lagged relationships, 320–321
　　PCA of, 320
　　prospective relationships, 320–321
　overview of, 322–324
　for single symptoms, dynamical properties of, 313–316
　　central tendencies of, 313
　　dispersion scores, 313–314
　　spread measures, 313–316
　theoretical approach to, 312–313
intensive longitudinally nested data, 441–442
intention-to-treat (ITT) analysis, 420
inter-individual variation, in subject-specific data analysis, 327–330
　data boxes, 328
　definition of, 327–329
　intra-individual variation and, 328–330
　R-technique for, 329
　unified framework for, 334
internal structure validity, of psychometrics, 59–61
internalizing psychopathology dimension, 223–224
interventions. See specific interventions
interventions, as science. See also Multiphase Optimization Strategy
　theoretical approach to, 267
interviews
　Classical Test Theory and, 49–50
　construct specification for, 50–51
　　continuum specification as part of, 51
　Item Response Theory and, 49–50
　measurement of constructs for, 49–50
　　principles of scale, 49–50
　　validity criteria in, 50
　methodological approach to, 45–46
　scale development for, practical considerations in, 51–52
　selection criteria for, 46–49
　　advantages of interviews, 47–48
　　disadvantages of interviews, 48–49
　surveys compared to, 46
　　in stimuli presentation, 46
　usefulness of, 45–46
intra-individual variation, in subject-specific data analysis, 327–330

data boxes, 328
definition of, 327–328
inter-individual variation and, 328–330
P-technique for, 329
intrinsic functional connectivity, fMRI for, 159–161
inversion RL models, 174–176
Item Response Theory (IRT), 26, 49–50
　in longitudinal studies, 233
ITT analysis. See intention-to-treat analysis

James, William, 122
James–Lange emotion theory, 122
Journal of Abnormal Psychology, 94
justification context, in data analysis, 97–98

Kenward–Roger method for calculating degrees of freedom, 440
k-means clustering approach, 469, 481
Kraepelin, Emile, 8

laboratory task performance heterogeneity, 99
lagged relationships, in intensive longitudinal data modeling, 320–321
latent change score models, in process research, 357–361
　auto-regressive models, 359–360
　DFA, 362
　time series analysis in, 357–359
latent growth curve models
　in process research, 355–357
　　common fate affects, 355
　　limitations of, 356
　　measurement invariance tests, 355–356
　in quantitative genetic research, classical twin studies, 224
latent variable models, 67
　in behavioral neuroscience, 169
　categorical, 75–77
　CFA of, 68, 71–72, 76–77
　　in exploratory-confirmatory spectrum, 72–73
　dimensional, 75–77
　EFA of, 70–71
　　atheoretical analytic approach, 70
　　in exploratory-confirmatory spectrum, 72–73
　　function and purpose of, 71
　　retention of factors in, 70
　　rotation of factors in, 71
　ESEM procedures, 72–73
　　advantages of, 72
　　estimation of, 72
　　for personality disorder syndromes, 72–73
　estimation of, 67–69
　evaluation of, 68–69
　formative, 68
　mapping of, 70
　pseudo-realism and, 67
　reflective, 68
　structural comparisons between, 75–77

law, legal systems and, adolescence and, methodological applications of, 214–215
learned action values, in RL models, 173
learning. See also machine learning
　definition of, 481
learning rules, in RL models, 172–173
leptin, 210
lesbian, gay, bisexual, transgender (LGBT) community, adolescence in, psychopathology of, 212–213
LH. See luteinizing hormone
life-span perspective, in developmental psychopathology, 199
linear moderation model, 401
Linnaeus, Carl, 33
Locke, John, 87
The Logic of Modern Physics, 5–6
longitudinal data, 428
longitudinal studies. See also comorbidity models; intensive longitudinal data; measurement models
　assessment protocols, 236
　authorship issues in, 235
　causal dynamics in, 232
　compensation for, 236
　critical design decisions in, 230–234
　　data wave collection criteria, 231–232
　　measurement strategies, 232–233
　　purpose and function issues, 231
　　for sample sizes, 233–234
　IRT methods in, 233
　management issues in, 234–237
　　for attrition, 236–237
　　codebooks and, 235
　　for documents, 235
　　through open communication, 235
　　through project roles, 235
　　variable labeling conventions and, 235
　in Open Science era, 237–238
　openness of, 237
　participant feedback in, 236
　purpose and function of, 230
　sampling in, 231
　self-report surveys in, 232
　short forms in, 236
　transparency in, 237
longitudinal transactional models of development, 196–197
　ECHO Program, 197
longitudinally nested data, 441
　intensive, 441–442
luteinizing hormone (LH), 210

Mach, Ernst, 9–10
machine learning
　classifiers from, 471, 475–476, 481
　　in supervised learning, 471–472
　clinical neuroscience influenced by, 470–471
　clinical theory influenced by, 469–470
　cluster identification through, 478
　　hierarchical clustering, 478
　cross-validation in, 468
　data analysis from, 476
　　cross-validation through, 475, 481
　　feature selection in, 474
　　overfitting, 473, 482

PCA in, 474
  in regression models, 475–476
  regularization and, 474, 482
 definition of, 467–468
 feature ablation approach, 469–470, 481
 fingerprints in, 468, 481
 folds and, 481
 function and purpose of, 468
 future applications of, 479
 generalization as result of, 468, 481
 holding-out, 468–482
 interpretation of, 476
 k-means clustering approach, 469, 481
 learning phase, 467
 multidimensionality of, 468–469
 outputs from, 476
 permutation testing, 476, 482
 pitfalls of, 478–479
 precision medicine, 467, 482
 prediction through, 467–468
  features of, 481
 study design for, 473
  confusion matrix in, 473, 477, 481
  group membership predictions, 472, 473
  outcome predictions, 472–473
 supervised, 471, 482
  classifiers from, 471–472
 testing phase, 467, 482
 unsupervised, 471–472, 482
magnetoencephalography (MEG), 153
Maher, Brendan, 93–94
major depressive disorder (MDD). *See also* depression
 MOST for, 268
males
 hypochondriasis in, 6
 hysteria phenotype and, 6
 pubertal timing for, 208
maltreatment, of children
 developmental neurobiology influenced by, 200
 MOST for, 267
  proximal mediators for, 270–271
  PTSD as result of, 267
MANOVA. *See* multivariate analysis of variance
MAP test. *See* Minimum Average Partial test
MAR mechanism. *See* missing at random mechanism
Markov decision process (MDP), in RL models, 172
Marr, David, 169
MASD. *See* mean absolute successive difference
maximum likelihood estimation, for missing data, 454
 analysis example, 452–454
 correlates in, 453
 with incomplete data, 451–452
 multiple imputation compared to, 454–456
 multiple regression model, 453
 with multivariate normality, 451
MCAR mechanism. *See* missing completely at random mechanism
MDD. *See* major depressive disorder

MDP. *See* Markov decision process
mean absolute successive difference (MASD), 316
mean squared successive difference (MSSD), 316–317
measurement invariance tests, 355–356
 for dynamic factor analysis, 331–335
measurement models, for longitudinal data
 alternative, 243–244
 analyses of, 248
 growth curve, 248
  commonly used, 246–247
   Factor Mean Shift model, 246–247
   Free Curve Slope Intercept factor model, 247
 improper, 243–244
 multiple, 242–243
 oblique simple structure factors, 246
 purpose and goals of, 241
mediation analysis. *See* statistical mediation analysis
mediator variables, in statistical mediation analysis, 397
medical marijuana laws (MMLs), 215
medicine. *See* behavioral medicine
Mednick, Sarnoff, 94
Meehl, Paul, 14–15
MEG. *See* magnetoencephalography
memory, connectionist models of, 170–171
menarche, 205–206
Mendelian disorders, in twin studies, 141
mental health, during adolescence, 212
 biological basis of, 213
 ethnic disparities in, 212
 prevalence estimates of, 211–212
 racial disparities in, 212
 sexual minority status and, disparities as result of, 212–213
 socioeconomic status factors for, 212
meta-analysis, through quantitative modeling
 areas for development in, 392–393
  through Bayesian estimates, 393
  through non-parametric approaches, 392–393
 coding of variables in, 391
 conceptual approach to, 383
 early applications of, 383
 elements of, 389–392
  estimate scaling, 390
  multivariate analysis, 390–391
  sampling/selection of studies, 390
  univariate analysis, 390–391
 estimation errors, 386–389
  bias in, 388–389
  heterogeneity in, 387–388
  variance in, 386–388
 forms of, 383–384
  higher-order modeling, 384
  inferences through, 384
  summative modeling, 384
 models, 384–389
  estimation of, 385–395
  fixed effects, 385, 387
  general features of, 384–385

  higher-order, 384
  random effects, 385, 387–388
  reparameterized, 385–386
  summative, 384
 overview of, 393–394
 PRISMA, 389–390
 reliability of variables in, 391
 replicability of results in, 383
 replications in, 391–392
  conceptual, 392
  direct, 392
  of results, 383
 as statistical analysis, 383–384
methodological education, 294–295
MFT model. *See* Multifactorial Threshold model
mHealth. *See* mobile health
mindful exploration, in NHST, 97
Minimum Average Partial (MAP) test, 55–56
missing at random (MAR) mechanism, 446
missing completely at random (MCAR) mechanism, 445–446
missing data
 maximum likelihood estimation, 454
  analysis example, 452–454
  correlates in, 453
  with incomplete data, 451–452
  multiple imputation compared to, 454–456
  multiple regression model, 453
  with multivariate normality, 451
 mechanisms of, 445–447
  MAR, 446
  MCAR, 445–446
  NMAR, 446–447
 motivating examples of, 445
  correlations among study variables, 445
  variables in, 445
 multiple imputation, 450
  analysis phase, 449
  examples of, 450, 454
  fully conditional specifications, 448
  imputation phase, 447
  maximum likelihood estimation compared to, 454–456
  pooling phase, 449
 multiple regression models for, 446
 patterns in, 445
 in randomized studies, 421
 theoretical approach to, 444
 variables in
  auxiliary, 447, 453–454
  in motivating examples, 445
missing heritability, 143–145
 CNVs, 145
 insufficient GWAS, 145
 non-additivity of genetic effects, 145
MITML scripts for multiple imputation, 463
mixed VAR models, 334–335
MLM frameworks. *See* multilevel modeling frameworks
MMLs. *See* medical marijuana laws
MMPI-2 depression scale, 14
mobile health (mHealth), 307

model-based causal inferences, 419
model-based cognitive neuroscience, 180
modeling. *See specific types of modeling*
moderated mediation model, 404–405
    components of, 406–407
    conditional indirect effects, 406–408
    in equation form, 405
    estimations of, 405–406
    first-stage parallel multiple, 405
    index of, 404
    probing of, 401–404
    research questions in, 405
    substantive conclusions of, 408
    in visual form, 405
moderation analysis, 400–403
    linear moderation model, 401
    partial effects in, 402
    plotting of moderation effects, 402–403
    probing of moderation effects, 402–403
    regression coefficients in, interpretation of, 401–402
    unconditional effects in, 402
    utility of, in clinical research, 401
modern descriptive psychiatry, 36–37
    assessment approaches in, 36
    BPD and, 36–37
    defining features of, 36
    theoretical approach to, 38–39
molecular genetics
    historical development of, 136
    risk factor determinations in, 136
monitoring, in AA, 304–305
    compliance factors, 304–305
    reactivity in, 305
monozygotic twins, 137–138
    DeFries–Fulker extreme analysis approach, 222–223
MOST. *See* Multiphase Optimization Strategy
MSSD. *See* mean squared successive difference
multi-component treatment packages, 281–282
    ATDs, 282
    SCED, 282
Multifactorial Threshold (MFT) model, 137–138
multifinality principle, 196, 199
multilevel modeling (MLM) frameworks, 179
Multiphase Optimization Strategy (MOST), for behavioral interventions, 269
    applications for, 267–268
        real world, 275
    for child maltreatment, 267
        proximal mediators for, 270–271
        PTSD as result of, 267
    clinical psychology benefits of, 268, 274–275
    in cognitive therapy, 268
    for MDD, 268, 276
    clustering in, 276
    cognitive processing component, 269–270
    conceptual model of, 269–271
        intervention components in, 269–270

optimization criterion, 271
    proximal mediators in, 270–271
    for PTSD, 269–270
continual optimization principle, 274
emotional regulation component, 269–270
evaluation of, 272–274
experimental design of, 271–273
    factorial designs, 272
    management of, 274–275
    resource management principle, 271–272
exposure and response prevention in, 269–270
failure of, 275–276
overview of, 268
preparation of, 268–269
for PTSD, 267, 276
    from child maltreatment, 267
    in conceptual model, 269–270
    proximal mediators for, 270–271
    in RCTs, 268
    relaxation training component, 269–270
multiple imputation phase, of missing data, 450
    analysis phase, 449
    examples of, 450, 454
    fully conditional specifications, 448
    imputation phase, 447
    maximum likelihood estimation compared to, 454–456
    pooling phase, 449
multiple regression models, for missing data, 446
multiple symptoms, intensive longitudinal data modeling for, 313–314, 319–322
    advanced techniques for, 321–322
    concurrent relationships, 319–320
    descriptive techniques for, 321
    lagged relationships, 320–321
    PCA of, 320
    prospective relationships, 320–321
multivariate analysis of variance (MANOVA), 428–429

National Epidemiologic Survey on Alcohol and Related Conditions (NESARC), 88
NEO Personality Inventory, 10–12
NESARC. *See* National Epidemiologic Survey on Alcohol and Related Conditions
nested data
    on alliance strength, 429–431
    alternative analytic approaches, 438–440
        with cluster-robust standard errors implementation, 439
        through fixed effect models, 440
        with GEE, 439–440
    in behavioral activation, 427
    in clinical psychology research, 426–427
    in cognitive therapy, 427
    definition of, 426
    dependence in, 427–429
        ANOVA and, 428–429
        in hierarchical data, 428

    in longitudinal data, 428
    MANOVA and, 428–429
discrete dependent variables, 441
dyadic, 441–442
groupings of, 429–430
    small numbers of, 440–441
as hierarchical, 427, 441
    dependence in, 428
    levels of effects, 431
        for ADHD, 429
        total effects, 429–431
in longitudinal data structures, 427
    dependence in, 428
    longitudinally, 441
        intensive, 441–442
multilevel models, 429, 438
    anxiety among children in families, 432–435
    extensions of, 438
    negative effects in, over time, 436–438
partially, 440
pre-post design for, 441
random factors in, 427
sources of, 426
theoretical approach to, 426
network modeling, in fMRI, 160
neural networks, 169
    ANNs, 171
    computation in, 170–171
    PDP, 171
    perceptrons, 170
        limitations of, 170–171
    pyramidal hippocampal neurons, 170
neurobehavioral traits
    ESI for, 21
    in HiTOP model, 21
    as liabilities, in clinical models, 25–26
    multi-modal assessment approaches, 21
    in psychoneurometric research strategy, 18–21
    threat sensitivity, 25–26
neuroeconomics, 168
neurometrics, 157
neuroscience. *See also* behavioral neuroscience; cognitive neuroscience
    machine learning as influence on, 470–471
neuroticism, endophenotypes for, 14
Neurovisceral Integration (NVI) Model, 119–120
    attractor basins and, 120
Neymanian randomized-based inference, 418–419
NHST. *See* null hypothesis statistical testing
NMAR mechanism. *See* not missing at random mechanism
nomothetic levels, of subject-specific data analysis, 327
non-compliance, in randomized studies, 420–421
normality interface, in developmental psychopathology, 195–196
not missing at random (NMAR) mechanism, 446–447
null hypothesis statistical testing (NHST), 96–103

data analysis in, 96–97
  discovery context in, 97–98
  heterogeneity factors in, 98–99
  justification context in, 97–98
  levels of, 99–100
  post-hoc, 97–98
data cleaning in, 97
in *DSM* approach classifications, 101–102
  operational definitions in, 101
equifinality principle in, 99
etiology in, 100
  heterogeneity and, 99
  qualitative factors, 100
  uniformly most potent factors in, 100
heterogeneity in
  in data analysis, 98–99
  etiological, 99
  laboratory task performance, 99
mindful exploration in, 97
predictors in, 98
as quasi-nearly always false, 96
reliability in, 100–101
reproducibility and, 373
validity of, 100–101
NVI Model. *See* Neurovisceral Integration Model

observational coding, in dyadic relationships, 342–343
observational studies, causal inferences in, 423
  analysis of, 422
  assumptions for identification in, 422
  covariate adjustments in, 423
  design of, 422
observations, in scientific method, 371–372
oculometry, 129–130
  biological mechanisms, 129
  early history of, 129
  methodological guidelines for, 129–130
  point of regard measures, 129–130
  smooth pursuit eye movements, 129–130
open concepts, 7–9, 14–15. *See also* endophenotypes
  defining properties in, 7
  hysteria phenotype as, 6–7
    explanatory hypotheses for, 6
    gender and, 6
    redefinition of, 6–7
    social stigma from, 6
    symptoms of, 6
  ontologies in, 15
  open textures of, 8–9
  in operational definitions, 5
  operationalism and, 14–15
  Pap on, 7–8
  partial definitions in, 7
  poor prognosis syndrome, 8
  scientific conventions in, 9–10
    Einstein as influence on, 9
    validators in, 8
Open Science era
  longitudinal studies in, 237–238
  reproducibility problem in, 374

open textures, of open concepts, 8–9
operational definitions
  as concept, development of, 5
  of depression, 5
  in NHST, 101
  open concepts in, 5
  phenotypes in, 5
operationalism, 14–15
operationalist conscience, 15
Opportunity social experiment, 83–86
optimal control of psychological processes, 332–333
  application of, 332–333
ordering, 399
organizational hypothesis, for puberty, 209
orthogenetic principle, 193–194
overfitting, 473, 482

PAMR. *See* post-auricular muscle response
Pap, Arthur, 7
  on open concepts, 7–8
Parallel Distributed Processing (PDP), 171
parallel models, in statistical mediation analysis, 400
parasympathetic nervous system (PSNS), 118
  respiratory sinus arrhythmia, 118
PARCHISY. *See* Parent–Child Interaction System
Parent–Child Interaction System (PARCHISY), 342–343
parsimonious factor pattern models, 245–246
parsimony, 175
partial effects, in moderation analysis, 402
partially nested data, 440
partner–partner attachment, 339–340
  anxious attachment styles, 339
  dyadic processes in, 339–340
passive rGE, 146–147, 225
Paul, Gordon, 467
Pavlov, Ivan, 168
payoff matrix, in game theory, 182
PCA. *See* principal component analysis
PDP. *See* Parallel Distributed Processing
PDS. *See* Pubertal Development Scale
peer–peer attachment, 338–339
  emotional dysregulation in, 339
  insecure, 339
  interpersonal emotional regulation, 339
  positive relationships in, 339
  risk distribution in, 338–339
  social baseline theory, 338–339
peer–peer designs
  with adult peers, 341
  in dyadic relationships, 340–341
    during adolescence, 341
perceptrons, 170
  limitations of, 170–171
peripheral psychophysiology
  ANS functioning in, 118–119
    enteric system in, 118
    PSNS in, 118
    SNS in, 118
  definition of, 118
  in dyadic relationships, 342

homeostasis and, 118–119
measures for, 118. *See also* cardiovascular measures; electrodermal activity; facial measures
methodological considerations in, 121–122
  baselines in, 121
  phasic measures, 121
  for population levels, 121–122
  tonic measures, 121
purpose and function of, 118
RDoC initiatives and, 118
theoretical models of, 118–120
  allostatic, 119
  Gray's Motivational Theory, 120
  heterostatic, 119
  homeorhetic, 119
  NVI Model, 119–120
  Polyvagal Theory, 119–120
permutation testing, 476, 482
personal sensing, AA and, 305–306
personality disorder syndromes. *See also specific syndromes*
  latent variable models for, 72–73
personalized interventions, 283–284
  capitalization models, 283
  through SDM, 283
personalized medicine procedures, 214
person-oriented theory, 330–331
  dynamic factor models in, 330–331
  methods in, 330–331
  subject-specific data analysis compared to, 327
PET. *See* positron emission tomography
phenotypes. *See also* endophenotypes
  in operational definitions, 5
photo-oculography (POG) methods, 129
placebo effects, in randomized studies, 421
planned missingness approach, in cognitive studies, 109
PNM research strategy. *See* psychoneurometric research strategy
POG methods. *See* photo-oculography methods
Poincaré, Henri, 9
point of regard measures, 129–130
Poldrack, Russel, 15
policy hypothesis, for RL model, 180
polynomial growth curve models, 249–250
Polyvagal Theory, 119–120
poor prognosis syndrome, 8
positron emission tomography (PET), 153
  fMRI influenced by, 157–159
post-auricular muscle response (PAMR), 127–129
posterior predictive causal inferences, 419
post-hoc data analysis, 97–98
post-traumatic stress disorder (PTSD), MOST for, 267, 276
  child maltreatment as source of, 267
  conceptual model, 269–270
  proximal mediators in, 270–271

precision medicine, 467, 482
prediction errors, neural signatures of, 184–185
predictive bias, in psychometrics, 62
pre-ejection period, in impedance cardiography, 124–125
Preferred Reporting Items for Systematic Reviews and Meta-Analyses (PRISMA), 389–390
prevention science, 199–200
primitives, in causal inferences, 415–416
principal component analysis (PCA), 320
  of machine learning, 474
Principles of psychopathology: An experimental approach (Maher), 94
PRISMA. *See* Preferred Reporting Items for Systematic Reviews and Meta-Analyses
problematic alcohol consumption, longitudinal assessment of, 259
  Factor Mean-Shift model, 261
  gender factors and
    factor correlations, 259
    in growth curve models, 255–258
    psychological distress and, 256
    variance differences in, 255–256
  growth curve models for, 255–257, 261–262
    alcohol use in, summary of, 254–255
    gender factors in, 255–258
    item-specific general method factors, 256–257
  manifest variable intercepts in, 252
  measurement errors in, 258–260
  multigroup factor loadings, 252
  observed means for, 253
  predicted means for, 253
  psychological distress as influence on, 255
    gender factors for, 256
  subfactor loadings, 260
PROCESS macro for conditional processes, 416
process research, in dyadic data, 354–364
  differential equation models for, 361–364
  dynamic processes, 354–355
  latent change score models in, 357–361
    auto-regressive models, 359–360
    DFA, 362
    time series analysis in, 357–359
  latent growth curve models in, 355–357
    common fate affects, 355
    limitations of, 356
    measurement invariance tests, 355–356
prodromal psychosis, 6–7
proneness. *See* externalizing proneness
prospective cohort studies, 80
prospective comorbidity models, 241–242
prospective relationships, in intensive longitudinal data modeling, 320–321
pseudo-realism, 67
PSNS. *See* parasympathetic nervous system
psychiatric epidemiology methods
  analyses of, 85–86

cause identification in, 87
  with DAGs, 87
  fourfold table in, 85–86
  incidence rate ratios in, 86–87
  incidence rates in, 86–87
case-control studies, 83–85
  advantages of, 84
  implementation challenges of, 84–85
  source populations in, 83–84
clinical psychology considerations, 87–88
in complex sampling designs, 88
cohort studies, 80–82
  exposure definitions and measurements in, 81–82
  identification of outcomes in, 80–82
  planning strategies for, 80–81
  prospective, 80
  retrospective, 80
  target populations in, 81
definition of, 80
epidemiologic experiments, 82–83
  Opportunity experiment, 83–86
  RCTs, 82–83
functions and goals of, 80
study designs in, 80–85. *See also specific designs*
psychiatry, psychology as distinct from, 168
psychobiological influence on disease, 289–293
  assessment measures, 291
  epidemiology of, 289–291
  implacable experimenter, 292
  mechanisms of, 291–293
  risk factors, 289–291
psychological distress, longitudinal assessment of, 248–253
  BSI, 248–249
    General Severity Index, 249
    growth curve models for, 251
  dimensionality in, 250
  early analyses of, 249
  growth curve models for, 251
    for BSI, 251
  observations of, 251–253
  parsimonious structure of, 250
  polynomial growth curve model, 249–250
  problematic alcohol consumption influenced by, 255
    gender factors for, 256
  Random Intercept model for, 250
psychological sweating, 119–125
psychology, psychiatry and as distinct from, 168
psychometrics. *See also* validity
  clinical measures in, 62–63
  dimensionality in, 54–57
    CFA evaluation of, 56–57
    EFA evaluation of, 55–56
    importance of, 55
    meaning of, 54
    statistical software for, 56–57
  in GAD Questionnaire, 54–55
  MAP test in, 55–56
  predictive bias in, 62
  purpose and function of, 54
  quality differences in, 62

reliability of, 57–60
  evaluation of, 58–59
  importance of, 58
  IRT and, 59–60
  meaning of, 57–58
  VSS method in, 55–56
psychoneurometric (PNM) research strategy, 18, 20–21
  future challenges for, 24–26
  IRT and, 26
  neurobehavioral trait constructs, 18–21
  prediction purpose of, 26
  score estimation in, 26
psychopathology. *See also* descriptive psychopathology; developmental psychopathology; experimental psychopathology
  behavioral measures in, integration of, 20
  biological reductionism issues, 26
  clinical assessments in, 20
    method variance in, 20
    modalities in, varieties of, 20
  clinical conceptualizations in, 20
  definition of, 93
  endophenotypes in, 13
  externalization of, 21
  of externalizing proneness, 21–24
    CFA approach to, 23–24
    EF factors for, 21–22
    EFA approach to, 22–24
    inhibition–disinhibition models, 22–24
    multi-method multivariate model of, 22–24
  inhibition–disinhibition models
    of externalizing proneness, 22–24
    liabilities of, 25–26
  inhibitory control in, 21
  multi-modal assessment approaches, 19–20
    for neurobehavioral traits, 21
  neural measures in, integration of, 20
    method variance in, 20
    modalities in, varieties of, 20
  neurobehavioral traits
    ESI for, 21
    in HiTOP model, 21
    as liabilities, in clinical models, 25–26
    multi-modal assessment approaches, 21
    in psychoneurometric research strategy, 18–21
    threat sensitivity, 25–26
  neuroscientific findings in, rigor and replicability of, 18
  psychoneurometric research strategy in, 18, 20–21
    future challenges for, 24–26
    IRT and, 26
    neurobehavioral trait constructs, 18–21
    prediction purpose of, 26
    score estimation in, 26
  puberty and, 208–210
    adrenarche and, 208
    in animal models, 209
    dual systems model for, 209

organizational hypothesis for, 209
    timing of, for boys, 208
    Research Domain Criteria initiative in, 18
psychophysiology. *See* peripheral psychophysiology
psychosis spectrum disorder, schizophrenia redefined as, 8
P-technique, for intra-individual variation analysis, 329
PTSD. *See* post-traumatic stress disorder
Pubertal Development Scale (PDS), 210
puberty, 205–210
    adrenarche, 206–208
    age of onset for, 207
        gender differences in, 205
        trends in, earlier onset, 206
    CPP, 207, 214
    hormonal influences on, 206–207
        adrenarche, 206–207
        on brain development, 207
        DHEA, 206–207, 210
        estradiol, 206–207, 210
        FSH, 210
        GH, 210
        GnRH, 206–207, 210
        gonadarche, 206–207
        leptin, 210
        LH, 210
        testosterone, 206–207, 210
    menarche, 205–206
    psychopathology and, 208–210
        adrenarche and, 208
        in animal models, 209
        dual systems model for, 209
        organizational hypothesis for, 209
        pubertal timing for boys, 208
    tempo of, 205–206, 208–210
        assessment of, 210
    timing of, 208–210
        for boys, 208
        measurement of, 210
        meta-analysis of, 208
        PDS for, 210
        single-item perceptual measures of, 210
pupillometry, 131–135
    biological mechanisms, 130
    clinical findings for, 131
    early history of, 130
    methodological guidelines for, 130–131
pyramidal hippocampal neurons, 170

quality of life issues, 288
quantitative genetic research, for child and adolescent psychopathology. *See also* cognitive studies
    for ADHD
        genetic factors for, 221–222
        heritability estimations for, 221–222
    ASDs
        etiologic factors for, 224–225
        genetic factors for, 221–222
        change variables in, 224–225
        Cholesky decomposition approach, 224
    in classical twin studies, 224
    in classical twin studies
        additive genetics in, 221
        Cholesky decomposition approach in, 224
        common/shared environments, 221
        DeFries–Fulker extreme analytic method, 222–223
        dominance genetic effects, 221
        Falconer Formula in, 221
        heritability estimations, 221
        latent growth curve modeling in, 224
        methodological considerations in, 221
        non-shared environments, 221
    etiologic factors in, 222–223
        for ASDs, 224–225
    gene-environment interplay in, 225–227. *See also* genotype-environment correlation
    genetic factors in, 221–222
        for ADHD, 221–222
        for ASDs, 221–222
    GxE in, 225
    heritability estimations
        for ADHD, 221–222
        for ASDs, 221–222
        in classical twin studies, 221
        environmental factors in, 222
    historic development of, 219
    latent growth curve modeling in, 224
        in classical twin studies, 224
    methodological considerations in, 219–221
        through adoption studies, 219–220
        through classical twin studies, 221
        through family studies, 219
    rGE in, 225–227
        active, 225–226
        detection of, 225–226
        evocative, 225–226
        long-term implications for, 226–227
        passive, 225
    in sibling-comparison studies, 227
    stability variables in, 224–225
    trait correlations in, 222–223. *See also* genotype-environment correlation
        common pathway model, 223–224
        CTCT measures, 223
        externalizing psychopathology dimension, 223–224
        independent pathways model, 223–224
        internalizing psychopathology dimension, 223–224
quantitative models. *See also* meta-analysis
    in descriptive psychopathology, 37
        assessment strategies in, 37
        for BPD, 37
        validity criterion in, 37

race, behavioral medicine and, health disparities influenced by, 294
Random Intercept model, 245–246, 250
randomized controlled trials (RCTs)
    in behavioral medicine, 293
    developmental psychopathology and, 199–200
    in health psychology, 293
    MOST in, 268
    as psychiatric epidemiology method, 82–83
    randomized studies, causal inferences in, 417–418
        Bayesian approach to, 419–420
        complications in, 420
            expectancy effects, 421
            missing data, 421
            non-compliance, 420–421
            placebo effects, 421
        covariates in, 420
        Fisherian, 418
        Hawthorne effect in, 420
        ITT analysis of, 420
        model-based, 419
        Neymanian, 418–419
        posterior predictive, 419
RCM. *See* Rubin Causal Model
RCTs. *See* randomized controlled trials
RDoC initiative. *See* Research Domain Criteria initiative
reactivity, in AA monitoring, 305
real time (RT) variables, 110–111
recruitment strategies, in dyadic relationship design strategies, 344–345
reflective latent variable models, 68
"refrigerator mothers," 371–372
regression coefficients, in moderation analysis, 401–402
regression models, for intensive longitudinal data, 313
regression-based approaches, for clinical research. *See* causal effects; conditional process analysis; moderation analysis; statistical mediation analysis
regularization, 474, 482
Reichenbach, Hans, 97
reinforcement learning (RL) models, of behavioral adaptation, 171–186
    in behavioral neuroscience, 168
    Bush–Mosteller model, 184
    challenges for, 169–170
    comparisons between, using behavioral data, 183
        through imaging data, 183
        for parameter identifiability, 183
    components of, 171
    connectionist models of, 170–171
    economic subjective value in, 185–186
    exceedance probability for, 175
    expected utility of, 185
    exploration/exploitation dilemma, 173
    forward modeling, 176
    historical development of, 168
    human behavior captured in, 173–175
        for individual differences, 178–179
        individual likelihood function, 174
        through model inversion, 174–176
        parameter estimations for, 174
        prediction of, 177–178
        interpretation of, 177
            with groundtruth parameters, 177
    inversion of, 174–176
    learned value in, 185–186

reinforcement learning (RL) models, of behavioral adaptation (cont.)
  as mathematical models, 183–184
    sampling theory in, 183–184
  MLM frameworks, 179
  neural correlates of, 185–186
  neurobiological links, 179–180
    experience through social exchange, 180
    HRFs, 179–180
  in neuroeconomics, 168
  overview of, 171
  parsimony in, 175
  payoff matrix and, in game theory, 182
  plausible alternatives to, 175–176
  policy hypothesis and, 180
  prediction errors and, neural signatures of, 184–185
  RW model, 184
  specification of, 171–173
    choice rules in, 173
    learned action values in, 173
    learning rules in, 172–173
    MDP in, 172
  subjective rationality of, 185
  task solving through, 176–177
  TD models, 184
  for testing alternative hypotheses, 181–183
  thought experiments in, 171
  trust game experiments and, 180–181
    reputation effects in, 181
  validation of, 175–176
relationships. *See* dyadic relationships
relativity, theory of, 9
reliability
  of meta-analytic variables, 391
  in NHST, 100–101
  of psychometrics, 57–60
    evaluation of, 58–59
    importance of, 58
    IRT and, 59–60
    meaning of, 57–58
reparameterized meta-analytic models, 385–386
replication
  in meta-analysis, 391–392
    conceptual, 392
    direct, 392
  of results, 383
  in scientific method, 372
    conceptual, 372
    direct, 372
    reproducibility in clinical psychology and, 379–380
reproducibility, in clinical psychology
  Bayesian inference and, 378–379
  best practices for, 374–380
    pre-registering of studies, 375
    replication methods in, 379–380
    reporting guidelines, 375–376
    transparency approach, 375
  confirmatory hypotheses for, 376–377
  exploratory hypotheses as distinct from, 376–377
  in multivariate modeling, 377
  effect sizes and, 377–378
  exploratory hypotheses for, 376–377

confirmatory hypotheses as distinct from, 376–377
  in multivariate modeling, 377
as problem
  HARKing effects, 373
  long-term impacts of, 373–374
  negative impacts of, 374
  NHST, 373
  in Open Science era, 374
  origins of, 373
  positive impacts of, 374
of reported effects, 371
as scientific concept, 371
statistical power and, 378
reputation effects, 181
Rescorla and Wagner (RW) model, 184
research assistants, training of, in cognitive studies, 112
Research Domain Criteria (RDoC) initiative, 18
  in cognitive studies, 105–106
  in descriptive psychopathology, 39–40
  in biological models, 38
  in *DSM*, 280
  peripheral psychophysiology and, 118
resilience
  developmental neurobiology of, 200
  developmental psychopathology and, 198–199
resource management principle, in MOST, 271–272
respiratory sinus arrhythmia (RSA), 123, 124
  PSNS, 118
response process validity, of psychometrics, 61
resting state designs, for fMRI, 156–157
retrospective cohort studies, 80
reverse inference, of fMRI, 163
rGE. *See* genotype-environment correlation
right-sizing growth comorbidity models, 244–246
  number of variables in, 244
RL models. *See* reinforcement learning models
Robins, Eli, 8
  poor prognosis syndrome, 8
romantic partner designs, in dyadic relationships, 341–342
  self-report methodologies, 341
RSA software, 342
RT variables. *See* real time variables
R-technique, for inter-individual variation analysis, 329
Rubin Causal Model (RCM), 415
  assignment mechanism, 415–417
    strongly ignorable, 417
    unconfounded, 417
RW model. *See* Rescorla and Wagner model

sampling theory
  in longitudinal studies, size of samples, 231
  in RL models, 183–184
Satterthwaite method, 440
SCED. *See* single-case experimental design

schizophrenia
  in adoption and twin studies, 143
  decline of, as phenotypic syndrome, 8
  experimental psychopathology and, 93–94
  in GWAS, 143
  as psychosis spectrum disorder, 8
scientific method, 371–373
  explanations in, 371–372
  observations in, 371–372
  replication in, 372
    conceptual, 372
    direct, 372
  transparency in, 372–373
SDM. *See* shared decision-making
selection bias
  Berkson's bias, 106
  clinician's illusion bias, 106
self-report surveys, 232
SEM. *See* structural equation modeling
sensitivity analyses, of causal inferences, 423
serial models, in statistical mediation analysis, 400
sexual minority status, during adolescence, 212–213
Shakow, David, 94
shared decision-making (SDM), 283
sibling-comparison studies, 227
simple conditional process model, 403–404
simple models, in statistical mediation analysis, 397–398
simultaneity measurement, 9
single nucleotide polymorphisms (SNPs), 142–143
single symptoms, in intensive longitudinal data modeling, 313–316
  central tendencies of, 313
  dispersion scores, 313–314
  spread measures, 313–316
single units, causal inferences on, 415–416
single-case experimental design (SCED), 282
single-gene disorders, 141
single-item perceptual measures, of pubertal timing, 210
skin conductance responses, 118
slow event-related designs, for fMRI, 154–155
smooth pursuit eye movements, 129–130
SNPs. *See* single nucleotide polymorphisms
SNS. *See* sympathetic nervous system
social baseline theory, 338–339
social relationships. *See* dyadic relationships
social stigma, from hysteria diagnosis, 6
Society for Research in Psychopathology (SRP), 94
socioeconomic status
  behavioral health during adolescence and, 212
  behavioral medicine influenced by, 294
  health psychology influenced by, 294
  mental health during adolescence and, 212

Specific Affect Coding System (SPAFF), 343
spread measures, 313–316
spread measures, in intensive longitudinal data modeling, 313–316
SRP. *See* Society for Research in Psychopathology
Stable Unit Treatment Value Assumption, 416
startle response, 127–129
　biological mechanisms, 127
　FPS, 127, 129
　identification of, 127
　methodological guidelines for, 127–128
　PAMR, 127–129
　startle-eye blink measures, 127–128
startle-eye blink measures, 127–128
statistical analysis, meta-analysis as, 383–384
statistical mediation analysis, 397–400
　causal inference in, 399
　　of indirect effects, 399
　　ordering in, 399
　direct effect inferences in, 400
　indirect effect inferences in, 398, 400
　　bootstrap confidence intervals, 398
　　causal elements of, 399
　mediator variables, 397
　multiple mechanisms in, 399–400
　parallel models, 400
　serial models, 400
　simple models, 397–398
　total effects in, 398–399
　suppression in, 399
statistical power, reproducibility and, 378
Stevens, Stanley Smith, 5, 14
strange situation tasks, 340
stress, allostatic load, 292
strongly ignorable assignment mechanism, 417
structural equation modeling (SEM), for dyadic data, 364
studies. *See specific studies*; *specific topics*
subject-specific data analysis, 327
　behavior space in, 329
　between-person structures in, 331
　dynamic factor analysis, 330–332
　　measurement invariance tests for, 331–335
　　person-oriented theory and, 330–331
　　in process research, for dyadic data, 362
　ergodic theorem in, 329, 333
　　consequences of, 330
　　non-ergodic processes, 329
　GIMME model, 333–334
　idiographic levels of, 333–335
　inter-individual variation in, 327–330
　　data boxes, 328
　　definition of, 327–329
　　intra-individual variation and, 328–330
　　R-technique for, 329
　　unified framework for, 334
　intra-individual variation in, 327–330
　　data aboxes, 328
　　definition of, 327–328

　　inter-individual variation and, 328–330
　　P-technique for, 329
　　unified framework for, 334
　nomothetic levels of, 327, 333–335
　person-oriented approaches compared to, 327
　VAR models, 334–335
　　mixed, 334–335
　within-person structures in, 331
substance abuse. *See also* problematic alcohol consumption
　during adolescence, psychopathology of, 206
subtraction analysis, of fMRI, 157–159
Sullivan, Jackie, 15
summative meta-analytic models, 384
supervised machine learning, 471, 482
　classifiers from, 471–472
suppression, 399
surveys
　Classical Test Theory and, 49–50
　construct specification for, 50–51
　　continuum specification as part of, 51
　interviews compared to, 46
　　in stimuli presentation, 46
　Item Response Theory and, 49–50
　measurement of constructs for, 49–50
　　principles of scale, 49–50
　　validity criteria in, 50
　methodological approach to, 45–46
　scale development for, practical considerations in, 51–52
　selection criteria for, 46–49
　　advantages of surveys, 46–47
　　disadvantages of surveys, 47
　usefulness of, 45–46
sweating. *See* psychological sweating
Sydenham, Thomas, 6–7
sympathetic nervous system (SNS), 118
skin conductance responses, 118
System for Coding Interactions in Family Functioning (SCIFF), 342–343

target populations
　in case-control studies, 83–84
　in cognitive studies, selection of, 106–107
　　Berkson's bias and, 106
　　clinician's illusion bias and, 106
　　sampling strategies, 106–107
　in cohort studies, 81
task solving, through RL models, 176–177
TD models. *See* temporal difference models
tempo, of puberty, 205–206, 208–210
　assessment of, 210
temporal difference (TD) models, 184
testosterone, 206–207, 210
theory of relativity, 9
theory of the intervention, 197
Thorndike, Edward, 168
TMM. *See* two method measurement
total effects, in statistical mediation analysis, 398–399
　suppression in, 399
training, 482. *See also* learning
trait comorbidity models, 242

transdiagnostic interventions, in *DSM*, 284
　capitalization model in, 283
　in CBT, 281
　mechanism-focused treatment development and, 281
　targets in, 280
　testing of, 282–283
　UP in, 281
transparency
　in longitudinal studies, 237
　reproducibility in clinical psychology and, 375
Triangular Decomposition comorbidity model, 246
trust game experiments, 180–181
　reputation effects in, 181
Turkheimer, Eric, 139
twin studies, 137–138. *See also* quantitative genetic research; sibling-comparison studies
　biometric modeling in, 136–137
　Cholesky decomposition approach in, 224
　clinical findings from, 139, 143
　　individual analyses from, 140–141
　　meta-analyses from, 140–141
　for dizygotic twins, 137–138
　　DeFries–Fulker extreme analysis approach, 222–223
　epigenetics and, 148
　　modifications in, 148
　Falconer Formula in, 221
　genetic architecture in, 145–147
　　environmental factors, 146
　　GxE, 147–148
　　rGE, 146–147
　genetic risk factor determination in, 136
　genetic variant identification in, 139–141
　GWAS and, 142–144
　　CDCV model, 142–143
　　clinical findings from, 143
　　missing heritability and, 145
　　for schizophrenia, 143
　　SNPs, 142–143
　heritability factors in, 136–137
　　definition of, 137
　　misinterpretation of, 137
　latent growth curve modeling, 224
　Mendelian disorders in, 141
　missing heritability in, 143–145
　　CNVs, 145
　　insufficient GWAS, 145
　　non-additivity of genetic effects, 145
　for monozygotic twins, 137–138
　　DeFries–Fulker extreme analysis approach, 222–223
　overview of, 136
　single-gene disorders in, 141
two method measurement (TMM), 109

unconditional effects, in moderation analysis, 402
unconfounded assignment mechanism, 417
Unified Protocol (UP), for the transdiagnostic treatment of emotional disorders, 281

uniformly most potent factors, in NHST, 100
unsupervised machine learning, 471–472, 482
UP. *See* Unified Protocol

vagotonia hypothesis, 122
validators
    endophenotypes and, 13
    in open concepts, 8
validity
    of NHST, 100–101
    of psychometrics, 60–62
        associative, 61–62
        consequential, 62
        content, 60–61
        evaluation of, 60–61
        importance of, 60
        internal structure, 59–61
        meaning of, 60
        response process, 61
VAR models. *See* vector-autoregressive models

variable models. *See also* latent variable models
    bifactor models, 73–75
        conceptual issues with, 75
        estimation of, 73–74
        features and structures of, 73
        hierarchical, 74
    common factor, 69
    conceptual foundations of, 66–68
    definitions of, 66–68
        sample realization, 67
    estimation of, 68–69
        of latent variables, 67–68
    evaluation of, 68–69
    factor mixture modeling, 75–77
        complexity of, 77
    hypothetical, 66
    RT, 110–111
    scope of, 66
variables
    in meta-analysis, reliability of, 391
    in missing data
        auxiliary variables, 447, 453–454

    in motivating examples, 445
vector-autoregressive (VAR) models, 320–321, 334–335
    mixed, 334–335
Very Simple Structure (VSS) method, 55–56
Vigouroux, Romain, 125
vocal arousal patterns, 343
VSS method. *See* Very Simple Structure method

Waismann, Friedrich, 8–9
within-person structures, in subject-specific data analysis, 331
women, menarche in, 205–206

Yarkoni, Tal, 15

Zubin, Joseph, 94